WARDLAW'S

# Perspectives in
# Nutrition

## 12e

Carol Byrd-Bredbenner
Jacqueline Berning
Danita Kelley
Jaclyn M. Abbot

McGraw Hill

WARDLAW'S PERSPECTIVES IN NUTRITION, TWELFTH EDITION

Published by McGraw Hill LLC, 1325 Avenue of the Americas, New York, NY 10121. Copyright ©2022 by McGraw Hill LLC. All rights reserved. Printed in the United States of America. Previous editions ©2019, 2016, and 2013. No part of this publication may be reproduced or distributed in any form or by any means, or stored in a database or retrieval system, without the prior written consent of McGraw Hill LLC, including, but not limited to, in any network or other electronic storage or transmission, or broadcast for distance learning.

Some ancillaries, including electronic and print components, may not be available to customers outside the United States.

This book is printed on acid-free paper.

3 4 5 6 7 8 9 LWI 26 25 24 23 22

ISBN 978-1-260-69559-5
MHID 1-260-69559-X

Portfolio Manager: *Lauren Vondra*
Product Developer: *Erin DeHeck*
Marketing Manager: *Tami Hodge*
Content Project Managers: *Jessica Portz/Rachael Hillebrand*
Buyer: *Susan K. Culbertson*
Design: *David W. Hash*
Content Licensing Specialist: *Abbey Jones*
Cover Image: *Lukas Gojda/Shutterstock*
Compositor: *MPS Limited*

**Library of Congress Cataloging-in-Publication Data**

Names: Byrd-Bredbenner, Carol, author. | Berning, Jacqueline R., author. |
  Kelley, Danita S., author. | Abbot, Jaclyn M., author.
  Title: Wardlaw's perspectives in nutrition/Carol Byrd-Bredbenner, Jacqueline
  Berning, Danita Kelley, Jaclyn Abbot.
  Description: Twelfth edition. | New York, NY : McGraw Hill LLC, [2022] |
  Includes bibliographical references and index.
  Identifiers: LCCN 2020027092 | ISBN 9781260695595 (hardcover) |
  ISBN 9781260788594 (spiral bound)
  Subjects: LCSH: Nutrition.
  Classification: LCC QP141 .W38 2022 | DDC 612.3—dc23
  LC record available at https://lccn.loc.gov/2020027092

The Internet addresses listed in the text were accurate at the time of publication. The inclusion of a website does not indicate an endorsement by the authors or McGraw Hill LLC, and McGraw Hill LLC does not guarantee the accuracy of the information presented at these sites.

mheducation.com/highered

# Brief Contents

©Brand X Pictures/Getty Images RF

# Meet the Author Team

Carol Byrd-Bredbenner, Ph.D., R.D., FAND, received her doctorate from Pennsylvania State University. Currently, she is Distinguished Professor in the Nutritional Sciences Department at Rutgers, The State University of New Jersey. She teaches a wide range of undergraduate and graduate nutrition courses. Her research interests focus on investigating environmental factors that affect dietary choices and health outcomes. Dr. Byrd-Bredbenner has authored numerous nutrition texts, journal articles, and computer software packages. She has received teaching awards from the American Dietetic Association (now called the Academy of Nutrition and Dietetics), Society for Nutrition Education, and U.S. Department of Agriculture. She was the recipient of the American Dietetic Association's Anita Owen Award for Innovative Nutrition Education Programs, American Society for Nutrition's Excellence in Nutrition Education Award, and Society for Nutrition Education and Behavior's Helen Denning Ullrich Award for Lifetime Excellence in Nutrition Education. She also was a Fellow of the United Nations, World Health Organization at the WHO Collaborating Center for Nutrition Education, University of Athens, Greece. She enjoys exploring food and culinary customs, traveling, diving, and gardening.

Jacqueline R. Berning, Ph.D., R.D., CSSD, earned her doctorate in nutrition from Colorado State University in Fort Collins, Colorado. She is currently Professor and Chair of the Health Science Department at the University of Colorado at Colorado Springs (UCCS), where she has won numerous teaching awards. Dr. Berning is published in the area of sports dietetics and was the sport dietitian for the Denver Broncos for over 25 years, Cleveland Indians for 18 years, and Colorado Rockies for 10 years. Currently, she is the sport dietitian for UCCS athletics and U.S. Lacrosse. She is active in the Academy of Nutrition and Dietetics, where she served as Chair of  the Program Planning Committee for FNCE and is currently Chair of the Appeals Committee. In 2014, Dr. Berning was awarded the Mary Abbot Hess Award for Culinary Events for teaching the University of Colorado football team how to grocery shop and cook. Additionally, she served 6 years as an ADA spokesperson and is former Chair of the Sports, Cardiovascular, and Wellness Nutritionists dietetics practice group. She enjoys walking, hiking, and gardening.

©Clinton Lewis

Danita Saxon Kelley, Ph.D., R.D., earned her doctorate in nutritional sciences from the University of Kentucky. She serves as Associate Dean of the College of Health and Human Services and is a Professor in the Family and Consumer Sciences Department at Western Kentucky University. Previously, Dr. Kelley was Director of the Didactic Program in Dietetics at Western Kentucky University. She is a Past President of the Board of Directors for the Kentucky Academy of Nutrition and Dietetics. Her scholarly work has focused on adolescents' eating patterns, the communication skills of dietetic students, histaminergic activity and regulation of food intake, and dietary restriction effects on the antioxidant defense system. She has received awards for teaching from the Kentucky Academy of Nutrition and Dietetics and the Dietetic Educators of Practitioners of the Academy of Nutrition and Dietetics. She enjoys singing, walking her dog, cheering for her family in water-ski competitions, and watching her children participate in athletic and musical endeavors.

Jaclyn Maurer Abbot, Ph.D., R.D., earned her doctorate in nutritional sciences at the University of Arizona. She is a Registered Dietitian Nutritionist and adjunct lecturer in the Nutritional Sciences Department at Rutgers, The State University of New Jersey. She teaches online undergraduate courses in nutrition and health and introductory sports nutrition. Her research focuses on nutrition communication and health promotion on an array of topics, including safe food handling, nutrition for optimizing fitness performance, nutrition knowledge and behavior, and disease prevention. She has delivered her research findings via formal classroom teaching, outreach programming, and peer-reviewed journals. She enjoys running, coaching youth sports, and spending time with her husband and 3 young children.

# *Preface*

## Welcome to the *Twelfth Edition of Wardlaw's Perspectives in Nutrition*

*Wardlaw's Perspectives in Nutrition* has the richly deserved reputation of providing an accurate, current, in-depth, and thoughtful introduction to the dynamic field of nutrition. We have endeavored to build upon this tradition of excellence by enriching this edition for both students and instructors. Our passion for nutrition, our genuine desire to promote student learning, and our commitment to scientific accuracy, coupled with constructive comments from instructors and students, guided us in this effort. Our primary goal has been to maintain the strengths and philosophy that have been the hallmark of this book yet continue to enhance the accessibility of the science content and the application of materials for today's students.

Nutrition profoundly affects all of our lives every day. For the authors, as well as many other educators, researchers, and clinicians, this is the compelling reason for devoting our careers to this dynamic field. The rapid pace of nutrition research and provocative (and sometimes controversial) findings challenge us all to stay abreast of the latest research and understand its implications for health. We invite you to share with us topics that you believe deserve greater or less attention in the next edition.

To your health!

Carol Byrd-Bredbenner

Jacqueline Berning

Danita Kelley

Jaclyn Maurer Abbot

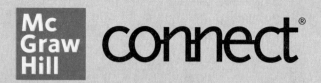

# Instructors: Student Success Starts with You

## Tools to enhance your unique voice

Want to build your own course? No problem. Prefer to use our turnkey, prebuilt course? Easy. Want to make changes throughout the semester? Sure. And you'll save time with Connect's auto-grading too.

**65%**
**Less Time Grading**

Laptop: McGraw Hill; Woman/dog: George Doyle/Getty Images

## Study made personal

Incorporate adaptive study resources like SmartBook® 2.0 into your course and help your students be better prepared in less time. Learn more about the powerful personalized learning experience available in SmartBook 2.0 at **www.mheducation.com/highered/connect/smartbook**

## Affordable solutions, added value

Make technology work for you with LMS integration for single sign-on access, mobile access to the digital textbook, and reports to quickly show you how each of your students is doing. And with our Inclusive Access program you can provide all these tools at a discount to your students. Ask your McGraw Hill representative for more information.

Padlock: Jobalou/Getty Images

## Solutions for your challenges

A product isn't a solution. Real solutions are affordable, reliable, and come with training and ongoing support when you need it and how you want it. Visit **www .supportateverystep.com** for videos and resources both you and your students can use throughout the semester.

Checkmark: Jobalou/Getty Images

SUPPORT AT every step

# Students: Get Learning that Fits You

## Effective tools for efficient studying

Connect is designed to make you more productive with simple, flexible, intuitive tools that maximize your study time and meet your individual learning needs. Get learning that works for you with Connect.

## Study anytime, anywhere

Download the free ReadAnywhere app and access your online eBook or SmartBook 2.0 assignments when it's convenient, even if you're offline. And since the app automatically syncs with your eBook and SmartBook 2.0 assignments in Connect, all of your work is available every time you open it. Find out more at **www.mheducation.com/readanywhere**

> *"I really liked this app—it made it easy to study when you don't have your textbook in front of you."*
>
> — Jordan Cunningham, Eastern Washington University

Calendar: owattaphotos/Getty Images

## Everything you need in one place

Your Connect course has everything you need—whether reading on your digital eBook or completing assignments for class, Connect makes it easy to get your work done.

## Learning for everyone

McGraw Hill works directly with Accessibility Services Departments and faculty to meet the learning needs of all students. Please contact your Accessibility Services Office and ask them to email accessibility@mheducation.com, or visit **www.mheducation.com/about/accessibility** for more information.

# *Connecting Instructors and Students to Additional Digital Resources*

## Saves students and instructors time while improving performance

 **Campus**

**McGraw Hill Campus** integrates all of your digital products from McGraw Hill with your school's Learning Management System for quick and easy access to best-in-class content and learning tools.

## Dietary Analysis Tools

**NutritionCalc Plus** is a powerful dietary analysis tool featuring more than 30,000 foods from the ESHA Research nutrient database, which is comprised of data from the latest USDA Standard Reference database, manufacturers' data, restaurant data, and data from literature sources. NutritionCalc Plus allows users to track food and activities, then analyze their choices with a robust selection of intuitive reports. The interface was updated to accommodate ADA requirements and modern mobile experience native to today's students. This tool is provided complimentary in Connect with *Perspectives in Nutrition*.

NutritionCalc **Plus**
5.0 Online

**Dietary Analysis Case Studies** One of the challenges instructors face with teaching nutrition classes is having time to grade individual dietary analysis projects. To help overcome this challenge, assign auto-graded dietary analysis case studies. These tools require students to use NutritionCalc Plus to analyze dietary data, generate reports, and answer questions to apply their nutrition knowledge to real-world situations. These assignments were developed and reviewed by faculty who use such assignments in their own teaching. They are designed to be relevant, current, and interesting!

**Assess My Diet** Students are using NutritionCalc Plus to analyze their own dietary patterns. But how can instructors integrate that information into a meaningful learning experience? With Assess My Diet, instructors can now assign auto-graded, personalized dietary analysis questions within Connect. These questions refresh their memory on the functions and food sources of each nutrient and prompt the students to evaluate their own eating behaviors. Students can evaluate their own nutrient intakes compared to current Dietary Reference Intakes and demonstrate their ability to perform calculations on their own data, such as percent of calories from saturated fat. They can compare the nutrient density of their own food selections to see which of their food choices provides the most fiber or iron. A benefit of the Assess My Diet question bank is that it offers assignable content that is personalized to the students' data, yet it is still auto-graded. It saves time and keeps all assignments in one place.

## Presentation tools allow you to customize your lectures

**Enhanced Lecture Presentations** Contain lecture outlines, art, photos, and tables. Fully customizable, adapted for ADA compliance, complete, and ready to use—these presentations will streamline your work and let you spend less time preparing for lecture!

**Editable Art** Fully editable (labels and leaders) line art from the text

**Animations** Over 50 animations bring key concepts to life, available for instructors *and* students.

## Digital Lecture Capture

**Tegrity®** is a fully automated lecture capture solution used in traditional, hybrid, "flipped classes" and online courses to record lessons, lectures, and skills.

## Virtual Labs and Lab Simulations    *Virtual* **Labs**

While the sciences are hands-on disciplines, instructors are now often being asked to deliver some of their lab components online, as full online replacements, supplements to prepare for in-person labs, or make-up labs.

These simulations help each student learn the practical and conceptual skills needed, then check for understanding and provide feedback. With adaptive pre-lab and post-lab assessment available, instructors can customize each assignment.

From the instructor's perspective, these simulations may be used in the lecture environment to help students visualize processes, such as digestion of starch and emulsification of lipids.

# Connecting Students to Today's Nutrition

## Our Intended Audience

This textbook was developed for students pursuing nutrition and health science careers as well as those wanting a better understanding of how nutrition affects their lives. Because this course often attracts students from a broad range of majors, we have been careful to include examples and explanations that are relevant to them and to include sufficient scientific background to make the science accessible to them. The appendices help students who wish to learn more or need assistance with the science involved in human physiology, chemistry, and metabolism.

To better bridge the span of differing science backgrounds and to enhance student interest and achievement of course objectives, we organized the presentation of the material within chapters to flow seamlessly from concrete to abstract learning. In chapters focusing on nutrients, for example, concrete concepts, such as food sources of the nutrients and recommended intakes, are introduced early in the chapter to create a framework for more abstract concepts, such as functions, digestion, and absorption.

## Accurate, Current Science That Engages Students

The twelfth edition continues the tradition of presenting scientific content that is reliable, accurate, and up-to-date. This edition incorporates coverage of recent nutrition research, as well as the recent updates to consumer guidelines and tools—Dietary Guidelines for Americans, MyPlate, *Healthy People,* and the new Nutrition Facts panel. It also retains the in-depth coverage students need to fully understand and appreciate the role of nutrition in overall health and to build the scientific knowledge base needed to pursue health-related careers or simply live healthier lives. To enhance these strengths and promote greater comprehension, new research findings and peer-reviewed references are incorporated and artwork is enhanced to further complement the discussions. The presentation of complex concepts was scrutinized to increase clarity through the use of clear, streamlined, precise, and student-friendly language. Timely and intriguing examples, illustrative analogies, clinical insights, culinary perspectives, historical notes, future perspectives, and thought-provoking photos make the text enjoyable and interesting to students and instructors alike.

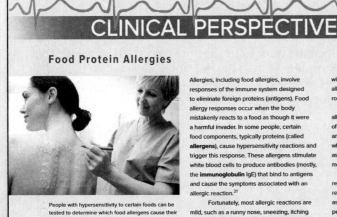

### CLINICAL PERSPECTIVE

#### Food Protein Allergies

People with hypersensitivity to certain foods can be tested to determine which food allergens cause their symptoms.

Science Photo Library/Getty Images

Allergies, including food allergies, involve responses of the immune system designed to eliminate foreign proteins (antigens). Food allergy responses occur when the body mistakenly reacts to a food as though it were a harmful invader. In some people, certain food components, typically proteins (called **allergens**), cause hypersensitivity reactions and trigger this response. These allergens stimulate white blood cells to produce antibodies (mostly, the **immunoglobulin** IgE) that bind to antigens and cause the symptoms associated with an allergic reaction.[21]

Fortunately, most allergic reactions are mild, such as a runny nose, sneezing, itching skin, hives, or digestive upset (indigestion, nausea, vomiting, diarrhea). For those who are severely allergic, exposure to the allergenic food may cause a generalized, life-threatening reaction involving all body systems (known as **anaphylaxis** or anaphylactic shock). Anaphylaxis causes decreased blood pressure

without immediate medical help. In the U.S., allergic reactions result in 200,000 emergency room visits and 150 to 200 deaths per year.

The protein in any food can trigger an allergic reaction. However, 8 foods account for 90% of all food allergies: peanuts, tree nuts (e.g., walnuts and cashews), milk, eggs, fish, shellfish, soy, and wheat (Fig. 7-16). Other foods frequently identified as causing allergic reactions are sesame seeds, meat and meat products, fruits, and cheese.

The only way to prevent allergic reactions is to avoid foods known to trigger reactions. Carefully reading food labels and asking questions when eating out are essential, perhaps life-saving, steps for those with food allergies.[21] In addition, individuals preparing foods at home or in restaurants need to know their menu ingredients and take steps to ensure that foods that cause an allergic reaction in a person do not come in contact with the food to be served to that individual. Even trace amounts of an allergen can cause a reaction. To prevent

# Connecting with a Personal Focus

## Applying Nutrition on a Personal Level

A key objective in nearly all introductory courses is for students to apply their new knowledge of nutrition to their own lives. Practical applications clearly linked to nutritional science concepts are woven throughout each chapter to help students apply their knowledge to improving and maintaining their own health and that of others for whom they are responsible, such as future patients or offspring.

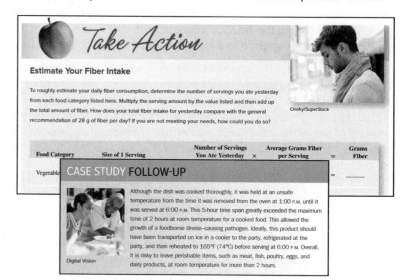

- **Take Action** features in each chapter allow students to examine their own diets and health issues.
- Updated **case studies** showcase realistic scenarios and thought-provoking questions.
- New discussion of the Nutrition Facts panel outlines the innovative changes to this important consumer tool.

## Applying Nutrition to Career and More

- ***Expert Perspective from the Field*** features examine cutting-edge topics and demonstrate how emerging, and sometimes controversial, research results affect nutrition knowledge and practice.
- ***Clinical Perspectives*** highlight the role of nutrition in the prevention and treatment of disease. These topics will be especially interesting to students planning careers in dietetics or health-related fields.
- ***Global Perspectives*** discuss concepts related to critical health and nutrition issues around the world. These timely features also aim to engage students with thought-provoking challenges.
- ***Historical Perspectives*** heighten awareness of critical discoveries and events that have affected our understanding of nutritional science.
- ***Perspective on the Future*** features address emerging trends affecting nutrition science and practice.
- ***Culinary Perspectives*** focus on interesting food trends and their impact on health.
- Each major heading in the chapters is numbered and cross-referenced to the end-of-chapter summary and study questions to make it easy to locate and prioritize important concepts.

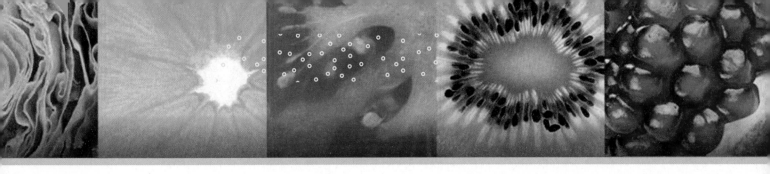

## HISTORICAL PERSPECTIVE
### Photographing Atoms

Discovering the molecular layout of biologically important molecules is critical to understanding their function and treating disease. The biochemist and crystallographer Dorothy Crowfoot Hodgkin developed X-ray techniques that permitted her to determine the structure of over 10 molecules, including insulin, vitamin B-12, vitamin D, and penicillin. Her work with insulin improved the treatment of diabete[s]. Knowing the structure of vitamin B-12 advanced our knowledge of its role in blo[od] health. Learn more about this Nobel Prize winner at www.nobelprize.org/prizes/chemistry/1964/hodgkin/biographical.

Digital Vision/Getty Images

## Perspective on the Future

The common wisdom that eating 3500 kcal less than you need will result in the loss of 1 pound has come under great scrutiny. Weight loss research models based on thermodynamics, mathematics, physics, and chemistry indicate that many more than 3500 calories may be stored in a pound of body fat. Researchers have developed a body weight planner that allows users to make personalized calorie and physical activity plans to reach a goal weight.[24] Learn more at www.pbrc.edu/research-and-faculty/calculators/weight-loss-predictor.

## NUTRITION
### Expert Perspective from the Field

**Tailoring a Healthy Eating Plan to Fit Your Lifestyle**

According to Dr. Judith Rodriguez,* find[ing] your lifestyle is the key to controlling w[eight and] diet. Dr. Rodriguez groups diets based [on] consumers match their lifestyles with p[...] you match what you like to eat or the c[...]

## CLINICAL PERSPECTIVE

**Foodborne Illness Can Be De[adly]**

Foodborne illness often means a few hours or even a few days of discomfort and then the illness resolves on its own. In some cases, though, foodborne illness causes more serious medical problems, which can have lifelong

inte[...]
mor[...]
the [...]
• *Liste[...]
caus[...]

## GL🌐BAL PERSPECTIVE

**How Big Is Your Food Print?**

Growing evidence indicates that what we eat may not only affect our personal health but also that of the environment. The world population is projected to increase to over 9 billion by 2050. The Food and Agricultural Organization (FAO) projects that food and feed production will need to increase by 70% to adequately feed the world's population. Many scientists believe that meat rich diets and the agricultural practices that support the production of food for these diets negatively affect the environment. For instance, producing food for nonvegetarian diets (especially beef-based diets) uses more water, fossil fuel energy, and acres of farmland than producing food for vegetarian diets.[25] Meat rich diets also cause greater emissions of greenhouse gases, such as carbon dioxide, methane, and nitrous oxide, which are associated with global warming.[26] Scientists are concerned that continued population growth may, in turn, decrease agricultural productivity, reduce farmers' incomes, and increase global food insecurity.[27]

Not all scientists agree with these findings and concerns, however. Some believe that consuming a low-fat vegetarian diet with some dairy products and/or meat may actually increase land use efficiency, thereby protecting environmental resources and promoting food security.[28] They point out that high quality farmland is required to grow fruits, vegetables, and grains, whereas meat and dairy products can be produced on the more widely available, lower quality land. Even though diets containing meat use more land, they can feed more people because of the greater availability of lower quality farmland. It appears that diets have different "agricultural land footprints," depending on the amount of plant-based and animal-based food they contain. Supporters of mixed animal/vegetable–based diets point out that vegetarian diets often include tofu and other meat substitutes produced from soy, chickpeas, and lentils. Many meat substitutes are highly processed and require energy-intensive production methods. Thus, including small amounts of meat may offer both environmental and nutritional benefits.

# Making Visual Connections

## Dynamic, Accurate Artwork

More than 1000 drawings, photographs, and tables in the text were critically analyzed to identify how each could be enhanced and refined to help students more easily master complex scientific concepts.

- Many images were updated or replaced to inspire student inquiry and comprehension and to promote interest and retention of information.
- Many illustrations were redesigned to use brighter colors and a more attractive, contemporary style. Others were fine-tuned to make them clearer and easier to follow. Navigational aids show where a function occurs and put it in perspective of the whole body.
- Coordinated color schemes and drawing styles keep presentations consistent and strengthen the educational value of the artwork. Color-coding and directional arrows in figures make it easier to follow events and reinforce interrelationships.

**MyPlate.gov**

**KEY**
- Protein
- Vegetables
- Fruits
- Grains
- Dairy
- Oils
- Other

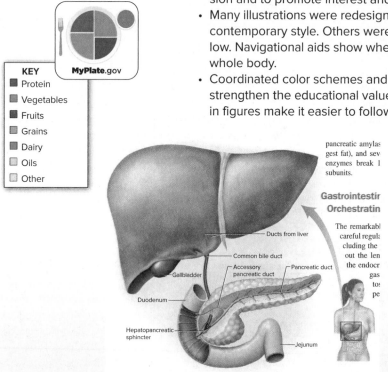

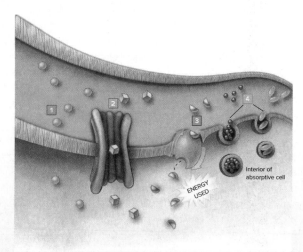

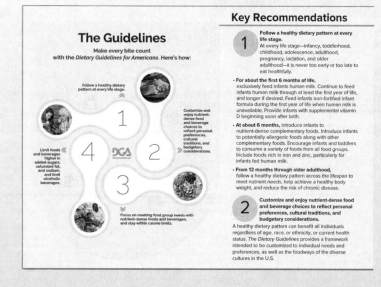

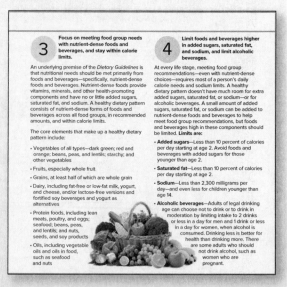

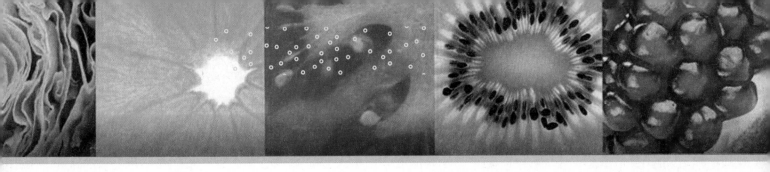

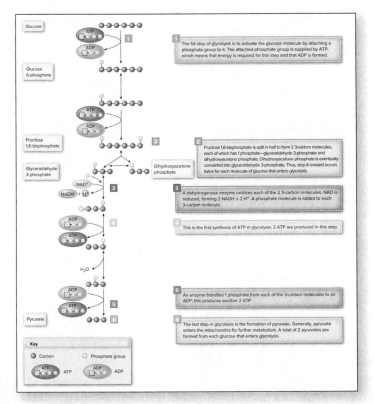

- In many figures, process descriptions appear in the body of the figure. This pairing of the action and an explanation walks students step-by-step through the process and increases the teaching effectiveness of these figures.
- Intriguing chapter opening photos pique students' curiosity by featuring seemingly unrelated topics that draw connections between the photo and nutrition.
- Finally, a careful comparison of artwork with its corresponding text was done to ensure that they are completely coordinated and consistent. The final result is a striking visual program that holds readers' attention and supports the goals of clarity, ease of comprehension, and critical thinking. The attractive layout and design of this edition are clean, bright, and inviting. This creative presentation of the material is geared toward engaging today's visually oriented students.

## Illustrative Chapter Summary

The visual chapter summary continues to reinforce key concepts and promote student engagement and comprehension.

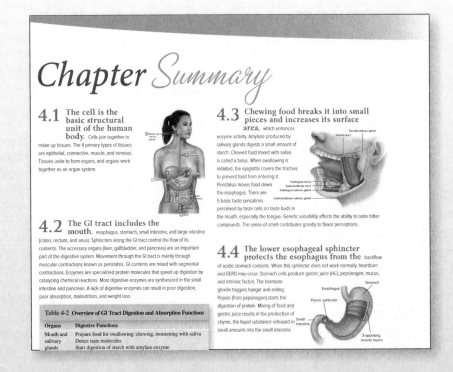

# Connecting with the Latest Updates

## Global Updates and Changes

- The entire twelfth edition updated, refined, and streamlined to enhance student learning
- Complete Dietary Guidelines update to include 2020–2025 recommendations
- Nutrition Facts panels updated to latest FDA regulations
- Latest Daily Values incorporated in nutrient content charts
- New *Culinary Perspective* features throughout
- Fresh, new art for visual engagement
- People-first language used throughout the text to put the person before diagnosis, such as "a person with alcoholism" rather than "an alcoholic"

## Chapter 1, *The Science of Nutrition*

- Updated statistics on leading causes of death
- *Culinary Perspective* featuring fermented foods
- Latest regulatory changes on *trans* fats introduced
- Streamlined and enhanced discussion of functional foods
- Expanded discussion of environmental factors affecting food choices
- Extensive revision of *Healthy People* goals and objectives
- Introduction of the concept of nutrition-focused physical exams
- New discussion on systematic reviews and meta-analyses
- New *Historical Perspective* on Joseph Goldberger
- New figure explaining human genome components

## Chapter 2, *Tools of a Healthy Diet*

- Expansion of summary of nutrient claims on food labels table to include omega-3 fatty acid claims
- Enhanced menu labeling *Expert Perspective from the Field*
- *Take Action* updated to include the latest dietary intake recommendations
- Streamlined discussion of MyPlate and international dietary guidance graphic symbols

## Chapter 3, *The Food Supply*

- Updated domestic and international food insecurity statistics highlighting the worldwide burden of malnutrition and hunger
- Updated food insecurity map
- Expanded discussion of food sustainability and agrobiodiversity
- New image depicting food sustainability from farm to table
- New *Culinary Perspective* on reducing food waste at the grocery store and home
- Extensive revision of discussion on amending agricultural plant and animal traits via selective breeding, mutagenesis, genetic (transgenetic) modification, genome editing, and safety and other concerns

- Expanded discussion of gene editing and illustration to increase comprehension
- Expanded discussion of food nanotechnology
- Latest BPA regulations added
- New non-nutritive sweetener, advantame, introduced
- Latest CDC foodborne illness statistics included
- Enhanced discussion of seafood toxins
- Fully updated discussion of water contamination in Flint, Michigan
- Enhanced *Expert Perspective from the Field* on sustainability in university food service

## Chapter 4, *Human Digestion and Absorption*

- Enhanced discussion on structure and function of nasal lining
- Updated procedure for treating choking to new Red Cross recommendations
- Updated *Global Perspective* to include latest global data on child death from diarrhea
- Expanded discussion of probiotics and prebiotics
- Expanded discussion of erosive and non-erosive gastroesophageal reflux disease (GERD) and management
- Fully updated discussion of sugar's role in nonalcoholic fatty liver disease
- New discussion of effects of opioids on intestinal mobility and constipation
- Irritable bowel disease presentation refined to incorporate probiotics and FODMAP dietary protocol
- New *Take Action* on comparing breads on gluten content
- Celiac disease and non-celiac gluten sensitivity prevalence statistics update

## Chapter 5, *Carbohydrates*

- Expanded content on function of pectin
- Typical sources of sweeteners (Table 5-1) expanded to include advantame
- Enhanced discussion of total sugar and added sugar declarations on Nutrition Facts panels
- Updated discussion on 100% fruit juice recommendations
- Addition of health concerns associated with high fructose corn syrup
- Streamlined discussion of non-nutritive (alternative) sweeteners
- Added discussion on advantame
- New *Culinary Perspective* on nutritive sweeteners
- Fully updated *Healthy People* carbohydrate intake goals
- Apps for managing diabetes mellitus introduced

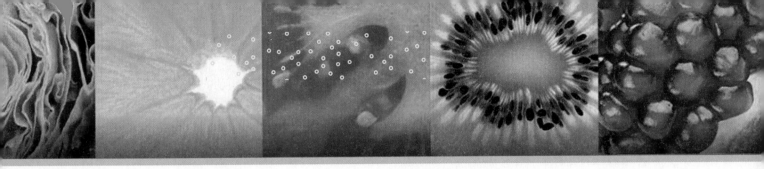

## Chapter 6, *Lipids*

- New FDA *trans* fats regulations incorporated
- Enhanced presentation of main sources of fatty acids (Table 6-1)
- Refined *Take Action* on dietary fat content
- New *Culinary Perspective* on phospholipids in food
- Revised discussion of phospholipids to reflect recent research findings on functions
- Table 6-2 enhanced and updated to reflect latest recommendations for fat intake
- New discussion on foods that affect blood cholesterol
- Streamlined *Expert Perspective from the Field* on a healthier approach to eating fats
- Refined fat content of foods chart (Figure 6-10)
- Refined fat absorption illustration to increase comprehension (Figure 6-16)

## Chapter 7, *Proteins*

- Enhanced discussion on pulses and legumes
- New *Culinary Perspective* on entomophagy
- Enhanced discussion on high protein diets
- Latest statistics on protein-energy malnutrition incorporated
- New feature on meat sweats
- Revised transaminase enzyme pathway to improve understanding (Figure 7-3)

## Chapter 8, *Alcohol*

- Enhanced feature on powdered alcohol
- Streamlined discussion of alcohol metabolism
- Revised *Healthy People* goals regarding alcohol use
- Updated alcohol consumption trends and statistics
- Refined discussion of potential benefits of alcohol intake
- Enhanced discussion of the effects of alcohol abuse on nutritional status
- Extensive revision of alcohol intake around the world
- Dangers of combining alcohol and caffeine added
- Updated cirrhosis section to reflect newest research

## Chapter 9, *Energy Metabolism*

- Improved clarity of caption explaining ATP stores and energy (Figure 9-4)
- Extensive revision of aerobic carbohydrate metabolism figure and caption to increase student comprehension (Figure 9-5)
- Increased clarity of ATP production sections for carbohydrates and lipids
- Streamlined discussion of ketosis in diabetes
- Modified disposal of excess amino groups figure and caption to enhance student understanding (Figure 9-17)
- Extensive revision of international incidence of cancer figure (Figure 9-18)

- Revised discussion on ATP concentrations to promote learning
- New *Take Action* on intermittent fasting and metabolism
- Recommendations added from the Advisory Committee on Heritable Disorders in Newborns and Children regarding inborn errors of metabolism
- New discussion of trimethylaminuria in inborn errors of metabolism section

## Chapter 10, *Energy Balance, Weight Control, and Eating Disorders*

- Most up-to-date map of obesity rates in the U.S.
- Enhanced discussion of estimated energy requirements
- Revised discussion on measuring body fat content
- Weight control objectives from *Healthy People* updated
- Extensive revision to popular diet approaches to weight control (Table 10-7)
- New *Take Action* on how to spot a fad diet
- Eating disorders section streamlined and updated
- Section on binge eating disorder refined

## Chapter 11, *Nutrition, Exercise, and Sports*

- Extensive revision of benefits of exercise section
- Refined discussion of section addressing calorie restriction and protein needs of wrestlers
- Enhanced discussion of boosting glycogen stores
- Expanded discussion of fat needs of athletes
- Enhanced section on ketogenic diets and athletic performance
- Streamlined discussion of calcium intake and relative energy deficiency in sports (REDS)
- Refined discussion of fluid intake and replacement strategies
- New *Culinary Perspective* on sports nutrition in the home kitchen
- Extensive revision of gene doping and editing in sports section

## Chapter 12, *The Fat-Soluble Vitamins*

- Amsler grid for macular degeneration added
- Role of lutein in brain development and cognitive function added to carotenoid section
- New figure depicting bioconcentration and vitamin A content
- *Historical Perspective* on rickets added
- Fitzpatrick sun-reactive scale added to discussion of skin type and vitamin D deficiency risk
- Enhanced and updated section on current vitamin D concerns and additional functions
- New *Culinary Perspective* on plant-based milk alternatives
- Expansion of vitamin K functions section
- Refined discussion of dietary supplements prevalence

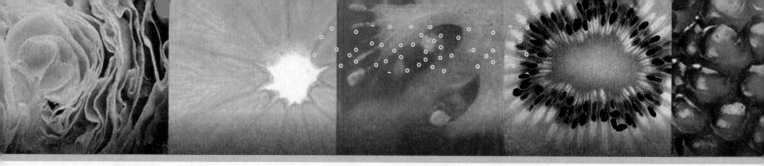

### Chapter 13, *The Water-Soluble Vitamins*

- New *Culinary Perspective* on preserving vitamins in fruits and vegetables
- Updated prevalence of thiamin deficiency in older adults
- Expanded section on riboflavin and plant-based milk alternatives
- Streamlined discussion on thiamin absorption and transport
- New *Culinary Perspective* on cooking methods for enhancing niacin bioavailability
- Updated discussion on pharmacologic use of niacin
- Figure added depicting biotinidase deficiency manifested as hypotonia in infants
- Refined discussion of B-6 metabolism and functions
- New image depicting vitamin B-6 deficiency manifested as seborrheic dermatitis
- Update of neural tube defect prevalence and maternal folate status
- New *Take Action* on energy drinks and B-vitamins
- Expanded discussion of vitamin C and cancer

### Chapter 14, *Water and Major Minerals*

- Expanded discussion of medical therapies used to slow bone loss
- Latest bottled water statistics
- Enhanced *Take Action* on calcium intake
- Image of uses of phosphorus beyond nutrient functions
- Figure added to depict the structure of chlorophyll and contributions to magnesium intake
- Art added to illustrate biological sources of sulfur

### Chapter 15, *Trace Minerals*

- Streamlined discussion of iron
- New feature on disease-causing bacteria and the need for iron
- Enhanced discussion of zinc
- New *Historical Perspective* on unleavened bread and zinc deficiency
- Streamlined discussion on zinc transport
- Menkes disease, a genetic condition impairing copper transport and utilization, pathology image added
- Extensive revision of iodine deficiency disorders
- Refined *Take Action* on local water supply fluoridation
- Extensive revision of the *Global Perspective* on nutrition
- Expanded discussion of dairy and calcium and cancer risk
- Updated iodine status worldwide map
- Enhanced illustration of heme and nonheme absorption

### Chapter 16, *Nutritional Aspects of Pregnancy and Breastfeeding*

- Expanded discussion of folate and vitamin B-12 needs during pregnancy
- Refined section on maternal factors increasing the risk of neural tube defects
- Streamlined discussion of maternal prepregnancy weight
- Refined section on recommendations for maternal weight gain during pregnancy
- Added section on postpartum weight loss
- Expanded discussion of maternal age to include older, first-time mothers
- Expanded discussion of breastfeeding links to reduced diabetes risk

### Chapter 17, *Nutrition during the Growing Years*

- Streamlined section on tracking child growth
- Extensive refinement of *Global Perspective* on autism
- Expanded discussion on energy needs during growth
- Expanded discussion on water needs during fever, diarrhea, and vomiting
- Expanded discussion of iron deficiency anemia during the growing years
- Updated American Academy of Pediatrics's vitamin D supplementation for exclusively breastfed infants recommendations
- Extensive revision of nutritional qualities of breast milk section
- New *Culinary Perspective* on homemade baby food added
- Contribution of snacks to children's diets added

### Chapter 18, *Nutrition during the Adult Years*

- Updated statistics and figure (Figure 18-1) summarizing life expectancy
- Vitamin D links with Alzheimer disease and other types of dementia added
- Strength training recommendations for older adults expanded and updated
- Expanded exercise guidelines for adults
- Added discussion on effects of dysphagia (trouble swallowing) on dietary status
- Revised *Clinical Perspective* to address drug-nutrient interactions
- Expanded discussion on Alzheimer disease
- New illustration depicting body composition changes with sarcopenia

# Acknowledgments

We offer a hearty and profound thank you to the many individuals who have supported and guided us along the way.

*To our loved ones:* Without your patience, understanding, assistance, and encouragement, this work would not have been possible.

*To our wonderful students—past, present, and future:* The lessons you have taught us over the years have enlightened us and sustained our desire to provide newer, better opportunities to help you successfully launch your careers and promote healthful lifelong living. Thank you, in particular, to the students who have used SmartBook®, as your feedback was instrumental in the revisions for this edition.

*To our amazing team at McGraw Hill:* Portfolio Manager Lauren Vondra and Product Developer Erin DeHeck—we thank you most of all for your confidence in us! We deeply appreciate your endless encouragement and patience as you expertly shepherded us along the way. A special thanks to Director Michelle Vogler, Marketing Manager Tami Hodge, and Content Project Managers Jessica Portz and Rachael Hillebrand for all their hard work. An additional thank you to Designer David Hash, Copy Editor Debra DeBord for her meticulous attention to detail, Content Licensing Specialist Shawntel Schmitt, and the many talented illustrators and photographers for their expert assistance.

To your health!

*Carol Byrd-Bredbenner*
*Jacqueline Berning*
*Danita Kelley*
*Jaclyn Maurer Abbot*

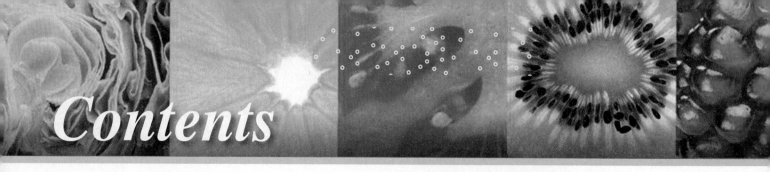

# Contents

©Stockbyte/Getty images RF

# 3  THE FOOD SUPPLY  75

# 4  HUMAN DIGESTION AND ABSORPTION  123

©D. Hurst / Alamy

©Brand X Pictures/Getty Images RF

# 7 PROTEINS 241

# 8 ALCOHOL 277

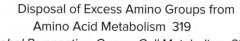

## Part 3  Metabolism and Energy Balance  303

## 9  ENERGY METABOLISM  303

©Michael Simons/123RF

## 10  ENERGY BALANCE, WEIGHT CONTROL, AND EATING DISORDERS  337

©Jules Frazier/Getty Images RF

# 13 THE WATER-SOLUBLE VITAMINS 461

©C Squared Studios/Getty Images

©C Squared Studios/Getty Images

# Part 5 Nutrition Applications in the Life Cycle 601

# 16 NUTRITIONAL ASPECTS OF PREGNANCY AND BREASTFEEDING 601

©D. Hurst/Alamy RF

A nutritious, delicious, and varied diet is key to good health and longevity. To learn more, carefully study this text and visit **nutrition.gov.** Letterberry/Shutterstock

# 1 The Science of Nutrition

## Learning Objectives

**After studying this chapter, you will be able to**

1. Define the terms *nutrition, carbohydrates, proteins, lipids* (fats and oils), *vitamins, minerals, water,* and *calories.*

2. Use the physiological fuel values of energy-yielding nutrients to determine the total energy content (calories) in a food or diet.

3. Describe the major characteristics of the North American diet and the food behaviors that often need improvement.

4. Describe the factors that affect our food choices.

5. Discuss the components and limitations of nutritional assessment.

6. List the attributes of lifestyles that are consistent with *Healthy People* goals and those that contribute to the leading causes of death in North America.

7. Describe the role of genetics in the development of nutrition-related diseases.

8. Explain how the scientific method is used in developing hypotheses and theories in the field of nutrition.

9. Identify reliable sources of nutrition information.

## Chapter Outline

**1.1**  Nutrition Overview

**Culinary Perspective: Fermented Foods**

**1.2**  Energy Sources and Uses

**1.3**  The North American Diet

**1.4**  Nutritional Health Status

**Global Perspective: The Price of Food**

**Clinical Perspective: Genetics and Nutrition**

**1.5**  Using Scientific Research to Determine Nutrient Needs

**1.6**  Evaluating Nutrition Claims and Products

IN OUR LIFETIMES, WE WILL eat about 60 tons of food served at 70,000 meals and countless snacks. Research over the last 50 years has shown that the foods we eat have a profound impact on our health and longevity. A healthy diet—especially one rich in fruits and vegetables—coupled with frequent exercise can prevent and treat many age-related diseases.[1] In contrast, eating a poor diet and getting too little exercise are **risk factors** for many common life-threatening, chronic diseases, such as cardiovascular (heart) disease, diabetes, and certain forms of cancer.[2,3] Another diet-related problem, drinking too much alcohol, can impair nutritional status and is associated with liver disease, some forms of cancer, accidents, and suicides. As you can see in Figure 1-1, diet plays a role in the development of most of the leading causes of death in the U.S. The combination of poor diet and too little physical activity contributes to well over half of these deaths.[3,4]

We live longer than our ancestors did, so preventing age-related diseases is more important now than ever before. Today, many people want to know more about how nutritious dietary choices can bring the goal of a long, healthy life within reach.[5] They may wonder what the best dietary choices are, how nutrients contribute to health, or if multivitamin and mineral supplements are needed. How can people know if they are eating too much saturated fat, *trans* fat, or cholesterol? Why are carbohydrates important? Is it possible to get too much protein?

*Figure 1-1* Leading causes of death in the U.S. The major health problems in North America are largely caused by a poor diet, excessive energy intake, and not enough physical activity. (Percentages do not total 100% due to rounding.)

Source: From Centers for Disease Control and Prevention, National vital Statistics Report, Canadian Statistics are quite similar.

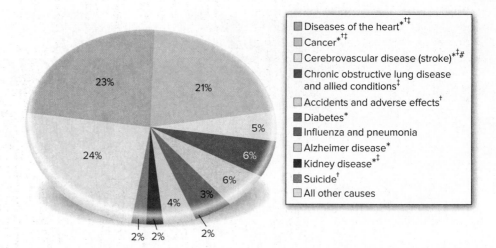

* Causes of death in which diet plays a part
† Causes of death in which excessive alcohol consumption plays a part
‡ Causes of death in which tobacco use plays a part
# Diseases of the heart and cerebrovascular disease are included in the more global term *cardiovascular disease.*

▶ Bold terms in the book are defined in the Glossary. Bold terms also are defined in the text and/or nearby when first presented.

Is the food supply safe to eat? Would a vegetarian diet lead to better health? This book, beginning with this chapter, will help you build the nutrition knowledge base needed to answer these questions (and many more!) and apply this knowledge to safeguard your health, as well as the health of others.

As you begin your study of nutrition, keep in mind that this field draws heavily on chemistry, biology, and other sciences. For the greatest understanding of nutrition principles, you may want to review human physiology (Appendix A), basic chemistry concepts (Appendix B), and the metric system (Appendix H).

## 1.1 Nutrition Overview

The Council on Food and Nutrition of the American Medical Association defines **nutrition** as the "science of food; the nutrients and the substances therein; their action, interaction, and balance in relation to health and disease; and the process by which the organism (e.g., human body) ingests, digests, absorbs, transports, utilizes, and excretes food substances." Food provides the nutrients needed to fuel, build, and maintain all body cells.

### Nutrients

You probably are already familiar with the terms *carbohydrates, lipids* (fats and oils), *proteins, vitamins,* and *minerals* (Table 1-1). These, plus water, make up the 6 classes of nutrients in food. **Nutrients** are substances essential for health that the body cannot make or that it makes in quantities too small to support health.

To be considered an essential nutrient, a substance must have these characteristics:

- Have a specific biological function
- Cause a decline in normal human biological function, such as the normal functions of the blood cells or nervous system, if removed from the diet
- Restore normal human biological function that was impaired by its absence if returned to the diet before permanent damage occurs

**Table 1-1 Nutrients in the Human Diet\***

| Energy-Yielding Nutrients | | | | | |
|---|---|---|---|---|---|
| **Carbohydrate** | **Lipids (Fats and Oils)** | | | **Protein (Amino Acids)** | |
| **Glucose** (or a carbohydrate that yields glucose) | Linoleic acid (omega-6) Alpha-linolenic acid (omega-3) | | Histidine Isoleucine Leucine | Lysine Methionine Phenylalanine | Threonine Tryptophan Valine |
| **Non-Energy-Yielding Nutrients** | | | | | |
| **Vitamins** | | | **Minerals** | | |
| **Water-Soluble** | **Fat-Soluble** | **Major** | **Trace** | **Questionable** | **Water** |
| Thiamin | A | Calcium | Chromium | Arsenic | Water |
| Riboflavin | D | Chloride | Copper | Boron | |
| Niacin | E | Magnesium | Fluoride | Nickel | |
| Pantothenic acid | K | Phosphorus | Iodide | Silicon | |
| Biotin | | Potassium | Iron | Vanadium | |
| B-6 | | Sodium | Manganese | | |
| B-12 | | Sulfur | Molybdenum | | |
| Folate | | | Selenium | | |
| C | | | Zinc | | |

\*This table includes nutrients that the current *Dietary Reference Intakes* and related publications list for humans. There is some disagreement about whether the questionable minerals and certain other minerals not listed in the table are required for human health. Fiber could be added to the list of required substances, but it is not a nutrient (see Chapter 5). The vitamin-like compound choline plays vital roles in the body but is not listed under the vitamin category at this time. Alcohol is a source of energy, but it is not a nutrient.

Nutrients can be assigned to 3 functional categories (Table 1-2):

1. Those that primarily provide energy (typically expressed in kilocalories [kcal])
2. Those that are important for growth and development (and later maintenance)
3. Those that regulate body processes and keep body functions running smoothly

Some overlap exists among these categories. The energy-yielding nutrients and water make up a major portion of most foods.[6]

Because carbohydrates, proteins, lipids, and water are needed in large amounts, they are called **macronutrients**. In contrast, vitamins and minerals are needed in such small amounts in the diet that they are called **micronutrients**. Let's now look more closely at the classes of nutrients.

**Table 1-2 Functional Categories of Nutrients**

| Provide Energy | Promote Growth and Development | Regulate Body Processes |
|---|---|---|
| Most carbohydrates | Proteins | Proteins |
| Proteins | Lipids | Some lipids |
| Most lipids (fats and oils) | Some vitamins | Some vitamins |
| | Some minerals | Some minerals |
| | Water | Water |

Alcoholic beverages are rich in energy (calories), but alcohol is not a nutrient.

foodiepics/Shutterstock

Many foods are rich sources of the nutrients we recognize today as essential for health.

©JGI/Blend Images LLC

## Carbohydrates

Carbohydrates are composed mainly of the **elements** carbon, hydrogen, and oxygen. Fruits, vegetables, grains, beans, and sugars are the primary dietary sources of carbohydrate. The main types of carbohydrates are simple and complex. Small carbohydrate structures are called sugars or simple carbohydrates—table sugar (sucrose) and blood sugar (glucose) are examples. Some sugars, such as glucose, can chemically bond together to form large carbohydrates, called polysaccharides or complex carbohydrates (Fig. 1-2). Examples of complex carbohydrates include the starch in grains and the glycogen stored in our muscles. Fiber, another type of complex carbohydrate, forms the structure of plants.

Glucose, which comes from simple carbohydrates and starch, is a major source of energy in most cells. It and most other carbohydrates provide an average of 4 calories per gram (kcal/g).[7] (Fiber provides little energy because it cannot be broken down by digestive processes.) When too little carbohydrate is eaten to supply sufficient glucose, the body is forced to make glucose from proteins. (Chapter 5 focuses on carbohydrates.)

## Lipids

Like carbohydrates, lipids (e.g., fats, oils, cholesterol) are **compounds** composed mostly of the elements carbon, hydrogen, and oxygen (Fig. 1-3). Note that the term *fats* refer to lipids that are solid at room temperature, whereas oils are those that are liquid at room temperature.

**macronutrient** Nutrient needed in gram quantities in the diet.

**micronutrient** Nutrient needed in milligram or microgram quantities in a diet.

**element** Substance that cannot be separated into simpler substances by chemical processes. Common elements in nutrition include carbon, oxygen, hydrogen, nitrogen, calcium, phosphorus, and iron.

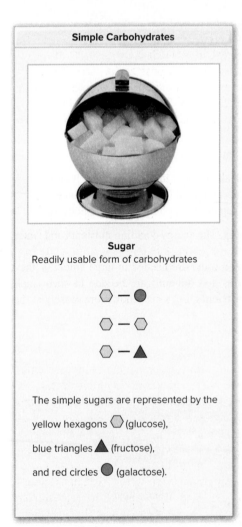

**Simple Carbohydrates**

**Sugar**
Readily usable form of carbohydrates

The simple sugars are represented by the
yellow hexagons ⬡ (glucose),
blue triangles ▲ (fructose),
and red circles ⬤ (galactose).

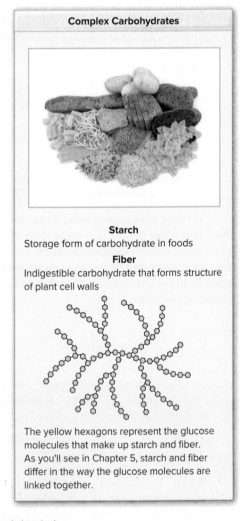

**Complex Carbohydrates**

**Starch**
Storage form of carbohydrate in foods

**Fiber**
Indigestible carbohydrate that forms structure of plant cell walls

The yellow hexagons represent the glucose molecules that make up starch and fiber. As you'll see in Chapter 5, starch and fiber differ in the way the glucose molecules are linked together.

*Figure 1-2* Two views of carbohydrates—dietary and chemical.

Photodisc/PunchStock; Shutterstock/Evgenia Sh.

Lipids yield more energy per gram than carbohydrates—on average, 9 calories per gram. (See Chapter 9 for details about the high energy yield of lipids.) Lipids are insoluble in water but can dissolve in certain organic solvents (e.g., ether and benzene).

The lipid type called a **triglyceride** is the major form of fat in foods and a key energy source for the body. Triglycerides also are the major form of energy stored in the body. They are composed of 3 fatty acids attached to a glycerol **molecule**. Fatty acids are long chains of carbon flanked by hydrogen with an acid group attached to the end opposite glycerol.

Most lipids can be separated into 2 basic types—saturated and unsaturated—based on the chemical structure of their dominant fatty acids. This difference helps determine whether a lipid is solid or liquid at room temperature, as well as its effect on health. Although almost all foods contain a variety of saturated and unsaturated fatty acids, plant oils tend to contain mostly unsaturated fatty acids, which make them liquid at room temperature. Many animal fats are rich in saturated fatty acids, which make them solid at room temperature. Unsaturated fats tend to be healthier than saturated fats—saturated fat raises blood cholesterol, which can clog arteries and eventually lead to cardiovascular disease.

Two specific unsaturated fatty acids—linoleic acid and alpha-linolenic acid—are essential nutrients. They must be supplied by our diets. These essential fatty acids have many roles, including being structural components of cell membranes and helping regulate blood pressure and nerve transmissions. A few tablespoons of vegetable oil daily and eating fish at least twice weekly supply sufficient amounts of essential fatty acids.[7]

Some foods also contain *trans* fatty acids—unsaturated fats that have been processed to change their structure from the more typical *cis* form to the *trans* form (see Chapter 6). Partially hydrogenated oils are the primary source of *trans* fats. In the U.S., partially hydrogenated oils have been banned for use in foods because they pose health risks. These oils will be phased out of the food supply in the next few years.[8] Partially hydrogenated foods are used to prepare many deep-fried foods (e.g., donuts, french fries), baked snack foods (e.g., cookies, crackers), and solid fats (e.g., stick margarine, shortening). (Chapter 6 focuses on lipids.)

## Proteins

Proteins, like carbohydrates and fats, are composed of the elements carbon, oxygen, and hydrogen (Fig. 1-4). Proteins also contain another element—nitrogen. Proteins are the main structural material in the body. For example, they are a major part of bone and muscle; they also are important components in blood, cell membranes, **enzymes**, and immune factors.[7] Proteins can provide energy for the body—on average, 4 calories per gram; however, the body typically uses little protein to meet its daily energy needs.

Proteins form when amino acids bond together. Twenty common amino acids are found in food; 9 of these are essential nutrients for adults, and 1 additional amino acid is essential for infants. (Chapter 7 focuses on proteins.)

## Vitamins

Vitamins have a wide variety of chemical structures and can contain the elements such as carbon, hydrogen, nitrogen, oxygen, phosphorus, sulfur, and others. The main function of vitamins is to enable many **chemical reactions** to occur in the body. Some of these reactions help release the energy trapped in carbohydrates, lipids, and proteins. Vitamins themselves provide no usable energy for the body.

The 13 vitamins are divided into 2 groups. Fat-soluble vitamins (A, D, E, and K) dissolve in fat. Vitamin C and the B-vitamins (thiamin, riboflavin, niacin, vitamin B-6, pantothenic acid, biotin, folate, and vitamin B-12) are water-soluble vitamins. The vitamin groups often act quite differently. For example, cooking is more likely to destroy water-soluble vitamins than fat-soluble vitamins. Water-soluble vitamins are excreted from the body much more readily than fat-soluble vitamins. As a result, fat-soluble vitamins, especially vitamin A, are much more likely to accumulate in excessive amounts in the body, which then can cause toxicity. (Vitamins are the focus of Part 4.)

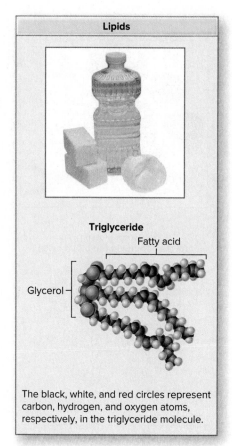

**Lipids**

**Triglyceride**

Fatty acid

Glycerol

The black, white, and red circles represent carbon, hydrogen, and oxygen atoms, respectively, in the triglyceride molecule.

*Figure 1-3* Dietary and chemical views of lipids.

Getty Images

**atom** Smallest unit of an element that still has all the properties of the element. An atom contains protons, neutrons, and electrons.

**molecule** Atoms linked (bonded) together; the smallest part of a compound that still has all the properties of a compound.

**compound** Atoms of 2 or more elements bonded together in specific proportions.

**enzyme** Compound that speeds the rate of a chemical process but is not altered by the process. Almost all enzymes are proteins (some are made of nucleic acids).

**chemical reaction** Interaction between 2 or more chemicals that changes both chemicals.

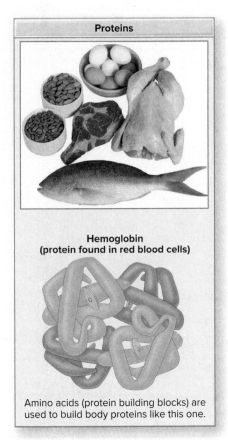

**Proteins**

**Hemoglobin**
**(protein found in red blood cells)**

Amino acids (protein building blocks) are used to build body proteins like this one.

*Figure 1-4* Dietary and chemical views of proteins.

Comstock Images/Getty Images

---

**organic compound** Substance that contains carbon atoms bonded to hydrogen atoms in the chemical structure.

**inorganic substance** Substance lacking carbon atoms bonded to hydrogen atoms in the chemical structure.

**metabolism** Chemical processes in the body that provide energy in useful forms and sustain vital activities.

**phytochemical** Physiologically active compound in plants that may provide health benefits.

**zoochemical** Physiologically active compounds in foods of animal origin that may provide health benefits.

## Minerals

The nutrients discussed so far are all complex organic compounds, whereas minerals are structurally very simple, inorganic substances. The chemical structure of an **organic compound** contains carbon atoms bonded to hydrogen atoms, whereas an **inorganic substance** generally does not. In this case, the term *organic* is not related to the farming practices that produce organic foods (these are described in Chapter 3).

Minerals typically function in the body as groups of one or more of the same atoms (e.g., sodium or potassium) or as parts of mineral combinations, such as the calcium- and phosphorus-containing compound called hydroxyapatite, found in bones. Because they are elements, minerals are not destroyed during cooking. (However, they can leak into cooking water and get discarded when food is drained.) Minerals yield no energy for the body but are required for normal body function. For instance, minerals play key roles in the nervous system, the skeletal system, and water balance.

Minerals are divided into 2 groups: major minerals and trace minerals. Major minerals are needed daily in gram amounts. Sodium, potassium, chloride, calcium, and phosphorus are examples of major minerals. Trace minerals are those that we need in amounts of less than 100 milligrams (mg) daily. Examples of trace minerals are iron, zinc, copper, and selenium. (Minerals are the focus of Part 4.)

## Water

Water is the 6th class of nutrients. Like minerals, water also is inorganic. Although sometimes overlooked as a nutrient, water is the nutrient needed in the largest quantity. Water ($H_2O$) has numerous vital functions in the body. It acts as a solvent and lubricant and is a medium for transporting nutrients to cells. It also helps regulate body temperature. Beverages, as well as many foods, supply water. The body even makes some water as a by-product of **metabolism**. (Water is examined in detail in Part 4.)

## Phytochemicals and Zoochemicals

**Phytochemicals** (plant components in fruits, vegetables, legumes, and whole grains) and **zoochemicals** (components in animals) are physiologically active compounds. They are not considered essential nutrients in the diet. Still, many of these substances provide significant health benefits.[9] For instance, numerous studies show reduced cancer risk among people who regularly consume fruits and vegetables. Researchers surmise that some phytochemicals in fruits and vegetables block the development of cancer (see Part 4).[10] Some phytochemicals and zoochemicals also have been linked to a reduced risk of cardiovascular disease.[11]

It will likely take many years for scientists to unravel the important effects of the many different phytochemicals and zoochemicals in foods. Multivitamin and mineral supplements currently contain few or none of these beneficial chemicals.

Tomatoes contain the phytochemical lycopene; thus, they can be called a functional food.

David R. Frazier Photolibrary, Inc./Alamy Stock Photo

**Table 1-3 Examples of Phytochemicals and Zoochemicals under Study**

| Phytochemicals | Food Sources |
| --- | --- |
| Allyl sulfides/organosulfides | Garlic, onions, leeks |
| Saponins | Garlic, onions, licorice, legumes |
| Carotenoids (e.g., lycopene) | Orange, red, and yellow fruits and vegetables; egg yolks |
| Monoterpenes | Oranges, lemons, grapefruit |
| Capsaicin | Chili peppers |
| Lignans | Flaxseed, berries, whole grains |
| Indoles | Cruciferous vegetables (broccoli, cabbage, kale) |
| Isothiocyanates | Cruciferous vegetables, especially broccoli |
| Phytosterols | Soybeans, other legumes, cucumbers, other fruits and vegetables |
| Flavonoids | Citrus fruit, onions, apples, grapes, red wine, tea, chocolate, tomatoes |
| Isoflavones | Soybeans, fava beans, other legumes |
| Catechins | Tea |
| Ellagic acid | Strawberries, raspberries, grapes, apples, bananas, nuts |
| Anthocyanosides | Red, blue, and purple produce (eggplant, blueberries) |
| Fructooligosaccharides | Onions, bananas, oranges |
| Stilbenoids (e.g., resveratrol) | Blueberries, grapes, peanuts, red wine |
| **Zoochemicals** | |
| Sphingolipids | Meat, dairy products |
| Conjugated linoleic acid | Meat, cheese |

Thus, nutrition and health experts suggest that a diet rich in fruits, vegetables, legumes, and whole-grain breads and cereals is the most reliable way to obtain the potential benefits of phytochemicals.[12] In addition, foods of animal origin, such as fatty fish, can provide the beneficial zoochemical omega-3 fatty acids (see Chapter 6), and fermented dairy products provide probiotics (see Chapter 4). Table 1-3 lists some phytochemicals and zoochemicals under study, with their common food sources.

### Functional Foods

Foods rich in phytochemicals (chemicals from plants) and zoochemicals (chemicals from foods of animal origin) are sometimes referred to as functional foods. A **functional food** provides health benefits beyond those supplied by the traditional nutrients it contains—the food or food ingredient offers additional components that may decrease disease risk and/or promote optimal health due to the physiologically active compounds they contain. Table 1-4 describes the categories of functional foods.[8]

The phytochemicals and zoochemicals that are present naturally in unmodified whole foods like fruits and vegetables are thought to provide many health benefits (see Table 1-3). Foods modified by adding nutrients, phytochemicals, zoochemicals, or herbs (see Chapter 18) also may provide health benefits. For instance, orange juice fortified with calcium may help prevent osteoporosis. Medical foods are designed to help enhance the management of health conditions. An example is phenylalanine-restricted formula fed to infants born with the inborn error of metabolism condition called phenylketonuria (PKU) (see Chapter 9). This formula helps them develop normally. An important trend in the food industry is the addition of nutrients, phytochemicals, and other components in hopes of boosting the healthfulness of the food supply.

Shutterstock/Ivonne Wierink

**Table 1-4  Functional Food Categories[9]**

**Conventional Foods: Unmodified Whole Foods**

| Fruits | Spices | Dairy products |
|---|---|---|
| Vegetables | Nuts | Fish |
| Herbs | | |

**Modified Foods: Fortified, Enriched, or Enhanced Foods**

Calcium-fortified orange juice

Omega-3-enriched bread

Breakfast bars enhanced with ginkgo biloba

Cheese made with plant sterols

**Medical Foods: Food, Formula, or Supplement Used under Medical Supervision to Manage a Health Condition**

Phenylalanine-free formulas for phenylketonuria (PKU)

Limbrel® for osteoarthritis

Axona® for Alzheimer disease

VSL#3® for ulcerative colitis

GlycemX™ 360 for diabetes management

**Special Dietary Use Foods: Foods That Help Meet a Special Dietary Need**

Infant formula for infants

Lactose-free foods for lactose intolerance

Sugar-free foods for weight loss

Gluten-free foods for celiac disease

Measuring Spoons: Stockbrokerxtra Images/Photolibrary; Orange Juice: ©Stockbyte/Getty Images; Baby Bottle: Photodisc; Lacctaid Carton: ©McGraw-Hill Education/Jill Braaten, photographer

*Knowledge Check*

1. What are the 6 classes of nutrients?
2. What characteristics do the macronutrients share?
3. How are vitamins categorized?
4. How are minerals different from carbohydrates, lipids, protein, and vitamins?
5. What are phytochemicals and zoochemicals?

### Fermented Foods

LapailrKrapai/Shutterstock

This man from Kyrgyzstan is drinking fermented horse milk, which is called kumis.

©Ronald Wixman

Fermented foods have been enjoyed around the world for centuries. Initially, people fermented soybeans, vegetables, milk, fruits, and even fish to preserve them. Fermentation was especially important when refrigeration or canning was not available. Fermented foods rely on bacteria, yeast, or fungi to convert natural sugars and starches to lactic and other acids, which in turn preserve the food. Research supports that there is a potential to use fermentation to help reduce short-chain fermentable carbohydrates (FODMAPs—see the Clinical Perspective in Chapter 4) during food preparation (e.g., proofing bread), potentially expanding food options for people with digestive disorders[13] (see Chapter 4). Today, interest focuses on fermented foods' flavor and probiotic and nutrient contents.

A very common fermented food is German sauerkraut (literally, "sour or pickled cabbage"). Sauerkraut is made by packing chopped cabbage and spices into a crock with rock salt. Over a few weeks, *Lactobacillus* bacteria on the cabbage multiply and cause fermentation, converting the raw cabbage to flavorful sauerkraut. Yogurt, another common fermented food, is made by adding a bacterial culture to warm milk. As the bacteria multiply, the pH drops, causing the formation of the smooth, soft, tangy curd that we know as yogurt. Some pickles and the traditional Korean cabbage dish called kimchi also are fermented. Other fermented foods include tempeh, a cakelike product made from slightly fermented soybeans, and miso, a rich paste made from soybeans, seasonings, and sometimes grains such as rice and barley. Both tempeh and miso are fermented with fungi. (Tofu, another soybean product, is not fermented.) Kombucha, a popular fermented tangy and slightly fizzy beverage, is made from green or black tea. Eating fermented foods may offer benefits to an individual's intestinal microbiota. Which fermented foods have you tried? How would you describe their flavors?

## 1.2 Energy Sources and Uses

Humans obtain the energy needed to perform body functions and do work from carbohydrates, fats, and proteins. Alcohol also is a source of energy, supplying about 7 calories per gram. As mentioned previously, it is not considered a nutrient, however, because alcohol has no required function. After digesting and absorbing energy-producing nutrients, the body transforms the energy trapped in carbohydrate, protein, fat, and alcohol into other forms of energy in order to

- Build new compounds
- Move muscles
- Transmit nerve impulses
- Balance **ions** within cells

(See Chapter 4 for more on digestion and absorption. Chapter 9 describes how energy is released from chemical bonds and then used by body cells to support the processes just described.)

**ion** Atom with an unequal number of electrons and protons. Negative ions (anions) have more electrons than protons; positive ions (cations) have more protons than electrons.

Physiological fuel values.

**Carbohydrate**
4 kcal per gram

**Protein**
4 kcal per gram

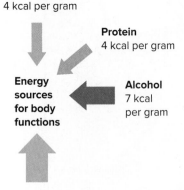

**Energy sources for body functions**

**Alcohol**
7 kcal per gram

**Fat**
9 kcal per gram

▶ The word *calorie* typically is used to mean kilocalorie; thus, this book uses the term *calorie* and its abbreviation, *kcal*.

▶ Many scientific journals express energy content of food as kilojoules (kJ), rather than kilocalories. A mass of 1 gram moving at a velocity of 1 meter/second possesses the energy of 1 joule (J); 1000 J = 1 kJ. Heat and work are 2 forms of energy; thus, measurements expressed in terms of kilocalories (a heat measure) are interchangeable with measurements expressed in terms of kilojoules (a work measure): 1 kcal = 4.18 kJ.

*Calorie* is often the term used to express the amount of energy in foods. Technically, a **calorie** is the amount of heat energy it takes to raise the temperature of 1 gram of water 1 degree Celsius (1°C). Because a calorie is such a tiny measure of heat, food energy is more accurately expressed in terms of the kilocalorie (kcal), which equals 1000 calories. (If the *c* in *calories* is capitalized, this also signifies kilocalories.) A **kilocalorie (kcal)** is the amount of heat energy it takes to raise the temperature of 1000 g (1 liter) of water 1°C. In everyday usage, the word *calorie* (without a capital *c*) also is used to mean kilocalorie. Thus, the term *calorie* and its abbreviation, *kcal,* are used throughout this book. Any values given in calories on food labels are actually in kilocalories (Fig. 1-5).

The calories in food can be measured using a bomb calorimeter (see Chapter 10). Or they can be estimated by multiplying the amount of carbohydrates, proteins, fats, and alcohol in a food by their physiological fuel values. The physiological fuel values are 4, 9, 4, and 7 for carbohydrate, fat, protein, and alcohol, respectively. These values are adjusted to account for the extent to which foods can be digested and for substances (e.g., waxes and fibers) that humans cannot digest. Thus, they should be considered estimates.

Physiological fuel values can be used to determine the calories in food. Consider these foods:

### 1 Large Hamburger

| | | |
|---|---|---|
| Carbohydrate | 39 grams × 4 = | 156 kcal |
| Fat | 32 grams × 9 = | 288 kcal |
| Protein | 30 grams × 4 = | 120 kcal |
| Alcohol | 0 grams × 7 = | 0 kcal |
| **Total** | | **564 kcal** |

### 8-ounce Pina Colada

| | | |
|---|---|---|
| Carbohydrate | 57 grams × 4 = | 228 kcal |
| Fat | 5 grams × 9 = | 45 kcal |
| Protein | 1 gram × 4 = | 4 kcal |
| Alcohol | 23 grams × 7 = | 161 kcal |
| **Total** | | **438 kcal** |

**HONEY WHEAT BREAD**

## Nutrition Facts

19 servings per container

| Serving size | 1 slice (36g) |
|---|---|

**Amount Per Serving**

**Calories** **80**

| | % Daily Value* |
|---|---|
| **Total Fat** 1g | 2% |
| Saturated Fat 0g | 0% |
| *Trans* Fat 0g | |
| **Cholesterol** 0mg | 0% |
| **Sodium** 200mg | 8% |
| **Total Carbohydrate** 15g | 5% |
| Dietary Fiber 2g | 8% |
| Total Sugars 1g | |
| Includes 0g added sugars | 0% |
| **Protein** 3g | |
| Vitamin D 0mcg | 0% |
| Calcium 0mg | 0% |
| Iron 1mg | 4% |
| Potassium 0mg | 0% |

\* The % Daily Value (DV) tells you how much a nutrient in a serving of food contributes to a daily diet. 2,000 calories a day is used for general nutrition advice.

**Figure 1-5** Use the nutrient values on the Nutrition Facts label to calculate the energy content of a food. Based on carbohydrate, fat, and protein content, a serving of this food (honey wheat bread) contains 81 kcal ([15 × 4] + [1 × 9] + [3 × 4] = 81). The label lists 80 because Nutrition Facts labels round values.

**kilocalorie (kcal)** Heat energy needed to raise the temperature of 1000 grams (1 L) of water 1 degree Celsius; also written as *Calorie*.

hamburger: ©Burke/Triolo/Brand X Pictures RF;
pina colada: C Squared Studios/Getty Images

These values also can be used to determine the portion of total energy intake that carbohydrate, fat, protein, and alcohol provide to your diet. Assume that one day you consume 283 g of carbohydrates, 60 g of fat, 75 g of protein, and 9 g of alcohol. This consumption yields a total of 2035 kcal ([283 × 4] + [60 × 9] + [75 × 4] + [9 × 7] = 2035). The percentage of your total energy intake derived from each nutrient can then be determined:

$$\% \text{ of energy intake as carbohydrate} = (283 \times 4) / 2035 = 0.56 \times 100 = 56\%$$

$$\% \text{ of energy intake as fat} = (60 \times 9) / 2035 = 0.27 \times 100 = 27\%$$

$$\% \text{ of energy intake as protein} = (75 \times 4) / 2035 = 0.15 \times 100 = 15\%$$

$$\% \text{ of energy intake as alcohol} = (9 \times 7) / 2035 = 0.03 \times 100 = 3\%$$

### Knowledge Check

1. What does the term *calorie* mean?
2. How do calories, kilocalories, and kilojoules differ?
3. How many calories are in a food that has 8 g carbohydrate, 2 g alcohol, 4 g fat, and 2 g protein?

## 1.3 The North American Diet

In the U.S. and Canada, large surveys are conducted to determine what people are eating. The U.S. government uses the National Health and Nutrition Examination Survey (NHANES) administered by the U.S. Department of Health and Human Services. In Canada, this information is gathered by Health Canada in conjunction with Agriculture and Agrifood Canada. Results from these surveys and others show that North American adults consume, on average, 16% of their energy intake as proteins, 50% as carbohydrates, and 33% as fats. These percentages are estimates, and they vary slightly from year to year and to some extent from person to person. Although these percentages fall within a healthy range[14] (see Chapter 2), many people are eating more than they need to maintain a healthy weight.[15]

Animal sources, such as meat, seafood, dairy products, and eggs, supply about two-thirds of the protein intake for most North Americans; plant sources provide only about a third. In many other parts of the world, it is just the opposite: plant proteins—from rice, beans, corn, and other vegetables—dominate protein intake. About half the carbohydrates in North American diets are simple carbohydrates (sugars); the other half are starches (e.g., pastas, breads, and potatoes). Most North Americans need to reduce their sugar intake and increase their intake of starch and fiber. Because approximately 60% of dietary fat comes from animal sources and only 40% from plant sources, many North Americans are consuming far more saturated fat and cholesterol than is recommended.

These surveys also indicate that most of us could improve our diets by focusing on rich food sources of vitamin A, vitamin E, iron, potassium, and calcium and reducing sodium intake. Nutrient intake varies in some demographic groups, which means these individuals need to pay special attention to certain nutrients. For example, older adults often get too little vitamin D, and many women of childbearing years have inadequate iron intake.

Many North Americans could improve their nutrient intake by moderating intake of sugared soft drinks and fatty foods and eating more fruits, vegetables, whole-grain breads, and reduced-fat dairy products. Vitamin and mineral supplements also can help meet nutrient needs but, as you'll see in Part 4, they cannot fully make up for a poor diet in all respects.[16]

Increasing vegetable intake, such as a daily salad, is one strategy to boost intake of important nutrients.

D. Hurst/Alamy Stock Photo

**hunger** Primarily physiological (internal) drive for food.

**appetite** Primarily psychological (external) influences that encourage us to find and eat food, often in the absence of obvious hunger.

## What Influences Our Food Choices?

Although we have to eat to obtain the nutrients we need to survive, many factors other than health and nutrition affect our food choices. Daily food intake is a complicated mix of the need to satisfy **hunger** (physical need for food) and social and psychological needs (Fig. 1-6).[17] In areas of the world where food is plentiful and fairly easy to access (e.g., the U.S., Canada, Europe, Australia, and Japan), food selection is largely guided by **appetite**—the psychological desire to eat certain foods and reject others. Appetite and food choice depend on many factors:

- *Food flavor, texture, and appearance preferences*—for many people, these are the most important factors affecting food choices. Creating more flavorful foods that are both healthy and profitable is a major focus of the food industry. Food preferences are affected by many factors, including genetics, culture, and lifestyle.
- *Culture* (knowledge, beliefs, religion, and traditions shared by a group or social network of people) teaches individuals which foods are considered proper or appropriate to eat and which are not. For example, many people in North America believe it is proper to eat beef; however, people in some cultures never consider eating beef. Some cultures savor foods such as blood, mice, and insects; even though these foods are packed with nutrients and safe to eat, few North Americans feel they are proper to eat. Early experiences with people, places, and situations influence lifelong food choices. New immigrants often retain the diet patterns of their country of origin until they become acculturated (i.e., they have adopted the cultural traits or social patterns of the new country).

*Figure 1-6* Food choices are affected by many factors. Which have the greatest impact on your food choices?

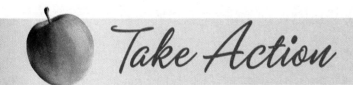

## Take Action

©Eric Audras/Photoalto/PictureQuest RF

### Why You Eat What You Do

Choose a day of the week that is typical of your eating pattern. List all the foods and drinks you consume for 24 hours. Using the factors that influence food choices discussed in Section 1.3, indicate why you consumed each item. Note that there can be more than 1 reason for choosing a particular food or drink.

Now ask yourself, What was the most frequent reason for eating or drinking? To what extent are health or nutrition concerns the reason for your food choices? Should you make these reasons higher priorities?

- *Lifestyle* includes the way we spend our resources and assign priorities. People with very busy lives often have limited time and energy to buy and prepare foods, so they opt for convenience or fast food. For some, it may be more important to spend extra time working rather than making it a priority to exercise and eat healthfully.

- *Routines and habits* related to food and eating affect *what* as well as *when* we eat. Most of us eat primarily from a core group of foods—only about 100 basic items account for 75% of a person's total food intake.

- *Food cost and availability* are important but play only moderate roles in food choices for many of us because food is relatively inexpensive and widely available in North America.[1] In fact, we spend less than 10% of our income on food—about half on food at home and half on food away from home.[18] However, as described in the Global Perspective, this is not the case in many other areas of the world. The type of food available also is affected by the environment.

- *Environment* includes your surroundings and experiences. In North America, the environment is filled with opportunities to obtain affordable, delicious, high-calorie food—vending machines, bake sales, food courts in shopping areas, and candy displays in bookstores—and encourages (via marketing) the consumption of these foods. Experiences with friends, family, and others also can influence food choices, as can mood and psychological needs. Geographic location can affect the food environment, too. Think about the regional cuisine in the area where you live. For example, in the southwestern U.S., barbecue is popular. In New England, seafood is readily available.

- *Food marketing* is any type of action a company takes to create a desire in consumers to buy its food; *advertising* is a type of food marketing. Food influencers, or thought leaders, are another type of food marketing. These individuals often are paid by companies to promote their products or services by recommending them on social media. Before following the advice of someone on social media, be sure to find out if that person is being paid to promote the product, and think about how this type of marketing affects your choices. The food industry in the U.S. spends billions of dollars annually on marketing. Some of this marketing is helpful, such as when it promotes the importance of calcium and fiber intake. However, the food industry more frequently advertises fast food, candy, cookies, cakes, and pastries because such products generate the greatest profits.

- *Health and nutrition concerns, knowledge, beliefs, and values* also can affect food choices. Those most concerned about health and who have the greatest nutrition knowledge tend to be well-educated, middle-income professionals. The same people are generally health oriented, have active lifestyles, and work hard to keep their bodies at a healthy weight. Values guide the food choices of many people. For instance, people who are concerned about the environment may opt to eat mostly locally produced food or organic foods. Those who value low-fat, high-fiber diets or animal rights may choose a vegan diet.

The major health problems in North America are largely caused by a poor diet, excessive energy intake, and not enough physical activity.

LiliGraphie/Shutterstock

romastudio © 123RF.com

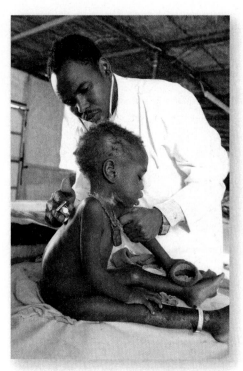

An assessment of this child's nutritional health status indicates stunted growth and edema due to limited protein intake caused by an inability to purchase enough protein-rich foods. Learn more about protein deficiency in Chapter 7.

©Jean-Marc Giboux/Getty Images

*Knowledge Check*

1. What type of food provides most of the protein in the diets of North Americans?
2. Which types of carbohydrates do most North Americans need to increase in their diets?
3. Which vitamins and minerals do many North Americans need to increase in their diets?
4. What factors affect food choices?

## 1.4 Nutritional Health Status

In a well-nourished person, the total daily intake of protein, fat, and carbohydrate weighs about 450 g (about 1 pound). In contrast, the typical daily mineral intake weighs about 20 g (about 4 teaspoons) and the daily vitamin intake weighs less than 300 mg (1/15th of a teaspoon). These nutrients can come from a variety of sources—fruits, vegetables, meats, dairy products, or other foods. Our body cells are not concerned with which food has supplied a nutrient—what is important, however, is that each nutrient be available in the amounts needed for the body to function normally.

The body's nutritional health is determined by the sum of its status with respect to each nutrient. There are 3 general categories of nutritional status: desirable nutrition, undernutrition, and overnutrition. The common term *malnutrition* can refer to either overnutrition or undernutrition, neither of which is conducive to good health.

Optimal, or **desirable nutritional status** for a particular nutrient is the state in which the body tissues have enough of the nutrient to support normal functions, as well as to build and maintain surplus stores that can be used in times of increased need. A desirable nutritional status can be achieved by obtaining essential nutrients from a variety of foods.

**Undernutrition** occurs when nutrient intake does not meet nutrient needs, causing surplus stores to be used. Once nutrient stores are depleted and tissue concentrations of an essential nutrient fall sufficiently low, the body's metabolic processes eventually slow down or even stop. The early stage of a nutrient deficiency is termed **subclinical** because there are no overt signs or symptoms that can be detected or diagnosed. If a deficiency becomes severe, clinical signs and symptoms eventually develop and become outwardly apparent. A **sign** is a feature that can be observed, such as flaky skin. A **symptom** is a change in body function that is not necessarily apparent to a health-care provider, such as feeling tired or achy. Table 1-4 describes the signs and symptoms associated with iron status.

Consumption of more nutrients than the body needs can lead to **overnutrition**. In the short run—for instance, a week or so—overnutrition may cause only a few symptoms, such as intestinal distress from excessive fiber intake. If an excess intake continues, the levels of some nutrients in the body may increase to toxic amounts. For example, too much vitamin A can have negative effects, particularly in children, pregnant women, and older adults. The most common type of overnutrition in industrialized nations—excess intake of energy-yielding nutrients—often leads to obesity. Obesity, in turn, can lead to other serious chronic diseases, such as type 2 diabetes and certain forms of cancer.

**sign** Physical attribute that can be observed by others, such as bruises.

**symptom** Change in health status noted by the person with the problem, such as a stomach pain.

| Table 1-4  **Nutritional Status, Using Iron as an Example** | |
| --- | --- |
| **Condition** | **Signs and Symptoms Related to Iron** |
| **Undernutrition:** nutrient intake does not meet needs | Decline in iron-related compounds in the blood, which reduces the ability of the red blood cells to carry oxygen to body tissues and, in turn, causes fatigue on exertion, poor body temperature regulation, and eventually pale complexion |
| **Desirable nutrition:** nutrient intake supports body function and permits storage of nutrients to be used in times of increased need | Adequate liver stores of iron, adequate blood levels of iron-related compounds, and normal functioning of red blood cells |
| **Overnutrition:** nutrient intake exceeds needs | Excess liver stores of iron, which damage liver cells |

# GL BAL PERSPECTIVE

## The Price of Food

The cost of food as a percentage of income varies widely around the world. In the U.S., for instance, food consumed at home by the average household is less than 6% of total income, whereas families in Nigeria spend almost two-thirds of their income on food eaten at home. Accounting partly for this difference is that a substantial amount is spent in the U.S. for food eaten away from home. As shown in the line chart, over the past century, the proportions of money spent in the U.S. on food eaten at home and away from home have gotten closer with each passing year.

Another reason for this differential among countries is that food production, packaging, and distribution have become more efficient in some parts of the world but less so in others. Greater efficiency means we can buy more for less. In developed nations, nearly all households have a refrigerator, which greatly reduces food spoilage and therefore food expenditures. Food packaging innovations also help keep food fresh longer. Pest management in homes, food processing plants, and farming further minimize food losses. In addition, U.S. government policies and legislation, primarily the Farm Bill, affect how food is produced and how the prices of some foods are set.

It is important to realize that the percentage of income spent on food in the countries shown in the bar graph is for the "average" household. The proportion of a family's budget spent on food tends to be higher for those with lower incomes. Regardless of why these proportional differences occur in the U.S. or in other nations, as the percentage of family income spent on food rises, the typical family has less to devote to other components that contribute to overall health and quality of life, such as health care, shelter, clothing, and education. In addition, these families may be unable to purchase sufficient amounts of nutritious food because they need to pay rent or other essential expenses. The cost of food can affect both food choices and health.

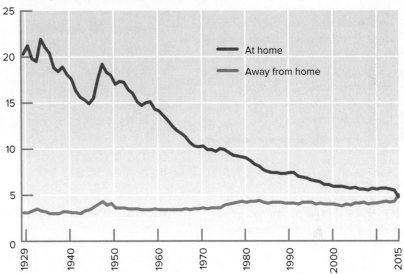

**Disposable Income Spent on Food in the U.S.**

Source: Calculated by the Economic Research Service, USDA, from various data sets from the U.S. Census Bureau and the Bureau of Labor Statistics.

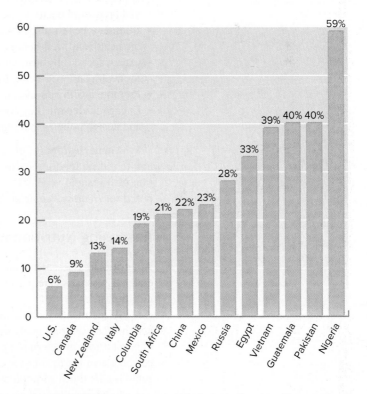

Source: USDA, ERS, Percent of household final consumption expenditures spent on food, alcoholic beverages, and tobacco that were consumed at home, by selected countries, 2016; www.ers.usda.gov/data-products/food-expenditures.aspx.

*Critical* Thinking

Ying loves to eat hamburgers, fries, and lots of pizza with double amounts of cheese. He rarely eats any vegetables and fruits but, instead, snacks on cookies and ice cream. He insists that he has no problems with his health, is rarely ill, and doesn't see how his diet could cause him any health risks. How would you explain to Ying that, despite his current good health, his diet could predispose him to future health problems?

**Table 1-5  Nutrition and Weight Status Objectives from *Healthy People 2030***

Reduce proportion of adults with obesity

Reduce household food insecurity and hunger

Increase fruit, vegetable, and whole grain consumption by people aged 2 years and over

Reduce consumption of saturated fat, sugar, and added sugar by people aged 2 years and over

Increase calcium, potassium, and vitamin D consumption by people aged 2 years and over

Reduce iron deficiency in children aged 1 to 2 years and females aged 12 to 49 years

Increase proportion of students participating in the School Breakfast Program

Increase proportion of worksites that offer employee nutrition programs

Note: Related objectives include those addressing osteoporosis, cancer, diabetes, foodborne illness, heart disease, nutrition during pregnancy, physical activity, and alcohol use.

▶ To review all of the *Healthy People* objectives, visit www.healthypeople.gov.

## National Health Objectives

Health promotion and disease prevention have been public health strategies in the U.S. and Canada since the late 1970s. One part of this strategy is *Healthy People*, a report issued by the U.S. Department of Health and Human Services, Public Health Service.[19] This is a national effort to set  science-based, 10-year national goals and objectives for improving the health and well-being of all Americans. The overarching goals of *Healthy People* are

- Attain healthy, thriving lives and well-being, free of preventable disease, disability, injury and premature death
- Eliminate health disparities, achieve health equity, and attain health literacy to improve the health and well-being of all
- Create social, physical, and economic environments that promote attaining full potential for health and well-being for all
- Promote healthy development, healthy behaviors, and well-being across all life stages
- Engage leadership, key constituents, and the public across multiple sectors to take action and design policies that improve the health and well-being of all[20]

Many *Healthy People* objectives are related to nutrition. For example, the objectives aim to reduce food insecurity and hunger, reduce intake of sodium, and increase consumption of fruits and vegetables. Table 1-5 lists the nutrition and weight status objectives proposed for *Healthy People 2030*. Learn more at health.gov/healthypeople.

## Assessing Nutritional Status

A nutritional assessment can help determine how nutritionally fit you are (Table 1-6). Generally, assessments are performed by a physician, often with the aid of a registered dietitian nutritionist.[5]

Assessments include an analysis of numerous background factors known to affect health. For example, many diseases have a genetic component, so family history plays an important role in determining nutritional and health status. Another background factor is a person's own medical history, especially any health conditions, diseases, or treatments that could hinder the absorption or use of a nutrient.

In addition to background factors, parameters that complete the picture of nutritional status are anthropometric, biochemical, clinical, dietary, and environmental assessments. **Anthropometric assessment** involves measuring various aspects of the body, including height, weight (and weight changes), body circumferences (e.g., waist, hips, arm), and skinfold thickness (an indicator of body fatness and body composition). Anthropometric measurements are easy to obtain and are generally reliable.

**Table 1-6  Conducting an Evaluation of Nutritional Health**

| Factors | Examples |
|---|---|
| **Background** | **Medical history** (e.g., current and past diseases and surgeries, body weight history, current medications)<br>**Family medical history** |
| **Nutritional** | **Anthropometric assessment** (e.g., height, weight, skinfold thickness, arm muscle circumference, body composition)<br>**Biochemical (laboratory) assessment** (e.g., compounds in blood and urine)<br>**Clinical assessment** (e.g., physical examination of skin, eyes, and tongue; ability to walk)<br>**Dietary assessment** (e.g., usual food intake, food allergies, supplements used)<br>**Environmental assessment** (e.g., education and economic background, marital status, housing condition) |

▶ A practical example using the ABCDEs for evaluating nutritional state can be illustrated in a person who chronically abuses alcohol. On evaluation, the physician notes the following:

**Anthropometric:** Low weight-for-height, recent 10 lb weight loss, muscle wasting in the upper body

**Biochemical:** Low amounts of the vitamins thiamin and folate in the blood

**Clinical:** Psychological confusion, skin sores, and uncoordinated movement

**Dietary:** Consumed mostly alcohol-fortified wine and hamburgers for the last week

**Environmental:** Currently residing in a homeless shelter, $35.00 in his wallet, unemployed

**Assessment:** This person needs professional medical attention, including nutrient repletion.

**Biochemical assessments** include the measurement of the concentrations of nutrients and nutrient by-products in the blood, urine, and feces and of specific blood enzyme activities. For example, the status of the vitamin thiamin is measured, in part, by determining the activity of an enzyme called transketolase used to metabolize glucose (see Part 4). To test for this, cells (e.g., red blood cells) are broken open and thiamin is added to see how it affects the rate of the transketolase enzyme activity.

During a **clinical assessment**, health-care providers search for any physical evidence of diet-related diseases (e.g., high blood pressure, skin conditions). The health-care provider tends to focus the clinical assessment on potential problem areas identified from a dietary assessment. Clinical assessments related to nutritional issues are referred to as **nutrition-focused physical exams**. These head-to-toe examinations of a person's physical appearance and functions search for evidence of nutrient deficiencies and toxicities. **Dietary assessment** examines how often a person eats certain types of foods (called a food frequency); the types of foods eaten over a long period of time, perhaps as far back as childhood (called a food history); and typical intake, such as foods eaten in the last 24 hours or several days (e.g., a 24-hour recall or a 3-day recall). Finally, an **environmental assessment** (based on background data) provides information on the person's education and economic background. This information is important because people who have inadequate education, income, and housing and/or live alone often have a greater risk of poor health. Those with limited education may have a reduced ability to follow instructions given by health-care providers and/or an income that hinders their ability to purchase, store, and prepare nutritious food. Taken together, these 5 parameters form the ABCDEs of nutritional assessment: anthropometric, biochemical, clinical, dietary, and environmental (Fig. 1-7).

## Limitations of Nutritional Assessment

Nutritional assessments can be helpful in improving one's health. However, it is important to recognize the limitations of these assessments. First, many signs and symptoms of nutritional deficiencies—diarrhea, skin conditions, and fatigue—are not very specific. They may

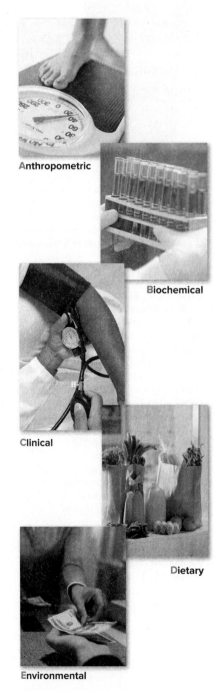

**Anthropometric**

**Biochemical**

**Clinical**

**Dietary**

**Environmental**

*Figure 1-7* The ABCDEs of nutritional assessment: anthropometric, biochemical, clinical, dietary, and environmental status.

Photodisc/Getty Images; © Adam Gault/age fotostock; Comstock Images/Jupiterimages; Janis Christie/Getty Images; Ryan McVay/Getty Images RF

be caused by poor nutrition or by other factors unrelated to nutrition. Second, it can take a long time for the signs and symptoms of nutritional deficiencies to develop and, because they can be vague, it is often difficult to establish a link between an individual's current diet and his or her nutritional status.

Third, a long time may elapse between the initial development of poor nutritional health and the first clinical evidence of a problem. For instance, a diet high in saturated fat often increases blood cholesterol, but it does not produce any clinical evidence for years. Nonetheless, the cholesterol is building up in blood vessels and may lead to a heart attack. Another example of a serious nutrition-related health condition with signs and symptoms that often don't appear until later in life is low bone density (osteoporosis) resulting from insufficient calcium intake, which may have begun in the teen years. Currently, a great deal of nutrition research is trying to identify better methods for detecting nutrition-related problems early—before they damage the body.

## Importance of Being Concerned about Nutritional Status

Regardless of the limitations of nutritional assessment, people who focus on maintaining desirable nutritional health are apt to enjoy a long, vigorous life and are less likely to develop health problems, such as those in Figure 1-8. For example, women who followed a healthy lifestyle experienced an 80% reduction in risk of heart attacks, compared with women without such healthy practices.[21] Here is what these healthy women did:

- Consumed a healthy diet that was varied, rich in fiber, and low in animal fat and *trans* fat and that included some fish
- Avoided becoming overweight
- Regularly drank a small amount of alcohol
- Exercised for at least 30 minutes daily
- Did not smoke

Soft drinks account for about 10% of the energy intake of teenagers in North America and, in turn, contribute to generally poor calcium intakes seen in this age group. Consuming insufficient calcium increases their risk of osteoporosis in future years.

Santirat Praeknokkaew/Shutterstock

*Figure 1-8* Examples of health problems associated with poor dietary habits. An upward arrow (↑) indicates that a high intake and a downward arrow (↓) indicates that a low intake contributes to the health problem. In addition to the habits in the figure, no illicit drug use, adequate sleep (7–8 hours), adequate water intake, regular physical activity, minimal emotional stress, a positive outlook on life, and close friendships provide a more complete approach to good health. Also important is regular consultation with health-care professionals—early diagnosis is especially useful for controlling the damaging effects of many diseases.

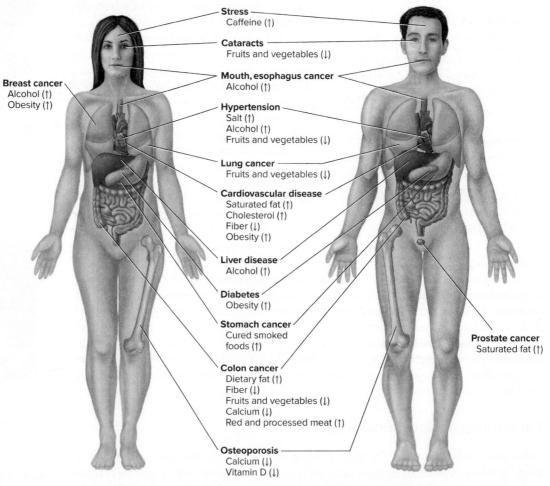

## CASE STUDY

Lane Oatey / Blue Jean Images/Getty Images

While Allen was driving to campus last week, he heard a radio advertisement for a nutrient supplement containing a plant substance that was discovered recently. It supposedly gives people more energy and helps them cope with the stress of daily life. This advertisement caught Allen's attention because he has been feeling run-down lately. He is taking a full course load and has been working 30 hours a week at a local restaurant. Allen doesn't have a lot of extra money to spare. Still, he likes to try new things, and this recent breakthrough sounded almost too good to be true. After searching for more information on the Internet, he discovered that the recommended dose would cost $60 per month. Because Allen is looking for some help with his low energy level, he decides to order a 30-day supply. Does this extra expense make sense to you?

## Getting Nutrition-Related Advice: The Nutrition Care Process

For those who feel they need to improve their diets and health, a safe approach is to consult a physician or registered dietitian (R.D.). A person with the credentials "R.D." after his or her name (or "R.D.N.," registered dietitian nutritionist) has completed a rigorous baccalaureate degree program approved by the Academy of Nutrition and Dietetics, has performed hundreds of hours of supervised professional practice, and has passed a registration examination. Starting in 2024, registered dietitian nutritionists will be required to also have a master's degree. Registered dietitian nutritionists are trained to provide scientifically valid nutrition advice. To find a registered dietitian nutritionist, visit the website of the Academy of Nutrition and Dietetics (www.eatright.org) or Dietitians of Canada (www.dietitians.ca) or call the dietary department of a local hospital.

When you meet with a registered dietitian nutritionist, you should expect that he or she will follow the **Nutrition Care Process**:[22]

Registered dietitians (also called registered dietitian nutritionists) are a reliable source of nutrition advice.

liquidlibrary/PictureQuest

- Conduct a nutritional assessment: asks questions about your food and nutrition history and anthropometric, biochemical, clinical, dietary, and environmental assessment data. The registered dietitian nutritionist also may conduct a nutrition-focused physical exam.
- Diagnose nutrition-related problem: uses your nutrition history and assessment data to determine your specific nutrition-related problem
- Create an intervention: formulates a diet plan tailored to your needs, as opposed to simply tearing a form from a tablet that could apply to almost anyone, that addresses the root cause of your nutrition problem with the goal of relieving the signs and symptoms of your diagnosis
- Monitor and evaluate progress: schedules follow-up visits to track your progress, answer questions, help keep you motivated, and perhaps reassess, rediagnose, and modify your intervention. Family members may be involved in the diet plan, when appropriate. The dietitian consults directly with your physician and readily refers you back to your physician for those health problems a nutrition professional is not trained to treat.

Be skeptical of health practitioners who prescribe very large doses of vitamin, mineral, or protein supplements for everyone.

### Knowledge Check

1. What is the difference between a sign and a symptom?
2. How does undernutrition differ from overnutrition?
3. What are the ABCDEs of nutritional assessment?
4. What are 3 limitations of nutritional assessment?
5. What should you expect when you meet with a registered dietitian nutritionist?

**Nutrition Care Process** Systematic approach used by registered dietitian nutritionists to ensure that patients receive high-quality, individualized nutrition care; process involves nutrition assessment, diagnosis, intervention, monitoring, and evaluation.

## Genetics and Nutrition

Your **genome**, in addition to lifestyle and diet, affects almost every medical condition. During digestion, the nutrients supplied by food are broken down, absorbed into the bloodstream, and transported to cells. There, genetic material, called **deoxyribonucleic acid (DNA)**, inside the nucleus of the cells directs how the body uses the nutrients consumed. As you can see in Figure 1-9, foods and humans contain the same nutrients, but the proportions differ—**genes** in body cells dictate the type and amount of nutrients in food that will be transformed and reassembled into body structures and compounds.

Genes direct the growth, development, and maintenance of cells and, ultimately, of the entire organism. Genes contain the codes that control the expression of individual traits, such as height, eye color, and susceptibility to many diseases. An individual's genetic risk for a given disease is an important factor, although often not the only factor, in determining whether he or she develops that disease.

Each year, new links between specific genes and diseases are reported. It is likely that soon it will be relatively easy to screen a person's DNA for genes that increase the risk of disease. Currently, there are about 1000 tests that can determine whether a person has genetic **mutations** that increase the risk of certain illnesses. For example, a woman can be tested for certain gene mutations that elevate her chances of developing breast cancer.

### Nutritional Diseases with a Genetic Link

Most chronic nutrition-related diseases (e.g., diabetes, cancer, osteoporosis, cardiovascular disease, hypertension, and obesity) are influenced by interactions among genetic, nutritional, and other lifestyle factors. Studies of families, including those with twins and

**deoxyribonucleic acid (DNA)** Site of hereditary information in cells. DNA directs the synthesis of cell proteins.

**genes** Hereditary material on chromosomes that makes up DNA. Genes provide the blueprint for the production of cell proteins.

**mutation** Change in the chemistry of a gene that is perpetuated in subsequent divisions of the cell in which it occurred; a change in the sequence of the DNA.

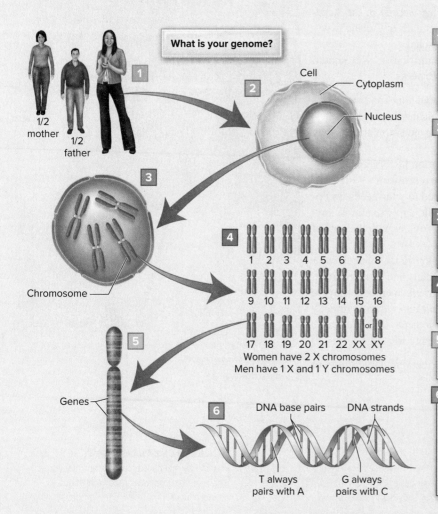

**What is your genome?**

1/2 mother
1/2 father

Cell
Cytoplasm
Nucleus

Chromosome

1 2 3 4 5 6 7 8
9 10 11 12 13 14 15 16
17 18 19 20 21 22 XX XY or
Women have 2 X chromosomes
Men have 1 X and 1 Y chromosomes

Genes

DNA base pairs   DNA strands

T always pairs with A
G always pairs with C

**1** A genome is an organism's complete set of DNA, including all of its genes. DNA contains all the information needed to build and maintain an organism throughout life. DNA is passed down from parents. The human genome has more than 3 billion DNA base pairs.

**2** Humans have billions of cells. We have a copy of our entire genome in every cell that has a nucleus. In humans, all cells, except red blood cells, contain a nucleus. The nucleus is a cell's "command center"—it controls cell growth and reproduction.

**3** Chromosomes are structures within the nucleus that contain hundreds to thousands of genes. Humans have a total of about 20,000 to 25,000 genes.

**4** Each parent provides 23 chromosomes, which pair up in their offspring.

**5** Each chromosome has genes that consist of DNA, which provides the code, or "recipe," for making proteins.

**6** Each DNA molecule is a long double helix—it looks like a spiral staircase with millions of steps. The two long strands are made of phosphate and sugar (deoxyribose) molecules. The strands are connected by pairs of molecules called base pairs. All base pairs are made of a combination of base molecules: adenine (A) and thymine (T) or guanine (G) and cytosine (C).

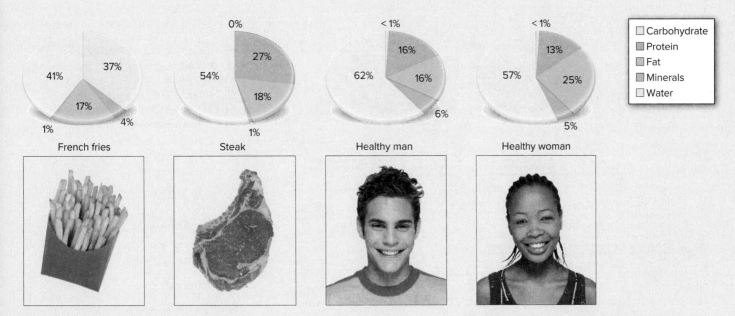

| | French fries | Steak | Healthy man | Healthy woman |
|---|---|---|---|---|

French fries: 37%, 41%, 17%, 1%, 4%, 0%

Steak: 0%, 27%, 18%, 54%, 1%

Healthy man: < 1%, 16%, 16%, 62%, 6%

Healthy woman: < 1%, 13%, 25%, 57%, 5%

Legend:
- Carbohydrate
- Protein
- Fat
- Minerals
- Water

*Figure 1-9* Proportions of nutrients in the human body, compared with those in typical foods—animal or vegetable. Note that the amount of vitamins found in the body is extremely small and, so, is not shown.

Comstock Images/Jupiterimages; ©FoodCollection/StockFood; MSPhotographic/Shutterstock; Stockbyte/Getty Images; Stockbyte/Getty Images

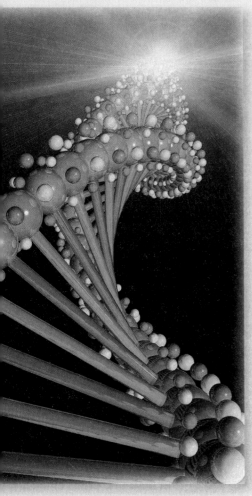

Genes are present on DNA—a double helix. The cell nucleus contains most of the DNA in the body.

©Brand X Pictures/age fotostock RF

adoptees, provide strong support for the effect of genetics in these disorders. In fact, family history is considered one of the most important risk factors in the development of many nutrition-related diseases.[23,24]

For example, both of the common types of diabetes, certain cancers (e.g., colon, prostate, and breast cancer), and osteoporosis have genetic links. In addition, about 1 in every 500 people in North America has a defective gene that greatly delays cholesterol removal from the bloodstream—this defective gene increases the risk of cardiovascular disease. Another example is hypertension (high blood pressure). Numerous North Americans are very sensitive to salt intake. When these individuals consume too much salt, their blood pressure climbs above the desirable range. The fact that more of these people with sensitivity are Black suggests that at least some cases of hypertension have a genetic component. Obesity also has genetic links. A variety of genes (likely 1000 or more) are involved in the regulation of body weight.

Although some individuals may be genetically predisposed to chronic disease, whether they actually develop the disease depends on lifestyle choices and environmental factors that influence the disease. It's important to realize that

▶ Epigenetics is the study of how environmental factors affect the way genetic potential is expressed.

predisposition to chronic disease is not the same type of genetic characteristic as being born with blue eyes or larger ears. With chronic disease, heredity is not necessarily destiny—individuals can exert some control over the expression of their genetic potential. For instance, those with a predisposition to premature heart disease can take steps to delay its onset by eating a nutritious diet, getting regular exercise, keeping weight under control, and getting medical treatment to lower blood cholesterol levels and control blood pressure. Likewise, those who did not inherit the potential for heart disease put themselves at risk of this disease by gaining excess body fat, smoking, abusing alcohol, and not getting medical treatment to keep blood cholesterol, blood pressure, and type 2 diabetes under control.

### Your Genetic Profile

By recognizing your potential for developing a particular disease, you can avoid behaviors that contribute to it. For example, women with a family history of breast cancer should avoid gaining excess body fat, minimize alcohol

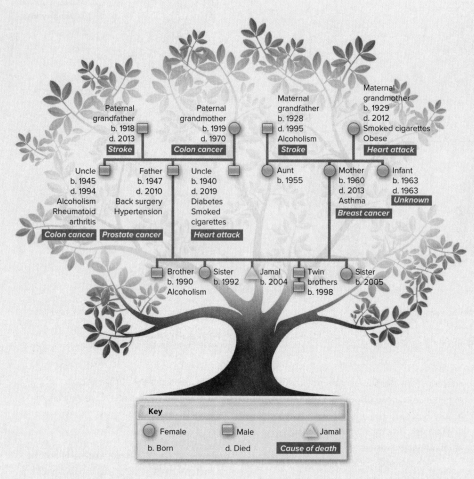

**Key**

○ Female    ▢ Male    △ Jamal

b. Born    d. Died    *Cause of death*

*Figure 1-10*  Example of a family tree for Jamal. If a person is deceased, the cause of death is shown. In addition to causes of death, medical conditions family members experienced are noted.

use, and get mammograms regularly. In general, the more relatives with a genetically transmitted disease and the more closely they are related to you, the greater your risk. One way to assess your risk is to create a family tree of illnesses and deaths (a genogram) by compiling a few key facts on your primary biological relatives: siblings, parents, aunts, uncles, and grandparents (Fig. 1-10).

High-risk conditions include having more than 1 first-degree relative (i.e., biological parents, siblings, and offspring) with a specific disease, especially if the disease occurred before age 50 to 60 years.[24] In the family

in Figure 1-10, prostate cancer killed Jamal's father. Knowing this, his physician likely would recommend that he be tested for prostate cancer more frequently or starting at an earlier age than men without a family history of the disease. Because Jamal's mother died of breast cancer, his sisters' doctor may recommend they consider having their 1st regular mammograms at a younger age than typical, as well as adopting other preventive practices. Heart attack and stroke also are common in the family, so all the children should adopt a lifestyle that minimizes the risk of developing these conditions, such as moderating their intake of animal fat and

**virus**  One of the smallest known types of infectious agents, many of which cause disease in humans. They do not metabolize, grow, or move by themselves. They reproduce by the aid of a living cellular host. A virus is essentially a piece of genetic material surrounded by a coat of protein.

sodium. Colon cancer is evident in the family, which makes it important for them to have careful screening throughout life.

### Gene Therapy

Scientists are developing therapies to correct DNA that causes some genetic disorders. Currently, gene therapy research focuses on diseases that have no other cures. Gene therapy may involve replacing a mutated gene with a healthy copy of the gene, inactivating a mutated gene that does not function normally, or introducing a new gene to help the body fight disease. When replacing a mutated gene, scientists typically isolate normal DNA, package it into a molecular delivery vehicle (usually a disabled **virus**), and inject it into the cells affected by the disease—such as liver cells. Inside the cell, the normal genetic material begins functioning and restores the cells to normal.

Hundreds of research studies are under way and success with gene therapy is growing, yet many obstacles remain to be overcome before gene therapy can become a safe, effective treatment. Government regulatory agencies in the U.S. (especially the National Institutes of Health and Food and Drug Administration) and Europe are actively overseeing human gene therapy research. Approved gene therapies are increasing; for instance, several are available for certain blood diseases and cancers. This ongoing research could lead to gene-based treatments for other types of cancer, heart disease, cystic fibrosis, and other diseases.

▶ These websites can help you gather more information about genetic conditions and testing.

www.geneticalliance.org
www.kumc.edu/gec/support
www.genome.gov
ghr.nlm.nih.gov/primer/therapy/genetherapy
www.dnalc.org
www.ncgr.org

## *Critical* Thinking

At family gatherings, Wesley notices that his parents, uncles, aunts, and older siblings tend to be overweight. His father has had a heart attack, as has his aunt. Two of his uncles died before the age of 60 from diabetes. His grandfather died of prostate cancer. Wesley wonders if he is destined to have obesity and develop heart disease, cancer, or diabetes. What advice would you give Wesley?

### Genetic Testing

Genetic tests analyze a person's genes to determine the likelihood of developing certain diseases. These tests are especially valuable for families afflicted by certain illnesses. In addition, they can help people who are healthy now predict the illnesses they will probably develop. Advance knowledge that a disease is likely to develop may provide opportunities to replace genes that encourage diseases, such as cancer and Alzheimer disease, with those that do not. Advance knowledge also could help couples wanting to have children make more informed choices (e.g., consider alternatives, such as adoption) and help health-care providers develop health and nutrition care plans that delay the onset of the disease. Additionally, this knowledge could help health-care providers diagnose diseases earlier and more accurately and prescribe individualized medical and nutrition therapies, instead of giving the same treatment to all patients with the same disease. It is likely that many medications may be more appropriate in people with certain genetic traits.

Some experts recommend that anyone considering genetic testing first have a genetic counselor analyze his or her family history, evaluate the risk of developing or passing along an inherited disease, and help determine whether testing is worth the time and effort. If you want to know if you are at risk of a specific genetic disease, it is a good idea to ask your physician about the possibility and likely usefulness of testing you. Some consumers opt to purchase at-home genetic test kits directly without involving a doctor or an insurance company. Typically, the test kit is mailed to the consumer, who collects the DNA sample (often by swabbing inside the cheek) and returns it to the lab. Results are provided by mail, fax, phone consultation, or web posting.

Genetic testing can help consumers take a more proactive role in protecting their health. However, given the limit on resources allocated to medical care in North America, it is not possible to identify all the people at genetic risk of the major chronic diseases and other health problems. In addition, in many cases, genetic susceptibility does not guarantee development of the disease. And, in almost all cases, there is no way to cure a specific gene alteration—only the resulting health problems can be treated. Researchers also are concerned that people who are found to have genetic alterations that increase disease risk may face job and medical insurance discrimination. Testing positive also could lead to unnecessary treatment. As well, a seemingly hopeless diagnosis could result in depression when a cure is out of reach.[25] Users of at-home genetic tests may face additional risks, such as receiving misleading results if unproven or invalid tests are used, making

Genetic analysis for disease susceptibility is becoming more common as the genes that increase the risk of developing various diseases are isolated and decoded.

Rob Melnychuk/Getty Images

unsafe health decisions if they do not receive guidance from a health-care professional, or finding that the testing company has not kept their genetic information confidential.

The wisdom of genetic testing is an open question. Perhaps preventive measures and careful scrutiny of the specific genetically linked diseases using one's family tree would suffice. Be aware that, throughout this book, discussions will point out how to avoid "controllable" risk factors that contribute to the development of genetically linked nutrition-related diseases present in your family tree.

### Knowledge Check

1. What is the role of genes?
2. What are 3 chronic nutrition-related diseases with a genetic link?
3. What is a genogram?

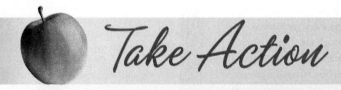

# Take Action

## Create Your Family Tree for Health-Related Concerns

Adapt this diagram to your own family tree. Under each heading, list year born, year died (if applicable), major diseases that developed during the person's lifetime, and cause of death (if applicable). Figure 1-10 provides an example.

Note that you are likely to be at risk of developing any diseases listed. Creating a plan for preventing such diseases when possible, especially those that developed in your family members before age 50 to 60 years, is advised. An online version of the family tree is available at www.hhs.gov/familyhistory. Speak with your health-care provider about any concerns arising from this activity.

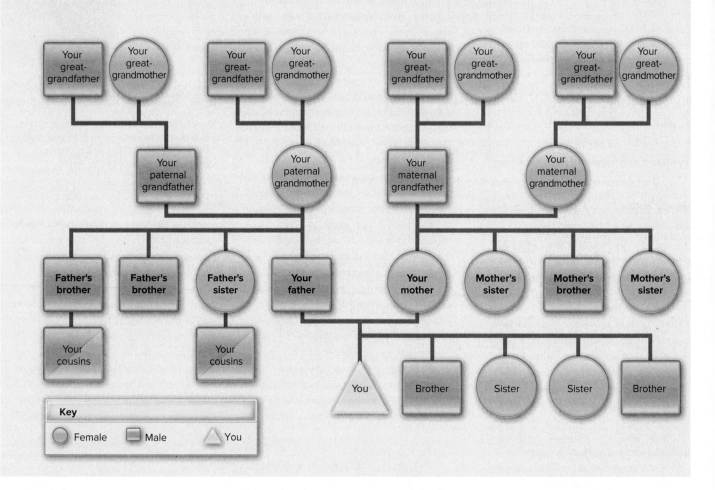

## 1.5  Using Scientific Research to Determine Nutrient Needs

*Perspective on the Future*

Nutritional genomics (nutrigenomics) examines how the foods you eat affect your genes and how your unique genetic profile affects the way your body uses nutrients, phytochemicals, and zoochemicals. In the future, nutrigenetic testing may help you get the most out of your diet and prevent chronic diet-related diseases.

How do we know what we know about nutrition? How has this knowledge been gained? In a word, research. Like other sciences, the research that sets the foundation for nutrition has developed through the use of the scientific method—a testing procedure designed to uncover facts and detect and eliminate error. The first step is the observation of a natural phenomenon. Scientists then suggest possible explanations, called hypotheses, about the causes of the phenomenon. Distinguishing a true cause-and-effect relationship from mere coincidence can be difficult. For instance, early in the 20th century, many people who suffered from the disease pellagra lived in orphanages, prisons, and mental hospitals, which suggested this disease was caused by germs that spread among people living close together. In time, however, it became clear that this connection was simply coincidental—the real cause of pellagra is a poor diet that contains too little of the B-vitamin niacin.

To test hypotheses and eliminate coincidental or erroneous explanations, scientists perform controlled experiments to gather data that either support or refute a hypothesis (Fig. 1-11). Very often, the results of an experiment lead to a new set of questions to be answered. If the results of many well-designed experiments provide valid evidence that supports a hypothesis, the hypothesis becomes generally accepted by scientists as a well-documented explanation for the phenomenon. As valid, extensive evidence about a related set of phenomena accumulates, a scientific theory or scientific law may be proposed. A scientific theory or law is a scientifically acceptable explanation of phenomena and how the phenomena are related to each other.

Sound scientific research requires the following:

1. Phenomena are observed.
2. Questions are asked and hypotheses are generated to explain the phenomena.
3. Research is conducted with **scientific integrity**.
4. Incorrect explanations are rejected and the most likely explanation is proposed.
5. Research results are scrutinized and evaluated by other scientists. Research conducted in an unbiased, scientific manner is published in a scientific journal.
6. The results are confirmed by other scientists and by more experiments and studies.

The scientific method requires an open, curious mind and a questioning, skeptical attitude. Scientists (as well as students) must not accept proposed hypotheses until they are supported by considerable evidence, and they must reject hypotheses that fail to pass critical analysis. An example of this need for skepticism involves stomach ulcers. For many years, it was generally accepted that stomach ulcers were caused mostly by a stressful lifestyle and a poor diet. Then, in 1983, Australian physicians Barry Marshall and Robin Warren reported in a respected medical journal that ulcers usually are caused by a common microorganism called *Helicobacter pylori* and can be cured with antibiotics. Initially, other physicians were skeptical about this finding and continued to prescribe medications, such as antacids, that reduce stomach acid. As more studies were published, however, and patients were cured of ulcers using antibiotics, the medical profession eventually accepted the findings. (Marshall and Warren were given the Nobel Prize for Medicine in 2005 for this discovery.) Scientific theories, laws, and discoveries always should be subjected to challenge and change.

## Making Observations and Generating Hypotheses

Historical observations have provided clues to important relationships in nutrition science. In the 15th and 16th centuries, for example, many European sailors on long voyages developed the often fatal disease scurvy. A British naval surgeon, James Lind, observed that the diet eaten while at sea differed from usual diets. Specifically, few fruits and vegetables were available onboard ships. He hypothesized that a missing dietary component caused scurvy. He set up an experiment in which he supplied sailors with a ration of salt water, vinegar, cider, citrus juice, or other liquid. The results of this experiment indicated that citrus (lemons, limes) prevents and cures scurvy. After this, British sailors were given a ration of lime juice, earning them

**scientific integrity** Integrity demonstrated by adhering to professional values and practices when conducting, interpreting, reporting, and using scientific research. These values and practices help ensure that research activities are objective, clear, reproducible, and useful. These values and practices also help prevent bias, fabrication or falsification of data, plagiarism, inappropriate interference by others who might want to influence the way the research is conducted or reported, censorship of scientific findings, and inadequate research activity security.

Careful research contributes to scientifically valid nutrition knowledge.

JGI/Daniel Grill/Blend Images/Getty Images

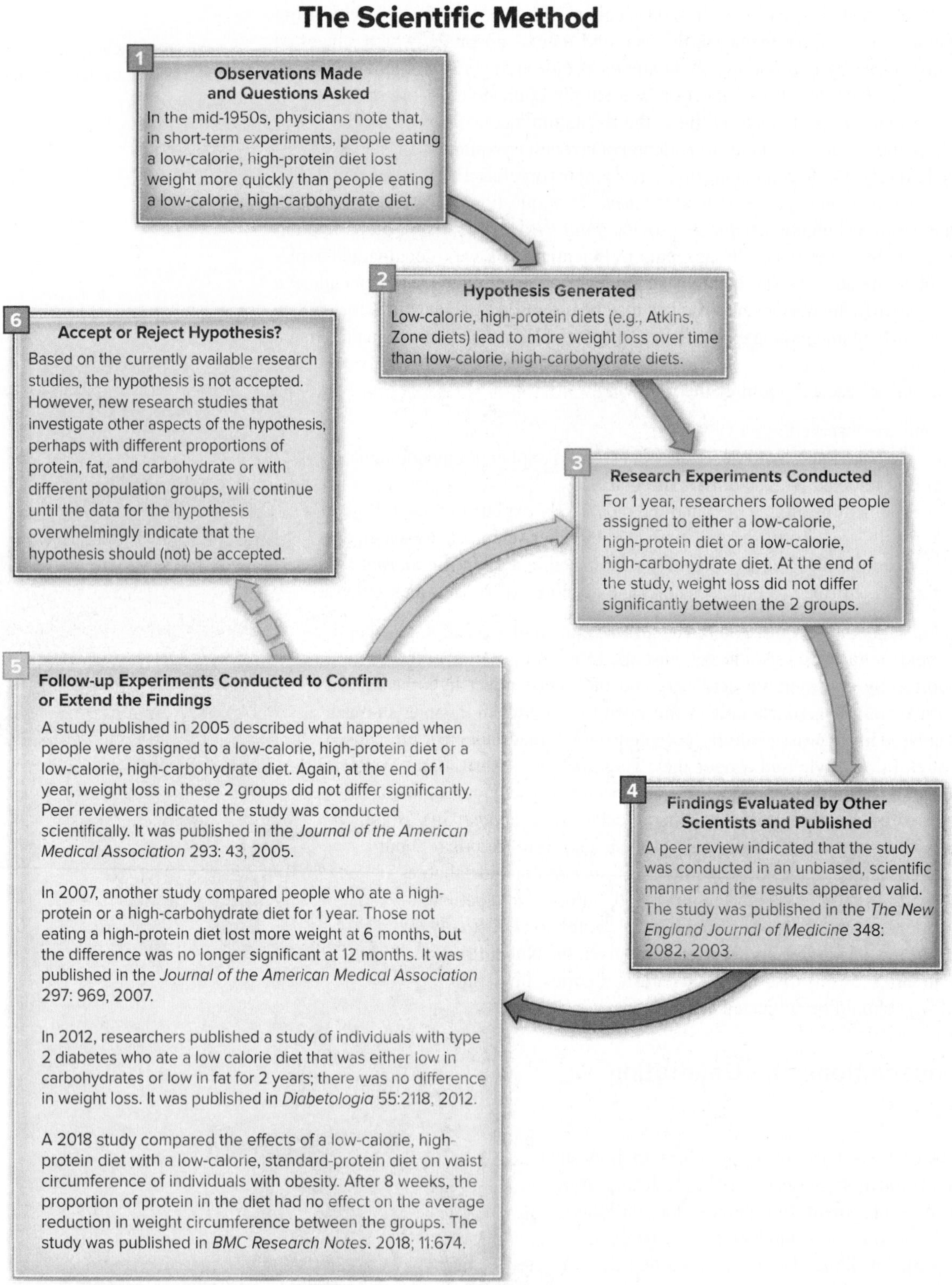

## The Scientific Method

**1　Observations Made and Questions Asked**

In the mid-1950s, physicians note that, in short-term experiments, people eating a low-calorie, high-protein diet lost weight more quickly than people eating a low-calorie, high-carbohydrate diet.

**2　Hypothesis Generated**

Low-calorie, high-protein diets (e.g., Atkins, Zone diets) lead to more weight loss over time than low-calorie, high-carbohydrate diets.

**3　Research Experiments Conducted**

For 1 year, researchers followed people assigned to either a low-calorie, high-protein diet or a low-calorie, high-carbohydrate diet. At the end of the study, weight loss did not differ significantly between the 2 groups.

**4　Findings Evaluated by Other Scientists and Published**

A peer review indicated that the study was conducted in an unbiased, scientific manner and the results appeared valid. The study was published in the *The New England Journal of Medicine* 348: 2082, 2003.

**5　Follow-up Experiments Conducted to Confirm or Extend the Findings**

A study published in 2005 described what happened when people were assigned to a low-calorie, high-protein diet or a low-calorie, high-carbohydrate diet. Again, at the end of 1 year, weight loss in these 2 groups did not differ significantly. Peer reviewers indicated the study was conducted scientifically. It was published in the *Journal of the American Medical Association* 293: 43, 2005.

In 2007, another study compared people who ate a high-protein or a high-carbohydrate diet for 1 year. Those not eating a high-protein diet lost more weight at 6 months, but the difference was no longer significant at 12 months. It was published in the *Journal of the American Medical Association* 297: 969, 2007.

In 2012, researchers published a study of individuals with type 2 diabetes who ate a low calorie diet that was either low in carbohydrates or low in fat for 2 years; there was no difference in weight loss. It was published in *Diabetologia* 55:2118, 2012.

A 2018 study compared the effects of a low-calorie, high-protein diet with a low-calorie, standard-protein diet on waist circumference of individuals with obesity. After 8 weeks, the proportion of protein in the diet had no effect on the average reduction in weight circumference between the groups. The study was published in *BMC Research Notes*. 2018; 11:674.

**6　Accept or Reject Hypothesis?**

Based on the currently available research studies, the hypothesis is not accepted. However, new research studies that investigate other aspects of the hypothesis, perhaps with different proportions of protein, fat, and carbohydrate or with different population groups, will continue until the data for the hypothesis overwhelmingly indicate that the hypothesis should (not) be accepted.

*Figure 1-11* This example shows how the scientific method was used to test a hypothesis about the effects of low-calorie, high-protein diets on weight loss. Scientists consistently follow these steps when testing all types of hypotheses. Scientists do not accept a nutrition or another scientific hypothesis until it has been thoroughly tested using the scientific method.

the nickname "limeys." About 200 years later, science had advanced to the point that researchers were able to identify vitamin C as the component in citrus juice that prevents and cures scurvy.

Observations of differences in dietary and disease patterns among various populations also have suggested important relationships in nutrition science. If a group tends to develop a certain disease but another group does not, scientists can speculate about how diet causes this difference. The study of diseases in populations, called epidemiology, ultimately forms the basis for many laboratory studies. Examples of the use of epidemiology are the links noticed between dietary components and health conditions such as scurvy, pellagra, and heart disease.

## Laboratory Animal Experiments

Human experiments are the most convincing to scientists; however, when scientists cannot test their hypotheses in experiments with humans, they often use laboratory animals. In fact, much of what is known about human nutritional needs and functions has been generated from laboratory animal experiments.

Experiments may be conducted with laboratory animals when the study would be unethical to conduct with humans. Although some people argue that laboratory animal experiments also are unethical, most believe that the careful, humane use of animals is an acceptable alternative. For example, most people think it is reasonable to feed rats a low-copper diet to study the importance of this mineral in the formation of blood vessels. Almost everyone, however, would object to a similar study in infants.

The use of laboratory animal experiments to study the role of nutrition in human diseases depends on the availability of an **animal model**—a disease in laboratory animals that closely mimics a human disease. For example, in the early 1900s, scientists showed that thiamin (a B-vitamin) cures a beriberi-like disease in chickens. As a result, chickens could be used to study this vitamin deficiency disease. If no animal model is available and human experiments are ruled out, scientific knowledge often cannot advance beyond what can be learned from epidemiological studies. Most human chronic diseases do not occur in laboratory animals.

Research using laboratory animals contributes to knowledge of nutrition.

G.K. & Vikki Hart/Getty Images

## Human Experiments

Before any research study can be conducted with humans (or laboratory animals), researchers must obtain approval from the research review board at their university, hospital, or company. The review board approves only studies that have a valid experimental protocol, are expected to produce important knowledge, and treat study participants fairly and ethically. The review board also assesses the potential treatment's risks and benefits to the study participants. In human studies, the review board also requires researchers to inform the participants of the study's purpose, procedures, known risks, and benefits so that they can make informed decisions about whether to participate. This process, called informed consent, is the voluntary, documented confirmation of the participants' agreement to participate in the study.

A variety of experimental approaches are used to test research hypotheses in humans. Migrant studies, for example, look at changes in the health of people who move from one country to another. Cohort studies start with a healthy population and follow it, looking for the development of disease. Other experimental approaches are case-control and double-blind studies.

### Case-Control Study

In a case-control study, scientists compare individuals who have the condition in question ("cases"), such as lung cancer, with individuals who do not have the condition ("controls"). The strongest case-control studies compare groups that are matched for other major characteristics (e.g., age, race, and biological sex) not being studied. You can think of a case-control study as a "mini" epidemiological study. This type of study may identify factors

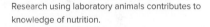

**animal model** Laboratory animal useful in medical research because it can develop a health condition (e.g., disease or disorder) that closely mimics a human disease and, thus, can be utilized to learn more about causes of a condition and its diagnosis in humans, as well as assess the usefulness and safety of new treatments or preventive actions.

other than the disease being studied, such as fruit and vegetable intake, that differ between the groups, thereby providing researchers with clues about the cause, progression, and prevention of the disease. However, without a controlled experiment, researchers cannot definitely claim cause and effect.

### Blinded Study

An important approach for more definitive testing of hypotheses is a controlled experiment using a blinded protocol. In a single-blind study, one group of participants—the experimental group—follows a specific protocol (e.g., receives a treatment, such as consuming a certain food or nutrient) while the participants in a corresponding control group follow their usual habits. The control group also usually receives a **placebo** (fake treatment). The placebo camouflages (blinds) who is in the experimental group and who is in the control group. Study participants are assigned randomly to the control or experimental group, such as by the flip of a coin. Scientists then observe both groups over time to identify any changes that occur in the experimental and the control groups. Sometimes individuals are used as their own control. First, they are observed for a period of time while they follow their usual habits. Then, they follow the experimental protocol and their responses are observed.

Double-blind studies have two features that help reduce the risk of bias (prejudice), which can easily affect the outcome of an experiment. First, during the course of the experiment, neither the study participants nor the researchers know who is getting the real treatment (experimental group) and who is getting the placebo (control group). An independent 3rd party holds the key to the study group assignment and the data until the study is completed. Second, the expected effects of the experimental protocol are not disclosed to the participants or researchers collecting the data until after the entire study is completed. These features reduce the possibility that the researchers will misperceive or overemphasize changes they hoped to see in the participants to prove a certain hypothesis they believe is true or will unconsciously ignore or minimize the importance of changes that disprove their hypothesis. These features also reduce the chance that study participants begin to feel better simply because they are involved in a research study or are receiving a new treatment, a phenomenon called the **placebo effect**. Double-blind studies improve the chances that any differences observed between the experimental and control groups really are due to the experimental treatment.

Sometimes only a single-blind study protocol is possible—in this case, only the study participants are kept uninformed about which participants are assigned to the experimental and control groups. Conducting blinded studies in nutrition is very challenging because it is difficult to create placebo foods and menus.

A recent example illustrates the need to test hypotheses based on epidemiological observations in double-blind studies.[25] Epidemiologists using primarily case-control studies found that smokers who regularly consumed fruits and vegetables had a lower risk of lung cancer than smokers who ate few of these foods. Some scientists proposed that beta-carotene (a yellow-orange pigment that is a precursor to vitamin A), present in many fruits and vegetables, reduced the lung damage that is caused by tobacco smoke and leads to lung cancer. However, in double-blind studies involving heavy smokers, the risk of lung cancer was found to be higher in those who took beta-carotene supplements than in those who did not. Some researchers criticized these studies, arguing that the beta-carotene was given too late in the smokers' lives to prevent lung cancer, but even these critics did not suspect that the supplement would increase cancer risk. Soon after these results were reported, the U.S. federal agency supporting other large studies of beta-carotene supplements halted the research, stating that these supplements are ineffective in preventing lung cancer. Although it appears that beta-carotene does not protect against lung cancer, other components in fruits and vegetables might offer protection and may become the focus of future studies.

## HISTORICAL PERSPECTIVE
### War on Pellagra

Early in the 20th century, pellagra killed millions in the U.S. Dr. Joseph Goldberger hypothesized that it was caused by poor diets, not infectious agents, and his experiments included feeding healthy people poor diets—they developed pellagra. He fed pellagra patients good diets—they improved. He and his assistant also experimented on themselves. They injected themselves with pellagra patients' blood and took capsules filled with their scabs—they didn't "catch" pellagra. Years after his death, others determined that the specific cause of pellagra is a deficiency of niacin (a B-vitamin). Learn more at history.nih.gov/exhibits/Goldberger/.

©Stockbyte/Punchstock

**placebo** Fake treatment (e.g., a sham medicine, supplement, or procedure) that seems like the experimental treatment; used to disguise whether a study participant is in the experimental or control group.

**placebo effect** *Placebo* is derived from the Latin word that means "I shall please." The placebo effect occurs when control group participants experience changes that cannot be explained by the action of the placebo they received. These changes may be linked to a treatment that is working, or a desire to help researchers achieve their goals. Overall, it is critical for researchers to take the placebo effect into consideration when interpreting research results.

Overall, health and nutrition advice provided by family, friends, and other well-meaning individuals cannot be accepted as valid until studied in a rigorous, scientific manner. Until that is done, it isn't possible to know whether a substance or procedure is truly effective. When people say, "I get fewer colds now that I take vitamin C," they overlook the facts that many cold symptoms disappear quickly with no treatment, that they want the supplement to work, and that the apparent curative effect of vitamin C or any other remedy is often coincidental rather than causal. Failure to understand the scientific method, as well as the accepted standards of evidence and current limitations of science, leads many people to believe erroneous information about health, nutrition, and disease. Using remedies that are not supported by credible scientific evidence can damage health and delay treatment that could preserve health.

## Peer Review of Experimental Results

Once an experiment is complete, scientists summarize the findings and seek to publish the results in scientific journals. It can seem challenging when you first start reading scientific journals. Table 1-7 lists what you can expect to find in research articles published in these journals.

Generally, before such research results are published in scientific journals, they are critically reviewed by other scientists familiar with the subject. A key objective of this peer review is to ensure that only the most unbiased, objective findings from carefully designed and executed research studies are published.

Peer review is an important step because most scientific research is funded by the government, nonprofit foundations, drug companies, and other private industries—all of which have strong expectations about research outcomes. It is important that the scientists conducting research studies will be fair in evaluating their results and will not be influenced by the funding agency. Peer review helps ensure that the researchers are as objective as possible. This then helps ensure that the results published in **peer-reviewed journals**, such as the *American Journal of Clinical Nutrition, New England Journal of Medicine,* and *Journal of the Academy of Nutrition and Dietetics,* are much more reliable than those found in popular magazines or promoted on television talk shows or websites.

Press releases from reputable journals and major universities are the main sources for the information presented in the popular media. Unfortunately, these press releases often oversimplify study findings, which may get misinterpreted or overextended in the popular press. Thus, when you hear or see a news report that cites a journal, it is best to review the journal article so that you can judge for yourself whether the research findings are valid.

## Follow-Up Studies

Even when a study has followed a well-designed protocol and the results have been published in a peer-reviewed journal, a single experiment is never enough to accept a particular hypothesis or provide a basis for nutrition recommendations. Rather, the results obtained in one laboratory must be confirmed by experiments conducted in other laboratories and possibly under varying circumstances. Only then can we really trust and use the results. The more lines of evidence available to support a hypothesis or an idea, the more likely it is to be true (Fig. 1-12). It is important to avoid rushing to accept new ideas as fact or incorporating them into your health habits until they are proved by several lines of evidence.

---

*Critical* **Thinking**

For thousands of years, early humans consumed a diet rich in vegetable products and low in animal products. These diets were generally lower in fat and higher in dietary fiber than modern diets. Do the differences in human diets throughout history necessarily tell us which diet is better—that of early humans or of modern humans? If not, what is a more reliable way to pursue this question of potential diet superiority?

► For a list of more nutrition- and health-related peer-reviewed journals, see Appendix L.

► These are examples of websites that provide reliable health and nutrition information.

www.nutrition.gov

www.eatright.org

www.webmd.com

**peer-reviewed journal** Journal that publishes research only after researchers who were not part of the study agree that the study was carefully designed and executed and the results are presented in an unbiased, objective manner. Thus, the research has been approved by peers of the research team.

---

**Table 1-7  Navigating a Research Article**

**Abstract:** summarizes the study

**Introduction:** provides brief background and rationale for the study; states the aims of the study

**Methods:** describes who participated in the study (also called the sample), how the study was conducted (also called study design), and how study variables were measured, collected, and analyzed

**Results:** explains what the study found; often includes tables and figures

**Discussion:** compares the study's results to previous studies; describes the limitations and strengths of the study; proposes possible conclusions and directions for future research

**References:** studies used to write the study background, establish the rationale, create the methods, and discuss the findings

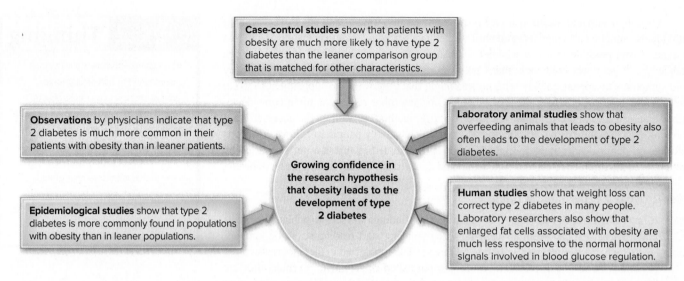

*Figure 1-12* Data from a variety of sources can come together to support a research hypothesis. This diagram shows how various types of research data support the hypothesis that obesity leads to the development of type 2 diabetes (see Chapter 5).

**systematic review** Critical evaluation and synthesis of research studies focusing on a specific topic or research question.

**Evidence Analysis Library** Source of systematic reviews conducted by the Academy of Nutrition and Dietetics to guide clinical decisions.

**USDA Nutrition Evidence Library** Source of systematic reviews to guide U.S. nutrition programs and policies.

**e-Library of Evidence for Nutrition Actions (eLENA)** Source of systematic reviews to help countries implement effective nutrition interventions, policies, and programs.

**Cochrane Collaboration** Source of systematic reviews to inform health-care decisions.

 Review nutrition-related systematic reviews at these sites.

Evidence Analysis Library (EAL) andeal.com

USDA Nutrition Evidence Library www.fns .usda.gov/nutrition-evidence-library-about

Library of Evidence for Nutrition Actions (eLENA) www.who.int/elena

Cochrane Collaboration www.cochrane.org

## Systematic Reviews

With many new studies being published every day, it is challenging to stay up-to-date with the latest research findings. Comparing the results of different studies and deciding which studies present the most useful data can be time consuming and difficult. However, this time and effort investment is critical if health professionals are to make well-informed clinical decisions based on current research. Fortunately, several organizations conduct and publish systematic reviews.

To create a **systematic review**, trained experts thoroughly search the literature for research focusing on a particular topic or research question, summarize the major findings of the studies located, rigorously grade the quality of the studies, compare and contrast the evidence across the studies, and write conclusions based on the strongest, most relevant research. For example, a systematic review by the World Health Organization (WHO) concluded that, to control blood pressure in children, potassium should be increased and sodium decreased in their diets. Systematic reviews that also use statistical analysis are called **meta-analyses**. Dietitians and other health-care providers can use systematic reviews and meta-analyses to determine which clinical methods are likely to help patients the most. Nutrition and health researchers use these studies to develop evidence-based policies and programs. Systematic reviews and meta-analyses are considered the best quality evidence about a research question because, by combining the findings of many studies, they reduce the risk of bias and generate more reliable results.

Systematic reviews that are especially useful to nutrition professionals include the **Evidence Analysis Library** (EAL) published by the Academy of Nutrition and Dietetics. The EAL's purpose is to guide clinical nutrition-related decisions to improve patient outcomes. The **USDA Nutrition Evidence Library** provides the evidence needed to develop and implement the most effective nutrition programs and policy, especially the Dietary Guidelines for Americans (see Chapter 2). At the international level, the WHO's **e-Library of Evidence for Nutrition Actions (eLENA)** provides evidence-informed guidance to help countries create and implement effective nutrition interventions, policies, and programs. The **Cochrane Collaboration** is an international network that conducts systematic reviews designed to help health practitioners, policymakers, and patients and their caregivers make more informed health-care decisions.

### Knowledge Check

1. What elements are required for scientific research to be considered valid?
2. What is the difference between single- and double-blind studies?
3. What is an animal model?
4. What is a peer-reviewed journal?
5. How are systematic reviews used?

# 1.6 Evaluating Nutrition Claims and Products

Nutrition claims often appear in media stories and in advertisements. Sometimes it is difficult to discern whether the claims are true. The following suggestions can help you make healthful and logical nutrition decisions.

1. Apply the basic principles of nutrition, as outlined in this book, to any nutrition claim. Do you note any inconsistencies?
2. Be wary if the answer is yes to any of the following questions about a health-related nutrition claim.
   - Are only advantages discussed and possible disadvantages ignored?
   - Is this a new or "secret" scientific breakthrough?
   - Are claims made about "curing" disease?
   - Do the claims sound too good to be true?
   - Is extreme bias against the medical community or traditional medical treatments evident? Physicians, nurses, dietitians, and other health-care providers as a group strive to cure diseases in their patients using proven techniques—they do not ignore reliable treatments.
3. Examine the scientific credentials of the individual, organization, or publication making the nutrition claim. Usually, a reputable author's educational background or present employer is affiliated with a nationally recognized university, research institute, or medical center that offers programs or courses in nutrition, medicine, or closely related fields.
4. If research is cited to support a claim, note the size and duration of any study. The larger the study and the longer it went on, the more dependable its findings. Also consider the type of study: epidemiology versus case-control versus blinded. Keep in mind that "contributes to," "is linked to," or "is associated with" does not mean "causes." Check to see if there are systematic reviews that evaluate the claim. Do reliable, peer-reviewed journal articles support the claims? Beware of testimonials about personal experience, disreputable publication sources, dramatic results (rarely true), and lack of evidence from supporting studies conducted by other scientists.
5. Be wary of press conferences and other hype regarding the latest findings. Much of this will not survive rigorous scientific evaluation.

## Buying Nutrition-Related Products

Popular nutrition-related products claim to increase muscle growth, enhance sexuality, boost energy, reduce body fat, increase strength, supply missing nutrients, increase longevity, and even improve brain function. Although some of us are willing to try these products and believe they can cause the miraculous effects advertised, few of these products have been thoroughly evaluated by reputable scientists. They may not be effective, and the amount and potency of the product may not match what is on the package.

A cautious approach to nutrition-related products is important because of sweeping changes in U.S. law in 1994. The Dietary Supplement Health and Education Act (DSHEA) of 1994 classified vitamins, minerals, amino acids, and herbal remedies as "foods," effectively restraining the U.S. Food and Drug Administration (FDA) from regulating them as rigorously as food additives and drugs. According to this act, rather than the manufacturer having to prove a dietary supplement is safe, the FDA must prove it is unsafe before preventing its sale. In contrast, the safety of food additives and drugs must be rigorously demonstrated before the FDA allows them to be sold (see Chapter 3). However, in recent years the FDA has increased efforts to prevent unsafe ingredients in dietary supplements and is working to improve safety by requiring supplement companies to notify the FDA before a new supplement ingredient is sold.[26]

Currently, a product labeled as a dietary supplement (or an herbal product) can be marketed in the U.S. without FDA approval if there is a history of its use or other evidence that it is reasonably safe when used under the conditions indicated in its labeling. (Note that the FDA can act if the product turns out to be dangerous, as with the ban on the sale of the supplement ephedra and DMAA after numerous reports of associated illnesses and deaths.) It is important to note that the "evidence" used to support a claim often is vague or unsubstantiated. Given its budget and regulatory constraints, the FDA is able to challenge only a few of these claims. However, some

▶ **10 Red Flags That Signal Poor Nutrition Advice**

1. Promise of a quick fix
2. Dire warnings of dangers from a single product or regimen
3. Claims that sound too good to be true
4. Simplistic conclusions drawn from a complex study
5. Recommendations based on a single study
6. Dramatic statements that are refuted by reputable scientific organizations or systematic reviews
7. Lists of "good" and "bad" foods
8. Recommendations made to help sell a product
9. Recommendations based on studies published without peer review
10. Recommendations from studies that ignore differences among individuals or groups

▶ To check for supplement recalls, visit www.fda.gov/safety/recalls-market-withdrawals-safety-alerts.

▶ These websites can help you evaluate ongoing nutrition and health claims.

www.acsh.org
www.quackwatch.com
ods.od.nih.gov
www.fda.gov
www.eatright.org

Appendix L also lists many reputable sources of nutrition advice for your use.

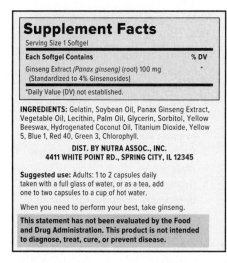

*Figure 1-13* The FDA requires the orange-highlighted disclaimer to appear on supplement labels.

other means for challenging these claims are emerging. For example, the Federal Trade Commission (FTC), which is responsible for ensuring that advertising is not deceptive, may investigate dubious claims made in advertisements. In addition, the supplement industry itself is trying to develop self-policing procedures.

To protect your health, it is important to scrutinize nutrition-related product labels carefully and check with reputable sources to be certain there is scientific proof that the product is likely to perform as described on the label. Be especially skeptical of using the product for purposes not stated on the label—a product is unlikely to perform a function that is not specifically stated on its label or package insert (legally part of the label). The labels on dietary supplements and herbal products are allowed to claim that general well-being results from consumption of the ingredients, to explain how the product provides a benefit related to a classic nutrient deficiency disease, and to describe how a nutrient affects human body structure or function (called structure/function claims). Structure/function claim examples include "maintains bone health" and "improves blood circulation." The labels of products bearing such claims also must prominently display a disclaimer regarding a lack of FDA review (Fig. 1-13). Despite this warning, many consumers mistakenly assume that the FDA has carefully evaluated the products. See Chapter 18 for more details on dietary supplements.

## Knowledge Check

1. What are 5 tips for determining whether nutrition claims are true?
2. Why does DSHEA make it wise to be cautious about dietary supplements?

## CASE STUDY FOLLOW-UP

Allen should be cautious about taking any supplement, especially if advertised as a "recent breakthrough." There likely have not been enough studies of the supplement to confirm its effectiveness. Also, as you have read, dietary supplements are not closely regulated by the FDA; a general phrase, such as "increases energy," is considered a structure/function claim, and such product labeling does not require prior approval by the FDA. Furthermore, the FDA

will not have evaluated either the safety or the effectiveness of such a product. Even harmful dietary supplements are difficult for the FDA to recall. There also is a chance that the supplement contains little or none of the advertised ingredient. Unfortunately, Allen will find all this out the hard way and will be out $60. His hard-earned money would be better spent on a nutritious diet and a medical checkup at the student health center. All consumers need to be cautious about nutrition information, especially regarding dietary supplements marketed as cure-alls and breakthroughs— let the buyer beware!

# Chapter Summary

**1.1** *Nutrition* **is defined as the "science of food;** the nutrients and the substances therein; their action, interaction, and balance in relation to health and disease; and the process by which the organism (e.g., human body) ingests, digests, absorbs, transports, utilizes, and excretes food substances." Nutrients are substances essential for health that the body cannot make or makes in quantities too small to support health. Nutrients primarily provide energy, support growth and development, and/or keep body functions running smoothly. Carbohydrates, proteins, lipids, and water are macronutrients. Vitamins and minerals are micronutrients. Phytochemicals are plant components, and zoochemicals are components in animals that may provide significant health benefits.

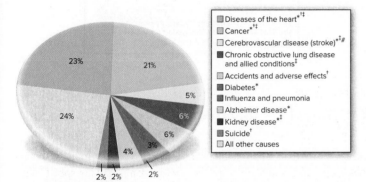

- ☐ Diseases of the heart*[†‡]
- ☐ Cancer*[†‡]
- ☐ Cerebrovascular disease (stroke)*[†‡#]
- ■ Chronic obstructive lung disease and allied conditions[‡]
- ☐ Accidents and adverse effects[†]
- ■ Diabetes*
- ■ Influenza and pneumonia
- ☐ Alzheimer disease*
- ■ Kidney disease*[‡]
- ■ Suicide[†]
- ☐ All other causes

23%  21%  5%  6%  6%  3%  4%  2%  2%  2%  24%

*Causes of death in which diet plays a part
[†]Causes of death in which excessive alcohol consumption plays a part
[‡]Causes of death in which tobacco use plays a part
[#]Diseases of the heart and cerebrovascular disease are included in the more global term *cardiovascular disease*.

Source: From Centers for Disease Control and Prevention, National vital Statistics Report, Canadian Statistics are quite similar.

**Simple Carbohydrates**

**Protein**

**Lipid**

**Complex Carbohydrates**

Photodisc/PunchStock; Comstock Images/Getty Images; Shutterstock/Evgenia Sh.; Getty Images

**1.2** **Humans obtain the energy** needed to perform body functions and do work from carbohydrates, fats, and proteins. Alcohol also provides energy but is not a nutrient. A kilocalorie is the amount of heat energy it takes to raise the temperature of 1000 g (1 liter) of water 1°C. The physiological fuel values are 4, 9, 4, and 7 for carbohydrate, fat, protein, and alcohol, respectively.

**1.3** **North American adults consume,** on average, 16% of their energy intake as proteins, 50% as carbohydrates, and 33% as fats. Animal sources, such as meat, seafood, dairy products, and eggs, are the main protein sources for North Americans. About half the carbohydrates in North American diets come from simple carbohydrates; the other half come from starches. Many North Americans are consuming more saturated fat, cholesterol, and sodium and less vitamin A, vitamin E, iron, potassium, and calcium than recommended. Daily food intake satisfies hunger (physical need for food) and social and emotional needs. Appetite and food choice depend on many factors.

romastudio © 123RF.com

**1.4** **Nutritional health is determined by the sum of the status of each nutrient.** Optimal, or desirable, nutritional status for a nutrient is the state in which body tissues have enough of the nutrient to support normal functions and to build and maintain surplus stores. Undernutrition occurs when nutrient intake does not meet needs, causing surplus stores to be used. The consumption of more nutrients than needed leads to overnutrition. *Healthy People* sets national health promotion and disease-prevention goals, many of which promote desirable nutrition that supports healthful lifestyles and reduces preventable death and disability. A nutritional assessment considers background factors, as well as anthropometric, biochemical, clinical, dietary, and environmental assessments.

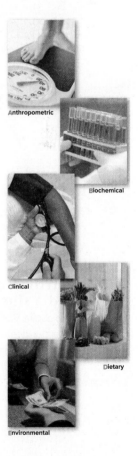

Anthropometric

Biochemical

Clinical

Dietary

Environmental

Photodisc/Getty Images; © Adam Gault / age fotostock; Comstock Images/Jupiterimages; Janis Christie/Getty Images; Ryan McVay/Getty Images RF

# 1.5 Research that creates the foundation for nutrition

has developed through the use of the scientific method. To test hypotheses and eliminate coincidental or erroneous hypotheses, scientists perform controlled experiments. The scientific method requires an open, curious mind and a questioning, skeptical attitude. Scientists must not accept hypotheses until they are supported by considerable research evidence. Scientific theories, laws, and discoveries always should be subjected to challenge and change. Experimental approaches used to test research hypotheses in humans include migrant, cohort, case-control, and blinded studies. Once an experiment is complete, scientists summarize the findings and seek to publish the results in a scientific, peer-reviewed journal. The objective of peer review is to ensure that only the most unbiased, objective findings from carefully designed and executed research studies are published. Systematic reviews are critical evaluations and syntheses of research studies focusing on a specific topic or research question, which are used to guide decisions about health care, nutrition programs, and nutrition policies.

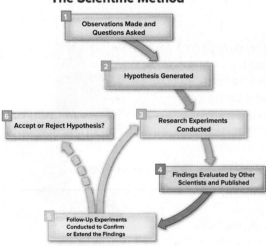

**The Scientific Method**

# 1.6 Nutrition claims often appear in media stories and in advertisements.

It can be difficult to discern whether claims are true. A cautious approach to nutrition-related claims and products is important. The Dietary Supplement Health and Education Act classified vitamins, minerals, amino acids, and herbal remedies as "foods," effectively restraining the FDA from regulating them as rigorously as food additives and drugs. When selecting nutrition-related products, carefully scrutinize product labels. For those who feel they need to improve their diets and health, a safe approach is to consult a physician or registered dietitian before purchasing dietary supplements.

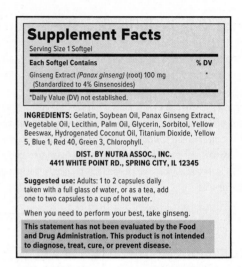

## Clinical Perspective

Genetic endowment affects almost every medical condition. Genes direct the growth, development, and maintenance of cells and, ultimately, of the entire organism. Most chronic nutrition-related diseases are influenced by genetic, nutritional, and lifestyle factors. Although some individuals may be genetically predisposed to develop chronic disease, the actual development of the disease depends on lifestyle and environmental factors. Scientists are developing therapies to correct some genetic disorders. Experts recommend that anyone considering genetic testing first undergo genetic counseling.

## Global Perspective

The cost of food as a percentage of income varies widely around the world. In the U.S., for instance, food consumed at home by the average household is less than 6% of total income, whereas families in Nigeria spend nearly two-thirds of their income on food eaten at home.

As the percentage of family income spent on food rises, the typical family has less to devote to other components that contribute to overall health and quality of life. The cost of food can affect both food choices and health.

# Study Questions

1. Which nutrient does not provide energy?

   a. carbohydrates
   b. vitamins
   c. protein
   d. lipids

2. Which element is found in protein but is not found in carbohydrates and lipids?

   a. nitrogen
   b. carbon
   c. hydrogen
   d. oxygen

3. How many calories per gram are supplied by fats and oils?

   a. 9
   b. 4
   c. 5
   d. 7

4. A kilocalorie is the amount of heat energy it takes to raise the temperature of 1 gram of water 1 degree Celsius (1°C).

   a. true
   b. false

5. How many calories are in a food that contains 12 g carbohydrate, 3 g alcohol, 0 g fat, and 3 g protein?

   a. 112
   b. 81
   c. 75
   d. 124

6. Animal sources, such as meat, seafood, dairy products, and eggs, supply about two-thirds of the protein intake for most people around the world.

   a. true
   b. false

7. Foods are selected based mainly on _____.

   a. hunger
   b. appetite
   c. custom
   d. cost

8. Anthropometric assessment involves measuring _____.

   a. blood pressure
   b. enzyme activity
   c. skinfold thickness
   d. all of the above

9. Biochemical assessment involves measuring _____.

   a. concentrations of nutrients in the blood
   b. nutrient content in the diet
   c. body circumferences
   d. education and economic levels

10. Genes direct the growth, development, and maintenance of cells.

    a. true
    b. false

11. Which disease has a genetic link?

    a. diabetes
    b. osteoporosis
    c. cancer
    d. all of the above

12. The risk of developing a specific genetic disease is high if you have 2 or more 1st-degree relatives with the disease, especially if the disease occurred before age 50 to 60 years.

    a. true
    b. false

13. Which type of study examines disease in populations?

    a. epidemiological study
    b. double-blind study
    c. single-blind study
    d. animal modeling

14. Which type of study follows a healthy population, looking for the development of diseases?

    a. cohort study
    b. case-control study
    c. migrant study
    d. double-blind study

15. Which statement is false about the Dietary Supplement Health and Education Act (DSHEA)?

    a. DSHEA classified vitamins as foods.
    b. DSHEA causes the FDA to regulate herbal supplements as rigorously as it does drugs.
    c. DSHEA effectively restrains the FDA from regulating mineral supplements.
    d. DSHEA requires the FDA to prove a dietary supplement is unsafe before preventing its sale.

16. Compare and contrast the characteristics of each macronutrient and micronutrient.

**17.** Define the term *functional food,* and give 3 examples.

**18.** Explain the concept of energy as it relates to foods. What are the physiological fuel (energy) values used for a gram of carbohydrate, fat, protein, and alcohol?

**19.** According to national nutrition surveys, which nutrients tend to be underconsumed by many adult North Americans? Why is this the case?

**20.** List 4 health objectives for the U.S. for the year 2020. How would you rate yourself in each area? Why?

**21.** What nutrition-related disease is common in your family? What step or steps could you take to minimize your risk?

**22.** Describe the types of studies that are conducted to test hypotheses in nutrition.

**23.** Describe how systematic reviews can be used.

**24.** Describe a nutrition claim you have heard recently that sounds too good to be true. What is the probable motive of the person or company providing the advice?

**25.** Discuss why the cost of food as a percentage of household income varies across countries.

**26.** Describe how knowing about your genetic background could help you take steps to avoid diet-related diseases.

©C Squared Studios/Getty Images

# References

1. U.S. Department of Agriculture, U.S. Department of Health and Human Services. Dietary Guidelines for Americans, 2020–2025, 9th Ed. 2020. dietaryguidelines.gov.

2. GBD 2017 Diet Collaborators. Health effects of dietary risks in 195 countries, 1990-2017: a systematic analysis for the Global Burden of Disease Study 2017; *Lancet*. 2019;393:1950.

3. Centers for Disease Control and Prevention. Leading causes of death. 2020; cdc.gov/nchs/fastats/leading-causes-of-death.htm.

4. Keadle SK and others. Causes of death associated with prolonged TV viewing: NIH-AARP diet and health study. *Am J Prev Med*. 2015;49:811.

5. Slawson DL and others. Position of the Academy of Nutrition and Dietetics: The role of nutrition and health promotion and chronic disease prevention. *J Acad Nutr Diet*. 2013;113:972.

6. USDA National Nutrient Database for Standard Reference 29. 2019; www.ars.usda.gov/main/site_main.htm?modecode=12-35-45-00.

7. Food and Nutrition Board. *Dietary Reference Intakes for energy, carbohydrate, fiber, fat, fatty acids, cholesterol, protein, and amino acids.* Washington, DC: Food and Nutrition Board; 2002.

8. U.S. Food & Drug Administration. Trans fat. 2018; https://www.fda.gov/food/food-additives-petitions/trans-fat

9. Crowe KM, Francis C. Position of the Academy of Nutrition and Dietetics: Functional foods. *J Acad Nutr Diet*. 2013;113:1096.

10. Zhang YJ and others. Antioxidant phytochemicals for the prevention and treatment of chronic diseases. *Molecules*. 2015;12:21138.

11. Cheng Y-C and others. Polyphenols and oxidative stress in atherosclerosis-related ischemic heart disease and stroke. *Oxid Med Cell Long*. 2017;8526438.

12. Volpe S. Fruit and vegetable intake and prevention of disease. *ASCM Health Fit J*. 2019;23:31.

13. Melini F and others. Health-promoting components in fermented foods: An up-to-date systemic review. *Nutrients*. 2019;11:1189.

14. Institute of Medicine. *Dietary Reference Intakes for energy, carbohydrate, fiber, fat, fatty acids, cholesterol, protein, and amino acids (macronutrients).* Washington, DC: National Academies Press; 2005.

15. Centers for Disease Control and Prevention. Adult obesity facts. 2020; cdc.gov/obesity/data/adult.html.

16. Freeland-Graves JH, Nitzke S. Position of the Academy of Nutrition and Dietetics: Total diet approach to healthy living. *J Acad Nutr Diet*. 2013; 113:307.

17. Tholin S and others. Genetic and environmental influences on eating behavior: The Swedish Young Male Twins Study. *Am J Clin Nutr*. 2005;81:564.

18. USDA, Economic Research Service. Food prices and spending. 2020; ers.usda.gov/data-products/ag-and-food-statistics-charting-the-essentials/food-prices-and-spending.aspx.

19. U.S. Department of Health and Human Services. Development of the National Health Promotion and Disease Prevention Objectives for 2030. 2020; www.healthypeople.gov/2020/About-Healthy-People/Development-Healthy-People-2030.

20. U.S. Department of Health and Human Services. Healthy People 2030 Framework. 2021; health.gov/healthypeople.

21. Centers for Disease Control and Prevention. Life expectancy. 2017; www.cdc.gov/nchs/fastats/life-expectancy.htm.

22. Academy of Nutrition and Dietetics. Evidence Analysis Library. 2020; andeal.org/ncp.

23. Camp KM, Trujillo E. Position of the Academy of Nutrition and Dietetics: Nutritional genomics. *J Acad Diet Nutr*. 2014;114:299.

24. Murgia C and Adamski MM. Translation of nutritional genomics into nutrition practice: the next step. *Nutr*. 2017;9:4.

25. Vitti JJ and others. Human evolutionary genomics: Ethical and interpretive issues. *Trends Genet*. 2012;28:137.

26. Goralczyk R. Beta-carotene and lung cancer in smokers: Review of hypotheses and status of research. *Nutr Cancer*. 2009;61:767.

27. United States Food and Drug Administration. Dietary supplements, 2019; www.fda.gov/food/dietary-supplements.

15 Kcal

27 Kcal

56 Kcal

Analyzing the calorie and nutrient content of food is an important step in creating dietary standards and guidance. Learn more about the nutrient content of food at **fdc.nal.usda.gov/**. Prostock-studio/Shutterstock

# 2 Tools of a Healthy Diet

## Learning Objectives

**After studying this chapter, you will be able to**

1. Explain the purpose of the Dietary Reference Intake (DRI) and its components (Estimated Average Requirements, Recommended Dietary Allowances, Adequate Intakes, Upper Levels, Estimated Energy Requirements, and Acceptable Macronutrient Distribution Ranges).

2. Compare the Daily Values to the Dietary Reference Intakes and explain how they are used on Nutrition Facts panels.

3. Describe Nutrition Facts panels and the claims permitted on food packages.

4. Describe the uses and limitations of the data in nutrient databases.

5. Discuss the Dietary Guidelines for Americans and the diseases they are intended to prevent or minimize.

6. Discuss the MyPlate food groupings and plan a diet using this tool.

7. Develop a healthy eating plan based on the concepts of variety, balance, moderation, nutrient density, and energy density.

## Chapter Outline

**2.1** Dietary Reference Intakes (DRIs)

**2.2** Daily Values (DVs)

**Global Perspective: Front-of-Package Nutrition Labeling**

**2.3** Nutrient Composition of Foods

**Expert Perspective from the Field: Menu Labeling: How Many Calories Are in That?**

**2.4** Dietary Guidelines for Americans

**2.5** MyPlate

---

**NUTRITION IS A POPULAR TOPIC** in the media. News stories often highlight up-to-the-minute research results. Magazine articles, websites, and books tout the "latest" way to lose weight or improve your diet. Deciding whether to incorporate advice given by the media can be a challenge. Relying on peer-reviewed research and recommendations by experts can help you make informed decisions about whether to follow any of this nutrition advice. There also are a number of other useful tools based on nutrition research and assessment methods that can assist you in deciding what advice to follow as well as in planning a dietary pattern that helps you live as healthfully as possible now while minimizing the risk of developing nutrition-related diseases later on.

One tool for planning diets that support overall health is the Dietary Reference Intakes—they provide guidance on the quantities of nutrients that are most likely to result in optimal health. Using Daily Values, food labels, nutrient databases, nutrient density, and energy density also can help facilitate your efforts to identify foods that contain the array of nutrients you need in the recommended amounts. The Dietary Guidelines outline key steps that support good health and reduce risk of chronic nutrition-related diseases. Finally, MyPlate is a handy tool you can use to create a dietary pattern that promotes excellent health and lets you eat foods you enjoy.

A key to healthy living is gaining a firm knowledge of these basic diet planning tools. With this understanding, you'll know why scientists believe that optimal nutritional health can be accomplished by doing what you've heard many times before: eat a balanced diet, consume a variety of foods, moderate the amount you eat, and stay physically active.

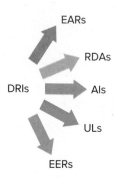

The DRIs are composed of Estimated Average Requirements (EARs), Recommended Dietary Allowances (RDAs), Adequate Intakes (AIs), Tolerable Upper Intake Levels (Upper Levels, or ULs), and Estimated Energy Requirements (EERs).

---

**Dietary Reference Intakes (DRIs)** Nutrient recommendations made by the Food and Nutrition Board, a part of the Institute of Medicine, and the National Academy of Science. These include EARs, RDAs, AIs, EERs, and ULs.

**Estimated Average Requirements (EAR)** Nutrient intake amounts estimated to meet the needs of 50% of the individuals in a specific life stage.

**Recommended Dietary Allowance (RDAs)** Nutrient intake amounts sufficient to meet the needs of 97 to 98% of the individuals in a specific life stage

**Adequate Intake (AI)** Nutrient intake amount set for any nutrient for which insufficient research is available to establish an RDA. AIs are based on estimates of intakes that appear to maintain a defined nutritional state in a specific life stage.

**Tolerable Upper Intake Level (UL)** Maximum chronic daily intake of a nutrient that is unlikely to cause adverse health effects in almost all people in a specific life stage.

**Estimated Energy Requirement (EER)** Estimate of the energy (kcal) intake needed to match the energy use of an average person in a specific life stage.

**Acceptable Macronutrient Distribution Range (AMDR)** Range of macronutrient intake, as percent of energy, associated with reduced risk of chronic diseases while providing for recommended intake of essential nutrients.

# 2.1 Dietary Reference Intakes (DRIs)

The first identification and documentation of nutrient deficiency diseases in the 1930s and 1940s, along with the rejection of many young men from military service in World War II due to the effects of poor nutrition on their health, made scientists realize the need for dietary intake recommendations. As a result, in 1941, a group of scientists formed the Food and Nutrition Board with the purpose of reviewing existing research and establishing the first official dietary standards. These standards were designed to evaluate nutrient intakes of large populations and to plan agricultural production. Since they were first published in 1943, these standards have been periodically reviewed and revised to reflect up-to-date scientific research.

The latest recommendations from the Food and Nutrition Board are called **Dietary Reference Intakes (DRIs)**.[1,2] The DRIs apply to people in both the U.S. and Canada because scientists from both countries worked together to establish them. DRIs include 5 sets of standards: Estimated Average Requirements (EARs), Recommended Dietary Allowances (RDAs), Adequate Intakes (AIs), Tolerable Upper Intake Levels (Upper Levels, or ULs), and Estimated Energy Requirements (EERs).[1] The DRIs are set for almost 40 nutrients; more may be added in the future.[3] Although not a DRI, Acceptable Macronutrient Distribution Ranges (AMDRs) were established for guidance on intake levels of carbohydrates, protein, and fat to help reduce the risk of nutrition-related chronic diseases.[4,5] As you can see from charts in Appendix J, the DRIs differ by life stage (i.e., age group, biological sex after age 9 years, pregnancy, lactation). All of the recommendations should be applied to dietary intake averaged over a number of days, not a single day.

## Estimated Average Requirements (EARs)

**Estimated Average Requirements (EARs)** are daily nutrient intake amounts that are estimated to meet the needs of half of the people in a certain life stage (Fig. 2-1). EARs are set for 17 nutrients. An EAR for a nutrient is set only when the Food and Nutrition Board agrees that there is an accurate method for measuring whether intake is adequate. These measures, called functional markers, typically evaluate the activity of an enzyme in the body or the ability of a cell or an organ to maintain normal physiological function.[1] If no measurable functional marker is available, an EAR cannot be set. Each EAR is adjusted to account

DRIs vary by life stage because nutrient needs differ with age and, after age 9 years, by biological sex. Pregnancy and lactation also affect nutrient needs; thus, there is a set of DRIs specially designed for these women.

Terry Vine/Blend Images LLC

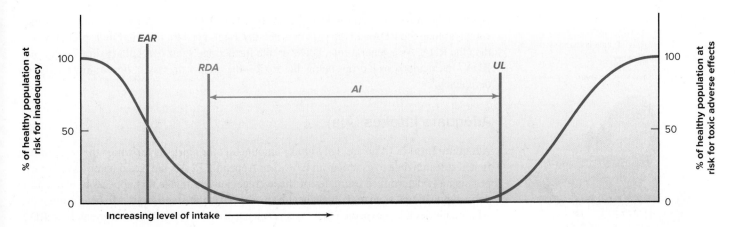

*Figure 2-1* This figure shows the relationship of the Dietary Reference Intakes (DRIs) to each other and the percent of the population covered by each.

**Estimated Average Requirement (EAR):** 50% of healthy North Americans would have an inadequate intake if they consumed the EAR, whereas 50% would meet their needs.

**Recommended Dietary Allowance (RDA):** 2 to 3% of healthy North Americans would have an inadequate intake if they met the RDA, whereas 97 to 98% would meet their needs.

**Upper Level (UL):** highest nutrient intake level that is likely to pose no risks of adverse health effects in almost all healthy individuals. At intakes above the UL, the margin of safety to protect against adverse effects is reduced. At intakes between the RDA and UL (Upper Level), the risk of either an inadequate diet or adverse effects from the nutrient is close to 0%.

**Adequate Intake (AI):** set for some nutrients instead of an RDA; lies somewhere between the RDA and UL. Thus, the AI should cover the needs of more than 97 to 98% of individuals.

for the amount of the nutrient that passes through the digestive tract unabsorbed. Because EARs meet the needs of only 50% of those in a life stage, they can be used to evaluate only the dietary adequacy of groups, not of individuals.[1] Specific EARs are listed in Appendix J.

To illustrate how EARs are determined, let's take a look at vitamin C. The amount of vitamin C needed daily to prevent scurvy is about 10 mg. However, as you will learn in Part 4 of this book, vitamin C has other functions as well, including some related to the immune system (see Appendix A for details on the immune system). In fact, the concentration of vitamin C in a component of the immune system—notably, white blood cells (specifically, neutrophils)—can be used as a functional marker for vitamin C. The Food and Nutrition Board concluded that nearly maximal saturation of these white blood cells with vitamin C is the best functional marker for optimal vitamin C status. It takes a daily vitamin C intake of about 75 mg for men and 60 mg for women to nearly saturate these white blood cells. These average amounts became the EARs for young adult men and women.

## Recommended Dietary Allowances (RDAs)

**Recommended Dietary Allowances (RDAs)** are daily nutrient intake amounts sufficient to meet the needs of nearly all individuals (97 to 98%) in a life stage (see Appendix J). RDAs are based on a multiple of the EARs (generally, the RDA = EAR × 1.2). Because of this relationship, an RDA can be set only for nutrients that have an EAR. (Recall that a measurable functional marker is required to set an EAR.) An additional consideration made when setting an RDA is the nutrient's ability to prevent chronic disease rather than just prevent deficiency.[1]

For example, to determine the RDA for vitamin C, its EAR (75 mg for men and 60 mg for women) was multiplied by 1.2. In this case, the RDA was set at 90 mg for men and 75 mg for women. The RDA for other life-stage groups was set similarly. Because smokers break down vitamin C more rapidly, the Food and Nutrition Board recommended that these individuals add 35 mg/day to the RDA set for their life stage.

The RDA is the goal for usual intake. To assess whether vitamin C intake meets the RDA, total the amount of vitamin C consumed in a week and divide by 7 to get an average daily intake. Keep in mind that the RDA is higher than the average human needs, so not

▶ The National Nutrition Conference for Defense, held in May 1941, examined the problem of poor nutritional status of many World War II military recruits. This conference led to many nutrition-related efforts, such as victory gardens and War Food Order No. 1. This order mandated that all flour sold for interstate commerce be enriched with thiamin, niacin, and iron. Riboflavin enrichment quickly followed. In 1998, flour was enriched with folic acid. Learn more about fortification and enrichment at www.foodinsight.org.

U.S. National Archives and Records Administration

▶ This online tool displays DRIs based on age and biological sex: nal.usda.gov/fnic /dri-calculator/.

Although Estimated Energy Requirements (EERs) provide a guide for energy needs, the best estimate of energy need is the amount needed to maintain a healthy weight.

everyone needs to have an intake equal to the RDA. Thus, even if average intake is somewhat less than the RDA and the person is healthy, one's need for this vitamin is probably less than the RDA. As a general rule, however, the further one's intake regularly drops below the RDA—particularly as it drops below the EAR—the greater the risk of developing a nutrient deficiency.[1]

## Adequate Intakes (AIs)

**Adequate Intakes (AIs)** are daily intake amounts set for nutrients for which there are insufficient research data to establish an EAR (see Appendix J). AIs are based on observed or experimentally determined estimates of the average nutrient intake that appears to maintain a defined nutritional state (e.g., bone health) in a specific life-stage group.[1] In determining the AI for a nutrient, it is expected that the amount exceeds the RDA for that nutrient, if an RDA is known. Thus, the AI should cover the needs of more than 97 to 98% of the individuals in a specific life-stage group. The actual degree to which the AI exceeds the RDA likely differs among the various nutrients and life-stage groups. Like the RDA, the AI can be used as the goal for usual intake of that nutrient by an individual. Currently, essential fatty acids, fiber, and certain vitamins and minerals, including some B-vitamins, the vitamin-like compound choline, and fluoride, have AIs.

## Tolerable Upper Intake Levels (Upper Levels, or ULs)

**Tolerable Upper Intake Levels,** or Upper Levels (**ULs**), are the maximum daily intake amounts of nutrients that are not likely to cause adverse health effects in almost all individuals (97 to 98%) in a life-stage group (see Appendix J).[1] The amount applies to chronic daily use and is set to protect even those who are very susceptible in the healthy general population. For example, the UL for vitamin C is 2000 mg/day. Intakes greater than this amount can cause diarrhea and inflammation of the stomach lining.

The UL for most nutrients is based on the combined intake of food, water, supplements, and fortified foods. The exceptions are the vitamin niacin and the minerals magnesium, zinc, and nickel—the UL for each refers only to nonfood sources, such as medicines and supplements. This is because niacin, magnesium, zinc, or nickel toxicity from food sources is unlikely.[6]

The UL is not a nutrient intake goal; instead, it is a ceiling below which nutrient intake should remain. Still, for most individuals, there is a margin of safety above the UL before any adverse effects are likely to occur. Too little information is available to set a UL for all nutrients, but this does not mean that toxicity from these nutrients is impossible. Plus, there is no clear-cut evidence that intakes above the RDA or AI confer any additional health benefits for most of us.

## Estimated Energy Requirements (EERs)

RDAs and AIs for nutrients are set high enough to meet the needs of almost all healthy individuals. In contrast, **Estimated Energy Requirements (EERs)** are set at the average daily energy (kcal) need for each life-stage group. Unlike most vitamins and minerals, energy (carbohydrate, fat, protein, alcohol) consumed in amounts above that needed is not excreted but is stored as body fat. Thus, to promote healthy weight, a more conservative standard is used to set EERs.[5] Overall, EERs are estimates because energy needs depend on energy expenditure and, in some cases, the energy needed to support growth or human milk production. For most adults, the best estimate of energy need is the amount required to achieve and maintain a healthy weight (see Chapter 10 for details).

## Acceptable Macronutrient Distribution Ranges (AMDRs)

In addition to EERs, the Food and Nutrition Board established **Acceptable Macronutrient Distribution Ranges (AMDRs)** for intake of carbohydrate, protein, fat, and essential fatty acids (see Appendix J). For each macronutrient, the AMDRs provide a range of intake, as a percent of energy, associated with good health and a reduced risk of chronic diseases while providing for recommended intakes of essential nutrients. The AMDRs complement the DRIs.[1] For example, the AMDR for fat is 20 to 35% of calories. For an average energy intake of 2000 kcal/day, this is equal to 400 to 700 kcal per day from fat. To translate this to grams of fat per day, divide by 9 kcal/g. Thus, a healthy amount of fat for a 2000 kcal diet is 44 to 78 g fat/day.

## Appropriate Uses of the DRIs

The DRIs are intended mainly for guiding nutrition programs (e.g., Supplemental Nutrition Assistance Program [SNAP], National School Lunch Program), nutrition labeling, U.S. military feeding practices, nutrition policies (e.g., Dietary Guidelines for Americans), nutrition research, monitoring of nutritional health, setting of nutrient standards, and diet planning (Table 2-1).[6] Specifically, a diet plan should aim to meet any RDAs or AIs set. Finally, when planning diets, it is important not to exceed the Upper Level for a nutrient (Fig. 2-2).[1,6] Keep in mind also that DRIs apply to healthy people—none are necessarily appropriate amounts for undernourished individuals or those with diseases or other health conditions that require higher intakes. Amounts of vitamins and minerals needed for good health will be discussed in Chapters 12 through 15.

## Putting the DRIs into Action to Determine the Nutrient Density of Foods

Nutrient density has gained acceptance in recent years as a tool for assessing the nutritional quality of an individual food.[7] To determine the **nutrient density** of a food, divide the amount of a nutrient (protein, vitamin, mineral) in a serving of the food by your daily recommended intake (e.g., RDA, AI). Next, divide the calories in a serving of the food by your daily calorie need (EER). Then, compare these values—a food is said to be nutrient dense if it provides a greater contribution to your nutrient need than to your calorie need. The higher a food's nutrient density, the better it is as a source for a particular nutrient. For example, the 70 mg of vitamin C and 65 calories provided by an orange supplies 108% of the RDA for a

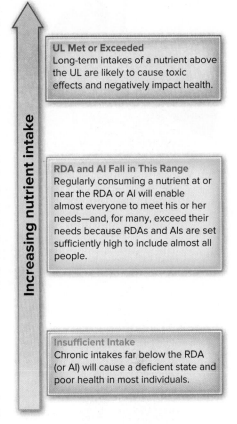

**UL Met or Exceeded**
Long-term intakes of a nutrient above the UL are likely to cause toxic effects and negatively impact health.

**RDA and AI Fall in This Range**
Regularly consuming a nutrient at or near the RDA or AI will enable almost everyone to meet his or her needs—and, for many, exceed their needs because RDAs and AIs are set sufficiently high to include almost all people.

**Insufficient Intake**
Chronic intakes far below the RDA (or AI) will cause a deficient state and poor health in most individuals.

Increasing nutrient intake

*Figure 2-2* Think of the nutrient standards that are part of DRIs as points along a line ranging from an insufficient intake to a healthy intake level to an excessive intake.

| Table 2-1 | **Putting the DRIs for Nutrient Needs to Use** |
|---|---|
| EAR | **Estimated Average Requirement.** Represents average nutrient need. Use only to evaluate adequacy of diets of groups, not individuals. |
| RDA | **Recommended Dietary Allowance.** Use to evaluate current intake for a specific nutrient. The more intake strays above or below this value, the greater the likelihood a person will develop nutrition-related problems. |
| AI | **Adequate Intake.** Use to evaluate current intake for a specific nutrient, realizing that an AI implies that further research is required before scientists can establish a more definitive intake amount needed to set an RDA. |
| UL | **Upper Level.** Use to evaluate the highest amount of daily nutrient intake that is unlikely to cause adverse health effects in the long run. This value applies to chronic use and is set to protect even very susceptible people in the healthy general population. As intake rises higher than the UL, the potential for adverse effects generally increases. |
| EER | **Estimated Energy Requirement.** Use to estimate energy needs according to height, weight, biological sex, age, and physical activity pattern. |
| AMDR | **Acceptable Macronutrient Distribution Range.** Use to determine whether percent of calories from each macronutrient falls within suggested range. The greater the discrepancy with AMDR, the greater the risk for nutrition-related chronic diseases. |

teenage girl (65 mg vitamin C) and only 4% of her 1800 daily calorie need. It is considered a nutrient-dense food for vitamin C. In contrast, the 52 mg of calcium in an orange provides only 4% of the teenage girl's calcium RDA (1300 mg).

On a nutrient-by-nutrient basis, comparing the nutrient density of different foods is an easy way to identify the more nutritious choice. It's more difficult to obtain an overall picture of nutritional quality. Some experts recommend averaging the nutrient density for key nutrients and comparing the average with the percent of daily calorie need provided. For example, as Figure 2-3 shows, an average of the nutrients in the fat-free milk is about 15% and supplies only 4% of calories, whereas the nutrients in the cola average approximately 0% while supplying 5% of calories. The fat-free milk is much more nutrient dense than a sugar-sweetened soft drink for many nutrients. Sugar-sweetened soft drinks and other foods that are not nutrient dense (e.g., chips, cookies, and candy) often are called **empty calorie foods** because they tend to be high in sugar and/or fat but few other nutrients—that is, the calories are "empty" of nutrients.

## Knowledge Check

1. Which dietary standard is set based on Estimated Average Requirements (EARs)?
2. Which dietary standard is set when an Estimated Average Requirement (EAR) cannot be set?
3. Which of the Dietary Reference Intakes (DRIs) is set at the maximum daily intake amount?
4. Determine the nutrient density for vitamin A and vitamin C of a baked sweet potato and french fries for someone your age and biological sex.

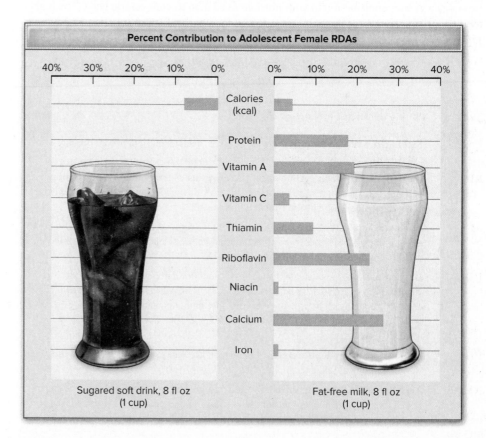

*Figure 2-3* Comparison of the nutrient density of a sugar-sweetened soft drink with that of fat-free (skim) milk. The milk provides a significantly greater contribution to nutrient intake per calorie than the soft drink. Compare the lengths of the bars indicating vitamin or mineral contribution with the bar that represents energy content. For the soft drink, no nutrient surpasses energy content. Fat-free milk, in contrast, has longer nutrient bars for protein, vitamin A, the vitamins thiamin and riboflavin, and calcium.

# 2.2 Daily Values (DVs)

The Nutrition Facts panel on a food label compares the amount of nutrients in the food with a set of standards called Daily Values (DVs). DVs are generic standards that were developed by the U.S. Food and Drug Administration (FDA) because the DRIs are age- and biological sex and it isn't practical to have different food labels for men and women or for teens and adults.

DVs have been set for 4 groups: infants, toddlers, pregnant or lactating women, and people over 4 years of age. The DVs that appear on all food labels—except those specially marketed to infants, toddlers, or pregnant or lactating women—are those set for people over age 4 years. This book will focus on the DVs for those over age 4 years.

DVs are based on 2 sets of dietary standards: Reference Daily Intakes and Daily Reference Values. These terms—*Reference Daily Intakes* and *Daily Reference Values*—do not appear on food labels. Instead, the term Daily Value is used to represent the combination of these 2 sets of dietary standards. Table 2-2 displays the DVs. Even though the term DV is used on Nutrition Facts panels, it is important for nutrition and health professionals to understand how Reference Daily Intakes and Daily Reference Values differ.

## Reference Daily Intakes (RDIs)

Reference Daily Intakes (RDIs) are set for vitamins and most minerals—these nutrients all have established nutrient standards. RDI values for people over age 4 years are set at the highest RDA value (or AI for nutrients that do not have RDAs) for any life-stage group (see Table 2-2). RDIs are used to calculate percent DV on Nutrition Facts panels.

## Daily Reference Values (DRVs)

Daily Reference Values (DRVs) are standards for energy-producing nutrients (fat, saturated fat, carbohydrate, protein, fiber), cholesterol, and sodium. Many of these nutrients do not have an established RDA or other nutrient standard (fat, saturated fat, carbohydrate). Like RDIs, DRVs are used to calculate percent DV on Nutrition Facts panels.

The DRVs for the energy-producing nutrients are based on daily calorie intake. The FDA selected 2000 calories as the reference for calculating percent DVs for energy-producing nutrients. The DRVs for energy-producing nutrients are equal to about

- 35% of calories for fat
- 10% of calories for saturated fat
- 60% of calories for carbohydrate
- 10% of calories for protein

## Putting the Daily Values into Action on Nutrition Facts Panels

With few exceptions, information related to the Daily Values is found on almost every food and beverage sold in the supermarket today. Their labels include the product name, name and address of the manufacturer, amount of product in the package, ingredients listed in descending order by weight, ingredients that are common allergens[8] (milk, eggs, fish, shellfish, peanuts, tree nuts, wheat, and soy—see Chapter 7 for details), and Nutrition Facts panel. The Nutrition Facts panel lists the amounts of certain food components and reports many of them as % Daily Value. Labels also must indicate the country of origin for certain products (i.e., chicken, lamb, goat, fish, fresh and frozen fruits and vegetables, peanuts, pecans, macadamia nuts, and ginseng).[9] This required labeling is monitored in North America by government agencies, such as the FDA in the U.S.

As you can see in Figure 2-4, Nutrition Facts panels present information for a single serving of food. Serving sizes are specified by the FDA so that they are consistent among similar foods. This means that all brands of ice cream, for example, must use the same

▶ DV: RDI and DRV

**Daily Value (DV):** Generic nutrient standard used on Nutrition Facts labels; it comprises both Reference Daily Intakes (RDIs) and Daily Reference Values (DRVs)

**Reference Daily Intakes (RDIs):** Part of the DV; generic nutrient standards set for vitamins and minerals (except sodium)

**Daily Reference Values (DRVs):** Part of the DV; generic nutrient standards set for energy-producing nutrients (fat, carbohydrate, protein, fiber), cholesterol, and sodium

▶ Canada also has a set of Daily Values for use on food labels (see Appendix D).

Fresh fruits, vegetables, and fish are not required to have Nutrition Facts panels. However, many grocers distribute leaflets or display posters like these that include Nutrition Facts panels for these foods. Raw meat and poultry products now have Nutrition Facts panels on their packages.

Source: FDA

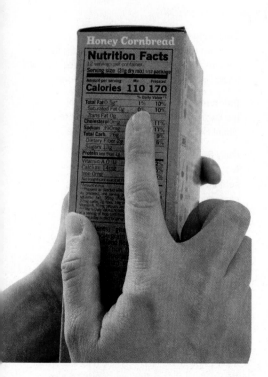

Use the Nutrition Facts panel to learn more about the nutrient content of the foods you eat. Nutrient content is expressed as % Daily Value. Canadian food laws and related food labels have a slightly different format (see Appendix D).

©McGraw-Hill EducationTara McDermott, photographer

### Table 2-2  Daily Values

| Dietary Constituent | Unit of Measure* | Current Daily Values for People over 4 Years of Age |
|---|---|---|
| **Daily Reference Values (DRVs)** | | |
| Total fat | g | 78 |
| Saturated fatty acids | g | 20 |
| Protein | g | 50 |
| Cholesterol | mg | 300 |
| Carbohydrate | g | 275 |
| Added sugar | g | 50 |
| Fiber | g | 28 |
| Sodium | mg | 2300 |
| **Reference Daily Intakes (RDIs)** | | |
| Vitamin A | µg Retinol Activity Equivalents | 900 |
| Vitamin D | µg | 20 |
| Vitamin E | mg | 15 |
| Vitamin K | µg | 120 |
| Vitamin C | mg | 90 |
| Folate | µg Dietary Folate Equivalents | 400 |
| Thiamin | mg | 1.2 |
| Riboflavin | mg | 1.3 |
| Niacin | mg | 16 |
| Vitamin B-6 | mg | 1.7 |
| Vitamin B-12 | µg | 2.4 |
| Biotin | µg | 30 |
| Pantothenic acid | mg | 5 |
| Calcium | mg | 1300 |
| Phosphorus | mg | 1250 |
| Iodine | µg | 150 |
| Iron | mg | 18 |
| Magnesium | mg | 420 |
| Copper | mg | 0.9 |
| Zinc | mg | 11 |
| Chloride | mg | 2300 |
| Potassium | mg | 4700 |
| Manganese | mg | 2.3 |
| Selenium | µg | 55 |
| Chromium | µg | 35 |
| Molybdenum | µg | 45 |
| Choline | mg | 550 |

*Abbreviations: g = gram; mg = milligram; µg = micrograms

U.S. government. Food labeling: Revision of the nutrition and supplement facts labels. 2016; www.regulations.gov /document?D=FDA-2012-N-1210-0875.

*Figure 2-4* Food packages must list product name, name and address of the manufacturer, amount of product in the package, and ingredients. The Nutrition Facts panel is required on virtually all packaged food products. The % Daily Value listed on the label is the percent of the amount of a nutrient needed daily that is provided by a single serving of the product.

**Servings per Container**

The number of servings of the size given in the serving size below that are in one package of the food.

**Serving Size**

Serving size is listed in household units (and grams). Pay careful attention to serving size to know how many servings you are eating: e.g., if you eat double the serving size, you must double the % Daily Values and calories.

Nutrients claims, such as "Good source," and health claims, such as "Reduce the risk of osteoporosis," must follow legal definitions.

**% Daily Value**

This shows how a single serving compares to the DV. Recall that the DVs for fat, saturated fat, cholesterol, protein, carbohydrate, added sugar, and fiber are based on a 2000-calorie diet.

**Protein DV**

% Daily Value for protein is generally not included due to expensive testing required to determine protein quality.

## Nutrition Facts

| 6 servings per container | |
|---|---|
| **Serving size** | **1 pouch (61g)** |
| **Amount Per Serving** | |
| **Calories** | **250** |
| | **% Daily Value\*** |
| **Total Fat** 7g | **8%** |
| Saturated Fat 2.5g | **10%** |
| *Trans* Fat 1g | |
| **Cholesterol** 5mg | **2%** |
| **Sodium** 400mg | **15%** |
| **Total Carbohydrate** 38g | **15%** |
| Dietary Fiber less than 1g | **4%** |
| Total sugars 6g | |
| Includes 0g added sugars | **0%** |
| **Protein** 7g | |
| Vitamin D 0mg | 0% |
| Calcium 120mg | 8% |
| Iron 1mg | 8% |
| Potassium 10mg | 0% |

\* The % Daily Value (DV) tells you how much a nutrient in a serving of food contributes to a daily diet. 2,000 calories a day is used for general nutrition advice.

INGREDIENTS: ENRICHED MACARONI PRODUCT (DURHAM WHEAT FLOUR, GLYCERYL MONO-STEARATE, SALT, NIACIN, FERROUS SULPHATE, THIAMIN MONONITRATE (VITAMIN B1), RIBOFLAVIN (VITAMIN B2), FOLIC ACID), CHEESE SAUCE MIX (WHEY, PARTIALLY HYDROGENATED SOYBEAN OIL, MALTODEXTRIN, WHEY PROTEIN CONCENTRATE, CORN SYRUP SOLIDS, SALT, MILKFAT, SUGAR, SODIUM, NATURAL FLAVOR, CITRIC ACID, MONOSODIUM GLUTAMATE, MODIFIED FOOD STARCH, LACTIC ACID, YELLOW 5)

**Daily Value Footnote**

The footnote explains what the DV indicates and that a 2000 calorie diet is the basis used.

**Nutrients**

These nutrients must appear on most labels. Labels of foods that contain few nutrients, such as candy and soft drinks, may omit some nutrients. Some manufacturers list more nutrients. Other nutrients must be listed if manufacturers make a claim about them or if the food is fortified with them.

Name and address of the food manufacturer

Ingredients are listed in descending order by weight.

**A Quick Guide to Nutrient Sources**

% Daily Value
20% or more = Rich source
10%–19% = Good source

serving size on their labels. The serving sizes on Nutrition Facts panels are based on typical serving sizes eaten by Americans; as a result, they may differ from the serving sizes recommended by MyPlate (see Section 2.5).

Starting in 2018, the following components must be listed on most Nutrition Facts panels: total calories (kcal), total fat, saturated fat, *trans* fat, cholesterol, sodium, total carbohydrate, total sugars, added sugar, fiber, protein, vitamin D, potassium, calcium, and iron. Labels of foods that contain few nutrients, such as candy and soft drinks, may omit some nutrients. In addition to the components required on Nutrition Facts panels, manufacturers can choose to list other nutrients, such as polyunsaturated fat or vitamin C. Manufacturers are required to include a nutrient on the Nutrition Facts panel if they make a claim about its health benefits (see the subsection Claims on Food Labels later in this section) or if the food is fortified with that nutrient.

Notice in Figure 2-4 that the amounts of fats, cholesterol, sodium, carbohydrates, protein, vitamins, and minerals in a food are listed. Most of these nutrients also are given as % Daily Value. Because protein deficiency is not a public health concern in the U.S., listing % Daily Value for protein is not mandatory on foods for people over 4 years of age. If the % Daily Value is given on a label, the FDA requires that the product be analyzed for protein quality (see Chapter 7). This procedure is expensive and time consuming; thus, many companies opt not to list % Daily Value for protein. However, labels on food for infants and children under 4 years of age must include the % Daily Value for protein, as must labels on any food carrying a claim about protein content.

Recall that all the values shown on Nutrition Facts panels are for a single serving of the food. Thus, to determine the total amount of calories or a nutrient in more than 1 serving, the value on the label must be multiplied by the number of servings consumed. For instance, let's say you ate the entire box of MicroMac shown in Figure 2-4—that would be 6 servings. The entire package would provide 1500 calories (250 kcal per serving × 6 servings per container = 1500 calories), 78% of total carbohydrate (13% per serving × 6 servings per container = 78%), 36 grams of sugar, and so on.

You can use the DVs to determine how a particular food fits into an overall diet (Fig. 2-5). If, for example, a single food provides 50% of the DV for fat, then it is a good idea to either select a different food that is lower in fat or be sure other choices that day are low in fat. The DVs also can help you determine how close your overall diet comes to meeting recommendations. For instance, if you consume 2000 calories per day, your total fat intake for the day should be 78 g or less. If you consume 10 g of fat at breakfast, you have 68 grams, or 87%, of your DV for fat left for the rest of the day. If you eat more or less than 2000 calories per day, you can still use the Nutrition Facts panel. For example, if you consume only 1600 calories per day, the total percent of DV you eat for fat, saturated fat, carbohydrate, and protein should add up to 80% DV because 1600/2000 = 0.8, or 80%. If you eat 3000 calories daily, the total percent of DV you eat of fat, saturated fat, carbohydrate, and protein in all the foods you eat in 1 day can add up to 150% DV because 3000/2000 = 1.5, or

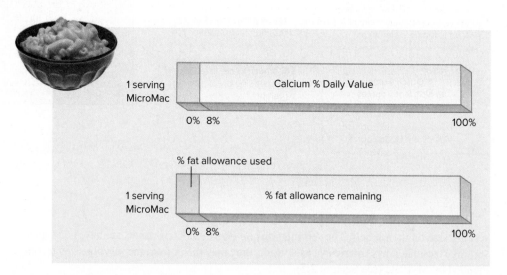

***Figure 2-5*** Nutrition Facts panels can help you track intake. If you needed 2000 calories and ate a single serving of MicroMac, you would still have 92% Daily Value of your fat allowance left. In addition, you would have met 8% Daily Value for calcium.

Brooke Becker/Shutterstock

150%. Remember, you need to make adjustments only for the nutrients that are based on calorie intake: carbohydrate, protein, fat, and saturated fat. For nutrients not based on calorie intake, such as vitamin D and cholesterol, just add percent of DVs in all the foods you eat to determine how close your diet comes to meeting recommendations.

As you may have noticed, the nutrients listed on Nutrition Facts panels tend to be the nutrients of greatest health concern in North America. Many people eat too much fat, saturated fat, *trans* fat, cholesterol, sodium, and sugar. Many also are concerned that they don't get enough fiber, calcium, iron, vitamin D, and potassium. Thus, for the best health, most people should aim to keep their intake of the following nutrients at or below 100% of the DV: total fat, saturated fat, cholesterol, total and added sugar, and sodium. Most people also should plan their diets to achieve 100% of the DV for fiber, vitamin D, potassium, iron, and calcium.

Nutrition Facts panels often include a footnote that explains that the DV shows how much a serving of the food contributes to a daily diet and that a 2000-calorie diet is the basis used. This footnote helps label readers see how the DVs are calculated for nutrients on the label.

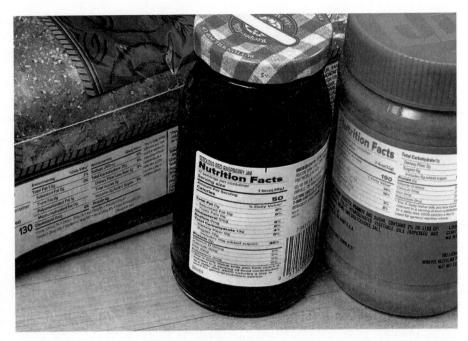

The nutrition information on the Nutrition Facts panels on these products can be combined to determine the nutrient intake for a peanut butter and jelly sandwich. For example, 2 slices of bread, 1 tablespoon of jam, and 2 tablespoons of peanut butter contain 500 calories ([130 × 2] + 50 + 190), which is 25% of the total calories needed on a 2000-calorie diet ([500/2000] × 100).

Jill Braaten/McGraw-Hill Education

### Claims on Food Labels

As a marketing tool directed toward health-conscious consumers, food manufacturers like to assert that some of their products have certain nutrient levels or health benefits. After reviewing hundreds of comments on the proposed rule allowing nutrient and health claims, the FDA, which has legal oversight for most food products, decided to permit certain specific claims. Although these claims must comply with FDA regulations, you can use Daily Value information on Nutrition Facts panels to verify the nutrient content and health claims made on food packages.[10]

**Nutrient content claims** are those that describe the nutrients in a food. Examples are "low in fat," "rich in vitamin A," and "zero calories." All nutrient content claims must comply with regulations set by the FDA. Table 2-3 summarizes the legal definitions of nutrient content claims permitted to appear on food packages. For example, if a product claims to be "low sodium," it must have 140 mg or less of sodium per serving.

**Health claims** describe a relationship between a disease and a nutrient, food, or food constituent.[11] All permitted health claims have significant scientific agreement that they are true. All health claims must use a *may* or *might* qualifier in the statement. The following are some of the permitted health claims.

- A diet with enough calcium may reduce risk of osteoporosis.
- A diet low in sodium may reduce risk of hypertension.
- A diet low in total fat may reduce risk of some cancers.
- A diet low in fat and rich in fiber-containing grain products, fruits, and vegetables may reduce risk of some cancers.
- A diet low in saturated fat and cholesterol may reduce risk of cardiovascular disease (typically referred to as *heart disease* on the label).
- A diet low in saturated fat and cholesterol and rich in fruits, vegetables, and grains that contain soluble fiber may reduce risk of heart disease.

▶ As you saw in Figure 1-13 (Chapter 1), the labels on nutrient and herbal supplements have a different layout than those on foods. These labels include a "Supplement Facts" heading.

▶ Claims on foods fall into these categories.

- Nutrient content claims—closely regulated by the FDA
- Health claims—closely regulated by the FDA
- Qualified health claims—regulated by the FDA, but there is only limited scientific evidence for the claims
- Structure/function claims—not FDA approved; manufacturer is responsible for their accuracy

▶ Canada has established a set of health claims for nutrition labels (see Appendix D).

- A diet low in saturated fat and cholesterol that also includes 25 g/day of soy protein may reduce risk of heart disease.
- Foods that contain plant stanol or sterol esters may reduce risk of heart disease (see Chapter 6).
- A diet adequate in folate may reduce a woman's risk of having a child with a brain or spinal cord defect (see Part 4).
- Sugar alcohols do not promote tooth decay (see Chapter 5).

Only food products that meet the following requirements can bear a health claim. First, the food must be a "good source" (before any fortification) of fiber, protein, vitamin A, vitamin C, calcium, or iron—it must provide at least 10% of the Daily Value for at least 1 of these nutrients. Second, a single serving of the food must contain no more than specified

**Table 2-3  Summary of Nutrient Claims on Food Labels**

| **Calories** | |
|---|---|
| Calorie free | Less than 5 kcal per serving |
| Low calorie | 40 kcal or less per serving (if the serving is small,* per 50 g of the food) |
| Reduced or fewer calories | At least 25% less kcal per serving than reference food |
| Light or lite | 50% less fat if half or more of the food's kcal are from fat; 50% less fat or 33% less kcal if less than half of the food's kcal are from fat |
| **Total Fat** | |
| Fat free | Less than 0.5 g fat per serving |
| Low fat | 3 g or less per serving (if the serving is small, per 50 g of the food) |
| Reduced or less fat | At least 25% less per serving than reference food |
| Lean | Seafood, poultry, or meat with less than 10 g total fat, 4.5 g or less saturated fat, and less than 95 mg cholesterol per reference amount |
| Extra lean | Seafood, poultry, or meat with less than 5 g total fat, less than 2 g saturated fat, and less than 95 mg cholesterol per reference amount |
| **Saturated Fat** | |
| Saturated fat free | Less than 0.5 g saturated fat and less than 0.5 g *trans* fatty acids per serving |
| Low saturated fat | 1 g or less per serving and 15% or less of kcal from saturated fat |
| Reduced or less saturated fat | At least 25% less per serving than reference food |
| **Alpha-Linolenic Acid (ALA) Omega-3 Fatty Acid** | |
| High in ALA | At least 320 mg ALA per serving |
| Medium in ALA | At least 160 mg ALA per serving |
| More | At least 160 mg more ALA than reference food |
| **Cholesterol** | |
| Cholesterol free | Less than 2 mg cholesterol and 2 g or less saturated fat per serving |
| Low cholesterol | 20 mg or less cholesterol and 2 g or less saturated fat per serving (if the serving is small, per 50 g of the food) |
| Reduced or less cholesterol | At least 25% less cholesterol per serving than reference food and 2 g or less saturated fat |
| **Sugar** | |
| Sugar free | Less than 0.5 g sugar per serving |
| No added sugar or without added sugar | No sugars or sugar-containing ingredient (e.g., jam, applesauce) added during processing or packing |
| Reduced sugar | At least 25% less sugar per serving than reference food |

*Small serving size or small reference amount = reference amount of 30 g or less or 2 tbsp or less.

amounts of fat, saturated fat, cholesterol, and sodium. If a food does not meet all the requirements, no health claim can be made for it, despite its other nutritional qualities. For example, even though whole milk is high in calcium, its label can't make a health claim about calcium and a reduced risk of osteoporosis because whole milk contains more saturated fat per serving than permitted in the FDA regulations. Third, the product must meet criteria specific to the health claim being made. For example, a health claim regarding fat and cancer can be made only if the product contains 3 g or less of fat per serving, which is the standard for low-fat foods.

The FDA permits qualified health claims based on incomplete scientific evidence as long as the label qualifies it with a disclaimer such as "this evidence is not conclusive" and the food meets the definition of being healthy (see Table 2-3).[11] So far, few qualified health claims have appeared on food packages (walnuts, fish and shellfish, and whole grains are some examples).[12,13]

Nutrition Facts panels can help you locate foods that will provide a nutrient-rich diet.

Fuse/Getty Images RF

## Table 2-3  continued

### Sodium

| | |
|---|---|
| Sodium free or salt free | Less than 5 mg sodium per serving |
| Very low sodium | 35 mg or less sodium per serving (if the serving is small, per 50 g of the food) |
| Low sodium | 140 mg or less sodium per serving (if the serving is small, per 50 g of the food) |
| Light (for sodium-reduced products) | If food meets definition of low calorie and low fat, and sodium is reduced by at least 50% |
| Light in sodium | At least 50% less sodium per serving than reference food |
| Reduced or less sodium | At least 25% less sodium per serving than reference food |
| Lightly salted | At least 50% less sodium than normally added to reference food; if it doesn't meet definition of low sodium, this must be stated on the label |
| No salt added, unsalted | If not sodium free, must declare "This Is Not a Sodium-Free Food" |

### Fiber

| | |
|---|---|
| Any claim | If food is not low in total fat, must state total fat in conjunction with fiber claim |

### Other Claims

| | |
|---|---|
| High, rich in, or excellent source | 20% or more of the DV per reference amount; may be used to describe protein, vitamins, minerals, dietary fiber |
| Good source, contains, or provides | 10 to 19% of the DV per reference amount; may be used to describe protein, vitamins, minerals, dietary fiber |
| More, added, extra, or plus | 10% or more of the DV per reference amount; may be used for vitamins, minerals, protein, dietary fiber, potassium |
| High potency | May be used to describe individual vitamins or minerals present at 100% or more of the DV per reference amount |
| Fortified or enriched | Vitamins and/or minerals added to the product in amounts at least 10% above levels normally present in food; *enriched* generally refers to replacing nutrients lost in processing, whereas *fortified* refers to adding nutrients not originally present in the specific food |
| Healthy | Varies with food type; generally is a food that is low fat and low saturated fat, has no more than 480 to 600 mg of sodium or 95 mg of cholesterol per serving, and provides at least 10% of the DV for vitamin A, vitamin C, protein, calcium, iron, or fiber |
| Light or lite | Used with calories and sodium (see above); also may be used to describe texture and color, as long as the label explains the intent—for example, *light brown sugar* and *light and fluffy* |

## Applying the Nutrition Facts Label to Your Daily Food Choices

Imagine that you are at the supermarket, looking for a quick meal before a busy evening. In the frozen food section, you find 2 brands of frozen cheese manicotti (see labels a and b). Which of the 2 brands would you choose? What information on the Nutrition Facts labels contributed to this decision?

### (a)

**Nutrition Facts**

1 serving per container
Serving size                           1 pkg (260g)

**Amount Per Serving**
**Calories**                                  **390**

| | % Daily Value* |
|---|---|
| **Total Fat** 18g | **25**% |
| Saturated Fat 9g | **45**% |
| *Trans* Fat 2g | |
| **Cholesterol** 45mg | **15**% |
| **Sodium** 880mg | **40**% |
| **Total Carbohydrate** 38g | **15**% |
| Dietary Fiber 4g | **15**% |
| Total sugars 12g | |
| Includes 0g added sugars | **0**% |
| **Protein** 17g | |
| Vitamin D 0mg | 0% |
| Calcium 400mg | 30% |
| Iron 8mg | 45% |
| Potassium 300mg | 6% |

* The % Daily Value (DV) tells you how much a nutrient in a serving of food contributes to a daily diet. 2,000 calories a day is used for general nutrition advice.

### (b)

**Nutrition Facts**

1 serving per container
Serving size                           1 pkg (260g)

**Amount Per Serving**
**Calories**                                  **230**

| | % Daily Value* |
|---|---|
| **Total Fat** 4g | **4**% |
| Saturated Fat 2g | **10**% |
| *Trans* Fat 1g | |
| **Cholesterol** 15mg | **4**% |
| **Sodium** 590mg | **25**% |
| **Total Carbohydrate** 28g | **10**% |
| Dietary Fiber 3g | **10**% |
| Total sugars 10g | |
| Includes 0g added sugars | **0**% |
| **Protein** 19g | |
| Vitamin D 0mg | 0% |
| Calcium 350mg | 25% |
| Iron 1mg | 4% |
| Potassium 240mg | 4% |

* The % Daily Value (DV) tells you how much a nutrient in a serving of food contributes to a daily diet. 2,000 calories a day is used for general nutrition advice.

▶ To learn more about nutrient content claims, visit www.fda.gov and search for nutrient content claims.

Recall from Chapter 1 that another type of claim, a structure/function claim, can appear on food labels. **Structure/function claims** describe how a nutrient affects human body structure or function such as "iron builds strong blood." They do not focus on disease risk reduction, as health claims do. The FDA does not approve or authorize structure/function claims; however, manufacturers are responsible for ensuring that these claims are accurate and not misleading.

# GL◯BAL PERSPECTIVE

### Front-of-Package Nutrition Labeling

The Nutrition Facts panel provides U.S. consumers with important, accurate information about a food's nutrient and calorie content. Similar labels appear on foods sold in many other countries.[15] To help busy consumers quickly make healthy food choices, the food industry has tried to condense the nutrition labels into nutrition symbols (e.g., check marks, traffic lights, stars, ratings) that are placed on the front of food packages or supermarket shelves. A variety of nutrition symbols have been used on food packages. For instance, a pink heart might indicate that a food is a good source of phytonutrients. A green star on a grocery shelf tag might indicate that a food is rich in vitamins and low in calories.

To prevent consumer confusion, the FDA invited the food industry to help it develop uniform eligibility criteria for front-of-package food labels. The goal is to create easy-to-understand labels that consumers can trust and use to choose healthier diets.[16] Several other countries already have regulated front-of-package labeling or are working to develop it. For instance, Denmark, Sweden, and Norway use a keyhole symbol to indicate foods that meet certain nutritional requirements. Australia uses stars to convey nutritional value. Many other nations are working to standardize and regulate front-of-package labeling. Although not mandated by the FDA, many food packages in the U.S. now display nutrient information on package fronts.

©Jill Braaten

## Knowledge Check

1. How do Reference Daily Intakes and Daily Reference Values differ?
2. Which nutrients on Nutrition Facts panels should most people aim to keep below 100% Daily Value?
3. What requirements must a food meet before a health claim can be made about it?

# 2.3 Nutrient Composition of Foods

Nutrient databases make it possible to estimate quickly the amount of calories and many nutrients in the foods we eat. With this information, it is possible to see how closely intake matches dietary standards such as the RDA and DV. These databases also can be used to determine the nutrient density and energy density of foods.

The data in nutrient databases are the results of thousands of analytical chemistry studies conducted in laboratories around the world. They are easy to use; however, generating the data requires years of research to develop laboratory methods that produce accurate and reliable data and years more to analyze samples of many different foods and then construct the data tables. Keep in mind that there are many nutrients, and all require a unique laboratory analysis method. To get an idea of the enormity of this task, multiply the number of nutrients by all the different foods (plant and animal species) people eat. As you might guess, countless foods have not been analyzed yet, and some nutrients have been measured in only a limited number of foods.

Nutrient values in the databases are average amounts found in the analyzed samples of the food. Currently, these values cannot account for the many factors that affect nutrient levels in the food we eat, factors such as farming conditions (e.g., soil type, fertilizers, weather, season, geographic region, genetic differences in plant varieties and animal breeds, and animal feed), maturity and ripeness of plants when harvested, food processing, shipping conditions, storage time, and cooking processes. For instance, the vitamin C content of an orange is influenced by where it was grown, the variety of the orange, and how ripe it was

Computerized nutrient data tables provide a quick and easy way to discover just how nutrient and energy dense the foods you eat are. Visit this website: fdc.nal.usda.gov/.

Evan Lorne/Shutterstock

when picked. It also is affected by how long it took to get the orange to the store where you bought it, the temperature of the truck that delivered it there, and how long it stayed in your refrigerator before you ate it. Nutrient databases also cannot account for how nutrients are handled in the body—as you'll see in later chapters, the ability to absorb nutrients, especially minerals, can be affected by factors such as medications, compounds in foods, and digestive disorders.

The variations in nutrient content do not mean that nutrient databases are unreliable or that you cannot depend on food to supply nutrients in amounts that support optimal health. But it is wise to view nutrient databases as tools that approximate nutrient content, rather than precise measurements. Even with these limitations, nutrient databases are important tools for estimating calories and nutrients.

## Putting Nutrient Databases into Action to Determine Energy Density and Dietary Intake

Nutrient databases can be used in many ways, including calculating a food's energy density. **Energy density** is determined by comparing a food's calorie content per gram weight of the food. Energy-dense foods are high in calories but weigh very little. Examples include nuts, cookies, most fried foods, and snack foods. For example, there are more than 5.5 kcal in 1 gram of bacon. Foods low in energy density contain large amounts of water, which makes them weigh a lot, but they contain few calories (keep in mind that water is calorie free). Low-energy-dense foods include fruits, vegetables, and any other food that incorporates lots of water during cooking, such as stews, casseroles, and oatmeal (Table 2-4). Lettuce, for instance, has about 0.1 calorie in a gram. As you'll see in Chapter 10, foods that are low in energy density help a person feel full, whereas foods with high energy density must be eaten in greater amounts in order to contribute to fullness.[7,17] Thus, low-energy-dense foods can help keep calorie intake under control.[18,19] And foods with high energy density can help people with poor appetites, such as some older people, maintain or gain weight.

You also can use nutrient databases to find out the amounts of nutrients and calories you consume. This involves locating a food you ate and noting the quantity of each nutrient. If you ate more or less than the serving size stated, you will need to adjust the values. For instance, if you ate 4 ounces of cheese and the database values are for 2 ounces, you'll need to double the values in the table. Because not every food has been analyzed, you may need to select a food that is similar to the one you actually ate. If you ate Roquefort cheese or Bob's Pizza, for example, you may need to use the values for blue cheese or Tom's Pizza. Many combination foods (e.g., tuna salad, bean burritos) may not be included in the tables; for these foods, you will need to identify the ingredients used, estimate the amounts in the recipe (e.g., 2 oz tuna, 2 tbsp mayonnaise), and look up the nutrient values for each ingredient. Becoming aware of the amounts of nutrients and calories in the foods you eat can help you improve the healthfulness of your diet.

**Table 2-4　Energy Density of Common Foods (Listed in Relative Order)**

| Very Low Energy Density (Less Than 0.6 kcal/g) | Low Energy Density (0.6 to 1.5 kcal/g) | Medium Energy Density (1.5 to 4 kcal/g) | High Energy Density (Greater Than 4 kcal/g) |
|---|---|---|---|
| Lettuce | Whole milk | Eggs | Graham crackers |
| Tomatoes | Oatmeal | Ham | Fat-free sandwich cookies |
| Strawberries | Cottage cheese | Pumpkin pie | Chocolate |
| Broccoli | Beans | Whole-wheat bread | Chocolate chip cookies |
| Salsa | Bananas | Bagels | Tortilla chips |
| Grapefruit | Broiled fish | White bread | Bacon |
| Fat-free milk | Non-fat yogurt | Raisins | Potato chips |
| Carrots | Ready-to-eat breakfast cereals with 1% low-fat milk | Cream cheese | Peanuts |
| Vegetable soup | Plain baked potato | Cake with frosting | Peanut butter |
| | Cooked rice | Pretzels | Mayonnaise |
| | Spaghetti noodles | Rice cakes | Butter or margarine |
| | | | Vegetable oils |

Source: Data adapted from Rolls B, Barnett RA. *Volumetrics.* New York: HarperCollins; 2000.

## Expert Perspective from the Field

### Menu Labeling: How Many Calories Are in That?

When eating out, have you ever wondered how many calories are in the food you are ordering? Or if it is high in fat or sodium? Keeping track of calorie and nutrient intake when eating out is challenging because, until recently, few restaurants, delis, and cafeterias provided nutrition information about menu items. In addition, many consumers and nutrition experts greatly underestimate the calories in restaurant meals.[20] You might be surprised to learn that a small milkshake has more calories than a large fries or that the calorie count of a fast-food fish sandwich surpasses that of a cheeseburger.

In 1990, the U.S. Congress required the use of Nutrition Facts panels on packaged foods because they believed that consumers have the right to know the nutrition content of these foods. However, these regulations did not include foods sold in restaurants. Now, restaurants and other businesses serving foods (e.g., movie theaters, amusement parks) that have 20 or more locations with the same name and similar menus must post calorie information for standard menu items on menus and menu boards next to the price or name of the food. Temporary menu items (e.g., daily specials) and condiments served on counters or on the table are exempt from labeling. If requested, restaurants also must provide additional written nutrition information about total calories, total fat, calories from fat, saturated fat, *trans* fat, cholesterol, sodium, total carbohydrates, fiber, sugars, and protein.[21] Vending machine operators with at least 20 machines must post calories in the foods sold in the machine.

Dr. Margo Wootan* pointed out that "having access to nutrition information is important because restaurant food represents a significant portion of the food and calories we eat. In fact, about half of the money Americans spend on food is used to buy foods prepared outside the home."[22] About one-third of all calories people in the U.S. eat are from restaurant food. Eating out can have a negative effect on health.[23] For instance, those who frequently eat at fast-food restaurants have an increased risk of obesity—probably because restaurant foods tend to be calorie rich and their large portion sizes promote overeating.

Menu labeling is an especially important tool for helping consumers choose meals that meet their health goals. However, studies of the effects of menu labeling on total calories ordered have inconsistent results.

Some studies show a positive effect of menu labeling on consumer choices; however, others find little or no effect.[24,25] Experts point out that drawing firm conclusions from the available published research is challenging because many studies were small, measured different variables, and used different research methods. Currently, researchers are developing and testing strategies to make it quicker and easier for consumers to use the menu labeling to make better food choices.

Jill Braaten/McGraw-Hill Education

An important outcome of menu labeling is the effect it has had on the calories in menu items. For example, the MenuStat project reports that two-thirds of the largest restaurant chains in the U.S. now offer new meal and beverage choices that have 12 to 20% fewer calories.[26] These lower-calorie items make it easier to make healthy choices when eating out. To learn more about menu labeling, visit www.menulabeling.org.

*Margo G. Wootan, DSc, is Vice President for Nutrition at the Center for Science in the Public Interest (CSPI), a consumer advocacy organization focusing on food, nutrition, and health. She cofounded the National Alliance for Nutrition and Activity and serves on the National Fruit and Vegetable Alliance steering committee. Dr. Wootan has testified before Congress and state legislatures about nutrition and health issues. She has received awards from the American Public Health Association, Association of State and Territorial Public Health Nutrition Directors, and Society for Nutrition Education and Behavior.*

### Knowledge Check

1. What are some factors that affect nutrient levels in food?
2. What is energy density?
3. What are some examples of high-energy-density and low-energy-density foods?

## *Critical* | Thinking

Margit would benefit from more variety in her diet. What are some practical tips she can use to increase her fruit and vegetable intake?

*Figure 2-6* Key recommendations from the Dietary Guidelines for Americans 2020–2025.

## 2.4  Dietary Guidelines for Americans

The diets of many people in the U.S. and Canada are too high in calories, fat, saturated fat, cholesterol, sugar, salt, and alcohol.[27] Many consume insufficient amounts of whole grains, fruits, and vegetables. These dietary patterns put many of us at risk of major chronic "killer" diseases, such as cardiovascular disease and cancer. In response to concerns about the prevalence of these killer disease patterns, every 5 years since 1980, the U.S. Department of Agriculture (USDA) and U.S. Department of Health and Human Services (DHHS) have published the Dietary Guidelines for Americans (Dietary Guidelines, for short).

The Dietary Guidelines are the foundation of the U.S. government's nutrition policy and education. They reflect what scientific experts believe is the most accurate and up-to-date scientific knowledge about nutritious diets, physical activity, and related healthy lifestyle choices. The Dietary Guidelines are designed to meet nutrient needs while reducing the risk of obesity, hypertension, cardiovascular disease, type 2 diabetes, osteoporosis, alcoholism, and foodborne illness. The Dietary Guidelines also guide government nutrition programs, research, food labeling, and nutrition education and promotion. For example, the Dietary Guidelines provide the scientific basis for the design of federal nutrition assistance programs, such as the USDA's school breakfast and lunch programs, SNAP (Supplemental Nutrition Assistance Program), and the WIC Program (Special Supplemental Nutrition Program for Women, Infants, and Children). In addition, MyPlate is based on the recommendations of the Dietary Guidelines (see the next section in this chapter). Figure 2-6 summarizes the Dietary Guidelines.

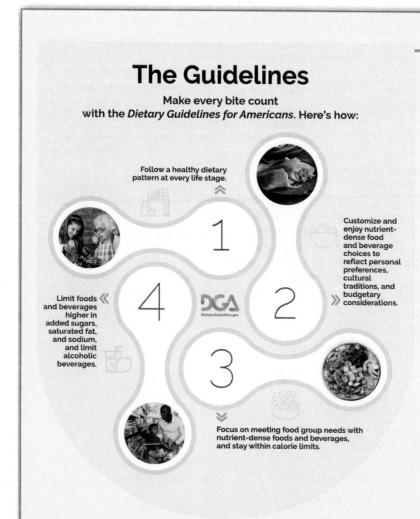

## The Guidelines

**Make every bite count**
with the *Dietary Guidelines for Americans*. Here's how:

Follow a healthy dietary pattern at every life stage.

Customize and enjoy nutrient-dense food and beverage choices to reflect personal preferences, cultural traditions, and budgetary considerations.

Limit foods and beverages higher in added sugars, saturated fat, and sodium, and limit alcoholic beverages.

Focus on meeting food group needs with nutrient-dense foods and beverages, and stay within calorie limits.

DGA
DietaryGuidelines.gov

## Key Recommendations

**1** **Follow a healthy dietary pattern at every life stage.**
At every life stage—infancy, toddlerhood, childhood, adolescence, adulthood, pregnancy, lactation, and older adulthood—it is never too early or too late to eat healthfully.

- **For about the first 6 months of life,** exclusively feed infants human milk. Continue to feed infants human milk through at least the first year of life, and longer if desired. Feed infants iron-fortified infant formula during the first year of life when human milk is unavailable. Provide infants with supplemental vitamin D beginning soon after birth.

- **At about 6 months,** introduce infants to nutrient-dense complementary foods. Introduce infants to potentially allergenic foods along with other complementary foods. Encourage infants and toddlers to consume a variety of foods from all food groups. Include foods rich in iron and zinc, particularly for infants fed human milk.

- **From 12 months through older adulthood,** follow a healthy dietary pattern across the lifespan to meet nutrient needs, help achieve a healthy body weight, and reduce the risk of chronic disease.

**2** **Customize and enjoy nutrient-dense food and beverage choices to reflect personal preferences, cultural traditions, and budgetary considerations.**
A healthy dietary pattern can benefit all individuals regardless of age, race, or ethnicity, or current health status. *The Dietary Guidelines* provides a framework intended to be customized to individual needs and preferences, as well as the foodways of the diverse cultures in the U.S.

Source: U.S. Department of Health and Human Services and U.S. Department of Agriculture. 2020 -2025 Dietary Guidelines for Americans. 9th Edition, 2020. Available at dietaryguidelines.gov.

To be sure the Dietary Guidelines are practical, relevant, and achievable, the experts who develop the Dietary Guidelines depend on high-quality research data. For example, they use data from national studies to understand the current health and dietary intake of Americans. They also look at food pattern research and use modeling techniques to understand how changing the types and amounts of food eaten would affect nutrient intake and health status. Systematic reviews of diet and health peer-reviewed research guide the decisions the experts make as they revise the Dietary Guidelines.

A basic premise of the Dietary Guidelines is that nutrient needs should be met primarily by consuming foods.[27] Foods provide an array of nutrients and other compounds that may have beneficial effects on health. In certain cases, fortified foods and dietary supplements may be useful sources of 1 or more nutrients that otherwise might be consumed in less than recommended amounts. These practices are especially important for people whose typical food choices lead to a diet that cannot meet nutrient recommendations such as for calcium. However, dietary supplements are not a substitute for a healthful diet.

The Dietary Guidelines for Americans 2020–2025 have key recommendations intended to help people of all ages achieve healthy eating patterns. The key recommendations have interconnected relationships; thus, all need to be implemented to achieve the best health.

Other scientific groups, such as the American Heart Association, American Cancer Society, Canadian Ministries of Health, and World Health Organization, also have issued dietary recommendations. All are consistent with the spirit of the Dietary Guidelines for Americans. These scientific groups, like the Dietary Guidelines, encourage people to modify their eating behavior in ways that are both healthful and pleasurable.

Source: U.S. Department of Health and Human Services and U.S. Department of Agriculture. 2020–2025 Dietary Guidelines for Americans. 9th Edition, 2020. Available at dietaryguidelines.gov.

**3** Focus on meeting food group needs with nutrient-dense foods and beverages, and stay within calorie limits.

An underlying premise of the *Dietary Guidelines* is that nutritional needs should be met primarily from foods and beverages—specifically, nutrient-dense foods and beverages. Nutrient-dense foods provide vitamins, minerals, and other health-promoting components and have no or little added sugars, saturated fat, and sodium. A healthy dietary pattern consists of nutrient-dense forms of foods and beverages across all food groups, in recommended amounts, and within calorie limits.

The core elements that make up a healthy dietary pattern include:

- Vegetables of all types—dark green; red and orange; beans, peas, and lentils; starchy; and other vegetables
- Fruits, especially whole fruit
- Grains, at least half of which are whole grain
- Dairy, including fat-free or low-fat milk, yogurt, and cheese, and/or lactose-free versions and fortified soy beverages and yogurt as alternatives
- Protein foods, including lean meats, poultry, and eggs; seafood; beans, peas, and lentils; and nuts, seeds, and soy products
- Oils, including vegetable oils and oils in food, such as seafood and nuts

monticello/Shutterstock

**4** Limit foods and beverages higher in added sugars, saturated fat, and sodium, and limit alcoholic beverages.

At every life stage, meeting food group recommendations—even with nutrient-dense choices—requires most of a person's daily calorie needs and sodium limits. A healthy dietary pattern doesn't have much room for extra added sugars, saturated fat, or sodium—or for alcoholic beverages. A small amount of added sugars, saturated fat, or sodium can be added to nutrient-dense foods and beverages to help meet food group recommendations, but foods and beverages high in these components should be limited. **Limits are:**

- **Added sugars**—Less than 10 percent of calories per day starting at age 2. Avoid foods and beverages with added sugars for those younger than age 2.
- **Saturated fat**—Less than 10 percent of calories per day starting at age 2.
- **Sodium**—Less than 2,300 milligrams per day—and even less for children younger than age 14.
- **Alcoholic beverages**—Adults of legal drinking age can choose not to drink or to drink in moderation by limiting intake to 2 drinks or less in a day for men and 1 drink or less in a day for women, when alcohol is consumed. Drinking less is better for health than drinking more. There are some adults who should not drink alcohol, such as women who are pregnant.

Daily Goal: Meet calorie needs with nutrient-dense foods and beverages from the food groups.

**Total Calories Per Day**

**85%**  **15%**

of calories are needed per day to meet food group recommendations healthfully, in nutrient-dense forms.

of remaining calories are available for other uses (including added sugars and saturated fat).

Follow these 3 steps.

1 Meet nutritional needs primarily from nutrient-dense foods and beverages.

2 Choose a variety of options from each food group.

3 Pay attention to portion size.

Everyone has a role in shifting from current eating patterns to those that align with the Dietary Guidelines—including individuals, families, food producers and retailers, schools, worksites, business and industry, communities, and government.

John Lund/Blend Images LLC

▶ The Academy of Nutrition and Dietetics suggests 5 basic principles with regard to diet and health.

- Be realistic; make small changes over time.
- Be adventurous; try new foods regularly.
- Be flexible; balance some sweet and fatty foods with physical activity.
- Be sensible; include favorite foods in smaller portions.
- Be active; include physical activity in daily life.

## The Science Underlying the *Dietary Guidelines* Demonstrates That Healthy Eating Across the Lifespan Can Promote Health and Reduce Risk of Chronic Disease

**Birth Through 23 Months**
- Lower risk of overweight and obesity
- Lower risk of type 1 diabetes
- Adequate iron status and lower risk of iron deficiency
- Lower risk of peanut allergy
- Lower risk of asthma

**Women Who Are Pregnant or Lactating**
- Favorable cognitive development in the child
- Favorable folate status in women during pregnancy and lactation

**Children and Adolescents**
- Lower adiposity
- Lower total and low-density lipoprotein (LDL) cholesterol

**Adults, Including Older Adults**
- Lower risk of all-cause mortality
- Lower risk of cardiovascular disease
- Lower risk of cardiovascular disease mortality
- Lower total and LDL cholesterol
- Lower blood pressure
- Lower risk of obesity
- Lower body mass index, waist circumference, and body fat
- Lower risk of type 2 diabetes
- Lower risk of cancers of the breast, colon, and rectum
- Favorable bone health, including lower risk of hip fracture

## Putting the Dietary Guidelines into Action

The Dietary Guidelines can be easily incorporated into our diets.[28,29] Table 2-5 provides a variety of easy-to-implement suggestions that can improve any diet. Despite popular misconceptions, the healthy diet recommended by the Dietary Guidelines is not especially expensive. Fruits, vegetables, and low-fat and fat-free milk often are similar in price to the chips, cookies, and sugared soft drinks they should replace. Plus, there are many lower-cost options, including canned and frozen fruits and vegetables and non-fat dry milk.

When applying the Dietary Guidelines to yourself, start by taking into account your current health status and family history for specific diseases. Then, identify specific changes you need to make and develop a plan for incorporating the changes into your lifestyle. My-Plate (see the next section in this chapter) can help you design a nutritious diet that meets your needs. When your plan is ready, make a couple of changes. As changes become part of your usual routine, add another change.[30] Continue making changes until your diet is healthful and reflects the Dietary Guidelines.

When making changes, it is a good idea to see whether they are effective. Keep in mind that results of dietary changes sometimes take a while to occur. Also, note that sometimes

**Table 2-5 Recommended Diet Changes Based on the Dietary Guidelines**

| If You Usually Eat This | Try This Instead | Benefit |
|---|---|---|
| White bread | Whole-wheat bread | • Higher nutrient density<br>• More fiber |
| Sugary breakfast cereal | Low-sugar, high-fiber cereal with fresh fruit | • Higher nutrient density<br>• More fiber<br>• More phytochemicals |
| Cheeseburger with french fries | Hamburger and baked beans | • Less saturated fat and *trans* fat<br>• Less cholesterol<br>• More fiber<br>• More phytochemicals |
| Potato salad | Three-bean salad | • More fiber<br>• More phytochemicals |
| Donuts | Bran muffin or bagel with light cream cheese | • More fiber<br>• Less fat |
| Regular soft drinks | Diet soft drinks | • Fewer kcal<br>• Less sugar |
| Fruit canned in syrup | Fresh or frozen fruit<br>Fruit canned in water or juice | • Less sugar<br>• Fewer kcal |
| Boiled vegetables | Steamed or sauteed vegetables | • Higher nutrient density due to reduced loss of water-soluble vitamins |
| Canned vegetables | Fresh or frozen vegetables<br>Low-sodium or rinsed canned vegetables | • Lower in sodium |
| Fried meats | Broiled meats | • Less saturated fat |
| Fatty meats such as ribs or bacon | Lean meats, such as ground round, chicken, or fish | • Less saturated fat |
| Whole milk | Low-fat or fat-free milk | • Less saturated fat<br>• Fewer kcal |
| Ice cream | Frozen yogurt | • Less saturated fat<br>• Fewer kcal |
| Mayonnaise or sour cream salad dressing | Oil-and-vinegar dressings or light creamy dressings | • Less saturated fat<br>• Less cholesterol<br>• Fewer kcal |
| Cookies | Air-popped popcorn with minimal margarine or butter | • Fewer kcal |
| Heavily salted foods | Foods flavored primarily with herbs, spices, or lemon juice | • Less sodium |
| Chips | Pretzels | • Less fat |

## Knowledge Check

1. Which government agencies publish the Dietary Guidelines for Americans?
2. What are the Dietary Guidelines?
3. How are the Dietary Guidelines used?

*Critical* **Thinking**

Cheng has grown up eating the typical American diet. Having recently read and heard many media reports about the relationship between nutrition and health, she is beginning to look critically at her diet and is considering making changes. However, she doesn't know where to begin. What advice would you give her?

## Take Action

### Are You Putting the Dietary Guidelines into Practice?

The advice provided by the Dietary Guidelines for Americans can help you determine the healthfulness of your diet and identify changes to make. This checklist includes the major points to consider. How closely are you following the basic intent of the Dietary Guidelines?

| Yes | No | |
|---|---|---|
| | | **Do you consume a variety of nutrient-dense foods and beverages within and among the basic food groups of MyPlate?** |
| | | **Do you choose foods that limit the intake of** |
| | | Saturated fat? |
| | | Cholesterol? |
| | | Added sugars? |
| | | Salt? |
| | | Alcohol (if used)? |
| | | **Do you emphasize in your food choices** |
| | | Vegetables? |
| | | Fruits? |
| | | Legumes (beans)? |
| | | Whole-grain breads and cereals? |
| | | Fat-free or low-fat milk or equivalent milk products? |
| | | **Do you keep your body weight in a healthy range by balancing energy intake from foods and beverages with energy expended?** |
| | | **Do you engage in at least 60 minutes of physical activity (above usual activity) at work or home on most days of the week?** |
| | | **Do you wash your hands, food contact surfaces, and fruits and vegetables before preparation?** |
| | | **Do you cook foods to a safe temperature to kill harmful microorganisms?** |

Pixtal/AGE Fotostock

changes don't result in the outcome you anticipated. Some people, for instance, who eat a diet low in saturated fat may not see a decrease in blood cholesterol because of genetic background.[31] If the changes are not leading to the health improvements you anticipated, it is a good idea to see a registered dietitian nutritionist or physician.

## 2.5 MyPlate

Since the early 20th century, researchers have worked to translate the science of nutrition into practical terms so that consumers could estimate whether their nutritional needs were being met. A plan with 7 food groups, based on foods traditionally eaten by North Americans, was an early format designed by the USDA. Daily food choices were to include items from each group. This plan was simplified in the mid-1950s to a 4-food-group plan: milk, meat, fruit and vegetable, and bread and cereal groups. In 1992, this plan was depicted using a pyramid shape. It was updated in 2005 to reflect new scientific knowledge and was called MyPyramid, Steps to a Healthier You. In 2011, the plan was simplified to make it easier for consumers to make healthy food choices. The most current healthy eating plan is called MyPlate (Fig. 2-7).[32]

Aimee M Lee/Shutterstock

MyPlate depicts the key elements of a healthy diet. It emphasizes the fruit, vegetable, grain, protein, and dairy food groups. The goal of MyPlate is to remind consumers to think about building a healthy plate at mealtimes and to visit the www.myplate.gov website to learn more about healthy eating. MyPlate recommendations are consistent with the Dietary Guidelines for Americans and focus on these key behaviors:

*Balancing Calories*
- Enjoy your food, but eat less.
- Avoid oversized portions.

*Foods to Increase*
- Make half your plate fruits and vegetables.
- Make at least half your grains whole grains.
- Switch to fat-free (skim) or low-fat (1%) milk.

*Foods to Reduce*
- Compare sodium in foods like soup, bread, and frozen meals and choose the foods with lower numbers.
- Drink water instead of sugary drinks.

The www.myplate.gov website has many tips and resources to help consumers plan food choices to meet MyPlate goals. This website provides in-depth information for every food group, including recommended daily intake amounts expressed in commonly used measures, such as cups and ounces, with examples and everyday tips. The website also includes recommendations for choosing healthy oils and physical activity.

## Putting MyPlate into Action

To put MyPlate into action, begin by estimating your energy needs (Fig. 2-8 or visit www.myplate.gov). Next, use Table 2-6 to discover how your energy needs correspond to the recommended number of servings from each food group. The servings are based on the sizes listed in Table 2-7.

**Figure 2-7** Evolution of USDA food guides.

Source: U.S. Department of Agriculture

▶ Nations around the world have created graphics to symbolize the components of a healthy diet. In the U.S., we have a plate. Many of the designs use important cultural symbols. In Japan, they have a spinning top. A ger (tent) is used in Mongolia. Benin's food guide uses a traditional African house design. A cedar tree is the image used in Lebanon. Honduras's food guide design is a cooking pot. Qatar uses a scallop shell. Visit this website at the Food and Agriculture Organization of the United Nations to compare the food guides used around the globe: www.fao.org/nutrition/education/food-dietary-guidelines. Think about how the foods are organized in each country. How are the components of a healthy diet the same? How do they differ?

**Table 2-6 MyPlate Recommendations for Daily Amounts of Foods to Consume from the Food Groups Based on Energy Needs**

| Energy Intake | 1000 | 1200 | 1400 | 1600 | 1800 | 2000 | 2200 | 2400 | 2600 | 2800 | 3000 | 3200 |
|---|---|---|---|---|---|---|---|---|---|---|---|---|
| Grains[a] | 3 oz-eq | 4 oz-eq | 5 oz-eq | 5 oz-eq | 6 oz-eq | 6 oz-eq | 7 oz-eq | 8 oz-eq | 9 oz-eq | 10 oz-eq | 10 oz-eq | 10 oz-eq |
| Vegetables[b,c] | 1 c | 1.5 c | 1.5 c | 2 c | 2.5 c | 2.5 c | 3 c | 3 c | 3.5 c | 3.5 c | 4 c | 4 c |
| Fruits | 1 c | 1 c | 1.5 c | 1.5 c | 1.5 c | 2 c | 2 c | 2 c | 2 c | 2.5 c | 2.5 c | 2.5 c |
| Dairy[d] | 2 c | 2.5 c | 2.5 c | 2.5–3 c | 2.5–3 c | 2.5–3 c | 3 c | 3 c | 3 c | 3 c | 3 c | 3 c |
| Protein foods[e] | 2 oz-eq | 3 oz-eq | 4 oz-eq | 5 oz-eq | 5 oz-eq | 5.5 oz-eq | 6 oz-eq | 6.5 oz-eq | 6.5 oz-eq | 7 oz-eq | 7 oz-eq | 7 oz-eq |
| Saturated fat maximum | 11 g | 13 g | 16 g | 18 g | 20 g | 22 g | 24 g | 27 g | 29 g | 31 g | 33 g | 36 g |
| Added sugar maximum | 25 g | 30 g | 35 g | 40 g | 45 g | 50 g | 55 g | 60 g | 65 g | 70 g | 75 g | 80 g |

Abbreviations: c = cup or cups; oz-eq = ounce equivalents; tsp = teaspoon; g=grams.
[a]At least half of these servings should be whole-grain varieties.
[b]Vegetables are divided into 5 subgroups [dark-green vegetables, orange vegetables, beans and peas (legumes), starchy vegetables, and other vegetables]. Over a week's time a variety of vegetables should be eaten.
[c]Beans and peas (legumes) can be counted either as vegetables (beans and peas subgroup) or in the protein foods group.
[d]Most dairy servings should be fat free or low fat.
[e]Oils are not a food group, but they provide essential nutrients. Fish, nuts, and vegetable oils (e.g., olive, sunflower, and Canola oils) are healthiest. Limit solid fats, such as butter, stick margarine, shortening, and meat fat, as well as foods that contain these.

| | Calorie Range (kcal) | |
|---|---|---|
| Children | Sedentary ⟶ | Active |
| 2–3 years | 1000 ⟶ | 1400 |
| **Females** | | |
| 4–8 years | 1200 ⟶ | 1800 |
| 9–13 | 1600 ⟶ | 2200 |
| 14–18 | 1800 ⟶ | 2400 |
| 19–30 | 2000 ⟶ | 2400 |
| 31–50 | 1800 ⟶ | 2200 |
| 51+ | 1600 ⟶ | 2200 |
| **Males** | | |
| 4–8 years | 1400 ⟶ | 2000 |
| 9–13 | 1800 ⟶ | 2600 |
| 14–18 | 2200 ⟶ | 3200 |
| 19–30 | 2400 ⟶ | 3000 |
| 31–50 | 2200 ⟶ | 3000 |
| 51+ | 2000 ⟶ | 2800 |

*Figure 2-8* Estimates of calorie needs by age and activity levels. *Sedentary* means a lifestyle that includes only the light physical activity associated with typical day-to-day life. *Active* means a lifestyle that includes physical activity equivalent to walking more than 3 miles per day at 3 to 4 miles per hour in addition to the light physical activity associated with typical day-to-day life.

### Table 2-7  MyPlate Food Serving Sizes

**Grains Group**

| 1-ounce equivalent = | 1 slice of bread<br>1 cup of ready-to-eat breakfast cereal<br>½ cup of cooked cereal, rice, pasta, or bulgur<br>1 mini bagel or small tortilla<br>½ muffin<br>3 cups of popcorn |
|---|---|

**Vegetable Group**

| 1 cup = | 1 cup of raw or cooked vegetables<br>1 cup of vegetable juice<br>2 cups of raw leafy greens |
|---|---|

**Fruits Group**

| 1 cup = | 1 cup of fruit<br>1 cup of 100% fruit juice<br>½ cup of dried fruit |
|---|---|

**Dairy Group**

| 1 cup = | 1 cup of milk, yogurt, or calcium-fortified soymilk<br>1 cup of frozen yogurt or pudding made with milk<br>1½ cups of ice cream<br>1½ ounces of natural cheese<br>2 ounces of processed cheese |
|---|---|

**Protein Group**

| 1-ounce equivalent = | 1 ounce of meat, poultry, fish, or cooked tempeh<br>1 egg<br>1 tablespoon of peanut butter or hummus<br>¼ cup of cooked beans<br>½ ounce of nuts or seeds |
|---|---|

**Oils**

| 1 teaspoon | 1 teaspoon of vegetable oil, fish oil, oil-rich foods (e.g., mayonnaise, soft margarine) |
|---|---|

Oils are not a food group, but they provide essential nutrients. Fish, nuts, and vegetable oils (e.g., olive, sunflower, and Canola oils) are healthiest. Limit solid fats, such as butter, stick margarine, shortening, and meat fat, as well as foods that contain these. Children 2–8 years should limit oil to 3–4 teaspoons/day; Older children and teens 9–19 years should limit oils to 5–6 teaspoons/day; adults should limit oils to 5–7 teaspoons/day.

Mizina/Getty Images

When planning menus using MyPlate, keep these points in mind.

1. No specific food is required for good nutrition. Every food supplies some nutrients but provides insufficient amounts of at least 1 essential nutrient.
2. No individual food group provides all essential nutrients in adequate amounts (Table 2-8). Each food group makes an important, distinctive contribution to nutritional intake.
3. The foods within a group may vary widely with respect to nutrients and energy content. For example, the energy content of 3 ounces of baked potato is 98 calories, whereas that of 3 ounces of potato chips is 470 calories.
4. To keep calories under control, pay close attention to the serving size of each choice when following MyPlate. Figure 2-9 provides a convenient guide for estimating portion sizes. Note that serving sizes listed for 1 serving in a MyPlate group are often less than individuals typically serve themselves or the sizes of portions served in many restaurants.[32]
5. Variety is the key to getting the array of nutrients offered by each food group. Variety starts with including foods from every food group and then continues by consuming a variety of foods within each group. The nutritional adequacy of diets planned using MyPlate depends greatly on the selection of a variety of foods (Table 2-9).

**Table 2-8 Major Nutrient Contributions of Groups in the MyPlate Food Guide Plan**

| Grains | Vegetables | Fruits | Dairy | Protein | Oils |
|---|---|---|---|---|---|
| Jules Frazier/Getty Images | Stockbyte/Stockdisc/Getty Images | ©Stockdisc/ PunchStock RF | Ingram Publishing / Alamy | ©Ingram Publishing/ Alamy RF | C Squared Studios/ Getty Images |
| Carbohydrate | Carbohydrate | Carbohydrate | Calcium | Protein | Fat |
| Thiamin‡ | Vitamin A | Vitamin A | Phosphorus | Thiamin | Essential fatty acids |
| Riboflavin‡ | Vitamin C | Vitamin C | Carbohydrate | Riboflavin | Vitamin E |
| Niacin‡ | Folate | Folate | Protein | Niacin | |
| Folate‡ | Magnesium | Magnesium | Riboflavin | Vitamin B-6 | |
| Magnesium§ | Potassium | Potassium | Vitamin A | Folate* | |
| Iron‡ | Fiber | Fiber | Vitamin D | Vitamin B-12† | |
| Zinc§ | | | Magnesium | Phosphorus | |
| Fiber§ | | | Zinc | Magnesium* | |
| | | | | Iron | |
| | | | | Zinc | |

*Primarily in plant protein sources; †Only in animal foods; ‡Both enriched or whole grain; §Whole grains. Bread:
Jules Frazier/Getty Images RF; Broccoli: ©Stockbyte/Getty Images RF; Strawberry: ©Stockbyte/Getty Images RF; Cheese: ©Ingram Publishing/Alamy RF; Meat: ©Ingram Publishing/Alamy RF; Bottles of oil:
©C Squared Studios/Getty Images RF

Here are some points that will help you choose the most nutritious diet.

- *Grains group:* Make at least half of your grain choices those that are whole grain. Whole-grain varieties of breads, cereals, rice, and pasta have the greatest array of nutrients and more fiber than other foods in this group. A daily serving of a whole-grain, ready-to-eat breakfast cereal is an excellent choice because the vitamins and minerals typically added to it, along with fiber it naturally contains, help fill in potential nutrient gaps. Although cakes, pies, cookies, and pastries are made from grains, these foods are higher in calories, fat, and sugar and lower in fiber, vitamins, and minerals than other foods in this group. The most nutritious diets limit the number of grain products with added fat or sugar.
- *Vegetables group:* Variety within the vegetables group (Table 2-10) is especially important because different types of vegetables are rich in different nutrients and phytochemicals (see Table 1-3 in Chapter 1).[33] For instance, dark-green vegetables (e.g., kale, bok

▶ Some research suggests that increasing variety in a diet can lead to overeating. Thus, as you include a wide variety of foods in your diet, pay attention to total energy intake as well.

▶ There are no "good" or "bad" foods—to make all foods fit into a nutritious dietary pattern and reduce the risk of many chronic diseases, balance calories eaten with needs, eat a variety of foods, and limit empty calorie foods.

**Portion sizes**

2 tbsp salad dressing, peanut butter, margarine, etc.

= 2 tbsp

Baked potato
Small/medium fruit
Ground or chopped food
Bagel
English muffin

= ½ to ⅔ cup

3 oz meat, poultry, or fish

= ½ to ¾ cup

Large apple or orange
1 cup ready-to-eat breakfast cereal

= 1 cup

*Figure 2-9* A golf ball, tennis ball, deck of cards, and baseball are standard-size objects that make convenient guides for judging MyPlate serving sizes. Your hand provides an additional guide (for the greatest accuracy, compare your fist with a baseball and adjust the estimates shown here accordingly).

Golf Ball: ©Ryan McVay/Getty Images RF; Tennis Ball: ©C Squared Studios/Getty Images RF; Cards: ©Stockbyte/Getty Images RF; Baseball: ©C Squared Studios/Getty Images; Hand/Serving Sizes, All: ©McGraw-Hill Education/Mark Dierker, photographer

**Table 2-9  Putting MyPlate into Practice: How Many MyPlate Servings from Each Food Group Does This Menu Provide?**

| Breakfast | | Lunch | | Study Break Snack | |
|---|---|---|---|---|---|

| Breakfast | | Lunch | | Study Break Snack | |
|---|---|---|---|---|---|
| 1 orange | Fruits | 8-inch pizza | Grains | 1 cup non-fat yogurt | Dairy |
| ¾ cup low-fat granola | Grains | *topped with* ⅓ cup | Vegetables | *topped with* ½ cup fresh | Fruits |
| *topped with* 2 tbsp dried | | chopped vegetables and | | fruit | |
| cranberries | Fruits | 2 oz low-fat cheese | Dairy | | |
| 1 cup fat-free milk | Dairy | 2 cups green salad | Vegetables | | |
| Optional: coffee or tea | | *topped with* ¾ oz nuts | Protein | | |
| | | 5 tsp salad dressing | Oils | | |
| | | Optional: diet soft drink or | | | |
| | | iced tea | | | |

| Dinner | | Late-Night Snack | | Nutrient Breakdown |
|---|---|---|---|---|

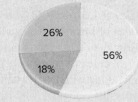

Calories: 1800
Carbohydrate: 56% of kcal
Protein: 18% of kcal
Fat: 26% of kcal

Ermin Gutenberger/iStock/Getty
Images

Tetra Images/Getty Images

| Dinner | | Late-Night Snack |
|---|---|---|
| 3½ oz salmon | Protein | 3 small chocolate chip |
| ½ cup asparagus | Vegetables | cookies as added sugar |
| 1¼ cups salsa | | |
| (½ cup fresh fruit and | Fruits | |
| ¾ cup vegetables) | Vegetables | |
| Sparkling water | | |

26%
56%
18%

Granola: ©Mitch Hrclicka/Getty Images RF; Pizza: ©Ingram Publishing RF; Yogurt: ©Jonelle Weaver/Getty Images RF; Salmon: © Tetra Images/Getty Images RF; Cookies:©ermingut/Getty Images

▶ The Exchange System is another menu planning tool. It organizes foods based on energy, protein, carbohydrate, and fat content. The result is a framework for designing diets, especially for the treatment of diabetes. For more information on the Exchange System, see Appendix E.

▶ For more suggestions on how to increase fruit, vegetable, and phytochemical intake, visit fruitsandveggies.org.

▶ A meal consisting of a bean burrito, a lettuce and tomato salad with oil-and-vinegar dressing, a glass of milk, and an apple covers all groups.

choy) tend to be good sources of iron, calcium, folate, and vitamins A and C. Vegetables with orange flesh (e.g., carrots, acorn squash) are rich in beta-carotene, the precursor to vitamin A. Starchy vegetables (e.g., corn) provide B-vitamins and carbohydrates. Legumes (beans and peas) also are in the protein group because they are rich in protein. Other vegetables, including celery, onions, and radishes, provide a wide array of phytochemicals, vitamins, and minerals. It is important to eat a variety of vegetables each week from all 5 vegetable subgroups. A goal of MyPlate is for the vegetables group and fruits group to make up half your plate.

- *Fruits group:* Like vegetables, fruits also vary in the nutrients and phytochemicals they contain. To be sure you get the fiber that fruits have to offer, keep the amount of fruit juice to less than half of total fruit intake. Select 100% fruit juice—punches, ades, fruit-flavored soft drinks, and most fruit drinks contain little or no juice but do have substantial amounts of added sugar and do not count toward fruit servings.
- *Dairy group:* Choose primarily low-fat (1%) and fat-free items from the dairy group, such as part skim cheese, fat-free milk, and low-fat yogurt. These foods contain all the nutrients in other milk products, except they are lower in fat, saturated fat, and cholesterol. In addition, go easy on dairy desserts (e.g., pudding, ice cream) and chocolate milk because

**Table 2-10  Vegetable Subgroup Recommendations per Week***

| | Dark-Green Vegetables  | Orange Vegetables  | Beans and Peas  | Starchy Vegetables 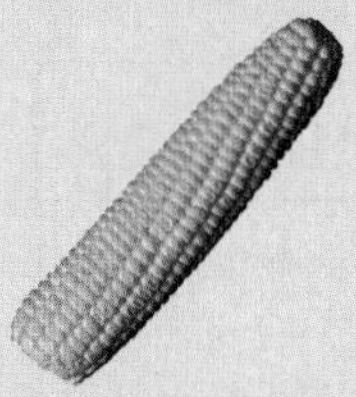 | Other Vegetables 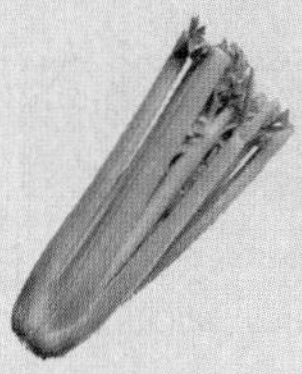 |
|---|---|---|---|---|---|
| **Life-Span Group** | | | | | |
| **Children** | | | | | |
| 2–3 years old | ½ cup | 2½ cups | ½ cup | 2 cups | 1½ cups |
| 4–8 years old | 1 cup | 3 cups | ½ cup | 3½ cups | 2½ cups |
| **Girls** | | | | | |
| 9–13 years old | 1½ cups | 4 cups | 1 cup | 4 cups | 3½ cups |
| 14–18 years old | 1½ cups | 5½ cups | 1½ cups | 5 cups | 4 cups |
| **Boys** | | | | | |
| 9–13 years old | 1½ cups | 5½ cups | 1½ cups | 5 cups | 4 cups |
| 14–18 years old | 2 cups | 6 cups | 2 cups | 6 cups | 5 cups |
| **Women** | | | | | |
| 19–30 years old | 1½ cups | 5½ cups | 1½ cups | 5 cups | 4 cups |
| 31–50 years old | 1½ cups | 5½ cups | 1½ cups | 5 cups | 4 cups |
| 51+ years old | 1½ cups | 4 cups | 1½ cups | 4 cups | 3½ cups |
| **Men** | | | | | |
| 19–30 years old | 2 cups | 6 cups | 2 cups | 6 cups | 5 cups |
| 19–30 years old | 2 cups | 6 cups | 2 cups | 6 cups | 5 cups |
| 51+ years old | 1½ cups | 5½ cups | 1½ cups | 5 cups | 4 cups |

*It is not necessary to eat vegetables from each subgroup daily; however, over a week they should be varied, as shown in this table.

Spinach: ©Florea Marius Catalin/E+/Getty Images; Carrots: ©Ingram Publishing/Alamy RF; Kidney Beans: ©MRS.Siwaporn/Shutterstock RF; Corn: ©Stockdisc/PunchStock RF; Celery: ©Stockbyte/Getty Images RF

of the added sugar. By reducing energy intake in this way, you free up calories that can be used to select more items from other food groups.

- *Protein group:* Except for beans and seafood, most foods in this group are high in fat. Meat, poultry, seafood, and eggs also supply cholesterol. When selecting foods from the protein group, focus on seafood, lean meat, poultry without skin, and beans—these foods are lower in fat than others in this group. To further reduce fat, avoid fried foods and trim away any fat you see on meat. Include protein-rich plant foods, such as beans and nuts, at least several times a week because many are rich in vitamins (e.g., vitamin E), minerals (e.g., magnesium), and fiber and contain less saturated fat than meat.

- *Oils:* Although not a food group, small amounts of oils are needed to supply you with health-promoting fats, called essential fatty acids (see Chapter 6). Oils are the fats from fish and plants that are liquid at room temperature. Include some plant oils on a daily basis, such as those in salad dressing and olive oil, and eat fish at least twice a week.

- *Empty calorie foods:* These are foods from the grains, vegetables, fruits, dairy, and protein groups that are high in solid fats and/or added sugars. The calories from solid fats and

Typical restaurant portions contain numerous servings from the individual groups in MyPlate.

Ingram Publishing/SuperStock

added sugars in a food are called empty calories because they add calories to the food but few or no nutrients (these calories are "empty" of nutrients). Most people eat far more empty calories than is considered healthy. Empty calories should be kept at the level that matches your calorie needs (see Table 2-6). These provide the most empty calories for Americans:

- Cakes, cookies, pastries, donuts, and ice cream (contain both solid fat and added sugars)
- Beverages (soft, energy, sports, and fruit drinks contain added sugars)
- Cheese, pizza, sausages, hot dogs, bacon, and ribs (contain solid fat)

Remember, a healthy diet includes foods from the grains, vegetables, fruits, dairy, and protein food groups in the recommended amounts. Variety means eating many different foods from each of these food groups. Variety makes meals more interesting and helps ensure that a diet contains sufficient nutrients. For example, carrots—a rich source of a pigment that forms vitamin A in our bodies—may be your favorite vegetable; however, if you choose carrots every day as your only vegetable source, you may miss out on other important vitamins supplied by other types of vegetables. This concept is true for all groups of foods.

It also is important to keep portion sizes under control so that you can eat a balanced and varied diet without consuming more calories, fat, cholesterol, sugar, and sodium than you need. Portion control requires only some simple planning and doesn't have to mean deprivation and misery. For example, if you eat a food that is relatively high in fat, salt, and energy, such as a bacon cheeseburger, it is a good idea to choose foods the rest of the day that are less concentrated sources of these nutrients, such as fruits and salad greens. If you prefer reduced-fat milk to fat-free milk, decrease the fat in other foods you eat. You could use low-fat salad dressings, choose a baked potato instead of french fries, or opt for jam instead of butter on toast. You also could choose smaller servings of high-fat or high-sugar foods you enjoy, such as regular soft drinks or chocolate. Overall, it's best to strive for smaller serving sizes of some foods (rather than eliminate these foods altogether) and include mostly nutrient-dense foods.

## Rating Your Current Diet

Regularly comparing your daily food intake with MyPlate recommendations for your age, biological sex, and physical activity level is a relatively simple way to evaluate your overall diet. The diets of many adults don't match the recommendations—many eat too few servings of whole grains, vegetables, fruits, and dairy products and go overboard on meat, oil, and empty calorie intake. Knowing how your diet stacks up can help you determine which nutrients likely are lacking and how you can take steps to improve. For example, if you do not consume enough servings from the dairy group, your calcium intake is most likely too low, so you'll need to find calcium-rich foods you enjoy, such as calcium-fortified orange juice or non-fat yogurt.

Customizing MyPlate to accommodate your own food habits may seem a daunting task now, but it is not difficult once you start using it. Try a few of the recommended diet changes in Table 2-5. After a few weeks, add a few more. Implementing even small diet changes can have positive results. Better health will likely follow as you strive to meet your nutrient needs and balance calorie intake with your needs. In addition, the guidance from the Dietary Guidelines for Americans regarding alcohol and sodium intake and safe food preparation can help you adopt other important changes to safeguard your health.

**Critical** | **Thinking**

Aiden, described in this chapter's Case Study, would benefit from more variety in his diet. What are some practical tips he can use to increase his fruit and vegetable intake?

Fruits are a rich source of nutrients and phytochemicals.

magone/123RF

Choosing a variety of foods every day helps meet all your nutrient needs.

cobraphotography/Shutterstock.com

*Knowledge Check*

1. What are examples of foods in each MyPlate food group?
2. What are empty calories?
3. What types of vegetables should be selected over a week's time?

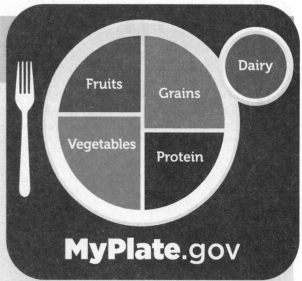

Source: U.S. Department of Agriculture

## Does Your Diet Meet MyPlate Recommendations?

In the accompanying chart, list all the foods you ate in the past 24 hours. For each food, indicate how many servings it contributes to each group based on the amount you ate (see Table 2-7 for serving sizes). Note that many of your food choices may contribute to more than 1 group. For example, toast with jam contributes to the grains group and added sugar. After entering all the values, add the number of servings consumed in each group. Finally, compare your total in each food group with the recommended number of servings shown in Table 2-6 or at the www.choosemyplate.gov website. Enter a minus sign (−) if your total falls below the recommendation, a zero (0) if it matches the recommendation, or a plus sign (+) if it exceeds the recommendation.

| Food or Beverage | Amount Eaten | Grains | Vegetables | Fruits | Dairy | Protein |
|---|---|---|---|---|---|---|
| | | | | | | |
| | | | | | | |
| | | | | | | |
| | | | | | | |
| | | | | | | |
| | | | | | | |
| | | | | | | |
| | | | | | | |
| | | | | | | |
| | | | | | | |
| **Total Eaten** | | | | | | |
| **Recommended Amount** | | | | | | |
| **Comparison of Recommended with Amount Eaten** | | | | | | |

## CASE STUDY FOLLOW-UP

Westend61/Getty Images

The most positive aspect of Aiden's diet is that it contains adequate protein, zinc, and iron because it is rich in animal protein. On the downside, his diet is low in calcium, some B-vitamins (such as folate), and vitamin C. This is because it is low in dairy products, fruits, and vegetables. It is also low in many of the phytochemical substances discussed in Chapter 1. In addition, his fiber intake is low because fast-food restaurants primarily use refined-grain products rather than whole-grain products. His diet is likely excessive in fat and sugar, too.

He could alternate between tacos and bean burritos to gain the benefits of plant proteins in his diet. He could choose a low-fat granola bar instead of the candy bar for breakfast, or he could take the time to eat a bowl of whole-grain breakfast cereal with low-fat or fat-free milk to increase his fiber and calcium intake. He also could order milk at least half the time at his restaurant visits and substitute diet soft drinks for the regular variety. This would help moderate his sugar intake. Overall, Aiden could improve his intake of fruits, vegetables, and dairy products if he focused more on variety in food choice and balance among the food groups.

# Chapter *Summary*

## 2.1 The Dietary Reference Intakes (DRIs)

differ by life stage and include Estimated Average Requirements (EARs), Recommended Dietary Allowances (RDAs), Adequate Intakes (AIs), Tolerable Upper Intake Levels (Upper Levels, or ULs), and Estimated Energy

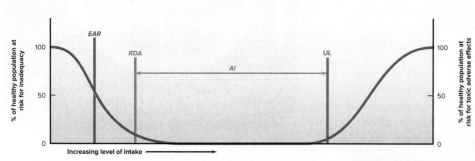

Requirements (EERs). EARs are daily nutrient intake amounts estimated to meet the needs of half of the people in a life stage. EARs are set only if a method exists for accurately measuring whether intake is adequate. RDAs are daily nutrient intake amounts sufficient to meet the needs of nearly all individuals (97 to 98%) in a life stage. RDAs are based on a multiple of the EAR. AIs are daily intake amounts set for nutrients for which there are insufficient data to establish an EAR. AIs should cover the needs of virtually all individuals in a specific life stage. A UL is the maximum daily intake amount of a nutrient that is not likely to cause adverse health effects in almost everyone. EERs are average daily energy needs. For each macronutrient, the Acceptable Macronutrient Distribution Ranges (AMDRs) provide a range of recommended intake, as a percent of energy. DRIs are intended mainly for diet planning. Nutrient density is a tool for assessing the nutritional quality of individual foods.

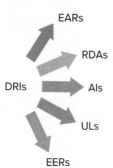

## 2.2 Daily Values (DVs)

are generic standards developed by the FDA for Nutrition Facts panels. DVs are based on Reference Daily Intakes and Daily Reference Values. Nutrition Facts panels present information for a single serving of food using serving sizes specified by the FDA. These components must be listed on most Nutrition Facts panels: total calories (kcal), total fat, saturated fat, *trans* fat, cholesterol, sodium, total carbohydrate, fiber, total and added sugars, protein, vitamin D, potassium, calcium, and iron. Food labels may include nutrient content claims, health claims, qualified health claims, and structure/function claims.

## 2.3 Nutrient databases make it possible to estimate quickly

the amount of calories and many nutrients in the foods we eat. The data in nutrient databases are the results of thousands of analytical chemistry studies. Nutrient values in the nutrient databases are average amounts found in the analyzed samples of the food. It is wise to view nutrient composition databases as tools that approximate nutrient intake, rather than precise measurements. Energy density is determined by comparing a food's calorie content with the weight of the food.

**Table 2-4 Energy Density of Common Foods (Listed in Relative Order)**

| Very Low Energy Density (Less Than 0.6 kcal/g) | Low Energy Density (0.6 to 1.5 kcal/g) | Medium Energy Density (1.5 to 4 kcal/g) | High Energy Density (Greater Than 4 kcal/g) |
|---|---|---|---|
| Lettuce | Whole milk | Eggs | Graham crackers |
| Tomatoes | Oatmeal | Ham | Fat-free sandwich cookies |
| Strawberries | Cottage cheese | Pumpkin pie | Chocolate |
| Broccoli | Beans | Whole-wheat bread | Chocolate chip cookies |
| Salsa | Bananas | Bagels | Tortilla chips |
| Grapefruit | Broiled fish | White bread | Bacon |
| Fat-free milk | Non-fat yogurt | Raisins | Potato chips |
| Carrots | Ready-to-eat breakfast cereals with 1% low-fat milk | Cream cheese | Peanuts |
| Vegetable soup | Plain baked potato | Cake with frosting | Peanut butter |
|  | Cooked rice | Pretzels | Mayonnaise |
|  | Spaghetti noodles | Rice cakes | Butter or margarine |
|  |  |  | Vegetable oils |

Source: Data adapted from Rolls B, Barnett RA. *Volumetrics*. New York: HarperCollins; 2000.

## 2.4 The Dietary Guidelines

are the foundation of the U.S. government's nutrition policy and education. They reflect what experts believe is the most accurate and up-to-date scientific knowledge about nutritious diets and related lifestyle choices. Dietary Guideline recommendations have 5 overarching recommendations: follow a healthy eating pattern; focus on variety, nutrient-dense

Source: U.S. Department of Health and Human Services and U.S. Department of Agriculture. 2015–2020 Dietary Guidelines for Americans. 8th Edition. December 2015. Available at http://health. gov/dietaryguidelines/2015 /guidelines/.

foods, and amount; limit calories from added sugars and saturated fats, and reduce sodium intake; shift to healthier food and beverage choices; and support healthy eating patterns for all.

## 2.5 MyPlate depicts the key elements of a healthy diet.

It emphasizes the fruits, vegetables, grains, protein, and dairy food groups. The goal of MyPlate is to remind consumers to think about building a healthy plate at mealtimes and to visit www.choosemyplate.gov to learn more about healthy eating. MyPlate recommendations are consistent with the Dietary Guidelines for Americans. The nutritional adequacy of diets planned using MyPlate depends on selecting a variety of foods, including grains, vegetables, fruits, dairy, protein, and oils in the recommended amounts, and keeping portion sizes under control.

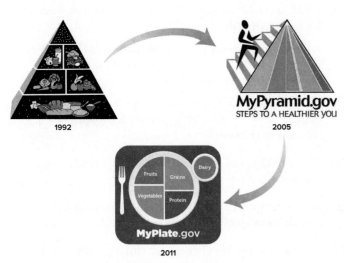

Source: U.S. Department of Agriculture

MyPlate focuses on these behaviors.

*Balancing Calories*
- Enjoy your food, but eat less.
- Avoid oversized portions.

*Foods to Increase*
- Make half your plate fruits and vegetables.
- Make at least half your grains whole grains.
- Switch to fat-free (skim) or low-fat (1%) milk.

©Stockbyte/Getty Images RF

*Foods to Reduce*
- Compare sodium in foods like soup, bread, and frozen meals and choose the foods with lower numbers.
- Drink water instead of sugary drinks.

## Expert Perspective

Having access to nutrition information when eating out is important because restaurant food represents a significant portion of the food we eat. Studies show that calorie labeling encourages companies to introduce healthier, lower calorie options.

### Nutrition Facts

1 serving per container

| Serving size | 1 pkg (260g) |
|---|---|

**Amount Per Serving**

**Calories** **390**

| | % Daily Value* |
|---|---|
| **Total Fat** 18g | **25%** |
| Saturated Fat 9g | **45%** |
| *Trans* Fat 2g | |
| **Cholesterol** 45mg | **15%** |
| **Sodium** 880mg | **40%** |
| **Total Carbohydrate** 38g | **15%** |
| Dietary Fiber 4g | **15%** |
| Total sugars 12g | |
| Includes 0g added sugars | **0%** |
| **Protein** 17g | |
| Vitamin D 0mg | 0% |
| Calcium 400mg | 30% |
| Iron 8mg | 45% |
| Potassium 300mg | 6% |

\* The % Daily Value (DV) tells you how much a nutrient in a serving of food contributes to a daily diet. 2,000 calories a day is used for general nutrition advice.

©Stockdisc/PunchStock RF

## Global Perspective

Front-of-package nutrition labeling can help consumers quickly make healthy food choices. To prevent consumer confusion and support the selection of healthier diets, easy-to-understand labels consumers can trust are needed.

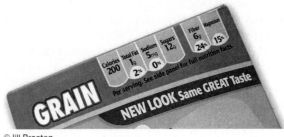

©Jill Braaten

## Study Questions

1. Which dietary standard is set at a level that meets the needs of practically all healthy people?

   a. RDA  c. UL
   b. DRI  d. EER

2. Which dietary standard is set at a level that meets the needs of about half of all healthy people?

   a. RDA  d. EER
   b. AI  e. both c and d
   c. EAR

3. Reference Daily Intakes are standards established for Nutrition Facts panels for energy-producing nutrients.

   a. true  b. false

4. Most people should aim to keep their intake of which nutrient at or below 100% Daily Value?

   a. total fat  c. vitamin A
   b. fiber  d. calcium

5. Foods that are a "good source" of a nutrient must contain at least _____% Daily Value of that nutrient.

   a. 5  c. 25
   b. 10  d. 50

6. Which factor affects nutrient levels in food?

   a. food processing  c. ripeness when harvested
   b. plant variety  d. all of the above

7. A food's energy density is determined by comparing its calorie content with the weight of the food.

   a. true  b. false

8. The FDA publishes the Dietary Guidelines for Americans.

   a. true  b. false

9. Which is true about the Dietary Guidelines for Americans?

   a. They are the foundation of the U.S. government's nutrition policy.
   b. They are designed to reduce the risk of obesity and hypertension.
   c. They guide government programs such as the USDA's school lunch program.
   d. All of the above are true.

10. Which food group is missing from this meal: cheese sandwich, macaroni salad, and orange juice?

    a. dairy group
    b. vegetables group

    c. fruits group
    d. protein group
    e. b and d
    f. a, b, and c

11. MyPlate recommends that at least 75% of the foods from the grains group should be whole grains.

    a. true  b. false

12. MyPlate recommends making at least half your plate fruits and vegetables.

    a. true  b. false

13. Which vegetables tend to be good sources of iron, calcium, folate, and vitamins A and C?

    a. starchy vegetables  d. dark-green vegetables
    b. legumes  e. all of the above
    c. vegetables with orange flesh

14. Describe the relationship between Estimated Average Requirements (EARs) and Recommended Dietary Allowances (RDAs).

15. How do RDAs and Adequate Intakes differ from Daily Values in their intention and application?

16. Why should values in nutrient composition databases be considered as approximate, not precise, values?

17. Based on the Dietary Guidelines, what are 2 changes the typical adult in the U.S. should consider making?

18. What changes would you need to make to meet the MyPlate guidelines on a regular basis?

19. What are empty calories? Which empty calorie foods do you eat most often?

20. What concerns have government agencies expressed about front-of-package nutrition labels?

21. How has menu labeling affected consumer purchasing in restaurants? How has it affected you?

Answer Key: 1-a; 2-e; 3-b; 4-a; 5-b; 6-d; 7-a; 8-b; 9-d; 10-e; 11-b; 12-a; 13-d; 14-refer to Section 2.1; 15-refer to Section 2.2; 16-refer to Section 2.3; 17-refer to Section 2.4; 18-refer to Section 2-5; 19-refer to Section 2-5; 20-refer to Global Perspective; 21-refer to Expert Perspective

# References

1. Murphy SP and others. History of nutrition: The long road leading to the Dietary Reference Intakes for the United States and Canada. *Adv Nutr*. 2016:157.

2. Chung M and others. Systematic review to support the development of nutrient reference intake values: Challenges and solutions. *Am J Clin Nutr*. 2010;92:273.

3. Flock MR and others. Long-chain omega-3 fatty acids: Time to establish a Dietary Reference Intake. *Nutr Rev*. 2013;71:692.

4. Food and Nutrition Board. *Dietary Reference Intakes: Guiding principles for nutrition labeling and fortification*. Washington, DC: National Academies Press; 2003.

5. Institute of Medicine. *Dietary Reference Intakes for energy, carbohydrate, fiber, fat, fatty acids, cholesterol, protein, and amino acids (macronutrients)*. Washington, DC: National Academies Press; 2005.

6. Institute of Medicine. Dietary Reference Intakes Tables and Application. 2019; nationalacademies.org/hmd/Activities/Nutrition/SummaryDRIs/DRI-Tables.aspx.

7. Hingle M and others. Practice paper of the Academy of Nutrition and Dietetics. Nutrient density: Foods for good health. *J Acad Nutr Diet*. 2016;116:1473.

8. Food Allergen Labeling and Consumer Protection Act of 2004, Public Law 108-282; 2004.

9. USDA, Agricultural Marketing Service. Country of origin labeling (COOL). 2019; www.ams.usda.gov/rules-regulations/cool.

10. U.S. Food and Drug Administration. CFR-Code of Federal Redulations Title 21. Part 101-Food Labeling. 2019; www.accessdata.fda.gov/scripts/cdrh/cfdocs/cfcfr/CFRSearch.cfm?fr=101.65

11. Food and Drug Administration. Title 21, Chapter I, Part 101, Subpart A-General Provisions. 2019; www.customsmobile.com/regulations/expand/title21_chapterI_part101_subpartA_section101.13#title21_chapterI_part101_subpartA_section101.13

12. Berhaupt-Glickstein A, Hallman W. Communicating scientific evidence in qualified health claims. *Crit Rev Food Sci Nutr*. 2015;57:2811.

13. U.S. FDA. FDA Announces New Qualified Health Claims for EPA and DHA Omega-3 Consumption and the Risk of Hypertension and Coronary Heart Disease. 2019; www.fda.gov/food/cfsan-constituent-updates/fda-announces-new-qualified-health-claims-epa-and-dha-omega-3-consumption-and-risk-hypertension-and.

14. European Food Information Council. Global update on nutrition labelling. 2018; www.eufic.org/en/healthy-living/article/global-update-on-nutrition-labelling.

15. U.S. Food and Drug Administration. Front-of-package labeling initiative questions & answers. 2018; www.fda.gov/food/food-labeling-nutrition/front-package-labeling-initiative-questions-answers-2009.

16. Crino M and others. The influence on population weight gain and obesity of the macronutrient composition. *Curr Obes Rep*. 2015;4:1.

17. Duffey KJ, Popkin BM. Energy density, portion size, and eating occasions: Contributions to increased energy intake in the United States, 1977–2006. *PLoS Med*. 2011;8(6):e1001050.

18. Poole SA and others. Relationship between dietary energy density and dietary quality in overweight young children: A cross-sectional analysis. *Pediatric Obes*, 2016;11:128.

19. Block JP and others. Consumers' estimation of calorie content at fast food restaurants: cross sectional observational study. *BMJ*.2013;346:f2907.

20. U.S. Food and Drug Administration. A labeling guide for restaurants and retail establishments selling away-from-home foods. 2016; www.fda.gov/food/guidanceregulation/guidancedocumentsregulatoryinformation/ucm461934.htm.

21. USDA, Economic Research Service. Food prices and spending. 2016; ers.usda.gov/data-products/ag-and-food-statistics-charting-the-essentials/food-prices-and-spending.aspx.

22. Center for Science in the Public Interest. *Literature review: Influence on nutritional information provision*. Washington, DC: Center for Science in the Public Interest; 2008.

23. Fernandez AC and others. Influence of menu labeling on food choices in real-life settings: a systematic review. *Nutr Rev*. 2016;74:534.

24. Sacco J and others. The influence of menu labeling on food choices among children and adolescents: a systematic review of the literature. *Perspect Public Health*. 2017;137:173.

25. Bleich SN and others. Calorie Changes in Chain Restaurant Menu Items: Implications for Obesity and Evaluations of Menu Labeling. *Am J Prev Med*. 2015;48:70

26. U.S. Department of Agriculture, U.S. Department of Health and Human Services. Dietary Guidelines for Americans, 2020-2025, 9th Ed. 2020. dietaryguidelines.gov.

27. Freeland-Graves JH, Nitzke S. Position of the Academy of Nutrition and Dietetics: Total diet approach to healthy living. *J Acad Nutr Diet*. 2013;113:307.

28. Gopinath B and others. Adherence to Dietary Guidelines positively affects quality of life and functional status of older adults. *J Acad Nutr Diet*. 2014;114:220.

29. Bickley P. Start by picking low hanging fruit. *BMJ*. 2012;344:e869.

30. Go V and others. Nutrient-gene interaction: Metabolic genotype-phenotype relationship. *J Nutr*. 2005;135:3016S.

31. USDA. MyPlate. 2019; www.choosemyplate.gov.

32. Keim NL and others. Vegetable variety is a key to improved diet quality in low-income women in California. *J Acad Nutr Diet*. 2014;114:430.

wavebreakmedia/Shutterstock

To achieve food sustainability (locally, nationally, and globally), we need access to adequate fertile farm land, clean water, safe and effective fertilizers, climate stability (and/or plants that grow efficiently in current climate conditions), and clean and sustainable energy sources. Learn more at **https://www.hsph .harvard.edu/nutritionsource/sustainability/**. kwest/Shutterstock

# 3

# The Food Supply

## Learning Objectives

**After studying this chapter, you will be able to**

1. Describe the health consequences of malnutrition and food insecurity for children and adults.
2. Differentiate food security from food insecurity in the U.S.
3. Describe the major U.S. government programs designed to increase food security.
4. Discuss factors that affect food sustainability.
5. Describe how organic foods differ from conventionally produced foods and their potential benefits.
6. Discuss how genetic engineering is used in foods and the potential risks and benefits of genetically modifying foods.
7. Describe why food additives are used in most processed foods and potential concerns about their use.
8. List important bacterial, viral, and parasitic causes of foodborne illness in the U.S.
9. Describe food handling practices that reduce the risk of foodborne illness.
10. Describe common environmental contaminants (heavy metals, industrial chemicals, pesticides, and antibiotics), their potential harmful effects, and how to reduce exposure to them.

A BOUNTIFUL, VARIED, NUTRITIOUS, SUSTAINABLE, and safe food supply is available to many individuals, especially those in developed countries. For example, most Americans have ready access to generous food supplies—more than 38,000 supermarkets in the U.S. stock, on average, nearly 42,000 items.[1] In addition to purchases from food stores, Americans are buying more ready-to-eat food than ever from restaurants, cafeterias, food trucks, vending machines, and the like. Despite ample food supplies and the efforts of private and government programs to help at-risk individuals obtain enough healthy foods, malnutrition and poor health from poor diets still plague people throughout the world.

Food preservation and processing methods (e.g., refrigeration, canning, and irradiation) and food additives, along with food production practices (e.g., conventional and organic farming and biotechnology), continue to expand the variety, availability, and sustainability of food. These processing and production methods offer many benefits; however, there are still concerns about the safety of food and water. For example, common foods such as fresh produce and ground beef sometimes are contaminated with harmful bacteria that cause foodborne illness. The safety of and need for pesticides, antibiotics in food animals (animals intended for consumption), biotechnology processes that genetically modify plants and animals, and food additives listed on food labels continue to be debated. This chapter addresses the complex issues of food access, food constituents, food sustainability, and food safety, as well as steps you can take to keep food safe in your home.

Poverty aggravates the problem of hunger in the developing world. One in 9 do not have enough food to eat.

brians101/iStock/Getty Images

▶ Often thought of as a problem of low- and middle-income countries, moderate to severe food insecurity affects 8% of the populations of upper-middle and higher-income nations like Europe and North America. A predictor of obesity in school-age children, adolescents, and adults is simply living in food-insecure households in these more affluent nations.[2]

Fresh produce may be too expensive for some families; however, frozen and canned vegetables provide similar amounts of nutrients and are a good alternative.

JGI/Jamie Grill/Getty Images

# 3.1 Food Availability and Access

Good nutritional status and health for each of us require access to a safe and healthy food supply. Worldwide, agriculture produces enough food to provide each person with 2940 kcal daily—more than enough to meet the energy requirements of each of the nearly 7.5 billion persons on earth. Even with this abundance, the Food and Agriculture Organization (FAO) of the United Nations estimates that, in recent years, more than 1 in every 9 people throughout the world, totaling 820 million people, experienced chronic food shortages and were undernourished. Another 2 billion people suffer from micronutrient (vitamin and mineral) deficiencies, sometimes called "hidden hunger."[2] The serious problems caused by hunger and malnutrition have decreased in recent decades, but the prevalence of undernourishment and hunger has slowly increased.[2] Hunger exists in virtually every nation and is on the rise in almost all regions of Africa, with undernourishment affecting nearly 20% of the population. Hunger has increased at slower rates in Asia where almost 15% of the population is undernourished[2]. Approximately 7% of the population experience hunger in Latin America and the Caribbean nations. Nearly all people suffering from hunger or malnutrition are poor.

According to the FAO, the problems of malnutrition (including overweight and obesity) and hunger account for over half the world's disease burden.[2] Overnutrition that results in overweight and obesity is the primary problem in industrialized countries, such as U.S., Canada, and the countries of Western Europe. However, overnutrition now is a global problem. The World Health Organization (WHO) estimates that 2 billion individuals are overweight or obese—nearly 39% of the world's adult population.[2] As developing countries become Westernized, their diets contain more meat, dairy, sugar, fat, processed foods, and alcohol but fewer whole grains, vegetables, and fruits. This phenomenon, known as the *nutrition transition*, is especially common in urban areas.[3] The "double burden" of food insecurity and malnutrition coupled with overnutrition poses new challenges for nutritionists and other health-care providers in developing countries.[4]

## Health Consequences of Malnutrition and Food Insecurity

When individuals don't get enough to eat or have access to only a few foods, problems related to hunger and malnutrition arise. The United Nations estimates that adults require a minimum of 2100 kcal per day to support a normal, healthy life (children require less). When energy intake falls below needs, a variety of problems arise—physical and mental activity declines; growth slows or ceases altogether; muscle and fat wasting occurs; the immune system weakens, increasing susceptibility to disease; and death rates rise (Fig. 3-1). The consequences of micronutrient deficiencies can be equally devastating. For example, vitamin A deficiency, found in approximately 30% of children under age 5 in developing countries, damages the eyes and sometimes causes blindness.[5] Vitamin A deficiency also increases vulnerability to common diseases, such as measles, diarrhea, and respiratory infections. Iodine deficiency is the world's leading cause of preventable mental retardation and brain damage. Vitamin A, iodine, iron, zinc, and folate are the micronutrients most likely to be in short supply in developing countries.[2] (See Part 4 for more details on micronutrient deficiencies.)

In the U.S., the Department of Agriculture (USDA) monitors **food security**, defined as "access by all people at all times to enough food for an active, healthy life."[6] **Food insecurity**, or lack of this access (Table 3-1), contributes to serious health and nutritional problems for millions of people. About 17% of the world's population has experienced moderate food insecurity, meaning that they do not have consistent access to sufficient, nutritious foods.[2] Worldwide, food insecurity is higher among women than men.[2] Food-insecure individuals often eat fewer servings of nutrient-dense foods, such as vegetables, milk, and meat, and consume poorer-quality diets in general.[7] These nutrient-poor diets can impair physical and mental health status and ability to learn.[8,9,10] Food-insecure children are more likely to have poorer general health and report more asthma, stomachaches, headaches, and colds, and they may not grow normally. Behavioral problems in school, lower educational achievement, higher rates of depression and suicidal symptoms, and increased levels of psychological

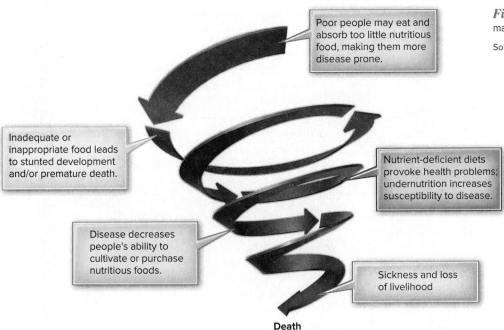

Poor people may eat and absorb too little nutritious food, making them more disease prone.

Inadequate or inappropriate food leads to stunted development and/or premature death.

Nutrient-deficient diets provoke health problems; undernutrition increases susceptibility to disease.

Disease decreases people's ability to cultivate or purchase nutritious foods.

Sickness and loss of livelihood

**Death**

*Figure 3-1* The spiral of poverty, malnutrition, and disease.

Source: Based on World Food Program graphic

distress also have been linked to food insecurity. Parents may compromise their own diets to allow children to have better diets. Food-insecure adults have a higher risk of chronic diseases, such as diabetes, and poorer management of these diseases. Health-care professionals who screen households to identify those at risk for food insecurity can target their services and create treatment plans to improve health and reduce health-associated risks related to food insecurity.

The relationships between food insecurity, poverty, and obesity have been studied intensely. Food insecurity may predispose individuals to overeat when food is more plentiful, or they may purchase mostly inexpensive, high-energy-density foods. Such diets tend to supply food-insecure individuals with enough (or too many) calories but limited amounts of some vitamins and minerals.[12] However, as noted in Chapter 10, many factors, including age, biological sex, race, and ethnicity, influence the development of obesity. In the U.S., food-insecure women may have a higher risk for obesity.[13]

## Food Insecurity in the U.S.

The USDA has monitored the food security of U.S. households since 1995. There are 4 levels of household food security: high, marginal, low, and very low (see Table 3-1). In food-secure households, food needs are met at all times, whereas food-insecure households may run out

### HISTORICAL PERSPECTIVE
### The Minnesota Semi-Starvation Experiment

In the 1940s, Dr. Ancel Keys and his research team examined the effects of undernutrition on 32 healthy male volunteers. The volunteers ate about 1800 calories daily for 6 months, losing an average of 24% of their body weight. The men experienced profound physiological and psychological symptoms, including fatigue, muscle soreness, cold intolerance, decreased heart rate and muscle tone, fluid retention, poor concentration abilities, moodiness, apathy, and depression. When the men were permitted to eat normally again, feelings of fatigue, recurrent hunger, and food cravings persisted, even after 12 weeks of rehabilitation. Full recovery required about 18 months. This study tells us much about the general state of undernourished adults worldwide.[14]

▶ **Food Insecurity Screener[11]**

Think about your household food situation over the past 12 months. For each statement, is your answer often true, sometimes true, or never true?

1. We worried whether our food would run out before we got money to buy more.
2. The food that we bought just didn't last, and we didn't have money to get more.

*Responses of "often true" or "sometimes true" to either question indicate food insecurity.*

| Table 3-1 USDA Descriptions for Food Security and Food Insecurity | | |
|---|---|---|
| **Food security** | **High food security** | No indications of food-access problems or limitations |
| | **Marginal food security** | 1 or 2 indications of food-access problems—typically, anxiety over food sufficiency or shortage of food in the house; little or no change in diets or food intake |
| **Food insecurity** | **Low food security** | Reduced quality, variety, or desirability of diet; little or no reduced food intakes |
| | **Very low food security** | Multiple indications of disrupted eating patterns and reduced food intake |

*Figure 3-2* Food insecurity in the U.S., average 2015–2017.

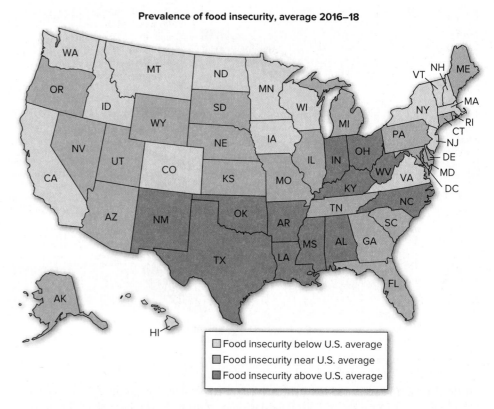

**Prevalence of food insecurity, average 2016–18**

☐ Food insecurity below U.S. average
☐ Food insecurity near U.S. average
☐ Food insecurity above U.S. average

**poverty guidelines** Federal poverty level; income level calculated each year by the U.S. Census Bureau. Guidelines are used to determine eligibility for many federal food and nutrition assistance programs.

▶ The federal minimum wage is $7.25 per hour. A full-time worker earning this wage makes $15,080 per year. If this person is the sole wage earner in a family of 2, they will live below the poverty guideline of $15,510. If rent is $800 per month, $9600 of the $15,080 goes to housing, leaving just $5480 for all other expenses, including food, in the year.

of food or manage food supplies by skipping meals, reducing meal size, or not eating when hungry. In 2017, 40 million Americans (1 in 8) experienced food insecurity, with greater than 12 million of these being children. Four and a half percent of households reported very low food security.[15] Figure 3-2 shows how food security varies by state in the U.S.

Food insecurity is closely linked to poverty. In the U.S., more than 38 million people (12% of the population)—one-eighth of whom are children—live at or below the **poverty guidelines**, currently estimated at about $25,750 annually for a family of 4.[16] Poverty rates are even greater for children living in single-parent households and for certain racial and ethnic groups. More than a quarter of African-American, Native American, and Hispanic individuals live in poverty, compared with approximately 13% of those who are Asian and Caucasian. Low-paying jobs and unemployment, coupled with a lack of health-care benefits, high housing costs, family break-ups, and catastrophic illness, contribute to economic hardship and poverty. Living in a food-insecure household in an upper-middle or higher-income country, like the U.S., increases the risk of obesity in school-age children, teens, and adults. This increased risk may be the result of substituting less expensive foods (higher in added sugar and fat) for their more expensive, nutritious counterparts, stress related to food insecurity, and possible physiological adaptations related to restricting food intake.[2]

Living in a **food desert** can limit access to healthy foods. Food deserts are geographic areas, both rural and urban, with the problems of low income and low access to supermarkets or large grocery stores and limited availability of many fresh and perishable foods. However, recent research indicates that improving access to healthy foods does not guarantee their consumption.[17] Rather, a person's income (and food prices), knowledge of nutrition, and food preferences may be the most important determinants of what is purchased.

## Programs to Increase Food Security in the U.S.

In the U.S., government programs and charitable programs help reduce food insecurity. Table 3-2 lists some ways individuals can help fight hunger in their communities.

### Government Programs

Since the 1930s, the U.S. government has provided food assistance to individuals and families in need. Today, the 15 food and nutrition assistance programs administered by the

**Table 3-2 Ways Individuals Can Help Fight Hunger in Their Communities**

- Donate food—high-protein foods (e.g., tuna and peanut butter), baby food, and culturally appropriate foods are often in high demand.

- Start or participate in a campus or community program that targets hunger. Community gardens and community kitchens are examples.

- Organize or work at a food drive.

- Donate money or time to organizations that fight hunger and food insecurity. Examples include Meals on Wheels, No Kid Hungry, Share our Strength, and UNICEF.

- Volunteer at a local food bank or pantry.

- Reduce your own food waste. This can stretch your food dollars.

- Advocate for programs that promote food security by writing letters to newspapers, contacting state and national legislators, or joining an organization that advocates politically for hungry people.

- Attend events targeted at achieving food security.

- Stay informed about hunger- and food-security-related issues and legislation.

- Pay special attention to World Food Day, October 16 each year.

Food pantries and soup kitchens provide food to a growing number of people in the U.S. Consider volunteering some of your time at a local program.

Ariel Skelley/Blend Images LLC

USDA account for more than two-thirds of the USDA budget, and about 25% of Americans participate in at least 1 of the programs.[18] Food and nutrition assistance programs increase access to food and help reduce food insecurity. However, not everyone in need of assistance receives it—individuals may not know about the programs, may find the application process too difficult, may not have transportation to the program site, or may feel uncomfortable about participating. The following are the largest government assistance programs.

- *Supplemental Nutrition Assistance Program (SNAP).* SNAP (formerly called the Food Stamp Program) is regarded as the cornerstone of the food assistance programs. It provides monthly benefits in the form of an Electronic Benefit Transfer card, which works as a debit card. The average benefit per person is $126 per month. About 44.2 million persons per month, or 13.5% of all Americans, participate in the program. Many of those receiving benefits are children: 31% of all children ages 5 to 17 and 13% of children less than 5 years old. Benefits can be used to purchase food, as well as seeds to grow food; they cannot be used to buy tobacco, alcoholic beverages, and nonedible products. SNAP Nutrition Education, intended to help SNAP participants make healthier food choices, is available in almost all states.

- *Special Supplemental Nutrition Program for Women, Infants, and Children (WIC).* This program provides low-income pregnant, breastfeeding, and postpartum women, infants, and children up to age 5 who are at nutritional risk with vouchers to purchase specific nutrient-dense foods. These individuals also receive nutrition education and referrals to health-care and social services. Currently, half of all U.S. infants, more than 25% of pregnant and postpartum women, and almost 25% of children under 5 years old participate in the program. Overall participation has declined in recent years due to improved economic conditions.

- *National School Lunch Program.* This program helps schools provide nutritious lunches to children, thereby helping them concentrate and learn. Each day, about 30 million children take part in the program. It is the second largest food and nutrition assistance program in the U.S. The program subsidizes lunches by providing schools with cash and food. All children can participate, but children from families with qualifying incomes receive either free or reduced-price lunches, which must meet federal nutrition guidelines. Summer food service programs operate in some locations across the country to meet the need for food assistance during the summer break.

- *School Breakfast Program.* This program began in response to concerns about children attending school hungry. It operates similarly to the National School Lunch Program. Breakfasts must meet federal nutrition guidelines.

▶ Many who are food insecure live in food deserts (locations with few supermarkets). They often have to rely on fast-food restaurants and corner stores—which usually means their access to healthy foods is limited. Mobile food pantries and inner-city farmers' markets are helping bring fresh foods to food deserts.

▶ One of the many challenges faced by food banks and pantries is to provide more nutrient-rich, fresh foods, such as fresh produce, whole grains, low-fat meat, and dairy products, instead of refined grains and highly processed foods.

▶ As of 2016, 39% of college undergraduates lived at or below the federal poverty line. Between tuition and cost of living, many of these students find themselves food insecure. In response to this, many college campuses have opened food pantries.[19]

- *Child and Adult Care Food Program.* Reimbursement is provided to eligible child-care and nonresidential adult day-care centers that provide meals and snacks. Like the School Lunch and School Breakfast programs, the meals must meet certain nutrition criteria.
- *Programs for seniors.* The Older Americans Act provides funding for nutrition programs targeted at older adults. Many communities offer congregate meal programs (often, lunch served daily at a variety of sites in a community) and home-delivered meals—popularly known as Meals on Wheels. Meals must meet nutrition guidelines and are available at little or no cost. Senior Farmers' Market Nutrition programs also are available in many states.
- *Food distribution programs.* Commodity foods are agricultural products purchased by the government. They typically include canned foods (fruits, vegetables, juices, meat, tuna), dry food (ready-to-eat cereal, non fat dry milk, beans, dehydrated potatoes, pasta, rice, infant cereal), peanut butter, and limited amounts of fresh foods, such as cheese, fruits, and vegetables. These programs distribute commodity foods and provide nutrition assistance to low-income households, emergency feeding programs, disaster relief programs, Indian reservations, and older adults. Many food banks and pantries distribute commodity foods to their clients.

### Hunger Relief Charitable Programs

In addition to the government programs, many private programs provide food assistance to individuals at food banks and pantries, soup kitchens, and homeless shelters. Each year, about 1 in 7 people in the U.S. obtain food from these programs.[20] The programs rely on support from individuals, faith-based organizations, businesses, foundations, and grants. Many of these programs greatly depend on volunteer workers.

The largest contributor of food to these emergency food programs is Feeding America. This organization distributes food it receives from private and corporate donations and government commodities to large, regional food banks, which then distribute it to individual food pantries, soup kitchens, emergency shelters, and other food programs. In 2018, Feeding America provided 4.3 billion meals across the U.S.[20]

## Food Insecurity and Malnutrition in the World's Developing Regions

According to the FAO, undernutrition in developing regions is decreasing. The proportion of underweight children worldwide has declined by almost half in the past 20 years.[2] However, malnutrition still disproportionately affects young children and women. Every year, approximately 2.6 million children under the age of 5 in low- and middle-income countries die of causes directly related to undernutrition; these deaths account for 45% of all deaths among infants and preschool children.[21] About 99 million of the world's children are underweight, and more than 160 million have stunted growth. Such children are more likely to suffer from infectious diseases and to have problems learning. In many households, women have less access to food than men because social customs dictate that women eat last. When a woman is poorly nourished, her developing fetus or breastfed infant also may suffer malnutrition.

Most hungry people in developing countries live in rural areas where they are unemployed or work as **subsistence farmers**—those who are able to grow only enough food for their families, with little extra to sell for income. Farming is difficult in many regions because of poor-quality farmland; lack of fertilizer, seeds, and farming equipment; and droughts or flooding. The poor health caused by food insecurity, concerns over agricultural sustainability, and malnutrition limits farmers' physical capabilities and ability to work. Natural disasters, war, and political unrest worsen conditions and can trigger food crises and famine.

By the year 2050, the earth's population will have increased to around 9.5 billion—think of this as adding 2 more countries about the size of India, the world's second most populous country. This population growth and the anticipated increases in urbanization and economic growth will drive an enormous increase in world food demand. Thus, reducing malnutrition and improving food security and sustainability in the future will present ever increasing challenges.

Food security is fostered by communities raising and distributing locally grown food.

Ariel Skelley/Blend Images LLC

Most experts agree that economic development to reduce poverty and improve local food systems and agricultural productivity are key. Raising agricultural productivity is especially important because most poor people live in rural areas and many are farmers.[22,23] With agricultural improvements, households can grow more crops, eat healthier diets, and earn an income from the extra food and crops they grow. Increasing agricultural productivity is costly and complex, though, and requires infrastructure (e.g., roads, irrigation, electricity, and banks), agricultural research, education, and a healthy population. Improvements in health, especially by eliminating micronutrient deficiencies, providing clean food and water, and offering family-planning methods, are critical because economic growth requires healthy people who are able to work, learn, and make improvements.

## Food Waste

Many experts believe that another way to improve global food security is to reduce the amount of food that goes to waste around the world. According to the FAO, about one-third of all food produced is lost instead of being eaten by humans.[24] That amount is even higher in the U.S., where 40% of food goes to waste annually. Food loss and waste can occur anywhere along the food production chain. In low-income countries, food waste and losses occur mainly at the early stages of food production (e.g., problems with harvesting, storage, and transportation). In higher income nations, food is wasted and lost at later stages in the chain, with consumer behavior being the largest contributing factor to food waste.[24] For example, every year, each American throws away 300 pounds of food.[25] The following are some examples of how food gets wasted.

- Farms may lose foods to pests, birds, insects, or disease. Some food may not be harvested because of labor shortages or is discarded because it is deemed too unattractive for purchase (think of a misshapen or spotted piece of fruit).
- During processing, foods may be discarded because of imperfections or because many consumers do not eat the food (e.g., North Americans typically eat limited amounts of animal organs).
- Food stores throw away unsold perishable foods, including fresh produce, meats, and dairy products.
- Restaurants and other food service establishments may prepare too much food. This uneaten food often ends up in a landfill.
- Many consumers waste food because they purchase and prepare too much food, let uneaten food spoil in the refrigerator, use leftovers poorly, and are confused about "best by," "use by," and "sell by" dates on foods. This uneaten food also ends up in the local landfill.

The FAO estimates that recovering half of the food that is wasted could substantially reduce hunger and shrink landfills, reduce greenhouse gas production, and decrease use of agricultural inputs (e.g., fresh water, fertilizers, pesticides, and fossil fuels). Reducing food loss can be tackled throughout the food chain. For example, many universities have eliminated trays in their dining services to decrease food waste (students can't simply load up a tray with food that looks appealing). Also, consumers can take many actions to reduce wasted food in the home. Instead of throwing out safe and nutritious uneaten food, donate it to a local food bank and compost scraps of food versus discarding them. Read more about food waste at www.lovefoodhatewaste.com, www.thinkeatsave.org, and www.un.org/zerohunger. Figure 3-3 describes strategies for reducing food waste throughout the food supply chain. Even more tips on how to reduce food waste at home are in this chapter's Culinary Perspective.

## Food Sustainability

A major factor that will impact the ability to achieve global food security is food sustainability. **Food sustainability** is the ability to produce enough food to maintain the human population. Concerns exist as to whether the food supply chain will be able to supply the food needed to nourish the global population.[26] To achieve food sustainability (locally, nationally, and globally), we need access to adequate, fertile farmland, clean water, safe and effective fertilizers, and clean and sustainable energy[2,26,27] sources. We also need climate stability and/or plants that grow efficiently in current and emerging climate conditions.

A variety of specialized nutritious foods are in use to treat malnutrition and eliminate the need for hospitalization. They are typically made from vegetable oil, dry milk powder, sugar, vitamins and minerals, and peanuts, chickpeas, corn, soy, rice, or wheat.

Mike Goldwater/Alamy Stock Photo

▶ Humanitarian crises such as those occurring in Syria, Nigeria, and Mozambique rely on organizations like the World Food Program (WFP) to provide support. The WFP provides nearly 90 million meals to the most vulnerable people in more than 80 countries annually. Nearly two-thirds of all people served are children.

*Figure 3-3* Food sustainability from farm to table.

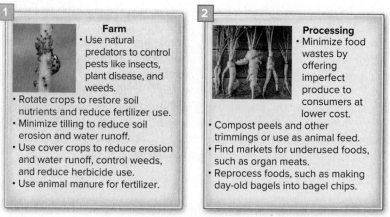

**1 Farm**
- Use natural predators to control pests like insects, plant disease, and weeds.
- Rotate crops to restore soil nutrients and reduce fertilizer use.
- Minimize tilling to reduce soil erosion and water runoff.
- Use cover crops to reduce erosion and water runoff, control weeds, and reduce herbicide use.
- Use animal manure for fertilizer.

**2 Processing**
- Minimize food wastes by offering imperfect produce to consumers at lower cost.
- Compost peels and other trimmings or use as animal feed.
- Find markets for underused foods, such as organ meats.
- Reprocess foods, such as making day-old bagels into bagel chips.

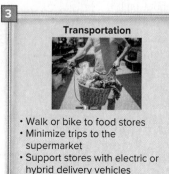

**3 Transportation**
- Walk or bike to food stores
- Minimize trips to the supermarket
- Support stores with electric or hybrid delivery vehicles

**4 Retail**
- Offer food with less packaging (e.g., in bins, unwrapped fresh fruit).
- Buy locally grown foods when possible.
- Donate food that is still safe to eat to soup kitchens.

**5 Table**
- Eat more plant-based foods.
- Minimize use of ultraprocessed foods.
- Match calorie intake to needs.
- Eat fewer animal-based foods.
- Recycle food containers.
- Compost food waste.

**agrobiodiversity** Variety and variability of animals, plants, and microorganisms that are used directly or indirectly for food and agriculture.

Achieving global food sustainability also requires serious evaluation and adjustment of what we eat and how we produce food. The global food supply is dominated by only 12 plant and 5 animal species—together they account for 75% of the food supply.[28] Increasing **agrobiodiversity** by diversifying the species we eat will increase variety in the food supply and may lead to a more nutrient-dense, higher quality diet. Agriculture is a major contributor to changes in the global environment—food production alone is responsible for creating nearly one-third of the global greenhouse gas emissions associated with agriculture and it accounts for 70% of the fresh water[29] used in agriculture. Livestock and dairy production use more water and land than plants and account for close to half of food production greenhouse gas emissions.[29] A shift toward a more plant-centered diet will help achieve food security goals.[28,29] Keep in mind that moving toward a more plant-centered diet does not mean having to follow a vegan diet. Research-based diets that take a more plant-centered approach include the Dietary Approaches to Stop Hypertension (DASH, see Chapter 14 Clinical Perspective) and the Mediterranean Diet (see Chapter 6). Table 3-3 describes strategies that researchers have proposed to help achieve global food sustainability.

---

**Table 3-3 Key Areas of Focus for Achieving Global Food Sustainability[28,30]**

1. *Promote dietary diversity to optimize nutrient density and lower environmental impact.* Eat a diet with a greater variety of vegetables, fruits, whole grains, legumes, nuts, and unsaturated oils; low to moderate amounts of sustainably farmed seafood and poultry; and limited red and/or processed meat, added sugar, refined grains, and starchy vegetables.

2. *Target food production sustainability.* Increase agrobiodiversity and improve current agricultural practices to reduce environmental impact.

3. *Reduce food loss and food waste.* Promote strategies to reduce food waste all along the food production chain—from field to plate.

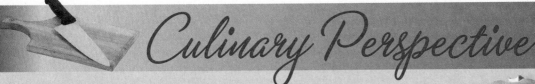

# Culinary Perspective

## Reducing Food Waste at the Grocery Store and Home

Reducing food waste is an important factor in achieving global food sustainability. U.S. consumers throw away 300 pounds of food every year—that's like discarding 300 loaves of sandwich bread! Fruits and vegetables top the list of foods wasted, followed by dairy and bread, then meat. Food waste not only wastes money but also negatively affects the environment by wasting all the resources used to produce it, like water, fertilizer, and pesticides. All this waste adds up to the equivalent of 33 million cars' worth of greenhouse gases. Everyone has a part in reducing food waste at the grocery store and in his or her own home kitchen.

There are a variety of ways to help reduce food waste—from creative cooking strategies to planning meals that reduce the chances leftovers will end up in the trash. Try these tips to reduce food waste, trim your grocery budget, and enjoy new recipes.

Don't throw out food scraps! Get creative—find ways to incorporate them into meals and snacks.

highwaystarz/123RF

### Tips to Reduce Food Waste at the Grocery Store

- *Plan ahead.* Use meal planning apps or websites to create a weekly menu plan. Not a planner? Start by creating a grocery list and sticking to it.  Shopping more often can help you avoid overbuying and wasting the excess that you don't use.

- *Be label savvy.* Understand that best-by and use-by dates on food packages are there to assure ideal quality, but, after these dates, foods are still nutritious and can be safe to eat. Sell-by dates indicate when a store should remove a product from the shelf, but close to one-third of a product's shelf life remains after this date.

- *Don't seek perfection.* Blemished or imperfect produce doesn't sell well and is often tossed before it gets to grocery store shelves. Blemished produce is safe and nutritious and can even save you money. Budget grocery stores often offer blemished produce at prices much lower than supermarkets. Companies like Imperfect Produce sell produce not "acceptable" to grocery stores for 30% less and deliver it straight to the consumer's door. Learn more at www.imperfectproduce.com.

### Tips to Reduce Food Waste in the Kitchen

- *Get creative with leftovers.*

- Toss leftover veggies and meat
  - With rice and seasonings for a quick skillet stir-fry
  - Onto a tortilla and top with cheese
  - With cooked pasta for a flavorful pasta salad
  - With eggs to make an omelet

- *Repurpose food "scraps."*
  - Blend trimmed portions from fruits into a smoothie.
  - Store vegetable scraps in the freezer until you have enough to make vegetable broth.
  - Use fruit peels to make fruit-flavored vinegar for salad dressing.
  - Dry citrus peels and add to vinegar to create a household cleanser.
  - Grind hard cheese rinds and sprinkle on soup or pasta.
  - Compost unusable food scraps and spread on soil.

- *Find new uses for foods past their prime.*
  - Shred wilting carrots, parsnips, or zucchini to add to muffin, bread, or pancake batter.
  - Use stale bread to make french toast or croutons, or pulverize it into breadcrumbs.
  - Arrange slices of very ripe fruit in a single layer on a pan and freeze. When frozen, store in a tightly closed container.
    - Use fruit as a cereal topping.
    - Blend to create a fruity salad dressing.
    - Add to smoothies.

Learn more creative ways to use leftovers and food scraps at zerowastechef.com.

---

The concept of food sustainability means different things to different people. Some view sustainability as buying local. Others look at sustainability from an agricultural perspective focusing on best farming practices to limit impact on climate change. Still others view food sustainability as achievable if we reduce food waste. Ultimately, the optimal approach to addressing and achieving global food sustainability likely will be a multifactorial approach that addresses all aspects of the food supply chain. What steps will you take to contribute to achieving global food sustainability?

**Knowledge Check**

1. What are the main causes of food insecurity?
2. What is hidden hunger?
3. How do food insecurity and hunger in developing countries differ from those seen in the U.S.?
4. What are some of the U.S. government food and nutrition programs that aim to reduce food insecurity?
5. What key areas of focus can help achieve global food sustainability?

The USDA organic seal identifies organic foods grown on USDA-certified organic farms.

## 3.2 Food Production

Agriculture, the production of food and livestock, has supplied humans with food for millennia. At one time, nearly everyone was involved in food production. Only about 1 in 3 people around the globe, and far fewer in the U.S. (less than 1%), is now involved in farming. Today, numerous advances in agricultural sciences are affecting our food supply; of particular note are organic food production and biotechnology.

### Organic Foods

Organic foods are widely available in supermarkets, specialty stores, farmers' markets, and restaurants. Consumers can select organic fruits, vegetables, grains, dairy products, meats, eggs, and many processed foods, including sauces and condiments, breakfast cereals, cookies, and snack chips. Interest in personal and environmental health has contributed to the increasing availability and sales of organic foods. Almost 13% of the fruit and vegetables sold are organic; overall, organic foods account for about 5% of foods sold.[31] Organic foods, because they often cost more to grow and produce, are typically more expensive than comparable conventional foods.

The term *organic* refers to the way agricultural products are produced. Organic production relies on farming practices such as **biological pest management**, composting, manure applications, and crop rotation to maintain healthy soil, water, crops, and animals. Synthetic pesticides, fertilizers, and hormones; antibiotics; sewage sludge (used as fertilizer); genetic engineering; and irradiation are not permitted in the production of organic foods (pesticides, antibiotics, and genetic engineering are discussed later in the chapter). Additionally, organic meat, poultry, eggs, and dairy products must come from animals allowed to graze outdoors and fed only organic feed.[32]

Foods labeled and marketed as organic must have at least 95% of their ingredients (by weight) meet USDA organic standards. The phrase "made with organic" can be used if at least 70% of the ingredients are organic. Small organic producers and farmers with sales less than $5000 per year are exempt from the certification regulation. Some farmers use organic production methods but choose not to be USDA certified. Their foods cannot be labeled as organic, but many of these farmers market and sell to those seeking organic foods.

### Organic Foods and Health

Consumers may choose to eat organic foods to reduce their pesticide intake, to support **sustainable agriculture** and protect the environment, and to improve the nutritional quality of their diets. Those who consume organic produce do ingest lower amounts of pesticides (about 1 in 4 organically grown fruits and vegetables contains pesticides and in lower

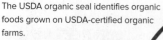

Farmers' markets are a good place to shop for organic foods.

Pixtal/age fotostock

**biological pest management** Control of agricultural pests by using natural predators, parasites, or pathogens. For example, ladybugs can be used to control an aphid infestation.

**sustainable agriculture** Agricultural system that provides a secure living for farm families; maintains the natural environment and resources; supports the rural community; and offers respect and fair treatment to all involved, from farmworkers to consumers to the animals raised for food.

amounts than conventional produce), but it's still not known whether or how this affects the health of most consumers. However, organic foods may be a wise choice for infants and young children because pesticide exposure may pose a greater risk to them (see Section 3.5 for more about pesticides). Compared to conventional agriculture, organic farming can reduce fossil fuel consumption, improve soil health, and reduce contamination with pesticides and herbicides.

The nutritional benefits of consuming organic foods continue to be hotly debated. Complicating this debate are the myriad research challenges that come into play when comparing the nutrient content of organic and conventional foods. Many factors can influence this comparison, including sampling methods, laboratory techniques, and statistical analyses. Recent meta-analyses show few differences between organically and conventionally grown foods for most nutrients.[33,34,35,36] However, minor variations have been found. Some organic foods have higher amounts of antioxidant compounds (e.g., some fruits, vegetables, and cereals) and omega-3 fats (e.g., beef and milk) and lower amounts of iodine and selenium (e.g., milk). A healthy dose of common sense is important here—as you will learn in Chapter 6, beef and milk are not important sources of omega-3 fats. Consider, too, that an "organic" label does not change a less healthy food into a healthier one. A cookie made with organic ingredients will have the same calorie and sugar content as a cookie made with identical conventional ingredients. At this point, it's not possible to recommend organic foods over conventional foods based on nutrient content—both can meet nutritional needs.

## Amending Agricultural Plant and Animal Traits

Plant and animal variations are amended by manipulating the genes of these organisms. The goal of this manipulation is to create plants or animals that have certain desired traits, such as improved disease resistance, drought tolerance, yields, or nutrient density. There are four main processes used in agriculture to manipulate genes: selective breeding, mutagenesis, genetic modification, and gene editing.

### Selective Breeding

Genetic manipulation is almost as old as agriculture. The first farmers used traditional selective breeding techniques. For instance, to improve his livestock, a farmer might selectively breed the best bull with the best cows. The offspring of selective breeding has a mix of genes from the two parents as well as normal random genetic mutations. The same is true for plants. By crossing two plants, a new variety that has traits of both of the original plants is produced. By the 1920s, these practices had made possible the selective breeding of better plant hybrids. As a result, corn production in the U.S. quickly doubled. Through similar methods, agricultural wheat was crossed with wild grasses to confer more desirable properties, such as greater yield, increased resistance to disease, and better performance in adverse climatic conditions. Farmers also began to crossbreed different types of plants to create new hybrids, such as citrons and Mandarin oranges to produce Meyer lemons.

### Mutagenesis

A second type of plant genetic manipulation, called mutagenesis, began in the 1950s. This process involves treating plant cells with radiation or chemicals to induce gene mutations that result in desired traits. This process leaves no harmful components in food. Foods modified in this way include red grapefruit, wheat, rice, barley, peanuts, peas, and cacao.[37]

Both selective breeding and mutagenesis affect many to all of a plant's genes. The traits that result in the new varieties sometimes improve the plant, but sometimes they do not. It take years of trial and error using these processes to produce new varieties that have the desired characteristics. To speed the development of new varieties, scientists developed biotechnology processes.

If you are on a budget that limits your ability to purchase organic produce, don't forgo buying produce simply because it's not organic. Washing with water, rubbing, and if desired, peeling conventionally grown produce will significantly reduce the limited pesticide residue that may be on it. Learn more at npic.orst.edu.

Fancy Collection/SuperStock

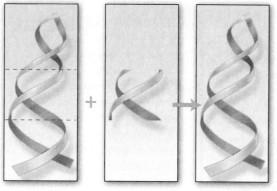

DNA of the host plant, corn.

Gene from bacteria (Bt gene) that produces a protein toxic to the European corn borer.

Bt gene inserted into DNA of corn plant. Now the corn plant is genetically modified. It makes the Bt toxin and, so, is resistant to the European corn borer.

*Figure 3-4* In this diagram, a gene from the bacterium *Bacillus thuringiensis* (Bt) is spliced into the DNA of a host corn plant. The corn plant, now referred to as a genetically modified plant, is resistant to the European corn borer. This genetic modification reduced the amount of pesticides used on corn by nearly 100%.

**recombinant DNA technology** Test tube technology that rearranges DNA sequences in an organism by cutting the DNA, adding or deleting a DNA sequence, and rejoining DNA molecules with a series of enzymes.

The genetically modified eggplant, Bt Brinjal, has been engineered to resist pests that cause up to 80% crop loss. Since its introduction in Bangladesh, farmers have been able to significantly reduce pesticide usage, improving crop yield and farmer incomes while reducing farmer illness and environmental impacts due to heavy pesticide spraying and exposure.

Lluis Real/age fotostock

## Genetic (Transgenetic) Modification

One of the biotechnology processes used to manipulate genes is known as transgenesis or genetic modification. It allows scientists to directly alter the genetic makeup of an organism. Using **recombinant DNA technology**, scientists can transfer genes (sometimes called transgenes) that confer specific traits, such as disease resistance, from almost any plant, animal, or microorganism to another (Fig. 3-4). The resulting organism is commonly referred to as a genetically modified or transgenic plant, animal, or organism. Foods contain thousands of genes—a genetically modified food differs from the original food by 1 to 8 genes.

Genetically modified foods have been available for more than 40 years. The first commercial genetically modified food—the Flavr Savr tomato, was engineered to remain firm after ripening on the vine. Bacteria and yeast have been genetically modified to produce substances used in food production. Chymosin, also known as rennet, is one such substance. Chymosin is the clotting agent needed to make cheese from milk. Traditionally, chymosin was harvested from the stomachs of calves, an expensive process. Now 80 to 90% of the cheese we eat is made using chymosin produced by genetically engineered bacteria or yeast. Chymosin produced in this manner is less costly and much purer than that extracted from calves. Another compound, recombinant bovine growth hormone (rBGH), produced by genetically engineered bacteria, increases milk production by 10 to 15%. Papaya and some varieties of squash are genetically engineered to be resistant to plant viruses. Scientists developed nonbrowning apples by silencing the genes that produce the enzyme that causes cut apples to turn brown.

Genetic modification is used primarily to confer herbicide tolerance and/or insect resistance to soy, corn, cotton, sugar beets, canola, and alfalfa.[37,38] Herbicide-tolerant crops, commonly referred to as "Roundup Ready®" crops, are genetically modified to be resistant to the herbicide glyphosate. Farmers can apply the herbicide to control weeds without harming the crop itself. This is intended to increase crop yields, decrease use of the most toxic herbicides, and reduce tilling to decrease weeds, minimize soil erosion, and save tractor fuel (all of which can be favorable for improving food sustainability, discussed in more detail later in this chapter). To aid pest control, the gene for a protein made by the soil bacterium *Bacillus thuringiensis* (Bt) was introduced into the genetic makeup of corn and cotton. The Bt protein is a naturally occurring insecticide that kills caterpillars, a major threat to these crops. Before Bt corn and cotton were introduced, farmers often applied toxic pesticides to protect these crops.

The U.S., followed by Brazil, Argentina, India, and Canada, is the leading user of genetically modified crops. Worldwide, 70 to 80% of soybeans and cotton and 24 to 32% of canola and corn are grown from genetically modified seeds.[38] Many countries, including most European countries, choose not to grow or import genetically modified foods or may require food labels to declare the presence of these ingredients.[39]

In addition to genetically modified plants, scientists also have developed genetically modified food animals with various traits. One example is transgenic farmed Atlantic salmon, AquaBounty. These salmon, grown in pens, have been engineered to produce extra amounts of growth hormone. The new genes permit the modified fish to reach full size in 18 months rather than 3 years.[40]

In the U.S., genetically modified foods are regulated by the FDA, USDA, and Environmental Protection Agency (EPA). The FDA's role is to ensure that the food is safe for humans and animals to eat (e.g., no toxins or allergens are present). It is the responsibility of the USDA to make sure genetically modified crops are safe to grow. The EPA ensures that pesticides introduced into foods (e.g., Bt corn) are safe for consumption and for the environment.

The labeling of genetically modified foods or ingredients is a contentious issue for many consumers and legislative bodies. Many individuals assert that consumers have the right to information about the presence or absence of genetically modified ingredients in foods. Because corn, soy, and sugar beets (to make sweeteners and oils and as cattle feed) are so widely used in the U.S. food supply, most people have consumed genetically modified ingredients. In 2016, the U.S. Congress approved a mandatory labeling law for genetically modified foods. Approximately 64 countries, including New Zealand, Australia, Japan, South Korea, and many European countries, require these foods to be labeled.

Numerous food regulation agencies, including the FDA, the National Academy of Sciences, and various international agencies, have found no evidence that genetically modified foods are harmful to humans. Despite this, consumers often question if eating these foods poses risks to human health or the environment.

Compared with selective breeding and mutagenesis, genetic modification processes allow access to a wider gene pool and faster, more accurate transfer of genes. Note, however, that developing a transgenic organism often takes millions of dollars and years of careful research. Genetically modified foods also are heavily regulated and it can take many years before approval is granted. The high cost, lengthy development process, and stringent government regulations, along with public concern about genetically modified foods, have led many companies to investigate other biotechnology processes.

### Genome Editing

The newest biotechnology process is called genome editing or gene editing. Figure 3-5 describes how gene editing works. **TALEN** and **CRISPR** are two editing tools scientists use to edit an organism's genome.[41,42]

**CRISPR** A gene-editing technology that permits scientists to precisely cut out, modify, or add to DNA in human, animal, plant, and other cells. *CRISPR* refers to a segment of regularly repeating DNA found in some bacteria; similar to TALEN.

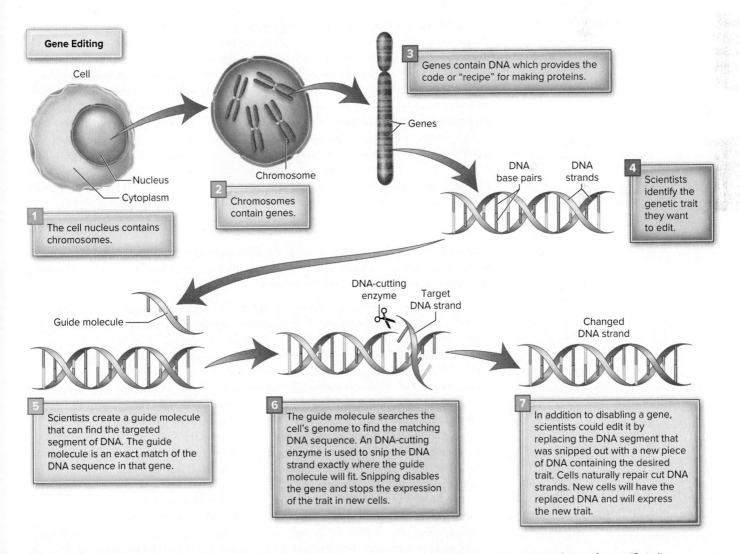

**Gene Editing**

Cell

Nucleus

Cytoplasm

1 The cell nucleus contains chromosomes.

Chromosome

2 Chromosomes contain genes.

3 Genes contain DNA which provides the code or "recipe" for making proteins.

Genes

DNA base pairs

DNA strands

4 Scientists identify the genetic trait they want to edit.

Guide molecule

DNA-cutting enzyme

Target DNA strand

Changed DNA strand

5 Scientists create a guide molecule that can find the targeted segment of DNA. The guide molecule is an exact match of the DNA sequence in that gene.

6 The guide molecule searches the cell's genome to find the matching DNA sequence. An DNA-cutting enzyme is used to snip the DNA strand exactly where the guide molecule will fit. Snipping disables the gene and stops the expression of the trait in new cells.

7 In addition to disabling a gene, scientists could edit it by replacing the DNA segment that was snipped out with a new piece of DNA containing the desired trait. Cells naturally repair cut DNA strands. New cells will have the replaced DNA and will express the new trait.

*Figure 3-5* Gene editing processes allow scientists to alter the genetic makeup of an organism by transferring genes that confer specific traits.

The gene editing process allows scientists to make precise changes to the DNA of individual animal and plant cells. These changes are like those that might occur naturally when two plants or animals breed, but the changes are precisely controlled. The precision of the edits makes it faster and less costly to change traits and produce new varieties than the processes of selective breeding, mutagenesis, and transgenetic modification. Because the changes are indistinguishable from those that occur through traditional breeding methods, the USDA does not regulate gene-edited foods.

Many scientists believe gene editing can make crops more resistant to disease and more able to withstand climatic shifts causing heat stress and droughts, increase food sustainability and global food security by yielding more nutritious and better-tasting food,[43] and improve the shelf life of food. This technology has been used to improve an array of plants, including citrus, rice, corn, and wheat. Calyno™ is a new soybean oil that has been gene edited to have a fat composition like that of olive oil.

Gene-editing technology also has the potential to reduce greenhouse gases and water pollution by reducing the amount of fertilizer used to supply plants with the nitrogen they need for high yields. In addition, gene editing promises to eradicate diseases in humans and food animals. For example, gene editing may be used to cure cystic fibrosis, hemophilia, sickle-cell disease, cancer, and other disorders. Gene editing also may permit the development of pigs resistant to swine fever and others with more muscle and less fat content. The citrus industry is endangered by citrus greening disease; scientists are working to develop genetically engineered citrus trees resistant to the disease. Additionally, the proteins responsible for food allergies could be edited to remove components that cause allergic reactions.

Although carefully applied biotechnology can have great benefits, there are some obstacles to its use. One frequently cited barrier concerns the ability of poor farmers to access this technology.[44] Research often is conducted by large corporations that patent their products (e.g., seeds of new plant varieties) and sell them to farmers at higher prices than conventional seed. Poor farmers who cannot afford the seeds may suffer economic losses. Another closely related concern is that most of the crops developed by corporations (corn, cotton, sugar beets, soybeans) have industrial uses, whereas far less research has focused on foods that are the dietary staples of millions of people, such as rice, wheat, yams, chickpeas, and peanuts. One positive development is that researchers at public universities are developing a variety of genetically engineered crops, including cowpeas resistant to insect pests and cassava resistant to root rot. Both cowpeas and cassava are important African staple crops. Like genetically modified food, a potential obstacle for gene-edited food is that consumers may not accept them.

### Safety and Other Concerns

Although food regulatory agencies, like the FDA and experts at the National Academy of Sciences, report there is no evidence that genetically modified foods are harmful to humans, concerns still exist.[45] and include the following.[37,46]

- *Production of new allergens or toxins in GM foods.* Evidence of allergen contamination was seen in an instance in soybeans but, in more than 20 years of the use of GM crops, there has been no evidence of any harmful reactions or effects in humans. Other harmful compounds can increase in GM plants, just as they can with traditional plant breeding. Again, there is no evidence that this has occurred, but new GM plants should be scrutinized.
- *Development of glyphosate-resistant "superweeds."* These superweeds have been found in different areas of the U.S. as well as several other countries. These weeds can reduce crop yields, and killing them requires larger applications of pesticides, raising concerns about environmental damage and higher costs for farmers.
- *Crossed species lines, such as inserting a gene from a fish into a grain.* Consumers may find it unnatural to introduce the genes from an unrelated species into another, especially the transfer of animal genes into plants.
- *"Gene flow" from GM crops to plants not intended to be modified.*
- *Development of Bt-resistant insects.* This is of particular concern to organic farmers who, although they do not grow GM crops, apply *Bacillus thuringiensis* bacteria to crops as a biological pest management method. Bt resistance also may lead to increased pesticide use by nonorganic farmers.

▶ One very successful application of genetic engineering is the production of insulin, a drug used to treat diabetes. Genetically engineered insulin, known as humulin, was approved by the FDA in 1982. It is produced by introducing the human gene for insulin into *Escherichia coli* bacteria and letting the bacteria grow in the laboratory and produce insulin. Today, almost all insulin produced in the U.S. is made this way.

- *Loss of genetic diversity.* If GM crops are widely adopted, the use of local, indigenous, conventional seeds may decline and these plants may disappear.
- *Loss of cultural heritage with older, native foods.* For example, in Mexico, traditional maize (corn) has a spiritual and cultural significance that is not fulfilled with GM corn.
- *Insufficient regulation and oversight of GM plants and animals.* Some argue that more rigorous regulation, testing, and oversight are needed to assure safety and benefits for consumers.
- *Development of trade barriers.* Regulations about GM foods vary around the world; thus, limitations of both imports and exports of foodstuffs can arise.
- *Many of these concerns are speculative;* however, finding a balance between using technological advances to improve the food supply and reducing the risk of unintended adverse consequences (and a fear of the unknown) is key in bringing scientists and consumers together on how best to use this technology.

### Meat and Milk from Cloned Animals

In 1997, the famous sheep Dolly introduced the world to animal cloning—the process of making genetically identical animals by nonsexual reproduction. Because their DNA is not altered, cloned animals are not genetically modified. Cloning is done by extracting the genetic material from a donor adult cell and transferring it to an egg that has had its own genetic material removed. The cloned embryo that results is transferred to the uterus of a female, where it continues to grow and develop until birth.[47] The biological process of cloning is not new—plants have been cloned for centuries, and some animals, such as worms and frogs, can clone on their own.

Some ranchers and farmers are interested in cloning as a way to reproduce their best-growing, best milk-producing, or best egg-laying animal for economic gain. After several years of study, the FDA announced in 2008 that it had determined that both meat and milk from cloned cattle, swine, and goats are safe to eat. However, these foods are still not in the marketplace. Although meat and milk from cloned animals appear safe for human consumption, many consumers are uncomfortable with cloning for religious and ethical reasons, and others question the need to add food produced from cloned animals to the food supply.

*Knowledge Check*

1. What substances and practices are not allowed in organic food production?
2. What are the potential advantages and disadvantages of eating organic foods?
3. What is sustainable agriculture?
4. What are 4 main processes used in agriculture to manipulate genes to produce new varieties?
5. How might gene editing help improve the food supply in the future?

## 3.3 Food Preservation and Processing

The vast majority of foods we purchase have been preserved or processed—frozen, refrigerated, canned, dehydrated, or milled, to name a few methods. Food preservation methods extend a food's shelf life by slowing the rate at which microorganisms (e.g., bacteria, mold, yeast) and enzymes in food cause spoilage. Food preservation permits a wide variety of good-quality, nutritious, and safe foods to be available year round. The oldest food preservation methods, some in use for thousands of years, are drying, salting, sugaring, smoking, and fermenting. Over the last 200 years, scientific discoveries and technological innovations have added pasteurization, sterilization, canning, aseptic processing, refrigeration, freezing, nitrogen packing, food irradiation, and preservative food additives to the list of food preservation techniques.

---

*Critical* **Thinking**

Labeling of foods with genetically modified ingredients varies around the world. Some countries require no labeling of ingredients, whereas others require clear information on the food label. Do you think consumers should know if their food contains genetically modified ingredients? What are some of the advantages and disadvantages of labeling foods as either containing these ingredients or "genetically modified free"? How would you use this information?

---

*Food Preservation Methods*

**Methods That Decrease Water Content to Deter Microbial Growth**

Drying (raisins)

Salting (salted fish)

Sugaring (candied fruit)

Smoking (smoked fish)

**Methods That Increase Acidity or Alcohol to Deter Microbial Growth**

Fermentation and pickling (sauerkraut, kimchi, pickles, cheese, yogurt, wine)

**Methods That Use Heat to Eradicate or Reduce Number of Microbes**

Pasteurization (milk)

Sterilization (aseptic cartons of milk, soup)

Canning (beef stew)

**Methods That Slow Rate of Microbial Growth**

Refrigeration (eggs)

Freezing (meat)

**Methods That Inhibit Microbial Growth**

Food additives: chemical preservatives (sodium nitrate in cured meat)

Irradiation (raspberries)

This Radura symbol indicates that a food has been irradiated.

Source: www.fda.gov/

**radiation** Energy that is emitted from a center in all directions. Various forms of radiation energy include X rays and ultraviolet rays from the sun.

**Nanotechnology** is the study of controlling matter at the atomic or molecular level. Nanoparticle properties differ significantly from the properties of larger particles. For instance, gold nanoparticles created the ruby red color in stained glass windows during the Middle Ages. This color depended on both the element (gold) and its particle size—gold atoms allow red light to pass through and block blue and yellow.

DeAgostini/Getty Images

## Food Irradiation

Food irradiation, sometimes known as cold or electronic pasteurization, uses controlled doses of radiant energy, or **radiation**, from gamma rays, X rays, or electron beams to extend the shelf life of food and to control the growth of insects and pathogens (bacteria, fungi, parasites) in foods.[48] Just as an airport scanner or dental X rays do not make your luggage or teeth radioactive, irradiated food is not radioactive. Irradiated foods are safe, in the opinion of the FDA and many other health authorities, including the American Academy of Pediatrics.

Foods approved for irradiation in the U.S. include fresh meat and poultry, shellfish, wheat and wheat powder, white potatoes, spices and dry vegetable seasonings, fresh shell eggs, and fresh produce. Irradiated food, except for dried seasonings, must be labeled with the international food irradiation symbol, the Radura, and a statement that the product has been treated by irradiation. Although the demand for irradiated foods still is low in the U.S., other countries, including Canada, Japan, Belgium, and Mexico, use food irradiation technology more widely. Barriers to its use include consumers' lack of familiarity with the technology, the potentially higher cost of irradiated foods, and concerns about the taste and safety of irradiated foods.[49]

## Food Nanotechnology

**Nanotechnology** is the study of controlling matter at the atomic or molecular level. A nanometer is one-billionth of a meter. The use of nanoparticles in food science and microbiology has many potential applications on the consumer level.[50] Today, nanoparticles are used to detoxify hazardous waste, clean polluted water, and preserve foods and drugs. They also can be used to brighten food and cosmetic colors, keep foods fresh longer, decrease fat content, and increase vitamin bioavailability. Food nanostructured ingredients can be used in a variety of ways, such as food additives, flavoring and fragance agents, and food packaging enhancement. Food nanosensors embedded in food packaging may enhance food safety by enabling the detection of food contaminants, microorganisms, and microtoxins that may be present in food.[51] The potential to use nanotechnology to enhance nutrient density, improve food safety and food processing techniques, and improve food packaging makes it a very promising area for food scientists to pursue.

Concerns arise whenever technology is applied to food production. A safety concern is that nanoparticles used in food production may seep into food.[52] Another concern is that nanoparticles are so small they can penetrate cells and the blood-brain barrier, making it possible to target the delivery of drugs and nutrients; however, these tiny particles may accumulate in body tissues and cause cell damage. For example, anticaking agents used to keep powdered ingredients from forming lumps may contain inorganic silica nanoparticles, which are toxic to human lung cells. Both organic and inorganic nanoparticles are used in food, with toxicity concerns being greatest for inorganic particles because they are not digested or metabolized by the body as organic nanoparticles are.[52]

The field of nanotechnology is a trillion-dollar business with much potential for use in food processing. The food industry is responsible for ensuring that products using nanotechnology meet all applicable safety requirements. The FDA does not universally state that foods using nanomaterials are safe or harmful. Strong safety standards to help guarantee safety, protect consumer health, and reduce the environmental impact of nanoparticle use are needed. Considerable research is under way to determine how to capture the benefits and minimize the risks of nanotechnology.

## Food Additives

If you are a food label reader, you may have seen terms like *ethoxylated monoglycerides* and *diglycerides* (used to improve dough characteristics in bread) or *xanthan gum* (used to thicken salad dressings and other foods) in the ingredient list. These ingredients are food additives. Some, like salt, vinegar, and alcohol, that preserve and flavor foods have been in use for thousands of years. Today, there are more than 10,000 substances classified as food additives.[53,54] These additives are found mainly in processed foods, which may be transported long distances and held for extended periods before purchase. The food additives help keep such foods nutritious, fresh, safe, and appealing.

### Intentional vs. Incidental Food Additives

Food additives can be classified as either intentional or incidental. Intentional food additives are purposely added to achieve a goal, such as a longer shelf life (preservative), greater nutritional value, or a more appealing color or flavor. Flavors and flavor enhancers are the most commonly used group of food additives—over 2600 are in use. Examples of common food additives and their functions are given in Table 3-4.

To illustrate typical uses of intentional food additives, consider a lunch menu of a cheeseburger, cucumber salad with dressing, and sports drink. You might find the following types of additives listed on the ingredient labels of the various foods in this meal.

- Nutrients and a preservative for freshness in the hamburger bun
- Natural food coloring in the cheddar cheese
- Wax on the cucumber skin to extend its shelf life
- Emulsifier (to prevent separating) and preservative in the dressing
- Additives that add sweetness, flavor, color, and nutrition; improve texture; and extend the shelf life of the sports drink

Incidental additives, also called indirect additives, are not intentionally added but become part of a food through some aspect of food cultivation, processing, packaging, transport, or storage. They have no function in finished products and their presence is not indicated on food ingredient labels.

There are safety concerns about some of these incidental additives. The presence of pesticide residue and arsenic in some foods is an example (both of these environmental contaminants are discussed in Section 3.5). Another incidental additive of concern is bisphenol A (BPA), a compound used in the lining of some metal food and beverage cans, plastic food storage containers, water bottles, and a host of other products. Almost all individuals in the U.S. consume BPA and excrete it in their urine. BPA is an **endocrine disrupter** that alters normal metabolic mechanisms and may be associated with adverse health effects, including liver and pancreatic damage, thyroid dysfunction, and increased risk of obesity, cardiovascular disease, and diabetes.[55] BPA exposure is of greatest concern during fetal development.[56] The FDA has recently stated that BPA as currently used is safe, but not all scientists agree with this assessment. Concerns exist that even low levels of BPA exposure are toxic and damages DNA in a way that increases cancer risk.[56] More research is needed on the sources of BPA exposure, how it is metabolized and excreted, and its role as an endocrine disrupter in humans. Many BPA-free products are now available, some as a result of an FDA ban on its use in baby bottles, sippy cups, and infant formula packaging.

### Synthetic vs. Natural Additives

Although most food additives are synthetic compounds, this does not necessarily make them less safe than natural compounds. The toxicity of a substance is determined by its effects in the body, not whether it is synthesized in a laboratory or a plant. The dose of the substance also is critical. Even a common substance, such as table salt, can cause illness or even death when ingested in large amounts. Further, many plants contain natural toxins that are even more potent and prevalent than the additives intentionally added to foods. Some cancer researchers suggest that we ingest at least 10,000 times more (by weight) natural toxins produced by plants than synthetic additives or pesticides.[57]

**endocrine disrupter**  Substance that interferes with the normal function of hormones produced in the body.

BPA, a breakdown product of coatings that line some food cans, can leach into foods.

Mark Dierker/McGraw-Hill Education

Many candies and soft drinks contain several intentional food additives, including colors, flavors, and sweeteners.

James Trice/iStock/Getty Images

**Table 3-4** Functions and Examples of Common Food Additives

| Type of Food Additive | Examples of Additives | Examples of Use |
|---|---|---|
| **Improve Freshness and Safety** | | |
| Antimicrobial agents | Sodium benzoate, sorbic acid, calcium propionate | Inhibit growth of molds, fungi, and bacteria in beverages, baked goods, jams, jellies, salad dressings, processed meats |
| Antioxidants | Butylated hydroxyanisole (BHA), butylated hydroxytoulene (BHT), ascorbic acid, erythorbic acid, alpha-tocopherol, sulfites | Control adverse effects of oxygen and/or prevent fats from spoiling; used in breakfast cereals, chewing gums, nuts, processed meats; prevent light-colored foods (sliced potatoes, white wine, fruit) from discoloring |
| Curing agents | Sodium nitrate, sodium nitrite | Prevent growth of *Clostridium botulinum* in bacon, ham, salami, hot dogs, other cured meats; contribute to pink color of cured meats |
| Acidic agents | Acetic acid, ascorbic acid, phosphoric acid, lactic acid | Add tartness and inhibit growth of microorganisms in foods such as beverages, salad dressings, candies, frozen desserts, salsas, pickles, processed meats |
| **Alter Nutritional Value** | | |
| Vitamins, minerals, protein | Thiamin, vitamin A, protein | Fortification (add nutrients): iodine in salt; enrichment (replace nutrients lost in processing): thiamin, riboflavin, niacin, folic acid, iron in cereal and grain products |
| Alternative sweeteners | Aspartame, sucralose | Sweeten products such as beverages, baked goods, yogurt |
| Fat replacers | Modified food starch, cellulose | Reduced-fat ice cream and low-fat salad dressings |
| **Enhance Flavor or Color** | | |
| Flavors and spices | Salts, sugars, herbs, spices, flavors | Grape flavor in popsicles |
| Flavor enhancers | Monosodium glutamate (MSG), guanosine monophosphate (GMP) | Enhance existing flavor or contribute savory flavor to foods, such as soups, rice, noodle mixes |
| Color additives | Beta-carotene, annatto, beet coloring, cochineal, caramel coloring | Natural colors obtained from plant, animal, or mineral sources; used in many foods; exempt from FDA certification |
| Certifiable color additives | FD&C Blue #1, FD&C Blue #2, FD&C Green #3, FD&C Red #3, FD&C Red #40, FD&C Yellow #5, FD&C Yellow #6, Citrus Red #2, Orange B | The only human-made dyes currently certified by the FDA for use in foods; found in a variety of foods |
| **Enhance Functional Characteristics** | | |
| Emulsifiers | Egg yolks, soy lecithin, mono- and diglycerides | Salad dressings, peanut butter, frozen desserts, baking mixes, margarine |
| Anticaking agents | Calcium silicate, ammonium citrate, magnesium stearate | Keep foods, especially powdered mixes, free flowing |
| Humectants | Glycerol, sorbitol | Retain moisture, flavor, and texture in foods, such as marshmallows, soft candies, energy bars |
| Stabilizers, thickeners | Pectin, gums (guar, carrageenan, xanthan), gelatin | Add creaminess and thickness to foods, such as frozen desserts, yogurt, dairy products, salad dressings, pudding, gelatin mixes |
| Enzymes | Lactase, rennet, chymosin, pectinase | Act on proteins, fats, or carbohydrates in foods; lactase makes milk more digestible; rennet and chymosin are required for cheese making; pectinase improves the clarity of some jellies and fruit juices |
| Leavening agents | Yeast, baking soda, baking powder | Contribute leavening gases (mainly $CO_2$) to improve texture of baked products, such as breads, cookies, cakes, baking mixes |

Burke/Triolo Productions/ Brand X Pictures/ Getty Images

## Regulation and Safety of Food Additives

Assuring the safety of food additives and regulating their use are complex tasks. The responsibility for these tasks lies with the FDA, as set out by the 1958 Food Additives Amendment of the Federal Food, Drug, and Cosmetics Act. Another amendment, enacted in 1960, specifies regulations for color additives. These laws require food manufacturers to carefully test and prove the safety of new additives and obtain FDA approval before they can be used in food (a recently approved food additive is the non-nutritive sweetener Advantame).

Some additives, however, are exempt from this approval and testing process. Examples are additives designated as prior-sanctioned substances and those on the Generally Recognized as Safe (GRAS) list. Prior-sanctioned substances are additives approved by the FDA and USDA as being safe to use prior to 1958; examples include sodium nitrate and sodium nitrite, used as preservatives of processed meats. A GRAS additive is one that scientific data, expert knowledge, and a history of safe use indicate can be used safely. Since 1958, some food additives such as safrole, a natural flavoring once used in root beer, and several color additives have been removed from the food supply due to safety concerns.

Starting in 1998, the FDA permitted food companies, using a panel of qualified experts, to make their own GRAS determinations.[54,58] The companies are encouraged, but not required, to report their findings to the FDA. There are thousands of GRAS additives, including many seemingly safe substances, such as spices and vitamins, that have a long history of safe use. However, approximately 3000 GRAS additives (including 1000 determined to be safe by industry) have not received any type of FDA review. Several organizations have identified this lack of oversight as a public health concern.[58,59]

Many people continue to wonder about the long-term safety of some food additives, including artificial sweeteners, food colors, sodium nitrite and sodium nitrate, and some preservatives. For instance, preliminary data suggest that consuming artificial sweeteners may alter gut microbiota (discussed in Chapter 4) and increase the risk of hyperglycemia and Metabolic Syndrome.[60] Some rodent studies indicate that long-term exposure to aspartame, an artificial sweetener, is linked to a higher risk of various cancers. However, research to confirm these effects in humans is lacking.[61] Sodium nitrite, added to cured meats to prevent growth of the deadly bacterium *Clostridium botulinum*, can be converted to carcinogenic nitrosamines in the stomach. Adding ascorbic acid or erythorbic acid to cured–meats limits nitrosamine production. Many deem the benefit of minimizing deadly botulism infections greater than the small risk of nitrosamine formation. Some artificial colors have been reported to cause allergic-type reactions in children and to increase hyperactive behavior;[62] others have been linked to cancer in animals. Food additives are inherently difficult to study. However, if credible future research confirms these effects, the amount permitted in food might be reduced or banned altogether.

A few food additives cause adverse symptoms in sensitive individuals. Sulfites, a group of sulfur-based chemicals, are used as antioxidants and preservatives in foods. About 1 in 100 persons, particularly those with asthma, experiences shortness of breath or gastrointestinal symptoms after ingesting sulfites. Because of this, sulfites are prohibited on salad bars and other raw vegetables. However, sulfites are found in a variety of foods, such as frozen or dehydrated potatoes, wine, and beer. Food labels indicate their presence. Monosodium glutamate, a flavor enhancer, also can cause problems; some individuals report flushing, chest pain, dizziness, rapid heartbeat, high blood pressure, headache, and/or nausea after consuming monosodium glutamate.

Many food producers are making foods with fewer additives to meet consumer preferences for foods with fewer ingredients and additives. Reading ingredient lists will help you identify these foods. Also, keep in mind that, the more processed a food is, the more additives it is likely to contain. Many prepackaged, precooked, frozen, canned, and instant foods, mixes, and snack foods contain additives. To lower your intake of additives, read food labels and eat fewer highly processed foods. Although no evidence shows that limiting additives will make you healthier, replacing highly processed foods with fruits, vegetables, whole grains, meats, and dairy products is a healthy practice.

▶ Many additives are ingredients you know, such as sodium chloride (salt), sucrose (table sugar), and sodium bicarbonate (baking soda). Many of the food additives used in the U.S. are listed at www.fda.gov/food/ingredientspackaginglabeling.

(a)

(b)

Depending on food choices, a diet can be either (a) essentially devoid of food additives or (b) high in food additives.

(a): C Squared Studios/Getty Images (b): Bob Coyle/McGraw-Hill Education

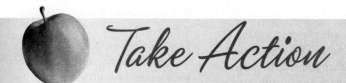

## Take Action

### A Closer Look at Food Additives

Evaluate the food label of a food item, either one in the supermarket or one you have available.

1. Write out the list of ingredients.

2. Identify the ingredients you think are food additives.

3. Based on the information available in this chapter, what are the functions and relative safety of these food additives?

Photodisc/Getty Images

### Knowledge Check

1. How can irradiated foods be identified?
2. How do intentional food additives differ from incidental food additives?
3. What are the broad functions of intentional food additives?
4. What concerns are voiced about food additives?

## HISTORICAL PERSPECTIVE

### A Safe Food Supply

Finding ways to preserve food to keep it safe and delicious has been a challenge for millennia. Starting in the early 1900s, Lloyd Augustus Hall found many new ways to preserve food. For example, he discovered that a mixture of sodium nitrate and nitrite helped prevent botulism in cured meats, like bologna and hot dogs (these preservatives give these meats a pinkish color). He also developed ways to treat spices and cereals to make them safer to eat. Many of the food preservative compounds used today are the direct result of Dr. Hall's research.

Learn more at webfiles.uci.edu/mcbrown /display/hall.html.

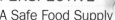

**foodborne illness** Sickness caused by the ingestion of food containing pathogenic microorganisms or toxins made by these pathogens.

## 3.4 Food and Water Safety

In addition to having access to abundant, varied, and nutritious foods, we must have safe food and water supplies to support good health. Scientific knowledge of the pathogens in food and of safe food handling practices, technological developments (e.g., refrigeration, water purification, and milk pasteurization), and laws and regulations have greatly improved the safety of the food and water supplies and have contributed to a steep decline in foodborne and waterborne illness.

Scientists and health authorities agree that North Americans enjoy relatively safe water and food supplies. Nonetheless, pathogens and certain chemicals in foods and water still pose a health risk. The following sections of the chapter examine these hazards and how you can minimize your exposure.

### Foodborne Illness Overview

**Foodborne illness** caused by microbial pathogens remains a significant public health problem in the 21st century. According to the U.S. Centers for Disease Control and Prevention (CDC), foodborne pathogens cause about 48 million illnesses each year. These illnesses result in an estimated 128,000 hospitalizations and over 3000 deaths.[63] In 80% of these illnesses, the specific microbial pathogen is unknown. Although it's impossible to know the true economic burden of these illnesses, experts estimate the annual cost in the U.S. to be about $78 billion when medical costs, loss of productivity, quality-of-life losses, and lost life expectancy are considered.[64]

Most cases of foodborne illness go undiagnosed because the symptoms are mild enough that the ill persons do not seek medical care. These symptoms typically include gastrointestinal effects, such as nausea, vomiting, diarrhea, and intestinal cramping. However, some bouts of foodborne illness, especially when coupled with ongoing health problems, are lengthy and lead to food allergies, seizures, blood poisoning (from microorganisms or their toxins in the bloodstream), organ failure, chronic complications (e.g., arthritis), and even death.

### Individuals at High Risk for Foodborne Illness

Individuals at increased risk of foodborne illness include those with immune systems weakened by disease or medical treatments (e.g., cancer, diabetes, HIV/AIDS); pregnant women and their fetuses; infants and young children; and elderly persons. Others who may be at disproportionately greater risk are those living in institutional settings and homeless persons.

### Foodborne Disease Outbreaks

A foodborne illness outbreak occurs when 2 or more people acquire the same illness from the same contaminated food or beverage. Most foodborne illness outbreaks occur among small groups in a localized area; however, our food system affects the spread of foodborne disease. Much of the food we eat is grown on large farms in the U.S. (e.g., lettuce from Arizona, tomatoes from California, beef from Colorado, eggs from Iowa) and is transported to either food processing plants or supermarkets across the country. Large-scale production means that, if contamination occurs at any point, many people can be affected. This was evident in an outbreak associated with cucumbers contaminated with *Salmonella* Poona, which over a period of 8 months sickened 907 people in 40 states, causing 204 hospitalizations and 6 deaths.[65] Similar multistate outbreaks have occurred with frozen berries, spinach, salad mix, peanuts, pistachios, poultry, ground meat, ice cream, and chocolate chip cookie dough.

Food mishandling in food service establishments, such as cafeterias and restaurants, can cause illness in many people, too. Although mishandling by food processors and food service establishments can lead to widespread problems, it is important to remember that food mishandling in home kitchens also causes foodborne illness.

## Microbial Pathogens

The greatest health risk from food and water today is contamination by viruses and bacteria and, to a lesser extent, by various forms of fungi and parasites. Foodborne illness occurs when microorganisms either directly infect the cells of the gastrointestinal tract, and sometimes other organs in the body, or secrete a toxin into food, which harms us when we eat it (called food intoxication). Unlike foodborne infections, live pathogens need not be present in food for a foodborne intoxication to occur. Toxin-producing bacteria need only to have, at some point, infused the food with their toxin.

Most of the pathogenic bacteria and viruses that cause foodborne illness originate in an infected human or animal and reach food by these fairly well-defined routes:

• *Contamination by feces.* Many foodborne illness-causing bacteria and viruses are excreted profusely in the feces of infected humans and animals. In countries with inadequate sanitation, the water used for drinking, cooking, washing dishes and produce, irrigating crops, and

Food contaminated in a central plant can go on to produce illness in people wherever that food is distributed and sold. According to the CDC, contaminated poultry is responsible for 15% of foodborne illness outbreaks in the U.S. Thorough cooking of poultry kills *Salmonella* and other pathogens, reducing foodborne illness.

Glow Images

Spinach, lettuce, unpasteurized milk and juice, and undercooked ground beef have been implicated in outbreaks of *E. coli* 0157:H7.

Florea Marius Catalin/E+/Getty Images

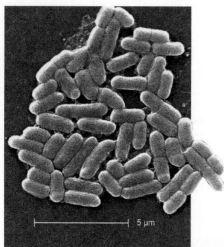

Most types of *E. coli* are not dangerous, but *E. coli* 0157:H7 is the leading cause of bloody diarrhea in the U.S. In children and the elderly, it can cause kidney failure and death.

Source: Janice Haney Carr/CDC

fishing is frequently contaminated with raw sewage and is a major source of illness. Even in the U.S., fecal contamination of soil and irrigation water from farm and wild animals contributes to the spread of pathogens. Fecal contamination also occurs when food is handled by a person who has come in contact with feces or sewage (as in using the bathroom or changing diapers) and has not thoroughly washed his or her hands. Insects, such as houseflies, also may carry bacteria from sewage to food. Foodborne illnesses that are acquired from fecal matter in one of these ways are said to be transmitted by the fecal-oral route.

- *Contamination by an infected individual.* Some pathogenic bacteria and viruses can be transferred to food directly by an infected individual. For example, a food handler who has an open wound or who coughs or sneezes onto food may contaminate the food. Pets also may be a source of foodborne pathogens that can contaminate food via the unwashed hands of food preparers.

- *Cross-contamination.* Cross-contamination occurs when an uncontaminated food touches a pathogen-contaminated food or object, such as a plate, knife, or cutting board, that has come in contact with contaminated food. For instance, let's say a person cuts up a raw chicken contaminated with pathogenic bacteria and then chops lettuce for a salad. When the person is cutting the chicken, the cutting board, the knife, and the food preparer's hands become contaminated with bacteria. If these items are not thoroughly washed before chopping the lettuce, they will cross-contaminate the lettuce with bacteria. Although the bacteria on the chicken will be killed with thorough cooking, the lettuce is not cooked and can cause foodborne illness.

### Bacteria

Bacteria are single-cell organisms found in the food we eat, the water we drink, and the air we breathe. They live in our intestines, on our skin, in our refrigerators, and on kitchen countertops. Luckily, most are harmless, but some are pathogenic and can cause illness. Any food can transmit pathogenic bacteria; however, the most common sources are raw or inadequately cooked meats, poultry, eggs, fish, and shellfish; unpasteurized dairy products; nuts; and fresh produce.

Table 3-5 lists many of the bacteria that cause foodborne illness and describes typical food sources and the symptoms of the illnesses they cause. *Salmonella* spp., *Clostridium perfringens*, *Campylobacter* spp., and *Staphylococcus aureus* are responsible for most bacterial foodborne illness in the U.S.63 Others, such as some strains of *Escherichia coli*, *Clostridium botulinum*, *Listeria monocytogenes*, and *Vibrio vulnificus*, cause fewer cases but are more likely to result in serious illness and death. In developing countries, pathogens such as *Vibrio cholerae* are more important.

To proliferate, bacteria require nutrients, water, and warmth. Most grow best in **danger zone** temperatures of 41° to 135°F (5° to 57°C) (Fig. 3-6). Pathogenic bacteria typically do not multiply in perishable foods when the foods are held at temperatures above 135°F (57°C) or stored at safe refrigeration temperatures, 32° to 40°F (0° to 4.4°C). One important exception is *Listeria* bacteria, which can multiply at refrigeration temperatures. Also note that high temperatures can kill toxin-producing bacteria, but high temperatures may not inactivate toxins the bacteria have produced in the food. Most pathogenic bacteria also require oxygen for growth, but *Clostridium botulinum* and *Clostridium perfringens* grow only in anaerobic (oxygen-free) environments, such as those found in tightly sealed cans and jars. Food acidity can affect bacterial growth, too. Although most bacteria do not grow well in acidic environments, some, such as pathogenic *E. coli*, can grow in acidic foods, such as fruit juice.

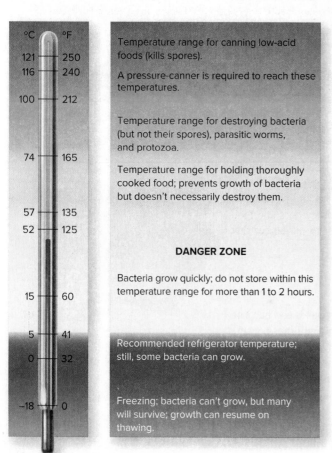

*Figure 3-6* Effects of temperature on microbes that cause foodborne illness.

**Table 3-5  Bacterial Causes of Foodborne Illness**

| Bacteria | Typical Food Sources | Symptoms | Additional Information |
|---|---|---|---|
| *Salmonella* species  | Raw and undercooked meats, poultry, eggs, and fish; produce, especially raw sprouts; peanut butter; unpasteurized milk | Onset: 6–72 hours; nausea, fever, headache, abdominal cramps, diarrhea, and vomiting; can be fatal in infants, the elderly, and those with impaired immune systems; lasts 4–7 days | Estimated 1.03 million infections/year; bacteria live in the intestines of animals and humans; food is contaminated by infected water and feces; about 2000 strains of *Salmonella* bacteria can cause disease, but 3 strains account for almost 50% of cases; *Salmonella enteritidis* infects the ovaries of healthy hens and contaminates eggs; almost 20% of cases are from eating undercooked eggs or egg-containing dishes; reptiles, such as turtles, also spread the disease |
| *Campylobacter jejuni* | Raw and undercooked meat and poultry, unpasteurized milk and cheese, contaminated water | Onset: 2–5 days; muscle pain, abdominal cramping, diarrhea (sometimes bloody), fever; lasts 2–7 days | Estimated 845,000 infections/year; produces a toxin that destroys intestinal mucosal surfaces; can cause Guillain-Barré syndrome, a rare neurological disorder that causes paralysis |
| *Escherichia coli* (0157:H7 and other strains) | Undercooked ground beef; produce—lettuce, spinach, sprouts; unpasteurized juice and milk | Onset: 1–9 days; bloody diarrhea, abdominal cramps; in children under age 5 and the elderly, hemolytic uremic syndrome (HUS) is a serious complication; red blood cells are destroyed and kidneys fail; can be fatal; lasts 2–10 days or longer in more serious illness | Most strains of *E. coli* are harmless, but those that cause disease produce a powerful toxin and are the leading cause of bloody diarrhea in the U.S.; estimated 265,000 cases/year; cattle and cattle manure are chief sources; petting zoos, lakes, and swimming pools can contain pathogenic *E. coli* |
| *Shigella* species | Fecal-oral transmission; water supplies, produce, and other foods contaminated by infected food handlers with poor hygiene | Onset: 1–3 days; abdominal cramps, fever, diarrhea (often bloody); lasts 5–7 days | Estimated 131,200 cases/year; humans and primates are the only sources; common in day-care centers and custodial institutions from poor hygiene; traveler's diarrhea often caused by *Shigella dysenteriae* |
| *Staphylococcus aureus* | Ham, chicken, tuna, egg, and potato salads; cream-filled pastries, custards; whipped cream | Onset: 1–6 hours; diarrhea, vomiting, nausea, abdominal cramps; lasts 1–3 days | Bacteria on skin and nasal passages of up to 50% of people and common in soil, water, air, everyday objects; can be passed to foods; multiplies rapidly when contaminated foods are held for extended time at room temperature; illness caused by a heat-resistant toxin that cannot be destroyed by cooking; estimated 241,000 cases/year |

**Table 3-5  Bacterial Causes of Foodborne Illness (continued)**

| Bacteria | Typical Food Sources | Symptoms | Additional Information |
|---|---|---|---|
| *Clostridium perfringens* | Beef, poultry, gravy, Mexican food | Onset: 8–24 hours; abdominal pain and diarrhea, usually mild; can be more serious in elderly or ill persons; lasts 1 day or less but may be longer in the elderly or infants | Estimated 966,000 cases/year; anaerobic, spore-forming bacteria widespread in soil and water; multiplies rapidly in prepared foods, such as meats, casseroles, and gravies, held for extended time at room temperature |
| *Listeria monocytogenes* | Unpasteurized milk, ice cream, and soft cheeses; raw meats and vegetables; ready-to-eat deli meats and hot dogs; refrigerated smoked fish | Onset: 9–48 hours for early symptoms, 3 days to 3 months for severe symptoms; fever, muscle aches, headache, vomiting; can spread to nervous system, resulting in stiff neck, confusion, loss of balance, or convulsion; can cause premature birth and stillbirth | Estimated 1600 cases with 255 fatalities/year; widespread in soil and water and can be carried in healthy animals; grows at refrigeration temperatures; about one-third of cases occur during pregnancy; high-risk persons should avoid uncooked deli meats, raw milk, soft cheeses (e.g., feta, Brie, and Camembert), blue-veined cheeses, Mexican-style cheeses (e.g., queso blanco made from unpasteurized milk), refrigerated meat spreads or pâtés, uncooked refrigerated smoked fish |
| *Clostridium botulinum* | Incorrectly home-canned vegetables, meats, and fish; incorrectly canned commercial foods; herb-infused oils; bottled garlic; potatoes baked in foil and held at room temperature; honey | Onset: 18–36 hours but can be 4 hours to 10 days; neurological symptoms—double and blurred vision, drooping eyelids, slurred speech, difficulty swallowing, muscle weakness, and paralysis of face, arms, respiratory muscles, trunk, and legs; can be fatal; lasts days to weeks | Estimated 55 cases/year, fatal in 5–10% of cases; caused by a neurotoxin; *C. botulinum* grows only in the absence of air in nonacidic foods; incorrect home canning causes most botulism, but in 2007 commercially canned chili sauce caused an outbreak; infant botulism is the most common type—honey should not be given to infants younger than 1 year of age because it can contain botulism spores |
| *Vibrio* | *V. parahemolyticus:* raw and undercooked shellfish, especially oysters | Onset: 4 hours–4 days; diarrhea (sometimes bloody), nausea, vomiting, fever, chills; lasts 2–6 days | Found in coastal waters; more infections in summer; number of infections hard to determine because it is difficult to isolate in the lab |
| | *V. vulnificus:* raw and undercooked shellfish, especially oysters | Onset: 1–2 days; vomiting, diarrhea, abdominal pain; in more severe cases, bloodstream infection with fever, chills, decreased blood pressure, blistering skin lesions; lasts 3 or more days | Estimated 95 cases/year; found in coastal waters; more infections in summer; those with impaired immune systems and liver disease at higher risk of infection; fatality rate of 35% with bloodstream infection |
| | *V. cholerae:* contaminated water and food, human carriers | Onset: 2–3 days; severe, dehydrating diarrhea, vomiting; dehydration, cardiovascular collapse, and death can occur | Occurs mainly in countries without adequate water purification and sewage treatment |
| *Yersinia enterocolitica* | Raw or undercooked pork, particularly pork intestines (chitterlings); tofu; water; unpasteurized milk; oysters | Onset: 1–11 days; fever, abdominal pain, diarrhea (often bloody); lasts 1–3 weeks or longer | Yersinosis most common in children under age 5 years; relatively rare; bacteria live mainly in pigs but can be found in other animals; can cause reactive arthritis, kidney and heart infections |

Table 3-6 **Viral Causes of Foodborne Illness**

| Viruses | Typical Food Sources | Symptoms | Additional Information |
|---|---|---|---|
| Norovirus (Norwalk and Norwalk-like viruses), human rotavirus | Foods prepared by infected food handlers; shellfish from contaminated waters; vegetables and fruits contaminated during growing, harvesting, and processing | Onset: 1–2 days; "stomach flu"—severe diarrhea, nausea, vomiting, stomach cramping, low-grade fever, chills, muscle aches; lasts 1–2 days or longer | Estimated 5.5 million infections per year; viruses found in stool and vomit of infected persons; food handlers can contaminate foods or work surfaces; noroviruses are very infectious—as few as 10–100 particles can lead to infection; workers with norovirus symptoms should not work until 2 or 3 days after they feel better |
| Hepatitis A virus | Foods prepared by infected food handlers, especially uncooked foods or those handled after cooking, such as sandwiches, pastries, and salads; shellfish from contaminated waters; vegetables and fruits contaminated during growing, harvesting, and processing | Onset: 15–50 days; anorexia, nausea, vomiting, fever, jaundice, dark urine, fatigue; may cause liver damage and death; lasts several weeks up to 6 months | Estimated 1600 infections/year from contaminated foods; children and young adults are more susceptible; a vaccine is available, decreasing the number of infections dramatically; immunoglobulin given within 1 week to those exposed to hepatitis A virus also can decrease infection |

As you can see, different types of pathogenic bacteria can thrive in a variety of environmental conditions. Some can even survive in harsh environmental conditions (e.g., dry conditions or very hot or cold temperatures) through spore formation. In the spore state, bacteria can remain viable for months or years—then, when environmental conditions improve, they begin proliferating. For example, uncooked rice contains little water, which prevents bacterial growth. It is safe to store dry, uncooked rice at room temperature. However, because cooked rice has a high water content, leaving it on the kitchen counter (at danger zone temperatures) instead of refrigerating it provides a hospitable environment (moisture, nutrients, warmth) for disease-causing bacteria and their spores to multiply rapidly. Steps to prevent bacterial and other foodborne illness are described later in this chapter.

▶ Some sources list the temperature range for the danger zone as 41° to 140°F (5° to 60°C). The FDA has lowered the upper end of the range to 135°F (57°C) because the risk posed by holding food between 135° and 140°F (57° and 60°C) is minimal.

### Viruses

Viruses, like bacteria, are widely dispersed in nature. Unlike bacteria, however, viruses can reproduce only after invading body cells, such as those that line the intestines. Thus, the key to preventing foodborne viral illnesses is to use sanitary food preparation practices to keep viruses from contaminating food and to cook food thoroughly to kill any that have found their way into food via contamination from infected food handlers, other foods, and feces.

Table 3-6 describes the 2 most common viral causes of foodborne illness, their typical food sources, and symptoms of the illnesses they cause. The highly contagious noroviruses are thought to account for almost half of all foodborne illness. If you have had "stomach flu," you may well have experienced a norovirus infection. There are many reports of norovirus outbreaks on cruise ships, in hotels and restaurants, and even in hospitals. Rotaviruses, a type of norovirus, are an important cause of diarrhea in children (see Chapter 4). The hepatitis A virus causes liver disease and is spread by contaminated food or water. A vaccine to prevent hepatitis A infection is now available.

### Parasites

Parasites live in or on another organism, known as the host, from which they absorb nutrients. Humans can serve as a host to parasites. These tiny ravagers rob millions of people around the globe of their health and, in some cases, their lives. Those hardest hit live in tropical countries where poor sanitation fosters the growth of parasites. However, epidemiologists report that parasitic infections seem to be on the increase in the U.S. and other industrialized countries.[66] For example, in 1993 a *Cryptosporidium* outbreak involving more than 400,000 people occurred in Milwaukee due to contamination of the water supply. A 2015 *Cyclospora* outbreak affected nearly 550 people in 31 states; although no single food was identified as the infection source, fresh produce was implicated.[67]

Botox is a drug made from *Clostridium botulinum*, the same bacterium that can cause life-threatening paralytic foodborne illness called botulism. Botox works by temporarily paralyzing specific muscles to help alleviate numerous health problems, such as excessive sweating, uncontrollable blinking, severe muscle contractions, and an overactive bladder. Cosmetically, it can be used to help smooth facial wrinkles.

Science Photo Library/Getty Images

Outbreaks of the highly contagious norovirus are reported often on cruise ships because passengers are in close contact with each other and because illnesses on cruise ships are monitored by health officials. About 20% of people exposed to norovirus are resistant to infections. When exposed to the viruses, these individuals do not develop the usual symptoms of abdominal pain, nausea, vomiting, and diarrhea. Resistance results from a heritable genetic mutation that prevents the release of an enzyme needed to transport the virus into gastrointestinal cells.

NAN/Alamy Stock Photo

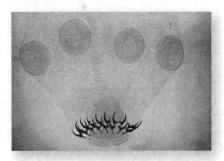

This micrograph shows the "suckers" (4 circles) and 2 rows of hooks of a tapeworm. Tapeworms can survive for years inside humans, using the suckers and hooks to attach to the small intestine.

Source: Centers for Disease Control and Prevention

▶ To find information about the microbial pathogens in food and water, visit www.cdc .gov/DiseasesConditions.

Paralytic shellfish poison is a natural toxin found in marine environments. It is produced by algae that are eaten by shellfish. When water conditions are favorable, such as when water temperatures are higher than normal, the algae "bloom," or reproduce in large numbers. This bloom is sometimes called a "red tide" because the water turns red, although the algae may be other colors. When toxin-producing algae are abundant, the toxin accumulates in shellfish, making them dangerous to eat.

Don Paulson Photography/Purestock/Alamy Stock Photo

The more than 80 foodborne parasites known to affect humans include mainly **protozoa** (1-celled animals), such as *Cryptosporidium* and *Cyclospora,* and **helminths**, such as tapeworms and the roundworm *Trichinella spiralis.* Table 3-7 describes common parasites, their typical food sources, and the symptoms of the illnesses they cause. Parasitic infections spread via person-to-person contact and contaminated food, water, and soil.

### Prions

Prions are infectious protein particles, found in the brain and spinal cord, that are thought to cause the progressive and fatal neurodegenerative diseases known collectively as transmissible spongiform encephalopathies (TSEs).[68] TSEs are characterized by tiny holes in the brain that give it a spongy appearance. They are extremely rare in humans.

Following a TSE outbreak in cattle in the United Kingdom in the 1980s and 1990s, doctors detected a new TSE, now known as variant Creutzfeldt-Jakob disease (vCJD), in a small number (<224 to date) of people, mostly in the United Kingdom. After years of research, scientists have concluded that vCJD is caused by eating prion-contaminated meat from cattle afflicted by bovine spongiform encephalopathy (BSE), the TSE found in cattle. Today, strict guidelines for monitoring cattle and preventing prion contamination of their feed have sharply reduced BSE worldwide. Because of this and other safeguards to the food supply, experts affirm that eating beef in the U.S. is not a risk factor for contracting vCJD.

### Toxins

A number of toxins produced by molds, algae, and plants can cause serious illness (Table 3-8). **Molds** are a type of fungus that can be scattered by the wind or carried by animals. Molds grow best in moist, dark places where air circulates. When conditions are right, the mold grows by sending rootlike "threads" deep into the food it lives on and forming endospores on the outside of the food. These endospores give mold its fuzzy, colorful look and are the form in which the mold travels to new locations. The foods most likely to mold in U.S. homes are cheese, breads, and fresh produce.

Thousands of types of mold grow on foods. Most just alter the color, texture, taste, and/or odor of foods, making them unpalatable and inedible. Some destroy crops and shorten the length of time a food will remain fresh and safe to eat. Others cause allergies or respiratory problems. A few fungi produce toxins known as **mycotoxins** (*myco* means "mold"), which cause blood diseases, nervous system disorders, and kidney and liver damage. The most important mycotoxins are aflatoxin, ergot, and those produced by the *Fusaria* fungi. Liver cancer–causing **aflatoxin** is produced by a mold that attacks peanuts, tree nuts (e.g., walnuts and pecans), corn, and oilseeds (e.g., cottonseed). Ergot is produced by a dark purple mold that grows on inappropriately stored grains, especially rye. Several types of *Fusaria* fungi can grow on grains stored for long periods and produce deadly mycotoxins.

Mycotoxins are rarely a problem in most industrialized nations because food production practices are designed to minimize mold growth. In addition, food producers and government inspectors closely monitor foods to detect molds and destroy any foods found to contain them. Unfortunately, mycotoxin poisonings are frequent in other parts of the world. For example, in Kenya in 2004 an aflatoxin poisoning caused liver damage and death in many people.

**Toxins in seafoods** Toxin-containing algae ingested by some fish and shellfish also cause foodborne disease. One example is the **ciguatera toxin** found in some large tropical and subtropical fish. Another example is **shellfish poisoning** from shellfish harvested from waters experiencing an algae population explosion called a red tide (sometimes the algae are so thick they color the water red). The fish and shellfish are not harmed by the toxins, and the toxins are not destroyed by cooking or freezing. To avoid ciguatera, do not eat large fish that may contain a higher amount of this toxin; instead, choose the smaller specimens and don't eat fish heads or organs where toxins concentrate. The only protection against shellfish poisoning is to avoid shellfish from affected waters until nature thins the algae population. Both

**Table 3-7  Parasitic Causes of Foodborne Illness**

| Parasite | Typical Food Sources | Symptoms | Additional Information |
|---|---|---|---|
| *Trichinella spiralis* | Pork, wild game | Onset: 1–4 weeks; GI symptoms followed by muscle weakness, fluid retention in the face, fever, flulike symptoms; larvae can live in muscles for years | Estimated 40–340 illnesses/year; the number of trichinosis infections has decreased greatly because pigs are now less likely to harbor this parasite; cooking pork to 145°F (63°C) plus 3 minutes of rest time before carving will kill *Trichinella*, as will freezing it for 3 days at −4°F (−20°C) |
| Anisakis | Raw or undercooked fish | Onset: 12 hours or less; violent stomach pain, nausea, vomiting when worms burrow into wall of stomach or intestine | Caused by eating the larvae of roundworms; the infection is more common where raw fish is routinely consumed; number of cases is unknown |
| Tapeworms | Raw or undercooked beef, pork, fish from regions with poor sanitary facilities | Onset: 2–4 months after ingesting live cysts; abdominal discomfort, diarrhea, general malaise | Common worldwide, rare in U.S.; caused by eating tapeworm cysts in undercooked or raw meat; in the human intestine, the cyst develops into a mature tapeworm, usually 6 to 23 feet in length; infected persons often pass segments of the tapeworm in the stool; rarely, the tapeworm infects muscle and central nervous tissues |
| *Toxoplasma gondii* | Raw or undercooked seafood and meat (pork, beef, lamb, or venison), unwashed fruits and vegetables | Onset: 5–20 days; most people are asymptomatic; those with symptoms have fever, headache, sore muscles, diarrhea; can spread to brain, eyes, heart, or other muscles; can be fatal to the fetus of pregnant women; lasts several weeks | The parasite is spread to humans from animals, including cats, the main reservoir of the disease; humans acquire the disease from ingesting contaminated meat or from fecal contamination from handling cat litter; an estimated 86,700 cases/year |
| *Cyclospora cayetanensis* | Water, contaminated food (e.g., imported fresh produce, raspberries, basil, snow peas) | Onset: 1 week; watery diarrhea, vomiting, muscle aches, fatigue, anorexia, weight loss; lasts 10–12 weeks | Most common in tropical and subtropical areas, but approximately 100–200 illnesses occur each year in the U.S. |
| *Cryptosporidium* | Water, any food contaminated by food handler or environmental source | Onset: 2–10 days; profuse, watery diarrhea, abdominal pain, fever, nausea, vomiting, weight loss; those with impaired immune systems become more ill; lasts 2 days–2 weeks in otherwise healthy persons | Estimated 748,000 cases/year; one of the most common waterborne diseases worldwide; also can be spread in water parks, community swimming pools, lakes, rivers, and streams |

**Table 3-8**  Toxins in Food

| Toxin | Typical Food Sources | Symptoms | Additional Information |
|---|---|---|---|
| **Mycotoxins** | | | |
| **Aflatoxin** (from *Aspergillus flavus and Aspergillus parasiticus*) | Corn, peanuts, rice, wheat, spices, nuts, especially when these foods are stressed due to drought or disease | *Acute* toxicity: liver damage or failure, malnutrition, malaise, impaired immune function; can be fatal *Chronic* toxicity: vomiting, abdominal pain, liver failure, liver cancer; can be fatal | Aflatoxin B-1 is the most common fungal toxin to contaminate grains and nuts; causes significant crop losses around the world and disease when consumed by humans; there have been no human illness outbreaks caused by aflatoxins in the U.S. |
| **Ergot** (from *Claviceps purpurea*) | Inappropriately stored grains, especially rye | Hallucinations, spontaneous abortion, severely constricted blood flow to limbs (which can lead to gangrene), tingling and burning sensations, involuntary muscle twitching and contractions | Ergot poisoning thought to be the cause of the odd physical behaviors attributed to women tried at the Salem witch trials; common throughout history but rare today due to inspections and agricultural practices |
| **Fish and Shellfish Toxins** | | | |
| **Ciguatera toxin** | Tropical and subtropical fish (e.g., amberjack, barracuda, grouper, hogfish, moray eel, snapper, scorpion fish, surgeonfish, triggerfish) that have ingested toxin-producing algae | Onset: 6 hours; nausea, vomiting, neurological symptoms (weakness, temperature reversal—hot feels cold and vice versa); symptoms can last days, months, or years | Number of cases not known |
| **Shellfish poisoning** (paralytic, diarrheic, neurotoxic, and amnesic) | Mussels, cockles, clams, scallops, oysters, crabs, lobster | *Paralytic* onset: 15 minutes to 10 hours; numb and tingling skin, respiratory paralysis, death *Diarrheic* onset: few minutes to a few hours; nausea, vomiting, diarrhea, abdominal pain, chills, headache, fever *Neurotoxic* onset: 30 minutes to 3 hours; tingling and numbness of mouth and throat, muscle aches, dizziness, reversal of sensations of hot and cold, diarrhea, vomiting *Amnesic* onset: 24 to 48 hours; vomiting, diarrhea, abdominal pain, confusion, memory loss, disorientation, seizure, coma, death | Toxins produced by algae on which the shellfish feed; the toxins accumulate in shellfish; not inactivated by cooking or freezing |
| **Scombroid poisoning** | Tuna, mackerel, bluefish, mahi-mahi, amberjack, sardine, anchovy | Onset: 2 to several hours; rash, diarrhea, flushing, sweating, headache, vomiting, difficulty breathing; symptoms resolve quickly | Bacterial decomposition of fish results in histamine production, which causes symptoms; one of the most common illnesses caused by seafood consumption |
| **Tetrodotoxin** | Pufferfish (fugu) liver, intestines, gonads, and skin | Onset: 20 minutes to 3 hours; mouth numbness, headache, nausea, diarrhea, vomiting, paralysis, respiratory distress | Fugu is a traditional delicacy in Japan, where chefs must be licensed to prepare and serve this fish |

**Table 3-8  Continued**

| Toxin | Typical Food Sources | Symptoms | Additional Information |
|---|---|---|---|
| **Plant Toxins** | | | |
| **Safrole** | Sassafras, mace, nutmeg | Cancer when consumed in high doses | Previously used as a food additive but now banned |
| **Solanine** | Potato sprouts, green spots on potato skins | Onset: 8–12 hours; nausea, diarrhea, vomiting, hallucinations, loss of sensation, paralysis | Can be prevented by storing potatoes in a dark area and discarding sprouts, peel, bruised or cut areas, and green-tinged spots |
| **Mushroom toxins** | Several mushroom species, such as amanita | Onset: minutes to 3 days; stomach upset, dizziness, hallucinations, and other neurological symptoms; more lethal varieties can cause liver and kidney failure, coma, death | Illness almost always caused by wild mushrooms picked by nonexperts; occurs worldwide, probably affects hundreds of people/year |
| **Herbal teas** | Teas containing senna or comfrey | Onset: depends on dose and it may be weeks before symptoms occur; abdominal pain, nausea, vomiting, diarrhea, liver damage | Teas are used in folk medicines and are not considered safe for internal use |
| **Lectins** | Raw or undercooked legumes, usually kidney beans | Onset: 1–3 hours; nausea, vomiting, abdominal pain, diarrhea | Caused by lectin proteins; as few as 4 or 5 raw beans can cause symptoms; outbreaks have occurred when beans were insufficiently cooked in crockpots and casseroles |

Aflatoxin: ©C Squared Studios/Getty Images RF; Ergot: ©Brand X Pictures/Getty Images RF; Ciguatera Toxin: ©Davies and Starr/Getty Images RF; Shellfish Poisoning: ©Ingram Publishing/Alamy RF; Scombroid Poisoning: Source: NOAA/Department of Commerce; Tetrodotoxin: ©Stephen Frink/Getty Images RF; Safrole: ©Comstock/Getty Images RF; Solanine: ©StockPhotosArt-Food/Alamy RF; Mushroom Toxins: ©Stockbyte/Getty Images RF; Herbal Teas: ©FoodCollection RF; Lectins: ©MRS.Siwaporn/Shutterstock RF

the U.S. and Canada quarantine waters experiencing an algae surge and prohibit shellfish harvesting until the shellfish are safe to eat. However, some countries are not as rigid in their regulations regarding shellfish and red tides. The frequency that red tides occur is increasing as ocean water warms with global climate change.

Scombroid poisoning is caused by eating certain fish left at room temperature for several hours after being caught. The toxin is not destroyed by freezing or cooking but can be prevented by refrigerating or freezing the fish immediately after they're caught.

**Toxins in Plants**

Plants contain a variety of **natural toxins** that can cause illness (see Table 3-8) but, in actuality, rarely do.[69] For instance, licorice contains a natural toxicant that can elevate blood

---

*Critical* **Thinking**

Recognizing that José is taking a nutrition class, his roommate asks him, "What is more risky: the bacteria that can be present in food or the additives listed on the label of my favorite cookie?" How should José respond? On what information should he base his conclusions?

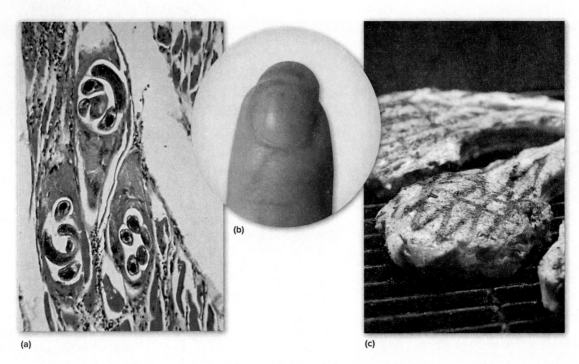

(a) Muscle tissue with *Trichinella spiralis* roundworm cysts. Eating meat that contains *Trichinella* cysts causes the disease trichinosis. (b) Splinter hemorrhages under fingernails are common in those with trichinosis. (c) Grill pork to an internal temperature of 145°F (63°C) plus 3 minutes of rest time before carving. This eliminates the risk of trichinosis and produces a desirable product. Trichinosis cases from pork are rare today because of more sanitary hog-feeding practices.

(a): Source: Centers for Disease Control and Prevention; (b): Centers for Disease Control and Prevention/Dr. Thomas F. Sellers, Emroy University, Atlanta, GA; (c): ORLIO/Shutterstock

Raw shellfish, especially bivalves (e.g., oysters and clams), present a particular risk related to foodborne viral disease. These animals filter feed, a process that concentrates viruses, bacteria, and toxins present in the water. Adequate cooking of shellfish will kill viruses and bacteria, but toxins may not be affected. It's important to buy shellfish from reliable sources who have harvested these foods from safe areas.

lynx/iconotec.com/Glow Images

Hunting wild mushrooms should be left to experts. Many varieties contain deadly toxins.

Brand X Pictures/PunchStock

pressure and cause heart failure. There is cyanide in lima beans and almonds. Nutmeg, bananas, and some herbal teas contain substances that can cause hallucinations. Plants produce and concentrate toxins to compete with their neighbors and protect themselves from plant-eating molds, bacteria, insects, and other predators, including people. When stressed by environmental conditions or damaged, plants tend to produce even greater quantities of toxins. An example is **solanine**, a powerful, narcotic-like toxin, produced by potatoes. The amount produced is normally small, but it increases when potatoes sprout and when they are stored in a brightly lit place.

Humans have coped with natural toxins for thousands of years by learning to avoid some of them and to limit intake of others. Farmers know potatoes must be stored in the dark so that solanine won't be synthesized. Cooking limits the potency of certain natural toxins. Spices are used in such small amounts that health risks from any toxins are unlikely to result. Another important way to cope with these toxins is to eat a wide variety of foods to minimize the chance that any single toxin is eaten in amounts that exceed the body's ability to detoxify it. Natural toxins are so widespread in foods that it would be unrealistic to try to avoid them totally—and doing so would likely limit food choices so severely that nutrient deficiency diseases would occur. Nevertheless, it's important to remember that some potentially harmful chemicals in foods occur naturally.

## Water Safety

Clean water is vital for good health. The disinfection of water is widely regarded as one of the major advances in public health in the last century, responsible for substantially reducing infectious diseases, especially the deadly typhoid fever. Today, the EPA regulates public water supplies, but the actual delivery and safety of water are under the jurisdiction of local municipal water departments. Under the U.S. Safe Drinking Water Act, all public drinking water suppliers are required to rigorously test for contaminants, such as bacteria, various chemicals, and toxic metals (e.g., lead and arsenic) and submit test results to the EPA.[70] Water treatments vary, depending on the water source, but all water is disinfected, usually with a chlorine-based chemical. Private water supplies, such as wells, are not regulated by the EPA but still should be tested for chemical and microbial contaminants. Local health departments can give advice on testing and keeping well water safe.

### Bottled Water

Bottled water is a popular beverage. Many consumers are attracted to its convenience, perceived health value, or taste. All bottled waters sold in the U.S. must list the source of the water on the label. Sources include wells, spas, springs, geysers, and quite often the public water supply. Some bottled water contains minerals, such as calcium, magnesium, and potassium, that either occur naturally or are added by the bottling company to improve the water's taste. The water is carbonated when carbon dioxide gas naturally occurs in the water source (called *naturally sparkling water*) or is added. Additives such as flavors and vitamins also are common. Bottled waters are regulated by the FDA, which sets high standards for their purity. It periodically collects and analyzes samples, but not to the extent that municipal water supplies are monitored. Water that is bottled in a sanitary manner and kept sealed will not spoil, although off-flavors may develop over time.

### Threats to Safe Water

The U.S. enjoys one of the cleanest water supplies in the world, but there are numerous threats to the safety of our water—agricultural runoff (animal waste, pesticides, fertilizer), inappropriate disposal of chemicals, municipal solid waste (containing bacteria, viruses, nitrates, synthetic detergents, household chemicals) leaking into waterways, inadequate treatment of human wastes, and pollution from boats and ships (contains solvents, gas, detergents, raw sewage), to name just a few. This makes regular testing of the water supply critical.

Bottled water is a convenient but relatively expensive source of water. In most cases, tap water is just as healthy a choice to meet water needs.

Stockbyte/Stockbyte/Getty Images

The EPA requires that the public be notified if water contamination is a danger to public health. For instance, nitrate contamination from fertilizer runoff is particularly dangerous to infants because it prevents oxygen from circulating in the body. As related earlier, *Cryptosporidium* can contaminate water supplies (it is not affected by normal chlorination procedures). Boiling tap water for a minimum of 1 minute is the best way to kill *Cryptosporidium*. Alternatively, individuals can purchase a water filter that screens out this parasite.

Even though the U.S. has one of the cleanest water supplies in the world, illnesses from contaminated drinking water do occur. The CDC monitors water-related outbreaks, which average about 30 per year.[71] Water safety experts note that these data likely underestimate the true number of illnesses caused by contaminated water.

## Preventing Foodborne and Waterborne Illnesses

Safe food and water supplies require a "farm-to-fork" approach. All those who grow our food, along with processors, distributors, and consumers, are responsible for food and water safety.[72] Several government agencies regulate and coordinate these efforts, monitor food and water, conduct research, enforce wholesomeness and quality standards and laws, and educate consumers (Table 3-9).

To do their part, consumers need to know how to handle food safely at home. In general, foodborne illness prevention focuses on using good personal and kitchen hygiene; handling food safely by using appropriate thawing, cooking (Fig. 3-7), chilling, and storage procedures; and knowing which foods pose extra risk to those more susceptible to foodborne illness. Complete the Take Action: Check Your Food Safety Skills to assess your own food safety practices.

▶ Raw milk has not undergone pasteurization, the heat-treatment process that kills pathogenic bacteria and makes milk safe to drink. Pasteurization does not alter milk's nutrient content or make milk less healthy or more allergenic. According to the FDA and the CDC, raw milk and products made from it, such as soft cheese, ice cream, and yogurt, can harbor dangerous bacteria and should not be consumed. Regulation of raw milk production and sales varies by state, but interstate commerce is prohibited. From 2007 to 2016, raw milk products caused 144 outbreaks, with 46% of these outbreaks affecting children under 5 years old. Learn more at www.cdc.gov/foodsafety/raw milk.

4 Simple Steps to Food Safety

CLEAN

SEPARATE

COOK

CHILL

Check your food safety steps.

CLEAN

Wash hands and surfaces often.

SEPARATE

Don't cross-contaminate.

COOK

Cook to safe temperatures.

CHILL

Refrigerate promptly.

Source: Courtesy of USDA, HHS, and the Ad Council;

**Table 3-9  U.S. Agencies Responsible for Monitoring the Food Supply**

| Agency Name | Responsibilities | Methods | How to Contact |
|---|---|---|---|
| Food and Drug Administration (FDA)  | • Ensures safety and wholesomeness of foods in interstate commerce (except meat, poultry, and processed egg products) <br> • Regulates seafood <br> • Controls product labels | • Conducts inspections <br> • Conducts food sample studies <br> • Sets standards for specific foods | www.fda.gov or call 1-888-463-6332 |
| U.S. Department of Agriculture (USDA) Food Safety and Inspection Service (FSIS)  | • Enforces whole-someness and quality standards for grains and produce (while in the field), meat, poultry, milk, eggs, and egg products | • Conducts inspections <br> • Monitors imported meat and poultry <br> • Administers "Safe Handling Label" | www.usda.gov /fsis or www.cdc .gov/foodsafety /outbreaks or call 1-800-535-4555 |
| Centers for Disease Control and Prevention (CDC) | • Promotes food safety | • Responds to emergencies concerning foodborne illness <br> • Surveys and studies environmental health problems <br> • Conducts research on foodborne illness <br> • Directs/enforces quarantines <br> • Conducts national programs for prevention and control of foodborne and other diseases | www.cdc.gov |
| Environmental Protection Agency (EPA) | • Regulates pesticides <br> • Establishes water quality standards | • Approves all U.S. pesticides <br> • Sets pesticide residue limits in food | www.epa.gov |
| National Marine Fisheries Service or NOAA Fisheries | • Monitors domestic and international conservation and management of living marine resources | • Conducts voluntary seafood inspection program <br> • Can use mark to show federal inspection | www.fisheries .noaa.gov |
| Bureau of Alcohol, Tobacco, Firearms and Explosives (ATF)  | • Enforces laws on alcoholic beverages | • Conducts inspections | www.atf.gov /alcohol-tobacco |
| State and local governments | • Promote milk safety <br> • Monitor food industry within their borders | • Conduct inspections of food-related establishments | Government pages of telephone book, Internet |

Digital Vision

Aaron and his wife attended an international potluck on a warm July afternoon. Their contribution was Argentine beef, a stewlike dish. They followed the recipe and the cooking time carefully, removed the dish from the oven at 1 P.M., and kept it warm by wrapping the pan in a towel. They drove to the party and set the dish out on the buffet table at 3 P.M. Dinner was to be served at 4 P.M., but the guests were enjoying themselves so much that no one began to eat until 6 P.M. Aaron made sure he sampled the Argentine beef they had prepared, but his wife did not. He also had some salads, garlic bread, and a sweet coconut dessert. The couple returned home at 11 P.M. and went to bed. At 2 A.M., Aaron knew something was wrong. He had severe abdominal pain and had to make a mad dash to the toilet. He spent most of the next 3 hours in the bathroom with diarrhea. By dawn, the diarrhea had subsided and he was feeling better. He ate a light breakfast and felt fine by noon. It's very likely that Aaron contracted foodborne illness from the Argentine beef. What precautions for avoiding foodborne illness were ignored by Aaron and the rest of the people at the party? How might this case study be rewritten to substantially reduce the risk of foodborne illness?

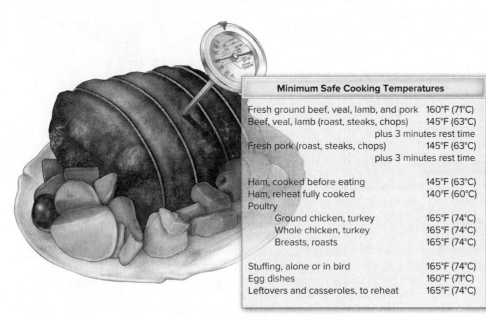

| Minimum Safe Cooking Temperatures | |
|---|---|
| Fresh ground beef, veal, lamb, and pork | 160°F (71°C) |
| Beef, veal, lamb (roast, steaks, chops) | 145°F (63°C) |
| | plus 3 minutes rest time |
| Fresh pork (roast, steaks, chops) | 145°F (63°C) |
| | plus 3 minutes rest time |
| Ham, cooked before eating | 145°F (63°C) |
| Ham, reheat fully cooked | 140°F (60°C) |
| Poultry | |
| Ground chicken, turkey | 165°F (74°C) |
| Whole chicken, turkey | 165°F (74°C) |
| Breasts, roasts | 165°F (74°C) |
| Stuffing, alone or in bird | 165°F (74°C) |
| Egg dishes | 160°F (71°C) |
| Leftovers and casseroles, to reheat | 165°F (74°C) |

*Figure 3-7* Minimum internal temperatures for cooking or reheating foods.

▶ Food recalls occur when food products are found to be contaminated. Examples include

- Frozen fruits and vegetables, sunflower kernels, sliced apples, avocado pulp, and sliced turkey for *Listeria*
- Bagged spinach and ground beef for *E. coli*
- Spices, pot pies, peanut butter, and hazelnuts for *Salmonella*

Recent food recall notifications are listed at www.foodsafety.gov.

▶ To find information about the purity of the water provided by municipal water departments, visit www.EPA.gov/water.

▶ A website providing information on efforts to promote U.S. food safety is www.foodsafety.gov.

▶ As the old adage says, when in doubt, throw the food out!

## Knowledge Check

1. What groups of people are particularly susceptible to foodborne illness?
2. Which bacterial and viral pathogens cause most foodborne illnesses?
3. How do foods become contaminated with viruses and bacteria?
4. What are the threats to safe water?
5. What is the temperature danger zone, and why is it important?

*Take Action*

## Check Your Food Safety Skills

Check yes if the item applies to you <u>100%</u> of the time.

| Yes | No | **I avoid unsafe food and water.** |
|-----|-----|-----|

1. I do not eat foods from cans that are leaking, bulging, or severely dented.
2. I do not eat raw or undercooked meat, fish, shellfish, poultry, or eggs (including raw eggs in homemade ice cream, eggnog, mayonnaise, and cookie dough).
3. I drink only milk and juices that have been pasteurized.
4. I avoid soft cheeses, cold deli salads, and cold smoked fish, and I heat hot dogs and deli meats to 165°F (75°C) if I am at increased risk of foodborne illness.
5. I use water for drinking and food preparation that is from a safe public water source or from a tested private source.

| Yes | No | **I prevent transmitting pathogens.** |
|-----|-----|-----|

6. I wash my hands for at least 20 seconds with warm water and soap before preparing foods and after handling unwashed produce or raw meat, fish, poultry, or eggs; using the bathroom; changing diapers; playing with pets; coughing; sneezing; or smoking.
7. I cover cuts, burns, sores, or infected areas when preparing foods.
8. I do not prepare food when sick with diarrhea or vomiting.

Photodisc/PunchStock

**I keep a sanitary kitchen.**

9. I prevent cross-contamination by washing counters, sinks, cutting boards, and other equipment thoroughly, especially after contact with raw meat, fish, poultry, or eggs.
10. I clean spills in my refrigerator promptly and discard refrigerated food that is past its expiration date or has spoiled.
11. I replace sponges and wash kitchen towels frequently. (Microwaving wet sponges for 1 minute helps kill bacteria.)
12. I use cutting boards that are made of hard plastic, marble, glass, or hardwood (oak, maple) and that are not deeply scratched.

**I handle and prepare foods safely.**

13. I take groceries home promptly and refrigerate or freeze perishable foods right away.
14. I wash fresh fruits and vegetables under running water and scrub firm produce (including melons) with a vegetable brush.
15. I discard soft or liquid foods (e.g., jam, syrups) that are moldy. For firm-textured foods, such as cheese, I trim off at least 1 inch around the mold.
16. I store raw meats and poultry below other foods in the refrigerator to prevent cross-contamination by drippings from leaky packages.
17. I marinate foods in the refrigerator, not on the countertop.
18. I thaw foods only in the refrigerator, under cold running water, or in a microwave oven—never on the countertop.
19. I use a refrigerator thermometer to be sure my refrigerator is at a safe temperature range (32° to 40°F [0° to 4.4°C]).
20. I cook foods to safe internal temperatures (see Fig. 3-7) and use a food thermometer to measure the temperatures.
21. I cook eggs until yolks and whites are firm, not runny.
22. I eat cooked food right away or I refrigerate or freeze it within 2 hours.
23. I discard leftover foods after 7 days.
24. I use coolers and ice to keep perishable food cool on picnics.

Now, count your yes responses and give yourself a grade.

| Score | Grade | Evaluation |
|-------|-------|------------|
| 0–5 | F | **FRIGHTENING!** Don't make another meal until you do some serious cleaning and improvement of your behaviors! The many food safety violations in your kitchen are hazardous to your health. |

| | | |
|---|---|---|
| 6–14 | **D** | **DANGER!** Your kitchen and food habits pose many food safety dangers. Review the items you **did not** check and make changes right away! |
| 15–18 | **C** | **CAUTION!** Some parts of your kitchen and your behaviors are food safe, but not all. Review the items you **did not** check and make changes to get your **whole** kitchen and self in top food safety shape. |
| 19–23 | **B** | **BETTER!** Overall, your kitchen and habits are in good shape, but there are areas where you can improve. Take steps to get your **whole** kitchen and self in awesome food safety shape. |
| 24 | **A** | **AWESOME! KEEP UP THE GOOD WORK!** |

# GL🌐BAL PERSPECTIVE

## Traveler's Diarrhea

Traveler's diarrhea afflicts 30 to 50% of those who travel to areas that tend to be hot and lack advanced water treatment systems and refrigeration, such as most of Central and South America, Mexico, Africa, Asia, and the Middle East. Traveler's diarrhea usually occurs abruptly and lasts for 3 to 4 days, or longer. According to the Centers for Disease Control and Prevention, most traveler's diarrhea is caused by bacterial infections, especially enterotoxigenic *Escherichia coli*, spread through contaminated food and water. The following guidelines can help reduce the risk of traveler's diarrhea.

- Wash your hands often, especially after using the bathroom and before eating.
- Eat foods that are freshly cooked and served piping hot.
- Avoid food from street vendors and buffets.
- Avoid salads and raw fruits and vegetables.
- Avoid raw or undercooked meat and seafood.
- Avoid tap water and beverages reconstituted with tap water (including ice and possibly fruit juice and milk).
- Bottled and sealed beverages, including soft drinks, water, beer, and wine, are generally safe.
- Beverages made with boiled water, such as coffee and tea, are generally safe.
- Travelers also can treat tap water by boiling, chemical disinfection, or filtering. To learn how, visit wwwnc.cdc.gov/travel.

Even when following these guidelines, traveler's diarrhea can be hard to avoid. Pepto-Bismol™, an over-the-counter (OTC) drug used to treat indigestion, can reduce traveler's diarrhea substantially when it is taken throughout the stay. However, before traveling to a high-risk area, it is wise to consult with a physician about using any medication, getting needed vaccinations, and taking other health precautions.

C Squared Studio/Getty Images

▶ A phrase familiar to those traveling to developing countries is "boil it, peel it, or don't eat it." Of course, this advice is simplistic—many foods that may be safe to eat can't be boiled or peeled.

# Foodborne Illness Can Be Deadly

Foodborne illness often means a few hours or even a few days of discomfort and then the illness resolves on its own. In some cases, though, foodborne illness causes more serious medical problems, which can have lifelong effects. High-risk populations—infants and young children, the elderly, pregnant women and their fetuses, and those with impaired immune systems—have the greatest risk of serious complications like these:

- *Hemolytic uremic syndrome (HUS).* Most cases of HUS are caused by the toxin produced by *Escherichia coli* 0157:H7. The toxin attacks red blood cells, causing them to break apart (called hemolysis), and the kidneys, causing waste products to build up (called uremia). Early symptoms of HUS include bloody diarrhea, vomiting, sleepiness, and low urine output. In the worst cases, the toxin damages multiple organs, causing seizures, permanent kidney failure, stroke, heart damage, liver failure, and even death. Fortunately, most individuals recover completely, although usually after weeks of intensive medical care. HUS afflicts children more often than adults, but adults experience the most severe infections.

- *Listeriosis. Listeria monocytogenes* bacteria cause listeriosis, a rare but serious disease. Listeriosis begins with muscle aches, fever, and nausea. It can spread to the nervous system, causing severe headache, stiff neck, loss of balance, and confusion. Pregnant women and their fetuses are particularly vulnerable—listeriosis can cause miscarriage, premature delivery, infection in the fetus, and fetal death. During pregnancy, women are 20 times more likely to develop the disease. The elderly also are susceptible.

- *Guillain-Barré syndrome (GBS). Campylobacter jejuni* is a cause of this rare nervous system disorder. In GBS, peripheral nerves (those that connect the spinal cord and brain to the rest of the body) are damaged by the body's own immune system. Early symptoms of GBS include tingling and pain in the legs, followed by severe muscular weakness. Paralysis can occur, and some individuals need help breathing with a ventilator. Recovery can take weeks to months, and about 30% of those with GBS do not fully recover, experiencing lifelong pain, weakness, and/or paralysis.

- *Reactive arthritis.* Foodborne illness caused by *Salmonella, Shigella, Campylobacter,* and others can cause reactive arthritis. This condition usually develops 2 to 6 weeks after the initial infection and causes inflammation throughout the body, but especially in the joints and eyes. Pain and swelling of the knees, ankles, and feet are common. Inflammation of the urinary tract and blistering of the palms of the hands and soles of the feet also are common. Genetic factors play an important role in determining who develops the disease. Most people with reactive arthritis require medical treatment and will recover after 2 to 6 months, but about 20% experience mild arthritis for a much longer time.

These diseases highlight the need for a safe food supply and safe food handling all along the food chain.

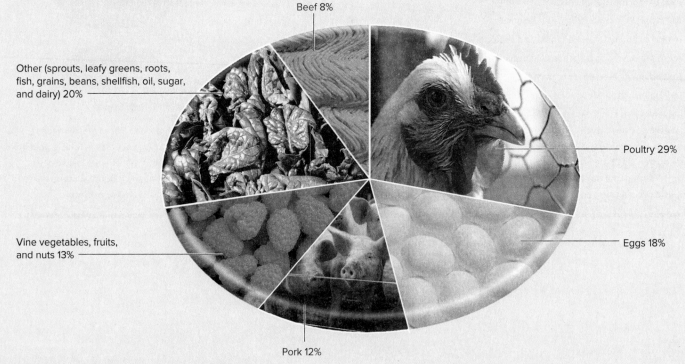

Beef 8%

Other (sprouts, leafy greens, roots, fish, grains, beans, shellfish, oil, sugar, and dairy) 20%

Poultry 29%

Vine vegetables, fruits, and nuts 13%

Eggs 18%

Pork 12%

Percentage of *Salmonella* infections associated with food types. beef: Brand X Pictures/PunchStock; poultry: Ingram Publishing/SuperStock; eggs: Brand X Pictures/PunchStock; pork: Shutterstock/DN1988; fruits: Digital Vision/Getty Images; greens: Author's Image/Glow Images

# 3.5 Environmental Contaminants in Foods

In addition to pathogens and natural toxins, a number of environmental toxins can contaminate food and cause health problems. Common environmental contaminants include heavy metals (lead, mercury), industrial chemicals (dioxins, polychlorinated biphenyls), and agricultural chemicals (pesticides, antibiotics).

## Lead

Lead can damage every organ system in the body—especially the nervous system and kidneys—and it impairs the synthesis of hemoglobin, the oxygen-carrying protein in the blood. Lead is particularly toxic to the developing nervous system of children; even low amounts in the body can lower IQ, cause behavior disorders, and impair coordination. It also can impair growth and hearing and predispose children to high blood pressure and kidney disease later in life. No safe amount of lead in the blood has been identified. In the U.S., 3% of children under the age of 6 have blood lead levels that are of concern.[73] In developing countries, many more children are affected.

Sources of lead include home plumbing and lead-based paints; both are more likely to be found in buildings constructed prior to 1986. In homes with lead pipes or solder, running cold water for 1–2 minutes before using it allows lead leached from pipes to go down the drain. Hot tap water should not be used for food preparation or consumption because more lead leaches into hot than cold tap water. Water filters that remove lead also are available. Dust and chips from lead-based paint are the most hazardous sources of lead for U.S. children. Both can be inadvertently ingested, especially by children. Keeping the home and hands clean can help reduce exposure. Also, never serve or store foods or beverages in lead-containing containers, such as leaded crystal, some pottery, and older or imported dishes.

Other documented lead sources include certain candies from Mexico, some dietary supplements, and toys painted with lead-containing paints (children may place toys in their mouths, making them particularly dangerous). Lead is no longer permitted in food cans in the U.S., but food cans from other countries may contain lead solder.

Preventing lead poisoning is best done by removing lead from the environment. Good nutrition plays a role, too. Children with iron deficiency absorb more lead, so preventing iron deficiency (see Part 4) may help limit lead absorption.[74]

▶ For more information about lead, visit www.epa.gov/lead and www.cdc.gov /nceh/lead/ or call the National Lead Information Center and Clearinghouse at 1-800-424-LEAD.

▶ Serious lead contamination of the water in Flint, Michigan, occurred when the city began to draw its water from the Flint River in 2014. Drinking water problems led to the discovery that corrosive river water was causing lead to leach from the old supply line pipes into the water that flowed through taps and into homes across the city. As a result, the number of children with elevated lead levels doubled in 1 year. These events triggered a massive federal response that included distribution of clean bottled water and water filters; expanded access to medical, educational, and behavioral services for children; expanded testing of blood lead levels; and funding for nutrition programs to increase calcium, iron, and vitamin C in diets to limit lead absorption. Flint now has a new water source, which recent testing has shown is within the federal and state standards for lead. While repair of the damaged water distribution systems is well underay, it will be years before the true extent of the damage of increased lead exposure on the cognitive and behavioral development of young children living in Flint is known.

## Arsenic

Arsenic is a toxic element found in soil and water throughout the world. Consumption of inorganic arsenic, a toxin and carcinogen, is linked to poor fetal growth and adverse immune and neurodevelopmental outcomes in infants and children, as well as cancer in adults.[75] Rice, which is especially efficient at taking up arsenic from soil and water, is the major source of inorganic arsenic in our diets.[76,77] Brown rice tends to have the highest concentrations of arsenic, as it accumulates in the husk; when white rice is produced, the husk is removed. Scientists are especially concerned about exposure to inorganic arsenic in infants and toddlers because they are fed infant rice cereal and other rice snacks frequently.[77] To address this, the FDA has proposed limits for inorganic arsenic content in infant cereal. Other recommendations for limiting dietary inorganic arsenic exposure include

- Feeding babies and toddlers a variety of grains and cereals, not just rice
- Consuming white rice in place of brown rice occasionally
- Cooking rice in an excess of water; draining this water after cooking reduces the arsenic content, but there also is some vitamin loss

## Dioxins and Polychlorinated Biphenyls (PCBs)

Dioxins (chlorine and benzene-containing chemicals) and PCBs are persistent industrial chemicals that have been linked to an increased risk of health problems, including

Eating a healthy diet with plenty of iron may help prevent lead poisoning in children.

White Rock/Getty Images

cancer, nerve damage, type 2 diabetes, and reproductive problems. Dietary exposure is primarily from fish, especially freshwater fish from contaminated lakes and rivers. The EPA and many state health departments provide guidance on fish species and waterways that are contaminated.[78] A key guideline for fish consumption is variety and moderation when local sources have the potential for contamination. Eating commercially caught fish is usually safe because these fish come from a variety of sources, and most rivers, streams, and lakes are not contaminated.

## Mercury

Mercury is abundant in the environment; in aquatic environments, bacteria convert mercury to the neurotoxin methylmercury. The FDA first limited the amount of mercury in foods in 1969, after 120 people in Japan became ill from eating mercury-contaminated fish. Birth defects in the offspring of some of these people also were traced to mercury exposure. Methylmercury can cause nerve damage, fatigue, and poor learning abilities. As with dioxins, fish are the primary source of the toxic chemical. Most at risk are children and pregnant and breastfeeding women. Guidelines aimed at limiting exposure to mercury, while encouraging healthy fish intake, are given in Table 3-10.

These guidelines are controversial, however. Fish is an important source of fatty acids that promote brain and nervous system development in fetuses and infants. Ongoing research suggests that the benefits from these fatty acids outweigh the risk of mercury from fish. As scientists learn more about the risks and benefits of eating fish, these guidelines may be relaxed to encourage fish intake.

## Pesticides and Antibiotics

Pesticides—products used to eliminate pests—include insecticides, herbicides (weed killers), fungicides, and rodenticides. Farmers have used them since the 1940s to limit crop-damaging pests and thereby increase agricultural production. Pesticides also can help improve the appearance of fruits and vegetables. For example, a fungicide helps prevent apple scab fungus, which makes apples look unappealing (and less likely to sell), but the fruit is still tasty and nutritious. Fungicides also help prevent carcinogenic aflatoxin from forming on some crops.

| Table 3-10  Guidelines for Children and Women Who May Become Pregnant or Are Pregnant or Breastfeeding to Limit Mercury in the Diet |
| --- |
| **Eat 8–12 ounces of a variety of fish a week.** |
| • Divide this amount over 2 or 3 servings. |
| • For young children, provide serving size right for the child's age and calorie needs. |
| **Choose fish lower in mercury.** |
| • Commonly eaten fish lower in mercury include salmon, shrimp, pollock, tuna (light canned), tilapia, catfish, and cod. |
| **Avoid 4 types of fish: tilefish from the Gulf of Mexico, shark, swordfish, and king mackerel.** |
| • These 4 types of fish are highest in mercury. |
| • Limit white (albacore) tuna to 6 ounces a week. |
| **When eating fish you or others have caught, pay attention to fish advisories for the stream, river, or lake where the fish were caught.** |
| • If advisories are not available, adults should limit such fish to 6 ounces a week, and young children should consume only to 1 to 3 ounces a week and not eat other fish that week. |

Source: *Fish: What pregnant women and parents should know. 2020;* www.fda.gov/Food/FoodborneIllnessContaminants/Metals/ucm393070.htm.

The EPA permits the use of about 10,000 pesticides, containing over 1000 active ingredients. About 1.2 billion pounds of pesticides (about half are herbicides) are used each year in the U.S., with much of this being applied to agricultural crops. However, pesticide use is not limited to agriculture—many pesticides are used in homes, businesses, schools, and health-care facilities for insect and rodent control. Pesticides also are widely applied to lawns, golf courses, and home gardens.

The 2 major classes of pesticides are synthetic pesticides and biopesticides. Many of the earliest synthetic pesticides were persistent chemicals—they did not break down easily and remained in water, soil, and plants for decades, thus posing a risk to the humans and animals that ingested contaminated water and plants. Many of these early pesticides have been banned. The synthetic compounds in use today break down much more quickly (they are less persistent), but they also are more powerful chemicals. Organophosphates, carbamates, organochlorine insecticides, and pyrethroid pesticides are the main classes of synthetic pesticides.[71]

Three types of biopesticides were developed to provide safer pesticide alternatives. The first type consists of microbial pesticides, such as the soil bacterium *Bacillus thuringiensis* (Bt), which produces proteins that are toxic to various insects. The second type is plants that are genetically modified to produce their own pesticides, such as the Bt protein. The third type is biochemical pesticides that don't directly kill an organism but limit its reproduction or growth.

A number of problems are associated with pesticide use. One is that the organisms intended for elimination can become resistant to the pesticide action. This means that either more of the pesticide must be applied or a new pesticide must be applied. Another problem is pesticide drift; once a pesticide is applied to a field, it can be carried by wind currents to nontarget sites. Pesticides that remain in the soil may be taken up by nontarget organisms or enter groundwater and aquatic habitats. Each of these paths is a route to the food chain. Still another problem is the unintended effects; pesticides can harm nontarget species, such as bees, frogs, and fish; decrease biological diversity by acting on the lower levels of the food chain; have harmful effects on water quality; disrupt natural wildlife habitats; damage soil and the nutrients in it; and contribute to erosion.

## Regulating Pesticides

The EPA, FDA, and USDA share responsibility for regulating pesticides. The EPA is responsible for determining that a pesticide is beneficial and will not pose unreasonable health or environmental risks. The EPA sets limits on how much of a pesticide may be used on food during growth and processing and how much may remain on the food you buy—this is known as pesticide tolerance. Because children are thought to be more vulnerable to any adverse effects of pesticides and their diets are their primary source of pesticide exposure, the Food Quality Protection Act of 1996 requires the EPA to consider children's pesticide

▶ The environmental damage caused by the pesticide DDT (dichlorodiphenyltrichloroethane) was chronicled by biologist, writer, and ecologist Rachel Carson in her book *Silent Spring*, published in 1962. DDT was widely used around the world because it increased crop yields and was inexpensive. But scientists soon learned that DDT and its metabolites are toxic to humans and animals, especially young fish and birds, and that it persists for years in the environment. DDT can no longer be used in the U.S. and many other countries, but it is still used in some low-income countries to kill malaria-carrying mosquitoes. Read more about Rachel Carson at www.rachelcarson.org.

Swordfish is a common source of mercury in our diets. Children and pregnant and lactating women should avoid swordfish and other fish that contain high levels of mercury.

rudisill/Getty Images

***Common Types of Pesticides***

**Organophosphates:** compounds toxic to the nervous system of insects and animals; some were used in World War II as nerve agents; they usually are not persistent in the environment

**Carbamates:** compounds that act similarly to the organophosphates but are less toxic

**Organochlorine insectides:** commonly used in the past, but many (e.g., DDT and chlordane) have been removed from the market due to their health and environmental effects and their persistence

**Pyrethroid pesticides:** compounds that mimic naturally occurring pesticides found in chrysanthemums, some of which are toxic to the nervous system

Applying pesticides to crops carries risks and benefits. Rural communities, where exposure is more direct, experience the greatest short-term risk.

Photo by Jeff Vanuga, USDA Natural Resources Conservation Service

**Parkinson's or Parkinson Disease?**

Many health conditions are named for the person who discovered them or the location where they were first observed. The name can be possessive (Parkinson's) or not (Parkinson)—both are correct. However, in a move to improve simplicity and consistency, international biomedical organizations have been endorsing the use of the nonpossessive form since the 1970s. Throughout this text, we use the nonpossessive form.[82]

exposure from foods they normally eat, such as apples and apple juice, potatoes, sugar, eggs, chicken, and beef.[79]

The FDA and USDA are jointly responsible for testing foods for pesticides and enforcing the EPA pesticide tolerances. The USDA reports indicate that about 59% of fresh fruits and vegetables have detectable pesticide residues, but only 0.36% exceed pesticide tolerances.[80] Imported fruits and vegetables tend to have higher amounts of pesticides than those grown domestically. The low level of residues that exceed tolerance confirms that farmers are using pesticides according to EPA regulations.

### Minimizing Exposure to Pesticides

There is no doubt that pesticides are toxic. Their purpose, after all, is to eliminate pests. Accidental pesticide poisonings occur each year, often related to careless use or storage of these chemicals. Studies also link people who work with pesticides, such as farmers and those who apply pesticides for a living, with higher rates of asthma, Parkinson disease (a neurological disorder), prostate cancer, leukemia, and other cancers.[81] However, there is much less certainty about the effects of long-term exposure to much lower doses, as occurs when we eat our usual diets. In this regard, infants (including before birth) and children deserve special attention. Young animals exposed to chemicals such as pesticides can experience damage to their developing reproductive, nervous, and immune systems. Risk may increase in young children because they often consume relatively higher doses of pesticides, when their lower body weight is considered. Infants and young children also may not metabolize the pesticides as readily as adults because the liver, the main organ that breaks down drugs and toxins for excretion, is still immature.

Even though government agencies work to keep pesticide residues in food to a minimum, the EPA suggests the following steps to further reduce pesticide exposure.

- Wash and scrub all fruits and vegetables under running water.
- Peel fruits and vegetables and discard outer leaves of vegetables.
- Trim fat from meat and skin from poultry, as some pesticide residues accumulate in fat.
- Select a variety of foods. This reduces the likelihood of exposure to a single pesticide.

Certified organic foods are grown without synthetic pesticides and can further minimize exposure to pesticides. However, these foods may contain very small amounts of pesticide residues because of background contamination (pesticides persisting in soil and water) and pesticide drift from nearby conventional farms. Several studies have documented that eating organic diets decreases pesticide excretion in the urine.

### Antibiotics

Farmers often give low-dose antibiotics to food animals to promote their growth and to prevent and treat disease. This accounts for approximately 80% of the antibiotics sold in the U.S. Many of these antibiotics are the same medications used to treat human infections. Scientists now know that this practice fosters the growth and spread of antibiotic-resistant bacteria strains in the animals. These bacteria can be readily transmitted to humans via the food supply, direct contact with animals, and environmental contamination (water, soil, and air around large-scale feeding areas). The development of antibiotic-resistant bacteria is a major public health concern, as infections from these bacteria are very difficult to treat and may be deadly. Further, the arsenal of antibiotics available to health-care providers is limited. This issue is receiving considerable attention from health-care professionals and scientists.[83] Note that the use of antibiotics is prohibited in organically produced animals.

Antibiotics are added to animal feed and water to promote growth and prevent disease. Scientists are increasingly questioning this practice because it may lead to the development of antibiotic-resistant bacteria.

Glow Images

Contamination of the food supply by nuclear power plant accidents poses a real, yet poorly understood, threat. Radioactive cesium and iodine released into the air after the 1986 Chernobyl (Ukraine) and 2011 Fukushima Daiichi (Japan) nuclear disasters settled on plants, soil, and water. Radioactive particles (radionuclides) entered our food supply from the milk and meat of animals grazing in contaminated areas. After nuclear accidents, radioactivity of the food supply is monitored carefully because consuming radionuclides may increase the risk for thyroid and other cancers. However, there is much to be learned about what constitutes an unsafe exposure to radionuclide-contaminated foods. Those exposed to radioactive iodine often are prescribed potassium iodine pills to prevent the thyroid gland from absorbing radioactive iodine.

## Knowledge Check

1. What are the sources of lead in the environment?
2. Which environmental contaminants are likely to be found in fish?
3. Which food is the major source of inorganic arsenic in our diets?
4. What is a pesticide tolerance?
5. What methods can you use to reduce the amount of pesticides you ingest?

## CASE STUDY FOLLOW-UP

Digital Vision

Although the dish was cooked thoroughly, it was held at an unsafe temperature from the time it was removed from the oven at 1:00 P.M. until it was served at 6:00 P.M. This 5-hour time span greatly exceeded the maximum time of 2 hours at room temperature for a cooked food. This allowed the growth of a foodborne illness–causing pathogen. Ideally, this product should have been transported on ice in a cooler to the party, refrigerated at the party, and then reheated to 165°F (74°C) before serving at 6:00 P.M. Overall, it is risky to leave perishable items, such as meat, fish, poultry, eggs, and dairy products, at room temperature for more than 2 hours.

# Expert Perspective from the Field

## Sustainability in University Food Service

The interest in sustainability on university campuses continues to grow. You probably are familiar with some of the practices associated with sustainability such as eating local and organic foods and recycling. The word *sustainability* has many meanings. A commonly used definition comes from the United Nations Brundtlund Report: "sustainable development is development that meets the needs of the present without compromising the ability of future generations to meet their own needs." According to Kristi Theisen,* sustainability expert, the sustainability concepts important in many college communities are the following.

*Food transparency.* Knowing which ingredients are used on menus is very important to many eating on college campuses today. Some want to know because they have allergies or special dietary needs. Others want to know where their food was grown. Many campus food service operators work closely with their supply chain to get the types of products and detailed information customers want.

*More local and fresh foods.* To protect the environment, support the local economy, and reap health benefits, many consumers want to eat more locally produced, fresh foods. To achieve these goals, college-based food service professionals are designing menus and thoughtfully arranging dining spaces to make it easy to make healthy choices. Menus often are planned around seasonal, local produce. For example, autumn and winter menus might feature squash, onions, carrots, and other root vegetables. Many universities now have their own gardens and use this produce in campus dining operations.

*Less waste.* Aligning with the EPA waste hierarchy (www.epa .gov/smm/sustainable-materials-management-non-hazardous-materials -and-waste-management-hierarchy), the best way to manage waste is to reduce it before it happens. One way university food service is responding to this is by "cooking to order." That means options are customized to the individual diner's preferences, and there's less waste because only what is ordered is prepared. Trayless dining in the cafeteria means individuals are less likely to take more food than they need, which results in less food waste, and the cost (energy, chemical, and water) of cleaning trays is eliminated. At some universities, food waste is composted and used frying oil is recycled to make biodiesel fuel. Leftover food may be delivered to community food programs. Reusable, recyclable, and compostable packaging reduces inorganic waste.

*Reduced energy use and lower carbon footprint.* Food service sustainability initiatives often are part of larger campus efforts to reduce energy use and carbon footprints. Nearly 1000 university presidents signed on to the American College and University President Climate Commitment (ACUPCC), pledging to make their campuses carbon neutral by 2020 and contribute 50% to decarbonization in the U.S. economy by 2030. So far, the ACUPCC has helped make campus operations greener by reducing greenhouse gas emissions and providing educational opportunities on climate and sustainability. As part of this commitment to eliminating carbon pollution from their campus operations, colleges and universities have embedded climate action into the curriculum and student experiences and expanded climate-related research activity. Thousands more use the Association for the Advancement of Sustainability in Higher Education's STARS (Sustainability Tracking, Assessment & Rating System™) sustainability dashboard to measure and report their progress.

*More social justice.* This area of sustainability demonstrates the close connection between the environment and people. Certifications, such as Fair Trade, that help communities around the world, along with local initiatives (e.g., on-campus food pantries for students struggling to afford food), are examples of how some campuses are addressing social justice.

*Extending education.* Extending learning to spaces outside classroom walls, such as the dining hall, is a goal of many campuses. Student organizations, academic departments, and even individual students partner with university sustainability offices and campus dining professionals to evaluate and improve food service practices and promote sustainability. Similarly, campus gardens and hydroponic greenhouses depend on students for their design and operation.

Barriers to the implementation of sustainability initiatives can include municipal infrastructure, investment costs, and logistics. If an urban campus wants to start composting, for example, but the facilities don't exist within its city and there isn't space on campus for this activity, having that option becomes very difficult. Another barrier may be an initial investment for food service equipment that is more energy efficient, as well as food safety, quality control, and transportation logistics when sourcing from small, local farms. If the barriers for some changes are too great, there likely are many other positive changes toward sustainability with fewer barriers that can be made.

Despite barriers, the benefits of advancing sustainability on campus are many. They include creating a healthier campus environment and gaining the knowledge and skills needed to improve the quality of life on campus and for people around the globe. What opportunities exist on your campus to increase sustainability?

*Kristi Theisen is a certified US Zero Waste Business Council Associate. She works with students and hundreds of college campuses across the country to advance sustainability and quality-of-life initiatives.*

zoryanchik/Shutterstock

# Chapter *Summary*

## 3.1 Food insecurity and hunger occur in virtually every country, but undernutrition is declining around the world.
About 795 million people (1 in 9) worldwide do not get enough food to meet their requirements, and another 2 billion suffer micronutrient deficiencies, especially of vitamin A, folate, zinc, iron, and iodine. Overweight and obesity are another form of malnutrition that is increasing around the world. Food insecurity is linked to poverty. Food-insecure people tend to have poorer diets and suffer more health problems. Children without enough food do not grow normally and are more likely to suffer diseases and death. In the U.S., the USDA monitors food insecurity. Food insecurity is a problem for 11.8% of U.S. households, with 4.5% of these households experiencing very low food security. The Supplemental Nutrition Assistance Program is the most important food-assistance program offered by the U.S. government. Other programs include WIC and the National School Lunch and Breakfast Programs. Emergency food programs also play an important role.

**Prevalence of food insecurity, average 2016–18**

☐ Food insecurity below U.S. average
☐ Food insecurity near U.S. average
☐ Food insecurity above U.S. average

www.ers.usda.gov

## 3.2 Organic foods are grown in ways that promote healthy soils, waterways, crops, and animals.
Many substances and processes cannot be used in organic food production. The USDA certifies foods as organic. Organic foods contain fewer pesticides, but their nutritional value may not differ from that of conventionally grown food. There are four main processes used in agriculture to manipulate genes to amend plant and animal traits: selective breeding, mutagenesis, genetic (transgenic) modification, and gene editing. Genetic modification and gene editing are biotechnologies that allow scientists to directly alter the genetic makeup of an organism. Genetic modification uses recombinant DNA technology to transfer genes (sometimes called transgenes) from almost any plant, animal, or microorganism to another. The most common genetically modified foods in the U.S. are soybeans, corn, sugar beets, canola, alfalfa, and cotton altered either to be herbicide resistant or to produce their own pesticides. Transgenic salmon reach maturity earlier than other salmon. Genetically modified foods are regulated by the FDA, USDA, and EPA. The newest biotechnology process, called genome editing or gene editing, allows scientists to make precise changes to the DNA of individual animal and plant cells. These changes are like those that might occur naturally when two plants or animals breed, but the changes are precisely controlled. The precision of the edits makes it faster and less costly to change traits and produce new varieties than the processes of selective breeding, mutagenesis, and transgenetic modification. Because the changes are indistinguishable from those that occur through traditional breeding methods, the USDA does not regulate gene-edited foods.

Source: USDA organic seal

## 3.3 Food spoilage results from microorganism and enzyme action.
Food preservation methods stop or slow the rate of spoilage. Food irradiation is approved for some foods but is not widely used. Nanotechnology is the study of controlling matter at the atomic or molecular level. Nanoparticles are used to preserve foods, brighten food, keep food fresh longer, decrease fat content, and increase vitamin bioavailability. Food additives are used to improve freshness and safety, alter nutritional value, and enhance the color, flavor, and functional characteristics of foods. Intentional food additives are used for a specific purpose, whereas incidental food additives, such as pesticides, arsenic, and BPA, become a part of food because of some aspect of production. They are regulated by the FDA, but there is concern about how well this is done. Some food additives are considered GRAS and have not had formal testing. Some people are concerned about the safety of food additives; however, no evidence shows that limiting additives will make you healthier. To lower your intake of additives, read food labels and eat fewer highly processed foods.

Source: www.fda.gov/

# 3.4 Foodborne illness is a significant cause of illness, hospitalization, and death in the U.S.

Individuals at increased risk of foodborne illness include those with immune systems weakened by disease or medical treatments, pregnant women and their fetuses, infants and young children, and elderly persons. Foodborne illness usually causes gastrointestinal effects but can have more serious, lasting effects. A host of pathogens, including viruses, bacteria, parasites, and toxins, can cause foodborne illness, but most are caused by viruses (norovirus) and bacteria (*Salmonella, Campylobacter, Clostridium perfringens,* and *Staphylococcus aureus*). Safe food handling practices can reduce the risk of foodborne illnesses. Raw or inadequately cooked meats, poultry, eggs, and shellfish; unpasteurized dairy products; and fresh produce often are implicated in outbreaks of foodborne illness. Clean water is vital to good health. Public water supplies are regulated by the EPA and municipal water systems. Bottled water is regulated by the FDA. There are numerous threats to safe water, including chemical and microbial contamination. Several government agencies, including the USDA, FDA, and CDC, are responsible for coordinating food safety efforts, but everyone has a responsibility for keeping food safe to eat. The risk of foodborne illness can be reduced by using good personal and kitchen hygiene, handling food safely, and avoiding foods that present extra risk. Washing hands, preventing cross-contamination, washing produce, keeping foods out of danger zone temperatures, and cooking meat, poultry, eggs, fish, and casseroles to safe temperatures are especially important. Cooked foods should be either consumed right away or refrigerated within 2 hours.

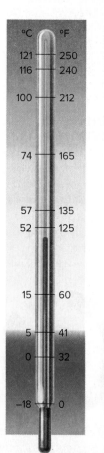

| Minimum Safe Cooking Temperatures | |
|---|---|
| Fresh ground beef, veal, lamb, and pork | 160°F (71°C) |
| Beef, veal, lamb (roast, steaks, chops) | 145°F (63°C) |
| | plus 3 minutes rest time |
| Fresh pork (roast, steaks, chops) | 145°F (63°C) |
| | plus 3 minutes rest time |
| Ham, cooked before eating | 145°F (63°C) |
| Ham, reheat fully cooked | 140°F (60°C) |
| Poultry | |
| Ground chicken, turkey | 165°F (74°C) |
| Whole chicken, turkey | 165°F (74°C) |
| Breasts, roasts | 165°F (74°C) |
| Stuffing, alone or in bird | 165°F (74°C) |
| Egg dishes | 160°F (71°C) |
| Leftovers and casseroles, to reheat | 165°F (74°C) |

| °C | °F | |
|---|---|---|
| 121 | 250 | Temperature range for canning low-acid foods (kills spores). |
| 116 | 240 | A pressure-canner is required to reach these temperatures. |
| 100 | 212 | |
| 74 | 165 | Temperature range for destroying bacteria (but not their spores), parasitic worms, and protozoa. |
| | | Temperature range for holding thoroughly cooked food; prevents growth of bacteria but doesn't necessarily destroy them. |
| 57 | 135 | |
| 52 | 125 | |
| | | **DANGER ZONE** |
| 15 | 60 | Bacteria grow quickly; do not store within this temperature range for more than 1 to 2 hours. |
| 5 | 41 | |
| 0 | 32 | Recommended refrigerator temperature; still, some bacteria can grow. |
| −18 | 0 | Freezing; bacteria can't grow, but many will survive; growth can resume on thawing. |

# 3.5 Environmental contaminants in food include lead,

arsenic, mercury, industrial contaminants (e.g., dioxins and polychlorinated biphenyls [PCBs]), pesticides, and antibiotics. Lead can damage the developing nervous system; children are most at risk. Iron-deficient children may be at more risk of lead toxicity. Rice is the major source of inorganic arsenic, a toxin and carcinogen. To reduce exposure to arsenic, consume a variety of grains and cereals (not just rice), replace brown rice with white rice occasionally, and cook rice in an excess of water. Mercury is found in fish—especially large fish—such as shark, swordfish, king mackerel, and tilefish. The FDA and EPA recommend that children and pregnant and breastfeeding

Photo by Jeff Vanuga, USDA Natural Resources Conservation Service

women limit their exposure to high-mercury fish. Farmers use pesticides to increase agricultural productivity. Pesticides are regulated by the EPA, USDA, and FDA. Pesticides in foods are a special concern for young children. Less than 0.5% of fruits and vegetables exceed pesticide tolerances. The widespread use of antibiotics in animal feed is a concern because they foster the growth of antibiotic-resistant bacteria.

## Clinical Perspective

Foodborne illness often is mild. However, infants, young children, elderly adults, and pregnant women and their fetuses are at high risk of serious complications from foodborne illness.

## Expert Perspective

The sustainability concepts important on many college campuses are food transparency, increasing local and fresh foods in food service operations, reducing waste, reducing energy use and lowering carbon footprint, increasing social justice, and promoting educational opportunities to build sustainability knowledge and skills.

## Global Perspective

Traveler's diarrhea is common in visitors to developing countries. Following guidelines about water and produce consumption can help reduce the likelihood of contracting traveler's diarrhea.

# Study Questions

1. Which of the following is a sign of malnutrition caused by insufficient food and nutrient intake?

   a. stunted growth
   b. wasting (loss of fat and muscle tissue)
   c. increased susceptibility to infection
   d. all of the above

2. About _____ of U.S. households are food insecure.

   a. 20%              c. 35%
   b. 14%              d. 1%

3. The Supplemental Nutrition Assistance Program benefits _____.

   a. only low-income women, infants, and children
   b. poor households that meet eligibility guidelines
   c. food banks and pantries in communities across the country
   d. seniors in adult day-care settings

4. Populations disproportionately affected by hunger and malnutrition include _____.

   a. preschool children and women
   b. working adults
   c. teenagers
   d. all of the above

5. Which of the following statements about organic foods is *not* true?

   a. They are always more nutritious than conventionally raised foods.
   b. Synthetic fertilizers, pesticides, antibiotics, synthetic hormones, and sewage sludge are prohibited in their production.
   c. A food labeled organic must have 95% of its ingredients by weight meet organic standards.
   d. The USDA is responsible for organic certification of farms and foods.

6. The main use of genetically modified foods is to _____.

   a. improve nutritional quality by increasing the amount of vitamins in foods
   b. eliminate potential allergens by altering the proteins synthesized in the plant or animal
   c. produce drought-resistant plants
   d. confer herbicide resistance to selected crops

7. In the U.S., almost all fruits, vegetables, and grains are genetically modified using modern techniques of biotechnology.

   a. true
   b. false

8. Which of the following statements about food irradiation is true?

   a. Irradiated food is radioactive.
   b. Irradiation can be used to destroy pathogens in food, such as *Salmonella* bacteria.
   c. All foods in the U.S. legally can be irradiated.
   d. Irradiation is used in only 2 countries around the world.

9. Food additive use and safety are regulated mainly by the _____.

   a. FDA              c. CDC
   b. USDA             d. EPA

Match the pathogen with the correct phrase.

10. multiplies at refrigerator temperatures; deli meats are a typical source
11. anaerobic; meats, casseroles, and gravies are typical sources
12. leading bacterial foodborne pathogen; raw and undercooked meat and poultry are typical sources
13. cause of millions of cases of "stomach flu" each year

   a. *Salmonella* spp.
   b. noroviruses
   c. *Listeria monocytogenes*
   d. *Clostridium perfringens*

14. Aflatoxin is produced by a mold that grows most often on _____.

   a. cheese              c. peanuts and corn
   b. bread products      d. fruit and other produce

15. The temperature danger zone is _____.

   a. 41° to 135°F (5° to 57°C)
   b. 32° to 40°F (0° to 4°C)
   c. 120° to 160°F (48° to 71°C)
   d. 0° to 212°F (0° to 100°C)

16. Thawing foods, such as chicken, on the counter overnight is a safe food handling practice.

   a. true
   b. false

17. Which of the following statements about pesticides is *not* true?

   a. Organic foods sometimes contain very low amounts of pesticides.
   b. Some pesticides can persist in the environment for many years.
   c. Washing produce removes all pesticide residues.
   d. The EPA regulates the type and amount of pesticides that may be applied to food.

18. The food most likely to contain mercury, dioxins, or PCBs is
_____.

    a. water from plumbing in older homes
    b. vegetables and fruits grown with pesticides
    c. fish from rivers, streams, and lakes
    d. milk from transgenic dairy cattle

19. Describe how the U.S. food security "safety net" protects people against food insecurity.

20. List the potential advantages and disadvantages of genetically modified foods.

21. Describe intentional, incidental, and GRAS food additives.

22. Describe the types of microbial pathogens that cause foodborne illness.

23. Create a checklist you could use at home for keeping foods safe.

24. Explain how you can minimize exposure to foods contaminated with environmental toxins.

25. Who is at greatest risk for serious health complications from foodborne illness?

26. What steps can travelers take to avoid traveler's diarrhea?

27. Why do some consumers choose organic foods over conventionally grown foods?

Answer Key: 1-d; 2-b; 3-b; 4-a; 5-a; 6-d; 7-b; 8-b; 9-a; 10-c; 11-d; 12-a; 13-b; 14-c; 15-a; 16-b; 17-c; 18-c; 19-refer to Section 3.1; 20-refer to Section 3.2; 21-refer to Section 3.3; 22-refer to Section 3.4; 23-refer to Section 3.4; 24-refer to Section 3.5; 25-refer to Clinical Perspective; 26-refer to Global Perspective; 27-refer to Expert Perspective

# References

1. Food Marketing Institute. Supermarket facts. 2020; www.fmi.org/research-resources/supermarket-facts.

2. FAO, IFAD and WFP. *The state of food insecurity in the world 2019. Meeting the 2019 international hunger targets: Taking stock of uneven progress.* Rome: FAO; 2019. www.fao.org.

3. Popkin BM and others. Global nutrition transition and the pandemic of obesity in developing countries. *Nutr Rev.* 2012;70:3.

4. Kimani-Murage EW and others. Evidence of a double burden of malnutrition in urban poor settings in Nairobi, Kenya. *PLoS ONE.* 2015;10:e0129943.

5. Stevens GA and others. Trends and mortality effects of vitamin A deficiency in children in 138 low-income and middle-income countries between 1991 and 2013: A pooled analysis of population-based surveys. *Lancet Global Health.* 2015;3:e528.

6. USDA. Definitions of food security. 2020; www.ers.usda.gov/topics/food-nutrition-assistance/food-security-in-the-us/definitions-of-food-security.aspx.

7. Leung CW and others. Food insecurity is inversely associated with diet quality of lower income adults. *J Acad Nutr Diet.* 2014;114:1943.

8. Gundersen C, Ziliak JP. Food insecurity and health outcomes. *Health Aff.* 2015;34:1830.

9. Johnson AD and Markowitz AD. Associations between household food insecurity in early childhood and children's kindergarten skills. *Child Devel.* 2017; DOI: 10.1111/cdev.12764.

10. Burke MP and others. Severity of household food insecurity is positively associated with mental disorders among children and adolescents in the United States. *J Nutr.* 2016;146:2019.

11. Hager ER and others. Development and validity of a 2-item screen to identify families at risk for food insecurity. *Pediatrics,* 2010;126:e26.

12. Aggarwal A and others. Nutrient intakes linked to better health outcomes are associated with higher diet costs in the US. *PLoS ONE.* 2012;7:e37533.

13. Pan L and others. Food insecurity is associated with obesity among US adults in 12 states. *J Acad Nutr Diet.* 2012;112:1403.

14. Keys A. *The biology of human starvation.* Minneapolis: University of Minnesota Press; 1950.

15. Coleman-Jensen A and others. *Household food security in the United States in 2018.* ERR-270, U.S. Department of Agriculture, Economic Research Service; 2020.

16. U.S. Department of Health and Human Services. Poverty guidelines. 2019. www.census.gov/newsroom/press-releases/2019/income-poverty.html

17. Ver Ploeg M, Rahkosky I. Recent evidence on the effects of food store access on food choice and diet quality. *Amber Waves.* 2016;May 2.

18. Oliveira V. The food assistance landscape: FY 2019 annual report. EIB-207. 2020; www.ers.usda.gov/publications/pub-details/?pubid=92895.

19. Levitan, M. Food pantries at college campuses across the U.S. tackle students food insecurity. 2019; diverseeducation.com/article/138359.

20. Feeding America. 2018 annual report. Solving hunger. www.feedingamerica.org/about-us/financials.

21. Black RE and others. Maternal and child undernutrition and overweight in low-income and middle-income countries. *Lancet.* 2014;382:427.

22. International Food Policy Research Institute (IFPRI). Global nutrition report 2016: From promise to impact: Ending malnutrition by 2030. Washington, DC: IFPRI; 2016.

23. Nordin SM and others. Position of the Academy of Nutrition and Dietetics: Nutrition security in developing nations: Sustainable food, water and health. *J Acad Nutr Diet.* 2013;113:581.

24. Food and Agriculture Organization of the United Nations. Food loss and food waste. 2019; www.fao.org/food-loss-and-food-waste/en/.

25. Gunders D. Wasted: How America is losing up to 40 percent of its food from farm to fork to landfill. 2012; www.nrdc.org/sites/default/files/wasted-food-IP.pdf.

26. Willett W and others. Food in the Anthropocene: the EAT-Lancet Commission on healthy diets from sustainable food systems. *Lancet.* 2019;S0140-6736(18)31788-4.

27. Morawicki RO and Díaz González DJ. Food sustainability in the context of human behavior. *Yale J Biol Med.* 2018;91:191.

28. Knorr and World Wildlife Fund. 50 foods for healthier people and a healthier planet. 2019; www.wwf.org

29. Harvard T.H. Chan. School of Public Health. The Nutrition Source. Sustainability. 2019. www.hsph.harvard.edu/nutritionsource/sustainability.

30. EAT Lancet Commission. Healthy Diets From Sustainable Food Systems. Food Planet Health. www.eatforum.org.

31. Organic Trade Association. 2019 organic industry survey. 2020; ota.com/resources/organic-industry-survey.

32. Agriculture Market Service, U.S. Department of Agriculture. National Organic Program. 2020; www.ams.usda.gov/about-ams/programs-offices/national-organic-program.

33. Smith-Spangler C and others. Are organic foods safer or healthier than conventional alternatives? *Ann Intern Med.* 2012;157:348.

34. Średnicka-Tober D and others. Higher PUFA and n-3 PUFA, conjugated linoleic acid, α-tocopherol and iron, but lower iodine and selenium concentrations in organic milk: A systematic literature review and meta- and redundancy analyses. *Br J Nutr.* 2016;115;1043.

35. Średnicka-Tober D and others. Composition differences between organic and conventional meat: A systematic literature review and meta-analysis. *Br J Nutr.* 2016;115:994.

36. Baranski M and others. Higher antioxidant and lower cadmium concentrations and lower incidence of pesticide residues in organically grown crops: A systematic literature review and meta-analyses. *Br J Nutr.* 2014;112:794.

37. Center for Science in the Public Interest. Straight talk on genetically engineered foods. Answers to frequently asked questions. 2015; cspinet.org/new/pdf/biotech -faq.pdf.

38. James C. *20th anniversary (1996 to 2015) of the global commercialization of biotech crops and biotech crop highlights in 2015.* ISAAA Brief No. 51. Ithaca, NY: ISAAA; 2015.

39. Center for Food Safety. Genetically engineered food labeling laws. 2016; www.centerforfoodsafety.org/ge-map/#.

40. U.S. Food &Drug Administration. AquaAdvantage salmon. 2019; www.fda .gov/AnimalVeterinary/DevelopmentApprovalProcess/GeneticEngineering /GeneticallyEngineeredAnimals/ucm280853.htm.

41. Specter M. How the DNA revolution is changing us. *National Geographic.* 2016;230:31.

42. Savac N, Schwank G. Advances in CRISPR/Cas9 gene editing. *Translational Res.* 2016;168:15.

43. Lassoued R and others. Benefits of genome-edited crops: expert opinion. *Transgenic Res.* 2019;28:247.

44. Whitty CJM. Africa and Asia need a rational debate on GM crops. *Nature.* 2013;497:31.

45. Zhang C and others. Genetically modified foods: A critical review of their promise and problems. *Food Sci Human Wellness.* 2016;5:116.

46. Gilbert N. A hard look at GM crops. *Nature.* 2013;497:24.

47. U.S. Food and Drug Administration, Center for Veterinary Medicine. Animal cloning. 2020; www.fda.gov/animal-veterinary/safety-health/animal-cloning.

48. U.S. Food and Drug Administration. Food irradiation: What you need to know. 2018; www.fda.gov/food/buy-store-serve-safe-food/food-irradiation-what-you -need-know.

49. Crowley OV and others. Factors predicting likelihood of eating irradiated meat. *J Appl Soc Psy.* 2013;43:95.

50. Samal D. Use of Nanotechnology in Food Industry: A review. *Int J Environ Ag Biotech.* 2017;2:2270.

51. Pathakotia K and others. Nanostructures: Current uses and future applications in food science. *J Food Drug Anal.* 2017;25:245.

52. McClements DJ and Xiao H. Is nano safe in foods? Establishing the factors impacting the gastrointestinal fate and toxicity of organic and inorganic food-grade nanoparticles. *NPJ Sci Food.* 2017;1: 6.

53. Nicole W. Secret ingredients: Who knows what's in your food? *Environ Health Perspect.* 2013;121:A126.

54. Neltner TG and others. Navigating the U.S. food additive regulatory program. *Comp Rev Food Sci Safety.* 2011;10:342.

55. Rancière F. Bisphenol A and the risk of cardiometabolic disorders: A systematic review with meta-analysis of the epidemiological evidence. *Environ Health.* 2015;14:46.

56. Jalal N and others. Bisphenol A (BPA) the mighty and the mutagenic. *Toxicol Rep.* 2018;5:76.

57. Ames B and others. Ranking possible carcinogenic hazards. *Science.* 1987;236:271.

58. The Pew Charitable Trusts. Fixing the oversight of chemicals added to our food. 2013; www.pewtrusts.org/en/research-and-analysis/reports/2013/11/07/fixing-the -oversight-of-chemicals-added-to-our-food.

59. Maffini MV and others. Looking back to look forward: A review of FDA's food safety assessment and recommendations for modernizing its programs. *Comp Rev Food Sci Safety.* 2013;12:439.

60. Suez J and others. Artificial sweeteners induce glucose intolerance by altering the gut microbiota. *Nature.* 2014;514(7521):181.

61. Position of the Academy of Nutrition and Dietetics. Use of nutritive and nonnutritive sweeteners. *J Acad Nutr Diet.* 2012;112:739.

62. Nigg JT and others. Meta-analysis of attention-deficit/hyperactivity disorder or attention-deficit/hyperactivity disorder symptoms, restriction diet, and synthetic food color additives. *J Am Acad Child Adolesc Psychiatry.* 2012;51:86.

63. Centers for Disease Control and Prevention. Estimates of foodborne illness in the United States. 2018; www.cdc.gov/foodborneburden.

64. Scharff RL. Economic burden from health losses due to foodborne illness in the United States. *J Food Protect.* 2012;75:123.

65. Centers for Disease Control and Prevention. Multistate outbreak of *Salmonella* Poona infections linked to imported cucumbers (final update). 2016; www.cdc .gov/salmonella/poona-09-15.

66. Karanis P and others. Waterborne transmission of protozoan parasites: A worldwide review of outbreaks and lessons learnt. *J Water Health.* 2007;5:1.

67. Centers for Disease Control and Prevention. Cyclosporiasis outbreak investigations—United States, 2019. 2019; www.cdc.gov/parasites /cyclosporiasis/outbreaks/2019/a-050119/index.html.

68. Centers for Disease Control and Prevention. Prion diseases. 2018; www.cdc.gov /prions.

69. Taylor SL. Food additives, contaminants and natural toxicants: Maintaining a safe food supply. In: Ross AC and others, eds. *Modern nutrition in health and disease.* 11th ed. Baltimore: Lippincott Williams & Wilkins; 2014.

70. Environmental Protection Agency. Ground water and drinking water. 2020; www.epa.gov/ground-water-and-drinking-water.

71. Beer KD and others. Surveillance for waterborne disease outbreaks associated with drinking water—United States, 2011–2012. *MMWR.* 2015;64:842.

72. Cody MM, Stretch T. Position of the Academy of Nutrition and Dietetics: Food and water safety. *J Acad Nutr Diet.* 2014;114:1819.

73. McClure LF and others. Blood lead levels in young children: US, 2009–2015. *J Pediatrics.* 2016;175:173.

74. Wright RO and others. Association between iron deficiency and blood level in a longitudinal analysis of children followed in an urban primary care clinic. *J Pediatr.* 2003;142:9.

75. U.S. Environmental Protection Agency. *Integrated Risk Information System.* Washington, DC: U.S. EPA; 2019. www.epa.gov/iris.

76. Sohn E. The toxic side of rice. *Nature.* 2014;514:S62.

77. Karagas MR and others. Association of rice and rice-product consumption with arsenic exposure early in life. *JAMA Pediatr.* 2016;170:609.

78. U.S. Environmental Protection Agency. Choose fish and shellfish wisely. 2019; www.epa.gov/choose-fish-and-shellfish-wisely.

79. U.S. Environmental Protection Agency. Pesticides. 2020; www.epa.gov /pesticides.

80. U.S. Department of Agriculture. Pesticide Data Program. 2017 Annual calendar summary. 2018; www.ams.usda.gov/AMSv1.0/pdp.

81. National Cancer Institute. Agricultural Health Study. 2019; aghealth.nih.gov.

82. Ayesa K and others. The case for consistent use of medical eponyms by eliminating possessive forms. *J Med Libr Assoc.* 2018;106:127.

83. Paulson JA and others. Nontherapeutic use of antimicrobial agents in animal agriculture: Implications for pediatrics. *Pediatr.* 2015;136:e1670.

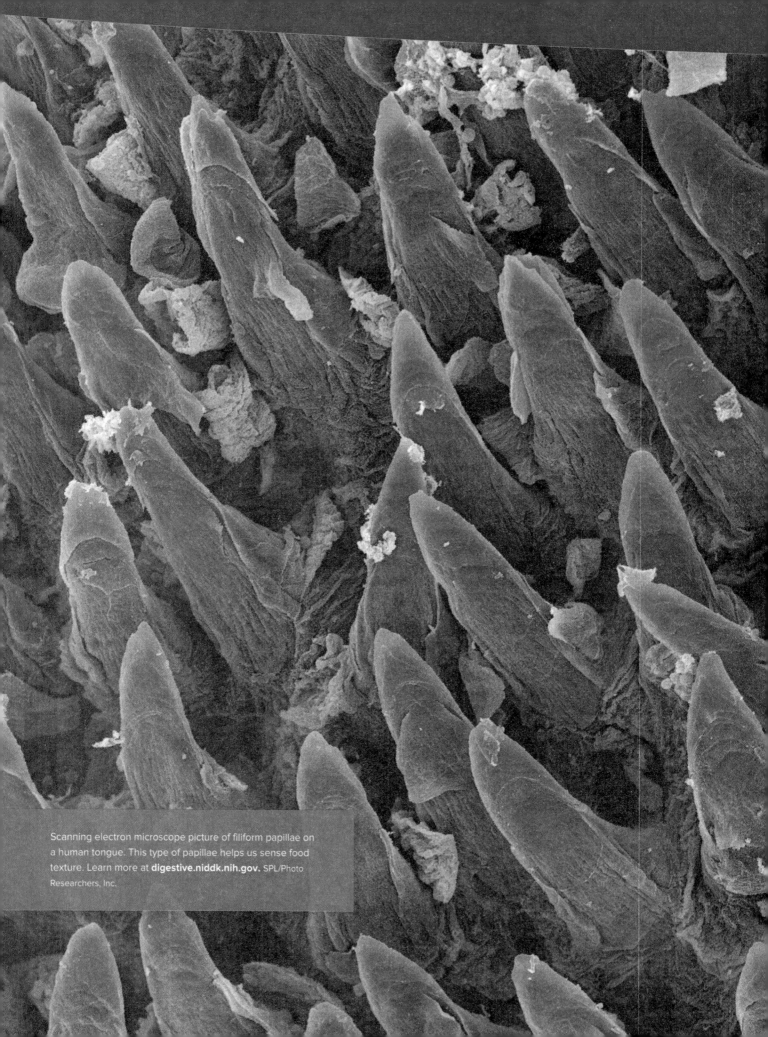

Scanning electron microscope picture of filiform papillae on a human tongue. This type of papillae helps us sense food texture. Learn more at **digestive.niddk.nih.gov.** SPL/Photo Researchers, Inc.

# 4 Human Digestion and Absorption

## Learning Objectives

**After studying this chapter, you will be able to**

1. Outline the basic anatomy and functions of digestive system organs.
2. Describe how the digestive tract processes foods and propels its contents from mouth to anus.
3. Describe the function of key enzymes and hormones required for digestion and absorption.
4. Explain the processes of nutrition absorption and how nutrients enter the circulatory system.
5. Identify major nutrition-related gastrointestinal disorders and typical approaches to prevention and treatment.
6. Explain why diarrhea represents a serious health challenge to infants and young children around the world and how it can be prevented and treated.

## Chapter Outline

**EARLY KNOWLEDGE OF DIGESTIVE PHYSIOLOGY** came from a surprising source—"the man with a hole in his stomach." In 1822, a fur trapper, Alexis St. Martin, was accidentally hit by a shotgun blast. The blast wound to his stomach never closed completely, allowing an opportunity for William Beaumont, a U.S. Army physician, to study for several years how foods are digested. For example, he lowered food tied to string through the hole in St. Martin's stomach and then periodically removed it to observe changes. Beaumont learned that the stomach releases its secretions in response to food in the stomach, rather than building up secretions between meals, as was commonly believed. He also discovered that the stomach secretions contain not only acid but also a substance that allows meat to be digested. We now know this substance as the digestive enzyme pepsin. Beaumont also observed that, when his subject was distressed or angry, digestion was impaired. Throughout his life, St. Martin remained in poor health, but he lived almost 60 years after the shooting accident.

Since the time of Beaumont and St. Martin, scientists have continued to study how the digestive system functions and the many digestive system disorders and diseases. This chapter will explore the processes of digestion and absorption and the related aspects of human physiology that support nutritional health. You will become acquainted with the basic anatomy (structure) and physiology (function) of the digestive system. You also will learn the causes of some common digestive system disorders, along with ways to prevent and manage them.

**adenosine triphosphate (ATP)** Chemical that supplies energy for many cellular processes and reactions.

**tissue** Collection of cells adapted to perform a specific function.

# 4.1 Organization of the Human Body

The cell is the smallest functional unit of the human body. (Appendix A reviews the parts of a cell.) The body's 10 trillion cells have the ability to grow, absorb nutrients and other substances, use energy, conduct countless metabolic and physiological functions, and excrete waste. Cellular processes and chemical reactions, which occur constantly in every living cell, require a continuous supply of energy in the form of dietary carbohydrate, protein, and fat. Almost all cells need oxygen to transform the energy in nutrients into the form the body can use—**adenosine triphosphate (ATP)** (see Chapter 9 for more on ATP). Cells also need water, building materials (e.g., amino acids and minerals), and chemical regulators (e.g., vitamins). Of course, adequately supplying all nutrients to the body's cells begins with a healthful diet.

Cells of the same type join together to form tissue (Fig. 4-1). **Tissue** is made of groups of similar cells working together to perform a specific task. Humans have 4 primary types of tissue: epithelial, connective, muscle, and nervous.

- **Epithelial tissue** is composed of cells that cover surfaces outside and inside the body. The skin and linings of the gastrointestinal (GI) tract are examples. Epithelial cells absorb nutrients, secrete important substances, excrete waste, and protect underlying tissues.
- **Connective tissue** supports and protects the body by holding structures (e.g., cells and cell parts) together, stores fat, and produces blood cells. Tendons, cartilage, and parts of bone, arteries, and veins are made of connective tissue.
- **Muscle tissue** can contract and relax and permits movement.
- **Nervous tissue**, found in the brain and spinal cord, transmits nerve impulses from 1 part of the body to another.

Tissues combine in a specific way to form structures, known as **organs**, which perform specific functions. All organs play a role in nutritional health, and nutrient intake

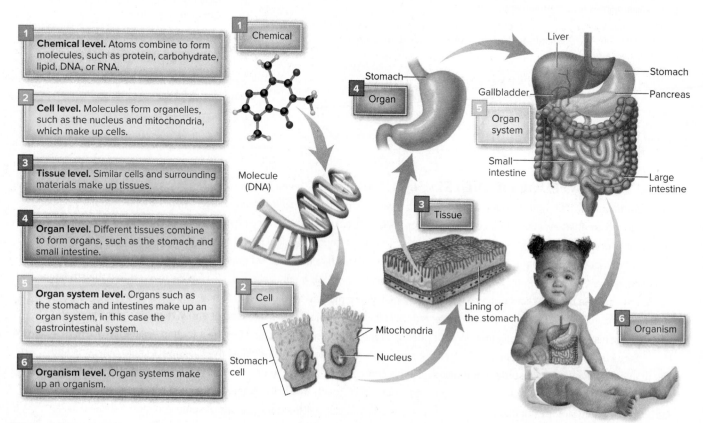

1 **Chemical level.** Atoms combine to form molecules, such as protein, carbohydrate, lipid, DNA, or RNA.

2 **Cell level.** Molecules form organelles, such as the nucleus and mitochondria, which make up cells.

3 **Tissue level.** Similar cells and surrounding materials make up tissues.

4 **Organ level.** Different tissues combine to form organs, such as the stomach and small intestine.

5 **Organ system level.** Organs such as the stomach and intestines make up an organ system, in this case the gastrointestinal system.

6 **Organism level.** Organ systems make up an organism.

*Figure 4-1* The levels of organization of the human body are chemical, cell, tissue, organ, organ system, and organism. Each level is more complex than the previous level. The organ system shown is the digestive system.

John Lund/Annabelle Breakey/Getty Images.

affects how well each organ functions. An **organ system** is formed when several organs work together to perform a specific function. For example, the digestive system includes the GI tract (mouth, esophagus, stomach, small intestine, and large intestine, which terminates with the rectum and anus), liver, pancreas, and gallbladder (Fig. 4-2). The coordinated work of all organ systems allows the entire body to function normally. Table 4-1 summarizes the components and functions of the body's organ systems.

The primary theme of human nutrition is to understand how nutrients affect different cells, tissues, organs, organ systems, and overall health. This chapter focuses on the digestive system. Learning how the digestive system makes nutrients in foods available to body organs, tissues, and cells is critical to understanding human nutrition.

**organs** Structure (e.g., heart, kidney, or eye) consisting of cells and tissues that perform a specific function in the organism.

**organ system** Group of organs classified as a unit because they work together to perform a function or set of functions.

## Knowledge Check

1. What is the form of energy that can be used by almost all cells?
2. Name the 4 primary types of tissues. Which of these tissue types covers the surfaces that are both outside and inside the body?
3. What organs make up the digestive system?

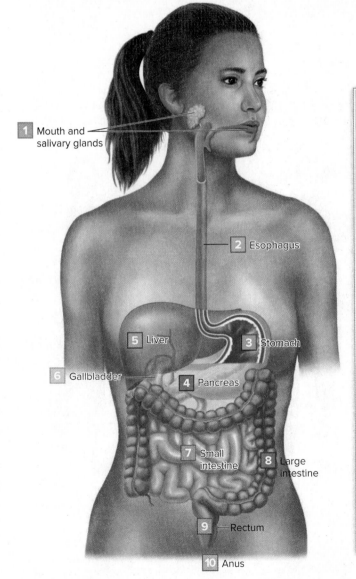

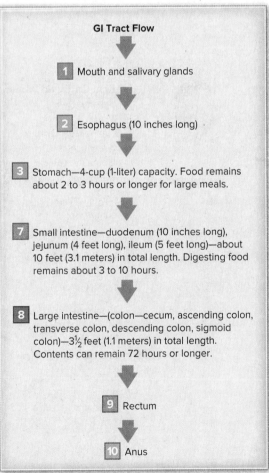

**GI Tract Flow**

↓

1 Mouth and salivary glands

↓

2 Esophagus (10 inches long)

↓

3 Stomach—4-cup (1-liter) capacity. Food remains about 2 to 3 hours or longer for large meals.

↓

7 Small intestine—duodenum (10 inches long), jejunum (4 feet long), ileum (5 feet long)—about 10 feet (3.1 meters) in total length. Digesting food remains about 3 to 10 hours.

↓

8 Large intestine—(colon—cecum, ascending colon, transverse colon, descending colon, sigmoid colon)—3½ feet (1.1 meters) in total length. Contents can remain 72 hours or longer.

↓

9 Rectum

↓

10 Anus

*Figure 4-2* Major organs of the gastrointestinal (GI) tract (1, 2, 3, 7, 8, and 9) and accessory organs (4, 5, and 6) used in digestion and the absorption of nutrients.

**Table 4-1  Organ Systems of the Body**

**Digestive System**

Major components: mouth, esophagus, stomach, intestines, and accessory organs (liver, gallbladder, and pancreas)

Functions: performs the mechanical and chemical processes of digestion of food, absorption of nutrients, and elimination of wastes

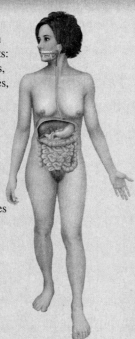

**Nervous System**

Major components: brain, spinal cord, nerves, and sensory receptors

Functions: detects and interprets sensation; controls movements and physiological and intellectual functions

**Cardiovascular System**

Major components: heart, blood vessels, and blood

Functions: carries blood and regulates blood supply; transports nutrients, waste products, hormones, and gases (oxygen and carbon dioxide) throughout the body; and regulates blood pressure

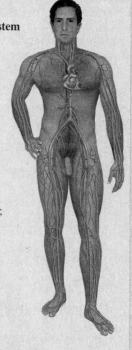

**Endocrine System**

Major components: endocrine glands, such as the pituitary, thyroid, and adrenal glands; hypothalamus; and pancreas

Functions: regulates metabolism, growth, reproduction, and many other functions by producing and releasing hormones

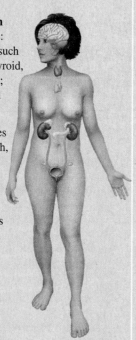

**Lymphatic and Immune Systems**

Major lymphatic components: lymph, lymphocytes, lymphatic vessels, and lymph nodes

Major immune components: mechanical (e.g., skin), chemical (e.g., lysozyme), and cellular (e.g., white blood cells)

Lymphatic functions: aids in fluid balance, fat absorption and transport, and immune functions

Immune functions: protects against microorganisms and other foreign substances

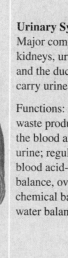

**Urinary System**

Major components: kidneys, urinary bladder, and the ducts that carry urine

Functions: removes waste products from the blood and forms urine; regulates blood acid-base (pH) balance, overall chemical balance, and water balance

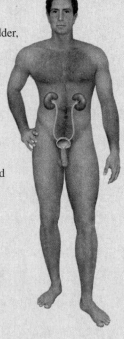

**Table 4-1** **Continued**

**Integumentary System**

Major components: skin, hair, nails, and sweat glands

Functions: protects the body, regulates body temperature, prevents water loss, and produces vitamin D

**Skeletal System**

Major components: bones, cartilage, ligaments, and joints

Functions: protects organs, supports body weight, allows body movement, produces blood cells, and stores minerals

**Muscular System**

Major components: smooth, cardiac, and skeletal muscle

Functions: produces body movement, heartbeat, and body heat; propels food in the digestive tract; and maintains posture

**Respiratory System**

Major components: lungs and respiratory passages

Functions: exchanges gases (oxygen and carbon dioxide) between the blood and the air; regulates blood acid-base (pH) balance

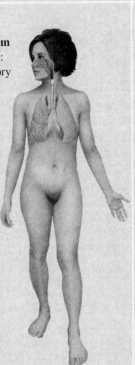

**Reproductive System**

Major components: gonads (ovaries and testes) and genitals

Functions: performs the processes of sexual maturation and reproduction; influences sexual functions and behaviors

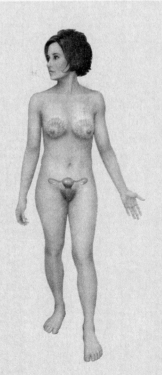

The cardiovascular and lymphatic organ systems together make up the circulatory system and, so, contribute to circulatory functions in the body. The lymphatic system is part of the immune system. The endocrine and nervous organ systems contribute to the regulatory functions. The digestive, urinary, integumentary, and respiratory organ systems contribute to the excretory functions, whereas the muscular and skeletal organ systems contribute to storage abilities in the body.

## 4.2 Digestive System Overview

The digestive system accomplishes the tasks of digestion, the mechanical and chemical processes of breaking down foods into smaller components, and absorption, the uptake of these components from the GI tract into either the blood or the lymph. All the nutrients found in foods—proteins, fats, carbohydrates, vitamins, minerals, and water—are made ready for use in the body's cells by the digestive system. An adult secretes about 29 cups (7 liters) of fluid containing water, mucus, acid, digestive enzymes, bile, and hormones into the GI tract each day to assist with the processes of digestion and absorption.[1] Finally, at the end of the GI tract, the excretion of waste matter can be initiated at an appropriate time. Like many other processes, such as breathing and the beating of the heart, digestion and absorption are carefully controlled by hormones and the nervous system. Table 4-2 provides an overview of the major functions of each digestive system organ.

In addition to its main functions of digestion and absorption, the GI tract also has an important role in the body's immune system and regulation of food intake. A heavy load of bacteria, viruses, and other microorganisms is continually introduced into the body with the food we eat. The GI tract is a physical barrier to the entry of these microorganisms into the body, and it produces hydrochloric acid (HCl) along with a host of immune components, such as antibodies, lymphocytes, and macrophages, that destroy microorganisms in the gut. Further, bacteria reside throughout the GI tract, but especially in the large intestine. These resident bacteria are referred to as the gut **microbiota**; they are part of the body's overall **microbiome**—the microorganisms that live throughout the body. Healthy intestinal bacteria help keep pathogenic (disease-causing) bacteria under control. Gut microbiota also have other roles, discussed in Section 4.7.

The regulation of food intake, discussed in Chapter 10, depends on many organ systems, neurotransmitters, and hormones. One such substance is **ghrelin**, a hormone released by the stomach, which acts to increase appetite and food intake.[2] Ghrelin levels are highest during fasting.

**microbiota** Microorganisms that inhabit a particular region, such as the gastrointestinal tract.

**microbiome** Microorganisms in a specific environment, such as our bodies. Learn more at commonfund.nih.gov/hmp.

**ghrelin** Hormone, made by the stomach, that increases food intake.

**Table 4-2  Overview of GI Tract Digestion and Absorption Functions**

| Organs | Digestive Functions |
|---|---|
| Mouth and salivary glands | Prepare food for swallowing: chewing, moistening with saliva<br>Detect taste molecules<br>Start digestion of starch with amylase enzyme<br>Start digestion of fat with lingual lipase |
| Esophagus | Moves food to stomach by peristaltic waves initiated by swallowing |
| Stomach | Secretes gastric juice containing acid, enzymes, and hormones<br>Mixes food with gastric juice, converting it to liquid chyme<br>Starts digestion of protein and fat<br>Kills microorganisms with acid<br>Secretes intrinsic factor, a protein required for vitamin B-12 absorption<br>Slowly releases chyme to the small intestine |
| Liver | Produces bile to aid fat digestion and absorption |
| Gallbladder | Stores and concentrates bile and releases it to the small intestine |
| Pancreas | Secretes pancreatic juice containing digestive enzymes and bicarbonate into the small intestine |
| Small intestine | Mixes chyme with bile and pancreatic juice to complete digestion<br>Secretes hormones that help regulate digestive processes<br>Secretes digestive enzymes<br>Absorbs nutrients and other compounds in foods<br>Transports remaining residue to large intestine |
| Large intestine (colon) | Absorbs water and electrolytes (sodium and potassium)<br>Forms and stores feces<br>Houses most of the gut microbiota |
| Rectum | Holds and expels feces via the anus |

## Anatomy of the GI Tract

The GI tract, also known as the **alimentary canal**, is a long, hollow, muscular tube that extends almost 15 feet from mouth to anus. Nutrients must pass through the wall of this tube to be absorbed into the body. The wall consists of 4 layers (Fig. 4-3).

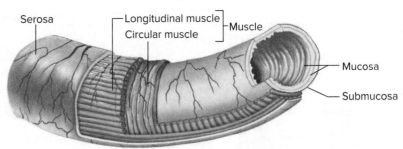

*Figure 4-3* The wall of the intestinal tract consists of 4 layers: mucosa, submucosa, muscle, and serosa.

- Mucosa, the innermost layer, is lined with epithelial cells and glands. The mucosa is not smooth and in some areas has tiny, fingerlike structures that project into the hollow interior of the tube known as the **lumen**. These projections increase the surface area of the mucosa.
- Submucosa, the second layer, consists of loose connective tissue, glands, blood vessels, and nerves. The blood vessels carry substances, including nutrients, both to and from the GI tract.
- Muscle, the next layer, occurs as double layers in most parts of the GI tract: an inner layer of circular smooth muscle that encircles the tube and an outer layer of longitudinal muscle fibers that runs up and down the tube. These muscles move food forward through the GI tract. The stomach has a third layer of muscle fiber running diagonally around it.
- Serosa, the outermost layer, protects the GI tract. The serosa secretes fluid that cushions the GI tract and reduces friction as it and other organs move.

Along the GI tract are **sphincters**, ringlike muscles that open and close like valves to control the flow of the contents (Fig. 4-4). The sphincters prevent contents from moving through the GI tract too quickly and allow thorough mixing with digestive system secretions. The sphincters also help propel food through the GI tract.

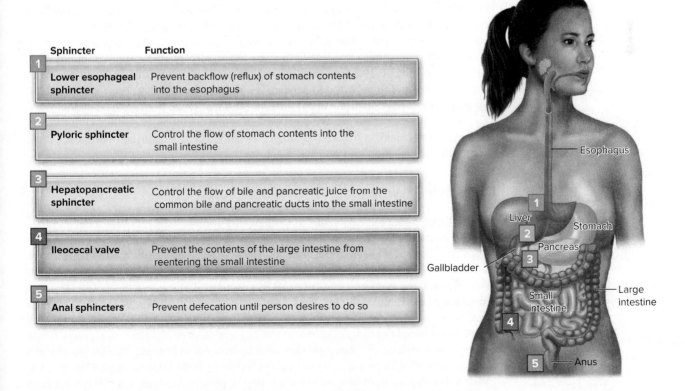

| | Sphincter | Function |
|---|---|---|
| 1 | **Lower esophageal sphincter** | Prevent backflow (reflux) of stomach contents into the esophagus |
| 2 | **Pyloric sphincter** | Control the flow of stomach contents into the small intestine |
| 3 | **Hepatopancreatic sphincter** | Control the flow of bile and pancreatic juice from the common bile and pancreatic ducts into the small intestine |
| 4 | **Ileocecal valve** | Prevent the contents of the large intestine from reentering the small intestine |
| 5 | **Anal sphincters** | Prevent defecation until person desires to do so |

*Figure 4-4* Sphincters of the GI tract. These circular muscles control the flow of contents through the GI tract. They open and close in response to stimuli from nerves, hormones, hormonelike compounds, and pressure that builds up around the sphincters.

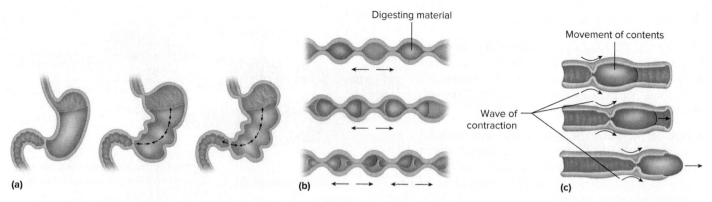

***Figure 4-5*** Mixing, segmentation, and peristalsis. (*a*) Strong contractions of the stomach muscles mix food and digestive juices. (*b*) Segmentation, a back-and-forth action in the small intestine, breaks apart contents of the small intestine into increasingly smaller pieces and mixes them with digestive juices. (*c*) Peristalsis, rhythmic waves of contraction and relaxation, moves the contents through the intestinal tract toward the anus.

▶ Hunger pangs are strong, somewhat uncomfortable peristaltic contractions that usually occur several hours after the last meal.

▶ The naming system for many enzymes is quite simple. The first part of the enzyme name usually indicates the target, followed by the suffix -ase. For instance, sucrase is the enzyme that digests the sugar sucrose; similarly, lactase digests lactose.

**digestive enzymes** Compounds that aid in the breakdown of carbohydrates, fats, and proteins.

**hydrolysis reactions** Chemical reaction that breaks down a compound by adding water. One product receives a hydrogen ion (H⁺); the other product receives a hydroxyl ion (—OH). Hydrolytic enzymes break down compounds using water in this manner.

## GI Motility: Mixing and Propulsion

Food is mixed with digestive secretions and propelled down the GI tract by a process called peristalsis. A snake swallowing its prey graphically illustrates the process. Recall that most of the GI tract has 2 layers of muscles—circular and longitudinal. Peristalsis consists of a coordinated wave of contraction (squeezing and shortening) and relaxation of these muscles. This process begins in the esophagus as 2 waves of muscle action closely following each other. The thickest and strongest muscles of the GI tract are in the stomach, where the contraction of 3 opposing muscle layers promotes complete mixing and churning of food and gastric juices. This contraction occurs as often as 3 times per minute after a meal.

The most frequent peristalsis takes place in the small intestine, where contractions occur about every 4 to 5 seconds. The small intestine also experiences segmental contractions (segmentation), which move the intestinal contents back and forth, causing the contents to break apart and mix with digestive juices. Figure 4-5 illustrates these mixing and propulsion processes. The large intestine has comparatively sluggish peristaltic waves that occur only 2 or 3 times per day, usually after a meal. These lead to mass movements that propel fecal matter from 1 part of the large intestine to the next and finally into the rectum for elimination. The amount of time it takes for a meal to make its way through the digestive system varies from day to day and from person to person. The size of a meal and the nutrients it contains also play a role. Studies indicate that normal transit times range from 24 to 60 hours.[3]

Vomiting reverses the normal digestive tract flow. Vomiting, controlled by the vomiting center in the brain, is triggered by toxins or irritants in the gastrointestinal tract, rapid changes in body position, and stomach distension.

## Digestive Enzymes and Other Secretions

**Digestive enzymes**, produced in the salivary glands, stomach, pancreas, and small intestine, are protein molecules that speed up digestion by catalyzing chemical reactions. Catalysis brings certain molecules close together and then creates a favorable environment for the chemical reaction. (Appendix B provides details on enzyme action.) Digestive enzymes catalyze chemical reactions known as **hydrolysis reactions**. In these reactions, water (*hydro-*) breaks apart (*-lysis*) molecules that are too large to pass through the GI tract wall. Hydrolysis reactions eventually yield simple molecules that are small enough to be absorbed through the intestinal wall. For example, sucrose (table sugar) cannot be absorbed. In Figure 4-6, you can see how a molecule of the sugar sucrose is hydrolyzed to form the smaller glucose and fructose molecules, both of which can be absorbed through the intestinal wall. Digestive enzymes aid mainly in the hydrolysis of carbohydrates, proteins, and fats. Each enzyme acts on a specific substance; for example, an enzyme that recognizes sucrose ignores lactose (milk sugar). Notice in Figure 4-6 that the sucrase enzyme hydrolyzes sucrose.

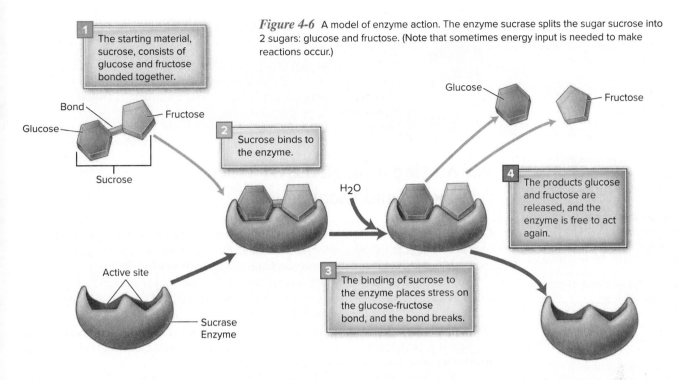

*Figure 4-6* A model of enzyme action. The enzyme sucrase splits the sugar sucrose into 2 sugars: glucose and fructose. (Note that sometimes energy input is needed to make reactions occur.)

**1** The starting material, sucrose, consists of glucose and fructose bonded together.

**2** Sucrose binds to the enzyme.

**3** The binding of sucrose to the enzyme places stress on the glucose-fructose bond, and the bond breaks.

**4** The products glucose and fructose are released, and the enzyme is free to act again.

The salivary glands and stomach produce relatively small amounts of the digestive enzymes. Most are synthesized by the pancreas and small intestine. The pancreas adjusts its enzyme production to match the macronutrient content of the diet. Increased protein intake results in an increase in protein-digesting enzymes and a low-fat diet will result in a decrease in the enzymes that aid fat digestion. Inadequate amounts of digestive enzymes may be produced when the small intestine or the pancreas is diseased or when an individual is malnourished. This scarcity can result in incomplete digestion and limited absorption. If food is not completely digested, bacteria in the large intestine convert some of it into gases and acids. The gases may distend (bloat) the abdomen. In addition, the feces may be foamy and greasy because of trapped gases and the presence of undigested fat. Chapters 5, 6, and 7 present a detailed review of the digestive enzymes used to process each macronutrient.

The digestive enzymes are just a single component of the 7 liters of secretions that enter the digestive tract each day. Other secretions include saliva from salivary glands, mucus produced along the entire GI tract, hydrochloric acid from the stomach, bicarbonate from the pancreas, bile from the liver, and numerous **hormones** (Table 4-3). The functions of these hormones and secretions are described in Sections 4.3, 4.4, and 4.5.

**Table 4-3  Important Secretions of the Digestive System**

| Secretion | Sites of Production | Functions |
|---|---|---|
| Saliva | Mouth | Dissolves taste-forming compounds; contains many compounds that aid swallowing, digestion, and protection of teeth |
| Mucus | Mouth, stomach, small and large intestines | Protects GI tract cells, lubricates digesting food |
| Enzymes (amylases, lipases, proteases) | Mouth, stomach, small intestine, pancreas | Break down carbohydrates, fats, and protein into forms small enough for absorption |
| Acid (HCl) | Stomach | Promotes digestion of protein, destroys microorganisms, increases solubility of minerals |
| Bile | Liver (stored in gallbladder) | Aids in fat digestion (emulsifies fat) |
| Bicarbonate | Pancreas, small intestine | Neutralizes stomach acid when it reaches small intestine |
| Hormones | Stomach, small intestine, pancreas | Regulate food intake, digestion, and absorption |

The stomach has millions of tiny gastric pits like this. The pits produce mucus to protect the stomach, and glands at the bottom of the pits secrete digestive juices (see Fig. 4-11). After a person eats, the muscles in the stomach contract and cause the pits to squirt their contents into the stomach and begin digesting food.

Image Source/Getty Images

**hormones** Chemical substance produced in the body that controls or regulates the activity of certain cells or organs. Hormones can be amino acid–like (epinephrine), proteinlike (insulin), or fatlike (estrogen).

**bolus** Mass of food that is swallowed.

**lysozyme** Set of enzyme substances produced by a variety of cells; can destroy bacteria by rupturing cell membranes.

**amylase** Starch-digesting enzyme from the salivary glands or pancreas.

**olfactory** Related to the sense of smell.

**epiglottis** Flap that folds down over the trachea during swallowing.

**trachea** Airway leading from the larynx to the bronchi.

**larynx** Structure located between the pharynx and trachea; contains the vocal cords.

▶ When we talk about liking the taste of a food, we mean we like its flavor. That's because aromatic and taste compounds, combined with the physical effect caused by food textures and certain chemicals in foods (e.g., the hot and irritating capsaicin in chili peppers), create flavor sensations.

## Knowledge Check

1. What is the difference between digestion and absorption?
2. What are the 4 layers of the GI tract?
3. What process propels the contents of the GI tract along its length?
4. How do enzymes aid digestion?
5. Where are most digestive enzymes synthesized?

# 4.3 Moving through the GI Tract: Mouth and Esophagus

Before we eat a bite of most foods, the work of digestion has already started. Food preparation, such as cooking, marinating, pounding, and dicing, often begins the process. Starch granules in food swell as they soak up water during cooking, making them much easier to digest. Cooking also softens tough connective tissues in meats and fibrous tissue of plants, such as that in broccoli stalks. As a result, the food is easier to chew, swallow, and break down during digestion.

In the body, digestion begins in the mouth, or oral cavity (Fig. 4-7). The teeth tear and grind solid food into smaller pieces, which increases the surface area exposed to saliva. During chewing, the tongue presses morsels of food against the hard palate and helps mix the food with saliva. The food is now referred to as a **bolus**.

The salivary glands produce about 4 cups (1 liter) of saliva each day. Saliva is a dilute, watery fluid that contains several substances, including mucus to lubricate the bolus and hold it together, **lysozyme** to kill bacteria, and **amylase** to break down starch into simple sugars. However, food remains in the mouth such a short time that only about 5% of the starch gets broken down by salivary amylase. Lingual **lipase**, also released from salivary glands, is a fat-digesting enzyme that initiates fat digestion. Saliva also helps prevent tooth decay because it contains antibacterial agents, minerals to repair teeth, and substances that neutralize acid.

## Taste and Smell

Saliva enhances our perception of the flavor of foods by dissolving the taste-forming compounds in foods. Taste buds, found on the papillae of the tongue and soft palate, contain specialized taste-receptor cells that can detect taste compounds in foods. The ability to detect the following 5 basic tastes is present in all areas of the tongue.

- Salty, from a variety of salts, such as NaCl (table salt) or KCl (potassium chloride)
- Sour, from acids, such as citric acid (think about how sour—and acidic—a lemon is)
- Sweet, from certain organic compounds, such as sugars: humans have an innate preference for sweetness
- Bitter, from a diverse group of compounds, including caffeine and quinine, and numerous other compounds in vegetables, fruits, and whole grains: many bitter compounds are toxic, but others are beneficial phytochemicals (see Chapter 1 and Part 4)
- Umami, a savory, brothy, or meaty taste, from amino acids (primarily glutamate): foods such as mushrooms, cooked tomatoes, Parmesan cheese, and seaweed cause the umami taste sensation, and the seasoning monosodium glutamate (MSG) may be added to processed and restaurant foods to enhance the umami sensation

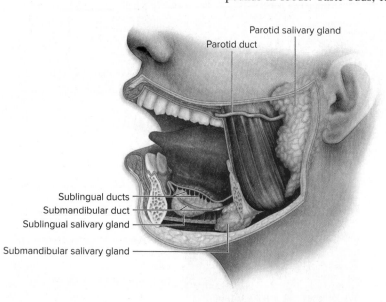

Parotid salivary gland
Parotid duct
Sublingual ducts
Submandibular duct
Sublingual salivary gland
Submandibular salivary gland

*Figure 4-7* The oral cavity (mouth) is the beginning of the GI tract. Teeth can tear and grind food into smaller pieces. The salivary glands near the oral cavity produce saliva, which lubricates and holds the chewed food together in a bolus. Saliva also contains enzymes and substances that help protect the teeth.

There also is evidence that we have taste receptors for calcium, magnesium, and other minerals.[4] The sense of taste is enhanced by input from approximately 6 million **olfactory** cells in the nose, which are stimulated when we chew. Thus, it makes perfect sense that, with nasal congestion, even strong-tasting foods may have little flavor. A variety of diseases and drugs, as well as the effects of aging, can alter taste and smell sensations.

Taste perception also is affected by human genetic variation in both taste and olfactory sensations. For example, the ability to taste bitter compounds is a heritable trait that can be measured by a person's ability to taste the compound 6-n-propylthiouracil (PROP).[5] This ability varies from person to person and between racial and ethnic groups. Super-tasters (7–40% of any given population) immediately detect a very bitter, unpleasant taste in response to PROP. They often dislike a number of bitter foods, such as coffee, broccoli, and kale. Other individuals are classified as medium tasters and non-tasters. It is not yet known how taste acuity influences food choices and health.[6]

## Swallowing

Swallowing moves food from the mouth into the esophagus, the 10-inch-long muscular tube that extends to the stomach. At its entrance is the **epiglottis**, a flaplike structure that prevents food from entering the **trachea** (windpipe). When food is swallowed, the epiglottis closes over the **larynx** (the opening of the trachea). The food bolus drops onto the epiglottis, and the esophagus relaxes and opens (Fig. 4-8). These involuntary responses ensure that the swallowed bolus, aided by peristalsis of the esophagus and gravity, travels down the esophagus, not into the trachea. Small pieces of food that enter the trachea may end up in the lungs and cause a serious infection. Larger pieces of food entering the trachea may cause choking and prevent speaking or breathing. To treat choking, bend the person forward at the waist and give 5 back blows between the shoulder blades with the heel of 1 hand, followed by 5 quick, upward abdominal thrusts. Thrusts are performed by placing a fist just above the choking person's navel, covering it with your other hand. Learn more at RedCross.org.

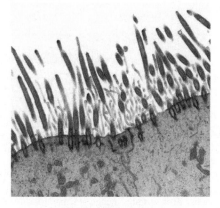

Nasal lining. Transmission electron micrograph (TEM) of a section through the cilia (red) covering the epithelial lining of the nasopharyx (upper part of the cavity behind the nose and mouth). Cilia are microscopic, hairlike structures covered with sticky mucus that traps dusts and other inhaled particles. The cilia move in waves to propel the mucus backward toward the throat, where it travels to the stomach and is digested and eventually excreted.

Steve Gschmeissner/Science Photo Library/Alamy Stock Photo

▶ "Dry mouth," or a decrease in saliva production, is a common complaint. It is a side effect of many medications, including some frequently used to treat depression, high blood pressure, pain, allergies, and cancer. When severe, dry mouth can cause sores in the mouth and tooth decay. Foods may taste bland and be hard to chew and swallow. Good oral hygiene, dietary changes, and artificial saliva products can help individuals manage dry mouth.

▶ Being a super-taster is probably related to having an abundance of fungiform papillae on the tongue. Fungiform papillae, irregularly scattered across the surface of the tongue, appear as small, red dots embedded in the even smaller and much more numerous filiform papillae.

### Knowledge Check

1. What substances are found in saliva?
2. For each of the 5 basic taste sensations, name at least 1 chemical compound that elicits the sensation.
3. How does the swallowing process prevent food from entering the trachea?

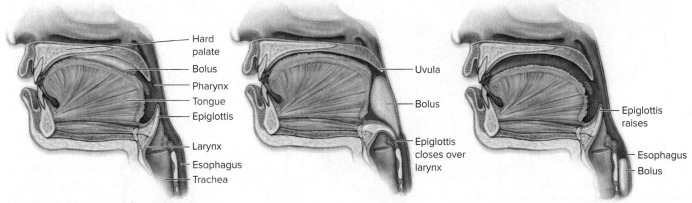

**(a)** Bolus of food is pushed by tongue against hard palate and then moves toward pharynx.

Hard palate
Bolus
Pharynx
Tongue
Epiglottis
Larynx
Esophagus
Trachea

**(b)** As bolus moves into pharynx, epiglottis closes over larynx.

Uvula
Bolus
Epiglottis closes over larynx

**(c)** Esophageal muscle contractions push bolus toward stomach. Epiglottis then returns to its normal position.

Epiglottis raises
Esophagus
Bolus

*Figure 4-8*  The process of swallowing. Swallowing occurs as the food bolus is forced (*a*) into the pharynx (throat) from the oral cavity, (*b*) through the pharynx, and (*c*) into the esophagus on the way to the stomach. Choking occurs when the bolus becomes lodged in the trachea (windpipe), blocking air to the lungs, instead of passing into the esophagus.

# 4.4 Moving through the GI Tract: Stomach

**parietal cells** Gastric gland cell that secretes hydrochloric acid and intrinsic factor.

**pepsinogen** Inactive protein precursor to the protein-digesting enzyme pepsin; produced in the stomach.

**chief cells** Gastric gland cell that secretes pepsinogen.

**gastrin** Hormone that stimulates HCl and pepsinogen secretion by the stomach.

**prostaglandins** Potent compounds that are synthesized from polyunsaturated fatty acid that produce diverse effects in the body.

**nonsteroidal anti-inflammatory drugs (NSAIDs)** Class of medications that reduce inflammation, fever, and pain but are not steroids. Aspirin, ibuprofen (Advil®), and naproxen (Aleve®) are some examples.

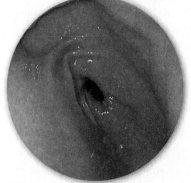

The heavy use of aspirin, an NSAID, may result in ulcer formation.

Comstock/Alamy Stock Photo

**(a)** Normal

*Figure 4-9* The esophagus can be seen opening into the stomach. (*a*) A healthy gastric mucosa; the small, white spots are reflections of light. (*b*) A bleeding peptic ulcer. An ulcer is a small erosion of the top layer of cells. A peptic ulcer typically has an oval shape and yellow-white color. Here the yellowish floor of the ulcer is partially obscured by black blood clots, and fresh blood is visible around the margin of the ulcer.

(a) Gastrolab/Science Source; (b) CNRI/SPL/ Science Source

The stomach often is regarded as primarily a mixing and holding tank for everything we ingest, but it also provides biochemical, hormonal, and protective functions vital to good health. The unique muscle structure of the stomach and the secretion of a variety of substances by specialized stomach cells permit the remarkably efficient work of the stomach.

The entry of food into the stomach is through the lower esophageal sphincter (sometimes called the cardiac sphincter due to its proximity to the heart), located between the esophagus and the stomach (Fig. 4-9). It prevents backflow (reflux) of the highly acidic stomach contents into the esophagus. If the sphincter malfunctions, causing reflux, the pain commonly known as **heartburn** occurs.

The average adult stomach holds about 2 ounces (50 ml) when empty and expands to 4 to 6 cups (1–1.5 liters) after a typical meal, but it can hold up to 16 cups (4 liters) when extremely full. Little digestion occurs in the stomach, and only water, a few forms of fats, and about 20% of any alcohol consumed can be absorbed there.

Each day, the stomach secretes about 8 cups (2 liters) of "gastric juice" that aids in the digestive process. Gastric juices are released when we see, smell, taste, or even think about food. These secretions include a very strong acid, called hydrochloric acid (HCl) (Figs. 4-10 and 4-11), from the **parietal cells**; **pepsinogen**, an inactive protein-digesting enzyme; and gastric lipase from the **chief cells**. **Gastrin**, a hormone made in the stomach, controls the release of HCl and pepsinogen. Gastrin secretion is highest at the beginning of a meal and declines as the meal progresses. The decline in gastrin secretion causes the release of HCl and pepsinogen to taper off.

The HCl produced by the stomach is very important. It inactivates the biological activity of ingested proteins, such as certain plant and animal hormones. This prevents them from affecting human functions. HCl also destroys most harmful bacteria and viruses (pathogens) in foods; dissolves dietary minerals (e.g., calcium) so that they can be more easily absorbed; and converts pepsinogen into the active protein-digesting enzyme pepsin.

The stomach also secretes mucus from mucous cells found on the gastric mucosa. Mucus lubricates and protects the stomach from being digested by HCl and pepsin. Mucus production relies on the presence of hormonelike compounds called **prostaglandins**. Heavy use of aspirin and other **nonsteroidal anti-inflammatory drugs (NSAIDs)** (e.g., ibuprofen, naproxen) can damage the stomach wall because they inhibit prostaglandin production. The reduced mucus barrier in the stomach means stomach acid may damage the stomach wall.

**(b)** Peptic ulcer

Contraction of the 3 muscle layers in the stomach thoroughly mixes food with gastric secretions. Mixing transforms solid food into **chyme** (pronounced kime), a soupy, acidic mixture. The pyloric sphincter, located between the stomach and the duodenum (the first part of the small intestine), controls the flow of chyme into the small intestine. Only 1 teaspoon of chyme is released at a time into the small intestine. **Glucose-dependent insulinotropic peptide** (formerly called gastric inhibitory peptide), a hormone, helps slow the release of chyme into the small intestine, giving the small intestine time to neutralize the acid and digest the nutrients. The pyloric sphincter also prevents the backflow of bile into the stomach (bile, discussed later in the chapter, can damage the stomach lining). It typically takes 1 to 4 hours for meals to move out of the stomach into the small intestine; less time is needed when meals are mostly liquid, more time when meals are large and high in fat.

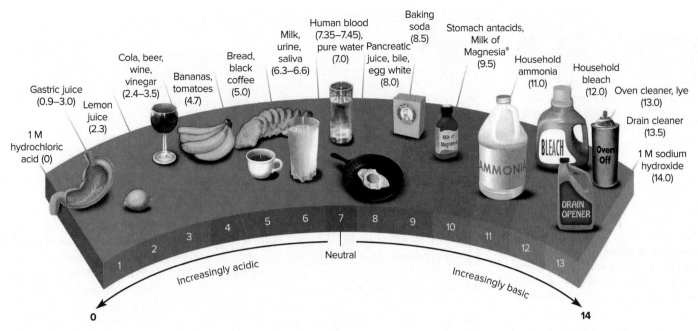

*Figure 4-10*  The approximate pH of various substances. The pH scale ranges from 0 (most acidic) to 14 (most basic).

Another important function of the stomach is the production of intrinsic factor (IF) and ghrelin. IF is required for the absorption of vitamin B-12 in the small intestine (discussed further in the Historical Perspective and Part 4). The hormone ghrelin plays a role in the short-term regulation of food intake by increasing appetite and food intake. Ghrelin concentrations increase before meals and decline after eating.

**chyme**  Liquid mixture of stomach secretions and partially digested food.

**glucose-dependent insulinotropic peptide**  Hormone that slows gastric motility and stimulates insulin release from the pancreas; formerly known as gastric inhibitory peptide.

*Figure 4-11*  Physiology of the stomach. Surface mucous cells produce mucus for protection from stomach acid and enzymes. Parietal cells produce the hydrochloric acid (HCl), and chief cells produce the enzymes. Mucous neck cells, scattered among the cells in the gastric pits, also produce mucus.

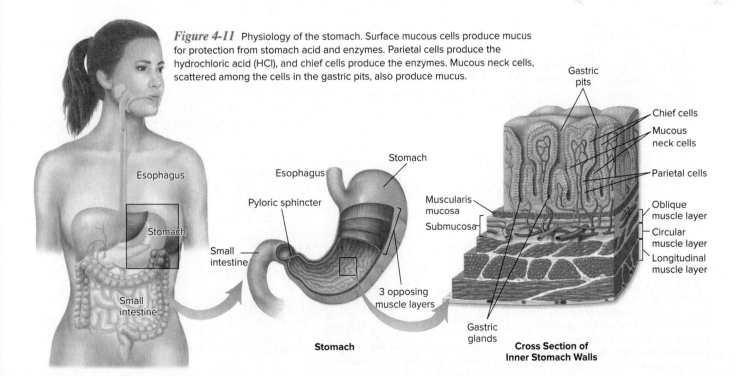

**HISTORICAL PERSPECTIVE**

**William Castle and Intrinsic Factor**

Pernicious anemia was a painful and nearly always fatal disease until a series of discoveries illuminated how poor absorption of vitamin B-12 causes the disease. One of these discoveries was in 1927 by a young physician, William Castle, who hypothesized that individuals with pernicious anemia lacked a vital stomach secretion. Castle tested his theory in an unusual way.[7] He consumed 300 g (2/3 lb) of raw ground beef (a source of vitamin B-12) each morning, waited an hour for the beef to mix with gastric juices, regurgitated the now semi-liquid mixture, and incubated it for several hours. He then infused it via a tube into the stomachs of patients suffering from pernicious anemia. In his experiments, Castle discovered that only the beef and gastric juice mixture cured the pernicious anemia—gastric juice and ground meat each given alone (another part of his experiment) did not improve the condition. Castle called the unknown essential gastric secretion "intrinsic factor." Today, it's known that vitamin B-12 absorption depends on the intrinsic factor protein secreted by parietal cells in the stomach.

---

**microvilli** Microscopic, hairlike projections of cell membranes of certain epithelial cells.

**glycocalyx** Projections of proteins on the microvilli. They contain enzymes to digest protein and carbohydrate.

---

### Knowledge Check

1. What are the components of gastric juice?
2. How do HCl and the enzyme pepsin aid in digestion?
3. Describe the location and function of the pyloric sphincter.

## 4.5  Moving through the GI Tract: Small Intestine and Accessory Organs

The small intestine is the major site of digestion and absorption of food. It is coiled below the stomach in the abdomen (Fig. 4-12). The small intestine is divided into 3 sections: the first part, the **duodenum**, is about 10 inches (25 cm) long; the middle segment, the **jejunum**, is about 4 feet (122 cm) long; and the last section, the **ileum**, is about 5 feet (152 cm) long. The small intestine is considered small because of its narrow, 1-inch (2.5 cm) diameter, not its length.

The interior of the small intestine has circular folds and fingerlike projections (**villi** and **microvilli**) that increase its surface area 600 times over that of a smooth tube. This large surface area contributes to the thoroughness and efficiency of digestion and absorption. The **circular folds** make the chyme flow slowly, following a spiral path as it travels through the small intestine. Slow spiraling and segmentation movements completely mix the chyme with digestive juices and bring it in contact with the villi that extend into the lumen (Fig. 4-13). Villi are lined with goblet cells that make mucus, endocrine cells that produce hormones and hormonelike substances, and cells that produce digestive enzymes and absorb nutrients (**enterocytes**). Each enterocyte has a brush border made up of microvilli that are covered with the digestive enzyme–containing **glycocalyx**. The villi and microvilli make the small intestine's interior look fuzzy, like terrycloth or velvet (Fig. 4-14).

Most digestion in the small intestine occurs in the duodenum and upper part of the jejunum and requires many secretions from the small intestine itself, as well as the pancreas, liver, and gallbladder. (Table 4-3 in Section 4.2 reviews these secretions and their functions.) Each day, the small intestine secretes about 6 cups (1.5 liters) of mucus, enzyme, and hormone-containing fluid. Enzymes produced in the small intestine, also known as brush border enzymes,

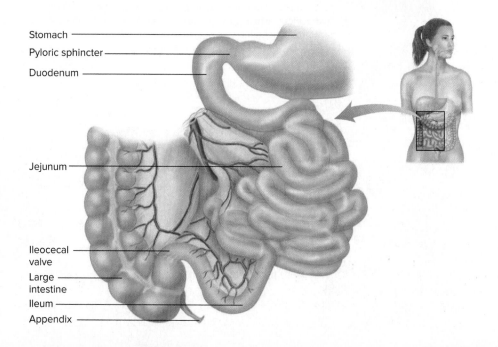

*Figure 4-12*  The small intestine and beginning of the large intestine. The 3 parts of the small intestine are the duodenum, jejunum, and ileum. Notice the smaller diameter of the small intestine, compared with the large intestine.

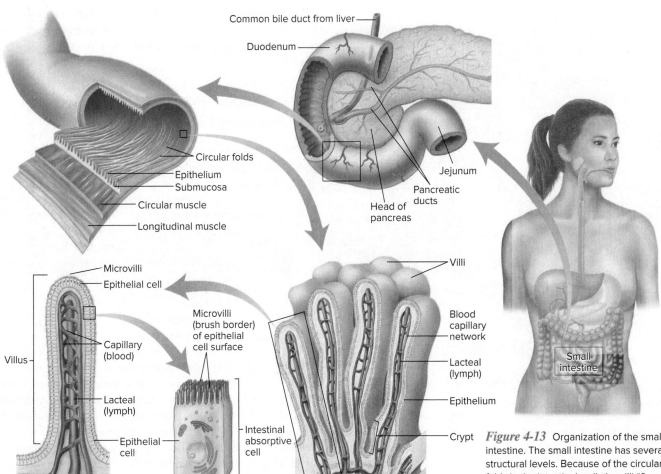

**Figure 4-13** Organization of the small intestine. The small intestine has several structural levels. Because of the circular folds in the intestinal wall, the villi "fingers" that project into the intestine, and the microvilli (brush border) on each absorptive cell that makes up the villi, the surface area for absorption is up to 600 times that of a smooth tube.

are responsible for the chemical digestion of the macronutrients. They typically complete the last steps of digestion, resulting in compounds that are small enough to be absorbed.

## Liver, Gallbladder, and Pancreas

The liver, gallbladder, and pancreas, known as the accessory organs of the digestive system, work with the small intestine but are not a physical part of it. Secretions from these organs are delivered through the common bile duct and the pancreatic duct. These ducts come together at the hepatopancreatic sphincter (also called the sphincter of Oddi) and empty into the duodenum (Fig. 4-15).

The liver produces bile, a cholesterol-containing, yellow-green fluid that aids in fat digestion and absorption. Bile emulsifies fat—it breaks the large fat globules into micelles, tiny fat droplets that are suspended in the watery chyme. The liver secretes about 2 to 4 cups (500 to 1000 ml) of bile per day. Bile is concentrated and stored in the gallbladder until needed. Bile released into the duodenum is reabsorbed in the ileum and returned to the liver. During a meal, bile is recirculated 2 or more times. This system of bile recycling is called the **enterohepatic circulation**. A small amount of bile is not reabsorbed and is excreted in feces—this is how cholesterol, 1 of the components of bile, is removed from the body.

The pancreas produces about 5 to 6 cups (1.5 liters) of pancreatic juice per day. This juice is an alkaline (basic) mixture of sodium bicarbonate ($NaHCO_3$) and enzymes. The sodium bicarbonate neutralizes the acidic chyme arriving from the stomach, thereby protecting the small intestine from damage by acid. Digestive enzymes from the pancreas include

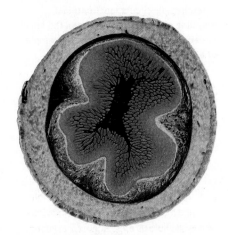

**Figure 4-14** This enhanced photo shows the entire width of the small intestine. The black center is the lumen. Protruding into the lumen are the fingerlike villi (colored in red). The villi greatly increase the surface area for absorption. The thick, outer, yellow layer is the circular and longitudinal muscle that pushes the chyme through the intestinal tract.

Steve Gschmeissner/Science Source

pancreatic amylase (to digest starch), pancreatic lipase (to digest fat), and several proteases (to digest protein). Pancreatic enzymes break large macronutrient molecules into smaller subunits.

## Gastrointestinal Hormones: A Key to Orchestrating Digestion

The remarkable work of the digestive system requires the careful regulation and coordination of several processes, including the production and release of hormones throughout the length of the GI tract. Five hormones, part of the endocrine system, play key roles in this regulation: gastrin, secretin, cholecystokinin (CCK), somatostatin, and glucose-dependent insulinotropic peptide (Table 4-4). To illustrate their functions, let's follow a turkey sandwich through the digestive system.

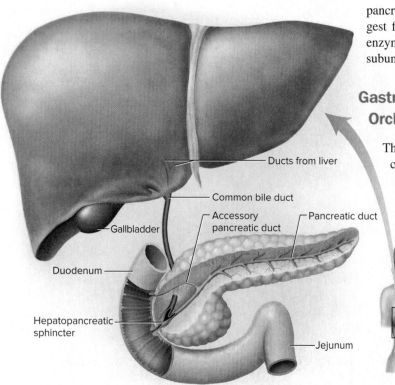

*Figure 4-15* The common bile duct from the liver and gallbladder and the pancreatic duct join together at the hepatopancreatic sphincter to deliver bile, pancreatic enzymes, and bicarbonate to the duodenum.

1. As you eat a turkey sandwich (or even just think about it), gastrin is produced by cells in the stomach. Gastrin signals other stomach cells to release HCl and pepsinogen (for protein digestion). After thorough mixing, the turkey sandwich, now liquid chyme, is released in small amounts into the small intestine.

2. As chyme is gradually released from the stomach into the small intestine, gastrin production slows and the small intestine secretes secretin and CCK. Both hormones trigger the release of enzyme- and bicarbonate-containing pancreatic juices that digest carbohydrate, fat, and protein and reduce the acidity of the intestinal contents. Fat in the small intestine (from the mayonnaise and the turkey) further stimulates the secretion of CCK by the small intestine. CCK promotes contraction of the gallbladder, which releases the stored bile that aids fat digestion. Relaxation of the hepatopancreatic sphincter allows bile and pancreatic juices to flow into the small intestine. CCK also slows GI motility to give digestive enzymes from the small intestine and pancreas enough time to do their work.

**enterohepatic circulation**   Continual recycling of compounds between the small intestine and the liver. Bile is an example of a recycled compound.

▶ In cystic fibrosis, the most common lethal genetic disease in White populations, thick, sticky mucus builds up in organs, especially the lungs and pancreas. This buildup prevents pancreatic enzymes from reaching the small intestine. If the condition is not treated, malabsorption of nutrients, weight loss, and malnutrition occur. Most individuals with cystic fibrosis are diagnosed in infancy. Most are prescribed pancreatic enzyme replacements, which are taken right before eating. An enteric coating protects the enzymes from destruction by stomach acid. Maintaining good nutritional status helps achieve normal growth and development and prevent the respiratory infections common in those with cystic fibrosis.

## CASE STUDY

Photodisc/Getty Images

Elise is a 20-year-old college sophomore. Over the last few months, she has been experiencing regular bouts of esophageal burning, pain, and a sour taste in the back of her mouth. This usually happens after a large lunch or dinner. Elise often takes an over-the-counter antacid to relieve these unpleasant symptoms. However, the symptoms have worsened and Elise has decided to visit the university health center.

The nurse practitioner at the center tells Elise it is good she came in for a checkup because she suspects Elise is experiencing heartburn and acid indigestion, but she might also be experiencing gastroesophageal reflux disease (GERD). What types of lifestyle and dietary changes may help reduce or prevent heartburn and GERD? What types of medications are especially helpful in treating this problem?

3. The sandwich becomes progressively digested and absorbed. The small intestine now releases somatostatin and glucose-dependent insulinotropic peptide. These hormones signal the stomach to slow motility and decrease the release of gastric juice.

Many other hormones, synthesized throughout the GI tract and in the brain and pancreas, contribute to the regulation of digestion and absorption. Some of these hormones are listed in Table 4-4.

**Table 4-4 Major Regulatory Hormones of the GI Tract\***

| Hormone | Released By | Functions |
|---------|-------------|-----------|
| Ghrelin | Stomach | Increases appetite and food intake |
| Gastrin | Stomach and duodenum in response to food reaching the stomach | Triggers the stomach to release HCl and pepsinogen; stimulates gastric and intestinal motility |
| Cholecystokinin (CCK) | Small intestine in response to dietary fat in chyme | Stimulates release of pancreatic enzymes and bile from the gallbladder |
| Secretin | Small intestine in response to acidic chyme<br>Small intestine as digestion progresses | Stimulates release of pancreatic bicarbonate |
| Motilin | Small intestine in response to gastric distension and dietary fat | Regulates motility of the gastrointestinal tract |
| Glucose-dependent insulinotropic peptide (GIP) | Small intestine in response to glucose, amino acids, and fat | Inhibits gastric acid secretion, stimulates insulin release |
| Peptide YY | Ileum and large intestine in response to fat in the large intestine | Inhibits gastric and pancreatic secretions |
| Somatostatin | Stomach, small intestine, and pancreas | Inhibits release of GI hormones; slows gastric emptying, GI motility, and blood flow to the intestine |

*\*At least 20 hormones are synthesized and released in the gastrointestinal tract.*

## Absorption in the Small Intestine

The absorptive cells of the small intestine originate in crypts (open-ended pits) located at the base of the villi. The absorptive cells migrate from the crypts to the villi. As they migrate, absorptive cells mature and their absorptive capabilities increase. By the time they reach the tips of the villi, they have been partially destroyed by digestive enzymes and have shed into the lumen. The body's entire supply of absorptive cells is replaced every 2 to 5 days.

The digestive capabilities and health of the small intestine rapidly deteriorate during a nutrient deficiency or in semistarvation. This is because cells that turn over rapidly, such as absorptive cells, depend on a constant supply of nutrients. These nutrients are provided by the diet, as well as from broken-down cell parts that are recycled.

Most nutrient absorption occurs in the small intestine (Fig. 4-16). The small intestine absorbs about 95% of the digested protein, carbohydrate (not including dietary fiber), fat, and alcohol consumed. Nutrients move from the lumen of the small intestine into the absorptive cells in the ways illustrated in Figure 4-17.

- **Passive diffusion**: When the concentration of a nutrient is higher in the lumen of the small intestine than in the absorptive cells, the difference in concentration, known as the concentration gradient, forces the nutrient into the absorptive cells. Fats, water, and some minerals are absorbed by passive diffusion.
- **Facilitated diffusion**: A higher concentration of a nutrient in the lumen than in the absorptive cells is not enough to move some nutrients into the absorptive cells. They need carrier proteins to shuttle them from the lumen into absorptive cells. For instance, the sugar fructose is absorbed by facilitated diffusion.
- **Active absorption**: In addition to the need for a carrier protein, some nutrients also require energy (ATP) for absorption. Active absorption, also known as active transport, allows the cell to concentrate nutrients on either side of the cell membrane. Amino acids and some sugars, such as glucose, are actively absorbed.
- **Endocytosis (phagocytosis/pinocytosis)**: In this type of active absorption, absorptive cells engulf compounds (in phagocytosis) or liquids (in pinocytosis). In both these processes, an absorptive cell forms an invagination in its cell membrane that engulfs the particles or fluid to form a vesicle. The vesicle is finally pinched off from the cell membrane and taken into the cell. This process allows immune substances (large protein particles) in human breast milk to be absorbed by infants.

*Critical* **Thinking**

Cancer treatments often involve the use of chemotherapy medications to prevent rapid cell production and growth. Cancer cells are the intended target. Diarrhea is a common side effect of chemotherapy. Why does chemotherapy often cause diarrhea?

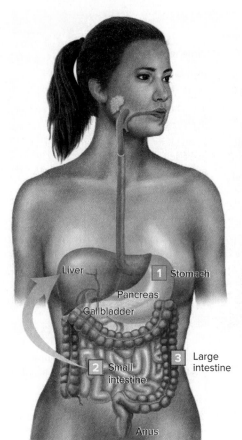

**Figure 4-16** Major sites of absorption along the GI tract. Note that some synthesis and absorption of vitamin K and biotin take place in the large intestine.

| Organ | Primary Nutrients Absorbed |
|---|---|
| **1 Stomach** | Alcohol (20% of total)<br>Water (minor amount) |
| **2 Small intestine** | Calcium, magnesium, iron, and other minerals<br>Glucose<br>Amino acids<br>Fats<br>Vitamins<br>Water (70 to 90% of total)<br>Alcohol (80% of total)<br>Bile acids |
| **3 Large intestine** | Sodium<br>Potassium<br>Some fatty acids<br>Vitamin K and biotin (synthesized by microorganisms in the large intestine)<br>Gases<br>Water (10 to 30% of total) |

**Figure 4-17** Nutrient absorption relies on 4 major absorptive processes. **1** Passive diffusion (in green) is diffusion of nutrients across the absorptive cell membranes. **2** Facilitated diffusion (in blue) uses a carrier protein to move nutrients down a concentration gradient. **3** Active absorption (in red) involves a carrier protein as well as energy to move nutrients (against a concentration gradient) into absorptive cells. **4** Phagocytosis and pinocytosis (in gray and orange) are forms of active transport in which the absorptive cell membrane forms an invagination that engulfs a nutrient to bring it into the cell.

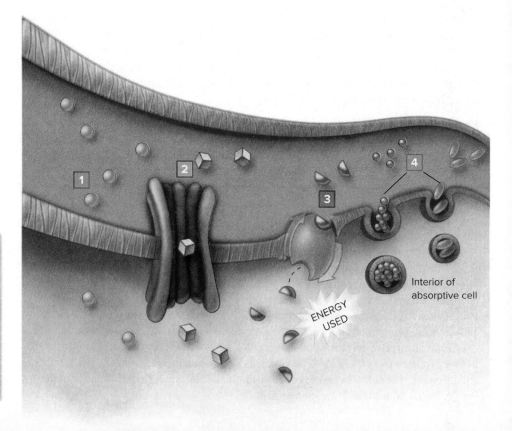

## Critical Thinking

The medical history of a young girl who is greatly underweight shows that she had three-quarters of her small intestine removed after she was injured in a car accident. Explain how this accounts for her underweight condition, even though her medical chart shows that she eats well.

# GLBAL PERSPECTIVE

## Diarrhea in Infants and Children

Diarrhea is rarely considered a serious threat to young children in countries such as the U.S. and Canada. However, in developing countries, diarrhea is a leading killer of children—it is the second leading cause of death in children less than age 5 years. In fact, more children die from diarrhea each year in developing countries than from malaria, measles, or HIV/AIDS.[8] Diarrhea in young children is typically caused by pathogenic microorganisms— viruses, bacteria, and parasites—found in water, food, and human and animal waste. One of the most common causes of severe diarrhea in young children around the world is rotavirus.[9] Like many other microbial pathogens, rotavirus replicates rapidly in the epithelial cells of the intestinal mucosa. Toxins produced by the virus cause the epithelial cells to slough off faster than they can be replaced. Fluid and **electrolytes**, which normally would have been absorbed in the intestine, are excreted rapidly. Infants and young children can become dangerously dehydrated very quickly. Death occurs if fluids are not replaced.

Friedrich Stark/Alamy Stock Photo

As discussed in Chapter 3, malnutrition afflicts many children around the world. Malnutrition increases susceptibility to diarrhea in several ways. The intestinal mucosa of a malnourished child can become thin, damaged, and leaky, allowing pathogens to invade more easily. Additionally, immune function declines in malnourished children. Repeated bouts of diarrhea can worsen malnutrition due to decreased food intake and poor absorption during illness.

**electrolytes** Compounds that separate into ions in water and, in turn, are able to conduct an electrical current. These include sodium, chloride, and potassium.

To prevent severe illness and death, it is vital that children with diarrhea be treated with oral rehydration therapy that provides oral rehydration salts (small amounts of the electrolytes sodium, chloride, and potassium) and the sugar glucose dissolved in water.[10] This simple recipe helped decrease the number of diarrhea-related deaths from 4.5 million in 1979 to 578,000 in 2013. Supplemental zinc (an essential mineral) can shorten the duration and severity of a bout with diarrhea.[11] Children treated with supplemental zinc are less likely to suffer from diarrhea and, if they do have diarrhea, it is less severe. Another promising advance in the prevention of diarrhea is the rotavirus vaccine, which substantially reduces rotaviral disease.[9]

Although life-threatening diarrhea can be successfully treated in many cases, prevention of diarrhea is critical. Keys to prevention include safe water and improved sanitation, handwashing with soap, breastfeeding, and immunization, along with the ready availability of affordable, nutritious foods that promote good health. Unfortunately, access to these basic necessities is limited for many.

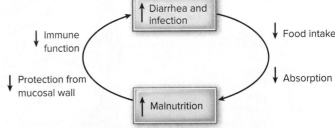

## Knowledge Check

1. What are the 3 sections of the small intestine?
2. Where is bile synthesized, and what is its function?
3. What is the role of the pancreas in digestion?
4. Which type of absorption requires energy?
5. Why is diarrhea life threatening for many young children in developing countries?

The GI tract digests the foods eaten. Despite what you may have heard, the order in which foods are eaten has no effect on digestive processes.

Jules Frazier/Getty Images

## 4.6 Moving Nutrients around the Body: Circulatory Systems

Nutrients absorbed in the small intestine are delivered to 1 of the body's 2 circulatory systems: the cardiovascular (blood) system and the **lymphatic system** (Fig. 4-18). The choice of system used to transport nutrients is based primarily on whether the nutrients are water or fat soluble.

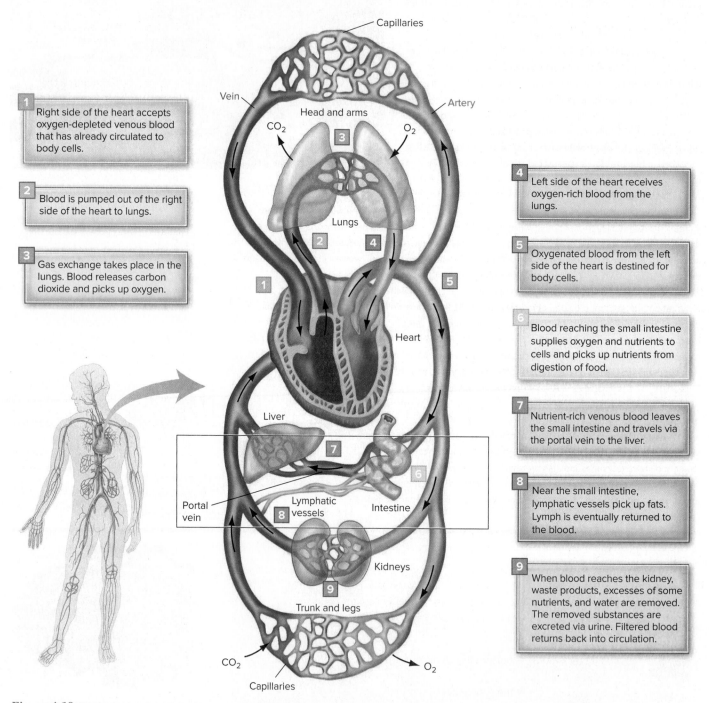

**1** Right side of the heart accepts oxygen-depleted venous blood that has already circulated to body cells.

**2** Blood is pumped out of the right side of the heart to lungs.

**3** Gas exchange takes place in the lungs. Blood releases carbon dioxide and picks up oxygen.

**4** Left side of the heart receives oxygen-rich blood from the lungs.

**5** Oxygenated blood from the left side of the heart is destined for body cells.

**6** Blood reaching the small intestine supplies oxygen and nutrients to cells and picks up nutrients from digestion of food.

**7** Nutrient-rich venous blood leaves the small intestine and travels via the portal vein to the liver.

**8** Near the small intestine, lymphatic vessels pick up fats. Lymph is eventually returned to the blood.

**9** When blood reaches the kidney, waste products, excesses of some nutrients, and water are removed. The removed substances are excreted via urine. Filtered blood returns back into circulation.

*Figure 4-18* Blood circulation through the body. This figure shows the paths that blood takes from the heart to the lungs (**1**–**7**), back to the heart (**4**), and through the rest of the body (**5**–**9**). The reddish-orange color indicates blood that is richer in oxygen; blue is for blood carrying more carbon dioxide. Keep in mind that arteries and veins go to all parts of the body. Pay particular attention to sites **7** and **8**. These sites are key parts of the process of nutrient absorption.

## Cardiovascular System

The cardiovascular system includes the heart, blood vessels (arteries, capillaries, veins), and blood. Water-soluble nutrients (proteins, carbohydrates, B-vitamins, and vitamin C) and **short-** and **medium-chain fatty acids** are transported by the cardiovascular system. These nutrients are absorbed directly into the bloodstream in the **capillary** beds inside the villi (see Fig. 4-13). Blood flows from the capillary beds into the **hepatic portal vein system** and collects in the large hepatic portal vein, which leads directly to the liver. The liver metabolizes or stores a portion of the absorbed nutrients, especially protein, lipids, glucose, and several vitamins and minerals. Nutrients not utilized or stored in the liver enter the general circulation. This nutrient-rich blood circulates throughout the body, delivering nutrients to all cells, where they are used for energy, growth, development, maintenance of tissues, and regulation of body processes. The carbon dioxide and other waste products produced by these processes are released into the blood and excreted by the lungs and kidneys.

## Lymphatic System

The lymphatic system contains lymph, which flows throughout the body in lymphatic vessels, which are similar to veins. Unlike blood, lymph is not pumped through the vessels. Instead, it slowly flows as muscles contract and squeeze the lymphatic vessels.

The lymphatic system provides an alternative route into the bloodstream for large molecules that cannot be absorbed by the capillary beds. Fat-soluble nutrients (most fats and the fat-soluble vitamins A, D, E, and K) and other substances, such as some proteins, are transported in lymph. Usually a clear fluid, lymph looks milky when it leaves the small intestine because of its fat content. Special lymphatic vessels (**lacteals**) in the villi transport nutrients to larger lymphatic vessels that connect to the thoracic duct. The thoracic duct extends from the abdomen to the neck, where it connects to the bloodstream at a large vein called the left subclavian vein. Once in the blood, nutrients originally absorbed by the lymphatic system are transported to body tissues in the cardiovascular system.

**short-chain fatty acid** Fatty acid that contains fewer than 6 carbon atoms.

**medium-chain fatty acid** Fatty acid that contains 6 to 10 carbons.

**capillary** Smallest blood vessel; the major site for the exchange of substances between the blood and the tissues.

**hepatic portal vein system** Veins leaving from the stomach, intestines, spleen, and pancreas that drain into the hepatic portal vein, which flows into the liver.

**lacteal** Tiny vessels in the small intestine villi that absorb dietary fat.

### Knowledge Check

1. What are 3 nutrients that are transported by the cardiovascular system?
2. What are 3 nutrients that are transported first in the lymphatic system?
3. Which organ first receives nutrients from the cardiovascular system?

## 4.7 Moving through the GI Tract: Large Intestine

The small intestine empties into the large intestine through the ileocecal valve, the sphincter between the ileum and the colon. After digestion and absorption in the small intestine, normally only water, some minerals, and undigested food fibers and starches are left. About 5% of carbohydrate, protein, and fat escapes absorption in the small intestine.

The large intestine, so called because its $2^1/_2$-inch (6-centimeter) lumen diameter is larger than that of the small intestine, is about 5 feet (1.5 meters) long. It has 3 main parts: the colon, rectum, and anus. The colon, the largest portion of the large intestine, has 5 sections: cecum, ascending colon, transverse colon, descending colon, and sigmoid colon (Fig. 4-19).

The large intestine performs 3 main functions. It houses gut microbiota that keep the GI tract healthy; it absorbs water and electrolytes, such as sodium and potassium; and it forms and expels feces.

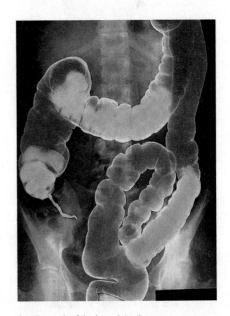

A radiograph of the large intestine.

CNRI/Science Source

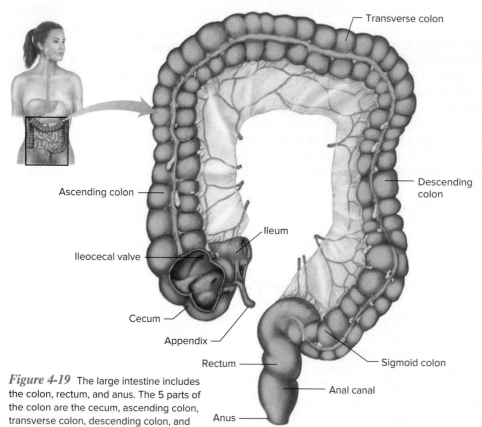

***Figure 4-19*** The large intestine includes the colon, rectum, and anus. The 5 parts of the colon are the cecum, ascending colon, transverse colon, descending colon, and sigmoid colon.

Yogurt is a convenient source of probiotic bacteria, which contribute to GI tract health.

## Gut Microbiota

The GI tract is home to hundreds of microbial species (mostly bacteria), collectively numbering more than 100 trillion microbial cells, or more than 10 times the number of cells (and 100 times the number of genes) in the human body![12] Most of these microorganisms reside in the large intestine, but some are found in the mouth, stomach, and small intestine. The ileocecal valve prevents microbes in the large intestine from migrating into the small intestine (where they could disrupt normal function and compete with the body for nutrients). Like other internal organs, the pancreas, gallbladder, and liver are microbe free.

Scientists now realize that the body's microorganisms are not merely idle passengers, but that they contribute significantly to health. The bacteria in the large intestine do this by

- Protecting against infection by pathogens by
  - Producing antimicrobial substances
  - Crowding out pathogens
  - Contributing to the health of the epithelial barrier by providing nutrients for cells
- Synthesizing vitamins, especially vitamin K and biotin, that can be absorbed in the colon
- Digesting and metabolizing complex carbohydrates, fibers, and starches and forming compounds such as beneficial short-chain fatty acids
- Modulating inflammation (positively and negatively) with their metabolic products

The bacterial composition of the large intestine varies from person to person and changes throughout the life span.[12,13] Factors that influence an individual's large intestinal microbiota are summarized in Figure 4-20. In healthy individuals, diet has the largest influence on the gut microbial composition and metabolic pathways. High-fiber, plant-based diets are thought to increase microbial diversity and numbers. Such diets also promote the production of beneficial short-chain fatty acids and may protect against colorectal cancer.[14]

Antibiotic treatment, radiation therapy, surgery, and some diseases often reduce the number of beneficial bacteria cells, which can allow pathogenic bacteria to multiply quickly. Disrupting the normal balance between beneficial and pathogenic bacteria, termed **dysbiosis**, can cause gastrointestinal infections (and vomiting, cramping, pain, diarrhea, and dehydration).[15] Researchers are intensely studying how intestinal microbiota and dysbiosis, fiber, and metabolites of beneficial intestinal bacteria, such as short-chain fatty acids, may prevent or treat disorders (e.g., gastrointestinal disorders, such as gastrointestinal infections, diarrhea, irritable bowel syndrome, colon cancer, inflammatory bowel disease, and liver disease).[12,13,16] Other disorders, including obesity, type 2 diabetes, rheumatoid arthritis, eczema, and psychiatric disorders, also are being studied. Treatments already include **fecal transplants** to treat colitis and other diseases for which patients have received high doses of antibiotics that greatly reduced their intestinal microbiota. Future therapies may include banking one's microbiota before undergoing chemotherapy.

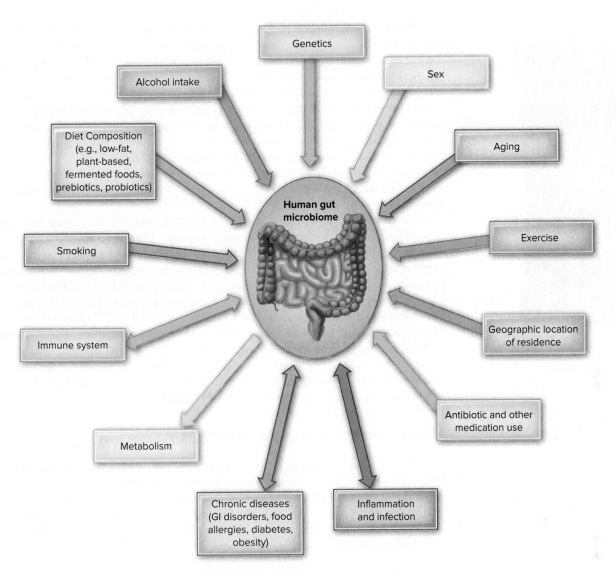

*Figure 4-20* The adult human gut microbiome composition is remarkably complex and variable. Its composition can be influenced by many genetic, environmental, and health-related factors. Dietary factors that appear to increase microbial diversity include the consumption of probiotics, such as yogurt, plant foods (sources of dietary fiber and prebiotics), coffee, tea, and red wine. Dietary factors that appear to decrease microbial diversity include the consumption of sugar-sweetened beverages and refined carbohydrates, a high calorie intake, and frequent snacking.

## Probiotics and Prebiotics

One proposed strategy for achieving a healthy balance of intestinal bacteria is the consumption of probiotics and prebiotics. **Probiotics** are live microorganisms that, when consumed in sufficient amounts, colonize the large intestine and provide health benefits.[17] Probiotics are found in fermented foods, such as yogurt and miso, and are sold as dietary supplements in powder, liquid, and capsule form. Common probiotic preparations include various strains of the bacteria *Lactobacilli* and *Bifidobacteria* and the yeast *Saccharomyces*; others may contain strains of *Enterococcus*, *Bacillus*, and *Escherichia* bacteria.[17]

Probiotic foods and supplements are thought to provide certain health benefits. Research supporting many probiotic claims is lacking because there are so many types and doses of these microorganisms to test.[18] A lack of well-controlled, randomized clinical trials also makes it challenging to clearly delineate which strains and dosages of probiotics are best for treating conditions.[19] Another challenge is that an individual's preexisting gut microbiome may determine whether probiotics are able to colonize the lining of the intestine, the area where the benefits of probiotics are conferred.

**probiotic** Live microorganisms that, when ingested in adequate amounts, confer a health benefit on the host.

**prebiotics** Substance that stimulates bacterial growth in the large intestine.

Inulin, a prebiotic, is found in asparagus.

Brand X Pictures/Getty Images

Recent research indicates that probiotic dietary supplements may prevent the bowel damage common in premature infants.[20] Probiotics also may help prevent and/or treat digestive disorders, such as antibiotic-associated diarrhea, acute infectious diarrhea (e.g., traveler's diarrhea), ulcerative colitis, and irritable bowel syndrome (see Section 4.8);[17,21,22,23] however, some research shows that using probiotics after antibiotic treatment (currently a common practice) may actually delay the normal recovery of intestinal microbiota. Thus, support for the universal use of probiotics to treat intestinal conditions is lacking.[18] Currently, scientists hypothesize that the most effective use of probiotics to positively manipulate the gut microbiome likely will be for personalized therapy.[24,25] Until more is clearly known, the American Gastroenterological Association advises against the indiscriminate use of probiotics and recommends that individuals speak with their physicians before trying these supplements.[26]

**Prebiotics** are nondigestible food ingredients that promote the growth of beneficial bacteria in the large intestine.[27] One example is **inulin**, a carbohydrate known as a fructan because it is made of several units of the sugar fructose. Inulin is found in many foods, including chicory, wheat, onions, garlic, asparagus, and bananas. Inulin also is added to some processed foods to add texture, bulk, and potential health benefits. **Resistant starch**, found in unprocessed whole grains, seeds, legumes, unripe fruit (e.g., bananas), and cooked and chilled pasta, potatoes, and rice, also functions as a prebiotic.[27] Resistant starch is not digested in the small intestine; thus, bacteria in the large intestine can ferment it. Prebiotics fermented in the large intestine produce short-chain fatty acids and other organic acids. In studies of prebiotics, participants typically ingest 10 to 20 grams per day; such large amounts can cause flatulence, bloating, and other GI distress. As with probiotics, the research that prebiotics improve health is not yet conclusive.

## Absorption of Water and Electrolytes

The GI tract receives a total of 10 liters of water per day (3 liters from the diet and 7 liters from intestinal secretions). The small intestine absorbs about 90% of the water, and the large intestine completes the job. Just 1% (less than 1/2 cup, or 120 ml) of the water in the GI tract remains in excreted feces. The large intestine also is the main site where electrolytes, especially sodium and potassium, are absorbed (see Fig. 4-16). Electrolyte absorption occurs mostly in the first half of the large intestine.

## Elimination of Feces

It takes 12 to 24 hours for the residue of a meal to travel through the large intestine. By the time the contents have passed through the first two-thirds of its length, a semi-solid mass has been formed. This mass remains in the large intestine until peristaltic waves and mass movements, usually greatest following the consumption of a meal, push it into the rectum. Feces in the rectum are a powerful stimulation for defecation, the expulsion of feces. This process involves muscular reflexes in the sigmoid colon and rectum, as well as relaxation of the internal and external anal sphincters. Only the external sphincter is under voluntary control. Once toilet-trained, a person can determine when to relax the sphincter for defecation, as well as when to keep it constricted.

When excreted, feces are normally about 75% water and 25% solids. The solids are primarily indigestible plant fibers, tough connective tissue from animal foods, and bacteria from the large intestine. During episodes of diarrhea, the percentage of water in feces rises.

*Knowledge Check*

1. What are the 3 main functions of the large intestine?
2. What are some of the beneficial actions of bacteria in the large intestine?
3. What is the difference between a prebiotic and a probiotic? Where can they be found in the diet?

# 4.8 When Digestive Processes Go Awry

The fine-tuned organ system we call the digestive system can develop problems. Knowing about common problems can help you avoid or lessen them.

## Heartburn and Gastroesophageal Reflux Disease

About half of U.S. adults occasionally experience heartburn (acid indigestion), making this the most commonly diagnosed GI disorder in U.S. adults. Heartburn has nothing to do with the heart; it occurs when stomach acid backs up into the esophagus (Fig. 4-21), causing a burning sensation or sour taste in the back of the mouth. Experiencing heartburn 2 or more times per week may signal the more serious gastroesophageal reflux disease (GERD).[28] GERD occurs when the lower esophageal sphincter relaxes and lets stomach contents backflow into the esophagus. (Normally, this sphincter relaxes only with swallowing.) Not everyone with GERD has heartburn—other symptoms include hoarseness, trouble swallowing, coughing, gagging, and nausea. In addition to the uncomfortable physical symptoms of GERD, more serious complications can occur. These include weight loss, esophageal ulceration, narrowing of the esophagus, bleeding in the esophagus, **anemia**, and a higher risk of adenocarcinoma of the esophagus, a cancer that has a poor prognosis. Based on observations using an **endoscope**, there are 2 types of GERD: erosive reflux disease (ERD) and non-erosive reflux disease (NERD).[29] Reflux and GERD can occur in infants and children, too, usually due to an immature digestive system. It can cause frequent spitting up, or vomiting, and coughing. Most children outgrow this by 1 year of age.

The cause of GERD is not known, but factors that may contribute to it include **hiatal hernia**, alcohol use, overweight, smoking, certain medications, and even pregnancy. Studies have shown that obesity slows stomach emptying and relaxes the lower esophageal sphincter.[30] Large meals and foods such as citrus fruits, chocolate, caffeinated drinks (e.g., coffee), fatty and fried foods, garlic, onion, spicy foods, and tomato-based foods (e.g., spaghetti sauce and pizza) may increase reflux. There also can be a close relationship between the brain and the gastrointestinal system; studies have demonstrated that higher levels of anxiety and depression are linked with increased risk of GERD, especially NERD.[29,31] Anxiety may increase acid reflux by reducing the pressure of the lower esophageal sphincter, altering esophageal motility, and/or increasing secretion of gastric acid.[32]

Heartburn and GERD are treated with both lifestyle modification and medications.[34] Lifestyle change recommendations include managing stress/anxiety (e.g., meditation, deep breathing, adequate sleep, daily exercise), eating small meals instead of large ones, avoiding foods that cause reflux, waiting several hours before lying down after eating (remaining upright limits reflux), losing weight, stopping smoking, and limiting alcohol intake. The following medications are used to treat GERD.

- Antacids (Tums®, Maalox®) are over-the-counter medications that neutralize stomach acid. Excessive intake of those that contain magnesium can cause diarrhea, and those that contain aluminum or calcium may cause constipation.
- $H_2$ blockers (cimetidine [Tagamet®], famotidine [Pepcid AC®]) block the increase of stomach acid production caused by histamine. Histamine, a breakdown product of the amino acid histidine, stimulates acid secretion by the stomach and has many other effects on the body. $H_2$ blockers are available in both prescription and less potent, nonprescription forms.
- Proton pump inhibitors (esomeprazole [Nexium®], lansoprazole [Prevacid®], rabeprazole [Aciphex®], pantoprazole) are the most potent acid-suppressing medications. They inhibit the ability of gastric cells to secrete hydrogen ions and make acid. Low doses of this class of medication also are available without prescription, such as omeprazole (Prilosec-OTC®).
- Prokinetic drugs (metoclopromide [Reglan®]) strengthen the lower esophageal sphincter and promote more rapid peristalsis in the small intestine.

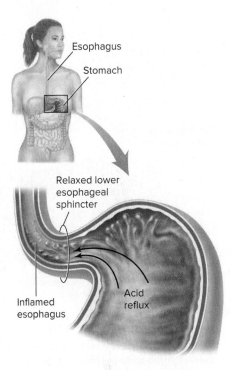

*Figure 4-21* Heartburn results from stomach acid refluxing into the esophagus.

▶ Many individuals with heartburn take a class of drugs that suppress HCl production in the stomach. Long-term use of these medications may increase the risk of osteoporosis and bone fracture. With a decrease in stomach acidity, both calcium and magnesium are less well dissolved and their absorption can decrease, reducing bone mineralization processes.[33]

**anemia** Decreased oxygen-carrying capacity of the blood. Can be caused by many factors, such as iron deficiency or blood loss.

**hiatal hernia** Protrusion of part of the stomach upward through the diaphragm into the chest cavity.

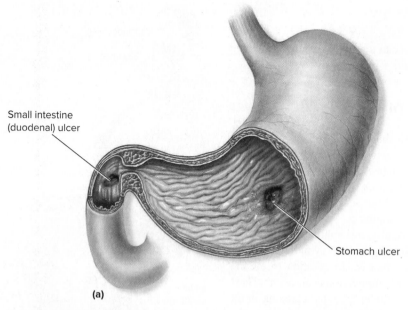

Small intestine
(duodenal) ulcer

Stomach ulcer

**(a)**

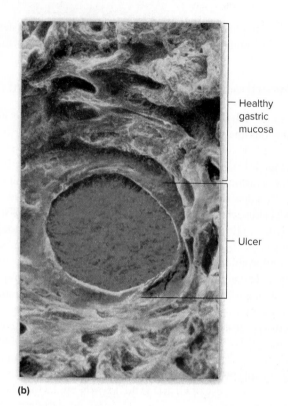

Healthy
gastric
mucosa

Ulcer

**(b)**

*Figure 4-22* (*a*) A peptic ulcer in the stomach or small intestine. *H. pylori* bacteria and NSAIDs (e.g., aspirin) cause ulcers by impairing mucosal defense, especially in the stomach. In the same way, smoking, genetics, and stress can impair mucosal defense, as well as cause an increase in the release of pepsin and stomach acid. All these factors can contribute to ulcers. (*b*) Close-up of a stomach ulcer. This needs to be treated, or eventual perforation of the stomach is possible.

J. James/Science Source

Surgery to strengthen the lower esophageal sphincter may be needed when lifestyle modifications and medications do not work.

## Ulcers

An **ulcer** is a very small (usually no larger than a pencil eraser) erosion of the top layer of cells in the stomach or duodenum (Fig. 4-22). The general term for this condition is **peptic ulcer**. About 10 to 20% of North Americans develop ulcers during their lifetimes.[33] Ulcers in younger people tend to develop in the small intestine, whereas in older people they occur in the stomach.

There are 2 main causes of peptic ulcers, infection with the bacterium *Helicobacter pylori* (*H. pylori*) and heavy use of NSAID medications, such as aspirin and ibuprofen. Excessive alcohol use, smoking, and emotional stress are not considered to be primary causes of ulcers but can contribute to ulcer progression and poor healing. Eating spicy foods does not cause ulcers.

Being infected with *H. pylori* does not necessarily lead to an ulcer. In fact, over half of people over the age of 60 are infected with *H. pylori* and most do not have an ulcer.[35] *H. pylori* causes ulcer by weakening the mucous coating that protects the stomach and duodenum. This allows HCl and digestive juices to attack and erode stomach and duodenal cells. *H. pylori* itself also irritates these cells. Recall that aspirin and other NSAID medications can cause ulcers by suppressing the synthesis of prostaglandins, compounds that promote the formation of the protective mucus.

The most common symptom of ulcer is a gnawing or burning pain in the stomach region between meals or during the night. This pain often can be relieved by eating or taking antacids. Other, less common symptoms are nausea, vomiting, loss of appetite, and weight loss. The primary complications of ulcers are bleeding and perforation. Slow bleeding eventually can cause anemia and fatigue. Rapid bleeding makes the feces tarry and black from the digested blood, or the person may vomit what looks like coffee grounds. Perforated

**peptic ulcer** Hole in the lining of the stomach or duodenum.

ulcers (those that eat through the stomach or intestinal wall) allow chyme to escape and enter the abdomen, where it may cause a major, even deadly, infection. It is important to pay attention to the early warning signs of an ulcer.

A combination of approaches is used for ulcer therapy (Table 4-5). Those infected with *H. pylori* are treated with antibiotics and either proton pump inhibitors or an H2 blocker to suppress acid production. Bismuth subsalicylate (a component of Pepto-Bismol®) is taken to protect the stomach lining from acid. Most people (80–95%) treated with these drugs heal their ulcers. Smoking can slow ulcer healing. A dietary recommendation is to avoid foods that increase ulcer symptoms, but a bland diet is not necessary.

## Nonalcoholic Fatty Liver Disease

**Nonalcoholic fatty liver disease** (NAFLD) is an increasingly common chronic liver disorder. It occurs when liver cells store excess fat, which may cause liver swelling, inflammation, and scarring.[36] Excess alcohol intake also can cause these changes, but individuals with NAFLD do not abuse alcohol. When inflammation or scarring is present, the condition is known as **nonalcoholic steatohepatitis** (NASH). NAFLD and NASH can be hard to diagnose because many individuals have no symptoms. When symptoms are present, they may include fatigue, nausea, poor appetite, weight loss, jaundice, and diffuse pain under the right rib cage (where the liver is located). Most cases are diagnosed when abnormal liver blood tests are discovered as a part of regular medical care.

NAFLD occurs in both adults and children. In Western countries, 17 to 46% of adults and 10 to 15% of children are thought to have abnormal amounts of fat in the liver, and it is even higher in those with obesity, diabetes, or **insulin resistance**. In these conditions, the liver stores fat instead of metabolizing it (fat metabolism is described in Chapter 9). Diets high in saturated fat and sugars also may contribute to fat storage in the liver. Fat in the liver is a trigger for abnormal oxidation and toxin formation, which damage liver cells. Sugar-sweetened beverages are linked with an increased risk for NAFLD, even in the absence of obesity (another factor increasing the risk for NAFLD).[37] Fructose, commonly found as high fructose corn syrup (see Chapter 5) and used to sweeten a majority of sugar-sweetened beverages, promotes the development of NAFLD.[37]

Changes in lifestyle that promote gradual weight loss—a reduction in calorie intake and daily physical activity—are the cornerstone for treatment of NAFLD and NASH.[38] With weight loss, liver fat and damage can be somewhat reversed. Avoiding all alcohol, limiting sugar-sweetened beverages, and eating a nutritious diet are also recommended. Scientists are studying if vitamin E, an antioxidant (described in Part 4), can reverse some of the liver changes.

## Gallstones

Gallstones, a frequent cause of illness and surgery, affect 10 to 20% of U.S. adults. The stones develop in the gallbladder when substances in the bile—mainly cholesterol (80% of gallstones) and bile pigments (20%)—form crystal-like particles.[39] Gallstones can be as small as a grain of sand or as large as a golf ball (Fig. 4-23). Gallstone formation is related to slow gallbladder motility and bile composition. Too little bile and phospholipids and too much cholesterol allow cholesterol to crystallize into stones. Factors that increase the risk of gallstone formation are listed in Table 4-6.[39,40]

**Table 4-5 Recommendations to Prevent Ulcers and Heartburn from Occurring or Recurring**

**Ulcers**

1 Stop smoking if you smoke.

2 Avoid large doses of aspirin, ibuprofen, and other NSAID compounds unless a physician advises otherwise. For people who must use these medications, the FDA has approved taking an NSAID along with a medication that reduces gastric damage.

3 Limit intake of coffee, tea, and alcohol (especially wine), if this helps.

4 Limit consumption of pepper, chili powder, and other strong spices, if this helps.

5 Eat nutritious meals on a regular schedule; include enough fiber (see Chapter 5 for sources of fiber).

6 Chew foods well.

7 Lose weight if you are currently overweight.

**Heartburn**

1 Follow ulcer prevention recommendations.

2 Wait about 2 hours after a meal before lying down.

3 Don't overeat. Eat smaller meals that are low in fat.

4 Elevate the head of the bed at least 6 inches.

*Figure 4-23* Gallbladder and gallstones after surgical removal from the body. Size and composition of the stones vary from case to case.

Pthawatc/Shutterstock

**Table 4-6  Factors Associated with Gallstone Formation**

- Overweight and obesity
- Prolonged fasting
- Rapid weight loss (more than 3 lb per week) and weight loss surgery
- High-calorie, low-fiber diet
- Other diseases: type 2 diabetes, inflammatory bowel disease, cystic fibrosis
- Sedentary lifestyle
- Some medications, especially estrogen replacement therapy and birth control pills
- Female biological sex
- Pregnancy
- Age over 60 years
- Family history of gallstones
- Ethnicity—especially Native American and Mexican-American

Preventing gallstones includes maintaining a healthy weight, especially for women. Avoiding rapid weight loss and eating a high-fiber diet (with plenty of fruits, vegetables, and whole grains) may help prevent stone formation. Regular physical activity also is important.

Most individuals with gallstones do not have symptoms; stones are usually detected during an examination for another illness. Symptoms, known as gallbladder "attacks," can include intermittent pain in the right upper abdomen, pain between the shoulder blades or near the right shoulder, nausea, vomiting, gas, and bloating. Attacks occur when stones block the bile ducts and stop the free flow of bile. They may last from 20 minutes to several hours.

For those with symptoms, surgical removal of the gallbladder is the most common treatment. Fortunately, removal of the gallbladder does not have serious consequences; after surgery, instead of being stored in the gallbladder, bile flows directly from the liver through the common bile duct to the small intestine.

## Food Intolerances

Food intolerances are caused by an individual's inability to digest certain food components, usually due to low amounts of specific enzymes. Note that food intolerances are not the same as food allergies. Food allergies cause an immune response as a result of exposure to certain food proteins (allergens) (see Chapter 7). Food intolerances afflict many individuals, and the symptoms vary widely, depending on the cause of the food intolerance. Common causes include

- Deficiencies in digestive enzymes, such as lactase (see Chapter 5)
- Sensitivities to food components, such as gluten (see Expert Perspective from the Field) and short-chain fermentable carbohydrates (see Clinical Perspective).
- Certain synthetic compounds added to foods, such as food coloring agents, sulfites, and monosodium glutamate (MSG). The food coloring tartrazine, for instance, causes airway spasms, itching, and reddening skin in some people. Sulfites, which often are added to protect the color of wine, dried foods (fruit, potatoes, soup mixes), and salad greens, may cause flushing, airway spasms, and a drop in blood pressure in susceptible people. MSG, a flavor enhancer frequently added to restaurant and processed foods, may increase blood pressure and cause numbness, sweating, vomiting, and headache in certain people.
- Residues of medications (e.g., antibiotics) and other chemicals used in the production of livestock and crops, as well as insect parts not removed during processing (see Chapter 3)
- Toxic contaminants, such as mold or bacteria (see Chapter 3)

## Intestinal Gas

Everyone has gas. In fact, we produce about 1 to 4 pints of gas each day and pass gas about 13 to 21 times a day, although there is considerable variability from person to person.[47] Gas is eliminated by burping and passing it through the rectum. Intestinal gas (also known as **flatulence**) is a mixture of carbon dioxide, oxygen, nitrogen, hydrogen, methane, and small amounts of sulfur-containing gas. The sulfur is responsible for the unpleasant odor associated with flatulence. Large quantities of intestinal gas can cause bloating and abdominal pain.

Gas comes primarily from the fermentation of undigested carbohydrates by bacteria in the large intestine. The bacteria produce gas as they metabolize the carbohydrates. Swallowed air also contributes to gas. Some people are particularly sensitive to certain carbohydrates (Table 4-7), whereas others can eat them with little problem. Enzyme preparations, such as Beano®, and lactase may help prevent gas by limiting the amount of undigested carbohydrate available to the bacteria in the large intestine. The enzyme in Beano® breaks down raffinose and other similar carbohydrates, whereas the lactase enzyme digests the lactose in milk. Because many gas-forming foods such as beans, vegetables, dairy products, and fruits are healthy foods, eliminating them from the diet is not recommended. Slowly increasing the amount of these foods in the diet can result in adaptation and less gas formation.[48] A rapid increase in fiber intake can worsen abdominal distension and flatulence.

*Critical* | Thinking

Joci is considering going on a new diet that emphasizes eating only fruits before noon, meat at lunchtime, and starch and vegetables at dinner. In addition, the diet recommends "cleansing" the intestines with laxatives and enemas every other week. What reasons would you give Joci to steer clear of this regimen? What are some possible harmful effects?

# Expert Perspective from the Field

## Gluten-Related Disorders: Celiac Disease and Non-celiac Gluten Sensitivity

For some individuals, consuming gluten causes damage to the GI tract, a multitude of disagreeable symptoms, and even serious health problems. The term *gluten* refers to specific proteins found in wheat and related grains, such as rye, barley, spelt, and triticale. Celiac disease is caused by an autoimmune response to the gluten proteins. In persons with celiac disease, these proteins trigger the immune system to damage the villi of the small intestine. The damaged, flattened villi are unable to adequately absorb nutrients, often leading to malnutrition. Currently, health experts believe that celiac disease results from both genetic and immunological factors.[41]

Celiac disease affects approximately 0.7% of the U.S. population.[42] As public and professional attention on gluten-related disorders has increased, undiagnosed celiac disease has declined, but the number of people without celiac disease who are avoiding gluten has increased (from 0.5% in 2009–2010 to 1.7% in 2013–14).[42] Celiac disease can affect a number of body systems. Classic symptoms include intestinal gas, bloating, diarrhea, constipation, abdominal pain, and weight loss or gain. Nongastrointestinal symptoms include anemia, early bone disease, fatigue, slower than normal growth in children, ataxia (impaired coordination) and other neurological conditions, a skin condition called dermatitis herpetiformis (shown in the accompanying photo), and infertility.[41] In the long term, untreated celiac disease may lead to severe malnutrition and an increased risk of aggressive GI tract cancers. Celiac disease is readily diagnosed by a physician with a blood test, followed by a biopsy of the small intestine to confirm the condition.

Although research is under way to develop drug therapies, Cynthia Kupper,* Executive Director of the Gluten Intolerance Group of North America and a celiac disease sufferer, points out that currently the only treatment is a lifetime of eating a gluten-free diet. A healthy diet that includes all food groups is important. However, only gluten-free grains such as corn, rice, quinoa, and buckwheat should be consumed.[41,43] A registered dietitian can help a person plan a healthy gluten-free diet. Ms. Kupper notes that food labels can help consumers avoid foods with ingredients that contain gluten because any food that contains wheat must state this clearly on the label. To further assist those following a gluten-free diet, the U.S. Food and Drug Administration (FDA)

**Small Intestine Villi**

Healthy, long villi

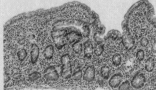

Damaged, blunted villi resulting from celiac disease

David Litman/Shutterstock

has issued regulations for voluntary labeling by food manufacturers of gluten-free products.[44]

Another gluten-related disorder is non-celiac gluten sensitivity (NCGS), which may afflict as many as 18 to 20 million individuals in the U.S. (though no one knows for certain how many people are sensitive to gluten). The symptoms are similar to celiac disease, but the small intestine is not damaged.[41,45] This condition is not well understood. Unfortunately, there is not an official diagnostic test for non-celiac gluten sensitivity, making it more difficult to diagnose. Even so, Ms. Kupper recommends getting a diagnosis from a physician rather than just eliminating gluten from the diet. The symptoms might be caused by another condition, such as wheat allergy, celiac disease, or inflammatory bowel disease, discussed later in this section. With current research demonstrating that symptoms of NCGS are related more to intolerance of short-chain fermentable sugars (e.g., FODMAPs) and not specifically gluten, some experts have proposed changing the condition name from NCGS to non-celiac wheat sensitivity (NCWS) and focusing dietary treatment strategies on the low-FODMAP diet[46] (see Clinical Perspective).

Gluten-free diets are sometimes promoted as being more healthful than gluten-containing diets. However, except for treating gluten-related disorders, a gluten-free diet is not recommended by health professionals. Claims that gluten-free diets promote weight loss, increase energy, cure other disorders, or are more nutritious than gluten-containing diets are not supported by scientific studies. To learn more about gluten-related disorders and the gluten-free diet, check out **www.gluten.net**, **www.celiac.org**, and **www.eatright.org**.

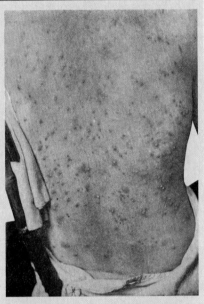

People with gluten-related disorders should avoid foods, such as bread, that are made with wheat.

DoubleVision/Science Source

*The Gluten Intolerance Group of North America reviews scientific research and translates it into practical information to help individuals with gluten intolerance disorders manage their diseases. Ms. Kupper often provides expert input to the food industry and government agencies, such as the National Institutes of Health (NIH) and the FDA.

## FODMAPs

Researchers in Australia have found that a diet low in short-chain fermentable carbohydrates, also known as FODMAPs, can reduce IBS symptoms.

F = Fermentable

O = Oligosaccharides, found in foods such as wheat, onions, beans, and artichokes

D = Disaccharides, such as lactose in dairy products

M = Monosaccharides, especially fructose found in foods such as honey, watermelon, apples, pears, and foods containing high fructose corn syrup

A = And

P = Polyols or sugar alcohols, such as xylitol, mannitol, and sorbitol, found in many fruits and food products sweetened with sugar alcohols

In some individuals, FODMAP carbohydrates are poorly digested and absorbed. This means that they are carried to the large intestine, where bacteria ferment them, also producing intestinal gas. This fermentation process can contribute to intestinal pain and bloating. Low-FODMAP diets may be a good option for individuals with IBS. Read more at www.med.monash.edu/cecs/gastro/fodmap/.

Strictly avoiding FODMAPs in the diet can lead to nutrient deficiencies; thus, supplementation may be warranted.[46] Individuals interested in following a low-FODMAP diet should consult with a registered dietitian nutritionist to help ensure nutritional adequacy of their diet.

Oleg Golovnev/Shutterstock

**Table 4-7  Carbohydrates that May Contribute to Intestinal Gas Formation**

| Carbohydrate(s) | | Description and Food Sources |
|---|---|---|
| Raffinose and stachyose | | Complex sugars found in beans and vegetables, such as cabbage, Brussels sprouts, and broccoli, that are poorly absorbed |
| Lactose | | Sugar found in milk and milk products (lactose intolerance is discussed in Chapter 5) |
| Fructose | | Sugar found in fruit, onions, artichokes, and wheat |
| Sorbitol | | Sugar alcohol that is poorly absorbed, found in many fruits (apples, pears, prunes) and used to sweeten some sugar-free products |
| Starches | | Some of the starch found in potatoes, corn, noodles, and wheat that is not fully digested |
| Fiber | | Soluble fiber found in beans, oat bran, and fruits; fiber added to processed foods |

# Constipation

Constipation is defined as difficult or infrequent (fewer than 3 times per week) bowel movements. Slow movement of fecal material through the large intestine causes constipation. As fluid is increasingly absorbed during the extended time the feces stay in the large intestine, they become dry and hard. Stool Types 1 and 2 of the Bristol Stool Scale (Fig. 4-24) represent constipation. Constipation is commonly reported by older adults because the colon becomes more sluggish as we age.

Constipation is caused by many factors.[49] It can occur when people regularly ignore normal urges to defecate for long periods. Constipation also can result from conditions such as diabetes mellitus, irritable bowel syndrome, and depression. Pregnant women frequently experience constipation because hormones released in pregnancy slow GI motility. Antacids, antidepressants, and calcium and iron supplements are examples of medications that can cause constipation. In addition, prolonged use or abuse of opioids for pain management or recreational drug use (e.g., heroin and fentanyl) can contribute to constipation.[50] Given the current opioid abuse epidemic—nearly 70% of drug-overdose deaths are attributed to opioid use and abuse[51]—understanding how this drug impacts nutritional health is essential.

Low-fiber diets also contribute to constipation; thus, eating foods with plenty of fiber, such as whole-grain breads and cereals, beans, fruits, and vegetables, and drinking more fluid help treat typical cases of mild constipation.[52,53] The recommended fiber intake for most adults is 25 to 38 grams per day. Fiber stimulates peristalsis by drawing water into the large intestine and helping form bulky, soft feces. The bulky fecal material stretches the peristaltic muscles; the muscles respond by constricting, causing the feces to be propelled forward.

People with constipation may need to develop more regular bowel habits—setting the same time each day for a bowel movement (usually on awakening or shortly after a meal) can help train the large intestine to respond routinely. Additionally, relaxation and daily exercise promote regular bowel movements.

Laxatives, medications that stimulate emptying of the intestines, can lessen more serious cases of constipation. There are several types of laxatives.

- Bulk-forming laxatives (Metamucil® and Citrucel®) contain different types of fiber (e.g., psyllium fiber, methylcellulose). Like fiber in food, bulk-forming laxatives draw water into the intestine and increase fecal volume.
- Osmotic laxatives (Milk of Magnesia®) keep fluid in the intestine, which helps keep the fecal matter soft and bulky.
- Stimulant laxatives (Dulcolax® and ExLax®) agitate intestinal nerves to stimulate the peristaltic muscles.
- Stool softeners (Colace®) allow water to enter the bowel more readily.
- Lubricant laxatives (mineral oil) are not recommended because they may block absorption of fat-soluble vitamins.

| Bristol Stool Scale | | |
|---|---|---|
| Type 1 | | Small, hard individual nuggets (hard to pass) |
| Type 2 | | Small, hard nuggets fused together |
| Type 3 | | Hard, dry logs |
| Type 4 | | Soft, smooth logs |
| Type 5 | | Soft, formed globs |
| Type 6 | | Very soft, mushy globs |
| Type 7 | | Runny, liquid |

*Figure 4-24* Although perhaps a bit embarrassing to talk about, the condition of your stools can provide information about your overall digestive health and wellness. The Bristol Stool Scale can help individuals better describe the condition of their stools. Additionally, a doctor may ask about the color, frequency, and odor of stools.

▶ The FDA recently issued guidelines to help those more easily find gluten-free foods. Foods labeled as "gluten free" must contain fewer than 20 parts per million of gluten. Learn more at fda.gov.

For most people, the bulk-forming and osmotic laxatives are the safest to use. However, the regular use of laxatives, especially stimulant laxatives, may lead to dependence and damage the intestine. Consult a physician before using laxatives longer than a week.

Physicians infrequently treat more severe cases of constipation with an **enema**. An enema is the insertion of fluid into the rectum and colon via the anus. The fluid stimulates the bowel, and the liquid and feces are expelled. Some individuals believe enemas and colonic cleansing remove toxins from the colon and the body, but toxins are removed by the liver and kidney, not the colon. There is little evidence that this practice is beneficial.

## Diarrhea

**Diarrhea**—loose, watery stools occurring more than 3 times per day—is a common GI tract problem. It usually lasts only a few days and goes away on its own. Most cases of diarrhea result from bacterial or viral infection, often from contaminated food or water (see Chapter 3). Stool Types 6 and 7 of the Bristol Stool Scale (see Fig. 4-24) represent diarrhea. These infections cause the intestinal tract to secrete fluid instead of absorbing it. Diarrhea also can be caused by parasites, food intolerances, medications (e.g., magnesium-containing antacids and certain antibiotics), megadoses of vitamin C supplements, intestinal diseases, and irritable bowel syndrome. Consuming substances that are not readily absorbed, such as the sugar alcohol sorbitol found in sugarless gum and candy, can cause diarrhea as well (see Chapter 5). When ingested in large amounts, unabsorbed substances draw excess water into the intestine, causing diarrhea.

The treatment of diarrhea generally requires consuming plenty of fluid (beverages, soups, broths) to replace lost fluid and electrolytes. Prompt treatment is vital for infants and older people because they are more susceptible to the effects of dehydration associated with diarrhea. Most infants and children with diarrhea can be treated at home. Special fluids, such as Pedialyte®, can be given for fluid and electrolyte replacement. Most children can continue to eat a normal diet. A health-care provider should evaluate children with diarrhea who are less than 6 months of age or who have blood in the stool, frequent vomiting, high fever, and signs of dehydration (fewer than 6 wet diapers per day, weight loss, extreme thirst, or dry, sticky mouth). Diarrhea in adults that lasts more than 3 days, especially if accompanied by fever, blood in the stool, or severe abdominal pain, also warrants investigation by a health-care provider. When recovering from diarrhea, it is best to avoid greasy, high-fiber, and very sweet foods because they can aggravate diarrhea.

## Irritable Bowel Syndrome

About 10 to 15% of the U.S. population suffers from irritable bowel syndrome (IBS), a chronic disorder that is more common in women than in men.[54] IBS symptoms include irregular bowel function (diarrhea, constipation, or alternating episodes of both), abdominal pain, and abdominal distension. Symptoms often worsen after eating and may be relieved by a bowel movement. Individuals with IBS may experience significantly decreased work productivity and quality of life. Diagnosis of IBS should be made by a physician.

Multiple related factors are thought to contribute to the development of IBS, including abnormal intestinal motility (leading to constipation or diarrhea), abnormal pain sensations, psychosocial distress, altered gut immunity, increased intestinal permeability, and intestinal dysbiosis.[54] Food has long been regarded by many with IBS as a trigger to their symptoms. Dietary constituents that appear to aggravate IBS symptoms include incompletely absorbed fermentable carbohydrates, such as fructose (in honey, fruit, and high fructose corn syrup added to many processed foods), lactose (dairy products), sugar alcohols such as sorbitol (see Chapter 5), and gas-forming foods (e.g., onions, cabbage, beans, broccoli).[56] The ingestion of wheat, gluten, and large, high-fat meals may trigger symptoms, too.[57]

▶ You may have heard that taking laxatives after overeating prevents the body from storing the excess calories as body fat. This erroneous and dangerous idea has gained popularity among some dieters. Laxatives may cause you to feel less full temporarily because they speed up the emptying of the large intestine and increase fluid loss. Most laxatives, however, do not speed the passage of food through the small intestine, where most digestion and most nutrient absorption take place. As a result, laxatives won't prevent fat gain from excess energy intake.

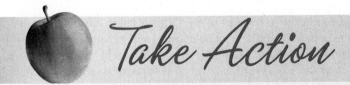

## *Take Action*

### Comparing Breads: With and Without Gluten

Mark Steinmetz

How does 100% whole-wheat bread (containing gluten) compare to gluten-free multigrain bread? Take a look at the accompanying chart and see how the 2 options compare nutrient-wise.

**Further Considerations:** Processed gluten-free foods (e.g., baked goods, snack foods,

breads) tend to be lower in dietary fiber, B-vitamins, vitamin D, and the minerals calcium, iron, zinc, and magnesium and higher in added sugar and saturated fat.[55] Think about how eating processed gluten-free foods could contribute to nutrient inadequacies and excesses and impact overall health.

| Calories and Nutrients | Gluten-Free Multigrain Bread (2 Slices) | 100% Whole-Wheat Bread (2 Slices) |
|---|---|---|
| Calories | 140 | 120 |
| Total fat (g) | 5 | 4 |
| Saturated fat (g) | 0 | 0 |
| Sodium (mg) | 260 | 220 |
| Total carbohydrate (g) | 23 | 22 |
| Sugar (g) | 2 | 1 |
| Dietary fiber (g) | 1 | 4 |
| Protein (g) | 2 | 8 |

The management of IBS is highly individualized. Increasing dietary fiber (either via diet or supplements) has long been touted as a way to manage IBS, but research does not consistently support this practice, except for increasing soluble fiber to help with constipation.[58] Peppermint oil, which relaxes the smooth muscle of the gastrointestinal tract, and low-fat, frequent, small meals may improve IBS symptoms. A gluten-free diet may improve symptoms in some individuals. Some probiotics may have positive effects on reducing IBS symptoms, especially abdominal pain, flatulence, and bloating.[16,59] Following a FODMAP (see Clinical Perspective) protocol of restriction, reintroduction, and personalization under the direction of a knowledgeable health professional holds promise as an effective treatment strategy for IBS.[60] A combination of prebiotics, probiotics, and a FODMAP dietary protocol to manipulate gut microbiota and treat and manage IBS is a promising therapeutic approach. A registered dietitian nutritionist can help those with IBS explore dietary management approaches to IBS while ensuring a nutritionally adequate diet.

### Inflammatory Bowel Disease

Inflammatory bowel disease (IBD) is a group of serious, chronic intestinal diseases that afflict about 1 in every 225 people in the U.S.[61] IBD is not related to IBS. The most common forms are **ulcerative colitis** and **Crohn disease**. With ulcerative colitis, recurring inflammation and ulceration occur in the innermost layer of the large intestine. However, with Crohn disease, the inflammation and ulceration can extend through all layers of the

**ulcerative colitis** Inflammation of the colon that can lead to ulcers (ulcerative colitis).

**Crohn disease** Inflammatory disease of the GI tract that often limits the absorptive capacity of the small intestine. Family history is a major risk factor.

## CASE STUDY FOLLOW-UP

Photodisc/Getty Images

Elise's GERD can be treated, but it may be a lifelong condition. To reduce the risk of GERD, Elise should eat small, frequent meals that are low in fat, not overeat at mealtime, wait about 2 hours after meals before lying down, and elevate the head of her bed about 6 inches (see Table 4-5). Additionally, she should limit her intake of chili powder, onions, garlic, peppermint, caffeine, alcohol, and chocolate; lose excess weight; and not smoke. If this advice does not control Elise's symptoms, her physician may turn to medications. The primary medications used to control GERD inhibit acid production in the stomach. If this and other medical therapy fail to control the problem, surgery to strengthen the lower esophageal sphincter is possible but generally does not cure the problem. Lifetime diet and lifestyle management, and most likely medications, will still be needed to manage the problem. Such management is important because long-standing GERD increases the risk of esophageal cancer.

GI tract. Crohn disease can occur in any part of the GI tract but is most common in the ileum and the ascending colon. The inflammation causes swelling and scar tissue, which can narrow the GI tract, creating a stricture. The ulcers can form fistulae, or deep tunnels, that extend from 1 area of the intestine to another or even to another organ outside the GI tract. IBD is most commonly diagnosed at around age 20; symptoms include rectal bleeding, diarrhea, abdominal pain, weight loss, and fever. Complications of IBD are caused by nutritional deficiencies and inflammation in other parts of the body. Nutritional problems include anemia due to blood loss; osteoporosis from lack of bone-forming nutrients; and protein-energy malnutrition from low food intake, malabsorption, and high nutrient requirements for healing the damaged gastrointestinal tract.[62] Poor growth and delayed puberty may occur in those who develop IBD in childhood. Individuals with IBD also are at a higher risk of colon cancer.

The cause of IBD is not yet known, but an overactive inflammatory response to antigens (foreign substances, such as bacteria or toxins) in the GI tract is suspected. There is a strong genetic association for IBD. Medical treatment for IBD includes medications to suppress the immune system, antibiotics, and surgery to remove the diseased area. Dietary intake and nutritional health should be closely monitored in those suffering from IBD.

## Hemorrhoids

**Hemorrhoids**, also called piles, are swollen veins of the rectum and anus (like varicose veins in the legs). The blood vessels in this area are subject to intense pressure, especially during bowel movements. Obesity, prolonged sitting, and violent coughing or sneezing add stress to the vessels. Many pregnant women also develop hemorrhoids (see Chapter 16). Hemorrhoids develop unnoticed until a strained bowel movement triggers symptoms, which may include itching, pain, and bleeding.

Itching is caused by moisture, swelling, or other irritation in the anal canal (an approximately 2-inch-long section between the rectum and anus). Pain, if present, is usually a steady ache. Bleeding from a hemorrhoid may appear in the toilet as a bright red streak in the feces. The sensation of a mass in the anal canal after a bowel movement is a symptom of an internal hemorrhoid that protrudes through the anus.

Anyone can develop a hemorrhoid—about 75% of adults do.[63] Pressure from prolonged sitting or exertion is often enough to bring on symptoms, although diet, lifestyle, and possibly heredity play a role. For example, a low-fiber diet can lead to hemorrhoids as a result of constipation and straining during bowel movements. If you think you have a hemorrhoid, you should consult your physician. Rectal bleeding, although usually caused by hemorrhoids, may also indicate other problems, such as cancer.

A physician may suggest a variety of self-care measures for hemorrhoids. Pain can be reduced by applying warm, soft compresses or sitting in a tub of warm water for 15 to 20 minutes. Dietary recommendations are the same as those for treating mild constipation, emphasizing the need to exercise every day and consume adequate fiber (25–35 grams daily) and fluid. Over-the-counter remedies, such as Preparation H®, offer symptom relief. Some hemorrhoids require removal procedures, usually done in a surgeon's office.

▶ Lactose intolerance and diverticulosis are 2 other common GI tract disorders (they are discussed in Chapter 5).

### Knowledge Check

1. What is the most common cause of peptic ulcers?
2. What are 2 nutritional factors that can increase the risk of gallstones?
3. What are the names of the disorders that may arise when abnormal amounts of fat are stored in the liver?
4. What are 3 factors that increase the chances of developing constipation?

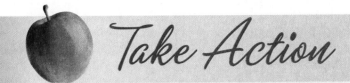

## Are You Eating for a Healthy Digestive System?

Our daily dietary patterns can help us maintain a healthy GI tract. Place a check in the box that best matches your practice for each item. For the items marked "I do this some of the time" or "I do this rarely," consider how you could move to the "I do this most of the time" box.

Design Pics/Monkey Business

| Action | I do this most of the time | I do this some of the time | I rarely do this |
|---|---|---|---|
| Drink plenty (8 cups or more) of water daily to stay hydrated. Adequate water helps keep fiber moving through the digestive system and prevents constipation. | | | |
| Avoid overeating. Large, high-fat meals can be slow to digest and lead to uncomfortable feelings of fullness and bloating. Large meals also trigger excess HCl production. | | | |
| Eat several fiber-rich foods daily. Following MyPlate recommendations for fruits, vegetables, and whole grains helps provide the fiber needed to prevent constipation. These foods also provide prebiotic compounds that help keep the gut microbiota flourishing. | | | |
| Eat foods containing probiotics. Regular intake of these healthy bacteria-containing foods (e.g., yogurt, miso, unpasteurized sauerkraut) help keep a healthy microbial population in the GI tract. | | | |
| Chew food thoroughly, eat slowly, and focus on eating during meals instead of getting distracted by things like TV. These actions help prevent overeating and may improve digestion. | | | |
| Avoid too many foods with sugar alcohols, such as maltitol, xylitol, and sorbitol (see Chapter 5 for more on sugar alcohols). These substances can cause diarrhea, gas, and bloating when consumed in large amounts. | | | |
| Avoid excessive intake of beer, wine, and other alcoholic beverages. Too much alcohol can irritate the stomach lining and damage the liver and pancreas (see Chapter 8). | | | |
| Maintain a healthy weight. Digestion may be impaired when an individual is underweight. Overweight persons are at risk for GERD. | | | |
| Practice safe food handling (see Chapter 3) to prevent foodborne illness and the conditions it causes (e.g., vomiting, diarrhea, mucosal damage) and associated increased risk for GI disorders (e.g., irritable bowel syndrome, indigestion, constipation, and GERD). | | | |
| Exercise at least 30 minutes per day. Exercise helps prevent constipation and prevents weight gain. | | | |

# Chapter Summary

## 4.1 The cell is the basic structural unit of the human body.
Cells join together to make up tissues. The 4 primary types of tissues are epithelial, connective, muscle, and nervous. Tissues unite to form organs, and organs work together as an organ system.

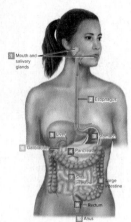

## 4.2 The GI tract includes the mouth,
esophagus, stomach, small intestine, and large intestine (colon, rectum, and anus). Sphincters along the GI tract control the flow of its contents. The accessory organs (liver, gallbladder, and pancreas) are an important part of the digestive system. Movement through the GI tract is mainly through muscular contractions known as peristalsis. GI contents are mixed with segmental contractions. Enzymes are specialized protein molecules that speed up digestion by catalyzing chemical reactions. Most digestive enzymes are synthesized in the small intestine and pancreas. A lack of digestive enzymes can result in poor digestion, poor absorption, malnutrition, and weight loss.

### Table 4-2  Overview of GI Tract Digestion and Absorption Functions

| Organs | Digestive Functions |
|---|---|
| Mouth and salivary glands | Prepare food for swallowing: chewing, moistening with saliva<br>Detect taste molecules<br>Start digestion of starch with amylase enzyme<br>Start digestion of fat with lingual lipase |
| Esophagus | Moves food to stomach by peristaltic waves initiated by swallowing |
| Stomach | Secretes gastric juice containing acid, enzymes, and hormones<br>Mixes food with gastric juice, converting it to liquid chyme<br>Starts digestion of protein and fat<br>Kills microorganisms with acid<br>Secretes intrinsic factor, a protein required for vitamin B-12 absorption<br>Slowly releases chyme to the small intestine |
| Liver | Produces bile to aid fat digestion and absorption |
| Gallbladder | Stores and concentrates bile and releases it to the small intestine |
| Pancreas | Secretes pancreatic juice containing digestive enzymes and bicarbonate into the small intestine |
| Small intestine | Mixes chyme with bile and pancreatic juice to complete digestion<br>Secretes hormones that help regulate digestive processes<br>Secretes digestive enzymes<br>Absorbs nutrients and other compounds in foods<br>Transports remaining residue to large intestine |
| Large intestine (colon) | Absorbs water and electrolytes (sodium and potassium)<br>Forms and stores feces<br>Houses most of the gut microbiota |
| Rectum | Holds and expels feces via the anus |

## 4.3 Chewing food breaks it into small pieces and increases its surface area,
which enhances enzyme activity. Amylase produced by salivary glands digests a small amount of starch. Chewed food mixed with saliva is called a bolus. When swallowing is initiated, the epiglottis covers the trachea to prevent food from entering it. Peristalsis moves food down the esophagus. There are 5 basic taste sensations perceived by taste cells on taste buds in the mouth, especially the tongue. Genetic variability affects the ability to taste bitter compounds. The sense of smell contributes greatly to flavor perceptions.

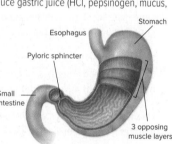

## 4.4 The lower esophageal sphincter protects the esophagus from the
backflow of acidic stomach contents. When this sphincter does not work normally, heartburn and GERD may occur. Stomach cells produce gastric juice (HCl, pepsinogen, mucus, and intrinsic factor). The hormone ghrelin triggers hunger and eating. Pepsin (from pepsinogen) starts the digestion of protein. Mixing of food and gastric juice results in the production of chyme, the liquid substance released in small amounts into the small intestine.

## 4.5 The small intestine has 3 sections: duodenum, jejunum, and ileum.
Most digestion occurs in the small intestine. Secretions from the liver, gallbladder, and pancreas are released into the small intestine. These secretions contain enzymes, bile, and sodium bicarbonate. Villi in the small intestine greatly increase its surface area, which enhances absorption. Villi are lined by enterocytes that release enzymes. Enterocytes are constantly broken down and replaced. Diseases, such as celiac disease, damage the villi and enterocytes. The liver, gallbladder, and pancreas aid digestion and absorption. The liver produces bile, which is stored in the gallbladder and used to emulsify fat. Pancreatic juices contain the alkaline sodium bicarbonate and

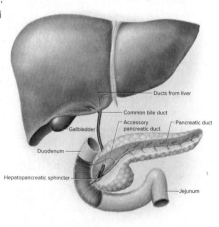

digestive enzymes. Bile and pancreatic juice are released into the small intestine via the pancreatic bile duct. Most nutrients are absorbed primarily in the small intestine. There are 4 main types of absorption: passive diffusion, facilitated diffusion, active absorption, and endocytosis. Hormones regulate digestion and absorption. Five hormones, part of the endocrine system, play key roles in this regulation: gastrin, secretin, cholecystokinin (CCK), somatostatin, and glucose-dependent insulinotropic peptide.

## 4.6 Nutrients absorbed into the absorptive cells are transported in the body via either the cardiovascular or the lymphatic

circulation. Water-soluble nutrients entering the cardiovascular system from absorptive cells travel via the hepatic portal vein to the liver, then to the general circulation and body tissues. Fat-soluble and large particles enter the lymphatic system from absorptive cells. Lymphatic vessels drain into the thoracic duct, which releases its contents to the bloodstream.

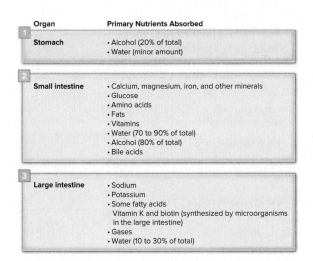

| Organ | Primary Nutrients Absorbed |
|-------|----------------------------|
| **1 Stomach** | • Alcohol (20% of total)<br>• Water (minor amount) |
| **2 Small intestine** | • Calcium, magnesium, iron, and other minerals<br>• Glucose<br>• Amino acids<br>• Fats<br>• Vitamins<br>• Water (70 to 90% of total)<br>• Alcohol (80% of total)<br>• Bile acids |
| **3 Large intestine** | • Sodium<br>• Potassium<br>• Some fatty acids<br>  Vitamin K and biotin (synthesized by microorganisms in the large intestine)<br>• Gases<br>• Water (10 to 30% of total) |

## 4.7 The large intestine is the last part of the GI tract. It houses over 1000 species of bacteria,

absorbs water and electrolytes, and forms and eliminates feces. GI contents entering the large intestine are mainly water, some minerals, fiber, and some starch. Carbohydrates (fiber and starch) can be digested to some extent by bacteria in the large intestine and form short-chain fatty acids, which serve as an energy

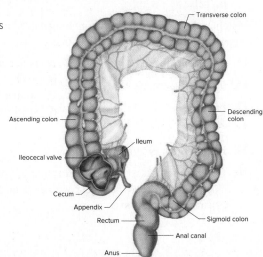

Transverse colon
Ascending colon
Ileocecal valve
Ileum
Cecum
Appendix
Rectum
Descending colon
Sigmoid colon
Anal canal
Anus

source for the large intestine and may help prevent and treat diseases. Bacteria in the large intestine also synthesize vitamin K and biotin and produce intestinal gas. Probiotics are live microorganisms found in fermented foods and supplements. They may promote intestinal health, such as preventing diarrhea in children. Prebiotics are nondigestible carbohydrates that promote the growth of beneficial bacteria in the large intestine. It takes 12 to 24 hours for contents to pass through the large intestine. Feces defecated through the rectum contain about 75% water and 25% solids—indigestible plant fibers, tough connective tissues from animal foods, and bacteria.

## 4.8 Disorders of the digestive system include heartburn, GERD, ulcers, nonalcoholic fatty liver disease, gallstones,

food intolerances, gluten-related disorders, intestinal gas, constipation, diarrhea, irritable bowel syndrome, inflammatory bowel disease, and hemorrhoids. These disorders often can be prevented or managed with healthy nutrition and lifestyle habits.

Pthawatc/Shutterstock

### Clinical Perspective

In some individuals, FODMAP carbohydrates are poorly digested and absorbed. Low-FODMAP diets may be a good option for individuals with IBS. Strictly avoiding FODMAPs in the diet can lead to nutrient deficiencies; thus, supplementation may be warranted.

### Expert Perspective

For some individuals, consuming gluten causes damage to the GI tract and serious health problems. *Gluten* refers to specific proteins found in wheat and related grains. Celiac disease is caused by an autoimmune response to gluten that results in damage to the villi of the small intestine. Non-celiac gluten sensitivity has symptoms that are similar to those of celiac disease, but the small intestine is not damaged. Except for treating gluten-related disorders, a gluten-free diet is not recommended by health professionals.

Mark Steinmetz

### Global Perspective

Diarrhea is a leading killer of children in developing countries. It is typically caused by pathogenic microorganisms—viruses, bacteria, and parasites—found in water, food, and human and animal waste. Being well nourished helps lower the risk of developing diarrhea. Breastfeeding also can prevent diarrhea in young children. Also key to prevention are improved sanitation, which keeps food and water pathogen free, and the ready availability of affordable, nutritious foods that promote good health. Unfortunately, access to these basic needs is limited for many.

# Study Questions

1. What is the smallest functional unit of the human body?

   a. organ
   b. organ system
   c. cell
   d. epithelial tissue

2. Most digestive enzymes are produced in the _____.

   a. mouth and esophagus
   b. esophagus and stomach
   c. small intestine and pancreas
   d. liver and gallbladder

3. The coordinated squeezing and shortening of the muscles of the GI tract is called _____.

   a. enzyme hydrolysis
   b. peristalsis
   c. olfaction
   d. diarrhea

4. Which part of the digestive system normally houses the largest number of bacteria?

   a. large intestine
   b. small intestine
   c. stomach
   d. pancreas

5. The main role of the stomach in digestion and absorption is to _____.

   a. absorb proteins and carbohydrates
   b. digest fats
   c. mix ingested foods to form chyme
   d. produce enzymes that digest carbohydrates and fats

6. Villi are found mainly in the _____.

   a. large intestine
   b. small intestine
   c. esophagus
   d. stomach

7. Which of the following is known to prevent or treat diarrhea in developing countries?

   a. oral rehydration therapy
   b. supplemental zinc
   c. prevention of malnutrition
   d. all of the above

8. Serious liver disease is most likely to result in malabsorption of _____.

   a. carbohydrate
   b. protein
   c. fat
   d. alcohol

9. The hormone _____ stimulates the release of pancreatic enzymes and bile from the gallbladder.

   a. gastrin
   b. lipase
   c. cholecystokinin
   d. secretin

10. The last section of the small intestine, known as the _____, is where _____ is absorbed and recirculated to the liver.

    a. jejunum; fat
    b. ileum; bile
    c. cecum; water
    d. duodenum; carbohydrate

11. _____ are absorbed in the large intestine.

    a. Vitamins and minerals
    b. Vitamins and water
    c. Fatty acids and minerals
    d. Water and electrolytes

12. Probiotics may be most useful in treating _____.

    a. diarrhea
    b. constipation
    c. celiac disease
    d. food intolerance

13. Which of the following digestive disorders is caused by the bacterium *Helicobacter pylori*?

    a. excessive intestinal gas
    b. constipation
    c. diarrhea
    d. peptic ulcer

14. Nonalcoholic fatty liver disease is common in adults and can cause liver swelling, inflammation, and scarring.

    a. true
    b. false

Match each secretion with the organ that produces it. Some organs may be selected more than once.

15. hydrochloric acid
16. sodium bicarbonate
17. bile
18. CCK
19. brush border enzyme

    a. pancreas
    b. liver
    c. stomach
    d. small intestine

20. Consider the digestion and absorption of your evening meal. Write out the steps that occur with each of the digestive organs and secretions.

21. Explain why, after a period of semistarvation, normal amounts of food can be difficult to digest and absorb.

22. Describe how the absorption and transport of water-soluble and fat-soluble nutrients differ.

23. There are more microbes in the gastrointestinal tract than there are cells in the body. Explain their importance to a healthy gastrointestinal tract.

24. Describe the dietary and lifestyle practices that may help prevent or treat the following gastrointestinal problems: gastroesophageal reflux disease, gallstones, constipation, and diarrhea.

25. Differentiate irritable bowel syndrome from inflammatory bowel disease. Which of these conditions is generally regarded as more serious?

26. Explain why diarrhea is the leading cause of death in young children living in developing countries.

27. Who should consume a gluten-free diet?

Answer Key: 1-c; 2-c; 3-b; 4-a; 5-c; 6-b; 7-d; 8-c; 9-c; 10-b; 11-d; 12-a; 13-d; 14-b; 15-c; 16-a and d; 17-b; 18-d; 19-d; 20-refer to Sections 4.3, 4.4, 4.5, 4.7; 21-refer to Section 4.5; 22-refer to Section 4.8; 23-refer to Section 4.7; 24-refer to Section 4.8; 25-refer to Section 4.8; 26-refer to Global Perspective; 27-refer to Expert Perspective

# References

1. Sullivan S and others. Nutritional physiology of the alimentary tract. In: Ross C and others, eds. *Modern nutrition in health and disease.* 11th ed. Baltimore: Lippincott, Williams & Wilkins; 2014.

2. Hunt RH and others. The stomach in health and disease. *Gut.* 2015;64:1650.

3. Degen LP and Phillips SF. Variability of gastrointestinal transit in healthy women and men. *Gut.* 1996;39:299.

4. Tordoff MG and others. Involvement of T1R3 in calcium-magnesium taste. *Physiol Genom.* 2008;34:338.

5. Tepper B. Nutritional implications of genetic taste variation: The role of PROP sensitivity and other taste phenotypes. *Annu Rev Nutr.* 2008;28:367.

6. Tepper B and others. Genetic sensitivity to the bitter taste of 6-n-propylthiouracil (PROP) and its association with physiological mechanisms controlling body mass index (BMI). *Nutrients.* 2014;6:3363.

7. Jandle JH. *William B. Castle 1897–1990. A biographical memoir.* Washington, DC: National Academies Press; 1995.

8. Schroder K and others. Increasing coverage of pediatric diarrhea treatment in high-burden countries. *J Glob Health.* 2019;9:0010503.

9. World Health Organization. Rotavirus vaccines. WHO position paper—January 2013. *Week Epidemiol Rec.* 2013;88:49.

10. Munos MK and others. The effect of oral rehydration solution and recommended home fluids on diarrhoea mortality. *Int J Epidemiol.* 2010;39:i75.

11. Black RE. Progress in the use of ORS and zinc for the treatment of childhood diarrhea. *J Glob Health.* 2019;9:010101.

12. Marchesi JR and others. The gut microbiota and host health: A new clinical frontier. *Gut.* 2016;65:330.

13. Tuddenham S and Sears CL. The intestinal microbiome and health. *Curr Opin Infect Dis.* 2015;28:464.

14. O'Keefe SJD and others. Fat, fibre, and cancer risk in African Americans and rural Africans. *Nature Commun.* 2015;6:6342.

15. McKenney PT and Pamer EG. From hype to hope: The gut microbiota in enteric infectious disease. *Cell.* 2015;163:1326.

16. Rodiño-Janeiro BK and others. A Review of Microbiota and Irritable Bowel Syndrome: Future in Therapies. *Adv Ther.* 2018;35: 289.

17. Schneiderhan J and others. Targeting gut flora to treat and prevent disease. *J Fam Prac.* 2016;65:33.

18. Abbasi J. Are Probiotics money down the toilet? Or Worse? *JAMA.* 2019;32:33.

19. Bafeta A and others. Harms Reporting in Randomized Controlled Trials of Interventions Aimed at Modifying Microbiota: A Systematic Review. *Ann Intern Med.* 2018;169:240.

20. AlFaleh K and Anabrees J. Probiotics for prevention of necrotizing enterocolitis in preterm infants. *Cochrane DB Syst Rev.* 2014;4:Art. No. CD005496.

21. Allen S and others. Probiotics for treating infectious diarrhoea. *Cochrane DB Syst Rev.* 2010;11:Art No. CD003048.

22. Goldenberg JZ and others. Probiotics for the prevention of pediatric antibiotic-associated diarrhea. *Cochrane DB Syst Rev.* 2015;12:Art. No. CD004827.

23. Goldenberg JZ and others. Probiotics for the prevention of *Clostridium difficile*-associated diarrhea in adults and children. *Cochrane DB Syst Rev.* 2013;5:Art No. CD006095.

24. Zmora N and others. Personalized Gut Mucosal Colonization Resistance to Empiric Probiotics Is Associated with Unique Host and Microbiome Features. *Cell.* 2018;174:1388.

25. Suez J and others. Post-Antibiotic Gut Mucosal Microbiome Reconstitution Is Impaired by Probiotics and Improved by Autologous FMT. *Cell.* 2018;174:1406.

26. American Gastrointestinal Association. 2020; gastro.org/news/aga-does-not-recommend-the-use-of-probiotics-for-most-digestive-conditions-2/.

27. Slavin J. Fiber and prebiotics: Mechanisms and health benefits. *Nutrients.* 2013;5:1417.

28. National Institute of Diabetes and Digestive and Kidney Diseases. Acid reflux (GER and GERD) in adults. 2020; www.niddk.nih.gov/health-information/digestive-diseases/acid-reflux-ger-gerd-adults.

29. Choi JM and others. Association Between Anxiety and Depression and Gastroesophageal Reflux Disease: Results From a Large Cross-sectional Study. *J Neurogastroenterol Motil.* 2018;24:593.

30. de Bortoli N and others. Voluntary and controlled weight loss can reduce symptoms and proton pump inhibitor use and dosage in patients with gastroesophageal reflux disease: A comparative study. *Dis Esophagus.* 2016;29:197.

31. Javadi SAHS and Shafikhani AA. Anxiety and depression in patients with gastroesophageal reflux disorder. *Electron Physician.* 2017;9:5107.

32. Avidan B and others. Reflux symptoms are associated with psychiatric disease. *Aliment Pharmacol Ther.* 2001;15:1907.

33. Ozdil K and others. Bone density in proton pump inhibitors users: A prospective study. *Rheumatol Int.* 2013;33:2255.

34. Ness-Jensen E and others. Lifestyle intervention in gastroesophageal reflux disease. *Clin Gastroenterol Hepatol.* 2016;14:175.

35. Anand BS and others. Peptic ulcer disease. 2015. emedicine.medscape.com/article/181753-overview.

36. Rinella ME. Nonalcoholic fatty liver disease: A systematic review. *JAMA.* 2015;313:2263.

37. Wijarnpreecha K and others Associations of sugar- and artificially sweetened soda with nonalcoholic fatty liver disease: a systematic review and meta-analysis. *QJM: Int J Med.* 2016;109:461.

38. Pimental CF and Lai M. Nutrition intervention for chronic liver disease and nonalcoholic fatty liver disease. *Med Clin North Am.* 2016;100:1303.

39. Marks JM. Gallstones. 2016; www.medicinenet.com/gallstones/article.htm.

40. Stokes CS and others. Gallstones: Environment, lifestyle and genes. *Dig Dis.* 2011;29:191.

41. Lebwohl B. Celiac disease and non-celiac gluten sensitivity. *BMJ.* 2015;351:h437.

42. Choung RS and others. Less hidden celiac disease but increased gluten avoidance without a diagnosis in the United States: Findings from the National Health and Nutrition Examination Surveys from 2009 to 2014. *Mayo Clin Proc.* 2017;92:30.

43. Rubio-Tapia A and others. ACG clinical guidelines: Diagnosis and management of celiac disease. *Am J Gastroenterol.* 2013;108:656.

44. Thompson T. The gluten-free labeling rule: What registered dietitian nutritionists need to know to help clients with gluten-related disorders. *J Acad Nutr Diet.* 2015;115:13.

45. Bardella MT and others. Non-celiac gluten sensitivity. *Curr Gastroenterol Rep.* 2016;18:63.

46. Roszkowska A and others. Non-Celiac Gluten Sensitivity: A Review. *Medicina.* 2019;55:E222.

47. National Digestive Diseases Information Clearinghouse. Constipation. 2014; www.niddk.nih.gov/health-information/health-topics/digestive-diseases/constipation.

48. Winham DM, Hutchins AM. Perceptions of flatulence from bean consumption among adults in 3 feeding studies. *Nutr J.* 2011;10:128.

49. National Digestive Diseases Information Clearinghouse. Gas in the digestive tract. Digestive diseases. 2016; digestive.niddk.nih.gov/ddiseases/pubs/gas.

50. Li J. Combining opioids and non-opioids for pain management: Current status. *Neuropharmacol.* 2019;158:3107619.

51. Volkow ND and Wargo EN. Overdose prevention through medical treatment of opioid use disorders. *Ann Intern Med.* 2018;169:190.

52. Liu LW. Chronic constipation: Current treatment options. *Can J Gastroenterol.* 2011;25 Suppl B:22B.

53. Eswaran S and others. Fiber and functional gastrointestinal disorders. *Am J Gastroenterol.* 2013;108:117.

54. Chey WD. Irritable bowel syndrome: A clinical review. *JAMA.* 2015;313:949.

55. Lerner BA and others. Going Against the Grains: Gluten-Free Diets in Patients Without Celiac Disease-Worthwhile or Not? *Dig Dis Sci.* 2019;64:1740.

56. Marsh A and others. Does a diet low in FODMAPs reduce symptoms associated with functional gastrointestinal disorders? A comprehensive systematic review and meta-analysis. *Eur J Nutr*. 2016;55:897.

57. Aziz I and others. Efficacy of a gluten-free diet in subjects with irritable bowel syndrome-diarrhea unaware of their HLA-DQ2/8 genotype. *Clin Gastroenterol Hepatol*. 2016;14:696.

58. Moayyedi P and others. The effect of fiber supplementation on irritable bowel syndrome: A systematic review and meta-analysis. *Am J Gastroenterol*. 2014;109:1367.

59. Melini F and others. Health-promoting components in fermented foods: an up-to-date systematic review. *Nutrients* 2019;11:1189.

60. Halmos EP and Gibson PR. Controversies and reality of the FODMAP diet for patients with irritable bowel syndrome. *J Gastroenterol Hepatol*. 2019;34:1134.

61. Crohn's and Colitis Foundation of America. What are Crohn's & colitis? 2020; www.ccfa.org/what-are-crohns-and-colitis/.

62. Lomer MC. Dietary and nutritional considerations for inflammatory bowel disease. *Proc Nutr Soc*. 2011;70:329.

63. National Institute of Diabetes and Digestive and Kidney Disease. Hemorrhoids. 2020; www.niddk.nih.gov/health-information/digestive-diseases/hemorrhoids.

Brand X Pictures/Getty Images

Grains are the chief source of carbohydrates for most people around the world. Typically, they are ground into flour or meal and made into bread—bread is so important that it is referred to as "the staff of life." For thousands of years, corn has been a staple grain of people living in Central and South America. In fact, more corn is harvested each year than any other grain. Learn more about the ancestry of corn at **learn.genetics.utah .edu/content/evolution/corn/**. Photo by Keith Weller, USDA-ARS

# 5 Carbohydrates

## Learning Objectives

**After studying this chapter, you will be able to**

1. Identify the major types of carbohydrates and give examples of food sources for each.
2. List alternative sweeteners that can be used to reduce sugar intake.
3. Describe recommendations for carbohydrate intake and health risks caused by low or excessive intakes.
4. List the functions of carbohydrates in the body.
5. Explain how carbohydrates are digested and absorbed.
6. Explain the cause of, effects of, and dietary treatment for lactose intolerance.
7. Describe the regulation of blood glucose, conditions caused by blood glucose imbalance, types of diabetes, and dietary treatments for diabetes.
8. Describe dietary measures to reduce the risk of developing type 2 diabetes.

## Chapter Outline

**FRUITS, VEGETABLES, DAIRY PRODUCTS, CEREALS,** breads, pasta, and desserts—all of these supply carbohydrates (i.e., sugar, starch, and fiber) (Fig. 5-1). Maybe you've avoided many of these foods in an attempt to reduce gastrointestinal symptoms, lose weight, or "bulk up" with muscle. Unfortunately, the benefits of carbohydrates are frequently misunderstood.[1] People often mistakenly think carbohydrate rich foods are fattening or cause diabetes. However, high carbohydrate foods—especially fiber rich foods, such as fruits, vegetables, legumes, and whole-grain breads and cereals—provide essential nutrients that help nourish our bodies and should constitute about 45 to 65% of our daily energy intake.[2] They also add interest to our diets—consider the vivid colors of fruits and vegetables, the crunchiness of cereals, and the delicious flavors of desserts.

Carbohydrates are a primary fuel source for cells, especially the cells of the central nervous system and red blood cells.[3] Muscle cells also rely on carbohydrates to fuel intense physical activity. Yielding an average of 4 kcal/g, carbohydrates are a readily available fuel for all cells in the form of glucose (a sugar) in the blood and glycogen (a starch) in the liver and muscles. Glycogen can be broken down to glucose and released into the blood to maintain blood glucose levels when the diet does not supply enough. Regular intake of carbohydrate is important because glycogen stores in the liver and muscles are exhausted in about 18 hours if no carbohydrate is consumed.[3] After that point, the body is forced to produce glucose from protein or to use fat as the primary source of energy; as you will learn later in this chapter, this eventually leads to health problems.

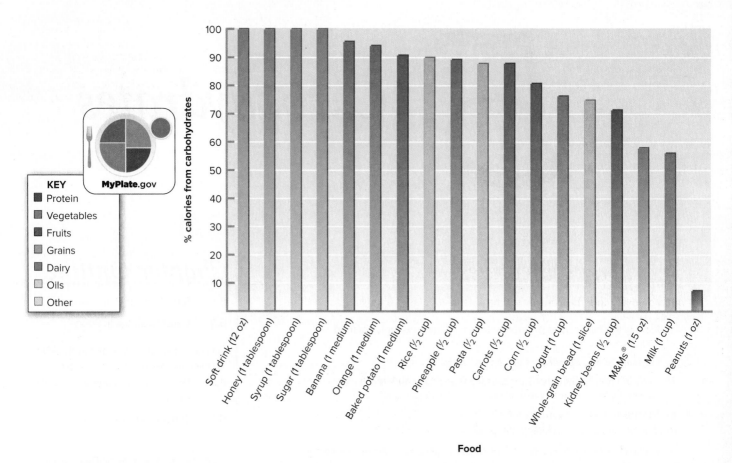

*Figure 5-1* Food sources of carbohydrates and the percentage of calories from carbohydrates. In addition to colors representing the MyPlate food groups, yellow is used for oils and pink is used for substances that do not fit easily into food groups (e.g., honey and soft drinks).

Source: ChooseMyPlate.gov, U.S. Department of Agriculture

# 5.1 Structures of Carbohydrates

**monosaccharide** Simple sugar, such as glucose, that is not broken down further during digestion.

**disaccharide** Class of sugars formed by the chemical bonding of 2 monosaccharides.

**polysaccharide** Large carbohydrate containing from 10 to 1000 or more monosaccharide units; also known as complex carbohydrate.

The carbohydrate family includes sugar, starch, and fiber. Most forms of carbohydrates are composed of carbon, hydrogen, and oxygen. Plants are the main source of carbohydrates. During photosynthesis, plants produce glucose by using carbon and oxygen from carbon dioxide in the air, hydrogen from water, and energy from the sun (Fig. 5-2). Plants either store the glucose or transform it into starch, fiber, fat, or protein.

The general formula for carbohydrates is $(CH_2O)n$ or $Cn(H_2O)n$, where *n* represents the number of times the formula is repeated. For example, the chemical formula for glucose is $C_6H_{12}O_6$, or $(CH_2O)_6$. The simpler forms of carbohydrates are called **monosaccharides** and **disaccharides.** Monosaccharides are single sugars with the general formula of $(CH_2O)_6$. Disaccharides are double sugars, made of 2 monosaccharide sugars, with the general formula of $(CH_2O)_{12}$. The more complex forms of carbohydrates (i.e., glycogen, starch, and fiber) are called **polysaccharides** and typically contain many glucose molecules linked together.

$$6CO_2 + 6H_2O \xrightarrow{light} C_6H_{12}O_6 + 6O_2$$

carbon dioxide    water    glucose    oxygen

*Figure 5-2* A summary of photosynthesis. Plants use carbon dioxide, water, and energy to produce glucose. Glucose is then stored in the leaves but also can undergo further metabolism to form starch and fiber in the plant. With the addition of nitrogen from the soil or air, glucose also can be transformed into protein.

## Monosaccharides: Glucose, Fructose, Galactose, Sugar Alcohols, and Pentoses

The common monosaccharides (*mono* means "1"; *saccharide* means "sugar") are glucose, fructose, and galactose. The structures of these monosaccharides are shown in Figure 5-3. Notice that each of these monosaccharides contains 6 carbon, 12 hydrogen, and 6 oxygen

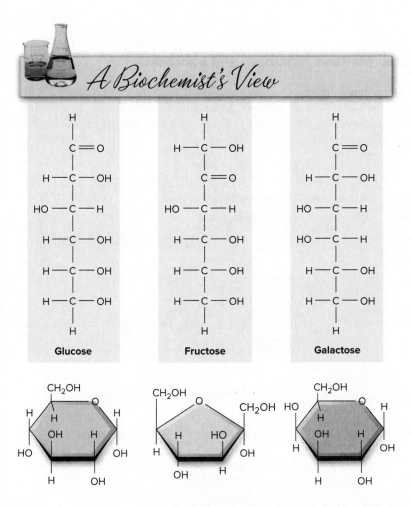

*A Biochemist's View*

molecules, but in slightly different configurations. Because each is a 6-carbon sugar, it is classified as a **hexose** (*hex* means "6"; *ose* refers to "sugar" or "carbohydrate").

**Glucose** is the most abundant monosaccharide, although we eat very little of it as a monosaccharide. Much of the glucose in our diets is linked together with additional sugars to form disaccharides or polysaccharides. In the bloodstream, glucose is sometimes called "blood sugar."

The monosaccharide **fructose** is found in fruits, vegetables, honey (which is about 50% fructose and 50% glucose), and high fructose corn syrup. Because high fructose corn syrup is sweeter and less expensive than table sugar, it is used to sweeten many food products, especially beverages. The presence of fructose in these products makes it a common sugar in our diets. In most North American diets, fructose accounts for about 9 to 11% of total energy intake.[4]

**Galactose** is the third major monosaccharide of nutritional importance. A comparison of the structure of this sugar with that of glucose shows that they are almost identical (see Fig. 5-3). Most of the galactose in our diets is found in combination with glucose. When galactose combines with glucose, it forms a disaccharide called **lactose,** which is found in milk and other dairy products.

The sugar alcohols, which are derivatives of monosaccharides, include sorbitol, mannitol, and xylitol. These are used primarily as sweeteners in sugarless gum and dietetic foods.

Two additional monosaccharides found in nature are ribose and deoxyribose. These are classified as "pentoses" because they contain 5 carbons (*penta* means "5"). Although these sugars do not need to be supplied by the diet, they are very important in the body because they are an essential part of the cell's genetic material. Ribose is part of ribonucleic acid (RNA), and deoxyribose is part of deoxyribonucleic acid (DNA).

**Simple Forms of Carbohydrates**

Monosaccharides: glucose, fructose, galactose

Disaccharides: sucrose, lactose, maltose

**Complex Forms of Carbohydrates**

Oligosaccharides: raffinose, stachyose

Polysaccharides: starches (amylose and amylopectin), glycogen, fiber

**hexose** Carbohydrate containing 6 carbons.

**glucose** Monosaccharide with 6 carbons; also called dextrose; a primary source of energy in the body; found in table sugar (sucrose) bound to fructose.

**fructose** Monosaccharide with 6 carbons that forms a 5-membered or 6-membered ring with oxygen in the ring; found in fruits and honey.

**galactose** Six-carbon monosaccharide that forms a 6-membered ring with oxygen in the ring; an isomer of glucose.

**condensation reaction** Chemical reaction in which a bond is formed between 2 molecules by the elimination of a small molecule, such as water.

## Disaccharides: Maltose, Sucrose, and Lactose

Carbohydrates containing 2 monosaccharides are called disaccharides (*di* means "2"). The linking of 2 monosaccharides occurs in a **condensation reaction.** During this reaction, 1 molecule of water is formed (and released) by taking a hydroxyl group (OH) from 1 sugar and a hydrogen (H) from the other sugar (Fig. 5-4).

### A Biochemist's View

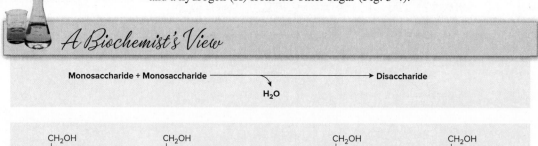

Monosaccharide + Monosaccharide ⟶ Disaccharide
$H_2O$

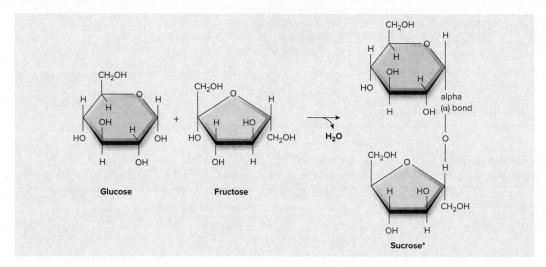

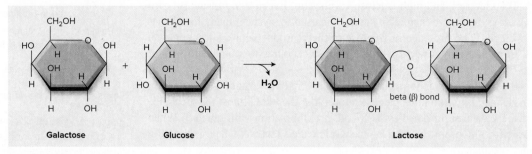

*Figure 5-4*  Two monosaccharides combine to form a disaccharide.

• Maltose is made up of 2 glucose molecules.

• Sucrose, or common table sugar, is made up of glucose and fructose.

• Lactose, or milk sugar, is made up of glucose and galactose. Note that lactose contains a different type of bond (beta, or β) than maltose and sucrose (alpha, or α); individuals who produce little of the enzyme lactase, which breaks this beta bond, have difficulty digesting lactose.

*This is the preferred representation of sucrose because both glucose and fructose are presented as they are usually drawn. Note: Typical horizontal representations of sucrose flip the fructose structure horizontally.

Source: International Union of Pure and Applied Chemistry; D. Blackman, Representing the Structure of Sucrose; *Biochem Ed.*, 1975; 3:77.

One carbon on each monosaccharide participating in the condensation reaction chemically bonds with a single oxygen. Two forms of this C—O—C bond exist in nature: alpha (α) bonds and beta (β) bonds. As shown in Figure 5-4, maltose and sucrose contain the alpha form, whereas lactose contains the beta form. Many carbohydrates contain long chains of glucose with the individual monosaccharides bonded together by either alpha or beta bonds.

Beta bonds differ from alpha bonds in that they cannot be easily broken down by digestive enzymes for absorption in the small intestine. Thus, foods that contain saccharide molecules linked together by beta bonds (e.g., in milk and dietary fiber) are often difficult or impossible for individuals to digest because they lack the enzymes necessary for breaking beta bonds apart.

The disaccharide **maltose** contains 2 glucose molecules joined by an alpha bond. When seeds sprout, they produce enzymes that break down polysaccharides stored in the seed to sugars, such as maltose and glucose. These sugars provide the energy for the plant to grow. Malting, the first step in the production of alcoholic beverages, such as beer, lets grain seeds sprout. Few other food products or beverages contain maltose. In fact, most of the maltose we ultimately digest in the small intestine is produced when we break down longer-chain polysaccharides.

**Sucrose,** common table sugar, is composed of glucose and fructose linked by an alpha bond. Large amounts of sucrose are found naturally in plants, such as sugarcane, sugar beets, and maple tree sap. The sucrose from these sources can be purified to various degrees. Brown, white, and powdered sugars are common forms of sucrose sold in grocery stores.

**Lactose,** the primary sugar in milk and milk products, consists of glucose joined to galactose by a beta bond. As discussed later in this chapter, many people are unable to digest large amounts of lactose because they don't produce enough of the enzyme lactase, which is needed to break this beta bond. This can cause intestinal gas, bloating, cramping, and discomfort as the unabsorbed lactose is metabolized into acids and gases by bacteria in the large intestine.[5]

Many terms are used to refer to monosaccharides and disaccharides and products containing these sugars. Monosaccharides and disaccharides often are referred to as *simple sugars* because they contain only 1 or 2 sugar units.

## Oligosaccharides: Raffinose and Stachyose

Oligosaccharides are complex carbohydrates that contain 3 to 10 single sugar units (*oligo* means "few"). Two oligosaccharides of nutritional importance are **raffinose** and **stachyose,** which are found in onions, cabbage, broccoli, whole wheat, and legumes such as kidney beans and soybeans. The beta bonds in oligosaccharides cannot be broken down by our digestive enzymes. Thus, when we eat foods with raffinose and stachyose, these oligosaccharides pass undigested into the large intestine, where bacteria metabolize them, producing gas and other by-products.[6]

Although many people have no symptoms after eating legumes, others experience unpleasant side effects from intestinal gas. An enzyme preparation, such as Beano®, can help prevent these side effects if taken right before a meal. This enzyme preparation works in the digestive tract to break down many of the indigestible oligosaccharides.

## Polysaccharides: Starch, Glycogen, and Fiber

Polysaccharides are complex carbohydrates that often contain hundreds to thousands of glucose molecules. The polysaccharides include some that are digestible, such as **starch,** and some that are largely indigestible, such as **fiber.** The digestibility of these polysaccharides is determined mainly by whether the glucose units are linked together by alpha or beta bonds.

### Digestible Polysaccharides: Starch and Glycogen

Starch, the major digestible polysaccharide in our diets, is the storage form of glucose in plants. There are 2 types of plant starch—**amylose** and **amylopectin**—both of which are a source of energy for plants and for animals that eat plants. Amylose and amylopectin are found in potatoes, beans, breads, pasta, rice, and other starchy products, typically in a ratio of about 1:4.

Amylose and amylopectin contain many glucose units linked by alpha bonds. The primary difference is that amylose is a linear, unbranched chain of glucose molecules that contains

Some over-the-counter products, such as Beano®, contain an enzyme, called alpha-galactosidase, that can break apart bonds in oligosaccharides. This helps decrease the amount of intact oligosaccharides reaching the large intestine and, thus, reduces the intestinal gas produced when beans and other legumes are eaten.

McGraw-Hill Education/Mark Dierker, photographer

## Critical | Thinking

Keith enjoys Mexican food, especially with a generous portion of black or pinto beans. However, he often develops gas and intestinal cramps after these meals. His friends suggest that he try using a product called Beano® to reduce his symptoms. How might it help Keith?

**raffinose** Indigestible oligosaccharide made of 3 monosaccharides (galactose-glucose-fructose).

**stachyose** Indigestible oligosaccharide made of 4 monosaccharides (galactose-galactose-glucose-fructose).

**starch** Carbohydrate made of multiple units of glucose attached together in a form the body can digest; also known as complex carbohydrate.

**fiber** Substance in plant foods that is not broken down by the digestive processes of the stomach or small intestine. Fiber adds bulk to feces. Fiber naturally found in foods is called dietary fiber.

## A Biochemist's View

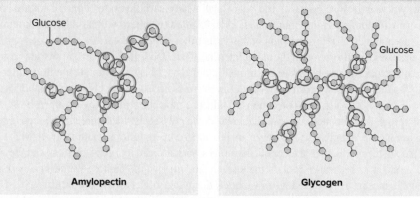

**Amylose**

**Amylopectin**

**Glycogen**

All bonds between glucose molecules are 1-4 alpha bonds. The bonds in the green boxes below are 1-4 alpha bonds. Notice how they are in a straight row.

For amylopectin and glycogen, all bonds between glucose molecules are 1-4 alpha bonds except the ones where the molecule branches. The circled areas represent branch points; an alpha 1-6 bond occurs where the chain branches (circled). The bond in the circle below is an alpha 1-6 bond. Notice how it links the upper and lower row of glucose chains in the green boxes below.

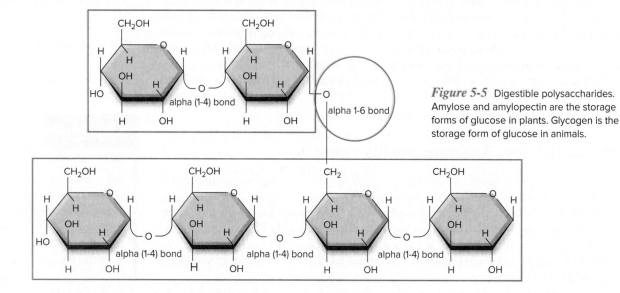

*Figure 5-5* Digestible polysaccharides. Amylose and amylopectin are the storage forms of glucose in plants. Glycogen is the storage form of glucose in animals.

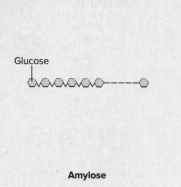

As some vegetables age, their sugars are converted to starches, making them taste less sweet.

Ingram Publishing

only 1 type of alpha bond (called a 1-4 bond), whereas amylopectin is a highly branched–chain structure that links glucose molecules using 2 types of alpha bonds (1-4 bonds link straight chains of glucose and 1-6 bonds link glucose at the branching points) (Fig. 5-5). Alpha 1-4 bonds are broken by amylase enzymes produced in the mouth and pancreas. Alpha 1-6 bonds are broken by an intestinal enzyme called alpha-dextrinase. The more numerous the branches in a starch, the more sites (ends) available for enzyme action. This explains why amylopectin causes blood glucose levels to increase more quickly than amylose.

The properties of amylopectin and amylose make them useful in food manufacturing. The branches in amylopectin allow it to retain water to form a very stable starch gel. Thus, food manufacturers commonly use starches rich in amylopectin to thicken sauces and gravies. Amylopectin also is used in many frozen foods because it remains stable over a wide temperature range. Amylose-rich molecules can be bonded to each other to produce modified food starch, a thickener used in baby foods, salad dressings, and instant puddings.

**Glycogen,** the storage form of carbohydrate in humans and other animals, also contains many glucose units linked together with alpha bonds. The structure of glycogen is similar to

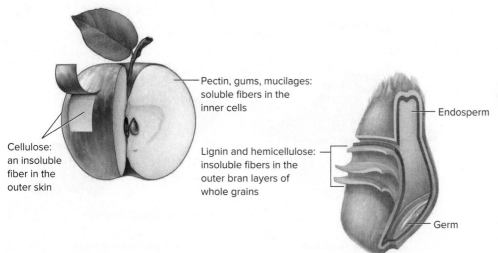

*Figure 5-6* Types of dietary fiber. The skin of an apple consists of the insoluble fiber cellulose, which provides structure for the fruit. The soluble fiber pectin "glues" the fruit cells together. The outside covering of a wheat kernel is made of layers of bran—insoluble fibers.

Pectin, gums, mucilages: soluble fibers in the inner cells

Cellulose: an insoluble fiber in the outer skin

Lignin and hemicellulose: insoluble fibers in the outer bran layers of whole grains

Endosperm

Germ

▶ The Food and Nutrition Board has recommended that the terms *soluble* and *insoluble fibers* gradually be replaced by other terms, such as *viscosity* and *fermentability,* which more clearly describe the properties of fibers. The actual terms used may change as scientific knowledge expands.

**total fiber** Combination of dietary fiber and functional fiber in a food; also called fiber.

**dietary fiber** Fiber in food.

**functional fiber** Fiber added to foods that has shown to provide health benefits.

**insoluble fibers** Fibers that mostly do not dissolve in water and are not metabolized by bacteria in the large intestine. These include cellulose, some hemicelluloses, and lignins; more formally called nonfermentable fibers.

**soluble fibers** Fibers that either dissolve or swell in water and are metabolized (fermented) by bacteria in the large intestine; include pectins, gums, and mucilages; more formally called viscous fibers.

that of amylopectin, but it is even more highly branched. The branched structure of glycogen allows it to be broken down quickly by enzymes in the body cells where it is stored.

Liver and muscle cells are the major storage sites for glycogen. The amount stored in these cells is influenced by the amount of carbohydrate in the diet. Although the amount of glycogen that can be stored is limited, glycogen storage is extremely important.[3] The approximately 90 grams (360 kcal) of glycogen stored in the liver can be converted into blood glucose to supply the body with energy, whereas the 300 grams (1200 kcal) of glycogen stored in muscles supply glucose for muscle use, especially during high-intensity and endurance exercise. (See Chapter 11 for a detailed discussion of carbohydrate use during physical activity.)

### Indigestible Polysaccharides: Dietary and Functional Fiber

Folklore surrounding fiber, or "roughage," has been a part of American culture since the 1800s, when a minister named Sylvester Graham traveled up and down the East Coast, extolling the virtues of fiber. He left us a legacy—the graham cracker. Although today's graham cracker bears little resemblance to the whole-grain product he promoted, present-day scientific evidence supports this early promotion of fiber as part of a healthy diet.

**Total fiber** (or just the term *fiber*) refers to the **dietary fiber** that occurs naturally in foods, as well as the **functional fiber** that may be added to food to provide health benefits.[2] Currently, Nutrition Facts labels include only dietary fiber and do not reflect any added functional fiber.

Fibers are composed primarily of the nonstarch polysaccharides **cellulose, hemicelluloses, pectins, gums,** and **mucilages. Lignins** are the only noncarbohydrate components of dietary fibers. Unlike the digestible polysaccharides that contain alpha bonds, the monosaccharide units in fibers are linked by beta bonds. As noted earlier, monosaccharide molecules joined by beta bonds are not broken down by human digestive enzymes. Thus, these undigested fibers pass through the small intestine into the large intestine, where bacteria metabolize some and form short-chain fatty acids and gas. These short-chain fatty acids provide fuel for cells in the large intestine and enhance intestinal health.[7] Pectins, gums, and mucilages are most readily digested by the intestinal bacteria, yielding about 1.5 to 2.5 kcal/g. Cellulose, hemicellulose, and lignins are more resistant to being broken down by bacteria. The body tends to adapt over time to a high fiber intake, leading to fewer symptoms of bloating, gas, and discomfort.[6]

Cellulose, hemicelluloses, and lignins form the structural part of the plant cell wall in vegetables and whole grains. Bran layers form the outer covering of all seeds; thus, whole grains (those in which the bran and outer layers have not been removed in processing) are good sources of fiber (Fig. 5-6). Because of their chemical structure, these fibers do not dissolve in water. Therefore, they are often referred to as **insoluble fibers.**

In contrast to the insoluble fibers, pectins, gums, mucilages, and some hemicelluloses dissolve easily in water and are classified as **soluble fibers.** In water, they become viscous (gel-like) in consistency. This property makes them useful for thickening jam, jelly, yogurt, and other food products. They also occur naturally inside and around plant cells in oat bran, many fruits, legumes, and psyllium.

Fruit is a good source of dietary fiber. Pectin, a naturally occurring fiber in fruit, helps jellies and jams gel.

sarsmis/123RF

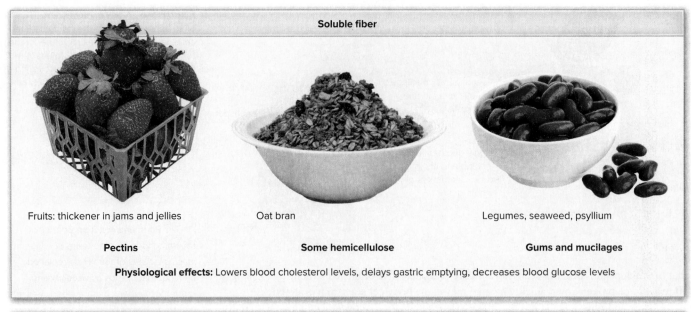

Soluble fiber

Fruits: thickener in jams and jellies

**Pectins**

Oat bran

**Some hemicellulose**

Legumes, seaweed, psyllium

**Gums and mucilages**

**Physiological effects:** Lowers blood cholesterol levels, delays gastric emptying, decreases blood glucose levels

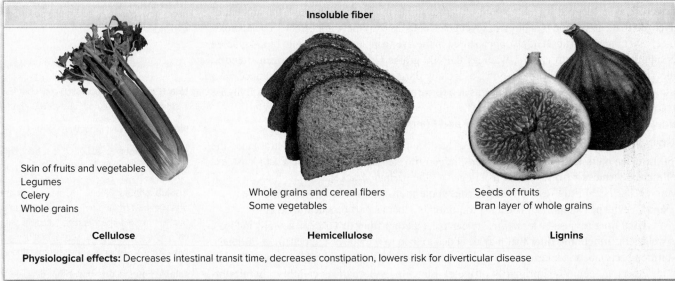

Insoluble fiber

Skin of fruits and vegetables
Legumes
Celery
Whole grains

**Cellulose**

Whole grains and cereal fibers
Some vegetables

**Hemicellulose**

Seeds of fruits
Bran layer of whole grains

**Lignins**

**Physiological effects:** Decreases intestinal transit time, decreases constipation, lowers risk for diverticular disease

*Figure 5-7* Soluble and insoluble fibers. Fibers can be classified as either soluble or insoluble based on their properties. Soluble fibers dissolve in water, whereas insoluble fibers do not dissolve in water.

Dynamic Graphics/PunchStock; Stockbytes/Getty Images; Moving Moment/Shutterstock; Ingram Publishing/Alamy Stock Photo; Photodisc/Getty Images; Mahirart/Shutterstock

The physical properties of soluble and insoluble fibers provide health benefits (Fig. 5-7). When consumed in adequate quantities, these fibers can lower blood cholesterol levels and blood glucose levels, thereby reducing risks of cardiovascular disease and diabetes.[8,9] In addition, fiber can decrease intestinal transit time, thus reducing risks of constipation and diverticular disease.[8,10,11,12] The health benefits of fiber are discussed in detail later in the chapter.

*Knowledge Check*

1. Which sugars are classified as monosaccharides? As disaccharides?
2. Why are foods that contain saccharide units linked by beta bonds difficult to digest?
3. What types of carbohydrates are classified as polysaccharides?

## 5.2 Carbohydrates in Foods

Carbohydrates are found in a wide variety of foods. Foods such as table sugar, jam, jelly, fruit, fruit juices, soft drinks, baked potatoes, rice, pasta, cereals, and breads are predominantly carbohydrates. Other foods, such as dried beans, lentils, corn, peas, and dairy products (milk and yogurt), also are good sources of carbohydrate, although they contribute protein and, in some cases, fat to our diets as well. Foods with little or no carbohydrate include meats, fish, poultry, eggs, vegetable oils, butter, and margarine.

Breads are a rich source of carbohydrate.

PhotoAlto/Alamy Stock Photo

### Starch

Starches contribute much of the carbohydrate in our diets. Recall that plants store glucose as polysaccharides in the form of starches. Thus, plant-based foods—such as legumes, tubers, and grains (wheat, rye, corn, oats, barley, and rice) used to make breads, cereals, and pasta—are the best sources of starch. A diet rich in these starches provides ample carbohydrate, as well as many micronutrients.

### Fiber

Fiber can be found in many of the same foods as starch, so a diet rich in grains, legumes, and tubers also can provide significant amounts of dietary fiber (especially insoluble cellulose, hemicellulose, and lignins). Because much of the fiber in whole grains is found in the outer layers, which are removed in processing, highly processed grains are low in fiber. Soluble fibers (pectin, gums, mucilages) are found in the skins and flesh of many fruits and berries; as thickeners and stabilizers in jams, yogurts, sauces, and fillings; and in products that contain psyllium and seaweed.

For individuals who have difficulties consuming adequate dietary fiber, fiber is available as a supplement (e.g., psyllium fiber in Metamucil®) or as an additive to certain foods (functional fiber). In this way, individuals with relatively low dietary fiber intakes can still obtain the health benefits of fiber.

### Nutritive Sweeteners

The various substances that impart sweetness to foods fall into 2 broad classes: nutritive sweeteners, which can be metabolized to yield energy, and non-nutritive (alternative) sweeteners, which provide no food energy (Table 5-1). The sweetness of sucrose (table sugar) makes it the benchmark against which all other sweeteners are measured. As shown in Figure 5-8, the alternative sweeteners are much sweeter than the nutritive sweeteners on a per gram basis.[1]

There are many forms of sugar in our diets.

C Squared Studios/Getty Images

| Table 5-1 Typical Sources of Sweeteners | |
|---|---|
| **Type of Sweetener** | **Typical Sources** |
| **Nutritive Sweeteners** | |
| **Sugars** | |
| Lactose | Dairy products |
| Maltose | Sprouted seeds, some alcoholic beverages |
| Glucose | Corn syrup, honey |
| Sucrose* | Table sugar, most sweets |
| Invert sugar | Some candies, honey |
| Fructose | Fruit, honey, some soft drinks, corn syrup |
| **Sugar Alcohols** | |
| Sorbitol | Sugarless candies, sugarless gum |
| Mannitol | Sugarless candies |
| Xylitol | Sugarless gum |
| **Non-nutritive (Alternative) Sweeteners†** | |
| Tagatose (Naturlose®) | Ready-to-eat cereals, diet soft drinks, health bars, frozen yogurt, fat-free ice cream, candies, frosting, sugarless gum |
| Advantame | Partial tabletop sugar replacer; used by the food industry to sweeten and enhance food flavors |
| Aspartame (Equal®) | Diet soft drinks, diet fruit drinks, sugarless gum, powdered diet sweetener |
| Acesulfame-K (Sunette®) | Sugarless gum, diet drink mixes, powdered diet sweeteners, puddings, gelatin desserts |
| Saccharin (Sweet'N Low®) | Diet soft drinks, tabletop sweetener |
| Sucralose (Splenda®) | Diet soft drinks, tabletop use, sugarless gum, jams, frozen desserts |
| Neotame | Tabletop sweetener, baked goods, frozen desserts, diet soft drinks, jams and jellies |
| Stevia (Truvia®) | Diet soft drinks |
| Monk fruit | Tabletop sweetener |

*Sucrose is broken down into glucose and fructose.
†Cyclamate is a non-nutritive sweetened used in Europe.

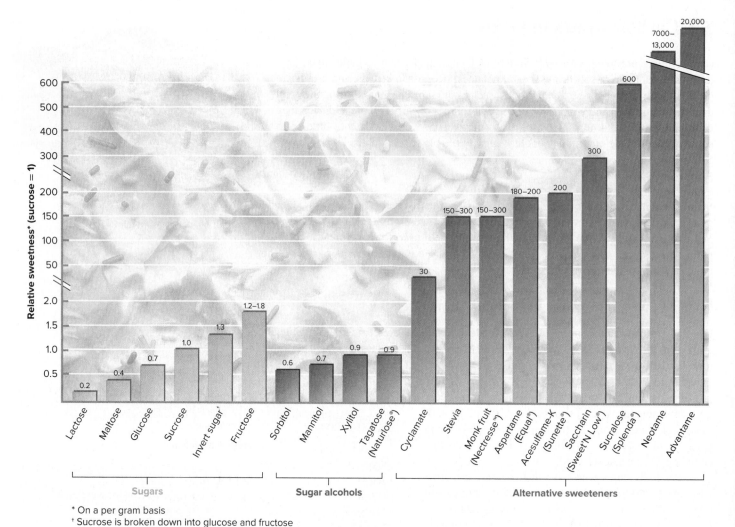

*Figure 5-8* Sweetness of sugars (nutritive sweeteners) and alternative (non-nutritive) sweeteners compared with sucrose.

Jules Frazier/Getty Images

The monosaccharides (glucose, fructose, and galactose) and disaccharides (sucrose, lactose, and maltose) are classified as nutritive sweeteners (Table 5-2).[1] Sucrose is obtained from sugarcane and sugar beet plants. Most of the sucrose and the other sugars we eat come from foods and beverages to which sugar has been added during processing and/or manufacturing. Nutrition Facts food labels list the amount of total sugars and how much of the total sugar is **added sugars** in a food (See Chapter 2, Section 2.2, for more details about the Nutrition Facts label). Added sugars include those sugars *added* during food manufacturing, such as honey, syrup, and concentrated fruit and vegetables juices. Added sugars do not include naturally occurring sugars, such as those in dairy and fruit. The major food sources of added sugars are soft drinks, candy, cakes, cookies, pies, fruit drinks, and dairy desserts such as ice cream. In general, the more processed the food, the more simple sugar it contains. The rest of the sugar in our diets is present naturally in foods such as fruits and juices.

Although the sugar in 100% fruit juice is naturally occurring, the biological response to this sugar is virtually identical to added sugars.[13,14] It is a good idea to limit fruit juice intake to 4 to 6 ounces daily because research suggests that consumption of fruit juice can contribute to an increase in all-cause mortality.

A nutritive sweetener used frequently by the food industry is high fructose corn syrup. High fructose corn syrup is made by treating cornstarch with acid and enzymes to break down much of the starch into glucose. Then enzymes convert some of the glucose into fructose. The final syrup is about 55% fructose, although it can range from 40 to 90% fructose. High fructose

**Table 5-2  Nutritive Sweeteners Used in Foods**

| | |
|---|---|
| Sugar | Honey |
| Sucrose | Molasses |
| Brown sugar | Date sugar |
| Turbinado sugar | Maple syrup |
| Invert sugar | Dextrin |
| Glucose | Fructose |
| Sorbitol | Maltose |
| Levulose | Caramel |
| Mannitol | Fruit sugar |
| Confectioner's sugar (powdered sugar) | Polydextrose |
| Corn syrup or sweeteners | Lactose |
| High fructose corn syrup | Agave nectar |

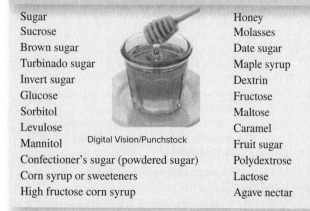

Digital Vision/Punchstock

corn syrup is similar in sweetness to sucrose, but it is much cheaper to use in food products. High fructose corn syrup is used in soft drinks, candies, jam, jelly, and desserts (e.g., packaged cookies).[4]

There are some health concerns associated with the consumption of high fructose corn syrup, including an increased risk for nonalcoholic fatty liver disease (see Chapter 4). Overall, however, the impact of the consumption of high fructose corn syrup on health is a controversial topic needing more research.[15]

### Sugar Alcohols

The sugar alcohols, sorbitol, mannitol, and xylitol, are nutritive sweeteners used in sugarless gum and candies. Sugar alcohols are not easily metabolized by bacteria in the mouth and thus do not promote dental caries as readily as do sugars such as sucrose. Sugar alcohols do contribute energy (about 1.5–3 kcal/g), but they are absorbed and metabolized to glucose more slowly than sugars. In large quantities, sugar alcohols can cause diarrhea, so labels must include this warning.

Sugar alcohols are listed individually on ingredient labels if only 1 sugar alcohol is used in a product; they are grouped together under the heading "sugar alcohols" if 2 or more are used. The calories listed on Nutrition Facts labels account for the calories in each sugar alcohol combined in the food products.

## Non-nutritive (Alternative) Sweeteners

Non-nutritive (alternative or artificial) sweeteners provide very low calorie or non-caloric sugar substitutes for people with diabetes and those trying to lose (or control) body weight. Alternative sweeteners include saccharin, cyclamate, aspartame, neotame, sucralose, acesulfame-K, tagatose, stevia, monk fruit, and advantame.[1,16] Alternative sweeteners yield little or no energy when consumed in amounts typically used in food products and do not promote dental caries.

The safety of sweeteners is determined by the FDA and is indicated by an **Acceptable Daily Intake (ADI)** guideline.[16] The ADI is the amount of alternative sweetener considered safe for daily use over one's lifetime. ADIs are based on studies in laboratory animals and are set at a level 100 times less than the level at which no harmful effects were noted in animal studies. Alternative sweeteners can be used safely by adults and children. Although general use is considered safe during pregnancy, pregnant women may want to discuss this issue with their health-care providers.[1]

### Saccharin

The oldest alternative sweetener, saccharin, is approximately 300 times sweeter than sucrose. Saccharin was once thought to pose a risk of bladder cancer based on studies using laboratory animals. It is no longer listed as a potential cause of cancer in humans because the earlier research is now considered weak and inconclusive.[1] The Joint FAO/WHO Expert Committee on Food Additives has set the ADI for saccharin at 5 mg/kg body weight per day. For a 154-pound (70 kg) adult, this equates to approximately 3 12-ounce diet soft drinks or 9 packets of the sweetener (e.g., Sweet'N Low®) daily.[16] Saccharin is used as a tabletop sweetener and in a variety of foods and beverages. It is not useful in cooking because heating causes it to develop a bitter taste.

### Aspartame

Aspartame is used throughout the world to sweeten beverages, gelatin desserts, chewing gum, cookies, and the toppings and fillings of prepared bakery goods. Aspartame breaks down and loses its sweetness when foods are cooked or heated. NutraSweet® and Equal® are brand names for aspartame.

Although aspartame yields about 4 kcal/g (the same calories as sucrose), it is 160 to 220 times sweeter than sucrose. Thus, because only a small amount of aspartame is needed to sweeten a food or beverage, it does not contribute calories to foods. The ADI for aspartame for an adult is 50 mg/kg body weight/day. This is equivalent to about 18 cans of diet soft drink or about 80 packets of Equal®.[16]

Soft drinks are common sources of sugars and alternative sweeteners.

Shutterstock/M. Unal Ozmen

INGREDIENTS: SORBITOL, GUM BASE, MANNITOL, GLYCEROL, HYDROGENATED GLUCOSE SYRUP, XYLITOL, ARTIFICIAL AND NATURAL FLAVORS, ASPARTAME, RED 40, YELLOW 6 AND BHT (TO MAINTAIN FRESHNESS). PHENYLKETONURICS: CONTAINS PHENYLALANINE.

*Sugarless Gum*

Sugar alcohols and the alternative sweetener aspartame are used to sweeten this product. Note the warning for people with phenylketonuria (PKU) that this product is made with aspartame and, thus, contains phenylalanine.

Nancy R. Cohen/Getty Images

**Acceptable Daily Intake (ADI)** Amount of a food additive considered safe for daily consumption over one's lifetime.

Scientific evidence has shown that the use of aspartame is safe for most individuals. However, the FDA has received reports of adverse reactions (headaches, dizziness, seizures, nausea, and other side effects) to aspartame. Although the percentage of people affected is very small, it is important for people who are sensitive to aspartame to avoid it. Those with the genetic disease phenylketonuria (PKU), which interferes with the metabolism of the amino acid phenylalanine, also should avoid aspartame because of its high phenylalanine content.

### Neotame

Neotame is approved by the FDA for use as a general purpose sweetener in a wide variety of food products, such as baked goods, nonalcoholic beverages (including soft drinks), chewing gum, confections and frostings, frozen desserts, gelatins and puddings, jams and jellies, processed fruits and fruit juices, toppings, and syrups. Neotame is heat stable and can be used in cooking and as a tabletop sweetener. Neotame is approximately 7000 to 13,000 times sweeter than sucrose. Thus, the small amounts needed to sweeten products do not contribute calories. Although neotame also contains phenylalanine, its bonding to other amino acids differs from that of aspartame and prevents it from being broken down. Therefore, it does not cause a problem for individuals with PKU. The ADI for neotame is 18 mg/kg body weight/day.[16]

### Acesulfame-K

The alternative sweetener acesulfame-K (the *K* stands for potassium) is sold for use in the U.S. as Sunette®. Acesulfame-K is 200 times sweeter than sucrose. It contributes no energy to the diet because it is not digested by the body.[1] Acesulfame-K can be used in baking because it does not lose its sweetness when heated. In the U.S., it is currently approved for use in chewing gum, powdered drink mixes, gelatins, puddings, baked goods, tabletop sweeteners, candy, throat lozenges, yogurt, and non-dairy creamers. The ADI for acesulfame-K is 15 mg/kg body weight/day.[16]

### Sucralose

Sucralose, sold as Splenda®, is 600 times sweeter than sucrose. It is the only artificial sweetener made from sucrose. It is made by substituting 3 chlorines (Cl) for 3 hydroxyl groups (–OH) on sucrose.[1] This substitution prevents it from being digested and absorbed. Sucralose is used as a tabletop sweetener and in soft drinks, chewing gum, baked goods, syrups, gelatins, frozen dairy desserts (e.g., ice cream), jams, and processed fruits and fruit juices. Sucralose is heat stable; thus, it can be used in cooking and baking. The ADI for sucralose is 5 mg/kg body weight/day.[16]

Extracts from the leaves of the stevia plant are used to make some alternative sweeteners.

BasieB/E+/Getty Images

### Tagatose

Tagatose, sold as Naturlose®, is an isomer of fructose. It is almost as sweet as sucrose and can be used in cooking and baking. Because it is poorly absorbed by the body, tagatose yields only 1.5 kcal/g. It has a prebiotic effect because it is fermented by bacteria in the large intestine (see Chapter 4). Tagatose is approved for use in ready-to-eat cereals, diet soft drinks, health bars, frozen yogurt, fat-free ice cream, soft and hard confectionary products, frosting, and chewing gum. It is metabolized like fructose, so individuals with disorders of fructose metabolism should avoid using it.

### Stevia

Stevia (also called rebiana) is an alternative sweetener derived from a plant from the Amazon rain forest.[17] It is 250 times sweeter than sucrose but provides no energy. Although stevia has been used in teas and as a sweetener in Japan since the 1970s, the FDA has only recently approved its use in beverages. Stevia can be purchased as a dietary supplement in natural and health food stores. In the U.S., it is combined with a sugar alcohol called erythritol and marketed as PureVia™ and Truvia®. Stevia also is blended with cane sugar and sold as Sun Crystals®. The ADI for stevia is 4 mg/kg body weight/day.

A variety of alternative sweeteners are available.

Mark Dierker/McGraw-Hill Education

### Luo han guo

Monk fruit (Luo han guo) is an intensely sweet, small, green fruit from Asia that has been cultivated for about 800 years. When concentrated, the fruit's juice is 150 to 300 times sweeter than sugar. This sweetener is heat stable and can be used in cooking and baking. Currently, it is being used mostly as a tabletop sweetener sold as Monk Fruit in the Raw® and Nectresse™. An ADI has not been set; however, the FDA has classified monk fruit sweeteners as Generally Recognized as Safe (GRAS).

### Advantame

Advantame is 1 of the newest approved alternative sweeteners. It is derived from aspartame and vanillin—a component of the extract of vanilla bean. It is 20,000 times sweeter than sugar, making it the highest intensity sweetener on the market. It is approved as a general use sweetener and flavor enhancer for food and beverage manufacturers. Advantame is heat stable and can be used in cooking and baking. It is recommended as a partial sugar replacement and currently available to the commercial and food ingredients market. The FDA has set an ADI of 32.8 mg per kilogram body weight (or the equivalent of about 800 cans of soda).

## Culinary Perspective

### Comparing Nutritive Sweeteners—What's the Best Sugar?

The idea that there is a more nutritious nutritive sweetener is appealing and helps explain why there are so many opinions about what the "best," "healthiest" sweetener is. The accompanying table shows a calorie comparison of popular nutritive sweeteners; notice the similar calorie values for 1 serving. Some sweeteners, such as honey, brown sugar, molasses, and coconut sugar, contain a few vitamins and minerals; however, the amount of these nutrients in a serving is very small. To get appreciable amounts of vitamins and minerals, you would need to eat very large amounts of the sweeteners, which also would result in a high intake of sugar and calories. It is better to focus on getting vitamins and minerals from more nutrient dense, lower energy dense foods such as fruits, vegetables, legumes, and whole grains and keep sugar intake low.

Sugar comes in a variety of forms. White granulated sugar, commonly known as table or "regular" sugar, is the form most used in baking and as a tabletop sweetener. Other forms of sugar, such brown sugars (e.g., turbinado and brown sugar), are made by mixing white sugar with various amounts of molasses.

MaraZe/Shutterstock

| Sweetener | Calories per Tablespoon |
|---|---|
| Table sugar | 46 |
| High fructose corn syrup (HFCS) | 53 |
| Honey | 64 |
| Agave nectar | 60 |
| Maple syrup | 52 |
| Blackstrap molasses | 58 |
| Brown sugar, packed | 53 |
| Raw sugar (turbinado) | 45 |
| Coconut sugar | 54 |

Regardless of the origin of a nutritive sweetener (i.e., processed from sugar cane, sugar beets, or agave plant or produced by honeybees), once digested and metabolized, the biological response to sugar is virtually the same. Therefore, the "best" or "healthiest" nutritive sweetener is that which you can enjoy in moderation.

▶ The *Healthy People* 2030 goals related to carbohydrate intake for those aged 2 years and older include

- Increase consumption of whole grains.
- Increase consumption of fruits.
- Increase consumption of vegetables.
- Increase contribution of dark-green vegetables, red and orange vegetables, and peas and beans.
- Reduce consumption of calories from added sugars.

Many of the foods we enjoy contain simple sugars. To improve nutrient intake, limit the consumption of sweets.

Getty Images/Digital Vision

# 5.3 Recommended Intake of Carbohydrates

According to the RDA, adults need about 130 g/day of digestible carbohydrate to supply adequate glucose for the brain and central nervous system to prevent the partial replacement of glucose by ketone bodies as an energy source (see Section 5.4). The Food and Nutrition Board recommends that, to provide for total body energy needs, carbohydrate intake should be considerably higher, ranging from 45 to 65% of total energy intake.[2] However, not all diet programs follow the Food and Nutrition Board recommendations. Some diet programs (e.g., keto-based, Atkins®, and South Beach® Diets) promote very low carbohydrate intakes, whereas others (e.g., the Pritikin® and Eat More, Weigh Less® Diets) promote very high carbohydrate intakes. Despite these differences in opinion, most scientists and nonscientists agree that carbohydrates in our diets should include mostly fiber rich fruits, vegetables, and whole grains and little added sugars and caloric sweeteners.[2]

North Americans obtain about 50% of their energy intakes from carbohydrates. The leading carbohydrate sources for U.S. adults are white bread, soft drinks, cookies, cakes, donuts, sugars, syrups, jams, and potatoes. Worldwide, carbohydrates account for about 70 to 80% of energy consumed, with much greater intakes of whole grains, fruits, vegetables, and legumes than is typical in North American diets.

The Dietary Guidelines for Americans recommend limiting added sugars to no more than 10% of daily total energy intake for those age 2 and older.[18] The World Health Organization also suggests that sugars added to foods during processing and preparation ("added sugars") should provide no more than about 10% of total daily energy intake.

The Institute of Medicine's Food and Nutrition Board set an Upper Limit of 25% of energy intake for added sugar consumption.[2]

The Adequate Intake for fiber is based on a goal of 14 g/1000 kcal consumed. For adults up to age 50, the Adequate Intake is set at 25 g for women and 38 g for men. After age 50, the Adequate Intake falls to 21 g/day and 30 g/day, respectively.[2] The Adequate Intake for fiber is aimed at reducing the risk of diverticular disease, cardiovascular disease, and other chronic diseases. The Daily Value used for fiber on food and supplement labels is 28 g for a 2000 kcal diet.

A nutritious daily diet with ample sources of carbohydrate should include approximately 6 ounces of grains (with 3 ounces as whole grains), 2 ½ cups of vegetables, 2 cups of fruit, and 3 cups of milk. As a protein alternative to meat, include more dried beans and lentils in your diet to increase fiber and total carbohydrate intake. Table 5-3 shows a diet containing the recommended intakes of carbohydrates.

## Our Carbohydrate Intake

Carbohydrates supply about 50% of the energy intakes of adults in North America. Although the proportion of total energy provided by carbohydrates is in line with recommendations, the types of carbohydrates consumed are not. Added sugars account for over 14.5% of total energy intake—more than the 10% total energy intake maximum recommended by the Dietary Guidelines for Americans.[18] High sugar intakes are, in large part, due to the popularity of sugar-sweetened beverages (Fig. 5-9).[19] Intake of caloric sweeteners,

**Table 5-3 Sample Menus Containing 1600 kcal with 25 g of Fiber and 2000 kcal with 38 g of Fiber\***

| Menu | 25 g Fiber Plan | | | 38 g Fiber Plan | | |
|---|---|---|---|---|---|---|
| | Serving Size | Carbohydrate Content (g) | Fiber Content (g) | Serving Size | Carbohydrate Content (g) | Fiber Content (g) |
| **Breakfast** | | | | | | |
| Muesli cereal | 1 cup | 60 | 6 | 1 cup | 60 | 6 |
| Raspberries | ½ cup | 11 | 2 | ½ cup | 11 | 2 |
| Whole-wheat toast | 1 slice | 13 | 2 | 2 slices | 26 | 4 |
| Margarine | 1 tsp | 0 | 0 | 1 tsp | 0 | 0 |
| Orange juice | 1 cup | 28 | 0 | 1 cup | 28 | 0 |
| 1% milk | 1 cup | 24 | 0 | 1 cup | 24 | 0 |
| Coffee | 1 cup | 0 | 0 | 1 cup | 0 | 0 |
| **Lunch** | | | | | | |
| Bean and vegetable burrito | 2 small | 50 | 4.5 | 3 small | 75 | 7 |
| Guacamole | ¼ cup | 5 | 4 | ¼ cup | 5 | 4 |
| Monterey Jack cheese | 1 oz | 0 | 0 | 1 oz | 0 | 0 |
| Pear (with skin) | 1 | 25 | 4 | 1 | 25 | 4 |
| Carrot sticks | — | — | — | ¾ cup | 6 | 3 |
| Sparkling water | 2 cups | 0 | 0 | 2 cups | 0 | 0 |
| **Dinner** | | | | | | |
| Grilled chicken (no skin) | 3 oz | 0 | 0 | 3 oz | 0 | 0 |
| Salad | ½ cup red cabbage ½ cup romaine ¼ cup peach slices | 7 | 3 | ½ cup red cabbage ½ cup romaine 1 cup peach slices | 19 | 6 |
| Toasted almonds | — | — | — | ½ oz | 3 | 2 |
| Fat-free salad dressing | 2 tbsp | 0 | 0 | 2 tbsp | 0 | 0 |
| 1% milk | 1 cup | 24 | 0 | 1 cup | 24 | 0 |
| Total | | 247 | 25 | | 306 | 38 |

\*The overall diet is based on MyPlate. Breakdown of approximate energy content: carbohydrate 58%; protein 12%; fat 30%.

Annabelle Breakey/Getty Images; BananaStock/PunchStock; Ingram Publishing/SuperStock

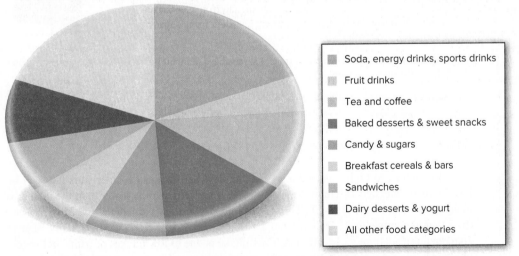

*Figure 5-9* Sources of added sugars in the diets of the U.S. population, ages 2 years and older.

Source: USDHHS/USDA. *Dietary Guidelines for Americans.* 9th ed., 2020; diearyguidelines.gov.

- Soda, energy drinks, sports drinks
- Fruit drinks
- Tea and coffee
- Baked desserts & sweet snacks
- Candy & sugars
- Breakfast cereals & bars
- Sandwiches
- Dairy desserts & yogurt
- All other food categories

including white sugar (sucrose) and high fructose corn syrup, adds approximately 270 calories daily to the diets of Americans ages 1 year and older.[18,20] Table 5-4 provides suggestions for reducing sugar intake.

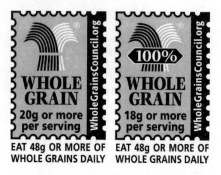

The Whole Grains Council has created stamps to help shoppers quickly locate foods containing whole grains. Both stamps indicate the total grams of whole grain in the food. To display the basic stamp, a food must have at least 8 grams of a whole grain in 1 serving. The 100% stamp indicates that all the grain ingredients in a food are whole grains and that each serving contains at least 16 grams of whole grains.

Whole Grain Stamps are a trademark of Oldways Preservation Trust and the Whole Grains Council, www.wholegrainscouncil.org. Courtesy of Oldways Preservation Trust and the Whole Grains Council, www.wholegrainscouncil.org

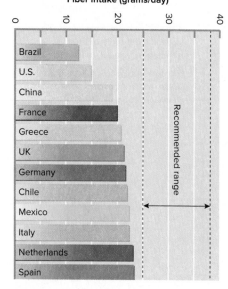

*Figure 5-10* Adult fiber intakes by country.[23]

## Table 5-4 Suggestions for Reducing Simple-Sugar Intake

Many foods we enjoy are sweet. These should be eaten in moderation.

**At the Supermarket**

- Read ingredient labels. Identify all the added sugars in a product. Select items lower in total sugar when possible.

- Buy fresh fruits or fruits packed in water, juice, or light syrup rather than those packed in heavy syrup.

- Buy fewer foods that are high in sugar, such as prepared baked goods, candies, sugared cereals, sweet desserts, soft drinks, and fruit-flavored punches. Substitute plain graham crackers, bagels, English muffins, diet soft drinks, and other low sugar alternatives.

- Buy reduced fat microwave popcorn to replace candy for snacks.

- Buy unsweetened or no-sugar-added versions of yogurt and fruit sauces (e.g., applesauce).

**In the Kitchen**

- Reduce the sugar in foods prepared at home. Try new low sugar recipes or adjust your own. Start by reducing the sugar gradually until you've decreased it by one-third or more.

- Experiment with spices, such as cinnamon, cardamom, coriander, nutmeg, ginger, and mace, to enhance the flavor of foods.

- Use home-prepared items with less sugar instead of commercially prepared products that are higher in sugar.

**At the Table**

- Reduce your use of white and brown sugars, honey, molasses, syrups, jams, and jellies.

- Choose fewer foods high in sugar, such as prepared baked goods, candies, and sweet desserts.

- Reach for fresh fruit instead of cookies or candy for dessert and between-meal snacks.

- Add less sugar to foods—coffee, tea, cereal, and fruit. Cut back gradually to a quarter or half the amount. Consider using sugar alternatives to substitute for some sugar.

- Reduce the number of sugared soft drinks, punches, and fruit juices you drink. Substitute water, diet soft drinks, and whole fruits.

- Reduce serving sizes of high sugar foods.

Mike Kemp/Rubberball/Getty Images; Martin Poole/Getty Images; Jack Holtel/McGraw-Hill Education

In contrast to high sugar intakes, the dietary fiber intakes of Americans fall well below the recommended 14 grams per 1000 calories and is generally lower than the intakes of many other countries (Fig. 5-10). Throughout life, both males and females eat 25 to 50% less fiber than recommended.[8] Only around 5% of the U.S. population meets the daily fiber intake recommendations.[21] Insufficient fiber intake is due to low intakes of fruits and vegetables and high consumption levels of refined grains, such as pasta, corn chips, white rice, and white bread. Dietary surveys indicate that Americans ages 1 and over eat 1½ cups or less of fruit daily, less than 1 or 2 cups of vegetables daily, and only 1 serving or less of whole grain daily, often in the form of breakfast cereals and breads.[18,22]

Many individuals lack knowledge about fiber rich food sources and their benefits. Also, food ingredient labels can be confusing. For example, manufacturers list enriched white (refined) flour as "wheat flour" on food labels. Most people think that. if "wheat flour" or "wheat bread" is on the label, they are buying a whole-wheat product. However, if the label does not list "whole-wheat flour" first, the product is not truly whole-wheat bread and does not contain as much fiber as it could. Careful label reading is important in the search for more fiber—especially for whole grains. Meeting fiber recommendations is possible if you include whole-wheat bread, fruits, vegetables, and legumes as a regular part of your diet. Eating a high fiber cereal topped with fruit for breakfast is a way to increase your fiber intake (Fig. 5-11). Use the Take Action activity to estimate the fiber content of your diet. What is *your* fiber score?

# Nutrition Facts

10 servings per container

**Serving size** 1 cup (55g)

| | Cereal | | Cereal with 1/2 Cup Skim Milk | |
|---|---|---|---|---|
| **Calories** | **170** | | **210** | |
| | | % DV* | | % DV* |
| **Total Fat** | 1g | **2%** | 1g | **2%** |
| Saturated Fat | 0g | **0%** | 0g | **0%** |
| *Trans* Fat | 0g | | 0g | |
| **Cholesterol** | 0mg | **0%** | 0mg | **0%** |
| **Sodium** | 300mg | **15%** | 350mg | **15%** |
| **Total Carb.** | 36g | **15%** | 42g | **15%** |
| Dietary Fiber | 7g | **25%** | 7g | **25%** |
| Total Sugars | 16g | | 16g | |
| Incl. Added Sugars | 11g | **22%** | 11g | **22%** |
| **Protein** | 4g | | 8g | |
| Vitamin D | 1mcg | 6% | 2mcg | 10% |
| Calcium | 20mg | 2% | 150mg | 10% |
| Iron | 10mg | 60% | 10mg | 60% |
| Potassium | 340mg | 8% | 560mg | 10% |
| Vitamin A | | 15% | | 20% |
| Vitamin C | | 20% | | 20% |
| Thiamin | | 25% | | 30% |
| Riboflavin | | 25% | | 35% |
| Niacin | | 25% | | 25% |
| Vitamin B₆ | | 25% | | 25% |
| Folic acid | | 30% | | 30% |
| Vitamin B₁₂ | | 25% | | 35% |
| Phosphorus | | 20% | | 30% |
| Magnesium | | 20% | | 25% |
| Zinc | | 25% | | 25% |
| Copper | | 10% | | 10% |

\* The % Daily Value (DV) tells you how much a nutrient in a serving of food contributes to a daily diet. 2,000 calories a day is used for general nutrition advice.

# Nutrition Facts

17 servings per container

**Serving size** 3/4 cup (30g)

| | Cereal | | Cereal with 1/2 Cup Skim Milk | |
|---|---|---|---|---|
| **Calories** | **170** | | **210** | |
| | | % DV* | | % DV* |
| **Total Fat** | 0g | **0%** | 0g | **0%** |
| Saturated Fat | 0g | **0%** | 0g | **0%** |
| *Trans* Fat | 0g | | 0g | |
| **Cholesterol** | 0mg | **0%** | 0mg | **0%** |
| **Sodium** | 100mg | **4%** | 150mg | **6%** |
| **Total Carb.** | 35g | **15%** | 41g | **15%** |
| Dietary Fiber | 1g | **4%** | 1g | **4%** |
| Total Sugars | 20g | | 20g | |
| Incl. Added Sugars | 14g | **28%** | 14g | **28%** |
| **Protein** | 7g | | 11g | |
| Vitamin D | 1mcg | 6% | 2mcg | 10% |
| Calcium | 0mg | 0% | 150mg | 10% |
| Iron | 2mg | 10% | 2mg | 10% |
| Potassium | 80mg | 2% | 280mg | 6% |
| Vitamin A | | 25% | | 30% |
| Vitamin C | | 0% | | 2% |
| Thiamin | | 25% | | 25% |
| Riboflavin | | 25% | | 35% |
| Niacin | | 25% | | 25% |
| Vitamin B₆ | | 25% | | 25% |
| Folic acid | | 25% | | 25% |
| Vitamin B₁₂ | | 25% | | 30% |
| Phosphorus | | 4% | | 15% |
| Magnesium | | 4% | | 8% |
| Zinc | | 10% | | 10% |
| Copper | | 2% | | 2% |

\* The % Daily Value (DV) tells you how much a nutrient in a serving of food contributes to a daily diet. 2,000 calories a day is used for general nutrition advice.

*Figure 5-11* The Nutrition Facts panel on food labels can help you choose more nutritious foods. Based on the information from these panels, note which cereal is the better choice for breakfast. When choosing a breakfast cereal, it is generally wise to focus on those that are rich sources of fiber. Added sugar content also can be used for evaluation. Note that added sugar does not include sugars naturally present in foods (e.g., raisins). The amount of sugar listed on the food label as Added Sugars is included in the amount displayed in the line above it, Total Sugars, *not* in addition to it.

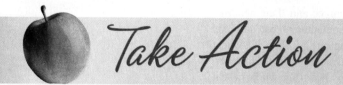

Onoky/SuperStock

## Estimate Your Fiber Intake

To roughly estimate your daily fiber consumption, determine the number of servings you ate yesterday from each food category listed here. Multiply the serving amount by the value listed and then add up the total amount of fiber. How does your total fiber intake for yesterday compare with the general recommendation of 28 g of fiber per day? If you are not meeting your needs, how could you do so?

| Food Category | Size of 1 Serving | Number of Servings You Ate Yesterday | × | Average Grams Fiber per Serving | = | Grams Fiber |
|---|---|---|---|---|---|---|
| Vegetables | 1 cup leafy greens or ½ cup other vegetables | _____ | × | 2 | = | _____ |
| Fruits | 1 whole fruit, ½ cup berries or chopped fruit, or ¼ cup dried fruit | _____ | × | 2.5 | = | _____ |
| Beans, split peas, lentils | ½ cup cooked | _____ | × | 7 | = | _____ |
| Nuts, seeds | ¼ cup or 2 Tbsp peanut butter | _____ | × | 2.5 | = | _____ |
| Whole-grain breads, pasta, rice | 1 slice bread, ½ cup pasta or rice, ½ muffin or bagel | _____ | × | 2.5 | = | _____ |
| Refined grain bread, pasta, rice | 1 slice bread, ½ cup pasta or rice, ½ muffin | _____ | × | 1 | = | _____ |
| Cereal | See Nutrition Facts Panel | _____ | × | _____ (see Nutrition Facts Panel) | = | _____ |
| | | | | **Total Fiber Grams** | | _____ |

Increasing intake of whole grains and vegetables is a healthful way to include carbohydrates in your diet.

Foodcollection

### Knowledge Check

1. Why is the RDA for carbohydrate intake set at 130 g/day? Is this an optimal intake?
2. What is the Adequate Intake for dietary fiber?
3. Why are the dietary fiber intakes of many North Americans far below the recommended level?

# 5.4 Functions of Carbohydrates in the Body

The digestible and indigestible carbohydrates in our diets have vital functions in our bodies.[3,8] These diverse functions are critical to normal metabolism and overall health.

## Digestible Carbohydrates

Most of the digestible carbohydrates in our diets are broken down to glucose. As glucose, they provide a primary source of energy, spare protein from use as an energy source, and prevent ketosis.

### Providing Energy

The main function of glucose is to act as a source of energy for body cells. In fact, red blood cells and cells of the central nervous system derive almost all their energy from glucose. Glucose also fuels muscle cells and other body cells, although many of these cells rely on fatty acids to meet energy needs, especially during rest and light activity. Recall that glucose provides 4 kcal of energy per gram.

### Sparing Protein from Use as an Energy Source

The amino acids that make up dietary protein are used to build body tissues and to perform other vital functions only when carbohydrate intake provides enough glucose for energy needs. If you do not consume enough carbohydrate to yield glucose, your body is forced to break down amino acids in your muscle tissue and other organs to make glucose. This process is termed **gluconeogenesis,** which means the production of new glucose (see Chapter 9 for details). However, when dietary carbohydrate intake is adequate to maintain blood glucose levels, protein is "spared" from use as energy. Generally, North Americans consume ample protein, so sparing protein is not an important role of carbohydrate in the diet. It does become important in some carbohydrate- and energy-reduced diets and in starvation. (Chapter 7 discusses the specific effects of starvation.)

### Preventing Ketosis

A minimal intake of carbohydrates—at least 50 to 100 g/day—is necessary for the complete breakdown of fats to carbon dioxide ($CO_2$) and water ($H_2O$) in the body.[3] When carbohydrate intake falls below this level, the release of the hormone **insulin** decreases, resulting in the release of a large amount of fatty acids from adipose tissue to provide energy for body cells. These fatty acids travel in the bloodstream to the liver. The subsequent incomplete breakdown of these fatty acids in the liver results in the formation of acidic compounds called ketone bodies, or keto-acids, and a condition called ketosis, or ketoacidosis (see Chapter 9). Ketone bodies include acetoacetic acid and its derivatives.

Although the brain and other cells of the central nervous system normally cannot utilize energy from fats, these cells can adapt to use ketones for energy when carbohydrate intake is inadequate. This is an important adaptive mechanism for survival during starvation. If the brain could not use ketone bodies, the body would be forced to produce much more glucose from protein to support the brain's energy needs. The resulting breakdown of muscles, heart, and other organs to provide protein for gluconeogenesis would severely limit our ability to tolerate starvation.

Excessive ketone production also can occur in untreated diabetes. This is not usually the result of low carbohydrate intake, however. Diabetic ketosis develops when insulin production is inadequate or cells resist insulin action, thereby preventing glucose from entering body cells. Cells then rely on ketone bodies from the breakdown of fats for energy. The accumulation of these ketones in the blood results in a more acidic pH. This condition, called diabetic ketoacidosis, is a very serious complication of untreated or poorly controlled diabetes.

Many low carbohydrate/high fat weight reduction diets (e.g., keto-based, Atkins®, and South Beach® Diets) and fasting regimens promote ketosis as a beneficial state for successful weight loss. Ketosis can suppress one's appetite, resulting in a lower calorie intake. It also can

lynx/iconotec.com/Glowimages

**gluconeogenesis** Generation (*genesis*) of new (*neo*) glucose from certain (glucogenic) amino acids.

**insulin** Hormone produced by beta cells of the pancreas. Among other processes, insulin increases the synthesis of glycogen in the liver and the movement of glucose from the bloodstream into muscle and adipose cells.

cause increased loss of water from the body, which may be reflected in lower body weight. However, over time, ketosis can lead to serious consequences, such as dehydration, loss of lean body mass, and electrolyte imbalances. If severe, ketosis can even cause coma and death. Learn more about these low carbohydrate/high fat weight reduction diets in Chapter 9.

## Indigestible Carbohydrates

Although fiber is indigestible, it plays an important role in maintaining the integrity of the GI tract and overall health.[8] Fiber helps prevent constipation and diverticular disease and enhances the management of body weight, blood glucose levels, and blood cholesterol levels.[8,9,10,11,12,24,25]

### Promoting Bowel Health

Fiber adds bulk to the feces, making bowel movements easier. When adequate fiber and fluid are consumed, the stool is large and soft because many types of plant fibers absorb water. The larger size stimulates the intestinal muscles, which aids elimination. Consequently, less force is necessary to expel the feces.

When too little fiber is eaten, the opposite can occur: the stool may be small and hard. Constipation may result, causing one to exert excessive force during defecation. Over time, excessive exertion can lead to the development of hemorrhoids. This high pressure from exertion also can cause parts of the large intestine wall to protrude through the surrounding bands of muscle, forming small pouches called **diverticula**.[10] Fibrous material, feces, and bacteria can become trapped in diverticula and lead to inflammation (Fig. 5-12).

Diverticular disease is asymptomatic (without noticeable symptoms) in about 80% of affected people. The asymptomatic form of this condition is called **diverticulosis** and is one of the most common conditions in Western countries.[10] If the diverticula become inflamed and symptomatic, the condition is known as **diverticulitis.** Intake of fiber then should be reduced to limit further bacterial activity and inflammation. Once the inflammation subsides, a high fiber and high fluid diet, along with regular physical activity, is advised to restore GI tract motility and reduce the risk of a future attack. Studies indicate that nuts, corn, and popcorn can be included in the diet because they do not increase the risk of diverticulitis or diverticular complications, as previously believed.

Over the past 30 years, epidemiological studies have shown an association between increased fiber intake and decreased risk of colon cancer. However, more recently, scientists have questioned these findings.[11,12] Current studies of diet and colon cancer are focusing on the potential preventive effects of increased intakes of fruits, vegetables, legumes, and whole-grain breads and cereals (rather than fiber per se); regular exercise; and adequate vitamin D, folate, magnesium, selenium, and calcium intakes. Overall, it appears that the potential cancer prevention benefits are, for the most part, due to the nutrients and other components that are commonly part of these foods, such as vitamins, minerals, and phytochemicals. Thus, it is more advisable to increase fiber intake using fiber rich foods than to rely on fiber supplements.[8]

### Reducing Obesity Risk

A diet high in fiber likely aids weight control and reduces the risk of accumulating body fat and becoming obese.[23] The bulky nature of high fiber foods fills us up without yielding much energy. Fibrous foods also absorb water and expand in the GI tract, which may result in a sense of fullness and contribute to satiety. Studies also support the intake of whole grains to achieve a lower body weight.[25]

### Enhancing Blood Glucose Control

When consumed in recommended amounts, soluble fibers slow glucose absorption from the small intestine and decrease insulin release from the pancreas. This contributes to better blood glucose regulation, which can be helpful in the treatment of diabetes. In fact, adults with high fiber diets and with greater whole-grain intakes are less likely to develop diabetes than are those with low fiber diets or lower whole-grain intakes, respectively.[8,9,26]

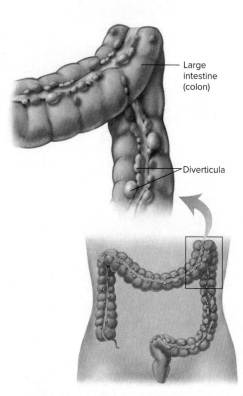

*Figure 5-12* Diverticula in the large intestine. A low fiber diet increases the risk of developing diverticula. About one-third of people over age 45 have this condition, whereas two-thirds of people over 85 do.

**diverticula** Pouches that protrude through the exterior wall of the large intestine.

**diverticulosis** Condition of having many diverticula in the large intestine.

**diverticulitis** Inflammation of the diverticula caused by acids produced by bacterial metabolism inside the diverticula.

## Reducing Cholesterol Absorption

A high intake of soluble fiber inhibits the absorption of cholesterol and the reabsorption of bile acids from the small intestine, thereby reducing the risk of cardiovascular disease and gallstones. The short-chain fatty acids resulting from the bacterial degradation of soluble fiber in the large intestine also reduce cholesterol synthesis in the liver. Overall, a fiber rich diet containing fruits, vegetables, legumes, and whole-grain breads and cereals is advocated as part of a strategy to reduce the risk of cardiovascular disease.[26] Recent studies have linked increased whole-grain intake with a decreased risk of dying from cardiovascular disease.[27] Recall from Chapter 2 that the FDA has approved the claims that diets rich in whole-grain foods and other plant foods and low in total fat, saturated fat, and cholesterol may decrease the risk of cardiovascular disease and certain cancers.

Because oatmeal is rich in soluble fiber, the FDA allows oatmeal package labels to list the benefits of oatmeal in lowering blood cholesterol as a part of a low fat diet.

John A. Rizzo/Photodisc/Getty Images

### Knowledge Check

1. What are 3 functions of digestible carbohydrates?
2. How does carbohydrate spare protein from use as an energy source?
3. Why are indigestible carbohydrates an important component of our diets?

## NUTRITION

# Expert Perspective from the Field

## Taxing Sugar-Sweetened Beverages

Many nutrition and health experts are concerned about the large increase in consumption of sugar-sweetened beverages that has occurred over the last few decades. For example, in the U.S., intake of these beverages adds an extra 155 calories to our daily diets![29]

Sugar-sweetened beverages are associated with an increased risk of obesity, diabetes, and heart disease.[30] Thus, to reduce intake and raise money for nutrition and health programs, researchers, such as Dr. Kelly Brownell,* have proposed taxing these beverages.[31] Dr. Brownell and colleagues estimated that a tax of 1 cent per ounce of sugar-sweetened beverages could generate about $15 million each year.[32] In addition, research indicates that taxing less healthy foods, such as sugar-sweetened beverages, reduces intake[33] and calorie intake[34] and is a cost-effective way to positively impact health-care costs.[35]

Many individuals and groups, especially in the beverage industry, strongly oppose taxing sugar-sweetened beverages. They argue that it would unfairly burden the poor and would not help solve the obesity epidemic. They believe that the increase in obesity rate is not related to sugar-sweetened beverages but is the result of inactivity and poor overall dietary habits. In response, Dr. Brownell points out that "sweetened beverages are the single largest source of added sugar in the American diet. Discouraging soft drink purchases could improve health by reducing empty calorie intake." Moreover, an impact study on the health cost effectiveness of a national sugar-sweetened beverage tax found that health gains and reduced health-care costs

tied to the taxation could offset any increased tax burden put on lower-income purchasers of these beverages.[35]

Dr. Brownell notes that public health advocates encountered similar industry resistance when they proposed increasing taxes on tobacco. Thus, he and other public health advocates plan to continue campaigning for a sugar-sweetened beverage tax as a means of helping reduce the incidence of obesity and related diseases.

*Kelly Brownell, Ph.D., is Director of the World Food Policy Center and Robert L. Flowers Professor of Public Policy at Duke University. Time magazine named him 1 of the "World's 100 Most Influential People." He is a member of the Institute of Medicine. He has received the James McKeen Cattell Award from the New York Academy of Sciences, the Outstanding Contribution to Health Psychology Award from the American Psychological Association, the Lifetime Achievement Award from Rutgers University, and the Distinguished Alumni Award from Purdue University.

Comstock/PunchStock

# 5.5 Carbohydrate Digestion and Absorption

The goal of carbohydrate digestion is to break down starch and sugars into monosaccharide units that are small enough to be absorbed. Food preparation can be viewed as the start of carbohydrate digestion because cooking softens the tough, fibrous tissues of vegetables, fruits, and grains. When starches are heated, the starch granules swell as they soak up water, making them much easier to digest. All these effects of cooking generally make these foods easier to chew, swallow, and break down during digestion.

## Digestion

The enzymatic digestion of some carbohydrates begins in the mouth. Saliva contains an enzyme called salivary **amylase,** which mixes with starch containing amylose when the food is chewed. Amylase breaks down the starch into smaller polysaccharides (called dextrins) and disaccharides (Fig. 5-13). Because food is in the mouth for such a short amount of time, this phase of digestion is only a minor part of the overall digestive process.

When food reaches the stomach, the salivary enzyme is inactivated by the acidity of the stomach. Thus, the digestion of carbohydrate stops until it passes into the small intestine. In the small intestine, the polysaccharides in the food that were first acted on in the mouth now are digested further by pancreatic amylase and dextrinase. Disaccharides are digested to their monosaccharide units by specialized enzymes in the absorptive cells of the small

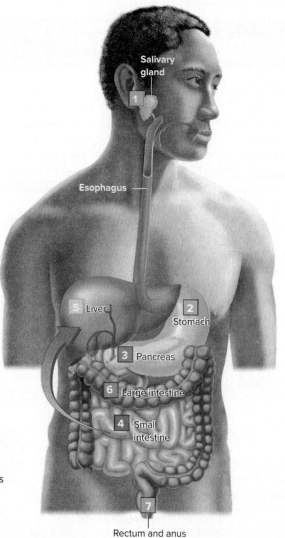

## Carbohydrates

**1** **Mouth:** Some starch is broken down to polysaccharide and disaccharide units by salivary amylase.

**2** **Stomach:** Salivary amylase is inactivated by the acidity in the stomach. No further digestion occurs in the stomach.

**3** **Pancreas:** Pancreatic amylase and dextrinase are secreted into the small intestine to break polysaccharides from starch into disaccharides.

**4** **Small intestine:** Enzymes in the wall of the small intestine break down the disaccharides into monosaccharides.

**5** **Liver:** The absorbed monosaccharides are transported to the liver by the portal vein.

**6** **Large intestine:** Some soluble fiber is metabolized into acids and gases by bacteria in the large intestine.

**7** **Rectum and anus:** Insoluble fiber escapes digestion and is excreted in feces.

*Figure 5-13* Carbohydrate digestion and absorption. Enzymes made by the salivary glands, pancreas, and small intestine participate in the process of digestion. Most carbohydrate digestion and absorption take place in the small intestine (see Chapter 4 for details).

intestine. The disaccharides include maltose from starch breakdown, lactose mainly from dairy products, and sucrose from sweetened foods. The enzyme **maltase** acts on maltose to produce 2 glucose molecules. **Sucrase** breaks down sucrose to produce glucose and fructose. **Lactase** digests lactose to produce glucose and galactose. Monosaccharides that occur in food (usually as glucose or fructose) do not require further digestion in the small intestine. The indigestible carbohydrates (dietary fibers and a small portion of starch in whole grains and some fruits, called resistant starch) cannot be broken down by the digestive enzymes of the small intestine. As discussed previously, they pass into the large intestine, where they are fermented by bacteria into acids and gases or are excreted in fecal waste.[6,7]

Intestinal diseases can interfere with the digestion of carbohydrates, such as lactose, and prevent their breakdown and absorption. When unabsorbed carbohydrates reach the large intestine, bacteria there digest them, producing acids and gases as by-products (see Fig. 5-13). If produced in large amounts, these gases can cause abdominal discomfort. People recovering from intestinal disorders, such as diarrhea, may need to avoid lactose for a few weeks or more because of temporary lactose maldigestion and malabsorption. A few weeks is often sufficient time for the small intestine to resume producing enough lactase enzyme to allow for more complete lactose digestion (see the discussion of lactose intolerance in Section 5.6).[28]

## Absorption

With the exception of fructose, monosaccharides are absorbed by an active absorption process. Recall from Chapter 4 that this process requires a specific carrier and energy input for the substance to be taken up by the absorptive cells in the small intestine. Following digestion, glucose and galactose are pumped into the absorptive cells, along with sodium (Fig. 5-14). The ATP energy used in the process pumps sodium back out of the absorptive cell.

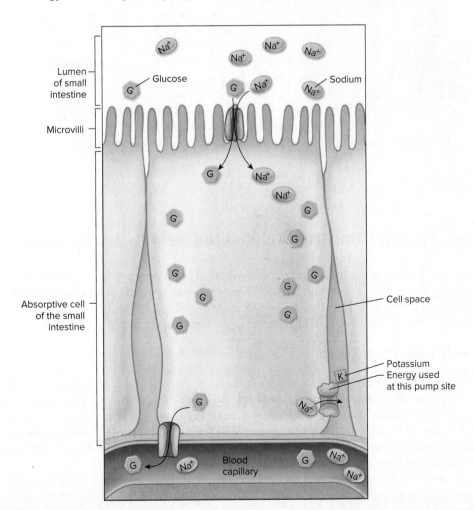

*Figure 5-14* Active absorption of glucose in the absorptive cells that line the villi in the small intestine (see Fig. 4-13 for a diagram of villi). Glucose and sodium pass across the absorptive cell membrane in a carrier-dependent, energy-requiring process. Once inside the absorptive cell, glucose can exit by facilitated diffusion and enter the bloodstream. Sodium is pumped out of the absorptive cell to maintain a low concentration in the absorptive cell and a high concentration in the extracellular fluid.

Fructose is taken up by the absorptive cells via facilitated diffusion. In this case, a carrier is used, but no energy input is needed. This absorptive process is slower than that of glucose or galactose. Once glucose, galactose, and fructose enter the intestinal cells, glucose and galactose remain in that form, whereas some fructose is converted to glucose. These monosaccharides are then transported via the portal vein to the liver. In the liver, fructose and galactose are converted to glucose.

Glucose is transported through the bloodstream for use by the cells of the body. If blood glucose levels are adequate to meet the energy needs of body cells, the liver stores additional glucose as glycogen. (Muscle cells also can store glycogen.) Although the liver's capacity to store glycogen is limited, glycogen storage provides an important reserve of energy to maintain blood glucose levels and cellular function. When carbohydrates are consumed in very high amounts, the glycogen storage capacity of the liver (and muscles) often is exceeded. The liver then converts the excess glucose to fat for storage in adipose tissue.

### Knowledge Check

1. What enzymes are involved in the digestion of carbohydrates?
2. Why do some individuals with intestinal diseases need to temporarily restrict their consumption of foods containing lactose?
3. How are monosaccharides absorbed?

---

## *Critical* Thinking

What foods and beverages contribute to your daily intake of added sugars? Identify two or three items you can consume less of or replace with other choices to decrease your consumption of sugars.

High carbohydrate beverages can contribute to tooth decay.

Amble Design/Shutterstock

---

## CASE STUDY

Ingram Publishing/
SuperStock

Myeshia, a 19-year-old female, recently read about the health benefits of calcium and decided to increase her intake of dairy products by drinking milk. Not long afterward, she experienced bloating, cramping, and gassiness. She suspected that the source of this discomfort was the milk she consumed, especially because her parents and sister had complained of the same problem. She wanted to determine if the milk was, in fact, the cause of her gastrointestinal discomfort, so the next day she substituted yogurt for the milk in her diet. Subsequently, she did not have any symptoms. What component of milk likely caused the problem? Why was she able to tolerate yogurt but not milk? What other foods might she include in her diet to supply calcium?

---

## 5.6 Health Concerns Related to Carbohydrate Intake

As a part of a nutritious diet, adequate carbohydrate intake is important for maintaining health and decreasing the risk of chronic disease. However, as with many nutrients, excessive intakes of different forms of carbohydrate can be harmful to overall health. The following sections will help you understand the risks of high intakes of different types of carbohydrates.

### Very High Fiber Diets

Adequate fiber intake provides many health benefits. However, very high intakes of fiber (i.e., above 50 to 60 g/day) can cause health risks. For example, high fiber consumption combined with low fluid intake can result in hard, dry stools that are painful to eliminate. Over time, this may cause hemorrhoids, as well as rectal bleeding, from increased exertion and pressure. In severe cases, the combination of excess fiber and insufficient fluid may contribute to blockages in the intestine, requiring surgery.

Very high fiber diets also may decrease the absorption of certain minerals and increase the risk of deficiencies. This occurs because some minerals can bind to fiber, which prevents them from being absorbed. In countries where fiber intake is often greater than 60 g/day, deficiencies of zinc and iron have been reported.

High fiber diets can be of concern in young children, elderly persons, and malnourished individuals, all of whom may not eat adequate amounts of foods and nutrients. For these individuals, high fiber intakes may cause a sense of fullness and reduce their overall intake of foods, energy, and nutrients.

## High Sugar Diets

Sugars constitute a large part of the diets of many Americans. In fact, on average, 13% of the calories consumed by U.S. adults are from added sugars. Daily intake is about 335 kcal for men and 239 kcal for women.[17] Teens consume about 360 kcal of added sugars per day.[20] Recall that most of the sugar we eat comes from foods and beverages to which sugar has been added during processing and/or manufacturing. Major sources of added sugar are soft drinks, cakes, cookies, fruit punch, and dairy desserts such as ice cream. Although sugars supply calories, they usually provide little else and often replace the intake of more nutritious foods. Children and adolescents are typically at greatest risk of overconsuming sugar and empty calories. Dietary surveys indicate that many children and adolescents are drinking an excess of sugar-sweetened beverages and too little milk. Milk contains calcium and vitamin D, both of which are essential for bone health. Thus, substituting sugared drinks for milk can compromise bone development and health.

High intakes of sugars also can increase the risk of weight gain and obesity. Recent evidence also links increased intakes of added sugars, particularly sugar-sweetened beverages, with an increased risk for type 2 diabetes in adults. The "supersizing" trend noted in food and beverage promotions is contributing to this concern. For example, in the 1950s, a typical soft drink serving was a 6.5-ounce bottle. Today, a 20-ounce bottle is a typical serving. This change alone contributes 170 extra calories of sugar to the diet. Drinking 1 bottle per day for a year amounts to 62,050 extra calories and a 17- to 18-pound (7.75 to 8.25 kg) weight gain.

The sugar in cakes, cookies, and ice cream also supplies extra energy, which promotes weight gain. Although dieters may be choosing more low fat and fat-free snack products, these usually are made with substantial amounts of added sugar in order to produce a dessert with an acceptable taste and texture. The resulting product often is a high calorie food that equals or even exceeds the energy content of the original high fat food product it was designed to replace.

High intakes of energy and sugar (especially fructose) have been associated with conditions that increase the risk of cardiovascular disease—namely, increased blood levels of triglycerides and LDL-cholesterol and decreased levels of HDL-cholesterol. Studies also have suggested a potential link between the increased consumption of sugar-sweetened beverages and energy with an increased risk of type 2 diabetes and Metabolic Syndrome.[36] However, scientists do not have enough evidence to conclude that increased sugar intake is a risk factor for cardiovascular disease, diabetes, or Metabolic Syndrome, although it is a good idea to limit sugar intake to the amounts suggested in MyPlate (see Chapter 2).

Diets high in sugar have been reported to cause hyperactivity in children. However, scientists have determined that hyperactivity and other behavioral problems are likely due to a variety of non-nutritional factors. Although eating a nutritious diet is important for a child's overall health and well-being, it will not prevent hyperactivity, behavioral problems, or learning disabilities (see Chapter 17).

High sugar diets increase the risk of developing dental caries (cavities), which can develop when bacteria in the mouth metabolize sugars into acids (Fig. 5-15). The acids gradually dissolve the tooth enamel and the underlying structure, causing decay, discomfort, and even nerve damage. Sugar from any source can lead to caries. Sticky and gummy foods that are high in sugar and adhere to teeth, such as caramels, licorice, and gummy bears, are the most likely to promote dental caries. Starches that are readily fermented in the mouth, such as crackers and white bread, also increase the risk of dental

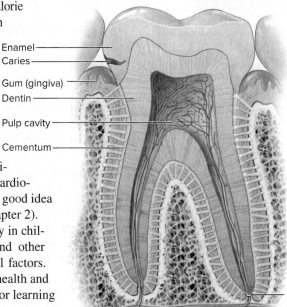

Enamel

Caries

Gum (gingiva)

Dentin

Pulp cavity

Cementum

Blood vessels and nerves

*Figure 5-15* Dental caries. Bacteria in the mouth metabolize sugars in food and create acids that can dissolve tooth enamel. This leads to the development of caries. If the caries progress into the pulp cavity, damage to the nerve and pain are likely.

Yogurt helps those with lactose intolerance meet their calcium needs.

Wavebreak Media/Getty Images

caries. Sipping fruit juices, soft drinks, and milk (which contains lactose) throughout the day bathes the teeth in sugar and can increase the risk of caries. Thus, parents should be cautioned against serving infants and young children these beverages to sip on between meals.

## Lactose Intolerance

The amount of the enzyme lactase produced in the small intestine often begins to decrease after early childhood. The insufficiency of lactase, referred to as *primary* lactose intolerance, can cause symptoms of abdominal pain, bloating, gas, and diarrhea after consuming lactose, especially in large amounts. The bloating and gas are caused by the bacterial fermentation of undigested lactose in the large intestine. The undigested lactose also draws water into the large intestine, causing diarrhea.

Primary lactose intolerance may occur in up to 75% of the world's population. In North America, approximately 25% of adults show signs of decreased lactose digestion. Those who have Asian, African, or Latino/Hispanic backgrounds are more likely to experience lactose intolerance than Caucasians. Some people with primary lactose intolerance do not experience symptoms. In addition, many are able to consume moderate amounts of lactose with little or no intestinal discomfort because bacteria in the large intestine break down the lactose. In fact, recent studies have shown that nearly all individuals with decreased lactase production can tolerate ½ to 1 cup of milk with meals. Hard cheese, yogurt, and acidophilus milk also are well tolerated because much of the lactose in these foods has been converted to lactic acid. Thus, it is unnecessary for many with lactose intolerance to greatly restrict their intakes of lactose-containing foods, such as milk and other dairy products.

Another type of lactose intolerance, called *secondary* lactose intolerance, occurs when conditions of the small intestine, such as Crohn disease and severe diarrhea, damage the cells that produce lactase. Secondary lactose intolerance also can cause gastrointestinal symptoms, but the symptoms are usually temporary and cease when the intestine recovers and lactase production normalizes.

## Glucose Intolerance

Maintaining blood glucose levels within normal ranges is important for providing adequate glucose for body functions and for preventing the symptoms associated with changes in blood glucose levels. The abnormal regulation of blood glucose can lead to either **hyperglycemia** (high blood glucose) or **hypoglycemia** (low blood glucose). Hyperglycemia is a more common condition than hypoglycemia and is most commonly associated with diabetes (technically, *diabetes mellitus*) and Metabolic Syndrome.

### Regulation of Blood Glucose

Under fasting conditions (at least several hours after eating), blood glucose normally varies between about 70 and 100 mg/dl of blood. However, if a fasting blood glucose level is equal to or above 126 mg/dl, it is classified as diabetes. The symptoms of diabetes include hunger, thirst, frequent urination, and weight loss. When blood glucose falls below 50 mg/dl, the condition is classified as hypoglycemia. A person with hypoglycemia may experience hunger, shakiness, irritability, weakness, and headache as energy availability decreases.

The liver is important in controlling the amount of glucose in the bloodstream. As the 1st organ to screen the sugars absorbed from the small intestine, the liver helps determine the amount of glucose that enters the bloodstream after a meal (see Fig. 5-13) and the amount that is stored as glycogen for later use.[3]

The pancreas also is important in blood glucose control. The pancreas releases small amounts of insulin as soon as a person starts to eat. Following carbohydrate digestion and absorption, blood glucose levels rise, signaling the pancreas to release large amounts of insulin. Insulin promotes increased glucose uptake by muscle and adipose cells. In addition,

**hyperglycemia** High blood glucose, above 125 mg/100 ml (dl) of blood.

**hypoglycemia** Low blood glucose, below 50 mg/100 ml (dl) of blood.

insulin promotes the use of glucose for energy and storage of excess glucose as glycogen. These actions lower blood glucose to the normal fasting range within a few hours after a person eats.

Other hormones in the body counteract the effects of insulin. When a person has not eaten carbohydrates for a few hours, the amount of glucose in the blood decreases. Another pancreatic hormone, called glucagon, is secreted in response to a decrease in blood glucose. It prompts the breakdown of glycogen in the liver and promotes gluconeogenesis, resulting in the release of glucose to the bloodstream and the normalization of blood glucose levels (Fig. 5-16). The hormones epinephrine (adrenaline) and norepinephrine, from the adrenal glands, also trigger the breakdown of glycogen in the liver and result in glucose release into the bloodstream. These hormones are responsible for the "fight-or-flight" reaction that we sometimes experience. They are released in large amounts in response to a perceived threat, such as a car approaching head-on. The resulting rapid release of glucose into the bloodstream promotes quick mental and physical reactions. The hormones cortisol and growth hormone also help regulate blood glucose by decreasing glucose use by muscle (Table 5-5).

In essence, the actions of insulin on blood glucose are balanced by the actions of glucagon, epinephrine, norepinephrine, cortisol, and growth hormone. If hormonal balance is not maintained, as during overproduction or underproduction of insulin or glucagon, major changes in blood glucose concentrations occur. This system of checks and balances allows blood glucose to be maintained within fairly narrow ranges.

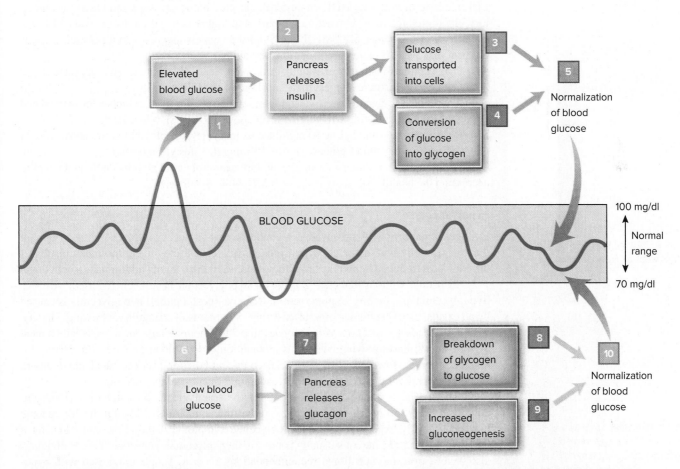

*Figure 5-16* Regulation of blood glucose. Insulin and glucagon are key factors in controlling blood glucose. Steps 1 to 5 show what happens when blood glucose rises above the normal range: when glucose is above normal 1, insulin is released 2 to lower it 3 and 4. Blood glucose then falls back into the normal range 5. Steps 6 to 10 show how the body responds when blood glucose drops below normal. When blood glucose falls below the normal range 6, glucagon is released 7, which has the opposite effect of insulin 8 and 9. This then restores blood glucose to the normal range 10. Other hormones, such as epinephrine, norepinephrine, cortisol, and growth hormone, also contribute to blood glucose regulation (see Table 5-5 for details).

**Table 5-5  Roles of Various Hormones in the Regulation of Blood Glucose**

| Hormone | Source | Target Organ or Tissue | Overall Effects on Organ or Tissue | Effect on Blood Glucose |
|---|---|---|---|---|
| Insulin | Pancreas | Liver, muscle, adipose tissue | Increases glucose uptake by muscles and adipose tissue, increases glycogen synthesis, suppresses gluconeogenesis | Decrease |
| Glucagon | Pancreas | Liver | Increases glycogen breakdown and release of glucose by the liver, increases gluconeogenesis | Increase |
| Epinephrine, norepinephrine | Adrenal glands | Liver, muscle | Increase glycogen breakdown and release of glucose by the liver, increase gluconeogenesis | Increase |
| Cortisol | Adrenal glands | Liver, muscle | Increases gluconeogenesis by the liver, decreases glucose use by muscles and other organs | Increase |
| Growth hormone | Pituitary gland | Liver, muscle, adipose tissue | Decreases glucose uptake by muscles, increases fat mobilization and utilization, increases glucose output by the liver | Increase |

## Metabolic Syndrome

More than one-third[37] of American adults have a condition known as Metabolic Syndrome. **Metabolic Syndrome** is characterized by a group of factors that increase the risk of type 2 diabetes and cardiovascular disease, including insulin resistance or glucose intolerance (causing high blood glucose), abdominal obesity, high blood triglycerides and LDL-cholesterol with low HDL-cholesterol, elevated blood pressure, increased inflammatory blood proteins (e.g., C-reactive protein), and higher concentrations of oxidized LDL-cholesterol (see Chapter 6).[38] Metabolic Syndrome also is associated with overall obesity, physical inactivity, genetic predisposition, and aging.

The National Institutes of Health, American Heart Association, and National Heart, Lung, and Blood Institute suggest that 3 or more of the following criteria be present for diagnosing Metabolic Syndrome: a waist circumference greater than 35 inches for women and 40 inches for men, a fasting triglyceride level above 150 mg/dl, blood HDL-cholesterol below 40 mg/dl for men and below 50 mg/dl for women, elevated blood pressure above 130/85 mm Hg, and fasting blood glucose above 100 mg/dl. Lifestyle modification (focusing on weight loss, decreased dietary fat intake, and increased physical activity) is fundamental to decreasing the health risks associated with Metabolic Syndrome.

## Hypoglycemia

▶ For more information on diabetes, consult this website: www.diabetes.org.

Hypoglycemia, or low blood sugar, is a condition that can occur in people with or without diabetes. In those with diabetes, hypoglycemia can occur if they inject too much insulin, if they don't eat frequently enough, and if they exercise without eating additional carbohydrate.

In people without diabetes, 2 types of hypoglycemia have been reported: **reactive hypoglycemia** and **fasting hypoglycemia.** Reactive (postprandial) hypoglycemia is caused by an exaggerated insulin response after eating. Symptoms of irritability, sweating, anxiety, weakness, headache, and confusion may develop 2 to 5 hours after a meal, especially a meal high in sugars. Fasting hypoglycemia is a condition of low blood glucose after fasting for 8 hours or more. However, it usually is caused by an underlying serious medical condition, such as cancer, liver disease, or renal disease, rather than by simply fasting.

The diagnosis of hypoglycemia requires the simultaneous presence of a blood glucose level below 50 mg/dl and classic hypoglycemic symptoms. Although healthy people may occasionally experience some hypoglycemic symptoms if they have not eaten for a prolonged period of time, this usually is not true hypoglycemia. However, these individuals also benefit from the nutritional recommendations given to people diagnosed with hypoglycemia. Regular meals consisting of a balance of protein, fat, and low glycemic load carbohydrates, plus ample soluble fiber, help prevent hypoglycemia. Individuals also should substitute protein-containing snacks for those that contain mostly sugar and aim to spread carbohydrate intake throughout the day. Finally, limiting caffeine and alcohol intake can be beneficial in preventing symptoms of hypoglycemia.

**reactive hypoglycemia** Low blood glucose that may follow a meal high in simple sugars, with corresponding symptoms of irritability, headache, nervousness, sweating, and confusion; also called postprandial hypoglycemia.

**fasting hypoglycemia** Low blood glucose that follows about a day of fasting.

# Diabetes Mellitus

As mentioned previously, an inability to regulate glucose metabolism can result in diabetes. Diagnosis of diabetes is based on a **fasting blood glucose** level equal to or above 126 mg/dl or a hemoglobin A1c level at or above 6.5%. Diabetes affects over 8% of North Americans and leads to over 200,000 deaths each year. An estimated 35% of our adult population shows evidence of pre-diabetes (indicated by a borderline high blood glucose level from 100 to 125 mg/dl or a borderline A1c level from 5.7 to 6.4%).

There are 2 major forms of diabetes: **type 1 diabetes** (formerly called insulin-dependent or juvenile-onset diabetes) and **type 2 diabetes** (formerly called non-insulin-dependent or adult-onset diabetes) (Table 5-6). The change in names to type 1 and type 2 diabetes stems from the fact that many with type 2 diabetes eventually must rely on insulin injections as a part of their treatment.[39] Approximately 90% of individuals with diabetes have type 2 diabetes.

In type 1 diabetes, individuals develop the classic symptoms of hyperglycemia (increased hunger, thirst, urination, and weight loss). No 1 symptom is diagnostic of diabetes. Other symptoms—such as unexplained weight loss, exhaustion, and blurred vision—may accompany these symptoms.[39]

In type 2 diabetes, 30 to 50% of individuals may not have any symptoms and are not aware that they have diabetes until diagnosed in routine health screening tests. Thus, new guidelines promote testing fasting blood glucose levels in adults over age 45 every 3 years to avoid missing cases and to prevent related morbidity and mortality.[39]

A 3rd form of diabetes is called gestational diabetes. Gestational diabetes occurs in approximately 2 to 10% of all pregnancies. It is usually treated with insulin and dietary modification, and it resolves after delivery of the baby. However, pregnant women who develop gestational diabetes are at high risk of developing type 2 diabetes later in life (see Chapter 16).

## Type 1 Diabetes

Although type 1 diabetes can occur at any age, it often begins in late childhood, between 8 and 12 years of age. The disease runs in families, suggesting a genetic link. Thus, children and siblings of those with diabetes are at increased risk. Most cases of type 1 diabetes begin as an autoimmune disorder that destroys the insulin-producing cells in the pancreas. As the pancreas loses its ability to synthesize insulin and thus regulate blood glucose levels, the clinical symptoms of the disease develop.

The onset of type 1 diabetes is associated with decreased release of insulin from the pancreas and increased blood glucose, especially after eating. When blood glucose exceeds the kidney's threshold for returning it to the bloodstream, the excess glucose ends up in the urine—hence the term *diabetes mellitus,* which means "flow of much urine" (*diabetes*) that is "sweet" (*mellitus*). Figure 5-17 shows a typical glucose tolerance curve observed in a patient

▶ A common clinical method to determine a person's success in controlling blood glucose is to measure glycated (or glycosylated) hemoglobin (hemoglobin A1c). Over time, blood glucose attaches to (glycates) hemoglobin in red blood cells, especially when blood glucose remains elevated.

## Table 5-6  Comparison of Type 1 and Type 2 Diabetes

|  | Type 1 Diabetes | Type 2 Diabetes |
|---|---|---|
| **Occurrence** | 5% of cases of diabetes | 90% of cases of diabetes |
| **Cause** | Autoimmune attack on the pancreas | Insulin resistance |
| **Risk Factors** | Moderate genetic predisposition | Strong genetic predisposition<br>Obesity and physical inactivity<br>Ethnicity<br>Metabolic Syndrome<br>Pre-diabetes |
| **Characteristics** | Distinct symptoms (frequent thirst, hunger, and urination)<br>Ketosis<br>Weight loss | Mild symptoms, especially in early phases of the disease (fatigue and nighttime urination)<br>Ketosis does not generally occur. |
| **Treatment** | Insulin<br>Diet<br>Exercise | Diet<br>Exercise<br>Oral medications to lower blood glucose<br>Insulin (in advanced cases) |
| **Complications** | Cardiovascular disease<br>Kidney disease<br>Nerve disease<br>Blindness<br>Infections | Cardiovascular disease<br>Kidney disease<br>Nerve damage<br>Blindness<br>Infections |
| **Monitoring** | Blood glucose<br>Urine ketones<br>Hemoglobin A1c | Blood glucose<br>Hemoglobin A1c |

**fasting blood glucose** Measurement of blood glucose levels after a period of 8 hours or more without food or beverages; also called fasting blood sugar.

**type 1 diabetes** Form of diabetes in which the person is prone to ketosis and requires insulin therapy.

**type 2 diabetes** Most common form of diabetes, in which ketosis is not commonly seen. Insulin therapy may be used but often is not required. This form of the disease is often associated with obesity.

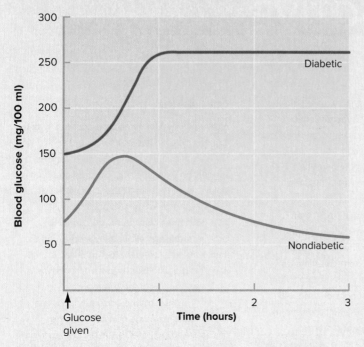

*Figure 5-17* Glucose tolerance test: a comparison of blood glucose concentrations in untreated diabetic and healthy, nondiabetic persons after consuming a 75 g test load of glucose.

Checking blood glucose regularly is an important part of diabetes therapy.

Nick Rowe/Getty Images

with type 1 diabetes, after eating a test load of about 20 teaspoons (75 grams) of glucose.

## Management

Diabetes care focuses on self-management, medication use, and lifestyle behaviors. Type 1 diabetes is treated by insulin therapy, either with injections several times per day or with an insulin pump. The pump dispenses insulin at a steady rate into the body, with greater amounts delivered after a meal. Although there is no single recommended eating pattern, nutrition therapy often includes 3 regular meals and 1 or more snacks (including a bedtime snack), as well as an individualized plan for carbohydrate, protein, and fat intake to maximize insulin action and minimize swings in blood glucose. The diet should supply energy in balance with expenditure, be low in

saturated fats and cholesterol, and meet overall nutritional needs.[39,40]

**Carbohydrate counting** and the food lists for diabetes are useful tools for balancing carbohydrate intake and improving blood glucose control while eating a variety of foods, including vegetables, fruits, whole grains, legumes, and dairy. The carbohydrate counting method awards 1 point to approximately 12 to 15 g of carbohydrate. The food list system is described in Appendix E.

Poorly controlled diabetes can lead to short-term and long-term health problems. The hormone imbalances that occur in people with uncontrolled type 1 diabetes lead to the breakdown of body fat for energy. Ketosis develops as fat is converted to ketone bodies. Ketones can increase to high levels in the blood, eventually ending up in the urine. Ketones also pull sodium and potassium ions with them into the urine, leading to dehydration, ion imbalance, coma, and even death. Treatment includes insulin and fluids, as well as sodium, potassium, and chloride.[39]

Over time, poorly controlled diabetes can cause degenerative conditions, such as blindness, cardiovascular disease, and kidney disease.[39,41] Nerves also can deteriorate, resulting in decreased nerve stimulation throughout the body (called neuropathy).[42] When this occurs in the intestinal tract, intermittent diarrhea and constipation result. Because of nerve deterioration in the arms, hands, legs, and feet, many people with

diabetes lose the sensation of pain associated with injuries and infections. Without normal pain sensations, they often delay treatment. This delay, combined with an environment rich in glucose that readily supports bacterial growth, sets the stage for damage to and the death of tissues in the extremities, sometimes even leading to the need for amputation. Poorly controlled diabetes also contributes to a rapid buildup of fats in blood vessel walls, which increases the risk of cardiovascular disease.[41]

The Diabetes Control and Complications Trial (DCCT) and other recent studies have shown that the development of diabetes-related cardiovascular disease and nerve damage can be delayed with aggressive treatment directed at keeping blood glucose within the normal range.[41,42] The therapy poses some risks of its own, however, such as hypoglycemia, so it must be implemented under the close supervision of a physician.

A person with diabetes should work regularly with a physician and dietitian to monitor and adjust (when necessary) diet, medications, and physical activity. Physical activity enhances glucose uptake by muscles independent of insulin action, which in turn can lower blood glucose.[43] This outcome is beneficial, but people with diabetes need to be aware of their own blood glucose response to physical activity and plan appropriately to avoid hypoglycemia.

Apps help individuals self-manage their diabetes by tracking their diets, blood glucose levels, insulin dosage, and other health

## Critical Thinking

Marc and Dan are twins who like the same activities and foods. At a recent doctor's appointment, Dan was told that he has type 2 diabetes. He has been feeling good and has not noticed any changes in his health. He does not understand why he has diabetes but his brother does not and why he has not had any noticeable symptoms. How would you explain this to him?

One Touch Reveal® mobile and web apps. Apps are currently available that connect   directly to a glucometer, allowing for enhanced diabetes monitoring and tracking.

Julia Pankin/Shutterstock

parameters (e.g., blood pressure, body weight, hemoglobin A1c). Some digital apps allow users to set reminders that can help manage their diabetes and make it easier to share health data with their health-care team. Apps that connect with a glucometer can alert individuals when blood glucose falls into an unhealthy range.

## Type 2 Diabetes

Type 2 diabetes is a progressive disease characterized by insulin resistance or loss of responsiveness by body cells to insulin. As a result, glucose is not readily transferred into cells and builds up in the bloodstream, causing hyperglycemia. In type 2 diabetes, insulin production may be low, normal, or at times even elevated. However, regardless of the amount of insulin produced, cells are less responsive to its actions.

Type 2 diabetes is the most common type, accounting for about 90% of the cases diagnosed in North America. Those over age 45 and with Latino/Hispanic, African, Asian, Native American, or Pacific Island backgrounds are at particular risk. The number of people with type 2 diabetes is on the rise, primarily because of widespread inactivity and obesity. There also has been a substantial increase in type 2 diabetes in children, mainly due to an increase in overweight coupled with limited physical activity in this population. Type 2 diabetes is genetically linked, so family history is a very important risk factor. Because of this genetic link, those with a family history should be careful to avoid other risk factors, such as obesity, inactivity, and diets rich in saturated fats, cholesterol, and high glycemic load foods.[39,40]

## Treatment

Treatments for type 2 diabetes are aimed at maintaining normal ranges of blood glucose through lifestyle modification and medication use. Adhering to a nutritious diet plan and a

regular physical activity program is an important component of therapy. Consistent exercise and an energy-controlled, nutritious diet eaten at regular mealtimes promote a healthy body weight, enhance the uptake of glucose by muscle cells, lower blood lipids and cardiovascular risk, and help achieve normal blood sugar levels. A Mediterranean-style eating pattern, emphasizing monounsaturated over saturated fats and food sources of omega-3 fatty acids, is recommended. For overweight or obese individuals, even a modest weight loss can improve blood glucose control.[39,43]

Many with type 2 diabetes need medications in addition to diet modification and regular activity to control blood glucose. Oral medications that reduce glucose production by the liver, increase insulin synthesis by the pancreas, slow the intestinal absorption of glucose, or decrease cellular resistance to insulin are used by many with type 2 diabetes to regulate blood glucose. However, when oral medications fail to normalize blood glucose levels, insulin injections are necessary.[39]

Moderate amounts of alcohol (1 serving/day) can be allowed in the diets of those with both type 1 and type 2 diabetes. In fact, for some individuals, small amounts of alcohol may help increase HDL-cholesterol and reduce cardiovascular disease risk. However, alcohol intake, especially without adequate food intake, can lead to severe hypoglycemia. Thus, those with diabetes need to use alcohol cautiously and monitor blood glucose levels closely to avoid hypoglycemia.

## Decreasing the Risk of Diabetes and Protecting Health If Diabetes Is Diagnosed

There are many lifestyle modifications that individuals with an increased risk of type 2 diabetes can adopt to decrease their likelihood of developing the disease. Obesity and inactivity are common risk factors associated with type 2 diabetes. Thus, maintaining a healthy weight, staying physically active, and following the Dietary Guidelines can decrease one's risk. For those with a family history of diabetes and for women with a history of gestational diabetes, regular testing of fasting blood glucose l evels or screening through glucose tolerance testing is an important part of personal health care.

Although diabetes is not yet a curable disease, it can be controlled through diet, exercise, and medications. Maintaining good glucose control is critical to prevent

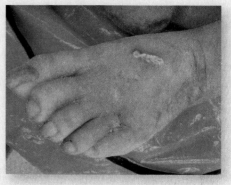

Over the years, poorly controlled blood glucose causes poor blood circulation and nerve damage. This damage causes a loss of sensation and ulcerations that heal very slowly or do not heal at all.

McGraw-Hill Education

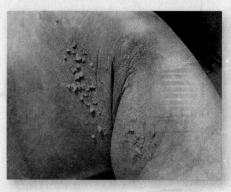

Individuals with type 2 diabetes sometimes have a condition called *acanthosis nigricans*. This darkening (hyperpigmentation) of the skin is found in skin folds of the neck, forehead, navel, armpits, and/or groin.

McGraw-Hill Education

the long-term, diabetes-related complications of cardiovascular disease, renal disease, blindness, and nerve damage. Diabetes education, lifestyle modification, medication management, and self-monitoring of blood glucose levels are essential in maintaining overall health for those with all types of diabetes.

Regular exercise has a key role in decreasing the risk of type 2 diabetes. For individuals with type 2 diabetes, exercise also can be an important part of managing the disease.

©Stockbyte/PunchStock RF

Carrots, criticized in the popular press for having a high glycemic index, actually contribute a low glycemic load to a diet.

Ingram Publishing/Alamy Stock Photo

**glycemic index (GI)** Ratio of the blood glucose response to a given food, compared with a standard (typically, glucose or white bread).

**glycemic load (GL)** Amount of carbohydrate in a food multiplied by the glycemic index of that carbohydrate. The result is then divided by 100.

▶ You might wonder why the glycemic index and glycemic load of white bread and whole-wheat bread are similar. This is because whole-wheat flour typically is ground so finely that it is quickly digested and absorbed. Thus, experts suggest we focus on more minimally processed grains, such as coarsely ground whole-wheat flour and steel-cut oats, to get the full benefits of these fiber sources in reducing blood glucose levels.

## Glycemic Index and Glycemic Load

Our bodies react uniquely to different sources of carbohydrates. For example, a serving of high fiber brown rice results in lower blood glucose levels compared with the same-size serving of mashed potatoes. As researchers investigated the glucose response to various foods, they noted that it was not always as predicted. Thus, they developed 2 tools, the glycemic index and the glycemic load, to indicate how blood glucose responds to various foods (Table 5-7).[44]

The **glycemic index (GI)** is a ratio of the blood glucose response of a given food compared with a standard (typically, glucose or white bread).[44] Glycemic index is influenced by a food's starch structure (amylose vs. amylopectin), fiber content, food processing, physical structure (small vs. large surface area), and temperature, as well as the amount of protein and fat in a meal.[44] Foods with particularly high glycemic index values are potatoes, breads, Gatorade®, short-grain white rice, honey, and jelly beans. A major shortcoming of the glycemic index is that it is based on a serving of food that would provide 50 grams of carbohydrate. However, this amount of food may not reflect the amount typically consumed.

**Glycemic load (GL)** takes into account the glycemic index and the amount of carbohydrate consumed, so it better reflects a food's effect on blood glucose than does the glycemic index alone. To calculate the glycemic load of a food, the number of grams of carbohydrate in 1 serving is multiplied by the food's glycemic index, then divided by 100 (because the glycemic index is a percentage). For example, vanilla wafers have a glycemic index of 77, and a serving of 5 wafers contains 15 g of carbohydrate. This yields a glycemic load of approximately 12:

$$(77 \times 15) / 100 = 12$$

Even though the glycemic index of vanilla wafers is considered high, the glycemic load calculation shows that the impact of this food on blood glucose levels is fairly low.

Why should we be concerned with the effects of various foods on blood glucose? Foods with a high glycemic load elicit an increased insulin response from the pancreas and a resulting drop in blood glucose. These dramatic fluctuations in blood glucose can cause short- and long-term consequences in individuals with diabetes. Chronically high insulin output leads to many harmful effects on the body, such as high blood triglycerides, increased fat deposition in the adipose tissue, increased fat synthesis in the liver, and a more rapid return of hunger after a meal. Thus, increasing intakes of lower glycemic load foods is often recommended as part of a healthful diet.[43] Because many foods with a low glycemic load contain higher amounts of dietary fiber, increasing one's intake of these foods will, in turn, increase fiber intake and may help reduce risk of cardiovascular disease, Metabolic Syndrome, and certain cancers.[45,46]

Use of the glycemic index and glycemic load remains somewhat controversial. Many researchers question their benefits. Nutritionally, neither tool indicates blood glucose responses when individual foods are eaten as a part of mixed meals. As most high glycemic foods are eaten in combination with low glycemic foods (e.g., rice cereals with milk, macaroni with cheese, bread with peanut butter), the glycemic index and glycemic load are often lower than the value given for these foods individually.

lynx/iconotec.com/Glowimages

## CASE STUDY FOLLOW-UP

Ingram Publishing/ SuperStock

Myeshia suspected she had a problem with milk because, when she consumed it, she developed bloating and gas. She was successful in reducing these symptoms by replacing milk with yogurt. As you have learned, yogurt is tolerated better than milk by people with lactose intolerance because the bacteria in yogurt digest much of the lactose. Hard cheeses also supply calcium and have lower levels of lactose. Note, however, that many people with lactose intolerance can consume small to moderate amounts of milk with few or no symptoms from the lactose that is present. Before eliminating foods from her diet, Myeshia would be wise to visit her physician to confirm her self-diagnosis.

**Table 5-7 Glycemic Index (GI) and Glycemic Load (GL) of Common Foods**

Reference food glucose = 100
Low GI foods—below 55
Intermediate GI foods—between 55 and 69
High GI foods—more than 70

Low GL foods—below 10
Intermediate GL foods—between 11 and 19
High GL foods—more than 20

| | Serving Size | Glycemic Index (GI)* | Carbohydrate (grams) | Glycemic Load (GL) |
|---|---|---|---|---|
| **Pastas/Grains** | | | | |
| Brown rice | 1 cup | 55 | 46 | 25 |
| White rice, short-grain | 1 cup | 72 | 53 | 38 |
| **Vegetables** | | | | |
| Carrots, boiled | 1 cup | 49 | 16 | 8 |
| Sweet corn | 1 cup | 55 | 39 | 21 |
| Potato, baked | 1 cup | 85 | 57 | 48 |
| **Dairy Foods** | | | | |
| Milk, fat-free | 1 cup | 32 | 12 | 4 |
| Yogurt, low fat | 1 cup | 33 | 17 | 6 |
| Ice cream | 1 cup | 61 | 31 | 19 |
| **Legumes** | | | | |
| Baked beans | 1 cup | 48 | 54 | 26 |
| Kidney beans | 1 cup | 27 | 38 | 10 |
| Lentils | 1 cup | 30 | 40 | 12 |
| **Sugars** | | | | |
| Honey | 1 tsp | 73 | 6 | 4 |
| Sucrose | 1 tsp | 65 | 5 | 3 |
| Lactose | 1 tsp | 46 | 5 | 2 |
| **Breads and Muffins** | | | | |
| Whole-wheat bread | 1 slice | 69 | 13 | 9 |
| White bread | 1 slice | 70 | 10 | 7 |
| **Fruits** | | | | |
| Apple | 1 medium | 38 | 22 | 8 |
| Banana | 1 medium | 55 | 29 | 16 |
| Orange | 1 medium | 44 | 15 | 7 |
| Peach | 1 medium | 42 | 11 | 5 |
| **Beverages** | | | | |
| Orange juice | 1 cup | 46 | 26 | 13 |
| Gatorade | 1 cup | 78 | 15 | 12 |
| Coca-Cola | 1 cup | 63 | 26 | 16 |
| **Snack Foods** | | | | |
| Potato chips | 1 oz | 54 | 15 | 8 |
| Chocolate | 1 oz | 49 | 18 | 9 |
| Jelly beans | 1 oz | 80 | 26 | 21 |

*Based on a comparison with glucose.

Source: Foster-Powell, K et al, "International table of glycemic index and glycemic load," *The American Journal of Clinical Nutrition*, Vol. 76 no. 1, January 2002, pp. 5–56.
Jacques Cornell/McGraw-Hill Education; Stockdisc/PunchStock; Judith Collins/Alamy Stock Photo; Comstock/Stockbyte/Getty Images; D. Hurst/Alamy Stock Photo; Jules Frazier/Getty Images; baibaz/Shutterstock; Stockbyte/Getty Images; Brand X Pictures/Getty Images

*Knowledge Check*

1. How do insulin and glucagon regulate blood glucose levels?
2. How does type 1 diabetes differ from type 2 diabetes?
3. What are the health risks associated with poorly controlled diabetes?
4. How does the glycemic index differ from the glycemic load?

► A term you might see on food labels is *net carbs*. Although this term is not FDA approved, sometimes it is used to describe the carbohydrates that increase blood glucose. Fiber and sugar alcohol content are subtracted from the total carbohydrate content to yield net carbs because they have a negligible effect on blood glucose.

# Chapter *Summary*

## 5.1 The general formula for carbohydrates is $(CH_2O)n$,
where *n* represents the number of times the ratio is repeated. The common monosaccharides are glucose, fructose, and galactose. Sugar alcohols are derivatives of monosaccharides. Additional monosaccharides found in nature are ribose and deoxyribose. Carbohydrates containing 2 monosaccharides are called disaccharides. Disaccharides include maltose, sucrose, and lactose. Oligosaccharides are complex carbohydrates that contain 3 to 10 single sugar units. Polysaccharides are complex carbohydrates that often contain hundreds to thousands of glucose molecules. Digestible polysaccharides are starch and glycogen. Dietary and functional fibers are indigestible polysaccharides.

---

**Simple Forms of Carbohydrates**

Monosaccharides: glucose, fructose, galactose

Disaccharides: sucrose, lactose, maltose

**Complex Forms of Carbohydrates**

Oligosaccharides: raffinose, stachyose

Polysaccharides: starches (amylose and amylopectin), glycogen, fiber

---

## 5.2 Carbohydrates are found in a wide variety of foods,
including table sugar, jam, jelly, fruits, soft drinks, rice, pasta, cereals, breads, dried beans, lentils, corn, peas, and dairy products. Starches contribute much of the carbohydrate in our diets. A diet rich in grains, legumes, and tubers also can provide significant amounts of dietary fiber (especially insoluble cellulose, hemicellulose, and lignins). Substances that impart sweetness to foods fall into 2 broad classes: nutritive sweeteners, which can be metabolized to yield energy, and non-nutritive (alternative or artificial) sweeteners, which provide no food energy. The sugar alcohols, sorbitol, mannitol, and xylitol, are nutritive sweeteners used in sugarless gum and candies. Non-nutritive sweeteners provide noncaloric or very-low-calorie sugar substitutes.

## 5.3 Adults need about 130 g/day of digestible carbohydrate
to supply adequate glucose for the brain and central nervous system, without having to rely on partial replacement of glucose by ketone bodies as an energy source. In North America, carbohydrates supply about 50% of energy intakes in adults. The Dietary Guidelines for Americans recommend limiting added sugars to a maximum of 10% of total energy intake. The Institute of Medicine's Food and Nutrition Board set an Upper Limit of 25% of energy intake for added sugar consumption.

Brand X Pictures/Stockbyte/Getty Images

The Adequate Intake for fiber is based on a goal of 14 g/1000 kcal consumed. For adults up to age 50 years, the Adequate Intake is set at 25 g for women and 38 g for men. After age 50, the Adequate Intake falls to 21 g/day and 30 g/day, respectively. In North America, carbohydrates supply about 50% of the energy intakes of adults. Sugar intake tends to be higher than recommended and fiber intake lower than recommended.

## 5.4 Most of the digestible carbohydrates
in our diets are broken down to glucose. As glucose, they provide a primary source of energy, spare protein for vital processes, and prevent ketosis. Fiber is indigestible; it helps prevent constipation and diverticular disease and enhances the management of body weight, blood glucose levels, and blood cholesterol levels.

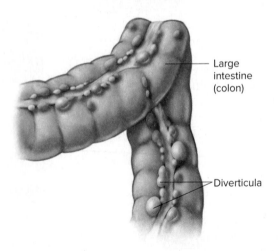

Large intestine (colon)

Diverticula

## 5.5 During digestion, starch and sugars are broken
into monosaccharide units that are small enough to be absorbed. The enzymatic digestion of some carbohydrates begins in the mouth with the action of an enzyme called salivary amylase. Salivary amylase is inactivated by the acidity of the stomach. In the small intestine, polysaccharides are digested further by pancreatic amylase and specialized enzymes in the absorptive cells of the small intestine. Glucose and galactose are absorbed by an active absorption process. Fructose is taken up by the absorptive cells via facilitated diffusion. Monosaccharides are transported via the portal vein to the liver. Within the liver, fructose and galactose are converted to glucose. Glucose is transported through the bloodstream for use by the cells of the body.

## Carbohydrate Digestion & Absorption

**1** **Mouth:** Some starch is broken down to polysaccharide and disaccharide units by salivary amylase.

**2** **Stomach:** Salivary amylase is inactivated by the acidity in the stomach. No further digestion occurs in the stomach.

**3** **Pancreas:** Pancreatic amylase and dextrinase are secreted into the small intestine to break polysaccharides from starch into disaccharides.

**4** **Small intestine:** Enzymes in the wall of the small intestine break down the disaccharides into monosaccharides.

**5** **Liver:** The absorbed monosaccharides are transported to the liver by the portal vein.

**6** **Large intestine:** Some soluble fiber is metabolized into acids and gases by bacteria in the large intestine.

**7** **Rectum and anus:** Insoluble fiber escapes digestion and is excreted in feces.

## Clinical Perspective

An inability to regulate glucose metabolism can result in diabetes, the major forms of which are type 1 diabetes and type 2 diabetes. Type 1 diabetes often begins as an autoimmune disorder in late childhood and runs in families, suggesting a genetic link. The onset of type 1 diabetes is caused by insufficient insulin release by the pancreas, which results in increased blood glucose levels. Type 1 diabetes is treated by insulin and nutrition therapy. Poorly controlled diabetes can cause blindness, cardiovascular disease, kidney disease, and nerve deterioration. A person with diabetes should work regularly with a physician and dietitian to monitor and adjust his or her diet, medications, and physical activity. Type 2 diabetes is a progressive disease characterized by insulin resistance or loss of responsiveness by body cells to insulin. As a result, glucose is not readily transferred into cells and builds up in the bloodstream, causing hyperglycemia. Treatments for type 2 diabetes are aimed at maintaining normal ranges of blood glucose through lifestyle modification and medication use.

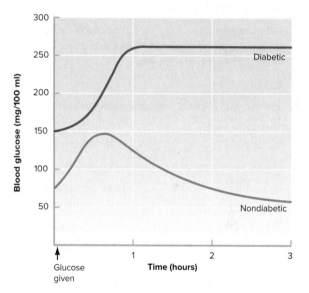

## 5.6 Adequate carbohydrate intake is important for maintaining health and decreasing the risk of chronic disease.

Very high intakes of fiber (i.e., above 50 to 60 g/day) combined with low fluid intake can result in hard, dry stools that are painful to eliminate. Very high fiber intakes also may decrease the absorption of several trace minerals. High intakes of sugars can displace more nutritious foods and increase the risk of weight gain, obesity, and dental caries. Lactose intolerance can cause symptoms of abdominal pain, bloating, gas, and diarrhea after consuming lactose, especially in large amounts. Glycemic index is a ratio of the blood glucose response of a given food compared with a standard, such as white bread. Glycemic load takes into account the glycemic index and the amount of carbohydrate consumed.

## Expert Perspective

Many experts are concerned about the high intake of sugar-sweetened beverages because these drinks are associated with an increased risk of obesity, diabetes, and heart disease. To reduce intake and raise money for nutrition and health programs, some have proposed taxing these beverages, whereas others oppose taxing sugar-sweetened beverages on the grounds that a tax would unfairly burden the poor and would not help solve the obesity epidemic.

Martin Poole/Getty Images

## Study Questions

1. Which of the following is a monosaccharide?

   a. lactose
   b. raffinose
   c. fructose
   d. maltose

2. Which of the following is classified as a digestible form of polysaccharide?

   a. cellulose
   b. raffinose
   c. lignin
   d. pectin

3. Carbohydrates are involved in all of the following functions *except* _____.

   a. providing energy
   b. preventing ketosis
   c. promoting bowel health
   d. promoting cell differentiation

4. Individuals with lactose intolerance have difficulty digesting milk products because they lack the enzyme needed to break apart the beta bond linkages.

   a. true
   b. false

5. The Adequate Intake for dietary fiber is set at 14 g/1000 kcal.

   a. true
   b. false

6. Which sweetener is classified as a non-nutritive sweetener?

   a. sorbitol
   b. honey
   c. aspartame
   d. mannitol

7. Which of the following is a poor source of dietary fiber?

   a. whole-grain oats
   b. fresh blueberries
   c. low fat yogurt
   d. dried lentils

8. Which of the following is a good source of starch?

   a. citrus fruits
   b. dark, leafy greens
   c. enriched grains
   d. barbequed chicken

9. Which term is used to describe an elevated blood sugar level?

   a. glucosuria
   b. hypoglycemia
   c. hyperlipidemia
   d. hyperglycemia

10. Which of the following is *not* a classic symptom of type 1 diabetes?

    a. polyuria
    b. polydipsia
    c. hunger
    d. rapid weight gain

11. Which population group is at lowest risk of diabetes?

    a. college athletes
    b. individuals with obesity
    c. individuals of Hispanic/Latino heritage
    d. adults over 45 years of age

12. High fiber diets can impair the absorption of trace minerals.

    a. true
    b. false

13. Which of the following is associated with low fiber diets?

    a. diverticulosis
    b. dental caries
    c. diarrhea
    d. lactose intolerance

14. Carbohydrate counting is an effective way for individuals with diabetes to monitor their daily carbohydrate intake.

    a. true
    b. false

15. The glycemic index is a ratio of the blood glucose response of a given food compared with a standard.

    a. true
    b. false

16. Describe how each of the 3 major dissaccharides plays a part in the human diet.

17. How do amylose, amylopectin, and glycogen differ from each other?

18. Compare the sweetness and food uses of nutritive and non-nutritive (alternative) sweeteners.

19. Compare the recommended intakes of carbohydrate with the typical intakes in North America.

20. What health benefits do indigestible carbohydrates offer?

21. Explain the main steps in carbohydrate digestion and absorption.

22. What are the possible effects of a diet too high in fiber?

23. How do type 1 and type 2 diabetes differ?

24. Why do some support taxing sugar-sweetened beverages? Why are others opposed?

Answer Key: 1-c; 2-d; 3-d; 4-a; 5-a; 6-c; 7-c; 8-c; 9-d; 10-d; 11-a; 12-a; 13-a; 14-a; 15-a; 16-refer to Section 5.1; 17-refer to Section 5.1; 18-refer to Section 5.2; 19-refer to Section 5.3; 20-refer to Section 5.4; 21-refer to Section 5.5; 22-refer to Section 5.6; 23-refer to Clinical Perspective; 24-refer to Expert Perspective

# References

1. Fitch C, Keim KS. Position of the Academy of Nutrition and Dietetics: Use of nutritive and nonnutritive sweeteners. *J Acad Nutr Diet.* 2012;112:739.

2. Food and Nutrition Board. *Dietary Reference Intakes for energy, carbohydrate, fiber, fat, fatty acids, cholesterol, protein, and amino acids.* Washington, DC: National Academies Press; 2005.

3. Keim NL and others. Carbohydrates. In: Shils ME and others, eds. *Modern nutrition in health and disease.* 11th ed. Philadelphia: Lippincott Williams & Wilkins; 2013.

4. Marriott BP and others. National estimates of dietary fructose intake increase from 1977 to 2004 in the United States. *J Nutr.* 2009;139:1228S.

5. Skypala I. Adverse food reactions—An emerging issue for adults. *J Am Diet Assoc.* 2011;111:1877.

6. Grabitske H, Slavin J. Gastrointestinal effects of low-digestible carbohydrates. *Crit Rev Food Sci Nutr.* 2009;49:327.

7. Lockyer S, Nugent AP. Health effects of resistant starch. *Nutr Bull.* 2017; doi.org/10.1111/nbu.1224.

8. Evans CE. Dietary fibre and cardiovascular health: A review of current evidence and policy. *Proc Nutr Soc.* 2019; doi:10.1017/S0029665119000673.

9. McRae MP. Dietary fiber intake and type 2 diabetes mellitus: An umbrella review of meta-analyses. *J Chiropr Med.* 2018;17:44.

10. Copeland E, Jones AS. Diverticular disease and diverticulitis: Causes, symptoms, and treatment. *Pharmaceu J.* 2019;article20206352.

11. McRae MP. Health benefits of dietary whole grains: An umbrella review of meta-analyses. *J Chiropr Med.* 2017;16:10.

12. Aune D and others. Dietary fibre, whole grains and risk of colorectal cancer: Systematic review and dose-response meta-analysis of prospective studies. *BMJ.* 2011;343:d6617.

13. Collin LJ and others. Association of sugary beverage consumption with mortality risk in US adults. *JAMA Network Open.* 2019;2:e193121.

14. Guasch-Ferre M, Hu FB. Are fruit juices just as unhealthy as sugar-sweetened beverages? *JAMA Network Open.* 2019;2:e193109.

15. Stanhope KL. Sugar consumption, metabolic disease and obesity: The state of the controversy. *Crit Rev Clin Lab Sci.* 2016;53:52.

16. Lohner S and others. Health outcomes of non-nutritive sweeteners: Analysis of the research landscape. *Nutr J.* 2017;16:55.

17. Samuel R and others. Stevia leaf to stevia sweetener: Exploring its science, benefits, and future potential. *J Nutr.* 2018;148:1186S.

18. U.S. Department of Health and Human Services and U.S. Department of Agriculture. Dietary Guidelines for Americans, 2020–2025, 9th ed. 2020. dietaryguidelines.gov.

19. Ervin RB, Ogden CL. *Consumption of added sugars among U.S. adults, 2005–2010.* NCHS Data Brief, #122. Hyattsville, MD: National Center for Health Statistics; 2013.

20. Welsh JA and others. Consumption of added sugars is decreasing in the United States. *Am J Clin Nutr.* 2011;94:726.

21. Quagliani D, Felt-Gunderson P. Closing America's fiber intake gap. *Am J Lifestyle Med.* 2017;11:80.

22. Murphy MM and others. Global assessment of select phytonutrient intakes by level of fruit and vegetable consumption. *Br J Nutr.* 2014;112:1004.

23. Murphy N and others. Dietary fibre intake and risks of cancers of the colon and rectum in the European Prospective Investigation into Cancer and Nutrition (EPIC). *PLOS ONE.* 2012;7(6):e39361.

24. Miketinas DC and others. Fiber intake predicts weight loss and dietary adherence in adults consuming calorie-restricted diets: The POUNDS lost (Preventing Overweight Using Novel Dietary Strategies) study. 2019;149:1742.

25. Dong D and others. Consumption of specific foods and beverages and excess weight gain among children and adolescents. *Health Affairs.* 2015;34:1940.

26. Ye E and others. Greater whole-grain intake is associated with lower risk of type 2 diabetes, cardiovascular disease, and weight gain. *J Nutr.* 2012;142:1304.

27. Chen G and others. Whole-grain intake and total, cardiovascular, and cancer mortality: A systematic review and meta-analysis of prospective studies. *Am J Clin Nutr.* 2016;104:164.

28. Szilagyi A, Ishayek N. Lactose intolerance, dietary avoidance, and treatment options. *Nutr.* 2018;10:1994.

29. Kit BK and others. Trends in sugar-sweetened beverage consumption among youth and adults in the United States: 1999–2010. *Am J Clin Nutr.* 2013;98:180.

30. Malik VS and others. Long-term consumption of sugar-sweetened and artificially sweetened beverages and risk of mortality in adults. *Circ.* 139:2113.

31. Brownell KD and others. The public health and economic benefits of taxing sugar-sweetened beverages. *New Eng J Med.* 2009;361:1599.

32. Brownell KD, Frieden TR. Ounces of prevention—The public policy case for taxes on sugared beverages. *N Engl J Med.* 2009;60:18.

33. World Health Organization. Taxes on sugary drinks: Why do it? 2017; apps.who .int/iris/bitstream/handle/10665/260253/WHO-NMH-PND-16.5Rev.1-eng.pdf;jse ssionid=F02C260701A5B52AAFF08456F52E92C2?sequence=1.

34. Epstein L and others. The influence of taxes and subsidies on energy purchased in an experimental purchasing study. *Psychol Sci.* 2010;21:406.

35. Wilde P and others. Cost-effectiveness of a US national sugar-sweetened beverage tax with a multistakeholder approach: Who pays and who benefits. *Am J Public Health.* 2019;109:276.

36. Bray GA. Potential health risks for beverages containing fructose found in sugar or high-fructose corn syrup. *Diabetes Care.* 2013;36:11.

37. Moore JX and others. Metabolic Syndrome prevalence by race/ethnicity and sex in the United States, National Health and Nutrition Examination Survey, 1988–2012. *Prev Chronic Dis.* 2017;14:E24.

38. Holvoet P and others. Association between circulating oxidized low-density lipoprotein and incidence of the Metabolic Syndrome. *JAMA.* 2008;299:2287.

39. American Diabetes Association. Standards of medical care in diabetes. *Diabetes Care.* 2019;37:11.

40. American Diabetes Association. Foundations of care and comprehensive medical evaluation. *Diabetes Care.* 2016;39:S23.

41. Papatheodorou K and others. Complications of diabetes 2017. *J Diabetes Res.* 2018;3086167.

42. Chawla A and others. Microvascular and macrovascular complications in diabetes mellitus: Distinct or continuum? *Indian J Endocrinol Metab.* 2016;20:546.

43. Colberg SR and others. Physical activity/exercise and diabetes: A position statement of the American Diabetes Association. *Diabetes Care.* 2016;39:2065.

44. Epstein L and others. The influence of taxes and subsidies on energy purchased in an experimental purchasing study. *J Am Diet Assoc.* 2008;108:S34.

45. Murphy N and others. Dietary fibre intake and risks of cancers of the colon and rectum in the European Prospective Investigation into Cancer and Nutrition (EPIC). *IUBMB Life.* 2011;63:7.

46. Ye E and others. Greater whole-grain intake is associated with lower risk of type 2 diabetes, cardiovascular disease, and weight gain. *Nutr Metab.* 2015;12:6.

Holly Curry/McGraw-Hill

The importance of consuming omega-3 fatty acids was discovered many years ago in studies of Greenland Eskimos. Their diet is very high in fish oils, and they exhibit diminished blood clotting and lower risks of heart disease. Two servings of cold-water fish, such as salmon, each week can meet omega-3 fatty acid needs. Learn more about these fats at **heart.org**.
Doug Allan/Getty Images

# 6 Lipids

## Learning Objectives

**After studying this chapter, you will be able to**

1. Recognize the basic chemical structure of fatty acids and describe how they are named.

2. Explain the functions of triglycerides, fatty acids, phospholipids, and sterols in the body.

3. Classify and evaluate the different fatty acids based on their health benefits or consequences.

4. Identify and explain the differences between the two essential fatty acids.

5. Identify food sources of triglycerides, fatty acids, phospholipids, and sterols.

6. Discuss the recommended intake of lipids.

7. Identify strategies for modifying total fat, saturated fat, and *trans* fat intake.

8. Summarize the digestion, absorption, and transport of lipids in the body.

9. Explain the relationship of dietary lipids, including *trans* fat, to chronic diseases.

10. Describe dietary measures to reduce the risk of developing cardiovascular disease.

## Chapter Outline

**LIPIDS (FATS AND OILS) GIVE** food a creamy, velvety mouthfeel. They also add a great deal of flavor to foods—think about the buttery taste of croissants or the savory flavor of beef. Most foods we eat contain at least some fat. Foods rich in fat and oils are vegetable oils, margarine, butter, avocado, and nuts (Fig. 6-1). All contain close to 100% of energy as fat. Many protein rich foods, such as meat, cheese, and peanut butter, are high in fat, too. Cakes, pies, cookies, muffins, chocolate, ice cream, and snack foods such as chips and crackers also contain sizable amounts of fat. In addition to providing flavor, texture, and energy, dietary fats supply fat-soluble vitamins (vitamins A, D, E, and K). They also are a compact source of calories—gram for gram, fats supply more than twice as many calories as both carbohydrates and proteins. Fat insulates the body and pads its organs to protect them from injuries. We also use fat to make hormones.

As you can see, some fats are essential for good health, so why do fats have such a bad reputation? It's because not all fats are created equal from a health perspective. Exploring the characteristics of the lipid family members will clarify this often misunderstood nutrient.

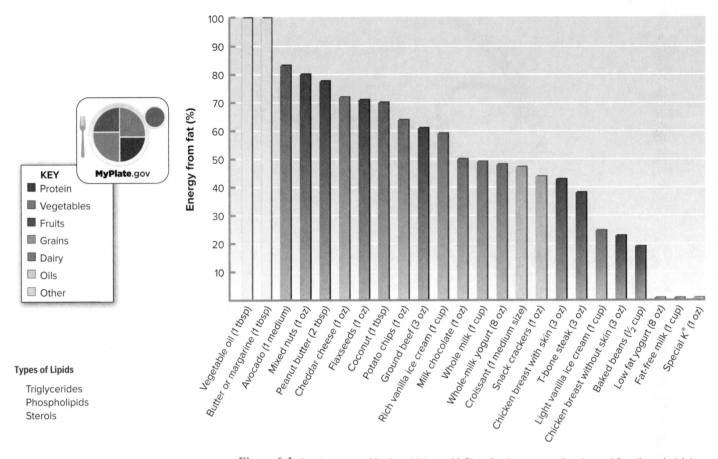

**KEY**
- Protein
- Vegetables
- Fruits
- Grains
- Dairy
- Oils
- Other

**Types of Lipids**

Triglycerides
Phospholipids
Sterols

*Figure 6-1* Food sources of fat. In addition to MyPlate food groups, yellow is used for oils and pink is used for substances that do not fit easily into food groups (e.g., candy, salty snacks).

Source: ChooseMyPlate.gov, U.S. Department of Agriculture

▶ Fat in foods has been considered the most satiating of all the macronutrients. However, studies show that protein and carbohydrate probably provide more satiety (gram for gram). High fat meals provide satiety because they are high in calories.

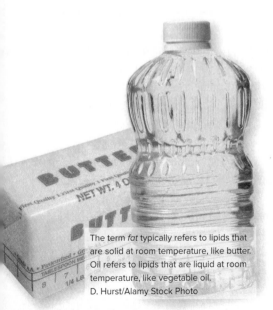

The term *fat* typically refers to lipids that are solid at room temperature, like butter. Oil refers to lipids that are liquid at room temperature, like vegetable oil.
D. Hurst/Alamy Stock Photo

# 6.1 Triglycerides

When you hear the word *lipid,* you may think of butter, lard, olive oil, and margarine. We usually refer to lipids simply as *fats and oils;* however, the lipid family can be further divided into triglycerides, phospholipids, and sterols. Although these diverse family members differ in their structures and functions, they all contain carbon, hydrogen, and oxygen, and none of them dissolve in water. However, they do dissolve in organic solvents, such as chloroform, benzene, and ether. Think of oil and vinegar salad dressing. No matter how long or hard you shake the dressing, when you stop shaking, the vinegar and oil quickly separate into layers, with the oil floating on the vinegar. This insolubility property sets lipids apart from carbohydrates and proteins.

Triglycerides are the most common type of lipid found in foods and in the body. About 95% of the fats we eat and 95% of the fat stored in the body are in the form of triglycerides.

## Structure

Each **triglyceride** molecule consists of 3 fatty acids attached (bonded) to 1 glycerol, which serves as a backbone for the fatty acids (Fig. 6-2). A triglyceride is built by attaching a fatty acid to each of glycerol's 3 hydroxyl groups (–OH). The fatty acids can be all the same fatty acid or they can be different. One water molecule is released when each fatty acid bonds to glycerol. The process of attaching fatty acids to glycerol is called **esterification.** The release of fatty acids from glycerol is called hydrolysis. Fatty acids released from the glycerol backbone are called **free fatty acids** to emphasize that they are unattached. A triglyceride that loses a fatty acid is a **diglyceride.** A **monoglyceride** results when 2 fatty acids are lost. The process of reattaching a fatty acid to glycerol that has lost a fatty acid is known as **re-esterification.**

All free fatty acids have the same basic structure: long chains of carbon atoms linked together and surrounded by hydrogen atoms. Free fatty acids have an acid (carboxyl) group at 1 end of the chain and a methyl group at the opposite end (Fig. 6-3). Fatty acids can vary in 3 ways: the number of carbons in the chain, the extent to which the chain is saturated with hydrogen, and the shape of the chain (straight or bent).

### Number of Carbons in the Fatty Acid Chain (Length)

Fatty acid chains usually have between 4 and 24 carbons. **Long-chain fatty acids** have 12 or more carbon atoms. Fats from beef, pork, and lamb and most plant oils are long chain. Long chains of carbon atoms take the most time to digest and are transported via the lymphatic system. **Medium-chain fatty acids** are 6 to 10 carbons in length, are digested almost as rapidly as glucose, and are transported via the portal system. Coconut and palm kernel oils are examples of medium-chain fatty acids. **Short-chain fatty acids** are usually less than 6 carbons in length. They are rapidly digested and transported via the portal system. About 3% of the fat in butter is short chain.

### Number of Double Bonds in the Fatty Acid (Saturation)

Fatty acids can be saturated, monounsaturated, or polyunsaturated. To understand saturation, remember, a carbon atom can form 4 chemical bonds, an oxygen atom can form 2 bonds, and a hydrogen atom can form only 1 bond.

Figure 6-3 shows a **saturated fatty acid (SFA).** Notice that every carbon in the chain has formed the maximum of 4 bonds. It is a saturated fatty acid because all the bonds between the carbons are single connections and the other carbon bonds are filled with hydrogens. To understand this concept, picture a school bus with a child in every seat. The school bus is "saturated" with children—there are no empty seats.

A **monounsaturated fatty acid (MUFA)** is shown in Figure 6-4. Notice how the carbons in the green shaded box in the chain are each missing 1 hydrogen. These carbons formed a double bond between each other by each giving up 1 hydrogen. (Remember, carbons can form only 4 bonds.) Fatty acids that have 1 double bond in the carbon chain are called monounsaturated fatty acids. They have 1 (*mono*) location in the carbon chain that is not saturated with hydrogen. Using the school bus example, a MUFA is like having 1 empty seat.

A **polyunsaturated fatty acid (PUFA)** has at least 2 double bonds in its carbon chain (Fig. 6-5). If the school bus were a PUFA, it would have 2 or more empty seats.

### Shape of the Fatty Acid Chain

The shape of the carbon chain varies with saturation. Saturated and *trans* fatty acids have straight carbon chains, and unsaturated *cis* fatty acids have bent or kinked carbon chains. In *cis* **fatty acids,** the hydrogens attached to the double-bonded carbons are on the same side of the carbon chain (see Fig. 6-5). In *trans* **fatty acids** (also called *trans* fats), the hydrogens attached to the double-bonded carbons zigzag back and forth across the carbon chain (Fig. 6-6). In Figure 6-7, notice how the *cis* fatty acid, which has the hydrogens next to the double bonds on the same side of the carbon chain, bends. The *trans* fatty acid, which has the hydrogens next to the double bonds on opposite sides of the carbon chain, is straight and resembles a saturated fatty acid.

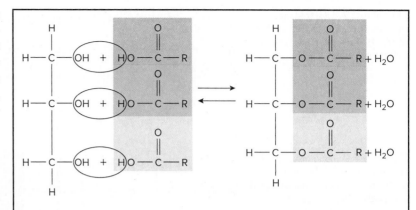

Triglycerides are built from a glycerol backbone and 3 fatty acids. Glycerol (in white) has 3 carbons in its chain. A triglyceride forms when each hydroxyl (−OH) group on the glycerol backbone bonds with the hydrogen atom from the acid (carboxyl) end of a fatty acid.

The bond between a fatty acid and glycerol is called an ester bond. One molecule of water ($H_2O$) forms each time an ester bond forms (this is called **esterification**). Thus, when a diglyceride (2 fatty acids attached to a glycerol backbone) forms, 2 molecules of water form. Similarly, forming a triglyceride will generate 3 water molecules. Esterification is a condensation reaction.

A molecule of water is used when a fatty acid breaks away from a glycerol backbone (this hydrolysis reaction can be thought of as "de-esterification"). Reattaching the fatty acid to a glycerol backbone (called **re-esterification**) will produce a water molecule.

*Figure 6-2* Esterification, hydrolysis, and re-esterification of glycerol and fatty acids to form triglycerides.

► Chemists classify triglycerides as esters. Tri*acyl*glyceride is the chemical name for triglyceride. *Acyl* refers to a fatty acid that has lost its hydroxyl group (−OH). A fatty acid loses its hydroxyl group when it attaches to glycerol.

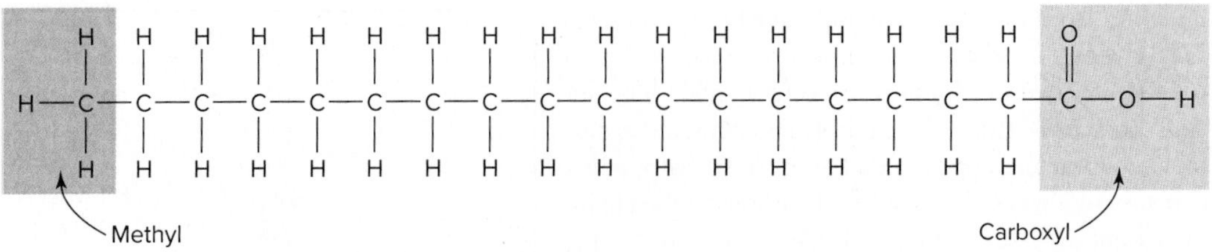

*Figure 6-3*  A saturated fatty acid has no carbon-carbon double bonds. This saturated fatty acid is called stearic acid.

*Figure 6-4*  A monounsaturated fatty acid has 1 carbon-carbon double bond. This monounsaturated fatty acid is called oleic acid.

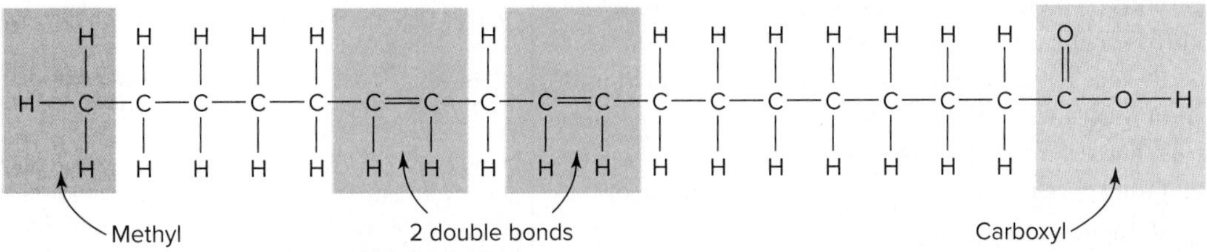

*Figure 6-5*  A polyunsaturated fatty acid has 2 or more carbon-carbon double bonds. This polyunsaturated fatty acid is called linoleic acid.

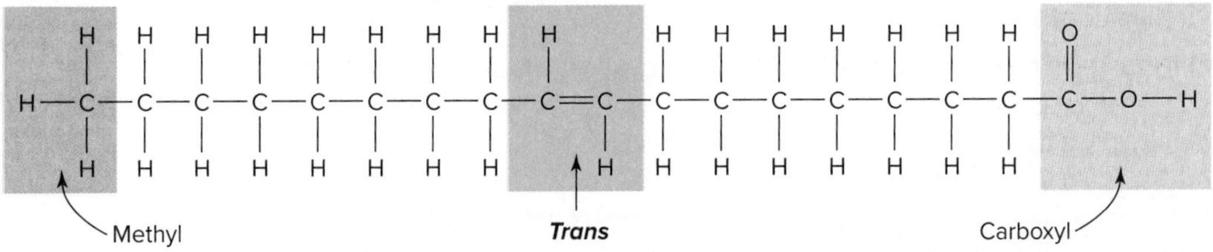

*Figure 6-6*  This monounsaturated fatty acid is a *trans* fat. It is called elaidic acid, which is the major *trans* fatty acid found in processed fats.

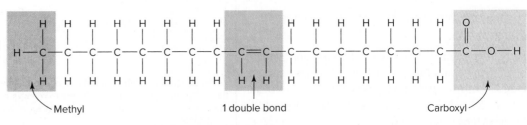

► Why are some fats solid at room temperature and others liquid? The fat's carbon chain shape and length determine this. Like crumpled paper, the kinked carbon chains of unsaturated fatty acids do not pack tightly together. This "loose" packing causes them to be soft or liquid at room temperature. In contrast, saturated fatty acids, with their straight carbon chain, like unfolded paper, pack tightly together. This tight packing helps them stay firm (not melt) at room temperature. However, the effect of the straight chain shape on saturated fats (but not unsaturated fats) can be overridden by chain length. That is, only saturated fatty acids with a long carbon chain, such as beef fat, are solid at room temperature. Medium- and short-chain saturated fats are soft or liquid at room temperature.

Most unprocessed unsaturated fatty acids, such as oils freshly pressed from nuts and seeds, are in the *cis* form. *Trans* fatty acids are found mostly in the polyunsaturated oils modified by food manufacturers using a process called hydrogenation.

**Hydrogenation** adds hydrogen to the carbon chain of unsaturated fats. As the amount of added hydrogen increases, the unsaturated fat becomes more and more saturated (until it is totally saturated) and increasingly solid. For instance, corn oil, which is polyunsaturated and liquid at room temperature, can be hydrogenated: a little to make squeeze margarine, some to make tub margarine, and a lot to make stick margarine.

Hydrogenation is like putting children in some of the empty bus seats except, when the children are added to the bus, it changes the shape of the bus. The shape change occurs because hydrogenation creates *trans* fatty acids, which have a straighter shape than *cis* fatty acids.

## Naming Fatty Acids

Two systems are commonly used to name fatty acids. Both are based on the numbers of carbon atoms and the location of double bonds in a fatty acid's carbon chain. The omega ($\omega$ or $n$) system indicates where the first double bond closest to the methyl (omega) end of the chain occurs. For example, the fatty acid linoleic acid is named 18:2 $\omega$6 (18:2 $n$6). This

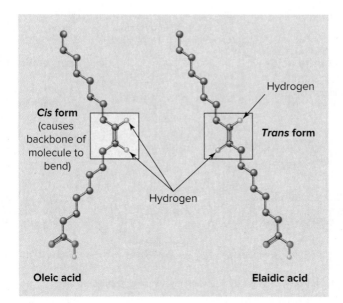

*Cis* form
(causes backbone of molecule to bend)

Hydrogen

*Trans* form

Hydrogen

**Oleic acid**

**Elaidic acid**

▶ Ball-and-stick models (e.g., see Fig. 6-7) show the spatial arrangement of atoms in a molecule. The blue balls are carbon, white hydrogen, and red oxygen. The lines between the balls represent bonds. For simplicity, the white hydrogen balls are shown only at the double-bond area.

*Figure 6-7* *Cis* and *trans* fatty acid comparison. *Cis* fatty acids have hydrogens on the same side of the double bond, whereas *trans* fatty acids have hydrogens on the opposite side of the double bonds.

means that linoleic acid, shown on the right in Figure 6-8, has 18 carbons in its carbon chain and 2 double bonds, and the first double bond starts at the 6th carbon from the omega end (orange boxed area of Fig. 6-8). The delta ($\Delta$) system describes fatty acids in relation to the carboxyl end of the carbon chain (blue boxed area of Fig. 6-8) and indicates the location of all double bonds. Thus, in the delta system, linoleic acid is written 18:2 $\Delta$9,12. Whereas the scientific community uses both of these systems nearly equally, the popular media use the omega system, as does this text.

**hydrogenation** Addition of hydrogen to some carbon-carbon double bonds and producing some trans fatty acids. This process is used to convert liquid oils into more solid fats.

## Essential Fatty Acids

Humans can synthesize a wide variety of fatty acids, but we *cannot* make 2 PUFAs: alpha-linolenic acid (the major omega-3 fatty acid in food) and linoleic acid (the major omega-6 fatty acid in food). Alpha-linolenic acid and linoleic acid are **essential fatty acids (EFAs).** We must get EFAs from foods because our bodies are unable to synthesize essential fatty acids with a double bond before the 9th carbon in the chain, counting from the omega end (see Fig. 6-8).

The location of the double bond closest to the omega carbon (methyl end) of the fatty acid identifies the fatty acid's family. If the 1st double bond of a polyunsaturated fatty acid

▶ Alpha is the first letter of the Greek alphabet, and omega is the last letter. Alpha looks like this: α. Omega looks like this in Greek: ω. In English, a lowercase *n* is sometimes used in place of the Greek letter for omega.

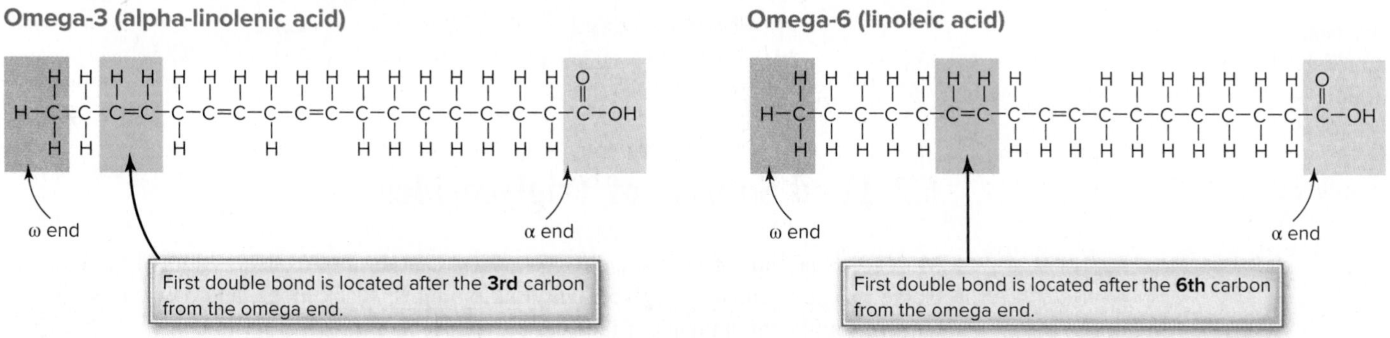

**Omega-3 (alpha-linolenic acid)**

ω end

α end

First double bond is located after the **3rd** carbon from the omega end.

**Omega-6 (linoleic acid)**

ω end

α end

First double bond is located after the **6th** carbon from the omega end.

*Figure 6-8* Omega-3 fatty acids have their first double bond 3 carbons in from the methyl (omega) end of the carbon chain. Likewise, omega-6 fatty acids have their first double bond 6 carbons in from the omega end. How would alpha-linolenic acid be named using the omega and delta systems?

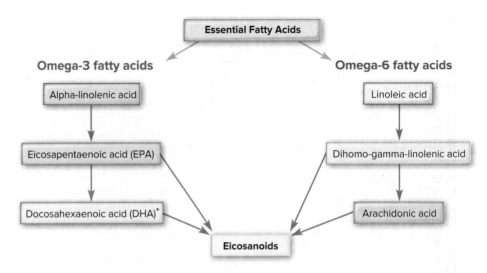

*Technically, DHA yields docosanoids, which are similar to eicosanoids.

***Figure 6-9*** The essential fatty acids alpha-linolenic acid and linoleic acid are used to make other important fatty acids.

The FDA has banned *trans* fats in food. Many food companies have eliminated *trans* fats, whereas others have fully hydrogenated polyunsaturated fats to eliminate *trans* fats. On food packages, hydrogenated fats appear in the ingredient list as hydrogenated fat. Food manufacturers can call a food "*trans* fat free" if it contains 0.5 gram or less of *trans* fats.
©McGraw-Hill Education

occurs between the 3rd and 4th carbons from the methyl end, it is called an omega-3 fatty acid ($\omega$−3). If the 1st double bond occurs between the 6th and 7th carbons on a polyunsaturated fatty acid, it is called an omega-6 fatty acid ($\omega$−6).

As you can see in Figure 6-9, the fatty acids eicosapentaenoic acid (EPA) and, subsequently, docosahexaenoic acid (DHA) are made from alpha-linolenic acid. Likewise, the fatty acids dihomo-gamma-linolenic acid and, subsequently, arachidonic acid are made from linoleic acid.

Different eicosanoids are produced from dihomo-gamma-linolenic acid, arachidonic acid, and eicosapentaenoic acid. **Eicosanoids** are hormonelike compounds, such as **prostaglandins,** prostacyclins, thromboxanes, leukotrienes, and lipoxins, that affect the body in the region where they are produced. (They are called *local* hormones because, unlike typical hormones, they are made *and* used in the same area of the body.)

**eicosanoids** Hormonelike compounds synthesized from polyunsaturated fatty acids, such as omega-3 fatty acids and omega-6 fatty acids.

**prostaglandins** Potent compounds that are synthesized from polyunsaturated fatty acid that produce diverse effects in the body.

### Knowledge Check

1. What characteristics do all lipids have?
2. How do saturated, monounsaturated, and polyunsaturated fats differ?
3. What are the differences between *cis* and *trans* fats?
4. How do the omega and delta systems for naming fatty acids differ?
5. Why must essential fatty acids be provided by the diet?

## 6.2 Food Sources of Triglycerides

Most foods provide at least some triglycerides. Certain foods, such as animal fat and vegetable oils, are primarily triglycerides. Bakery items, snack foods, and dairy desserts also contain significant amounts of fat. In contrast, fat-free milk and fat-free yogurt, as well as many breakfast cereals and yeast breads, contain little or no fat. Other than coconuts and avocados, fruits and vegetables are low in fat.

Table 6-1 lists the main sources of each type of fatty acid. For instance, fats from animal sources and tropical oils (coconut, palm, palm kernel) are rich in saturated fatty acids. Omega-3 fatty acid sources include cold-water fish (salmon, tuna, sardines, mackerel), walnuts, and flaxseed. Fish oil and flaxseed oil supplements are another source of omega-3 fatty acids.

Most triglyceride rich foods contain a mixture of fatty acids. As you can see in Figure 6-10, butter contains saturated, monounsaturated, and polyunsaturated fatty acids. Saturated fats are the predominant fatty acid in butter, so it is referred to as a

| Table 6-1 Main Sources of Fatty Acids and Their State at Room Temperature | | | |
|---|---|---|---|
| **Type and Health Effects** | **Double Bonds** | **Main Sources** | **State at Room Temperature** |
| **Saturated Fatty Acids** *Increase blood levels of LDL-cholesterol* | 0 | | |
| Long-chain | 0 | Lard; fat in beef, pork, and lamb | Solid |
| Medium- and short-chain | 0 | Milk fat (butter), coconut oil, palm oil, palm kernel oil | Soft or liquid |
| **Monounsaturated Fatty Acids** *Decrease blood levels of LDL-cholesterol* | 1 | Olive oil, canola oil, peanut oil | Liquid |
| **Polyunsaturated Fatty Acids** *Decrease blood levels of LDL-cholesterol* | 2 or more | Sunflower oil, corn oil, safflower oil, fish oil | Liquid |
| **Essential Fatty Acids** Omega-3: Linolenic acid | 3 | | Liquid |
| Eicosapentaenoic acid (EPA)/ docosahexaenoic acid (DHA): *improves cognition, behavior, and mood; decreases risk of macular degeneration; normalizes blood lipids* | | Cold-water fish (salmon, tuna, halibut, sardines, mackerel) | |
| Alpha-linolenic acid: *reduces inflammation responses, blood clotting, and plasma triglycerides* | | Walnuts, flaxseed, hemp oil, canola oil, soybean oil, chia seeds | |
| Omega-6: linoleic acid, arachidonic acid *Regulates blood pressure and increases blood clotting* | 2 or more | Beef, chicken, eggs, safflower oil, sunflower oil, corn oil | Solid to liquid |
| **Naturally Occurring Trans Fatty Acids\*** *Naturally occuring trans fatty acids, like those found in meat and dairy from ruminant animals (cows, goats, sheep), are not associated with an increase risk of cardiovascular disease* | Usually 2 | Milk and dairy products, Meat | Soft to very solid |

\*Note that artificially made trans fats, once found in foods like vegetable shortening and margarine, are now banned because they increased LDL-cholesterol and lowered HDL-cholesterol.

Flax is a common plant that produces seeds high in omega-3 fats. The seeds can be ground and used as a meal in baked goods. They also can be pressed to extract the oil, which is sold as a nutritional supplement. Seed: LAMB/Alamy Stock Photo; Plant: Purestock/Alamy Stock Photo

Cold-water fish are high in essential fatty acids, but are fish safe to eat? Some people worry about potential health risks related to fish and fish oil supplements, such as carcinogens (e.g., DDT, dieldrin, heptachlor, PCBs, dioxin) and toxins (e.g., methylmercury). Although these contaminants are present in low levels in fresh and salt water, they can be concentrated in fish. To minimize exposure, select small, nonpredatory fish; vary the type of fish you eat; buy fish from reputable markets; and discard fatty portions of fish, where toxins concentrate. If you catch your own fish, always check local advisories to be sure that the water you plan to fish in is safe. You can call your local health department or visit your state government's website to learn about fishing advisories.

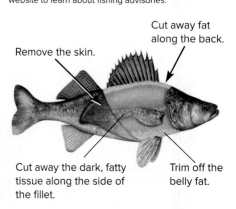

Cut away fat along the back.

Remove the skin.

Cut away the dark, fatty tissue along the side of the fillet.

Trim off the belly fat.

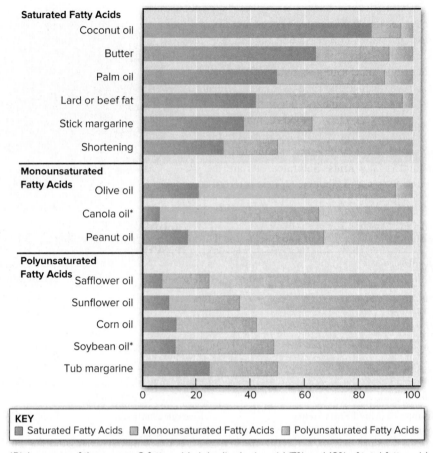

*Rich source of the omega-3 fatty acid alpha-linolenic acid (7% and 12% of total fatty acid content for soybean oil and canola oil, respectively).

*Figure 6-10*  Saturated, monounsaturated, polyunsaturated, and *trans* fatty acid composition of common fats and oils (expressed as % of all fatty acids in the product).

saturated fat. Similarly, olive oil contains saturated, monounsaturated, and polyunsaturated fatty acids, but monounsaturated fat predominates. Thus, it is referred to as a monounsaturated fat.

## Hidden Fats

The fat in some foods is visible: butter on bread, mayonnaise in potato salad, and marbling in raw meat. In many foods, however, fat is hidden, as is the fat in whole milk, cheese, pastries, cookies, cake, hot dogs, crackers, french fries, and ice cream. Nutrition Facts labels can help you learn more about the quantity of fat in the foods you eat (Fig. 6-11).

## Fat Replacements

To help consumers trim their fat intake and still enjoy the mouthfeel sensations fat provides, food companies offer low fat versions of many foods. To lower the fat in foods, manufacturers may replace some of the fat with water, protein (Dairy-Lo®), or forms of carbohydrates such as starch derivatives (Z-trim®), fiber (Maltrin®, Stellar™, Oatrim), and gums. Manufacturers also may use engineered fats, such as salatrim (Benefat®), which are made with fat and sucrose (table sugar) but provide few or no calories because they cannot be digested and/or absorbed well.

Fat replacements, such as gum fiber, are often used in soft serve ice cream.
Shutterstock/pixelliebe

*Take Action*

## Is Your Diet High in Saturated and Total Fat?

Instructions: In each row of the following list, circle your typical food selection from column A or B.

| Column A | | Column B |
|----------|----|----------|
| Bacon and egg sandwich | or | Avocado toast on whole grain bread |
| Donut or sweet roll | or | Whole-wheat roll, bagel, or bread |
| Breakfast sausage | or | Prosciutto |
| Whole milk | or | Reduced fat, low fat, or fat-free milk |
| Cheeseburger | or | Beyond Burger© or other plant-based burger |
| Potato salad | or | Potatoes roasted with olive oil |
| Ground chuck | or | Ground round |
| Soup with a cream base | or | Soup with a broth base |
| Macaroni and cheese | or | Macaroni with marinara sauce |
| Cheese and crackers | or | Peanut butter and whole-grain crackers |
| Lemon, pumpkin, or other breads made with butter | or | Lemon, pumpkin, or other breads made with oil |
| Ice cream | or | Frozen yogurt, sherbet, or reduced fat ice cream |
| Butter or stick margarine | or | Vegetable oils or soft margarine in a tub |

LAMB/Alamy Stock Photo

Purestock/Alamy Stock Photo

### Interpretation

The foods listed in column A tend to be high in saturated fat, cholesterol, and total fat. Those in column B generally are low in 1 of these dietary components. If you want to help reduce your risk of cardiovascular disease, choose more foods from column B and fewer from column A.

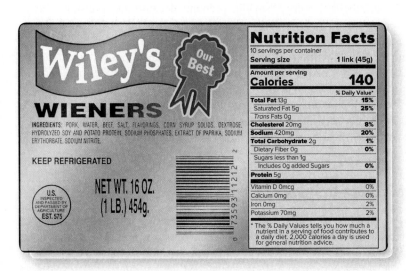

*Figure 6-11* Reading labels helps you locate hidden fat. Who would think that wieners (hot dogs) can contain about 86% of energy content as fat? Looking at the hot dog itself does not suggest that almost all its energy content comes from fat, but the label shows otherwise. Do the math: 120 kcal from fat/140 kcal = 0.86, or 86%.

*Critical* Thinking

Martin wants to cut back on the saturated fat in his diet. He mentions to you that he is still going to eat hamburgers and fries, but now he will order the burger without cheese and dip his fries in ketchup instead of Ranch dressing. Will he really reduce his fat intake?

► A "reduced fat" food may not be lower in calories than its full fat counterpart. That's because, when fat is removed from a product, something must be added—commonly, sugars—in its place.

► Table 2-3 in Chapter 2 defines the fat claims that are permitted on food labels, such as "low fat," "fat free," and "reduced fat."

So far, fat replacements have had little impact on our diets, partly because the currently approved forms are either not very versatile or not used extensively by manufacturers. In addition, fat replacements are not practical for use in the foods that provide the most fat in our diets—beef, cheese, whole milk, and pastries.

### Knowledge Check

1. How do the 3 classes of fatty acid affect blood levels of cholesterol?
2. What are examples of foods that contain hidden fats?
3. Name 3 foods that contain monounsaturated fat.

## 6.3 Functions of Triglycerides

Triglycerides are essential for optimal health. They provide a concentrated source of energy, insulate and cushion vital organs, and help transport essential nutrients in the bloodstream. However, high intakes, especially of saturated and *trans* fat, and imbalances of EFAs can present health challenges.

### Provide Energy

Triglycerides in food and body fat cells are a rich source of energy, with each gram providing about 9 calories. Triglycerides are the main fuel source for all body cells, except the nervous system and red blood cells. When you are resting or engaging in light physical activity, triglycerides provide 30 to 70% of the energy you burn. The exact amount depends on your glycogen stores, how physically fit you are, and the intensity and duration of the exercise.

► Only the glycerol backbone from a triglyceride can be used as a fuel for the nervous system. Another brain fuel that originates with triglycerides is ketones—compounds formed when fatty acids do not metabolize completely; large amounts of ketones form when carbohydrate (glucose) intake is restricted or insulin is low. (See Chapter 9.)

### Provide Compact Energy Storage

Triglycerides are the body's main storage form of energy. Excess calories from carbohydrate, fat, protein, and alcohol can be converted to fatty acids and then to triglycerides. Triglycerides make an excellent energy "savings account" because they are stable (don't react with other cell parts) and calorie dense. Fat cells contain about 80% lipid and only 20% water and protein. Muscle cells also contain fat and protein but are 73% water. This difference means that the lipid rich fat cells can deliver much more energy than the water rich muscle cells. Another reason triglycerides make an excellent storage form of energy is that they can expand to an average of 2 to 3 times their normal size. A single adipose (fat) cell can increase in weight about 50 times. Although it is important to have some body fat stores, very small and large stores can pose numerous health risks. (Chapter 10 discusses the health issues associated with underweight, overweight, and obesity.)

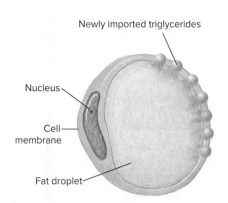

Newly imported triglycerides

Nucleus

Cell membrane

Fat droplet

**Adipose cell**

Newly imported triglycerides (blue dots), eventually merge with the larger central fat droplet. With prolonged overnutrition, adipose cells are stimulated to grow and become mature fat cells.

### Insulate and Protect the Body

The insulating layer of fat just beneath the skin (called subcutaneous fat) is made mostly of triglycerides. Subcutaneous fat insulates the body, keeping body temperature at a constant level. Visceral fat is the fat surrounding some organs—kidneys, for example. This fat cushions vital organs and keeps them from jostling around and getting injured. We usually do not notice the insulating function of subcutaneous fat because we wear clothes and add more when needed. However, people who are starving lose most of their body fat and, as a result, feel chilled even when the environment is warm.

## Aid Fat-Soluble Vitamin Absorption and Transport

Fats found in food carry fat-soluble vitamins (vitamins A, D, E, and K) to the small intestine, where dietary fat assists in the absorption of these vitamins. Fat-soluble vitamins are transported in the bloodstream in the same manner as dietary fat (see Chapter 4). Those who eat an extremely low fat diet, use mineral oil as a laxative, take certain medications (e.g., the weight loss medication orlistat), or have diseases that affect fat absorption (e.g., cystic fibrosis) may be unable to absorb sufficient amounts of fat-soluble vitamins.

## Essential Fatty Acid Functions

Essential fatty acids, along with phospholipids and cholesterol, are important structural components of cell membranes.[1] They also keep the cell membrane fluid and flexible so that substances can flow into and out of the cell membrane. The omega-3 fatty acid docosahexaenoic acid (DHA) is needed during fetal life and infancy for normal development and function of the retina (the part of the eye that senses light). Starting in the 1st few weeks of embryonic life, DHA is vital for normal development and maturation of the nervous system. Throughout life, DHA helps regulate nerve transmission and communication.

Eicosanoids, which are made from essential fatty acids, have over 100 different actions, such as regulation of blood pressure, blood clotting, sleep/wake cycles, body temperature, inflammation or hypersensitivity reactions (e.g., asthma), stomach secretions, labor during childbirth, and immune and allergic responses. For example, some types of eicosanoids cause inflammation, whereas other types prevent the inflammation associated with inflammatory diseases and allergic reactions. Other eicosanoids help the blood form clots, whereas other types help prevent clots. Still other eicosanoids from omega-6 fats influence blood vessels to constrict and raise blood pressure, yet other omega-6 eicosanoids, along with the omega-3 eicosanoids, influence the body to lower blood pressure by dilating blood vessels.

Eicosanoids also have other important roles in the body, many of which are only just being discovered. For example, they assist in

- Regulating cell division rates, which may help prevent certain cancers or slow the growth of existing tumors and help prevent cancer from spreading to other parts of the body
- Maintaining normal kidney function and fluid balance
- Directing hormones to their target cells
- Regulating the flow of substances into and out of cells
- Regulating ovulation, body temperature, immune system function, and hormone synthesis

When at rest or during low-intensity activity (<30% $VO_{2max}$), fatty acids are the primary fuel for the body.
PunchStock

*Knowledge Check*

1. What are 3 functions of triglycerides?
2. Under what conditions is fat used as the main fuel in the body?
3. What roles do essential fatty acids play in the body?

*Critical* **Thinking**

You tell your roommate that you have lost 5 pounds of fat weight. She responds by asking "How does your body 'excrete' fat?" What would be your answer?

*A Biochemist's View*

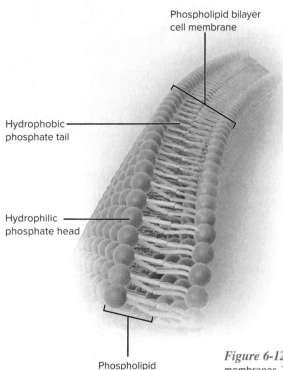

Lecithin—Phospholipid

Lecithin is a type of phospholipid. It is similar to a triglyceride. Instead of the 3 fatty acids found in a trigylceride, it contains 2 fatty acids and a phosphate group.

## 6.4 Phospholipids

Many types of phospholipids are found in food and the body, especially the brain. The structure of these lipids is very similar to that of triglycerides—except that 1 fatty acid is replaced with a compound (phosphate) that contains the mineral phosphorus and often has nitrogen attached. Phosphate gives phospholipids an important quality—making them soluble in both water and fat, thus keeping fats suspended in the blood and body fluids.

Here's how phospholipids work: The phosphate end (head) of the phospholipid is hydrophilic ("water loving") and will mix with water. The fatty acid end (tail) of the phospholipid is hydrophobic ("water fearing") and is attracted to fats. When placed in water, phospholipids cluster together, with their hydrophilic phosphate heads facing outward in contact with the water and their hydrophobic tails extending into the cluster away from the water.

### Phospholipid Functions

In the body, phospholipids have 2 major roles: as a cell membrane component and an emulsifier. Phospholipids, along with fatty acids and cholesterol, are a primary component of cell membranes (Fig. 6-12). A cell membrane is the double-layered outer covering of a cell that corrals the cell's contents and regulates the movement of substances into and out of the cell. Imagine the cell membrane as being like corrugated cardboard. The cardboard has a smooth outside edge and a smooth inside edge, and the corrugated area fills in the space between the 2 edges. The hydrophilic phosphate heads of phospholipids orient themselves to form the cell membrane's outside edge (the part that is exposed to the blood) or inside edge (the part that is exposed to the watery cell components). Regardless of whether the hydrophilic heads are facing toward the inside or outside of the cell, their hydrophobic tails point away from the heads—so they form the "corrugation." Because the heads and tails orient themselves in this way, the cell membrane remains fluid so that compounds can move into and out of the cell.

Phospholipids also serve as emulsifiers in the body. Bile and lecithins are the body's main emulsifiers. An **emulsifier** is a compound that forms a shell around fat droplets so that the droplets can be suspended in water and not clump together (Fig. 6-13). The hydrophobic tails of the phospholipids reach toward fat droplets and form the inside of the shell. The outside of the shell is made of the hydrophilic heads that extend away from the fat droplet. With the hydrophilic heads on the outside, fat droplets mix with water (stay suspended) and "repel" other fat droplets (don't clump together). Emulsifiers are essential for fat to be digested and transported through the bloodstream.

### Sources of Phospholipids

Phospholipids can be synthesized by the body or supplied by the diet. For example, lecithins are found in foods such as egg yolks, wheat

Phospholipid bilayer cell membrane

Hydrophobic phosphate tail

Hydrophilic phosphate head

Phospholipid

*Figure 6-12* Hydrophilic phosphate heads form the outer and inside edges of cell membranes. The hydrophobic tails point away from the heads.

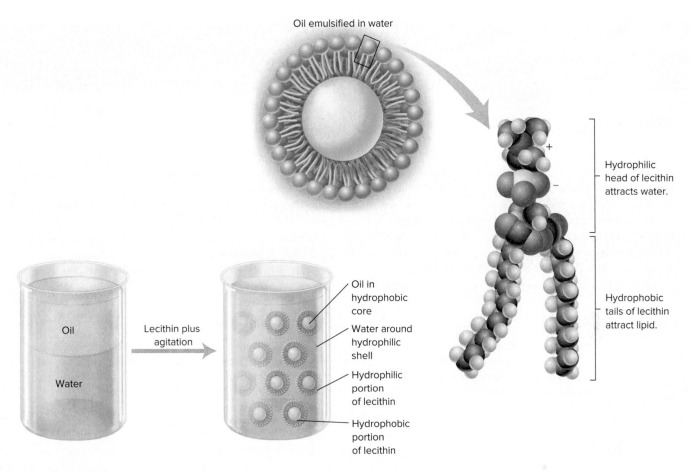

Oil emulsified in water

Hydrophilic head of lecithin attracts water.

Hydrophobic tails of lecithin attract lipid.

Oil in hydrophobic core

Water around hydrophilic shell

Hydrophilic portion of lecithin

Hydrophobic portion of lecithin

Lecithin plus agitation

Oil

Water

*Figure 6-13* When an emulsifier, such as lecithin, is added to water and oil and agitated, the 2 solutions (oil and water) form an emulsion.

germ, and peanuts. Although lecithin supplements are available, they are not needed because the liver can produce sufficient amounts of phospholipids. Lecithin supplements have been promoted as a way to lose weight; however, studies indicate that lecithin has no effect on weight loss. Some studies show that lecithin-derived phosphatidylserine may lower cholesterol and increase cognition and mood in Alzheimer patients.[2,3] It is important to note that high doses of lecithin can cause intestinal gas, diarrhea, and weight gain.

**emulsifier** Compound that can suspend fat in water by isolating individual fat droplets using a shell of water molecules or other substances to prevent the fat from coalescing.

Lecithins are a family of phospholipids synthesized by the body and found in foods such as peanuts, wheat germ, soybeans, egg yolks, and liver.

*Knowledge Check*

1. What is the main structural difference between phospholipids and triglycerides?
2. What are the main functions of phospholipids?
3. How are phospholipids used in the food industry?

*Culinary Perspective*

## Phospholipids in Food

The phospholipids in food often are used as an emulsifier in food preparation and manufacturing. Their ability to emulsify fats works the same in foods as it does in the body. For example, eggs are used in many muffin recipes. The lecithins in yolks emulsify fat in muffin batter and keep it suspended in the other ingredients. Mayonnaise is thick because phospholipids in egg yolks and mustard emulsify the oil and vinegar used to make this food. Food manufacturers add emulsifiers to keep the fat and watery compounds in them from separating. Emulsifying fats in foods such as cakes, muffins, and salad dressings gives them body and a smooth texture. Without emulsifiers, these foods would seem oily and have a sandy or rough texture.

Food emulsions can be temporary, semi-permanent, or permanent. As shown in the accompanying table, the difference depends on how the ingredients are blended and whether an emulsifier is added.

| Emulsion Type | How It's Made | Examples | Tips for Success |
|---|---|---|---|
| Temporary | Agitating (stirring, whisking, blending, shaking) immiscible liquids | Vinaigrette (oil and vinegar) | Liquid ingredients stay emulsified longer with vigorous agitation; adding paprika and/or dried mustard also helps keep it mixed. Usually, it separates in an hour or less. Agitate again to re-mix. |
| Semi-permanent | Agitating immiscible liquids and an emulsifier (usually egg) | Hollandaise or Béarnaise sauce | Ingredients combine to have a semi-thick, creamy texture that will stay mixed for a few hours. It's best to serve as soon as it is made. |
| Permanent | Slow agitation and gradual addition of immiscible liquids along with an emulsifier (usually egg) | Mayonnaise | Adding ingredients slowly and agitating together at a steady speed forms a thick, permanent, uniform mixture. |

Iconotec/Glow Images

anopdesignstock/iStock/Getty Images

Emulsions are found in more than just sauces and dressings. Chocolate is an emulsion of milk and cocoa butter. Ice cream is an emulsion of fat, water, sugar, ice, and air bubbles. Bologna, hot dogs, soups, gravies, and cakes are emulsions, too. Learn more about emulsions at www.aocs.org/stay-informed/inform-magazine/featured-articles/emulsions-making-oil-and-water-mix-april-2014.

# 6.5 Sterols

Sterols are the last type of lipid. Sterols are a type of steroid. The structure of sterols is very different from that of the long carbon chains seen in fatty acids and phospholipids. Instead, the carbons are mostly arranged in multi-ringed structures (see A Biochemist's View).

## Sterol Functions

From a nutrition perspective, cholesterol is the most well-known sterol. This waxy substance is required to synthesize many compounds. For instance, our bodies use cholesterol to make sex hormones such as testosterone and estrogens, the active form of vitamin D, and adrenal hormones such as cortisone. Cholesterol also is used to make bile, which is required to emulsify fats so that they can be digested normally.

In addition, cholesterol, along with phospholipids, forms cell membranes and allows fat-soluble substances to move into and out of the cell. Cholesterol, along with phospholipids and proteins, also forms the shell covering chylomicrons (droplets that transport lipids). This shell is what allows fat droplets to float through the water-based bloodstream (see Section 6.8).

## Sources of Sterols

Cholesterol is found in foods of animal origin, such as meat, fish, poultry, eggs, and dairy products. (Foods of plant origin do not contain cholesterol.) Most people get about one-third of their cholesterol from the foods they eat, and the rest is manufactured by their bodies. Of the approximately 875 mg of cholesterol produced daily by our bodies, about 400 mg is used to replenish bile stores and about 50 mg is used to make steroid hormones. The rest of the cholesterol serves as components of plasma lipoproteins that carry cholesterol to tissues. On average, American diets supply about 180 to 325 mg of cholesterol per day.[4] Of that, we absorb about 40 to 60%. It is not necessary to consume exogenous (dietary) cholesterol because the body can synthesize all the cholesterol it needs. In humans, the balance between cholesterol intake and outflow is not precise, resulting in cholesterol deposits in tissues.

Although plants do not contain or produce cholesterol, they do make other sterols, stanols, and sitostanols, such as ergosterol (a form of vitamin D) and beta sitosterols (added to some margarines, such as Take Control®). Eating margarine that contains sitostanol can reduce the body's absorption of cholesterol and bile, which is made from cholesterol, thereby reducing blood cholesterol levels, which decreases the risk of heart disease.

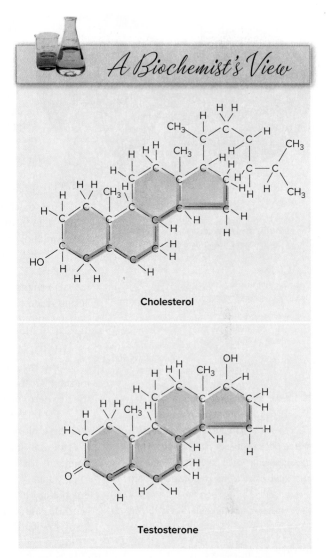

*A Biochemist's View*

**Cholesterol**

**Testosterone**

The carbons in sterols are arranged in rings. There is a carbon on each corner of this structural drawing of cholesterol and testosterone.

Egg yolks are the main source of cholesterol in the North American diet. The Food and Nutrition Board suggests limiting intake of high cholesterol foods. Photodisc/Getty Images

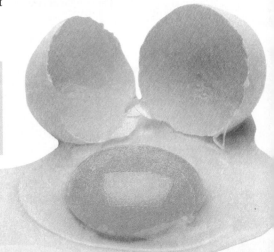

## Knowledge Check

1. What is the main structural difference between sterols and triglycerides?
2. Why do we need sterols?
3. What are 2 dietary sources of cholesterol?

The main saturated fat in coconut oil is lauric acid, a medium-chain fatty acid. Lauric acid increases levels of HDL and LDL in the blood but is not thought to negatively affect the overall ratio of these 2 lipoproteins.
Dynamic Graphics/PunchStock

▶ Infants and children younger than 2 years need to get about half their total calories from fat to meet calorie needs and to obtain sufficient fat for normal brain development. For children 2 to 3 years of age, keep total fat intake between 30 and 35% of calories. Between the ages of 4 and 18 years, keep fat intake between 25 and 35% of calories.

# 6.6  Recommended Fat Intakes

Fats are an essential part of a healthful diet but, for optimal health, the type and total amount of fat consumed need careful attention. There is no RDA for fat, but there is an Adequate Intake for infants. As you can see in Table 6-2, the Institute of Medicine's Acceptable Macronutrient Distribution Range for total fat is 20 to 35% of calories for most age groups. A total fat intake that exceeds 35% of calories often means saturated fat intake is too high. A low intake of total fat (less than 20% of calories) increases the chances of getting too little vitamin E and essential fatty acids and may adversely affect blood levels of triglycerides and a type of cholesterol called high-density lipoprotein (HDL) cholesterol, which is sometimes called "good" cholesterol (see Section 6.7).

The Institute of Medicine also recommends that saturated fat intake, including *trans* fats, and cholesterol levels be kept as low as possible while still consuming a nutritionally adequate diet. *Healthy People* and other expert groups, such as the Academy of Nutrition and Dietetics, American Heart Association, and Dietary Guidelines for Americans committee, also suggest that healthy people limit saturated fats and *trans* fat intake.[5,6,7,8] Currently, strong research evidence shows that replacing saturated fat with unsaturated fat, especially polyunsaturated fat, lowers total blood cholesterol and LDL. Replacing saturated fat with carbohydrates also lowers total blood cholesterol and LDL, but it increases triglycerides and lowers HDL.[9] Replacing saturated fat with polyunsaturated fat can be achieved by eating a healthy diet composed of fruits, vegetables, whole grains, low-fat dairy products, lean meats, plant-based proteins (e.g., nuts, seeds, beans, and peas), and seafood (twice a week). (See Expert Perspective from the Field in this chapter.)

Fat intake recommendations are lower for those at risk of heart disease, such as people with high blood levels of low-density lipoprotein (LDL) cholesterol, the so-called bad cholesterol, and individuals with diabetes. For example, the American Heart Association recommends that these individuals restrict dietary fat to 20% of total calories and saturated fat to less than 7% of total calories. Even more stringent is Dr. Dean Ornish's recommendation that dietary fat be limited to 10% of total calories.[9] Low fat diets can help lower the risk of heart disease and, in some cases, partly reverse damage already done to arteries. However, many low fat diets are high in carbohydrates, which may increase blood triglyceride levels

## Table 6-2  Recommendations for Daily Intake for Adults

| Fat Component | DRI* | Dietary Guidelines* | AHA* | TLC* |
|---|---|---|---|---|
| % fat of calories | 20–35% | 20–35% (based on DRI) | Consume a diet of fruits, vegetables, legumes, whole grains, and fish. | 25–35% |
| Saturated fat | Eat as little as possible | <10% of calories starting at age 2 | Replace with mono- and polyunsaturated fats. | <7% of calories |
| *Trans* fat | Eat as little as possible. | As of 2018, artificial *trans* fat is no longer Generally Recognized as Safe (GRAS) and no longer added to food in the U.S. | Minimize. | Lower intake. |
| Unsaturated fat | | A healthy eating pattern includes oils. | | Monounsaturated: up to 20% Polyunsaturated: up to 10% |
| Omega-6 | 5–10% | Based on Adequate Intake (DRI); 11–12g women, 14–17g men | | |
| Omega-3 | 1–2% of calories | Based on Adequate Intake (DRI); 1.1g women, 1.6g men | | |
| Dietary cholesterol | As low as possible while consuming a healthy diet | Minimize intake while maintaining a healthy eating pattern. | Minimize intake. | <200 mg/day |

*DRI = Dietary Reference Intake; Dietary Guidelines = Dietary Guidelines for Americans, 2020; AHA = American Heart Association; TLC = Therapeutic Lifestyle Change Diet. Blank areas indicate that no recommendation was given.

and raise the risk of heart disease. Thus, it's best for those getting less than 20% of total calories from fat to be monitored by a physician. Elevated blood triglycerides often decline over several months, especially if carbohydrate choices are high in fiber, weight is kept at a healthy level, and regular exercise is undertaken.included.

## Mediterranean Diet

There is some evidence that up to 40% of calories from fat can be healthy if monounsaturated fats account for most of the fat. The 1st evidence that a diet high in monounsaturated fats can be heart healthy was reported over 60 years ago in the 7 Countries Study conducted by Ancel Keys and colleagues.[11] This study led to today's popular Mediterranean Diet. Those who follow the traditional Mediterranean Diet enjoy some of the lowest recorded rates of chronic disease in the world. In years gone by, Greek farmers in Crete drank a glass of monounsaturated fat rich olive oil for breakfast! Even today, the average consumption of olive oil in Greece is 20 liters per person per year.

The traditional Mediterranean Diet features the following:

- Olive oil as the main fat
- Abundant daily intake of fruits, vegetables (especially leafy greens such as purslane, which is high in omega-3 fatty acids), whole grains, beans, nuts, and seeds
- An emphasis on minimally processed and, wherever possible, seasonally fresh and locally grown foods
- Daily intake of small amounts of cheese and yogurt
- Weekly intake of low to moderate amounts of fish
- Limited use of eggs and red meat
- Regular exercise
- Moderate drinking of wine at mealtime

## Essential Fatty Acid Needs

The Institute of Medicine has set Adequate Intakes for essential fatty acids. These recommendations add up to less than 120 calories daily for women and 170 calories for men—that's about 2 to 4 tablespoons daily of oils rich in these fatty acids. A deficiency of essential fatty acids is very unlikely, but insufficient intake for many weeks can lead to diarrhea, slowed growth, delayed healing of wounds and infections, and flaky, itchy skin. Although the Institute of Medicine has not yet set an Upper Level for the safe intake of omega-3 fats, Greenland Eskimos safely consume about 6.5 g/day, which is 3 to 5 times higher than the Adequate Intake.

## Our Fat Intake

Most North Americans get more than enough total dietary fat. In fact, during the last century, our fat intake has doubled. Added fats are those that we add to food, such as butter on bread and shortenings used to make cookies, pastries, and fried foods. In terms of types of fat, many people get too much saturated fat and too little monounsaturated and polyunsaturated fat.

Dairy products (whole milk, cheese, ice cream, butter), beef, chicken, mayonnaise, and margarine are the main contributors of saturated fat. Before they were banned, the major *trans* fat sources were margarine and baked goods made with shortening, such as cakes, cookies, crackers, pies, and breads. Vegetable oils are the prime contributors of polyunsaturated fat. Figure 6-14 compares the fat content of a high fat meal with a meal lower in fat. What changes can you make to bring your fat intake under control?

Omega-6 fatty acid intakes are usually plentiful, but omega-3 intakes often are lower than optimal. Omega-3 fat needs can be met by eating at least 2 portions of cold-water fish each week. For individuals who do not eat fish regularly, walnuts, flaxseeds, chia seeds, and canola, soybean, and flaxseed oils also supply omega-3 fatty acids. Supplements also are an option but should be discussed with a health-care provider

▶ **Foods That Affect Blood Cholesterol**

For many years, it was thought that foods high in cholesterol, like eggs, would significantly increase blood cholesterol in the body and increase the risk of heart disease. However, research shows that most individuals consuming cholesterol-containing foods will not increase their risk of cardiovascular disease.[10] Certain saturated fat rich foods—such as fried foods; processed meats such as sausage, hot dogs, and bacon; and desserts such as cakes, cookies, ice cream, and pastries—can negatively impact blood cholesterol levels as well as body weight. To find out how much cholesterol is in the foods you eat, check their food labels or use your favorite search engine to find out.

In late autumn, workers spread nets under olive trees, then shake the trees. The olives that fall are taken to a mill, where they are pressed to remove the oil. Learn more at www.explorecrete.com/nature/olive-oil-history.html.
Patricia Fenn/Flickr Open/Getty Images

Foods of animal origin, such as meat and butter, as well as hydrogenated fats, are the primary contributors of saturated fat to our diets.
Jill Braaten/McGraw-Hill Education

**Adequate Intake**

**Linoleic acid**

Men: 14–17 g/day

Women: 11–12 g/day

**Alpha-linolenic acid**

Men: 1.6 g/day

Women: 1.1 g/day

**1/4 lb cheeseburger = 31 g fat**

**Sandwich with 2 slices ham = 6 g fat**

*Figure 6-14*   Knowing the fat content of foods can help you plan a healthy diet.
Left: Dynamic Graphics/PunchStock; Right: Ingram Publishing/Alamy Stock Photo

Omega-3 fatty acids from animal sources, like fish, are more bioavailable than plant sources. Two servings of cold-water fish, such as salmon, each week can meet omega-3 needs.
Olga Nayashkova/Shutterstock

prior to use. The National Institutes of Health recommends choosing a fish oil supplement that has 650 mg of EPA and 650 mg of DHA. Those who have bleeding disorders, are scheduling surgery, or are taking anticoagulants (e.g., aspirin, warfarin [Coumadin®], or the herb ginkgo biloba) should check with their physician to minimize the risk of harmful side effects from an omega-3 supplement, which can prolong bleeding time.

*Knowledge Check*

1. What are the recommendations regarding total fat intake?
2. What happens to blood LDL and triglyceride levels when saturated fat intake is replaced with polyunsaturated fats or with carbohydrates?
3. How does our fat intake compare with recommendations?
4. What are some steps you can take to ensure a sufficient intake of omega-3 fatty acids?

## 6.7 Fat Digestion and Absorption

The body is very efficient at digesting and absorbing dietary fat (Fig. 6-15). (For a complete review of the digestive process, see Chapter 4.)

### Digestion

Fat digestion begins in the mouth, where lingual lipase is secreted. This enzyme helps break down triglycerides with short- and medium-chain fatty acids that are found in milk fat.

In the stomach, gastric lipase helps break triglycerides into monoglycerides, diglycerides, and free fatty acids. Fat floats on top of the watery contents of the stomach, which limits the extent of lipid digestion in the stomach.

## CASE STUDY

UpperCut Images/
SuperStock

William is a college senior; he will be graduating in a few months and has begun interviewing for jobs. William's college eating habits have not been great. He states that he has not cooked a single meal in his entire 4 years of college and has gained nearly 50 pounds. He decides to get a checkup and talk to the campus registered dietitian nutritionist to see if he can lose a few pounds. He finds out that his total cholesterol is 240 mg/dl, his LDL-cholesterol is 150 mg/dl, and his blood pressure is on the verge of hypertension. William's dad had his first heart attack at age 52 and currently is on medication to lower his blood pressure. What factors put William at risk for heart problems? How did his diet influence this risk? Regarding diet, what recommendations would you make?

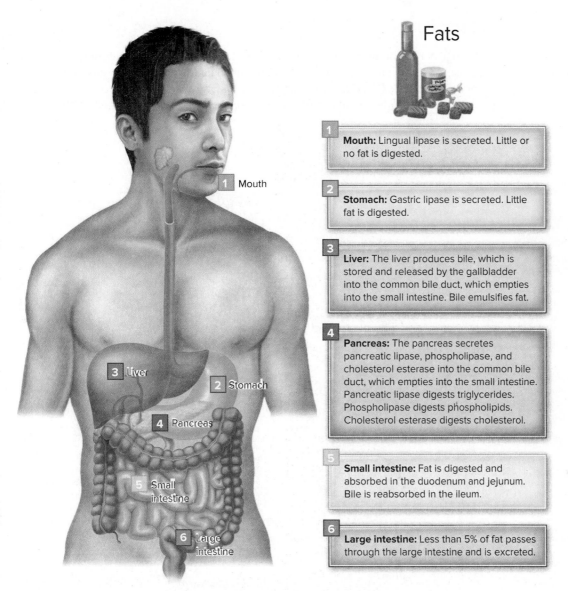

## Fats

**1 Mouth:** Lingual lipase is secreted. Little or no fat is digested.

**2 Stomach:** Gastric lipase is secreted. Little fat is digested.

**3 Liver:** The liver produces bile, which is stored and released by the gallbladder into the common bile duct, which empties into the small intestine. Bile emulsifies fat.

**4 Pancreas:** The pancreas secretes pancreatic lipase, phospholipase, and cholesterol esterase into the common bile duct, which empties into the small intestine. Pancreatic lipase digests triglycerides. Phospholipase digests phospholipids. Cholesterol esterase digests cholesterol.

**5 Small intestine:** Fat is digested and absorbed in the duodenum and jejunum. Bile is reabsorbed in the ileum.

**6 Large intestine:** Less than 5% of fat passes through the large intestine and is excreted.

1 Mouth
3 Liver
2 Stomach
4 Pancreas
5 Small intestine
6 Large intestine

*Figure 6-15* Lipid digestion and absorption. Enzymes made by the mouth, stomach, pancreas, and small intestine, as well as bile from the liver, participate in the process of digestion. Lipid digestion and absorption take place mostly in the small intestine (see Chapter 4 for details).

Fat digestion occurs mostly in the small intestine. Recall that the presence of fat in the small intestine triggers the release of the hormone cholecystokinin (CCK) from intestinal cells. CCK stimulates the release of bile from the gallbladder and lipase and colipase from the pancreas, all of which are delivered to the small intestine by the common bile duct. Bile emulsifies fats. That is, it breaks fat into many tiny droplets, called **micelles,** and forms a shell around the micelles that keeps the fat droplets suspended in the water-based intestinal contents. Emulsification increases the surface area of lipids and allows pancreatic lipase to efficiently break triglycerides into monoglycerides and free fatty acids. Fat digestion is very rapid and thorough because the amount of pancreatic lipase released usually is much greater than the amount needed. In addition, colipase helps lipase latch onto micelles.

Phospholipids and cholesterol also are digested mostly in the small intestine. Phospholipase enzymes from the pancreas and enzymes from the small intestine mucosa break phospholipids into their basic parts: glycerol, fatty acids, phosphoric acid, and other components (e.g., choline). Cholesterol esters (cholesterol with a fatty acid attached) are broken down to cholesterol and free fatty acids by a pancreatic enzyme called cholesterol esterase.

**micelle** Water-soluble, spherical structure formed by lecithin and bile acids in which the hydrophobic parts of the molecules face inward and the hydrophilic parts face outward.

## Absorption

The lipid portion of the micelles is absorbed by the brush border of the absorptive cells lining the duodenum and jejunum sections of the small intestine (Fig. 6-16). About 95% of dietary fat is absorbed. The carbon chain length of a fatty acid or monoglyceride determines whether it is absorbed by the cardiovascular or the lymphatic system. After absorption,

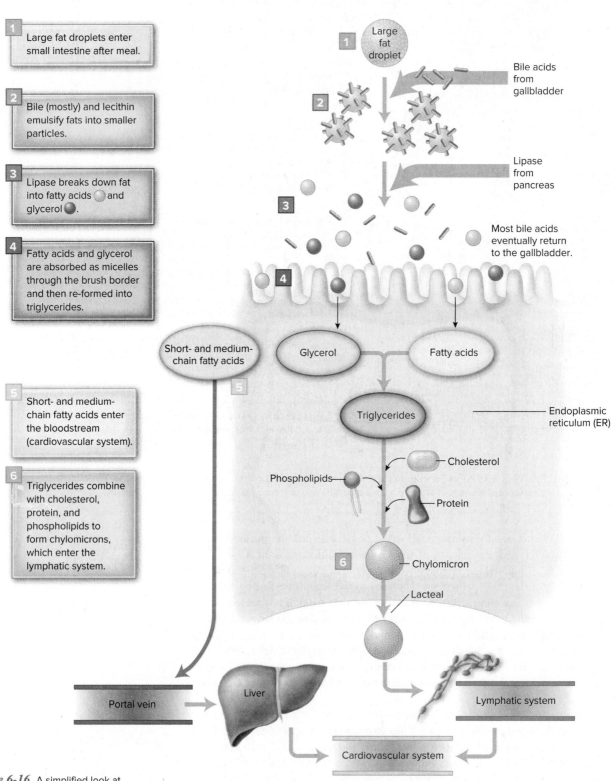

1  Large fat droplets enter small intestine after meal.

2  Bile (mostly) and lecithin emulsify fats into smaller particles.

3  Lipase breaks down fat into fatty acids ◯ and glycerol ●.

4  Fatty acids and glycerol are absorbed as micelles through the brush border and then re-formed into triglycerides.

5  Short- and medium-chain fatty acids enter the bloodstream (cardiovascular system).

6  Triglycerides combine with cholesterol, protein, and phospholipids to form chylomicrons, which enter the lymphatic system.

Large fat droplet

Bile acids from gallbladder

Lipase from pancreas

Most bile acids eventually return to the gallbladder.

Short- and medium-chain fatty acids

Glycerol

Fatty acids

Triglycerides

Endoplasmic reticulum (ER)

Cholesterol

Phospholipids

Protein

Chylomicron

Lacteal

Portal vein

Liver

Lymphatic system

Cardiovascular system

*Figure 6-16*  A simplified look at absorption of triglycerides.

short- and medium-chain fatty acids (<12 carbons) mostly enter the cardiovascular system via the portal vein, which leads directly to the liver. The vast majority of fatty acids consumed are long-chain fatty acids (≥12 carbons). Long-chain fatty acids are re-esterified into triglycerides in the absorptive cell. After further packaging (described in Section 6.8), they enter the lymphatic circulation, along with fat-soluble vitamins and dietary cholesterol.

Recall that bile, and the cholesterol it contains, is recycled by enterohepatic circulation. That is, bile is reabsorbed in the ileum and returned to the liver (via the portal vein) to be used again in fat digestion. About 98% of bile is recycled, and the rest is eliminated in the feces. Increasing the amount of bile that is excreted can help lower blood cholesterol levels because, when less is recycled, the liver takes more cholesterol out of the blood to restore the bile supply. Certain medications and diets rich in soluble fiber, which binds bile and carries it out with the feces, reduce the amount of recycled bile, thus lowering blood cholesterol.

### Knowledge Check

1. What is the role of cholecystokinin in the digestion of lipids?
2. What is the role of bile in fat digestion?
3. How does the chain length of a fatty acid affect its absorption?

## 6.8 Transporting Lipids in the Blood

Transporting lipids through the water-based blood and lymphatic system presents a challenge because water and fat do not mix. Lipids are transported in the blood as lipoproteins called chylomicrons, very-low-density lipoproteins, intermediate-density lipoproteins, low-density lipoproteins, and high-density lipoproteins. **Lipoproteins** have a core, made of lipids, that is covered with a shell composed of protein, phospholipid, and cholesterol. The shell lets the lipoprotein circulate in the blood. Figure 6-17 and Table 6-3 show the composition role of these lipoproteins.

### Transporting Dietary Lipids Utilizes Chylomicrons

Triglycerides that are re-formed in the absorptive cells of the intestine are packaged with other lipids, such as cholesterol and phospholipids, into lipoproteins called **chylomicrons.** These large lipid droplets are surrounded by a thin shell of phospholipid, cholesterol, and protein, which allows the chylomicrons to float freely in the blood (Fig. 6-18). The protein portion of the shell lipoproteins contains **apolipoproteins.** A series of letters (*A* through *E*) with subclasses are used to identify apolipoproteins. For convenience, they are abbreviated

**lipoprotein** Compound, found in the bloodstream, containing a core of lipids with a shell composed of protein, phospholipid, and cholesterol.

**chylomicron** Lipoprotein made of dietary fats that are surrounded by a shell of cholesterol, phospholipids, and protein. Chylomicrons are formed in the absorptive cells (enterocytes) in the small intestine after fat absorption and travel through the lymphatic system to the bloodstream.

**apolipoprotein** Protein attached to the surface of a lipoprotein or embedded in its outer shell. Apolipoproteins can help enzymes function, act as lipid-transfer proteins, or assist in the binding of a lipoprotein to a cell-surface receptor.

### *Critical* Thinking

As part of his annual health checkup, Juan has a blood sample drawn for the measurement of cholesterol values. The results of the test indicate that his total cholesterol is 210 mg/dl, his HDL-cholesterol is 65 mg/dl, and his triglycerides are 100 mg/dl. Juan has read that total cholesterol should be less than 200 mg/dl to minimize cardiovascular problems. However, he is happy with the result of the blood test. How would Juan explain his satisfaction to his parents?

**Table 6-3 Composition and Roles of the Major Lipoproteins in the Blood**

| Lipoprotein | Primary Component | Key Role |
|---|---|---|
| Chylomicron | Triglyceride | Carries dietary fat from the small intestine to cells |
| VLDL | Triglyceride | Carries lipids both taken up and made by the liver to cells |
| LDL | Cholesterol | Carries cholesterol made by the liver and from other sources to cells |
| HDL | Protein | Helps remove cholesterol from cells and, in turn, excrete cholesterol from the body |

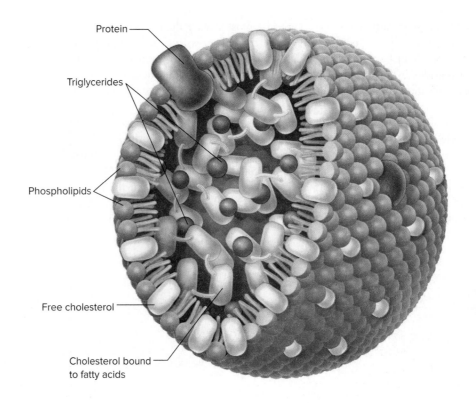

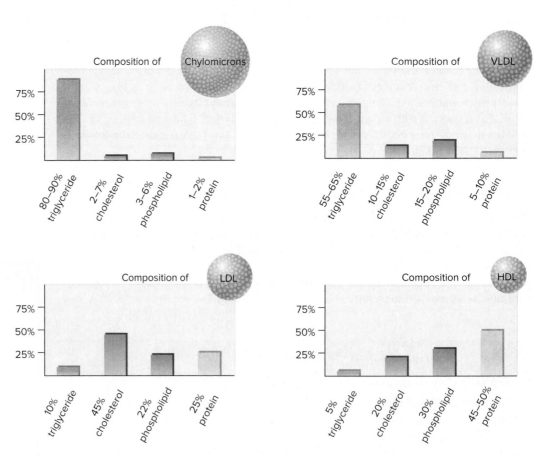

***Figure 6-17*** Structure and composition of lipoproteins. This lipoprotein structure allows fats to circulate in the bloodstream. Compare the amount of triglyceride and protein in each lipoprotein. Notice that lipoproteins with more triglycerides (orange bars) are lower in density, whereas those with more protein (yellow bars) are higher in density.

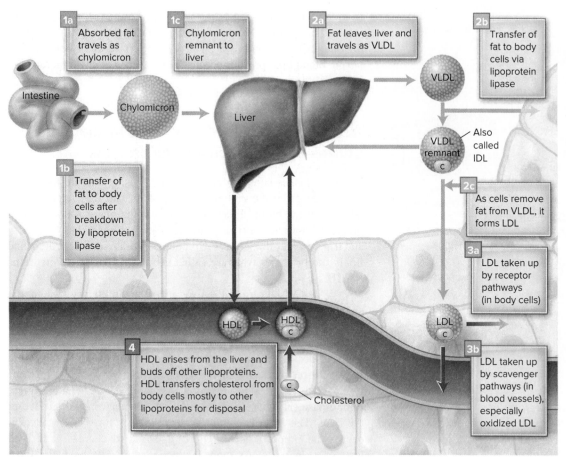

**1a** Absorbed fat travels as chylomicron

**1c** Chylomicron remnant to liver

**2a** Fat leaves liver and travels as VLDL

**2b** Transfer of fat to body cells via lipoprotein lipase

**1b** Transfer of fat to body cells after breakdown by lipoprotein lipase

Intestine

Chylomicron

Liver

VLDL

VLDL remnant c — Also called IDL

**2c** As cells remove fat from VLDL, it forms LDL

**3a** LDL taken up by receptor pathways (in body cells)

HDL → HDL c

LDL c

**4** HDL arises from the liver and buds off other lipoproteins. HDL transfers cholesterol from body cells mostly to other lipoproteins for disposal

c — Cholesterol

**3b** LDL taken up by scavenger pathways (in blood vessels), especially oxidized LDL

*Figure 6-18* Lipoprotein interactions. **1** Chylomicrons carry absorbed fat to body cells. **2** VLDL carries fat taken up from the bloodstream by the liver, as well as any fat made by the liver, to body cells. **3** LDL arises from VLDL and carries mostly cholesterol to cells. **4** HDL arises from body cells, mostly in the liver and intestine, as well as from particles that bud off other lipoproteins. HDL carries cholesterol from cells to other lipoproteins and to the liver for excretion.

"apo," followed by an identifying letter—apo A, apo B-48, apo C-II, and so on. Apolipoproteins can turn on a lipid-transfer enzyme (e.g., apo C-II turns on lipoprotein lipase), assist in binding a lipoprotein to a receptor on cell surfaces (e.g., apo B-48 binds chylomicrons to the liver), or assist enzymes (e.g., apo A-I activates lecithin:cholesterol acyltransferase).

Chylomicrons are secreted from the intestinal cells into the lymphatic system via the lacteals (specialized lymphatic vessels) in the intestinal villi. Recall that lacteals connect to larger lymphatic vessels, which then connect to the thoracic duct. The thoracic duct extends from the abdomen to the neck, where it connects to the bloodstream at a large vein called the left subclavian vein. Once in the blood, nutrients originally absorbed by the lymphatic system are transported to body tissues in the vascular system.

The enzyme lipoprotein lipase (LPL) is attached to the inside of most cell membranes, including those in blood vessels, muscles, fat tissue, and other cells. When LPL is activated by apo C-II, it hydrolyzes triglycerides from the chylomicrons, resulting in free fatty acids, which get taken up by cells and repackaged into triglycerides and either used for energy or stored for later use. Certain cells, such as muscles, tend to use the triglycerides as energy, whereas adipose cells tend to store them. As more cells take up the free fatty acids, chylomicrons decrease in size and become more dense and are then called chylomicron remnants.

After a meal, the whole process of removing chylomicrons from the blood via LPL activity takes

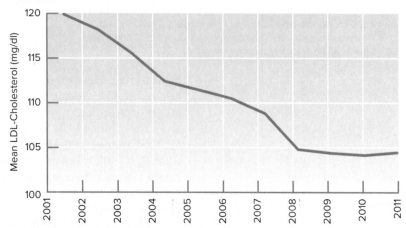

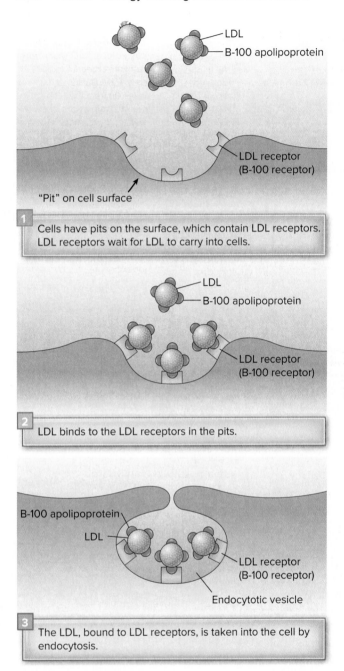

**1** Cells have pits on the surface, which contain LDL receptors. LDL receptors wait for LDL to carry into cells.

**2** LDL binds to the LDL receptors in the pits.

**3** The LDL, bound to LDL receptors, is taken into the cell by endocytosis.

*Figure 6-19* Receptor pathway for cholesterol uptake. **1** LDL receptors wait for LDL to carry into cells. **2** LDL receptors capture circulating LDL and **3** release it inside the cell to be metabolized. Once free of their load, LDL receptors return to the cell surface to await new LDL.

from 2 to 10 hours, depending partly on how much fat was in the meal. After 12 to 14 hours of fasting, no chylomicrons should be in the blood. It is a good idea for people to fast 12 to 14 hours before having blood lipid profiles because the presence of chylomicrons can affect the results.

## Transporting Lipids Mostly Made by the Body Utilizes Very-Low-Density Lipoproteins

The liver makes some fat and cholesterol using carbon, hydrogen, and energy from the carbohydrate, protein, and free fatty acids it takes from the blood. Free fatty acids are the major source of "ingredients" for triglyceride synthesis. The liver coats the cholesterol and triglycerides that collect in that organ with a shell of protein and lipids and produces what are called **very-low-density lipoproteins (VLDLs).**

When VLDLs from the liver enter the circulatory system, the enzyme LPL in the lining of blood vessels hydrolyzes the triglycerides in VLDLs, which allows them to enter body cells, including adipose tissue for fat storage and muscle tissue for energy. As triglycerides are released, VLDLs get more and more dense and become **intermediate-density lipoproteins (IDLs).** IDLs lose additional triglycerides by activating an enzyme called hepatic triglyceride lipase (HTGL), which is found on the endothelial surface of the liver. The hydrolysis of IDL triglycerides by HTGL and LPL causes the proportions of IDL triglyceride to decrease and cholesterol to increase. As more triglycerides are removed, the IDLs become **low-density lipoproteins (LDLs).** LDLs are composed primarily of cholesterol.

## LDL Removal from the Blood

The **receptor pathway for cholesterol uptake** removes LDL from the blood by cells that have the LDL receptor called B-100. The liver, as well as other cells, has this receptor (Fig. 6-19). Once inside a cell, LDL is broken down to protein and free cholesterol. These LDL components are used for maintaining the cell membrane or synthesizing specialized compounds, such as estrogen, testosterone, and vitamin D. When the free cholesterol concentration inside the cell increases to the point at which the cell can no longer take up any more LDL, the B-100 receptor stops taking LDL from the blood. When this occurs, the concentration of LDL increases in the blood. Some of the LDL that remains in the blood becomes damaged (oxidized) by free radicals, although a diet rich in **antioxidants** can help reduce LDL oxidation. Oxidized LDL increases the risk of cardiovascular disease and Metabolic Syndrome.[12]

The oxidized LDL, along with the non-oxidized LDL, can be taken up and degraded by the **scavenger pathway for cholesterol uptake.** In this pathway, certain "scavenger" white blood cells called macrophages leave the bloodstream and embed themselves in blood vessels. The macrophages detect LDL, then engulf and digest it. Once engulfed, LDL generally is prevented from reentering the bloodstream. Macrophages are able to pick up enormous amounts of LDL.

Cholesterol builds up in the macrophages and eventually kills them. They are replaced with new macrophages, which also eventually die. Over time, the cholesterol-filled macrophages build up on the inner blood vessel walls—especially in the arteries—and plaque

**antioxidant** Compound that protects other compounds, such as unsaturated fats, and body tissues from the damaging effects of oxygen (*anti* means "against"; oxidant means "oxygen").

develops. Diets rich in saturated fat, *trans* fat, and cholesterol encourage this process. The plaque eventually mixes with connective tissue (collagen) and is covered with a cap of smooth, fibrous muscle cells and calcium. **Atherosclerosis,** also referred to as hardening of the arteries, develops as plaque thickens in the vessel (Fig. 6-20). This thickening eventually chokes off the blood supply to organs, setting the stage for a heart attack and other problems, or it breaks apart and causes a clot to form in an artery.

## HDL's Role in Removing Blood LDL

A final critical participant in the extensive process of fat transport is **high-density lipoprotein (HDL).** Its high proportion of protein makes it the heaviest (most dense) lipoprotein. The liver and intestine produce most of the HDL found in the blood. HDL roams the bloodstream, picking up cholesterol from dying cells and other sources. HDL donates the cholesterol to other lipoproteins for transport back to the liver to be excreted. Some HDL travels directly back to the liver. Another beneficial function of HDL is that it blocks the oxidation of LDL.

Many studies have demonstrated that the amount of HDL in the blood can closely predict the risk of cardiovascular disease. Risk increases with low HDL levels because little blood cholesterol is transported back to the liver and excreted. Women tend to have high amounts of HDL, especially before menopause, whereas low amounts are more common in men.

Because high amounts of HDL slow the development of cardiovascular disease, any cholesterol carried by HDL is considered "good" cholesterol. In contrast, cholesterol carried by LDL is termed "bad" cholesterol because high amounts of LDL speed the development of cardiovascular disease. Still, some LDL is needed for normal body functions; LDL is a problem only when there is too much in the blood.

### Knowledge Check

1. How are lipids carried in the blood?
2. How do chylomicrons, LDL, IDL, VLDL, and HDL differ in their composition?
3. What role do apolipoproteins play in lipid metabolism?
4. What determines if LDL is metabolized by the receptor pathway or the scavenger pathway?
5. What role does HDL play in cardiovascular disease?

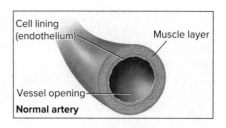

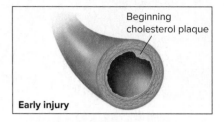

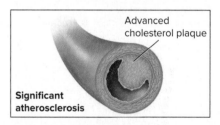

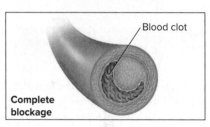

*Figure 6-20* Progression of atherosclerosis.

# 6.9 Health Concerns Related to Fat Intake

Dietary fat is essential for good health. However, high intakes can adversely affect health status.

## High Polyunsaturated Fat Intake

Intakes of polyunsaturated fats greater than 10% of total calorie intake seem to increase the amount of cholesterol deposited in arteries, which raises the chances of developing cardiovascular disease. High intakes also may impair the immune system's ability to fight disease and may promote weight gain.

## Excessive Omega-3 Fatty Acid Intake

Diets that include fish rich in omega-3 fatty acids twice a week (8 ounces/week) can reduce blood clotting abilities and may favorably affect heart rhythm in some people—both of these effects help lower the chances of having a heart attack. Larger intakes of fish (4 to 8 ounces/day)

**atherosclerosis** Buildup of fatty material (plaque) in the arteries, including those surrounding the heart.

Choosing low-fat dairy products can help keep fat intake under control.
Christopher Kerrigan/McGraw-Hill Education

## HISTORICAL PERSPECTIVE

Heart disease is the leading cause of death in the U.S. Recently, scientists discovered that heart disease was common in ancient times, too. MRIs from one-third of the 137 mummies from Egypt, Peru, Alaska, and the southwestern U.S. have revealed obvious signs of atherosclerosis. Heart disease is often thought of as a "modern disease" resulting from fatty diets and limited physical exercise. Although the mummies probably ate relatively lean diets and got plenty of exercise, infections were a major cause of death. Infection and inflammation likely were the cause of the atherosclerosis in the mummies, just as they are linked to heart disease in modern times.

Source: http://www.nbcnews.com/health /mummy-study-showsclogged-arteries-are -nothing-new-1C8790657;
Mummy: webking/iStock/Getty Image

further reduce the risk of heart disease by lowering blood triglyceride levels in those whose levels are high. However, an excessive intake of omega-3 fatty acids may impair the function of the immune system, allow uncontrolled bleeding, and cause hemorrhagic stroke (bleeding in the brain that damages it). Excessive levels of omega-3s are usually the result of supplement use.

## Imbalances in Omega-3 and Omega-6 Fatty Acids

On average, Americans consume 20 times more omega-6 fatty acids than omega-3s. Both fatty acids use the same metabolic pathways; as a result, they compete with each other. Thus, the body may not have enough of some compounds and too much of others. For example, as you saw in Figure 6-9, omega-6 fatty acids can be converted to arachidonic acid, which then can be used to make inflammation-causing eicosanoids called prostaglandins. In contrast, the omega-3 fatty acids EPA and DHA can be made into substances that help decrease inflammation, pain, and blood triglycerides.[13] Low intakes of omega-3s may worsen inflammatory diseases, such as arthritis. Although it is not known what causes or cures arthritis, an imbalance in the intakes of omega-3 and omega-6 fatty acids may play a role.[13,14,15,16] (See Expert Perspective from the Field .)

## Intake of Rancid Fats

Rancid (spoiled) fats smell and taste bad. They also contain compounds (peroxides and aldehydes) that can damage cells. Polyunsaturated fats go rancid fairly easily because their double bonds are easily damaged (broken) by oxygen, heat, metals, or light (sunlight or artificial light). The broken double bonds cause the polyunsaturated fats to decompose. (Saturated and *trans* fats are less susceptible to rancidity because they have no or few double bonds in their carbon chains.)

The foods most likely to become rancid are those high in polyunsaturated fats (e.g., fish and vegetable oils), packaged fried foods (e.g., potato chips), and fatty foods with a large surface area (e.g., powdered egg yolks). To prevent rancidity, food manufacturers can break the double bonds and add hydrogen (hydrogenate them). Or they can protect the double bonds in fats by sealing foods in airtight packages or adding antioxidants, such as certain nutrients (vitamin E, vitamin C), or additives, such as butylated hydroxyanisol (BHA) and butylated hydroxytolune (BHT). (See Part 4 to learn more about the antioxidant functions of vitamins and Chapter 3 to learn about additives.)

## Diets High in *Trans* Fat

*Trans* fatty acids from hydrogenated fats have harmful health effects. Hydrogenated fats were popular for many years because they helped food manufacturers produce high quality baked and fried products. For example, some foods are more pleasing when made with solid fats. Pastries and pies made with oil tend to be oily and mealy, whereas those made with solid fats are flaky and crispy. Although solid fat from animals, such as butter or lard, could be used instead of hydrogenated fat, hydrogenated fat is cholesterol free. Another advantage of hydrogenation is that it delays fat decomposition and spoilage (rancidity) in packaged foods.

Despite any advantages of hydrogenation, consuming *trans* fatty acids raises blood cholesterol levels, which increases the risk of heart disease. In addition, *trans* fats lower HDL- (good) cholesterol and increase inflammation in the body. Studies with monkeys have indicated that diets rich in *trans* fats raise body weight and the amount of body fat stored in the abdomen, even when calories are at levels that should only maintain weight. Much of the stored abdominal fat is visceral fat, which increases the risk of type 2 diabetes.[19]

# Expert Perspective from the Field

## A Healthier Approach to Eating Fats

Recommendations for dietary fat consumption and lowering of heart disease risk have changed considerably over the years. Dr. Penny Kris-Etherton* and other experts recommend keeping saturated fat intake under control in the context of an overall healthy diet. Healthy diets are low in refined carbohydrates and added sugars, contain little or no *trans* fat, and replace most saturated fat with unsaturated fat, especially polyunsaturated fats. It is important to include both omega-6 and omega-3 fatty acids (Table 6-4) as substitutions for saturated fat because both have been found to reduce CVD mortality and major risk factors.[16]

Dr. Kris-Etherton also points out that several studies have reported that low omega-6 intakes were associated with an increased risk of heart disease and that replacing saturated fat with omega-6 fatty acids reduced that risk. She stated, "Omega-6 fatty acids have independent cholesterol-lowering properties beyond the simple removal of saturated fats." Omega-6 fatty acids clearly provide health benefits. For instance, replacing saturated fatty acids with omega-6s reduces heart disease

risk.[17] For optimal heart health, the American Heart Association recommends that omega-6 fatty acid intake account for at least 5 to 10% of calorie intake. Reducing omega-6 intake below this level likely would increase the risk of heart disease.[18]

*Penny Kris-Etherton, Ph.D., R.D., is Distinguished Professor of Nutrition in the Department of Nutritional Sciences at Pennsylvania State University and a Fellow of the American Heart Association. She is the recipient of the Lederle Award for Human Nutrition Research from the American Society for Nutritional Sciences, the Foundation Award for Excellence in Research, and the Marjorie Hulsizer Copher Award from the Academy of Nutrition and Dietetics. She has served on the National Academies Panel on Macronutrients, American Heart Association Nutrition Committee, National Cholesterol Education Program Second Adult Treatment Panel, and 2005 Dietary Guidelines for Americans Advisory Committee.*

### Table 6-4 Major Sources of Polyunsaturated Fatty Acids

| Monounsaturated | Omega-3 | Omega-6 |
|---|---|---|
| Olives | Fatty fish, such as salmon, mackerel, tuna, herring, sardines | Meat, poultry, eggs, mayonnaise |
| Peanut butter | Seeds, such as flaxseed, chia seeds | Seeds, such as pumpkin, sunflower |
| Avocados | Yeast, marine algae | Salad dressing |
| Oils, such as olive, peanut, canola, sesame, and safflower oil with high oleic | Oils, such as flaxseed, soybean, and canola oil | Oils, such as corn, cottonseed, safflower, and soybean oil |
| Nuts, such as almonds, cashews, hazelnuts, macadamia nuts, peanuts, pecans, pistachios | Nuts, such as walnuts | Nuts, such as pine nuts, walnuts |

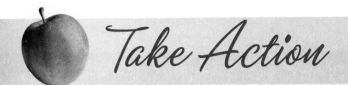

*Take Action*

## Understanding Your Lipid Blood Test  (Lipid Panel)

A lipid blood test, or lipid panel, measures the specific lipids in your blood to help you and your health-care provider understand your risk for CVD. When no other risk factors are present, this test is first conducted between the ages of 9 and 11 and again between the ages of 17 and 21. If risks are present, then the test is conducted regularly.

### Preparing for a Lipid Panel Test

The most accurate results require fasting 12 to 14 hours before the test. Usually, the test is done in the morning, which means you can fast overnight. A small amount of blood is drawn from a vein using a syringe. Sometimes a finger stick can be used to collect blood for the test.

### Lipids Measured

A lipid panel measures the lipid levels in the blood sample that are linked to cardiovascular disease risk. It typically includes the following 4 measurements.

*Total cholesterol:* all the cholesterol in all the lipoprotein particles combined

*High-density lipoproteins (HDL-C):* cholesterol in HDL particles ("good" cholesterol)

*Low-density lipoproteins (LDL-C):* cholesterol in LDL particles ("bad" cholesterol)

*Triglycerides:* triglycerides in all the lipoprotein particles combined; most is in the very-low-density lipoproteins (VLDLs)

The following information may also appear on the report.

*Very-low-density lipoprotein (VLDL-C):* calculated by dividing the triglyceride value by 5 (based on the typical composition of VLDL particles)

*Non-high-density lipoprotein (non-HDL-C):* calculated by subtracting HDL-C from total cholesterol

*Lipoprotein particle number/concentration (LDL-P):* total LDL particles; this measurement may more accurately predict CVD risks

### Interpreting the Results

### Total Cholesterol

*Desirable:* <200 mg/dL
*Borderline high:* 200–239 mg/dL
*High:* ≥240 mg/dL

It is important to know how much of your total cholesterol is made up of LDL-C or HDL-C. For most people, about 70% of their total cholesterol is LDL-C and 30% is HDL-C. This balance between LDL-C

Image Source/Andrew Brookes

and HDL-C is important; when the balance is off, especially when LDL-C is high and HDL-C is low, the risk of developing heart disease increases.

### HDL-C

*Less than average risk of CVD:* ≥60 mg/dl
*Average CVD risk:* 40–50 mg/dL
*Increased CVD risk:* <40 mg/dL for men and <50 mg/dL for women

Regular and consistent exercise increases HDL-C and lowers LDL-C.

### LDL-C

*Optimal:* <100 mg/dL
For those with CVD risk factors or family history: <70 mg/dL
*Near optimal:* 100–129 mg/dL
*Borderline high:* 130–159 mg/dL
*High:* 160–189 mg.dL
*Very high:* >190 mg/dL

Strategies to lower LDL-C include reducing saturated fat and increasing polyunsaturated fats intake, engaging in exercise, and, when appropriate, taking cholesterol-lowering drugs.

### Triglycerides

*Desirable:* <150 mg/dL
*Borderline high:* 150–199 mg/dL
*High:* 200–499 mg/dL
*Very high:* >500 mg/dL

Smoking, excessive alcohol consumption, uncontrollable diabetes, and some medications can contribute to elevated triglyceride levels.

**VLDL-C**

*Normal:* 5–40 mg/dL

The higher the VLDL number, the greater the risk for a heart attack or stroke.

**Non-HDL-C**

*Optimal:* <130 mg/dL
*Near/above optimal:*130–159 mg/dL
*Borderline High:* 160–189 mg/dL
*High:* 190–219 mg/dL
*Very high:* >220 mg/dL

**Non-HDL-P** represents the atherogenic cholesterol, which can build up and narrow arteries.

**LDL-P** test evaluates the number of particles of LDL in the blood. An increase in small, dense LDL-P may explain why some people have heart attacks even when their total cholesterol and LDL-C are not very high.

**What Is Your Goal?**

Keep in mind that your lipid panel report is only a guideline. What is normal for you may not be for others. Review the results with your health-care provider and develop a strategy to manage your CVD risk.

Your goals should be based on your risk factors (see **Clinical Perspective** in this chapter). Adherence to a healthy lifestyle that includes a diet with more polyunsaturated fat than saturated fat, little to no *trans* fat, and regular, consistent exercise is an important part of maintaining heart health.

To help people control *trans* fat intake, the FDA requires *trans* fats to be on Nutrition Facts labels (food labels in Canada also must list *trans* fats in foods, as well as their negative health effects). Both the U.S. and Canada have banned the use of partially hydrogenated oils. To comply with the new regulation, food manufacturers have reformulated many products to make them *trans* fat free (less than 0.5 g/serving is defined by the FDA as *trans* fat free). The reformulated products often use interesterified fats, which are made by interchanging the fatty acids in solid fats and liquid oils. This interchange creates a fat with properties similar to those of *trans* fats—that is, it is solid at room temperature, stands up to high-temperature cooking methods, and stays fresh a long time. Interesterified fats appear to be healthier than *trans* fats, but more research is needed.

Both *trans* and interesterified fat intakes can be kept to a minimum when eating out by limiting fried (especially deep fat–fried) foods, pastries, flaky bread products (e.g., pie crusts, crackers, croissants, and biscuits), and cookies. At home, keep intake of these fats under control by using little or no stick margarine or shortening. Instead, substitute vegetable oils and softer tub or squeeze margarine. Applesauce and fruit purées can be substituted for shortening in many baked goods. Also, to avoid deep-fat frying in shortening, try baking, pan-frying, broiling, steaming, grilling, or stir frying. Most non-dairy creamers are rich in hydrogenated vegetable oils, so replace them with reduced fat milk or non-fat dry milk.

## Diets High in Total Fat

Diets high in total fat increase the risk of obesity (see Chapter 10), certain types of cancer (see Part 4), and cardiovascular disease. Diets high in fat, especially saturated fat, may increase the risk of colon, prostate, and breast cancer.[20,21,22] Although it isn't known how high fat diets increase risk, a theory related to colon cancer is that bile, which is secreted into the intestine to emulsify dietary fat, may irritate colon cells. As fat intake rises, more bile is secreted, which then irritates the cells more intensely and frequently, perhaps damaging them and causing them to become cancerous. In the case of breast and prostate cancer, the risk of both climbs as blood levels of estrogen hormones rise. High fat diets elevate blood lipid levels, which in turn raise blood estrogen levels. In contrast, low-fat

Naturally occurring *trans* fats found in animal products, like meat and milk, do not have the harmful properties of the *trans* fats formed from hydrogenation.
Frank Bean/Getty Images

## Cardiovascular Disease (CVD)

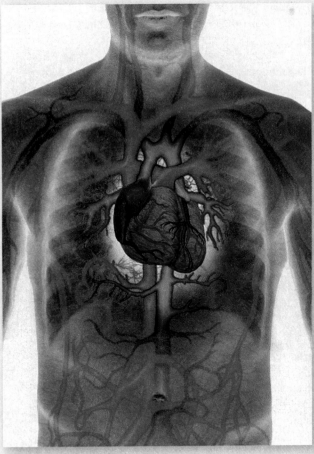

Don Farrall/Getty Images

Cardiovascular disease (CVD) is the major killer of North Americans. In the U.S., every 40 seconds someone has a heart attack. Every 80 seconds, someone dies from a heart attack. Each year, about 630,000 people die

▶ A *Healthy People* goal is to increase cardiovascular health and reduce death from coronary heart disease and stroke.

▶ CVD typically involves the coronary arteries and thus is frequently termed *coronary heart disease* (*CHD*) or *coronary artery disease* (*CAD*).

**homocysteine** Amino acid not used in protein synthesis. Instead, it arises during metabolism of the amino acid methionine. Homocysteine likely is toxic to many cells, such as those lining the blood vessels.

of CVD in the U.S., about 60% more than die of cancer.[23] The figure rises to almost 1 million if strokes and other circulatory diseases are included. Heart disease is the leading cause of death for both men and women. Women generally lag about 10 years behind men in developing the disease. Still, it eventually kills more women than any other disease—twice as many as cancer. And, for each person in North America who dies of CVD, 20 more (over 13 million people) have symptoms of the disease.

High fat diets, especially those rich in saturated and *trans* fats, increase the risk of CVD. (Recall that the vascular system includes the blood, heart, arteries, and veins.) The symptoms develop over many years and often do not become obvious until old age. Nonetheless, autopsies of those under 20 years of age have shown that many already had atherosclerotic plaque in their arteries.

### Development of CVD

Atherosclerotic plaque is probably 1st deposited to repair injuries in the lining in any artery. The damage that starts plaque formation can be caused by smoking, diabetes, hypertension, **homocysteine** (likely, but not a major factor), and LDL.[24,25,26] Viral and bacterial infections and ongoing blood vessel inflammation also may promote plaque formation.[27]

As atherosclerosis progresses, plaque thickens over time, causing arteries to harden, narrow, and become less elastic. This makes them unable to expand to accommodate the normal ups and downs of blood pressure.

Affected arteries are further damaged as blood pumps through them and pressure increases. In the final phase, a clot or spasm in a plaque-clogged artery blocks the flow of blood and leads to a heart attack (myocardial infarction) or stroke (cerebrovascular accident).

Recall that blood supplies the heart muscle and brain—and other body organs—with oxygen and nutrients. When blood flow via the coronary arteries surrounding the heart is interrupted, a heart attack may occur, which damages the heart muscle. If blood flow to parts of the brain is interrupted long enough, part of the brain dies, causing a stroke. Factors that typically bring on a heart attack in a person at risk include dehydration, severe emotional stress, strenuous physical activity if not physically fit, sudden awakening during the night or just getting up in the morning (linked to an abrupt increase in blood pressure and stress), and high fat meals, which increase blood clotting.

### Risk Factors for CVD

In addition to a high fat diet, the American Heart Association has identified several other factors that affect the risk of heart disease. The more risk factors a person has, the greater the risk of CVD. Some of the risk factors cannot be changed, but others can. The risk factors that cannot be changed are age, biological sex, genetics, and race.

- *Age.* The risk of CVD increases with age. Over 83% of people who die of CVD are at least 65 years old.
- *Biological Sex.* Men have a greater chance of having a heart attack than women do, and they have attacks earlier in life. Even after menopause, when women's death rate from heart disease increases,[28] it's not as great as the risk men face.
- *Genetics.* Having a close relative who died prematurely from CVD, especially before age 50, may increase the risk. Those with the highest risk of premature CVD have genetic defects that block the removal of chylomicrons and triglycerides from the blood, reduce the liver's ability to remove LDL-cholesterol

from the blood, limit the synthesis of HDL-cholesterol, or increase blood clotting.

- *Race.* Race may affect CVD risk. For example, those of African heritage have more severe high blood pressure levels compared with Caucasians, which puts them at higher risk of CVD. Heart disease risk also is higher among those of Hispanic/Latino, Native American, and native Hawaiian descent, as well as some Asian groups, which is partly due to higher rates of obesity and diabetes in these groups.

The risk factors that can be modified are blood cholesterol levels, blood triglyceride levels, hypertension, smoking, physical inactivity, obesity, diabetes, liver and kidney disease, and low thyroid hormone levels.

- *Blood cholesterol levels.* A total blood cholesterol level over 200 mg/dl (especially over 240 mg/dl), along with an LDL-cholesterol level of 160 mg/dl or higher, increases CVD risk. When high blood cholesterol levels accompany other risk factors (e.g., high blood pressure and smoking), CVD risk increases even more. Reducing dietary intakes of cholesterol, saturated fat, and total fat; keeping weight under control; and exercising can help lower blood cholesterol levels, as can prescription medications.
- *Blood triglyceride levels.* Fasting blood triglyceride levels should be below 150 mg/dl. Excess triglycerides in the blood is called *hypertriglyceridemia.* Blood triglycerides are derived from fats in the foods we eat. In some individuals, simple carbohydrates and alcohol raise plasma triglyceride levels. A high triglyceride level in combination with a low HDL and high LDL may speed up atherosclerosis. Individuals with a high

▶ **Systolic blood pressure** over 140 mm (millimeters of mercury) and **diastolic blood pressure** over 90 mm indicate hypertension. Normal systolic blood pressure value is less than 120 mm and less than 80 mm for diastolic. (Systolic blood pressure is the maximum pressure in the arteries when the heart beats. Diastolic blood pressure is the pressure in the arteries when the heart is between beats.)

▶ The values in Table 6-5 are for adults. Age-specific values for adolescents, which take their growth and maturation differences into account, have been published.[27,29]

triglyceride level should lower their saturated fat intake and increase monounsaturated fat and omega-3 fatty acids.

- *Hypertension.* Hypertension (high blood pressure) damages the heart muscle by making it thicker and stiffer. This damage causes the heart to work harder than normal. It also increases the risk of stroke, heart attack, kidney failure, and congestive heart failure. Hypertension accompanied by high blood cholesterol levels, smoking, obesity, or diabetes raises the risk of heart attack or stroke by several times. Reducing sodium intake, losing weight, and taking medication can help bring hypertension under control. Exercise also may help.
- *Smoking.* Smokers have a 2 to 4 times greater risk of CVD than nonsmokers. Even exposure to secondhand smoke can increase the risk of CVD. Smoking boosts a person's genetically linked risk of CVD, increases risk even when blood lipids are low, and makes blood more likely to clot. Smoking also tends to negate the lower CVD risk that females have compared with males. In fact, smoking is the main cause of about 20% of the CVD cases in women. In addition, women who smoke and take oral contraceptives are at even greater risk of CVD.
- *Physical inactivity.* Lack of exercise increases the CVD risk. Regular moderate to vigorous physical activity lowers CVD risk, helps control blood cholesterol levels, reduces the risk of diabetes and obesity, and may even reduce blood pressure.
- *Obesity.* Many adults gain weight as they grow older. This weight gain, especially if it is around the waist, is a chief contributor to the increases in LDL blood cholesterol levels common in older adults. Obesity increases inflammation in the body and reduces the adipose cells' production of the hormone adiponectin. The reduced level of this hormone in the blood elevates the risk of having a heart attack. Obesity also leads to insulin resistance in many people, creating a risk of diabetes.
- *Diabetes.* Diabetes greatly increases the risk of developing CVD. Even when blood

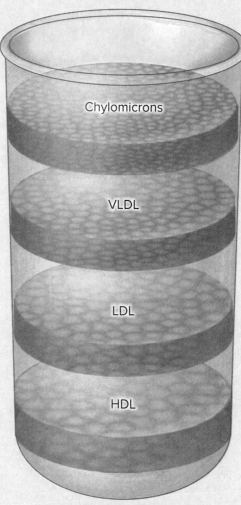

**Centrifuge tube**

One way to measure the amount of chylomicron, VLDL, LDL, and HDL particles in the bloodstream is to centrifuge the serum portion of the blood at high speed for about 24 hours in a sucrose rich solution. The lipoproteins settle out in the centrifuge tube based on their density, with chylomicrons at the top and HDLs at the bottom.

glucose (sugar) levels are well controlled, diabetes increases the risk of heart attack and stroke, but the risks are even greater when blood glucose levels are not well controlled. About 75% of those with diabetes die of some form of CVD. Diabetes also negates the female advantage of reduced CVD risk.

- *Liver and kidney disease and low thyroid hormone levels.* Certain forms of liver and kidney disease and low concentrations of thyroid hormone can increase blood LDL-cholesterol and thus increase the risk of CVD. Medical treatment can help control these conditions and lower CVD risk.

## Table 6-5  Ratings of Blood Lipoprotein Levels

| Lipoprotein (mg/dl) | Rating |
|---|---|
| **Total Cholesterol** | |
| <200 | Desirable |
| 200–239 | Borderline high |
| ≥240 | High |
| **LDL-Cholesterol** | |
| <100 | Optimal |
| 100–129 | Near optimal/above optimal |
| 130–159 | Borderline high |
| 160–189 | High |
| ≥190 | Very high |
| **HDL-Cholesterol** | |
| <40 | Low |
| ≥60 | High |
| **Triglycerides** | |
| <100 | Optimal |
| 100–149 | Near optimal |
| 150–199 | Borderline high |
| 200–499 | High |
| ≥500 | Very high |

### Assessing CVD Risk

The National Cholesterol Education Program (NCEP) suggests that all adults age 20 years or older have a blood lipoprotein profile done every 5 years. This profile is most useful when the person fasts for 12 to 14 hours before the test. (Only total cholesterol and HDL values are accurate if the person has not fasted.) Table 6-5 shows how blood lipoprotein levels are interpreted.

The NCEP also has developed tables to help you calculate your risk of developing heart disease in the next 10 years. The risk is based on age, total and HDL blood cholesterol levels, blood pressure, and whether you smoke. Knowing your score can help you and your health-care provider determine if you need to make lifestyle changes, go on medication, or both.

### Preventing CVD

The following are lifestyle changes that can lower LDL blood cholesterol levels and reduce health risks.

- Keep total fat intake between 20 and 35% of total calories.

- Keep saturated fat intake to less than 7% of total calories.
- Keep *trans* fat intake low.
- Keep polyunsaturated fat under 10% of total calories.
- Keep monounsaturated fat under 20% of total calories.
- Lower cholesterol intake to less than 200 mg per day.
- Include 2 grams of plant stanols/sterols in your daily diet to help reduce cholesterol absorption in the small intestine and lower its return to the liver.
- Increase soluble fiber intake to 20 to 30 grams per day.
- Moderate sugar intake.[30]
- Keep body weight at a healthy level.
- Increase physical activity.

Frequently eating fruits, vegetables, nuts, and plant oils also can help reduce plaque buildup in arteries and slow the progression of cardiovascular disease.[31] That's because these foods are rich in antioxidants, which likely reduce LDL oxidation and slow the need for macrophages to pick up oxidized LDL. Supplements of antioxidant nutrients, such as vitamins C and E, may help. However, large studies of people with CVD have shown no benefit from megadoses of vitamin E (200–400 mg/day, equaling about 400–800 IU/day).[32,33] Still, some experts suggest that vitamin E supplements (up to 200 mg [400 IU] per day) may be helpful for *preventing* CVD; these should be taken under a physician's guidance, however. Note that antioxidant supplements can be harmful to some, especially those taking certain medications that reduce blood clotting (anticoagulants) because vitamin E also reduces blood clotting. High intakes of iron probably speed LDL oxidation, making it unwise to take an iron supplement unless a physician prescribes it. To learn more, visit www.nih.gov and search for cholesterol guidelines.

### Heart Attack Symptoms

A heart attack can strike with the sudden force of a sledgehammer, with pain radiating up the neck or down the arm. It can sneak up at night, masquerading as indigestion, with slight pain or pressure in the chest. Many times, the symptoms are so subtle in women that it often is too late once she or health professionals realize that a heart attack is taking (or has recently taken) place. If there is any suspicion at all that a heart attack is occurring, the person should first chew an aspirin (325 mg) thoroughly and then call 911. Aspirin helps reduce the blood clotting that precipitates a heart attack. The typical warning signs are

- Intense, prolonged chest pain or pressure, sometimes radiating to other parts of the upper body (men and women)
- Shortness of breath (men and women)
- Sweating (men and women)
- Weakness (men and women)
- Nausea and vomiting (especially women)
- Dizziness (especially women)
- Jaw, neck, and shoulder pain (especially women)
- Irregular heartbeat (men and women)

### Symptoms of Stroke

Each year, 700,000 North Americans suffer strokes, and almost 25% of them die. Over 90% of strokes (i.e., ischemic strokes) occur when a blood clot blocks blood flow to the brain—think of a stroke as a "brain attack," similar to a heart attack. The other 10% are hemorrhagic strokes, occurring when a blood vessel bursts. The major risk factor for stroke is high blood pressure. Individuals experiencing any of the following symptoms of stroke should seek immediate treatment because physicians can administer drugs that can limit the further death of brain cells and reduce the extent of the damage caused by most ischemic strokes. The stroke warning signs are

- Sudden numbness or weakness of the face, arm, or leg, especially on 1 side of the body
- Sudden confusion and/or trouble speaking or understanding
- Sudden trouble seeing in 1 or both eyes
- Sudden trouble walking, dizziness, and/or loss of balance or coordination
- Sudden, severe headache with no known cause

▶ Benecol® and Take Control® margarines contain plant stanols/sterols.

▶ The soluble fiber in 1½ cups of oatmeal a day will reduce blood cholesterol levels by about 15%.

diets seem to lower blood estrogen levels. Another possible explanation of the relationship between high fat diets and cancer is that high fat diets usually are low in fiber and other phytonutrients—thus, high fat diets may lack the protective plant compounds that help prevent certain cancers.

Lowering the intake of dietary fat is 1 way to help control calorie intake and avoid being overfat. Obesity is linked with a greater risk of cancer of the colon, breast, and uterus. (The risks associated with obesity are discussed in Chapter 10.)

Many manufacturers offer products that are lower in fat than traditional products. Even though these products are lower in fat, portion size and total calories provided still must be considered.

McGraw-Hill Education/Jill Braaten

### Knowledge Check

1. What is the risk of eating a diet that is low or high in omega-3 fatty acids?
2. What risk factors for cardiovascular disease can be modified by lifestyle changes?
3. What are the risks and benefits of *trans* fatty acids?
4. What health conditions are associated with diets high in total fat?

## CASE STUDY FOLLOW-UP

UpperCut Images/SuperStock

William has a few factors that are putting him at risk for heart disease. He has a family history, he is male, his total cholesterol is over 200 mg/dl, and his LDL is approaching 160 mg/dl. Additionally, he has gained weight in his 4 years of college and is on the verge of hypertension. William's diet is high in saturated fat and *trans* fat. He could improve his eating habits by having a breakfast rich in fiber and nutrients, such as whole-grain cereal, fat-free milk, and fruit juice. Instead of fatty burgers and fast foods, he could choose lean meat, chicken, and fish. To round out meals, William needs several servings of fruits and vegetables. He could improve his intake of monounsaturated fats by eating a small handful of peanuts each day as a snack and having cold-water fish more often. He would benefit from cooking more meals at home and adding a consistent exercise program to his daily routine.

# Chapter *Summary*

## 6.1 Triglycerides are the most common type of lipid found in foods and in the body.
Each triglyceride molecule consists of 3 fatty acids attached to a glycerol. A triglyceride that loses a fatty acid is a diglyceride. A monoglyceride results when 2 fatty acids are lost. The carbon chains of fatty acids can vary in 3 ways: the number of carbons in the chain, the extent to which the chain is saturated with hydrogen, and the shape of the chain (straight or bent). Hydrogenation adds hydrogen to the carbon chain of unsaturated fats. The systems commonly used to name fatty acids, omega and delta, are based on the numbers of carbon atoms and the location of double bonds in a fatty acid's carbon chain. Essential fatty acids (alpha-linolenic acid and linoleic acid) must be obtained from the diet because humans cannot synthesize them.

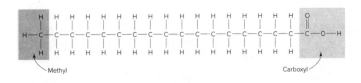

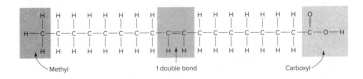

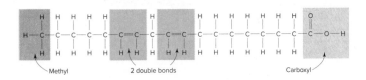

## 6.2 Triglycerides are the main fuel source for all body cells, except the nervous system and red blood cells.
Triglycerides are the body's main storage form of energy. The insulating layer of fat just beneath the skin is made mostly of triglycerides. Fats in food carry fat-soluble vitamins (vitamins A, D, E, and K) to the small intestine. Essential fatty acids, along with phospholipids and cholesterol, are important structural components of cell membranes. They also keep the cell membranes fluid and flexible so that substances can flow into and out of the cell. Eicosanoids, which are made from essential fatty acids, have over 100 different actions, such as regulating blood pressure, blood clotting, sleep/wake cycles, and body temperature.

## 6.3 Most foods provide at least some triglycerides.
Most triglyceride rich foods contain a mixture of fatty acids. The fat in some foods is visible; however, fat is hidden in many foods. Fat replacements help consumers trim fat intake and still enjoy the mouthfeel sensations fat provides.

## 6.4 The structure of phospholipids is very similar to that of triglycerides,
except a fatty acid is replaced with a compound that contains the mineral phosphorus and often has nitrogen attached. Phospholipids function in a watery environment without clumping together. The hydrophilic head of phosphate is attracted to water, and the fatty acid tail of phospholipids is attracted to fats. When placed in water, phospholipids cluster together, with their hydrophilic phosphate heads facing outward in contact with water and their hydrophobic tails extending into the cluster away from the water. In the body, phospholipids have 2 major roles as a cell membrane component and an emulsifier. Phospholipids can be synthesized by the body or supplied by the diet.

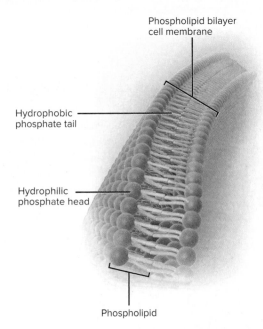

Phospholipid bilayer cell membrane

Hydrophobic phosphate tail

Hydrophilic phosphate head

Phospholipid

## 6.5 The carbons in the structure of sterols are mostly arranged in many rings.
Cholesterol, the most well-known sterol, is used in the body to make bile and steroid hormones, such as testosterone, estrogens, the active form of vitamin D hormone, and corticosteroids. Cholesterol is found in foods of animal origin, such as meat, fish, poultry, egg yolks, and dairy products. Foods of plant origin do not contain cholesterol.

Photodisc/Getty Images

## 6.6 The Acceptable Macronutrient Distribution Range for total fat is 20 to 35% of calories for most age groups.
Saturated fat intake, including *trans* fats, and cholesterol should be kept as low as possible while still consuming a nutritionally adequate diet. Cholesterol intake should be limited to about 300 mg daily. Fat intake recommendations are lower for those at risk of heart disease. Adequate Intakes for essential fatty acids equal less than 120 calories daily for women and 170 calories for men—that's about 2 to 4 tablespoons daily of oils rich in these fatty acids. Most North Americans get too much saturated and too little monounsaturated and polyunsaturated fat. Omega-6 fatty acid intake is usually plentiful, but omega-3 intakes often are lower than optimal.

**Table 6-2 Recommendations for Daily Intake for Adults**

| Fat Component | DRI* | Dietary Guidelines* | AHA* | TLC* |
|---|---|---|---|---|
| % fat of calories | 20–35% | 20–35% (based on DRI) | Consume a diet of fruits, vegetables, legumes, whole grains, and fish | 25–35% |
| Saturated fat | Eat as little as possible | <10% of calories starting at age 2 | Replace with mono and polyunsaturated fats | <7% of calories |
| Trans fat | Eat as little as possible | As of 2018, artificial *trans* fat is no longer Generally Recognized as Safe (GRAS) and no longer added to food in the U.S. | Minimize | Lower intake |
| Unsaturated fat | | A healthy eating pattern includes oils | | Monounsaturated: up to 20% Polyunsaturated: up to 10% |
| Omega-6 | 5–10% | Based on Adequate Intake (DRI); 11-12g women, 14-17g men | | |
| Omega-3 | 1–2% of calories | Based on Adequate Intake (DRI); 1.1g women, 1.6g men | | |
| Dietary Cholesterol | As low as possible while consuming a healthy diet | Minimize intake while consuming a healthy eating pattern | Minimize intake | <200 mg/day |

\* DRI=Dietary Reference Intake, Dietary Guidelines=Dietary Guidelines for Americans, 2020; AHA=American Heart Association, TLC=Therapeutic Lifestyle Change Diet. Blank areas indicate no recommendation was given.

## 6.7 Fat digestion occurs mostly in the small intestine.

The presence of fat in the small intestine triggers the release of cholecystokinin from intestinal cells. Cholecystokinin stimulates the release of bile and pancreatic enzymes. Bile emulsifies fats and allows enzymes to efficiently break triglycerides into monoglycerides and free fatty acids. Phospholipids and cholesterol are digested mostly in the small intestine. After absorption, short- and medium-chain fatty acids mostly enter the circulatory system. Long-chain fatty acids enter the lymphatic circulation.

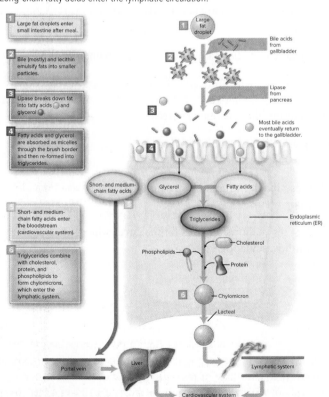

## 6.8 Fats are transported in the blood as lipoproteins called chylomicrons,

very-low-density lipoproteins (VLDLs), intermediate-density lipoproteins (IDLs), low-density lipoproteins (LDLs), and high-density lipoproteins (HDLs). Lipoproteins have a core, made of lipids, that is covered with a shell composed of protein, phospholipid, and cholesterol. The shell lets the lipoprotein circulate in the blood. The receptor pathway for cholesterol uptake removes LDL from the blood, breaks it down, and uses the component parts for maintaining the cell membrane or synthesizing compounds. Oxidized LDL is removed from the blood by the scavenger pathway for cholesterol uptake. Over time, cholesterol builds up in macrophages. When macrophages have collected and deposited cholesterol for many years at a heavy pace, plaque builds up on the inner blood vessel walls. HDL roams the bloodstream, picking up cholesterol from dying cells and other sources, and donates the cholesterol to other lipoproteins for transport back to the liver to be excreted.

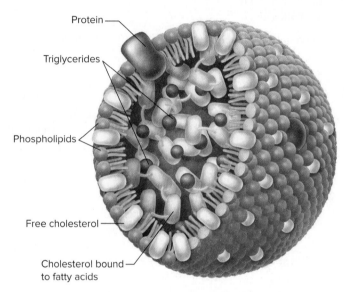

## 6.9 Intakes of polyunsaturated fats greater

than 10% of total calorie intake seem to increase the amount of cholesterol deposited in arteries. Diets that include fish rich in omega-3 twice a week can reduce blood clotting abilities and may favorably affect heart rhythm. Omega-6 and omega-3 fatty acids use the same metabolic pathways; as a result, imbalances in intake of these fatty acids may cause health problems. Rancid fats contain compounds that can damage cells. *Trans* fatty acids raise blood cholesterol levels, lower HDL-cholesterol levels, and increase inflammation in the body. Diets high in total fat increase the risk of obesity; colon, prostate, and breast cancer; and cardiovascular disease (CVD).

### Clinical Perspective

Atherosclerotic plaque is probably first deposited to repair injuries in the lining in any artery. As atherosclerosis progresses, plaque thickens over time, causing arteries to harden, narrow, and become less elastic. CVD risk factors are age, biological sex, genetics, race, blood cholesterol levels, blood triglyceride levels, hypertension, smoking, physical inactivity, obesity, diabetes, liver and kidney disease, and low thyroid hormone levels. All adults age 20 years or older should have a blood lipoprotein profile done every 5 years. Lifestyle changes can lower blood LDL-cholesterol levels and reduce health risks.

### Expert Perspective

In the diet, saturated fats should be replaced with unsaturated fats, with an emphasis on polyunsaturated fats .

## Study Questions

**1.** The lipids found in chylomicrons can be taken up by other cells with the help of ____________.

**a.** bile
**b.** pancreatic amylase
**c.** lipoprotein lipase (LPL)
**d.** low-density lipoprotein (LDL)

**2.** Triglycerides consist of ____________.

**a.** glycerol
**b.** cholesterol
**c.** 3 fatty acids
**d.** both a and b
**e.** both a and c

Match the fat-related terms on the right to their definitions on the left.

**3.** ____________ lipid that is solid at room temperature

**4.** ____________ chief form of fat in food

**5.** ____________ sterol manufactured in the body

**6.** ____________ similar to triglycerides, except a fatty acid has been replaced by a phosphorus

**7.** ____________ lipid that is liquid at room temperature

**a.** fat
**b.** cholesterol
**c.** oil
**d.** phospholipid
**e.** triglyceride

**8.** *Trans* fatty acids tend to ____________ blood cholesterol.

**a.** raise
**b.** lower
**c.** have no effect on

**9.** Chylomicrons are the principal transport vehicle for ____________.

**a.** glucose
**b.** triglycerides
**c.** cholesterol
**d.** free fatty acids

**10.** Which of the following lipoproteins is responsible for transporting cholesterol from the liver to tissues?

**a.** chylomicrons
**b.** low-density lipoprotein (LDL)
**c.** high-density lipoprotein (HDL)
**d.** very-low-density lipoprotein (VLDL)

**11.** Which essential fatty acid can help lower the risks of coronary heart disease?

**a.** omega-3
**b.** omega-6
**c.** omega-9
**d.** both a and b

**12.** Mike has been told to reduce his fat intake to less than 25% of his total calories (2500 per day). How many grams of fat should he consume?

**a.** 69 grams or less
**b.** 76 grams or less
**c.** 89 grams or less
**d.** 93 grams or less

**13.** Monounsaturated fatty acids ____________.

**a.** are liquid at room temperature
**b.** have 1 double bond in the fatty acid chain
**c.** are provided by plants
**d.** lower blood cholesterol levels
**e.** all of the above

**14.** Fats that are liquid at room temperature can be made more solid by the process of ____________.

**a.** esterification
**b.** hydrogenation
**c.** emulsification
**d.** calcification

**15.** For good health, experts now recommend that more of our dietary fat be polyunsaturated fatty acids than saturated fatty acids.

**a.** true
**b.** false

**16.** Describe the chemical structures of saturated and polyunsaturated fatty acids and their different effects in the human body.

**17.** What are the functions of lipids in the human body?

**18.** What are the recommendations of health-care professionals regarding fat intake? What do these recommendations mean in terms of actual food choices?

**19.** Trace the digestion of fat from the beginning to the end of the digestive tract.

**20.** Describe the structures, origins, and roles of the 4 major lipoproteins.

**21.** List the main risk factors for the development of cardiovascular disease.

Answer Key: 1-c; 2-e; 3-a; 4-e; 5-b; 6-d; 7-c; 8-a; 9-b; 10-b; 11-d; 12-a; 13-e; 14-b; 15-a; 16-refer to Sections 6.1 to 6.4; 17-refer to Sections 6.3 to 6.5; 18-refer to Section 6.6; 19-refer to Section 6.7; 20-refer to Section 6.8; 21-refer to Section 6.9, Clinical Perspective, and Expert Perspective

# References

1. Spector AA and Kim H. Emergence of omega 3 fatty acids as Biomedical Research. *PLEFA* 2019;32:47.

2. More MI and others. Positive effects of soy lecithin-derived phosphatidylserine plus phosphatidic acid on memory, cognition, daily functioning, and mood in elderly patients with Alzheimer's disease and dementia. *Adv Ther.* 2014;31:1247.

3. Ramdath, D and others. Beyond the Cholesterol-lowering Effect of Soy Protein. *Nutrients.* 2017:9;324.

4. Food and Nutrition Board. *Dietary Reference Intakes for energy, carbohydrate, fiber, fat, fatty acids, cholesterol, protein, and amino acids.* Washington, DC: National Academies Press; 2015.

5. Sacks, FM and others. Dietary Fast and Cardiovascular Disease: A Presidential Advisory from the American Heart Association. *Circ.* 2017;136:3.

6. Vannice G, Rasmussen H. Position of the Academy of Nutrition and Dietetics and Dietitians of Canada: Dietary fatty acids for healthy adults. *J Acad Nutr Diet.* 2014;114:1.

7. U.S. Department of Health and Human Services and U.S. Department of Agriculture. Dietary Guidelines for Americans, 2020–2025, 9th ed. 2020. dietaryguidelines.gov.

8. USDHHS. Healthy people 2020; HealthyPeople.gov.

9. Freeman A and others. Intensive Cardiac Rehabilitation: An underutilized Resource. *Curr Cardiol Rep.* 2019;21;19.

10. Soliman, GA. Dietary cholesterol and the lack of evidence in cardiovascular disease. *Nutrients.* 2018;10(6).

11. Vanitallie TB. Ancel Keys: A tribute. *Nutr Metab.* 2005;14:4.

12. Paredes S and others. Novel and traditional lipid profiles in metabolic syndrome reveal a high atherogencity. *Sci Rep.* 2019;9.

13. Bjorklund G and others. Has human diet a role in reducing nociception related to inflammation and chronic pain? *Nutr.* 2019;26:66;153.

14. Chehade L and others. Lifestyle modification in Rheumatoid Arthritis. Curr *Rheumato Rev* 2019;15:3;209.

15. van Zwol W and others. The future of lipid lowering therapy. *J Clin Med.* 2019;8;7.

16. Severson T and others. Roundtable discussion: Dietary fats in prevention of athersclerotic cardiovascular disease. *J Clin Lipidol.* 2018;12:574.

17. Briggs MA and others. Saturated fatty acids and cardiovascular disease: Replacements for saturated fat to reduce cardiovascular risks. *Healthcare.* 2017;5;2;29.

18. Hopper L and others. Reduction in saturated fat intake for cardiovascular disease. *Cochrane DB Syst Rev.* 2015:CD011737.

19. Wang M and others. Prediction of type 2 diabetes mellitus using non invasive MRI quantitation of visceral abdominal adiposity tissue volume. *Quant Imaging Med. Surg.* 2019;6:1076.

20. Zeng H and others. Secondary bile acids and short chain fatty acids in the colon. *Int J Mol Sci.* 2019;20:5.

21. Weigl J and others. Can nutrition lower the risk of recurrence in breast cancer. *Breast Care.* 2018;2;86.

22. Maly IV and Hofman WA. Fatty acid and calcium regulation in prostate cancer. *Nutrients.* 2018;10:6.

23. Centers for Disease Control and Prevention, National Center for Health Statistics. Multiple Cause of Death 1999-2015 on CDC WONDER Online Database, released December 2016. Data are from the Multiple Cause of Death Files, 1999-2015, as compiled from data provided by the 57 vital statistics jurisdictions through the Vital Statistics Cooperative Program. Accessed at wonder.cdc.gov/mcd-icd10.html.

24. Valensi P and others. Type 2 diabetes: Why should diabetologists and cardiologists work more closely together? *Diabetes Metab.* 2019;29:1.

25. Kjeldsen S. Hypertension and cardiovascular risk: general aspects. *Pharmacology Research.* 2018;129;95.

26. Lubin J. Risk of cardiovascular disease from cumulative cigarette use and the impact of smoking intensity. *Epidemiol.* 2016;3;395

27. Brack MC. Cardiovascular sequelae of pneumonia. *Curr Opin Pulm Med.* 2019;3;257.

28. Karim R and others. Relationship between serum levels of sex hormones and progression of subclinical atherosclerosis in postmenopausal women. *J Clin Endocrin Metab.* 2008;93:131.

29. Jolliffe C, Janssen I. Age-specific lipid and lipoprotein thresholds for adolescents. *J Cardio Nurs.* 2008;23:56.

30. Welsh JA and others. Caloric sweetener consumption and dyslipidemia among US adults. *JAMA.* 2010;303:1490.

31. Sabaté J and others. Nut consumption and blood lipid levels. A pooled analysis of 25 intervention trials. *Arch Intern Med.* 2010;170:821.

32. Kwon-Myung S and others. Efficacy of vitamin and antioxidant supplements in prevention of cardiovascular disease. *BMJ.* 2013;346:f10.

33. Loffredo L and others. Supplementation with vitamin E alone is associated with reduced myocardial infarction. *Nutr Metab Cardiovasc Dis.* 2015:75:354.

LAMB/Alamy Stock Photo

Many cultures around the world enjoy the flavors of insects and benefit from the protein they provide—3 ounces of the locusts in this stir-fry provide about 11 grams of protein! Learn more about entomophagy at **www.fao.org/docrep/018/i3253e /i3253e.pdf**.

# 7 Proteins

## Learning Objectives

**After studying this chapter, you will be able to**

1. Describe how amino acids form proteins.
2. Define *essential* and *nonessential amino acids* and explain why adequate amounts of each of the essential amino acids are required for protein synthesis.
3. Distinguish between high quality and low quality proteins and list sources of each.
4. Describe how 2 low quality proteins can be complementary to each other to provide the required amounts of essential amino acids.
5. Explain the methods used to measure the protein quality of foods.
6. Identify the factors that influence protein needs.
7. Calculate the RDA for protein for a healthy adult with a given body weight.
8. Explain positive nitrogen balance, negative nitrogen balance, and nitrogen equilibrium and list the conditions under which they occur.
9. Describe how protein is digested and absorbed in the body.
10. Explain the primary functions of protein in the body.
11. Describe the differences between the two types of protein-energy malnutrition.
12. Describe the symptoms and treatment of food allergies.
13. Discuss the advantages and disadvantages of a vegetarian diet plan.

## Chapter Outline

THE TERM *PROTEIN* COMES FROM the Greek word *protos,* which means "to come first." This is an appropriate name, given that proteins are a primary component of all cells throughout the body. In fact, aside from water, proteins form the major part of lean body tissue, totaling about 17% of body weight.[1] Many of our body proteins are found in muscle, connective tissue, and organs. Hemoglobin, antibodies, hormones, and enzymes are examples of proteins in the body.

Proteins are crucial to the regulation and maintenance of essential body functions. For example, maintenance of fluid balance, hormone and enzyme production, cell synthesis and repair, and vision each require specific proteins. The body synthesizes proteins in many sizes and configurations so that they can serve these greatly varied functions.[1]

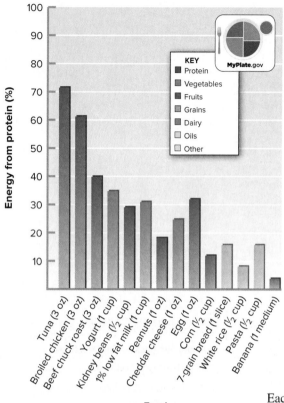

*Figure 7-1* Protein content of foods. In addition to MyPlate food groups, yellow is used for oils and pink is used for substances that do not fit easily into food groups (e.g., candy, salty snacks).

Source: ChooseMyPlate.gov, U.S. Department of Agriculture

**nonessential amino acids**  Amino acids that the human body can synthesize in sufficient amounts. There are 11 nonessential amino acids. These also are called dispensable amino acids.

In industrialized countries, such as the U.S. and Canada, most people consume diets rich in protein. In contrast, diets in low-income countries often contain insufficient amounts of protein. As you'll see, consuming inadequate amounts of protein can impair many metabolic processes because the body is unable to build the proteins it needs. For example, the immune system no longer functions efficiently when it lacks key proteins, which leads to an increased risk of infection, disease, and, if severe, even death. This chapter will examine the functions of protein, its metabolism, the sources of protein, and the consequences of eating diets that are too low or too high in protein.

# 7.1  Structure of Proteins

Like carbohydrates and lipids, proteins are made of the elements carbon, hydrogen, and oxygen. However, all proteins also contain the element nitrogen. Some proteins also contain the mineral sulfur. Together, these elements form various amino acids, which serve as the building blocks for protein synthesis.

## Amino Acids

The amino acids needed to make body proteins are supplied by the protein-containing foods we eat as well as through cell synthesis (Fig. 7-1).

Each amino acid is composed of a central carbon bonded to 4 groups of elements: a nitrogen (amino) group, an acid (carboxyl) group, hydrogen, and a side chain (often signified by the letter *R*). The basic, or "generic," model of an amino acid and the structures of 2 amino acids, glycine and alanine, are shown in Figure 7-2. (The chemical structures of the rest of the amino acids are shown in Appendix B.)

The side chain (R group) makes each amino acid unique and determines the structure, function, and name of the amino acid. For example, if R is a hydrogen, the amino acid is glycine; if R is a methyl group ($-CH_3$), the amino acid is alanine. Some amino acids have chemically similar side chains. These related amino acids form special classes, such as acidic amino acids, basic amino acids, and branched-chain amino acids. For instance, the acidic amino acids lose a hydrogen in reactions and become negatively charged, whereas the basic amino acids gain a hydrogen and become positively charged. This allows them to participate in different enzymatic reactions in the body.

The body needs 20 different amino acids to function. Although all amino acids are needed for life, 11 of them do not need to be obtained from the diet. They are classified as **nonessential** (or *dispensable*) **amino acids** because our bodies make them, using other amino acids we consume (Table 7-1). The 9 amino acids the body cannot make are

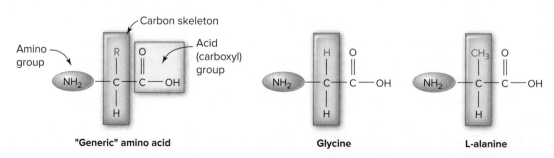

*Figure 7-2*  Amino acid structure. The side chain (R) differentiates glycine from alanine.

known as nutritionally **essential** (or *indispensable*) **amino acids** because they must be obtained from foods. Essential amino acids cannot be synthesized in the body because body cells cannot make the **carbon skeleton** of the amino acid, cannot attach an amino group to the carbon skeleton, or cannot do the whole process fast enough to meet the body's needs.

Several nonessential amino acids may be classified as "conditionally essential" amino acids during infancy, disease, or trauma.[1] For example, a person with the genetic disease phenylketonuria (PKU) has a limited ability to metabolize the essential amino acid phenylalanine due to a deficiency of the enzyme phenylalanine hydroxylase. This enzyme is needed to convert phenylalanine to the nonessential amino acid tyrosine. As a result, individuals with PKU cannot produce sufficient tyrosine, thereby making tyrosine a "conditionally" essential amino acid because it must be obtained from the diet. Following trauma and infection, the amino acids glutamine and arginine may be considered conditionally essential because supplemental amounts have been shown to promote recovery.

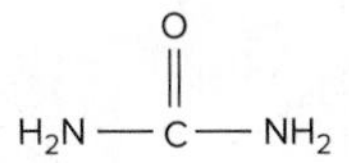

## Synthesis of Nonessential Amino Acids

Nonessential amino acids can be synthesized through a process called transamination. **Transamination** involves the transfer of an amino group from 1 amino acid to a carbon skeleton to form a new amino acid. As illustrated in Figure 7-3, alanine donates its amino group to the carbon skeleton of pyruvic acid to become the nonessential amino acid glutamic acid.

Glutamic acid (and several other amino acids) also can lose an amino group without transferring it to another carbon skeleton. This process is called **deamination**. The lost amino group (in the form of ammonia) is incorporated into **urea** in the liver, transported via the bloodstream to the kidneys, and excreted in the urine. Once an amino acid breaks down to its amino-free carbon skeleton, the carbon skeleton can be used for energy or synthesized into other compounds, such as glucose (see Chapter 9).

**essential amino acids** Amino acids that the human body cannot synthesize in sufficient amounts or at all and therefore must be included in the diet. There are 9 essential amino acids. They also are called indispensable amino acids.

**carbon skeleton** Amino acid without the amino group ($-NH_2$).

**urea** Nitrogenous waste product of protein metabolism and the major source of nitrogen in the urine; chemically,

$$H_2N - \underset{\underset{\displaystyle C}{\overset{\displaystyle O}{\|}}}{} - NH_2$$

**Table 7-1 Classification of Amino Acids**

| Essential Amino Acids | Nonessential Amino Acids |
|---|---|
| Histidine | Alanine |
| Isoleucine[*] | Arginine |
| Leucine[*] | Asparagine |
| Lysine | Aspartic acid |
| Methionine | Cysteine |
| Phenylalanine | Glutamic acid |
| Threonine | Glutamine |
| Tryptophan | Glycine |
| Valine[*] | Proline |
| | Serine |
| | Tyrosine |

[*]Branched-chain amino acids.

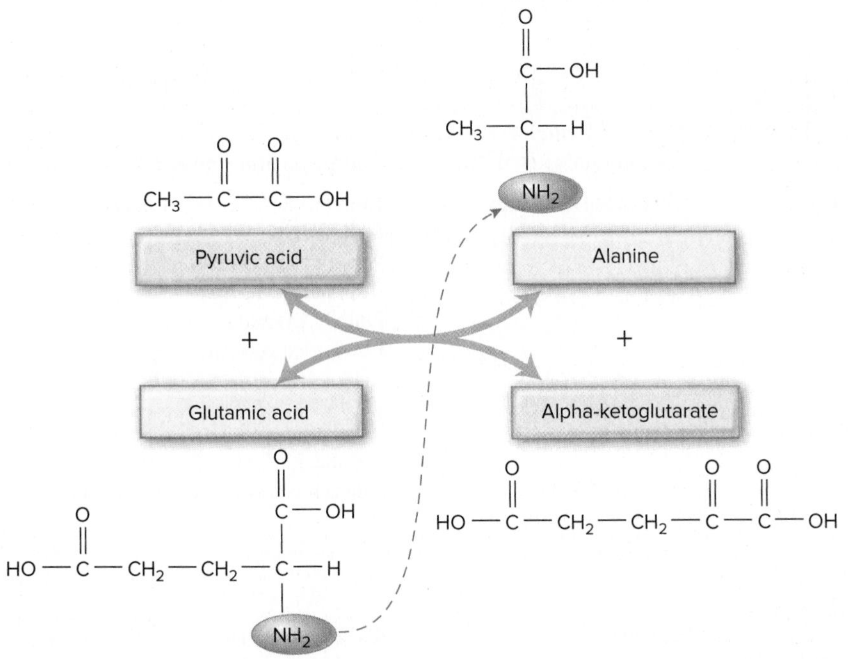

*Figure 7-3* In *transamination*, the pathway allows cells to synthesize nonessential amino acids. In this example, pyruvic acid gains an amino group from glutamic acid to form the amino acid alanine. In *deamination*, the pathway allows for loss of an amino group without transferring it to another carbon skeleton. In this example, glutamic acid loses its amino group to form alpha-ketoglutaric acid.

High protein foods can provide all the essential amino acids needed by our bodies.

Pixtal/age fotostock

**pool** Amount of a nutrient within the body that can be easily mobilized when needed.

## Amino Acid Composition: Complete and Incomplete Proteins

Animal and plant proteins can differ greatly in their proportions of essential and nonessential amino acids. Animal proteins, such as meat, poultry, fish, eggs, and milk, contain ample amounts of all 9 essential amino acids. (Gelatin—made from the animal protein collagen—is an exception because it loses an essential amino acid during processing and is low in several essential amino acids.) In contrast, plant proteins do not contain the needed amounts of essential amino acids. With the exception of quinoa, buckwheat, and soy protein, they are low in at least 1 of the 9 essential amino acids.

Dietary proteins are classified according to their amino acid composition. Because they contain sufficient amounts of all the essential amino acids, animal proteins (except gelatin) are classified as complete, or high quality, proteins. Plant proteins (except soybeans, buckwheat, and quinoa) are classified as incomplete, or low quality, proteins because they contain limited amounts of 1 or more of the essential amino acids.

Cells require a **pool** of essential amino acids for the synthesis of body proteins. Thus, a single plant protein, such as wheat (which is low in the amino acid lysine), cannot support the synthesis of body protein if it is the sole source of dietary protein. Even a variety of low quality proteins may not provide sufficient amounts of essential amino acids for protein synthesis if food choices are not carefully planned. When this occurs, proteins cannot be made and the remaining amino acids may be used for energy or converted to carbohydrate or fat.

The essential amino acid in smallest supply in a food or diet in relation to body needs is called the **limiting amino acid** because it limits the amount of protein the body can synthesize. For example, assume the letters of the alphabet represent the 20 different amino acids we need. If *A* represents an essential amino acid, we need 4 of these letters to spell the hypothetical protein *ALABAMA*. If the body had an *L*, a *B*, and an *M*, but only 3 *A*s, the "synthesis" of *ALABAMA* would not be possible. *A* would be the limiting amino acid, preventing the synthesis of the protein *ALABAMA*.

When 2 or more plant proteins are combined to compensate for deficiencies in essential amino acid content in each protein, the proteins are called **complementary proteins** (Table 7-2).

### Table 7-2  Limiting Amino Acids in Plant Sources of Protein

| Food | Primary Limiting Amino Acid | Create a Complete Protein By Combining It With | Complementary Food Protein Combinations |
|---|---|---|---|
| Legumes (peanuts, dry beans such as navy, black, and kidney beans) | Methionine | Grains, nuts, or seeds | Hummus and whole-wheat pita bread<br>Bean burrito<br>White beans and pasta<br>Bean and barley stew<br>Pinto beans and polenta<br>Black-eyed peas and rice |
| Nuts and seeds (cashews, walnuts, almonds, sunflower seeds) | Lysine | Legumes | Vegetarian chili with kidney beans and cashews<br>Sesame, buckwheat, and bean bread<br>Salads made with nuts, beans, and seeds |
| Grains (wheat, rice, oats, corn) | Lysine | Legumes | Red beans and rice<br>Lentil soup and cornbread<br>Barley and black beans<br>Peanut butter sandwich |

Bean: ©McGraw-Hill Education/Jacques Cornell, photographer; sunflower seeds: ©McGraw-Hill Education/Jacques Cornell, photographer; rice: ©McGraw-Hill Education/Jacques Cornell, photographer

When complementary protein sources are combined, the amino acids in 1 source can make up for the limiting amino acid in the other sources to yield a high quality (complete) protein for the diet. Mixed diets generally provide high quality protein because these diets often contain complementary proteins. Complementary proteins do not need to be consumed at the same meal but can be balanced over the course of a day to provide a sufficient supply of amino acids for body cells. For nonvegetarians, adding a small amount of animal protein to a plant-based dish (e.g., pizza with cheese or spaghetti with meatballs) is a way of providing adequate essential amino acids.

Pulses are an important component of most vegetarian diets. Pulses are a type of legume and include dried beans, dry peas, chickpeas, and lentils. Pulses are high in protein, fiber, and various minerals and are recognized as part of a healthy diet throughout the world. Pulse crops are among the most sustainable and "earth-friendly" crops farmers can grow. It takes just 43 gallons of water to produce 1 pound of pulses, compared to 216 gallons for soybeans and 368 for peanuts. They also help improve soil quality. Pulses will play a major role in meeting the food needs of the world's growing population. See the Culinary Perspective on beans, lentils, and dried peas in Part 4 to learn more.

### Knowledge Check

1. Why are some amino acids classified as essential and others as nonessential?
2. What are complementary proteins? Give 2 examples.
3. What does the term *limiting amino acid* mean?

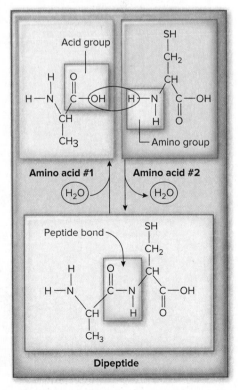

Small amounts of animal protein in a meal quickly add up to meet daily protein needs.

©Ingram Publishing/SuperStock

## 7.2 Synthesis of Proteins

Within body cells, amino acids can be linked together by a chemical bond, called a peptide bond, to form needed proteins (Fig. 7-4). **Peptide bonds** form between the amino group of 1 amino acid and the acid (carboxyl) group of another. Through peptide bonding of amino acids, cells can synthesize dipeptides (joining of 2 amino acids), tripeptides (joining of 3 amino acids), oligopeptides (joining of 4 to 9 amino acids), and polypeptides (joining of 10 or more amino acids). Most proteins are polypeptides, ranging from approximately 50 to 2000 amino acids. The body can synthesize many different proteins by joining different combinations of amino acids with peptide bonds.

**codon** Specific sequence of 3 nucleotide units within DNA that codes particular amino acids needed for protein synthesis.

### Transcription and Translation of Genetic Information

The synthesis of body proteins is determined through a process called gene expression. Gene expression begins when deoxyribonucleic acid (DNA) replicates, making an exact copy of the gene. Thus, each gene serves as a template to guide the duplication of genetic information carried by the DNA.

As you know, DNA is a double-stranded molecule in a helical form. Each strand of DNA is composed of 4 nucleotides (building blocks of DNA): adenine (A), guanine (G), cytosine (C), and thymine (T). Each of the nucleotides is complementary to (binds to) another nucleotide; A and T are complementary, as are C and G.

DNA-coded instructions for protein synthesis consist of a sequence of 3 nucleotides per unit of instruction (e.g., CTC), which dictate where each amino acid is to be placed in a protein and in which order. These nucleotide units are called **codons**, and each represents a specific amino acid. For example, the codon CTC represents the amino acid glutamic acid. Some amino acids have only 1 possible codon, whereas others have as many as 6. For instance, the amino acid glutamic acid actually has 2 codons: CTC and CTT. Having the correct codons in the right sequence is critical for producing the needed amino acid and a normal protein. DNA with mistakes in the order or types of amino acids can result in profound health consequences (see the discussion of sickle-cell disease later in this section).

Protein synthesis takes place in the ribosomes, which are located in the cytosol of the cell. Because DNA is in the cell nucleus, the DNA code used for the synthesis of a specific

*Figure 7-4* Peptide bonds link amino acids. This reaction also is reversible.

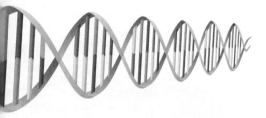

protein must be transferred from the nucleus to the cytosol to allow for such synthesis. This transfer is the job of messenger RNA (mRNA). To produce mRNA, the DNA unwinds from its super-coiled state. Enzymes read the code on the DNA and transcribe that code into a complementary single-stranded mRNA molecule, called the primary transcript (Fig. 7-5). This process is called **DNA transcription**. The primary transcript mRNA undergoes processing in the cell nucleus to remove any parts of the DNA code that do not code for protein synthesis. The mRNA then travels to the ribosomes. The ribosomes read the codons on the

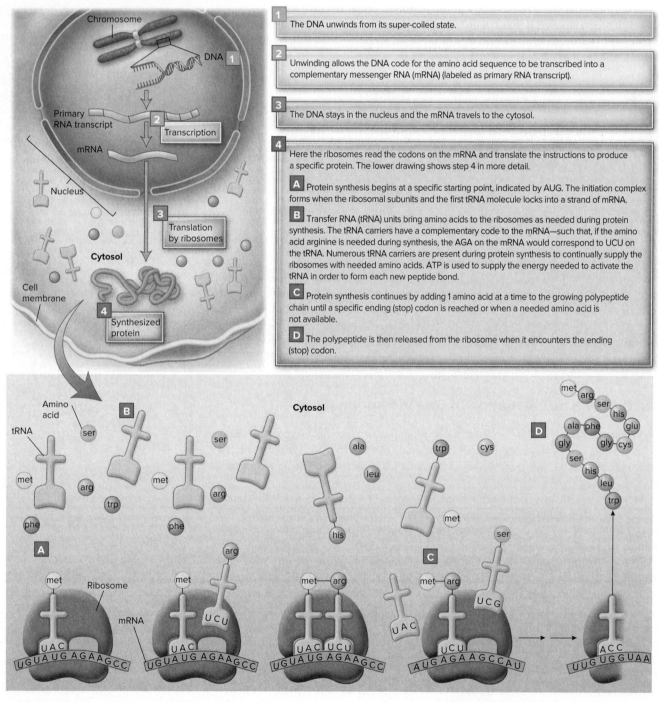

*Figure 7-5* Summary of protein synthesis. DNA present in the nucleus of the cell is composed of 4 nucleotides: adenine (A), guanine (G), cytosine (C), and thymine (T). The DNA code is read 3 nucleotides at a time, with each specific unit being called a codon. Each DNA codon represents a specific amino acid.

mRNA and translate those instructions to produce a specific protein. This is the **mRNA translation** phase of protein synthesis. Amino acids are added 1 at a time to the polypeptide chain as directed by the instructions on the mRNA. Protein synthesis begins at a specific starting point on the mRNA (indicated by AUG) and continues until a specific ending (stop) codon is reached, such as UAA, UAG, or UGA. Energy input from ATP is needed to add each amino acid to the growing polypeptide chain, making protein synthesis very "costly" to the body in terms of energy use.

One key participant in protein synthesis in the cytosol is transfer RNA (tRNA). The tRNA units take amino acids to the ribosomes as needed during protein synthesis. The tRNA carriers have a complementary code to the mRNA. Numerous tRNA carriers are present during protein synthesis to continually supply the ribosomes with needed amino acids.

Once synthesis of the polypeptide is completed, indicated by the ending codon, it is released from the ribosome, as is the mRNA. The polypeptide now twists and folds into a very complex 3-dimensional structure (see the section Protein Organization). The DNA code determines not only the shape but also the function of the protein. Thus, if the DNA contains errors, an incorrect mRNA will be produced. The ribosomes, in turn, will read this incorrect message and produce an abnormal polypeptide.

The genetic disorder sickle-cell anemia illustrates what can happen when amino acid sequencing errors occur.[2] In this disease, the amino acid valine replaces glutamic acid in the DNA sequence of half of the 4 polypeptide chains of hemoglobin. This error produces a profound change in hemoglobin structure (Fig. 7-6). Instead of forming normal, donut-shaped discs, the red blood cells collapse into crescent, or sickle, shapes. This limits their ability to carry oxygen to tissues, resulting in many serious health concerns. The disease can lead to severe bone and joint pain, abdominal pain, headache, convulsions, paralysis, and even death when the sickle cells clump in the capillary beds and impede blood flow. Treatment usually involves blood transfusions, medications to increase red blood cell synthesis, and bone marrow transplants.

## Protein Organization

The sequential order and strong peptide bonding of the amino acids in the polypeptide chain, called the *primary structure,* determine a protein's shape. Amino acids must be accurately positioned in order for the amino acids to interact and fold correctly into the intended shape for the protein. This, in turn, allows weaker chemical bonds to form between amino acids near each other and stabilizes the structure. This creates a spiral-like or pleated sheet shape called the *secondary structure.* The unique 3-dimensional folding of a protein, called *tertiary structure,* determines the protein's overall shape and physiological function. Thus, if a protein fails to form the appropriate configuration, it cannot function. In some cases, 2 or more separate polypeptides interact to form a large, new protein, with *quaternary structure* (Fig. 7-7). In this way, a protein may be active when the units are joined but inactive when the units are separate. Hemoglobin is an example of a protein with quaternary structure.

**DNA transcription** Process of forming messenger RNA (mRNA) from a portion of DNA.

**mRNA translation** Synthesis of polypeptide chains at the ribosome according to information contained in strands of messenger RNA (mRNA).

(a)

(b)

*Figure 7-6* An example of the consequences of errors in DNA coding of proteins. (*a*) Red blood cell from a person with sickle-cell anemia—note its abnormal crescent (sickle) shape; (*b*) normal red blood cell—note its rounded shape.

Callista Images/Getty Images

Alanine
|
Glycine
|
Serine
|
Valine
|
Leucine
|
Lysine
|
Glycine
|
Valine

**Primary**      **Secondary**      **Tertiary**      **Quaternary**

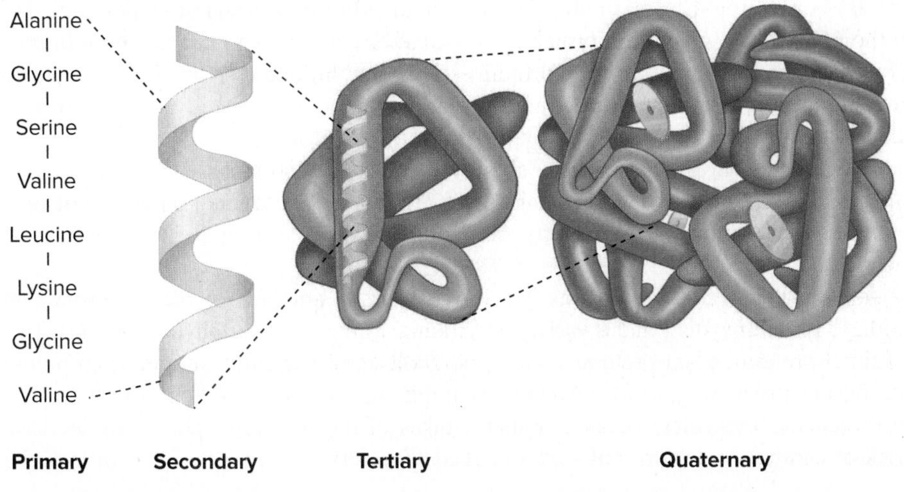

*Figure 7-7* Protein organization. Four levels of structure are found in proteins. The primary structure of a protein is the linear sequence of amino acids in the polypeptide chain. Secondary structure consists of areas in the polypeptide chain that have a specific shape stabilized by hydrogen and sulfur bonds. The 3-dimensional shape of proteins is called tertiary structure. It determines the function of the protein. Some proteins also show quaternary structure, where 2 or more protein units join together to form a larger protein, such as hemoglobin, depicted in the figure.

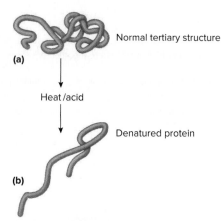

*Figure 7-8* Denaturation. (*a*) Protein in a typical, coiled state. (*b*) Protein is now partly uncoiled, exhibiting a denatured state. This uncoiling typically reduces or eliminates biological activity and is usually caused by treatment by heat, enzymes, acid or alkaline solutions, or agitation.

**Synopsis of the Steps in Protein Synthesis**

Part of DNA code (gene) is transcribed to mRNA in the cell nucleus.

mRNA leaves the nucleus and travels to cell cytosol.

Ribosomes in the cytosol read the mRNA code and translate it into directions for a specific sequence of amino acids in a polypeptide chain.

To produce the polypeptide, tRNA takes the appropriate amino acid to the ribosome as dictated by the mRNA code. The amino acid is added to the existing amino acid chain, which begins with the amino acid methionine.

When synthesis of the polypeptide is complete, it is released from the ribosome.

The polypeptide folds into its active 3-dimensional form.

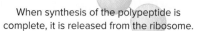

Sulfur-containing amino acids stabilize many compounds, such as the hormone insulin. Sulfur atoms can bond together (—S—S—), creating a bridge between 2 protein strands or 2 parts of the same strand. This stabilizes the structure of the molecule and helps create the *secondary structure*.

## Denaturation of Proteins

Exposure to acid or alkaline solutions, enzymes, heat, or agitation can change a protein's structure, leaving it in a denatured state (Fig. 7-8). Alteration of a protein's 3-dimensional structure is called **denaturation**. Although denaturation does not affect the protein's primary structure, unraveling a protein's shape often destroys its normal biological function.

Sometimes the denaturation of proteins is beneficial. For example, the secretion of hydrochloric acid in the stomach during digestion denatures food proteins, which increases their exposure to digestive enzymes and aids in the breakdown of polypeptide chains. The heat produced during cooking also can denature proteins, making them safer to eat (e.g., when harmful bacterial protein is denatured) and more pleasing to eat (e.g., when eggs solidify in cooking). However, denaturation also can be harmful to physiological function and overall health. During illness, changes in gastrointestinal acidity, body temperature, or body pH can cause essential proteins to denature and lose their function.

## Adaptation of Protein Synthesis to Changing Conditions

Most vital body proteins are in a constant state of breakdown, rebuilding, and repair. This process, called protein turnover, allows cells to adapt to changing circumstances. For example, when we eat more protein than necessary for health, the liver makes more enzymes to process the waste product from the resulting amino acid metabolism—namely, ammonia—into urea. Overall, protein turnover is a process by which a cell can respond to its changing environment by increasing the production of needed proteins while reducing the production of proteins not currently needed.[1]

> *Knowledge Check*
>
> 1. How are the amino acids in a protein linked together?
> 2. Why is the structure of a protein important?
> 3. What effects does denaturation have on a protein?

## 7.3 Sources of Protein

Proteins and amino acids are supplied by the diet as well as by the recycling and reutilization of amino acids released during the breakdown of body protein. For example, the intestinal tract lining is constantly sloughed off. The digestive tract treats sloughed cells just like food particles and absorbs their amino acids released during digestion. In fact, most protein breakdown products—amino acids—released throughout the body can be recycled and added to the pool of amino acids available for future protein synthesis. By comparing the 250 to 300 g of protein an adult makes and degrades each day with the 65 to 100 g of protein typically consumed by adults, you can see how important recycled amino acids are as a protein source for the body.[3] Nonetheless, dietary protein is needed to replenish and maintain an adequate amino acid pool for protein synthesis and repair.

In typical North American diets, about 70% of dietary protein is supplied by meat, poultry, fish, milk, cheese, legumes, and nuts (Fig. 7-9).[4] Worldwide, only 35% of protein comes from animal sources. Plants are the major source of protein in many areas of the world.

As shown in Table 7-3, plants can provide ample amounts of dietary protein in addition to providing fiber and a variety of vitamins, minerals, and phytochemicals. Unlike animal proteins, plant proteins contain no cholesterol and little saturated fat, unless added during processing. North Americans might benefit from adding soy and other plant proteins to their diets because higher intakes of these proteins may help decrease the risk of cardiovascular disease, certain cancers, obesity, and diabetes.[5,6,7,8] In fact, the

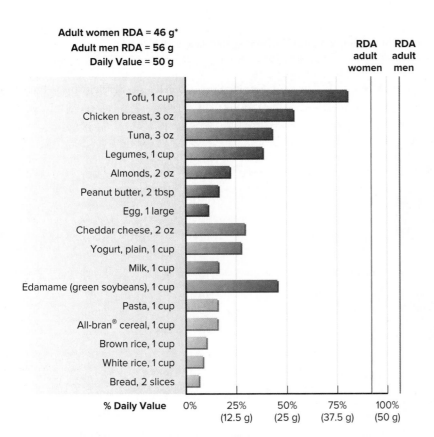

Adult women RDA = 46 g*
Adult men RDA = 56 g
Daily Value = 50 g

*RDA based on 0.8 g/kg body weight for 125 lb (57 kg) woman and 154 lb (70 kg) man.

*Figure 7-9* Food sources of protein.

FDA has approved a health claim regarding the benefits of soy protein in lowering blood cholesterol levels.

As a way to add more plant proteins to your diet, consider these suggestions.

- At your next cookout, try a veggie burger instead of a hamburger. These are available in the frozen foods section of the grocery store. Many restaurants, including fast food restaurants, have veggie burgers on their menus.
- Sprinkle sunflower seeds or chopped walnuts on top of a salad to add taste and texture.
- Mix chopped pecans or almonds into the batter of banana bread, muffins, or pancakes to boost your intake of monounsaturated fats and protein.
- Eat edamame (green soybeans) or roasted soy nuts as a snack.
- Spread peanut butter, instead of butter or cream cheese, on bagels.
- Consider using soy milk, especially if you have lactose intolerance. Look for varieties that are fortified with calcium.
- Substitute black beans or vegetarian refried beans for the meat or fish in your tacos.
- Make a stir-fry with tofu, cashew nuts, and a variety of vegetables.

## Evaluation of Food Protein Quality

Scientists use various measures to evaluate the protein quality of a food. These measures indicate a food protein's ability to support body growth and maintenance. Protein quality is determined primarily by the food's digestibility (amount of amino acids absorbed) and amino acid composition compared with a reference protein (e.g., egg white protein) that provides the essential amino acids in amounts needed to support growth. The digestibility of animal proteins is relatively high (90–100%), in contrast to that of plant proteins (70%).

D. Hurst/Alamy Stock Photo

**Table 7-3  Protein Content of Sample Menus Containing 1600 kcal and 2000 kcal**

|  | Menu | 1600 kcal | | 2000 kcal | |
|---|---|---|---|---|---|
|  |  | Serving Size | Protein (g) | Serving Size | Protein (g) |
| **Breakfast** | Low-fat granola | ⅔ cup | 5 | ⅔ cup | 5 |
|  | Blueberries | 1 cup | 1 | 1 cup | 1 |
|  | Fat-free (skim) milk | 1 cup | 8.5 | 1 cup | 8.5 |
|  | Coffee | 1 cup | 0 | 1 cup | 0 |
| **Lunch** | Grilled chicken breast | 3 oz | 25 | 4 oz | 33 |
|  | Salad greens | 3 cups | 5 | 3 cups | 5 |
|  | Baked taco shell strips | ½ cup | 2 | ½ cup | 2 |
|  | Low-fat salad dressing | 2 tbsp | 0 | 2 tbsp | 0 |
|  | Fat-free (skim) milk | 1 cup | 8.5 | 1 cup | 8.5 |
| **Dinner** | Yellow rice | 1¼ cups | 5 | 2½ cups | 10 |
|  | Shrimp | 4 large | 5 | 6 large | 7 |
|  | Mussels | 4 medium | 8 | 6 medium | 12 |
|  | Clams | 5 small | 12 | 10 small | 24 |
|  | Peas | ¼ cup | 4 | ½ cup | 4 |
|  | Sweet red pepper | ¼ cup | 0 | ½ cup | 0 |
| **Snack** | Whole-grain cracker | 6 small | 3 | 12 small | 6 |
|  | Cottage cheese | 2 oz | 7.5 | 2 oz | 7.5 |
|  | Banana | ½ small | 0.5 | ½ small | 0.5 |
|  | **Total** |  | 100 |  | 131 |

BreakFast: Floortje/Getty Images; Lunch: John A. Rizzo/Getty Images; Dinner: Kevin Sanchez/Cole Group/Photodisc/Getty Images; Snack:Mariia Boiko/Shutterstock

It is important to note that the concept of protein quality applies only under conditions in which protein intakes are equal to or less than the amount of protein needed to meet the requirement for essential amino acids. When protein intake exceeds this amount, the efficiency of protein use is decreased, even with the highest quality proteins. This occurs because, once essential amino acid needs have been met, the remaining amino acids (both essential and nonessential) are mostly broken down and used as energy.

### Biological Value (BV)

The **biological value (BV)** of a protein is a measure of how efficiently the absorbed food protein is converted into body tissue protein. If a food possesses adequate amounts of all 9 essential amino acids, it should allow a person to efficiently incorporate amino acids from food protein into body protein.

To determine the BV, nitrogen retention in the body is compared with the nitrogen content of the food protein. More nitrogen is retained when a food's amino acid pattern closely matches the amino acid pattern of body protein. The better the match, the higher the BV. In contrast, if the amino acid pattern in a food is quite unlike body tissue amino acid patterns, more nitrogen is excreted because many of the amino acids in the food will not be

Legumes are rich sources of protein. One-half cup meets about 10% of protein needs but contributes only about 5% of energy needs.

Moving Moment/Shutterstock

incorporated into body protein. The BV of such a food protein is low, as little of the nitrogen is retained in body tissues.

Egg white protein has a BV of 100, the highest BV of any single food protein. This means that essentially all the nitrogen absorbed from egg protein is retained and incorporated into body tissue protein. Most animal proteins have a high BV, reflecting an amino acid composition similar to that of human tissues. Plants have amino acid patterns that differ greatly from those of humans. Therefore, the BV of plant proteins is usually much lower than that of animal proteins.

## Protein Efficiency Ratio (PER)

**Protein efficiency ratio (PER)** is another method for assessing a food's protein quality. The PER compares the amount of weight gain by a growing laboratory animal consuming a specific amount of the protein being studied with the weight gain by an animal consuming a specific amount of a reference protein, such as casein (milk protein). The PER of a food reflects its BV because the weight gain and growth measured in the PER depend on the incorporation of food protein into body tissue. Thus, animal proteins with a high BV also have a high PER, whereas plant proteins generally have a lower BV and PER because they are incomplete proteins. The FDA uses the PER to set standards for the labeling of foods intended for infants.

## Chemical Score

The protein quality of a food also can be evaluated by its chemical score. To calculate a **chemical score**, the amount of each essential amino acid in a gram of the food protein being tested is divided by the "ideal" amount for that amino acid in a gram of the reference protein (usually egg protein). The lowest (limiting) amino acid ratio that is calculated for the essential amino acids of the test protein equals its chemical score. Chemical scores range from 0 to 1.0.

## Protein Digestibility Corrected Amino Acid Score (PDCAAS)

The most widely used measure of protein quality is called the **Protein Digestibility Corrected Amino Acid Score (PDCAAS).** This score is derived by multiplying a food's chemical score by its digestibility. For example, to determine the PDCAAS of wheat, multiply its chemical score (0.47) by its digestibility (0.90). This gives a PDCAAS of approximately 0.40. The highest PDCAAS is 1.0, which is the score for soy protein and most animal proteins. A protein missing any of the 9 essential amino acids (e.g., gelatin) has a PDCAAS of 0 because its chemical score is 0.

For nutrition labeling purposes, protein content (when listed as % Daily Value) is reduced if the PDCAAS is less than 1.0. For example, if the protein content of ½ cup of spaghetti noodles is 3 g, only 1.2 g are counted when calculating % Daily Value, since the PDCAAS of wheat is 0.40 (3 g × 0.40 = 1.2). Other PDCAAS values are egg white, 1.0; soy protein, 0.92 to 0.99; beef, 0.92; and black beans, 0.53. Currently, the Nutrition Facts panel rarely contains the % Daily Value for protein because manufacturers do not want to spend the money needed to determine the PDCAAS.

$$BV = \frac{\text{Nitrogen retained (g)}}{\text{Nitrogen absorbed (g)}} \times 100$$

$$PER = \frac{\text{Weight gain (g)}}{\text{Protein consumed}}$$

$$\text{Chemical score} = \frac{\begin{array}{c}\text{mg of limiting amino}\\\text{acid per g of protein}\end{array}}{\begin{array}{c}\text{mg of limiting amino acid}\\\text{per g of an "ideal" protein}\end{array}}$$

$$PDCAAS = \text{Chemical score} \times \text{Digestibility}$$

▶ The concept of biological value has clinical importance whenever protein intake must be limited. This is because it is important that the small amount of protein consumed be used efficiently by the body. For example, protein intake during liver disease and kidney disease may need to be controlled to lessen the effects of the disease. In these cases, most of the protein consumed should be of high biological value, such as eggs, milk, and meat.

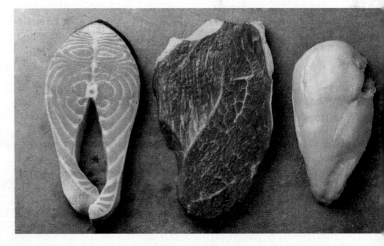

Meat, fish, and poultry are the main sources of high biological value protein.

Lisovskaya Natalia/Shutterstock

## Knowledge Check

1. What are 3 ways of assessing protein quality?
2. Why does egg protein have a high biological value (BV)?
3. Why is it important to use protein foods with high biological values with patients who have kidney disease?
4. What factors affect the protein quality of a food?

# *Culinary Perspective*

## Entomophagy

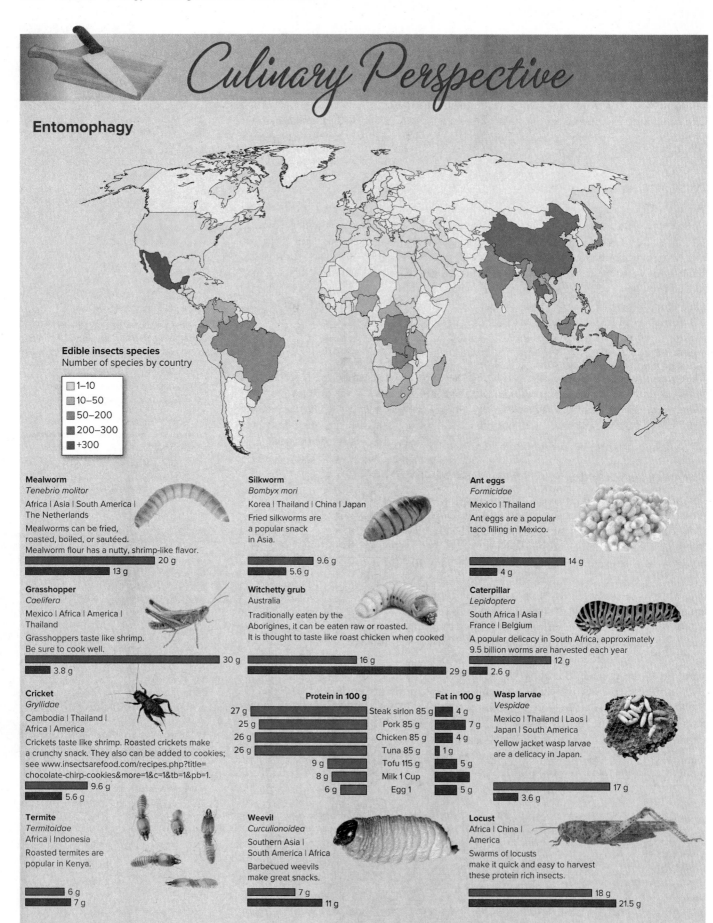

**Edible insects species**
Number of species by country

- ☐ 1–10
- ☐ 10–50
- ☐ 50–200
- ☐ 200–300
- ☐ +300

### Mealworm
*Tenebrio molitor*

Africa | Asia | South America | The Netherlands

Mealworms can be fried, roasted, boiled, or sautéed. Mealworm flour has a nutty, shrimp-like flavor.

20 g
13 g

### Silkworm
*Bombyx mori*

Korea | Thailand | China | Japan

Fried silkworms are a popular snack in Asia.

9.6 g
5.6 g

### Ant eggs
*Formicidae*

Mexico | Thailand

Ant eggs are a popular taco filling in Mexico.

14 g
4 g

### Grasshopper
*Caelifera*

Mexico | Africa | America | Thailand

Grasshoppers taste like shrimp. Be sure to cook well.

30 g
3.8 g

### Witchetty grub
Australia

Traditionally eaten by the Aborigines, it can be eaten raw or roasted. It is thought to taste like roast chicken when cooked

16 g
29 g

### Caterpillar
*Lepidoptera*

South Africa | Asia | France | Belgium

A popular delicacy in South Africa, approximately 9.5 billion worms are harvested each year

12 g
2.6 g

### Cricket
*Gryllidae*

Cambodia | Thailand | Africa | America

Crickets taste like shrimp. Roasted crickets make a crunchy snack. They also can be added to cookies; see www.insectsarefood.com/recipes.php?title=chocolate-chirp-cookies&more=1&c=1&tb=1&pb=1.

9.6 g
5.6 g

| Protein in 100 g | | Fat in 100 g |
|---|---|---|
| 27 g | Steak sirlon 85 g | 4 g |
| 25 g | Pork 85 g | 7 g |
| 26 g | Chicken 85 g | 4 g |
| 26 g | Tuna 85 g | 1 g |
| 9 g | Tofu 115 g | 5 g |
| 8 g | Milk 1 Cup | 5 g |
| 6 g | Egg 1 | 5 g |

### Wasp larvae
*Vespidae*

Mexico | Thailand | Laos | Japan | South America

Yellow jacket wasp larvae are a delicacy in Japan.

17 g
3.6 g

### Termite
*Termitoidae*

Africa | Indonesia

Roasted termites are popular in Kenya.

6 g
7 g

### Weevil
*Curculionoidea*

Southern Asia | South America | Africa

Barbecued weevils make great snacks.

7 g
11 g

### Locust
Africa | China | America

Swarms of locusts make it quick and easy to harvest these protein rich insects.

18 g
21.5 g

Find an insect in your kitchen and chances are you'll squash it, not eat it. But should you hold off on squashing that little bug and consider serving it for dinner instead? Entomophagy, or simply, the consumption of insects, isn't a new phenomenon, but it's not quite mainstream in Western culture, either. Almost 2000 species of insects are included in the traditional diets of 2.5 billion people worldwide.[9] Beetles, caterpillars, bees, wasps, and ants account for almost 75% of the insects eaten for meals and snacks.

With the growing concern about global food security, exploring the idea of adding edible insects to our menus should not be overlooked. Consider that edible insects raised for food have smaller environmental footprints than beef, pork, and poultry. Plus, they are very versatile. Edible insects can be consumed whole, chopped, or ground into flour to make breads and cookies.

Adding edible insects to your meal plan takes an open mind and adventurous palate but can offer some great nutritional value. Insects are nutrient dense. They are rich in protein, amino acids, fat, and carbohydrates, along with vitamins and minerals. Their crunchy exoskeletons are a good source of insoluble fiber in the form of chitin.[10]

You may be reluctant to order a cricket appetizer, but be aware that you probably already have eaten some insects this year. In fact, the FDA estimates that we eat close to 2 pounds of insects or insect parts each year—mainly (and allowed by FDA standards because they present no health hazards) in vegetables, rice, beer, pasta, spinach, and broccoli.

Interested in trying out edible insects? You can find them at some ethnic grocery stores, online, and even some adventurous restaurants. Learn more about edible insects and find recipes online at edibleinsects.com.

## CASE STUDY

©Fancy Collection/ SuperStock RF

Bethany is a college freshman. She lives in a campus residence hall and teaches aerobics in the afternoon. She eats 2 or 3 meals a day in the dining hall and snacks between meals. Bethany decided to become a vegetarian after reading an article describing the health benefits of a vegetarian diet. Yesterday, her diet consisted of a café latte and Danish pastry for breakfast; a vegetarian tomato-rice dish, pretzels, and a diet soft drink for lunch; 2 cookies in the afternoon after aerobics class; and a vegetarian sub sandwich with 2 glasses of fruit punch for dinner. In the evening, she had a bowl of popcorn. What is missing from Bethany's current diet? How can she improve her new diet to meet her nutrient needs? What foods would Bethany need to include in her diet to increase her protein intake?

▶ Recent evidence indicates that eating protein often during the day (about 6 times) decreases body and abdominal body fat and increases lean body mass and thermogenesis more than when protein is eaten at 3 meals per day.[11]

## 7.4 Nitrogen Balance

Nitrogen balance is a method to determine protein needs. Healthy individuals who are not in periods of growth or recovering from illness or injury need to consume protein in an amount that replaces the protein lost in urine, feces, sweat, skin cells, hair, and nails. When protein intake equals the amount lost, protein balance (equilibrium) is maintained as long as energy intake is adequate to prevent the use of protein for energy.

When protein intake is less than what is lost, an individual is in negative protein balance (negative nitrogen balance). Negative protein balance often develops in individuals eating inadequate protein accompanied by a serious, untreated illness or injury (see Section 7.7) and in those with diseases that increase protein breakdown. For example, individuals with untreated acquired immune deficiency syndrome (AIDS) synthesize protein at rates similar to those of healthy people, but they often break down body protein at much higher rates. Over time, the increased rates of protein breakdown result in wasting of lean body mass. Negative protein balance eventually causes loss of proteins in skeletal muscles, blood, the heart, the liver, and other organs.

When protein intake is greater than losses, a state of positive protein balance is attained. During periods of growth and recovery from injury, trauma, or illness, positive protein balance is required to supply sufficient materials for building and repairing tissues. In addition, the hormones insulin, growth hormone, and testosterone all stimulate

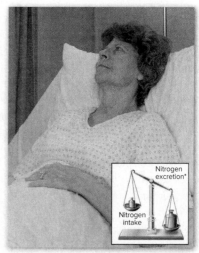

| **Positive Nitrogen Balance** | **Nitrogen Equilibrium** | **Negative Nitrogen Balance** |
|---|---|---|
| **Situations when positive nitrogen balance occur:**<br><br>Growth<br>Pregnancy<br>Recovery stage after illness/injury<br>Athletic training resulting in increased lean body mass<br>Increased secretion of certain hormones, such as insulin, growth hormone, and testosterone | **Situation when nitrogen equilibrium occurs:**<br><br>Healthy adult meeting protein and energy needs | **Situations when negative nitrogen balance occur:**<br><br>Inadequate intake of protein<br>Inadequate energy intake<br>Conditions such as fevers, burns, and infections<br>Bed rest (for several days)<br>Deficiency of essential amino acids (e.g., poor quality protein consumed)<br>Increased protein loss (as in some forms of disease)<br>Increased secretion of certain hormones, such as thyroid hormone and cortisol |

*Based on losses of urea and other nitrogen-containing compounds in the urine as well as protein losses in feces, skin, hair, nails, and other minor routes.

*Figure 7-10*   Determining nitrogen balance requires measuring nitrogen intake and loss.

Top, Left: ©Dave and Les Jacobs/Blend Images LLC; Middle: ©Rolf Bruderer/Getty Images RF

Legumes, such as cannellini beans, boost the protein content of meals.

©Corbis/Punchstock RF

protein synthesis for the building of new tissue. Merely eating more protein does not build additional body protein; an individual must be in a situation requiring positive nitrogen balance for this to occur.

Researchers and clinicians can measure dietary protein intake and body losses of protein to determine protein balance. Because nitrogen is a component of protein and can be more easily quantified, nitrogen, rather than protein, is measured. Figure 7-10 provides examples of the states of nitrogen (protein) balance. Nitrogen makes up approximately 16% of the weight of an amino acid (100/16 = 6.25). Therefore, nitrogen intake multiplied by 6.25 provides an estimate of protein intake:

$$\text{Nitrogen (g)} \times 6.25 = \text{Protein (g)}$$

Nitrogen balance studies are difficult to conduct because an accurate measure of all sources of nitrogen intake and loss over a 24-hour period is needed. Outside of hospital and research environments, this is not usually feasible. Thus, it is easier to calculate protein needs based on the RDA.

## Recommended Intakes of Protein

The RDAs for protein are listed in Appendix J. These guidelines provide recommendations for healthy individuals during periods of growth and development (infancy, childhood, pregnancy, lactation), as well as during normal adulthood. For most adults, the RDA for protein is 0.8 g/kg body weight. The recommended percentage of energy that should come from protein ranges from 10 to 35%. Healthy weight is used as a baseline for the RDA because excess fat storage doesn't contribute much to protein needs. As noted in the following equations, the RDAs for adults are 56 g/day for a 154 lb (70 kg) man and 46 g/day for a 125 lb (57 kg) woman.

Convert weight from pounds to kg:

$$\frac{154 \text{ pounds}}{2.2 \text{ pounds/kg}} = 70 \text{ kg (man)}$$

$$\frac{125 \text{ pounds}}{2.2 \text{ pounds/kg}} = 57 \text{ kg (woman)}$$

Calculate RDA:

70 kg × 0.8 g protein/kg = 56 g (man)
56 kg × 0.8 g protein/kg = 46 g (woman)

The RDA for protein does not include additional protein amounts needed during recovery from illness or injury or that might be required to support the needs of highly trained athletes.[3] During these conditions, protein needs can range from approximately 0.8 to 2.0 g/kg body weight.[12] Mental stress, physical labor, and routine weekend sports activities do not require an increase in the protein RDA.[3]

For many North Americans, protein needs are easily met through our typical diets. In fact, North Americans typically consume protein in amounts exceeding the RDA, equaling about 100 g of protein daily for men and 65 g daily for women.[3] Excess protein—whether from dietary sources of protein or amino acid supplements—cannot be stored as such, so the carbon skeletons from excess protein are metabolized for energy needs or used for other purposes.

The amount of protein fed to infants should closely follow RDA guidelines. At intakes in excess of the Adequate Intake, their kidneys may have difficulty excreting the large amounts of urea formed in metabolizing the protein. The quantity of protein in breast milk and formula is well matched to infant needs.

©Createas/PictureQuest RF

## Meeting Protein Needs When Dieting to Lose Weight

Your father has been gaining weight for the last 5 years. His physician has suggested that he lose 20 pounds to decrease his risk of heart disease and type 2 diabetes. You know that it will be important for your father to meet his protein needs as he tries to lose weight. Design a 1-day diet for him that contains about 1500 kcal with 15% of energy intake as protein. A nutrient analysis computer program, or a website (fdc.nal.usda.gov/) will provide some help. Does your diet plan meet the RDA for protein and follow MyPlate guidelines?

©David Buffington/Getty Images RF

*Perspective*
*on the Future*

Americans tend to get plenty of protein, but most of it is eaten at dinner. Emerging research indicates that distributing protein more evenly throughout the day increases satiety and maximizes muscle repair and synthesis. Aiming for 25 to 35 grams of protein at each meal is well within the protein AMDR and can be especially helpful for older adults who naturally lose muscle as part of aging, those on weight loss diets to help them preserve lean muscle during weight loss, and physically active individuals.[13]

**Knowledge Check**

1. When is it important to be in positive nitrogen balance?
2. What situations increase the risk of being in negative nitrogen balance?
3. During what stage of the life cycle are people generally in nitrogen equilibrium?

## 7.5  Protein Digestion and Absorption

For some foods, the first step in protein breakdown takes place during cooking. Cooking unfolds (denatures) proteins and softens the tough connective tissues in meat. This can make many protein rich foods easier to chew and aids in breakdown during digestion and absorption in the GI tract.

The enzymatic digestion of protein begins in the stomach with the secretion of hydrochloric acid. Once proteins are denatured by stomach acid, **pepsin**, a major enzyme produced by the stomach, begins to break the long polypeptide chains into shorter chains of amino acids through hydrolysis reactions (Fig. 7-11). Pepsin does not completely separate proteins into amino acids because it can break only a few of the many peptide bonds found in these large molecules.

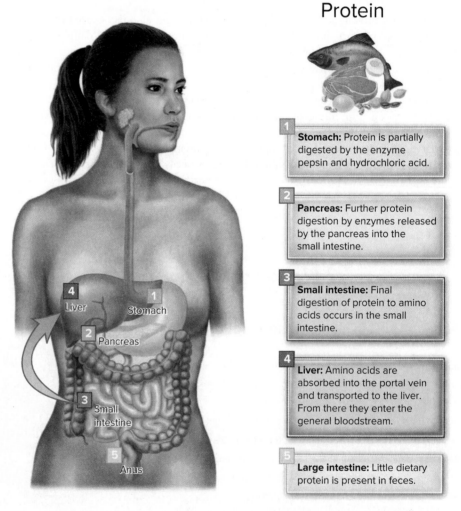

*Figure 7-11*  A summary of protein digestion and absorption. Enzymatic protein digestion begins in the stomach and ends in the absorptive cells of the small intestine, where the last peptides are broken down into single amino acids.

The release of pepsin is controlled by the hormone **gastrin**. Thinking about food or chewing food stimulates gastrin-producing cells of the stomach to release the hormone. Gastrin also strongly stimulates the stomach's parietal cells to produce acid, which aids in digestion and the activation of pepsin. Pepsin is actually stored as an inactive enzyme (called pepsinogen) to prevent it from digesting the stomach lining. Once pepsinogen enters the stomach's acidic environment (pH between 1 and 2), part of the molecule is split off, forming the active enzyme pepsin.

From the stomach, the partially digested proteins move with the rest of the nutrients and other substances from the meal (now called chyme) into the duodenum. In the small intestine, chyme triggers the release of the hormones secretin and cholecystokinin (CCK) from the walls of the small intestine. These, in turn, stimulate the pancreas to release the proteases (protein-splitting enzymes, including trypsin, chymotrypsin, and carboxypeptidase) into the small intestine. These enzymes digest the polypeptides into short peptides and amino acids that can be actively absorbed into the cells of the small intestine (Fig. 7-12). Any remaining short peptides are broken down to individual amino acids by peptidase enzymes. Amino acids then travel via the portal vein to the liver for use in protein synthesis, energy needs, conversion to carbohydrate or fat, or release into the bloodstream for transport to other cells.

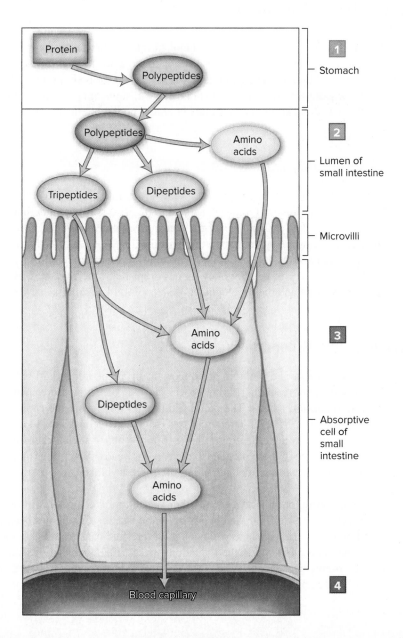

*Figure 7-12* Protein digestion takes place in the stomach 1, the lumen of the small intestine 2, and the absorptive cells of the small intestine 3. Absorption of amino acids from the lumen into the intestinal cells requires transporters. Most amino acids are transported by sodium-dependent transporters in an energy-requiring process (active absorption). Any remaining peptides get broken down to amino acids within the absorptive cell. The amino acids are then transported out of the cell and released into the bloodstream 4.

Except during infancy, it is uncommon for intact proteins to be absorbed from the digestive tract. However, in early infancy (up to 4 to 5 months of age), the gastrointestinal tract is somewhat permeable to small proteins, so some whole proteins can be absorbed. Because proteins from foods such as cow's milk and egg white may predispose an infant to food allergies, pediatricians and registered dietitians recommend waiting until an infant is 4 to 6 months of age or older before introducing common allergenic foods (see Clinical Perspective).[14]

### Knowledge Check

1. Name 4 enzymes that are involved in protein digestion and absorption.
2. What are the end products of polypeptide digestion?
3. How is the absorption of proteins different in early infancy?

## 7.6  Functions of Proteins

Proteins function in many crucial ways in metabolism and in the formation of essential compounds and structures (Fig. 7-13). Recall that the amino acids needed for the synthesis of proteins are supplied by the diet as well as by the recycling of body protein. However, if we do not eat adequate amounts of carbohydrate and fat, some amino acids will be used to produce energy, which makes them unavailable to build body proteins for other essential functions.

### Producing Vital Body Structures

One of the primary functions of protein is to provide structural support to body cells and tissues. The key structural proteins (collagen, actin, and myosin) make up more than a third of body protein and provide a matrix for muscle, connective tissue, and bone. During periods of growth, new proteins are synthesized to support the development of body tissues and structures. During malnutrition or disease, body proteins are often broken down to supply energy. Thus, the synthesis of protein for body tissues drops below normal rates, eventually resulting in protein wasting and the development of a condition known as kwashiorkor (see Section 7.7).

### Maintaining Fluid Balance

The blood proteins albumin and globulin are important in maintaining fluid balance between the blood and the surrounding tissue space. Normal blood pressure in the arteries forces blood into **capillary beds**. The blood fluid then moves from the capillary beds into the spaces between nearby cells (interstitial spaces) to provide nutrients to those cells (Fig. 7-14). Proteins such as albumin are too large to move out of the capillary beds into the tissues. But the presence of these proteins in the capillary beds attracts the right amount of fluid back to the blood, partially counteracting the force of blood pressure to maintain fluid balance.

When protein consumption is inadequate, the concentration of proteins in the blood eventually decreases. Excessive fluid then builds up in the surrounding tissues because the amount of blood protein is inadequate to pull enough of the fluid back from the tissues into the bloodstream. As fluid builds up in the interstitial spaces, the tissues swell with excess fluid, resulting in **edema**. This can be a sign of a serious medical condition, so it is important for physicians to determine the cause of edema.

---

### Critical | Thinking

Krista eats twice as much protein as her body needs. What happens to this extra protein?

---

**capillary beds** Minute blood vessels, 1 cell thick, that create a junction between arterial and venous circulation. Gas and nutrient exchange occurs here between body cells and the bloodstream. Figure A-6 in Appendix A provides a detailed view of a capillary bed.

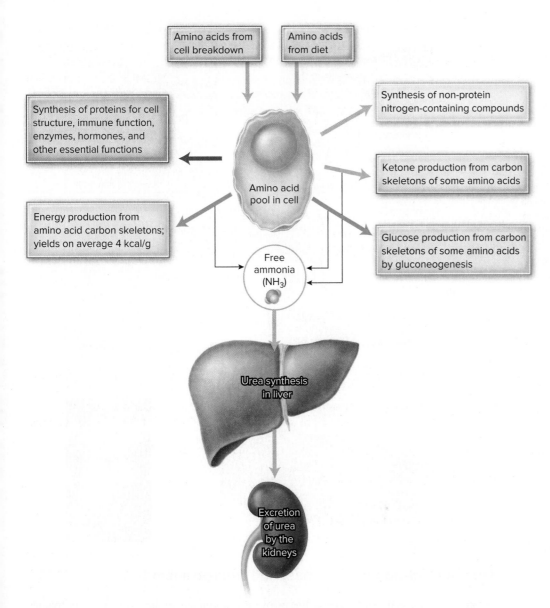

**Figure 7-13** The amino acid pool supplies amino acids for varied protein functions. The nitrogen-containing ammonia ($NH_3$) released during amino acid breakdown is converted to urea and excreted in the urine.

## Contributing to Acid-Base Balance

The acid-base balance in the body is expressed in terms of pH, which reflects the concentration of hydrogen ions [H]. A solution with a high hydrogen ion concentration has a low pH and is therefore more acidic, whereas a solution with a low hydrogen ion concentration has a high pH and is more alkaline (see Appendix B for additional information). Proteins play an important role in regulating acid-base balance and body pH. For example, proteins located in cell membranes pump chemical ions into and out of cells. The ion concentrations that result from the pumping action help keep the blood slightly alkaline (pH = 7.35–7.45). In this way, proteins act as **buffers**—compounds that help maintain acid-base balance within a narrow range. Proteins are especially good buffers for the body because they have negative charges, which attract positively charged hydrogen ions. This allows them to accept and release hydrogen ions as needed to prevent detrimental changes in pH.

**buffer** Compound that helps maintain acid-base balance within a narrow range.

*Figure 7-14* Role of protein in maintaining fluid balance. (*a*) Blood proteins help draw fluid forced into interstitial spaces by blood pressure back into the capillary bed. (*b*) Without sufficient protein in the bloodstream, there are too few blood proteins to counteract the force of blood pressure pushing fluid into the interstitial spaces between cells. When fluid between the cells builds up, it is called edema. Fluid then remains in the interstitial spaces between cells. (*c*) Examples of feet with edema. In some cases, applying pressure to the swollen area, such as in the photo on the right, causes an indentation that persists after the release of the pressure (sometimes called pitting edema).

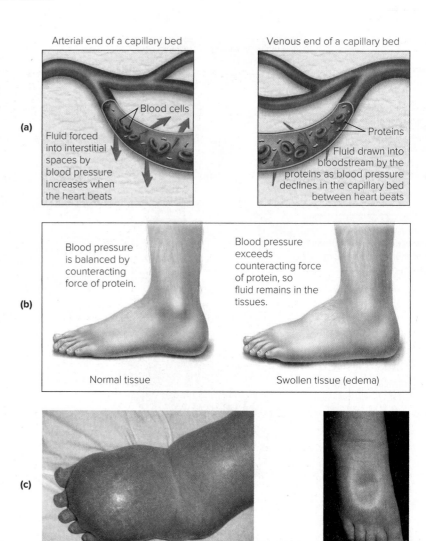

## Forming Hormones, Enzymes, and Neurotransmitters

Amino acids are required for the synthesis of most hormones in the body. Some hormones, such as the thyroid hormones, are made from only 1 amino acid, whereas others, such as insulin, are composed of many amino acids. Hormones act as messengers in the body and aid in regulatory functions, such as controlling the metabolic rate and the amount of glucose taken up from the bloodstream. Amino acids also are required for the synthesis of enzymes. Cells contain thousands of enzymes that facilitate chemical reactions fundamental to metabolism. Many neurotransmitters, released by nerve endings, also are derivatives of amino acids. This is true for dopamine (synthesized from the amino acid tyrosine), norepinephrine (synthesized from the amino acid tyrosine), and serotonin (synthesized from the amino acid tryptophan).

## Contributing to Immune Function

Antibody proteins are a key component of the immune system. Antibodies bind to foreign proteins (called antigens) that invade the body and prevent their attack on target cells. In a normal, healthy individual, antibodies are very efficient in combating these antigens to prevent infection and disease. However, without sufficient dietary protein, the immune system lacks the material needed to build this defense. Thus, immune incompetence (called **anergy**) develops and reduces the body's ability to fight infection. Anergy can turn measles into a fatal disease for a malnourished child. It also increases the risk of illness and infection in protein-deficient adults.

**anergy** Lack of an immune response to foreign compounds entering the body.

# Expert Perspective from the Field

## Nutrition and Immunity

The immune system is a complex network of organs, tissues, cells, and secretions that protects the body from foreign organisms (pathogens), such as bacteria, viruses, parasites, fungi, and toxins. When the body detects the presence of these "non-self" cells, or antigens, innate and acquired immune responses attempt to destroy the antigens.

**Innate (nonspecific) immunity** is present at birth and provides the first barrier of protection against invading antigens. Innate immunity includes *physical barriers,* provided by the skin and mucous membranes, that prevent access to the inside of the body; *chemical secretions,* such as the hydrochloric acid secreted by the stomach, that destroy antigens; *physiological barriers,* such as fever, that prevent the growth of antigens; and *phagocytic cells* that engulf and destroy antigens. Innate immunity provides a general, nonspecific response, as it has a limited ability to recognize antigens that have previously attacked the body.

**Acquired (specific) immunity** provides an immune response that is initiated by the recognition of a specific antigen. Acquired immunity develops over a person's lifetime. Acquired immunity also is referred to as *adaptive immunity* because, after exposure to an antigen, the immune system can recognize the antigen and adapt its response to it. When the acquired immune system is triggered, the bone marrow and thymus are stimulated to produce *antibodies (immunoglobulins)* and other specialized immune cells that destroy the specific antigens.

Because infants have limited acquired immunity at birth, nutrition experts recommend breastfeeding. Dr. Stephanie Atkinson* is one of many scientists whose research has shown that human milk has high concentrations of many protective immune components, such as immunoglobulins and lactoferrin (see Chapter 17). These immune components from the mother can be absorbed by infants and help protect them while their own immune systems are maturing. Although the immune components in human milk benefit all infants, for infants born in low-income countries where exposure to pathogens is much greater than in North America, it can be life saving.

Nutrition is an important part of maintaining both innate and acquired immunity. According to Dr. Atkinson, nutritional deficiencies can suppress the immune system's ability to prevent infection and disease. Malnutrition results in loss of immune tissue, decreased production of immune cells, decreased number and effectiveness of antibodies, and breakdown of physical barriers to antigens. This increases the risk of infection, disease, and death.

Severe protein-energy malnutrition (PEM) has profound effects on immune function. Dr. Atkinson indicated that this is of particular concern in infants and children with PEM because they have increased nutritional needs to support growth and often have simultaneous micronutrient deficiencies (e.g., zinc deficiency), infections, and

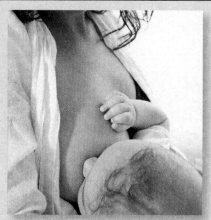

Compassionate Eye Foundation/Three Images/
Getty Images

diarrhea, which result in even greater impairment to immune responses. The severity of the nutrient deficiencies and the child's overall health determine whether the impairment can be reversed by supplementing the diet with the deficient nutrients.

Specific nutrients can increase immune protection during critical illness and trauma. For example, the amino acids arginine and glutamine are considered "immunomodulators" because they promote protein synthesis and immune responses during illness. Glutamine also is important in maintaining the integrity of the intestinal mucosa, thereby preventing bacteria in the GI tract from entering the bloodstream. Studies of omega-3 and omega-6 polyunsaturated fatty acids suggest that these essential fatty acids also may be immunomodulators. Specialized nutritional formulas can provide these key nutrients during times of increased nutritional need, such as during disease or injury.

The study of nutrition and immune function is a relatively new field. Thus, scientists' understanding of how nutritional deficiencies and interventions affect immune responses is far from complete. However, it is clear that maintaining optimal nutritional status is an important way to support immune function and reduce the risk of infection and disease.

*Stephanie Atkinson, Ph.D., is a Professor, Department of Pediatrics and Associate Member, Department of Biochemistry and Biomedical Sciences at McMaster University in Hamilton, Ontario, Canada. She directs an internationally recognized research program focused on pediatric nutrition. Among the many honors she has received are a Career Scientist Award from the Ministry of Health in Ontario, the McHenry Award from the Canadian Society for Nutritional Sciences, and a distinguished service award from the Dietitians of Canada.*

## Transporting Nutrients

Many proteins function as transporters for other nutrients, carrying them through the bloodstream to cells and across cell membranes to sites of action. For instance, the protein hemoglobin carries oxygen from the lungs to cells. Lipoproteins transport large lipid molecules from the small intestine through the lymph and blood to body cells. Some vitamins and minerals also have specific protein carriers that aid in their transport into and out of tissues and storage proteins. Examples include retinol-binding protein (a carrier protein for vitamin A), transferrin and ferritin (carrier and storage proteins, respectively, for iron), and ceruloplasmin (a carrier protein for copper).

## Forming Glucose

The body must maintain a fairly constant concentration of blood glucose to supply energy, especially for red blood cells, brain cells, and other nervous tissue cells that rely almost exclusively on glucose for energy. If carbohydrate intake is inadequate to maintain blood glucose levels, the liver (and kidneys, to a lesser extent) is forced to make glucose from the amino acids present in body tissues (see Fig. 7-13). This process is called gluconeogenesis (see Chapter 9).

Making glucose from amino acids is a normal backup system the body utilizes to supply needed glucose. For example, when you skip breakfast and haven't eaten since 7 P.M. the night before, glucose must be synthesized from amino acids. However, when this occurs chronically, as in starvation, the conversion of amino acids into glucose results in the development of widespread muscle wasting in the body (called cachexia).

## Providing Energy

Proteins supply very little energy for healthy individuals. Under most conditions, body cells use primarily fats and carbohydrates for energy. Although proteins and carbohydrates contain the same amount of usable energy—on average, 4 kcal/g—proteins are a very costly source of energy, considering the amount of metabolism and processing the liver and kidneys must perform to convert protein into an energy source (see Fig. 7-13).

### Knowledge Check

1. What are 3 functions of proteins?
2. How do proteins help maintain fluid balance?
3. How do proteins contribute to immune function?

# 7.7 Health Concerns Related to Protein Intake

Many people living in low-income countries suffer from malnutrition and disease because dietary protein supplies are limited.[15] In contrast, the residents of wealthier countries tend to eat more protein than they need and may boost their intake even higher by consuming protein or amino acid supplements.[3] As you know, getting sufficient amounts of protein is required for good health, but getting too little or too much can have serious health consequences.[15,16]

**HISTORICAL PERSPECTIVE**

Flaky Paint Skin and Bloated Bellies

For centuries, kwashiorkor has sickened and killed millions of young children. Its cause remained elusive until Cicely Williams carefully observed and listened to parents in western Africa, where she was working in the 1930s. She discovered that this condition is the result of severe malnutrition that occurs when toddlers are weaned from breast milk, which is rich in protein and other nutrients, to a nutrient-poor, starchy gruel. This physician's description of kwashiorkor is still valid today. Learn more at apps.who.int/iris/handle/10665/72346.

©Centers for Disease Control and Prevention/Dr. Lyle Conrad

## Protein-Energy Malnutrition

Protein deficiency rarely develops as an isolated condition. It most often occurs in combination with a deficiency of energy (and other nutrients) and results in a condition known as **protein-energy malnutrition (PEM)**, or protein-calorie malnutrition (PCM). In many low-income areas of the world where diets are often low in protein and energy, PEM is a very serious public health concern. Although PEM can affect people of all ages, its most devastating consequences are seen in children. Without adequate protein and energy, children fail to grow normally, and many develop diarrhea, infections, and diseases and die early in life. Of the 156 million children who are malnourished worldwide, 50 million experience wasting, and 3 million die each year.[15]

PEM usually occurs as either marasmus or kwashiorkor. These conditions differ in the severity of the overall energy and protein deficit and the related clinical characteristics (Fig. 7-15). **Marasmus** develops slowly from a severe deficiency of energy (and, in turn, protein and micronutrients). Over time, this leads to extreme weight loss, muscle and fat loss, and growth impairment. **Kwashiorkor** occurs more rapidly in response to a severe protein deficit, typically accompanied by underlying infections or disease. Kwashiorkor is characterized by edema, mild to moderate weight loss, growth impairment, and the development of a fatty liver (excess accumulation of fat in the liver).

PEM is most prevalent in parts of Africa, Southeast Asia, Central America, and South America.[1] However, it also is seen in some population groups in industrialized countries, such as the U.S. Those at greatest risk are individuals living in poverty and/or isolation and those with substance abuse problems, anorexia nervosa, or debilitating diseases (e.g., AIDS or cancer). Some hospitalized patients also are at increased risk of PEM because of poor prior health, low dietary intakes, and increased protein needs for recovery from surgery, trauma, and/or disease. Malnourished patients face a much greater risk of complications and even death. Consequently, hospitals have developed nutrition support teams to ensure appropriate nutritional care for at-risk patients.

**protein-energy malnutrition (PEM)** Condition resulting from insufficient amounts of energy and protein, which eventually results in body wasting and increased susceptibility to infections.

**marasmus** Condition that results from a severe deficit of energy and protein, which causes extreme loss of fat stores, muscle mass, and body weight.

**kwashiorkor** Condition occurring primarily in young children who have an existing disease and consume a marginal amount of energy and severely insufficient protein. It results in edema, poor growth, weakness, and an increased susceptibility to further infection and disease.

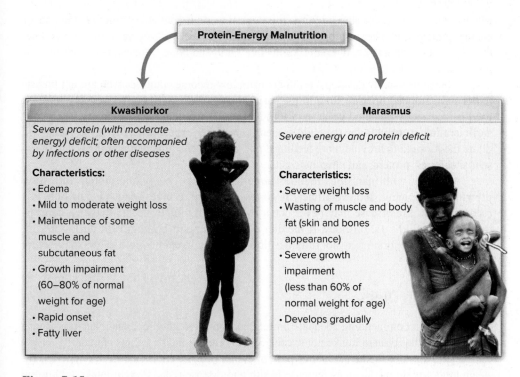

*Figure 7-15* Classification of undernutrition in children.
Left: Christine Osborne Pictures/Alamy Stock Photo; Right: ©Peter Turnley/Corbis/VCG via Getty Images

Some hospitalized patients are at risk of protein-energy malnutrition (PEM) because of poor dietary intakes and increased needs for recovery from surgery, trauma, or disease.

©McGraw-Hill Education

### Kwashiorkor

*Kwashiorkor* is a word from Ghana that means "the disease that the first child gets when the new child comes." From birth, an infant in low-income areas of the world is usually breastfed. Often, by the time the child is 12 to 18 months old, the mother is pregnant or has already given birth again. The mother's diet is usually so marginal that she cannot produce sufficient milk to continue breastfeeding the older child. This child's diet then abruptly changes from nutritious human milk to starchy roots and gruels. These foods have low protein densities compared with their energy content. Additionally, the foods are usually high in plant fibers and bulk, making it difficult for the child to consume enough to meet energy needs and nearly impossible to meet protein needs. Many children in these areas also have infections and parasites that elevate protein and energy needs and often precipitate the development of kwashiorkor.

The presence of edema in a child who still has some subcutaneous fat is the hallmark of kwashiorkor (see Fig. 7-15). Other major symptoms of kwashiorkor are apathy, diarrhea, listlessness, failure to grow and gain weight, infections, and withdrawal from the environment. These symptoms also complicate other diseases that may be present. For example, measles, a disease that normally makes a healthy child ill for only a week or so, can become severely debilitating and even fatal in a child with kwashiorkor.

Many symptoms of kwashiorkor can be explained based on our knowledge of proteins. Proteins play important roles in fluid balance, growth, immune function, and the transport of other nutrients. Thus, protein deficiency can severely compromise these functions.

If children with kwashiorkor are helped in time—infections are treated and a diet ample in protein, energy, and other essential nutrients is provided—the disease process often reverses and they begin to grow again. Unfortunately, by the time many of these children reach a hospital or care center, they already have severe infections. Thus, despite good medical care, many still die. Those who survive often continue to battle chronic infections and diseases.

### Marasmus

Marasmus is the result of chronic PEM. It is caused by diets containing minimal amounts of energy, protein, and other nutrients. The word *marasmus* means "to waste away." Over time, the severe lack of energy and protein results in a "skin and bones" appearance, with little or no subcutaneous fat (see Fig. 7-15).

Marasmus usually develops in infants from low-income countries who are not breastfed or have stopped breastfeeding in the early months of life. Often, the weaning formula used is incorrectly prepared because of unsafe water and because the parents cannot afford sufficient infant formula for the child's needs. The latter problem may lead the parents to dilute the formula to provide more feedings, not realizing that this deprives the infant of essential calories, protein, and other nutrients.

An infant with marasmus requires large amounts of energy and protein to restore growth, development, and overall health. Unless the child receives adequate nutrition, full recovery from the disease may never occur. Most brain growth occurs between conception and the child's 1st birthday. If the diet does not support brain growth during the 1st months of life, the brain may not fully develop, resulting in poor cognitive and intellectual growth.

## High Protein Diets

In addition to recommending adequate protein consumption, the Food and Nutrition Board also suggests that protein intake not exceed 35% of energy intake.[3] Diets containing an excessive or disproportionate amount of protein do not provide additional health benefits. Instead, high protein intakes may increase health and disease risks. Protein intake within the range of recommended intakes is linked with with normal kidney function in healthy individuals.[17,18] However, in those with impaired kidney function, high protein diets may overburden the kidneys' capacity to excrete excess nitrogen as urea. A lower protein diet with adequate fluid intake is recommended for these individuals to help preserve kidney health.[18]

When excess protein is primarily from a high intake of animal proteins, the overall diet is likely to be low in plant-based foods and consequently low in fiber, some vitamins (vitamins C and E and folate), minerals (magnesium and potassium), and beneficial phytochemicals. Animal proteins are often rich in saturated fat and cholesterol and may be cured with sodium-containing compounds to create hot dogs, ham, salami, and luncheon meats. As a result, these diets can increase the risk of cardiovascular disease and cancer.[5,6,8,19] Although very high intakes of meat, particularly processed meat, are associated with increased risk of cardiovascular disease, a more modest intake of meat consumed as part of a varied diet containing healthy amounts of unsaturated fatty acids appears to have little effect on heart disease.[20]

High protein diets also may increase urinary calcium loss and eventually lead to a loss of bone mass and an increased risk of osteoporosis.[5] These findings are somewhat controversial, however, and are less of a concern for individuals with adequate calcium intakes.

Other concerns, particularly with athletes, are the health risks associated with excess protein and amino acid supplementation. As described earlier, our bodies are designed to obtain amino acids from dietary sources of whole proteins. This assures a supply of amino acids in proportions needed for body functions and prevents amino acid toxicity, especially for methionine, cysteine, and histidine—the most toxic amino acids.[15] When individual amino acid supplements are taken, chemically similar amino acids can compete for absorption, resulting in amino acid imbalances and toxicity risk.

### Knowledge Check

1. How does kwashiorkor differ from marasmus?
2. Which individuals might be at greatest risk of PEM in the U.S.?
3. What areas of the world have the highest incidence of PEM?
4. Why might excessive protein intake be harmful?

## 7.8 Vegetarian Diets

Vegetarianism has evolved over the centuries from a necessity into an option. Today, approximately 2.5% of adults in the U.S. and 4% of adults in Canada follow a vegetarian diet. Additionally, 20 to 25% of Americans report that they eat at least 4 meatless meals a week.[5] Most people choose vegetarian diets for religious, philosophical, ecological, or health-related reasons. For example, Hindus, Seventh-Day Adventists, and Trappist monks follow vegetarian diets as a part of their religious practices. Others adopt vegetarian practices because they are concerned about the economic and ecological impact of eating meat-based diets. They recognize that meat is not an efficient way of obtaining protein because it requires the use of approximately 40% of the world's grain production to raise meat-producing animals. Diets rich in fruits, vegetables, legumes, and grains frequently result in increased intakes of antioxidant nutrients (e.g., vitamins C and E and carotenoids), dietary fiber, and healthful phytochemicals and decreased intakes of saturated fat and cholesterol. Vegetarianism also may offer protection against obesity. Thus, the American Cancer Society, the World Cancer Research Fund, the American Heart Association, and the Heart and Stroke Foundation of Canada encourage plant-based diets to promote health and reduce risk of chronic disease.[5,29,30,31,32,33]

The amino acids in legumes are best used when combined with nuts, seeds, or grains.

Pixtal/age fotostock

## Food Protein Allergies

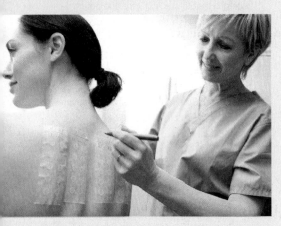

People with hypersensitivity to certain foods can be tested to determine which food allergens cause their symptoms.

Science Photo Library/Getty Images

Allergies, including food allergies, involve responses of the immune system designed to eliminate foreign proteins (antigens). Food allergy responses occur when the body mistakenly reacts to a food as though it were a harmful invader. In some people, certain food components, typically proteins (called **allergens**), cause hypersensitivity reactions and trigger this response. These allergens stimulate white blood cells to produce antibodies (mostly, the **immunoglobulin** IgE) that bind to antigens and cause the symptoms associated with an allergic reaction.[21]

Fortunately, most allergic reactions are mild, such as a runny nose, sneezing, itching skin, hives, or digestive upset (indigestion, nausea, vomiting, diarrhea). For those who are severely allergic, exposure to the allergenic food may cause a generalized, life-threatening reaction involving all body systems (known as **anaphylaxis** or anaphylactic shock). Anaphylaxis causes decreased blood pressure and respiratory distress so severe that the person cannot breathe—death will occur

without immediate medical help. In the U.S., allergic reactions result in 200,000 emergency room visits and 150 to 200 deaths per year.

The protein in any food can trigger an allergic reaction. However, 8 foods account for 90% of all food allergies: peanuts, tree nuts (e.g., walnuts and cashews), milk, eggs, fish, shellfish, soy, and wheat (Fig. 7-16). Other foods frequently identified as causing allergic reactions are sesame seeds, meat and meat products, fruits, and cheese.

The only way to prevent allergic reactions is to avoid foods known to trigger reactions. Carefully reading food labels and asking questions when eating out are essential, perhaps life-saving, steps for those with food allergies.[21] In addition, individuals preparing foods at home or in restaurants need to know their menu ingredients and take steps to ensure that foods that cause an allergic reaction in a person do not come in contact with the food to be served to that individual. Even trace amounts of an allergen can cause a reaction. To prevent cross-contact, anything that will be used to prepare an allergen-free meal (e.g., hands,

---

**allergen** Substance (e.g., a protein in food) that induces a hypersensitive response, with excess production of certain immune system antibodies. Subsequent exposure to the same protein leads to allergic symptoms.

**immunoglobulins** Proteins (also called antibodies) in the blood that are responsible for identifying and neutralizing antigens, as well as pathogens that bind specifically to antigens.

| Peanut/tree nuts | Milk products | Soy |
|:---:|:---:|:---:|
|  |  |  |

| Wheat | Eggs | Fish/shellfish |
|:---:|:---:|:---:|
|  |  |  |

**Figure 7-16** These foods account for the vast majority of all food allergies.

All: Photos courtesy of Dennis Gottlieb

workspace, pans, utensils, plates) should be washed thoroughly before preparing the allergen-free meal. Unlike foodborne illness pathogens, such as bacteria and viruses, cooking an allergenic food often does not render its allergens harmless.[22]

The prevalence of food allergies has increased in the last 30 years.[21,22] Although difficult to estimate, it appears that approximately 5 to 8% of children and 2 to 4% of adults have food allergies.[21,22] It is unclear why some people develop allergies and what steps might help decrease the risk of developing food allergies. Most research indicates that maternal dietary restrictions do not play a significant role in preventing food allergies in children.[21] After a child is born, the following steps may help prevent food allergies.[14] These guidelines are especially important for families with a history of any type of allergy.

• Feed babies only breast milk or infant formula until they are at least 4 to 6 months old.
• Consult a pediatrician before giving infants and children highly allergenic foods (e.g.,

thinned peanut butter, eggs, milk). If an allergy is diagnosed, children must avoid the problematic foods. For children not at high risk for food allergies, when they are able to tolerate some solid foods (between 4 and 6 months of age), introducing small amounts of potentially allergenic foods may help prevent allergies.[23]

Many young children with food allergies outgrow them.[21,24] Thus, parents should not assume that the allergy will be long-lasting. Allergies to certain foods (e.g., milk, egg, soy, wheat) are more likely to be outgrown than are allergies to other foods (e.g., peanuts, tree nuts, fish, shellfish).[21] Those with allergies may be tested by physicians periodically to determine whether they have outgrown the allergy. If so, the food(s) can be reintroduced into the diet and eaten safely.

▶ Food allergies and food intolerances are not the same. Food allergies cause an immune response as a result of exposure to certain food components, typically proteins. In contrast, food intolerances (see Chapter 4) are caused by an individual's inability to digest certain food components, usually due to low amounts of specific enzymes. Generally, larger amounts of an offending food are required to produce the symptoms of food intolerance than to trigger allergic symptoms. Food allergies tend to be far more life threatening than food intolerances.

▶ The American Academy of Allergy and Immunology has a toll-free number (800-822-2762) to answer questions about food allergies and help direct people to specialists who treat allergy problems. Free information on food allergies is available by contacting Food Allergy Research & Education at www .foodallergy.org.

## Knowledge Check

1. What are the symptoms of food allergies?
2. Which foods cause most food allergies?
3. What steps can parents take to help prevent food allergies in their children?

# GL BAL PERSPECTIVE

## How Big Is Your Food Print?

Growing evidence indicates that what we eat may affect not only our personal health but also that of the environment. The world population is projected to increase to over 9 billion by 2050. The Food and Agricultural Organization (FAO) projects that food and feed production will need to increase by 70% to adequately feed the world's population. Many scientists believe that meat rich diets and the agricultural practices that support the production of food for these diets negatively affect the environment. For instance, producing food for nonvegetarian diets (especially beef-based diets) uses more water, fossil fuel energy, and acres of farmland than producing food for vegetarian diets.[25] Meat rich diets also cause greater emissions of greenhouse gases, such as carbon dioxide, methane, and nitrous oxide, which are associated with global warming.[26] Scientists are concerned that continued population growth may, in turn, decrease agricultural productivity, reduce farmers' incomes, and increase global food insecurity.[27]

Not all scientists agree with these findings and concerns, however. Some believe that consuming a low-fat vegetarian diet with some dairy products and/or meat may actually increase land use efficiency, thereby protecting environmental resources and promoting food security.[28] They point out that high quality farmland is required to grow fruits, vegetables, and grains, whereas meat and dairy products can be produced on the more widely available, lower quality land. Even though diets containing meat use more land, they can feed more people because of the greater availability of lower quality farmland. It appears that diets have different "agricultural land footprints," depending on the amount of plant-based and animal-based food they contain. Supporters of mixed animal/vegetable–based diets point out that vegetarian diets often include tofu and other meat substitutes produced from soy, chickpeas, and lentils. Many meat substitutes are highly processed and require energy-intensive production methods. Thus, including small amounts of meat may offer both environmental and nutritional benefits.

**Table 7-4  Food Plan for Vegetarians Based on MyPlate**

| Food Group | MyPlate Servings | | Key Nutrients Supplied‡ |
| | Lacto-vegetarian* | Vegan† | |
| --- | --- | --- | --- |
| Grains | 5 | 6 | Protein, thiamin, niacin, folate, vitamin E, zinc, magnesium, iron, fiber |
| Protein foods (beans, nuts, seeds) | 5 | 5 | Protein, vitamin B-6, zinc, magnesium, fiber |
| Vegetables | 2 (include 1 dark green daily) | 2½ (include 1 dark green daily) | Vitamin A, vitamin C, folate, vitamin K, potassium, magnesium |
| Fruit | 1½ | 1½ | Vitamin A, vitamin C, folate |
| Milk | 3 | — | Protein, riboflavin, vitamin D, vitamin B-12, calcium |
| Fortified soy milk | — | 3 | Protein, riboflavin, vitamin D, vitamin B-12, calcium |

*This plan contains about 75 grams of protein in 1650 kcal.
†This plan contains about 79 grams of protein in 1800 kcal.
‡One serving of vitamin- and mineral-enriched ready-to-eat breakfast cereal is recommended to fill possible nutrient gaps. Alternatively, a balanced multivitamin and mineral supplement can be used.

Vegetarian versions of common foods are becoming more readily available in groceries and restaurants.

Pizza: Ingram Publishing/Alamy Stock Photo; Burger: Ingram Publishing/Alamy

The eating patterns of vegetarians can vary considerably, depending on the extent to which animal products are excluded. Vegans follow the most restrictive diet, as they eat only plant foods. Because they do not eat any animal foods, their diets may be low in high biological value protein, riboflavin, vitamin D, vitamin B-12, calcium, and zinc unless carefully planned.[5] Lacto-vegetarians are similar to vegans because their diets exclude meat, poultry, eggs, and fish but differ in that they include dairy products in their diets. Lacto-ovo-vegetarians include eggs in their diets but avoid meat, poultry, and fish. These last 2 groups eat some animal foods, so their diets often contain ample amounts of nutrients that may be low or missing in strictly plant-based diets. However, to reduce the risk of nutrient deficiencies, all vegetarians need to follow nutritional recommendations (Table 7-4) when making daily food choices.[7,34,35]

Vegetarian diets require knowledge and creative planning to yield high quality protein and other key nutrients without animal products. Earlier in this chapter, you learned about complementary proteins, whereby the essential amino acids deficient in 1 protein source are supplied by those of another consumed at the same meal or the next. Recall that many legumes are deficient in the

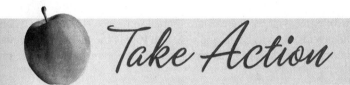

## Take Action

## Protein and the Vegan

Alana is excited about the possible health benefits of her new vegan diet. However, she is concerned that her diet may not contain enough protein, vitamins, and minerals. Use a nutrient analysis computer program, or a website (fdc.nal.usda.gov/) to calculate her protein intake and see if her concerns are valid.

andresr/Getty Images

| Breakfast | Protein (g) |
|---|---|
| Calcium-fortified orange juice, 1 cup | |
| Soy milk, 1 cup | |
| Fortified bran flakes, 1 cup | |
| Banana, medium | |

**Snack**

Calcium-enriched granola bar

**Lunch**

| | |
|---|---|
| Garden veggie burger, 4 oz | |
| Whole-wheat bun | |
| Mustard, 1 tbsp | |
| Soy cheese, 1 oz | |
| Apple, medium | |
| Green leaf lettuce, 1½ cups | |
| Peanuts, 1 oz | |
| Sunflower seeds, ¼ cup | |
| Tomato slices, 2 | |
| Mushrooms, 3 | |
| Vinaigrette salad dressing, 2 tbsp | |
| Iced tea | |

Alana's diet contained 2150 kcal, with _____ g (you fill in) of protein (is this adequate for her?), 57 g of total dietary fat (only 9 g of which came from saturated fat), and 50 g of fiber. Her vitamin and mineral intake with respect to those of concern to vegetarians—vitamin B-12, vitamin D, calcium, iron, and zinc—met her needs.

**Dinner**

Kidney beans, ½ cup
Brown rice, ¾ cup
Fortified margarine, 2 tbsp
Mixed vegetables, ¼ cup
Hot tea

**Snack**

Strawberries, ½ cup
Angel food cake, 1 small slice
Soy milk, ½ cup

Total protein: _____

## Critical Thinking

Landon, a new vegetarian, has been complaining of hair loss, stomachaches, and long healing time of cuts and bruises. Explain in metabolic terms why he might have these symptoms.

**Vegetarianism Websites**

www.ivu.org

www.vegetariannutrition.net

essential amino acid methionine, whereas cereals are limited in lysine. Thus, eating a combination of legumes and cereals, such as beans and rice, will supply the body with adequate amounts of all essential amino acids. Variety is an especially important characteristic of a nutritious vegan diet.[35]

At the forefront of nutritional concerns for vegetarians are riboflavin, vitamins D and B-12, calcium, iron, and zinc.[5,36,37] A major source of riboflavin, vitamin D, and calcium in the typical North American diet is milk, which is omitted from the vegan diet. However, riboflavin can be obtained from green leafy vegetables, whole-grain breads and cereals, yeast, and legumes—components of most vegan diets. Alternate sources of vitamin D include fortified foods (e.g., soy milk) and dietary supplements, as well as regular sun exposure (see Part 4).

Calcium-fortified foods are the vegan's best option for obtaining calcium. These include fortified soy milk, fortified orange juice, calcium rich tofu (check the label), and certain ready-to-eat breakfast cereals, breads, and snacks. Green leafy vegetables also contain calcium, but it is not well absorbed.[38] Dietary supplements provide another option for meeting calcium needs (see Part 4). It is important to read supplement labels and to plan supplement use carefully because a typical multivitamin and mineral supplement supplies only 25 to 45% of daily calcium needs.

Vitamin B-12 occurs naturally only in animal foods. Plants can contain soil or microbial contaminants that provide trace amounts of vitamin B-12, but these are negligible sources of the vitamin. Therefore, vegans need to eat food fortified with vitamin B-12 or take supplements to protect against deficiency.[36]

To obtain iron, vegans can consume whole-grain breads and cereals, dried fruits and nuts, and legumes.[5,35] The iron in these foods is not absorbed as well as the iron in animal foods, but eating a good source of vitamin C with these foods enhances iron absorption (see Part 4).

Vegans can obtain zinc from whole-grain breads and cereals, nuts, and legumes. However, phytic acid and other substances in these foods limit zinc absorption. Grains are most nutritious when consumed as breads because the leavening (rising of the bread dough) reduces the influence of phytic acid.[5]

## Special Concerns for Infants and Children

Infants and children are at highest risk of nutrient deficiencies as a result of poorly planned vegetarian diets.[5] However, with the use of complementary proteins and good sources of the problem nutrients discussed earlier, the energy, protein, vitamin, and mineral needs of vegetarian and vegan infants and children can be met. The most common nutritional concerns for infants and children following vegetarian and vegan diets are deficiencies of iron, vitamin B-12, vitamin D, zinc, and calcium.[5,7,37]

Vegetarian and vegan diets tend to be high in bulky, high fiber, low calorie foods that cause fullness. Although this side effect can be a welcome advantage for adults, children have a small stomach volume and relatively high nutrient needs for their size and may feel full before their energy needs are met. For this reason, the fiber content of a child's diet may need to be decreased by replacing high fiber sources with some refined grain products, fruit juices, and peeled fruit. Including concentrated sources of energy, such as fortified soy milk, nuts, dried fruits, and avocados, can help meet calorie and nutrient needs.

Overall, vegetarian and vegan diets can be appropriate during infancy and childhood. However, to achieve normal growth and ensure adequate intake of all nutrients, these diets must be implemented with knowledge and, ideally, professional guidance.[5,35]

*Knowledge Check*

1. How does a vegan diet differ from a lacto-vegetarian diet?
2. Which nutrients are likely to be low in a vegan diet?
3. What are 2 nutritional risks for children on vegetarian diets?

## CASE STUDY FOLLOW-UP

©Fancy Collection/
SuperStock RF

Bethany's dietary intake for this day is not as healthy as it could be because it does not come close to following recommendations. Many of the components of a healthy vegetarian diet—whole grains, nuts, soy products, beans, 2 to 4 servings of fruit, and 3 to 5 servings of vegetables per day—are missing. With so few fruits and vegetables, her diet also is low in the many phytochemicals that may provide numerous health benefits. It is apparent that Bethany has not yet learned to implement the concept of complementary proteins, so the quality of the protein in her diet is low. Unless she makes a more informed effort at diet planning, Bethany will not reap the health benefits she had hoped for when she chose to follow a vegetarian diet.

# Chapter Summary

## 7.1 Amino acids, the building blocks of proteins, contain a usable form of nitrogen for humans.
Of the 20 amino acids needed by the body, 9 must be provided in the diet (essential). The other 11 can be synthesized by the body (nonessential) from amino acids in the body's amino acid pool. High quality, also called complete, proteins contain ample amounts of all 9 essential amino acids. Foods derived from animal sources provide high quality protein. Lower quality, or incomplete, proteins lack sufficient amounts of 1 or more essential amino acids. This is typical of plant foods. Different types of plant foods eaten together often complement each other's amino acid deficits, thereby providing high quality protein in the diet.

## 7.2 Individual amino acids are linked together to form proteins.
The sequential order of amino acids determines the protein's ultimate shape and function. This order is directed by DNA in the cell nucleus. Diseases, such as sickle-cell anemia, can occur if the amino acids are incorrect in a polypeptide chain. When the 3-dimensional shape of the protein is unfolded (denatured) by treatment with heat, acid or alkaline solutions, or other processes, the protein also loses its biological activity.

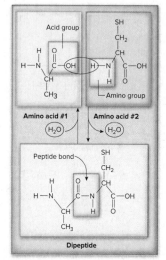

## 7.3 Almost all animal products are rich sources of protein.
The high quality of these proteins means that they can be easily converted into body proteins. Legumes, nuts, seeds, and grains are good sources of plant protein. Protein quality can be measured by determining the extent to which the body can

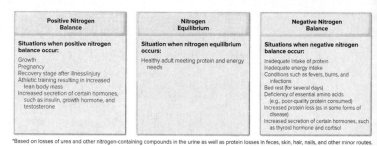

| Positive Nitrogen Balance | Nitrogen Equilibrium | Negative Nitrogen Balance |
| --- | --- | --- |
| **Situations when positive nitrogen balance occur:**<br>Growth<br>Pregnancy<br>Recovery stage after illness/injury<br>Athletic training resulting in increased lean body mass<br>Increased secretion of certain hormones, such as insulin, growth hormone, and testosterone | **Situation when nitrogen equilibrium occurs:**<br>Healthy adult meeting protein and energy needs | **Situations when negative nitrogen balance occur:**<br>Inadequate intake of protein<br>Inadequate energy intake<br>Conditions such as fevers, burns, and infections<br>Bed rest (for several days)<br>Deficiency of essential amino acids (e.g., poor-quality protein consumed)<br>Increased protein loss (as in some forms of disease)<br>Increased secretion of certain hormones, such as thyroid hormone and cortisol |

*Based on losses of urea and other nitrogen-containing compounds in the urine as well as protein losses in feces, skin, hair, nails, and other minor routes.

retain the nitrogen contained in the amino acids absorbed. This is called biological value (BV). In addition, the balance of essential amino acids in a food can be compared with an ideal pattern to determine the food's chemical score. When multiplied by the degree of digestibility, the chemical score yields the Protein Digestibility Corrected Amino Acid Score (PDCAAS).

## 7.4 The adult RDA for protein is 0.8 g per kg of healthy body weight.
For a typical 154 lb (70 kg) person, this corresponds to 56 g of protein daily; for a 125 lb (57 kg) person, this corresponds to 46 g/day. The North American diet generally supplies plenty of protein: men typically consume about 100 g of protein daily, and women consume about 65 g. These intakes also are of sufficient quality to support body functions.

## 7.5 Protein digestion begins in the stomach, where proteins are broken down into shorter polypeptide chains of amino acids.
In the small intestine, these polypeptide chains are digested into dipeptides and amino acids, which are absorbed by the small intestine, where any remaining

peptides are broken down into amino acids. Absorbed amino acids then travel via the portal vein to the liver.

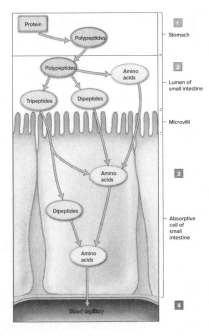

## 7.6 Important body components—such as muscles, connective tissue, transport

proteins, enzymes, hormones, and antibodies—are made of proteins. Protein is required for the maintenance of fluid and acid-base balance, normal immune function, and the transport of nutrients. Proteins also provide carbon skeletons, which can be used to synthesize glucose when necessary.

IMTSSA/CNRI/
Science Source

## 7.7 Undernutrition can lead to protein-energy malnutrition in the form of kwashiorkor or marasmus.

Kwashiorkor results primarily from an inadequate energy intake with a severe protein deficit, often accompanied by disease and infection. Kwashiorkor frequently occurs when a child is weaned from human milk and fed mostly starchy gruels. Marasmus results from extreme starvation—a negligible intake of both protein and energy. Marasmus commonly occurs during famine, especially in infants. These conditions have been noted in some North Americans with cancer, AIDS, malabsorption disease, anorexia nervosa, alcoholism, or limited income and resources to obtain food. High protein diets (above 35% of energy intake) do not provide additional health benefits. They are associated with dehydration, overburdening of the kidneys' capacity to excrete nitrogen wastes, increased risk of cardiovascular disease and certain cancers, increased urinary calcium loss, and risk of amino acid imbalance and toxicity.

## 7.8 Vegetarian diets are becoming more popular as individuals recognize the possible health benefits of plant-based diets. Low intakes of riboflavin, vitamins D and

B-12, calcium, iron, and zinc are of greatest concern in vegetarian diets. Vegetarian diets require knowledge and creative planning to obtain high quality protein and other key nutrients but can be nutritionally adequate when guidelines are followed.

## Clinical Perspective

Food allergies involve immune system responses designed to eliminate allergens that the body mistakenly reacts to as though they were harmful invaders. Symptoms range from mild to life threatening. The only way to prevent reactions is to avoid foods known to trigger allergic reactions.

All: Photos courtesy of Dennis Gottlieb

## Expert Perspective

The immune system is a complex network of organs, tissues, cells, and secretions that protects the body from foreign organisms. When the body detects the presence of "non-self" cells or antigens, innate and acquired immune responses attempt to destroy the antigens. Nutritional deficiencies can suppress the immune system's ability to prevent infection and disease.

## Global Perspective

What we eat may affect our personal health and the environment. Many scientists believe that meat rich diets and the agricultural practices that support the production of food for these diets negatively affect the environment. Other scientists believe that consuming a small amount of dairy and/or meat may actually increase land use efficiency because many meat substitutes are highly processed and require energy-intensive production methods. Including small amounts of meat in the diet may offer both environmental and nutritional benefits.

## Study Questions

1. Essential amino acids must be supplied by the diet because the body cannot synthesize them in adequate amounts.

   a. true      b. false

2. A process involved in the synthesis of nonessential amino acids is called _____.

   a. ketogenesis      c. transamination
   b. gluconeogenesis      d. supplementation

3. The carbon skeleton of an amino acid is the portion remaining after an amino group has been removed.

   a. true      b. false

4. Which of the following is classified as a complete protein?

   a. kidney beans      c. whole-grain bread
   b. fat-free milk      d. corn tortillas

5. The sequential order of amino acids in a polypeptide chain is called the _____.

   a. primary structure      c. tertiary structure
   b. secondary structure      d. quaternary structure

6. Which of the following is a rich source of protein?

   a. citrus fruits      c. enriched grains
   b. dark, leafy greens      d. barbequed chicken

7. Which of the following is *not* a means of determining the protein quality of a food?

   a. biological value      c. protein efficiency ratio
   b. chemical score      d. complementary score

8. Hospitalized patients recovering from illness or trauma usually need additional protein to attain positive nitrogen balance.

   a. true      b. false

9. Proteins are involved in all of the following functions *except* _____.

   a. providing energy      c. promoting bowel
   b. aiding in             function
       immune function      d. providing cell structure

10. Which of the following population groups is at increased risk of PEM?

    a. college athletes      c. the elderly
    b. obese individuals      d. adolescents

11. Many children with kwashiorkor maintain some muscle and subcutaneous fat.

    a. true      b. false

12. Which of the following is *not* a usual characteristic of marasmus?

    a. edema      c. impaired growth
    b. severe weight loss      d. muscle wasting

13. Which of the following is associated with excessive protein intakes?

    a. dehydration      c. diarrhea
    b. anemia      d. diabetes

14. Which of the following foods is a common cause of food allergies?

    a. peanuts      c. eggs
    b. shellfish      d. all of the above

15. Which of the following nutrients would most likely be low in a vegan diet?

    a. vitamin C      c. vitamin B-12
    b. thiamin      d. dietary fiber

16. Why is it important for essential amino acids lost from the body to be replaced by the diet?

17. Explain the process for synthesizing nonessential amino acids.

18. What is a limiting amino acid? Explain why this concept is a concern in a vegetarian diet.

19. Describe factors that determine whether a person is in nitrogen equilibrium.

20. Describe the process of protein digestion from ingestion to excretion.

21. Describe the functions of protein.

22. Outline the major differences between kwashiorkor and marasmus.

23. Plan a diet for a vegan using Table 7-4.

24. What is a food allergy?

Answer Key: 1-a; 2-c; 3-a; 4-b; 5-a; 6-d; 7-d; 8-a; 9-c; 10-c; 11-a; 12-a; 13-a; 14-d; 15-c; 16-refer to Section 7.1; 17-refer to Section 7.2; 18-refer to Sections 7.3 and 7.8; 19-refer to Section 7.4; 20-refer to Section 7.5; 21-refer to Section 7.6; 22-refer to Section 7.7; 23-refer to Section 7.8; 24-refer to Clinical Perspective

# References

1. Gropper SS and Smith JL. Protein. In: *Advanced nutrition and human metabolism.* 7th ed. Belmont, CA: Wadsworth, Cengage Learning; 2018.

2. Azar S and Wong TE. Sickle cell disease: A brief update. *Med Clin North Am.* 2017;101:375.

3. Food and Nutrition Board. *Dietary Reference Intakes for energy, carbohydrate, fiber, fat, fatty acids, cholesterol, protein, and amino acids.* Washington, DC: National Academies Press; 2005.

4. U.S. Department of Agriculture. *Nutrient content of the U.S. food supply.* Research Report 57. Washington, DC: U.S. Department of Agriculture; 2007.

5. Dinu M and others. A heart healthy diet: Recent insights and practical recommendations. *Curr Cardiol Rep.* 2017;19:95. doi.org/10.1007/s11886-017-0908-0.

6. Patel H and others. Plant based nutrition: An essential component of cardiovascular disease prevention and management. *Curr Cardiol Rep.* 2017;19:104.

7. Melina V and others. Position of the American Dietetic Association: Vegetarian diets. *J Acad Nutr Diet.* 2016;116:1970.

8. Appleby PN and Key TJ. The long term health of vegetarian and vegans. *Proc Nutr Soc.* 2016;75:287.

9. van Huis A and others. Edible insects: Future prospects for food and feed security. FAO Forestry, 2013; Paper 171.

10. Kouřimská L, Adámková A. Nutritional and sensory quality of edible insects. *NFS J.* 2016;4:22.

11. Arciero PJ and others. Increased protein intake and meal frequency reduces abdominal fat during energy balance and energy deficit. *Obesity.* 2013;21:1357.

12. Wise AK and others. Energy expenditure and protein requirements following burn injury. *Nutr Clin Pract.* 2019;34:673.

13. Memerow MM and others. Dietary protein distribution positive influences 24-h muscle protein synthesis in healthy adults. *J Nutr.* 2014;144:876.

14. Perkin MR and others. Enquiring about tolerance (EAT) study: Feasibility of an allergenic food introduction. *J Allerg Clin Immunol.* 2016;137:1477.

15. United Nations Children's Fund. UNICEF data: Monitoring the situation of children and women. 2016; data.unicef.org/resources/joint-child-malnutrition-estimates-2016-edition.

16. Ibrahim MD and others. Impact of childhood malnutrition on host defense and infection. *Clin Microbiol Rev.* 2017;30:919.

17. Van Elswyk ME and others. A systematic review of renal health in healthy individuals associated with protein intake above the US recommended daily allowance in randomized controlled trials and observational studies. *Adv Nutr.* 2018;9:404. .

18. Bilancio G and others. Dietary protein, kidney function and mortality: Review of evidence from epidemiological studies. *Nutrients.* 2019;11:196.

19. Fernandez de Jauregui and others. Common dietary patterns and risk of cancers of the colon and rectum: Analysis from the United Kingdom women's cohort study (UKWCS). *Int J Cancer.* 2018;143:773.

20. Salter AM. Impact of consumption of animal products on cardiovascular disease, diabetes, and cancer in developed countries. *Animal Frontiers.* 2013;3:20.

21. Wright BL and others. Clinical management of food allergy. *Pediatr Clin North Am.* 2015;62:1409.

22. Collins SC. Practice paper of the Academy of Nutrition and Dietetics: Role of the registered dietitian nutritionist in the diagnosis and management of food allergies. *J Acad Nutr Diet.* 2016;116:1621.

23. Togias A and others. Addendum guidelines for the prevention of peanut allergy in the United States. *J Allergy Clin Immunol.* 2017;139:29.

24. Savage J, Johns CD. Food allergy: Epidemiology and natural history. *Immunol Allergy Clin North Am.* 2016;35:45.

25. Marlow HJ and others. Diet and the environment: Does what you eat matter? *Am J Clin Nutr.* 2009;89:1699S.

26. Carlsson-Kanyama A, Gonzalez AD. Potential contributions of food consumption patterns to climate change. *Am J Clin Nutr.* 2009;89:1704S.

27. Battisti DS, Naylor RL. Historical warnings of future food insecurity with unprecedented seasonal heat. *Science.* 2009;323:240.

28. Peters CJ and others. Testing a complete-diet model for estimating the land resource requirements of food consumption and agricultural carrying capacity: The New York State example. *Renew Ag Food Sys.* 2007;22:145.

29. Petersen KS and others. Healthy dietary patterns for preventing cardiometabolic disease: The role of plant-based foods and animal products. *Curr Dev Nutr.* 2017;1:12.

30. Kahleova H and others. Cardio-metabolic benefits of plant-based diets. *Nutrients.* 2017;9:848.

31. McMacken S, Shah S. A plant-based diet for the prevention and treatment of type 2 diabetes. *J Geriatr Cardiol.* 2017;14:342.

32. Barnard ND and others. A systematic review and meta analysis of changes in body weight in clinical trials of vegetarian diets. *J Acad Nutr Diet.* 2015;115:954.

33. Kim H and others. Healthy plant-based diets are associated with lower risk of all-cause mortality in US adults. *J Nutr.* 2018;148:624.

34. Forestell CA. Flexitarian diet and weight control: Healthy or risky eating behavior? *Front Nutr.* 2018;5:59.

35. Rogerson D. Vegan diets: Practical advice for athletes and exercisers. *J Int Soc Sports Nutr.* 2017;14:36.

36. Herrmann W and others. Enhanced bone metabolism in vegetarians—The role of vitamin B12 deficiency. *Clin Chem Lab Med.* 2009;47:1381.

37. Sanders TA. DHA status of vegetarians. *Prostaglandins Leukot Essent Fatty Acids.* 2009;81:137.

38. Weaver CM. Should dairy be recommended as part of a healthy vegetarian diet? *Am J Clin Nutr.* 2009;89:1634S.

D. Hurst/Alamy Stock Photo

Wine grapes are grown in almost every country. Some studies suggest that compounds in grapes and wine may promote heart health. Learn more about viticulture (cultivation of grapes) and enology (science of wine) at **wineserver.ucdavis.edu**.
roycebair/RooM/Getty Images

# $8$ Alcohol

## Learning Objectives

**After studying this chapter, you will be able to**

1. Describe the sources of alcohol (ethanol) and the calories it provides.
2. Define the terms *alcoholic drink equivalent* and *moderate drinking*.
3. Outline the process of alcohol absorption, transport, and metabolism.
4. Explain how alcohol consumption affects blood alcohol concentration.
5. Define *binge drinking* and describe the problems associated with it.
6. Discuss potential health risks and benefits of alcohol consumption.
7. Describe the effects of chronic alcohol use on the body and nutritional status.
8. List the signs of an alcohol use disorder.

ON ANY GIVEN DAY, BEER, WINE, OR HARD LIQUOR, or spirits, is consumed by about 23% of the adult population in North America.[1] Although not an essential nutrient, the alcohol in these beverages is an energy-rich substance that supplies 7 kcal/g. Alcohol intake, when averaged across the population, contributes nearly 5% of total energy intake.[2] However, among alcohol consumers, alcoholic beverages provide, on average, 17% of calories, although many drinkers will exceed this amount.[1,2]

For some individuals, alcohol adds to the enjoyment of a meal or social times shared with friends and family. For others, moderate alcohol use may help relieve tensions and enhance relaxation. In middle-aged and older adults, moderate alcohol consumption may even reduce the risk of cardiovascular disease. Unfortunately, moderate use of alcohol can escalate to an alcohol use disorder in susceptible individuals.

Alcohol is a **narcotic**, an agent that reduces sensations and consciousness, and a central nervous system depressant. It is the most commonly abused drug in North America. The harmful effects of alcohol are well known. Alcohol abuse can cause motor vehicle accidents, destroy families and friendships, and encourage violence, suicide, rape, and other aggressive behaviors. It also can cause multiple nutrient deficiencies. Although too much alcohol damages almost every organ in the body, the liver and brain are especially vulnerable to its toxic effects. Alcohol abuse, behind smoking and obesity, is the third leading cause of preventable death in the U.S. Because alcohol is so widely consumed and its abuse touches many lives, this chapter will examine this substance in detail. It will explore the sources of alcohol, the production of alcohol, the metabolism of alcohol, and the health concerns related to intakes of alcohol.

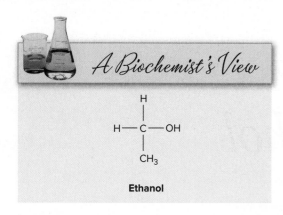

*A Biochemist's View*

H
|
H—C—OH
|
CH₃

**Ethanol**

▶ A survey of 80 restaurants and bars showed that wine, beer, and mixed drinks often were at least 50% larger than standard sizes.[3] This can make it difficult to keep alcohol intake at a safe level.

# 8.1 Sources of Alcohol

The form of alcohol we consume, chemically known as ethanol ($CH_3CH_2OH$), is supplied mostly by beverages such as beer, wine, distilled spirits (hard liquor, such as vodka, tequila, and rum), liqueurs, cordials, and hard cider. As discussed in this chapter's Culinary Perspective, it also is used as an ingredient in foods, such as chicken cooked in wine (coq au vin), flaming desserts, and chocolate candy filling.

As shown in Table 8-1, beverages vary in alcohol and calorie content. Most beers are about 5% alcohol by volume (ABV) or less, although some exceed 11% ABV. Wines generally range in alcohol content from approximately 5 to 14%. Fortified wines (wines that have spirits added to increase their alcohol content) typically contain 15 to 22% ABV. Distilled wine spirits, such as brandy, contain more than 22% ABV. For hard liquor (distilled spirits), alcohol content is listed by "proof" rather than by percentage. The proof is twice the percentage of alcohol content. Thus, an 80-proof vodka or gin is 40% ABV.

A standard drink is usually defined as the size that provides approximately 14 g of alcohol (known as an alcoholic drink equivalent). In general, this equates to a 12-ounce 5% ABV beer, 5-ounce glass of 12% ABV wine, or 1.5-ounce pour of 80-proof hard liquor (Fig. 8-1). Many individuals are not aware of these definitions of serving sizes—some may consider a 20-ounce glass of beer or an 8-ounce glass of wine to be a "drink," when, in fact, these servings are both closer to 2 drinks. Drinks served in bars and restaurants may vary considerably from the defined standard drink serving. Additionally, alcoholic drink equivalents also depend on the ABV amount. To calculate drink equivalents, multiply the volume in ounces by the alcohol content in percentage, and divide by 0.6 ounce of alcohol per drink equivalent. For example, using a 16 fl oz beer with 7% ABV,

**16 fl oz × (0.07 ABV/0.6 fl oz per drink equivalent) = 1.9 drink equivalents**

Both variability in portion size and alcohol content must be considered when calculating the number of drink equivalents.

| Table 8-1 Alcohol and Energy Content of Alcoholic Beverages | | | |
|---|---|---|---|
| **Beverage** | **Amount (fl oz)** | **Alcohol (g)** | **Energy (kcal)** |
| **Beer** | | | |
| Regular | 12 | 12 | 150 |
| Light | 12 | 10 | 75–100 |
| Ultra light | 12 | 7–8 | 55–66 |
| **Distilled Spirits** | | | |
| Gin, rum, vodka, bourbon, tequila, whiskey (80 proof) | 1.5 | 14 | 95 |
| Liqueurs | 1.5 | 14 | 160 |
| **Wine** | | | |
| Red | 5 | 14 | 100 |
| White | 5 | 14 | 100 |
| Dessert, sweet | 5 | 23 | 225 |
| Rose | 5 | 14 | 100 |
| **Mixed Drinks** | | | |
| Martini | 3.5 | 32 | 220 |
| Manhattan | 3.5 | 30 | 225 |
| Whiskey sour | 3.5 | 17 | 135 |
| Margarita (frozen) | 8 | 20 | 175 |
| Margarita (low calorie) | 5 | 14 | 100 |
| Rum and cola | 8 | 15 | 170 |

Beer: David Toase/Photodisc/Getty Images; Tequila: ©C Squared Studios/Photodisc/Getty Images; Wine: Ryan McVay/Photodisc/Getty Images; Mixed Drink: ©Ingram Publishing/Alamy RF

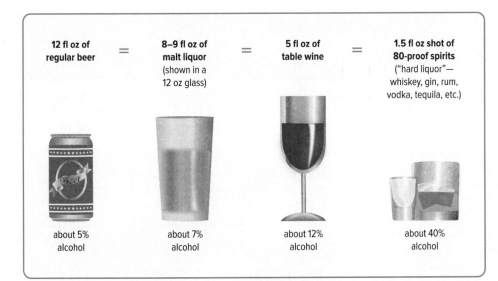

| 12 fl oz of regular beer | = | 8–9 fl oz of malt liquor (shown in a 12 oz glass) | = | 5 fl oz of table wine | = | 1.5 fl oz shot of 80-proof spirits ("hard liquor"— whiskey, gin, rum, vodka, tequila, etc.) |

about 5% alcohol    about 7% alcohol    about 12% alcohol    about 40% alcohol

*Figure 8-1* These are examples of 1 *standard drink equivalent.* In determining a safe level of intake, it is important to observe these standard serving sizes.

Source: National Institute on Alcohol Abuse and Alcoholism, http://www.niaaa.nih.gov/alcohol-health/overview-alcohol-consumption/standard-drink

Most guidelines for "moderate alcohol intake" suggest no more than 1 alcoholic drink equivalent per day for women and no more than 2 per day for men. This does not mean that one can abstain from drinking during the week and then safely consume 7 or more drinks on a single occasion. As discussed in a later section, this is defined as binge drinking, and it can have serious consequences.

## Production of Alcoholic Beverages

The alcohol we consume is produced by fermentation. The process of fermenting foods to produce mead (fermented honey), beer, wine, and other alcoholic products dates back thousands of years. Grains, cereals, fruits, honey, milk, potatoes, and other carbohydrate rich foods can be used to make alcoholic beverages.

Fermentation occurs when yeast, a microorganism, converts carbohydrates to alcohol and carbon dioxide. The carbohydrate must be in the form of simple sugars, such as maltose or glucose, for yeast to use it as food. If the carbohydrate is a starch, such as that found in cereal grain seeds (e.g., barley), it must be broken down to simpler forms, or "malted," before fermentation can occur. During malting, the grain seeds are allowed to sprout; the sprouting process produces enzymes in the seed that break starches into simple sugars. Each molecule of glucose that is fermented produces 2 molecules of ethanol and 2 molecules of carbon dioxide: $C_6H_{12}O_6$ (glucose) $\rightarrow 2\ CH_3CH_2OH$ (ethanol) $+ 2\ CO_2$. Additionally, the reaction yields energy the yeast can use.

Fermentation begins when yeast, water, and a food rich in carbohydrates are combined and left at room temperature. During the first stage, the yeast cells multiply, using the sugars for energy, and produce small amounts of alcohol. When the oxygen in the mixture is depleted, the second stage begins, during which the yeast ferments the remaining sugar to produce alcohol and carbon dioxide under anaerobic (without oxygen) conditions. After fermentation has ceased (when the sugar is used up or the alcohol content is high enough to inactivate the yeast), the product can be finished in a variety of ways, or the alcohol itself can be recovered from the product by **distilling** it into spirits, such as gin or whiskey.

Tequila is made by fermenting juice from agave plants, then distilling it.

Carlos S. Pereyra/age fotostock

Distilled spirits, such as tequila, vodka, and gin, have the highest alcohol content of the alcohol-containing beverages.

Shutterstock/Q77photo

**distill** To separate 2 or more liquids that have 2 different boiling points. Alcohol is boiled off and the vapors are collected and condensed. Distillation produces a high alcohol content in hard liquor.

## CASE STUDY

McGraw-Hill Education/Gary He, photographer

Charles, a college student, has noticed that his pants are getting hard to button. A quick check on the scale in the gym confirms a 7-pound weight gain over the last 12 weeks. The main change in Charles's diet is his alcohol intake—he now typically drinks 5 or 6 12-ounce beers on Friday and Saturday nights and drinks another 3 or 4 beers during the week. How many extra calories per week is Charles consuming? If each pound of weight gain results from a surplus of 3500 kcal, can Charles's weight gain be explained by his beer consumption?

# Culinary Perspective

## Cooking with Alcohol

Wine, beer, cider, spirits, and liqueurs are included in many recipes for their noticeable effects on flavor. Even the addition of a small amount of an alcoholic beverage contributes to distinctive flavors and new compounds that form during cooking. For example, ethanol itself can combine with acids in the food and oxygen to create new, flavorful compounds. Also, as alcohol-containing sauces and glazes cook, the flavor compounds become even more concentrated.

Red wine is a common addition to sauces for pasta. Spirits, wine, or beer may be included in glazes for meat—for example, whiskey-based glaze for chicken. Strongly flavored beers may be included in recipes for hearty stews and breads. Sometimes a higher alcohol wine or liqueur is used to finish a dish, such as adding a splash of sherry (a wine fortified with brandy) to a creamy squash soup. Many desserts include bourbon, rum, or sweet wines. Alcohol-based extracts, such as vanilla extract, also are widely used. The alcohol content of extracts may be as high as 90%, but typically only very small amounts are used.

Many people wonder how much ethanol remains after cooking. A common belief is that, because alcohol boils at a lower temperature (173°F [78°C]) than water (212°F [100°C]), it will evaporate during cooking. However, as the chart illustrates, 5 to 40% of the alcohol in baked and simmered dishes is retained. When alcohol is flamed (flambéed), a quick exposure to heat, 75% of it remains.

When added to a hot pan, alcohol creates a burst of flames. This cooking method, called flambé, is used to prepare bananas foster, cherries jubilee, and steak with brandy sauce.

UpperCut Images/SuperStock

### Alcohol Burn-Off Chart

| Preparation Method | % Retained |
| --- | --- |
| Alcohol added to boiling liquid and removed from heat | 85% |
| Alcohol flamed | 75 % |
| Alcohol stirred into baked/simmered dish | |
|     15 minutes cooking time | 40% |
|     30 minutes cooking time | 35% |
|     1 hour cooking time | 25% |
|     1.5 hours cooking time | 20% |
|     2 hours cooking time | 10% |
|     2.5 hours cooking time | 5% |

Source: USDA Table of Nutrient Retention Factors, Release 6, 2007; www.ars.usda.gov/SP2UserFiles/Place/80400525/Data/retn/retn06.pdf

▶ Powdered alcohol is made by combining alcohol with a carbohydrate, forming a dry, crystalline product that can be added to liquids. It has been approved for sale in the U.S. but is not currently commercially available. Many serious regulatory and safety concerns are associated with this product, including the ease of access by underage consumers and the potential for spiking drinks and overdosing. As a result, 35 states have banned its sale.[4]

### Knowledge Check

1. What are the chemical name and formula for the substance we know as alcohol?
2. How does the proof of an alcoholic beverage (e.g., vodka or tequila) relate to its alcohol content?
3. What is the alcoholic drink equivalent for 8 fl oz of white wine, 12% ABV?
4. What 3 ingredients are required for alcoholic fermentation?

# 8.2 Alcohol Absorption and Metabolism

Alcohol, unlike most carbohydrate, protein, and fat molecules, requires no digestion. It also does not need specific transport mechanisms or receptors to enter cells. Therefore, it is absorbed rapidly throughout the digestive tract by simple diffusion. The stomach absorbs about 20% of ingested alcohol, with the remainder being absorbed in the duodenum and jejunum. When food is consumed with alcohol, absorption is slowed. Larger meals with a high fat content leave the stomach more slowly, thereby slowing the absorption of any alcohol consumed. In contrast, alcohol consumed on an empty stomach is absorbed quickly from the stomach and small intestine into the bloodstream.

Alcohol is readily dispersed throughout the body because alcohol is found wherever water is distributed in the body. Alcohol moves easily through the cell membranes; however, as it does, it damages proteins in the membranes.[5]

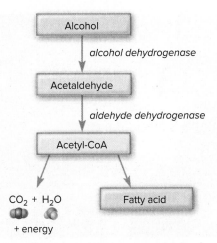

The alcohol dehydrogenase pathway of alcohol metabolism

## Alcohol Metabolism: 3 Pathways

Alcohol cannot be stored in the body, so it has absolute priority in metabolism as a fuel source, taking precedence over other energy sources, such as carbohydrate. At low to moderate intakes, alcohol is metabolized through a series of reactions called the alcohol dehydrogenase (ADH) pathway[5] (see Chapter 9). Although the cells lining the stomach metabolize 10 to 30% of alcohol via the ADH pathway, the liver is the chief site for alcohol metabolism. When a person drinks moderate to excessive amounts of alcohol, the ADH pathway cannot keep up with the demand to metabolize all the alcohol. Under these circumstances, the liver activates the microsomal ethanol oxidizing system (MEOS) to help metabolize alcohol. The MEOS pathway produces the same intermediates as the ADH pathway, but it requires energy to function (see Chapter 9). As a person's alcohol intake increases over time, the MEOS becomes increasingly active, allowing for more efficient metabolism of alcohol and a greater tolerance to alcohol. This means that increasing amounts of alcohol are necessary to produce the same effects.

The MEOS also metabolizes drugs and other substances foreign to the body. Activation of the MEOS by excessive alcohol intake reduces the liver's capacity for metabolizing drugs because the metabolism of alcohol takes priority.[5] Thus, MEOS activation by alcohol increases the potential for drug interactions and toxicities.

The 3rd metabolic pathway for metabolizing alcohol—the catalase pathway—in the liver and other cells makes a minor contribution to alcohol metabolism in comparison with the alcohol dehydrogenase pathway and the MEOS. Table 8-2 summarizes alcohol metabolism.

The 3 metabolic pathways (ADH, MEOS, and catalase) metabolize nearly all the alcohol consumed. Only a small percentage (2–10%) of alcohol intake is excreted unmetabolized through the lungs, urine, and sweat.[5]

Women are smaller, have less body water, and have less alcohol dehydrogenase in their stomachs. This makes them more susceptible than men to the detrimental effects of alcohol.

Image Source/SuperStock

| Table 8-2 Alcohol Metabolism Summary | | | |
|---|---|---|---|
| **Alcohol Metabolic Pathway** | **Main Location of Pathway Activity** | **Alcohol Intake That Activates Pathway** | **Extent of Participation in Alcohol Metabolism** |
| Alcohol dehydrogenase pathway (ADH) | Stomach Liver (mostly) | Low to moderate intake | Metabolizes about 90% of alcohol |
| Microsomal ethanol oxidizing system (MEOS) | Liver | Moderate to excessive intake | Increases with increasing alcohol intake |
| Catalase pathway | Liver Other cells | Moderate to excessive intake | Minor |

One is legally intoxicated at a blood alcohol concentration of 0.08% in the U.S. and Canada. However, for many individuals, driving is often noticeably impaired at blood alcohol concentrations of 0.02 to 0.05%.

piotr290/Getty Images

## Factors Affecting Alcohol Metabolism

The key to alcohol metabolism lies in one's ability to produce the enzymes used in the alcohol dehydrogenase pathway because this pathway metabolizes about 90% of the alcohol consumed.[5] Ethnicity, biological sex, and age affect the production and activity of enzymes in the alcohol dehydrogenase pathway.[6] For instance, many individuals of Asian descent have normal to high alcohol dehydrogenase enzyme activity, allowing a very rapid conversion of alcohol by the first enzyme used in this pathway (alcohol dehydrogenase), but they have very low activity of the second enzyme (aldehyde dehydrogenase) in this pathway that is needed to complete alcohol breakdown. The resulting buildup of acetaldehyde commonly causes flushing, dizziness, nausea, headaches, rapid heartbeat (tachycardia), and rapid breathing (hyperventilation). These reactions can be so severe that it's uncomfortable to drink more or even at all.[5,6]

Compared with men, women produce less of the alcohol dehydrogenase enzyme in the cells that line their stomachs—as a result, women absorb about 30 to 35% more unmetabolized alcohol from the stomach directly into the bloodstream. Another biological sex related factor is that, in comparison with men, women are generally smaller in body size, have more body fat, and have less total body water. As a result, alcohol becomes more concentrated in the blood and body tissues of a woman than in a similar-size man because alcohol can be diluted by water, but not by adipose tissue. For these reasons, a woman becomes intoxicated on less alcohol than a similar-size man.

Other factors that can affect alcohol metabolism include the alcohol content of the beverage, the amount of alcohol consumed, and the individual's usual alcohol intake. As described earlier, individuals who drink large amounts of alcohol regularly are able to metabolize alcohol more rapidly and have a greater tolerance to alcohol.

## Rate of Alcohol Metabolism

The body is fairly well equipped to metabolize moderate amounts of alcohol. A social drinker who weighs 150 pounds (about 70 kg) and has normal liver function metabolizes about 5 to 7 g of alcohol per hour (or about 1/2 alcoholic drink equivalent). Figure 8-2 shows the relationship among biological sex, number of drinks, and body weight. When the rate of alcohol

| | Effect on Women | | | | | | | | | Drinks | Effect on Men | | | | | | | | |
|---|---|---|---|---|---|---|---|---|---|---|---|---|---|---|---|---|---|---|---|
| **Body weight in pounds** | 90 | 100 | 120 | 140 | 160 | 180 | 200 | 220 | 240 | | 100 | 120 | 140 | 160 | 180 | 200 | 220 | 240 | **Body weight in pounds** |
| **ONLY SAFE DRIVING LIMIT** | .00 | .00 | .00 | .00 | .00 | .00 | .00 | .00 | .00 | 0 | .00 | .00 | .00 | .00 | .00 | .00 | .00 | .00 | **ONLY SAFE DRIVING LIMIT** |
| **DRIVING SKILLS SIGNIFICANTLY AFFECTED** | .05 | .05 | .04 | .03 | .03 | .03 | .02 | .02 | .02 | 1 | .04 | .03 | .03 | .02 | .02 | .02 | .02 | .02 | **DRIVING SKILLS SIGNIFICANTLY AFFECTED** |
| **LEGALLY INTOXICATED** | .10 | .09 | .08 | .07 | .06 | .05 | .05 | .04 | .04 | 2 | .08 | .06 | .05 | .05 | .04 | .04 | .03 | .03 | |
| | .15 | .14 | .11 | .11 | .09 | .08 | .07 | .06 | .06 | 3 | .11 | .09 | .08 | .07 | .06 | .06 | .05 | .05 | |
| | .20 | .18 | .15 | .13 | .11 | .10 | .09 | .08 | .08 | 4 | .15 | .12 | .11 | .09 | .08 | .08 | .07 | .06 | |
| | .25 | .23 | .19 | .16 | .14 | .13 | .11 | .10 | .09 | 5 | .19 | .16 | .13 | .12 | .11 | .09 | .09 | .08 | **LEGALLY INTOXICATED** |
| **DEATH POSSIBLE** | .30 | .27 | .23 | .19 | .17 | .15 | .14 | .12 | .11 | 6 | .23 | .19 | .16 | .14 | .13 | .11 | .10 | .09 | |
| | .35 | .32 | .27 | .23 | .20 | .18 | .16 | .14 | .13 | 7 | .26 | .22 | .19 | .16 | .15 | .13 | .12 | .11 | |
| | .40 | .36 | .30 | .26 | .23 | .20 | .18 | .17 | .15 | 8 | .30 | .25 | .21 | .19 | .17 | .15 | .14 | .13 | **DEATH POSSIBLE** |
| | .45 | .41 | .34 | .29 | .26 | .23 | .20 | .19 | .17 | 9 | .34 | .28 | .24 | .21 | .19 | .17 | .15 | .14 | |
| | .51 | .45 | .38 | .32 | .28 | .25 | .23 | .21 | .19 | 10 | .38 | .31 | .27 | .23 | .21 | .19 | .17 | .16 | |

Approximate blood alcohol percentage

Approximate blood alcohol percentage

***Figure 8-2*** Approximate relationship between alcohol consumption and blood alcohol concentration (units are % or mg of alcohol per 100 ml of blood). Note that effects can vary among people and whether food also is consumed. A blood alcohol concentration of 0.02 begins to impair driving. One is legally intoxicated at a blood alcohol concentration of 0.08 in the U.S. and Canada. Blood alcohol concentrations exceeding 0.30 can cause death.

(young women): ©Photodisc/Getty Images; (young man): ©RubberBall Productions RF

| Possible behaviors & feelings | BAC | Impairments |
|---|---|---|
| Relaxed, slight euphoria, sense of well-being, loss of inhibition, less alert. | .01–.06 | Slight impairment of judgment, reason, memory, concentration. |
| You probably believe you are functioning better than you really are! Still relaxed, less inhibited, and more extroverted. May appear intoxicated. | .06–.10 | Depth perception, reaction times, balance, speech, vision, hearing reduced. Judgment, reason, memory decline. |
| Obviously intoxicated. Not thinking straight. Mood swings. Vomiting (a sign of alcohol poisoning) is common. | .11–.15 | Motor coordination significantly impaired. Very slow reaction time. Slurred speech. Vision and hearing impairment. Judgment, reason, memory suffer. |
| Very incapacitated. May pass out and have memory blackout. Nausea and vomiting common. | .16–.20 | Worsening motor coordination, reaction time, memory, attention, balance, reason, judgment. |
| Extremely drunk; in a stupor. May not feel pain. Nausea, vomiting; may choke on vomit. May pass out. | .21–.29 | All mental, physical, and sensory functions are severely impaired. |
| You don't know where you are. Loss of consciousness may occur. Seizures. Danger of life-threatening alcohol poisoning. | .30–.39 | Many physiological functions are severely impaired: slowed heart rate and breathing, hypothermia, hypoglycemia. Loss of bladder control. |
| Life-threatening alcohol poisoning. Death likely. | >0.40 | |

*Figure 8-3* As blood alcohol content (BAC) increases, individuals become increasingly impaired. Alcohol poisoning, a dangerous condition, can occur with high alcohol intake.

▶ Binge drinking, or the rapid consumption of several standard drinks in 1 or 2 hours, overwhelms the liver. Blood alcohol content increases until drinking stops and may require several hours to return to normal. Learn more about the dangers of binge drinking and alcohol poisoning at awareawakealive.org.

Although many young adults do not recognize the true impact of binge drinking habits, it poses a significant risk to their overall health and safety. Nutritional status also can be adversely affected.

©Ryan McVay/Getty Images RF

consumption exceeds the liver's metabolic capacity, blood alcohol levels rise and the symptoms of intoxication appear as the brain and central nervous system are exposed to alcohol.[7]

Blood alcohol concentration (BAC) can be determined by measuring the amount of alcohol excreted ted through the lungs because the alcohol contents of exhaled air and blood are directly related. The constant relationship between the alcohol content of blood and that of exhaled air makes it possible to use breathalyzer tests as a legal basis for defining alcohol impairment and intoxication.

If blood alcohol levels rise high enough, the person experiences acute alcohol toxicity, also known as alcohol poisoning (Fig. 8-3 and Table 8-3). This dangerous condition requires immediate medical treatment—left untreated, alcohol poisoning can cause respiratory failure and death. Inhalation of vomit also can result in death and has occurred at levels lower than those that cause alcohol poisoning. The risk of consuming toxic levels of alcohol is greater when drinking distilled spirits because the higher alcohol content makes it easier to ingest more alcohol in less volume and less time than with beer or wine. **Binge drinking**, defined as having 4 or more drinks for females and 5 or more drinks for males on a single occasion, increases the risk of alcohol poisoning.

## *Critical* Thinking

Kevin went out with his friends to celebrate the end of the semester. Over an hour's time, he had 4 drinks. Kevin weighs approximately 160 lb. According to Figure 8-2, what would Kevin's blood alcohol level be? Is this within a legally safe limit to drive?

▶ *Healthy People* 2030 goals[10] regarding alcohol use include

- Reduce the proportion of adolescents using alcohol.
- Reduce the proportion of persons engaging in binge drinking of alcoholic beverages.
- Reduce the proportion of people with alcohol use disorder.
- Increase abstinence from alcohol among pregnant women.
- Reduce motor vehicle crash deaths involving an alcohol-impaired driver.

▶ The age group with the highest rate of impaired driving (32% BAC over 0.8) is 21- to 24-year olds. With the rapid increase in ride-sharing apps and services, these opportunities help reduce accident rates and save lives. Some popular ride-sharing companies offer free rides on major holidays to help reduce rates of impaired driving (check out SoberRide.com). Additionally, the SaferRide mobile app allows a user to share his or her location with a ride-sharing company, taxi, or friend, making it easier for the user to get a safe ride home if he or she is too impaired to drive (or even use a ride-sharing app).

Many colleges require students who break alcohol and drug rules to attend counseling or educational programs to reduce their alcohol intake and binge drinking episodes. Colleges and communities offer a variety of educational and treatment strategies for individuals with alcohol dependency.

Getty Images RF

### Table 8-3 Signs and Symptoms of Alcohol Poisoning

- Confusion, stupor
- Vomiting
- Low blood sugar (hypoglycemia)
- Severe dehydration
- Seizures
- Slow or irregular breathing and heartbeat
- Blue-tinged or pale skin
- Low body temperature (hypothermia)
- Loss of consciousness

Call 911 or your local emergency number if you suspect alcohol poisoning and the person cannot be roused or is unconscious. If the person is conscious, you can call 1-800-222-1222 to be routed to your local poison control center for further instructions.

### Knowledge Check

1. Which enzyme system is the most important for metabolizing low to moderate amounts of alcohol?
2. Name 3 factors that can affect alcohol metabolism.
3. How quickly can most people metabolize 1 alcoholic beverage?
4. Why does binge drinking increase the risk of alcohol poisoning?

## 8.3 Alcohol Consumption

Over the course of a year, approximately 70% of North American adults consume alcohol and most of them do not drink excessively. However, the amount consumed varies considerably, depending on biological sex, age, ethnicity, religious beliefs, and place of residence.[8,9] In general, research shows that excess alcohol consumption, as measured by more than 1 drink per day for women and 2 drinks per day for men, is on the rise.[9] Men use alcohol at higher rates than women.

### College and Underage Drinking

The largest drinking population in North America consists of young, White college students, many of whom are not yet of legal drinking age.[11] College students are drinking more heavily and more frequently than ever before. In fact, excessive alcohol consumption is a bigger problem than illicit drug use on most college campuses. Many young adults consider drinking alcohol to be a "rite of passage" into adulthood and may incorporate drinking competitions into initiations to clubs and social circles. Alcohol producers also target advertising and marketing efforts to college students, many of whom underestimate the risks of alcohol consumption.

Of the 56% of college students who report using alcohol, approximately 37% engage in

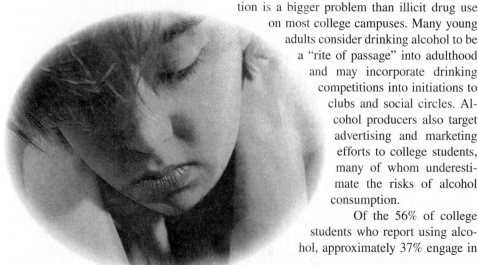

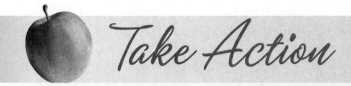

## Take Action

### Alcohol and Driving

Laura, Diane, Marc, and Jade attended an off-campus college graduation party over the weekend. The friends agreed to stay for 3 hours because Laura and Jade had to work the next day. When it came time to leave, Laura and Diane noticed that Marc and Jade seemed loud and boisterous and were slurring words. Marc also lost his balance and tripped several times as they walked to the car. Laura and Diane wondered if they should still ride home with their friends or if they should call a cab or take the bus instead. What would you do in their situation? Based on Marc's and Jade's behavior and symptoms, what do you think their blood alcohol concentrations (BACs) were? Were their BACs within a legally safe limit to drive?

Ryan McVay/Getty Images/RF

**Table 8-4  Impact of Harmful and Underage College Drinking**

**Death:** About 1825 college students between the ages of 18 and 24 die each year from alcohol-related unintentional injuries, including motor vehicle crashes.

**Assault:** About 696,000 students between the ages of 18 and 24 are assaulted each year by another student who has been drinking.

**Sexual abuse:** About 97,000 students between the ages of 18 and 24 report experiencing alcohol-related sexual assault or date rape each year.

**Academic problems:** About 25% of college students report academic consequences of their drinking, including missing class, falling behind, doing poorly on exams or papers, and receiving lower grades overall. Higher amounts of drinking lead to more serious academic problems.

**Alcohol use disorder:** Twenty percent of college students meet the criteria for a diagnosis of alcohol abuse and 6% for a diagnosis of alcohol dependence in the past 12 months, according to questionnaire-based self-reports about their drinking.

**Other consequences:** These include suicide attempts, injury, unsafe sex, driving under the influence, vandalism, property damage, and police involvement.

The consequences of excessive and underage drinking affect virtually all college campuses, college communities, and college students, whether they choose to drink or not.

Source: www.collegedrinkingprevention.gov

binge drinking, the most common form of alcohol abuse.[11] Similarly, 33% of high school students report drinking at least sometimes and, when they do consume alcohol, they often binge drink.[12] As noted in Table 8-4, the consequences of binge drinking are serious and can lead to major academic, health, social, and legal problems. According to the National Institute on Alcohol Abuse and Alcoholism, thousands of college students are treated in emergency rooms for alcohol poisoning (described in Table 8-3) each year.[13]

A variety of strategies to tackle the problem of excessive drinking on college campuses are available. Interventions can target individuals or whole campus communities. Read about these strategies at www.collegedrinkingprevention.gov/.

## 8.4 Health Effects of Alcohol

Low to moderate use of alcohol has been associated with several social and health-related benefits. However, despite the possible benefits of regular, moderate alcohol use, excessive alcohol intake has serious effects on health and nutritional status. This relationship between

Thirty-one percent of all motor vehicle fatalities are alcohol-related.

Ingram Publishing

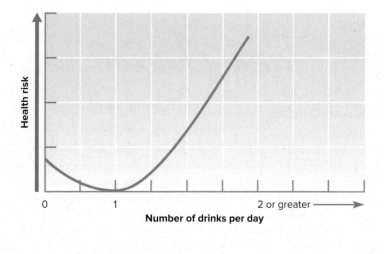

*Figure 8-4* The J-shaped curve illustrates the relationship between alcohol intake and health risks. Many studies show that a daily intake of ½ to 1 drink per day in women (1 to 2 drinks daily in men) is associated with the lowest health risk when compared with no alcohol intake or with higher alcohol intake. Despite this, health professionals do not recommend that those who currently abstain from alcohol should drink alcohol.

Of all the alcohol sources, red wine is often singled out as the best choice because of the added bonus of the many phytochemicals present, such as resveratrol. These leach out of the grape skins when red wine is made. Dark beer also contains some phytochemicals, but in lower amounts. Tea, chocolate, fruits, and vegetables also are good sources of phytochemicals.

Jenny Cundy/Image Source

the amount of alcohol consumed and its health effects is often described by alcohol researchers as a "J-shaped curve" (Fig. 8-4). Some researchers, however, continue to question if modest intakes are more healthful than no alcohol intake.[14]

## Guidance for Using Alcohol Safely

The U.S. Surgeon General's office, National Academy of Science, U.S. Department of Agriculture, and U.S. Department of Health and Human Services do not recommend that a nondrinker start consuming alcohol because risks often outweigh possible benefits. However, they do not specifically discourage moderate alcohol use. The following are suggestions for individuals who choose to drink alcohol.[15]

- Drinking alcohol should be done sensibly and in moderation—defined as the consumption of up to 1 drink per day for women and up to 2 drinks per day for men.
- Alcoholic beverages should not be consumed by some individuals, including those who cannot restrict their alcohol intake to moderate levels, women who are or may be pregnant, anyone younger than the legal drinking age, individuals taking medications that can interact with alcohol, those with specific medical conditions, and those who are driving or participating in activities requiring skill, coordination, and alertness.
- Mixing alcohol and caffeine is not regarded as safe. This can result in higher alcohol intakes and more intoxication. Caffeine does not change blood alcohol content levels.

## Potential Benefits of Alcohol Intake

Many people enjoy meeting a friend for a beer or having a glass of wine in the evening with dinner. Others report a reduced feeling of anxiety and stress after having a drink at the end of their workday. Moderate alcohol use can stimulate appetite and increase dietary intake, which may be helpful to those with depressed appetites such as some elderly individuals.[16,17] These behaviors are not considered harmful as long as they are practiced by individuals of a legal drinking age, continue in moderation, and cause no obvious harm.

In middle-aged and older adults, moderate alcohol use has been shown to lower the risk of cardiovascular disease and overall mortality in comparison with nondrinkers.[18,19,20] Alcohol can lower serum low-density lipoprotein (LDL) levels, increase protective high-density lipoprotein (HDL) levels, and decrease platelet aggregation (blood cell accumulation)—all of which may help reduce the risk of heart disease. Note that high alcohol intakes do not confer these benefits. Some studies suggest that the heart disease prevention benefits are associated with red wine consumption and a phytochemical it contains, called **resveratrol**. However, other compounds and types of alcoholic beverages may produce similar benefits.

*Figure 8-5* Alcohol affects virtually every organ.

1. Cognitive deficits: difficulties with memory, learning, and problem solving

2. Vasodilation and flushing of the skin

3. Cancer of the oral cavity, throat, larynx, and esophagus

4. Increased blood pressure, heart muscle damage, and resulting heart failure

5. Breast cancer

6. Irritation of the stomach lining (gastritis) and stomach cancer

7. Fatty infiltration of the liver, alcoholic hepatitis, cirrhosis, liver failure, and liver cancer

8. Impaired pancreatic function and related hypoglycemia, pancreatic cancer

9. Malabsorption of nutrients in the small intestine

10. Abdominal fat deposition and fluid accumulation (ascites)

11. Cancer of the colon and rectum

12. Decrease in bone mineral density and increased vulnerability to hip fracture

Although the heart health benefits of alcohol have received the most attention, a J-shaped relationship between alcohol use and risk of type 2 diabetes and dementia also occurs.[21,22] Alcohol intake also can stimulate the appetite and may cause those trying to control calorie intake to overeat.[17] Although moderate intakes may be protective, high alcohol intakes can injure the pancreas and brain. Additional research is required.

## Risks of Excessive Alcohol Intake

The excessive consumption of alcohol contributes significantly to 4 of the 10 leading causes of death in North America—heart failure, certain forms of cancer, motor vehicle and other accidents, and suicides. Excessive alcohol consumption is responsible for nearly 1 in 10 deaths among U.S. adults and causes approximately 88,000 deaths annually.[23]

As shown in Figure 8-5, excessive alcohol intake affects many organs and systems in the body. Heavy drinkers may develop heart damage that causes arrhythmias (abnormal heartbeats) and fluid retention in the lungs. Excess alcohol intake also can contribute to high blood pressure and the risk of stroke. Limiting alcohol intake to no more than 2 drinks per day (men) and 1 drink per day (women) may help prevent or improve these conditions.

Cancer also is related to alcohol consumption—acetaldehyde, a compound formed in the metabolism of ethanol, is a known carcinogen. Research suggests that 3.5% of all cancer deaths in the U.S. (about 19,500 deaths per year) are due to alcohol intake.[24] The cancers

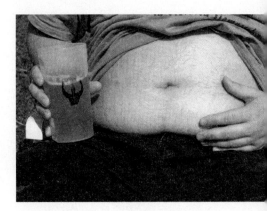

A "beer belly," common in many drinkers, occurs because alcohol promotes the synthesis of fat in the liver and promotes a positive energy balance that contributes to a risk of obesity, especially abdominal obesity.

©imageBROKER/Alamy

**fatty liver** Accumulation of triglycerides and other lipids inside liver cells; most often caused by excessive alcohol intake. Other causes include malnutrition and obesity.

**cirrhosis (see-ROH-sis)** Chronic degenerative disease, caused by poisons (e.g., alcohol) that damage liver cells, that results in a reduced ability to synthesize proteins and metabolize nutrients, drugs, and poisons.

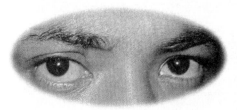

*Figure 8-6* *Jaundice* is a yellowish discoloration of the skin, the sclerae of the eyes, and other tissues caused by the buildup of bile pigments in the blood. Centers for Disease Control and Prevention (CDC)

most strongly related to alcohol consumption are those of the oral cavity, pharynx, larynx, esophagus, liver, breast, colon, and rectum.[25,26] Cancer risk increases with higher alcohol intake, but there is no safe consumption level. Abstaining from alcohol is a way to reduce cancer risk.

Prolonged, excessive alcohol intake can cause significant liver damage and lead to the development of cirrhosis of the liver.[27,28] Alcohol-related liver damage also can increase the risk of liver cancer. Other adverse effects of excessive alcohol intake are osteoporosis, brain damage and cognitive impairment, inflammation of the stomach lining, intestinal bleeding, pancreatitis, suppression of the immune system (with an increased risk of infections), sleep disturbances, impotence, hypoglycemia (an effect of acute excessive alcohol intake), hyperglycemia (an effect of chronic, excessive alcohol intake on pancreatic function), abdominal obesity, high blood triglyceride levels, and nutrient deficiencies.

### Cirrhosis of the Liver

The liver is 1 of the largest organs in the body. It has many functions, including nutrient storage, protein and enzyme synthesis, and the metabolism of protein, fats, and carbohydrates. It also is vital for removing toxins from the body and for metabolizing drugs. Because the liver is the main organ for alcohol metabolism, long-term alcohol abuse damages the liver.

The first change seen in the liver is fat accumulation, known as **fatty liver**, or steatosis. Fatty liver occurs in response to the increased synthesis of fat and trapping of fat in the liver. More than 90% of heavy drinkers develop fatty liver. This condition may be reversible, but only if the person abstains from drinking alcohol.[28]

If alcohol consumption persists, inflammation of the liver cells, known as alcoholic hepatitis, can develop. Alcoholic hepatitis produces symptoms of nausea, poor appetite, vomiting, fever, pain, and **jaundice** (yellow-orange coloration of the skin and whites of the eyes) (Fig. 8-6), resulting from the liver's inability to excrete bile pigments, which instead spill from damaged hepatocytes (liver cells) into the blood. Alcoholic hepatitis is a serious condition, which frequently progresses to the chronic, irreversible liver disease known as cirrhosis.

**Cirrhosis** is characterized by the loss of functioning hepatocytes (Fig. 8-7). As cirrhosis progresses, the synthesis of proteins, such as those required for normal blood clotting and

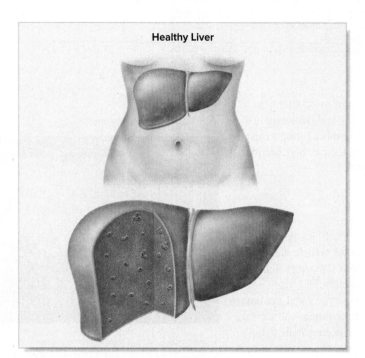

**Healthy Liver**

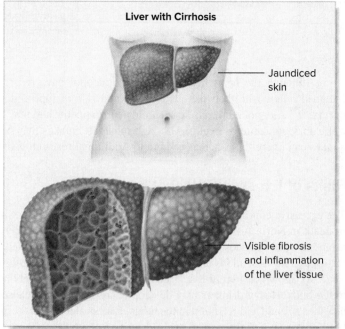

**Liver with Cirrhosis**

Jaundiced skin

Visible fibrosis and inflammation of the liver tissue

*Figure 8-7*  Notice the difference between the healthy liver on the left and the diseased liver with cirrhosis on the right. The bumpy, nodular surface of the cirrhotic liver is from excess scar tissue, which forms as healthy cells are destroyed. The cirrhotic liver can no longer function normally.

nutrient transport, decreases dramatically. Ascites, abnormal fluid retention in the abdomen, amounting to as much as 4 gallons (15 liters), is another common complication of cirrhosis. Nutritional status also is often poor.[27]

Whereas the early stages of alcoholic liver injury (fatty liver and alcoholic hepatitis) may be reversible with abstinence, cirrhosis is not, though symptoms may improve. Many individuals with cirrhosis develop liver failure. Once a person has cirrhosis, there is a 50% chance of death within 4 years. More than 36,000 people die from cirrhosis each year in the U.S.—most are between the ages of 40 and 65 years.[29] A liver transplant is necessary for long-term survival, and successful liver transplantation requires that patients abstain from consuming any alcohol.

Cirrhosis develops in about 10 to 15% of cases of alcoholism and affects about 2 million people in the U.S. Nonalcoholic fatty liver disease (discussed in Chapter 4) and hepatitis viruses are other leading causes of cirrhosis. The amount of alcohol that will damage the liver varies from person to person, but consuming less than 2 and 3 drink equivalents per day for women and men, respectively, will probably not injure the liver.[28] Drinking more than this over a period of 10 or more years can eventually lead to liver damage and cirrhosis. In addition to the amount and duration of alcohol consumption, genetic factors and individual factors, such as obesity, diabetes, exposure to hepatotoxins (e.g., acetaminophen [Tylenol®]), iron overload disorders, and infections causing hepatitis, determine the risk for the disease.

A number of possible mechanisms are thought to cause the liver damage that results from alcohol abuse. In chronic alcoholism, the increased concentration of acetaldehyde in the liver is thought to damage this organ. The accumulation of fat in liver cells causes inflammation and cell damage. Scientists also believe that the production of free radicals from alcohol metabolism contributes to liver damage. The highly reactive free radical molecules destroy cell membranes and lead to chronic inflammation. Additionally, disruption of the gut microbiota may contribute to liver damage.

A nutritious diet may help delay some of the complications associated with alcohol abuse and alcoholic liver disease. However, alcoholism usually brings about the serious destruction of vital tissues regardless of the quality of one's diet. In fact, studies in laboratory animals show that, even when a nutritious diet is consumed, alcohol abuse leads to cirrhosis. However, a poor diet often results in nutrient deficiencies that compound the problem of cirrhosis by making the liver more vulnerable to toxic substances produced by alcohol metabolism and by causing additional health concerns related to malnutrition.

Uric acid is formed from the breakdown of purines found in DNA and RNA. When uric acid production exceeds the kidneys' ability to excrete it, uric acid crystals form in joints, resulting in the very painful condition known as gout. Gout is well known as a disease of affluence—its risk factors include overweight, excessive alcohol intake, and high dietary purine intake (gravies, organ meats, fish eggs, anchovies, sardines, and many meats are high in purines), along with genetics, male sex, and older age.

©Science Photo Library/SuperStock RF

## Effects of Alcohol Abuse on Nutritional Status

Individuals who abuse alcohol often have poor nutritional status and are at risk of developing nutrient deficiencies for a variety of reasons. Alcohol, a poor source of nutrients, may replace food in the diet. For instance, if a person were to use beer as a nutrient source, he or she would need to consume daily 40 to 55 bottles (12 oz each) to meet protein needs and 65 bottles for thiamin needs.

When an individual relies on alcohol for the majority of his or her energy needs, protein-energy malnutrition develops. The symptoms of protein-energy malnutrition with alcoholism are similar to those seen in individuals with kwashiorkor (see Chapter 7). In addition to protein deficiencies, deficiencies of vitamins and minerals can result in further complications. These deficiencies are usually the result of decreased intake, greater loss from increased rates of urination, impaired absorption, altered metabolism, and alcohol-related tissue damage in the GI tract, liver, and pancreas.[27]

Alcohol abusers are at increased risk of protein, vitamin, and mineral deficiencies.

KatarzynaBialasiewicz/Getty Images

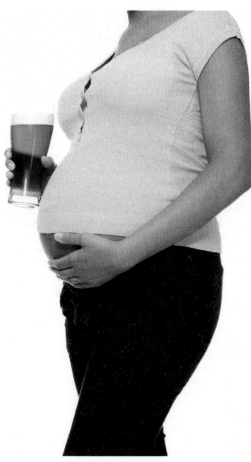

Drinking alcohol during pregnancy poses grave danger for the developing offspring.

PhotoMediaGroup/Shutterstock

**Wernicke-Korsakoff syndrome (ver-NIK-ee KOR-sah-koff)** Thiamin deficiency disease caused by excessive alcohol consumption. Symptoms include eye problems, difficulty walking, and deranged mental functions.

It is important for medical personnel to be aware of the severe consequences of alcohol abuse on nutritional health. Early intervention, guided by a dietitian and physician, is an essential part of the treatment for alcoholism to correct nutrient deficiencies, minimize tissue damage, and restore overall health.

### Water-Soluble Vitamins

Excessive alcohol intake can lead to deficiencies of the water-soluble vitamins thiamin, riboflavin, niacin, vitamin B-6, vitamin B-12, and folate. These deficiencies are due to poorer quality diets, increased nutrient need for alcohol metabolism, reduced absorption, and, in part, increased vitamin loss in the urine as a result of the more frequent urination caused by excessive alcohol intake (see Part 4).

Chronic alcohol abuse often leads to a severe form of thiamin deficiency called **Wernicke-Korsakoff syndrome**. This causes significant changes in brain and nervous system function, which, if untreated, result in irreversible paralysis of the eye muscles, loss of sensation in the lower extremities, loss of balance with abnormal gait, and memory loss.[27] Alcoholics are at increased risk of developing niacin deficiency because alcohol metabolism requires large quantities of this vitamin.[30] The metabolism of alcohol can increase the excretion of vitamin B-6 in the urine, which, if not offset with increased dietary intake, increases the risk of developing anemia and peripheral neuropathy (weakness or numbness in the arms and legs). Excessive alcohol intake also can impair the absorption of vitamin B-12, which also increases the risk of anemia and neuropathy.

### Fat-Soluble Vitamins

Excessive alcohol intake can result in deficiencies of the fat-soluble vitamins A, D, E, and K (see Part 4). Chronic alcohol abuse damages the liver and pancreas, which impairs the liver's ability to secrete bile and the pancreas's ability to secrete enzymes that digest fats. Decreased bile and pancreatic lipase secretions, in turn, lead to poor absorption of fat and fat-soluble vitamins.[25]

Vitamin A deficiency is common in alcoholics. Vitamin A is normally stored in specialized liver cells and, as these cells are damaged, vitamin A stores are lost.[27] The risk of vitamin A deficiency is compounded by the liver's inability to produce the protein needed to deliver vitamin A to all parts of the body. Alcohol also can decrease the amount of beta-carotene (a precursor of vitamin A) that the liver converts to vitamin A. Alcoholics often have trouble seeing in the dark (a condition called night blindness) because of vitamin A deficiency. Individuals with alcoholic liver disease are less able to synthesize vitamin K–containing compounds that are needed to help blood clot, in turn increasing the risk of bleeding. Additionally, vitamin D deficiency is common in alcoholics.[31] This may be due to poor intestinal absorption and/or activation of vitamin D to its biologically active form in the liver. Vitamin D is required for calcium absorption and bone health; thus, a deficiency can cause bone loss and increase the risk of osteoporosis.

### Minerals

Individuals who abuse alcohol are at increased risk of developing mineral deficiencies as well. Deficiencies of calcium, magnesium, zinc, and iron are most common (see Part 4). Poor absorption of calcium (because of low vitamin D status) contributes to a deficiency of this mineral. Increased urinary excretion of magnesium contributes to low blood concentrations of magnesium and deficiency symptoms. One of the classic symptoms of magnesium deficiency is tetany, a condition characterized by muscle twitches, cramps, spasms, and seizures. Decreased absorption and increased urinary excretion contribute to zinc deficiency, which leads to changes in taste and smell, loss of appetite, and impaired wound healing.

Iron deficiency commonly occurs in alcoholics. Excessive alcohol consumption can damage gastrointestinal tissues, causing bleeding, malabsorption, and the eventual development of iron deficiency.

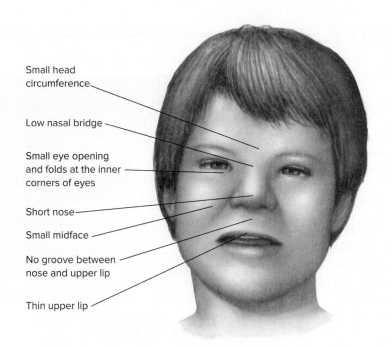

Small head circumference

Low nasal bridge

Small eye opening and folds at the inner corners of eyes

Short nose

Small midface

No groove between nose and upper lip

Thin upper lip

*Figure 8-8* The facial features shown are typical of children with fetal alcohol syndrome. Additional abnormalities in the brain and other internal organs accompany fetal alcohol syndrome but are not immediately apparent from simply looking at the child. Milder forms of alcohol-induced changes from a lower alcohol exposure to the fetus are known as fetal alcohol spectrum disorder.

## Alcohol Consumption during Pregnancy and Breastfeeding

More than half of all women in the U.S. of childbearing age drink alcohol. If a woman chooses to drink during pregnancy, she may cause serious harm to her developing offspring. Alcohol slows nutrient and oxygen delivery to the fetus, thereby retarding growth and development. Alcohol also may displace nutrient dense foods in the mother's diet. Although the most severe damage occurs during the first 12 to 16 weeks of pregnancy, when organs are undergoing major developmental steps, consuming alcohol at any time during pregnancy can cause lifelong damage.

The alcohol-related health consequences are known collectively as fetal alcohol spectrum disorders (FASD).[32] An individual with an FASD may experience a range of effects, including facial malformation (Fig. 8-8), growth retardation (including a smaller than normal brain), birth defects, lifelong learning disabilities, a short attention span, and hyperactivity. The most serious FASD is known as fetal alcohol syndrome (FAS). Individuals with FAS have facial malformations, growth deficits, and central nervous system problems. No one knows how many individuals have an FASD, but the Centers for Disease Control and Prevention reports that a range of 6–9 out of 1000 children suffer from FAS and at least 3 times as many more experience FASD.[32] It is not clear how alcohol causes physical malformations and disabilities. They may be the result of alcohol itself or compounds produced during alcohol metabolism. Within minutes after consumption, the alcohol travels through the mother's blood to the developing offspring. The effects of alcohol are intensified by the offspring's small size and are prolonged because the fetus is unable to metabolize the alcohol and must wait for maternal blood to carry it away.

FASD are 100% preventable simply by avoiding alcohol during pregnancy. Because no one is sure how much alcohol it takes to cause FASD, health experts agree that women should abstain from drinking alcohol during pregnancy. The more alcohol consumed during pregnancy, the worse the effects are likely to be. Also, experts recommend that women planning pregnancy or who are sexually active and without effective birth control should avoid alcohol. For more information about FASD and FAS, visit the website www.cdc.gov/ncbddd/fasd/index.html.

For centuries, new mothers were advised to drink a little wine or beer before breastfeeding to relax and allow babies to suckle longer. However, alcohol actually reduces the

# GL BAL PERSPECTIVE

### Alcohol Intake around the World

International alcohol intake patterns vary tremendously. According to World Health Organization data, over half of the populations in 3 major regions—the Americas, Europe, and the Western Pacific—consume alcohol. Worldwide, less than half of the population aged 15 years and older consume alcohol. Of those who consume alcohol, the average intake is almost 17 liters (about 575 ounces) of pure alcohol annually.[34]

Alcohol intake around the world is increasing, perhaps in response to Westernization and aggressive marketing by the alcohol industry. According to the World Health Organization, 1 in 20 of all deaths can be attributed to the harmful affects of alcohol. More than one-fourth of all 15- to 19-year-olds worldwide are drinkers, with rates highest in Europe (44%), the Americas (38%), and Western Pacific (38%). Almost half (45%) of all recorded alcohol consumed is in the form of spirits, followed by beer (34%) and wine (12%).[34]

An extreme example of alcohol consumption is in certain areas of Russia, where some consume as much as a bottle of vodka daily. Worldwide, about 4% of deaths are attributed to alcohol intake, yet in areas of Russia more than half of the deaths of men aged 15 to 54 years are due to

Mauro Pezzotta/Shutterstock

alcohol.[35] These deaths are caused by accidents and violence, acute heart disease, tuberculosis, pneumonia, liver and pancreatic disease, and cancers of the mouth, esophagus, and liver. Many are from alcohol poisoning, which kills more than 40,000 Russians each year.

Alcohol, typically wine, consumed with meals is a tradition enjoyed in many cultures. As you know, this type of moderate alcohol intake may offer some health benefits for those who choose to drink alcohol. Moderation is key to reaping benefits—risks rise quickly when women consume more than 1 drink per day and men consume more than 2 drinks daily. Risks are particularly high when many alcoholic beverages are consumed in rapid succession—this practice can result in alcohol poisoning and death.

---

mother's ability to produce milk and causes babies to drink less and have disrupted sleep patterns. In addition, infants are not efficient at breaking down alcohol, so its detrimental effects are much more pronounced than in adults. The safest plan for mothers who are breastfeeding is to avoid alcohol altogether.[33] However, breastfeeding women who want to drink alcohol are advised to limit their intake to no more than 1 to 2 drinks and then wait at least 2 hours before breastfeeding. The amount of alcohol in breast milk peaks about 30 to 60 minutes after the mother ingests it, then declines.

### Knowledge Check

1. What are the guidelines for using alcohol safely?
2. What are possible benefits of moderate alcohol intake by middle-aged and older adults?
3. Ingestion of alcohol is related to developing at least 7 different kinds of cancer. Name 5 of them.
4. Why does persistent alcohol abuse cause liver damage? What are the signs and symptoms of liver damage?
5. How does alcohol abuse impair nutritional status?
6. What are the dangers of consuming alcohol during pregnancy?

# 8.5 Alcohol Use Disorders

Alcohol use disorders (formerly known by the terms *alcohol abuse* and *alcohol dependence*) are common mental health disorders that pose serious risks for many individuals. In addition to the multiple health problems previously described, persons with an alcohol use disorder (AUD) may experience deteriorating personal relationships, financial burdens, and frequent absences or poor performance in the workplace, school, or other settings. The criteria for diagnosing an AUD are shown in Table 8-5.

AUDs afflict many individuals and families. Research shows that 29% of Americans experience an AUD in their lifetimes, with nearly 14% of adults (or about 1 in 7) meeting the criteria for an AUD diagnosis in a given year.[37] The factors linked to developing alcohol-related problems include genetics, biological sex, age of drinking onset, and ethnicity. Parental and peer attitudes favoring excessive drinking and the presence of mental health disorders, such as depression and anxiety disorder, are other significant risk factors.

▶ Swiss chemist Paracelsus (1493–1541) made the observation that "the dose determines the poison." This is true for alcohol because, the greater the dose of alcohol, the greater the risk of alcohol poisoning and death.

▶ The CAGE questionnaire is commonly used in routine health care to screen for an AUD.[36]

**C:** Have you ever felt you ought to *cut* down on drinking?

**A:** Have people *annoyed* you by criticizing your drinking?

**G:** Have you ever felt bad or *guilty* about your drinking?

**E:** Have you ever had a drink first thing in the morning (an *eye-opener*) to steady your nerves or get rid of a hangover?

More than a single positive response in the CAGE questionnaire suggests an alcohol problem. An easy-to-use online screening tool is found at www.alcoholscreening.org.

---

**Table 8-5  DSM-5 Diagnostic Criteria for an Alcohol Use Disorder**

**In the past year, have you . . .**

1. Had times when you ended up drinking more, or longer, than you intended?
2. More than once wanted to cut down or stop drinking or tried to but couldn't?
3. Spent a lot of time drinking? Or being sick or getting over other aftereffects?
4. Wanted a drink so badly you couldn't think of anything else?
5. Found that drinking—or being sick from drinking—often interfered with taking care of your home or family? Or caused troubles? Or school problems?
6. Continued to drink even though it was causing trouble with your family or friends?
7. Given up or cut back on activities that were important or interesting to you, or gave you pleasure, in order to drink?
8. More than once gotten into situations while or after drinking that increased your chances of getting hurt (such as driving, swimming, using machinery, walking in a dangerous area, or having unsafe sex)?
9. Continued to drink, even though it was making you feel depressed or anxious or adding to another health problem? Or having had a memory blackout?
10. Had to drink much more than you once did to get the effect you want? Or found that your usual number of drinks had much less effect than before?
11. Found that, when the effects of alcohol were wearing off, you had withdrawal symptoms, such as trouble sleeping, shakiness, restlessness, nausea, sweating, a racing heart, or a seizure? Or sensed things that were not there?

The presence of at least 2 of these symptoms indicates an alcohol use disorder (AUD).

The severity of the AUD is defined as

**Mild:** The presence of 2 to 3 symptoms

**Moderate:** The presence of 4 to 5 symptoms

**Severe:** The presence of 6 or more symptoms

Source: American Psychiatric Association. *Diagnostic and Statistical Manual of Mental Disorders*. 5th ed. Arlington, VA: American Psychiatric Publishing, 2013.

## Critical | Thinking

José is a well-liked 17-year-old. It always seems as if everything is going his way—an A on a test, a scholarship to college, you name it. Lately, however, José has experienced some disappointments. His grandfather has just passed away and he and his girlfriend broke up. When he arrived home late with the smell of alcohol on his breath, his parents started to worry. What signs and symptoms should they be aware of that indicate a problem with alcohol?

Drinking alcohol at a young age increases the risk of alcohol addiction.

Image Source/Getty Images

### Genetic Influences

Genetic factors account for approximately 40 to 50% of a person's risk of alcoholism.[6] Twins and 1st-degree relatives (parents, siblings, and offspring) share a tendency toward alcohol addiction. Children of alcoholics have a 4 times greater risk of developing alcoholism, even when adopted by a family with no history of alcoholism.

Scientists are actively studying the genetic basis for alcohol dependence. The genes regulating alcohol metabolism enzymes—alcohol dehydrogenase and aldehyde dehydrogenase—have been of particular interest to researchers.[6] These genes have several possible variants that can occur. Scientists are trying to understand how each variant alters alcohol metabolism and affects the risk of developing alcohol dependence or alcoholic liver disease. For example, as discussed earlier, the inability to completely metabolize alcohol quickly may cause a person to feel ill, causing him or her to be unlikely to drink large amounts of alcohol. Other genes, such as those that make antioxidant enzymes, neurotransmitters, and immune factors, are also under study.

Individuals with a family history of alcoholism need to be especially alert for evidence of the early signs of alcohol dependence. However, it is important to recognize that genetic risk is not destiny—not all children of alcoholic parents go on to develop alcohol use problems. Some alcoholics have no family history of alcohol problems at all, so genetic factors do not fully explain why some individuals develop alcohol-related problems but others do not.

### Effect of Biological Sex

Biological sex plays a key role in alcohol dependence and metabolism. The male:female ratio of alcohol use disorders is 4:1. However, women are more susceptible to the adverse effects of alcohol—liver disease, heart muscle damage, cancer, and brain injury. As previously noted, the recommended limit for alcohol use is lower for women than for men because an equivalent amount of alcohol is more concentrated in women—they are smaller and have more fat and less body water than men. Additionally, women cannot metabolize alcohol as quickly as men, due to lower activity of the alcohol dehydrogenase in the stomach, so women's blood alcohol concentrations remain elevated for longer periods.

### Age of Onset of Drinking

Not only does alcohol consumption in underage youths contribute to nearly 4400 deaths in the U.S. each year (mainly from homicides, motor vehicle accidents, and suicides), but drinking at a young age also is an important risk factor for later alcohol dependence.[38] Researchers have found that drinking before age 14 is especially problematic. This is particularly worrisome, considering that about 35% of high school students report that they currently consume alcohol.

### Ethnicity and Alcohol Use

Patterns of alcohol use vary among ethnic groups. In North America, non-Hispanic White individuals have the highest levels of consumption and those of Asian heritage have the lowest use.[39] However, alcohol use disorders are highest in Native American Indians. Health burdens related to alcohol consumption (e.g., alcohol-related motor vehicle accidents, liver disease) are greatest in American Indians and Black individuals.[39] Differences in alcohol use and effects may result from social factors, such as the availability of alcohol in communities, and biological factors that affect vulnerability to alcohol. As discussed previously, many Asians experience uncomfortable effects after drinking alcohol, which likely accounts for low alcohol use in this population.

## Mental Health and Alcohol Use

Mental health disorders, such as depression and generalized anxiety disorder, often go hand in hand with alcohol use disorders.[39] Alcohol dependence and abuse may aggravate or even cause mental health disorders. Conversely, individuals with depression or other disorders may use alcohol to self-medicate their conditions. It's important that both mental health disorders and alcohol abuse be identified and treated.

The majority of suicides and interfamily homicides are alcohol-related. Alcohol consumption appears to increase the risk of youth suicide—the younger the drinker, the more likely he or she is to commit suicide.

## The Economic Costs of Alcohol Abuse

The CDC estimates that the annual cost of alcohol abuse in the U.S. is $249 billion (more than 70% of which is from binge drinking) in lost productivity, health-care costs, automobile collisions, and criminal justice expenses.[41] For instance, a liver transplant costs several hundred thousand dollars (a recent estimate is $714,000) and may be needed in cases of excessive alcohol use.[42] In contrast, a typical outpatient counseling program to treat a person who is abusing alcohol may cost $5000 to $10,000 or more.

## CASE STUDY FOLLOW-UP

Charles is consuming 13 to 16 beers per week. Using the average of 14.5 beers per week, his beer consumption provides an additional 2175 calories each week and 26,100 calories in the 12-week period. Dividing 26,100 calories by 3500 calories per pound of fat yields an estimated weight gain of 7.5 pounds. It is very likely that Charles's weight gain is due to his new beer drinking habit.

McGraw-Hill
Education/Gary He,
photographer

### Knowledge Check

1. What factors predispose a person to alcohol dependence?
2. Why is the age at which a person started drinking alcohol of concern?
3. Why might women suffer more ill effects from alcohol consumption than men?
4. Which ethnic groups are at increased risk of alcohol-related problems?

# CLINICAL PERSPECTIVE

## Treatment of Alcohol Use Disorders

The treatment of an AUD typically involves behavioral therapy, along with medication.[43] Behavioral therapy is usually facilitated by a psychologist, social worker, or counselor. Important goals of counseling are to identify triggers for alcohol consumption and ways to compensate for the loss of pleasure from drinking. Total abstinence must be the ultimate objective because, for most alcoholics, there is no such thing as safe and controlled social drinking. Because other mental health disorders often go hand in hand with AUDs, these conditions must be treated as well for successful outcomes.

Three medications are approved in the U.S. to treat AUDs.[43]

- Naltrexone (ReVia®) blocks the craving for alcohol and the pleasure of intoxication (Fig. 8-9).
- Acamprosate (Campral®) is thought to act on neurotransmitter pathways in the brain to decrease the desire to drink.
- Disulfiram (Antabuse®) causes physical reactions, such as vomiting, when drinking alcohol. It does so by blocking the complete breakdown of alcohol in the liver in a way similar to that experienced naturally by some individuals of Asian decent (as described earlier in the chapter).

Mutual-help programs also facilitate the recovery from alcohol dependence. Alcoholics Anonymous® (AA), a 12-step program, is among the most well-known programs. As an informal society chartered in 1935, AA includes more than 2 million recovered alcoholics. The only requirement for membership is the desire to stop drinking. There are no rules, regulations, dues, or fees. Further information about AA is available at www.aa.org. Another organization, Al-Anon, helps family members and friends recover from the effects of living with an alcohol-dependent person; visit al-anon.org for more information.

Current research does not support the generally negative public opinion about the prognosis for recovery from alcoholism. In most job-related alcoholism treatment programs, where workers are socially stable and well motivated, recovery rates reach 60% or more. This remarkably high cure rate is probably accounted for by early detection. Once a person moves from problem drinking to an advanced stage of alcoholism, treatment success seldom exceeds 50%. Early identification and intervention remain the most important steps in the treatment of alcoholism.

▶ Alcoholics who stop drinking may substitute caffeine, nicotine, and/or sweets for alcohol. Because alcoholics usually have poor nutritional status, these substitutions often have a significant negative impact on their overall nutritional status. However, these substitutions are not as harmful as alcohol abuse.

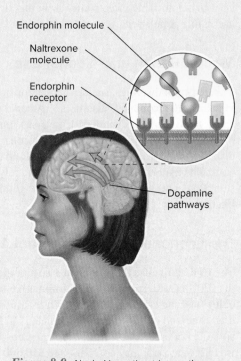

*Figure 8-9* Alcohol ingestion triggers the release of endorphins in the brain. Endorphins are chemical compounds produced in the brain that act as natural painkillers and elicit feelings of well-being. It is likely that this binding, in turn, causes a release of the neurotransmitter dopamine, which is thought to cause the characteristic high associated with alcohol use. Naltrexone (ReVia®) works by blocking alcohol's ability to bind to brain receptors. This, then, reduces dopamine release and blocks the pleasant feelings elicited by alcohol use.

# Take Action

## Do You Know Why These Are Alcohol Myths?

Myths about alcohol and its effects abound. See if you can correctly identify what makes the statements inaccurate. Read each alcohol myth, analyze it to determine why it is not true, and then check the facts!

| Myths | Facts |
|---|---|
| "Beer before liquor, you've never been sicker. Liquor before beer and you're in the clear." | The *amount* of alcohol consumed determines how sick a person will become, not the sequence in which different types are consumed. One effect of excessive alcohol consumption is a *hangover*, characterized by fatigue, headache, dizziness, sensitivity to light or noise, poor sleep, inability to concentrate, dehydration, shakiness, and gastrointestinal disturbances—nausea, vomiting, diarrhea, or stomach pain. Too much beer, wine, or hard liquor can cause a hangover, regardless of the combination or sequence they are consumed. |
| Light beer is healthier than regular beer. | Compared to regular beer, light beers have fewer calories because they have fewer carbohydrates and/or lower alcohol content. The composition of light beer varies from brand to brand. Many have a 4 to 5% alcohol concentration, compared to the typical 5 to 6% found in regular beer. This small reduction in alcohol and calories does not make it healthier than regular beer. Drinking light beer is not a green light for overconsumption—the guidelines for moderate consumption still apply. |
| A breastfeeding woman can produce more milk by having a drink just before feeding her baby. | The consumption of alcohol does not increase milk production. The mother's blood alcohol content and the concentration of alcohol in breast milk go hand in hand. Some research shows that babies drink less breast milk when it contains alcohol. The baby may become sleepy from the alcohol or may not like its flavor or odor. See Chapter 17 for more information on breastfeeding. |
| Eating greasy food after a night's drinking will reduce a hangover. | The best way to prevent a hangover is to drink less. Eating greasy food like pizza won't speed the metabolism of alcohol by the liver or prevent the toxic effects of too much alcohol. |
| Dark-colored beers have a higher alcohol content (and more calories) than pale beers. | Beer color comes from chemical reactions, known as browning reactions, in the malt (germinated cereal grains). Darker beer is made from darker malt that has undergone more of these browning reactions. The malt color has no effect on the alcohol or calorie content of the beer. Dark and light beers vary in their alcohol content. |
| Drinking hard liquor, such as whiskey and rum, is more harmful than drinking beer or wine. | Drinking too much of any alcoholic beverage is harmful. The harmful effects of consuming alcohol, such as liver disease and alcohol use disorders, are related to the amount of ethanol consumed, not its source. |
| Alcohol warms you up. | Drinking alcohol temporarily increases blood flow to the skin, making you feel warmer, but this actually cools the body. |
| Alcohol is a stimulant. | Early effects of drinking alcohol include a loss of inhibition and increased self-confidence, which makes some people think alcohol is a stimulant drug. In reality, alcohol is a central nervous system depressant and has many negative effects on brain function when consumed in greater than moderate amounts. |
| Beer is a good source of vitamins and minerals. | Beer is made from hops, grains, and yeast, which contribute small amounts of the B-vitamins, potassium, iron, and other minerals. But the main ingredient in beer is water, which dilutes these small amounts even further. The small amount of nutrients found in beer cannot compete with the more concentrated amounts found in nutrient dense foods. For example, beer is sometimes touted as being a good source of folate, but 1 pint of beer might contain 30 µg of folate, much less than the 140 µg found in ½ cup of asparagus. |
| Alcohol stimulates the appetite. | This depends on how the alcohol is consumed. Drinking small amounts of alcohol at a meal or in a social setting may help stimulate the appetite. However, drinking larger amounts of alcohol can reduce the appetite, and healthy foods may be displaced by alcoholic beverages. Heavy drinkers often have poor diets and nutritional status. |

Jodi Jacobson/E+/Getty Images

# Chapter *Summary*

## 8.1 Alcohol, chemically known as ethanol, is found in beer, wine,
distilled spirits (hard liquor, such as vodka and rum), liqueurs, cordials, and hard cider. These beverages range from 5% ABV (alcohol by volume) to more than 40% ABV. A standard drink is defined as 14 g of alcohol, and that equates to 12 oz of 5% ABV beer, 5 oz of 12% ABV wine, and 1.5 oz of distilled spirits (80 proof). Ethanol is produced by the chemical process known as fermentation. Grains, cereals, fruits, honey, milk, potatoes, and other carbohydrate rich foods are used to make alcoholic beverages.

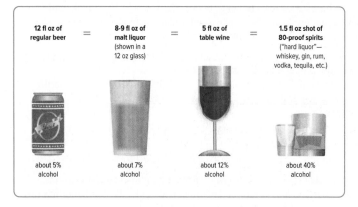

| 12 fl oz of regular beer | = | 8-9 fl oz of malt liquor (shown in a 12 oz glass) | = | 5 fl oz of table wine | = | 1.5 fl oz shot of 80-proof spirits ("hard liquor"— whiskey, gin, rum, vodka, tequila, etc.) |
|---|---|---|---|---|---|---|
| about 5% alcohol | | about 7% alcohol | | about 12% alcohol | | about 40% alcohol |

Source: National Institute on Alcohol Abuse and Alcoholism, http://www.niaaa.nih.gov/alcohol-health/overview-alcohol-consumption/standard-drink

## 8.2 Alcohol is readily absorbed in the GI tract because it does not require digestion.
The rate of absorption is affected by biological sex, ethnicity, body size and composition, and alcohol and food intake. Alcohol is primarily metabolized in the liver, although small amounts also are metabolized in the stomach. The body uses the alcohol dehydrogenase (ADH) pathway to metabolize most alcohol, but the microsomal ethanol oxidizing system (MEOS) becomes more active with chronically high alcohol intakes. Blood alcohol levels can be determined by measuring the amount of alcohol excreted through the lungs with a breathalyzer test. A blood alcohol content of 0.08% is the legal definition of intoxication and is associated with impaired judgment and coordination. Alcohol poisoning, a very dangerous condition, occurs when blood alcohol content continues to rise, causing vomiting, irregular breathing and heartbeat, low blood sugar, severe dehydration, seizures, confusion, coma, and death.

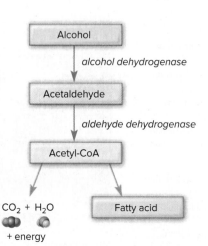

Alcohol
↓ *alcohol dehydrogenase*
Acetaldehyde
↓ *aldehyde dehydrogenase*
Acetyl-CoA
↓          ↓
$CO_2$ + $H_2O$     Fatty acid
+ energy

Table 8-2 Alcohol Metabolism Summary

| Alcohol Metabolic Pathway | Main Location of Pathway Activity | Alcohol Intake That Activates Pathway | Extent of Participation in Alcohol Metabolism |
|---|---|---|---|
| Alcohol dehydrogenase pathway (ADH) | Stomach Liver (mostly) | Low to moderate intake | Metabolizes about 90% of alcohol |
| Microsomal ethanol oxidizing system (MEOS) | Liver | Moderate to excessive intake | Increases with increasing alcohol intake |
| Catalase pathway | Liver Other cells | Moderate to excessive intake | Minor |

## 8.3 About 70% of North American adults consume alcohol;
many are light or moderate drinkers, but some consume excessive amounts. College students are frequent drinkers and many engage in binge drinking. Binge drinking is a dangerous practice associated with a variety of accidents, crimes, health risks, and even death.

Table 8-3 Signs and Symptoms of Alcohol Poisoning

- Confusion, stupor
- Vomiting
- Low blood sugar (hypoglycemia)
- Severe dehydration
- Seizures
- Slow or irregular breathing and heartbeat
- Blue-tinged or pale skin
- Low body temperature (hypothermia)
- Loss of consciousness

Call 911 or your local emergency number if you suspect alcohol poisoning and the person cannot be roused or is unconscious. If the person is conscious, you can call 1-800-222-1222 to be routed to your local poison control center for further instructions.

## 8.4 If alcohol is consumed, it should be consumed in moderation with meals.
Women are advised to drink no more than 1 drink per day, men no more than 2 drinks a day. The benefits of alcohol use are associated with low to moderate alcohol consumption. These benefits include the pleasurable and social aspects of alcohol use and possibly a reduced risk of cardiovascular disease. Excessive consumption of alcohol is responsible for nearly 10% of deaths in U.S. adults. Alcohol increases the risk of developing heart damage, inflammation of the pancreas, GI tract damage, certain forms of cancer, and hypertension. The liver is particularly vulnerable to the toxic effects of alcohol. Liver damage occurs as fatty liver, alcoholic hepatitis, and cirrhosis. Fatty liver and hepatitis may be relieved by abstention from alcohol, but cirrhosis cannot. Cirrhosis results in the impaired synthesis of vital proteins and abnormal fluid retention. Most people with cirrhosis are malnourished. Liver failure is the typical outcome of cirrhosis. Nutritional problems are common among alcoholics. Alcohol abuse can impair nutrient intake and absorption, alter nutrient metabolism, and increase nutrient excretion.

Deficiencies of protein, fat- and water-soluble vitamins, and the minerals calcium, magnesium, iron, and zinc are most common.

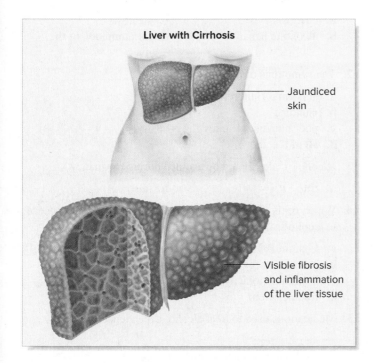

**Liver with Cirrhosis**

Jaundiced skin

Visible fibrosis and inflammation of the liver tissue

## Clinical Perspective

The treatment of an alcohol use disorder includes behavioral and pharmacologic therapy. Three medications are approved to treat alcoholism. Alcoholics Anonymous® is another avenue for help. Recovery rates can be high, especially with early intervention.

*CAGE Questionnaire*

C:  Have you ever felt you ought to *cut* down on drinking?

A:  Have people *annoyed* you by criticizing your drinking?

G:  Have you ever felt bad or *guilty* about your drinking?

E:  Have you ever had a drink first thing in the morning (an *eye-opener*) to steady your nerves or get rid of a hangover?

## Global Perspective

Alcohol intake around the world is increasing. Alcohol intake accounts for about 4% of deaths worldwide.

Jenny Cundy/Image Source

## 8.5 Alcohol use disorders affect 13% of the adult population in the U.S.

The DSM-5 diagnostic criteria and CAGE questionnaire can help identify individuals experiencing an alcohol problem. Many personal, social, and employment problems result from excessive alcohol use. Genetic factors account for 40 to 50% of a person's risk of alcoholism. Differences in the alcohol-metabolizing genes may account for some of these genetic risks. Males have higher rates of alcoholism, but women are at higher risk of alcohol-related damage throughout the body. Drinking at a young age puts one at risk of alcohol problems later in life. Native American Indians and Black individuals experience high rates of health burdens related to alcohol consumption. Mental health and alcohol use disorders occur together.

## Study Questions

1. An 80-proof alcohol is approximately 80% alcohol.

   **a.** true         **b.** false

2. A standard drink is defined as the amount that provides approximately 14 g of alcohol. Which of the following is considered a standard-size drink?

   **a.** 16-ounce beer (5% ABV)
   **b.** 20-ounce wine cooler (7% ABV)
   **c.** 5-ounce glass of wine (12% ABV)
   **d.** 4-ounce pour of hard liquor (80 proof)

3. Alcohol requires no digestion and can enter cells without specific transport mechanisms.

   **a.** true         **b.** false

4. Which of the following is the primary pathway used in the metabolism of alcohol?

   **a.** alcohol dehydrogenase pathway
   **b.** microsomal ethanol oxidizing system
   **c.** catalase pathway
   **d.** none of the above

5. Which of the following affects the metabolism of alcohol?

   **a.** biological sex      **c.** ethnicity
   **b.** diet composition    **d.** all of the above

6. What blood alcohol concentration denotes legal intoxication in the U.S. and Canada?

   **a.** 1.0%         **c.** 0.05%
   **b.** 0.08%       **d.** 0.10%

7. Compared with individuals who begin drinking alcohol at a legal age, those who begin drinking as teenagers have higher rates of alcohol use disorders in adulthood.

   **a.** true         **b.** false

8. Which of the following would be considered moderate drinking?

   **a.** 1 drink per day for women and 2 drinks per day for men
   **b.** 2 drinks per day for both men and women
   **c.** 2 drinks per day for women and 3 drinks per day for men
   **d.** 4 drinks on any single occasion for men and women

9. In moderation, alcohol may help raise HDL-cholesterol.

   **a.** true         **b.** false

10. Risk of cancer of the _____ increases greatly with high alcohol consumption.

    **a.** esophagus      **b.** bone
    **b.** lung           **d.** all of the above

11. The first stage of alcoholic liver disease is _____.

    **a.** cirrhosis        **c.** steatosis
    **b.** alcoholic hepatitis    **d.** inflammation of the liver

12. The symptoms of cirrhosis include _____.

    **a.** abnormal fluid retention
    **b.** jaundice
    **c.** poor nutritional status
    **d.** all of the above

13. Alcohol intake should be avoided during pregnancy.

    **a.** true         **b.** false

14. Which of the following is *not* a common nutritional concern in alcoholics?

    **a.** vitamin B-12 deficiency
    **b.** protein-energy malnutrition
    **c.** vitamin A toxicity
    **d.** iron deficiency

15. Medications used to treat alcohol dependence act on _____.

    **a.** the brain to reduce alcohol cravings
    **b.** the liver to block complete metabolism of alcohol
    **c.** the stomach to prevent alcohol absorption
    **d.** both a and b

16. What is a standard-size serving of alcohol? How many servings are considered moderate for men and women?

17. Describe how alcohol is metabolized. What is a toxic by-product of alcohol metabolism?

18. Define *binge drinking,* and list 4 problems associated with this practice.

19. Describe the health benefits and risks of alcohol use.

20. What criteria may indicate that someone is dependent on alcohol?

Answer Key: 1-b; 2-c; 3-a; 4-a; 5-d; 6-b; 7-a; 8-a; 9-a; 10-a; 11-c; 12-d; 13-a; 14-c; 15-d; 16-refer to Section 8.1; 17-refer to Section 8.2; 18-refer to Section 8.3; 19-refer to Section 8.4 and Global Perspective; 20-refer to Section 8.5 and Clinical Perspective.

# References

1. Butler LB and others. Trends in energy intake from alcoholic beverages among US adults by sociodemographic characteristics, 1989–2012. *J Acad Nutr Diet.* 2016;116:1087.

2. Nielsen SJ and others. *Calories consumed from alcoholic beverages by U.S. adults, 2007–2010.* NCHS data brief, #110. Hyattsville, MD: National Center for Health Statistics; 2012.

3. Kerr W and others. Alcohol content variation of bar and restaurant drinks in northern California. *Alcohol Clin Exp Res.* 2008;32:1623.

4. Naimi TS, Moscher JF. Powdered alcohol products new challenge in an era of needed regulation. *JAMA.* 2015;314:119.

5. National Institute on Alcohol Abuse and Alcoholism. Alcohol metabolism: An update. Alcohol alert 72. 2007; pubs.niaaa.nih.gov.

6. Wall TL and others. Biology, genetics and environment: Underlying factors influencing alcohol metabolism. *Alcohol Res.* 2016;38:59.

7. Schukit M. Alcohol and Alcohol Use Disorders. in Jameson, JL and others, eds. *Harrison's principles of internal medicine.* New York: McGraw-Hill; 2018.

8. Substance Abuse and Mental Health Services Administration. National Survey on Drug Use and Health (NSDUH). Table 2.41B—Alcohol use in lifetime, past year, and past month among persons aged 18 or older, by demographic characteristics: Percentages, 2013 and 2014. 2014; www.samhsa.gov/data/sites/default/files /NSDUH-DetTabs2014/NSDUH-DetTabs2014.pdf.

9. Dwyer-Lindgren L and others. Drinking patterns in US counties from 2002 to 2012. *Am J Pub Health.* 2015;105:1120.

10. U.S. Department of Health and Human Services. Healthy people 2030. 2020; www.healthypeople.gov.

11. Substance Abuse and Mental Health Services Administration. 2014 National Survey on Drug Use and Health (NSDUH). Table 6.88B—Alcohol use in the past month among persons aged 18 to 22, by college enrollment status and demographic characteristics: Percentages, 2013 and 2014. 2014; www.samhsa .gov/data/sites/default/files/NSDUH-DetTabs2014/NSDUH-DetTabs2014.pdf.

12. Kann L. Youth risk behavior surveillance—United States, 2015. *MMWR.* 2016;65(6):1.

13. National Institute on Alcohol Abuse and Alcoholism. College drinking. 2019; www.niaaa.nih.gov/alcohol-health/special-populations-co-occurring-disorders/ college-drinking.

14. Chokshi DA and others. J-shaped curves and public health. *JAMA.* 2015;314:1339.

15. U.S. Department of Health and Human Services and U.S. Department of Agriculture. Dietary Guidelines for Americans, 2020–2025, 9th Ed. 2020. dietaryguidelines.gov.

16. Ferreira M, Weems S. Alcohol consumption by aging adults in the United States: Health benefits and detriments. *J Am Diet Assoc.* 2008;108:1668.

17. Kase CA and others. The relationship of alcohol use to weight loss in the context of behavioral weight loss treatment. *Appetite.* 2016;99:105.

18. Constanzo S and others. Wine, beer or spirit drinking in relation to fatal and non-fatal cardiovascular events: A meta-analysis. *Eur J Epidemiol.* 2011;11:833.

19. Gemes K and others. Alcohol consumption is associated with a lower incidence of acute myocardial infarction: Results from a large prospective population-based study in Norway. *J Intern Med.* 2016;279:365.

20. Klatsky KL. Alcohol and cardiovascular disease: Where do we stand today? *J Intern Med.* 2015;278:238.

21. Knott C and others. Alcohol consumption and the risk of type 2 diabetes: A systematic review and dose-response meta-analysis of more than 1.9 million individuals from 38 observational studies. *Diabetes Care.* 2015;38:1804.

22. Kim JW. Alcohol and cognition in the elderly: A review. *Psychiatry Investig.* 2012;9:8.

23. Stahre M. Contribution of excessive alcohol consumption to deaths and years of potential life lost in the United States. *Prev Chronic Dis.* 2014;11:130293.

24. Nelson DE and others. Alcohol-attributable cancer deaths and years of potential life lost in the United States. *Am J Pub Health.* 2013;103:641.

25. American Institute for Cancer Research. The Continuous Update Project. Washington, DC: AICR; 2019; www.aicr.org/continuous-update-project /index.html.

26. Connor J. Alcohol consumption as a cause of cancer. *Addiction.* 2016; doi:10.1111/add.13477.

27. Beier JI and others. Nutrition in liver disorders and the role of alcohol. In: Ross C and others, eds. *Modern nutrition in health and disease.* Philadelphia: Lippincott Williams & Wilkins; 2014.

28. National Institute of Diabetes, Digestion and Kidney Diseases. Cirrhosis. 2014; www.niddk.nih.gov/health-information/health-topics/liver-disease/cirrhosis /Pages/facts.aspx.

29. Xu J and others. Deaths: Final data for 2013. *National Vital Statistics Report* .2016;64(2):1.

30. Badawy AA-B. Pellagra and alcoholism: A biochemical perspective. *Alcohol Alcohol.* 2014;49:238.

31. Quintero-Platt G and others. Vitamin D, vascular calcification and mortality among alcoholics. *Alcohol Alcohol.* 2015;50:18.

32. Centers for Disease Control and Prevention. Fetal alcohol spectrum disorders (FASD). 2020; www.cdc.gov/ncbddd/fasd/.

33. American Academy of Pediatrics. Breastfeeding and the use of human milk. *Pediatrics.* 2012;129:e827.

34. World Health Organization. Global status report on alcohol and health. 2018; www.who.int/substance_abuse/publications/global_alcohol_report/en/.

35. Zaridze D and others. Alcohol and mortality in Russia: prospective observational study of 151,000 adults. The *Lancet.* 2014;383:1465.

36. U.S. Preventive Services Task Force. Screening and behavioral counseling interventions in primary care to reduce alcohol misuse: U.S. preventive services task force recommendation statement. *Ann Intern Med.* 2013;159:210.

37. Grant BF and others. Epidemiology of DSM-5 alcohol use disorders: Results from the National Epidemiologic Survey on Alcohol and Related Conditions III. *JAMA Psychiatry.* 2015;72:757.

38. National Institute on Alcohol Abuse and Alcoholism. Underage drinking. 2019; pubs.niaaa.nih.gov/publications/UnderageDrinking/UnderageFact.htm.

39. Delker E and others. Alcohol consumption in demographic subpopulations: An epidemiologic overview. *Alcohol Res.* 2016;38:7.

40. Goldstein RB and others. Sex differences in prevalence and comorbidity of alcohol and drug use disorders: Results from wave 2 of the National Epidemiologic Survey on Alcohol and Related Conditions. *J Stud Alcohol Drugs.* 2012;73:938.

41. Sacks JJ and others. 2010 national and state costs of excessive alcohol consumption. *Am J Prev Med.* 2015;49:e73.

42. Bentley TS and Ortner N. 2020 U.S. organ and tissue transplants: Cost estimates, discussion, and emerging issues. 2020; www.milliman.com/en /insight/2020-us-organ-and-tissue-transplants.

43. National Institute on Alcohol Abuse and Alcoholism. Treatment for alcohol problems: Finding and getting help. 2014; pubs.niaaa.nih.gov/publications /Treatment/treatment.htm.

Life depends on energy from the sun. During photosynthesis, plants transform solar energy into chemical energy in the form of carbohydrates. During energy metabolism, we transform this chemical energy into ATP. Learn more at **health.nih.gov.**

# 9 Energy Metabolism

## Learning Objectives

**After studying this chapter, you will be able to**

1. Identify the properties of metabolism, catabolism, and anabolism.
2. Explain oxidation and reduction reactions.
3. Illustrate the metabolic pathways involved in ATP production from carbohydrates.
4. Illustrate the metabolic pathways involved in ATP production from fats.
5. Discuss how ATP production from protein differs from catabolism of carbohydrate and fat.
6. Describe the process of gluconeogenesis.
7. Describe how the body metabolizes alcohol.
8. Summarize how fed and fasted states affect metabolism.
9. Recognize the common inborn errors of metabolism.

## Chapter Outline

**THE MACRONUTRIENTS AND ALCOHOL ARE** rich sources of energy; however, the energy they provide is not in the form that cells can use. Thus, the body must have a process for breaking down energy-yielding compounds to release and convert their chemical energy to a form the body can use.[1] That process is energy metabolism—an elaborate, multistep series of energy-transforming chemical reactions. Energy metabolism occurs in all cells every moment of every day for our entire lifetime; it is slowest when we are resting and fastest when we are physically active.

Understanding energy metabolism clarifies how carbohydrates, proteins, fats, and alcohol are interrelated and how they serve as fuel for body cells. In this chapter, you will see how the macronutrients and alcohol are metabolized and discover why proteins can be converted to glucose but most fatty acids cannot. Studying energy metabolism pathways in the cell also sets the stage for examining the roles of vitamins and minerals. As you'll see in this and subsequent chapters, many micronutrients contribute to the enzyme activity that supports metabolic reactions in the cell.[2]

# 9.1 Metabolism: Chemical Reactions in the Body

**metabolic pathway** Series of chemical reactions occurring in a cell.

**intermediate** Chemical compound formed in 1 of many steps in a metabolic pathway.

**Metabolism** refers to the entire network of chemical processes involved in maintaining life. It encompasses all the sequences of chemical reactions that occur in the body. Some of these biochemical reactions enable us to release energy from carbohydrate, fat, protein, and alcohol. They also permit us to synthesize a new substance from another and prepare waste products for excretion.[1] A group of biochemical reactions that occur in a progression from beginning to end is called a **metabolic pathway**. Compounds formed in any of the many steps in a metabolic pathway are called **intermediates**.

All the pathways that take place within the body can be categorized as either anabolic or catabolic. **Anabolic** pathways use small, simpler compounds to build larger, more complex compounds (Fig. 9-1). The human body uses compounds, such as glucose, fatty acids, cholesterol, and amino acids, as building blocks to synthesize new compounds, such as glycogen, hormones, enzymes, and other proteins, that keep the body functioning, growing, and developing normally. For example, to make glycogen (a storage form of carbohydrate), we link many units of the simple sugar glucose. Energy must be used for anabolic pathways to take place.

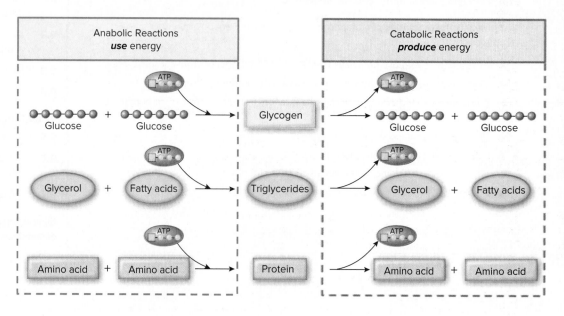

***Figure 9-1*** Anabolic reactions take simple compounds and build complex compounds and use energy. Catabolic reactions break down complex compounds into simpler compounds and produce energy.

Conversely, **catabolic** pathways break down compounds into small units. For instance, a glycogen molecule can be broken down into many glucose molecules when energy is needed. Later, the complete catabolism of this glucose results in the release of carbon dioxide ($CO_2$), water ($H_2O$), and energy in the form of ATP.

The body strives to balance anabolic and catabolic processes. However, sometimes 1 process is more prominent than the other. For example, during growth, anabolism is more prominent because more tissue is being synthesized than broken down. However, during weight loss or a wasting disease, such as cancer, catabolism predominates because more tissue is being broken down than synthesized.

## Converting Food into Energy

The energy used by all cells of the body initially comes from the sun. During photosynthesis, plants convert solar energy and carbon dioxide into glucose and other organic (carbon-containing) compounds (see Chapter 5). Virtually all organisms use the sun—either indirectly, as we do, or directly—as their source of energy.[1]

As shown in Figure 9-2, the series of catabolic reactions that produce energy for body cells begins with digestion and continues when monosaccharides, amino acids, fatty acids, glycerol, and alcohol are sent through a series of metabolic pathways, which capture energy in

## Catabolism

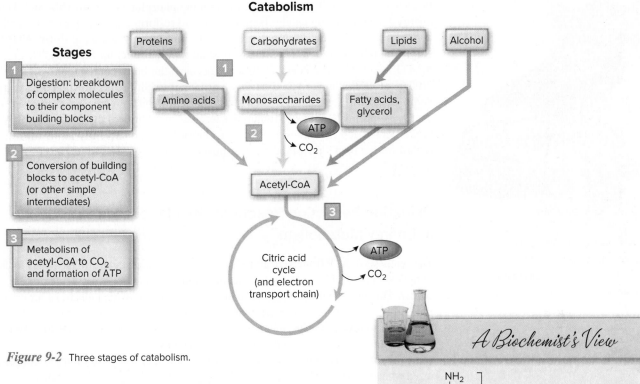

**Figure 9-2** Three stages of catabolism.

a compound called **adenosine triphosphate (ATP)**—the main form of energy the body uses. Heat, carbon dioxide, and water also result from these catabolic pathways. The heat produced helps maintain body temperature. Plants can use the carbon dioxide and water to produce glucose and oxygen via photosynthesis.

### Adenosine Triphosphate (ATP)

Only the energy in ATP and related compounds can be used directly by the cell to synthesize new compounds (anabolic pathways), contract muscles, conduct nerve impulses, and pump ions across membranes.[3] A molecule of ATP consists of the organic compound adenosine (comprised of the nucleotide adenine and the sugar ribose) bound to 3 phosphate groups (Fig. 9-3). The bonds between the phosphate groups contain energy and are called high-energy phosphate bonds. Hydrolysis (breaking) of the high-energy bonds releases this energy. To release the energy in ATP, cells break a high-energy phosphate bond, which creates **adenosine diphosphate (ADP)** plus $P_i$, a free (inorganic) phosphate group (Fig. 9-4). Hydrolysis of ADP results in the compound **adenosine monophosphate (AMP)**.

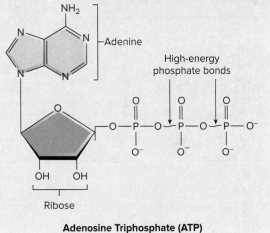

**Adenosine Triphosphate (ATP)**

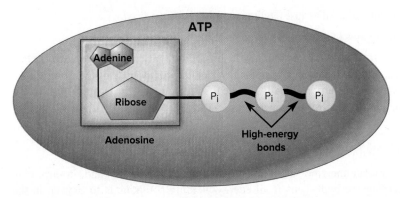

**Figure 9-3** ATP is a storage form of energy in the body. The energy is contained in the high-energy phosphate bonds. $P_i$ is the abbreviation for an inorganic phosphate group.

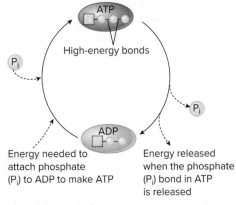

**Figure 9-4** When 1 high-energy phosphate group ($P_i$) is removed from ATP, ADP is formed and energy is released for use by the body. When energy is captured by ADP and inorganic phosphate is added, ATP can be regenerated.

Yuri Kevhiev/Alamy Stock Photo

▶ The mnemonic "LEO [loss of electrons is oxidation] the lion says GER [gain of electrons is reduction]" can help you differentiate between oxidation and reduction.

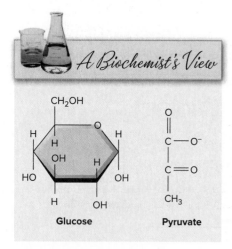

**Glucose**          **Pyruvate**

*A Biochemist's View*

▶ The term *antioxidant* is typically used to describe a compound that can donate electrons to oxidized compounds, putting them into a more reduced (stable) state. Oxidized compounds tend to be highly reactive; they seek electrons from other compounds to stabilize their chemical configuration. Dietary antioxidants, such as vitamin E, donate electrons to these highly reactive compounds, making them less reactive (see Part 4 for more about antioxidants).

**Recycling of ATP**  ATP is regenerated by adding phosphate back to AMP and ADP. The energy released during catabolism permits $P_i$ to re-form a high-energy bond with AMP and ADP, thereby regenerating ATP. A cell is constantly breaking down ATP for energy, then rebuilding it to maintain a constant supply of fuel for the body. This recycling of ATP is an essential survival strategy because the body contains only about 0.22 lb (100 g) of ATP at any given time, but a sedentary adult breaks down and resynthesizes about 88 lb (40 kg) of ATP each day. The requirement increases even more during exercise—during 1 hour of strenuous exercise, an additional 66 lb (30 kg) of ATP are broken down and resynthesized. In fact, the runner who currently holds the American record for the men's marathon was estimated to have used 132 lb (60 kg) of ATP to run the race.[4]

## Oxidation-Reduction Reactions: Key Processes in Energy Metabolism

The synthesis of ATP from ADP and $P_i$ involves the exchange of electrons, mostly in the form of hydrogen ions ($H^+$), from energy-yielding compounds (carbohydrate, fat, protein, and alcohol). This process uses oxidation-reduction reactions, in which electrons are transferred in a series of reactions from energy-yielding compounds eventually to oxygen. These reactions form water and release much energy, which is used to produce ATP. See Appendix B for more information on oxidation-reduction reactions.

A substance is *oxidized* when it loses 1 or more electrons. For example, copper is oxidized when it loses an electron:

$$Cu^+ \rightleftharpoons Cu^{2+} + e^-$$

A substance is *reduced* when it gains 1 or more electrons. For example, iron is reduced when it gains an electron:

$$Fe^{3+} + e^- \rightleftharpoons Fe^{2+}$$

The movement of electrons governs oxidation-reduction processes. If a substance loses electrons (is oxidized), another substance must gain electrons (is reduced). These processes go together; 1 process cannot occur without the other.[2] In the previous examples, the electron lost by copper can be gained by the iron, resulting in this overall reaction:

$$Cu^+ + Fe^{3+} \rightarrow Cu^{2+} + Fe^{2+}$$

Oxidation-reduction reactions involving organic (carbon-containing) compounds are somewhat more difficult to visualize. Two simple rules help identify whether these compounds are oxidized or reduced:

If the compound gains oxygen or loses hydrogen, it has been *oxidized*.

If it loses oxygen or gains hydrogen, the compound has been *reduced*.

Enzymes control oxidation-reduction reactions in the body. Dehydrogenases, a class of these enzymes, remove hydrogens from energy-yielding compounds or their breakdown products. These hydrogens are eventually donated to oxygen to form water ($H_2O$). In the process, large amounts of energy are converted to ATP.[1]

### Niacin and Riboflavin: Key Players in Energy Metabolism

Two B-vitamins, niacin and riboflavin, assist dehydrogenase enzymes and, in turn, play a role in transferring the hydrogens from energy-yielding compounds to oxygen in the

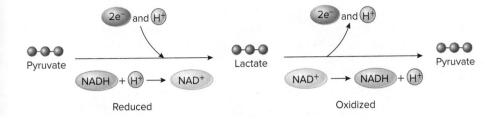

metabolic pathways of the cell.[2] In the following reaction, niacin functions as the **coenzyme** nicotinamide adenine dinucleotide (NAD). NAD is found in cells in both its oxidized form (NAD) and its reduced form (NADH). During intense (anaerobic) exercise, the enzyme lactate dehydrogenase helps reduce pyruvate (made from glucose) to form lactate. During reduction, 2 hydrogens, derived from NADH + H$^+$, are gained. Lactate is oxidized back to pyruvate by losing 2 hydrogens. NAD$^+$ is the hydrogen acceptor. That is, the oxidized form of niacin (NAD$^+$) can accept 1 hydrogen ion and 2 electrons to become the reduced form NADH + H$^+$. (The plus [+] on NAD$^+$ indicates it has 1 less electron than in its reduced form. The extra hydrogen ion [H$^+$] remains free in the cell.) By accepting 2 electrons and 1 hydrogen ion, NAD$^+$ becomes NADH + H$^+$, with no net charge on the coenzyme.

Riboflavin plays a similar role. In its oxidized form, the coenzyme form is known as flavin adenine dinucleotide (FAD). When it is reduced (gains 2 hydrogens, equivalent to 2 hydrogen ions and 2 electrons), it is known as FADH$_2$.

The reduction of oxygen (O) to form water (H$_2$O) is the ultimate driving force for life because it is vital to the way cells synthesize ATP. Thus, oxidation-reduction reactions are a key to life.

### Knowledge Check

1. What is the main form of energy used by the body?
2. Describe a time when the body may be more catabolic than anabolic.
3. What is the difference between oxidation and reduction reactions?
4. What is the function of a coenzyme?

▶ A new tool for understanding how individuals differ in the metabolic response to nutrients may lie in the ability to track the actual metabolic intermediates made during metabolism. This approach, called *metabolomics*, should be more accurate than looking for differences in DNA between individuals to predict dietary responses.

## 9.2 ATP Production from Carbohydrates

This section examines how ATP is produced from carbohydrates. Subsequent sections will explore how ATP is produced from the energy found in fats, protein, and alcohol. Along the way, you will see how these energy-yielding processes are interconnected.

ATP is generated through cellular respiration. The process of **cellular respiration** oxidizes (removes electrons) food molecules to obtain energy (ATP). Oxygen is the final electron acceptor. As you know, humans inhale oxygen and exhale carbon dioxide. When oxygen is readily available, cellular respiration may be **aerobic**. When oxygen is not present, **anaerobic** pathways are used. Aerobic respiration is far more efficient than anaerobic metabolism at producing ATP. As an example, the aerobic respiration of a single molecule of glucose will result in a net gain of 30 to 32 ATP. In contrast, the anaerobic metabolism of a single molecule of glucose is limited to a net gain of 2 ATP.

The 4 stages of aerobic cellular respiration of glucose are shown in Fig. 9-5.[1,5]

**coenzyme** Compound that combines with an inactive protein, called an apoenzyme, to form a catalytically active enzyme, called a holoenzyme. In this manner, coenzymes aid in enzyme function.

**cellular respiration** Oxidation (electron removal) of food molecules resulting in the eventual release of energy, CO$_2$, and water. See also *respiration*.

**aerobic** Requiring oxygen.

**anaerobic** Not requiring oxygen.

**cytosol** Water-based phase of a cell's cytoplasm; excludes organelles, such as mitochondria.

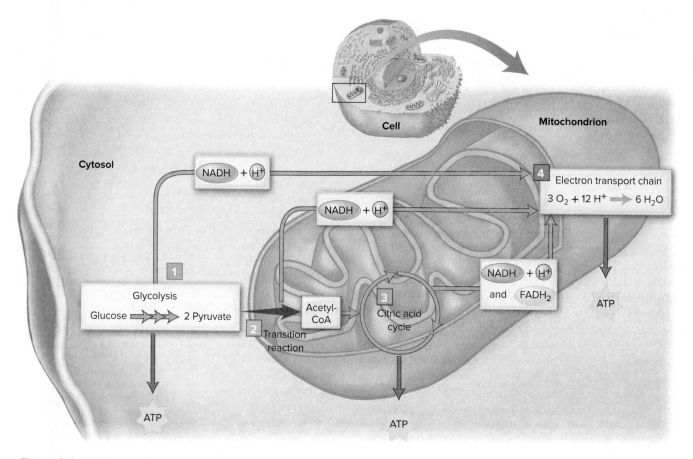

***Figure 9-5***   The 4 stages of aerobic carbohydrate metabolism.

Stage **1**, *glycolysis:* In this pathway, glucose (a 6-carbon compound) is oxidized and forms 2 molecules of the 3-carbon compound pyruvate, produces $NADH + H^+$, and generates a net of 2 molecules of ATP. Glycolysis occurs in the **cytosol** of cells.

Stage **2**, *transition reaction:* The transition reaction converts 2 molecules of the 3-carbon pyruvate from glycolysis into 2 molecules of the 2-carbon molecule acetyl-CoA. The transition reaction also produces $NADH + H^+$ and releases carbon dioxide ($CO_2$) as a waste product. The transition reaction is given this name because these molecules are transitioning from the cytosol into the **mitochondria** of cells where this reaction occurs.

Stage **3**, *citric acid cycle:* In this pathway, acetyl-CoA enters the citric acid cycle, resulting in the production of $NADH + H^+$, $FADH_2$, and ATP. Carbon dioxide is released as a waste product. Like the transition reaction, the citric acid cycle takes place within the mitochondria of cells.

Stage **4**, *electron transport chain:* The $NADH + H^+$ produced by stages 1 through 3 of cellular respiration and $FADH_2$ produced in stage 3 enter the electron transport chain, where $NADH + H^+$ is oxidized to $NAD^+$ and $FADH_2$ is oxidized to FAD. At the end of the electron transport chain, oxygen is combined with hydrogen ions ($H^+$) and electrons to form water. The electron transport chain takes place within the mitochondria of cells. Most ATP is produced in the electron transport chain; thus, the mitochondria are the cell's major energy-producing organelles.

## Glycolysis

Because glucose is the main carbohydrate involved in cell metabolism, we will track its step-by-step metabolism as an example of carbohydrate metabolism. Glycolysis has 2 roles: to break down carbohydrates to generate energy and to provide building blocks for synthesizing other needed compounds. During glycolysis, glucose is converted to 2 units of a 3-carbon compound called pyruvate. Initially, glycolysis requires 2 ATPs; however, it generates 4 ATP, thus yielding a net of 2 ATP. The details of glycolysis can be found in Figure 9-6.

## Transition Reaction: Synthesis of Acetyl-CoA

When oxygen is present, the pyruvate dehydrogenase enzyme complex converts pyruvate into a 2-carbon compound called acetyl-CoA. This conversion process takes place in the

▶ *CoA is short for coenzyme A. The A stands for acetylation because CoA provides the 2-carbon acetyl group to start the citric acid cycle.*

**Acetyl-CoA**

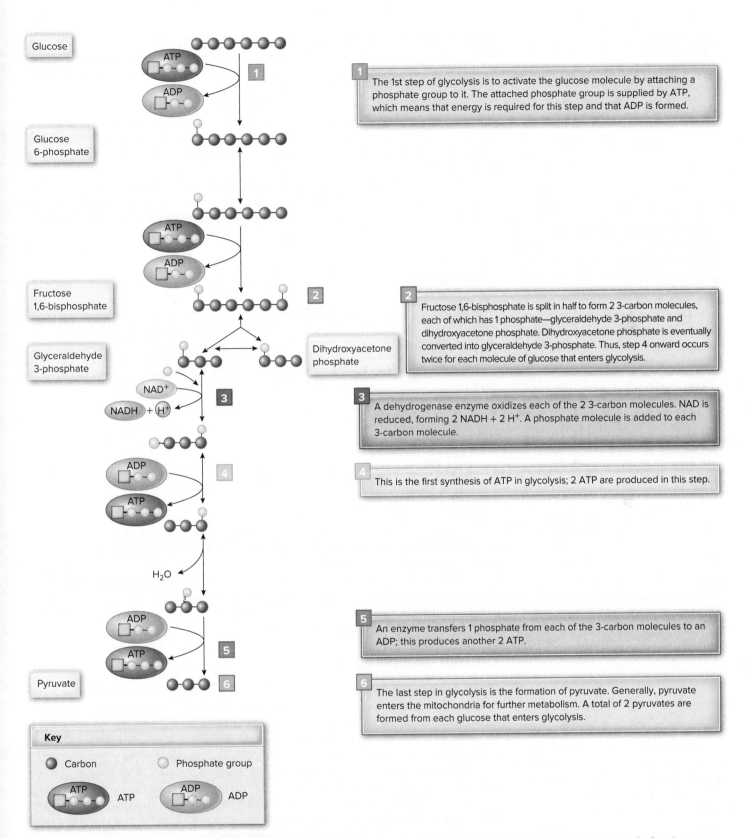

**Glucose**

**1** The 1st step of glycolysis is to activate the glucose molecule by attaching a phosphate group to it. The attached phosphate group is supplied by ATP, which means that energy is required for this step and that ADP is formed.

**Glucose 6-phosphate**

**Fructose 1,6-bisphosphate**

**2** Fructose 1,6-bisphosphate is split in half to form 2 3-carbon molecules, each of which has 1 phosphate—glyceraldehyde 3-phosphate and dihydroxyacetone phosphate. Dihydroxyacetone phosphate is eventually converted into glyceraldehyde 3-phosphate. Thus, step 4 onward occurs twice for each molecule of glucose that enters glycolysis.

**Glyceraldehyde 3-phosphate**

**Dihydroxyacetone phosphate**

$NAD^+$

$NADH$ + $H^+$

**3** A dehydrogenase enzyme oxidizes each of the 2 3-carbon molecules. NAD is reduced, forming 2 NADH + 2 H$^+$. A phosphate molecule is added to each 3-carbon molecule.

**4** This is the first synthesis of ATP in glycolysis; 2 ATP are produced in this step.

$H_2O$

**5** An enzyme transfers 1 phosphate from each of the 3-carbon molecules to an ADP; this produces another 2 ATP.

**Pyruvate**

**6** The last step in glycolysis is the formation of pyruvate. Generally, pyruvate enters the mitochondria for further metabolism. A total of 2 pyruvates are formed from each glucose that enters glycolysis.

**Key**

● Carbon    ○ Phosphate group

ATP    ADP

**Figure 9-6** Glycolysis takes place in the cytosol portion of the cell. This process breaks glucose (a 6-carbon compound) into 2 units of a 3-carbon compound called pyruvate. More details can be found in Appendix C.

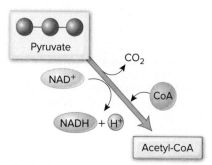

*Figure 9-7* During the transition reaction, pyruvate is metabolized to acetyl-CoA, which then enters the citric acid cycle. In the process, NADH + H⁺ is produced and $CO_2$ is lost. Pyruvate dehydrogenase enzyme assists in the transition reaction of pyruvate to acetyl-CoA.

▶ Intermediates of the citric acid cycle, such as oxaloacetate, can leave the cycle and go on to form other compounds, such as glucose. Thus, the citric acid cycle should be viewed as a traffic circle, rather than as a closed circle.

▶ How many ATP are produced by 1 molecule of glucose? The metabolism of 1 glucose molecule will yield

| Glycolysis | 2 NADH |
| | 2 ATP |
| Transition reaction | 2 NADH |
| Citric acid cycle | 6 NADH |
| | 2 FADH₂ |
| | 2 GTP |
| **Total** | **10 NADH** |
| | **2 FADH₂** |
| | **2 GTP** |
| | **2 ATP** |

The NADH, FADH₂, and GTP generated undergo oxidative phosphorylation in the electron transport chain to yield

  2.5 ATP molecules per NADH

  1.5 ATP molecules per FADH₂

  1.0 ATP molecule per GTP

Thus, 28 ATP molecules are synthesized in the electron transport chain. In addition to the 4 ATP synthesized via glycolysis and the citric acid cycle, each glucose molecule produces 32 ATP. (Including the 3.0 ATP per NADH and 2.0 ATP per FADH₂ produced, a total of 38 ATP are produced.)

| Net Production of ATP from Each Glucose Molecule | |
|---|---|
| Glycolysis | 2 ATP |
| Citric acid cycle | 2 ATP |
| Electron transport chain | 28 ATP |
| **Total** | **32 ATP** |

mitochondria and is known as the transition reaction[6] (see Fig. 9-7). The transition reaction also reduces NAD⁺ to 2 NADH + 2 H⁺, which will eventually enter the electron transport chain. Carbon dioxide is a waste product of the transition reaction and is eventually eliminated by way of the lungs.

The transition reaction is irreversible; thus, acetyl-CoA cannot be converted back to glucose. The significance of this irreversible step is illustrated in individuals who have a deficiency of pyruvate dehydrogenase. This condition results in lactic acidosis, which causes many tissues, including the central nervous system, to malfunction. Whereas glycolysis requires only the B-vitamin niacin as NAD, the pyruvate dehydrogenase complex that converts pyruvate to acetyl-CoA requires coenzymes from 4 B-vitamins—thiamin, riboflavin, niacin, and pantothenic acid. In fact, CoA is made from the B-vitamin pantothenic acid. For this reason, carbohydrate metabolism depends on an ample supply of these vitamins (see Part 4).[2]

## Knowledge Check

1. What is the goal of glycolysis?
2. How many 3-carbon compounds are made from a 6-carbon glucose molecule?
3. What is the end product of glycolysis?
4. Which vitamins are involved in the transition reaction?

## Citric Acid Cycle

The acetyl-CoA molecules produced by the transition reaction enter the citric acid cycle, which also is known as the tricarboxylic acid cycle (TCA cycle) and the Krebs cycle. The citric acid cycle is a series of chemical reactions that cells use to convert the carbons of an acetyl group to carbon dioxide while harvesting energy to produce ATP.[3]

It takes 2 turns of the citric acid cycle to process 1 glucose molecule because glycolysis and the transition reaction yield 2 acetyl-CoA. Each complete turn of the citric acid cycle produces 2 molecules of $CO_2$ and 1 potential ATP in the form of 1 molecule of guanosine triphosphate (GTP), as well as 3 molecules of NADH + H⁺ and 1 molecule of FADH₂. Oxygen does not participate in any of the steps in the citric acid cycle; however, it does participate in the electron transport chain. The details of the citric acid cycle can be found in Figure 9-8; further details are in Appendix C.

## Electron Transport Chain

The final pathway of aerobic respiration is the electron transport chain, located in the mitochondria. The electron transport chain functions in most cells in the body. Cells that need a lot of ATP, such as muscle cells, have thousands of mitochondria, whereas cells that need very little ATP, such as adipose cells, have fewer mitochondria. Almost 90% of the ATP produced from the catabolism of glucose is produced by the electron transport chain.

The electron transport chain involves the passage of electrons along a series of electron carriers. As electrons are passed from 1 carrier to the next, small amounts of energy are released. NADH + H⁺ and FADH₂, produced by glycolysis, the transition reaction, and the citric acid cycle, supply both hydrogen ions and electrons to the electron transport chain. The metabolic process, called **oxidative phosphorylation**, is the way in which energy derived from the NADH + H⁺ and FADH₂ is transferred to ADP + $P_i$ to form ATP (Fig. 9-9). Oxidative phosphorylation requires the minerals copper and iron. Copper is a component of an enzyme, whereas iron is a component of **cytochromes** (electron-transfer compounds) in the electron transport chain. In addition to ATP production, hydrogen ions, electrons, and oxygen combine to form water. The details of the electron transport chain are presented in Appendix C.

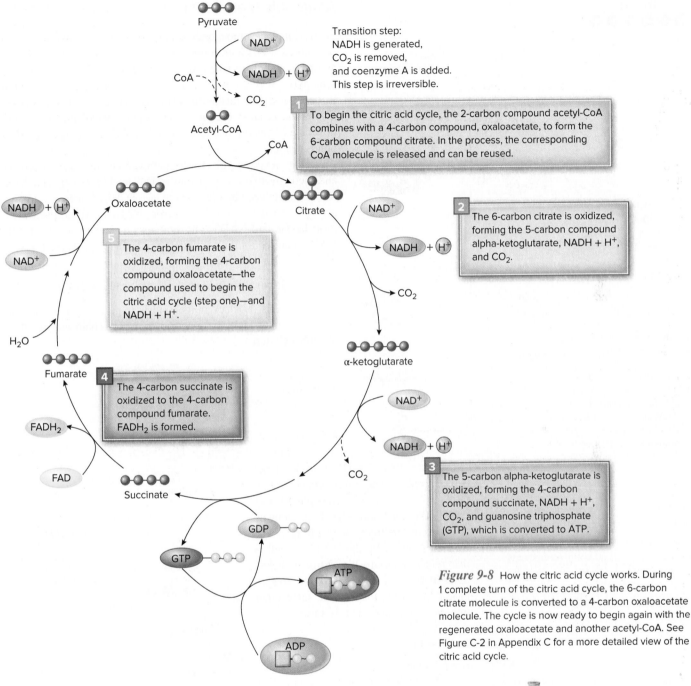

**Transition step:**
NADH is generated,
$CO_2$ is removed,
and coenzyme A is added.
This step is irreversible.

**1** To begin the citric acid cycle, the 2-carbon compound acetyl-CoA combines with a 4-carbon compound, oxaloacetate, to form the 6-carbon compound citrate. In the process, the corresponding CoA molecule is released and can be reused.

**2** The 6-carbon citrate is oxidized, forming the 5-carbon compound alpha-ketoglutarate, NADH + H⁺, and $CO_2$.

**5** The 4-carbon fumarate is oxidized, forming the 4-carbon compound oxaloacetate—the compound used to begin the citric acid cycle (step one)—and NADH + H⁺.

**4** The 4-carbon succinate is oxidized to the 4-carbon compound fumarate. $FADH_2$ is formed.

**3** The 5-carbon alpha-ketoglutarate is oxidized, forming the 4-carbon compound succinate, NADH + H⁺, $CO_2$, and guanosine triphosphate (GTP), which is converted to ATP.

*Figure 9-8* How the citric acid cycle works. During 1 complete turn of the citric acid cycle, the 6-carbon citrate molecule is converted to a 4-carbon oxaloacetate molecule. The cycle is now ready to begin again with the regenerated oxaloacetate and another acetyl-CoA. See Figure C-2 in Appendix C for a more detailed view of the citric acid cycle.

## The Importance of Oxygen

Most people can hold their breath for only a few minutes. That's because oxygen is essential for energy metabolism to continue—without oxygen, metabolism stops and death occurs. Here's why. NADH + H⁺ and $FADH_2$ produced during the citric acid cycle can be regenerated into NAD⁺ and FAD only by the eventual transfer of their electrons and hydrogen ions to oxygen, as occurs in the electron transport chain. The citric acid cycle has no ability to oxidize NADH + H⁺ and $FADH_2$ back to NAD⁺ and FAD. This is ultimately why oxygen is essential to many life forms—it is a final acceptor of the electrons and hydrogen ions generated from the breakdown of energy-yielding nutrients. Without oxygen, most of our cells are unable to extract enough energy from energy-yielding nutrients to sustain life.[1]

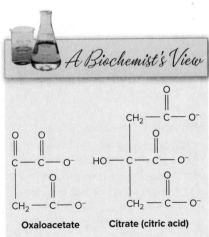

A Biochemist's View

**Oxaloacetate**          **Citrate (citric acid)**

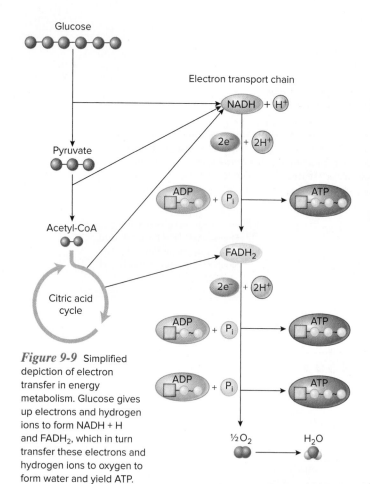

*Figure 9-9* Simplified depiction of electron transfer in energy metabolism. Glucose gives up electrons and hydrogen ions to form NADH + H and FADH$_2$, which in turn transfer these electrons and hydrogen ions to oxygen to form water and yield ATP.

▶ In Figure 9-9, the coenzymes NADH + H$^+$ and FADH$_2$ donate a pair of electrons to a specialized set of electron carriers called the electron transport chain. As electrons are passed down the chain, they lose part of their energy, which is then captured and stored by the production of ATP.

▶ In anaerobic environments, some microorganisms, such as yeast, produce ethanol, a type of alcohol, instead of lactate from glucose. Other microorganisms produce various forms of short-chain fatty acids. All this anaerobic metabolism is referred to as fermentation.

## Anaerobic Metabolism

Some cells lack mitochondria and, so, are not capable of aerobic respiration. Cells with mitochondria are capable of turning to anaerobic metabolism when oxygen is lacking. When oxygen is absent, pyruvate that is produced through glycolysis is converted into lactate. Anaerobic metabolism is not nearly as efficient as aerobic respiration because it converts only about 5% of the energy in a molecule of glucose to energy stored in the high-energy phosphate bonds of ATP.[1]

The anaerobic glycolysis pathway encompasses glycolysis and the conversion of pyruvate to lactate (Fig. 9-10). The 1-step reaction, catalyzed by the enzyme lactate dehydrogenase, involves a simple transfer of a hydrogen from NADH + H$^+$ to pyruvate to form lactate and NAD$^+$. The synthesis of lactate regenerates the NAD$^+$ required for the continued function of glycolysis. The reaction can be summarized as

$$\text{Pyruvate} + \text{NADH} + \text{H}^+ \rightarrow \text{Lactate} + \text{NAD}^+$$

For cells that lack mitochondria (e.g., red blood cells) and therefore cannot use the electron transport chain and oxidative phosphorylation pathways, anaerobic glycolysis is the only way to make ATP. Therefore, when red blood cells convert glucose to pyruvate, NADH + H$^+$ builds up in the cell. Eventually, the NAD$^+$ concentration falls too low to permit glycolysis to continue.[6] The anaerobic glycolysis pathway produces lactate to regenerate NAD$^+$. The lactate produced by the red blood cell is then released into the bloodstream, picked up primarily by the liver, and used to synthesize pyruvate, glucose, or some other intermediate in aerobic respiration.

### Cori Cycle

Even though muscle cells have mitochondria, during high intensity exercise they rely heavily on anaerobic glycolysis to quickly produce ATP. Anaerobic glycolysis causes lactate accumulations and NAD$^+$ regeneration, both of which allow anaerobic glycolysis to continue in the muscle. The lactate generated is transported from the muscles to the liver, where it is converted to glucose, which can then be returned to the muscles. This process is known as the Cori cycle (Fig. 9-11).

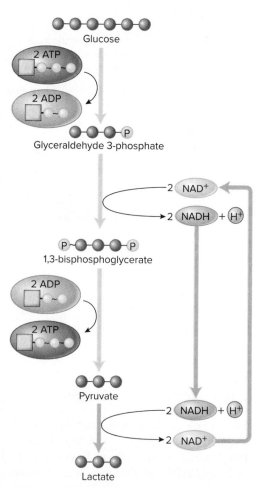

*Figure 9-10* Anaerobic glycolysis "frees" NAD$^+$ and it returns to the glycolysis pathway to pick up more hydrogen ions and electrons. Recall that the 6-carbon glucose molecule produces 2 of the 3-carbon pyruvate molecules.

*Figure 9-11* The Cori cycle. During strenuous exercise, skeletal muscle and other cells that lack mitochondria release lactate into the blood, which travels to the liver, where it is converted back into glucose and released back into circulation.

**4** Glucose returns to the muscle.

Glucose

NAD⁺

ATP

NADH

Pyruvate → Lactate

**1** In the absence of oxygen, muscle produces lactate from pyruvate.

Glucose

ADP

ATP

**3** Liver enzymes convert lactate to glucose using ATP.

Lactate

**2** Lactate leaves the muscle via blood and enters the liver.

**Liver**

**Muscle**

## Knowledge Check

1. How is citric acid formed during the citric acid cycle?
2. How many NADH + H⁺ are formed in the citric acid cycle?
3. Why is the citric acid cycle called a cycle?
4. What are the end products of the electron transport chain?
5. How does the Cori cycle help muscles continue to work during high intensity exercise?

Quick bursts of activity rely on the production of lactate to help meet the ATP energy demand.

Dave and Les Jacobs/Blend Images LLC

## CASE STUDY

RASimon/Getty Images

Melissa is a 45-year-old woman who has obesity. At her last physical, her doctor told her that she needs to lose weight. Melissa decided to follow a low carbohydrate, high protein diet she has read about. She knows it will be difficult to follow because many of the foods Melissa likes are rich in carbohydrates and the diet eliminates almost all carbohydrates during the first 2 weeks. Although she is ready to try the diet, she is confused about certain phases of the program, especially the part where the author talks about ketones. In the book, the author states that anyone going on this diet should purchase ketone strips to dip in his or her urine for the detection of ketones. The author strongly suggests these tests, especially during the extremely low carbohydrate phase of the diet. Melissa wonders if she should be considering this diet if the author is telling her to check something and she wonders what ketones are.

What are ketones, and why does a very low carbohydrate diet produce an increase in ketones in both the blood and the urine? In metabolic terms, explain how ketones are formed. Why do individuals with diabetes need to be concerned about high levels of ketones in the body?

# 9.3 ATP Production from Fats

▶ How many ATP are produced by a 16-carbon fatty acid? The metabolism of a 16-carbon fatty acid will yield

| Beta-oxidation | 7 FADH$_2$ |
| | 7 NADH |
| Citric acid cycle | 24 NADH |
| (8 turns; once for each | 8 FADH$_2$ |
| 2-carbon fragment) | 8 GTP |
| **Total** | **31 NADH** |
| | **15 FADH$_2$** |
| | **8 GTP** |

The NADH, FADH$_2$, and GTP undergo oxidative phosphorylation in the electron transport chain to yield

2.5 ATP molecules per NADH

1.5 ATP molecules per FADH$_2$

1.0 ATP molecule per GTP

Thus, 108 ATP molecules are synthesized in the electron transport chain. (Including the 3.0 ATP per NADH and 2.0 ATP per FADH$_2$ produced, a total of 131 ATP are produced.)

| Net ATP Production from a 16-Carbon Fatty Acid* | |
| --- | --- |
| Beta-oxidation | 0 ATP |
| Citric acid cycle | 8 ATP |
| Electron transport chain | 100 ATP |
| **Total** | **108 ATP** |
| **ATP used for activation** | **−2 ATP** |
| **Net ATP** | **106 ATP** |

*Only NADH and FADH$_2$ are formed during beta-oxidation, which are subsequently converted to ATP in the electron transport chain.

Just as cells release the energy in carbohydrates and capture it as ATP, they also release and capture energy in triglyceride molecules. This process begins with **lipolysis**, the breaking down of triglycerides into free fatty acids and glycerol. The further breakdown of fatty acids for energy production is called **fatty acid oxidation** because the donation of electrons from fatty acids to oxygen is the net reaction in the ATP-yielding process. This process takes place in the mitochondria.

The fatty acids used to generate energy can come from triglycerides in the diet or from stored triglycerides in adipose tissue. Following high fat meals, the body stores excess fat in adipose tissue. However, during periods of low calorie intake or fasting, triglycerides from fat cells are broken down into fatty acids by an enzyme called hormone-sensitive lipase and released in the blood. The activity of this enzyme is increased by hormones such as glucagon, growth hormone, and epinephrine and is decreased by the hormone insulin. The fatty acids are taken up from the bloodstream by cells throughout the body and are shuttled from the cell cytosol into the mitochondria using a carrier called **carnitine** (Fig. 9-12).[7]

## ATP Production from Fatty Acids

Almost all fatty acids in nature are composed of an even number of carbons, ranging from 2 to 26. The first step in transferring the energy in such a fatty acid to ATP is to cleave the carbons, 2 at a time, and convert the 2-carbon fragments to acetyl-CoA. Fatty acid oxidation is also called beta-oxidation because the process of converting a free fatty acid to multiple acetyl-CoA molecules begins with the beta carbon, the 2nd carbon on a fatty acid (counting after the carboxyl [acid] end).[1] (See Chapter 6.) During beta-oxidation, NADH + H$^+$ and FADH$_2$ are produced. Thus, as with glucose, a fatty acid is eventually degraded, 2 carbons at a time, into acetyl-CoA while some of the chemical energy contained in the fatty acids is transferred to NADH + H$^+$ and FADH$_2$ (Fig. 9-13).

The acetyl-CoA enters the citric acid cycle, and 2 carbon dioxides are released, just as with the acetyl-CoA produced from glucose. Thus, the breakdown product of both glucose and fatty acids—acetyl-CoA—enters the citric acid cycle. A big difference between glucose and fatty acids is that most fatty acids have far more carbons and, thus, can go around the citric acid cycle many more times than glucose. For instance, a 16-carbon fatty acid yields 8 2-carbon acetyl-CoAs, each of which can go around the citric acid cycle, whereas a 6-carbon glucose forms 2 acetyl-CoAs and thus can go around the citric acid cycle only twice. Additionally, each fatty acid carbon results in about 7 ATP, whereas glucose oxidation results

▶ Carnitine helps shuttle free fatty acids in the cytosol into mitochondria. Carnitine is a popular nutritional supplement. In healthy people, cells produce the carnitine needed, and carnitine supplements provide no benefit. In patients hospitalized with acute illnesses, however, carnitine synthesis may be inadequate. These patients may need to have carnitine added to their intravenous feeding (total parenteral nutrition) solutions.

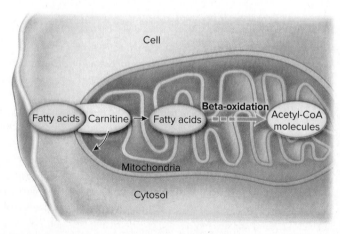

*Figure 9-12* Fatty acids are taken up from the bloodstream by various cells and shuttled by carnitine into the inner portion of the cell mitochondria. The fatty acid then undergoes beta-oxidation, in which 2 carbon fragments are removed from the fatty acid and produce acetyl-CoA, NADH + H$^+$, and FADH$_2$.

in only about 5 ATP per carbon. This difference in ATP is because fatty acids have relatively fewer oxygen atoms than oxygen rich glucose. Glucose carbons exist in a more oxidized state than fat; as a result, fats yield more energy than carbohydrates (9 kcal/g vs. 4 kcal/g).[1]

*Remember, cells cannot make glucose from the 2-carbon fragment from fatty acids. That is because there is no pathway that converts fatty acids to glucose.*

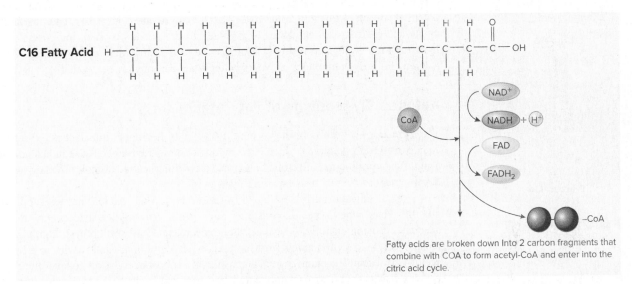

Fatty acids are broken down into 2 carbon fragments that combine with COA to form acetyl-CoA and enter into the citric acid cycle.

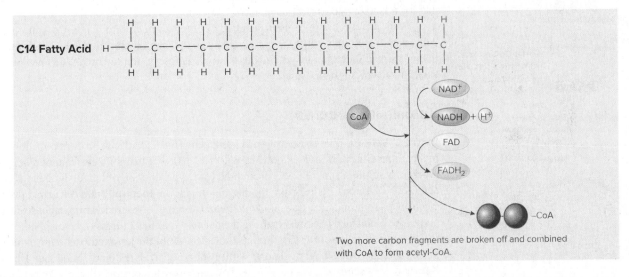

Two more carbon fragments are broken off and combined with CoA to form acetyl-CoA.

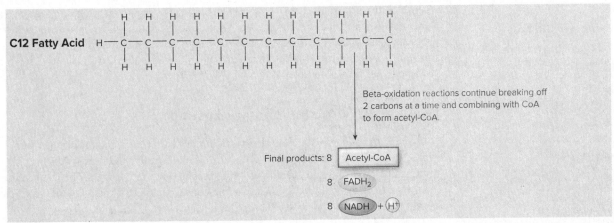

Beta-oxidation reactions continue breaking off 2 carbons at a time and combining with CoA to form acetyl-CoA.

Final products: 8 Acetyl-CoA

8 FADH$_2$

8 NADH + H$^+$

*Figure 9-13* Overview of beta-oxidation. A 2-carbon fragment is cleaved from a fatty acid to form 1 acetyl-CoA, 1 FADH$_2$, and 1 NADH$^+$. The 2-carbon acetyl-CoA then enters the citric acid cycle; see Fig. 9-14.

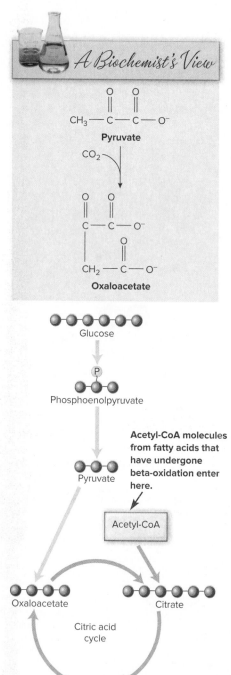

*A Biochemist's View*

**Figure 9-14** As acetyl-CoA concentrations increase due to beta-oxidation, oxaloacetate levels are maintained by pyruvate from carbohydrate metabolism. In this way, carbohydrates help oxidize fatty acids.

**ketone bodies** Incomplete breakdown products of fat, containing 3 or 4 carbons. Most contain a chemical group called a ketone. An example is acetoacetic acid.

**ketosis** Condition of having a high concentration of ketone bodies and related breakdown products in the bloodstream and tissues.

## Carbohydrate Aids Fat Metabolism

In addition to its role in energy production, the citric acid cycle provides compounds that leave the cycle and are used for other pathways. This results in a slowing of the cycle, as eventually not enough oxaloacetate is formed to combine with the acetyl-CoA entering the cycle. Cells are able to compensate for this by synthesizing additional oxaloacetate. One potential source of this additional oxaloacetate is pyruvate (Fig. 9-14). Thus, as fatty acids create acetyl-CoA, carbohydrates (e.g., glucose) are needed to keep the concentration of pyruvate high enough to resupply oxaloacetate to the citric acid cycle. Overall, the entire pathway for fatty acid oxidation works better when carbohydrate is available.

## Ketones: By-products of Fat Catabolism

The formation of **ketone bodies** occurs mainly with hormonal imbalances—chiefly, inadequate insulin production to balance glucagon action in the body.[8] These imbalances lead to a significant production of ketone bodies and a condition called **ketosis**. The key steps in the development of ketosis are shown in Figure 9-15.

Recall that the citric acid cycle functions best when oxaloacetate is supplied from glucose. Thus, when a person consumes a low carbohydrate diet, oxaloacetate may be inadequate, causing citric acid cycle activity to decrease. However, lipolysis continues, which means acetyl-CoA production from fatty acids also continues and results in a buildup of acetyl-CoA because oxaloacetate is not available to allow acetyl-CoA to enter the citric acid cycle. Because acetyl-CoA cannot enter the citric acid cycle, these molecules join together and form ketone bodies (i.e., acetoacetate, acetone, and beta-hydroxybutyrate).

Most ketone bodies are subsequently converted back into acetyl-CoA in other body cells, where they then enter the citric acid cycle and can be used for fuel. One of the ketone bodies formed (acetone) leaves the body via the lungs, giving the breath of a person in ketosis a characteristic, fruity smell.

## Ketosis in Diabetes

In type 1 diabetes, little to no insulin is produced. This lack of insulin does not allow for normal carbohydrate and fat metabolism. Without sufficient insulin, cells cannot readily utilize glucose, resulting in rapid lipolysis and the excess production of ketone bodies.[9] If the concentration of ketone bodies rises too high in the blood, they spill into the urine, pulling the electrolytes sodium and potassium with them. Eventually, severe ion imbalances occur in the body. The blood also becomes more acidic because 2 of the 3 forms of ketone bodies contain acid groups. The resulting condition, known as **diabetic ketoacidosis**, can induce coma or death if not treated immediately, with insulin, electrolytes, and fluids (see Chapter 5). Ketoacidosis usually occurs only in ketosis caused by uncontrolled type 1 diabetes; in fasting, blood concentrations of ketone bodies typically do not rise high enough to cause the problem. A ketogenic diet (high in fat, moderate in protein, and very low in carbohydrates) may help individuals with type 2 diabetes better maintain blood glucose level and keep hemoglobin A1c (a blood marker of average blood glucose levels over the past 2 to 3 months) at a healthier level.

## Ketosis in Semistarvation or Fasting

When a person is in a state of semistarvation or fasting, the amount of glucose in the body falls, so insulin production falls. This fall in blood insulin then causes fatty acids to flood into the bloodstream and eventually form ketone bodies in the liver. The heart, muscles, and some parts of the kidneys then use ketone bodies for fuel. After a few days of ketosis, the brain also begins to metabolize ketone bodies for energy.

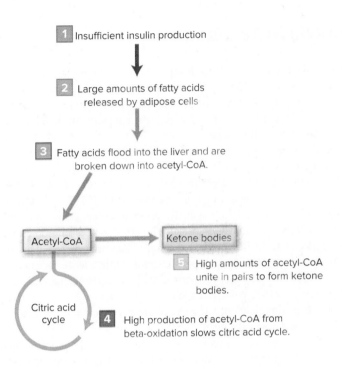

1. Insufficient insulin production

2. Large amounts of fatty acids released by adipose cells

3. Fatty acids flood into the liver and are broken down into acetyl-CoA.

Acetyl-CoA → Ketone bodies

Citric acid cycle

4. High production of acetyl-CoA from beta-oxidation slows citric acid cycle.

5. High amounts of acetyl-CoA unite in pairs to form ketone bodies.

*A Biochemist's View*

**Ketones**

**Acetoacetic acid**

**Beta-hydroxybutyric acid**

**Acetone**

*Figure 9-15* Key steps in ketosis. Any condition that limits insulin or glucose availability to cells results in some ketone body production.

This adaptive response is important to semistarvation or fasting. As more body cells begin to use ketone bodies for fuel, the need for glucose as a body fuel diminishes. This then reduces the need for the liver and kidneys to produce glucose from amino acids (and from the glycerol released from lipolysis), sparing much body protein from being used as a fuel source (see Section 9.4). The maintenance of body protein mass is a key to survival in semistarvation or fasting—death occurs when about half the body protein is depleted, usually after about 50 to 70 days of total fasting.[10]

**Critical Thinking**

The use of a very low carbohydrate diet to induce ketosis for weight loss is covered in Chapter 10. Why is careful physician monitoring needed if this type of diet is followed?

*Knowledge Check*

1. How are fatty acids shuttled into the mitochondria for energy production?
2. What is the end product of beta-oxidation?
3. How do fatty acids enter the citric acid cycle?
4. What conditions must exist in the body to promote the formation of ketones?

Metabolism is part of everyday life; metabolic activity increases when we increase physical activity and slows during fasting and semistarvation.

Paul Bradbury/age fotostock

# 9.4 Protein Metabolism

► Branched-chain amino acids are added to some liquid meal replacement supplements given to hospitalized patients. Some fluid replacement formulas marketed to athletes also contain branched-chain amino acids (see Chapter 11).

Protein metabolism takes place primarily in the liver. Only branched-chain amino acids—leucine, isoleucine, and valine—are metabolized mostly at other sites—in this case, the muscles.[2] Protein metabolism begins after proteins are degraded into amino acids. To use an amino acid for fuel, cells must first deaminate them (remove the amino group) (see Chapter 7). These pathways often require vitamin B-6 to function. Removal of the amino group produces carbon skeletons, most of which enter the citric acid cycle. Some carbon skeletons also yield acetyl-CoA or pyruvate.[6]

Some carbon skeletons enter the citric acid cycle as acetyl-CoA, whereas others form intermediates of the citric acid cycle or glycolysis (Fig. 9-16). Any part of the carbon skeleton that can form pyruvate (i.e., alanine, glycine, cysteine, serine, and threonine) or bypass acetyl-CoA and enter the citric acid cycle directly (such amino acids include asparagine, arginine, aspartic acid, histidine, glutamic acid, glutamine, isoleucine, methionine, proline, valine, and phenylalanine) are called **glucogenic amino acids** because these carbons can become the carbons of glucose. Any parts of carbon skeletons that become acetyl-CoA (leucine and lysine, as well as parts of isoleucine, phenylalanine, tryptophan, and tyrosine) are called **ketogenic amino acids** because these carbons become acetyl-CoA, and, if insulin levels are low, they can become ketones. The factor that determines whether an amino acid is glucogenic or ketogenic is whether part or all of the carbon skeleton of the amino acid can yield a "new" oxaloacetate molecule during metabolism, 2 of which are needed to form glucose.

*Figure 9-16* Amino acid metabolism. Amino acids that can yield glucose can be converted to pyruvate 1, directly enter the citric acid cycle 3, or be converted directly to oxaloacetate 4. Amino acids that cannot yield glucose are converted to acetyl-CoA and are metabolized in the citric acid cycle 2. The glycerol portion of triglycerides 5 can be converted to glucose. The alpha keto acid portion of amino acids can be used to make glucose 6. Fatty acids from beta-oxidation enter the citric acid cycle at acetyl-CoA.

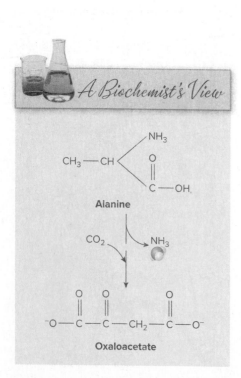

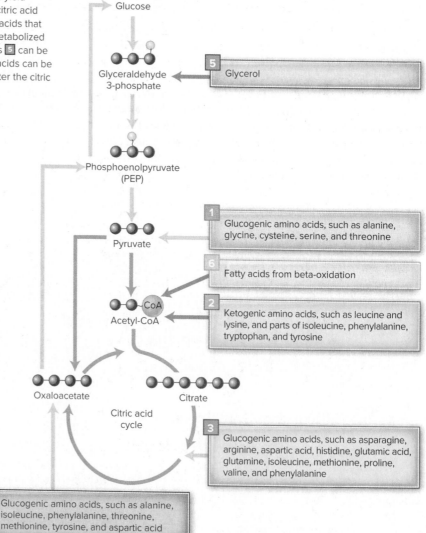

## Gluconeogenesis: Producing Glucose from Glucogenic Amino Acids and Other Compounds

The pathway to produce glucose from certain amino acids—**gluconeogenesis**—is present only in liver cells and certain kidney cells. The liver is the primary gluconeogenic organ. A typical starting material for this process is oxaloacetate, which is derived primarily from the carbon skeletons of some amino acids, usually the amino acid alanine. Pyruvate also can be converted to oxaloacetate as well as other gluconeogenic precursors, such as lactate and glycerol (see Fig. 9-14).

Gluconeogenesis begins in the mitochondria with the production of oxaloacetate. The 4-carbon oxaloacetate eventually returns to the cytosol, where it loses 1 carbon dioxide, forming the 3-carbon compound phosphoenolpyruvate, which then reverses the path back through glycolysis to glucose. It takes 2 of these 3-carbon compounds to produce the 6-carbon glucose. This entire process requires ATP, as well as coenzyme forms of the B-vitamins biotin, riboflavin, niacin, and B-6.[6]

To learn more about gluconeogenesis, examine Figure 9-16 and trace the pathway that converts the amino acid glutamine to glucose. Glutamine first loses its amino group to form its carbon skeleton, which enters the citric acid cycle directly and is converted by stages to oxaloacetate. Oxaloacetate loses 1 carbon as carbon dioxide, and the 3-carbon phosphoenolpyruvate produced then moves through the gluconeogenic pathway to form glucose. Eventually, 2 glutamine molecules are needed to form 1 glucose molecule.

## Gluconeogenesis from Typical Fatty Acids Is Not Possible

Typical fatty acids cannot be turned into glucose because those with an even number of carbons—the typical form in the body—break down into acetyl-CoA molecules. Acetyl-CoA can never re-form into pyruvate; the step between pyruvate and acetyl-CoA is irreversible. The options for acetyl-CoA are forming ketones and/or combining with oxaloacetate in the citric acid cycle. However, 2 carbons of acetyl-CoA are added to oxaloacetate at the beginning of the citric acid cycle, and 2 carbons are subsequently lost as carbon dioxide when citrate converts back to the starting material, oxaloacetate. Thus, at the end of 1 cycle, no carbons from acetyl-CoA are left to turn into glucose; it is impossible to convert typical fatty acids into glucose.[6]

The glycerol portion of a triglyceride is the part that can become glucose. Glycerol enters the glycolysis pathway and can follow the gluconeogenesis pathway from glyceraldehyde 3-phosphate to glucose. Glucose yield from glycerol is insignificant.[1]

## Disposal of Excess Amino Groups from Amino Acid Metabolism

The catabolism of amino acids, primarily from the liver, yields amino groups ($-NH_2$), which then are converted to ammonia ($NH_3$). The ammonia must be excreted because its buildup is toxic to the brain. The liver prepares the amino groups for excretion in the urine with the urea cycle. Some stages of the urea cycle occur in the cytosol and some in the mitochondria. During the urea cycle, 2 nitrogen groups—1 ammonia group and 1 amino group—react through a series of steps with carbon dioxide molecules to form urea and water. Eventually, urea is excreted in the urine (Fig. 9-17). In

**gluconeogenesis**  Generation (*genesis*) of new (*neo*) glucose from certain (glucogenic) amino acids.

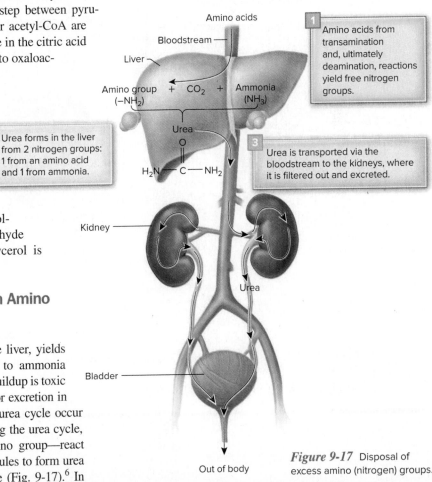

*Figure 9-17* Disposal of excess amino (nitrogen) groups.

# GL🌐BAL PERSPECTIVE

## Cancer Cell Metabolism

Cancer is characterized by abnormal, uncontrolled cell growth (see Clinical Perspective in Part 4 for more information). As you can see in Figure 9-18, this disease affects many people around the world.[11] Scientists have been working for years to find successful treatments for cancer. One area of investigation focuses on preventing cancer cell growth by disrupting the metabolism of these cells.[12]

Recall that, in normal metabolism, ATP is generated through glycolysis, the citric acid cycle, and the electron transport chain. During glycolysis, glucose is converted to pyruvate. When oxygen levels are low, pyruvate is converted to lactate and only the 2 ATP from glycolysis are generated. When oxygen is sufficient, pyruvate is routed to the citric acid cycle and electron transport chain, generating a total of 30 to 32 ATP. Things are different, however, in cancer cells.

Even when oxygen is plentiful, cancer cells use glycolysis and produce lactate. This alteration in metabolism during cancer, known as the Warburg effect, was first observed by Nobel Prize–winner Otto Warburg nearly a century ago. Researchers believe that this altered glucose metabolism ensures that cancer cells have the energy needed to support their rapid growth. Because cancer cells are burning glucose in this wasteful manner, they starve healthy cells of energy and nutrients. As the starving healthy cells weaken and die, cancer cells gain the space they need to proliferate. This wasteful use of glucose also may promote cancer cell growth—that's because their excessive nutrient use produces free radicals, which may increase DNA mutations and promote cancer (see Part 4).

Cancer cells also use protein and fats wastefully. In both healthy and cancer cells, the nitrogen component (amino group) of the amino acid glutamine is used to synthesize nonessential amino acids (via transamination), which are then used to build body proteins (see Chapter 7). Cancer cells grow rapidly and need a large supply of glutamine to support cell synthesis—the high rate of glutamine use by cancer cells impairs normal protein synthesis in the body. Cancer cells also burn glutamine for energy in the citric acid cycle (see Fig. 9-16). Cancer cells also can use fat for energy, but they use fat mostly to make the new lipids and phospholipids they need for their cell membranes (see Chapter 6).

Many cancer cells die when glucose is unavailable; however, blocking glycolysis has not been shown to be a useful cancer treatment. One potential treatment involves blocking the enzymes needed to convert pyruvate to lactate or lactate back to pyruvate. Another possibility is blocking the enzyme needed to convert glutamine to glutamate, the form required to enter the citric acid cycle and generate energy. Still another possibility involves blocking the enzymes needed to build new lipids from fatty acids. Drugs that alter the enzymes that cancer cells use during metabolism may prove to be effective cancer treatments.

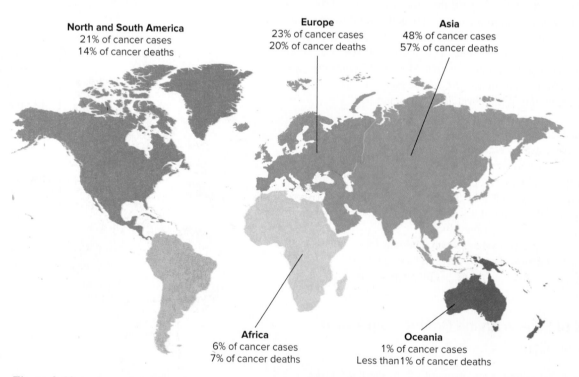

**North and South America**
21% of cancer cases
14% of cancer deaths

**Europe**
23% of cancer cases
20% of cancer deaths

**Asia**
48% of cancer cases
57% of cancer deaths

**Africa**
6% of cancer cases
7% of cancer deaths

**Oceania**
1% of cancer cases
Less than 1% of cancer deaths

*Figure 9-18* International incidence of cancer cases and deaths. Every year there are 18.1 million new cancer cases and 9.6 million cancer deaths worldwide.

Shutterstock/Pyty

## Knowledge Check

1. In order to use amino acids as a fuel, what must happen to the nitrogen attached to the amino acid?
2. Explain the difference between glucogenic and ketogenic amino acids.
3. What part of the amino acid is used in the metabolic pathways?
4. What is the name of the pathway that converts amino acids to glucose?
5. Can fat be used to synthesize glucose? Why or why not?

liver disease, ammonia can build up to toxic concentrations in the blood, whereas in kidney disease the toxic agent is excess amounts of urea. The form of nitrogen in the blood—ammonia or urea—is a diagnostic tool for detecting liver or kidney disease.

## 9.5 Alcohol Metabolism

The alcohol dehydrogenase (ADH) pathway is the main way in which alcohol is metabolized. First, alcohol is converted in the cytosol to acetaldehyde by the action of the alcohol dehydrogenase enzyme and the $NAD^+$ coenzyme. $NAD^+$ picks up 2 hydrogen ions and 2 electrons from the alcohol to form $NADH + H^+$ and produces the intermediate acetaldehyde (Fig. 9-19). Next, coenzyme A and aldehyde dehydrogenase convert acetaldehyde to acetyl-CoA and yield $NADH + H^+$.

Alcohol, carbohydrate, protein, and fat all contribute chemical energy to the body.

AlexPro9500/Getty Images

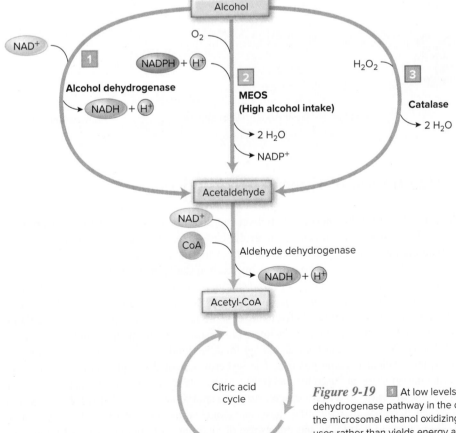

*Figure 9-19* ▣ At low levels of alcohol (ethanol) intake, the alcohol dehydrogenase pathway in the cytoplasm is used. ▣ At high levels of alcohol intake, the microsomal ethanol oxidizing system (MEOS) in the cytoplasm is used. The MEOS uses rather than yields energy and accounts in general for about 10% of alcohol metabolism. ▣ Catalase occurs in the peroxisomes of cells and is a minor pathway.

The metabolism of alcohol occurs predominantly in the liver, although approximately 10 to 30% of alcohol is metabolized in the stomach. Different forms (known as polymorphisms) of alcohol dehydrogenase and aldehyde dehydrogenase are found in the stomach and the liver.

The acetyl-CoA formed through the ADH pathway has several metabolic fates. Small amounts can enter the citric acid cycle to produce energy. However, the breakdown of alcohol in the ADH pathway utilizes $NAD^+$ and converts it to NADH. As $NAD^+$ supplies become limited and NADH levels build, the citric acid cycle slows and blocks the entry of acetyl-CoA. Because of the toxic effects of alcohol and acetaldehyde, the metabolism of alcohol takes priority over continuation of the citric acid cycle. Thus, most of the acetyl-CoA is directed toward fatty acid and triglyceride synthesis, resulting in the accumulation of fat in the liver (called steatosis). Clinicians are often alerted to this condition by high levels of triglycerides in the blood.

When a person drinks moderate to excessive amounts of alcohol, the ADH pathway cannot keep up with the demand to metabolize all the alcohol and acetaldehyde. To prevent the toxic effects of alcohol and acetaldehyde, the body utilizes a second pathway, called the microsomal ethanol oxidizing system (MEOS), to metabolize the excess alcohol. As shown in Figure 9-19, this system uses oxygen and a different niacin-containing coenzyme (NADP) and produces water and acetaldehyde. When excessive amounts of alcohol are consumed in a small amount of time, the ability of these enzyme systems to metabolize alcohol completely is exceeded, and the result is alcohol poisoning (see Chapter 8). In addition to metabolizing excess alcohol, the MEOS system metabolizes drugs, preventing them from being toxic in the body.

The MEOS differs in several ways from the ADH pathway. First, the MEOS uses potential energy (in the form of $NADPH + H^+$, another niacin coenzyme), rather than yielding potential energy (as $NADH + H^+$ in the ADH pathway) in the conversion of alcohol to acetaldehyde. This use of energy may, in part, explain why individuals who consume large amounts of alcohol do not gain as much weight as might be expected from the amount of alcohol-derived energy they consume.

The body has an additional pathway, called the catalase pathway, for metabolizing alcohol. However, this is a relatively minor pathway in comparison with the ADH and MEOS pathways.

### Knowledge Check

1. Where does alcohol dehydrogenase catabolize alcohol?
2. When is the ADH system used to metabolize alcohol versus the MEOS system?
3. In addition to the ADH and MEOS pathways, what other pathway allows the metabolism of alcohol?

## 9.6 Regulation of Energy Metabolism

As shown in Figure 9-20, energy metabolism can take many forms in the body. Carbohydrates can be used for fat synthesis—the acetyl-CoA from the breakdown of carbohydrate is the building block for fatty acid synthesis. By stringing together glycolysis and the citric acid cycle, cells can convert carbohydrates into carbon skeletons for the synthesis of certain amino acids and can use the energy in carbohydrates to form ATP. These pathways also can turn the carbon skeletons of some amino acids into the carbon skeletons of others. In addition, they can convert carbon skeletons from some amino acids to glucose or have them drive ATP synthesis by serving as substrates (precursors) for intermediates in the citric acid cycle. Finally, fatty acids can provide energy for ATP synthesis or produce ketone bodies; however, they cannot become glucose. The glycerol part of the triglyceride either can be converted into glucose and used for fuel or can contribute to ATP synthesis via participation in glycolysis, the citric acid cycle, and electron transport chain metabolism (Table 9-1).

When it comes to regulating these metabolic pathways, the liver plays the major role—it responds to hormones and makes use of vitamins. Additional means of regulating metabolism involve ATP concentrations, enzymes, hormones, vitamins, and minerals.[1]

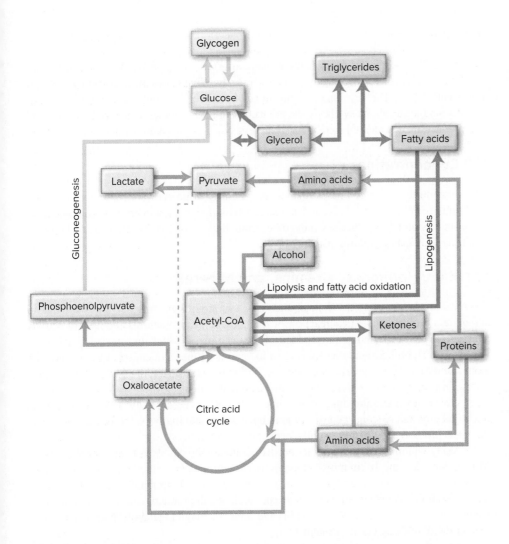

*Figure 9-20* Overview of cell metabolism. Note that acetyl-CoA forms a crossroads for many pathways and that the citric acid cycle also can be used to help build compounds. Anabolic and catabolic processes may appear to share the same pathways, but generally this is true for only a few steps. Because a specific set of enzymes must be activated to promote anabolism and a different set to activate catabolism, the cell has significant control over metabolism.

**Table 9-1  What Happens Where: A Review**

| Pathway | Location in Cell | Organs |
|---|---|---|
| Glycolysis (glucose → pyruvate) | Cytosol | All |
| Transition reaction (pyruvate → acetyl-CoA) | Mitochondria | All |
| Citric acid cycle (acetyl-CoA → $CO_2$) | Mitochondria | All except red blood cells, parts of the kidney, and brain |
| Gluconeogenesis | Begins in mitochondria, then moves to cytosol | Mostly liver and to a lesser extent in the kidneys |
| Beta-oxidation (fatty acid → acetyl-CoA) | Mitochondria | All |
| Glucogenic amino acid oxidation (amino acids → pyruvate) | Cytosol | Liver and to a lesser extent in the kidneys |
| Non-glucogenic amino acid oxidation (amino acids → acetyl-CoA) | Mitochondria | Liver and to a lesser extent in the kidneys |
| Alcohol oxidation (ethanol → acetaldehyde) (acetaldehyde → acetyl-CoA) | Cytosol Mitochondria | Liver |

## The Liver

The liver is the location of many nutrient interconversions (Fig. 9-21). Most nutrients must pass 1st through the liver after absorption into the body. What leaves the liver is often different from what entered. The key metabolic functions of the liver are conversions between various forms of simple sugars, fat synthesis, the production of ketone bodies, amino acid metabolism, urea production, and alcohol metabolism. Nutrient storage is an additional liver function.[2]

## ATP Concentrations

ATP concentration in a cell helps regulate metabolism. High ATP concentrations decrease ATP synthesis and energy-yielding reactions, such as glycolysis, and promote anabolic reactions, such as protein synthesis. On the other hand, high ADP concentrations stimulate ATP synthesis and energy-yielding pathways.[13]

## Enzymes, Hormones, Vitamins, and Minerals

Enzymes are key regulators of metabolic pathways; both their presence and their rate of activity are critical to chemical reactions in the body. Enzyme synthesis and rates of activity are controlled by cells and by the products of the reactions in which the enzymes participate. For example, a high protein diet leads to an increased synthesis of enzymes associated with amino acid catabolism and gluconeogenesis. Within hours of a shift to a low protein diet, the synthesis of enzymes associated with amino acid metabolism slows.[6]

Hormones, including insulin, regulate metabolic processes. Low levels of insulin in the blood promote gluconeogenesis, protein breakdown, and lipolysis. Increased blood insulin levels promote the synthesis of glycogen, fat, and protein.

Many vitamins and minerals are needed for metabolic pathways to operate (Fig. 9-22). Most notable are the B-vitamins—thiamin, riboflavin, niacin, pantothenic acid, biotin, vitamin B-6, folate, and vitamin B-12—as well as the minerals iron and copper. Because so many metabolic pathways depend on nutrient input, health problems can develop from nutrient deficiencies.[2] (The roles that vitamins and minerals play in metabolism are discussed in greater detail in Chapters 12 through 15.)

> ### *Critical* Thinking
>
> If you had unlimited resources to design a drug that inhibits fat synthesis, which type of metabolism (aerobic or anaerobic) would you look to affect? What unintended metabolic consequences might result from using such a drug? Figure 9-22 might help you answer this question.

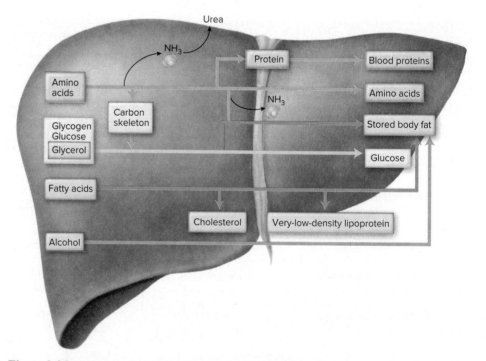

*Figure 9-21* Most nutrients must pass 1st through the liver after absorption into the body. What leaves the liver is often different from what entered.

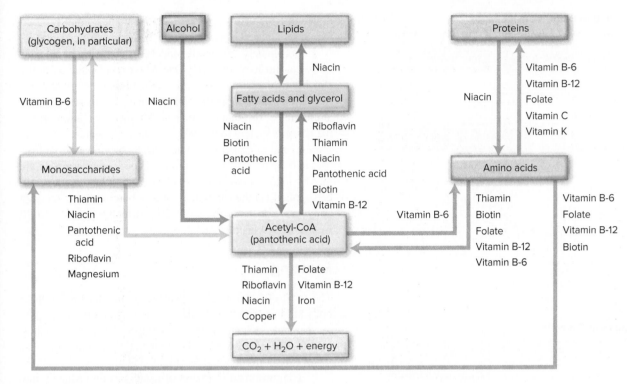

*Figure 9-22* Many vitamins and minerals participate in the metabolic pathways.

### Knowledge Check

1. Where does glycolysis take place in a cell?
2. What factors determine the regulation of glycolysis and citric acid cycle pathways?
3. How do ATP concentrations regulate metabolism?

## 9.7 Fasting and Feasting

Both fasting and feasting affect metabolism. The type of macronutrient and the rate at which it is used vary when the calorie supplies are insufficient or exceed needs.

### Fasting

In the 1st few hours of a fast, the body fuels itself with stored liver glycogen and fatty acids from adipose tissue. As the fast progresses, body fat continues to be broken down and liver glycogen becomes exhausted. Although most cells can use fatty acids for energy, the nervous system and red blood cells use only glucose for energy. To provide the needed glucose, the body begins breaking down lean body tissue and converts glucogenic amino acids, via gluconeogenesis, to glucose (Fig. 9-23).[7,8] During the 1st few days of a fast, body protein is broken down rapidly—in fact, it supplies about 90% of needed glucose, with the remaining 10% coming from glycerol. At this rate of breakdown, body protein would be quickly depleted and death would occur within 2 to 3 weeks. (Death would occur regardless of the amount of body fat a person had because fatty acids cannot be used for gluconeogenesis.) Sodium and potassium depletion also can result during fasting because these elements are

**Postprandial fasting (0 to 6 hours after eating)**

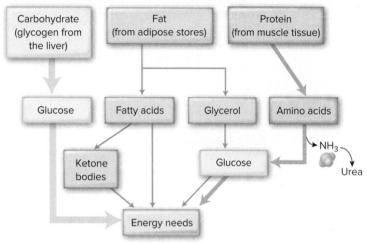

**Short-term fasting (3 to 5 days)**

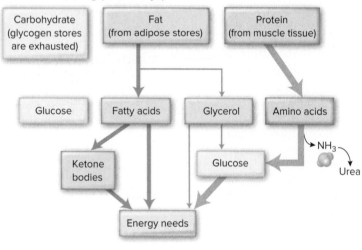

**Long-term fasting (5 to 7 days and beyond)**

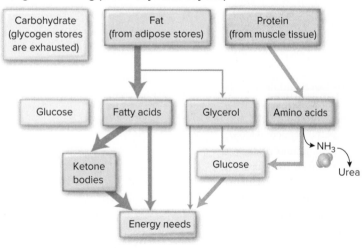

drawn into the urine along with ketone bodies. Finally, blood urea levels increase because of the breakdown of protein.

Fortunately, the body undergoes a series of adaptations that prolong survival. One of these adaptations is the slowing of metabolic rate and a reduction in energy requirements. This helps slow the breakdown of lean tissue to supply amino acids for gluconeogenesis. Another adaptation allows the nervous system to use less glucose (and, hence, less body protein) and more ketone bodies. After several weeks of fasting, half or more of the nervous system's energy needs are met by ketone bodies; nonetheless, some glucose must still be supplied via the catabolism of lean body mass. When lean body mass declines by about 50% (usually within 7 to 10 weeks of total fasting), death occurs.

## Feasting

The most obvious result of feasting is the accumulation of body fat. In addition, feasting increases insulin production by the pancreas, which in turn encourages the burning of glucose for energy, as well as the synthesis of glycogen and, to a lesser extent, protein and fat (Fig. 9-24).[7,14]

Fat consumed in excess of need goes immediately into storage in adipose cells.[2] Compared with the conversion of carbohydrate and protein, relatively little energy is required to convert dietary fat into body fat. Therefore, high fat, high energy diets promote the accumulation of body fat.

Protein consumed in excess of need—contrary to popular belief—does not promote muscle development. Some of the excess protein can reside in amino acid pools in the body, but the amount is not significant. Amino acids left over in the body after a large meal can be used to synthesize fatty acids, but this is typically of minor importance in humans.[14] The process of storing amino acids as fat requires ATP and the B-vitamins biotin, niacin, and pantothenic acid.[2] The energy cost of converting dietary protein to body fat is higher than it is for the conversion of dietary fat to body fat.

Carbohydrate consumed in excess of need is used 1st to maximize glycogen stores. Once glycogen stores are filled, carbohydrate consumption stimulates the use of carbohydrate as fuel and the storage of excess amounts as body fat. This then lessens the need for any fat catabolism. However, the

*Figure 9-23* Postprandial fasting encourages the use of mostly glucose, as well as some fatty acids and amino acids for energy needs. As the fast progresses, glycogen stores are depleted, which causes the rapid use of carbon skeletons of certain amino acids from body protein to produce glucose. This supplies glucose to glucose-dependent cells, such as red blood cells. Long-term fasting leads to the reduced breakdown of body protein and increased use of adipose stores, which are used to produce ketones. Ketones can provide a significant proportion of the fuel required by glucose-dependent cells, thereby sparing body protein and prolonging life. Note that the thickness of the arrows in the figures conveys the relative use of each energy source during the stages of fasting.

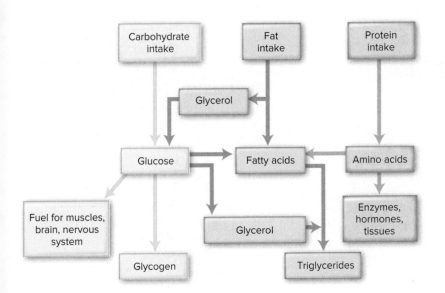

*Figure 9-24* Feasting encourages glycogen and triglyceride synthesis and storage and allows amino acids to participate in the synthesis of body proteins. Minimal synthesis of fatty acids using glucose or carbon skeletons of amino acids occurs unless intake is quite excessive in comparison with overall energy needs.

pathway for storing carbohydrate as body fat is not very active in humans.[14] In addition, it requires the B-vitamins biotin, niacin, and pantothenic acid, and it is energetically expensive to convert carbohydrate to body fat (Table 9-2). *Anyone who consumes more calories from any of the energy-yielding nutrients than what the body can use will gain weight.*

The pathways for the synthesis of fat from excess carbohydrate or protein intake, called lipogenesis, are found primarily in the cytosol of liver cells. Synthesis involves a series of steps that link the acetyl-CoA formed from either glucose or amino acids into a 16-carbon saturated fatty acid, palmitic acid. Insulin increases the activity of a key enzyme—fatty acid synthase—used in the pathway. Palmitic acid can later be lengthened to an 18- or 20-carbon chain either in the cytosol or in the mitochondria.[1] Ultimately, the fatty acids and glycerol (produced during glycolysis from glyceraldehyde 3-phosphate) are used to synthesize triglycerides, which are subsequently delivered by very-low-density lipoproteins (see Chapter 6) via the bloodstream to adipose tissues for storage.

Feasting especially encourages the synthesis of glycogen and the storage of fat.

©Bear Dancer Studios/Mark Dierker

**Table 9-2 Metabolism of ATP-Yielding Compounds**

| Nutrient | Yields Glucose? | Yields Amino Acids for Body Proteins? | Yields Fat for Adipose Tissue Stores? | Energy Cost of Conversion to Adipose Tissue Stores |
|----------|-----------------|---------------------------------------|---------------------------------------|----------------------------------------------------|
| Glucose | Yes | No | Yes | High |
| Fatty acids | No | No | Yes | Minimal |
| Glycerol | Yes but not a major pathway | No | Yes | High |
| Amino acids | Yes | Yes | Yes but is inefficient | High |
| Alcohol | No | No | Yes | High |

► Fasting encourages the following.

Glycogen breakdown

Fat breakdown

Gluconeogenesis

Synthesis of ketone bodies

► Feasting encourages the following.

Glycogen synthesis

Fat synthesis

Protein synthesis

Urea synthesis

## Take Action

### Intermittent Fasting and Metabolism

Your friend is very overweight and decides to go on an intermittent fast to lose weight. He eats on alternate days. On fasting days he eats no food or beverages with calories, and on nonfasting days he eats whatever he wants. His rationale for going on the diet is that, theoretically, over the course of a week, he can cut the total number of calories he typically consumes and, because he is eating fewer calories, the pounds will drop off.

Based on your knowledge of energy metabolism, answer the following questions.

1. During your friend's fasting days, what are the likely sources of energy for his body's cells? What metabolic adaptations occur to provide glucose for the nervous system?

2. After about a month on the program, your friend's weight loss starts to stabilize. Explain why and how weight stabilizes on this type of weight loss program. What are intermittent fasting's long-term effects on health?

Pgiam/Getty Images

### Knowledge Check

1. In the 1st few hours of a fast, what is the primary fuel for the body?
2. What adaptations occur that help slow the breakdown of lean body mass during prolonged fasting?
3. What happens to excess amounts of ingested fat, protein, and carbohydrate?

## CASE STUDY FOLLOW-UP

RASimon/Getty Images

A very low carbohydrate diet that produces ketones is not the best way to lose body fat. Although body weight may decline, the large production of ketones means that the body is not capable of oxidizing fatty acids; therefore, the ketones are excreted in the urine. The body is using protein (amino acids) as a fuel source for the brain and nervous system. This loss of amino acids, especially from muscle, is part of the loss of body mass. A better weight loss program would be to reduce total calories to create a calorie deficit and to maintain an exercise or fitness program that keeps muscle tissue metabolically active to oxidize fat.

The energy used to perform physical activity is in the form of ATP, which can be supplied by carbohydrate, fat, or protein. The proportion supplied by each macronutrient depends on the length of time after eating and the type and intensity of exercise.

U.S. Air Force photo by Staff Sgt. Desiree N. Palacios

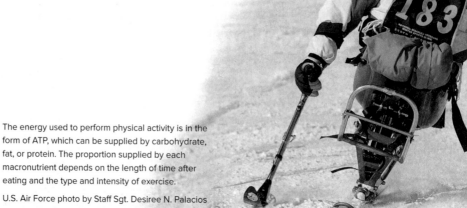

## Inborn Errors of Metabolism

Some people lack a specific enzyme to perform normal metabolic functions—they are said to have an inborn error of metabolism. The metabolic pathway in which the enzyme is supposed to participate does not function normally. Typically, this causes alternative metabolic products to be formed, some of which are toxic to the body.

Inborn errors of metabolism occur when a person inherits a defective gene coding for a specific enzyme from both parents. Both parents are likely to be carriers of the defective gene—that is, they have 1 healthy gene and 1 defective gene for the enzyme in their chromosomes. When each parent donates the defective form of the gene to the offspring, the offspring has 2 defective copies of the gene and therefore little or no activity of the enzyme that the gene normally would produce. If a person has a defective gene, he or she produces a defective protein based on the instructions contained in that defective gene. It also is possible that 1 or both parents have the disease and are not simply carriers. Generally, however, individuals who have an inborn error of metabolism are advised to see a genetic counselor to assess the risk of passing on the inborn error of metabolism to their offspring (see Chapter 1).

The following are some characteristics of inborn errors of metabolism.[15]

- They appear soon after birth. Such a disorder is suspected when otherwise physically well children develop a loss of appetite, vomiting, dehydration, physical weakness, or developmental delays soon after birth. For some of these conditions, infants are screened for the potential to have a specific inborn error of metabolism.
- They are very specific, involving only 1 or a few enzymes. These enzymes usually participate in catabolic pathways (in which compounds are degraded).
- No cure is possible, but typically the disorders can be controlled. The type of control depends on the inborn error—control might include reducing intake of the substance the individual is unable to metabolize normally, taking pharmacologic doses of vitamins, and replacing a compound that cannot be synthesized.

Some of the most common inborn errors of metabolism are phenylketonuria (PKU), galactosemia, glycogen storage disease, and trimethylaminuria. A number of other, very rare inborn errors of metabolism involve various amino acids, fatty acids, and the sugars fructose and sucrose. Typically, in large hospitals and state health departments, physicians, nurses, and registered dietitian nutritionists can help affected persons and their families cope with these and other inborn errors of metabolism.[15]

### Newborn Screening

Newborn screening is the process of testing newborn babies for treatable genetic errors of metabolism.[16] Newborn screening is a public health program that provides early identification and follow-up for the treatment of infants with genetic and metabolic disorders. There are no national mandates to test newborns; each state determines which newborn screening tests are required, so required tests vary widely among states. The Advisory Committee on Heritable Disorders in Newborns and Children recommends screening for 35 core conditions; however, some states test for fewer of these conditions and some test for more than 60. To learn more, visit www.babysfirsttest.org/newborn-screening/states.

### Phenylketonuria (PKU)

PKU is estimated to occur in about 1 per 10,000 to 15,000 births.[16] Most carriers can be detected with a simple blood test. People of Irish descent are especially affected.[17] Today, most infants in the U.S. are diagnosed within a few days of life because all states require them to be tested for this inborn error of metabolism.[18]

The majority of PKU cases occur because the enzyme phenylalanine hydroxylase does not function efficiently in the liver. Normally, phenylalanine hydroxylase converts the amino acid phenylalanine into the amino acid tyrosine. If this reaction does not take place, phenylalanine accumulates in the blood and tyrosine is deficient. If not corrected within 30 days of birth, this phenylalanine buildup leads to the production of toxic phenylalanine by-products, such as phenylpyruvic acid, which then can lead to severe, irreversible intellectual disability.[19]

Sufficient phenylalanine
hydroxylase activity

Normal: Phenylalanine ⟶ Tyrosine

Reduced phenylalanine
hydroxylase activity

PKU: Phenylalanine ⟶ Phenylpyruvic acid
Phenyllactic acid
Other related
products

As soon as they are diagnosed, infants are started on a phenylalanine-restricted diet.[20] Recall

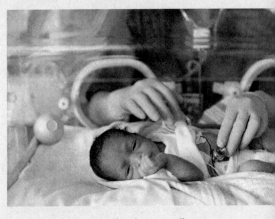

An infant who does not develop normally may have an inborn error of metabolism. A physician should investigate this possibility.

ERproductions Ltd./Blend Images LLC

that phenylalanine is an essential amino acid, which means that even someone with PKU has to obtain phenylalanine from his or her diet; however, the amount of phenylalanine consumed needs to be controlled carefully to prevent toxic amounts from building up.[19,21] During infancy, nutritional needs can change frequently, so these infants are monitored continually through blood phenylalanine testing.

Starting in infancy, special formulas are used to provide nutrients for individuals with PKU. Because infants have high protein needs, satisfying their protein requirements—without also having high intakes of phenylalanine—is impossible without these specially prepared formulas. For infants, formulas are designed to provide about 90% of protein needs and 80% of energy needs. Human milk or regular infant formula then can be used to make up the difference and supply small amounts of phenylalanine.[22]

Later in life, foods can be used to make up the difference, especially foods low in phenylalanine. Fruits and vegetables are naturally low in phenylalanine, and breads and cereals have a moderate amount. Dairy products, eggs, meats, and nuts are very high in amino acids, including phenylalanine, so they are not allowed in the diet. Foods and beverages that contain the alternative sweetener aspartame also are not allowed because aspartame contains phenylalanine (see Chapter 5). Older children and adults can use a formula (such as Phenyl-Free®) that is very low in phenylalanine, which allows them to consume more foods but limits phenylalanine intake. Overall, the majority of their nutrient intake throughout life will come from a special formula.

Ideally, the low phenylalanine diet is followed for life.[23] In the past, health-care professionals thought it was appropriate to end the diet after age 6 years because brain development was complete. However, it is now known that discontinuing this diet leads to decreased intelligence as well as behavior problems, such as aggressiveness, hyperactivity, and decreased attention span.[22]

If a woman with PKU has abandoned the diet, she needs to return to it at least 6 months before becoming pregnant.[24] Otherwise, the fetus—even though it does not have PKU—will be exposed to a high blood phenylalanine level and related toxic products from the mother. This can result in miscarriage or birth defects. All pregnancies for women with PKU are high risk and require close medical supervision.

### Galactosemia

Galactosemia is a rare genetic disease—the most common form occurs in 1 in 30,000 to 60,000 births.[16,25] It is more common in those of Italian and Irish descent.[26] In galactosemia, 2 specific enzyme defects lead to a reduction in metabolism of the monosaccharide galactose to glucose (a 3rd defect is very rare). Galactose then builds up in the bloodstream, which can lead to very serious bacterial infections, intellectual disability, and cataracts in the eyes.

An infant with galactosemia typically develops vomiting after a few days of consuming infant formula or breast milk. Both contain much galactose as part of the milk sugar lactose. The child is then switched to a soy formula. In addition, all dairy products and other lactose-containing products (e.g., butter, milk solids), organ meats, and some fruits and vegetables must be avoided. Strict label reading also is important for controlling the disease because lactose can be found in a variety of products. Even in well-controlled cases, slight intellectual disability (e.g., speech delays) and cataracts occur.

### Glycogen Storage Disease

Glycogen storage disease, a group of 13 diseases, occurs in 1 in 20,000 to 25,000 births. In glycogen storage disease, the liver is unable to convert glycogen to glucose. There are a number of possible enzyme defects along the pathway from glycogen to glucose. The most common forms cause poor physical growth, low blood glucose, and liver enlargement. Low blood glucose results because liver glycogen breakdown is typically used to maintain blood glucose between meals (see Chapter 5). People with glycogen storage disease typically have to consume frequent meals in order to regulate blood glucose. They also consume raw cornstarch between meals because it is slowly digested and helps maintain steady blood glucose. Careful monitoring of blood glucose is very important for these people so that low blood glucose levels can be detected and treated quickly.[22]

### Trimethylaminuria

Trimethylaminuria is a disorder in which a strong, fishy-smelling compound, trimethylamine (TMA), accumulates and is excreted in the urine, sweat, and breath. Due to the body odors and accompanying bad breath, many patients avoid social gatherings, which leads to isolation and depression. Primary trimethylaminuria is an inherited enzyme deficiency in which TMA is not efficiently converted to nonodorus trimethylamine-N-oxide (TMAO) in the liver. More than

Children with PKU must restrict their intake of high protein foods, such as milk and meat.

Shutterstock/komokvm

100 cases have been reported in the medical literature. Some clinicians believe that the disorder is underdiagnosed because many people with mild symptoms do not seek help. Symptoms can be improve by changes in the diet to avoid TMAO precursors, such as marine fish. Treatment with activated charcoal to lowe TMA may be beneficial.

▶ A drug, sapropterin dihydrochloride, 6-R-L-erythro 5,6,7,8-tetrahydrobiopterin (BH4), is used to treat mild forms of PKU. Although the drug cannot cure PKU, it appears to lower blood levels of phenylalanine.[27,28]

### Knowledge Check

1. What are the characteristics of inborn errors of metabolism?
2. What is the cause of PKU?
3. What dietary restrictions must those with galactosemia observe?

# Take Action

## Newborn Screening in Your State

There are no mandatory national newborn screening standards in the U.S., even though dozens of metabolic diseases are detectable by newborn screening tests. Each state has developed its own newborn screening program for infants born there. To learn which metabolism disorder tests are required for newborns in your state, visit the website: genes-r-us.uthscsa.edu. Why don't all states require the same tests? Once a newborn has tested positive, what resources and professionals can help the family manage the disorder?

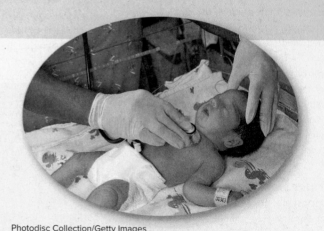

Photodisc Collection/Getty Images

# Chapter *Summary*

## 9.1 *Metabolism* refers to the entire network of chemical processes involved in

maintaining life. It encompasses all the sequences of chemical reactions that occur in the body. Some of these biochemical reactions enable us to release and use energy from carbohydrate, fat, protein, and alcohol. Metabolism is the sum total of all anabolic and catabolic reactions. ATP is the energy currency for the body. A molecule of ATP consists of the organic compound adenosine (comprised of the nucleotide adenine and the sugar ribose) bound to 3 phosphate groups. As ATP is broken down to ADP plus $P_i$, energy is released from the broken bond. Every cell contains catabolic pathways, which release energy to allow ADP to combine with $P_i$ to form ATP. The synthesis of ATP from ADP and $P_i$ involves the transfer of energy from energy-yielding compounds (carbohydrate, fat, protein, and alcohol). This process uses oxidation-reduction reactions, in which electrons (along with hydrogen ions) are transferred in a series of reactions from energy-yielding compounds eventually to oxygen. The process of cellular respiration oxidizes (removes electrons) to obtain energy (ATP). Oxygen is the final electron acceptor. When oxygen is readily available, cellular respiration may be aerobic. When oxygen is not present, anaerobic pathways are used.

glucose passes through several steps, which convert it to 2 units of a 3-carbon compound called pyruvate. Glycolysis nets 2 ATP. Pyruvate passes from the cytosol into the mitochondria, where the enzyme pyruvate dehydrogenase converts pyruvate into the compound acetyl-CoA in a process called a transition reaction. Acetyl-CoA molecules enter the citric acid cycle, which is a series of chemical reactions that convert carbons in the acetyl group to carbon dioxide while harvesting energy to produce ATP. In the citric acid cycle, acetyl-CoA undergoes many metabolic conversions, which result in the production of GTP, ATP, NADH, and $FADH_2$. NADH and $FADH_2$ enter the electron transport chain, which passes electrons along a series of electron carriers. As electrons pass from 1 carrier to the next, small amounts of energy are released. This metabolic process is called oxidative phosphorylation, and it is the pathway in which energy derived from glycolysis, the transition reaction, and the citric acid cycle is transferred to ADP + $P_i$ to form ATP and water.

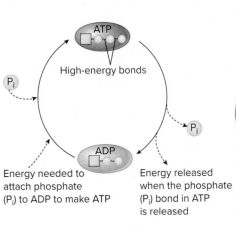

## 9.2 Glucose metabolism begins with glycolysis, which literally means "breaking down glucose."

Glycolysis has 2 roles: to break down carbohydrates to generate energy and to provide building blocks for synthesizing other needed compounds. During glycolysis,

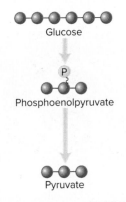

## 9.3 The first step in generating energy from a fatty acid is to cleave the carbons, 2 at a time, and convert the 2-carbon fragments to acetyl-CoA.

The process of converting a free fatty acid to multiple acetyl-CoA molecules is called beta-oxidation because it begins with the beta carbon. Fatty acids can be oxidized for energy but cannot be converted into glucose. During low carbohydrate intakes and uncontrolled diabetes, more acetyl-CoA is produced in the liver than can be metabolized. This excess acetyl-CoA is synthesized into ketone bodies, which can be used as an energy source by other tissues or excreted in the urine and breath.

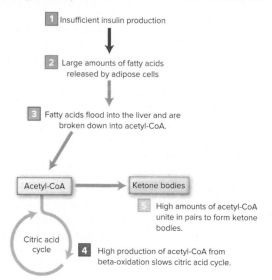

**1** Insufficient insulin production

**2** Large amounts of fatty acids released by adipose cells

**3** Fatty acids flood into the liver and are broken down into acetyl-CoA.

Acetyl-CoA → Ketone bodies

**5** High amounts of acetyl-CoA unite in pairs to form ketone bodies.

Citric acid cycle

**4** High production of acetyl-CoA from beta-oxidation slows citric acid cycle.

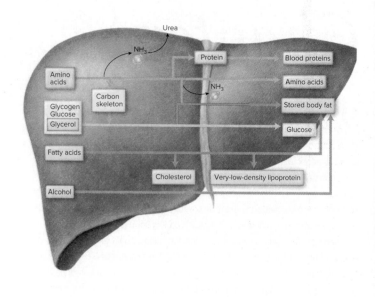

## 9.4 Protein metabolism begins after proteins are degraded into amino acids.

To use an amino acid for fuel, cells must first deaminate them (remove the amino group, –NH₂). Resulting carbon skeletons mostly enter the citric acid cycle. Some carbon skeletons also yield acetyl-CoA or pyruvate. The process of generating glucose from amino acids is called gluconeogenesis. Acetyl-CoA molecules cannot participate in gluconeogenesis; thus, ketogenic amino acids cannot participate in gluconeogenesis.

## 9.5 The alcohol dehydrogenase (ADH) pathway

is the main pathway for alcohol metabolism. Alcohol is converted in the cytosol to acetaldehyde by the action of the enzyme alcohol dehydrogenase and the coenzyme $NAD^+$. $NAD^+$ picks up 2 hydrogen ions and 2 electrons from the alcohol to form $NADH + H^+$ and produces the intermediate acetaldehyde. Acetaldehyde is converted to acetyl-CoA, again yielding $NADH + H^+$ with the aid of the enzyme aldehyde dehydrogenase and coenzyme A. Most of the acetyl-CoA is used to synthesize fatty acids and triglycerides, resulting in the accumulation of fat in the liver. When an individual consumes too much alcohol, a second pathway—called the microsomal ethanol oxidizing system (MEOS)—is activated to help metabolize the excess alcohol.

## 9.6 The liver plays the major role in regulating metabolism.

Additional means of regulating metabolism involve enzymes, ATP concentrations, and minerals. Many micronutrients (thiamin, niacin, riboflavin, biotin, pantothenic acid, vitamin B-6, magnesium, iron, and copper) play important roles in the metabolic pathway.

## 9.7 During fasting, the body breaks down both amino acids and fats for energy.

The body undergoes a series of adaptations that prolong survival. One of these adaptations is the slowing of metabolic rate and the reduction in energy requirements. This helps slow the breakdown of lean tissue to supply amino acids for gluconeogenesis. Another adaptation allows the nervous system to use less glucose and more ketone bodies. Fat consumed in excess of need goes into storage in adipose cells. Compared with the conversion of carbohydrate and protein, relatively little energy is required to convert dietary fat into body fat. Therefore, high fat diets promote the accumulation of body fat.

## Clinical Perspective

Inborn errors of metabolism occur when a person inherits a defective gene coding for a specific enzyme from 1 or both parents. Some of the most common inborn errors of metabolism are phenylketonuria (PKU), galactosemia, glycogen storage disease, and trimethylaminuria. Strict diets can help minimize many of the serious effects of these diseases.

## Global Perspective

Cancer cells use glycolysis for energy, even when oxygen is abundant. Altered glucose, glutamine, and fat metabolism are hallmarks of cancer cells.

ERproductions Ltd./Blend Images LLC

# Study Questions

1. The energy currency for the body is _____.

   a. NAD
   b. FAD
   c. TCA
   d. ATP

2. Glycolysis is a biochemical pathway that _____.

   a. breaks down glucose
   b. produces energy
   c. takes place in the cytosol
   d. all of the above

3. Glycolysis begins with _____ and ends with

   _____.

   a. pyruvate; water
   b. pyruvate; glucose
   c. glucose; pyruvate
   d. pyruvate; acetyl-CoA

4. When muscle tissue is exercising under anaerobic conditions, the production of _____ is important because it assures a continuous supply of NAD.

   a. glucose-6-phosphate
   b. pyruvate
   c. lactate
   d. glycogen

5. The net ATP production in glycolysis is _____.

   a. 1 ADP
   b. 2 ATP
   c. 4 FADH
   d. 2 GTP
   e. none of the above

6. The common pathway for the oxidation of glucose and fatty acid is _____.

   a. glycolysis
   b. the urea cycle
   c. the citric acid cycle
   d. ketosis

7. The oxidation of fatty acids occurs in the _____.

   a. cell membrane
   b. mitochondria
   c. nucleus
   d. cytosol

Match the definitions on the right with the terms on the left.

8. beta-oxidation

9. ketosis

   a. breakdown of glucose to pyruvate

   b. breakdown of fat to 2-carbon units called acetyl-CoA

10. electron transport chain

11. gluconeogenesis

12. glycolysis

   c. synthesis of glucose from noncarbohydrate sources

   d. formation of excess ketone bodies

   e. electrons transferred back and forth to make ATP

13. Metabolism is regulated by _____.

   a. hormones
   b. enzymes
   c. the energy status of the body
   d. all of the above

14. During periods of starvation, the body uses protein as a fuel source for the brain and central nervous system in a pathway called gluconeogenesis.

   a. true                    b. false

15. Insulin is _____.

   a. a coenzyme in the glycolytic pathway
   b. a cofactor needed for gluconeogenesis
   c. an anabolic hormone
   d. a catabolic hormone

16. The electron transport chain is a series of biochemical reactions that transfers the energy in NADH + H$^+$ and FADH$_2$ to _____.

   a. iron and copper
   b. oxygen
   c. ATP
   d. ketones

17. Which of the following is an end product of the electron transport chain?

   a. glucose
   b. ketones
   c. water
   d. amino acids

18. Before protein becomes an energy source, what must be removed from the molecule?

   a. acid
   b. nitrogen
   c. R group
   d. carbons

19. When a chemical compound becomes oxidized, it _____ electrons.

   a. gains
   b. loses
   c. converts
   d. inverts

**20.** What is the "common denominator" compound of many pathways of energy metabolism (citric acid cycle, glycolysis, beta-oxidation)? Why is it considered important in the body's metabolism?

**21.** Describe the process of glycolysis.

**22.** Explain why most fatty acids cannot become glucose.

**23.** Trace the steps of gluconeogenesis from body protein to the formation of glucose.

**24.** Describe the metabolism of alcohol.

**25.** Identify the vitamins and minerals used in ATP synthesis.

**26.** Describe how the fuels used by the body change as a fast progresses from a few hours to a week.

**27.** Explain how metabolism is altered in cancer cells.

**28.** Describe dietary restrictions for those with PKU.

Answer Key: 1-d; 2-d; 3-c; 4-c; 5-b; 6-c; 7-b; 8-b; 9-d; 10-e; 11-c; 12-a; 13-d; 14-a; 15-c; 16-c; 17-c; 18-b; 19-b; 20-refer to Sections 9.1 to 9.4; 21-refer to Section 9.2; 22-refer to Section 9.3; 23-refer to Section 9.4; 24-refer to Section 9.5; 25-refer to Section 9.6; 26-refer to Section 9.7; 27-refer to Global Perspective; 28-refer to Clinical Perspective

©Photodisc/Getty Images

# References

**1.** Berg J and others. *Biochemistry.* 9th ed. New York: WH Freeman; 2019.

**2.** Gropper SS and others. *Advanced nutrition and human metabolism.* 7th ed. Belmont, CA: Thomson/Wadsworth; 2018.

**3.** Bender D, Mayes P. The citric acid cycle: The central pathway of carbohydrates, lipid and amino acid metabolism. In: Rodwell V and others, eds. *Harper's illustrated biochemistry.* 31st ed. New York:Appleton & Lange Medical Books/McGraw-Hill; 2018.

**4.** Buono MJ, Kolkhorst FW. Illuminations: Estimating ATP synthesis during a marathon run: A method to introduce metabolism. *Adv Physiol Educ.* 2001;25:70.

**5.** Bender D, Mayes P. Overview of metabolism & the provision of metabolic fuels. In: Rodwell V and others, eds. *Harper's illustrated biochemistry.* 31st ed. New York:Appleton & Lange Medical Books/McGraw-Hill; 2018.

**6.** Harvey RA and others. *Biochemistry.* 7th ed. Philadelphia: Lippincott Williams & Wilkins; 2017.

**7.** Longo N and others. Carnitine transport and fatty acid oxidation. *Biochem Biophys Acta.* 2017;1863:2422.

**8.** Botham K, Mayes P. Oxidation of fatty acids: Ketogenesis. In: Rodwell V and others, eds. *Harper's illustrated biochemistry.* 31st ed. New York: Appleton & Lange Medical Books/McGraw-Hill; 2018.

**9.** Kanikarla-Marie P, Jain S. Hyperketonemia and ketosis increase with risk of complications in type 1 diabetes. *Free Radic Biol Med.* 2016;95:268.

**10.** Palmiere C and others. Postmortem biochemistry in suspected starvation induced ketoacidosis. *J Forensic Leg Med.* 2016;42:51.

**11.** World Health Organization, International Agency for Research on Cancer. Globocan. 2020; globocan.iarc.fr.

**12.** Mycielska M and others. Extracellular citrate affects critical elements of cancer cell metabolism and supports cancer development in vivo. *Cancer Res.* 2018;10:2513.

**13.** Botham K, Mayes P. Bioenergetics: The role of ATP. In: Rodwell V and others, eds. *Harper's illustrated biochemistry.* 31st ed. New York: Appleton & Lange Medical Books/McGraw-Hill; 2018.

**14.** Green CR and others. Branched-chain amino acid catabolism fuels adipocyte differentiation and lipogenesis. *Nat Chem Biol.* 2016;1:15.

15. Trahms C. Medical nutrition therapy for genetic metabolic disorders. In: Mahan L and others, eds. *Food and the nutrition care process.* 14th ed. St. Louis: Elsevier/Saunders; 2017.

16. Berry SA and others. Newborn screening 50 years later: Access issues faced by adults with PKU. *Genet Med.* 2013;15:591.

17. O'Donnell KA and others. The mutation spectrum of hyperphenylalaninaemia in the Republic of Ireland: The population history of the Irish revisited. *Eur J Human Genet.* 2002;10:530.

18. Champion MP. An approach to the diagnosis of inherited metabolic disease. *Arch Dis Child Educ Pract Ed.* 2010;95:40.

19. de Groot MJ and others. Pathogenesis of cognitive dysfunction in phenylketonuria: Review of hypotheses. *Mol Genet Metab.* 2010;99:S86.

20. Rocha J, MacDonald A. Dietary intervention in the management of phenylketonuria: Current perspectives. *Pediatric Health Med Ther.* 2016;7:155.

21. Al Hafid N. Phenylketonuria: A review of current and future treatments. *Transl Pediatr.* 2015;4:304.

22. Banta-Wright S and others. Breastfeeding success among infants with PKU. *J Pediatr Nurs.* 2012;4:319.

23. Ney DM and others. Dietary glycomacropeptide supports growth and reduces the concentrations of phenylalanine in plasma and brain in a murine model of phenylketonuria. *J Nutr.* 2008;138:316.

24. Koch R and others. Psychosocial issues and outcomes in maternal PKU. *Mol Genet Metab.* 2010;99:S68.

25. Coman DJ and others. Galactosemia, a single gene disorder with epigenetic consequences. *Pediatr Res.* 2010;67:286.

26. Murphy M and others. Genetic basis of transferase-deficient galactosemia in Ireland and the population history of the Irish Travellers. *Eur J Hum Genet.* 1999;7:549.

27. Muntau A and others. International best practice for the evaluation of responsiveness to sapropterin dihydrochloride in patients with phenylketonuria. *Molecul Genet Metab.* 2019;127:1.

28. Somaraju UR, Rani U. Sapropterin dihydrochloride for phenylketonuria. *Cochrane DB Syst Rev.* 2015;CD008005.

Sabine Scheckel/Getty Images

Lack of sleep is a risk factor for weight gain and obesity. Although staying awake instead of sleeping increases daily energy expenditure by about 5%, energy intake, especially after dinner, is often higher than needed for weight maintenance. Eating more when you are tired may be how the body adapts to help you stay awake. Researchers reported that, when those who were getting too little sleep started sleeping enough, they ate less, especially from fats and carbohydrates, and lost weight.[1] Learn more at **www.nhlbi.nih. gov/health-topics/education-and-awareness/sleep-health**.

# 10 Energy Balance, Weight Control, and Eating Disorders

## Learning Objectives

**After studying this chapter, you will be able to**

1. Describe energy balance and its relationship to energy intake and expenditure.

2. Evaluate the different techniques used to measure energy expenditure by the body.

3. Explain internal and external regulation of hunger, appetite, and satiety.

4. Describe the methods used for assessing body composition and determining whether body weight and composition are healthy.

5. Explain the impact of genetics and environment on body weight and composition.

6. Outline the key components of programs designed to treat overweight, obesity, and underweight.

7. Discuss the characteristics of fad diets.

8. Evaluate weight loss programs to determine whether they are safe and likely to result in long-term weight loss.

9. Identify common characteristics and health risks of different eating disorders.

THIS CHAPTER BEGINS WITH SOME good news and some bad news. The good news is that, if you stay at a healthy body weight, you increase your chances of living a long and healthy life. The bad news is that, in the last 30 years, there has been a dramatic increase in the percentage of individuals who have obesity. This problem is occurring not only in the U.S. but around the world, especially in low-income countries where Westernized dietary patterns (high fat, high calorie) are increasing in popularity. In 1990, no state in the U.S. had an obesity prevalence rate exceeding 15%. By 2015, no state had an obesity prevalence rate less than 20%, and 4 states had a prevalence rate of 35% or higher.[2] Currently in North America, more than 1 out of 3 of adults have obesity. The current trend is not likely to be reversed without a national commitment to weight maintenance and approaches that make our social environment more favorable to maintaining a healthy weight. These data suggest that there is a good chance that many of us will have significant weight gain in adulthood.[3] There is no quick cure for overweight, despite what advertisements claim. Any success comes from hard work and commitment. Unfortunately, for most people, weight reduction efforts fizzle before they achieve a healthy weight range (Fig. 10-1).

*Figure 10-1* Obesity rates among U.S. adults in 2018. To see other years, visit **www.cdc.gov/obesity/data/prevalence-maps.html**.

Source: CDC Behavioural Risk Factor Surveillance System.

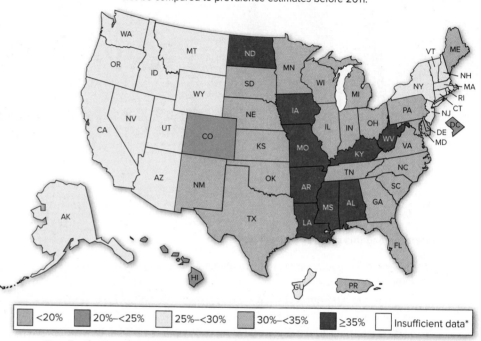

**Prevalence of Self-Reported Obesity among U.S. Adults by State and Territory, BRFSS, 2018**

Prevalence estimates reflect BRFSS methodological changes started in 2011. These estimates should not be compared to prevalence estimates before 2011.

| ▨ <20% | ▨ 20%–<25% | ▢ 25%–<30% | ▨ 30%–<35% | ▨ ≥35% | ▢ Insufficient data* |

*Sample size <50 or the relative standard error (dividing the standard error by the prevalence) ≥30%.

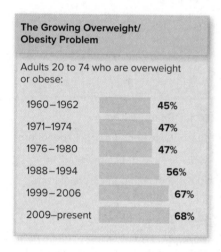

**The Growing Overweight/Obesity Problem**

Adults 20 to 74 who are overweight or obese:

| 1960–1962 | **45%** |
| 1971–1974 | **47%** |
| 1976–1980 | **47%** |
| 1988–1994 | **56%** |
| 1999–2006 | **67%** |
| 2009–present | **68%** |

Popular ("fad") diets are generally monotonous, ineffective, and confusing; they may even be dangerous for some population groups and individuals with health disorders. The relentless pursuit of thinness may drive some to develop eating disorders, which involve severe distortions of the eating process. The safest and most logical approach to maintaining a healthy weight is to watch calorie intake, exercise regularly, and get problem eating behaviors under control.[4] Preventing excess weight gain in the first place is the most successful approach of all.[5]

## 10.1 Energy Balance

**Energy balance** is the relationship between energy intake and energy expenditure. When the calories consumed from food and beverages (energy intake) match the amount of energy expended, **energy equilibrium** occurs. If energy intake exceeds energy expended, the result is a **positive energy balance.** The excess energy consumed is stored, resulting in weight gain (Fig. 10-2). There are some situations in which positive energy balance is desired, such as during the growth stages of the life cycle (pregnancy, infancy, childhood, adolescence) and to restore body weight to healthy levels after losses caused by starvation, disease, or injury. However, during other times, such as adulthood, positive energy balance over time can cause body weight to climb to unhealthy levels. The process of aging itself does not cause weight gain; rather, weight gain stems from a pattern of excess food intake coupled with limited physical activity and slower metabolism.[4]

     **Negative energy balance** results when energy intake is less than energy expenditure. Weight loss occurs because energy stored in the body—fat and muscle—is used to make up for the shortfall in energy intake. Negative energy balance is desired in adults when body fatness exceeds healthy levels. Negative energy balance during growth stages of the life cycle generally is not recommended because it can impair normal growth.

*Critical* Thinking

A 20-year-old classmate of yours has been watching her parents and grandparents gain weight over the years. How should she explain energy balance to them?

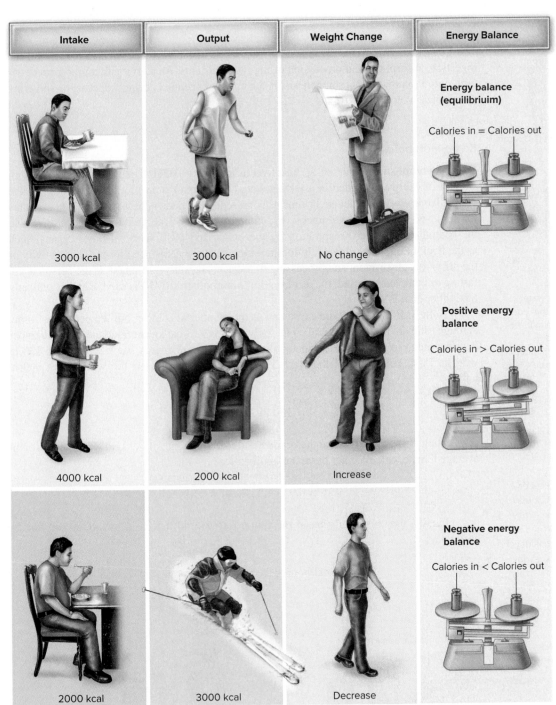

| Intake | Output | Weight Change | Energy Balance |
|---|---|---|---|
| 3000 kcal | 3000 kcal | No change | **Energy balance (equilibriuim)** Calories in = Calories out |
| 4000 kcal | 2000 kcal | Increase | **Positive energy balance** Calories in > Calories out |
| 2000 kcal | 3000 kcal | Decrease | **Negative energy balance** Calories in < Calories out |

*Figure 10-2*  States of energy balance.
- Energy equilibrium occurs when calories consumed equal calories expended.
- Positive energy balance is when intake of calories exceeds calories burned.
- Negative energy balance results when fewer calories are eaten than used.

## Energy Intake

The amount of energy in a food or beverage can be estimated using nutrient databases or nutrient analysis software. Calorie values in these tables and programs can be derived by directly measuring calorie content using a device called a **bomb calorimeter** (Fig. 10-3). Calorie content is most commonly calculated by determining the grams of carbohydrate, protein, and fat (and possibly alcohol) in a food and multiplying these compounds by their physiological fuel values. (Recall from Chapter 1 that the physiological fuel values are 4 kcal/g for carbohydrates and proteins, 9 kcal/g for fat, and 7 kcal/g for alcohol.)

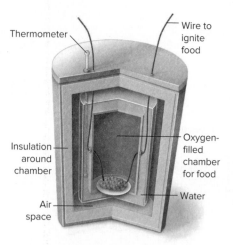

Thermometer

Wire to ignite food

Insulation around chamber

Oxygen-filled chamber for food

Air space

Water

*Figure 10-3* Bomb calorimeters measure calorie content by igniting and burning a dried portion of food. The burning food raises the temperature of the water surrounding the chamber holding the food. The increase in water temperature indicates the number of kilocalories in the food because 1 kilocalorie equals the amount of heat needed to raise the temperature of 1 kg of water by 1°C.

 When a person is resting, the following are the approximate percentages of total energy used by the body's organs.

| Liver | 27% | 380 kcal/day |
|---|---|---|
| Brain | 19% | 265 kcal/day |
| Skeletal muscle | 18% | 250 kcal/day |
| Kidney | 10% | 140 kcal/day |
| Heart | 7% | 100 kcal/day |
| Other | 19% | 265 kcal/day |

Classwork leads to mental activity but little physical activity. Hence, energy is burned at a rate of only about 1.5 kcal per minute.

Stockbyte/PunchStock RF

# Energy Expenditure

The body uses energy for 3 main purposes: basal metabolism; physical activity; and the digestion, absorption, and processing of ingested nutrients. An additional, minor form of energy output, known as thermogenesis, is the energy expended during fidgeting or shivering in response to cold (Fig. 10-4).[4]

## Basal Metabolism

**Basal metabolism** (expressed as **basal metabolic rate [BMR]**) represents the minimum amount of energy expended in a fasting state (12 hours or more) to keep a resting, awake body alive in a warm, quiet environment. For a sedentary person, basal metabolism accounts for about 60 to 70% of total energy expenditure. Some of the processes involved include the beating of the heart, respiration by the lungs, and the activity of other organs, such as the liver, brain, and kidneys.[4] It does not include energy expended for physical activity or the digestion, absorption, and processing of nutrients recently consumed. If the person is not fasting or completely rested, the term **resting metabolism** (RMR) is used. RMR is typically 6% higher than BMR.

Both BMR and RMR are expressed as the number of calories burned per unit of time. A rough estimate of basal metabolic rate for women is 0.9 kcal/kg per hour and 1.0 kcal/kg per hour for men. To see how basal metabolism contributes to energy needs, consider a 130 lb woman. First, knowing that there are 2.2 lb for every kg, convert her weight into metric units:

$$130 \text{ lb}/2.2 = 59 \text{ kg}$$

Then, using a rough estimate of the basal metabolic rate of 0.9 kcal/kg per hour for an average female, calculate her basal metabolic rate:

$$59 \text{ kg} \times 0.9 \text{ kcal/kg/hour} = 53 \text{ kcal/hour}$$

Finally, use this hourly basal metabolic rate to find her basal metabolic rate for an entire day:

$$53 \text{ kcal/hour} \times 24 \text{ hours} = 1272 \text{ kcal/day}$$

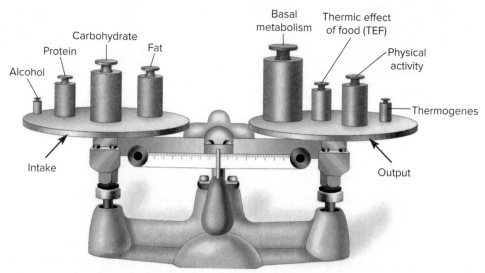

Protein
Carbohydrate
Fat
Alcohol
Intake

Basal metabolism
Thermic effect of food (TEF)
Physical activity
Thermogenes
Output

*Figure 10-4* Major components of energy intake and expenditure. The size of each component shows the relative contribution of that component to energy balance. Alcohol is an additional source of energy only for those who consume it.

These calculations are only an estimate of actual basal metabolism—it can vary 25 to 30% among individuals. The following factors increase basal metabolism.

- Greater muscle mass
- Larger body surface area
- Male biological sex (males typically have more body surface area and muscle mass than females)
- Body temperature (fever or cold environmental conditions)
- Higher than normal secretions of thyroid hormones (a key regulator of basal metabolism)
- Aspects of nervous system activity (e.g., the release of stress hormones)
- Growth stages of the life cycle
- Caffeine and tobacco increase metabolism; however, using tobacco to control body weight is not recommended because too many health risks are increased.
- Recent exercise

Of these factors, the amount of muscle mass a person has is the most important.

These factors decrease basal metabolism.

- Lower than normal secretions of thyroid hormones (hypothyroidism)
- Restricted calorie intake, as well as fasting and starvation
- Less body surface area and muscle mass
- Aging after age 30 years

Basal metabolism decreases by about 10 to 20% (about 150 to 300 kcal/day) when calorie intake declines and the body shifts into a conservation mode. This shift helps us survive during periods of famine and starvation, but it also is a barrier to sustained weight loss during dieting that involves extremely low calorie intake.[5] Basal metabolism drops 1 to 2% for each decade past the age of 30 years as a result of the lean body mass loss that typically occurs with advancing age. However, physical activity helps maintain lean body mass and helps preserve BMR throughout adulthood.[6]

### Energy for Physical Activity

Physical activity increases energy expenditure above and beyond basal energy needs by as much as 25 to 40%. In choosing to be active or inactive, we determine much of our total energy expenditure for a day. Climbing stairs rather than riding the elevator, walking rather than driving to the store, and standing in a bus rather than sitting increase physical activity and hence energy expenditure. The increased rate of obesity in North America is caused in part by our inactivity.[6]

### Thermic Effect of Food

The **thermic effect of food (TEF)** is the energy the body uses to digest, absorb, transport, store, and metabolize the nutrients consumed in the diet. The TEF accounts for about 5 to 10% of the energy consumed each day. If daily energy intake were 3000 calories, TEF would account for 150 to 300 calories. As with other components of energy output, the total amount varies somewhat among individuals.[4] In addition, food composition influences TEF. For example, the TEF value for a protein rich meal (20 to 30% of the energy consumed) is higher than that for a carbohydrate rich (5 to 10%) or fat rich (0 to 5%) meal because it takes more energy to metabolize amino acids into fat than to convert glucose into glycogen or transfer absorbed fat into adipose stores. The TEF of alcohol is 20%. In addition, large meals result in higher TEF values than the same amount of food eaten over many hours.[4]

### Adaptive Thermogenesis

**Thermogenesis,** the process of heat production by humans and other organisms, makes a fairly small contribution to overall energy expenditure. Thermogenesis goes by other names, such as thermoregulation and non-exercise activity thermogenesis (NEAT). Adaptive thermogenesis heat is produced when the body expends energy for nonvoluntary physical activity triggered by extreme cold conditions, overfeeding, trauma, or starvation. Examples of nonvoluntary activities include fidgeting, shivering when cold, maintaining muscle tone, and holding the body up when not lying down. This nonvoluntary production of heat increases energy expenditure, resulting in less storage of energy. In some cases, this can be a considerable amount of energy lost as heat.

**basal metabolism** Minimum amount of energy the body uses to support itself when fasting, resting, and awake in a warm, quiet environment.

**thermic effect of food (TEF)** Energy the body uses to digest, absorb, transport, store, and metabolize nutrients.

**thermogenesis** Heat production by humans.

▶ Resting Energy Expenditure (REE) is the amount of calories needed during a nonactive period. It is used to estimate calorie needs in clinical situations. The Harris-Benedict equation can be used to estimate Resting Energy Expenditure.

*Men*
$$REE = 66.5 + (13.8 \times WT) + (5 \times HT) - (6.8 + AGE)$$

*Women*
$$REE = 655.1 + (9.6 \times WT) + (1.9 \times HT) - (4.7 \times AGE)$$

The variables in the formulas correspond to the following.

REE = Resting Energy Expenditure
WT = Weight in kg (lb ÷ 2.2)
HT = Height in cm (inches × 2.54)
AGE = Age in years

▶ The TEF of alcohol is 20%. About 90% of the energy in food can be used by the body as an energy source, whereas only 80% of the energy in alcohol can be used by the body to make energy.[7]

▶ **High Fructose Corn Syrup (HFCS) and Your Waistline.** Many food manufacturers use HFCS instead of table sugar (sucrose) because high fructose corn syrup (HFCS) costs less—corn is abundant in the U.S. and is subsidized by the government. In addition, HFCS is easier to use because it is a liquid and dissolves easily.

HFCS entered the U.S. food supply in 1966 and, by 2005, the per capita availability of the sweetener was about 59 lb per year.[8] During the same time period, the per capita availability of table sugar (made from sugarcane and sugar beets) dropped to 63 lb, down from a peak of 102 lbs per capita in 1972.[8] The rate of HFCS use paralleled rising rates of obesity,[9] causing some to hypothesize that high intakes of HFCS lead to excess weight gain. By 2018, consumers had cut HFCS intake to 40 lbs per year, but sugar intake had risen to 69 lb per year. Most of the reduction in HFCS intake was due to decreased soda consumption, which is down 15% from its peak consumption in 1999.

Experts have proposed several hypotheses to explain how HFCS might be related to weight gain. One hypothesis is that HFCS triggers a desire to eat sweets, which leads to excessive energy intake. However, most of the HFCS used has the same sweetness as sucrose. Another hypothesis is that, as intake of HFCS increases and that of sucrose decreases, this change in the fructose-to-glucose ratio in our diets leads to adverse metabolic consequences, such as greater fat synthesis in the liver and diminished release of satiety hormones. However, the fructose-to-glucose ratio in the food supply has not changed.[10] There also is no evidence that the body metabolizes HFCS and sucrose differently. The impacts of HFCS and sucrose on fasting plasma glucose, insulin, and the appetite-regulating hormones ghrelin and leptin are the same.[11]

Currently, scientific evidence does not support the hypothesis that HFCS promotes weight gain more than sucrose or any other caloric sweetener. Evidence does indicate that we like sweet flavors and, when intake of HFCS goes up, sucrose intake declines, and vice versa.

Brown adipose tissue is a specialized form of fat tissue that participates in thermogenesis. The brown appearance results from the large number of capillaries and abundant mitochondria that it contains. It is found in small amounts in infants and hibernating animals. Brown fat cells' mitochondria are structurally different and have special pores in the inner mitochondrial membrane formed by a protein called thermogenin, or uncoupling protein 1 (UCP1). UCP1 allows electrons to flow back into the mitochondria rather than through the electron transport chain and results in heat generation rather than ATP synthesis. Adults have very little brown fat, and its role in adulthood is not known. It is thought to be important mostly for thermoregulation during infancy, when brown fat accounts for as much as 5% of body weight. Hibernating animals use brown fat to generate heat during cold winter months.

### Knowledge Check

1. What percentage of total energy expenditure is spent on basal metabolism?
2. What factors increase basal metabolism?
3. How much energy is expended via the thermic effect of food?
4. What is adaptive thermogenesis?
5. What is brown fat, and what function does it play in an infant?

## 10.2 Measuring Energy Expenditure

The amount of energy a body uses can be measured by both direct and indirect calorimetry or can be estimated based on height, weight, degree of physical activity, and age.

**Direct calorimetry** estimates energy expenditure by measuring the amount of heat released by the body. Direct calorimetry works because approximately 60% of the energy the body uses eventually leaves as heat. Heat release is measured by placing a person in an insulated chamber, often the size of a small bedroom, that is surrounded by a layer of water. The change in water temperature before and after the body releases heat is used to determine the amount of energy the person has expended. Recall that a calorie is related to the amount of heat required to raise the temperature of water. Direct calorimetry is expensive and complex to use.

**Indirect calorimetry,** the most commonly used method to determine energy use by the body, involves collecting expired air from an individual during a specified amount of time (Fig. 10-5). This method works because a predictable relationship exists between the body's use of energy and the amount of oxygen consumed and carbon dioxide produced. The procedure to collect the air can be done in a laboratory or with a handheld device that allows the individual to be mobile and not restricted to the lab. Data tables showing energy costs of different exercises are based on information from indirect calorimetry studies.

In another approach to indirect calorimetry, a person drinks doubly labeled water ($^2H_2O$ and $H_2^{18}O$); then, his or her urine and blood samples are analyzed to examine $^2H$ and $^{18}O$ excretion. The labeled oxygen is eliminated from the body as water and carbon dioxide, whereas the labeled hydrogen is eliminated only as water. Subtracting hydrogen losses from oxygen losses provides a measure of carbon dioxide output. This method is quite accurate but also very expensive. It is the basis for determining Estimated Energy Requirements for humans.

**Estimated Energy Requirements (EERs)** are measurements based on formulas, developed by the Food and Nutrition Board, that can estimate energy needs using a person's weight, height, biological sex, age, and physical activity level. To determine your estimated EER, use the appropriate equation, inserting your information. These formulas are for adults. (Do the calculations within the parentheses first and then the calculations in the brackets.)

*Men 19 Years and Older*

$$EER = 662 - [9.53 \times AGE] + PA \times [(15.91 \times WT) + (539.6 \times HT)]$$

*Women 19 Years and Older*

$$EER = 354 - [6.91 \times AGE] + PA \times [(9.36 \times WT) + (726 \times HT)]$$

The variables in the formulas correspond to the following.

EER = Estimated Energy Requirement
AGE = Age in years
PA = Physical Activity Estimate (see the following table)
WT = Weight in kg (lb ÷ 2.2)
HT = Height in meters (inches ÷ 39.4)

### Physical Activity (PA) Estimates

| Activity Level | PA (Men) | PA (Women) |
|---|---|---|
| Sedentary (e.g., no exercise) | 1.00 | 1.00 |
| Low activity (e.g., walks the equivalent of 2 miles per day at 3 to 4 mph) | 1.11 | 1.12 |
| Active (e.g., walks the equivalent of 7 miles per day at 3 to 4 mph) | 1.25 | 1.27 |
| Very active (e.g., walks the equivalent of 17 miles per day at 3 to 4 mph) | 1.48 | 1.45 |

Source: Food and Nutrition Board. Dietary reference intakes for energy, carbohydrate, fiber, fat, fatty acids, cholesterol, protein, and amino acids. Washington, DC: National Academy Press, 2005.

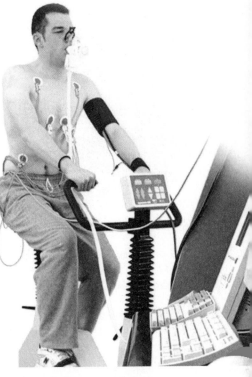

Consider a man who is 25 years old, is 5 feet 9 inches (1.75 meters) tall and 154 lb (70 kg), and has an active lifestyle. His EER is

$$EER = 662 - [9.53 \times 25] + 1.25 \times [(15.91 \times 70) + (539.6 \times 1.75)] = 2997$$

Remember, EERs are only estimates—many other factors, such as genetics and hormones, can affect actual energy needs.

A simple method of tracking energy expenditure, and thus energy needs, is to use the forms in Appendix M. Begin listing all the activities performed (including sleep) in a 24-hour period and recording the number of minutes spent in each activity; the total should be 1440 minutes (24 hours). Next, record the energy cost for each activity in kcal per minute, following the directions in Appendix M. Multiply the energy cost by the minutes to determine the energy expended for each activity. Finally, total all the kcal values to calculate your estimated energy expenditure for the day.

**Figure 10-5** Indirect calorimetry measures oxygen intake and carbon dioxide output to determine energy expended during daily activities. Handheld indirect calorimeters also are available.

Samuel Ashfield/SPL/Photo Researchers, Inc/ Science Source

### Knowledge Check

1. How do direct and indirect calorimetry differ?
2. Why is it possible to use direct calorimetry and indirect calorimetry to measure energy expenditure?
3. What is your Estimated Energy Requirement?

### CASE STUDY

Jamie Grill Photography/ Getty Images

Christy is a college freshman. She is excited about college and has good support from her family and friends to help her succeed. Her greatest concerns this first semester are to establish good study habits and time management, to get along with her new roommate, and to avoid the "Freshman 15." For the last 3 months, all she has heard about is the amount of weight that she will gain in her freshman year. She has never had a weight problem, but in high school she was on the track team and played basketball. She is not quite sure what she should be eating, so for now she has a salad for lunch and dinner and skips breakfast because she has an 8 A.M. class. She gets very hungry at around 10 P.M. and her roommate has been having pizza delivered to the dorm—she cannot resist having several slices. What advice would you give Christy? With these eating habits, is the Freshman 15 likely to happen to Christy? Why can't she resist eating pizza?

| Calorie Guidelines | | |
|---|---|---|
| **Children** | **Sedentary** ⟶ | **Active** |
| 2–3 years | 1000 ⟶ | 1400 |
| **Females** | **Sedentary** ⟶ | **Active** |
| 4–8 years | 1200 ⟶ | 1800 |
| 9–13 | 1600 ⟶ | 2200 |
| 14–18 | 1800 ⟶ | 2400 |
| 19–30 | 2000 ⟶ | 2400 |
| 31–50 | 1800 ⟶ | 2200 |
| 51+ | 1600 ⟶ | 2200 |
| **Males** | **Sedentary** ⟶ | **Active** |
| 4–8 years | 1200 ⟶ | 2000 |
| 9–13 | 1800 ⟶ | 2600 |
| 14–18 | 2200 ⟶ | 3200 |
| 19–30 | 2400 ⟶ | 3000 |
| 31–50 | 2200 ⟶ | 3000 |
| 51+ | 2000 ⟶ | 2800 |

Energy expenditure estimates by age and activity.

We have an innate taste for sweet and acquire a taste for fat.

Ingram Publishing/Fotosearch RF

## 10.3 Eating Behavior Regulation

Two factors drive our desire to eat: hunger and appetite (Fig. 10-6). **Hunger,** the physiological drive to find and eat food, is controlled primarily by internal body mechanisms, such as organs, hormones, hormonelike factors, and the nervous system.[12] **Appetite,** the psychological drive to eat, is affected mostly by external factors that encourage us to eat, such as social custom, time of day, mood (e.g., feeling sad or happy), memories of pleasant tastes, and the sight of a tempting dessert.

Internal and external signals that drive hunger and appetite generally operate simultaneously and lead us to decide whether to reject or eat a food. For example, external signals can cause cephalic phase responses by the body—that is, saliva flows and digestive hormones and insulin are released in response to seeing, smelling, and initially tasting food. These physiological responses encourage eating and prepare the body for the meal.[12] Although hunger and appetite are closely intertwined, they don't always coincide. Almost everyone has encountered a mouthwatering dessert and devoured it, even on a full stomach. Alternately, there are times when we are hungry but have no appetite for the food being served. Where food is ample, appetite—not hunger—mostly triggers eating.

Fulfilling either or both drives by eating sufficient food normally brings a state of **satiety,** in which we feel satisfaction and no longer have the drive to eat. The hypothalamus, a portion of the brain, is the key integration site for the regulation of satiety (Fig. 10-7). The hypothalamus communicates with the endocrine and nervous systems and integrates many internal cues, including blood glucose levels, hormone secretions, and **sympathetic nervous system** activity, that both inhibit and encourage food intake.[12] If these internal signals stimulate the satiety centers of the hypothalamus, we stop eating. If they stimulate the feeding centers in the hypothalamus, we eat more.[12] Surgery and some cancers and chemicals can harm the hypothalamus. Damage to the satiety center causes humans to become obese, whereas damage to the feeding center inhibits eating and eventually leads to weight loss.[12]

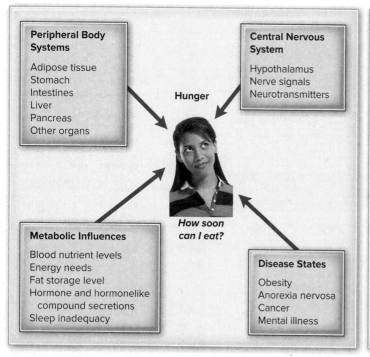

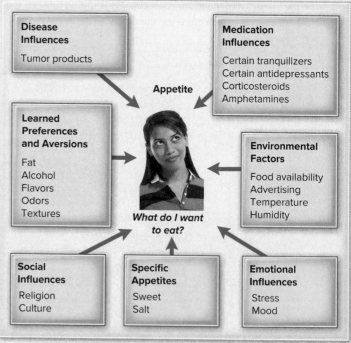

**Figure 10-6** Although some factors have an impact on both hunger and appetite, internal factors are mainly responsible for the physiological drive to eat (hunger), whereas external factors primarily influence the psychological drive to eat (appetite). These factors combine to play a role in the complex and interrelated processes that help determine when, what, and how much we eat.

Getty Images/Digital Vision RF

## Satiety Process

Feelings of satiety are elicited first by the sensory aspects of food (e.g., food flavor and smell, the size and shape of the portion served, and dietary variety) and the knowledge that a meal has been eaten (see Fig. 10-7). Chewing also seems to contribute to satiety, in part linked to the release of the neurotransmitter histamine, which affects the satiety center in the brain. Next, the stomach and intestines expand (called gastrointestinal distension) as they fill with digesting food and drink, which further contributes to satiety. Low-energy-dense foods (those high in water and/or fiber) promote satiety because they expand the stomach and intestines to a greater extent than lighter-weight foods (e.g., oils and snack foods).

Finally, the effects of digestion, absorption, and metabolism promote satiety. For example, the secretion of hormones, such as cholecystokinin, glucagon-like peptide-1 (GLP-1), and peptide YY$_{3-36}$, during digestion helps shut off hunger. Nutrient receptors in the small intestine also are thought to help promote satiety. (Infusing fats or carbohydrates directly into the small intestine causes feelings of satiety, whereas infusing these nutrients directly into the blood does not cause this effect.) Studies suggest that an apolipoprotein on chylomicrons absorbed into the blood after a meal also signals satiety to the brain. The metabolism of certain nutrients, especially carbohydrates, is linked to an increased production of serotonin (a brain neurotransmitter), which causes us to feel calm and may reduce our food intake.[11] The metabolism of protein may promote short-term satiety by decreasing the secretion of hormones, such as **ghrelin** (made by the stomach), that stimulate eating.[13] Nutrient use in the liver also signals satiety.

In addition to the short-term control of satiety depicted in Figure 10-7, food intake also is affected by body composition—specifically, the amount of body fat. **Leptin** is a protein made by adipose tissue; it functions as a hormone to reduce food intake, thus decreasing body fat. The obesity gene (ob gene) codes for the synthesis of leptin. When the ob gene is functioning normally, leptin is made and signals satiety. However, when there is a mutation in the ob gene and leptin is not made in sufficient quantities, the desire to eat is increased dramatically, metabolic rate declines (which reduces energy output), and weight gain occurs. Because treating people with leptin doesn't cause significant weight loss, experts suggest that, instead of protecting against obesity, leptin may be more important for signaling low body fat stores and setting in motion adaptations that promote energy conservation, delaying the effects of starvation.[12]

## Signals to Eat

Several hours after eating, concentrations of macronutrients in the blood begin to fall and the body starts using energy from body stores. This change causes feelings of satiety to diminish, and feeding signals begin to dominate again.[12] Endorphins, the body's natural painkillers, and hormones, such as cortisol and ghrelin, stimulate appetite and increase food intake.

As you can see, the regulation of food intake and satiety is complex and involves body cells (brain, adipose tissue, stomach, intestines, liver, and other organs), hormones (e.g., cholecystokinin and ghrelin), neurotransmitters (e.g., serotonin), dietary components, and social customs. This system is not perfect, however; your body weight can increase (or decrease) over time if you are not careful to balance your energy intake with your energy output.

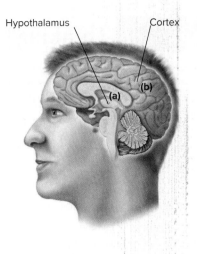

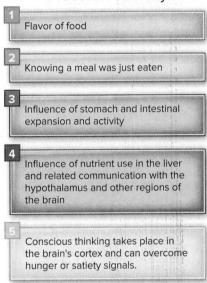

**Process of Satiety**

1. Flavor of food
2. Knowing a meal was just eaten
3. Influence of stomach and intestinal expansion and activity
4. Influence of nutrient use in the liver and related communication with the hypothalamus and other regions of the brain
5. Conscious thinking takes place in the brain's cortex and can overcome hunger or satiety signals.

*Figure 10-7* The hypothalamus and satiety. (a) The hypothalamus is the region in the brain that does most of the processing of signals regarding food intake. (b) The process of satiety starts with eating food and concludes with actions in the hypothalamus and other regions of the brain, such as the cortex.

**sympathetic nervous system** Part of the nervous system that regulates involuntary vital functions, including the activity of the heart muscle, smooth muscle, and adrenal glands.

**ghrelin** Hormone, made by the stomach, that increases food intake.

**leptin** Hormone made by adipose tissue that influences long-term regulation of fat mass. Leptin also influences reproductive functions, as well as other body processes, such as insulin release.

*Knowledge Check*

1. What factors affect hunger?
2. What factors affect appetite?
3. What factors affect satiety?
4. How do body fat stores affect food intake?

▶ One BMI unit equals about 6 to 7 lb.

▶ *Healthy weight* is currently the preferred term to use for weight recommendations. Older terms, such as *ideal weight* and *desirable weight,* are no longer used in medical literature, although you still may hear them used in clinical practice.

**Pounds per Inch Shortcut for Estimating Healthy Body Weight**

*Women:* Start with 100 lb; then add 5 lb for each inch of height above 5 feet.

*Men:* Start with 106 lb; then add 6 lb for each inch of height above 5 feet.

The weight estimate is given a ±10% range.

*Example:* A 6-foot-tall man's healthy weight is 178 lb (106 + [12 × 6] = 178).

The 10% range is about 18 lb (178 × 10%).

His healthy weight range = 160 to 196 (178±18).

Obese: BMI ≥30

Overweight: BMI 25–29.9

Healthy weight: BMI 18.5–24.9

Underweight: BMI <18.5

# 10.4 Estimating Body Weight and Composition

In the past, weight-for-height tables were the typical method for determining whether a person's weight was healthy. These tables consider biological sex and frame size, predicting the weight range at a specific height that is associated with the greatest longevity. The latest table (issued in 1983) and methods for determining frame size can be found in Appendix G.

Weight-for-height tables provide a good estimate of weights associated with health and longevity. However, the focus has shifted from using weight-for-height tables to considering the components of body weight (e.g., body fat and fat-free mass [muscle, bone, water]) and their relative proportions because of the increased health risks associated with excess body fatness. Instead of just assessing body weight, experts now recommend evaluating total amount of body fat, location of body fat, and weight-related medical problems.[5]

## Body Mass Index

Currently, body mass index (BMI) is the preferred weight-for-height standard because it is more closely related to body fat content (Fig. 10-8).[4] BMI is convenient to use because it is easier to measure height and weight than body fat and because BMI values apply to both men and women. Table 10-1 lists the BMI for various heights and weights. Either of the following equations can be used to calculate BMI.

$$\frac{\text{Body weight (in kg)}}{\text{Height}^2 \text{ (in meters)}} \quad \text{or} \quad \frac{\text{Body weight (in lb)} \times 703}{\text{Height}^2 \text{ (in inches)}}$$

A healthy weight-for-height is a BMI ranging from 18.5 to <25. Health risks from excess weight may begin when BMI is 25 or more. When interpreting BMI, it is important to remember that any weight-for-height standard is a crude estimate of body fatness—a BMI of 25 to <30 is a marker of overweight, not necessarily a marker of overfat. Even agreed-upon standards for BMI are not appropriate for everyone; they do not apply to children, teens, older adults, and pregnant and lactating women.[14] Many men (especially athletes) have a BMI greater than 25 because of extra muscle tissue. Adults less than 5 feet tall may have a high BMI but are not overweight or overfat. For this reason, BMI alone should not be used to diagnose overweight or obesity. Still, overfat and overweight conditions generally appear together.

## Measuring Body Fat Content

Body fat can range from 2 to 70% of body weight. Desirable amounts of body fat are about 8 to 24% of body weight for men and 21 to 35% for women. Men with over 24% body fat and

*Figure 10-8* Examples of body shapes associated with different BMI values.

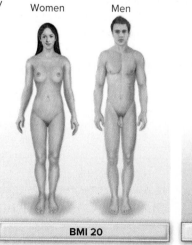
Women    Men
**BMI 20**

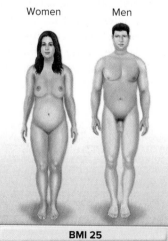
Women    Men
**BMI 25**

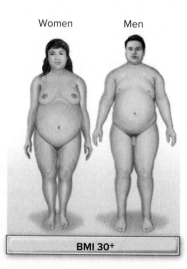
Women    Men
**BMI 30+**

**Table 10-1  Body Weight in Pounds According to Height and Body Mass Index (BMI)**

| Height (Inches)* | Healthy BMI | | | | | | Overweight BMI | | | | | Obese BMI | | |
| --- | --- | --- | --- | --- | --- | --- | --- | --- | --- | --- | --- | --- | --- | --- |
| | 19 | 20 | 21 | 22 | 23 | 24 | 25 | 26 | 27 | 28 | 29 | 30 | 35 | 40 |
| | Body Weight (Pounds) | | | | | | | | | | | | | |
| 58 | 91 | 96 | 100 | 105 | 110 | 115 | 119 | 124 | 129 | 134 | 138 | 143 | 167 | 191 |
| 59 | 94 | 99 | 104 | 109 | 114 | 119 | 124 | 128 | 133 | 138 | 143 | 148 | 173 | 198 |
| 60 | 97 | 102 | 107 | 112 | 118 | 123 | 128 | 133 | 138 | 143 | 148 | 153 | 179 | 204 |
| 61 | 100 | 106 | 111 | 116 | 122 | 127 | 132 | 137 | 143 | 148 | 153 | 158 | 185 | 211 |
| 62 | 104 | 109 | 115 | 120 | 126 | 131 | 136 | 142 | 147 | 153 | 158 | 164 | 191 | 218 |
| 63 | 107 | 113 | 118 | 124 | 130 | 135 | 141 | 146 | 152 | 158 | 163 | 169 | 197 | 225 |
| 64 | 110 | 116 | 122 | 128 | 134 | 140 | 145 | 151 | 157 | 163 | 169 | 174 | 204 | 232 |
| 65 | 114 | 120 | 126 | 132 | 138 | 144 | 150 | 156 | 162 | 168 | 174 | 180 | 210 | 240 |
| 66 | 118 | 124 | 130 | 136 | 142 | 148 | 155 | 161 | 167 | 173 | 179 | 186 | 216 | 247 |
| 67 | 121 | 127 | 134 | 140 | 146 | 153 | 159 | 166 | 172 | 178 | 185 | 191 | 223 | 255 |
| 68 | 125 | 131 | 138 | 144 | 151 | 158 | 164 | 171 | 177 | 184 | 190 | 197 | 230 | 262 |
| 69 | 128 | 135 | 142 | 149 | 155 | 162 | 169 | 176 | 182 | 189 | 196 | 203 | 236 | 270 |
| 70 | 132 | 139 | 146 | 153 | 160 | 167 | 174 | 181 | 188 | 195 | 202 | 207 | 243 | 278 |
| 71 | 136 | 143 | 150 | 157 | 165 | 172 | 179 | 186 | 193 | 200 | 208 | 215 | 250 | 286 |
| 72 | 140 | 147 | 154 | 162 | 169 | 177 | 184 | 191 | 199 | 206 | 213 | 221 | 258 | 294 |
| 73 | 144 | 151 | 159 | 166 | 174 | 182 | 189 | 197 | 204 | 212 | 219 | 227 | 265 | 302 |
| 74 | 148 | 155 | 163 | 171 | 179 | 186 | 194 | 202 | 210 | 218 | 225 | 233 | 272 | 311 |
| 75 | 152 | 160 | 168 | 176 | 184 | 192 | 200 | 208 | 216 | 224 | 232 | 240 | 279 | 319 |
| 76 | 156 | 164 | 172 | 180 | 189 | 197 | 205 | 213 | 221 | 230 | 238 | 246 | 287 | 328 |

Each entry gives the body weight in pounds for a person of a given height and BMI (kg/m²). Pounds have been rounded off. To use the table, find the appropriate height in the far left column. Move across the row to a weight. The number at the top of the column is the BMI for the height and weight.

*To state height in inches in feet and inches, divide by 12 (there are 12 inches in a foot) to determine the number of feet. The remainder is inches. For example (58 inches/12 inches) = 4 feet 10 inches.

women with over about 35% body fat have obesity. Women need more body fat because some "sex-specific" fat is associated with reproductive functions. This fat is normal and factored into calculations. The further body fatness rises above desirable levels, the greater the health risks are likely to be (Table 10-2).

To measure body fat content accurately using typical methods, both body weight and body volume must be measured. Body weight is easy to measure. One method used to estimate body volume is **underwater weighing.** This technique determines body volume by measuring body weight when under water and body weight in air and entering these values into a mathematical formula that accounts for the differences in the relative densities of fat tissue and lean tissue (Fig. 10-9). When done correctly, it can be very accurate, with a 2 to 3% error margin.[16] **Air displacement,** another method for determining body volume, measures the space a person takes up inside a small chamber, such as the BodPod® (Fig. 10-10). This method also has a 2 to 3% error margin and is an accurate alternative to underwater weighing.[17]

Once body weight and body volume are known, body density and body fat can be calculated:

$$\text{Body density} = \frac{\text{Body weight}}{\text{Body volume}}$$

% body fat = ([495/Body density] − 450)

***Figure 10-9*** During underwater weighing, the person exhales as much air as possible and then holds his or her breath and bends over at the waist. When the person is totally submerged, underwater weight is recorded. Body volume is calculated by entering this value and weight in air into a formula.

David Madison/Getty Images

| Table 10-2 Obesity-Related Health Conditions |
| --- |
| Surgical risk |
| Pulmonary disease and sleep disorders |
| Type 2 diabetes |
| Hypertension |
| Cardiovascular disease (e.g., coronary heart disease and stroke) |
| Bone and joint disorders (including gout) |
| Gallstones |
| Skin disorders |
| Various cancers, such as kidney, gallbladder, colon and rectum, and uterus (women) and prostate gland (men) |
| Shorter stature (in some cases of obesity) |
| Pregnancy risks* |
| Reduced physical agility and increased risk of accidents and falls |
| Menstrual irregularities and infertility* |
| Vision problems |
| Premature death |
| Infections[15] |
| Liver damage and eventual failure |
| Erectile dysfunction in men |

*Estrogen, the primary "female" hormone, is synthesized mainly by the ovaries but also by adipose tissue. Greater than normal amounts of body fat increase estrogen levels in women with obesity. Higher than normal estrogen levels can adversely affect pregnancy, menstruation, and fertility.

The greater the degree of obesity, the more likely and the more serious these health problems generally become. They are much more likely to appear among people who have excess upper body fat distribution and/or are greater than twice healthy body weight.

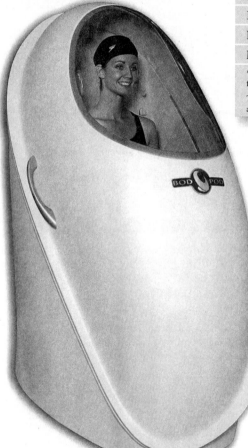

*Figure 10-10* A BodPod® determines body volume by measuring the volume of air displaced when a person sits in a sealed chamber for a few minutes.

BOD POD® Body Composition Tracking System photo provided courtesy of COSMED USA, Inc.

For example, if the person in the underwater weighing tank in Figure 10-9 has a body density of 1.06 g/cm³, she has 17% body fat; $[(500/1.06) - 450] = 22$. This method has been found to overestimate body fat in some African American men.

**Skinfold thickness** is a common anthropometric method to estimate total body fat content. Technicians use calipers to measure the fat layer directly under the skin at multiple sites and then enter those values into a mathematical formula (Fig. 10-11). The accuracy of this method is good (3 to 4% error margin) when performed by a trained technician.[16]

**Bioelectrical impedance** estimates body fat content by sending a painless, low-energy electrical current through the body. Researchers surmise that adipose tissue resists electrical flow more than lean tissue does because adipose tissue contains less electrolytes and water than lean tissue. Thus, greater electrical resistance is associated with more adipose tissue. Electrical resistance measurements can be used to estimate total body fat (3 to 4% error margin) if the individual has prepared by having normal body hydration, resting for 12 hours, fasting for 4 hours, and not drinking alcohol for 48 hours before the test (Fig. 10-12).

**Dual energy X-ray absorptiometry (DXA)** is considered the most accurate way to determine body fat (2 to 4% error margin), but the equipment is very expensive and not widely available. The usual whole-body scan requires about 5 to 20 minutes and a very low dose of radiation, which is less than a chest X ray. This method can estimate body fat, fat-free soft tissue, and bone minerals. Thus, obesity, osteoporosis, and other health conditions can be investigated using DXA (Fig. 10-13).

## Assessing Body Fat Distribution

Some people store fat in upper body areas. Others accumulate fat lower on the body. Excess fat in either place generally presents health risks, but each storage space also has

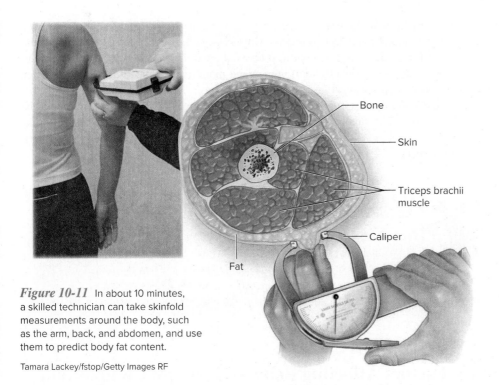

*Figure 10-11* In about 10 minutes, a skilled technician can take skinfold measurements around the body, such as the arm, back, and abdomen, and use them to predict body fat content.

Tamara Lackey/fstop/Getty Images RF

▶ Another method to assess body fat is to measure total-body electrical conductance (TOBEC) when placed in an electromagnetic field. Still another method exposes the bicep muscle to a beam of near-infrared light and assesses interactions of the light beam with fat and lean tissues. This inexpensive, flashlight-size device can quickly estimate body composition; however, this method is not very accurate.

▶ To determine if body weight is healthy, consider these factors.
- Weight
- Body composition
- Body fat distribution
- Age and physical development
- Health status
- Family history of obesity and weight-related diseases
- Personal feelings about one's body weight

its unique risks. Upper body (android) obesity is more often related to cardiovascular disease, hypertension, and type 2 diabetes.[18] Instead of emptying fat directly into general circulation, as other adipose cells do, abdominal adipose cells release fat directly to the liver by way of the portal vein. This direct delivery to the liver likely interferes with the liver's ability to clear insulin and alters the liver's lipoprotein metabolism. Abdominal adipose cells also make substances that increase inflammation, insulin resistance, blood clotting, and blood vessel constriction. All these changes can lead to long-term health problems.

High blood testosterone (primarily a male hormone) levels apparently encourage upper body fat storage, as does a diet with a high glycemic load, alcohol intake, and smoking.

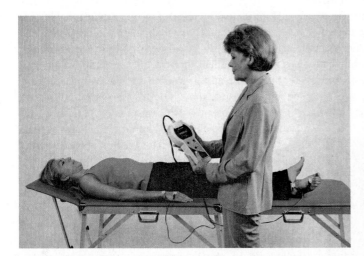

*Figure 10-12* Bioelectrical impedance estimates total body fat in less than 5 minutes and is based on the principle that body fat resists the flow of electricity because it is low in water and electrolytes. The degree of resistance to electrical flow is used to estimate body fatness.

Maltron International Ltd

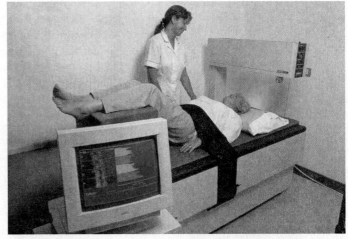

*Figure 10-13* Dual energy X-ray absorptiometry (DXA) measures body fat by releasing small doses of radiation through the body to assess body fat and bone density. DXA is considered the most accurate method for determining body fat.

SPL/Science Source

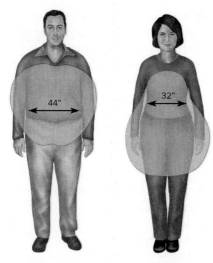

Upper body fat distribution    Lower body fat distribution
(android: apple shape)         (gynoid: pear shape)

*Figure 10-14* Body fat stored primarily in the upper body (android) form brings higher risks of ill health associated with obesity than does lower body (gynoid) fat. The woman's waist circumference of 32 inches and the man's waist circumference of 44 inches indicate that the man has upper body fat distribution but the woman does not, based on a cutoff of 35 inches for women and 40 inches for men. Another way you can state this is that waist circumference should be less than half a person's height.

**identical twins** Two offspring that develop from a single ovum and sperm and, consequently, have the same genetic makeup.

Does the difference in body fat between parents and children arise from nature, nurture, or both?

Lopolo/Shutterstock

This characteristic male pattern of fat storage appears in an apple shape (large abdomen [potbelly] and thinner buttocks and thighs). Upper body obesity is assessed by simply measuring the waist at the narrowest point just above the navel when relaxed. A waist circumference more than 40 inches (102 cm) in men and more than 35 inches (88 cm) in women indicates upper body obesity (Fig. 10-14).[4]

Estrogen and progesterone (primarily female hormones) encourage the storage of fat in the lower body. The small abdomen and much larger buttocks and thighs give a pearlike appearance. After menopause, blood estrogen levels fall, encouraging abdominal fat distribution in women.

> ### Knowledge Check
>
> 1. What is the difference between body weight and body composition?
> 2. What is body mass index, and how is it calculated?
> 3. Name 3 techniques used to assess body fat and how they work.
> 4. How is body fat distribution used to assess health risks?
> 5. Which body shape is at the greatest risk for health problems?

## 10.5 Factors Affecting Body Weight and Composition

Observations indicate that a child has only a 10% chance of becoming obese if neither parent has obesity. When a child has 1 parent with obesity, that risk rises to 40%; when both parents have obesity, it soars to 80%. This section explores whether this increasing risk is due to nature or nurture.

### Role of Genetics

Studies of **identical twins** provide some insight into the contribution of nature (genetics) to body weight. Even when identical twins are raised apart, they tend to show similar weight gain patterns, both in overall weight and in body fat distribution. Nurture—eating habits and nutrition, which vary between twins who are raised apart—seems to have less to do with weight gain patterns than nature does.[4]

Research suggests that genes account for up to 40 to 70% of weight differences between people. The genes may be those that determine body type, metabolic rate, and the factors that affect hunger and satiety. For example, basal metabolism increases as body surface increases, and therefore those who inherited genes that cause them to grow tall use more energy than shorter people, even at rest. As a result, tall people appear to have an inherently easier time maintaining healthy body weight. Some individuals are thought to have a genetic predisposition to obesity because they inherit a "thrifty metabolism"—which uses energy frugally. This metabolism enables them to store fat readily and use less energy to perform tasks than a typical individual. In earlier times when food supplies were scarce, a thrifty metabolism would have been a safeguard against starvation. Now, with the abundant food supply available in Westernized countries, people with a thrifty metabolism need to engage in physical activity and make wise food choices to prevent storing excess amounts of fat. If you think you are prone to weight gain, you may have inherited a thrifty metabolism.

### Set-Point Theory

The **set-point theory** proposes that humans have a genetically predetermined body weight or body fat

content, which the body closely regulates. It is not known what cells control this set point or how it actually functions in weight regulation. There is evidence, however, that mechanisms exist that help regulate weight. For example, some research suggests that the hypothalamus monitors the amount of body fat and tries to keep that amount constant over time. Recall from earlier in the chapter that the hormone leptin forms a communication link between adipose cells and the brain, which allows for some weight regulation.[12]

Numerous studies indicate that the body tries to maintain its weight and resist weight loss. For instance, when volunteers who lost weight through starvation had access to food, they tended to eat in such a way as to regain their original weight. Also, after an illness is resolved, a person generally regains lost weight. When energy intake is reduced, secretions of thyroid hormones fall, which slows basal metabolism and conserves body weight. When weight is lost, the body becomes more efficient at storing fat by increasing the activity of the enzyme lipoprotein lipase, which takes fat into cells.

Some evidence also supports the idea that a set point helps prevent weight gain. Studies of men with no history of obesity found it was hard for some to gain weight, even with high calorie intakes. If a person overeats, basal metabolism and thermogenesis tend to increase in the short run, which causes some resistance to weight gain. However, in the long run, evidence that the set point helps us resist weight gain is weaker than the evidence that the set point helps us resist weight loss. When a person gains weight and stays at that weight for a while, the body tends to establish a new set point.

Opponents of the set-point theory argue that weight does not remain constant throughout adulthood—the average person gains weight slowly, at least until old age. In addition, if an individual is placed in a different social, emotional, or physical environment, weight can be altered and maintained at markedly higher or lower levels. These arguments suggest that humans, rather than having a set point determined by genetics or amount of body fat, actually settle into a particular stable weight based on their circumstances, often referred to as a "settling point."

## Role of Environment

Some researchers argue that body weight similarities among family members stem more from learned behaviors than from genetic similarities. Even couples and friends (who generally have no genetic link) may behave similarly toward food and eventually assume similar degrees of leanness or fatness.[19] The effects of environment are supported by the fact that our gene pool has not changed much in the last 50 years, but the ranks of people with obesity recently have grown in what the U.S. Centers for Disease Control and Prevention describes as epidemic proportions.

Environmental factors have important effects on what we eat. These factors may define when eating is appropriate, what is preferable to eat, and how much food should be eaten. As you recall from Chapter 1 (see Fig. 1-6), our food choices are affected by numerous environmental factors, including food availability and preferences, food marketing, social networks, culture, education, lifestyle, health concerns, and income—all of which can affect calorie intake and weight gain. For example, those with limited incomes tend to have a greater risk of obesity. People who experience significant emotional stress, are members of a cultural or ethnic group that prefers higher body weight, have a social network of overweight friends, or get insufficient sleep are more likely to carry excess body fat. These patterns suggest the important role of nurture in determining body weight and composition.

Student life is often full of physical activity. This is not necessarily true for a person's later working life; hence, weight gain is a strong possibility.

Ryan McVay/Getty Images RF

**set-point theory**  Theory that humans have a genetically predetermined body weight, which is closely regulated. It is not known what cells control this set point or how it actually functions in weight regulation.

**Table 10-3** Factors That Encourage Excess Body Fat Storage and Obesity

| Factors | How They Promote Fat Storage |
| --- | --- |
| Aging | Adults tend to gain weight as they age due to the slowing of basal metabolism and increasingly sedentary lifestyles. |
| Female biological sex | Women naturally have greater fat stores than men. Women may not lose all weight gained in pregnancy. At menopause, abdominal fat deposition is favored. |
| High calorie diet | Excess energy intake, binge eating, and preference for high energy density foods favor weight gain. |
| Sedentary lifestyle | A low or decreasing amount of physical activity ("couch potato") favors weight gain. |
| Weight history | Overweight children and teens have an increased risk of being overweight in adulthood. |
| Social, cultural, and behavioral factors | Lower socioeconomic status, learned food preferences, overweight friends and family, a cultural/ethnic group that prefers higher body weight, a lifestyle that discourages healthy meals and adequate exercise, easy availability of inexpensive high calorie food, excessive television viewing, smoking cessation, lack of adequate sleep, emotional stress, and a greater number of meals eaten away from home are linked with increased fat storage. |
| Certain medications | Certain medications stimulate appetite, causing food intake to increase. |
| Geographic location | Regional differences, such as high-fat diets and sedentary lifestyles in the Midwest and areas of the South, lead to different rates of obesity in different places. |
| Genetic characteristics | These affect basal metabolic rate, the thermic effect of food, adaptive thermogenesis, the efficiency of storing body fat, the relative proportion of fat and carbohydrate used by the body, and possibly increased hunger sensations linked to the activity of various brain chemicals. |

**Marfan syndrome** Genetic disorder affecting muscles and skeleton, characterized by tallness, long arms, and little subcutaneous fat. Some medical historians speculate that Abraham Lincoln suffered from Marfan syndrome.

**Prader-Willi syndrome** Genetic disorder characterized by shortness, mental retardation, and uncontrolled appetite, caused by a dysfunction of the nervous system, leading to extreme obesity.

## Genetic and Environmental Synergy

Even though our genetic backgrounds have a strong influence on body weight and composition, genes are not destiny—both nature and nurture are involved (Table 10-3). Hereditary factors interact with environmental factors to determine actual body weight and composition. Even with a genetic potential for leanness, it is possible for a person who overeats to gain excess fat. Conversely, an individual genetically predisposed to obesity can avoid excess fat storage with a healthy diet and sufficient amounts of regular physical activity.

## Diseases and Disorders

Body weight and fatness can be affected by certain diseases, hormonal abnormalities, rare genetic disorders, and psychological disturbances. For instance, cancer, AIDS, hyperthyroidism, **Marfan syndrome,** and anorexia nervosa tend to cause a person to have limited fat stores. A very small percentage of obesity cases are caused by brain tumors, ovarian cysts, hypothyroidism, and congenital syndromes such as **Prader-Willi syndrome.**

### Knowledge Check

1. What evidence supports the role of genetics in determining body weight?
2. What evidence supports the role of the environment in determining body weight?
3. Why is it likely that both genetics and the environment determine body weight?

# 10.6 Treatment of Overweight and Obesity

The treatment of overweight and obesity should be considered similar to the treatment of any chronic disease: it requires long-term lifestyle changes.[5] Too often, however, people view a "diet" as something they go on temporarily and, once the weight is lost, they revert to their old dietary habits and physical activity routines. It is mostly for this reason that so many people regain lost weight (called weight cycling or yo-yo dieting). Instead, those who weigh more or less than is healthy should emphasize healthy, active lifestyles with lifelong dietary modifications (Fig. 10-15)—if started early, these modifications also can help prevent obesity.[18]

▶ Weight control objectives from *Healthy People* include

- Reduce the proportion of adults who have obesity.
- Increase the proportion of adults who are at a healthy weight.
- Reduce the proportion of children and adolescents who have obesity.
- Prevent inappropriate weight gain in youth and adults.
- Increase the proportion of physician office visits that include counseling or education related to nutrition or weight.

**RATE OF LOSS**

☐ Encourages slow and steady weight loss, rather than rapid weight loss, to promote lasting weight
☐ Sets goal of 1 to 2 lb of fat loss per week
☐ Includes a period of weight maintenance for a few months after 10% of body weight is lost
☐ Evaluates need for further dieting before more weight loss begins

**FLEXIBILITY**

☐ Supports participation in normal activities (e.g., parties, restaurants)
☐ Adapts to individual habits and tastes

**INTAKE**

☐ Meets nutrient needs (except for energy needs)
☐ Includes common foods, with no foods being promoted as magical or special
☐ Recommends a fortified ready-to-eat breakfast cereal or balanced multivitamin/mineral supplement, especially when intake is less than 1600 kcal per day
☐ Uses MyPlate as a pattern for food choices

**BEHAVIOR MODIFICATION**

☐ Focuses on maintenance of healthy lifestyle (and weight) for a lifetime
☐ Promotes reasonable changes that can be maintained
☐ Encourages social support
☐ Includes plans for relapse, so that one does not quit after a setback
☐ Promotes changes that control problem eating behaviors

**OVERALL HEALTH**

☐ Requires screening by a physician for people with existing health problems, those over 40 (men) to 50 (women) years of age who plan to increase physical activity substantially, and those who plan to lose weight rapidly
☐ Encourages regular physical activity, sufficient sleep, stress reduction, and other healthy changes in lifestyle
☐ Addresses underlying psychological weight issues, such as depression or marital stress

*Figure 10-15* Characteristics of a sound weight loss diet. If you are considering a weight loss diet, compare it to this checklist. The healthiest weight loss diets have all these characteristics. Keep in mind that, during times of underfeeding or overfeeding, the body makes numerous physiological adjustments that resist weight change. This compensation is most pronounced during times of underfeeding and is why slow, steady weight loss is advocated.

*Figure 10-16* The key to weight loss and maintenance can be thought of as a triangle, in which the 3 corners consist of 1 controlling energy intake, 2 performing regular physical activity, and 3 controlling problem behaviors. The 3 corners of the triangle support each other—without 1 corner, the triangle becomes incomplete.

1: Digital Vision/Getty Images RF; 2: Ingram Publishing RF; 3: Ryan McVay/Getty Images RF

A sound weight loss program should include 3 key components.[6,20]

1. Control of energy intake
2. Regular physical activity
3. Control of problem behaviors

A 1-sided approach that focuses only on restricting energy intake is a difficult plan of action. Instead, adding physical activity and the control of problem behaviors will contribute to success in weight loss. Accepting that these changes must be maintained for a lifetime improves the likelihood that lost weight will not be regained (Fig. 10-16).

A weight loss program should be considered successful only when those involved in the process remain at or close to their lower weights. Only about 5% of people who follow commercial diet programs actually lose weight and then remain close to that weight. Typically, one-third of the weight lost during dieting is regained within a year after the diet ends, and almost all weight lost is regained within 3 to 5 years. Some programs have success rates higher than 5%, as do some people who simply lose weight on their own without enrolling in a supervised plan. Overall, however, the statistics are grim. Currently, only the surgical approaches to obesity treatment routinely show success in maintaining the weight loss in most people (see the Clinical Perspective in this chapter).

Because lost weight is frequently regained, many dieters try a variety of weight loss diets. The negative health consequences associated with this weight cycling are an increased risk of upper body fat deposition, discouragement, diminished self-esteem, and possibly a decline in high-density lipoprotein (HDL) cholesterol and immune system function. Nevertheless, experts still encourage people with obesity to attempt weight loss, with a strong focus on maintaining that lower weight.

Foods, such as fruit, that are nutrient dense and have low energy density help weight control efforts.

Dennis Gray/Cole Group/Getty Images RF

## Control of Energy Intake

Adipose tissue, which is mostly fat, contains about 3500 kcal/lb. Therefore, to lose 1 pound of adipose tissue per week, energy intake must be decreased by approximately 500 kcal/day, or physical activity must be increased by 500 kcal per day. Alternately, a combination of both strategies can be used.[4] A goal of losing 1 lb or so of stored fat per week may require limiting energy intake to 1200 kcal/day for women and 1500 kcal/day for men. The energy allowance can be higher for very active people. Keep in mind that, in our very sedentary society, decreasing energy intake is a vital component of weight loss.

To reduce energy intake, some experts suggest consuming less fat (especially saturated fat and *trans* fat), whereas others suggest consuming less carbohydrate, especially from refined (high glycemic load) carbohydrate sources. Protein intakes in excess of what is typically needed by adults also are sometimes used as a weight loss strategy (especially plant protein sources).[21] All these approaches can be used simultaneously. It appears that low energy density (low fat, high fiber) approaches are the most successful in long-term studies. There is no long-term evidence for the effectiveness of the other approaches.[22]

Low energy density diets may be successful because foods low in energy density provide bigger portions for a given number of calories. Water is the dietary component that has the biggest impact on the energy density of foods; water adds weight but no calories and therefore decreases energy density. Increasing the water content of recipes (e.g., by adding vegetables) helps reduce energy intake and enhance satiety. A surprising finding in recent years is that people tend to eat a consistent weight or volume of food over a few days. Thus, when foods contain fewer calories per gram, people consume less energy but still report feeling just as full and satisfied. Using energy density as a guide to food selection helps individuals consume foods that health professionals recommend—fruits, vegetables, legumes, low-fat dairy products, and whole grains.[23]

Overall, it is best to consider healthy eating a lifestyle change, rather than simply a weight loss plan. Healthy eating choices that bring calories under control start with eating smaller portions and using MyPlate as a pattern for daily intake. Many people often underestimate portion size, so measuring cups and a food-weighing scale can help them learn appropriate portion sizes. Reading Nutrition Facts panels also can help individuals find low energy density foods. Learning to identify lower calorie versions of favorites also is helpful (Table 10-4). Note that liquids do not stimulate satiety mechanisms as strongly as solid foods; thus, experts advise choosing beverages that have few or no calories. Another method is to become aware of calorie and nutrient intake by writing down food intake for 24 hours, calculating energy intake by using nutrient analysis software, and then adjusting future food choices as needed.

## Regular Physical Activity

Regular physical activity is very important for everyone, especially people who are trying to lose weight or maintain a lower body weight.[6] Obviously, more energy is burned during physical activity than at rest. Even expending only 100 to 300 extra kcal/day above and beyond normal daily activity, while controlling energy intake, can lead to a steady weight loss. Physical activity also has so many other benefits, including a boost for overall self-esteem and maintenance of bone mass.

Adding any of the activities in Table 10-5 to one's lifestyle can increase energy expenditure. Duration and regular performance, rather than intensity, are the keys to success with this weight loss approach. Another key is finding activities that are enjoyable and can be continued throughout life. There is no single activity that is better than another—for example, walking vigorously 3 miles daily can be as helpful as aerobic dancing or jogging. Activities of lighter intensity are less likely to lead to injuries as well. Some resistance exercises, such

Slow, steady weight loss is 1 of the characteristics of a sound weight loss program.

Ryan McVay/Getty Images RF

▶ As you read brochures, articles, or research reports about specific diet plans, look beyond the weight loss promoted by the diet's advocate to see if the reported weight loss was maintained. If the weight maintenance aspect is missing, then the program is not successful.

▶ Meal replacement formulas to replace a meal or snack are appropriate to use once or twice a day, if desired. These are not a magic bullet for weight loss, but they have been shown to help some people control calorie intake.

▶ Learn how portion sizes have changed over the years. Visit www.nhlbi.nih.gov/ health/educational/wecan/eat-right/ distortion.htm.

Physical activity complements any diet plan.

Michael Simons/123RF

*Perspective*
*on the Future*

The common wisdom that eating 3500 kcal less than you need will result in the loss of 1 pound has come under great scrutiny. Weight loss research models based on thermodynamics, mathematics, physics, and chemistry indicate that many more than 3500 calories may be stored in a pound of body fat. Researchers have developed a body weight planner that allows users to make personalized calorie and physical activity plans to reach a goal weight.[24] Learn more at www.pbrc.edu/research-and-faculty/calculators/weight-loss-predictor.

▶ Spot-reducing by using diet and physical activity is not possible. "Problem" local fat deposits can be reduced in size, however, using lipectomy (surgical removal of fat). This procedure carries some risks and is designed to help a person lose about 4 to 8 lb per treatment.

▶ Many smartphones have built-in apps that monitor the number of steps taken. An often-stated goal for activity is to take 10,000 steps/day—typically, we take half that many or less.

**Table 10-4  Saving Calories: Ideas for Getting Started**

| | | | | | |
|---|---|---|---|---|---|
| Save | 140 kcal | by choosing | 3 oz lean beef | instead of | 3 oz well-marbled beef |
| Save | 175 kcal | by choosing | ½ broiled chicken | instead of | ½ batter-fried chicken |
| Save | 210 kcal | by choosing | 3 oz lean roast beef | instead of | ½ cup beef stroganoff |
| Save | 65 kcal | by choosing | ½ cup boiled potatoes | instead of | ½ cup fried potatoes |
| Save | 140 kcal | by choosing | 1 cup raw vegetables | instead of | ½ cup potato salad |
| Save | 150 kcal | by choosing | 2 tbsp low kcal salad dressing | instead of | 2 tbsp regular salad dressing |
| Save | 310 kcal | by choosing | 1 apple | instead of | 1 slice apple pie |
| Save | 150 kcal | by choosing | 1 English muffin | instead of | 1 Danish pastry |
| Save | 60 kcal | by choosing | 1 cup cornflakes | instead of | 1 cup sugar-coated cornflakes |
| Save | 45 kcal | by choosing | 1 cup 1% milk | instead of | 1 cup whole milk |
| Save | 150 kcal | by choosing | 6 oz wine cooler made with sparkling water | instead of | 6 oz gin and tonic |
| Save | 150 kcal | by choosing | 1 cup plain popcorn | instead of | 1 oz potato chips |
| Save | 185 kcal | by choosing | 1 slice angel food cake | instead of | 1 slice white iced cake |
| Save | 150 kcal | by choosing | 12 oz sugar free soft drink | instead of | 12 oz regular soft drink |
| Save | 140 kcal | by choosing | 12 oz light beer | instead of | 12 oz regular beer |

**Table 10-5  Approximate Energy Costs of Various Activities**

| Activity | kcal/kg per Hour | Activity | kcal/kg per Hour | Activity | kcal/kg per Hour |
|---|---|---|---|---|---|
| Aerobics—heavy | 8.0 | Dressing/showering | 1.6 | Running or jogging (10 mph) | 13.2 |
| Aerobics—medium | 5.0 | Driving | 1.7 | Downhill skiing (10 mph) | 8.8 |
| Aerobics—light | 3.0 | Eating (sitting) | 1.4 | Sleeping | 1.2 |
| Backpacking | 9.0 | Food shopping | 3.6 | Swimming (.25 mph) | 4.4 |
| Basketball—vigorous | 10.0 | Football—touch | 7.0 | Tennis | 6.1 |
| Cycling (5.5 mph) | 3.0 | Golf | 3.6 | Volleyball | 5.1 |
| Bowling | 3.9 | Horseback riding | 5.1 | Walking (2.5 mph) | 3.0 |
| Calisthenics—heavy | 8.0 | Jogging—medium | 9.0 | Walking (3.75 mph) | 4.4 |
| Calisthenics—light | 4.0 | Jogging—slow | 7.0 | Water skiing | 7.0 |
| Canoeing (2.5 mph) | 3.3 | Ice skating (10 mph) | 5.8 | Weight lifting—heavy | 9.0 |
| Cleaning | 3.6 | Lying down | 1.3 | Weight lifting—light | 4.0 |
| Cooking | 2.8 | Racquetball—social | 8.0 | Window cleaning | 3.5 |
| Cycling (13 mph) | 9.7 | Roller skating | 5.1 | Writing (sitting) | 1.7 |

The values refer to total energy expenditure, including that needed to perform the physical activity plus that needed for basal metabolism, the thermic effect of food, and thermogenesis.
Example: For a 150 lb person who played tennis for 1.5 hours,
　　150 lb/2.2 = 68 kg
　　68 kg × 6.1 kcal/hr × 1.5 hr = 622 kcal expenditure

as weight training, also should be added to increase lean body mass and, in turn, fat use (see Chapter 11). As lean muscle mass increases, so does overall metabolic rate.

Unfortunately, opportunities to expend energy in our daily lives continue to diminish as technology eliminates almost every reason to move our muscles. The easiest way to increase physical activity is to make it an enjoyable part of a daily routine. To start, one might wear walking shoes and walk between classes and to and from the parking lot. Some people find using physical activity apps on their smartphones a way to help motivate them to walk and be active. There also are plenty of other simple ways to increase the activity of daily living, such as parking the car farther away from the shopping mall entrance or getting up to change the channels on the television.

## Control of Problem Behaviors

Controlling energy intake and increasing physical activity also mean modifying problem behaviors.[6,25] Only the dieter can decide which behaviors derail weight loss efforts. What events start (or stop) the action of eating or exercising? Becoming aware of these events can help dieters change their behaviors and improve their habits. The following key behavior modification techniques help organize intervention strategies into manageable steps and bring problem behaviors under control.

- **Chain-breaking** separates the link between behaviors that tend to occur together—for example, snacking on chips while watching television (Fig. 10-17).
- **Stimulus control** alters the environment to minimize the stimuli for eating—for example, storing foods out of sight and avoiding the path by the vending machines. Positive stimulus control includes keeping low-fat snacks on hand to satisfy hunger/appetite or placing walking shoes in a convenient, visible location.
- **Cognitive restructuring** changes one's frame of mind regarding eating—for example, instead of using a difficult day as an excuse to overeat, substitute another pleasure or reward, such as a relaxing walk with a friend.
- **Contingency management** prepares one for situations that may trigger overeating (e.g., when snacks are within arm's reach at a party) or hinder physical activity (e.g., rain).
- **Self-monitoring** tracks which foods are eaten, when, why, how one feels (usually using a diary), which physical activities are completed, and body weight. Self-monitoring helps people understand more about their habits and reveals patterns—such as unconscious overeating—that may explain problem behaviors that lead to weight gain.

Controlling energy intake and boosting energy expenditure are critical for losing and maintaining lost weight. Many times, we know what we should do, but some eating and exercise behaviors keep us from reaching our goals. Table 10-6 lists steps for modifying behaviors that promote weight loss and maintenance.

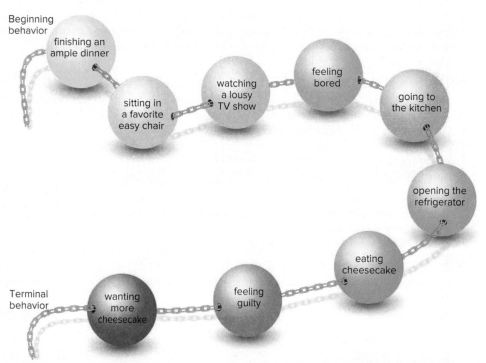

*Figure 10-17* Analyzing behavior chains is a good way to understand more about your behaviors and to pinpoint how to change unwanted habits. The earlier in the chain that you substitute a nonfood link, the easier it is to stop a chain reaction. Behaviors that can be substituted to break a behavior chain include

1. Fun activities (taking a walk, calling a friend)
2. Necessary activities (writing a paper, cleaning a room, taking a shower)
3. Urge-delaying activity (set a kitchen timer and wait 30 minutes before eating)

Using activities to interrupt behavior patterns that lead to inappropriate eating (or inactivity) can be a powerful means of changing behaviors.

## Table 10-6 Behavior Modification Principles for Weight Loss and Control

### Shopping

1. Shop for food after eating—buy nutrient dense foods.
2. Shop from a list; limit purchases of irresistible "problem" foods. It helps to shop first for fresh foods around the perimeter of the store. Shopping online also can help control purchases of problem foods.
3. Avoid ready-to-eat foods.

### Plans

1. Plan to limit food intake as needed.
2. Plan meals, snacks, and physical activity.
3. Eat meals and snacks at scheduled times; don't skip meals.
4. Exercise with a partner and schedule a time together.
5. Plan to take an exercise class at the YMCA or local recreation center.

### Activities

1. Store food out of sight, preferably in the freezer, to discourage impulsive eating.
2. Eat food only in a "dining" area.
3. Keep serving dishes off the table, especially dishes of sauces and gravies.
4. Use smaller dishes, glasses, and utensils.
5. Keep exercise equipment handy and visible.

### Holidays and Parties

1. Drink fewer alcoholic beverages.
2. Plan eating behavior before parties.
3. Eat a low calorie snack before parties.
4. Practice polite ways to decline food.
5. Take opportunities to dance, swim, or engage in other physical activities at parties.

### Eating Behavior

1. Put your fork down between mouthfuls and chew thoroughly before taking the next bite.
2. Leave some food on the plate.
3. Concentrate on eating; do nothing else while eating (e.g., do not watch TV).
4. Don't label certain foods "off limits"; this sets up an internal struggle that can keep you feeling deprived and defeated. Control problem foods by buying small amounts and eating a little at a time, such as 1 snack-size candy bar.
5. Limit eating out to once or twice a week.

### Portion Control

1. Make substitutions, such as a regular hamburger instead of a "quarter pounder" or cucumbers instead of croutons in salads.
2. Think small. Share an entrée with another person. Order a cup of soup instead of a bowl or an appetizer in place of an entrée.
3. Use a to-go container (doggie bag). Ask your server to put half the entrée in a to-go container before bringing it to the table.
4. Become aware of portion sizes and use guides (see Chapter 2, Fig. 2-9) to judge portions.
5. Learn to recognize feelings of satiety and stop eating.

### Reward and Social Support

1. Plan specific nonfood rewards for specific behavior (behavioral contracts).
2. Solicit help from family and friends and suggest how they can help you. Encourage family and friends to provide this help in the form of praise and nonfood rewards.
3. Use self-monitoring records as a basis for rewards.

### Self-monitoring

1. Keep a diet diary (note the time and place of eating, the type and amount of food eaten, who is present, and how you feel) and use it to identify problem areas.
2. Keep a physical activity diary (note which exercise is done, when, and how long) and use it to identify when more physical activity can be included.
3. Check body weight regularly.

### Cognitive Restructuring

1. Avoid setting unreasonable goals.
2. Think about progress, not shortcomings.
3. Avoid imperatives, such as *always* and *never*.
4. Counter negative thoughts with positive restatements.
5. Don't get discouraged by an occasional setback. Take charge immediately—think about how progress got disrupted and make a plan for avoiding it next time.
6. Eating a particular food doesn't make a person "bad" and shouldn't lead to feelings of guilt. Change responses such as "I ate a cookie, so I'm a failure" to "I ate a cookie and enjoyed it. Next time, I'll have a piece of fruit."
7. Seek professional help before weight, dietary intake, or sedentary behavior gets out of control.

# Expert Perspective from the Field

## Tailoring a Healthy Eating Plan to Fit Your Lifestyle

According to Dr. Judith Rodriguez,* finding a weight loss diet that suits your lifestyle is the key to controlling weight and eating a healthy diet. Dr. Rodriguez groups diets based on common principles to help consumers match their lifestyles with popular diets. These groupings help you match what you like to eat or the concerns you have about your diet and find a diet program that might work best for you. In the accompanying table, choose your eating pattern/lifestyle preferences in the left column. Then, consider the matching characteristics/descriptions in the center column. Finally, take a look at the diet resources listed in the last column to help you achieve the diet changes that are important to you.

| Eating Pattern/ Lifestyle | Characteristics/Descriptions | Resources |
|---|---|---|
| Enjoy eating a variety of foods from different food groups in a reasonable balance | These diets develop and manage healthy eating patterns. | • *200 Surefire Ways to Eat Well & Feel Better* by Dr. Judith Rodriguez<br>• *Intuitive Eating* by Evelyn Tribole, RD<br>• *Change One* by John Hastings, Peter Jaret, Mindy Hermann, RD<br>• *Food: What the Heck Should I Eat?* by Mark Hyman, MD |
| Would like to manage lifestyle diseases through diet | These diets have specific eating patterns and foods to consume based on certain diseases or risk of disease. | • *Diabetes Meal Planning and Nutrition for Dummies* by Toby Smithson, RDN, CDE<br>• *Blood Pressure Down: The 10 step plan to lower your blood pressure in 4 weeks without prescription drugs* by Janet Bond Brill, PhD, RD, LD<br>• *The Bloated Belly Diet* by Tamara Duker Freuman, MS, RD, CDN<br>• *The Low-FODMAP Diet* by Kate Scarlata, RDN, LDN, and Dede Wilson<br>• *The Mediterranean Diet Weight Loss Solution* by Julene Stassou, MS, RD |
| Would like to manage weight and healthy food choices in college | These diets teach students how to make better food choices for health and weight management. | • *The College Student's Guide to Eating Well on Campus* by Ann Litt, MS, RD<br>• *Dressing On The Side (and Other Diet Myths Debunked)* by Jaclyn London, MS, RD<br>• *Lose It For The Last Time* by Amy Newman Shapiro, RD, DCN, CPT |
| Enjoy eating ethnic foods and making them a part of a normal diet | These plans and diets help focus on international foods and eating patterns. | • Mediterranean Diet & Pyramid<br>• Latino Diet & Pyramid<br>• African Diet & Pyramid<br>• Asian Diet & Pyramid<br>All at *www.oldwayspt.org* |
| Enjoy eating more vegetables and whole grains compared to animal products | These diets encourage consumers to consume more fruits, vegetables, and whole grains to help with weight loss and prevention of lifestyle disease. | • *The Plant Powered Diet* by Sharon Palmer, RD<br>• *The Plant Based Solution* by Joel Kahn, MD<br>• *Vegan for Everybody: Foolproof Plant-Based Recipes* by Bonnie Taub Dix<br>• *Smart Meal Prep for Beginners* by Toby Amidor, MS, RD, CDN |
| Enjoy eating more animal products instead of cereals, breads, and pasta | These diets focus on a low carbohydrate and higher protein diet that includes lean meats and healthy fats. | • *Healthy Keto* by Prevention Magazine with Rachel Lustgarden, RD, CDN<br>• *Fill Your Plate, Lose the Weight* by Sarah Mirkin, RDN |
| Enjoy preparing quick meals and want nutrition tips for busy families with kids | These resources provide tips for teaching children healthy eating and making family food fun. | • *The Picky Eater Project: 6 Weeks to Happier Healthier Family Mealtimes* by Natalie Digate Muth, MD, RDN, and Sally Sampson<br>• *Obesity Prevention for Children: Before It's Too Late* by Avin Eden MD and Sari Greaves, RDN<br>• www.kidseatright.org website |

Fuse/Getty Images RF

*Judith Rodriguez, PhD, RD FADA, LDN is Professor Emeritus of Nutrition in the Brooks College of Health, Department of Nutrition & Dietetics, at the University of North Florida. She is author of The Diet Selector: From Atkins to the Zone, More Than 50 Ways to Help You Find the Best Diet for You, and Contemporary Nutrition for Latinos: Latino Lifestyle Guide to Nutrition and Health. Dr. Rodriguez has served as President of the American Dietetic Association, has received the Distinguished Dietitian Award from Florida Dietetic Association, and was named an Outstanding Dietetics Educator by the American Dietetic Association (now called The Academy of Nutrition and Dietetics).*

We are faced with many opportunities to overeat. It takes much perseverance to control dietary intake.

Pando Hall/Getty Images RF

## Weight Loss Maintenance

Losing excess weight might be easier than keeping it off. Several studies from the National Weight Registry have identified 4 behaviors for successful maintainers.[6,26]

1　Eat a low-fat, high carbohydrate diet. Individuals who have been successful at losing weight and keeping it off appear to eat about 25% of their total intake as fat and about 56% of their calories as carbohydrates, mainly in the form of fruits, vegetables, and whole grains.

2　Eat breakfast. About 90% of the participants in the National Weight Registry eat breakfast at least 4 days a week. Eating breakfast causes the body to burn more fat throughout the day and there is less tendency to overeat due to hunger.[27,28] Most of the participants in the studies eat whole-grain cereal, fat-free milk, and fruit for breakfast.

3　Self-monitor by regularly weighing oneself and keeping a food journal. Self-monitoring helps individuals know when weight is creeping up and signals them to pay more attention to their diets and exercise.

4　Have a physical activity plan. Participants in the National Weight Registry program exercise about 1 hour a day. A regular exercise program helps individuals maintain lost weight and feel better.

---

### Knowledge Check

1. What are the 3 components of a sound weight loss program?
2. What are key behavior modification techniques that can help bring problem behaviors under control?
3. What are 4 behaviors that help keep weight off?

---

▶ Some believe fasting is an effective weight loss method. However, many misconceptions are associated with fasting.

- Does fasting cause weight loss? Yes, in the form of water loss—with a trip to the water fountain, a faster's prunelike, dried-out body plumps up to juicy plum size again.

- Does fasting rid the body of accumulated toxins? Toxins are poisons—a "buildup" would cause illness and death quickly. The liver and kidneys work continuously to filter out toxins we ingest or produce (e.g., ammonia), detoxify them, and rapidly excrete them. In reality, fasting may produce more toxins than normal in the form of ammonia and ketones that need to be detoxified and excreted immediately.

## 10.7 Fad Diets

Fad diets claim miraculous weight loss or improved health—often by unhealthy or unrealistic eating plans and perhaps touting "miracle" foods, specific rituals (e.g., eating only fruit for breakfast or cabbage soup every day), or certain foods that people would not normally eat in large amounts. Some are so monotonous that they are hard to follow for more than a short time. Fad diets may lead to some immediate weight loss simply because daily energy intake is monitored. Fad diets rarely lead to lasting weight loss or help retrain eating and exercise habits. Plus, some can actually cause harm (Table 10-7).

Instead of adding to the over $30 billion a year Americans already spend on fad diets,[29] a better choice for losing weight and keeping it off is to follow an eating and exercise plan that you can live with every day for the rest of your life. The goal should be lifelong weight control, not immediate weight loss.

Fad diets often promise rapid weight loss. Unfortunately, quick weight loss cannot consist primarily of fat because a high energy deficit is needed to lose a large amount

**Table 10-7  A Summary of Popular Diet Approaches to Weight Control**

| Approach | Examples | Characteristics | Outcomes |
|---|---|---|---|
| **Moderate energy restriction** | • DASH Diet<br>• Weight Watcher's<br>• Volumetrics | Generally, 1200 to 1800 kcal/day, with moderate fat intake<br>Reasonable balance of macronutrients<br>Encourage exercise<br>May use behavioral approach | Generally, good with moderate weight loss |
| **Restricted carbohydrate** | • Keto Diet<br>• Paleo Diet<br>• Dr. Atkins Diet Revolution<br>• South Beach Diet | Generally, less than 50 g of carbohydrate per day<br>May eliminate dairy products<br>Usually have high protein and high fat consumption | Ketosis; reduced exercise capacity due to poor glycogen stores in the muscles; excessive animal fat and cholesterol intake; constipation, headaches, halitosis (bad breath), and muscle cramps |
| **Low fat** | • Macrobiotic Diet<br>• Pritikin Diet | Generally, less than 20% of energy intake from fat<br>Limited (or elimination of) animal protein sources; also limited fats, nuts, and seeds | Flatulence; possibly poor mineral absorption from excess fiber; limited food choices sometimes lead to deprivation; not necessarily to be avoided, but certain aspects of many of the plans possibly unacceptable |
| **Novelty diets** | • Intermittent Fasting<br>• Cabbage Soup Diet<br>• Whole 30 Diet | Restricted calorie intake due to the number of times you can eat or due to the elimination of certain food groups<br>Promotes certain nutrients, foods, or combinations of foods as having unique, magical, or previously undiscovered qualities<br>Eliminates certain food groups<br>May promote gimmicks that have no effect on weight loss | Malnutrition; no change in habits, which leads to relapse; unrealistic food choices lead to possible binge eating |

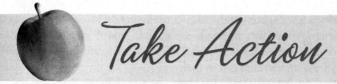

*Take Action*

## How to Spot a Fad Diet

Fad diets are built on gimmicks. Some emphasize a single food or food group and may exclude all other foods. Some tell you that certain foods get stuck in the body and never get digested. Fad diets can be costly to your pocketbook and your health. They may emphasize expensive foods. The foods they recommend may lower the nutrient stores in your body or increase fat levels in your blood. Some fad diet followers lose weight because they eat less food, but, when boredom sets in and they go back to their old eating habits, weight bounces back to pre-diet levels—and sometimes beyond.

How can you tell if a program or diet plan is a fad diet? The Academy of Nutrition and Dietetics published 10 red flags to help consumers determine if the diet or nutrition information they are receiving is credible.[9]

1. Recommendations that promise a quick fix

2. Dire warnings of danger from a single product or regimen

3. Claims that sound too good to be true

4. Simplistic conclusions drawn from a complex study

5. Recommendations based on a single study

6. Dramatic statements that are refuted by reputable scientific organizations

7. Lists of "good" and "bad" foods

8. Recommendations made to help sell a product; often, testimonials are used

9. Recommendations based on studies published without peer review

10. Recommendations from studies that ignore differences among individuals or groups

Now, put the red flags into action.

1. What popular diet have you heard about recently?

2. Gather information on this diet from the Internet.

3. Compare the information to the 10 red flags. Is this popular diet a fad diet? Why or why not?

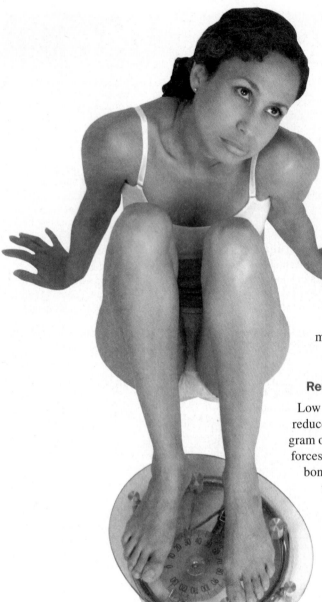

of adipose tissue. Diets that promise a weekly weight loss of 10 to 15 lb cannot ensure that the weight loss is from adipose tissue stores alone. Subtracting enough energy from one's daily intake to lose that amount of adipose tissue simply is not possible. Lean tissue and water, rather than adipose tissue, account for the major part of the weight lost when weight loss exceeds a few pounds weekly.

Probably the cruelest characteristic of these diets is that they essentially guarantee failure for the dieter. The diets are not designed for permanent weight loss. Habits are not changed, and the food selection is so limited that the person cannot follow the diet for long. Although dieters assume they have lost fat, they have actually lost mostly muscle and other lean tissue mass. As soon as they begin eating normally again, much of the lost weight returns in a matter of weeks. The dieter appears to have failed, when actually the diet has failed. The gain and loss cycle (yo-yo dieting) can cause blame, guilt, and negative health effects.

Health professionals can help dieters design and follow a healthy weight loss plan—unfortunately, current trends suggest that people are spending more time and money on fad diets and quick fixes than on professional help.[5]

### Restricted Carbohydrate Diets

Low carbohydrate diets lead to reduced glycogen synthesis and therefore reduced amounts of water in the body (about 3 g of water are stored per gram of glycogen). As discussed in Chapter 9, a very low carbohydrate intake forces the liver to produce glucose via gluconeogenesis. The source of carbons for this glucose is mostly body proteins. (Recall also from Chapter 9 that typical fatty acids cannot form glucose.) It appears that those who follow a low carbohydrate, high protein diet do lose weight; however, the amount of weight lost and potential improvements in health are no greater than for those who follow a moderate fat, higher carbohydrate, moderate protein diet.

Low carbohydrate diets work primarily in the short run because they limit total food intake. In long-term studies, these diets have not shown an advantage over diets that simply limit energy intake in general. Some research suggests that a low carbohydrate diet may increase LDL-cholesterol.[30]

When you see a new diet advertisement, look first to see how much carbohydrate it contains. If breads, cereals, fruits, and vegetables are extremely limited, you are probably looking at a low carbohydrate diet (see Table 10-7).

### Low-Fat Diets

Low-fat diets, especially those that are very low in fat, turn out to be very high carbohydrate diets. These diets contain approximately 5 to 10% of energy intake as fat. Low-fat diets are not harmful for healthy adults, but they are difficult to follow. These diets contain mostly grains, fruits, and vegetables. Eventually, many people get bored with this type of diet because they cannot eat favorite foods—they want some foods higher in fat or protein. Low-fat diets are very different from the typical North American diet, which makes it hard for many adults to follow them consistently.

People on diets often have a healthy BMI—rather than worrying about weight loss, these individuals should focus on a healthy lifestyle that promotes weight maintenance and acceptance of their body characteristics. For those at a healthy weight, the desire to lose weight may stem from unrealistic weight expectations (especially for women) and lack of appreciation for the natural variety in body shape and weight. Not everyone can look like a movie star, but all of us can strive for good health and a healthy lifestyle.

Ingram Publishing/SuperStock RF

## Novelty Diets

Novelty diets are built on gimmicks. Some emphasize a single food or food group and exclude almost all others. They might include only grapefruit, rice, or eggs. The rationale behind these diets is that you can eat only these foods for just so long before becoming bored and, in theory, reduce your energy intake. However, chances are that you will abandon the diet entirely before losing much weight.

The most questionable of the novelty diets propose that "food gets stuck in your body." The supposition is that food gets stuck in the intestine, putrefies, and creates toxins that cause disease. In response, these diets recommend not consuming certain foods or eating them only at certain times of the day. These recommendations make no physiological sense, however—as you know from Chapter 4, the digestive tract is efficient at digesting foods and eliminating waste.

Quack fad diets usually involve a costly product or service that doesn't lead to weight loss. Often, those offering the gimmick don't realize that they are promoting quackery because they have been victims themselves. For example, they tried the product and by pure coincidence lost weight, and they erroneously believe it worked for them, so they wish to sell it to others. Numerous weight loss gimmicks have come and gone and are likely to resurface.[31] If you hear that an important aid for weight loss is discovered, you can feel confident that, if it is legitimate, major peer-reviewed journals and authorities, such as the Surgeon General's Office or the National Institutes of Health, will make us aware of it. Information in diet books, websites, infomercials, and advertisements needs to be viewed with a scientist's skeptical, questioning eye.

In time, very low carbohydrate, high protein diets typically leave a person wanting more variety in meals, so the diets are abandoned. Dropout rates are very high on these diets.

Ingram Publishing/SuperStock RF

### Knowledge Check

1. What are the 10 red flags that can help you determine if nutrition information is credible?
2. Why can't rapid weight loss consist mainly of fat?
3. What are the characteristics of each of the main types of fad diets?

# CLINICAL PERSPECTIVE

## Professional Help for Weight Control

The first professional to see for weight control advice is one's family physician. Doctors are best equipped to assess overall health and the appropriateness of weight loss and weight gain. The physician then may recommend a registered dietitian nutritionist for a specific eating plan and answers to diet-related questions. Registered dietitian nutritionists are uniquely qualified to help design eating plans for weight control because they understand both food composition and the psychological importance of food. Exercise physiologists also can provide advice about physical activity. The expense for professional interventions is tax deductible in the U.S. in some cases (see a tax advisor) and often covered by health insurance if prescribed by a physician.

### Drug Treatment for Weight Loss

In skilled hands, prescription medications can aid weight loss in some instances. However, success with medications has been shown only in those who also modify their behavior, decrease their energy intake, and increase their physical activity.[5,32]

People who are candidates for medications for obesity include those with a BMI of 30 or more or a BMI of 27 to 29.9 with obesity-related (comorbid) conditions, such as type 2 diabetes, cardiovascular disease, hypertension, or excess waist circumference; those with no contraindications to the use of the medication; and those ready to undertake lifestyle changes that support weight loss. The following are the general classes of medications used. Significant research is under way to develop additional medications.[33]

- Medications that enhance norepinephrine and serotonin activity in the brain by reducing the re-uptake of these neurotransmitters by nerve cells (Belviq®). This effect causes the neurotransmitters to remain active in the brain longer, which prolongs a sense of reduced hunger. These medications appear to be effective in helping some people who eat healthy diets but who simply eat too much.[5]
- Amphetamine-like medication (phentermine [Fastin® or Ionamin®]), which prolongs epinephrine and norepinephrine activity in the brain. This therapy is effective for some people in the short run but has not yet been

proven effective in the long run. A new drug (Qsymia®) combines an appetite suppressant (phentermine) and an anti-seizure/migraine headache drug (topiramate). Topiramate helps induce feelings of fullness, makes foods taste less appealing, and boosts calorie burning.
- Medications that inhibit lipase enzyme action in the small intestine, thereby reducing fat digestion by about 30% (orlistat [Xenical®, Alli®]). Malabsorbed fat is deposited in the feces. Fat intake has to be controlled, however, because large amounts of fat in the feces can cause gas, bloating, and oily discharge. The malabsorbed fat also carries fat-soluble vitamins into the feces, so a multivitamin and mineral supplement is recommended.
- Medications that are not approved for weight loss per se that can have weight loss as a side effect. For example, certain antidepressants (e.g., bupropion [Wellbutrin®]) have this effect.[29] Using a medication in this way is termed off-label because the product label does not include weight loss as an FDA-approved use.

### Treatment of Severe Obesity

Severe (morbid) obesity—weighing at least 100 lb over healthy body weight (or twice one's healthy body weight)—requires professional treatment. Because of the serious health implications of severe obesity, drastic measures may be necessary. Such treatments are recommended only when traditional diets fail because these treatments have serious physical and psychological side effects that require careful monitoring by a physician. These practices include very-low-calorie diets and gastroplasty.

**Very-low-calorie diets (VLCDs),** or modified fasts, are used to treat severe obesity if more traditional dietary changes have failed. VLCDs provide 400 to 800 calories daily, often in liquid form, and tend to be used if a person has obesity-related diseases that are not well controlled.[5] About half the calories in these diets are carbohydrate, and the rest is high quality protein. The resulting ketosis from this low carbohydrate intake may help decrease hunger. However, the main reasons for weight loss are minimal calorie intake and absence of food choice. About 3 to 4 lb can be lost per week; men tend to lose at a faster rate than women. Careful monitoring by a physician is

crucial throughout this very restrictive form of weight loss. Major health risks include heart problems and gallstones. If behavioral therapy and physical activity supplement a long-term support program, maintenance of the weight loss is more likely but still difficult. Medications for obesity also may be included in the maintenance phase of a VLCD.

**Gastroplasty** (gastric bypass surgery, also called stomach stapling) may be recommended for those who have a BMI ≥40, have had obesity for at least 5 years with several nonsurgical attempts to lose weight, and have no history of alcoholism or major psychiatric disorders. Gastroplasty works by reducing the stomach capacity to about 30 ml (the volume of 1 egg or shot glass) and bypassing a short segment of the upper small intestine (Fig. 10-18). Another surgical approach is banded gastroplasty. In vertical-banded gastroplasty, a vertical staple line is made down the length of the stomach to create a small stomach pouch. At the pouch outlet, a band is placed to keep the opening from stretching. With gastric banding, a band is placed around the upper portion of the stomach to create a small stomach pouch. A salt solution can be injected into the band through a port to adjust the size of the pouch over time.

With gastroplasty, weight loss is achieved because most of the food bypasses the stomach and small intestine, which results in less digestion and absorption of nutrients. Gastroplasty requires major lifestyle changes, such as the need for frequent, small meals and the elimination of sugars from the diet, to avoid dumping syndrome (severe diarrhea that begins almost immediately after eating concentrated sugar, such as regular soft drinks, candy, and cookies).

The surgery is costly and may not be covered by health insurance. In addition, the patient faces months of difficult adjustments. However, with gastroplasty, about 75% of people who have severe obesity eventually lose half or more of their excess body weight. In addition, the surgery's success at long-term maintenance often leads to dramatic health improvements, such as reduced blood pressure and the elimination of type 2 diabetes. The risks of this very serious surgery include bleeding, blood clots, hernias, and severe infections. About 2% die from the surgery itself. In the long

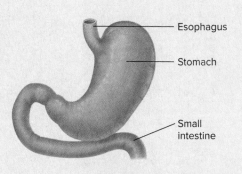

**Normal stomach**

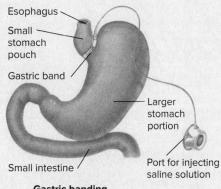

Esophagus

Small stomach pouch

Gastric band

Larger stomach portion

Port for injecting saline solution

Small intestine

**Gastric banding (e.g., LAP-BAND® procedure)**

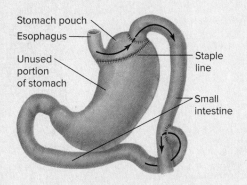

Stomach pouch
Esophagus
Unused portion of stomach
Staple line
Small intestine

**Gastric bypass**

Esophagus
Removed portion of stomach
Gastric sleeve

**Gastric sleeve**

*Figure 10-18* The most common forms of gastroplasty for treating severe obesity. The gastric bypass is the most effective method. In banded gastroplasty, the band prevents expansion of the outlet for the stomach pouch.

run, nutrient deficiencies can develop if the person is not adequately treated in the years following the surgery (chewable multivitamin and mineral supplements are often used). Follow-up surgery frequently is needed after weight loss to remove excess skin that was previously padded with fat.

### Treatment of Underweight

Being underweight (BMI below 18.5) also carries health risks, including loss of menstrual function, low bone mass, complications with pregnancy and surgery, and slow recovery after illness. In growing children and teens, underweight can interfere with normal growth and development. Significant underweight also is associated with increased death rates, especially when combined with cigarette smoking.

Underweight can be caused by excess physical activity, severely restricted calorie intake, and health conditions such as cancer, infectious disease (e.g., tuberculosis), digestive tract disorders (e.g., chronic inflammatory bowel disease), eating disorders, and mental stress or depression. Active children and teens who do not take the time to consume enough energy to support their needs may become underweight. Genetic background may confer characteristics that promote low body weight, such as a higher metabolic rate, a lean or petite body frame, or both.

Gaining weight can be a formidable task for an underweight person. An extra 500 calories per day may be required to gain weight, even at a slow pace, in part because of the increased expenditure of energy from thermogenesis. One approach for treating underweight is to

replace less energy dense foods with higher energy dense foods. For example, a cup of granola instead of bran flakes can add an extra 300 to 500 calories. Choosing bean soup over vegetable soup adds 50 calories or more. Portion sizes may need to be increased gradually. Having a regular meal and snack schedule also aids in weight gain and maintenance. Making time to eat regular meals can help those who are underweight attain an appropriate weight and help with digestive disorders, such as constipation, that are sometimes associated with irregular eating schedules. Excessively physically active people can reduce their activity. If their weight remains low, they can add muscle mass through resistance training (weight lifting), but to gain weight they must increase energy intake to support that physical activity. If these changes do not lead to weight gain within a few weeks, the person should seek medical intervention to identify the cause of this condition and get appropriate treatment.

Although the media and the fashion world promote a very thin physique, being underweight carries severe risks.

Karl Prouse/Catwalking/Getty Images

## Knowledge Check

1. Which weight loss medications may be used to treat obesity?
2. Who should use a modified fast for weight loss?
3. What is gastroplasty?
4. What health risks are associated with being underweight?

# Take Action

## Changing for the Better

Even if you are satisfied with your current weight, many people see their weight climb as they get older, so it is a good idea to be aware of how to keep weight under control. The following behavioral change method can help you do just that. It also can be applied to changing exercise habits, self-esteem, and many other behaviors (Fig. 10-19).

1. **Become aware of the problem.** Calculate your current weight status to determine if you have a weight problem. First, measure your height and weight. Then, using Table 10-1, record your BMI: _____. When BMI exceeds 25, health risks from overweight may start. It is especially advisable to consider weight loss if your BMI exceeds 30.

   Now, use a tape measure to measure the circumference of your waist (at the narrowest point just above the navel, with stomach muscles relaxed). Record the circumference: _____ inches.

   When BMI exceeds 25 and a waist circumference is more than 40 inches (102 cm) in men or 35 inches (88 cm) in women, health risks rise. Does your waist circumference exceed the standard for your biological sex? _____

   Whether you feel the need to pursue a program of weight loss now or in the future, it is important to find out more about how to make changes.

2. **Gather baseline data.** Look back at the food diary you completed in Chapter 1. What factors most influence your eating habits? Do you eat out of stress, boredom, or depression? Is eating too much food your problem, or do you mainly eat a poor diet? Next, think about whether it is worth changing these practices—a cost:benefit analysis can help you decide (Fig. 10-20).

3. **Set goals.** Setting realistic, achievable goals and allowing a reasonable amount of time to pursue them increase the likelihood of success. What final goal would you like to achieve? Why do you want to pursue this goal (e.g., improve health, lose weight, boost self-esteem)?

*Figure 10-20* Cost:benefit analysis applied to changing eating habits. This process helps put behavioral change into the context of total lifestyle.

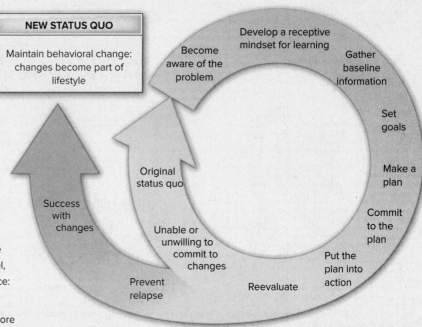

*Figure 10-19* A model for behavioral change. It starts with awareness of the problem and ends with the incorporation of new behaviors intended to address the problem.

| Benefits of changing eating habits |
| --- |
| What do you expect to get, now or later, that you want? |
| What may you avoid that would be unpleasant? |
| *feel better physically and psychologically* |
| *look better* |

| Costs involved in changing eating habits |
| --- |
| What do you have to do that you don't want to do? |
| What do you have to stop doing that you would rather continue doing? |
| *take time to plan meals and shop* |
| *must give up some food volume* |

| Benefits of not changing eating habits |
| --- |
| What do you get to do that you enjoy doing? |
| What do you avoid having to do? |
| *no need for planning* |
| *can eat without feeling guilty* |

| Costs of not changing eating habits |
| --- |
| What unpleasant or undesirable effects are you likely to experience now or in the future? |
| What are you likely to lose? |
| *creeping weight gain* |
| *low self-esteem and poor health* |

4. **Make a plan.** List several steps that will be necessary to achieve your goal. Changing only a few behaviors at a time increases the chances of success. You might choose to walk 60 minutes daily, eat less fat, eat more whole grains, or not eat after 8 P.M. What steps will you take to achieve the goal? If you are having trouble listing the steps, you may want to consult a health professional for assistance.

5. **Commit to the plan.** Next, ask yourself, "Can I do this?" Be honest with yourself. Commitment is an essential component in the success of behavioral change. Permanent change is not quick or easy. Drawing up a behavioral contract often adds incentive to follow through with a plan (Fig. 10-21). The contract can list goal behaviors and objectives, milestones for measuring progress, and regular rewards for meeting the terms of the contract. After finishing a contract, you should sign it in the presence of some friends. This formality encourages commitment.

6. **Put the plan into action.** Thinking of a lifetime commitment can be overwhelming, so start with a trial of 6 or 8 weeks. Aim for a total duration of 6 months of new activities before giving up. Keep your plan on track.

   - Focus on reducing, but not necessarily extinguishing, undesirable behaviors. For example, it's usually unrealistic to say that you will never eat a certain food again. It's better to say, "I won't eat that problem food as often as before."

   - Monitor progress. Note your progress in a diary and reward positive behaviors. While conquering some habits and seeing improvement, you may find yourself quite encouraged about your plan of action—that can motivate you to move ahead with the plan.

   - Control environments. In the early phases of behavioral change, try to avoid problem situations, such as parties, favorite restaurants, and people who try to derail your plans. Once new habits are firmly established, you can probably more successfully resist the temptations of these environments.

7. **Reevaluate and prevent relapse.** After practicing a program for several weeks to months, take a close and critical look at your

---

**Name** _Alan Young_

**Goal**
I agree to _ride my exercise bike_
(specify behavior)
under the following circumstances _for 30 minutes, 4 times per week_
(specify where, when, how much, etc.)
_in the evening._

Substitute behavior and/or reinforcement schedule _I will reinforce myself if I've achieved my goal after a month with a weekend off campus._

**Charting progress**
To keep track of my progress, _I will mark the days I exercise on a calendar._

**Environmental planning**
To help me do this, I am going to (1) arrange my physical and social environment by _downloading new music_

and (2) control my internal environment (thoughts, images) by _coordinating riding the bike with the first T.V. watching I do in the evening._

**Reinforcements**
Reinforcements provided by me daily or weekly (if contract is kept):
_I will buy myself a new piece of clothing for off-campus trip._

Reinforcements provided by others daily or weekly (if contract is kept):
_at the end of a month, if I've completed my goal, my parents will buy me a fitness club membership for winter._

**Social support**
Behavioral change is more likely to take place when other people support you. During the school term, please meet with the other person at least 3 times to discuss your progress. The name of my "significant helper" is _Dad_

**Figure 10-21** Completing such a contract can help generate commitment to behavioral change. What would your contract look like?

---

original plan. Does it actually lead to the goals you set? Are there any new steps toward your goal that you want to add? Do you need new reinforcements? Have you experienced relapses? What triggered these relapses? How can you avoid future relapses? You may have noticed a behavior chain in some of your relapses (see Fig. 10-19)—how can you break the chain?

8. **Maintain behavioral change.** If you have used the activities in this section, you are well on your way to permanent behavioral change. Change isn't easy, but the results can be worth the effort.

## 10.8 Eating Disorders

**Disordered eating** can be defined as mild and short-term changes in eating patterns that occur in response to a stressful event, an illness, or a desire to modify the diet for health and/or personal appearance reasons. The problem may be no more than a bad habit, a style of eating adapted from friends or family members, or an aspect of preparing for athletic competition. Although disordered eating can lead to changes in body weight and certain nutritional problems, it seldom requires in-depth professional attention. However, in today's world, given the common practice of dieting, skipping meals, eating at odd times, and having hectic jobs and schedules, it may not be obvious when disordered eating stops and an **eating disorder** begins (Fig. 10-22). It takes a skilled, multidisciplinary medical team and extensive assessments to accurately make the diagnosis of an eating disorder. A team approach also is needed to effectively treat eating disorders because these conditions have complex signs and symptoms, involve both physical and mental health, and must consider an array of eating and other health behaviors.

Although obesity is the most common eating disorder in our society, the eating disorders explored in this section involve much more severe distortions of the eating process, and they can develop into life-threatening conditions if left untreated.[34] The main types of eating disorders are anorexia nervosa, bulimia nervosa, and binge eating disorder. Two other categories are Other Specified Feeding or Eating Disorder (OSFED) and Unspecified Feeding and Eating Disorder (UFED).[35,36]

### Prevalence and Susceptibility

In the U.S., 20 million women and 10 million men suffer from an eating disorder. Additionally, half a million adolescents struggle with either an eating disorder or disordered eating. Eating disorders are 6 to 10 times more common in females than males. Currently, between 3 to 5% of women and about 1% of men in North America will develop some form of anorexia nervosa or bulimia nervosa in their lifetimes.[35] What is most alarming about these disorders is the increasing number of cases reported each year.[37]

Some people are more susceptible to these disorders for genetic, psycho-social, and physical reasons. Eating disorders are not restricted to any socioeconomic class, ethnicity, age group, or biological sex.

Many eating disorders start with a simple diet. Stress and a lack of appropriate coping mechanisms, dysfunctional family relationships, body size dissatisfaction, social media, influences of cultural values, and drug abuse may cause dieting to get out of control.[38] Stress may be caused by physical changes associated with entering puberty, having to maintain a certain weight to look attractive or competent on a job, having to maintain a lean profile for a sport, leaving home for college, or losing a friend. Eating disorders develop 85% of the time during adolescence or early adulthood, but some reports indicate that their onset can occur during childhood or later in adulthood.

Disordered eating can escalate into physiological changes associated with sustained food restriction, binge eating, purging, and fluctuations in weight that interfere with everyday activities. It also involves emotional and cognitive changes that affect how people perceive and experience their bodies, such as feelings of distress or extreme concern about body shape or weight.[34] Eating disorders frequently co-occur with other psychological disorders, such as depression, anxiety disorders, and addictions.[34]

Eating disorders are not due to a failure of will power or behavior; rather, they are real, treatable mental illnesses that require complex professional intervention that must go beyond nutritional therapy.[39] Without treatment, eating disorders can cause serious physical health complications, including heart conditions and kidney failure, which may even lead to death. Self-help groups for those with eating disorders, as well as their families and friends, represent nonthreatening first steps into treatment. People also can attend self-help group meetings to get a sense of whether they really do have an eating disorder.

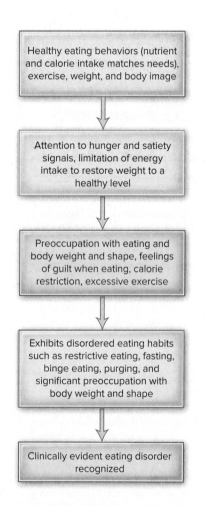

*Figure 10-22* Eating behaviors occur on a continuum. This figure illustrates how a person can go from ordered to disordered eating behaviors.

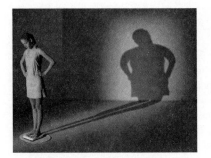

For people with eating disorders, the differences among real, perceived, and desired body images may be too difficult to accept.

Steve Niedorf Photography/Stone/Getty Images

Eating disorders are commonly seen in people who must maintain low body weight, such as jockeys, wrestlers, figure skaters, and ballet dancers.

Mikhail Pogosov/Shutterstock

## Anorexia Nervosa

The term *anorexia* implies a loss of appetite; however, a denial of appetite more accurately describes the behavior of people with anorexia nervosa. The term *nervosa* refers to disgust with one's body. **Anorexia nervosa** is characterized by extreme weight loss, a distorted body image, and an irrational, almost morbid fear of obesity and weight gain. These individuals believe they are fat, even though they are not and others tell them so. Some realize they are thin but continue to be haunted by certain areas of their bodies that they believe to be fat (e.g., thighs, buttocks, and stomach). The discrepancy between actual and perceived body shape is an important gauge of the severity of the disease.

Estimating the prevalence of this eating disorder is difficult because of underreporting, but approximately 1 in 200 (0.5%)[34] adolescent girls in North America eventually develops anorexia nervosa and 20% of those who develop anorexia nervosa will die prematurely from complications related to their disorder. This relatively high incidence may be due to girls' tendency to blame themselves for weight gain associated with puberty. It happens less commonly among adult women, specifically African-American women. Men account for approximately 10% of cases of anorexia nervosa, partly because the ideal image conveyed for men is big and muscular. Among males, some gay men as well as athletes are most prone to develop anorexia nervosa, especially athletes who participate in sports that require weight classes, such as boxers, wrestlers, and jockeys. Other activities that may foster eating disorders in men include cycling, swimming, dancing, and modeling.[35,40]

There are two main types of anorexia nervosa: restricting and binge eating/purging type. Patients who are classified as restricting usually limit their food intake, fast, and/or exercise excessively. Patients who are classified with binge eating/purging anorexia nervosa engage in binge eating, then purge themselves in response to severe food restriction. People with anorexia nervosa have very low body weights, fear weight gain,

Concern over self-image begins early in life—we develop images of "acceptable" and "unacceptable" body types. Of all the attributes that constitute attractiveness, many people view body weight as the most important. For many, fatness is the most dreaded deviation from our cultural ideals of body image.

Dave J. Anthony/Getty Images RF

**Table 10-8 Typical Characteristics of Those with Eating Disorders**

| Anorexia Nervosa | Bulimia Nervosa |
|---|---|
| • Rigid dieting, causing dramatic weight loss, generally to less than 85% of what would be expected for one's age (or BMI of 17.5 or less)<br>• False body perception—thinking "I'm too fat," even when extremely underweight; relentless pursuit of control<br>• Rituals involving food, excessive exercise, and other aspects of life<br>• Maintenance of rigid control in lifestyle; security found in control and order<br>• Feeling of panic after a small weight gain; intense fear of gaining weight<br>• Feelings of purity, power, and superiority through maintenance of strict discipline and self-denial<br>• Preoccupation with food, its preparation, and observing another person eat<br>• Helplessness in the presence of food<br>• Typically, lack of menstrual periods after what should be the age of puberty<br>• Some binge eat and purge. | • Secretive binge eating; generally not overeating in front of others<br>• Eating when depressed or under stress<br>• Bingeing on a large amount of food, followed by fasting, laxative or diuretic abuse, self-induced vomiting, or excessive exercise (at least weekly for 3 months)<br>• Shame, embarrassment, deceit, and depression; low self-esteem and guilt (especially after a binge)<br>• Fluctuating weight ($\pm$10 lb or 5 kg) resulting from alternate bingeing and fasting<br>• Loss of control; fear of not being able to stop eating<br>• Perfectionism, "people pleaser"; food as the only comfort/escape in an otherwise carefully controlled and regulated life<br>• Erosion of teeth; swollen glands from vomiting<br>• Purchase of syrup of ipecac, a compound sold in pharmacies that induces vomiting<br><br>**Binge Eating Disorder**<br>• Same characteristics as bulimia nervosa without purging<br>• Eats faster than normal<br>• Eats even when full or not hungry<br>• Feels unpleasantly full after binge eating<br>• Eats alone to hide binge eating from others |

People who exhibit only a few of these characteristics may be at risk but probably do not have a disorder. They should, however, reflect on their eating habits and related concerns and take appropriate action, such as seeking a careful evaluation by a physician.

**disordered eating** Mild to short-term abnormal changes in eating patterns that occur in relation to a stressful event, an illness, or a desire to modify one's diet for a variety of health and personal appearance reasons.

**eating disorder** Severe alterations in eating patterns linked to physiological changes; the alterations include food restricting, binge eating, purging, weight fluctuations, and emotional and cognitive changes in perceptions of one's body.

**anorexia nervosa** Eating disorder involving a psychological loss or denial of appetite followed by self-starvation; it is related, in part, to a distorted body image and to social pressures.

**binge eating disorder** Eating disorder characterized by recurrent binge eating and feelings of loss of control over eating.

**bulimia nervosa** Eating disorder in which large quantities of food are eaten on a single occasion (binge eating) and counteracted by purging food from the body, fasting, and/or excessive exercise.

have a distorted view of their body weight, and evaluate their self-worth almost entirely in terms of their weight.[39] They focus on controlling food, eating, and weight instead of human relationships. Some characteristics of those with anorexia nervosa are listed in Table 10-8.

Keep in mind that only a health professional can correctly evaluate the criteria required to make a diagnosis of eating disorders and exclude other possible diseases. If you think you know someone who is at risk for this or other eating disorders, suggest that the person seek professional evaluation because, the sooner treatment begins, the better the chances for recovery.[41]

### Physical Effects of Anorexia Nervosa

Anorexia nervosa produces profound physical effects.[42] The person often appears to be skin and bones. Body weight less than 85% of that expected is a clinical indicator of anorexia nervosa.[43] BMI is a more reliable indicator of the degree of malnourishment; generally, a BMI of 17.5 or less indicates a severe case. For children under age 18, growth charts should be used to assess weight status (see Chapter 17 and Appendix K).

This state of semistarvation forces the body to conserve as much energy as possible and results in most of the physical effects of anorexia nervosa (Fig. 10-23). Thus, many complications can be ended by returning to a healthy weight, provided the duration of anorexia nervosa has not been too long. These are the predictable effects caused by hormonal responses to and nutrient deficiencies from semistarvation:[34,35,39,44]

- Low body weight (15% or more below what is expected for age, height, and activity level)
- Lowered body temperature and cold intolerance caused by loss of an insulating fat layer
- Slower metabolic rate caused by decreased synthesis of thyroid hormones

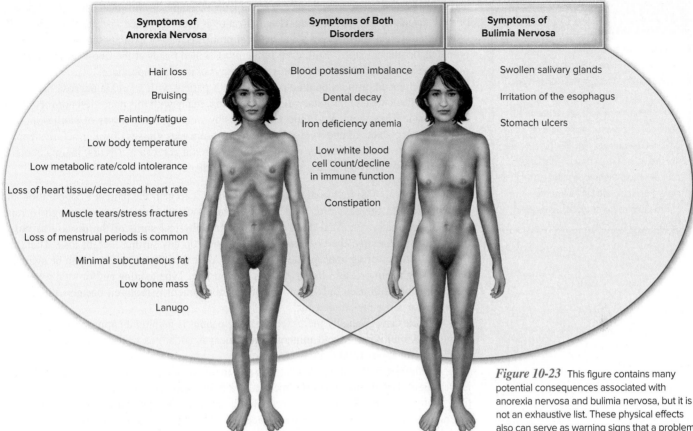

| Symptoms of Anorexia Nervosa | Symptoms of Both Disorders | Symptoms of Bulimia Nervosa |
|---|---|---|
| Hair loss | Blood potassium imbalance | Swollen salivary glands |
| Bruising | Dental decay | Irritation of the esophagus |
| Fainting/fatigue | Iron deficiency anemia | Stomach ulcers |
| Low body temperature | Low white blood cell count/decline in immune function | |
| Low metabolic rate/cold intolerance | Constipation | |
| Loss of heart tissue/decreased heart rate | | |
| Muscle tears/stress fractures | | |
| Loss of menstrual periods is common | | |
| Minimal subcutaneous fat | | |
| Low bone mass | | |
| Lanugo | | |

*Figure 10-23* This figure contains many potential consequences associated with anorexia nervosa and bulimia nervosa, but it is not an exhaustive list. These physical effects also can serve as warning signs that a problem exists and professional evaluation is needed.

- Decreased heart rate as metabolism slows, leading to easy fatigue, fainting, and an overwhelming need for sleep. Other changes in heart function also may occur, including loss of heart tissue and poor heart rhythm.
- Iron deficiency anemia, which leads to further weakness
- Rough, dry, scaly, and cold skin from a deficient nutrient intake, which may show multiple bruises because of the loss of protection from the fat layer normally present under the skin
- Low white blood cell count, which increases the risk of infection and potentially death
- Abnormal feeling of fullness or bloating, which can last for several hours after eating
- Loss of hair
- Appearance of **lanugo**—downy hairs that appear on the body after a person has lost much body fat through semistarvation—that helps trap air, reducing heat loss that occurs with the loss of fat tissue
- Constipation from semistarvation and laxative abuse
- Low blood potassium caused by a deficient nutrient intake, the use of some types of diuretics, and vomiting. Low blood potassium increases the risk of heart rhythm disturbances, another leading cause of death in those with anorexia nervosa.
- Loss of menstrual periods, common because of low body weight, low body fat content, and the stress of the disease. Accompanying hormonal changes cause a loss of bone mass and increase the risk of osteoporosis later in life.
- Changes in neurotransmitter function in the brain, leading to depression
- Eventual loss of teeth caused by acid erosion if frequent vomiting occurs
- In athletes, muscle tears and stress fractures caused by decreased bone and muscle mass

### Treatment of Anorexia Nervosa

With prompt, vigorous, and professional help, many people with anorexia nervosa can lead normal lives. Treatment requires a multidisciplinary team of experienced physicians, registered dietitian nutritionists, psychologists, and other health professionals.[34] An ideal setting is an

## HISTORICAL PERSPECTIVE

### The Golden Cage

The earliest criteria for diagnosing anorexia nervosa, described in 1870, were the first medical standards to recognize that societal influences can affect health. Through extensive observations and treatment of eating disorder patients starting in the 1930s, the psychoanalyst Hilde Bruch greatly advanced understanding of the interplay of culture and social structure of eating disorders. Her 1973 book *Eating Disorders: Obesity, Anorexia Nervosa, and the Person Within* brought the emotional aspects of eating disorders to the forefront of treatment. A later book, *The Golden Cage: The Enigma of Anorexia Nervosa*, helped consumers grasp the serious and complicated nature of eating disorders. Learn more at www.nimh.nih.gov/ health/topics/ eating-disorders/ index.shtml.

### *Critical* Thinking

Jennifer is an attractive 13-year-old. However, she's very compulsive. Everything has to be perfect—her hair, her clothes, even her room. Since her body started to mature, she's become quite obsessed with having perfect physical features as well. Her parents are worried about her behavior. The school counselor told them to look for certain signs that could indicate an eating disorder. What might those signs be?

Early treatment of eating disorders improves chances of success. Note that such help is commonly available at student health centers and student guidance/counseling facilities on college campuses.

ESB Professional/Shutterstock

eating disorders clinic in a medical center. Outpatient therapy, day hospitalization (6–12 hours), or total hospitalization may be used. Hospitalization is necessary once a person falls below 75% of expected weight, experiences acute medical problems, and/or exhibits severe psychological problems or suicidal risk.[34] Still, even in the most skilled hands at the finest facilities, efforts may fail. Thus, the prevention of anorexia nervosa is of utmost importance.

Experienced, professional help is the key. A patient with anorexia nervosa may be on the verge of suicide and near starvation. In addition, many with this condition are very clever and resistant. They may try to hide weight loss by wearing many layers of clothes, putting coins in their pockets or underwear, and drinking numerous glasses of water before stepping on a scale. The average time for recovery from anorexia nervosa is 7 years; many insurance companies cover only a fraction of the cost of treatment.

**Nutrition Therapy**    The first goal of nutrition therapy is to gain the patient's cooperation and trust in order to increase oral food intake. Ideally, weight gain must be enough to raise the metabolic rate to normal and reverse as many physical signs of the disease as possible.[37] Early in therapy, the goal is to minimize or stop any further weight loss. Then, the focus shifts to restoring appropriate food habits. After this, the expectation can be switched to slow weight gain—2 to 3 lb per week is appropriate. Tube feeding and/or total parenteral nutrition support is used only if immediate renourishment is required because this drastic measure can cause patient distrust.

Calories are gradually increased until the patient is gaining an appropriate amount of weight.[45] A multivitamin and mineral supplement is added, as well as enough calcium to raise intake to about 1500 mg/day.[34]

In addition to helping patients reach and maintain adequate nutritional status, the registered dietitian nutritionist on the medical team also provides accurate nutrition information, promotes a healthy attitude toward food, and helps the patient learn to eat based on natural hunger and satiety cues. Patients need considerable reassurance during the refeeding process because of uncomfortable effects, such as bloating, increase in body heat, and increase in body fat. The medical team also should assure patients that they will not be abandoned after gaining weight.

**Psychological and Related Therapy**    Once the physical problems of patients with anorexia nervosa are addressed, the treatment focus shifts to the underlying emotional problems of the disorder. To heal, these patients must reject the sense of accomplishment they associate with an emaciated body and begin to accept themselves at a healthy body weight. Establishing a strong relationship with either a therapist or another supportive person is especially important to recovery. A key aspect of psychological treatment is showing affected individuals how to regain control of other facets of their lives and cope with tough situations. As eating evolves into a normal routine, they can return to previously neglected activities.

Family therapy often is important, especially for younger patients who still live at home. Medications also are sometimes part of the therapy for patients with anorexia nervosa. However, their use is aimed primarily at preventing relapse in patients who have been treated but have an existing psychiatric disorder, such as depression, anxiety, or obsessive-compulsive disorder.

## Bulimia Nervosa

**Bulimia nervosa** (*bulimia* means "great [ox] hunger") is characterized by episodes of binge eating followed by attempts to purge the excess energy consumed by vomiting or misusing laxatives, diuretics, or enemas. Some people exercise excessively to try to burn off a binge's high energy intake.

A person with bulimia nervosa uses food as a way to cope with critical situations.[46] Unlike those with anorexia nervosa, people with bulimia nervosa recognize their behavior as abnormal.[43] These individuals often have very low self-esteem and are depressed—about 50% have major depression.

Up to 4% of adolescent and college-age women suffer from bulimia nervosa. About 10% of the cases occur in men.[35] However, many people with bulimic behavior are probably never diagnosed, perhaps because their symptoms are not obvious, and many with bulimia nervosa lead secret lives, hiding their abnormal eating habits.

Many susceptible people have genetic factors and lifestyle patterns that predispose them to becoming overweight, and many frequently try weight reduction diets as teenagers. Like people with anorexia nervosa, those with bulimia nervosa are usually female and successful. They are usually at or slightly above a normal weight.

### Binge-Purge Cycle

The typical characteristics of those with bulimia nervosa are described in Table 10-8. In addition to bingeing and purging, those with bulimia nervosa often have elaborate food rules, such as avoiding all sweets. Thus, eating just 1 cookie or donut may cause those with bulimia to feel guilty that they have broken a rule and proceed to binge. Usually, this action leads to significant overeating. A binge can be triggered by a combination of hunger from recent dieting, stress, boredom, loneliness, and depression. Bingeing often follows a period of strict dieting and thus can be linked to intense hunger. The binge is not like normal eating; once begun, it seems to propel itself. The person eats rapidly, loses control over eating, and may get little enjoyment of the flavors of the food being eaten. Binge-purge cycles may be practiced daily, weekly, or at longer intervals.[47] Binge eating often occurs at night, when others are less likely to interrupt them, and usually lasts from 30 minutes to 2 hours.

Most commonly, people with bulimia consume sweet, high carbohydrate convenience foods during binges because these foods can be purged relatively easily and comfortably by vomiting. In a single binge, foods supplying 3000 kcal or more may be eaten.[39] Purging follows, in hopes that no weight will be gained. However, even when vomiting follows the binge, 33 to 75% of the food energy taken in is still absorbed, which causes some weight gain. When laxatives or enemas are used, about 90% of the energy is absorbed because these products act in the large intestine, beyond the point of most nutrient absorption. Clearly, the belief that purging soon after bingeing will prevent excessive energy absorption and weight gain is a misconception.

Another way people with bulimia attempt to compensate for a binge is by engaging in excessive exercise to burn the excess calories. Exercise is considered excessive when it is done at inappropriate times or settings or when a person continues to exercise despite injury or medical complications. Some people try to estimate the amount of energy eaten in a binge and then exercise to counteract this energy intake. This practice, referred to as "debting," represents an effort to control their weight.

After bingeing and purging, those with bulimia nervosa usually feel guilty and depressed.[43] Over time, they experience low self-esteem and feel hopeless about their situation (Fig. 10-24). Sufferers gradually distance themselves from others, spending more and more time preoccupied by thoughts of food and body image and engaging in bingeing and purging.

### Physical Effects of Bulimia Nervosa

Most of the health problems associated with bulimia nervosa arise from vomiting.[35,48]

- Repeated exposure of teeth to the acid in vomit causes demineralization, making the teeth painful and sensitive to heat, cold, and acids (Fig. 10-25). Eventually, the teeth may decay severely, erode away from fillings, and finally fall out.
- Blood potassium can drop significantly because of regular vomiting or the use of certain diuretics. This drop can disturb the heart's rhythm and even cause sudden death.
- Salivary glands may swell as a result of infection and irritation from persistent vomiting.
- Stomach ulcers and tears in the esophagus develop, in some cases.

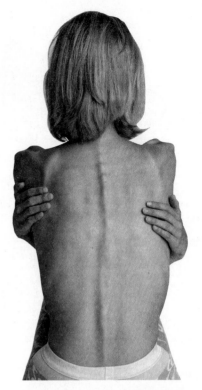

A young woman in a self-help group for those with anorexia nervosa explained her feelings to the other group members: "I have lost a specialness that I thought it gave me. I was different from everyone else. Now I know that I'm somebody who's overcome it, which not everybody does."

Tomas Houda/Alamy RF

Bulimia nervosa can lead to tragic consequences.

Fancy Collection/SuperStock RF

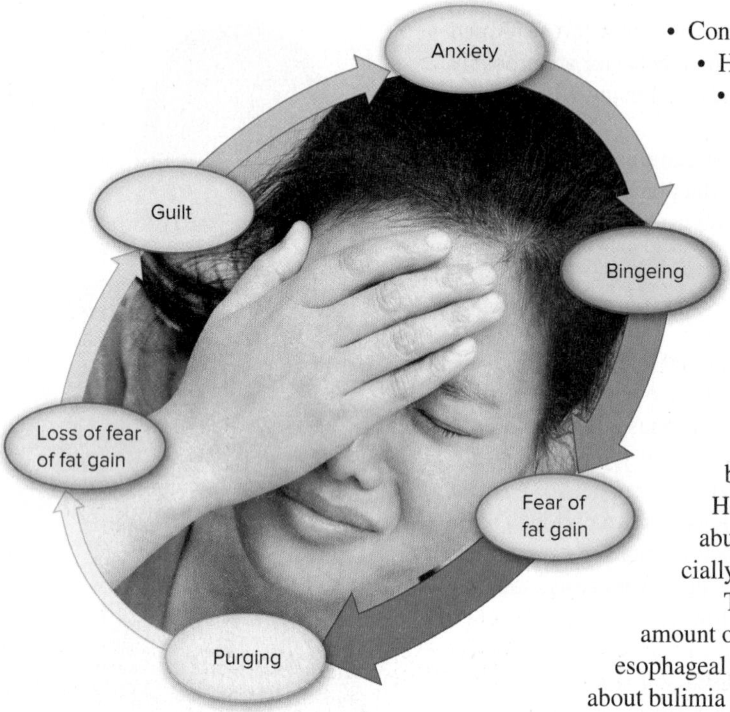

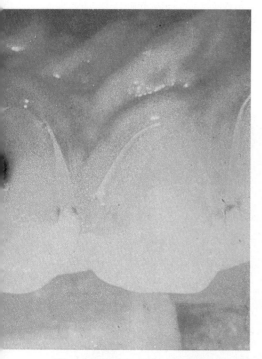

**Figure 10-24** Bulimia nervosa's vicious cycle of obsession.

fatchoi/Getty Images

▶ Bulimia nervosa is rare in low-income countries, which suggests that our culture is an important causal factor.

**Figure 10-25** Excessive tooth decay is common in patients with bulimia. Dental professionals are sometimes the first health professionals to notice signs of bulimia nervosa.

Paul Casamassimo, DDS, MS

- Constipation may result from frequent laxative use.
  - Hands may be calloused or cut by teeth while inducing vomiting
    - Ipecac syrup, sometimes used to induce vomiting, is toxic to the heart, liver, and kidneys and can cause accidental poisoning when taken repeatedly.

Overall, bulimia nervosa is a potentially debilitating disorder that can lead to death, usually from suicide, low blood potassium, or overwhelming infections.

### Treatment of Bulimia Nervosa

Therapy for bulimia nervosa requires a team of experienced psychotherapists and registered dietitian nutritionists.[47] If a patient has lost significant weight, this weight loss must be treated before psychological treatment and nutrition counseling begin. Hospitalization may be necessary in cases of extreme laxative abuse, regular vomiting, substance abuse, and depression, especially if physical harm is evident.

The first goal of treatment for bulimia nervosa is to decrease the amount of food consumed in a binge session in order to reduce the risk of esophageal tears from related purging by vomiting. Patients also are taught about bulimia nervosa and its consequences. They must recognize that they are dealing with a serious disorder that can have grave medical complications if not treated. Next follows nutrition counseling and psychotherapy.

Nutrition Therapy In general, the focus of nutrition therapy is to develop regular eating and exercise habits and correct misconceptions about food. To establish regular eating patterns, some specialists encourage patients to self-monitor by keeping a food diary, in which they record food intake, internal sensations of hunger, environmental factors that trigger binges, and thoughts and feelings that accompany binge-purge cycles. Avoiding binge foods and not constantly stepping on a scale may be recommended early in treatment. Patients also are discouraged from following strict rules about healthy food choices because such rules simply mimic the typical obsessive attitudes associated with bulimia nervosa. Rather, they should be encouraged to regularly consume moderate amounts of a variety of foods from each food group.[37]

Psychological and Related Therapy People with bulimia nervosa need psychological help because they can be very depressed and are at a high risk of suicide. The primary aims of psychotherapy are to improve patients' self-acceptance and to help them be less concerned about body weight. Psychotherapy helps correct the all-or-none thinking typical of those with bulimia: "If I eat 1 cookie, I'm a failure and might as well binge." In addition, the therapist guides the person in using methods other than bingeing and purging to cope with stressful situations. Group therapy often is useful in fostering strong social support. One goal of therapy is to help patients accept some depression and self-doubt as normal. Certain antidepressants may be used to treat bulimia nervosa; however, they should be used in conjunction with other therapies.

Because relapse is likely, therapy should be long-term. About 50% of people with bulimia nervosa recover completely from the disorder. Others continue to struggle with it, to varying degrees, for the rest of their lives. This fact underscores the need for prevention because treatment is difficult.

## Binge Eating Disorder

Binge eating disorder involves eating large amounts of food frequently; however, it is not usually followed by purging. During most binge episodes, individuals eat much more rapidly than usual, eat until feeling uncomfortably full, eat large amounts of food when not feeling physically hungry, eat alone because of being embarrassed by how much they are eating, and/or feel distressed, depressed, or very guilty after overeating. Those who binge eat consume food without regard to biological need and often in a recurrent, ritualized fashion. Some people with this disorder eat food continually over an extended period; others cycle episodes of bingeing with normal eating.[49]

### Prevalence and Susceptibility

Note that obesity and binge eating are not necessarily linked. Not all people with obesity engage in binge eating and, although obesity may result, it is not necessarily an outcome of binge eating. Nonetheless, about 8% of the population with obesity have binge eating disorder. It is most common among those with extremely high BMIs and those with a long history of frequent restrictive dieting, although obesity is not a criterion for having binge eating disorder. Approximately 30 to 50% of people in organized weight control programs have binge eating disorder, whereas about 2 to 5% of North Americans in general have this disorder. Many more people in the general population have less severe forms of the disease but do not meet the formal criteria for diagnosis. The number of cases of binge eating disorder is far greater than that of either anorexia nervosa or bulimia nervosa.

For some people, frequent dieting beginning in childhood or adolescence is a precursor to binge eating disorder. During periods when little food is eaten, they get very hungry and feel driven to eat in a compulsive, uncontrolled way. Many individuals with binge eating disorder (about 40% of whom are males) perceive themselves as hungry more often than normal. Almost half of those with severe binge eating disorder exhibit clinical depression symptoms and isolate themselves from others.

Stressful events and feelings of depression or anxiety can trigger binge eating. Giving themselves "permission" to eat a "forbidden" food also can precipitate a binge. Other triggers include loneliness, anxiety, self-pity, depression, anger, rage, alienation, and frustration.[49] In general, people engage in binge eating to induce a sense of well-being and perhaps even emotional numbness, usually in an attempt to avoid feeling and dealing with emotional pain and anxiety. For example, someone with a stressful or frustrating job might go home every night and eat until bedtime. Another person might eat normally most of the time but find comfort in consuming large quantities of food when an emotional setback occurs.

People with binge eating disorder exert some of the same characteristics as seen in people with bulimia nervosa. These characteristics include strong cravings, poor self-control, a diminished sensitivity to pleasure, and patterns of compulsive behavior.

### Treatment of Binge Eating Disorder

Overall, people who have binge eating disorder are usually unsuccessful in controlling it without professional help.[49] Depending on the specific symptoms, these individuals may need treatment as outlined earlier for anorexia nervosa or bulimia nervosa.

The focus of nutrition therapy for people with binge eating disorder mirrors that of controlling the binge eating associated with bulimia nervosa. Psychological therapy involves helping those with binge eating disorder identify personal emotional needs and express emotions. Because this problem is a common predisposing factor in binge eating, communication issues should be addressed during treatment. Those who engage in binge eating often must be helped to recognize their own buried emotions in anxiety-producing situations, and learning simple but appropriate phrases to say to oneself can help stop binge eating when the desire is strong. Self-help groups, such as Overeaters Anonymous, aim to help recovery from binge eating disorder. The treatment philosophy attempts to create an environment of encouragement and accountability to overcome this eating disorder. Antidepressants, as well as other medications, may be prescribed to help reduce binge eating in these individuals by decreasing depression.

## Other Specified Feeding and Eating Disorders (OSFED)

OSFED are subtypes of eating disorders—patients meet some, but not all, of the criteria for diagnosis with anorexia nervosa, bulimia nervosa, or binge eating disorder.[35] For example,

Binge eating episodes usually include foods that carry the social stigma of so-called junk foods—ice cream, cookies, sweets, potato chips, and similar snack foods.

Donna Day/Image Bank/Getty Images

Night eating syndrome is an eating disorder under study. In this disorder, people eat a lot in the late evening or eat food in order to fall asleep again once awakened in the night. This night eating can contribute to weight gain, so affected persons are urged to seek treatment.

Ryan McVay/Getty Images RF

## CASE STUDY FOLLOW-UP

There is a good chance that, if Christy keeps eating as she is, she will gain weight in her freshman year of college. Christy skips breakfast, eats a light lunch and dinner, and then becomes hungry late at night and cannot resist the high calorie, high fat pizza that is delivered to the dorm. Recent data show that women who skip breakfast are more likely to weigh more than those who eat breakfast.[27] This pattern of eating leads to weight gain. Christy needs to wake up and eat a simple breakfast, such as a bowl of cereal with fat-free milk and a banana. In addition, she needs to eat a balanced lunch and dinner and find ways to add exercise to her daily routine to avoid the Freshman 15.

an individual may have most of the characteristics of anorexia nervosa but have a normal weight. Or an individual may binge eat and purge, but less often than the criteria for bulimia nervosa. Some may engage only in purging or eat large amounts of food at night (night eating syndrome). The vast majority of people with eating disorders fall into this category.[50] A diagnosis of Unspecified Feeding or Eating Disorder (UFED) is given when medical professionals do not specify how a patient does not meet the criteria for anorexia nervosa, bulimia nervosa, or binge eating disorder.

### Other Related Conditions

Although not recognized as diagnosable eating or body image disorders, muscle dysmorphia, orthorexia, and pregorexia are appearing in research journal articles. Muscle dysmorphia, sometimes called bigorexia, is a condition characterized by an excessive concern that one has underdeveloped muscles.[51] Individuals with muscle dysmorphia, however, tend to have well-developed musculature. This condition appears to be more common among males, especially bodybuilders. Many individuals with muscle dysmorphia devote many hours each day to lifting weights and performing resistance exercises. Most sufferers continue to exercise even when injured and to the point that social relationships and performance at work and school are impaired. Some also use anabolic steroids or other muscle-enhancing drugs, take nutritional supplements, and follow specific dietary patterns they believe will enhance muscle development. Treatment from a sports medicine physician and a counselor trained to work with athletes may help individuals overcome this condition.

*Orthorexia* comes from the Greek word *orthos* (meaning "straight, proper") and *orexia* ("appetite"). Sometimes this condition is called the "health food eating disorder." Healthy eating is taken to such an extreme that eating the "right" foods dominates a person's life.[52] Individuals with orthorexia spend many hours each day searching for foods that are "pure" (e.g., free of herbicides, pesticides, and artificial ingredients or packaged in a specific manner, such as no plastic packaging). This excessive concern for "righteous eating" can progress to the point that it crowds out other activities, impairs relationships, and even becomes physically dangerous if the diet is severely restricted. Little is known about this condition or how to treat it.

*Pregorexia* is a term coined by the popular press to describe women who decrease calories and exercise excessively to control weight gain during pregnancy.[53] Whether this problem actually exists or not, eating a diet that is nutrient rich and sufficient in calories is essential for an optimal pregnancy outcome (see Chapter 16). Likely the best treatment for these individuals is similar to that for those with anorexia nervosa.

### Prevention of Eating Disorders

A key to developing and maintaining healthful eating behavior is to realize that some concern about diet, health, and weight is normal, as are variations in what we eat, how we feel, and even how much we weigh. For example, most people experience some minimal weight change (up to 2 to 3 lb) throughout the day and even more over the course of a week. A large weight fluctuation or an ongoing weight gain or weight loss is more likely to indicate that a problem is present. If you notice a large change in your eating habits, how you feel, or your body weight, it is a good idea to consult your physician. Treating physical and emotional problems early helps prevent eating disorders and promotes good health.

Not only is the treatment of eating disorders far more difficult than prevention, but these disorders also have devastating effects on the entire family. For this reason, parents, friends, and professionals working with children and teens must emphasize the importance of an overall healthful diet that focuses on moderation, as opposed to restriction and perfection. These caregivers also can help children and teens form positive habits and appropriate expectations, especially regarding body image.[54] The following are some ways that caregivers and health professionals can help growing children and adolescents avoid eating disorders.

• Discourage restrictive dieting, meal skipping, and fasting (except for religious reasons).
• Encourage children to eat only when hungry.

- Promote good nutrition and regular physical activity in school and at home.
- Promote regularly eating meals as a family unit.[55]
- Provide information about normal changes that occur during puberty.
- Correct misconceptions about nutrition, healthy body weight, and approaches to weight loss.
- Carefully phrase any weight-related recommendations and comments.
- Don't overemphasize numbers on a scale. Instead, primarily promote healthful eating irrespective of body weight.
- Increase self-acceptance and appreciation of the power and pleasure emerging from one's body.
- Encourage coaches to be sensitive to weight and body image issues among athletes.
- Emphasize that thinness is not necessarily associated with better athletic performance.
- Enhance tolerance for diversity in body weight and shape.
- Encourage normal expression of emotions.
- Build respectful environments and supportive relationships.
- Provide adolescents with an appropriate, but not unlimited, degree of independence, choice, responsibility, and self-accountability for their actions.

## Knowledge Check

1. What is the difference between eating disorders and disordered eating?
2. What are the characteristics of anorexia nervosa, bulimia nervosa, and binge eating disorder?
3. Why is a multidisciplinary health team recommended for dealing with eating disorders?
4. What type of advice should be given to growing children and adolescents to help them avoid eating disorders?

Disordered eating affects many college students. It is important to get help before disordered eating becomes an eating disorder. Counselors are aware of this and are available to help.

Ted Foxx/Alamy RF

# Take Action

## Assessing Risk of Developing an Eating Disorder

British investigators have developed a 5-question screening tool, called the SCOFF Questionnaire, for recognizing eating disorders:[56]

1. Do you make yourself **S**ick because you feel full?
2. Do you lose **C**ontrol over how much you eat?
3. Have you lost more than **O**ne stone (about 13 lb) recently?
4. Do you believe yourself to be **F**at when others say you are thin?
5. Does **F**ood dominate your life?

Two or more positive responses suggest an eating disorder.

1. After completing this questionnaire, do you feel that you might have an eating disorder or the potential to develop an eating disorder?
2. Do you think any of your friends might have an eating disorder?
3. What counseling and educational resources exist in your area or on your campus to help with a potential eating disorder?

4. If a friend has an eating disorder, what do you think is the best way to assist him or her in getting help?

Peter Cade/Photodisc/Getty Images

# Chapter Summary

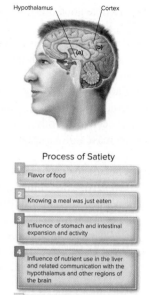

## 10.1 Energy balance considers energy intake and energy output.

Negative energy balance occurs when energy output surpasses energy intake, resulting in weight loss. Positive energy balance occurs when energy intake is greater than output, resulting in weight gain. Basal metabolism, the thermic effect of food, physical activity, and thermogenesis account for total energy use by the body. Basal metabolism, which represents the minimum amount of energy used to keep the resting, awake body alive, is primarily affected by lean body mass, body surface area, and thyroid hormone concentrations. Physical activity is energy expenditure above and beyond that expended for basal metabolism. The thermic effect of food is the increase in metabolism that facilitates the digestion, absorption, and processing of nutrients recently consumed. Thermogenesis is heat production caused by shivering when cold, fidgeting, and other stimuli. In a sedentary person, about 70 to 80% of energy use is accounted for by basal metabolism and the thermic effect of food.

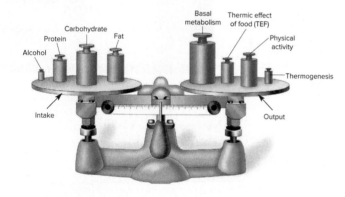

## 10.2 Direct calorimetry estimates energy expenditure by measuring the amount of body heat released by a person.

Indirect calorimetry, the most commonly used method to determine energy use by the body, involves collecting expired air from an individual during a specified amount of time. This method works because a predictable relationship exists between the body's use of energy and the amount of oxygen consumed and carbon dioxide produced. Estimated Energy Requirements (EERs), which are based on formulas developed by the Food and Nutrition Board, can be used to estimate energy needs.

Samuel Ashfield/SPL/Science Source

## 10.3 Groups of cells in the hypothalamus and other regions in the brain affect hunger,

the primarily internal desire to find and eat food. These cells monitor macronutrients and other substances in the blood and read low amounts as a signal to promote feeding. A variety of external (appetite-related) forces, such as food availability, affect satiety. Hunger cues combine with appetite cues to promote feeding. Numerous factors elicit satiety, such as flavor, smell, chewing, and the effects of digestion, absorption, and metabolism.

### Process of Satiety

1. Flavor of food
2. Knowing a meal was just eaten
3. Influence of stomach and intestinal expansion and activity
4. Influence of nutrient use in the liver and related communication with the hypothalamus and other regions of the brain
5. Conscious thinking takes place in the brain's cortex and can overcome hunger or satiety signals.

## 10.4 A person of healthy weight generally shows good health and performs daily activities

without weight-related problems. A body mass index (weight in kilograms/height$^2$ in meters) of 18.5 to <25 indicates healthy weight, although weight in excess of this value may not lead to ill health. A healthy weight is best determined in conjunction with a thorough health evaluation by a health-care provider. A body mass index of 25 to <30 represents overweight. Obesity is defined as a total body fat percentage over 25% (men) or 35% (women), or a body mass index of 30 or more. Fat distribution greatly determines health risks from obesity. Upper body fat storage, as measured by a waist circumference greater than 40 inches (102 cm) (men) or 35 inches (88 cm) (women), increases the risks of hypertension, cardiovascular disease, and type 2 diabetes more than does lower body fat storage.

## 10.5 Research suggests that genes account for up to 40 to 70% of weight differences between people.

The genes may be those that determine body type, metabolic rate, and the factors that affect hunger and satiety. Some individuals are thought to have a genetic predisposition to obesity because they inherit a thrifty metabolism. The set-point theory proposes that humans have a genetically predetermined body weight or body fat content, which the body closely regulates. Environmental factors have important effects on what we eat. These factors may define when eating is appropriate, what is preferable to eat, and how much food should be eaten. Even though our genetic backgrounds have a strong influence on body weight and composition, genes are not destiny—both nature and nurture are involved.

Lopolo/Shutterstock

# 10.6 A sound weight loss program emphasizes a wide variety of low energy density foods,

adapts to the dieter's habits, consists of readily obtainable foods, strives to change poor eating habits, stresses regular physical activity, and stipulates the participation of a physician if weight is to be lost rapidly or if the person is over the age of 40 (men) or 50 (women) years and plans to perform substantially greater physical activity than usual. A pound of adipose tissue contains about 3500 kcal. Thus, if energy output exceeds intake by about 500 kcal per day, a pound of adipose tissue can be lost per week. Physical activity as part of a weight loss program should be focused on duration, rather than intensity. Behavior modification is a vital part of a weight loss program because the dieter may have many habits that discourage weight maintenance.

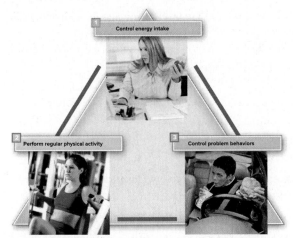

1: Digital Vision RF; 2: Ingram Publishing RF; 3: Ryan McVay/Getty Images RF

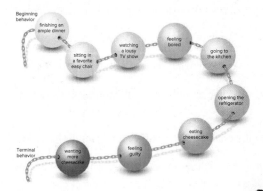

# 10.7 Many fad diets promise rapid weight loss; however, these diets are not designed for permanent weight loss.

Low carbohydrate diets work in the short run because they limit total food intake; however, long-term studies have shown that the weight generally returns in about a year. Weight loss drugs are reserved for those who have obesity or have weight-related problems, and they should be administered under close physician supervision.

Michael Simons/123RF

# 10.8 Anorexia nervosa usually starts with dieting in early puberty and proceeds to the near-total refusal to eat.

Early warning signs include intense concern about weight gain and dieting, as well as abnormal food habits. Eventually, anorexia nervosa can lead to numerous negative physical effects. The treatment of anorexia nervosa includes increasing food intake to support gradual weight gain. Psychological counseling attempts to help patients establish regular food habits and find means of coping with the life stresses that led to the disorder. Bulimia nervosa is characterized by secretive bingeing on large amounts of food within a short time span and then purging by vomiting or misusing laxatives, diuretics, or enemas. Alternately, fasting and excessive exercise may be used to offset calorie intake. Both men and women are at risk. Vomiting as a means of purging is especially destructive to the body. The treatment of bulimia nervosa includes psychological as well as nutritional counseling. Binge eating disorder is most common among people with a history of frequent, unsuccessful dieting. Binge eaters binge without purging. Emotional disturbances are often at the root of this disordered form of eating. Treatment addresses deeper emotional issues, discourages food deprivation and restrictive diets, and helps restore normal eating behaviors. The treatment of eating disorders may include certain medications. Other Specified Feeding and Eating Disorders (OSFED) are subtypes of eating disorders—patients meet some, but not all, of the criteria for a diagnosis of anorexia nervosa, bulimia nervosa, or binge eating disorder.

## Clinical Perspective

The treatments for severe obesity include surgery to reduce stomach volume to approximately 1 oz (30 ml) and very-low-calorie diets containing 400 to 800 kcal/day. Both of these measures should be reserved for people who have failed at more conservative approaches to weight loss. They also require close medical supervision. Underweight can be caused by a variety of factors, such as excessive physical activity and genetic background. Sometimes being underweight requires medical attention. A physician should be consulted 1st to rule out underlying disease. The underweight person may need to increase portion sizes and include energy-dense foods in the diet.

## Expert Perspective

Finding a weight loss diet that suits your lifestyle is key to weight control and eating a healthy diet. To find the healthy eating plan that is right for you, keep your eating pattern and lifestyle in mind.

## Study Questions

1. Positive energy balance occurs when energy output surpasses energy intake.

   **a.** true          **b.** false

2. What is the approximate basal metabolism of a 175-pound man?

   **a.** 3840 kcal/day          **c.** 1909 kcal/day
   **b.** 1227 kcal/day          **d.** 1745 kcal /day

3. All the following raise basal metabolism rate *except* _____.

   **a.** growing          **c.** fever
   **b.** muscle mass          **d.** starvation

4. Direct calorimetry estimates energy expenditure by collecting expired air from an individual during a specified amount of time.

   **a.** true          **b.** false

5. Which protein produced in fat tissue regulates body weight by signaling the brain about changes in body fat stores?

   **a.** ghrelin          **c.** thyroid hormone
   **b.** leptin          **d.** cholecystokinin

6. The most accurate method for diagnosing obesity is _____.

   **a.** underwater weighing
   **b.** dual energy X-ray absorptiometry (DEXA)
   **c.** skinfold thickness
   **d.** body mass index

7. Jane's waist measurement is 35 inches. Which of the following statements about Jane's body fat distribution is correct?

   **a.** Jane is at an increased risk of developing diabetes, hypertension, and heart disease.
   **b.** Jane's waist circumference indicates central body fat distribution.
   **c.** Jane has a "pear" shape.
   **d.** Both a and c are correct.
   **e.** Both a and b are correct.

8. The set-point theory proposes that humans have a genetically predetermined body weight or body fat content, which the body closely regulates.

   **a.** true          **b.** false

9. A sound weight loss program _____.

   **a.** includes a wide variety of low energy density foods
   **b.** stresses regular physical activity
   **c.** includes behavior modification to change problem behaviors
   **d.** all of the above

10. The danger of low carbohydrate diets is that they cause fast weight loss in the form of _____.

    **a.** fat tissue          **d.** both a and b
    **b.** water          **e.** both b and c
    **c.** muscles

11. Surgery to reduce stomach volume should be reserved for people who have failed at more conservative approaches to weight loss.

    **a.** true          **b.** false

12. Which eating disorder starts with dieting and proceeds to near-total refusal to eat?

    **a.** bulimia nervosa          **c.** anorexia nervosa
    **b.** binge eating          **d.** night eating syndrome

13. What is lanugo?

    **a.** downy, soft hair that develops on people with anorexia nervosa
    **b.** a sudden drop in blood pressure from excessive dieting
    **c.** a popular fad diet
    **d.** a code word for an eating disorder

14. Bulimia nervosa is characterized by all of the following *except* _____.

    **a.** refusal to eat          **c.** purging
    **b.** secretive bingeing          **d.** misuse of laxatives

15. Binge eating disorder is more widespread than anorexia nervosa or bulimia nervosa.

    **a.** true          **b.** false

16. Create a pie chart showing the relative proportion of the major components of energy expenditure.

17. How does direct calorimetry differ from indirect calorimetry?

18. Contrast the 2 drives that influence our desire to eat.

19. Make a chart that compares the methods for measuring body fat.

20. Why do both genetic and environmental factors affect body weight?

21. What are the characteristics of a sound weight loss diet?

22. Why is the claim for quick, effortless weight loss by any method always misleading?

23. What are the typical characteristics of individuals with anorexia nervosa and those with bulimia nervosa?

24. Describe your eating patterns and lifestyle. Then, identify the type of eating plan that fits your lifestyle.

25. What types of weight loss treatment are available for severe obesity?

Answer Key: 1-b; 2-c; 3-d; 4-b; 5-b; 6-b; 7-e; 8-a; 9-d; 10-e; 11-a; 12-c; 13-a; 14-a; 15-a; 16-refer to Section 10.1; 17-refer to Section 10.2; 18-refer to Section 10.3; 19-refer to Section 10.4; 20-refer to Section 10.5; 21-refer to Section 10.6; 22-refer to Section 10.7; 23-refer to Section 10.8; 24-refer to Expert Perspective; 25-refer to Clinical Perspective

# References

1. Noh J. The effects of circadian and sleep disruption on obesity. *J Obes Metab Syndr.* 2018;27:78.

2. Centers for Disease Control and Prevention. *U.S. obesity 1985–2018.* Atlanta: CDC; 2018.

3. Hales CM and others. Trends in obesity and severe obesity prevalence in US youth and adults by sex and age, 2007–2008 to 2015–2016. *JAMA.* 2018;319:1723.

4. Cheskin LH, Podda KH. Obesity: Management. In: Ross AC and others, eds. *Modern nutrition in health and disease.* 11th ed. Philadelphia: Lippincott Williams & Wilkins; 2012.

5. Polsky S and others. Obesity: Epidemiology, etiology, and prevention. In: Ross AC and others, eds. *Modern nutrition in health and disease.* 11th ed. Philadelphia: Lippincott Williams & Wilkins; 2012.

6. Pownall HJ and others. Changes in regional body composition over 8 years in a randomized lifestyle trial. *Obes.* 2016;9:1899.

7. Suter P and others. Effect of ethanol on energy expenditure. *Am J Physiol Regul Integr Comp Physiol.* 1994;266:4.

8. U.S. Department of Agriculture, Economic Research Service. Food availability. 2020; www.ers.usda.gov/data-products/food-availability-per-capita-data-system.

9. Bray G. Energy and fructose from beverages sweetened with sugar or high-fructose corn syrup pose a health risk for some people. *Adv Nutr.* 2013;1:220.

10. Campos VC, Tappy L. Physiological handling of dietary fructose containing sugars: Implications for health. *Int J Obes.* 2016;1:S6.

11. Tappy L, Le KA. Metabolic effects of fructose and the worldwide increase in obesity. *Physiol Rev.* 2010;90:23.

12. Hussain SS and others. Control of food intake and appetite. In: Ross AC and others, eds. *Modern nutrition in health and disease.* 11th ed. Philadelphia: Lippincott Williams & Wilkins; 2012.

13. Mollahosseini M and others. Effect of whey protein supplementation on long and short term appetite: A meta-analysis of randomized controlled trials. *Clin Nutr ESPEN.* 2017;20:34.

14. Flicker L and others. Body mass index and survival in men and women aged 70 to 75. *J Am Geriatr Soc.* 2010;58:234.

15. Wensveen FM and others. The "Big Bang" in obese fat: Events initiating obesity-induced adipose tissue inflammation. *Eur J Immunol.* 2015;9:244.

16. Heyward V, Wagner D. *Applied body composition assessment.* 2nd ed. Champaign, IL: Human Kinetics; 2004.

17. Lee SY, Gallagher D. Assessment methods in human body composition. *Curr Opin Clin Nutr Metab Care.* 2008;11:566.

18. Bajaj HS and others. Clinical utility of waist circumference in predicting all-cause mortality in preventive cardiology clinic population: A PReCIS database study. *Obes.* 2009;17:1615.

19. Powell K and others. The role of social networks in the development of overweight and obesity among adults: A scoping review. *BMC Public Health.* 2015;15:996.

20. Karkkainen U and others. Successful weight maintainers among young adults—A ten-year prospective population study. *Eat Behav.* 2018;29:91.

21. Teunissen-Beekman KFM and others. Effect of increased protein intake on renal acid load and renal hemodynamic responses. *Physiol Rep.* 2016;4:e12687.

22. Ornish D. Was Dr. Atkins right? *J Am Diet Assoc.* 2004;104:537.

23. Ello-Martin J and others. Dietary energy density in the treatment of obesity: A year-long trial comparing 2 weight-loss diets. *Am J Clin Nutr.* 2007;85:1465.

24. Thomas DM and others. Time to correctly predict the amount of weight loss with dieting. *J Acad Diet Nutr.* 2014;114:857.

25. Wilson TG. Behavioral treatment of obesity: Introduction. *Behav Res Ther.* 2010;May 23:705.

26. Raynor HA and others. Amount of food group variety consumed in the diet and long-term weight loss maintenance. *Obes Res.* 2005;13:883.

27. Astbury NM and others. Breakfast consumption affects appetite, energy intake, and the metabolic and endocrine responses to foods consumed later in the day in male habitual breakfast eaters. *J Nutr.* 2011;141:1381.

28. Sievert K and others. Effect of breakfast on weight and energy intake: Systematic review and meta-analysis of randomised controlled trials. *BMJ.* 2019;364:l42.

29. American Dietetic Association. Position of the American Dietetic Association: Food and nutrition misinformation. *J Am Diet Assoc.* 2006;106:601.

30. Brinkworth GD and others. Long-term effects of a very low carbohydrate weight loss diet compared with an isocaloric low fat diet after 12 mo. *Am J Clin Nutr.* 2009;90:23.

31. Pittler M, Ernst E. Dietary supplements for body-weight reduction: A systematic review. *Am J Clin Nutr.* 2004;79:529.

32. Moyers S. Medications as adjunct therapy for weight loss: Approved and off-label agents in use. *J Am Diet Assoc.* 2005;105:948.

33. Isidro ML, Cordido F. Approved and off-label uses of obesity medications, and potential new pharmacologic treatment options. *Pharmaceuticals.* 2010;3:125.

34. Kan C, Treasure J. Recent research and personlized treatment of anorexia nervosa. *Psychiar Clin North Am.* 2019;42:11.

35. American Psychiatric Association. *Diagnostic and statistical manual of mental disorders, fifth edition (DSM-5).* 5th ed. Washington, DC: APA; 2013.

36. Wonderlich SA and others. The validity and clinical utility of binge eating disorder. *Int J Eat Disord.* 2009;42:687.

37. Koletzko B and others. 3.22 Nutrition rehabilitation in eating disorders. *World Rev Nutr Diet.* 2015;113:259.

38. Courbasson C and others. Substance use disorders, anorexia, bulimia, and concurrent disorders. *Can J Public Health.* 2005;96:102.

39. Tchanturia K and others. Cognitive flexibility and clinical severity in eating disorders. *PLoS ONE.* 2011;6:e20462.

40. Riebl SK and others. The prevalence of subclinical eating disorders among male cyclists. *J Am Diet Assoc.* 2007;107:1214.

41. Le Grange D, Eisler I. Family interventions in adolescent anorexia nervosa. *Child Adol Psychiatr Clin N Am.* 2009;18:159.

42. Attia E. Anorexia nervosa: Current status and future directions. *Ann Rev Med.* 2010;61:425.

43. Walsh BT. Eating disorders. In: Fauci AS and others, eds. *Harrison's principles of internal medicine.* 20th ed. New York: McGraw-Hill; 2008.

44. Miller K and others. Medical findings in outpatients with anorexia nervosa. *Arch Intern Med.* 2018;165:561.

45. Van Wymelbeke V and others. Factors associated with the increase in resting energy expenditure during refeeding in malnourished anorexia nervosa patients. *Am J Clin Nutr.* 2004;80:1469.

46. Broussard B. Women's experiences of bulimia nervosa. *J Adv Nurs.* 2005;49:43.

47. Wilson GT, Sysko R. Frequency of binge eating episodes in bulimia nervosa and binge eating disorder: Diagnostic considerations. *Int J Eat Disord.* 2009;42:603.

48. Johannson AK. Eating disorders and oral health: A matched case-control study. *Eur J Oral Sci.* 2012;120:61.

49. Chiba FY and others. Peridontal condition, changes in salivary biochemical parameters, and oral health-related quality of life in patients with anorexia and bulimia nervosa. *J Peridontol.* 2019;90:1423.

50. Cossrow N and others. Estimating the prevalence of binge eating disorder in a community sample from the United States: Comparing DSM-IV-TR and DSM-5 criteria. *J Clin Psychiatry.* 2016;8e:968.

51. Corazza O and others. The emergence of exercise addiction, body dysmorphic disorders and other image-related psychopathological correlates in fitness settings: A cross sectional study. *PLoS ONE.* 2019;14:e0213060.

52. McComb SE, Mills JS. Orthorexia nervosa: A review of psychosocial risk factors. *Appetite.* 2019;140:50.

53. Mathieu J. What is pregorexia? *J Am Diet Assoc.* 2009;109:977.

54. Keel PK, Brown TA. Update on course and outcome in eating disorders. *Int J Eat Disord.* 2010;43:195.

55. Neumark-Sztainer D and others. Dieting and disordered eating behaviors from adolescence to young adulthood: Findings from a 10-year longitudinal study. *J Am Diet Assoc.* 2011;111:1004.

56. Kutz AM and others. Eating disorder screening: A systematic review and meta-analysis diagnostic test characteristics of SCOFF. *J Gen Intern Med.* 2019; 10.1007/s11606-019-05478-6.

**Design element:** Expert Perspective from the Field Magnifying glass: Aldo Murillo/Getty Images RF; Knowledge Check Puzzle piece: Image Source/Getty Images RF; Take Action Apple: lynx/iconotec.com/Glow Images; Global Perspective Globe: Corbis/VCG/Getty Images RF; Historical Perspective Beakers: Colin Anderson/Getty Images RF; Clinical Perspective (EKG): Tara McDermott/McGraw-Hill Education; Culinary Perspective Cutting board/knife: Ingram Publishing RF

The best exercise for you is one you want to continue.
Learn more at **www.mayoclinic.com/health/fitness/HQ00171.**
© Dudarev Mikhail/Shutterstock

# 11 Nutrition, Exercise, and Sports

**EXERCISE, LIKE HEALTHY EATING,** is essential to good health. However, many people get far less exercise than is needed for good health. In fact, the majority of health problems in North America are related to poor diet and insufficient physical activity. Only about half of adults report engaging in physical activity that lasts 30 minutes or longer on at least 5 days each week. Less than one-third of adults meet the guidelines for both aerobic and muscle-strengthening activity. Approximately 40% do not participate in any regular activity. And, when adults do start an exercise program, about half quit within 3 months.

The effects of diet and exercise on health status are closely related to each other. Recall that regular exercise helps chyme move through the intestinal tract, promotes calcium deposition in the bones, and strengthens the heart so that nutrients can be delivered to cells efficiently. Similarly, what you eat and drink affects exercise performance—whether you are a recreational athlete or an elite athlete or if you just want to maintain your health.

Athletes invest a lot of time and effort in training; their quest to find a competitive edge has spurred research studies designed to determine how diet affects exercise performance. Good eating habits can't substitute for physical training and genetic endowment, but healthy food and beverage choices are crucial for top-notch performance, contributing to endurance and helping speed the repair of injured tissues.[1] Unfortunately, there is much misinformation regarding the effect of diet and nutrients on exercise performance. A good working knowledge and understanding of sports nutrition can help individuals choose diets that let them perform as close to their potential as possible. In this chapter, you will discover how exercise benefits the entire body and how nutrition relates to fitness and sports performance.

*Figure 11-1* The benefits of regular, moderate physical activity and exercise.

Samuel Borges Photography/Shutterstock

Strengthens bones and joints

Reduces blood pressure

Improves blood glucose regulation

Increases cardiovascular function and improves blood lipid profile

Reduces stress and improves self-image

Aids in weight loss/weight control

Increases flexibility and balance

Increases muscle mass and strength

Improves immune function

Improves GI tract peristalsis

Reduces risk of colon cancer, prostate cancer, and likely breast cancer

Improves sleep (if activity is done in the morning or afternoon)

Slows aging process

## 11.1  Benefits of Fitness

The benefits of regular physical activity (and exercise) include enhanced heart function, improved balance, a reduced risk of falling, better sleep habits, healthier body composition (less body fat, more muscle mass), and reduced injury to muscles, tendons, and joints. Physical activity also can reduce stress and positively affect blood pressure, blood cholesterol levels, blood glucose regulation, and immune function. In addition, physical activity aids in weight control, both by raising Resting Energy Expenditure for a short period of time after exercise and by increasing overall energy expenditure (Fig. 11-1).[2,3,4,5,6,7]

Almost everyone can benefit from regular exercise. *Healthy People* objectives for U.S. adults include

▶ Recall how the terms *physical activity* and *exercise* are related. Exercise is physical activity done with the intent to gain health and fitness benefits, whereas physical activity is simply part of day-to-day activities.

- Reduce the proportion of adults engaging in no leisure-time physical activity.
- Increase the proportion of adults who meet  physical activity guidelines for aerobic and muscle-strengthening activity.

The Physical Activity Guidelines for Americans suggest that being active is 1 of the most important actions that people of all ages can do to improve their health. Evidence shows that health benefits start immediately after exercising, and even short bouts of exercise are beneficial. The key guidelines for adults include

- Adults should move more and sit less throughout the day and should know that some physical activity is better than none.
- For substantial health benefits, each week adults should engage in 150 to 300 minutes of moderate intensity or 75 to 150 minutes of vigorous aerobic activity, or an equivalent combination of  moderate- and vigorous-intensity aerobic physical activity. Preferably, aerobic activity  should be spread throughout the week.
- Additional health benefits can be gained by participating in physical activity beyond the equivalent of 300 minutes of moderate-intensity physical activity a week.

- Adults also should do muscle-strengthening activities of moderate or greater intensity that involve all major muscle groups on 2 or more days a week because these activities provide additional health benefits.

### Knowledge Check

1. Describe 3 benefits of physical activity.
2. When do the health benefits of exercise start?
3. How much resistance exercise should be performed during a week?

## 11.2 Characteristics of a Good Fitness Program

A good fitness program meets a person's needs—the ideal program for an individual may not be right for another. The first step in designing a fitness program is to define goals. Some may want a program that trains them for athletic competition; others may want to lose weight or just increase their stamina or improve their balance. Different goals mean different fitness programs. To reach goals, fitness program planning should consider the mode, duration, frequency, intensity, and progression of exercise, as well as consistency and variety. Also, a good fitness program must help individuals achieve and maintain fitness.[8]

### Mode

*Mode* refers to the type of exercise performed. The American College of Sports Medicine (ACSM) defines **aerobic exercise** as "any activity that uses large muscle groups, can be maintained continuously, and is rhythmic in nature." It is a mode of exercise that causes the heart and lungs to work harder than at rest. Aerobic exercise includes activities such as brisk walking, running, lap swimming, and cycling. **Resistance exercise,** or strength training, is defined as activities that use muscular strength to move a weight or work against a resistant load. A mode of exercise that increases the ability of a joint to move through its entire range of motion is called **flexibility exercise.**

### Duration

Duration is the amount of time spent in an exercise or physical activity session. Generally, exercise should last at least 30 minutes, not counting time for warm-up and cooldown. Ideally, exercise should be continuous (without stopping), but research has shown that 10-minute bouts of exercise done 3 times throughout the day can lower the risk of cardiovascular disease, cancer, and diabetes.

### Frequency

Frequency is the number of times an activity is performed weekly. For the best fitness level, daily aerobic activity is recommended. However, aerobic exercise performed 3 to 5 days a week appears to achieve cardiovascular fitness. To meet desired weight loss goals, the frequency of exercise may need to be 5 or 6 times per week. To achieve muscular fitness, 2 to 3 days of resistance training are needed weekly. Similarly, 2 to 3 days per week of flexibility exercises are recommended.

### Intensity

Intensity is the level of effort required, or how hard the exercise is to perform. Intensity can be described as low intensity (very mild increased heart rate exercise), moderate intensity (exercise that increases breathing, sweating, and heart rate but permits one to carry on a conversation), or vigorous intensity (exercise that significantly increases breathing, sweating, and heart rate, which makes it difficult to carry on a conversation).

Fran Polito/Getty Images

▶ To stick with an exercise program, experts recommend the following.

- Start slowly.
- Vary activities; make exercise fun.
- Include friends and others.
- Set specific attainable goals and monitor progress.
- Set aside a specific time each day for exercise; build it into daily routines, but make it convenient.
- Reward yourself for being successful in keeping up with your goals.
- Don't worry about occasional setbacks; focus on long-term benefits to your health.

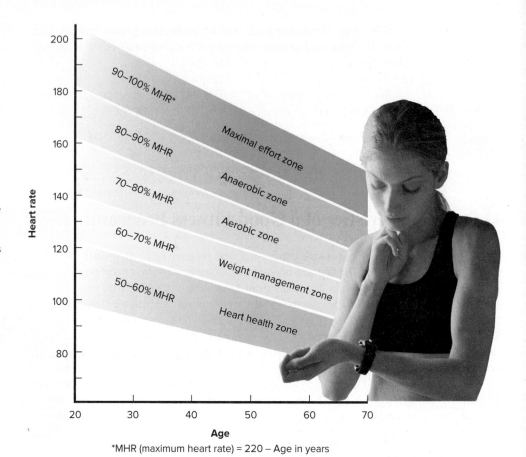

▶ The maximum heart rate for a 20-year-old person is 200 beats per minute.

220 − Age in years = Maximum heart rate

220 − 20 = 200

The target zone for a 20-year-old person is 120 to 180 heartbeats per minute.

Maximum heart rate × 0.6 = Low end of maximum heart rate target zone

200 × 0.6 = 120

Maximum heart rate × 0.9 = High end of maximum heart rate target zone

200 × 0.9 = 180

*MHR (maximum heart rate) = 220 − Age in years

*Figure 11-2* Heart rate training chart. This chart shows the number of heartbeats per minute that corresponds to various exercise intensities.

Duncan Smith/Getty Images

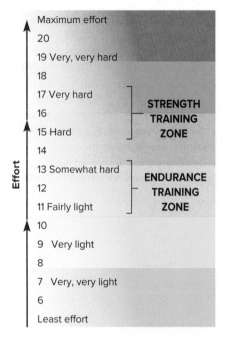

*Figure 11-3* Rating of Perceived Exertion (RPE) scale. This scale is used to estimate exercise intensity. A rating between 12 and 15 is recommended to achieve a high physical fitness level.

Traditionally, heart rate has been used to define intensity (Fig. 11-2). This popular and simple method uses a percentage of age-predicted maximum heart rate. To estimate maximum heart rate in beats per minute, subtract a person's age from 220. The range of heart rates between 60 and 90% of maximum is sometimes called the target zone. The lower end of this range is calculated by multiplying maximum heart rate by 0.6; the upper end is computed by multiplying maximum heart rate by 0.9. Medications, such as those for high blood pressure, may affect maximum heart rate. A physician can help those with health conditions personalize their target zone.

Another way to determine exercise intensity is the Borg Scale of Perceived Exertion, which often is called the **Rating of Perceived Exertion (RPE)** scale. The scale ranges from 6 to 20, with numbers corresponding to subjective feelings of exertion (Fig. 11-3). For example, the number 9 is rated as "very light" exertion, and the number 19 is considered close to maximal effort, or "very, very hard," as would occur in an all-out sprint. To achieve fitness, aim for an intensity of 12 to 15. At this level, you are working at a moderate intensity but still can talk to an exercise partner.

Individuals need to monitor how their bodies feel during exercise. For example, a jogger who wants to engage in moderate-intensity exercise should aim for an RPE scale rating of "somewhat hard" (12 to 14), which is of moderate intensity. If the jogger feels that muscle fatigue and breathing are "very light" (9), then he or she can increase the intensity. On the other hand, if the jogger feels the exertion is "extremely hard" (19), he or she should slow down to achieve the moderate-intensity range.

Because this rating is what the exerciser perceives, the actual exertion differs among those with different levels of fitness. That is, a very fit person will have a lower RPE scale rating when jogging than will a less fit person engaging in the same activity. Similarly, as a less fit person becomes more fit, his or her RPE will drop over time as the same exercise becomes easier. To keep increasing fitness levels, the exerciser should continue to exercise at a moderate intensity.

Energy needs dictate the amount of oxygen used by the body's cells (1.5 or 2.5 ATP molecules are produced from each molecule of oxygen). Thus, another way to determine exercise intensity is to measure oxygen consumption during exercise. A treadmill test is commonly used to determine the maximum amount of oxygen a person can consume in a unit of time (ml/min). In this test, oxygen consumption is measured as the treadmill speed and/or grade is gradually increased until the subject can no longer increase oxygen consumption as workload increases. The oxygen consumption measured at this point is **VO$_{2max}$**. Because of individual differences in VO$_{2max}$, it is generally best to express exercise intensity as a percentage of VO$_{2max}$.

Exercise intensity also is sometimes expressed in units called metabolic equivalents (METs). One MET is the expenditure of 1 kcal/kg/hour, or on average 3.5 ml O$_2$/kg/minute. This approximates Resting Energy Expenditure. A brisk walk represents about 4.5 METs of energy expenditure. Exercise prescriptions given to people recovering from a heart attack often are in MET units.[1]

## Progression

Progression describes how the duration, frequency, and intensity of exercise increase over time. The first 3 to 6 weeks of an exercise program make up the initiation phase. This phase corresponds to the time it takes for the body to adapt to the exercise program. The next 5 to 6 months of training are the improvement stage, during which intensity and duration increase to a point that no further physical gains are achieved. This plateau marks the beginning of the maintenance stage. At this stage, exercisers may want to evaluate their fitness goals. If they have achieved their goals, they can continue their fitness program in the same way to maintain their level of fitness. If they have not reached their goals, they can adjust their exercise duration, frequency, intensity, and/or mode.

## Consistency

The easiest way to have consistency in physical activity is to make it part of a daily routine, similar to other regular activities, such as eating. The best time to exercise is whenever it fits best into one's lifestyle—first thing in the morning, at lunchtime, before dinner, or later. Many people find that the best time to exercise is when they need an energy pick-me-up or a break from work or studying. When schedules are tight, exercise can be done in short segments, such as breaks between classes.

## Variety

Although some people enjoy doing the same activities day after day, boredom is a common reason many people abandon fitness programs. Just as a variety of foods helps ensure a nutritious diet, a varied fitness routine helps exercise different muscles for overall fitness, keeps exercising interesting and fun, and helps individuals stick with their fitness programs. Variety can be achieved in a number of ways, such as exercising indoors or outdoors or alternating aerobic exercise with resistance exercise. The Physical Activity Pyramid (Fig. 11-4) shows how to add variety to fitness programs and to increase fitness levels.

## Achievement and Maintenance of Fitness

Starting a new fitness program has 2 main activities. Begin by discussing the fitness program goals with a health-care provider. This is especially important for men over 40 and women over 50 years of age who have been inactive for many years or have existing health problems. Then, assess and record baseline fitness scores—these provide benchmarks against which to measure progress. Benchmarks should be based on fitness program goals. A benchmark for a person wanting to increase muscle strength might be the amount of weight that can be lifted with the arms or the number of push-ups that can

Taking your pulse (heart rate) helps you determine if your exercise output is in the target zone. To measure your pulse, count the number of heartbeats for 6 seconds and then multiply that number by 10 to determine heart rate per minute. There also are watches, apps for smartphones, and fitness trackers that contain heart rate monitors.

PeopleImages/Getty Images

### *Critical* Thinking

Dominica started working out about 10 weeks ago with the goal of losing weight. Initially, she lost about 8 pounds but now notices that the scale in the gym is not changing. In fact, this week she has gained a pound, yet her clothes are fitting better and she can now buy smaller-size pants. Why is Dominica maintaining weight but losing inches around her waist?

**VO$_{2max}$** Maximum volume of oxygen that can be consumed per unit of time.

*Figure 11-4* Each week, try to balance
your physical activity using this guide.

Man with Remote: C Squared Studios/Getty
Images; In-line Skater: ©Ingram Publishing
Fotosearch RF; Biker & Woman Stretching & Man
Using Weight Machine: ©Ingram Publishing/
Alamy RF; Ingram Publishing/Alamy Stock Photo;
Ingram Publishing/Alamy RF; Drinking Takeaway
Coffee: Dean Drobot/Shutterstock; Gardening:
©Comstock/Punchstock RF; Stroller: ©Ingram
Publishing/Fotosearch RF; Man Vacuum Cleaning:
Elnur/Shutterstock

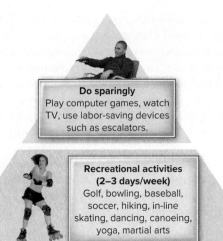

**Do sparingly**
Play computer games, watch
TV, use labor-saving devices
such as escalators.

**Recreational activities
(2–3 days/week)**
Golf, bowling, baseball,
soccer, hiking, in-line
skating, dancing, canoeing,
yoga, martial arts

**Aerobic exercise
(3–5 days/week
20–60 minutes)**
Running, cycling, cross-
country skiing, in-line
skating, stair stepping

**Flexibility exercise
(2–3 days/week)**
Static stretching of major
muscle groups. Hold each
pose 10–30 seconds.

**Strength exercise
(2–3 days/week
8–10 exercises
1 set of 8–12 reps)**
Bicep curl, tricep press,
squats, lunges, push-ups

**Physical activity
(most days of the week,
accumulate 30+ minutes)**
Take the stairs, garden, wash and wax
your car, rake leaves, mow the lawn,
walk to do your errands, walk the dog,
clean your house, play with your kids

be done without stopping. The amount of time it takes to walk a mile and the heart rate after the walk may be appropriate benchmarks for individuals wanting to build stamina. Determining how far a person can bend or stretch can be a benchmark for those wanting to improve flexibility.

Most new exercise programs should start with short intervals of exercise at the lower end of the maximum heart rate target zone and work up to a total of 30 minutes of activity incorporated into each day. If necessary, exercise can be broken into 3 sessions lasting 10 minutes each. When individuals are able to perform physical activity for 30 minutes daily, they can begin concentrating on the goals of their physical activity program. As fitness progresses, exercisers can work up to a higher level of their maximum heart rate target zone.

To prepare for and recover safely from an exercise session, warm-up and cooldown periods should be included. Warm-up includes 5 to 10 minutes of low-intensity exercises, such as walking, slow jogging, or stretching, and calisthenics that warm muscles and prepare them for exercise. The warm-up should be gradual and sufficient to increase muscle and body temperature, but it should not cause fatigue or deplete energy stores. The cooldown should include activities like those performed during the warm-up, but at a lower intensity to prevent blood from pooling in the extremities and a sudden drop in blood pressure. The cooldown helps the body recover slowly and return heart rate to normal. Generally, the cooldown aims for a heart rate of about 50 to 60% of a person's maximum heart rate (see Fig. 11-2).

A total fitness plan includes strength training, aerobic training, and flexibility exercise.

Weight Machines: ©Fuse/Getty Images RF; Jogging: Cultura/Alamy; Stretching: Sam Edwards/ age fotostock

## Knowledge Check

1. How do aerobic and resistance exercise differ?
2. What are the components of a comprehensive fitness program?
3. What is perceived exertion, and how is it measured?
4. What is your target heart rate zone?

# 11.3 Energy Sources for Muscle Use

Recall that cells cannot directly use the energy released from breaking down macronutrients. Rather, to use the chemical energy in foods, body cells must first convert the energy in foods to adenosine triphosphate (ATP).

## ATP: Immediately Usable Energy

When the body uses energy, 1 of the phosphates in ATP is cleaved off, releasing usable energy for cell functions, including muscle contractions. The product remaining is ADP and inorganic phosphate ($P_i$). A resting muscle cell contains a small amount of ATP—just enough to keep the muscle working maximally for about 1 to 2 seconds. To produce more ATP for muscle contraction over extended periods, the body uses phosphocreatine (PCr). In addition, dietary carbohydrates, fats, and proteins are used as energy sources (Fig. 11-5). The breakdown of all these compounds releases energy to make more ATP (Table 11-1).

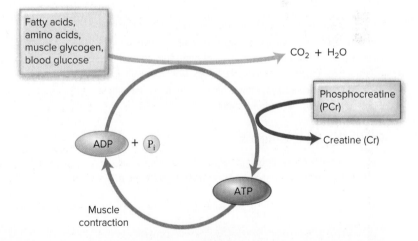

Fatty acids, amino acids, muscle glycogen, blood glucose

$CO_2 + H_2O$

Phosphocreatine (PCr)

Creatine (Cr)

ADP $+$ $P_i$

ATP

Muscle contraction

*Figure 11-5* Energy sources for muscular activity. Different fuels are used for ATP synthesis. As shown, ATP also can be synthesized rapidly using phosphocreatine.

## Phosphocreatine: Initial Resupply of Muscle ATP

**Phosphocreatine (PCr)** is a high-energy compound created from ATP and **creatine (Cr)** and is stored in small amounts in muscle cells. Creatine is an organic molecule in muscle cells that is synthesized from 3 amino acids: glycine, arginine, and methionine.[9] It also can be provided by supplements (see Table 11-11). As soon as ADP from the breakdown of ATP begins to accumulate in a contracting muscle, an enzyme is activated, transferring a high-energy $P_i$ from PCr to ADP—this transfer re-forms ATP (Fig. 11-6).

**phosphocreatine (PCr)** High-energy compound that can be used to re-form ATP from ADP.

**creatine (CR)** Organic molecule in muscle cells that serves as a part of the high-energy compound creatine phosphate, or phosphocreatine.

©Ingram Publishing RF

## How Physically Fit Are You?

The fitness assessments presented here are easy to do and require little equipment. Also included are charts to compare your results with those typical of your peers.

### Cardiovascular Fitness: 1-Mile Walk

Measure a mile on a running track or on a little-trafficked neighborhood street. With a stopwatch or watch with a second hand, walk the mile as fast as you can. Note the time it took.

### Strength: Push-Ups

Lie facedown on the floor. Get up on your toes and hands. Women can use the same position or can use knees, if necessary, instead of toes. Keep your back straight, with hands flat on the floor directly below your shoulders. Lower your body, bending your elbows, until your chin grazes the floor. Push back up until your arms are straight. Count the number of push-ups you can do (you can rest when in the up position).

### Strength: Curl-Ups

Lie on the floor on your back with your knees bent, feet flat. Rest your hands on your thighs. Now, squeeze your stomach muscles, push your back flat, and raise your upper body high enough for your hands to touch the tops of your knees. Don't pull with your neck or head, and keep your lower back on the floor. Count how many curl-ups you can do in 1 minute.

### Flexibility: Sit-and-Reach

Place a yardstick on the floor and apply a 2-foot piece of tape on the floor perpendicular to the yardstick, crossing at the 15-inch mark. Sit on the floor with your legs extended and the soles of your feet touching the tape at the 15-inch mark, the 0-inch mark facing you. Your feet should be about 12 inches apart. Put 1 hand on the other, exhale, and very slowly reach forward as far as you can along the yardstick, lowering your head between your arms. Don't bounce! Note the farthest inch mark you reach. Don't hurt yourself by reaching farther than your body wants to. Relax, and then repeat 2 more times.

Now, check your results. If you want to improve,

- Do aerobic exercise that makes you breathe hard for at least half an hour on almost or all days of the week.

- Lift weights that challenge you 2 or 3 times per week.

- Stretch after activity at least a couple of times per week.

- Walk more.

| Cardiovascular: 1-Mile Walk (Time, in Minutes) | | | | |
|---|---|---|---|---|
| | **Under 40 Years** | | **Over 40 Years** | |
| | **Men** | **Women** | **Men** | **Women** |
| Excellent | 13:00 or less | 13:30 or less | 14:00 or less | 14:30 or less |
| Good | 13:01–15:30 | 13:31–16:00 | 14:01–16:30 | 14:31–17:00 |
| Average | 15:31–18:00 | 16:01–18:30 | 16:31–19:00 | 17:01–19:30 |
| Below average | 18:01–19:30 | 18:31–20:00 | 19:01–21:30 | 19:31–22:00 |
| Poor | 19:31 or more | 20:01 or more | 21:31 or more | 22:01 or more |

Source: Cooper Institute.

## Strength: Push-Ups (Number Completed without Rest)

| | Ages 17–19 | | Ages 20–29 | | Ages 30–39 | | Ages 40–49 | | Ages 50–59 | | Ages 60–65 | |
|---|---|---|---|---|---|---|---|---|---|---|---|---|
| | Men | Women | Men | Women | Men | Women | Men | Women | Men | Women | Men | Women |
| Excellent | >56 | >35 | >47 | >36 | >41 | >37 | >34 | >31 | >31 | >25 | >30 | >23 |
| Good | 47–56 | 28–35 | 40–47 | 30–36 | 34–41 | 31–37 | 28–34 | 25–31 | 25–31 | 21–25 | 24–30 | 19–23 |
| Above average | 35–46 | 21–27 | 30–39 | 23–29 | 25–33 | 22–30 | 21–28 | 18–24 | 18–24 | 15–20 | 17–23 | 13–18 |
| Average | 19–34 | 11–20 | 17–29 | 12–22 | 13–24 | 10–21 | 11–20 | 8–17 | 9–17 | 7–14 | 6–16 | 5–12 |
| Below average | 11–18 | 6–10 | 10–16 | 7–11 | 8–12 | 5–9 | 6–10 | 4–7 | 5–8 | 3–6 | 3–5 | 2–4 |
| Poor | 4–10 | 2–5 | 4–9 | 2–6 | 2–7 | 1–4 | 1–5 | 1–3 | 1–4 | 1–2 | 1–2 | 1 |
| Very poor | <4 | <2 | <4 | <2 | <2 | 0 | 0 | 0 | 0 | 0 | 0 | 0 |

Source: adapted from Golding, Lawrence Arthur et al., *The Y's Way to Physical Fitness, 3rd ed.* Champaign, IL: Published for the YMCA of the USA by Human Kinetics Publishers, 1986.

## Strength: Curl-Ups (Number Completed in 60 Seconds)

| | Ages 18–25 | | Ages 26–35 | | Ages 36–45 | | Ages 46–55 | | Ages 56–65 | | Ages 65+ | |
|---|---|---|---|---|---|---|---|---|---|---|---|---|
| | Men | Women | Men | Women | Men | Women | Men | Women | Men | Women | Men | Women |
| Excellent | >49 | >43 | >45 | >39 | >41 | >33 | >35 | >27 | >31 | >24 | >28 | >23 |
| Good | 44–49 | 37–43 | 40–45 | 33–39 | 35–41 | 27–33 | 29–35 | 22–27 | 25–31 | 18–24 | 22–28 | 17–23 |
| Above average | 39–43 | 33–36 | 35–39 | 29–32 | 30–34 | 23–26 | 25–28 | 18–21 | 21–24 | 13–17 | 19–21 | 14–16 |
| Average | 35–38 | 29–32 | 31–34 | 25–28 | 27–29 | 19–22 | 22–24 | 14–17 | 17–20 | 10–12 | 15–18 | 11–13 |
| Below average | 31–34 | 25–28 | 29–30 | 21–24 | 23–26 | 15–18 | 18–21 | 10–13 | 13–16 | 7–9 | 11–14 | 5–10 |
| Poor | 25–30 | 18–24 | 22–28 | 13–20 | 17–22 | 7–14 | 13–17 | 5–9 | 9–12 | 3–6 | 7–10 | 2–4 |
| Very poor | <25 | <18 | <22 | <13 | <17 | <7 | <13 | <5 | <9 | <3 | <7 | <2 |

Source: adapted from Golding, Lawrence Arthur et al., *The Y's Way to Physical Fitness, 3rd ed.* Champaign, IL: Published for the YMCA of the USA by Human Kinetics Publishers, 1986.

## Flexibility: Sit-and-Reach (in Inches)

| | Men | Women |
|---|---|---|
| Super | >10.5 | >11.5 |
| Excellent | 6.5–10.5 | 8–11.5 |
| Good | 2.5–6.0 | 4.5–7.5 |
| Average | 0–2.0 | 0.5–4.0 |
| Fair | −3–−0.5 | −2.5–0 |
| Poor | −7.5–−3.5 | −6.0–−3.0 |
| Very poor | <−7.5 | <−6.0 |

Source: topendsports.com.

These charts are typical of those used by health and fitness experts. For a more thorough assessment of fitness or for development of an exercise plan appropriate for your fitness level, consult a certified personal trainer, an exercise physiologist, or another fitness professional.

**Table 11-1  Energy Stored in the Human Body**

| Energy Source* | Major Storage | When Used | Activity |
|---|---|---|---|
| ATP | All tissues | All the time | Sprinting (0–3 sec) |
| Phosphocreatine (PCr) | All tissues | Short bursts | Shot put, high jump, bench press |
| Carbohydrate (anaerobic) | Muscles | High-intensity exercise lasting 30 seconds to 2 minutes | 200-meter sprint |
| Carbohydrate (aerobic) | Muscles and liver | Exercise lasting 2 minutes to 3 hours or more | Jogging, soccer, basketball, swimming, gardening, car washing |
| Fat (aerobic) | Muscles and fat cells | Exercise lasting more than a few minutes; greater amounts are used at lower exercise intensities | Long-distance running, marathons, ultra endurance events, cycling, day-long hikes |

*All energy sources operate at the same time; however, the predominant source depends on the intensity and duration of the exercise.

If no other system for resupplying ATP were available, PCr could probably maintain maximal muscle contractions for about 10 seconds.[1] However, because the energy released from the metabolism of glucose and fatty acids also begins to contribute ATP, and thus spares some PCr use, PCr can function as the major source of energy for events lasting up to about 1 minute.

The main advantage of PCr is that it can be activated instantly and can replenish ATP quickly enough to meet the energy demands of the fastest and most powerful sports events, such as jumping, lifting, throwing, or sprinting. The disadvantage of PCr is that too little is made and stored in muscles to sustain a high rate of ATP resupply for more than a few minutes.

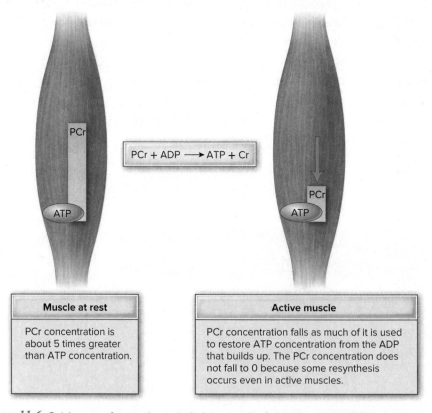

$$PCr + ADP \longrightarrow ATP + Cr$$

| **Muscle at rest** | **Active muscle** |
|---|---|
| PCr concentration is about 5 times greater than ATP concentration. | PCr concentration falls as much of it is used to restore ATP concentration from the ADP that builds up. The PCr concentration does not fall to 0 because some resynthesis occurs even in active muscles. |

*Figure 11-6* Quick energy for muscle use includes a supply of phosphocreatine (PCr). Phosphocreatine can rapidly replenish ATP stores as activity begins, but in less than 60 seconds it can become nearly depleted in maximally contracting human forearm muscles. It takes 4 minutes of rest to replenish half the PCr and 7 minutes to replenish 95% of the PCr. Similarly, it takes about 7 minutes of rest to replenish 95% of the PCr depleted with repeated knee extensions against resistance.

## Carbohydrate: Major Fuel for Short-Term, High-Intensity, and Medium-Term Exercise

As you know, glucose breaks down during glycolysis, producing a 3-carbon compound called pyruvate. Glycolysis does not require oxygen, but it yields only a small amount of ATP. If oxygen is present, pyruvate is metabolized further, yielding much more ATP.

### Anaerobic Pathway

When the oxygen supply in muscle is limited (anaerobic state) or when the physical activity is intense (e.g., running 200 meters or swimming 100 meters), pyruvate from glycolysis accumulates in the muscle and is converted to lactate (Fig. 11-7). Because the breakdown of 1 glucose molecule to 2 pyruvates yields 2 ATP, glycolysis can resupply some ATP depleted in muscle activity.[9] Carbohydrate is the only fuel that can be used for this process.

Glycolysis provides most of the energy for sports events in which energy production is near maximal for 30 to 120 seconds. The advantage of the anaerobic pathway is that, other than PCr breakdown, it is the fastest way to resupply ATP in muscle.[1] The anaerobic pathway has 2 major disadvantages.

- It cannot sustain ATP production for long.
- Only about 5% of the total ATP production from muscle glycogen can be released.

The rapid accumulation of lactate from anaerobic glycolysis may be associated with the onset of fatigue. A hypothesis suggests that anaerobic glycolysis releases hydrogen ions that increase the acidity within the muscle cell. This acidity disturbs the normal cell environment and activity of key enzymes in glycolysis, which causes anaerobic ATP production to slow and fatigue to set in. The acidity also leads to a net potassium loss from muscle cells, providing another cause of fatigue.[9] By trial and error, we learn an exercise pace that controls muscle lactate concentrations from anaerobic glycolysis.

During anaerobic glycolysis, the lactate that is produced accumulates in the muscles and is eventually released into the bloodstream. The heart can use lactate directly for its energy needs, as can less active muscle cells situated near active ones. The liver (and to some extent the kidneys) takes up some of the lactate from the blood and resynthesizes it into glucose, using an energy-requiring process. This glucose then can reenter the bloodstream and be used by cells for energy.

### Aerobic Pathway

If plenty of oxygen is available in muscle tissue (aerobic state) and physical activity is moderate to low intensity (e.g., jogging or distance swimming), most of the pyruvate produced by glycolysis in the cell cytoplasm is shuttled to the mitochondria and metabolized into carbon dioxide and water by a series of oxygen-requiring reactions (see Chapter 9). About 95% of the ATP produced from the complete metabolism of glucose is formed aerobically in the mitochondria (Fig. 11-8).

Bursts of muscle activity use a variety of energy sources, including PCr and ATP.

George Doyle/Stockbyte/Getty Images

▶ For many years, both scientists and athletes believed that the accumulation of lactate in exercising muscles caused muscle fatigue and soreness. Today, there is little evidence to show that lactate alone causes fatigue and soreness. Depletion of muscle glycogen and blood glucose along with the accumulation of other metabolic by-products in the muscle can contribute to muscle fatigue and soreness.

▶ When an acid loses a hydrogen ion, as typically happens at the pH level of the body, the name of that acid is given the suffix *-ate*. Thus, pyruvic acid is called pyruvate and lactic acid is called lactate when in the context of body metabolism.

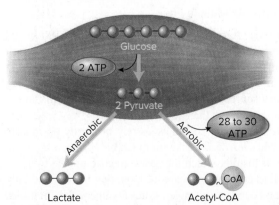

*Figure 11-7*  Both anaerobic and aerobic pathways can supply ATP. However, the aerobic pathway can supply more ATP but does it at a slower rate, whereas the anaerobic pathway supplies less ATP at a more rapid rate.

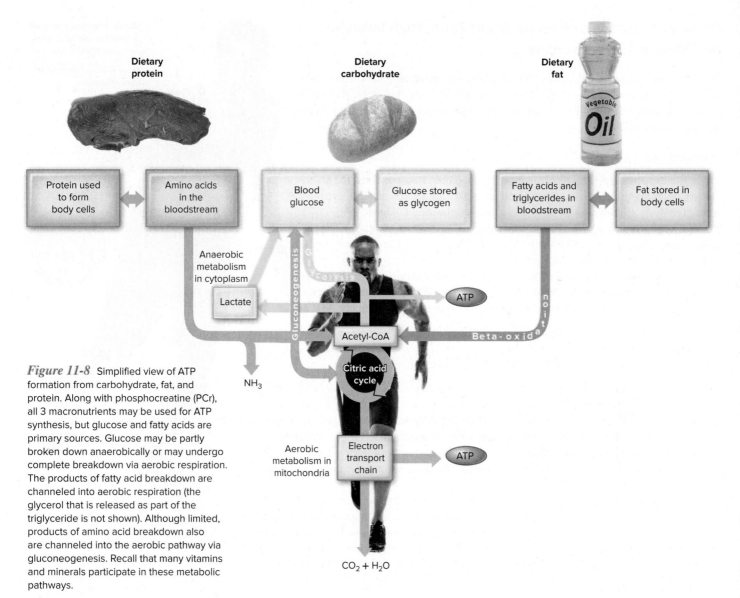

*Figure 11-8*  Simplified view of ATP formation from carbohydrate, fat, and protein. Along with phosphocreatine (PCr), all 3 macronutrients may be used for ATP synthesis, but glucose and fatty acids are primary sources. Glucose may be partly broken down anaerobically or may undergo complete breakdown via aerobic respiration. The products of fatty acid breakdown are channeled into aerobic respiration (the glycerol that is released as part of the triglyceride is not shown). Although limited, products of amino acid breakdown also are channeled into the aerobic pathway via gluconeogenesis. Recall that many vitamins and minerals participate in these metabolic pathways.

Protein: ©Stockbyte/Getty Images RF;
Carbohydrate: ©Ingram Publishing/Alamy;
Fat: McGraw-Hill Education/Jacques Cornell, photographer; Man Jogging: 4x6/Getty Images

The aerobic pathway supplies ATP more slowly than the anaerobic pathway, but it releases much more energy. Also, ATP production via the aerobic pathway can be sustained for hours. As a result, this pathway of glucose metabolism makes an important energy contribution to physical activity lasting from about 2 minutes to 3 hours or more (Fig. 11-9).[1]

### Muscle Glycogen versus Blood Glucose as Muscle Fuel

Recall that glycogen is the temporary storage form of glucose in the liver (about 100 g) and muscles (about 400 g in sedentary people). Glycogen is broken down to glucose, which can be metabolized by both the anaerobic and aerobic pathways. Liver glycogen is used to maintain blood glucose levels, whereas muscle glycogen supplies the glucose to the working muscle. Glycogen is, in fact, the primary source of glucose for ATP production in muscle cells during fairly intense activities that last for less than about 2 hours.

In short events (e.g., less than 30 minutes or so), muscles rely primarily on muscle glycogen stores for carbohydrate fuel. Muscles do not take up much blood glucose during short-term exercise because the action of insulin, which increases glucose uptake by muscles, is blunted by other hormones, such as epinephrine and glucagon, that increase initially during exercise.[1] As exercise time increases, muscle glycogen stores decline and the muscles

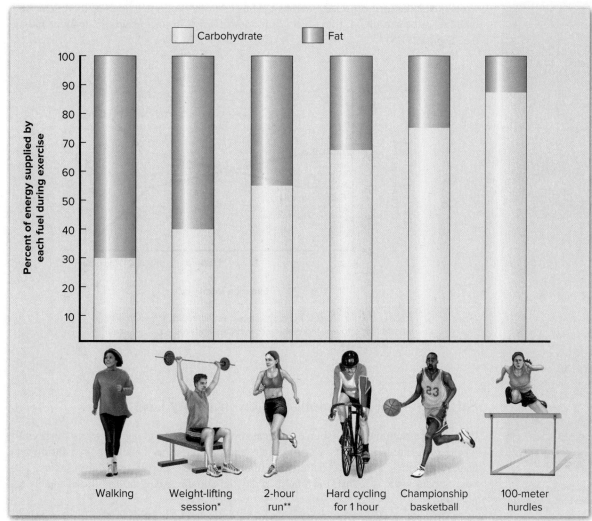

*Figure 11-9* Rough estimates of carbohydrate and fat use during various forms of exercise.

*With regard to weight lifting, carbohydrate use can be somewhat greater and fat use somewhat less if the session is intense and fast-paced (e.g., circuit training). Fat use generally is higher because much of the time spent weight lifting is for rest periods.
**With regard to endurance running, the balance of fat and carbohydrate used will vary somewhat depending on whether the athlete is consuming carbohydrate during the run. The values shown are for a runner consuming carbohydrate during the run; more fat and less carbohydrate would be used if carbohydrates were not consumed.

begin to take up blood glucose to use as an energy source. The depletion of glycogen in the muscles contributes to fatigue, whereas the depletion of glycogen in the liver leads to a fall in blood glucose.[9]

Once glycogen stores are exhausted, a person can continue working at only about 50% of maximal capacity. Athletes call this point of glycogen depletion "hitting the wall," because further exertion is hampered. Thus, for exercise that requires 70% or more of maximal effort for more than an hour or so, athletes (e.g., long-distance runners or cyclists) should consider increasing the amount of carbohydrate stored in their muscles. Diets high in carbohydrate can be used to increase muscle glycogen stores—up to double the typical amounts—in advance of competition, thereby delaying the onset of fatigue and improving endurance. This process is called carbohydrate or glycogen loading (see Section 11.5).

The maintenance of blood glucose becomes an increasingly important consideration as exercise duration increases beyond about 20 to 30 minutes. By maintaining blood glucose levels, the body saves muscle glycogen for muscle use during sudden bursts of effort, such as a sprint to the finish in a marathon race. Without the maintenance of blood glucose, irritability, sweating, anxiety, weakness, headache, and confusion may occur (cyclists call this "bonking"). A carbohydrate intake of 0.7 g/kg/hour (about 30–60 g/hour) during

The energy to perform comes from carbohydrate, fat, and protein. The relative mix of fuels depends on the pace and duration of exercise.

Sean Sullivan/Getty Images

strenuous endurance exercise, such as cycling, that lasts about 1 hour or more, can help maintain adequate blood glucose concentrations, which in turn results in delay of fatigue (see Section 11.5).[10]

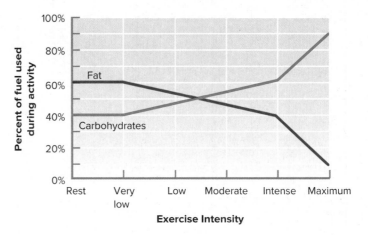

**Figure 11-10**   As the intensity of exercise increases, the exercising muscles tend to rely more on carbohydrates and less on fat. At lower intensities of exercise, fat is the predominant fuel.

## Fat: Main Fuel for Prolonged, Low-Intensity Exercise

Fat is the predominant fuel source when at rest and during prolonged exercise, especially when exercise remains at a low or moderate (aerobic) rate (Fig. 11-10). In fact, during very lengthy activities, such as a triathlon, an ultra-marathon, occupations requiring manual labor, or even work at a desk for 8 hours a day, fat supplies about 50 to 90% of the energy required.[1]

The rate at which muscles use fatty acids is affected by training level. The more trained a muscle, the greater its ability to use fat as a fuel. Training increases the size and number of mitochondria and the levels of enzymes involved with the aerobic synthesis of ATP. Training also increases muscle myoglobin, which enhances oxygen availability in muscles and is needed to metabolize fat (Table 11-2). Overall, training allows an athlete to use fat for fuel more readily, thereby conserving glycogen for when it is really needed—such as for a burst of speed at the end of a race.

**Table 11-2  Adaptations to Endurance Exercise Training in Skeletal Muscle**

| Change | Advantage |
|---|---|
| Increased ability of muscle to store glycogen (high carbohydrate diet increases this even further) | More glycogen fuel available for the final minutes of an event |
| Increased triglyceride storage in muscle | Conserves glycogen by allowing for increased fat use |
| Increased mitochondrial size and number | Conserves glycogen by allowing for increased fat use (even at high exercise outputs) |
| Increased myoglobin content | Increased oxygen delivery to muscles and increased ability to use fat for fuel |
| Increased cardiac output | Increased blood flow to promote adequate delivery of oxygen and nutrients to muscles |

As you know, most of the energy stored in the body is in fat, stored as triglycerides. Most of this energy resides in adipose tissue, although some is stored in the muscle itself. That stored in muscles is especially used as activity increases from a low to a moderate pace.

The advantage of fat over other energy sources is that it provides more than twice as much energy (9 kcal/g) and thus can provide more ATP. In addition, there is plenty of fat stored in the body, compared with the very limited carbohydrate stores. However, carbohydrate metabolism is more efficient than fat because it produces more ATP per unit of oxygen and it is the only fuel source that can support intense (anaerobic) activity. Fat utilization simply cannot occur fast enough to meet the ATP demands of short-duration, high-intensity physical activity—if fat were the only available fuel, we would be unable to carry out physical activity more intense than a fast walk or jog.[1]

## Protein: A Minor Fuel Source during Exercise

Most of the energy supplied from protein comes from metabolism of the branched-chain amino acids—leucine, isoleucine, and valine. These amino acids can be used to make glucose, or they can enter the citric acid cycle as precursors to glucose and provide energy during exercise.

Although amino acids derived from protein can fuel muscles, their contribution is relatively small compared with that of carbohydrate and fat. As a rough guide, only about 5% of the body's general energy needs, as well as the typical energy needs of exercising muscles, is supplied by amino acid metabolism.[9] However, proteins can contribute to energy needs in endurance exercise, perhaps as much as 15%, especially as glycogen stores in the muscle are exhausted.[9] Endurance exercise is when protein is most likely to make its most significant contribution as a fuel source—but, even then, it provides limited energy, with estimates ranging from 3 to 15%.[9,11] In contrast, protein is used least in resistance exercise (e.g., weight lifting).

Despite the fact that the primary muscle fuels for weight lifting are phosphocreatine (PCr) and carbohydrate, high protein supplements are marketed to weight lifters and bodybuilders and sold in nearly every health food and fitness store. Consuming more protein than the body needs or can use will not lead to greater muscle mass. Eating high carbohydrate, moderate protein foods immediately after a weight training workout can enhance the anabolic effect of the activity. This most likely increases blood concentrations of insulin and growth hormone and contributes to protein synthesis.[1] Remember, it is impossible to increase muscle mass simply by eating protein; putting physical strain on muscle through strength training or other physical activity is needed, as is adequate protein intake to support growth and recovery.

**ATP Yield from Aerobic Fatty Acid Metabolism and Aerobic Glucose Metabolism**

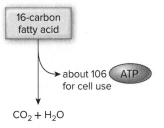

16-carbon fatty acid → about 106 ATP for cell use → $CO_2 + H_2O$

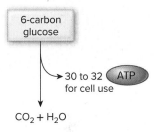

6-carbon glucose → 30 to 32 ATP for cell use → $CO_2 + H_2O$

## CASE STUDY

Ryan McVay/ Photodisc/Getty Images

Jake, a college junior, is 6 feet tall and weighs 175 pounds. He has been lifting weights since he was a college freshman. Although he has gotten significantly stronger over the last 2 years, several weeks ago he decided he wanted to have more muscle definition. He read (mainly on the Internet) a lot about nutrition and resistance training, especially about the role of protein and muscle growth, and decided to take a protein supplement to get bigger muscles and more definition. He has been taking the supplement for about 3 weeks. It consists of a whey protein powder, which he mixes with either water or milk. It contains about 60 grams of protein per serving, and he has 2 protein drinks per day.

His breakfasts consist of a protein shake; for lunch, he has a sandwich with extra meat and a small salad with fat-free dressing. He consumes another shake around 4 p.m. and isn't hungry again until about 7 P.M. Yesterday for dinner, he ate 2 chicken breasts with french fries, 4 hard-cooked eggs, and an iced tea.

Unfortunately, this past week he noticed that during his lifting he was tired and could not lift as much weight as the week before. Today while lifting, he started to feel fatigued 20 minutes into the training session. He is not sure why he can't finish his workout and thinks maybe he should eat more protein.

What role does protein have in resistance exercise? What is causing him to be so fatigued that he cannot finish his workout? Should he consume more protein?

## Table 11-3  Fuel Use* Estimate Based on Percent VO$_{2max}$

| VO$_{2max}$ | | Muscle Glycogen | Muscle Triglyceride | Blood Glucose | Free Fatty Acids in the Bloodstream |
|---|---|---|---|---|---|
| Low intensity (e.g., fast walk)—30 to 50% of VO$_{2max}$ | | 5% | 20% | 5% | 70% |
| Moderate intensity (e.g., fast jog)—50 to 65% of VO$_{2max}$ | | 30% | 30% | 10% | 30% |
| High intensity (e.g., 3-hour marathon pace)—70 to 80% of VO$_{2max}$ | | 55% | 15% | 15% | 15% |
| Very high intensity (e.g., sprints)—85 to 150% of VO$_{2max}$ | | 70% | 10% | 10% | 10% |

*The total amount of energy is approximate and may vary considerably between individuals.

Golfer: ©Ingram Publishing/Alamy RF; Soccer Player: Thomas Barwick /Getty Images; Male Runner: Squaredpixels/Getty Images; Female Runner: 4x6/Getty Images

## Fuel Use and VO$_{2max}$

Wrestlers who restrict calories to lose weight to qualify for a lower weight classification may need protein intakes as high as 1.8 to 2.0 g/kg to prevent loss of lean body mass.[12]

As you can see in Table 11-3, fuel sources for muscle cells can be estimated based on percent of VO$_{2max}$. For example, fat use drops as exercise intensity increases. Carbohydrate use then becomes more important for meeting energy needs. In very high-intensity activities, the ATP equivalent to the "extra" 50% above 100% of VO$_{2max}$ is produced anaerobically from PCr and glycolysis.

### Knowledge Check

1. What is the main form of energy that cells use?
2. What fuels anaerobic exercise?
3. What fuels aerobic exercise?
4. Why is creatine so important for fueling high-intensity, short-duration exercise?
5. How does fitness level affect the fuels burned for exercise?
6. When is protein used as a fuel source during exercise?

# 11.4 The Body's Response to Physical Activity

Physical activity has many effects on the body. The most pronounced effects are typically seen in the muscular, circulatory, and skeletal systems.

## Specialized Functions of Skeletal Muscle Fiber Types

The body contains 3 major types of muscle tissue: skeletal muscle (the type involved in locomotion); smooth muscle (the type found in internal organs, except the heart); and cardiac (heart) muscle. Skeletal muscle is composed of 3 main types of **muscle fibers,** which have distinct characteristics (Table 11-4).[9]

- *Type I (slow-twitch—oxidative).* These muscle fibers contract slowly and have a high capacity for oxidative metabolism. They also are called red fibers because of their high myoglobin content. Type I fibers are fueled by the aerobic respiration of fat.
- *Type IIA (fast-twitch—oxidative, glycolytic).* These muscle fibers have moderate oxidative capacity and are fueled by glycolysis using glucose (anaerobic) plus the aerobic respiration of both fat and glucose.
- *Type IIX (fast-twitch—glycolytic).* These muscle fibers have less oxidative capacity than other muscle fibers. They also are called white fibers (in rodents, type IIX fibers are called type IIB) because they have less mitochondria and myoglobin than other fibers. Type IIX fibers are fueled by glycolysis using glucose (anaerobic).

Prolonged, low-intensity exercise, such as walking, mainly uses type I muscle fibers, so the predominant fuel is fat. As exercise intensity increases, type IIA and type IIX fibers are gradually recruited; in turn, the contribution of glucose as a fuel increases. Type IIA and type IIX fibers also are important for rapid movements, such as a jump shot in basketball.

The relative proportions of the 3 fiber types throughout the muscles of the body vary from person to person and are constant throughout each person's life. However, some shifting in proportions of fiber types can occur with training and aging. The individual differences in fiber-type distribution are partially responsible for producing elite marathon runners who could never compete at the same level as sprinters, or elite gymnasts who could never be competitive as long-distance swimmers. Although the proportion of muscle fiber types is largely determined by genetics, appropriate training can develop muscles within limits. For example, aerobic training enhances the capacity of type IIA muscle fibers to produce ATP and may bring about a relative change in size. Overall, great athletes are born, but their genetic potential must be nurtured by training.[9]

## Adaptation of Muscles and Body Physiology to Exercise

With training, muscle strength becomes matched to the muscles' work demands. Muscles enlarge after being made to work repeatedly, a response called **hypertrophy.** Certain cells in the muscles gain bulk and improve their ability to work. Conversely, after several days without activity, muscles diminish in size and lose strength, a response called **atrophy.** Both

### Relative Distribution of Muscle Fiber Types

| Activity Level | Type I | Type IIA + Type IIX |
|---|---|---|
| Nonathlete | 45–50% | 50–55% |
| Sprinter | 20–35% | 65–85% |
| Marathoner | 80% | 20% |

The quick, powerful movements of a gymnast rely primarily on type IIA and type IIX muscle fibers. What sorts of physiological changes would you expect to occur in a gymnast who has trained diligently for many years?

View Stock/Getty Images

**muscle fiber** Essentially, a single muscle cell; an elongated cell, with contractile properties, that forms the muscles in the body.

**Table 11-4 Muscle Fiber Summary**

| Muscle Fiber | Description | Structure | Primary Fuel Source | Activities When Used |
|---|---|---|---|---|
| Type I | Slow-twitch; high oxidative metabolism capacity | High density of capillaries, mitochondria, myoglobin | Aerobic respiration of fat | Aerobic activity, such as endurance exercise |
| Type IIA | Fast-twitch; moderate oxidative metabolism capacity | Rich in capillaries and mitochondria | Anaerobic glycolysis; aerobic respiration of fat and glucose | Aerobic/anaerobic activities, such as middle-distance running, swimming |
| Type IIX | Fast-twitch; lower oxidative metabolism capacity | Less dense in mitochondria and myoglobin | Anaerobic glycolysis | Anaerobic activity, such as sprinting |

| Person | Typical VO$_{2max}$ (ml O$_2$/kg/min) |
|---|---|
| Sedentary elderly adult | <20 |
| Typical middle-aged adult | 35–45 |
| Elite athlete | 65–75 |

hypertrophy and atrophy are forms of adaptation to the workload applied. Thus, many marathon runners have well-developed leg muscles but little arm or chest muscle development.

Repeated aerobic exercise produces beneficial changes in the circulatory system. Because the body needs more oxygen during exercise, it responds to training by producing more red blood cells and expanding total blood volume. Training also leads to an increase in the number of capillaries in muscle tissue; as a result, oxygen can be delivered more easily to muscle cells. Finally, training causes the heart, a muscle itself, to strengthen. Then, each contraction empties the heart's chamber more efficiently, so more blood is pumped with each beat. As exercise increases the heart's efficiency, its rate of beating at rest and during submaximal exercise decreases.[1]

The more physically fit a person is, the more work the muscles and body can do and the more oxygen the person can consume. Typical VO$_{2max}$ values range from 20 to 65 ml O$_2$/kg/min, depending on age, biological sex, and fitness level. Most people can improve their VO$_{2max}$ by 15 to 20% or more with training.[9]

Another adaptation that occurs with exercise is increased bone density. By placing a mechanical stress on bone, exercise stimulates bone development by promoting the deposition of calcium in bones. Weight-bearing exercise, such as running, gymnastics, basketball, soccer, walking, and volleyball, is essential for the normal development and maintenance of a healthy skeleton.

### Knowledge Check

1. How do the functions of the muscle fiber types differ?
2. What is the predominant fuel used by each muscle fiber type?
3. How does repeated exercise affect the circulatory system?

► Review Table 10-5 in Chapter 10, which lists the energy costs of typical forms of physical activity.

## 11.5 Power Food: Dietary Advice for Athletes

Athletic training and genetic makeup are very important determinants of athletic performance. A good diet won't substitute for either factor, but making wise food choices will allow an athlete to maximize his or her athletic potential. On the other hand, poor food choices can seriously reduce performance.

### Energy Needs

Athletes need varying amounts of food energy, depending on their body size, their body composition, and the type of training or competition. A petite gymnast may need only 1800 kcal/day to sustain normal daily activities without losing body weight; a tall, muscular swimmer may need 4000 kcal/day. If an athlete experiences daily fatigue and/or weight loss, the first consideration should be whether that person is consuming enough food. Up to 6 meals per day may be needed, including 1 before each workout.

Monitoring body weight is an easy way to assess the adequacy of calorie intake. Athletes should strive to maintain weight during competition and training. Generally, if athletes are losing weight, then energy intake is inadequate; however, if athletes are gaining weight, then energy intake is too high. If an athlete needs to lose weight, his or her food intake should be lowered by 200 to 500 kcal per day. This slight reduction will allow the athlete to continue to train and compete yet will create a calorie deficit so that weight loss can be achieved. Reducing fat intake is the best way to cut calories and not affect performance. On the other hand, if an athlete needs to gain weight, increasing food intake by 500 to 700 kcal/day will eventually achieve that goal. The extra calories should come from a healthy balance of carbohydrate, protein, and fat; exercise needs to be maintained to make sure this gain is mostly in the form of lean muscle mass.

Athletes often expend much energy. An increase in food and beverage intake can easily provide ample carbohydrate, protein, and other nutrients to support activity.

Dave and Les Jacobs/Blend Images LLC

Some athletes compete in sports that require them to maintain a lean profile, whereas others must maintain a certain weight. Gymnasts, swimmers, figure skaters, and dancers are required to maintain a lean profile, whereas wrestlers, boxers, jockeys, judoists, and rowers are often weighed before matches to verify that they meet weight restrictions. Athletes in sports such as these tend to eat and drink less than needed to support training and competition needs—this puts them at risk of eating disorders and the effects of poor nutritional status, including osteoporosis, menstrual dysfunction, kidney failure, heat-related illness, dehydration, and even death.

## Carbohydrate Needs

Carbohydrates are the primary energy source for exercising muscles. Anyone who exercises vigorously, especially for more than an hour per day on a regular basis, needs to consume moderate to high amounts of carbohydrates (Fig. 11-11). Numerous servings of whole grains, starchy vegetables, and fruits provide enough carbohydrate to maintain adequate liver and muscle glycogen stores, especially when replacing glycogen losses from workouts on the previous day. Table 11-5 shows some nutritious, carbohydrate rich foods.

The Acceptable Macronutrient Distribution Range (AMDR) for carbohydrate is 45 to 65% of total energy intake; however, this percentage is poorly correlated to both the amount

To prevent deaths from unsafe weight loss practices in wrestlers, the National Collegiate Athletic Association and many states now establish a weight class at the beginning of the season to eliminate the severe weight loss practices that often happen at the end of the season. Each school must have a physician or an athletic trainer conduct an initial weight assessment during the 1st week of October using body weight, body composition (body fat), and specific gravity of urine (to determine level of hydration at time of weighing). Minimum wrestling weight is set at the athlete's lean body weight plus 5% body fat. Each wrestler has the option of modifying his weight over an 8-week period under the following guidelines: no more than 1.5% of body weight can be lost per week, and the final weight cannot fall below the calculated minimum wrestling weight. A national certification period is held in December; at that time, the process is repeated and a weight class is set that remains in place for the rest of the wrestling season.

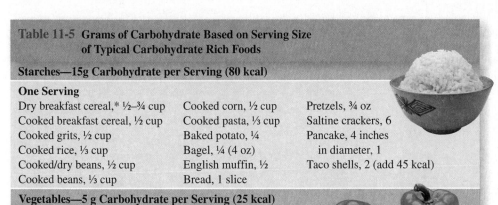

**Table 11-5 Grams of Carbohydrate Based on Serving Size of Typical Carbohydrate Rich Foods**

**Starches—15g Carbohydrate per Serving (80 kcal)**

**One Serving**

| | | |
|---|---|---|
| Dry breakfast cereal,* ½–¾ cup | Cooked corn, ½ cup | Pretzels, ¾ oz |
| Cooked breakfast cereal, ½ cup | Cooked pasta, ⅓ cup | Saltine crackers, 6 |
| Cooked grits, ½ cup | Baked potato, ¼ | Pancake, 4 inches |
| Cooked rice, ⅓ cup | Bagel, ¼ (4 oz) | in diameter, 1 |
| Cooked/dry beans, ½ cup | English muffin, ½ | Taco shells, 2 (add 45 kcal) |
| Cooked beans, ⅓ cup | Bread, 1 slice | |

**Vegetables—5 g Carbohydrate per Serving (25 kcal)**

**One Serving**

| | |
|---|---|
| Raw vegetables, 1 cup | Cooked or canned |
| Vegetable juice, ½ cup | vegetables, ½ cup |

**Fruits—15 g Carbohydrate per Serving (60 kcal)**

**One Serving**

| | | |
|---|---|---|
| Canned fruit, ½ cup | Apricots (dried), 8 | Grapes, 17 |
| Fruit juice, ½ cup | Grapefruit, ½ | Dates, 3 |
| Watermelon cubes, 1¼ cups | Banana, 1 small | Peach, 1 |
| Apple or orange, 1 small | Figs (dried), 1½ | |

**Milk—12 g Carbohydrate per Serving**

**One Serving**

| | |
|---|---|
| Milk, 1 cup | Soy milk, 1 cup |
| Plain low-fat yogurt, ⅔ cup | |

**Sweets—15 g Carbohydrate per Serving (Variable kcal)**

**One Serving**

| | |
|---|---|
| Cake, 2-inch square | Ice cream, ½ cup |
| Cookies, 2 small | Sherbet, ½ cup |

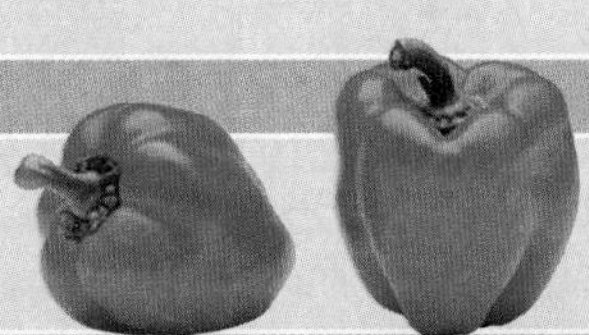

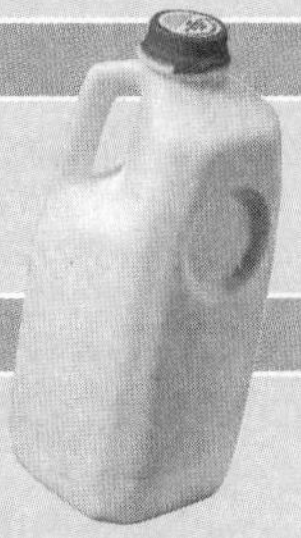

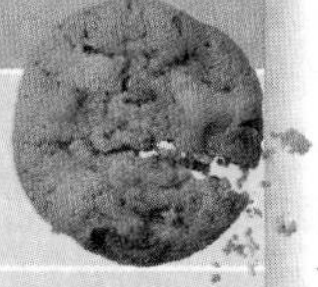

*The carbohydrate content of dry cereal varies widely. Check the labels of those you choose and adjust the serving size accordingly.

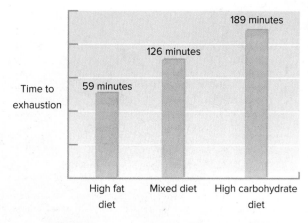

Time to exhaustion

59 minutes — High fat diet

126 minutes — Mixed diet

189 minutes — High carbohydrate diet

*Figure 11-11* In his classic study, Bergstrom and colleagues had participants consume 3 different diets: a high fat and high protein diet, a mixed diet, and a high carbohydrate diet. Subjects cycled at 75% of $VO_{2max}$ until exhaustion. The high carbohydrate diet group had the longest exercise time as well as the greatest glycogen content.[13]

of carbohydrate actually eaten and the fuel needed to support an athlete's training and competition. It is more important for the athlete's carbohydrate intake to match the fuel needs of training and glycogen restoration; this is defined as carbohydrate availability.[14]

Carbohydrate intake for low-intensity activity should range from 3 to 5 g/kg of body weight, whereas athletes engaged in moderate exercise lasting about an hour per day should consume 5 to 7 g/kg of body weight. When exercise duration approaches several hours per day, the carbohydrate recommendation increases to 6 to 10 g/kg of body weight. Individuals participating in extreme activity lasting 4 to 5 hours should consume 8 to 12 g/kg of body weight.[15] Attention to carbohydrate intake is especially important when performing multiple training bouts in a day, such as swim practices or track and field events, as well as tournament play, such as volleyball, basketball, or soccer. The depletion of carbohydrate ranks just behind the depletion of fluid and electrolytes as a major cause of fatigue.

Table 11-6 shows sample menus for diets providing food energy ranging from 1800 to 5000 kcal/day. In addition, the Exchange System (see Appendix E) is a useful tool for planning all types of diets, including high carbohydrate diets for athletes.

**Table 11-6  Sample Daily Menus, Based on MyPlate, That Provide Various Total Energy Intakes**

| 1800 kcal | 3000 kcal | 4000 kcal | 5000 kcal |
|---|---|---|---|
| **Breakfast** | **Breakfast** | **Breakfast** | **Breakfast** |
| 1 cup fat-free milk | 1 cup fat-free milk | 3 slices french toast | 4 slices whole-wheat toast |
| 1 cup Cheerios® | 2 cups Cheerios® | 2 tbsp syrup | 2 tsp margarine |
| ½ whole-wheat bagel | 1 whole-wheat bagel | 1 tsp margarine | 2 poached eggs |
| 1 tsp margarine | 1 tsp margarine | 1 banana | 1 cup low-fat yogurt |
| | ½ cup grapes | 1 cup low-fat yogurt | ½ cup granola |
| | 1 large muffin | | |
| **Lunch** | **Lunch** | **Lunch** | **Lunch** |
| 2 oz sliced turkey breast | 3 oz sliced turkey breast | 4 medium beef and bean tacos | 3 chicken enchiladas |
| 2 slices whole-wheat bread | 2 slices whole-wheat bread | topped with lettuce and shredded | 1 oz shredded cheese |
| 1 tsp mayonnaise | 1 oz cheese | cheese | 1 cup romaine lettuce with 1 cup |
| ½ cup carrots | 1 tsp mayonnaise | 1 cup Spanish rice | carrots, celery, and green peppers |
| 1 medium banana | 1 banana | 1 cup romaine lettuce | 2 tbsp salad dressing |
| 1 cup apple juice | ½ cup carrots | 2 tbsp salad dressing | 1 banana |
| 2 oz peanuts | 1 cup yogurt | 1½ cups orange juice | 1 cup apple juice |
| | 1 cup apple juice | | 1 fresh orange |
| **Snack** | **Snack** | **Snack** | **Snack** |
| 1 granola bar | 1 granola bar | 2 slices whole-wheat bread | 2 whole-grain bagels |
| 1 cup low-fat yogurt | 1 cup applesauce | 4 tbsp peanut butter | 3 tbsp almond butter |
| | | | 1 cup grapes |
| **Dinner** | **Dinner** | **Dinner** | **Dinner** |
| 3 oz roast beef | 1 cup spaghetti with meat sauce | 4 oz turkey breast | 5 oz grilled salmon (or other fish) |
| 1 medium baked potato | 1½ cups pasta | 2 cups mashed potatoes | 2 cups rice pilaf |
| 1 tsp margarine | 1 tbsp Parmesan cheese | ½ cup corn | 1 cup asparagus |
| 1 cup romaine lettuce | 1 cup romaine lettuce | 1 roll | 1 cup green beans |
| 2 tsp salad dressing | 2 tsp salad dressing | 1 tsp margarine | 2 tsp margarine |
| 1 cup green beans | 1 cup green beans | 1 cup vanilla pudding | 1 cup low-fat yogurt topped with |
| 1 cup fat-free milk | ½ cup fat-free milk | ½ cup sliced fruit | ½ cup melon and ½ cup granola |
| | | 1 cup fat-free milk | 1½ cups fat-free milk |
| **Nutrient Contribution** | **Nutrient Contribution** | **Nutrient Contribution** | **Nutrient Contribution** |
| **57% carbohydrate** | **64% carbohydrate** | **50% carbohydrate** | **48% carbohydrate** |
| **20% protein** | **16% protein** | **17% protein** | **18% protein** |
| **25% fat** | **21% fat** | **33% fat** | **34% fat** |

## Boosting Glycogen Stores

The first source of glucose for the exercising muscle is its own glycogen store. During endurance exercise that exceeds 90 minutes, such as marathon running, muscle glycogen stores progressively decline. When they drop to critically low levels, high-intensity exercise cannot be maintained. In practical terms, the athlete is exhausted and must either stop exercising or drastically reduce the pace.

Glycogen depletion also may be a gradual process, occurring over repeated days of heavy training in which muscle glycogen breakdown exceeds its replacement, as well as during high-intensity exercise that is repeated several times during competition or training. For example, a distance runner who averages 10 miles per day but does not take the time to consume enough carbohydrates or a swimmer who completes several interval sets at above maximal oxygen consumption can deplete his or her glycogen stores rapidly.

Because carbohydrates are such an important fuel for exercise and the body has limited capacity to store them, researchers have investigated ways to maximize the body's ability to store carbohydrates. The original regimen of getting the body to store more glycogen than typical is called **carbohydrate loading,** or **glycogen loading.** It involves altering both exercise and diet. There are a number of carbohydrate-loading strategies. The classic method of carbohydrate loading depleted muscle glycogen stores with 3 days of heavy training by eating a very low carbohydrate diet. This was followed by 3 days of high carbohydrate intake and rest to promote muscle glycogen synthesis. This classic method often left athletes exhausted and prone to injuries during the first 3 days. A modified method tapers off training intensity and duration on consecutive days after a bout of glycogen-depleting exercise 6 days before competition. During the first 3 days of tapering, the athlete consumes a normal mixed diet, followed by a high carbohydrate diet 3 days before competition. Today, most athletes who train daily have a higher proportion of carbohydrates in their diets to replenish glycogen stores daily.

High carbohydrate foods should form the basis of an athlete's diet.

Ingram Publishing/SuperStock

| Carbohydrate-Loading Regimen | | | | | | |
|---|---|---|---|---|---|---|
| **Days before Competition** | 6 | 5 | 4 | 3 | 2 | 1 |
| **Exercise Time (Minutes)** | 60 | 40 | 40 | 20 | 20 | Rest |
| **Carbohydrate Intake (g/kg Body Weight)** | 5 | 5 | 5 | 10 | 10 | 10 |

The classic and modified carbohydrate-loading regimens usually increase muscle glycogen stores 50 to 85% over typical conditions, when dietary carbohydrate intake is only about 50% of total intake. However, the process is relatively slow. It takes from 2 to 6 days to fully load the muscles with glycogen. These regimens pose problems for athletes who compete on back-to-back days or who do not want to alter their training before a competition. Other carbohydrate-loading strategies, some as short as 1 day in duration, can help athletes increase muscle glycogen levels above typical levels.[16]

Carbohydrate loading is for athletes who compete in continuous, intense aerobic events lasting more than 60 minutes or in shorter events with repeated bouts of exercise occurring more than once in a 24-hour period. Above normal glycogen stores do not allow an athlete to work harder during shorter exercise periods, such as 5 and 10 km races, and may harm performance because of muscle stiffness and heaviness. With each gram of glycogen stored in the muscle, 3 to 4 grams of water also are stored. Although this water helps maintain hydration, the additional water weight may make the muscles feel stiff and therefore make carbohydrate loading inappropriate. Athletes who want to try carbohydrate loading should do so during training, and long before an important competition, to experience its effects on performance.

## Fat Needs

A fat intake of 15 to 25% of energy is generally recommended for athletes. Rich sources of unsaturated fat, such as canola, soybean, and olive oils, should be emphasized, and saturated fat and *trans* fat intake should be limited.[15]

**Appropriate Activities for Carbohydrate Loading**

- Marathons
- Long-distance swimming
- Cross-country skiing
- 30-kilometer runs
- Triathlons
- Tournament-play basketball
- Soccer
- Cycling time trials
- Long-distance canoe racing

**Inappropriate Activities for Carbohydrate Loading**

- American football
- 10-kilometer or shorter runs
- Walking and hiking
- Most swimming events
- Single basketball games
- Weight lifting
- Most track and field events

**carbohydrate (glycogen) loading** Exercise and eating regimen that increases the amount of glycogen stored in muscles to levels higher than normal.

**Table 11-7  Current Recommendations for Protein Intake Based on Body Weight (kg)***

| Group | g/kg | Amount for a 70 kg Person (g) |
|---|---|---|
| RDA for adults | 0.8 | 56 |
| Strength-trained athletes, muscle mass maintenance phase | 1.0–1.2 | 70–84 |
| Strength-trained athletes, muscle mass gain phase | 1.5–1.8 | 105–126 |
| Moderate-intensity endurance athletes | 1.2–1.4 | 84–98 |
| High-intensity endurance athletes | 1.8 | 126 |

*Calculate kilograms by dividing pounds by 2.2.
Source: Adapted from Burke L and Deakin V., *Clinical Sports Nutrition*. Roseville, Australia: McGraw-Hill, 2015; Thomas, DT and others, "Position of the Academy of Nutrition and Dietetics, Dietitians of Canada, and the American College of Sports Medicine: Nutrition and Athletic performance," *J Acad Nutr Diet.* 2016;116:501.

Recall that during exercise you are always burning a combination of carbohydrate and fat. At rest, fewer carbohydrates and more fat are used. However, during high-intensity exercise the body relies on more carbohydrate and very little fat. If you exercise at moderate intensity, the mixture of fuel is close to half carbohydrate and half fat (see Fig. 11-10).

### Ketogenic Diet and Athletic Performance

Studies of the effects of ketogenic diets on athletic performance have mixed results. Athletes in high-intensity sports who follow a ketogenic diet tend to have lower glycogen and glycolytic enzymes that impair their ability to maintain sprint and high-intensity workouts.[17] On the other hand, a ketogenic diet may benefit endurance athletes, especially during the pre-season or whenever long bouts of low- to moderate-intensity exercise are performed.[17]

Knowledge of whether and how low carbohydrate (ketogenic or not) diets can enhance athletic performance is still evolving. Some evidence suggests that athletes wanting to follow a ketogenic diet may need to modify this diet to achieve performance goals. For instance, including some carbohydrate from sources that are lower in carbohydrate may be needed to provide enough carbohydrates for optimal performance.[18]

### Protein Needs

Typical recommendations for protein intake for most athletes range from 1.2 to 1.8 g of protein/kg of body weight, which is twice as high as the RDA of 0.8 g/kg of body weight for adults. As you can see in Table 11-7, recommended protein intake is at the lower end of

Weight-restricted athletes who feel they must significantly limit their energy intake and athletes who are vegetarians should be sure to consume at least 1.2–2.0 g of protein per kg of body weight each day, the upper recommendation for most athletes.
©Comstock/Getty Images RF

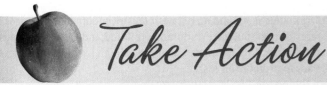

©Stockbyte/Getty Images RF

## Meeting the Protein Needs of an Athlete: A Case Study

Mark is a college student who has been lifting weights at the student recreation center. Mark's weight has been stable at 154 lb (70 kg). The total energy and protein content of Mark's current diet is 3470 kcal, 125 g of protein (14% of total energy intake supplied by protein). This diet is representative of the types and amounts of food that Mark chooses on a regular basis. The trainer at the center has recommended a protein drink to help Mark build muscle mass. Answer the following questions and determine whether a protein drink is needed to supplement Mark's diet.

1. Determine Mark's protein needs based on the RDA (0.8 g/kg).

   a. Mark's estimated protein RDA: _____

   b. What are the maximum recommendations for protein intake for athletes? _____

   c. Calculate the maximum protein recommendation for Mark. _____

2. Compare Mark's protein intake with the recommended intake amounts.

   a. What is the difference between Mark's estimated protein needs as an athlete (from question 1) and the amount of protein in his current diet? _____

   b. Is his current protein intake inadequate, adequate, or excessive? _____

3. Mark takes his trainer's advice and goes to the supermarket to purchase a protein drink to add to his diet. Four products are available; they contain the following label information.

|  | Amino Fuel | Sugar Free 90% Plus Protein | Dynamic Muscle Builder | Super Mega Mass 2000 |
|---|---|---|---|---|
| Serving size | 3 tbsp | 3 tbsp | 3 tbsp | ¼ scoop |
| Kcal | 104 | 110 | 103 | 104 |
| Protein (g) | 15 | 24 | 10 | 5 |

The trainer recommends that Mark add the supplement to his diet 2 times a day. Mark chooses Dynamic Muscle Builder.

   a. How much protein would be added to Mark's diet daily from 2 servings of the supplement alone? _____

   b. How much total protein would Mark now consume in 1 day? _____

   c. What is the difference between Mark's estimated protein needs as an athlete and his total protein intake with the supplement? _____

4. What is your conclusion—does Mark need the protein supplement? _____

### Answers to Calculations

1a. Mark's estimated protein RDA: 70 kg × 0.8 g/kg = 56 g

1b. Maximum recommendation for protein intake for athletes = 1.7 g/kg

1c. Applied to Mark: 1.7 × 70 = 119 g

2a. Difference between Mark's diet and the maximum recommended amount for athletes: 125 − 119 = 6 g

2b. Mark's current diet is adequate.

3a. Two servings of protein supplement alone = 20 g of protein

3b. Mark's total protein consumption: 125 g + 20 g = 145 g protein

3c. Difference between Mark's estimated maximum protein needs as an athlete and total protein consumption: 145 g − 119 g = 26 g of protein

High protein drinks, bars, and other products, which are often marketed to athletes, are unnecessary in most cases.

©imageBROKER/Alamy

the range for strength training maintenance and moderate-intensity endurance activities. The highest recommendations are for high-intensity endurance training and during the muscle mass gain phase of strength training. Protein intakes may go as high as 2.0 g/kg/day when athletes are restricting calories to promote fat loss.[19]

Energy needs are not the reason for the higher protein recommendations for athletes (recall that protein is not a major fuel for exercise). The extra protein is needed for the repair of tissue and the synthesis of the new muscle that results from training. The high level set for muscle mass gain during strength training, theoretically, is required for the synthesis of new muscle tissue brought on by the loading effect of this training. Once the desired muscle mass is achieved, protein intake need not exceed 1.2 g/kg of body weight.

Although it has long been a popular belief among athletes that additional protein increases strength and enhances performance, sports nutritionists and exercise physiologists generally agree that consuming protein at levels above recommendations does not build bigger or stronger muscles.[20] Protein intakes above recommendations result in an increased use of amino acids for energy needs and has disadvantages, such as insufficient carbohydrate intake and increased urine production, which may interfere with body hydration. No advantages, such as an increase in muscle protein synthesis, are seen. Despite marketing claims, protein supplements are an expensive and unnecessary part of a fitness plan.

Any athlete not specifically on a low calorie regimen can easily meet protein recommendations simply by eating a variety of foods (see Table 11-6). To illustrate, a 116 lb (53 kg) woman performing endurance activity can meet daily protein needs of 64 g (53 × 1.2) by eating a 3 oz chicken breast, a small hamburger (3 oz), and 2 glasses of milk. Similarly, a 170 lb (77 kg) man wanting to gain muscle mass through strength training needs to consume just 6 oz of chicken, ½ cup of cooked beans, 6 oz of canned tuna, and 3 glasses of milk to achieve an intake of 130 g of protein (77 × 1.7) in a day. Plus, for both athletes, these calculations do not include the protein they will get from the grains they eat. As you can see, simply by meeting their energy needs, many athletes consume much more protein than is required.

## Vitamin and Mineral Needs

Vitamin and mineral needs are the same or slightly higher for athletes, compared with those of sedentary adults. Still, because athletes usually have such high energy intakes, they tend to consume plenty of vitamins and minerals. An exception is athletes consuming low calorie diets (about 1200 kcal or less), such as some female athletes participating in events in which maintaining a low body weight is crucial. These diets may not meet B-vitamin and other micronutrient needs.[23] To meet vitamin and mineral needs, vegetarian athletes and athletes consuming low calorie diets should consume fortified foods (e.g., ready-to-eat breakfast cereals) or a balanced multivitamin and mineral supplement.

### Iron Deficiency and Impaired Performance

Because iron is involved in red blood cell production, oxygen transport, and energy production, a deficiency of this mineral can noticeably detract from optimal athletic performance.[24] The potential causes of iron deficiency in athletes vary. As in the general population, female athletes are most susceptible to low iron status due to monthly menstrual losses. Special diets followed by athletes, such as low energy and vegetarian (especially vegan) diets, are likely to be low in iron. Distance runners should pay special attention to iron intake because their intense workouts may lead to gastrointestinal bleeding.

Another concern is sports anemia, which occurs because exercise causes blood plasma volume to expand, particularly at the start of a training regimen before the synthesis of red blood cells increases. This expansion results in dilution of the blood. In sports anemia, even if iron stores are adequate, blood iron tests may appear low. Sports anemia is not detrimental to performance, but it is hard to differentiate between sports anemia and true anemia.

True anemia, noted as reduced blood hemoglobin and hematocrit levels (see Part 4), may affect up to about 15% of male and 30% of female athletes. It is a good idea, especially for women athletes, to have their iron status checked at the beginning of a training season and at least once during midseason, as well as to monitor their dietary iron intake. Once depleted, iron stores can take months to replenish. For this reason, athletes must be especially careful to meet their iron needs.

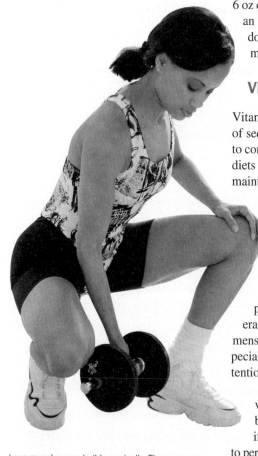

Lean muscle mass builds gradually. The average amount of muscle gained in a week after 16 weeks of resistance exercise and recommended protein intakes (see Table 11-7) is about 0.20 to 0.35 lb for women and 0.4 to 0.7 lb for men.[21,22]

Getty Images RF

Any blood test indicating low iron status—sports anemia or not—is cause for follow-up. For some athletes, the use of iron supplements may be advisable. However, the indiscriminate use of iron supplements is not advised because toxic effects are possible. It is important that physicians investigate the cause of the deficiency because iron deficiency can be caused by blood loss. If caught early, serious medical conditions often can be treated or prevented.

### Calcium Intake and Relative Energy Deficiency in Sport (REDS)

Athletes, especially women trying to maintain a lean profile, can have a restricted caloric intake, which leads to energy intake that is insufficient to support daily living, growth, and health. This condition is called Relative Energy Deficiency in Sport, or REDS. If the energy deficiency continues without being addressed, poor bone growth, stress fractures, and early osteoporosis can occur. Other potential effects include low estrogen levels, delayed menses, and increased risk of injury.[25]

Research has clearly documented the importance of regular menstruation to maintain bone mineral density. Disturbing reports show that female athletes who do not menstruate regularly have far less dense spinal bones than both nonathletes and female athletes who menstruate regularly. These female athletes are at increased risk of bone fractures during training and competition. If irregular menstrual cycles persist, severe bone loss, much of which is not reversible, and osteoporosis can result.[26] This combination of risks outweighs the benefits of weight-bearing exercise for bone density.

Women with symptoms of REDS are best treated by a team that includes a physician, a registered dietitian nutritionist, a psychologist, and an athletic trainer. The primary goals of treatment are to control and manage the athlete's diet, to restore normal hormone levels and menstruation, and to monitor and treat any injuries or other medical complications. Treatment strategies to reach these goals may include a slight reduction (10 to 20%) in the amount of training and a higher energy intake for a 2 to 5% increase in weight. Gaining weight, either by cutting back on training or by consuming more energy, helps some athletes with amenorrhea increase their chances of resuming normal menstrual activity. Most athletes experiencing amenorrhea fear weight gain and must be counseled that an increase in muscle weight can improve their stamina and performance. Extra calcium in the diet does not necessarily compensate for the effects of menstrual irregularities, but inadequate dietary calcium can make matters worse. Calcium supplementation should be implemented in all athletes presenting with amenorrhea.

> ▶ Paula Findlay, a competitive triathlete, revealed that she suffered from iron deficiency, a condition she believes played a role in her last-place finish during the 2012 London Olympics.

> ▶ Learn more about the nutrition and health needs of female athletes at www.olympic.org/hbi.

**_Critical_ Thinking**

Joe is a wrestler who qualified for the 125 lb weight classification in the annual state high school competition. After a few matches, he began to feel dizzy and faint. He was disqualified because he was unable to continue the match. Later, the coach found out that Joe had spent 2 hours in the sauna before weighing in, which had made him dehydrated. What are the consequences of dehydration? What can you suggest as a safer alternative for weight loss?

### Knowledge Check

1. What is the primary source of energy for an exercising muscle?
2. What is glycogen loading?
3. How does iron deficiency anemia affect athletic performance?
4. What is Relative Energy Deficiency in Sport (REDS)?

Marathon runners can lose 6 to 10% of their body weight during a race.

©Comstock/Stockbyte/Getty Images RF

## 11.6 Fluid Needs for Active Individuals

Exercise can raise muscle temperature 15 to 20 times above resting muscle temperature—this heat can be dissipated through the evaporation of sweat from the skin. During prolonged exercise, sweat loss ranges from 3 to 8 cups (750 to 2000 ml) per hour. Sweat losses tend to be greatest during hot weather and in endurance sports or those that require athletes to wear heavy equipment (e.g., football). As little as a 2% weight loss through sweating puts athletes at risk of dehydration and decreased performance.

Active individuals need more fluids than those who are sedentary to replace fluid lost in sweat and, thereby, maintain blood volume and allow the body to regulate internal temperature normally.[10] Insufficient fluid intake leads to dehydration, which causes a decline in endurance, strength, and overall performance and sets the stage for heat exhaustion, heat cramps, and potentially fatal heatstroke (Fig. 11-12).[27]

## HISTORICAL PERSPECTIVE

Gatorade®, a rehydration fluid, was born when a football coach asked Dr. Robert Cade at the University of Florida why players didn't need to urinate during a game. Cade's research discovered that the players lost so much sweat and were so dehydrated that no fluid was left to form urine. Learn more at www .gatorade.com.mx/company/heritage.

Rubberball/Getty Images

**heat exhaustion** First stage of heat-related illness that occurs because of depletion of blood volume from fluid loss by the body. This depletion may increase body temperature and can lead to headaches, dizziness, muscle weakness, and visual disturbances, among other effects.

**heat cramps** Frequent complication of heat exhaustion. Cramps usually occur in individuals who have experienced large sweat losses from exercising for several hours in a hot climate and have consumed a large volume of water. The cramps occur in skeletal muscles and consist of contractions for 1 to 3 minutes at a time.

**heatstroke** Condition in which the internal body temperature reaches 104°F or higher. Blood circulation is greatly reduced. Nervous system damage may ensue, and death is likely. Sweating generally ceases, which cause the skin of individuals who suffer heatstroke to feel hot and dry.

Heat exhaustion and heatstroke are on a continuum. **Heat exhaustion** is the 1st stage of heat-related illness caused by dehydration. Common symptoms of heat exhaustion include profuse sweating, headache, dizziness, nausea, vomiting, muscle weakness, visual disturbances, and flushing of the skin. A person with heat exhaustion should be taken to a cool environment immediately, and excess clothing should be removed. The body should be dunked in cool water or covered with cool towels. Fluid replacement, as tolerated, should be provided.[27] It is critical that the individual get immediate medical attention to prevent tissue damage and possible death.

**Heat cramps** are a frequent complication of heat exhaustion, but they may appear without other symptoms of dehydration. Heat cramps occur in skeletal muscles and consist of contractions lasting 1 to 3 minutes at a time. The cramp moves down the muscle and causes excruciating pain. It is important not to confuse heat cramps with other forms of muscle cramps, such as those caused by intestinal tract upset. Heat cramps usually occur in individuals who have experienced significant sweating from exercising for several hours in a hot climate and who have consumed a large volume of water without replacing sodium losses. The best way to prevent heat cramps is to exercise moderately at first, have adequate salt intake before engaging in long and strenuous activity in hot conditions, and avoid becoming dehydrated.[27]

Left unchecked, heat exhaustion can rapidly progress to **heatstroke.** Heatstroke can occur when the internal body temperature reaches 104°F or higher. Exertional heatstroke results from

| Relative humidity (%) | 70° | 75° | 80° | 85° | 90° | 95° | 100° | 105° | 110° |
|---|---|---|---|---|---|---|---|---|---|
| 100 | 72° | 80° | 91° | 108° | | | | | |
| 90 | 71° | 79° | 88° | 102° | 122° | | | | |
| 80 | 71° | 78° | 86° | 97° | 113° | 136° | | | |
| 70 | 70° | 77° | 85° | 93° | 106° | 124° | 144° | | |
| 60 | 69° | 76° | 82° | 90° | 100° | 114° | 132° | 149° | |
| 50 | 70° | 75° | 81° | 88° | 96° | 107° | 120° | 135° | 150° |
| 40 | 68° | 74° | 79° | 86° | 93° | 101° | 110° | 123° | 137° |
| 30 | 67° | 73° | 78° | 84° | 90° | 96° | 104° | 113° | 123° |
| 20 | 66° | 72° | 77° | 82° | 87° | 93° | 99° | 105° | 112° |
| 10 | 65° | 70° | 75° | 80° | 85° | 90° | 95° | 100° | 105° |
| 0 | 64° | 69° | 73° | 78° | 83° | 87° | 91° | 95° | 99° |

**Air temperature (°F)**

| Heat index | Heat disorders possible with prolonged exposure and/or physical activity |
|---|---|
| 80°–89° | Fatigue |
| 90°–104° | Sunstroke, heat cramps, and heat exhaustion |
| 105°–129° | Sunstroke, heat cramps, or heat exhaustion likely and heatstroke possible |
| 130° or higher | Heatstroke/sunstroke highly likely |

NOTE: Direct sunshine increases the heat index by up to 15°F.

*Figure 11-12* Heat index chart showing associated heat disorders.

high blood flow to exercising muscles, which overloads the body's cooling capacity. Symptoms include nausea, confusion, irritability, poor coordination, seizures, hot and dry skin, rapid heart rate, vomiting, diarrhea, and coma. If heatstroke is left untreated, circulatory collapse, nervous system damage, and death are likely. The death rate from heatstroke is approximately 10%.[27]

For heatstroke victims, cooling the skin with ice packs or cold water is the usually recommended immediate treatment until medical help arrives. To decrease the risk of developing heatstroke, athletes should watch for rapid changes in body weight (2% or more), replace lost fluids and sodium, and avoid exercising in extremely hot, humid conditions.

## Fluid Intake and Replacement Strategies

Paying attention to fluid intake before exercising can help ensure that athletes begin with optimal fluid levels. During exercise, the recommended fluid status goal is a loss of no more than 2% of body weight, especially in hot weather. Athletes should first calculate 2% of their body weight and then, by trial and error, determine how much fluid they must drink to avoid losing more than this amount of weight during exercise. This determination is most accurate if an athlete is weighed before and after a typical workout. For every pound (½ kg) lost, 3 cups (about ¾ liter) of fluid should be consumed during or immediately after exercise. If weight change can't be monitored, urine color is another measure of hydration status—it should be no darker than the color of lemonade (Fig. 11-13).[1]

Thirst is a late sign of dehydration and, so, is not a reliable indicator of an athlete's need to replace fluid during exercise. An athlete who drinks only when thirsty is likely to take 48 hours to replenish fluid loss. After several days of training, an athlete relying on thirst as an indicator can build up a fluid debt that will impair performance.

Fluid intake during exercise, when possible, can help minimize fluid loss and a drop in body weight. Drinking fluids during practice is a good idea, even when sweating can go unnoticed, such as when swimming or during the winter.[10] However, fluid replacement mostly has to take place after exercise because it is difficult to consume enough fluid during exercise to prevent weight loss.

The following guidelines can meet most athletes' fluid needs.[15]

### Before Exercise
- Freely drink beverages (e.g., water, sports drinks) during the 24-hour period before an event, even if not thirsty.
- Drink about 2 to 3 cups of fluid (0.2 to 0.3 fl oz/kg of body weight) about 2 to 3 hours before exercise. This allows time for both adequate hydration and the excretion of excess fluid.

### During Exercise
- Start drinking fluids early and at consistent intervals to prevent excessive dehydration and maintain weight during the exercise bout.
- Drink 1 to 1.5 cups (250 to 375 ml) of fluid every 10 to 15 minutes.
- Fluids that are flavored and cooler than the environmental temperature promote fluid replacement.[29]
- If exercise lasts more than 1 hour, fluid replacement beverages should contain 4 to 8% carbohydrates to maintain blood glucose levels. Sodium also should be included in the beverage in amounts of 0.5 to 0.7 gram of sodium per liter of water to replace sodium lost in sweat.

### After Exercise
- Drink 3 cups of fluid for each pound lost during exercise.
- Restore weight before the next exercise period.

## Water Intoxication

In athletes, water intoxication is most often caused by overdrinking before, during, or after exercise without also replacing sodium losses. To prevent water intoxication, athletes should drink beverages containing sodium and should consume enough fluid during exercise to minimize loss of body weight (i.e., to avoid significant dehydration), but they should avoid overdrinking. Sports drinks containing at least 100 mg of sodium per 8-ounce serving have been shown to help maintain blood sodium level better than plain water.[30]

▶ Sufficient fluid intake is important at any age, but it is particularly critical for young children participating in activities such as youth soccer, T-ball, and basketball. That's because, for any given level of dehydration, children's core body temperatures rise faster than those of adults. Children who participate in sports activities must be taught to prevent dehydration by drinking above and beyond thirst and drinking at frequent intervals—for example, every 20 minutes.[28]

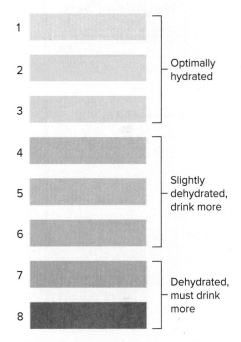

*Figure 11-13*   Urine color chart.

Adequate fluid intake is important before, during, and after exercise. Skipping fluids before or during events will almost certainly impair performance. For peak performance, it is important that weight be restored before the next exercise period.

©JGI/Blend Images LLC RF

## Sports Drinks

Water is the recommended beverage of choice for fluid replacement for physical activity lasting less than 1 hour. Using a sports drink during high-intensity, stop-and-go sports (lasting longer than 1 hour) such as basketball, volleyball, and sprint cycling can delay fatigue and maintain hydration.[31] Long-endurance athletes who consume only water as a fluid replacement risk diluting their blood (i.e., plasma sodium) and increasing their urine output, thus shutting off the drive to drink and leading to dehydration.

When exercise extends beyond 1 hour, sports drinks can offer several advantages over water alone—even more so in hot weather (Fig. 11-14).

- Carbohydrates in sports drinks supply glucose to muscles as they become depleted of glycogen, thus enhancing performance.
- Electrolytes in sports drinks help maintain blood volume, enhance the absorption of water and carbohydrates from the intestines, and stimulate thirst.

Beverages containing alcohol should be avoided because they increase urine output and reduce fluid retention. In addition, high intakes of caffeine (greater than 500 mg per day, or the amount in 4 to 5 cups of brewed coffee) can increase urine output. Carbonated beverages also should be avoided, as they reduce the desire to consume fluids because the carbonation leads to feelings of stomach fullness. Drinks with a sugar content above 10%, such as soft drinks or fruit juices, take longer to absorb, less efficiently contribute to hydration, and are not recommended.

Overall, the American College of Sports Medicine suggests that beverages containing electrolytes and carbohydrates (sports drinks) can provide benefits to athletes over water alone, especially when athletes have not eaten a pre-exercise meal or when they are participating in intense exercise, 2-a-day training, tournament play, or back-to-back competition.[15,32] If athletes have not consumed sports drinks and want to try them, they can experiment with them during practice, before using them during competition.

### Knowledge Check

1. What are the symptoms of heat exhaustion?
2. What are the symptoms of heatstroke?
3. How much fluid should an athlete drink after exercise?
4. How should athletes determine if they are dehydrated?
5. When would you recommend a sports drink over water?

*Figure 11-14* Sports drinks for fluid and electrolyte replacement typically contain simple carbohydrates plus sodium and potassium. The various sugars in this product total 14 g per 1-cup (240 ml) serving. In percentage terms based on weight, the sugar content is about 6% ([14 g sugar per serving / 240 g per serving] × 100 = 5.8%). (*Note:* 1 ml of water weighs 1 g.) Sports drinks typically contain about 6 to 8% sugar. This provides ample glucose and other monosaccharides to aid in fueling working muscles, and it is well tolerated. Sports drinks containing 0.5 to 0.7 g/L of sodium and 0.8 to 2.0 g/L of potassium are recommended for athletes.[12]

## 11.7 Food and Fluid Intake before, during, and after Exercise

The composition of food and fluids consumed before, during, and after athletic events or exercise training sessions can affect performance and the speed with which the athlete recovers from the exercise bout. Careful planning is needed to ensure that food intake meets the athlete's needs.

### Pre-exercise Meal

The pre-event or pre-exercise training meal keeps the athlete from feeling hungry before and during the exercise bout, and it maintains optimal levels of blood glucose for the exercising muscles. A pre-exercise meal has been shown to improve performance, compared with exercising in a fasted state. Athletes who train early in the morning before eating or drinking risk developing low liver glycogen stores, which can impair performance, particularly if the exercise regimen involves endurance training.

Allowing for personal preferences and psychological factors, a pre-exercise meal should be high in carbohydrate, nongreasy, non-gas-producing, and readily digested. Recall that eating carbohydrate before exercise can help restore suboptimal liver glycogen stores, which may be called on during prolonged training and high-intensity competition. Fat in pre-exercise meals should be limited because it delays stomach-emptying time and takes longer to digest.

A meal eaten 3.5 to 4 hours before exercising can have as much as 4 g of carbohydrate per kg of body weight and 26% of the calories from fat. To avoid indigestion, nausea, vomiting, and gastrointestinal distress, the carbohydrate and fat content of the meal should be reduced the closer the meal is to the exercise time (Table 11-8). For example, 1 hour before exercising, the athlete should consume only 1 g carbohydrate/kg body weight.[1] Likewise, fat in meals eaten closer to exercise start time should provide less than 26% of calories. Allowing time for partial digestion and absorption provides for a final addition to muscle glycogen, additional blood sugar, and relatively complete emptying of the stomach.

Commercial liquid formulas providing a high carbohydrate meal are popular with athletes because they leave the stomach rapidly. Other appropriate pregame meals are toast with jelly, a baked potato, spaghetti with tomato sauce, cereal with skim milk, and low-fat yogurt with fruit-sugar flavorings (Table 11-9).

| Table 11-8 | Rule of Thumb for Approximate Pre-event Carbohydrate Intake | |
|---|---|---|
| Hours Before | Grams per Kilogram Body Weight | For a 70 kg Person |
| 1 | 1 | 70 |
| 2 | 2 | 140 |
| 3 | 3 | 210 |
| 4 | 4 | 280 |

## Fueling during Exercise

For sporting events that are longer than 60 minutes, consuming carbohydrate during activity can improve athletic performance because prolonged exercise depletes muscle glycogen stores, and low levels of blood glucose lead to both physical and mental fatigue.[10,33] Recall that, when the supply of energy from carbohydrates runs low, athletes often complain of "hitting the wall," the point at which maintaining a competitive pace seems impossible. To avoid this situation, a general guideline for endurance events is to consume 30 to 60 g of carbohydrate per hour; however, an athlete should experiment during training sessions to establish the level that leads to optimal performance.[10]

Sports drinks that are 6 to 8% carbohydrate can provide fuel for endurance events. They supply the necessary fluid, electrolytes, and carbohydrate to keep athletes performing at their best. An alternative to sports drinks are carbohydrate gels (e.g., High5 Energy Gel™, Torq Gel™, and GU® energy gel and bars such as Clif Bars®). Gels contain about 25 to 28 g of carbohydrate per serving, and popular energy bars range from 2 to 45 g of carbohydrate per serving. Sports drinks, by comparison, contain about 14 g of carbohydrate per 8 oz serving. Overall, choose a bar with about 40 g of carbohydrate and no more than 10 g of protein, 4 g of fat, and 5 g of fiber. The bars also are typically fortified with vitamins and minerals, often to 100% of the Daily Values. Thus, these bars can be seen as a convenient, although somewhat expensive, source of nutrients. If the athlete prefers solid sources of carbohydrate, fig cookies, gummy bears, and jellybeans yield a quick source of glucose at a much lower cost. However, any carbohydrate-containing food, including energy bars and gels, must be accompanied by fluid to ensure adequate hydration.[1]

| Table 11-9 Convenient Pre-event Meals | |
|---|---|
| **Breakfast Options** | |
| Cornflakes, ¾ cup<br>Reduced fat milk, 1 cup<br>Blueberry muffin, 1<br>Orange juice, 4 oz | 450 kcal<br>82% carbohydrate<br>(92 g) |
| Low-fat fruit yogurt, 1 cup<br>Plain bagel, ½<br>Apple juice, 4 oz<br>Peanut butter (for bagel), 1 tbsp | 482 kcal<br>68% carbohydrate<br>(84 g) |
| Whole-wheat toast, 1 slice<br>Jam, 1 tsp<br>Apple, 1 large<br>Reduced fat milk, 1 cup<br>Oatmeal, ½ cup (with reduced fat milk, ½ cup) | 507 kcal<br>73% carbohydrate<br>(98 g) |
| **Lunch or Dinner Options** | |
| Chili with beans, 8 oz<br>Baked potato with sour cream and chives<br>Chocolate milk shake, 9 oz | 900 kcal<br>65% carbohydrate<br>(150 g) |
| Spaghetti noodles, 2 cups<br>Spaghetti sauce, 1 cup<br>Reduced fat milk, 1½ cups<br>Green beans, 1 cup | 761 kcal<br>66% carbohydrate<br>(129 g) |
| Orange, 1 large<br>Reduced fat milk, 1½ cups<br>Chicken noodle soup, 1 cup<br>Saltine crackers, 12<br>Buttered beans, 1 cup<br>Corn, 1 cup<br>Angelfood cake, 1 slice | 829 kcal<br>70% carbohydrate<br>(160 g) |

The rule of thumb when timing pre-activity meals is to allow 4 hours for a big meal (about 1200 kcal), 3 hours for a moderate meal (about 800 to 900 kcal), 2 hours for a light meal (about 400 to 600 kcal), and an hour or less for a snack (about 300 kcal).

Cereal: ©George Doyle/Getty Images RF; Spaghetti: ©Comstock/Getty Images RF

Elite athletes, such as Olympic beach volleyball gold medal winner Kerri Walsh, know that modifying their diet and training regimens to match the specific needs of their sports is key to optimum performance. Replenishing carbohydrates and fluids is especially important when training.

McGraw-Hill Education/Gary He, photographer

▶ Why isn't fat a way to improve athletic performance during an endurance event? Although it is true that fat is used along with carbohydrate as fuel during prolonged aerobic activity, the processes of digestion, absorption, and metabolism of fat are relatively slow. Therefore, the consumption of fat during activity is not likely to translate into better athletic performance.

▶ For more information on sports nutrition, visit the Gatorade Sport Science Institute website (www.gssiweb.com). For more information on sports medicine, visit the home page of the *Physician and Sports Medicine* journal at www.physsportsmed .com; this journal details current issues in sports medicine, including injury prevention, nutrition, and exercise. Also helpful are the websites of the American College of Sports Medicine (www.acsm.org), Centers for Disease Control and Prevention (www.cdc.gov/nccdphp/dnpa), and American Council on Exercise (www.acefitness.org).

## Recovery Meals

After strenuous exercise, recovery meals promote muscle protein synthesis and reloading of muscles with glycogen. Optimal recovery depends on the quantity and timing of nutrient intake.[15] Glycogen resynthesis is greatest immediately after exercise because the muscles are very insulin sensitive.[1] Thus, consuming 1 to 1.5 g of carbohydrate per kg of body weight up to 1 hour after exercise and at 2-hour intervals for up to 6 hours helps reload muscles with glycogen for the next day's exercise (Table 11-10). High glycemic load carbohydrates consumed immediately after exercise can contribute to rapid glycogen synthesis (see Table 5-7 in Chapter 5). Athletes can consume a simple sugar candy, a sugar-sweetened soft drink, fruit or fruit juice, or a sports or recovery drink right after training. Later, whole-grain bread, mashed potatoes, and brown rice can contribute additional carbohydrates during a meal. Eating a small amount of protein (10 to 20 g) along with carbohydrate can stimulate muscle repair and muscle protein synthesis during recovery from exercise. For a 154 lb (70 kg) athlete, the amount of carbohydrate and protein needed for recovery corresponds to about 70 g of carbohydrate and 15 g of protein, the amount found in 1 bagel and 16 ounces of Gatorade® recovery beverage or in a turkey sandwich with 1 cup of fruit-flavored yogurt.

In summary, the following are key factors for achieving the most rapid recovery of muscle glycogen after exercise.

- The availability of adequate carbohydrate
- The ingestion of carbohydrate as soon as possible after the completion of exercise
- The selection of high glycemic load carbohydrates

Fluid and electrolyte intake also is an essential component of an athlete's recovery diet.[15,32] Recall that replenishing body fluids as quickly as possible is especially important if more than 1 workout a day is performed or if the environment is hot and humid. Generally, if food and fluid intake is sufficient to restore weight loss, it also will supply enough electrolytes to meet the athlete's needs during recovery from endurance activities.

| Table 11-10  Sample Postexercise Meals for Rapid Muscle Glycogen Replacement |
| --- |
| **Option 1** |
| Bagel, 1 regular<br>Peanut butter, smooth, 2 tbsp<br>Low-fat chocolate milk, 8 oz<br>Banana, 1 medium<br>**600 kcal, 87 g carbohydrate, 23 g protein, 18 g fat** |
| **Option 2** |
| Carnation® Breakfast Essentials, 1 packet<br>Fat-free milk, 8 oz<br>Banana, 1 medium<br>Peanut butter, 1 tbsp<br>Blend until smooth.<br>**438 kcal, 70 g carbohydrate, 17 g protein, 10 g fat** |
| **Option 3** |
| Gatorade® recovery drink (16.9 oz)<br>**130 kcal, 14 g carbohydrate, 16 g protein** |

Bagel: Burke/Triolo Productions/Getty Images; Banana: ©Brand X Pictures/Getty Images RF; Milk: Burke/Triolo Productions/Getty Images

### Knowledge Check

1. What is the purpose of a pre-exercise meal?
2. What is the primary nutrient that should be consumed in a pre-exercise meal?
3. How important is timing in pre-exercise and recovery meals?
4. What effect does glycemic index have on a recovery meal?

# Culinary Perspective

## Sports Nutrition in the Home Kitchen

Commercial sports nutrition products range from drinks to even jelly beans. These products provide a convenient way to get tailored sports nutrition on the go, but they can be expensive. It's easy to make your own sports nutrition products at home or choose whole-food alternatives. Studies show that athletes perform equally well, whether they are using commercial sports products, whole foods, or homemade options. Check out the alternatives to commercial sports nutrition products in the chart.

| Commercial Sports Product (Calories/Carb g/ Pro g/Fat g/Sodium mg)* | Homemade Option | Whole-Food Option |
|---|---|---|
| Energy sports bar (240/42/5/10/200) | Banana bread, homemade granola bars (mix oats, seeds, dried fruit, cinnamon, honey, and peanut or nut butter; then press into a parchment paper–lined pan; refrigerate overnight) | Whole-grain toast or crackers with peanut butter |
| Sports gel/jellybeans (100/30/0/0/70) | Pitted dates stuffed with an almond (or another nut) | Raisins, apple sauce squeeze pack, banana, honey packet, maple syrup, blackstrap molasses |
| Sports drink (8 oz) (50–70/16/0/0/110) | Mix (yields 1 cup): ¼ cup sugar ¼ tsp salt ¼ cup hot water 1/3 cup 100% fruit juice 3½ cups cold water | Plain water, coconut water (eat with a snack containing carbohydrate and sodium, such as pretzels) |

Supakorn Sangpech/Shutterstock

*Average nutrient content: carb = carbohydrates, pro = protein, g = grams, mg = milligrams.

# GLOBAL PERSPECTIVE

### Gene Doping and Editing in the Wide World of Sports

In their quest for excellence, athletes are constantly looking for ways to gain an edge over the competition. Some of the methods they use, such as healthy eating and rigorous training programs, are acceptable means for becoming the best athlete possible. However, other substances and methods, such as using anabolic steroids or blood doping, are neither appropriate nor permitted by the International Olympic Committee or the national governing bodies. One of the newest inappropriate methods, considered an impending threat to the world of sports, is gene doping—the practice of using gene therapy to artificially enhance athletic performance.

With the advent of precise gene-editing tools such as CRISPR, scientists and sports officials have speculated that unscrupulous athletes may attempt to use them to gain an advantage over competitors. CRISPR is being tested in a few medical studies with humans, but not for performance-enhancing purposes.[34] In response to this rapidly developing technology, the World Anti-Doping Agency (WADA) has expanded the ban on gene doping to include gene editing designed to enhance athletic performance by altering genome sequences and/or the transcriptional or epigenetic regulation of gene expression.

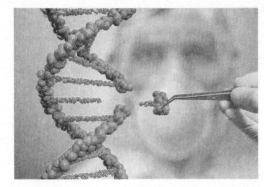

vchal/Shutterstock

Sports scientists who are advising WADA are considering how to best detect gene doping. One way would be to have a record of an athlete's entire genetic code. Another may be to have a record of just the sections of the genetic code that contains the genes associated with athletic performance.[35]

| Critical | Thinking |
|---|---|
| How would you advise someone who was planning to buy a purported "muscle-building" protein supplement? What risks are important to point out? | |

**ergogenic aid** Substance or treatment intended to directly improve exercise performance.

# 11.8 Ergogenic Aids to Enhance Athletic Performance

Today's athletes are as likely as their predecessors to seek ways to improve performance. Most don't want to miss out on any advantage, whether real or perceived, that might give them the winning edge. As a result, many experiment with diet composition, supplements, and other aids in hopes of gaining an ergogenic (work-producing) benefit. An **ergogenic aid** is a nutritional, psychological, pharmacologic, mechanical, or physiological substance or treatment intended to improve exercise performance. Most of these aids, such as artichoke hearts, bee pollen, dried adrenal glands from cattle, seaweed, freeze-dried liver flakes, gelatin, and ginseng, are ineffective. In fact, scientific support for ergogenic effectiveness exists for only a few dietary substances: sufficient water and electrolytes; abundant carbohydrates; a healthy, varied diet; and caffeine.[1,36] Protein and amino acid supplements are not among those aids because athletes can easily meet protein needs from foods, as Table 11-6 demonstrates. Nutrient supplements should be used only to meet specific dietary shortcomings, such as inadequate iron intake.

As summarized in Table 11-11, no scientific evidence supports the effectiveness of many substances touted as performance-enhancing aids. Many are useless; some are dangerous. The risk-benefit ratio of any ergogenic aid merits careful evaluation before use.[1] Athletes should be skeptical of any substance until its ergogenic effect is scientifically verified. The FDA has a limited ability to regulate dietary supplements (see Chapter 1), and the manufacturing processes for dietary supplements are not as tightly regulated as they are for medications. As a result, some supplements do not contain the substance and/or the amount listed on the label and may contain substances that will cause athletes to test positive for various banned substances. Even substances that have been supported by systematic scientific studies should be used with caution because the conditions under which they were tested may not match those of the intended use.

**Table 11-11  Evaluation of Popular Ergogenic Aids**

| Substance/Practice | Rationale for Use | Reality |
|---|---|---|
| **Sports Supplements That May Be Useful in Some Circumstances** | | |
| Creatine | Increase phosphocreatine (PCr) in muscles to keep ATP concentration high | Use of 20 g per day for 5 to 6 days and then a maintenance dose of 2 g per day may improve performance in athletes who undertake repeated bursts of activity, such as in sprinting and weight lifting. Vegetarian athletes may especially show benefits because creatine is low or nonexistent in their diets. Some of the muscle weight gain noted with use results from water contained in muscles. Endurance athletes do not benefit from use. Little is known about the safety of long-term creatine use. Continual use of high doses has led to kidney damage in a few cases. Cost: $25 to $65 per month. |
| Sodium bicarbonate (baking soda) | Counter lactic acid buildup | Partially effective in some circumstances in which lactate is rapidly produced, such as wrestling, but can induce nausea and diarrhea. The dose used is 300 mg/kg, 1 to 3 hours before exercise. Cost: nil. |
| Caffeine | Stimulate nervous system, heighten sense of awareness, and may enhance nerve conduction | Drinking 2 to 3 5-ounce cups of coffee (3–9 milligrams of caffeine per kg of body weight) about 1 hour before an event improves performance in most events except short, high-intensity exercise. It also may increase cognitive functioning during exercise. Intakes of more than about 600 milligrams (6 to 8 cups of coffee) elicit a urine concentration illegal under NCAA rules (greater than 15 mg per ml). Possible side effects are increased blood pressure, increased heart rate, intestinal distress, and insomnia. Cost: $0.08 per 300 mg. |

**Table 11-11  Continued**

| Substance/Practice | Rationale for Use | Reality |
|---|---|---|
| **Sports Supplements That Are Not Effective** | | |
| Beta-hydroxy-beta methylbutyric acid (HMB) | Decrease protein catabolism, causing a net growth-promoting effect | Research in livestock and humans suggests that supplementation with this substance may increase muscle mass. Still, safety and effectiveness of long-term HMB use in humans are unknown. Cost: $100 per month. |
| Ketone salts | Alternative fuel for muscle and glycogen sparing | Contrasting evidence exists regarding the value of exogenous ketones for endurance performance. Further studies are needed. |
| Coenzyme Q-10 | Part of the electron transport chain; improve VO2max and reduce fatigue | Research in athletes has found that coenzyme Q-10 has anti-inflammatory benefits but has found no increase in performance. |
| Glucosamine | Aid in repair of joint damage | Most of the positive evidence is for repair of knee damage in older people, but a large study showed no clear benefit for such use. It may be of use to athletes experiencing knee damage, but reliable evidence is lacking. Cost: $30 per month. |
| **Dangerous or Illegal Substances/Practices: U.S. Anti-Doping Agency Prohibits the Use of These Substances** | | |
| Anabolic steroids (and related substances, such as androstenedione and tetrahydrogestrinone [THG]) | Increase muscle mass and strength | Although effective for increasing protein synthesis, anabolic steroids are illegal in U.S. unless prescribed by a physician. They have numerous potential side effects, such as premature closure of growth plates in bones (possibly limiting the potential height of a teenage athlete), bloody cysts in the liver, increased risk of cardiovascular disease, increased blood pressure, and reproductive dysfunction. Possible psychological consequences include increased aggressiveness, drug dependence (addiction), withdrawal symptoms (e.g., depression), sleep disturbances, and mood swings (known as "roid rage"). Use of needles for injectable forms adds further health risk. They are banned by the International Olympic Committee, National Football League, Major League Baseball, and other sports organizations. |
| Growth hormone | Increase muscle mass | It may increase height; at critical ages, it also may cause uncontrolled growth of the heart and other internal organs and even death; it is potentially dangerous and requires careful monitoring by a physician. The use of needles for injections adds further health risk. It is banned by the International Olympic Committee. |
| Blood doping | Enhance aerobic capacity by injecting red blood cells harvested previously from the athlete or use the hormone erythropoietin (Epogen®) to increase red blood cell number | It may offer aerobic benefit, but very serious health consequences are possible, including thickening of the blood, which puts extra strain on the heart. It is an illegal practice under Olympic guidelines. |
| Gamma hydroxybutyric acid (GHB) | Promoted as a steroid alternative for body building | The FDA has never approved it for sale as a medical product; it is illegal to produce or sell GHB in the U.S. GHB-related symptoms include vomiting, dizziness, tremors, and seizures. Many victims require hospitalization, and some have died. Clandestine laboratories produce virtually all the chemical accounting for GHB abuse. The FDA is working with the U.S. Attorney's office to arrest, indict, and convict individuals responsible for the illegal operations. |

Substances that are promoted to athletes but have yet to show any clear ergogenic effects include pyruvic acid (pyruvate), ribose, chromium, coenzyme Q-10, medium-chain triglycerides, L-carnitine, conjugated linoleic acid (CLA), bovine colostrum, and insulin. Any use of these products is not recommended at this time. Some of these substances are defined in the glossary.

Finally, rather than waiting for a magic bullet to enhance performance, athletes should concentrate their efforts on improving their training routines and sports techniques while consuming healthy diets.

The National Collegiate Athletic Association's Committee on Competitive Safeguards and Medical Aspects of Sports has developed lists of supplements that are permissible and nonpermissible for athletic departments to dispense. The following are some key examples.

| *Permissible* | *Nonpermissible* |
|---|---|
| Vitamins and minerals | Amino acids |
| Energy bars (if no more than 30% protein) | Creatine |
| | Glycerol |
| Sports drinks | Beta-hydroxy-beta methylbutyric acid (HMB) |
| Meal replacement drinks (e.g., Ensure Plus®, Boost®) | L-carnitine |
| | Protein powders |

## Knowledge Check

1. What is an ergogenic aid?
2. How do you know if a supplement is safe?
3. Should all athletes take a supplement?

## CASE STUDY FOLLOW-UP

Ryan McVay/
Photodisc/Getty
Images

Jake decided to see the sports dietitian at his college. She asked him to keep a 3-day food record so that she could analyze his diet. When they met to go over his diet, she told him that he was overconsuming protein and underconsuming carbohydrate. Jake told her he needed more protein due to his lifting, but the registered dietitian nutritionist pointed out that he was consuming almost 300 g/day of protein. When the dietitian calculated his protein requirement using the recommended protein requirements for athletes, he actually needed 120 g/day. He was eating so much protein that he was not getting enough carbohydrate to fuel his workouts. To increase carbohydrate intake, they worked out a plan to add more whole-grain cereals for breakfast instead of the protein shake. He added rice, potatoes, or pasta and vegetables with his chicken for the evening meal and cut back his protein shake to once a day. A couple of weeks later, Jake's energy returned, he was lasting longer in the weight room, and he was lifting more weight.

# *Chapter Summary*

## 11.1 The benefits of regular physical activity include enhanced heart function, improved balance, reduced risk of falling, better

sleep habits, healthier body composition, and reduced injury to muscles, tendons, and joints. A gradual increase in regular physical activity is recommended for all healthy persons. A minimum plan includes 30 minutes of physical activity on most (or all) days; 60 to 90 minutes per day provides even more benefit, especially if weight control is an issue.

Samuel Borges Photography/Shutterstock

## 11.2 A good fitness program meets a person's needs. To reach goals, fitness program

planning should consider the mode, duration, frequency, intensity, and progression of exercise, as well as consistency and variety. Before starting a new fitness

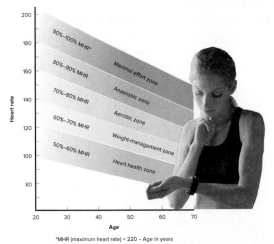

Duncan Smith/Getty Images

program, discuss program goals with a health-care provider. Also, assess and record baseline fitness scores. Most new exercise programs should start with short intervals of exercise at the lower end of the maximum heart rate target zone and work up to a total of 30 minutes of activity incorporated into each day. To prepare for and recover safely from an exercise session, warm-up and cooldown periods should be included.

## 11.3 At rest, muscle cells use mainly fat for fuel. For intense exercise of short duration, muscles

use mostly phosphocreatine (PCr) for energy. During more sustained, intense activity, muscle glycogen breaks down to lactic acid, providing a small amount of ATP. For endurance exercise, both fat and carbohydrate are used as fuels; carbohydrate is used increasingly as activity intensifies. Little protein is used to fuel muscles. Fuel sources for muscle cells can be estimated based on percent of $VO_{2max}$.

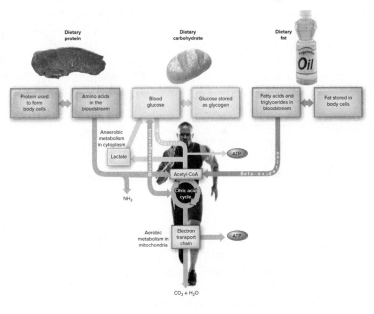

Protein: ©Stockbyte/Getty Images RF; Carbohydrate: ©Ingram Publishing/Alamy; Fat: McGraw-Hill Education/Jacques Cornell, photographer; Male Jogging: 4x6/Getty Images

## 11.4 Physical activity has many effects on the body. The most pronounced effects are

typically seen in the muscular, circulatory, and skeletal systems. The body contains 3 major types of muscle tissue: skeletal muscle, smooth muscle, and cardiac muscle. Skeletal muscle is composed of 3 main types of muscle fibers, which have distinct characteristics. Prolonged, low-intensity exercise, such as a slow jog, uses mainly type I muscle fibers, so the predominant fuel is fat. As exercise intensity increases, type IIA and type IIX fibers are gradually recruited; in turn, the contribution of glucose as a fuel increases. Type IIA and type IIX fibers also are important for rapid movements, such as a jump shot in basketball. The relative proportions of the 3 fiber types throughout the muscles of the body vary from person to person and are constant throughout each person's life. With training, muscle strength becomes matched to the muscles' work demands. Muscles enlarge after being made to work repeatedly. Repeated aerobic exercise strengthens the heart and increases the number of capillaries in muscle tissue; as a result, oxygen can be delivered more easily to muscle cells. Another adaptation that occurs with exercise is increased bone density.

View Stock/Getty Images

## 11.5 Athletic training and genetic makeup are very important determinants of athletic performance.

Monitoring body weight is an easy way to assess the adequacy of calorie intake. Athletes should strive to maintain weight during competition and training. Athletes should obtain at least 60% of total energy needs from carbohydrates. Carbohydrate-loading regimens usually increase muscle glycogen stores 50 to 85% over typical conditions. Carbohydrate loading is for athletes who compete in continuous, intense aerobic events lasting more than 60 to 90 minutes. A fat intake of 15 to 25% of energy is generally recommended for athletes. Typical recommendations for protein intake for most athletes range from 1.2 to 1.7 g of protein/kg of body weight. The extra protein is needed for the repair of tissue and the synthesis of new muscle that results from training. Vitamin and mineral needs are the same or slightly higher for athletes, compared with those of sedentary adults. Relative Energy Deficiency in Sport, or REDS, occurs when athletes have an energy deficiency insufficient to support daily living, growth, and health; if the deficiency continues, it can lead to poor bone growth, stress fractures, and early osteoporosis.

Ingram Publishing/SuperStock

## 11.6 To maintain the body's ability to regulate internal temperature,

athletes must consume sufficient fluids because dehydration leads to a decline in endurance, strength, and overall performance and sets the stage for heat exhaustion, heat cramps, and potentially fatal heatstroke. During exercise, the recommended fluid status goal is a loss of no more than 2% of body weight. To prevent water intoxication, athletes should drink beverages containing sodium and should consume enough fluid during exercise to minimize the loss of body weight. Long-endurance athletes should consider consuming sports drinks with electrolytes.

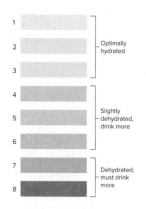

## 11.7 The composition of food eaten before, during,

and after athletic events or exercise training sessions can affect performance and the speed with which an athlete recovers from the exercise bout. Pre-exercise training meals keep the athlete from feeling hungry before and during the exercise bout and maintain optimal levels of blood glucose for the exercising muscles. A pre-exercise meal should be high in carbohydrate, low in fat, and readily digested. For sporting events lasting more than 60 minutes, consuming carbohydrate during activity can improve athletic performance. Carbohydrate rich foods and a small amount of protein should be consumed 30 minutes after exercise and again at 2-hour intervals for up to 6 hours.

| Table 11-8 Rule of Thumb for Approximate Pre-event Carbohydrate Intake | | |
|---|---|---|
| Hours Before | Grams per Kilogram Body Weight | For a 70 kg Person |
| 1 | 1 | 70 |
| 2 | 2 | 140 |
| 3 | 3 | 210 |
| 4 | 4 | 280 |

## 11.8 An ergogenic aid is a nutritional,

psychological, pharmacologic, mechanical, or physiological substance or treatment intended to improve exercise performance. Most of these aids are ineffective.

©Brand X Pictures/Getty Images RF

## Global Perspective

Gene doping and editing are the practice of using gene therapy to artificially enhance athletic performance. The same techniques used to alter a person's DNA to fight diseases could be used to boost an athlete's competitive capabilities.

# Study Questions

1. The benefits of regular physical activity include _____.

   a. a reduced risk of falling
   b. better sleep habits
   c. healthier body composition
   d. all of the above

2. Which mode of exercise is defined as any activity that uses large muscle groups, can be maintained continuously, and is rhythmic in nature?

   a. aerobic
   b. resistance
   c. flexibility
   d. none of the above

3. The predominant fuels for the 50-meter sprint are _____.

   a. fat and protein
   b. carbohydrate and protein
   c. protein and phosphocreatine
   d. ATP and phosphocreatine

4. The predominant fuel for a 2-hour marathon is _____.

   a. protein
   b. fat
   c. carbohydrate
   d. water

5. The amount of ATP stored in a muscle cell can keep a muscle active for about _____.

   a. 2 to 4 seconds
   b. 10 to 30 seconds
   c. 1 to 3 minutes
   d. 1 to 3 hours

6. There are 4 main types of muscle fibers.

   a. true
   b. false

7. Although genetics largely determines the proportion of muscle fiber type, training can develop muscle fibers within some limits.

   a. true
   b. false

8. Which of the following athletes would *not* benefit from carbohydrate loading?

   a. marathon runner
   b. long-distance cyclist
   c. triathlete
   d. football player

9. Athletes who are involved in endurance activities may need to consume _____ grams of carbohydrates per kilogram of body weight.

   a. 5 to 7
   b. 7 to 8
   c. up to 10
   d. 3 to 4

10. All athletes should consume at least 2.0 grams of protein/kg of body weight.

    a. true
    b. false

11. Iron deficiency can impair athletic performance.

    a. true
    b. false

12. Relative Energy Deficiency in Sport (REDS) can lead to _____.

    a. poor bone growth
    b. stress fractures
    c. early osteoporosis
    d. weight loss
    e. all of the above

13. Water intoxication is a condition that can occur when athletes drink too much _____.

    a. alcohol
    b. water
    c. sports drinks
    d. milk

14. Thirst is an accurate indicator of fluid needs.

    a. true
    b. false

15. Most ergogenic aids are effective and enhance athletic performance.

    a. true
    b. false

16. How does greater physical fitness contribute to greater overall health?

17. What are the characteristics of a good fitness program?

18. What is the difference between anaerobic and aerobic exercise? Explain why aerobic respiration is increased by regular exercise.

19. Describe how ATP is resupplied after the initiation of exercise and at various times thereafter.

20. Describe the types of muscle fibers and the predominant fuel used by each.

21. List 5 nutrients of special interest to athletes and appropriate food sources of each.

22. What are some measures athletes can take to determine if their fluid intake is adequate?

23. What advice would you give a friend planning to run a 5-kilometer race about food intake before, during, and after the event?

24. What advice would you give a friend about the general effectiveness of amino acid supplements?

25. Your roommate is determined to lose weight; he has committed to run about 20 miles a week and is restricting his carbohydrate intake. What advice would you give him about his diet?

26. Ryan is a defensive lineman on your university football team; he is 6 feet 5 inches tall and weighs 290 lb. Ryan eats about 5000 calories per day. Calculate how many grams of protein he needs to consume. Does his recommendation fall within the AMDR for protein?

27. How might gene therapy give athletes a competitive edge?

red peppers: Jules Frazier/Getty Images; grapes: ©Stocktrek/Getty Images RF

# References

1. Rawson E and others. *Williams' nutrition for health, fitness, and sport.* 12th ed. New York: McGraw-Hill; 2020.

2. Bacon S and others. Effects of exercise, diet and weight loss on high blood pressure. *Sports Med.* 2004;34:307.

3. Strengthen your muscles to reduce diabetes. *Harv. Health Lett.* 2014;39:9.

4. Ross R and others. Exercise induced reduction in obesity and insulin resistance in women: A randomized control trial. *Obesity Res.* 2004;12:789.

5. Farrell SW and others. Is there a gradient of mortality risk among all men with low cardiorespiratory fitness? *Med Sci Sport Exerc.* 2015;47:1825.

6. Warden SJ and others. Physical activity when young provides life-long benefits to cortical bone size and strength in men. *PNAS.* 2014;111:5337.

7. Chan B and others. Incident fall risk and physical activity and physical performance among older men: The osteoporotic fractures in men study. *Am J Epidemiol.* 2007;165:696.

8. American Heart Association and American Stroke Association. Recommendations for physical activity in adults. 2014; www.heart.org.

9. Jeukendrup A. *Sport nutrition.* 3rd ed. Champaign, IL: Human Kinetics; 2019.

10. Jeukendrup A. A step towards personalized sports nutrition: Carbohydrate intake during exercise. *Sports Med.* 2014;44:25.

11. Moore DR and others. Ingested protein dose response of muscle and albumin protein synthesis after resistance exercise in young men. *Am J Clin Nutr.* 2009;89:161.

12. Phillips SM. The impact of protein quality of resistance exercise induced changes in muscle mass. *Nutr Metab.* 2016;13:64.

13. Bergstrom J and others. Diet, muscle glycogen and physical performance. *Acta Physiol Scand.* 1967;71:140.

14. Hearris MA and others. Regulation of muscle glycogen metabolism during exercise: Implications for endurance performance and training adaptations. *Nutrients.* 2018;45:615.

15. Position of the Academy of Nutrition and Dietetics, Dietitians of Canada, and the American College of Sports Medicine. Nutrition and athletic performance. *JAND.* 2016;116:501.

16. Doering TM and others. Repeated muscle glycogen supercompensation with four days' recovery between exhaustive exercise. *J Sci Med Sport.* 2019;22:907.

17. Zajac A and others. The effects of a ketogenic diet on exercise metabolism and physical performance in off-road cyclists. *Nutrients.* 2014;6:2491.

18. Burke L. Re-examining high-fat diets for sports performance: Did we call the 'nail in the coffin' too soon? *Sports Med.* 2015;45:S33.

19. Phillips SM and others. Dietary protein for athletes: From requirements to optimum adaptation. *J Sports Sci.* 2011;29:S29.

20. Betts J and others. Recovery of endurance running capacity effect of carbohydrate-protein mixtures. *Int J Sport Nutr Exerc Metab.* 2005;15:590.

21. Phillips SM and others. Body composition and strength changes in women with milk and resistance exercise. *Med Sci Sport Exerc.* 2010;42:1122.

22. Abe T and others. Whole body muscle hypertrophy from resistance training: Distribution and total mass. *Br J Sports Med.* 2003;37:543.

23. Woolf K, Manore MM. B-vitamins and exercise: Does exercise alter requirements. *Int J Sport Nutr Exerc Metab.* 2006;16:453.

24. Hinton PS. Iron and the endurance athlete. *Appl Physiol Nutr Metab.* 2014;39:1012.

25. Daly JP, Stumbo JR. Female athlete triad. *Prim Care.* 2018;45:4.

26. Mountjoy M and others. The IOC consensus statement on relative energy deficiency in sport (RED-S): 2018 Update. *Br J Sports Med.* 2018;52:687.

27. DeFranco MJ and others. Environmental issues for team physicians. *Am J Sports Med.* 2008;36:2226.

28. AAP. Climatic heat stress and the exercising child and adolescent. *Pediatrics.* 2007;120:683.

29. Burdon CA and others. Influence of beverage temperature on exercise performance in the heat: A systematic review. *Int J Sport Nutr Exerc Metab.* 2010;20:166.

30. Sterns RH and others. Treatment of hyponatremia. *Semin Nephrol.* 2009;29:282.

31. Davis J and others. Carbohydrate drinks delay fatigue during intermittent, high-intensity cycling in active men and women. *Int J Sport Nutr Exerc Metab.* 1997;7:261.

32. Shirreffs SM, Sawka MN. Fluid and electrolyte needs for training, competition and recovery. *J Sports Sci.* 2011;29(Suppl 1):S39.

33. Nybo L. CNS fatigue provoked by prolonged exercise in the heat. *Front Biosci.* 2010;2:779.

34. Beiter T and others. Direct and long-term detection of gene doping in conventional blood samples. *Gene Ther Advance Online.* 2010;September 2:1.

35. Miller E. Future Olympic athletes could be required to have their entire DNA sequenced to test for gene doping. *Science.* 2018;82:85.

36. Foskett A and others. Caffeine enhances cognitive function and skill performance during simulated soccer activity. *Int J Sport Nutr Exerc Metab.* 2009;19:410.

Spacecraft are designed to block much of the radiation from the sun, which decreases the body's synthesis of vitamin D. Thus, astronauts receive vitamin D supplements. Learn more about space food at **www.nasa.gov/aeroresearch/resources/ artifact-opportunities/space-food/**. forplayday 123RF.com

# The Fat-Soluble Vitamins

## Learning Objectives

**After studying this chapter, you will be able to**

1. Define the term *vitamin* and list 3 characteristics of vitamins as a group.
2. Classify the vitamins according to whether they are fat soluble or water soluble.
3. List 3 important food sources for each fat-soluble vitamin.
4. List the major functions of each fat-soluble vitamin.
5. Describe the deficiency symptoms for each fat-soluble vitamin and state the conditions in which deficiencies are likely to occur.
6. Describe the toxicity symptoms caused by excess consumption of certain fat-soluble vitamins.
7. Evaluate the use of vitamin and mineral supplements with respect to their potential benefits and risks to health.

## Chapter Outline

**WHEN IT COMES TO VITAMINS,** we often hear "if a little is good, then more must be better." Some people believe that consuming vitamins far in excess of their needs provides them with extra energy, protection from disease, and prolonged youth. Actually, our total vitamin needs to prevent deficiency are quite small. In general, humans require about 1 oz (28 g) of vitamins for every 150 lb (70 kg) of food they consume. Although plants can synthesize all the vitamins they need, animals vary in their ability to synthesize vitamins. For example, guinea pigs and humans are among the few organisms that cannot make their own supply of vitamin C and must get it from their diets.

Long before any vitamins had been identified, certain foods were known to cure conditions brought on by what we now know are vitamin deficiencies. The ancient Greeks, for example, treated night blindness with beef liver, a rich source of vitamin A. As you'll see, vitamin A plays a critical role in vision. During the 15th and 16th centuries, many British sailors on long sea voyages died from the disease scurvy. After it was discovered that eating lemons and limes prevents scurvy, citrus was included as a routine part of British sailors' rations, and deaths from scurvy declined sharply. We now know that scurvy results from a deficiency of vitamin C.

Today, consumers in countries like the U.S. and Canada rarely consider vitamin deficiencies when making choices about diet and vitamin supplements. Instead, many focus on the ability of vitamins, and of diets rich in vitamins, to decrease the risk of developing chronic diseases, such as cancer, heart disease, and bone disease. However, some vitamin deficiencies are still a public health concern in specific groups of people in higher-income countries and in large populations in many low-income countries. Vitamin A deficiency, for example, is a primary cause of childhood blindness and disease in many low-income areas of the world. Vitamin D deficiency plays a role in bone disorders in all countries.

Vitamins are divided into fat-soluble vitamins and water-soluble vitamins. This chapter will provide an overview of vitamins and an in-depth discussion of fat-soluble vitamins.

**Fat-Soluble Vitamins**

    Vitamin A
    Vitamin D
    Vitamin E
    Vitamin K

**Water-Soluble Vitamins**

    B-vitamins
      Thiamin
      Riboflavin
      Niacin
      Pantothenic acid
      Biotin
      Folic acid
      Vitamin B-6 (pyridoxine)
      Vitamin B-12 (cobalamin)
    Vitamin C

▶ The synthetic and natural forms of most vitamins have similar characteristics. One exception, natural vitamin E, is about twice as potent as the synthetic form. In contrast, synthetic folic acid is almost twice as potent as the natural form.

# 12.1 Vitamins: Essential Dietary Components

**Vitamins** are essential, organic (containing carbon bonded to hydrogen) substances needed in small amounts in the diet. They are not a source of energy. Instead, they aid in energy metabolism as well as in the growth, development, and maintenance of body tissues.

During the 1st half of the 20th century, scientists identified each of the 13 vitamins now recognized as essential. For the most part, as the vitamins were discovered, they were named alphabetically: A, B, C, D, and E. Later, some substances originally classified as B-vitamins were dropped from the list because they were shown to be nonessential substances. The B-vitamins originally were thought to have a single chemical form but turned out to exist in many forms. Thus, the label "vitamin B" now comprises 8 B-vitamins. Vitamins A, D, E, and K dissolve in organic solvents, such as ether and benzene, and are referred to as **fat-soluble vitamins.** The B-vitamins and vitamin C, in contrast, dissolve in water and are classified as **water-soluble vitamins.**

Vitamins are indispensable in human diets because they either cannot be synthesized in the body at all or are synthesized in insufficient quantities. However, a substance does not qualify as a vitamin merely because the body can't make it. Evidence must suggest that health declines when the substance is not consumed.[1,2,3] In fact, when vitamin intake is insufficient to meet needs, a deficiency develops, accompanied by a measurable decline in health. If the deficiency is not in advanced stages, the deficiency and related symptoms can be alleviated by increased intakes of the vitamin.

In addition to preventing deficiency diseases, a few vitamins have been useful as pharmacologic agents (drugs) in treating several nondeficiency conditions. These treatments often require the administration of megadoses, amounts much higher than typical human needs for the vitamin. For example, megadoses of a form of niacin can be used as part of blood cholesterol–lowering treatment for certain individuals. Nevertheless, any claimed benefits for the use of vitamin supplements, especially intakes above the Upper Level (if set), should be viewed critically because many unproved claims are continually being made.[1,2,3]

Foods of both plant and animal origin supply vitamins in the diet. Dietary supplements also can provide needed vitamins. Whether vitamins in supplements are isolated from foods or synthesized in a laboratory, these vitamins are usually similar chemical compounds and generally work equally well in the body. Contrary to claims in the health food literature, "natural" vitamin supplements isolated from foods are, for the most part, no more healthful than those synthesized in a laboratory. However, vitamins consumed in foods as part of a varied diet may be more beneficial than vitamins taken separately as dietary supplements. Because some vitamins exist in several related forms that differ in chemical or physical properties, it is important to consume enough vitamins in the forms the body can use and in amounts that are in balance with other nutrients in the diet.

## Absorption of Vitamins

Fat-soluble vitamins are absorbed along with dietary fat. Thus, adequate absorption of fat-soluble vitamins depends on the efficient use of bile and pancreatic lipase in the small intestine to digest dietary fat and adequate intestinal absorption (Fig. 12-1). Under optimal conditions, about 40 to 90% of the fat-soluble vitamins are absorbed when they're consumed in recommended amounts. In contrast, absorption of B-vitamins and vitamin C typically ranges from 90 to 100% and occurs in the small intestine independent of dietary fat.

## Malabsorption of Vitamins

Vitamins consumed in food must be absorbed efficiently from the small intestine to meet body needs. If the absorption of a vitamin is decreased, a person must consume larger amounts of it to avoid deficiency symptoms. For example, fat malabsorption (resulting from GI tract and pancreatic disease) may cause poor absorption of fat-soluble vitamins.[1,2,3] Alcohol abuse and certain intestinal diseases also can lead to malabsorption of some B-vitamins (see Chapter 13). Individuals with these diseases usually require vitamin supplements to prevent deficiencies.

*Figure 12-1* An overview of the digestion and absorption of vitamins.

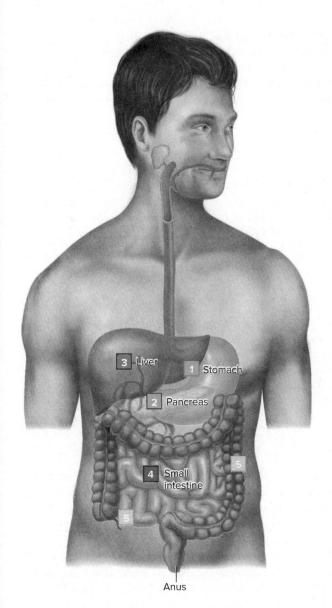

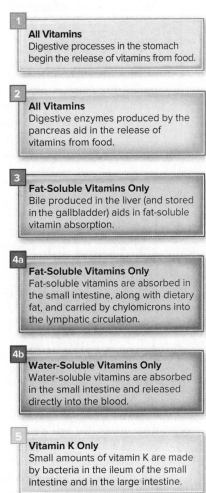

**1**

**All Vitamins**
Digestive processes in the stomach begin the release of vitamins from food.

**2**

**All Vitamins**
Digestive enzymes produced by the pancreas aid in the release of vitamins from food.

**3**

**Fat-Soluble Vitamins Only**
Bile produced in the liver (and stored in the gallbladder) aids in fat-soluble vitamin absorption.

**4a**

**Fat-Soluble Vitamins Only**
Fat-soluble vitamins are absorbed in the small intestine, along with dietary fat, and carried by chylomicrons into the lymphatic circulation.

**4b**

**Water-Soluble Vitamins Only**
Water-soluble vitamins are absorbed in the small intestine and released directly into the blood.

**5**

**Vitamin K Only**
Small amounts of vitamin K are made by bacteria in the ileum of the small intestine and in the large intestine.

▶ People with diseases that result in poor fat absorption, such as cystic fibrosis and Crohn disease, are at high risk of fat-soluble vitamin deficiencies. Medications, such as the weight loss drug orlistat (Xenical®, Alli®), also can interfere with fat absorption. Unabsorbed fat carries fat-soluble vitamins to the large intestine, where they are incorporated into the feces and excreted. A multivitamin and mineral supplement usually is prescribed as part of the treatment for preventing nutrient deficiencies associated with fat malabsorption.

## Transport of Vitamins

Once absorbed, fat-soluble vitamins are packaged for transport through the lymphatic system and delivered by the bloodstream to target cells throughout the body in a manner similar to that for dietary fats—namely, by way of chylomicrons and other blood lipoproteins.[1,2,3] Recall that, as a chylomicron circulates, much of its triglyceride content is removed by body cells. What remains—the remnant—is taken up by the liver. This remnant contains the fat-soluble vitamins absorbed from the diet. The liver then "repackages" fat-soluble vitamins with new proteins for transport in the blood, or it stores them in adipose tissue or the liver for future use. In contrast to the fat-soluble vitamins, water-soluble vitamins are delivered directly to the bloodstream and distributed throughout the body.

## Storage of Vitamins in the Body

With the exception of vitamin K, fat-soluble vitamins are not readily excreted from the body. Instead, they are often stored in the liver and/or adipose tissue. In contrast, most water-soluble vitamins are excreted from the body quite rapidly, resulting in limited stores. Two exceptions are vitamin B-12 and vitamin B-6, which

Foods provide a wide array of vitamins.

Ariel Skelley/Blend/Image Source

## Critical Thinking

Miguel is taking vitamin supplements that supply nutrients in amounts much higher than 100% of the Daily Value. How would you explain to him that the supplement he is taking is unsafe because it contains amounts that exceed the Daily Values by 10 times for many nutrients, including vitamin A?

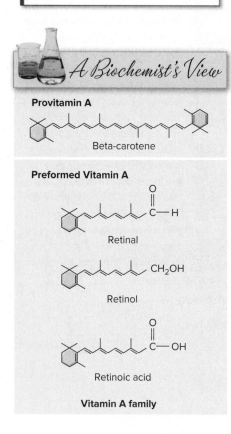

*A Biochemist's View*

**Provitamin A**

Beta-carotene

**Preformed Vitamin A**

Retinal

Retinol

Retinoic acid

**Vitamin A family**

are stored to a greater extent than the other water-soluble vitamins. Because of the limited storage of many vitamins, they should be consumed daily. However, the signs and symptoms of a deficiency usually do not occur until the vitamin is lacking in the diet for at least several weeks and body stores are essentially depleted. Thus, an occasional lapse in dietary intake of most vitamins is not a serious health concern in otherwise healthy individuals.

## Vitamin Toxicity

Although the toxic effects of an excessive intake of any vitamin is theoretically possible, toxicity from the fat-soluble vitamins A and D is the most likely to occur.[2,3] However, these vitamins are unlikely to cause toxicities unless taken in amounts at least 5 to 10 times greater than the DRI guidelines. Because the daily use of balanced multivitamin and mineral supplements usually supplies less than twice the Daily Value of the components, this practice is unlikely to cause toxic effects in adults.

### Knowledge Check

1. Which vitamins are classified as water soluble? Which are fat soluble?
2. How does the absorption of fat-soluble vitamins differ from that of water-soluble vitamins?
3. Why are large doses of certain fat-soluble vitamins more likely to cause toxic effects than large intakes of water-soluble vitamins?

## 12.2 Vitamin A

Although vitamins per se were not isolated and identified until the 20th century, vitamin A was known for more than 3500 years as a factor needed to prevent night blindness. Ancient Egyptians and the Greek physician Hippocrates recommended the consumption of beef liver, a cure that still works today. *Vitamin A* refers to the preformed retinoids and provitamin A carotenoids that can be converted to vitamin A activity.

**Retinoids** is a collective term for the biologically active forms of vitamin A. They are called preformed vitamin A because, unlike carotenoids, they do not need to be converted in the body to become biologically active. Retinoids exist in 3 forms: retinol (an alcohol), retinal (an aldehyde), and retinoic acid. The tail segment of the vitamin A structure terminates in 1 of these 3 chemical groups (alcohol, aldehyde, or acid) and determines the name or classification. To some extent, these forms can be interconverted (Fig. 12-2). However, retinoic acid cannot be converted back to the other forms. The ability to interconvert forms helps maintain adequate amounts of each retinoid form for its unique functions.

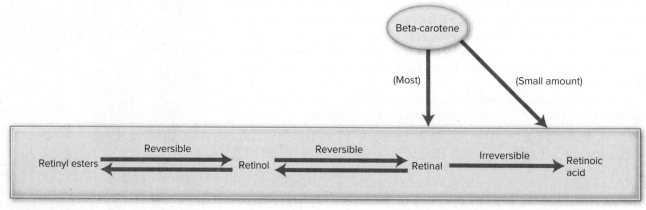

*Figure 12-2* Interconversions of beta-carotene and various retinoids. Retinyl esters do not have vitamin A activity until retinol and the attached fatty acid ester are separated in the intestinal tract.

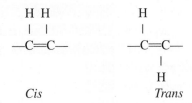

Retinol            Retinal            Retinoic acid

The tail of the vitamin A molecule can vary from *cis* to *trans* configuration. This orientation influences the function of the specific retinoid.

*Cis*                *Trans*

**Carotenoids** are yellow-orange pigmented materials in fruits and vegetables, some of which are **provitamins**—that is, they can be converted into vitamin A. Of the 600 or more known carotenoids, only alpha-carotene, beta-carotene, and beta-cryptoxanthin can be converted to biologically active forms of vitamin A. Other carotenoids, such as lycopene, lutein, and zeaxanthin, are not converted into vitamin A and, thus, do not have vitamin A activity in humans.[4]

## Vitamin A in Foods

Retinoids (preformed vitamin A) are found in liver, fish, fish oils, fortified milk, and eggs. Margarine is fortified with vitamin A, as are fat-free, low-fat, and reduced fat milks. The provitamin A carotenoids are found mainly in dark green vegetables, such as broccoli, spinach, and other greens as well as yellow-orange vegetables and fruits, such as carrots, winter squash, sweet potatoes, mangoes, cantaloupe, peaches, and apricots. About 70% of the vitamin A in the typical North American diet comes from animal (preformed vitamin A) sources, whereas plant-based carotenoids (provitamin A) provide most of the vitamin A in the diets of people in low-income areas of the world. Figure 12-3 displays the vitamin A content of various foods.

Beta-carotene has the greatest amount of provitamin A activity. It accounts for some of the orange color in carrots and other carotenoid rich foods. In dark green vegetables, this yellow-orange coloring is masked by the dark green pigment chlorophyll, although these vegetables do contain provitamin A. Therefore, consuming a varied diet rich in both dark green and yellow-orange vegetables can provide vitamin A.

In the past, the amounts of vitamin A (and most other nutrients) were expressed in International Units (IUs). Today, there are more sensitive means for measuring nutrients. Consequently, milligram (1/1000 of a gram) and microgram (1/1,000,000 of a gram) measurements have generally replaced IUs as the units of measure.

Dietary vitamin A activity is currently expressed in Retinol Activity Equivalents (RAE) to account for the different biological activities of retinol and the provitamin A carotenoids. One RAE is equal to 1 µg of retinol, 12 µg of beta-carotene, and 24 µg of the other 2 provitamin A carotenoids (alpha-carotene and beta-cryptoxanthin).[1,2] Table 12-1 is a tool for converting the amounts of vitamin A and carotenes expressed in a particular unit of measure into another unit of measure.

Retinol equivalent (RE), is another older unit of measurement for vitamin A activity. RE was based on the assumption that carotenoids made

**Critical Thinking**

Why do you think beef liver has such a high amount of vitamin A?

Many vegetables are rich in provitamin A carotenoids.

Photodisc Collection/Getty Images

*Figure 12-3*  Food sources of vitamin A. In addition to MyPlate food groups, yellow is used for oils, and pink is used for substances that do not fit easily into food groups (e.g., candy, salty snacks).

Ingram Publishing/Alamy Stock Photo

Source: ChooseMyPlate.gov, U.S. Department of Agriculture

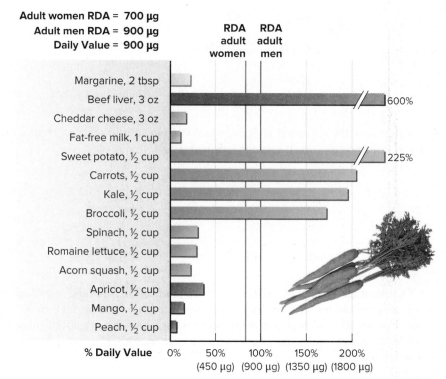

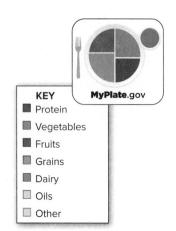

a greater contribution to vitamin A needs than is now known to be the case. Nutrient databases may contain this older RE standard because it will take some time to update these resources.

To compare the older RE or IU standards with current RAE recommendations, assume that, for any preformed vitamin A in a food or added to food, 1 RE (3.3 IU) = 1 RAE. There is no easy way to convert RE or IU units to RAE units for foods that naturally contain provitamin A carotenoids. A general rule of thumb is to divide the older values for foods containing carotenoids by 2 and then do the conversion from RE to IU to RAE, as shown in Table 12-1. There also is no easy way to do this calculation for food containing a mixture of preformed vitamin A and carotenoids. Generally speaking, these foods contain less vitamin A than the RE or IU values suggest.

## Vitamin A Needs

The RDA for vitamin A is 900 µg Retinol Activity Equivalents (RAE) per day for adult men and 700 µg RAE per day for adult women.[2] At this intake, adequate body stores of vitamin A are maintained in healthy adults. The Daily Value used on food packages and supplements is 900 µg. At present, there is no DRI for beta-carotene or any of the other provitamin A carotenoids.[1] The average intakes of adult men and women in North America currently meet DRI guidelines for vitamin A. Thus, vitamin A is no longer a mandatory nutrient to include on food and supplement labels.

**Table 12-1  Conversion Values for Retinol Activity Equivalents**

| 1 Retinol Activity Equivalent (RAE) | 1 IU Vitamin A Activity |
| --- | --- |
| = 1 µg retinol | = 0.3 µg retinol |
| = 12 µg beta-carotene | = 3.6 µg beta-carotene |
| = 24 µg alpha-carotene and beta-cryptoxanthin | = 7.2 µg alpha-carotene and beta-cryptoxanthin |

## Absorption, Transport, Storage, and Excretion of Vitamin A

Preformed vitamin A is found in foods of animal origin as retinol and retinyl ester-compounds (retinol attached to a fatty acid). Retinyl esters don't have vitamin A activity until the retinol and fatty acid are separated in the intestinal tract.[2] This process requires bile and pancreatic lipase enzymes. Up to 90% of retinol is absorbed into the cells of the small intestine via specific carrier proteins. After absorption, a fatty acid is attached to retinol to form a new retinyl ester. These retinyl esters are packaged into chylomicrons before entering the lymphatic circulation.

Dietary carotenoids often are attached to proteins that must be split off by digestive enzymes prior to absorption. Once freed, carotenoids are absorbed primarily by passive diffusion. Carotenoid absorption rates range from 5 to 60% of intake. Inside intestinal cells, provitamin A carotenoids can be cleaved to form retinal or, to a lesser extent, retinoic acid. Retinal is then converted to retinol. Retinol can have a fatty acid attached to it to become a retinyl ester and enter the lymphatic system as part of a chylomicron. The chylomicrons deliver vitamin A to tissues for storage or cellular use. Retinoic acid can enter the bloodstream directly for transport to the liver. Carotenoids also can enter the bloodstream directly; however, this occurs to a lesser extent.[4]

Over 90% of the body's vitamin A stores are found in the liver, with small amounts in adipose tissue, kidneys, bone marrow, testicles, and eyes. Normally, the liver stores enough vitamin A in the form of retinyl esters to last for several months to protect against vitamin A deficiency.[2]

When vitamin A (as a retinoid) is released from the liver into the bloodstream, it is bound to a retinol-binding protein (RBP) (Fig. 12-4). Synthesis of RBP depends on having adequate amounts of retinol, protein, and zinc. In the bloodstream, retinol-binding protein is bound to another protein called transthyretin (commonly known as prealbumin). In contrast, when carotenoids are released from the liver, they are carried by lipoproteins and are taken up into cells by specific apoprotein receptors.[4] Within the body's cells, retinoids are bound to specific RBPs, which direct them to functional sites in the cell. Nearly all cells contain 1 or more of these binding proteins. The distribution of the cellular RBPs differs among tissues, possibly reflecting their different functional needs for vitamin A.[4]

Although vitamin A is not readily excreted by the body, the primary means of excretion is via urine. Carotenoids are excreted via bile that is eliminated with feces.

▶ During protein-energy malnutrition, the synthesis of retinol-binding protein and transthyretin (prealbumin) is reduced by the lack of sufficient amino acids and energy. These proteins often are used as clinical indicators of a person's protein status because decreased concentrations in the blood can suggest inadequate protein intake.

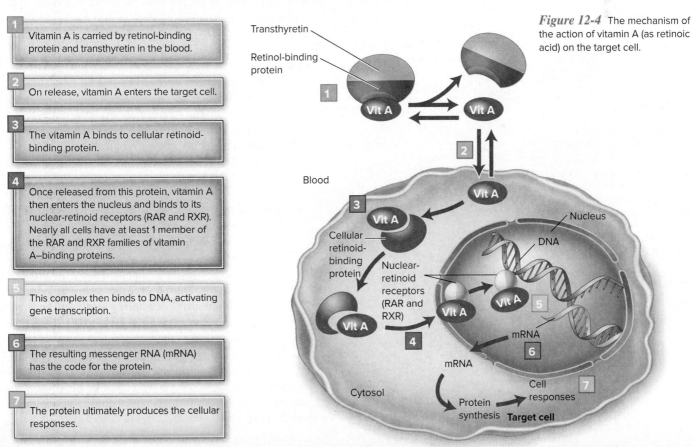

1. Vitamin A is carried by retinol-binding protein and transthyretin in the blood.

2. On release, vitamin A enters the target cell.

3. The vitamin A binds to cellular retinoid-binding protein.

4. Once released from this protein, vitamin A then enters the nucleus and binds to its nuclear-retinoid receptors (RAR and RXR). Nearly all cells have at least 1 member of the RAR and RXR families of vitamin A–binding proteins.

5. This complex then binds to DNA, activating gene transcription.

6. The resulting messenger RNA (mRNA) has the code for the protein.

7. The protein ultimately produces the cellular responses.

*Figure 12-4* The mechanism of the action of vitamin A (as retinoic acid) on the target cell.

## Functions of Vitamin A (Retinoids)

Retinoids perform different functions in the body. Key functions include growth and development, cell differentiation, vision, and immune function.

### Growth and Development

Retinoids play an important role in embryonic development. From studies of animals, scientists learned that vitamin A is involved in the development of the eyes, limbs, cardiovascular system, and nervous system. They also noted that a lack of vitamin A during early stages of pregnancy resulted in birth defects and fetal mortality (death). Retinoic acid also is necessary for the production, structure, and normal function of epithelial cells in the lungs, trachea, skin, and GI tract, as well as in many other systems. It is important for the formation and maintenance of mucous-forming cells in these organs.

### Cell Differentiation

In the cell nucleus, retinoids bind to 2 main families of retinoid receptors (see Fig. 12-4). These receptors (called **RXR** *and* **RAR**) bind to specific DNA sites (called retinoic acid response elements) that regulate the formation of messenger RNA (needed to copy genetic material from DNA) and the subsequent formation of proteins through gene expression. Gene expression directs **cell differentiation**—the process in which **stem cells** develop into specialized cells with unique functions in the body. Vitamin A is especially important in maintaining normal differentiation of the cells that make up the structural components of the eye, such as the cornea (clear lens) and the retina (rod and cone cells).[4]

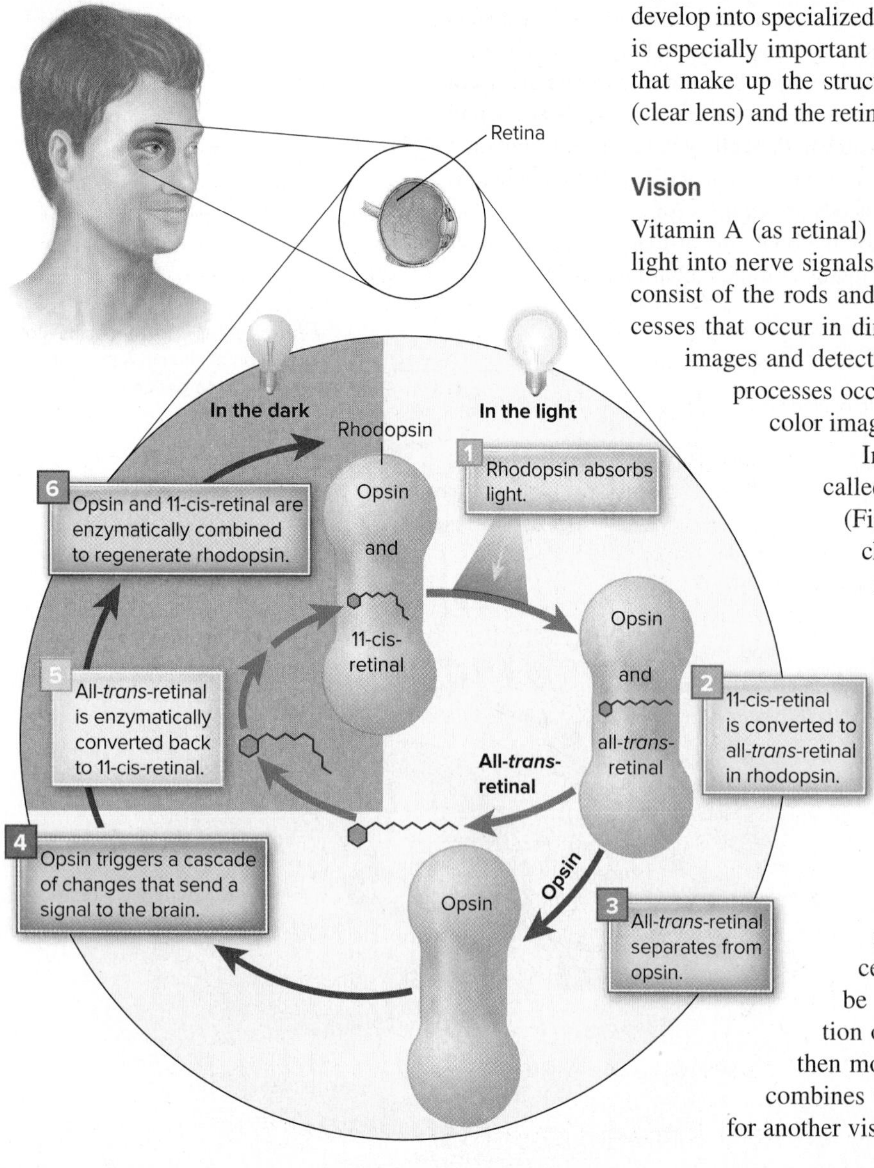

### Vision

Vitamin A (as retinal) is needed in the **retina** of the eye to turn visual light into nerve signals to the brain. The sensory elements of the retina consist of the rods and cones. **Rods** are responsible for the visual processes that occur in dim light, translating objects into black-and-white images and detecting motion. **Cones** are responsible for the visual processes occurring under bright light, translating objects into color images.

In the rods, 11-*cis*-retinal binds to a protein called **opsin** to form the visual pigment **rhodopsin** (Fig. 12-5). The absorption of light catalyzes a change in the shape of 11-*cis*-retinal to all-*trans*-retinal, causing opsin to separate from all-*trans*-retinal.[4] (The separation is a **bleaching process.**) This leads to a cascade of biochemical events that trigger a change in the ion permeability of the photoreceptor cells and initiate a signal to the nerve cells that communicate with the brain's visual center. Thousands of rod cells containing millions of molecules of rhodopsin are triggered simultaneously to produce this signal. During exposure to bright light, the rod's rhodopsin is completely activated and cannot respond to more light. To keep the visual process functioning, all-*trans*-retinal must eventually be converted back to 11-*cis*-retinal. This regeneration occurs within several minutes. The 11-*cis*-retinal then moves back to the photoreceptor cells, where it recombines with the opsin, forming rhodopsin, and is ready for another visual cycle.

*Figure 12-5* The bleaching and regeneration of rhodopsin, a photoreceptor pigment in rod cells of the retina composed of 11-cis-retinal and the protein opsin. The yellow background indicates the bleaching events that occur in the light; the gray background indicates the regenerative events that can occur in either light or dark conditions.

Not all retinal is used in each cycle. Some is stored in the eye to maintain vitamin A pools. If vitamin A pools become depleted, the process of **dark adaptation** is impaired, making it difficult to adjust to seeing in dim light, known as night blindness. You may have had a brief experience similar to night blindness when you walked into a dark movie theater or when a bright light was suddenly shined in your eyes. However, this brief difficulty in seeing is not related to vitamin A deficiency because your vision quickly returns once your eyes have adjusted to the change in light.

### Immune Function

As early as the 1920s, researchers recognized that vitamin A (mostly as retinoic acid) is important for immune system functions. They observed that increased incidence of infection was an early symptom of vitamin A deficiency. Many studies since have shown that vitamin A–deficient individuals have greater susceptibility to illness and infection. This may be, in part, because vitamin A helps maintain the **epithelium,** a barrier that protects the body against the entry of disease pathogens. In many regions of the world where vitamin A deficiency is common, vitamin A supplementation has been shown to reduce the severity of some infections, such as measles and diarrhea, in vitamin A–deficient children.

### Use of Vitamin A Analogs in Dermatology

Several synthetic compounds with a chemical makeup similar to that of vitamin A (called analogs) have been used in topical and oral medications (e.g., Retin-A® and Accutane®) to treat acne and **psoriasis.** Retinoid-based medications also have been used topically to lessen the damage from excess sun and UV-light exposure. It is important that a physician monitor the use of these medications because high doses of oral, and even topical, retinoids can cause serious toxic effects. A pregnancy test is required before Accutane® is prescribed to women because high doses of vitamin A can cause birth defects.

## Carotenoid Functions

Scientists have known for many years that several dietary carotenoids can be converted to vitamin A in the body. More recently, evidence from research studies suggests that carotenoids may have functions other than provitamin A activity. These studies indicate that diets high in carotenoid rich fruits and vegetables may decrease the risk of certain eye diseases, cancers, and cardiovascular disease.[5,6,7] This has sparked interest in the potential benefits of supplementing diets with specific carotenoids to lower disease risks.

The most familiar carotenoid is beta-carotene, the carotenoid with the most vitamin A activity. Because of its chemical structure, beta-carotene may act as an antioxidant within tissues, thereby protecting them from free radical damage. Evidence that beta-carotene might protect eye tissues was suggested by epidemiological studies showing a decreased risk of cataracts in people with high blood levels of antioxidant nutrients.[7] However, in follow-up studies, long-term supplementation of beta-carotene (and vitamin E and/or vitamin C) did not prevent or reduce the incidence of cataracts.[8]

Other studies have focused on a possible role for beta-carotene in the prevention of lung cancer. Although the consumption of fruits and vegetables rich in beta-carotene has been associated with a reduced risk of lung cancer, large studies examining the effectiveness of beta-carotene supplements in preventing lung cancer in high-risk populations (smokers and asbestos workers) revealed that these supplements actually increased the risk of lung cancer. Thus, beta-carotene supplements are not an effective means of reducing the risk of cancer and may even increase the risk in certain individuals.[9,10]

Data from the European Prospective Investigation into Cancer and Nutrition Cohort (EPIC) and the Nurses' Health Study suggest a possible link between blood levels of beta-carotene and alpha-carotene and a decreased risk of breast cancer.[11,12] Although more research is needed, high overall intakes of dietary antioxidants may help lower breast cancer occurrence.[13]

Studies of diets that contain high amounts of the carotenoids lutein and zeaxanthin suggest that they may help protect against age-related **macular degeneration** of the eye (Figs. 12-6 and 12-7).[7,14] To date, however, a direct link between increased intake of lutein and zeaxanthin and the prevention of age-related macular degeneration has not been

**RXR, RAR** Abbreviations for *retinoid X receptor* and *retinoic acid receptor*. These 2 subfamilies of retinoid receptors in the nucleus interact with retinoic acid and bind with specific sites on DNA, allowing for gene expression.

**stem cell** Unspecialized cell that can be transformed into a specialized cell.

**bleaching process** Process by which light depletes the rhodopsin concentration in the eye by separating opsin from all-*trans*-retinal. This fall in rhodopsin concentration allows the eye to become adapted to bright light.

**dark adaptation** Process by which the rhodopsin concentration in the eye increases in dark conditions, allowing improved vision in the dark.

**epithelium** Covering of internal and external surfaces of the body, such as the lungs, GI tract, blood vessel linings, and skin.

**psoriasis** Immune system disorder that causes a chronic, inflammatory skin condition (painful patches of red, scaly skin).

**macular degeneration** Chronic eye disease that occurs when tissue in the macula (the part of the retina responsible for central vision) deteriorates. It causes a blind spot or blurred vision in the center of the visual field.

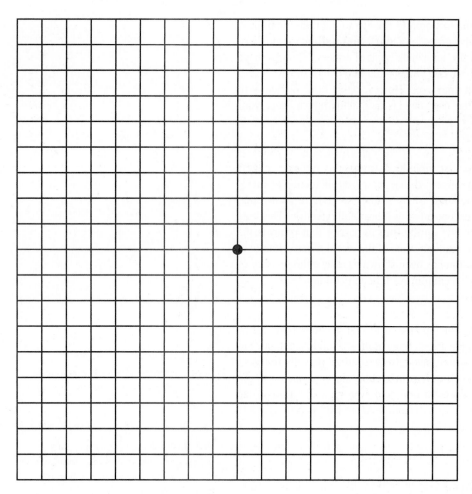

*Figure 12-6* The Amsler grid is used to detect and monitor damage from macular degeneration or other eye diseases. While in a well-lighted area and wearing reading glasses (if you use them), hold the grid 12 to 15 inches away from your eyes. Cover 1 eye and look directly at the center dot. Notice if the lines look straight or look blurry, wavy, dark, or missing. Repeat with your other eye. Seeing anything other than straight lines indicates a need to visit an ophthalmologist soon. Learn more at www.aao.org/eye-health/tips-prevention /facts-about-amsler-grid-daily-vision-test.

▶ Concentrations of lutein and zeaxanthin are 500 to 1000 times higher in the macula of the eye than in other tissues. Spinach and kale are 2 notable sources of these carotenoids.

*Figure 12-7* Age-related macular degeneration causes a blind spot in the center of the visual field. Although increased dietary lutein and zeaxanthin have been associated with decreased risk of macular degeneration in several studies, further research is needed to better understand the possible relationships between macular degeneration and carotenoids.

Source: National Eye Institute, National Institutes of Health

demonstrated. A mixture of antioxidant supplements (lutein, zeaxanthin) may slow the progression of age-related macular degeneration in certain populations; however, more research is needed.[15]

Over the last decade, there has been increased interest in a carotenoid pigment found in tomatoes, called lycopene.[6] Much of this interest developed when scientists observed that men with greater intakes of tomato products and higher blood levels of lycopene had a reduced risk of prostate cancer. However, tomatoes may contain many phytochemicals in addition to lycopene, some of which may provide protection against disease. Thus, increased blood lycopene levels may simply reflect increased tomato intake and may not be responsible for reduced cancer risk.

Beta-carotene, lycopene, and lutein have been studied for their possible role in reducing the risk of cardiovascular disease (CVD).[15,16] Because carotenoids are carried in the blood with lipoproteins, scientists have proposed that they may inhibit oxidation of the lipoprotein LDL. Beta-carotene supplementation alone has not been consistently shown to decrease risk of CVD. There is evidence that lycopene may reduce CVD risk by decreasing LDL oxidation and cholesterol synthesis, as well as by increasing LDL receptor activity in cells. Some studies have suggested a beneficial effect of blood lutein levels on coronary heart disease, stroke, and Metabolic Syndrome. However, the evidence for cardioprotective benefits is not consistent in all studies. Until these complex relationships are more clearly understood, nutritionists do not recommend carotenoid supplements. Instead, they advocate increased intakes of carotenoid rich fruits and vegetables.

An emerging area of interest is the role of lutein in brain development and cognitive function across the life span. Adults 50 years of age and older with higher measures of lutein have improved brain functions, such as processing speed and memory.[17] Supplementation with lutein shows improved verbal and memory performance in older adults. Potential benefits of increased lutein levels in children also are reported. It is not certain how lutein contributes to brain function, but it may be linked to its protective effects as an antioxidant. Further research is needed to expand the understanding of how dietary lutein affects brain function and cognitive development.

## Vitamin A Deficiency Diseases

North Americans have little risk of developing vitamin A deficiency because this vitamin is abundant in our food supply.[2] Deficiencies of vitamin A and vitamin E are the least common nutrient deficiencies in the U.S, occurring in less than 1% of the population.[18] Vitamin A deficiency, however, is a major public health problem in low-income countries. Worldwide, vitamin A deficiency is a leading cause of nonaccidental blindness. Children in low-income nations in Africa, Asia, and South America are especially susceptible because inadequate intakes and low stores of vitamin A fail to meet their needs for growth. In the poorest nations, approximately 500,000 children become blind each year because of vitamin A deficiency.

Although vitamin A deficiency is not commonly seen in North America, several population groups are considered at risk. Low-income and older adults, people with alcoholism or liver disease (which limits vitamin A storage), and individuals with severe fat malabsorption—which may develop in gluten-sensitive enteropathy (celiac disease), chronic diarrhea, pancreatic insufficiency, Crohn disease, cystic fibrosis, and AIDS—may develop vitamin A deficiency. Premature infants also are at risk of deficiency because they are born with low stores of vitamin A.[2,4]

Vitamin A deficiency results in many changes in the eye. When the retinol in the blood is insufficient to replace the retinal lost during the visual cycle, the rods in the retina regenerate rhodopsin more slowly. The resulting night blindness is a common early symptom of vitamin A deficiency, as discussed earlier. Without enough retinoic acid, mucous-forming cells deteriorate and are no longer able to synthesize mucus. The eye, especially the cornea, is adversely affected by the loss of mucus because mucus helps keep the eye surface moist and washes away dirt particles that settle on the eye. This leads to the development of conjunctival xerosis (abnormal dryness of the **conjunctiva** of the eye). Bitot's spots (foamy, gray spots on the eye consisting of hardened epithelial cells) also appear as vitamin A deficiency worsens (Fig. 12-8). These conditions often progress to keratomalacia (softening of the cornea) and scarring (Fig. 12-9). This sequence of changes in the eye—collectively known as **xerophthalmia**—causes irreversible blindness in millions of people worldwide.

**conjunctiva** Mucous membrane covering the front surface of the eye and the lining of the eyelids.

**xerophthalmia** Condition marked by dryness of the cornea and eye membranes that results from vitamin A deficiency and can lead to blindness.

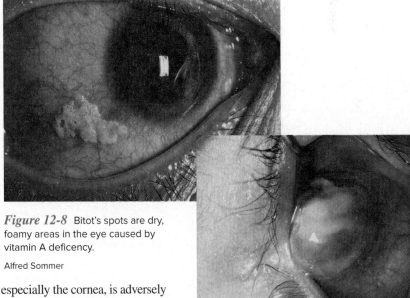

***Figure 12-8*** Bitot's spots are dry, foamy areas in the eye caused by vitamin A deficency.

Alfred Sommer

***Figure 12-9*** Vitamin A deficiency can have severe effects on the eye, progressing from night blindness to dry eye membranes (conjunctival xerosis), corneal dryness (corneal xerosis), softening of the cornea (keratomalacia), and eventually blindness.

Alfred Sommer

**follicular hyperkeratosis** Condition in which keratin, a protein, accumulates around hair follicles.

**wean** To accustom an infant to a diet containing foods other than just milk.

Vitamin A deficiency also produces skin changes, referred to as **follicular hyperkeratosis.** Keratin, the normal component in the outer layers of the skin, protects the inner layers and reduces water loss through the skin. During severe vitamin A deficiency, keratinized cells, which are normally present only in the outer layers, replace the normal mucous-forming epithelial cells in the underlying skin layers. Hair follicles become plugged with keratin, giving a dry, rough, sandy texture to the skin.

In infants and young children, vitamin A deficiency can impair growth. If adequate vitamin A stores are established before an infant is **weaned,** they can help protect against deficiency. Vitamin A supplementation for infants and young children at risk also may protect them. Finding suitable foods to improve vitamin A intake is essential as a long-term solution to vitamin A deficiency.[19]

## Vitamin A Toxicity

The signs and symptoms of toxicity from excessive vitamin A—called hypervitaminosis A—appear with long-term supplement use at 5 to 10 times the RDA for retinoids[2] (Fig. 12-10). Correspondingly, the Upper Level is set at 3000 µg/day of retinol to prevent harmful effects. No Upper Level is set for carotenoids because vitamin A toxicity results only from excess intakes of retinoids.[1]

Three kinds of vitamin A toxicity occur: acute, chronic, and teratogenic.[20,21] Acute toxicity is caused by the ingestion of 1 very large dose of vitamin A or several large doses taken over a few days (about 100 times the RDA). The effects of acute toxicity include GI tract upset, headache, blurred vision, and poor muscle coordination. Once the dosing is stopped, the signs disappear. Extraordinarily large doses of 500 mg in children and 10 g in adults can be fatal.

In chronic toxicity, infants and adults show a wide range of signs and symptoms: joint pain, loss of appetite, skin disorders, headache, reduced bone minerals, liver damage, double vision, hemorrhage, and coma. These symptoms occur with repeated intakes of at least 10 times the RDA guidelines. The treatment is simply to discontinue the supplement. The symptoms then decrease over the next few weeks as blood concentrations fall within a normal range. Permanent damage to the liver, bones, and eyes, however, can occur with the chronic ingestion of excessive amounts of the vitamin.

The most serious and tragic effects of hypervitaminosis A are **teratogenic** (causing birth defects). Vitamin A and its related analog forms (all-*trans*-retinoic acid [topical tretinoin, or Retin-A®] and 13-*cis*-retinoic acid [oral isotretinoin, or Accutane®]) are used to treat

*Critical* **Thinking**

Julie bought a new juicer and has been making carrot and mango juice, which she drinks 2 or 3 times daily. She has noticed recently that the palms of her hands are turning orange. What is likely causing this? Is she at risk of vitamin A toxicity?

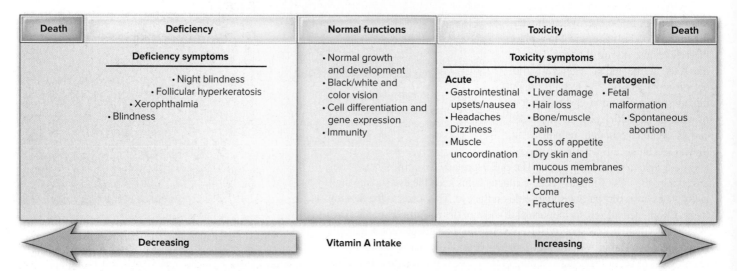

*Figure 12-10* Consuming the right amount of vitamin A is critical to overall health. A very low (deficient) or very high (toxic) vitamin A intake (as retinoids) can produce harmful symptoms and can even lead to death.

## CASE STUDY

Emmi has been under a lot of stress and is concerned about her overall health. She works nights at a local restaurant and takes a full course load during the day at the local community college. At lunch today, her friend Jessi suggested that she take Nutramega supplements to help prevent colds, flu, and other illnesses and reduce her stress.

Emmi was a little surprised to learn that a month's supply of Nutramega costs about $50, but she decided to buy the supplements, anyway. The Nutramega label recommends taking 2 or 3 tablets daily for health maintenance and 2 or 3 tablets every 3 hours at the first sign of illness. When Emmi read the label on the bottle, she noted that each tablet contained the following nutrients (listed as % Daily Value): 33% vitamin A (75% as preformed vitamin A), 500% vitamin C, 50% zinc, and 10% selenium.

If you were Emmi's registered dietitian nutritionist or health clinician, would you recommend that Emmi use this product? Are there any health risks associated with its use, considering the dosages recommended on the label for maintenance and illness? What alternative suggestions might you give Emmi to help her maintain her overall health?

Rachel Frank/Corbis/Glow Images

various skin disorders, such as acne and psoriasis. However, these analogs are teratogenic in humans and have caused spontaneous abortion and birth defects. Women of childbearing age should be cautioned against using these medications or should use reliable methods to prevent pregnancies that might result in fetal malformations.

It is even possible for pregnant women to get too much vitamin A from food if they frequently consume increased amounts of foods rich in vitamin A, such as liver or fortified ready-to-eat breakfast cereals. For this reason, pregnant women should limit their intake of these foods and, if taking supplements, should check that much of the supplemental vitamin A is in the form of beta-carotene. The FDA recommends that women of childbearing age limit their intake of preformed vitamin A to 100% of the Daily Value.[2]

Consuming carotenoids in large amounts from foods does not readily result in toxicity. The carotenoids' rate of conversion to vitamin A is relatively slow. In addition, the efficiency of carotenoid absorption from the small intestine decreases markedly as dietary intake increases.[1] If one consistently consumes large amounts of carrots, carrot juice, or winter squash, the resulting high carotenoid concentrations in the body can turn the skin a yellow-orange color, a condition termed hypercarotenemia, or carotenemia.[1] (*Hyper* means "high" and *emia* means "in the bloodstream.") This is not harmful to overall health.

▶ One pound of polar bear liver contains enough vitamin A to keep a human healthy for several years. A team of early Arctic explorers became gravely ill from vitamin A toxicity after eating polar bear liver. Using this simplified food chain pyramid, hypothesize why polar bear liver is so high in vitamin A.

### Knowledge Check

1. What are 3 sources of provitamin A and 3 sources of preformed vitamin A?
2. Why is the carotenoid beta-carotene classified as a provitamin?
3. How does vitamin A affect vision?
4. What are 2 symptoms of vitamin A deficiency?
5. What population groups are at highest risk of vitamin A deficiency?
6. What are the signs and symptoms of vitamin A toxicity?

## 12.3 Vitamin D

Bone deformities that were likely caused by the vitamin D deficiency disease rickets have been described since ancient times. It wasn't until 1918, however, when scientists cured rachitic dogs (dogs afflicted with rickets) by feeding them cod liver oil, that diet was linked with this disease. Soon after, vitamin D was discovered and cod liver oil became a daily supplement for millions of children.

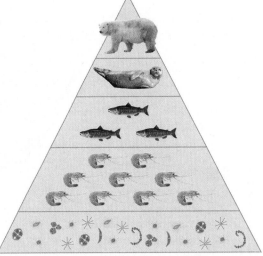

**SIMPLIFIED ECOLOGICAL PYRAMID**

alinabel/Shutterstock; JackF/iStock/Getty Images Plus/Getty Images; tony mills/Alamy Stock Photo; Foodcollection; Timothy Knepp/U.S. Fish and Wildlife Service; KittyVector/Shutterstock

# GL🌐BAL PERSPECTIVE

## Vitamin A Deficiency

In many low-income countries, vitamin A deficiency is a major public health concern (Fig. 12-11). According to the World Health Organization, an estimated 250 million preschool children are vitamin A deficient. Women of childbearing years also are at increased risk of deficiency, especially in impoverished areas of Africa and Southeast Asia. In many of these areas, where HIV infection also is prevalent, vitamin A deficiency in pregnancy increases the likelihood of the transmission of HIV to the developing fetus and increases the risk of maternal mortality. Approximately 600,000 women die each year from pregnancy- and childbirth-related causes. Many of these deaths result from complications secondary to poor vitamin A and overall nutritional status.[19]

As discussed in this chapter, vitamin A deficiency can lead to serious consequences, such as night blindness, total blindness, impaired growth, and an increased incidence of infections. Due to the prevalence of vitamin A deficiency worldwide, it is the leading cause of preventable blindness in children, resulting in 250,000 to 500,000 cases annually. Approximately half of these children die within a year from severe infections, measles, diarrhea, and anemia.[19]

Golden rice was genetically engineered to synthesize beta-carotene. This rice was developed for use as a fortified food in areas of the world that have limited access to vitamin A rich foods.

JIANG HONGYAN/Shutterstock

In 1998, a partnership was formed among the World Health Organization (WHO), the United Nations Children's Fund (UNICEF), the Canadian International Development Agency (CIDA), the U.S. Agency for International Development (U.S.AID), and the Micronutrient Initiative (MI) to combat vitamin A deficiency. This coalition of international agencies, called the Vitamin A Global Initiative, has worked to decrease vitamin A deficiency by promoting breastfeeding, fortification of foods (e.g., sugar fortification in Guatemala), and educational programs to increase home gardening of vitamin A rich fruits and vegetables in rural areas of Africa and Southeast Asia. WHO, UNICEF, and other international agencies also have provided vitamin A supplements (as a complement to immunization programs) to at-risk populations. Such strategies have been shown to decrease vitamin A–related mortality by 25% and greater in some of these areas. For children ages 6 months to under 5 years, the mortality from vitamin A deficiency has decreased by about 50% in low- and middle-income countries,[22] yet 95% of these deaths have been concentrated in sub-Saharan Africa and southern Asia, where vitamin A deficiency remains most prevalent. Ongoing programs to increase the production of and access to nutrient rich foods (e.g., fish) that are native to the diets and livelihoods of the individuals in many at-risk rural areas are food-based strategies that can help prevent nutrient deficiencies.

*Figure 12-11* Vitamin A deficiency affects many low-income countries.

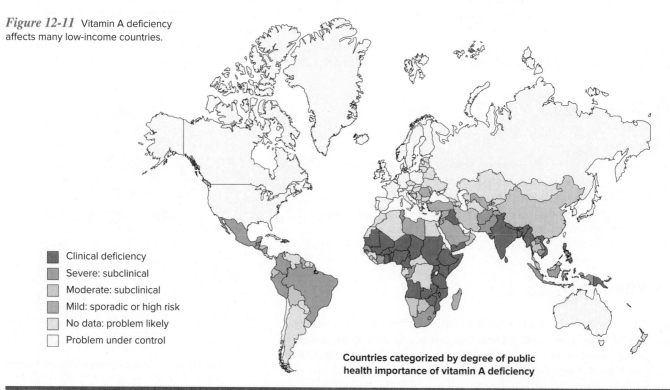

- ■ Clinical deficiency
- ■ Severe: subclinical
- ■ Moderate: subclinical
- ■ Mild: sporadic or high risk
- ■ No data: problem likely
- □ Problem under control

**Countries categorized by degree of public health importance of vitamin A deficiency**

Most scientists classify vitamin D as a vitamin. However, in the presence of sunlight, skin cells can synthesize a sufficient supply of vitamin D from a derivative of cholesterol. Because a dietary source is not required if synthesis is adequate to meet needs, the vitamin is more correctly classified as a "conditional" vitamin, or **prohormone** (a precursor of an active hormone). In the absence of UV-light exposure, an adequate dietary intake of vitamin D is essential to prevent the deficiency diseases rickets and osteomalacia and to provide for cellular needs.

## Vitamin D₂ in Foods

The best food sources of vitamin D are fatty fish (e.g., sardines, mackerel, herring, and salmon), cod liver oil, fortified milk, and some fortified breakfast cereals (Fig. 12-12). In North America, milk is generally fortified with 10 µg (400 IU) of vitamin D per quart. Food products made from milk, such as cheese and ice cream, are not usually fortified. Although eggs, butter, liver, and a few brands of margarine contain some vitamin D, large servings must be eaten to obtain an appreciable amount of the vitamin. Thus, these foods are not considered a significant source. Most fortified foods and supplements containing vitamin D are in the form of ergocalciferol, or vitamin D₂, the same form found naturally in foods. Ergocalciferol has vitamin D activity in humans, but in lower amounts than provided by the vitamin D₃ formed in the body.

## Vitamin D₃ Formation in the Skin

The synthesis of vitamin D₃ begins with a compound called 7-dehydrocholesterol, a precursor of cholesterol synthesis located in the skin. During exposure to sunlight, 1 ring on the molecule undergoes a chemical transformation, forming the more stable vitamin D₃ (cholecalciferol). This change allows vitamin D₃ to enter the bloodstream for transport to the liver and kidneys, where it undergoes conversion to its bioactive form (calcitriol).

For many individuals, sun exposure provides 80 to 100% of the vitamin D₃ required by the body.[3] The amount of sun exposure needed, however, depends on the time of day, the geographic location, the season of the year, one's age, one's skin color, and the use

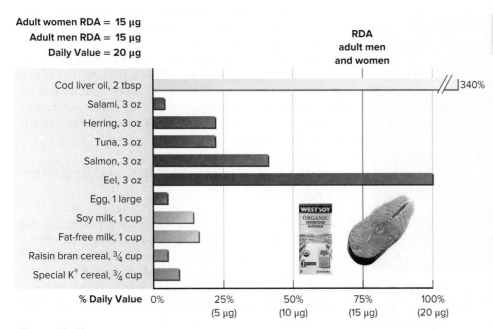

*Figure 12-12* Food sources of vitamin D.

Soy Milk: Andrew Resek/McGraw-Hill Education; Seafood: Digital Vision/Getty Images

### A Biochemist's View

Addition of hydroxyl groups (—OH) by the liver (carbon #25) and then by the kidney (carbon #1) yields the final product.

HO — Cholecalciferol (vitamin D₃)

OH — Carbon #25

Carbon #1

HO — OH

Active form of vitamin D: 1,25 dihydroxy D₃ (calcitriol)

The form produced by the body is called cholecalciferol (vitamin D₃). A form typically found in or added to foods is ergocalciferol (vitamin D₂). It has a double bond in the starred position in the top structure.

**Vitamin D family**

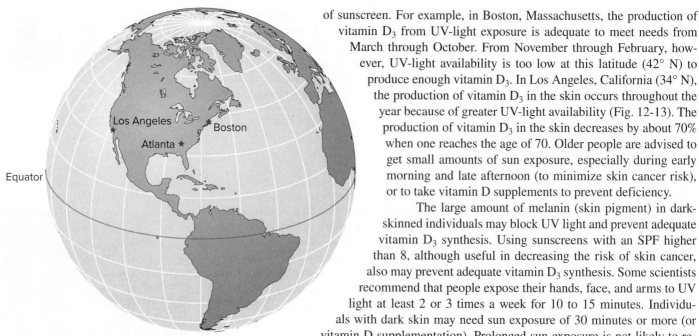

*Figure 12-13* As the distance from the equator increases, available uv light decreases.

of sunscreen. For example, in Boston, Massachusetts, the production of vitamin $D_3$ from UV-light exposure is adequate to meet needs from March through October. From November through February, however, UV-light availability is too low at this latitude (42° N) to produce enough vitamin $D_3$. In Los Angeles, California (34° N), the production of vitamin $D_3$ in the skin occurs throughout the year because of greater UV-light availability (Fig. 12-13). The production of vitamin $D_3$ in the skin decreases by about 70% when one reaches the age of 70. Older people are advised to get small amounts of sun exposure, especially during early morning and late afternoon (to minimize skin cancer risk), or to take vitamin D supplements to prevent deficiency.

The large amount of melanin (skin pigment) in dark-skinned individuals may block UV light and prevent adequate vitamin $D_3$ synthesis. Using sunscreens with an SPF higher than 8, although useful in decreasing the risk of skin cancer, also may prevent adequate vitamin $D_3$ synthesis. Some scientists recommend that people expose their hands, face, and arms to UV light at least 2 or 3 times a week for 10 to 15 minutes. Individuals with dark skin may need sun exposure of 30 minutes or more (or vitamin D supplementation). Prolonged sun exposure is not likely to result in vitamin D synthesis beyond needs or in toxic amounts because excess amounts of **previtamin $D_3$** in the skin are rapidly degraded. Overall, those who do not receive enough UV-light exposure to synthesize adequate amounts of vitamin $D_3$ should make certain that they have adequate vitamin D in their diets.

| Type | Characteristics | Reaction to Sun Exposure |
|---|---|---|
| | Pale to very fair skin; blond or red hair; light eye color | Always burns, never tans |
| | Fair skin; light hair and eye color | Almost always burns, rarely tans |
| | Fair skin; medium to dark hair; eye color varies | Burns, then tans |
| | Light brown to tan skin; dark hair and eyes | Less likely to burn, tans easily |
| | Dark skin; dark hair and eyes | Rarely burns, tans easily |
| | Very dark skin; dark eyes and hair | No burn, tans easily |

Assessment of skin type using the Fitzpatrick sun-reactive scale has been linked to vitamin D status, where rates of vitamin D deficiency may be higher in persons with darker skin types.[23,24]

## Vitamin D Needs

An Institute of Medicine (IOM) expert committee has determined that, on average, North Americans require approximately 10 µg (400 IU) of vitamin D daily. To meet these estimated needs, the IOM set the RDA to 15 µg/day (600 IU/day) for individuals 1 to 70 years old and 20 µg (800 IU) for adults over age 70 years. Although vitamin D is supplied by the diet and sun exposure, the IOM committee based its recommendations on minimal sun exposure.[25,26,27,28,29] The Daily Value used on food and supplement labels is 20 µg (see Chapter 2).

Although full-term infants are born with a supply of vitamin D, the American Academy of Pediatrics recommends that breastfed infants be given a vitamin D supplement to meet current recommendations of 10 µg/day (400 IU). This supplementation should continue until they are weaned to infant foods fortified with, or rich in, vitamin D.

## Absorption, Transport, Storage, and Excretion of Vitamin D

Following the consumption of foods containing vitamin $D_2$, about 80% of vitamin $D_2$ is incorporated (along with other dietary fats) into micelles in the small intestine, absorbed, and transported to the liver by chylomicrons through the lymphatic system (Fig. 12-14). Patients with diseases that may result in fat malabsorption syndromes (e.g., cystic fibrosis, Crohn disease, and gluten-sensitive enteropathy) are at increased risk of vitamin $D_2$ malabsorption and deficiency.

When vitamin D (either $D_3$ synthesized in the skin or $D_2$ consumed in food or supplements) enters the general circulation, it is bound to a vitamin D–binding protein for transport

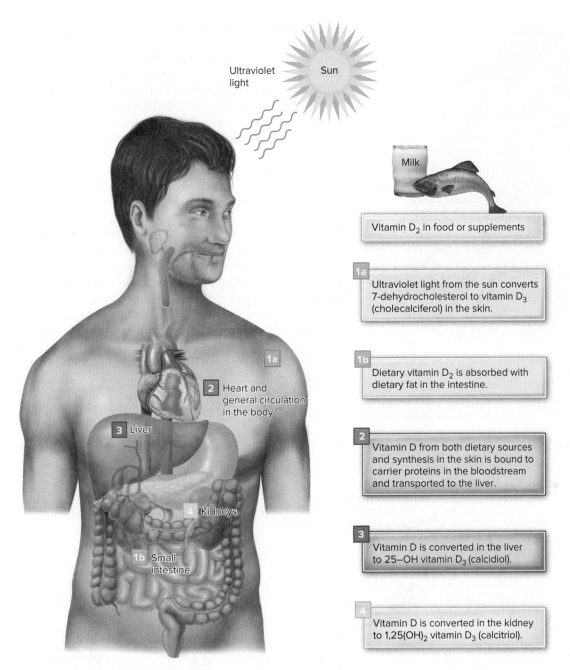

Ultraviolet light

Sun

Milk

Vitamin $D_2$ in food or supplements

**1a**
Ultraviolet light from the sun converts 7-dehydrocholesterol to vitamin $D_3$ (cholecalciferol) in the skin.

**1b**
Dietary vitamin $D_2$ is absorbed with dietary fat in the intestine.

**2**
Vitamin D from both dietary sources and synthesis in the skin is bound to carrier proteins in the bloodstream and transported to the liver.

**3**
Vitamin D is converted in the liver to 25–OH vitamin $D_3$ (calcidiol).

**4**
Vitamin D is converted in the kidney to $1,25(OH)_2$ vitamin $D_3$ (calcitriol).

**1a**
**2** Heart and general circulation in the body
**3** Liver
**4** Kidneys
**1b** Small intestine

*Figure 12-14* Whether synthesized in the skin or obtained from dietary sources, vitamin D ultimately functions as a hormone: $1,25(OH)_2$ vitamin $D_3$ (calcitriol)..

to muscle or adipose cells for storage or to the liver and kidneys. In the liver, vitamin D is hydroxylated (the addition of –OH) on carbon 25, converting it to 25–OH vitamin $D_3$ (calcidiol). This inactive form circulates in the blood for many weeks and serves as an additional storage pool of vitamin D. The next stop is the kidney, the principal site for the production of $1,25(OH)_2$ vitamin $D_3$, also known as calcitriol. This is the active form of the vitamin that, as needed, binds to specific vitamin D receptors in target tissues to induce vitamin D actions.

The synthesis of $1,25(OH)_2$ vitamin $D_3$ is tightly regulated by the parathyroid gland and the kidneys. When there's a shortage of calcium in the blood, the parathyroid gland increases the production of parathyroid hormone (PTH). PTH then increases the production of $1,25(OH)_2$ vitamin $D_3$ in the kidneys to restore calcium balance.

Vitamin D is excreted primarily through bile that is lost in the feces. Small amounts of vitamin D also are excreted in the urine.

**previtamin $D_3$** Precursor of 1 form of vitamin D, produced as a result of sunlight opening a ring on 7-dehydrocholesterol in the skin.

Ultraviolet light on the skin provides about 80 to 100% of the vitamin D humans use. Few foods provide significant amounts of vitamin D; thus, sun exposure often is the most reliable way of maintaining vitamin D status.

Purestock/Alamy Stock Photo

## Functions of Vitamin D

Calcitriol, the most active form of vitamin D, has several important functions.[25] Its most recognized function is its hormone-like role in maintaining the body's concentration of calcium and phosphorus (Fig. 12-15).[26] This helps maintain skeletal health, but it also can have the opposite effect and result in bone demineralization. When blood calcium levels are low, vitamin D promotes increased intestinal absorption of calcium and phosphorus from foods to maintain blood levels of these minerals. This makes calcium and phosphorus available for body cells and for incorporation into bones when there is more than needed for basic functions. However, when blood levels of calcium and phosphorus start to fall, vitamin D (with PTH from the parathyroid gland) can release these minerals from bone to restore blood levels. Although this action eventually can weaken the bones if it continues for a prolonged period of time, it helps provide the calcium and phosphorus needed for many basic life functions. If the bones did not supply calcium and phosphorus for these functions, a person could quickly have serious, even fatal, health consequences. Thus, vitamin D preserves the functions of calcium and phosphorus even if dietary intakes of these minerals are inadequate.[25]

### Current Vitamin D Concerns and Additional Functions

The fortification of milk with vitamin D and the prophylactic use of cod liver oil almost completely eradicated the epidemic of

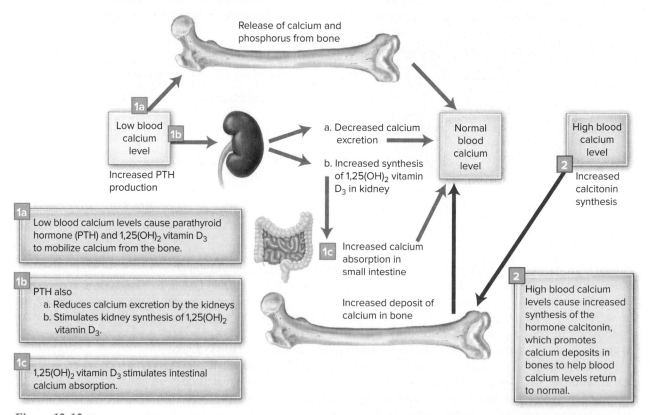

*Figure 12-15*  The active vitamin D hormone—1,25(OH)$_2$ vitamin D$_3$ (calcitriol)—and parathyroid hormone interact to control blood calcium concentration. Low blood calcium is a trigger for the actions shown in [1a], [1b], and [1c]—all of which raise blood calcium levels. Conversely, when calcium levels in the blood become too high, [2] the hormone calcitonin responds by promoting calcium deposition in the bone.

People who remain almost fully covered during the day produce little vitamin D₃

Goodshoot/Alamy Stock Photo

rickets in the 20th century. However, vitamin D deficiency has reemerged as a global health concern. Some experts feel that low intakes of vitamin D, coupled with behaviors that limit UV-light exposure (e.g., time spent indoors, the use of sunscreen, the use of clothing to fully cover the skin), have resulted in widespread inadequate vitamin D status. In the U.S. population, the prevalence of low blood levels of vitamin D that might be linked with a risk of vitamin D deficiency has remained fairly constant over the past several decades.[30]

The RDA is based on intakes shown to promote calcium homeostasis and prevent vitamin D–related bone disorders. However, some experts believe vitamin D also may have other important functions.[29] Some studies suggest that vitamin D helps regulate immune function and the secretion of several hormones (insulin, renin, PTH). Although the exact role of vitamin D in many tissues is not fully known, some scientists believe it may be involved in cell cycle regulation. Additionally, vitamin D may decrease the risk of certain types of infections and autoimmune diseases, such as multiple sclerosis, and offer protection against diabetes, hypertension, dementia, and certain cancers.[25,26,27,29,31,32,33] However, much of this evidence has come from observational studies with inconsistent results.

Recent reports summarizing data from numerous articles indicate that supplementation with vitamin D does not benefit cardiovascular outcomes and that additional intervention studies of vitamin D supplementation are needed to further explore vitamin D's role in health outcomes and diseases.[34,35] The large VITAL clinical trial, studying the effects of vitamin D and omega-3 fatty acid supplements, reported no positive effects of supplementation on the occurrence of cardiovascular disease or multiple cancers.[36] In the study, participants were given 2000 IU of vitamin D, 1 gram of omega-3 fatty acids, or a combination of both supplements every day.

Typical inputs from UV exposure (2000 IU), food sources (150–200 IU), and dietary supplements (200 IU) often provide 2400 IU daily or less for the average individual. Some researchers estimate that, to support the diverse functions of vitamin D, the body may require 3000 to 4000 IU (75–100 µg) of vitamin D daily to maintain optimal blood levels of 25–OH vitamin D₃. However, more research is needed to determine the daily intake necessary to maintain optimal vitamin D status, particularly for persons who have limited sun exposure.

In a recent clinical trial, vitamin D supplementation of 4000 or 10,000 IU did not provide benefits to bone status in healthy adults.[37] Instead, bone density and strength decreased

*Figure 12-16* The bone deformities and bowed legs of rickets, a vitamin D deficiency disease in children.

Jeff Rotman/Photolibrary/Getty Images

Milk is usually fortified with vitamin D, so even fat-free milk and low-fat milk contain vitamin D.

Royalty-Free/Corbis

when compared to supplementation with 400 IU. Reports from the VITAL study also indicate that vitamin D supplements of 2000 IU in vitamin D–sufficient people did not provide benefits to bone density or bone loss.[38] Further investigation is needed, as supplementation may be beneficial in persons with vitamin D insufficiency. Additional research may also help identify if toxic effects develop from vitamin D supplementation at these levels.

## Vitamin D Deficiency Diseases

The skeleton will not mineralize normally without adequate calcium and phosphorus in the blood for deposition in bone. This causes the bones to weaken and bow under pressure. When these effects occur in the growing bones of a child, this vitamin D deficiency disease is called **rickets** (Fig. 12-16). The signs of rickets include enlarged head, joints, and rib cage; a deformed pelvis; and bowed legs. In higher-income countries, rickets is most commonly associated with fat malabsorption, as seen in children with cystic fibrosis. However, an increase in cases has been seen in infants and children who have dark skin pigmentation, low milk intakes, and/or minimal sun exposure due to protective clothing, sunscreens, and/or limited outdoor activity.[31,39,40] In fact, some scientists are concerned about an epidemic of vitamin D deficiency and rickets reemerging in industrialized countries.[31,39,40]

Vitamin D deficiency in adults is called **osteomalacia,** which means "soft bones." It is characterized by poor calcification of newly synthesized bone, resulting in fractures in the hip, spine, and other bones. (Do not confuse this disease with osteoporosis.) Osteomalacia is most likely to occur in adults with kidney or liver disease (both of which impair the synthesis of calcitriol) or intestinal diseases that impair fat absorption (e.g., Crohn disease).[3] Other adults at risk of deficiency include those with dark skin and those with limited UV exposure.[31,39,40]

Many older adults who live in northern climates or reside in nursing homes also are at increased risk of vitamin D deficiency. Not only do many of these individuals have little sun exposure, but they also may have reduced vitamin D levels from low dietary intakes and/or impaired kidney function, which limits conversion to the active form of vitamin D.[31]

A person with a low circulating concentration of 25–OH vitamin $D_3$ should take at least 20 to 25 µg (800–1000 IU) of vitamin D each day until the concentration reaches the normal range.[3,31] After blood concentrations are adequate, 10 µg (400 IU/day) of supplemental vitamin D should be sufficient for most people.[3]

## Vitamin D Toxicity

Vitamin D toxicity can occur from excessive vitamin D supplementation, especially in the form of cholecalciferol.[41] It does not result from excess sun exposure (vitamin D in the skin is readily broken down) or from natural sources in the diet. The Upper Limit is 1000 to 3000 IUs per day for infants and children up to 8 years of age and 4000 IUs per day for individuals 9 years of age and older. Above these UL intakes, vitamin D can cause an overabsorption of calcium and hypercalcemia (increased calcium in the blood). Excess blood calcium, in turn, leads to deposits of calcium in the kidneys, heart, and lungs; anorexia; nausea; vomiting; bone demineralization; weakness; joint pain; and kidney dysfunction. In the early stages of toxicity, the symptoms often are treatable if vitamin D is withdrawn. However, if excess supplementation continues, vitamin D toxicity eventually can be fatal.

### Knowledge Check

1. Why is vitamin D often classified as a conditional vitamin, or prohormone?
2. What are the rich dietary sources of vitamin D?
3. What are 3 functions of vitamin D?
4. What are the consequences of vitamin D deficiency?
5. Why are those who live in northern latitudes at risk of vitamin D deficiency?
6. Why was an Upper Level of intake established for vitamin D?

# Culinary Perspective

## Plant-Based Milk Alternatives

Individuals' consumption of plant-based milk alternatives has increased over the past decade, whereas ingestion of cow's milk has decreased. Such alternative products are derived from soy, almond, peanut, rice, oat, coconut, and other plants. The basic process of producing some plant-based beverages involves grinding the plant, such as almonds or rice, and mixing it with water, followed by various treatments to increase its shelf life or modify other of its characteristics. Nutrients may be added to the product to increase its nutritional value, or sugars may be added to enhance its flavor.

Consumers may choose to include cow's milk in their diets for its protein, vitamin D, calcium, and other nutrients, or they may select alternatives to cow's milk because of an allergy, lactose intolerance, a vegan dietary pattern, religious dietary law restrictions, or other health-related interests. Soy and nut allergies, as well as taste, also may influence consumer decisions. Choosing a product for its potential benefits also typically has additional consequences. For example, coconut milk might be selected for its lower caloric value and plant composition, but the protein and riboflavin it supplies is much lower than the same volume of cow's milk.

R.Bordo/Shutterstock.

Nutritional value comparisons per cup (240 mL) for some of these alternative milk beverages are shown in the accompanying table. Their nutrient content can vary from brand to brand, based on the nutrients added in processing.

| Nutrient | Cow's Milk (Vitamin D Added) | Soy Milk (Unsweetened; Calcium, Vitamin A, and Vitamin D Added) | Almond Milk (Sweetened; Calcium, Vitamin A, and Vitamin D Added) | Coconut Milk (Unsweetened, Calcium and Vitamin D Added) |
|---|---|---|---|---|
| Energy (kcal) | 149 | 80 | 91 | 80 |
| Protein (g) | 8 | 7 | 1 | 1 |
| Carbohydrate (g) | 13 | 4 | 16 | 8 |
| Sugars (g) | 12 | 1 | 15 | 1 |
| Fiber (g) | 0 | 1 | 1 | 2 |
| Lipid (g) | 8 | 4 | 2.5 | 4.5 |
| Cholesterol (mg) | 36 | 0 | 0 | 0 |
| Vitamin D (IU) | 101 | 119 | 101 | 120 |
| Vitamin A (IU) | 300 | 503 | 499 | 0 |
| Riboflavin (mg) | 0.4 | 0.5 | 0.4 | 0 |
| Calcium (mg) | 300 | 301 | 451 | 350 |
| Sodium (mg) | 125 | 90 | 130 | 34 |

## 12.4 Vitamin E

The importance of vitamin E was first noted in 1922, when researchers discovered that a substance in vegetable oil was needed for normal reproduction in rats. Initially named "vitamin E" (it followed the discovery of vitamin D), researchers later named the substance tocopherol, from the Greek words *toco,* meaning "childbirth," and *pherein,* meaning "to bear." Vitamin E was not fully recognized as an essential nutrient in humans until the mid-1960s, when vitamin E deficiency was observed in children with fat malabsorption diseases.

The first RDA for vitamin E was established in 1968. Like RDAs for other nutrients, it has been subsequently revised as knowledge about vitamin E has grown.

Vitamin E is a family of 8 naturally occurring compounds—4 tocopherols (alpha, beta, gamma, delta) and 4 tocotrienols (alpha, beta, gamma, delta)—with widely varying degrees of biological activity. Vitamin E has a long carbon chain tail attached to a ringed structure. This tail exists in many possible isomer forms. The most active form of the vitamin is alpha-tocopherol. This is the form found in some foods and in varying amounts in vitamin supplements. Gamma-tocopherol is a potentially beneficial form of vitamin E found in many vegetable oils. However, it does not have as much biological activity as alpha-tocopherol.[1]

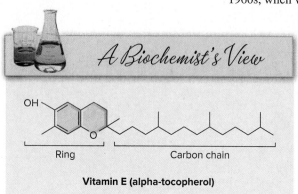

### A Biochemist's View

OH

| Ring | Carbon chain |

**Vitamin E (alpha-tocopherol)**

### Vitamin E in Foods

Good food sources of vitamin E include plant oils (e.g., cottonseed, canola, safflower, and sunflower oils), wheat germ, avocado, almonds, peanuts, and sunflower seeds (Fig. 12-17). Products made from the plant oils—margarine, shortenings, and salad dressings—also are good sources. Animal fats and dairy products contain little vitamin E.

The vitamin E content of a food depends on harvesting, processing, storage, and cooking because vitamin E is highly susceptible to destruction by oxygen, metals, light, and heat, as in deep-fat frying. Thus, foods that are highly processed and/or deep-fried are usually poor sources of vitamin E.

### Vitamin E Needs

The RDA for vitamin E is 15 mg/day of alpha-tocopherol for both men and women. The recommendation is based on the amount of vitamin E needed to prevent a breakdown of red

▶ When you look at food or supplement labels, the type of vitamin E they contain will be labeled as "d" or "dl." If you see "d" (now called RRR-alpha-tocopherol) next to vitamin E on a label, all of that vitamin E will be active in the body. If you see "dl" (now called all-rac-alpha-tocopherol) on a label, only about half of the vitamin E will be active in the body.

*Figure 12-17* Food sources of vitamin E.

D. Hurst/Alamy Stock Photo; Jacques Cornell/ McGraw-Hill Education

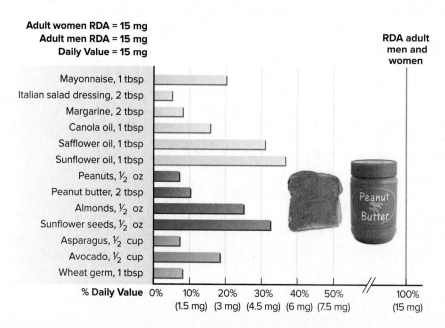

Adult women RDA = 15 mg
Adult men RDA = 15 mg
Daily Value = 15 mg

RDA adult men and women

blood cell membranes, a process called **hemolysis.**[1] The 15 mg allotment is approximately equivalent to 22 IUs of a natural source and 33 IUs of a synthetic source of vitamin E.

Adults consume, on average, only two-thirds of the RDA for vitamin E each day.[42] Increasing one's intake of vitamin E rich foods, eating ready-to-eat breakfast cereals containing vitamin E, or taking a vitamin E supplement daily could close the gap between typical vitamin E intakes and needs.

The Daily Value for vitamin E used on food and supplement labels is 15 mg. When converting IUs of vitamin E in synthetic form (as in most supplements) to milligrams, 1 IU equals about 0.45 mg. If the vitamin E is from a natural source, 1 IU equals 0.67 mg because the natural form of vitamin E is more potent than the synthetic form.

## Absorption, Transport, Storage, and Excretion of Vitamin E

The degree of vitamin E absorption depends on the amount consumed and the absorption of dietary fat. Absorption occurs by passive diffusion and can vary from 20 to 70% of dietary intake. As with other fat-soluble nutrients, vitamin E must be incorporated into micelles in the small intestine, a process dependent on bile and pancreatic enzymes. Once taken up by the intestinal cells, vitamin E is incorporated into chylomicrons for transport in the lymph and eventually the blood.

As chylomicrons are broken down, most of the vitamin E is carried to the liver as chylomicron remnants. A small amount is carried directly to other tissues. The liver repackages the vitamin E from the chylomicron remnants with other lipoproteins (VLDL, LDL, and HDL) for delivery to body tissues. Unlike other fat-soluble vitamins, vitamin E does not have a specific transport protein in the blood, so it is carried by these lipoproteins. Vitamin E also differs from other fat-soluble vitamins in that it does not accumulate in the liver; instead, approximately 90% of the vitamin E in the body is localized in adipose tissue.

Vitamin E can be excreted via the bile, urine, and skin. However, because vitamin E absorption is often low, most vitamin E is excreted via the small amount of bile that exits the body in the feces.

## Functions of Vitamin E

Vitamin E is an important part of the body's antioxidant network, which helps maintain the integrity of cell membranes by stopping chain reactions caused by free radicals. **Free radicals** are very unstable compounds that have 1 or more unpaired electrons. Normally, atoms left with an unpaired electron after oxidation reactions immediately pair with each another, creating more stable compounds. However, when this does not occur, free radicals remain and act as strong oxidizing (electron-seeking) agents. These can be very destructive to electron-dense cell components, such as cell membranes and DNA.

As damaging as free radicals may be, they do play important roles in the body. For example, as part of the immune system's arsenal against invading pathogens, white blood cells (leukocytes) generate free radicals to destroy bacteria, viruses, and other agents that cause infections. To protect itself, the body must be able to regulate free radical activity and avoid potentially damaging effects. This task is assigned to vitamin E and other antioxidants, such as glutathione peroxidase, catalase, and superoxide dismutase (Fig. 12-18).

Antioxidants function in a variety of ways to regulate free radicals and prevent the damage they cause. For instance, in lipid rich areas of the body, free radicals can initiate a chain of reactions known as peroxidation. Lipid peroxidation reactions break apart fatty acids and create lipid **peroxyl radicals** (also called reactive oxygen species because they contain oxygen radicals). The chain of reactions continues to break apart fatty acids until 2 free radicals pair and stabilize each other. However, many lipid peroxyl radicals may be produced by these reactions before stabilization occurs.

**free radical** Compound with an unpaired electron, which causes it to seek an electron from another compound. Free radicals are strong oxidizing agents and can be very destructive to electron-dense cell components, such as DNA and cell membranes.

**peroxyl radical** Peroxide compound containing a free radical; designated R-O-O•, where R is a carbon-hydrogen chain broken off a fatty acid and the dot is an unpaired electron.

Plant oils are rich sources of vitamin E.

C Squared Studios/Getty Images

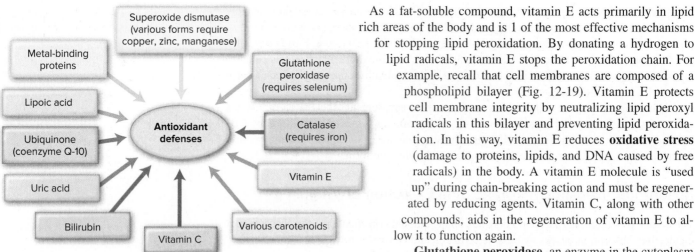

**Figure 12-18** The body does not rely solely on vitamin E for antioxidant protection. Such protection is a team effort, using a number of nutrients, metabolites, and enzyme systems.

As a fat-soluble compound, vitamin E acts primarily in lipid rich areas of the body and is 1 of the most effective mechanisms for stopping lipid peroxidation. By donating a hydrogen to lipid radicals, vitamin E stops the peroxidation chain. For example, recall that cell membranes are composed of a phospholipid bilayer (Fig. 12-19). Vitamin E protects cell membrane integrity by neutralizing lipid peroxyl radicals in this bilayer and preventing lipid peroxidation. In this way, vitamin E reduces **oxidative stress** (damage to proteins, lipids, and DNA caused by free radicals) in the body. A vitamin E molecule is "used up" during chain-breaking action and must be regenerated by reducing agents. Vitamin C, along with other compounds, aids in the regeneration of vitamin E to allow it to function again.

**Glutathione peroxidase,** an enzyme in the cytoplasm and mitochondria of cells, is important in catalyzing the breakdown of hydrogen peroxides and lipid peroxides. These compounds are reactive oxygen species that can form free radicals if they combine with other compounds. Thus, by eliminating these peroxides, glutathione peroxidase helps vitamin E reduce oxidative damage to cells. The activity of glutathione peroxidase depends on the mineral selenium (the functional part of this enzyme). **Catalase** also is important for the removal of hydrogen peroxide. This heme iron–dependent enzyme is found in the peroxisomes of the cell.

Another vital defense system to prevent the overproduction of reactive species in cells is provided by a family of enzymes known as superoxide dismutase enzymes. These enzymes are important in eliminating superoxide radicals that form when oxygen reacts with other compounds. Two superoxide dismutase enzymes require copper and zinc. One of these enzymes is located inside the cell (in the cytosol) and the other is outside the cell. The 3rd superoxide dismutase enzyme is found in the mitochondria and requires the mineral manganese.

Many individuals take supplements of vitamin E and other antioxidants in hopes of preventing cancer, cardiovascular disease, and other chronic diseases associated with free radical damage.[43,44,45,46,47] Two large studies (SELECT: Selenium and Vitamin E Cancer Prevention Trial and Physicians' Health Study II) reported that vitamin E and/or selenium did not decrease the risk of prostate or other cancers.[44,45] However, a meta-analysis suggests that the effects of vitamin E may vary with the stage of prostate cancer and blood levels of the alpha-tocopherol form of vitamin E.[48] Although early studies also suggested a decreased risk of atherosclerosis with vitamin E supplementation, later studies failed to show a benefit.[46] Inconsistent research findings indicate more research is needed before experts are able to recommend using vitamin E to prevent cancer or cardiovascular disease.

## Vitamin E Deficiency

Overt vitamin E deficiency is rare in humans. Individuals at greatest risk of deficiency are those with fat malabsorption conditions, such as cystic fibrosis or Crohn disease, smokers, and **preterm** infants.[1] Preterm infants are particularly susceptible because they are born with limited stores of vitamin E and often have insufficient intestinal absorption of this vitamin. Smokers are at increased risk of deficiency because the oxidative stress and lipid peroxidation caused by cigarette smoke increase vitamin E needs.[49]

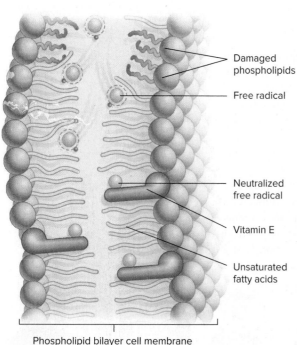

Phospholipid bilayer cell membrane

**Figure 12-19** Fat-soluble vitamin E can donate an electron to stop free radical chain reactions. If not interrupted, these reactions cause extensive oxidative damage to cell membranes.

Vitamin E deficiency is characterized by the premature breakdown of red blood cells (hemolysis) and the development of **hemolytic anemia.** Because of the serious risk of this, preterm infants are given supplemental vitamin E and specialized formulas containing vitamin E early in life.

Vitamin E deficiency also can impair immune function and cause neurological changes in the spinal cord and peripheral nervous system.[50,51] These symptoms have been noted in individuals with vitamin E deficiency resulting from a genetic abnormality in lipoprotein synthesis that decreases vitamin E transport and distribution in the body.

## Vitamin E Toxicity

Although vitamin E is relatively nontoxic, excessive amounts can interfere with the role of vitamin K in blood clotting. This causes insufficient clotting and a risk of **hemorrhage.** These risks are of particular concern in individuals taking daily aspirin or anticoagulation medications, such as warfarin (Coumadin®), to prevent blood clots. Megadoses of vitamin E can result in severe hemorrhaging in these individuals. To prevent toxicity-related problems, the UL for vitamin E is set at 1000 mg (1500 IUs) of alpha-tocopherol from natural sources or 1100 IUs from synthetic sources.[1]

Taking large amounts of alpha-tocopherol also may decrease the antioxidant activity of gamma-tocopherol in the body. To compensate, some experts recommend that vitamin E supplements contain a mixture of natural tocopherols. This combination is more expensive, however, than natural or synthetic alpha-tocopherol alone.

### Knowledge Check

1. How does vitamin E function as an antioxidant in the body?
2. What are 3 foods rich in vitamin E?
3. Why is excess supplementation of vitamin E of concern in individuals taking daily aspirin or anticoagulation medications?

▶ Oxidizing agents that cause cell damage include highly reactive oxygen species, such as the singlet oxygen ($^1O_2$), hydrogen peroxide ($H_2O_2$), hydroxyl radical (•OH), superoxide ($O_2$•.), ozone ($O_3$), and nitrogen-oxygen combinations that are typical of air pollutants (NO•).

▶ Because an antioxidant protects other compounds by becoming oxidized itself, in a chemical sense, antioxidants are more correctly termed **redox agents.** In other words, they can undergo both oxidation (loss of an electron) and later reduction (regaining an electron). Nevertheless, *antioxidant* is still the most common term, even in the scientific literature.

**preterm**   Born before 37 weeks of gestation (also referred to as premature).

**hemolytic anemia**   Disorder that causes red blood cells to break down faster than they can be replaced.

**hemorrhage**   Bleeding.

**coagulation**   Formation of a blood clot.

## 12.5 Vitamin K

The discovery of vitamin K centered on its role in blood clotting (Fig. 12-20). A Danish researcher first noted the relationship between vitamin K and blood clotting when he observed that chicks fed a diet with the fat extracted developed hemorrhages. Thus, he named this new lipid-soluble factor "vitamin K" after *koagulation,* the Danish spelling for **coagulation.**

The family of compounds known as vitamin K, or the quinones, includes **phylloquinones** (vitamin $K_1$) from plants and **menaquinones** (vitamin $K_2$) found in fish oils and meats. Menaquinones also are synthesized by bacteria in the human colon. A synthetic compound, called menadione, can be converted to menaquinone in body tissues. Phylloquinone, the main dietary form of the vitamin, is the most biologically active form.

## Vitamin K Sources

About 10% of the vitamin K absorbed each day comes from bacterial synthesis in the colon.[52] The remainder is provided by the diet. Although the vitamin K content of individual foods varies, green leafy vegetables (e.g., kale, turnip greens, parsley, salad greens, cabbage, and spinach), broccoli, peas, and green beans are the best sources (Fig. 12-21). Vegetable oils, such as soy and canola,

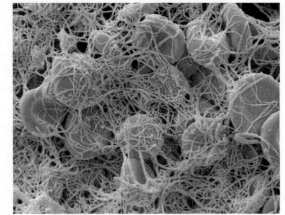

***Figure 12-20***   VVitamin K is essential for normal blood clotting.

Steve Gschmeissner/Science Photo Library/Getty Images

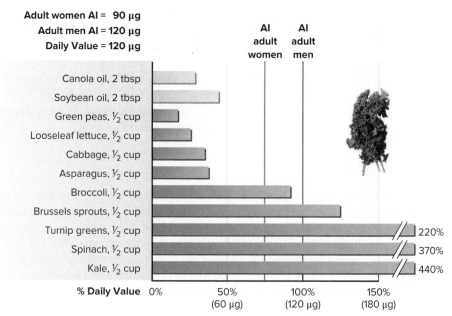

Adult women AI =  90 µg
Adult men AI = 120 µg
Daily Value = 120 µg

**Figure 12-21**  Food sources of vitamin K.

Stockdisc/Getty Images

also are good sources. Vitamin K is relatively stable to heat processing, but it can be destroyed by exposure to light.

## Vitamin K Needs

For women, the Adequate Intake for vitamin K is 90 µg/day; for men, it is 120 µg/day. These Adequate Intakes are based on the apparent adequacy of usual intakes and the lack of information to determine an EAR and RDA.[2] Although current intakes provide adequate vitamin K for blood clotting functions, it is not known whether increased intakes might be beneficial for other functions. The Daily Value for vitamin K is 120 µg/day.

## Absorption, Transport, Storage, and Excretion of Vitamin K

Approximately 80% of dietary vitamin K (primarily as phylloquinone) is taken up by the small intestine and incorporated into chylomicrons. This process requires bile and pancreatic enzymes. The menaquinones synthesized by bacteria in the colon are absorbed by passive diffusion. Vitamin K can be incorporated into the lipoprotein VLDL in the liver for storage or carried by HDL and LDL to cells throughout the body. Most vitamin K excretion occurs via the bile that passes out of the body in the feces. A small amount of excretion also occurs via the urine.[52]

A daily salad containing dark green vegetables provides abundant vitamin K.

C Squared Studios/Getty Images

## Functions of Vitamin K

Vitamin K is needed for the synthesis of blood clotting factors by the liver and the conversion of preprothrombin to the active blood clotting factor called prothrombin (Fig. 12-22). In these reactions, carbon dioxide ($CO_2$) is added to a glutamic acid in preprothrombin, yielding prothrombin. Because prothrombin contains the Gla amino acid gamma-carboxyglutamic acid, it is called a Gla protein. All vitamin K–dependent proteins contain Gla residues, which are needed to bind calcium and form blood clots.[52]

Vitamin K is converted to an inactive form once it has activated the clotting factors. It must then be reactivated for its biological action to persist. Drugs such as warfarin (Coumadin®), which strongly inhibit this reactivation process, act as powerful anticoagulants.

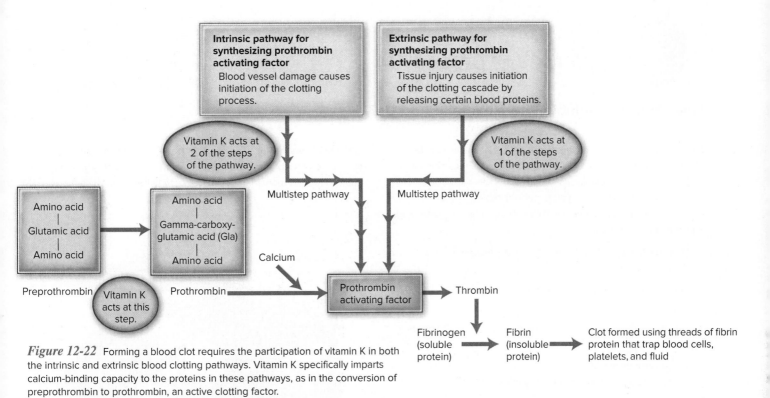

*Figure 12-22* Forming a blood clot requires the participation of vitamin K in both the intrinsic and extrinsic blood clotting pathways. Vitamin K specifically imparts calcium-binding capacity to the proteins in these pathways, as in the conversion of preprothrombin to prothrombin, an active clotting factor.

People taking warfarin to lessen blood clotting should maintain a consistent dietary vitamin K intake and avoid vitamin K supplementation.[52]

Vitamin K also plays a role in bone metabolism. Three additional vitamin K–dependent Gla proteins (called osteocalcin, matrix Gla protein, and protein S) are synthesized in bone. The functions of these proteins are not clearly understood. However, synthesis of these proteins is reduced in vitamin K–deficient animals and results in changes in bone health.[52]

Another function of vitamin K is linked with preventing calcification in blood vessels. This action may decrease the risk of cardiovascular disease.[53] Vitamin K also is being explored for its potential benefits related to kidney disease, diabetes mellitus, and obesity.[53]

## Vitamin K Deficiency

Vitamin K deficiency can occur in newborns because vitamin K stores are typically low at birth and the intestinal tracts of newborns do not yet have bacteria that can synthesize vitamin K. This increases the risk of bleeding due to defective blood clotting; thus, newborn infants in North America are given vitamin K injections within 6 hours of delivery.

A deficiency of vitamin K in older infants, children, teens, and adults is rare, although it can occur with prolonged use of antibiotics that disrupt vitamin K synthesis or with impaired fat absorption.[2] Because a deficiency results in poor clot formation and hemorrhaging, physicians always check a patient's vitamin K status prior to any surgical procedure.

Excessive intakes of vitamin A and vitamin E negatively affect the actions of vitamin K.[2,52,54] Vitamin A is thought to interfere with the absorption of vitamin K from the small intestine, whereas large doses of vitamin E can lead to a decrease in vitamin K–dependent clotting factors and increased bleeding tendency. In either case, megadose supplements of these vitamins increase the risk of vitamin K deficiency and bleeding, as noted in earlier discussions of the Upper Levels of the vitamins.

*A Biochemist's View*

Phylloquinone (K$_1$)

Menaquinone (K$_2$)

**Vitamin K family**

[ ] = Repeated section

## Take Action

### Does Your Fat-Soluble Vitamin Intake Add Up?

From NHANES and other dietary surveys of the American population, it's known that many individuals do not consume the recommended intakes for all the fat-soluble vitamins. Diets are often low in vitamins D and E and in carotenoid rich fruits and vegetables. The following questions can help you determine if your dietary intake of these foods and nutrients is adequate.

1. Do you eat at least 1 cup of yellow-orange vegetables or 2 cups of dark green, leafy vegetables each day?

2. Do you consume at least 1 cup of yellow-orange fruit or juice (100% juice) each day?

3. Do you consume 2 to 3 cups of milk or yogurt, or 2 to 3 ounces of cheese, each day?

4. Do you include at least 1 teaspoon of plant oils (cottonseed, canola, sunflower, corn, or olive) in your daily diet?

5. Do you include at least ¼ cup of plant seeds or nuts in your diet each day?

6. Do you eat at least 2 to 3 servings of salmon, tuna, herring, or fish oils each week?

If you answered no to any of these questions, your diet may be lacking in essential fat-soluble vitamins, helpful phytochemicals, and other important nutrients. Review the information on food sources for each of the fat-soluble vitamins in this chapter and the guidelines provided on choosemyplate.gov to help you plan a more healthful diet.

Digital Vision/Getty Images

---

▶ One means of detecting a vitamin K deficiency is an increase in blood clotting time, a measure of how quickly prothrombin in the blood can form a clot.

---

**bilirubin (bi-li-RUBE-in)** Bile pigment derived from hemoglobin during the destruction of red blood cells; excess in the blood causes skin and eyes to become yellow (jaundiced).

## Vitamin K Toxicity

No Upper Level has been set for vitamin K.[2] Although vitamin K can be stored in the liver and bone, storage is limited. Vitamin K also is more readily excreted than other fat-soluble vitamins. When used in its natural forms of phylloquinones or menaquinones, increased amounts of vitamin K have not caused harmful effects. In contrast, injections of menadione, a synthetic form of vitamin K, have caused hemolytic anemia, excess **bilirubin** in the blood (jaundice), and death in newborns. Thus, menadione is no longer used for the treatment of vitamin K deficiency.

As you can see, the fat-soluble vitamins have numerous important functions in the body. Table 12-2 provides a summary of these vitamins.

### Knowledge Check

1. What are 3 foods that are rich sources of vitamin K?
2. How does vitamin K help in the formation of blood clots?
3. Why should people taking the drug Coumadin® avoid taking vitamin K supplements?
4. What population groups are at increased risk of a vitamin K deficiency?

**Table 12-2 Summary of the Fat-Soluble Vitamins**

| Major Vitamin | Functions | Deficiency Symptoms | People at Risk | Sources | RDA or Adequate Intake | Toxicity Symptoms |
|---|---|---|---|---|---|---|
| **Vitamin A** | | | | | | |
| Preformed retinoids, provitamin A carotenoids | Vision in dim light and color vision, cell differentiation, bone growth, immunity, reproduction | Poor growth, night blindness, total blindness, dry skin, xerophthalmia, hyperkeratosis, impaired immune function | Rare in U.S. but common in preschool children living in poverty in low-income countries and patients with fat malabsorption syndromes | Preformed vitamin A (retinoids): liver, fortified milk, fish liver oils; provitamin A (carotenoids): red, orange, dark green, and yellow vegetables; orange fruits | 700 µg (RAE) for women, 900 µg for men | Headache, vomiting, double vision, dry mucous membranes, bone and joint pain, liver damage, hemorrhage, coma, spontaneous abortions, birth defects; Upper Level is 3000 µg of preformed vitamin A |
| **Vitamin D** | | | | | | |
| Cholecalciferol D$_3$, ergocalciferol D$_2$ | Maintenance of calcium and phosphorus concentrations, immune function, cell cycle regulation | Rickets in children, osteomalacia in older adults | Dark-skinned individuals, older adults with low intakes or low UV exposure, patients with fat malabsorption syndromes | Vitamin D–fortified milk, fish oils, oily fish | 15 µg for ages 1 to 70 years, 20 µg for above 70 years | Calcification of soft tissues, impaired growth, excess calcium in the blood, excretion in the urine; Upper Level is 100 µg |
| **Vitamin E** | | | | | | |
| Tocopherols, tocotrienols | Antioxidant, prevention of free radical damage | Hemolysis of red blood cells, degeneration of sensory neurons | Patients with fat malabsorption syndromes | Plant oils, seeds, nuts, products made from oils | 15 mg alpha-tocopherol for men and women | Inhibition of vitamin K metabolism; Upper Level is 1000 mg |
| **Vitamin K** | | | | | | |
| Phylloquinone, menaquinone | Synthesis of blood clotting factors and bone proteins | Hemorrhage due to poor blood clotting | Those taking antibiotics for a long period of time, adults with low green vegetable intake, patients with fat malabsorption syndromes | Green vegetables, synthesis by intestinal microorganisms | 90 µg for women, 120 µg for men | Rare, can cause hemolytic anemia; no Upper Level has been set |

## 12.6 Dietary Supplements: Healthful or Harmful?

About 50% of the U.S. population uses dietary supplements, spending over $46 billion annually—with over 30% of U.S. adults taking multivitamin and mineral supplements.[3,55,56] Opinions differ within the scientific and medical communities as to whether supplements should be widely prescribed. Some scientists and clinicians suggest that, because most Americans fail to meet the guidelines for the intake of fruits, vegetables, and whole grains, they may have inadequate intakes of several micronutrients. Although current evidence is insufficient to support the recommendation of multivitamin and mineral supplementation for the general population,[34,55,57] many people take them to enhance or maintain their overall well-being and/or for the health of specific sites, such as bones, heart, or eyes.[58] Among adults, more women than men use multivitamin and mineral supplements, and the incidence of use rises with increasing age and higher socioeconomic status.[59]

According to the Dietary Supplement Health and Education Act (DSHEA) of 1994, a supplement is defined as any product intended to supplement the diet that contains 1 or more of the following ingredients.

- A vitamin
- A mineral
- An amino acid
- An herb, a botanical, or a plant extract
- A combination of any of the above

Recall from Chapter 1 that dietary supplements are regulated differently than drugs and food additives, which the FDA tests extensively and regulates for safety, effectiveness, dose size, concerns about interactions with other substances, and health-related claims. In contrast, the FDA does not closely monitor dietary supplements, except folic acid, unless there is evidence that a supplement is dangerous or is marketed with an illegal claim. Thus, individuals need to seek advice from their physician, registered dietitian nutritionist, or pharmacist to understand the potential health benefits and risks related to the use of dietary supplements.

Due to the limited regulation of dietary supplements, some supplement manufacturers make broad claims about the benefits of their products. Without scientific evidence, they cannot claim that their products will prevent, treat, or cure diseases. However, current laws allow them to make structure or function claims and do not prevent them from making unproven claims with regard to conditions that are not diseases. Thus, many manufacturers market their products as a means to increase energy, enhance performance, lose weight, reduce body fat, eliminate signs of aging, and relieve symptoms of menopause, fatigue, and stress. For example, a product that claims to prevent premenstrual symptoms, increase energy, enhance mood, or promote bowel health can be sold without any evidence that the product effectively does so because these conditions are not diseases. However, a product that claims to decrease the risk of cardiovascular disease by reducing blood cholesterol levels must have scientific evidence to justify this claim. The FDA does not evaluate a supplement's effectiveness before it is marketed.

The quality, purity, and consistency of dietary supplement products also are not closely monitored by the FDA. Although the FDA has established standards to help ensure a product's quality, strength, and purity, studies of dietary supplements indicate that product quality can vary significantly. To aid consumers in purchasing supplements meeting acceptable standards, the U.S. Pharmacopeia (USP) designation may be listed on products that meet established USP standards for strength, quality, purity, packaging, labeling, solubility, and storage life. However, because the USP labeling of dietary supplements is voluntary, many manufacturers use their own standards for manufacturing and quality control.

Although dietary supplements can replace specific nutrients lacking in a diet, they cannot fully correct a nutritionally poor diet. For example, most supplements do not contain fiber or phytochemicals that have health-promoting benefits. Others contain limited amounts of calcium, needed for bone health and cellular functions. Supplements also may contain excess amounts of individual nutrients that can increase the risk of vitamin and mineral toxicities and nutrient interactions. For instance, a supplement with a high amount of zinc can interfere with the absorption and utilization of iron or copper; high intakes of folate can mask the symptoms of a vitamin B-12

Long-term supplementation with just 3 to 5 times the Daily Value for some fat-soluble vitamins—particularly preformed vitamin A (retinoids)—can cause toxic effects.

Glow Images

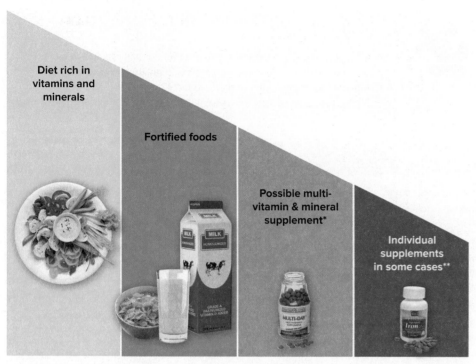

Diet rich in
vitamins and
minerals

Fortified foods

Possible multi-
vitamin & mineral
supplement*

Individual
supplements
in some cases**

*Men and older women generally should use iron-free formulations.
**Iron and calcium supplements for younger women are examples.

*Figure 12-23* A savvy approach to nutrient supplementation emphasizes nutrient rich foods that provide vitamins and minerals as well as fiber, numerous phytochemicals, and omega-3 fatty acids. Vitamin and/or mineral supplements may be important during certain life stages and when treating certain health conditions. To minimize potential toxic effects, avoid exceeding the Upper Limits for nutrients if also consuming fortified foods and beverages.

salad: C Squared Studios/Getty Images; cereal: Foodcollection; OJ: Sergei Vinogradov/seralexvi123; milk: PhotoSpin, Inc/Alamy Stock Photo; multi-vitamin: Ken Karp/McGraw-Hill Education; supplements: John Flournoy/McGraw-Hil

deficiency; and excessive intakes of vitamins A and D can result in toxicity. Thus, nutrition experts suggest that the best way to meet nutritional needs is to eat a diet that includes a variety of nutrient rich fruits, vegetables, whole grains, low-fat dairy products, fish, lean meat and meat alternatives, nuts, and seeds. The approach to supplementation shown in Figure 12-23 can help you plan a diet rich in vitamins and minerals.

Although vitamin and mineral requirements can be met by eating a nutritionally varied diet, there may be times when the use of vitamin and/or mineral supplements is necessary.

- Women with excessive bleeding during menstruation may need iron supplements to prevent anemia.
- Women who may become pregnant or are pregnant may require iron and folate supplements to meet their needs.
- Individuals with low calorie intakes may require a multivitamin and mineral supplement to correct for limited intakes.
- Vegans may require calcium, iron, zinc, vitamin D, and vitamin B-12 supplements to prevent deficiencies.
- Newborn infants may need a single dose of vitamin K (as directed by a physician) to prevent bleeding problems.
- Infants and young children may need fluoride supplements to prevent dental caries.
- Individuals with limited sunlight exposure and a low intake of dairy products may need vitamin D supplements.
- Individuals with lactose intolerance or milk allergies may need calcium and vitamin D supplements.
- Individuals with specific medical conditions or those using medications that alter nutrient metabolism or nutritional status may require specific vitamin/mineral supplements.

Registered dietitian nutritionists can assess the need for supplements in healthy people and those with disease or health risks. They also can provide guidance for choosing foods rich in specific nutrients, as well as an appropriate supplement. Registered dietitian nutritionists often suggest that you start by choosing a nationally recognized brand that contains no more than 100% of the Daily Value for the nutrients listed. Be careful that the total intake from your diet, including foods fortified with vitamins and minerals, plus your supplement does not exceed the Upper Level for any vitamin or mineral. When choosing a supplement, check for superfluous ingredients, such as bee pollen, lecithin, hesperidin complex, inositol, laetrile ("vitamin B-17"), pangamic acid, or para-aminobenzoic acid (PABA), which are not needed in our diets and often add significant expense to supplements.

▶ These websites can help you evaluate the safety and claims of various supplements.

www.acsh.org
www.quackwatch.com
www.ncahf.org
www.eatright.org
ods.od.nih.gov
www.consumerlab.com
www.integrativerd.org

### Knowledge Check

1. According to DSHEA, supplements to the diet contain which ingredients?
2. What types of claims do current laws allow supplement manufacturers to make about the benefits of their products?
3. Describe 3 circumstances when a vitamin and/or mineral supplement is needed.

## CASE STUDY FOLLOW-UP

Rachel Frank/Corbis/Glow Images

The use of Nutramega poses some health risks for Emmi. Taking 2 or 3 tablets every 3 hours would mean taking at least 16 tablets per day. This alone would provide an intake of vitamin A, vitamin C, and zinc well in excess of the Upper Levels for these nutrients. Her intake of preformed vitamin A would be 1.2 times the Upper Level; of vitamin C, 3.6 times the Upper Level; and of zinc, 2.2 times the Upper Level. Her intake of selenium, however, would fall well below the Upper Level set for that nutrient. This is how the math works out:

*Vitamin A:* 33% (0.33 times) the Daily Value of 900 µg RAE equals 297 µg RAE per tablet. Sixteen tablets yield 4752 µg RAE. The Upper Level is 3000 µg RAE for preformed vitamin A. Because 75% of the vitamin A is preformed vitamin A, this yields 3564 µg RAE of preformed vitamin A (4752 × 0.75 = 3564), or 1.2 times the Upper Level (3564/3000 = 1.2).

*Vitamin C:* 500% (5) times the Daily Value of 90 mg equals 450 mg per tablet. Sixteen tablets yield 7200 mg. The Upper Level is 2000 mg. This amount is 3.6 times the Upper Level (7200/2000 = 3.6).

*Zinc:* 50% (0.5) times the Daily Value of 11 mg equals 5.5 mg per tablet. Sixteen tablets yield 88 mg. The Upper Level is 40 mg. This amount is 2.2 times the Upper Level (88/40 = 2.2).

*Selenium:* 10% (0.1) times the Daily Value of 55 µg equals 5.5 µg per tablet. Sixteen tablets yield 88 µg. This is less than the Upper Level of 400 µg.

If Emmi takes the maintenance dose of 2 or 3 tablets per day, she will not be at risk of toxicities. However, Nutramega is very expensive, compared with the cost of a typical multivitamin and mineral supplement. Overall, Emmi is smart to be concerned about meeting her nutrient needs, but the stress she is under does not increase her nutrient needs. A healthy diet, as shown in Table 2-9 in Chapter 2, should be her primary strategy to manage her stress and stay healthy.

*Take Action*

## A Closer Look at Supplements

With the current popularity of vitamin and mineral supplements, it is more important than ever to understand how to evaluate a supplement. Study the label of a supplement you use or 1 readily available from a friend or the supermarket. Then answer the following questions.

1. Based on the recommended dosage, are there any individual vitamins or minerals for which the intake is greater than 100% of the Daily Value? List these vitamins and minerals.

2. How do the suggested intakes of the vitamins and minerals in the supplement compare with the current DRIs for these nutrients?

3. Are any suggested intakes above the Upper Levels for the nutrients? List these nutrients and the Upper Level for each.

4. Are there any non-nutrient ingredients, such as herbs or botanical extracts, in the supplement? You often can find these by looking for ingredients that do not have a % Daily Value.

5. Does at least 25 to 50% of the vitamin A in the product come from beta-carotene or other provitamin A carotenoids (to reduce the risk of vitamin A toxicity)?

6. Are there any warnings on the label for individuals who should not consume this product?

7. Are there any other signs that tip you off that this product may be more harmful than healthful?

Ryan McVay/Getty Images

# Chapter Summary

## 12.1 Vitamins are essential, organic compounds needed for important metabolic reactions in the body.

They are not a source of energy. Instead, they promote many energy-yielding and other reactions in the body, thereby aiding in the growth, development, and maintenance of various body tissues. Vitamins cannot be synthesized in the body at all or are synthesized in insufficient amounts. Vitamins A, D, E, and K are fat soluble, whereas the B-vitamins and vitamin C are water soluble. Fat-soluble vitamins are absorbed along with dietary fat. They travel by way of the lymphatic system into general circulation, carried by chylomicrons. In disease states that limit fat digestion, fat-soluble vitamin absorption may be compromised, thereby increasing the risk of deficiency in these individuals. Fat-soluble vitamins are excreted less readily from the body than water-soluble vitamins and thus pose a potential threat for toxicity, especially of vitamins A and D. Toxicities of these fat-soluble vitamins generally occur with high doses of supplements, rather than from foods.

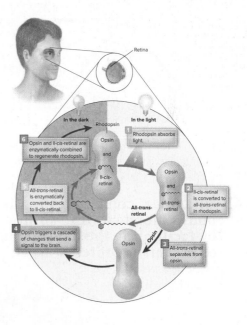

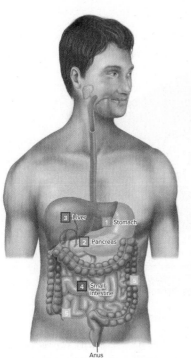

## 12.2 Vitamin A consists of a family of retinoid compounds:

retinal, retinol, and retinoic acid. A plant derivative, known as beta-carotene, along with 2 other carotenoids, yields vitamin A after metabolism by the small intestine or liver. Vitamin A is found in foods of animal origin, such as liver, fish oils, and fortified milk. Carotenoids are obtained from plants and are especially plentiful in dark green and yellow-orange vegetables and fruits. Vitamin A contributes to the maintenance of vision, the normal development of cells (especially mucous-forming cells), and immune function. Preformed vitamin A can be quite toxic when taken at doses 2 to 4 times or more the RDA. Use of vitamin A supplements is especially dangerous during pregnancy because it can lead to fetal malformations.

## 12.3 Vitamin D can be obtained from food and it is produced by the body.

The synthesis of vitamin $D_3$ begins with a precursor of cholesterol synthesis located in the skin and depends on ultraviolet light. With adequate sun exposure, no dietary intake of vitamin D is needed. Vitamin D food sources include fish oils and fortified milk. The provitamin, whether produced in the skin or obtained from the diet, is metabolized in the liver and kidneys to yield 1,25(OH)₂ vitamin $D_3$ (calcitriol), the active form of vitamin D. Calcitriol is important for calcium and phosphorus absorption from the small intestine and, along with other hormones, for the regulation of bone metabolism. It also is important in gene

455

expression and immune function. The risk of vitamin D deficiency may be greater than previously observed, especially in the elderly and individuals lacking regular sunlight exposure. Vitamin D deficiency results in harmful changes in bone, a condition known as rickets in children and osteomalacia in adults. Toxicity of vitamin D can occur with excess supplementation, causing the deposition of calcium in the kidneys, heart, and lungs.

## 12.4 Vitamin E functions as an antioxidant.

By donating electrons to electron-seeking, or oxidizing, compounds (e.g., free radicals), it neutralizes their action and prevents the widespread destruction of both cell membranes and DNA. Vitamin E is 1 of several components in the body's defense system against oxidizing agents. Vitamin E is plentiful in plant oils and food products that contain these oils. Overt vitamin E deficiency is rare. Toxicity from megadose therapy inhibits vitamin K activity and, in turn, increases the risk of hemorrhage.

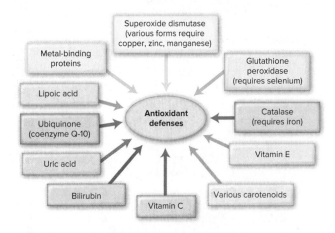

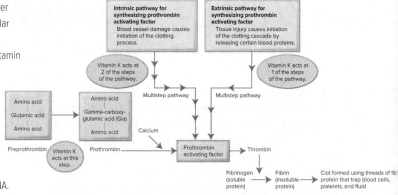

## 12.5 Vitamin K contributes to the body's blood clotting ability by facilitating the conversion of precursor proteins, such as

prothrombin, to active clotting factors that promote blood coagulation. Vitamin K also plays a role in bone metabolism. About 10% of the vitamin K absorbed each day comes from bacterial synthesis in the large intestine. Most vitamin K comes primarily from green leafy vegetables and vegetable oils in the diet. Vitamin K deficiency is rare, but it can occur in newborns. Thus, newborns are given vitamin K injections shortly after birth as a preventive measure.

C Squared Studios/Getty Images

## 12.6 Taking a multivitamin and mineral supplement to help meet nutrient needs is recommended by some experts,

but others suggest that only some people need them. Taking many nutrient supplements can lead to nutrient-related toxicity, so their use should be considered carefully. The clearest evidence for good nutrition supports a diet rich in fruits, vegetables, whole-grain breads and cereals, low-fat dairy products, fish, lean meat and meat alternatives, and nuts and seeds, rather than relying on supplements to meet nutritional needs.

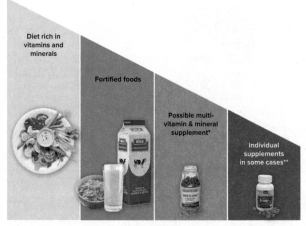

*Men and older women generally should use iron-free formulations.
**Iron and calcium supplements for younger women are examples.

salad: C Squared Studios/Getty Images; cereal: Foodcollection; OJ: Sergei Vinogradov/ seralexvi123; milk: PhotoSpin, Inc/Alamy Stock Photo; multi-vitamin: Ken Karp/McGraw-Hill Education; supplements: John Flournoy/ McGraw-Hil

## Global Perspective

Vitamin A deficiency is common in much of the low-income world. International groups promote breastfeeding, food fortification, supplement use, and educational programs to combat this problem.

JIANG HONGYAN/Shutterstock

# Study Questions

1. Which population group is at lowest risk of fat-soluble vitamin deficiencies?

   a. low-birth-weight, premature infants
   b. very-low-income families
   c. patients with malabsorption diseases
   d. pregnant women

2. Carotenoids are a precursor form of _____.

   a. vitamin K
   b. vitamin E
   c. vitamin D
   d. vitamin A

3. Which food provides very little vitamin A?

   a. mango
   b. spinach
   c. banana
   d. liver

4. Vitamin A is involved in all of the following functions *except* _____.

   a. vision and dark adaptation
   b. hemoglobin synthesis
   c. resistance to infection
   d. cell differentiation

5. Vitamin A deficiency is associated with the symptoms of night blindness, keratinization, and increased infections.

   a. true
   b. false

6. Which of the following vitamins also can be classified as a hormone because the body can synthesize it?

   a. vitamin A
   b. vitamin D
   c. vitamin E
   d. vitamin K

7. Which of the following is a good source of vitamin D?

   a. yellow-orange vegetables
   b. salmon and sardines
   c. dark, leafy greens
   d. enriched grains

8. Which of the following is a function of vitamin D?

   a. serves as an antioxidant to protect against lipid peroxidation
   b. serves as a coenzyme in energy metabolism
   c. regulates calcium homeostasis
   d. produces blood clotting factors

9. Vitamin D deficiency has been associated with an increased risk of diabetes, multiple sclerosis, and hypertension.

   a. true
   b. false

10. Vitamin D deficiency in children results in a condition called _____.

    a. osteomalacia
    b. beriberi
    c. rickets
    d. xerophthalmia

11. Wheat germ and vegetable oils are good sources of vitamin E.

    a. true
    b. false

12. Large doses of vitamin E have been shown to interfere with vitamin K activity and to increase the risk of bleeding.

    a. true
    b. false

13. Which of the following is the best source of vitamin K?

    a. citrus fruits
    b. dark, leafy greens
    c. enriched grains
    d. nuts and seeds

14. Which vitamin aids in blood clotting?

    a. vitamin A
    b. vitamin D
    c. vitamin E
    d. vitamin K

15. Vitamin and mineral supplements are tightly regulated by the FDA.

    a. true
    b. false

16. Describe 3 major differences between water-soluble and fat-soluble vitamins.

17. Describe 3 differences between retinoids and carotenoids.

18. Why is vitamin D also classified as a hormone or a conditional vitamin?

19. Explain how vitamin E helps protect the integrity of cell membranes.

**20.** Explain why newborn infants are given an injection of vitamin K.

**21.** What factors should you consider when deciding whether to use a nutrient supplement?

**22.** Describe the efforts to reduce vitamin A deficiency globally.

Photodisc Collection/Getty Images

# References

**1.** Food and Nutrition Board, Institute of Medicine. *Dietary Reference Intakes for vitamin C, vitamin E, selenium, and carotenoid*s. Washington, DC: National Academies Press; 2000.

**2.** Food and Nutrition Board, Institute of Medicine. *Dietary Reference Intakes for vitamin A, vitamin K, arsenic, boron, chromium, copper, iodine, iron, manganese, molybdenum, nickel, silicon, vanadium, and zinc.* Washington, DC: National Academies Press; 2001.

**3.** Food and Nutrition Board, Institute of Medicine. *Dietary Reference Intakes for vitamin D and calcium.* Washington, DC: National Academies Press; 2011.

**4.** Ross AC. Vitamin A and carotenoids. In: Shils ME and others, eds. *Modern nutrition in health and disease.* 10th ed. Philadelphia: Lippincott Williams & Wilkins; 2006.

**5.** Li J and others. Dietary inflammatory potential and risk of cardiovascular disease among men and women in the U.S. *J Am Coll Cardiol.* 2020;76:2181.

**6.** Ma L and others. Lutein and zeaxanthin and the risk of age-related macular degeneration: A systematic review and meta-analysis. *Br J Nutr.* 2012;107:360.

**7.** Bungau S and others. Health benefits of polyphenols and carotenoids in age-related eye diseases. *Oxidative Medicine and Cellular Longevity.* 2019; Article ID 9783429, 22 pages, 2019. https://doi.org/10.1155/2019/9783429

**8.** Matthew MC and others. Antioxidant vitamin supplementation for preventing and slowing the progression of age-related cataract. *Cochrane DB Syst Rev.* 2012;6:Art. No. CD0045.

**9.** Middha P and others. B-carotene supplementation and lung cancer incidence in the Alpha-tocopherol, Beta-carotene Cancer Prevention Study: The role of tar and nicotine. *Nicotine & Tobacco Res.* 2018; doi.org/10.1093/ntr/nty115.

**10.** Herrera E and others. Aspects of antioxidant foods and supplements in health and disease. *Nutr Rev.* 2009;67:S140.

**11.** Bakker MF and others. Plasma carotenoids, vitamin C, tocopherols, and retinol and the risk of breast cancer in the European Prospective Investigation into Cancer and Nutrition Cohort. *Am J Clin Nutr.* 2016;103:454.

**12.** Eliassen AH and others. Plasma carotenoids and risk of breast cancer over 20 y of follow-up. *Am J Clin Nutr.* 2015;101:1197.

**13.** Pantavos A and others. Total dietary antioxidant capacity, individual antioxidant intake and breast cancer risk: The Rotterdam study. *Int J Cancer.* 2014;173:2178.

**14.** Boyd K. Have AMD? Save your sight with an Amsler grid. 2016; *Am Acad Ophthalmol.* www.aao.org/eye-health/tips-prevention/facts-about-amsler-grid-daily-vision-test.

**15.** The Age-Related Eye Disease Study 2 Group. Lutein + zeaxanthin and omega-3 fatty acids for age-related macular degeneration: The Age-Related Eye Disease Study 2 (AREDS2) randomized clinical trial. *JAMA.* 2013;309:2005.

**16.** Leermakers ETM and others. The effects of lutein on cardiometabolic health across the life course: A systematic review and meta-analysis. *Am J Clin Nutr.* 2016;103:481.

**17.** Stringham JM and others. Lutein across the lifespan: From childhood cognitive performance to the aging eye and brain. *Curr Dev Nutr.* 2019;3:nzz066.

**18.** Centers for Disease Control and Prevention. Second national report on biochemical indicators of diet and nutrition in the U.S. population, executive summary. 2012; www.cdc.gov/nutritionreport/report.html.

**19.** World Health Organization. Vitamin A deficiency. 2019; www.who.int/nutrition/topics/vad/en.

**20.** Penniston KL, Tanumihardjo SA. The acute and chronic toxic effects of vitamin A. *Am J Clin Nutr.* 2006;83:191.

**21.** Greaves R and others. Vitamin A: The first vitamin. *Clin Chimica Acta.* 2010;411:907.

**22.** Stevens GA and others. Trends and mortality effects of vitamin A deficiency in children in 138 low-income and middle-income countries between 1991 and 2013: A pooled analysis of population-based surveys. *Lancet Glob Health.* 2015;3:e528.

**23.** Webb AR and others. Colour counts: Sunlight and skin type as drivers of vitamin D deficiency at UK latitudes. *Nutrients.* 2018;10:457.

**24.** Khalid AT and others. Utility of sun-reactive skin typing and melanin index for discerning vitamin D deficiency. *Pediatr Res.* 2017;82:444.

**25.** Food and Nutrition Board, Institute of Medicine. *Dietary Reference Intakes for calcium and vitamin D.* Washington, DC: National Academies Press; 2011.

**26.** Lappe JM, Heaney RP. The anticancer effect of vitamin D: What do the randomized trials show? In: Holick MF, ed. *Nutrition and health: Vitamin D.* New York: Springer Science and Business Media; 2010.

27. Buell JS and others. 25-hydroxyvitamin D, dementia and cerebrovascular pathology in elders receiving home services. *Neurol.* 2010;74:18.

28. Wimalawansa SJ. Vitamin D in the new millennium. *Curr Osteoporos Rep.* 2012;10:4.

29. Bair T and others. Association between vitamin D deficiency and prevalence of cardiovascular disease. *J Am Coll Cardiol.* 2010;55:A141.

30. Schleicher RL. The vitamin D status of the US population from 1988 to 2010 using standardized serum concentrations of 25-hydroxyvitamin D shows recent modest increases. *Am J Clin Nutr.* 2016;104:454.

31. Holick MF and others. Guidelines for preventing and treating vitamin D deficiency and insufficiency revisited. *J Clin Endocrinol Metab.* 2012;97:2011.

32. Ascherio A and others. Vitamin D as an early predictor of multiple sclerosis activity and progression. *JAMA Neurology.* 2014;71:306.

33. Lappe JM and others. Vitamin D and calcium supplementation reduces cancer risk: Results of a randomized trial. *Am J Clin Nutr.* 2007;85:1586.

34. Jenkins and others. Supplemental vitamins and minerals for CVD prevention and treatment. *J Am Coll Cardiol.* 2018;71:2570.

35. Newberry SJ and others. A systematic review of health outcomes (update). Evidence report/technology assessment No. 217. AHRQ Publication No. 14-E004-EF. Rockville, MD: Agency for Healthcare Research and Quality. 2014; www.effectivehealthcare.ahrq.gov/reports/final.cfm.

36. Manson JE and others. Vitamin D supplements and prevention of cancer and cardiovascular disease. *N Engl J Med.* 2019;380:33.

37. Burt LA and others. Effect of high-dose vitamin D supplementation on volumetric bone density and bone strength. A randomized clinical trial. *JAMA.* 2019;322:736.

38. Hlavink E. Vitamin D supplements flop again for bone health—But other data suggest no adverse cardiovascular effect. MedpageToday. 2019; www.medpagetoday.com/meetingcoverage/asbmr/82341.

39. Zittermann A and others. Serum 25-hydroxyvitamin D response to vitamin D supplementation in infants: A systematic review and meta-analysis of clinical intervention trials. *Eur J Nutr.* 2020;59:359.

40. Goldacre M and others. Hospitalisation for children with rickets in England: A historical perspective. *The Lancet.* 2014;383:597.

41. Stephenson DW, Peiris AN. The lack of vitamin D toxicity with megadoses of daily ergocalciferol (D₂) therapy: A case report and literature review. *S Med J.* 2009;102:765.

42. Talegawkar SA and others. Total alpha-tocopherol intakes are associated with serum alpha-tocopherol concentrations in African American adults. *J Nutr.* 2007;137:2297.

43. Gann PH. Randomized trials of antioxidant supplementation for cancer prevention: First bias, now chance next, cause. *JAMA.* 2009;301:102.

44. Lippman SC and others. Effect of selenium and vitamin E on risk of prostate cancer and other cancers: The Selenium and Vitamin E Cancer Prevention Trial (SELECT). *JAMA.* 2009;301:39.

45. Gaziano JM and others. Vitamins E and C in the prevention of prostate and total cancer in men: The Physicians' Health Study II randomized controlled trial. *JAMA.* 2009;301:52.

46. Saremi A, Arora R. Vitamin E and cardiovascular disease. *Am J Ther.* 2010;17:e56.

47. Traber MG. Heart disease and single vitamin supplementation. *Am J Clin Nutr.* 2007;85:293S.

48. Key TJ and others. Carotenoids, retinol, tocopherols, and prostate cancer risk: Pooled analysis of 15 studies. *Am J Clin Nutr.* 2015;102:1142.

49. Bruno RS, Traber MG. Cigarette smoke alters human vitamin E requirements. *J Nutr.* 2005;135:671.

50. De la Fuente M and others. Vitamin E ingestion improves several immune functions in elderly men and women. *Free Radical Res.* 2008;42:272.

51. Muller DPR. Vitamin E and neurological function. *Molec Nutr Food Res.* 2010;54:710.

52. Suttie JW. Vitamin K. In: Shils ME and others, eds. *Modern nutrition in health and disease.* 10th ed. Philadelphia: Lippincott Williams & Wilkins; 2006.

53. Halder M and others. Double bonds beyond coagulation insights into differences between vitamin K1 and K2 in health and disease. *Int J Mol Sci.* 2019;20:896.

54. Booth S and others. Effect of vitamin E supplementation on vitamin K status in adults with normal coagulation status. *Am J Clin Nutr.* 2004;80:143.

55. NIH State-of-the Science Panel. National Institutes of Health State-of-the-Science Conference statement: Multivitamin/mineral supplements and chronic disease prevention. *Am J Clin Nutr.* 2007;85:275S.

56. *Nutrition Business Journal. NBJ's supplement business report* 2015. New York: Penton Media; 2015.

57. Fortmann SP and others. Vitamin and mineral supplements in the primary prevention of cardiovascular disease and cancer: An updated systematic evidence review for the U.S. Prevention Services Task Force. *Ann Intern Med.* 2013;159:824.

58. Bailey RL and others. Why US adults use dietary supplements. *JAMA Intern Med.* 2013;173:355.

59. Cowan and others. Dietary supplement use differs by socioeconomic and health-related characteristics among U.S. adults, NHANES 2011-2014. *Nutrients.* 2018;10:1114.

C Squared Studios/Getty Images

Is niacin an extraterrestrial vitamin? One source of niacin could have been from meteorites landing on Earth. Learn more at **www.nasa.gov/feature/goddard/nasa-researchers -find-frozen-recipe-for-extraterrestrial-vitamin.**

Belish/Image Source/Shutterstock

# 13

# The Water-Soluble Vitamins

## Learning Objectives

**After studying this chapter, you will be able to**

1. Identify the water-soluble vitamins.
2. List important food sources for each water-soluble vitamin.
3. Describe how each water-soluble vitamin is absorbed, transported, stored, and excreted.
4. List the major functions of and deficiency symptoms for each water-soluble vitamin.
5. Describe the toxicity symptoms from the excess consumption of certain water-soluble vitamins.

## Chapter Outline

FOR CENTURIES, SCURVY, PELLAGRA, and other vitamin deficiency diseases caused enormous suffering and death. It wasn't until early in the 20th century that scientists discovered that these illnesses are caused by the absence of certain vital substances from the diet—now called vitamins.[1] Researchers discovered that restoring vitamins to the diet dramatically reversed deficiency diseases if they were supplied before significant deterioration of the body took place.

Today, outright vitamin deficiency diseases are rare in North America, although those with poor diets, those with intestinal conditions, smokers, alcohol abusers, the elderly, and those who take certain medications may be at risk. Depending on our food choices, typical diets contain ample and varied natural sources of vitamins, as well as many foods enriched or fortified with vitamins. In recent years, research in vitamin nutrition has aimed to identify the effects of subclinical vitamin deficiencies as well as the benefits of vitamin supplementation in both healthy adults and those with chronic diseases. It should be noted, however, that vitamin deficiency diseases have not been eradicated and still are significant public health problems in some low-income countries.

Recall from Chapter 12 that, as vitamins were discovered, they were named after letters of the alphabet. The second vitamin to be discovered is water soluble and was designated "vitamin B." Although this water-soluble substance was initially thought to be a single chemical compound, subsequent research showed that "vitamin B" actually is several compounds. Numbers were added to the letter *B* to distinguish these compounds. Of the 8 B-vitamins, only 2 are still commonly referred to by letter and number: vitamin B-6 and vitamin B-12. The others now are usually referred to by the following names: thiamin (previously B-1), riboflavin (previously B-2), niacin (previously B-3), pantothenic acid, biotin, and folate. The older designations, however, are sometimes used on vitamin supplement labels. Vitamin C also is a water-soluble vitamin.

**Water-Soluble Vitamins**

B-vitamins
    Thiamin
    Riboflavin
    Niacin
    Pantothenic acid
    Biotin
    Vitamin B-6
    Folate
    Vitamin B-12
Vitamin C

▶ Interestingly, the amount of vitamin C, folate, vitamin K, and some carotenoids can increase in green leafy vegetables in response to light exposure, as is found in supermarkets. The green leaves are still living and continue to synthesize vitamins and phytochemicals.[4]

As you'll see in this chapter, the water-soluble vitamins, like the fat-soluble vitamins, work together to maintain health. The water-soluble vitamins discussed in this chapter are the 8 B-vitamins and vitamin C. A newcomer to the list of important nutrients is choline.

# 13.1 Water-Soluble Vitamin Overview

Like fat-soluble vitamins, water-soluble vitamins are essential organic substances needed in small amounts for the normal function, growth, and maintenance of body tissues (Fig. 13-1). For example, thiamin, riboflavin, niacin, pantothenic acid, and biotin are especially important for energy metabolism. Vitamin B-6, folate, and vitamin B-12 are important for amino acid metabolism and red blood cell synthesis.[2] Vitamin C participates in the synthesis of numerous compounds, including collagen, and choline is required for nervous system function and aids amino acid and lipid metabolism.[3] In contrast to the fat-soluble vitamins, only small amounts of water-soluble vitamins are stored in the body. The risk of water-soluble vitamin toxicity tends to be low because these vitamins are readily removed by the kidneys and excreted in the urine. In fact, Tolerable Upper Intake Levels have been set for only 4 of the water-soluble vitamins and choline.

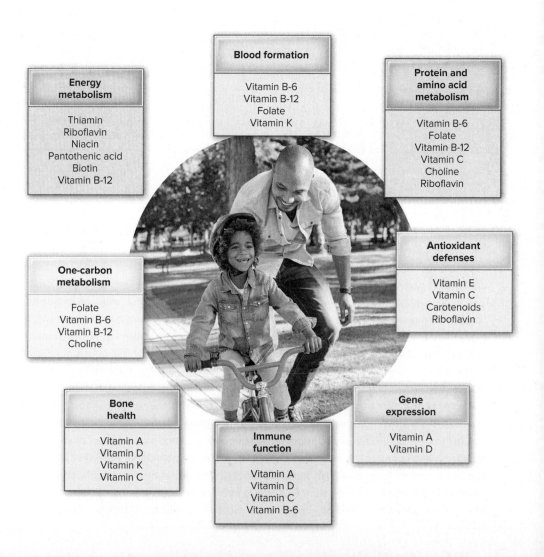

*Figure 13-1* Vitamins and choline work together to maintain health.
© Digital Vision/Getty Images RF

## Culinary Perspective

### Tips for Preserving Vitamins in Fruits and Vegetables

Compared with fat-soluble vitamins, water-soluble vitamins are more easily destroyed during cooking. A food's vitamin content can be decreased by exposure to heat, light, air, and alkaline substances. Water-soluble vitamins can leach into cooking water, whereas fat-soluble vitamins can leach into cooking fats and oils. Retention of the B-vitamins and vitamin C is greatest in foods that are prepared by steaming, stir-frying, and microwaving. These cooking methods limit exposure to heat and water. Fruits and vegetables are especially important sources of many vitamins.

C Squared Studios/Getty Images

### Tips for Preserving Vitamins in Fruits and Vegetables

| Preservation Tips | Why? |
|---|---|
| Keep fruits and vegetables cool until eaten. | Enzymes in fruits and vegetables begin to degrade vitamins once they are harvested. Chilling limits this process. |
| Refrigerate fruits and vegetables (except bananas, onions, potatoes, tomatoes, and fruit that is not fully ripened) in the vegetable crisper drawer or in plastic bags with tiny holes that allow food to breathe. | Nutrients keep best at temperatures near freezing, at high humidity, and away from air. |
| Trim, peel, and cut fruits and vegetables minimally just before serving. | Oxygen breaks down vitamins faster when more of the food surface is exposed. |
| Microwave, steam, stir-fry, sauté, or lightly cook most fruits and vegetables. | More nutrients are retained when there is less contact with water and a shorter cooking time. |
| Minimize cooking time and reheating. | Prolonged cooking (slow simmering) and heat speed vitamin breakdown. |
| Do not add baking soda to vegetables to enhance the green color. | Alkalinity destroys vitamin D, thiamin, and other vitamins. |
| Store canned foods in a cool, dry location. Store frozen foods at 0°F (−32°C) or colder. | Careful storage protects vitamins. |
| Eat canned and frozen foods within 12 months. | Vitamin content declines as storage time increases. |

## Coenzymes: A Common Role of B-Vitamins

All B-vitamins form **coenzymes** (see Chapter 9), which are small, organic molecules that are a type of **cofactor.** Metals (e.g., zinc, magnesium) are another type of cofactor. Cofactors combine with inactive enzymes (called apoenzymes) to form active enzymes (called holoenzymes) that are able to catalyze specific reactions (Fig. 13-2). Table 13-1 lists examples of the coenzymes formed from B-vitamins.

**Table 13-1  B-Vitamins and Coenzyme Examples**

| B-Vitamin | Coenzyme Examples* | Abbreviation | Function |
|---|---|---|---|
| Thiamin | Thiamin pyrophosphate (thiamin diphosphate) | TPP (TDP) | Decarboxylation |
| Riboflavin | Flavin adenine dinucleotide<br>Flavin mononucleotide | FAD<br>FMN | Electron (hydrogen) transfer<br>Electron (hydrogen) transfer |
| Niacin | Nicotinamide adenine dinucleotide<br>Nicotinamide adenine dinucleotide phosphate | NAD<br>NADP | Electron (hydrogen) transfer<br>Electron (hydrogen) transfer |
| Pantothenic acid | Coenzyme A | CoA | Acyl (2-C groups) transfer |
| Biotin | N-carboxylbiotinyl lysine | | Carboxylation; $CO_2$ transfer |
| Vitamin B-6 | Pyridoxal phosphate | PLP | Transamination: amino group transfer |
| Folic acid | Tetrahydrofolic acid | THFA | 1-carbon unit transfer |
| Vitamin B-12 | Methylcobalamin | | 1-carbon unit transfer |

*Some B-vitamins form more than 1 coenzyme.

All 8 B-vitamins participate in energy metabolism; some also have other roles within cells. Figure 13-3 shows where the B-vitamin coenzymes function in energy metabolism. Because of the role of B-vitamins in energy metabolism, the need for many of them increases somewhat with higher amounts of physical activity. Still, this is not a major concern because the higher food intake that usually accompanies an increase in energy expenditure contributes more B-vitamins to the diet.

In foods, B-vitamins are present as vitamins or as coenzymes, both of which are sometimes bound to protein. Digestion frees B-vitamins from coenzymes or protein. Unbound (free)

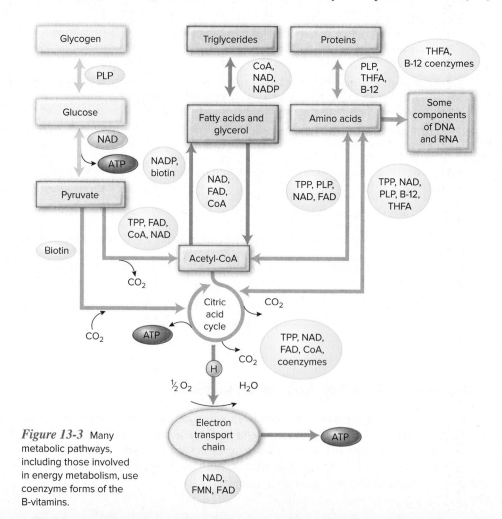

**Figure 13-2** The enzyme-coenzyme interaction. The B-vitamins form coenzymes, which are compounds that enable specific enzymes to function.

**Figure 13-3** Many metabolic pathways, including those involved in energy metabolism, use coenzyme forms of the B-vitamins.

vitamins are the main form absorbed in the small intestine. Typically, 50 to 90% of the B-vitamins in the diet are absorbed.[2] Once inside cells, the coenzyme forms of the vitamins are resynthesized. Vitamin supplements sold in the coenzyme form have no specific benefits to consumers because vitamins must be released from the coenzyme before they can be absorbed.

### The B-Vitamins and Epigenetics

**Epigenetics** is the study of heritable changes to gene expression (switching genes on and off) that are not due to changes in the underlying DNA sequence. One epigenetic mechanism is DNA methylation—the addition of a methyl ($CH_3$) group to cytosine, 1 of the bases in DNA (see Chapter 7). The B-vitamins folate, vitamin B-6, and vitamin B-12, along with choline, are key components in forming methyl (1 carbon) groups. Both too little methylation (hypomethylation) and excessive methylation (hypermethylation) are thought to cause some disease processes. Scientists are actively researching the connections among B-vitamin nutrition, methylation, and various diseases.[5]

## Enrichment and Fortification of Grains

Grains are an important source of many B-vitamins, minerals, and fiber. However, when grains are milled, the seeds are crushed and the germ, bran, and husk layers are removed. This refining process leaves just the starch-containing endosperm, which is the only portion of the grain used to make white flour, as well as many bread and cereal products (Fig. 13-4). Because the discarded parts are rich in nutrients, milling leads to a loss of vitamins, minerals, and fiber.

To counteract this nutrient loss, in the U.S. nearly all bread and cereal products made from milled grains are enriched with 4 B-vitamins—thiamin, riboflavin, niacin, and folic acid—and with the mineral iron. This enrichment program, begun in the 1940s, adds to the nutrient intakes of nearly all U.S. residents and helps protect them from the common deficiency diseases associated with a dietary lack of these nutrients. In fact, according to a national survey of dietary intakes, only about half of the folate and thiamin, and three-quarters of the niacin and riboflavin, we ingest occurs naturally in foods.[6] However, foods made with enriched and fortified grains have less vitamin B-6, potassium, magnesium, zinc, fiber, and phytochemicals than those from whole grains because refined grains lack the nutrient rich germ and bran (Fig. 13-5). Nutrition experts therefore recommend that at least half of the grains consumed daily be whole grain, such as brown rice, oatmeal, popcorn, and whole-wheat bread and pasta.

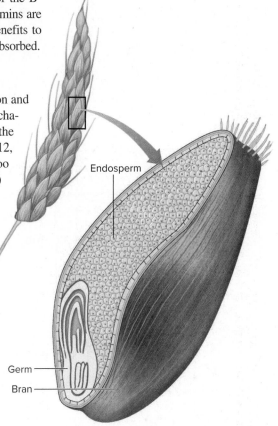

Endosperm

Germ

Bran

*Figure 13-4* When grains are milled, the bran and germ are removed and discarded, leaving only the starch rich endosperm.

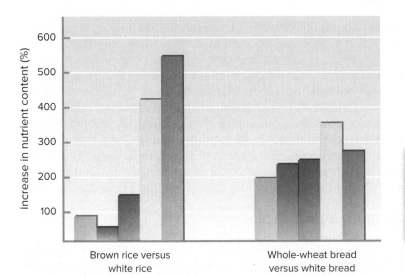

*Figure 13-5* When compared with white rice, brown rice has 93% more vitamin B-6, 50% more potassium, 160% more zinc, 435% more magnesium, and 550% more fiber. Similarly, compared with white bread, whole-wheat bread has 200% more vitamin B-6, 250% more potassium, 260% more zinc, 370% more magnesium, and 285% more fiber.

Traditionally, in tropical areas of the world, white rice has been favored over brown rice because it stays fresh longer. That's because, in warm climates and without refrigeration, the fat in the germ of brown rice goes rancid quickly.
C Squared Stuidos/Getty Images RF

### Knowledge Check

1. Describe how the ways in which fruits and vegetables are stored and prepared can impact their water-soluble vitamin retention.
2. A common role of B-vitamins is to form coenzymes. What is the role of coenzymes in the cell?
3. Which B-vitamins are added to enriched bread and cereal products?

*A Biochemist's View*

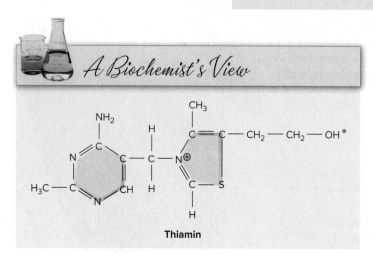

**Thiamin**

# 13.2 Thiamin

For centuries, the devastating effects of the disease beriberi were known in Asian countries where milled (polished) white rice was the main food. Those afflicted with beriberi developed extreme weakness, paralysis, and fatigue, often dying within several months. In the late 1800s, Kanehiro Takakaia, a Japanese naval physician, was able to rid the Japanese Imperial Navy of the scourge of beriberi by improving sailors' rations with the addition of nutrient rich barley, meat, and fish. However, scientists did not link the disease beriberi with a vitamin deficiency until early in the 1900s, when it was discovered that a vital factor in the rice germ that is removed in milling cures beriberi. That factor is thiamin, also known as vitamin B-1.

Thiamin consists of a central carbon attached to a 6-member, nitrogen-containing ring and a 5-member, sulfur-containing ring. Its name comes from *thio,* meaning "sulfur," and *amine,* referring to the nitrogen groups in the molecule. Two phosphate groups are added (at the red dot in thiamin's structure) to form this vitamin's coenzyme, thiamin pyrophosphate (TPP), also called thiamin diphosphate (TDP).

The chemical bond between each ring and the central carbon in thiamin (shown in red in the structure) is easily broken by prolonged exposure to heat, as can occur in cooking. When this happens, the vitamin can no longer function in the body. Thiamin also is susceptible to destruction in alkaline (basic) conditions. As mentioned in the Culinary Perspective: Tips for Preserving Vitamins in Fruits and Vegetables, baking soda (a base) to the cooking water of green vegetables helps keep them bright green, but this practice destroys thiamin and is not recommended.

## Thiamin in Foods

Thiamin is found in a wide variety of foods, although generally in small amounts. As can be seen in Figure 13-6, foods rich in thiamin are pork products, sunflower seeds, and legumes. Whole and enriched grains and cereals, green peas, asparagus, organ meats (e.g., liver), peanuts, and mushrooms also are good sources. In the U.S., major contributors of thiamin are bread and rolls, ready-to-eat cereals, bakery products, pasta, milk, white potatoes, pork, ham, and bacon.[7] Eating a variety of foods in accord with MyPlate is a reliable way to obtain sufficient thiamin.

A few foods contain compounds, called thiamin antagonists, that lower the bioavailability of thiamin. Some species of fresh fish and shellfish contain thiaminase enzymes that destroy thiamin. Cooking inactivates these enzymes. Other foods, including coffee, tea, blueberries, red cabbage, Brussels sprouts, and beets, contain compounds that oxidize thiamin and make it inactive.[8] However, eating these foods has not been linked to thiamin deficiency.

## Thiamin Needs and Upper Level

The RDAs for thiamin are 1.2 mg/day for adult men and 1.1 mg/day for women.[2] The Daily Value on food and supplement labels is 1.2 mg. In the U.S., for men, the average daily intake for thiamin from food is 1.9 mg/day. For women, it is 1.4 mg/day.[9] There appear to be no adverse effects with excess intake of thiamin from food or supplements because it is readily excreted in the urine. Thus, no Upper Level is established for this nutrient.[2]

## CASE STUDY

© Hill Street Studios/Blend Images LLC RF

Mr. Martin is a 72-year-old man admitted to the hospital with weight loss, mild anemia, "pins and needles" in his hands and feet, difficulty walking, and occasional memory lapses. He has been a vegan for 20 years. He has used drugs to prevent heartburn for the last 5 years. His doctor determines that Mr. Martin has a water-soluble vitamin deficiency. As you read through this chapter, identify the water-soluble vitamin deficiency that accounts for his problems. What dietary and lifestyle changes might help Mr. Martin prevent a reoccurrence of these problems?

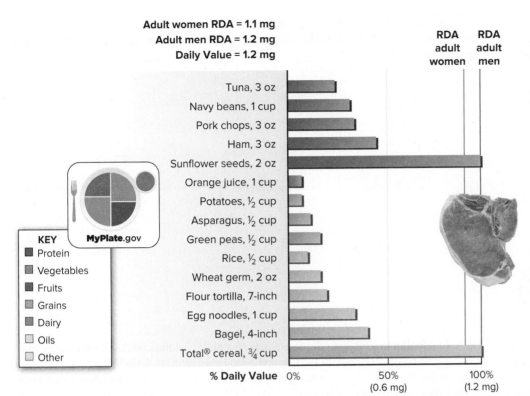

Adult women RDA = 1.1 mg
Adult men RDA = 1.2 mg
Daily Value = 1.2 mg

| | RDA adult women | RDA adult men |

Tuna, 3 oz
Navy beans, 1 cup
Pork chops, 3 oz
Ham, 3 oz
Sunflower seeds, 2 oz
Orange juice, 1 cup
Potatoes, ½ cup
Asparagus, ½ cup
Green peas, ½ cup
Rice, ½ cup
Wheat germ, 2 oz
Flour tortilla, 7-inch
Egg noodles, 1 cup
Bagel, 4-inch
Total® cereal, ¾ cup

**KEY**
■ Protein
■ Vegetables
■ Fruits
■ Grains
■ Dairy
□ Oils
□ Other

**% Daily Value** 0%  50% (0.6 mg)  100% (1.2 mg)

*Figure 13-6* Food sources of thiamin. In addition to MyPlate food groups, yellow is used for oils and pink is used for substances that do not fit easily into food groups (e.g., candy, salty snacks).
©Foodcollection RF

Source: ChooseMyPlate.gov, U.S. Department of Agriculture

## Absorption, Transport, Storage, and Excretion of Thiamin

Thiamin is readily absorbed in the small intestine by a sodium-dependent active absorption process. It is transported mainly by red blood cells in its coenzyme form (thiamin pyrophosphate). Little thiamin is stored; only a small reserve (25–30 mg) is found in the muscles, brain, liver, and kidneys.[8] Any excess intake is rapidly filtered out by the kidneys and excreted in the urine.[2]

## Functions of Thiamin

The coenzyme thiamin pyrophosphate (TPP) is required for the metabolism of carbohydrates and branched-chain amino acids.[8] TPP is necessary for 2 different types of reactions. First, it works with specific enzymes to remove carbon dioxide (known as **decarboxylation**) from certain compounds. The conversion of pyruvate to acetyl-CoA, a critical reaction in the aerobic respiration of glucose, is an example of the decarboxylation action of TPP.

Pork is a rich source of thiamin.
©Michael Lamotte/Cole Group/Getty Images RF

**decarboxylation** Removal of 1 molecule of carbon dioxide from a compound.

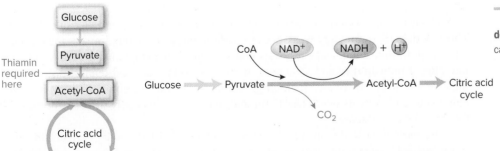

**transketolase** Enzyme whose functional component is thiamin pyrophospate (TPP). It converts glucose to other sugars.

**pentose phosphate pathway** Pathway for glucose breakdown that produces NADPH used in biosynthetic pathways and the 5-carbon sugar used to synthesize RNA and DNA.

**peripheral neuropathy** Problem caused by damage to the nerves outside the spinal cord and brain. Symptoms include numbness, weakness, tingling, and burning pain, often in the hands, arms, legs, and feet.

## HISTORICAL PERSPECTIVE

### Beriberi Cure

In 1886, the Dutch physician Christjaan Eijkman went to the Indonesian island of Java to discover the cause of the disease beriberi. Because Eijkman initially hypothesized that beriberi was caused by a bacterium, he started by inoculating rabbits, monkeys, and chickens with bacteria from people with beriberi. Only the chickens developed symptoms. After a short period, however, the chickens improved. While trying to understand these results, Eijkman noted that the chickens' diets were changed during the experiment. When fed polished white rice, they became ill, but, when fed brown rice, they improved. Eijkman tested the white and brown rice diets many times and was able to induce and cure beriberi.

These results led him to incorrectly hypothesize that beriberi was caused by a toxin in white rice. In fact, as other researchers discovered, white rice lacked the anti-beriberi factor, later named thiamin, present in brown rice. Eijkman was awarded the Nobel Prize in Physiology or Medicine in 1929 (shared with Frederick Hopkins, a British biochemist and vitamin researcher) for his beriberi research and for his new methods of investigation.

Learn more at www.nobelprize.org/prizes /medicine/1929/summary.

©GK Hart/Vikki Hart/Getty Images RF

A similar decarboxylation reaction occurs in the citric acid cycle. As shown in the following diagram, TPP aids in the conversion of the intermediate compound alpha-ketoglutarate to succinyl-CoA.

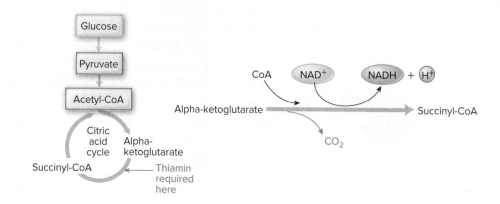

In addition to TPP, both of these reactions require 3 additional B-vitamin coenzymes: CoA (pantothenic acid), NAD (niacin), and FAD (riboflavin) (see Fig. 13-3). TPP functions in a similar manner (as a decarboxylase) in the metabolism of the branched-chain amino acids valine, leucine, and isoleucine.

TPP also functions as a coenzyme for **transketolase,** an enzyme in the **pentose phosphate pathway.** This pathway converts glucose to ribose, a 5-carbon sugar required in several metabolic pathways. The pentose phosphate pathway also forms a niacin-containing coenzyme required for biosynthesis of some compounds.

Thiamin is required for normal function of the nervous system. The brain and nerves rely on glucose for energy. When thiamin is lacking, glucose metabolism is severely disrupted because pyruvate cannot be converted to acetyl-CoA, the compound that enters the citric acid cycle (see Fig. 13-3). However, researchers believe that thiamin has additional critical noncoenzyme roles that protect brain and nerve function.[10]

## Thiamin Deficiency

As described previously, the thiamin deficiency disease beriberi was endemic throughout much of Asia due to poor diets. Although much less common today, beriberi is still a problem in parts of Asia, particularly among refugees and impoverished persons and infants.[11] Individuals who abuse alcohol may experience the thiamin deficiency disorder known as Wernicke-Korsakoff syndrome. Additionally, some individuals—including those with heart failure, gastrointestinal diseases, critical illness, eating disorders, and obesity—are at risk of moderate thiamin deficiency.[10,12] The prevalence of thiamin deficiency in older adults is much less common than once thought.[13]

### Beriberi

In Sinhalese, the language of Sri Lanka, the word *beriberi* means "I can't, I can't." Those with beriberi are very weak because a deficiency of thiamin impairs the nervous, muscle, gastrointestinal, and cardiovascular systems. The symptoms of beriberi include **peripheral neuropathy** and weakness, muscle wasting, pain and tenderness, enlargement of the heart, difficulty breathing, edema, anorexia, weight loss, poor memory, confusion, and convulsions.[8] The brain and nerves are especially affected because of their reliance on glucose.

Beriberi often is described as dry, wet, or infantile beriberi. In dry beriberi, the main symptoms are related to the nervous and muscular systems. In wet beriberi, in addition to the neurological symptoms, the cardiovascular system is affected. The heart is enlarged, breathing may be difficult, and **congestive heart failure** may occur. Infants develop beriberi when breast milk contains insufficient thiamin. Infantile beriberi causes heart problems, convulsions, and death if not treated promptly. Like most water-soluble vitamins, only small amounts of thiamin are stored in the body. Thus, some signs of thiamin deficiency can develop after only 14 days on a thiamin-free diet.[2]

### Wernicke-Korsakoff Syndrome

Wernicke-Korsakoff syndrome (also known as cerebral beriberi) is found mainly among heavy users of alcohol.[14] These individuals have a 3-pronged problem related to thiamin: alcohol decreases thiamin absorption, alcohol increases thiamin excretion in the urine, and those with alcoholism may consume a poor-quality diet lacking sufficient thiamin. Because thiamin is not readily stored in the body, the syndrome can occur rapidly. Damage to brain tissue occurs in individuals with Wernicke-Korsakoff syndrome. The symptoms can include changes in vision (double vision, crossed eyes, rapid eye movements), **ataxia,** apathy, confusion, hallucinations, and memory loss and distortions (confabulation). The symptoms, especially those of the eye, improve with high doses of thiamin.[8]

**congestive heart failure** Condition resulting from severely weakened heart muscle, resulting in ineffective pumping of blood. This leads to fluid retention, especially in the lungs. Symptoms include fatigue, difficulty breathing, and leg and ankle swelling.

**ataxia** Inability to coordinate muscle activity during voluntary movement; incoordination.

### Knowledge Check

1. How is the coenzyme TPP involved in energy metabolism? What is 1 critical reaction that requires TPP?
2. What are 3 foods that are rich sources of thiamin?
3. Why are those with alcoholism at risk of thiamin deficiency?
4. Why do individuals with beriberi feel tired?

## 13.3 Riboflavin

Riboflavin, also known as vitamin B-2, was once called "yellow enzyme" because it has a distinctive yellow-green fluorescence. In fact, its name comes from its color (*flavin* means "yellow" in Latin). Riboflavin contains 3 linked, 6-membered rings, with a sugar alcohol attached to the middle ring.

### Riboflavin in Foods

Almost one-quarter of the riboflavin in our diets comes from milk products. The rest typically is supplied by coffee, tea, and other nonalcoholic beverages; breads and rolls; ready-to-eat cereal; eggs; bakery products; and meat.[7] Foods rich in riboflavin are liver, mushrooms, spinach and other green leafy vegetables, broccoli, asparagus, milk, and cottage cheese (Fig. 13-7). Unless fortified, most plant-based milk alternatives contain limited amounts of riboflavin (see Culinary Perspective in Chapter 12). Exposure to light (ultraviolet radiation) causes riboflavin to break down rapidly. To prevent this light-induced breakdown, paper and plastic containers—not glass—should be used as packaging for riboflavin rich foods, such as milk, milk products, and cereals.

### Riboflavin Needs and Upper Level

The RDAs for riboflavin for adult men and women are 1.3 and 1.1 mg/day, respectively. The Daily Value on food and supplement labels is 1.3 mg. In the U.S., the average intake from food for riboflavin is 2.5 mg/day for men and 1.8 mg/day for women.[9] There appear to be no adverse effects from consuming large amounts of riboflavin because of its limited absorption and rapid excretion via the urine, so no Upper Level has been set.[2]

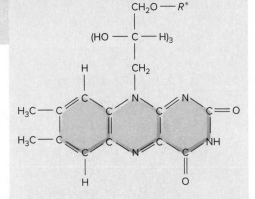

**Riboflavin (oxidized)**

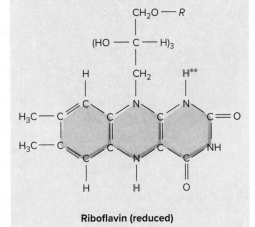

**Riboflavin (reduced)**

$R^*$ = H in free riboflavin; phosphate in the coenzyme FMN; adenine dinucleotide in the coenzyme FAD

$^{**}$ = Addition of 2 hydrogens (in red) in the reduced form

*Figure 13-7*  Food sources of riboflavin.
©Stockbyte/Getty Images RF

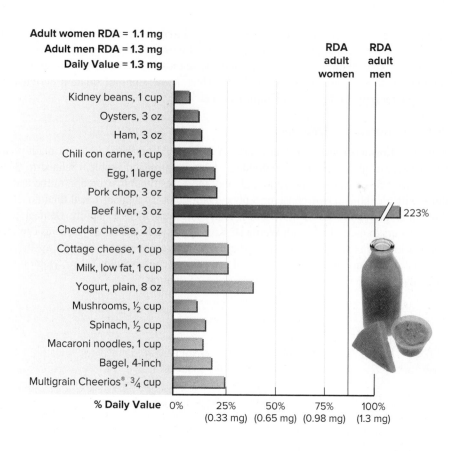

**Adult women RDA = 1.1 mg**
**Adult men RDA = 1.3 mg**
**Daily Value = 1.3 mg**

Kidney beans, 1 cup
Oysters, 3 oz
Ham, 3 oz
Chili con carne, 1 cup
Egg, 1 large
Pork chop, 3 oz
Beef liver, 3 oz — 223%
Cheddar cheese, 2 oz
Cottage cheese, 1 cup
Milk, low fat, 1 cup
Yogurt, plain, 8 oz
Mushrooms, ½ cup
Spinach, ½ cup
Macaroni noodles, 1 cup
Bagel, 4-inch
Multigrain Cheerios®, ¾ cup

RDA adult women | RDA adult men

**% Daily Value**  0%   25% (0.33 mg)   50% (0.65 mg)   75% (0.98 mg)   100% (1.3 mg)

Milk products are good sources of riboflavin. Plastic and cardboard containers protect the riboflavin from UV radiation, which causes riboflavin to break down.

© McGraw-Hill Education/Ken Cavanagh, Photographer

## Absorption, Transport, Storage, and Excretion of Riboflavin

In the stomach, hydrochloric acid (HCl) releases riboflavin from its bound forms. About 60 to 65% of the free riboflavin is absorbed, primarily via active transport or facilitated diffusion in the small intestine.[15] In the blood, riboflavin is transported by protein carriers. Riboflavin is converted to its coenzyme forms in most tissues, but this occurs mainly in the small intestine, liver, heart, and kidneys. A small amount of riboflavin is stored in the liver, kidneys, and heart. Any excess intake is excreted in the urine.[2,16] For people who take excessive amounts in supplement form, riboflavin imparts a bright yellow color to the urine, which glows under a black light.

## Functions of Riboflavin

Riboflavin is a component of 2 coenzymes that play key roles in energy metabolism: flavin mononucleotide (FMN) and flavin adenine dinucleotide (FAD).[16] These coenzymes, also referred to as flavins, have oxidation and reduction functions. (See Section 9.1 in Chapter 9.) FAD is the oxidized form of the coenzyme. When it is reduced (gains 2 hydrogens, equivalent to 2 hydrogen ions and 2 electrons), it is known as $FADH_2$.

The riboflavin coenzymes are involved in many reactions in various metabolic pathways. They are critical for energy metabolism and are involved in the formation of other compounds, including other B-vitamins and antioxidants.

### Energy Metabolism

- In the citric acid cycle, the oxidation of succinate to fumarate requires the FAD-containing enzyme *succinate dehydrogenase*. The $FADH_2$ formed in this reaction donates hydrogen to the electron transport chain.

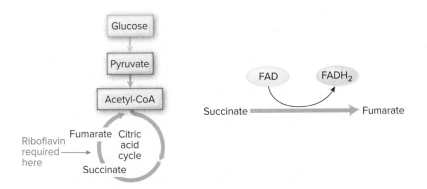

- In fatty acid breakdown (beta-oxidation) to acetyl-CoA, the enzyme *fatty acyl dehydrogenase* requires FAD (see Chapter 9).
- FMN shuttles hydrogen atoms into the electron transport chain.

### *Activation of Other B-Vitamins*

- The formation of niacin from the amino acid tryptophan requires FAD (see Section 13.4).
- The formation of the active vitamin B-6 coenzyme (pyridoxal phosphate) requires FMN.
- FAD is required for the synthesis of the folate metabolite 5-methyl-tetrahydrofolate; in this way, riboflavin participates indirectly in homocysteine metabolism (see Sections 13.8 and 13.9).

### *Antioxidant Function*

- The synthesis of the antioxidant compound glutathione depends on the FAD-containing enzyme *glutathione reductase*. Recall from Chapter 12 that glutathione is an important part of the cell's antioxidant defense network.

## Riboflavin Deficiency

Riboflavin deficiency, called **ariboflavinosis,** primarily affects the mouth, skin, and red blood cells. The symptoms include inflammation of the throat, mouth (stomatitis), and tongue (glossitis); cracking of the tissue around the corners of the mouth (angular cheilitis); and moist, red, scaly skin (seborrheic dermatitis) (Fig. 13-8). Poor growth, anemia, fatigue, confusion, and headaches also may occur. Some of the symptoms of ariboflavinosis may result from deficiencies of other B-vitamins because they work in the same metabolic pathways as riboflavin and are often supplied by the same foods.

▶ Several small clinical trials suggest that high doses of riboflavin supplements (200–400 mg/day) may reduce the frequency and duration of migraine headaches.[17] Researchers hypothesize that riboflavin may improve mitochondrial function in the brain. Check with your physician before taking riboflavin supplements.

(a)

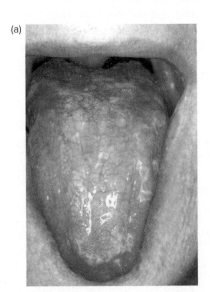

(b)

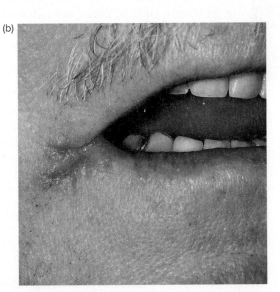

*Figure 13-8* (*a*) Glossitis is a painful, inflamed tongue that can signal a deficiency of riboflavin, niacin, vitamin B-6, folate, or vitamin B-12. (*b*) Angular cheilitis, also called cheilosis or angular stomatitis, is another result of a riboflavin deficiency. It causes painful cracks at the corners of the mouth. Both glossitis and angular cheilitis can be caused by other medical conditions; thus, further evaluation is required before diagnosing a nutrient deficiency.

(a) ©Medical-on-Line/Alamy; (b) ©Dr. P. Marazzi/Science Photo Library/Science Source

Ariboflavinosis develops after 2 months on a riboflavin-deficient diet and is rare in otherwise healthy people. However, biochemical evidence of deficiency (low riboflavin levels in red blood cells or reduced activity of the enzyme glutathione reductase) is sometimes seen in adolescent girls, women of reproductive age, and elderly people.[11,18] Correcting moderate riboflavin deficiency with supplementation improves hematologic status. Diseases such as cancer, certain forms of cardiovascular disease, and diabetes can lead to or worsen a riboflavin deficiency.[2] People with alcoholism, malabsorption disorders, or very poor diets may be at risk of riboflavin deficiency. The long-term use of phenobarbital also may adversely affect riboflavin status because this drug increases the breakdown of riboflavin and other nutrients in the liver. Marginal riboflavin intake may occur in those who do not consume milk or milk products. Presently, little is known about the effects of a marginal riboflavin deficiency.

### Knowledge Check

1. What foods are rich in riboflavin?
2. What are 3 general functions of riboflavin?
3. What are the 2 coenzymes formed from riboflavin?
4. Describe why a riboflavin deficiency may cause fatigue.

## 13.4 Niacin

Pellagra—the deficiency disease of the B-vitamin niacin—is the only dietary deficiency disease ever to reach epidemic proportions in the U.S.[19] In the early 1900s, pellagra affected thousands in the southeastern states before scientists discovered its link with niacin poor diets. Niacin, or vitamin B-3, exists in 2 forms—nicotinic acid (niacin) and nicotinamide (niacinamide). Both forms are used to synthesize the niacin coenzymes: nicotinamide adenine dinucleotide ($NAD^+$) and nicotinamide adenine dinucleotide phosphate ($NADP^+$).

### Niacin in Foods

Chicken is a good source of niacin. Also, the tryptophan that chicken contains can be used to synthesize niacin.
©Ingram Publishing/SuperStock RF

Niacin can be obtained from foods as the vitamin itself (preformed niacin) or synthesized in the body from the essential amino acid tryptophan.[2,20] Poultry, meat, and fish provide about 27% of the preformed niacin in North American diets. Another 28% is from foods made from enriched flours and grains, including breads, rolls, ready-to-eat cereals, cookies, cakes, and pancakes. Coffee and tea also contribute a little preformed niacin to the diet.[7] Figure 13-9 shows some rich sources of preformed niacin—mushrooms, wheat bran, fish, poultry, and peanuts. Protein rich foods also are good sources of niacin because they provide tryptophan. Unlike some other water-soluble vitamins, niacin is very heat stable, and little is lost in cooking.

In the synthesis of niacin from tryptophan, 60 mg of dietary tryptophan is needed to make about 1 mg of niacin.[2] Riboflavin and vitamin B-6 coenzymes also are required. Protein is about 1% tryptophan, so 1 g of protein provides 10 mg of tryptophan. The overall contribution of dietary protein to niacin can be roughly estimated as shown in the following example of a diet containing 90 g of protein.

$$1 \text{ g protein yields } 10 \text{ mg tryptophan}$$

$$60 \text{ mg tryptophan yields } 1 \text{ mg niacin}$$

$$90 \text{ g protein} \times 10 \text{ mg tryptophan/g protein} = 900 \text{ mg tryptophan}$$

$$\frac{900 \text{ mg tryptophan}}{60 \text{ mg tryptophan/mg niacin}} = 15 \text{ mg niacin}$$

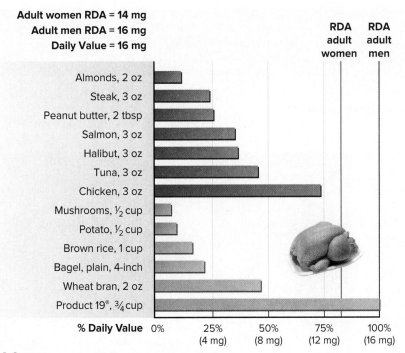

*Figure 13-9* Food sources of niacin. © Foodcollection RF

A "shortcut" for this calculation is to divide protein intake (in grams) by 6. In the previous example, 90 g protein/6 yields 15 mg niacin.

To account for the direct (preformed) and indirect (from tryptophan) sources of niacin, dietary requirements and the amounts in foods are expressed as niacin equivalents (NE).[2] Thus, a diet that provides 13 mg preformed niacin and 90 g protein supplies approximately 28 NE (13 mg preformed + 15 mg from tryptophan). Nutrient databases often underestimate niacin in the diet because the amount contributed by tryptophan is not included.

## Niacin Needs and Upper Level

The niacin RDA for adult men is 16 mg/day; for adult women, it is 14 mg/day. The RDA for niacin is expressed as niacin equivalents (NE) to account for preformed niacin in foods and niacin synthesized from tryptophan. Niacin intake is ample in the U.S., with intakes of preformed niacin from food averaging 31.8 mg/day for men and 21.4 mg/day for women.[9] These figures underestimate intake, however, because they do not include niacin synthesized from tryptophan, which supplies about half the NE in the diet. The Daily Value for niacin on food and supplement labels is 16 mg. The Upper Level for niacin, 35 mg/day, applies only to niacin supplements and fortified foods.[2]

## Absorption, Transport, Storage, and Excretion of Niacin

Nicotinic acid and nicotinamide are readily absorbed from the stomach and the small intestine by active transport and passive diffusion, so generally almost all niacin that is consumed is absorbed. However, the bioavailability of niacin is low in some grains, especially corn. This is because the niacin is tightly bound to protein, so less than 30% can be absorbed.

After being absorbed, niacin is transported via the portal vein to the liver, where it is stored or delivered to the body's cells. Niacin is converted to its coenzyme forms in all tissues. Any excess niacin is excreted in the urine.[20]

## Culinary Perspective

### Cooking Method Enhances Niacin Bioavailability

Niacin bioavailability can be improved in corn by soaking it in an alkaline solution of calcium hydroxide dissolved in water (known as lime water). This culinary practice, originating in Central America, releases the skin from the corn kernels so that the dough can be formed. The alkaline soak also releases niacin from protein, providing the added benefit of protecting against niacin deficiency.

Masa is a dough made from ground corn that has been soaked in an alkaline solution. This alkaline treatment is sometimes referred to as nixtamalization. The dough is used to prepare Latin American foods, such as tamales and tortillas.

BobNoah/Shutterstock

### Functions of Niacin

Like the coenzyme forms of riboflavin, the coenzyme forms of niacin, $NAD^+$ and $NADP^+$, are active participants in oxidation-reduction reactions.[20] The niacin coenzymes function in at least 200 reactions in cellular metabolic pathways, especially those that produce ATP. $NAD^+$ is required mainly for the catabolism of carbohydrates, proteins, and fats (Fig. 13-10). $NAD^+$ acts as an electron and hydrogen acceptor in glycolysis and the citric acid cycle. Under anaerobic conditions, $NAD^+$ is regenerated when pyruvate is converted to lactate. Under aerobic conditions, $NADH + H^+$ donates electrons and hydrogens to acceptor molecules in the electron transport chain, thereby contributing to ATP synthesis. Alcohol metabolism also requires niacin coenzymes (see Section 9.5 in Chapter 9).

The catabolic reactions just described start with an oxidized form of a niacin coenzyme. However, synthetic pathways in the cell—those that make new compounds—use $NADPH + H^+$, the reduced form of the coenzyme. This coenzyme is important in the biochemical pathway for fatty acid synthesis. Cells that synthesize a lot of fatty acids (e.g., those in the liver and female mammary glands) have higher concentrations of $NADPH + H^+$ than cells not involved in fatty acid synthesis (e.g., muscle cells).

### Niacin Deficiency

Because almost every metabolic pathway uses either $NAD^+$ or $NADPH + H^+$, it is not surprising that a niacin deficiency causes widespread damage in the body. The niacin deficiency disease pellagra, once a significant public health problem in the U.S., is now eradicated here, thanks to the enrichment of grains and protein rich diets. The discovery of how

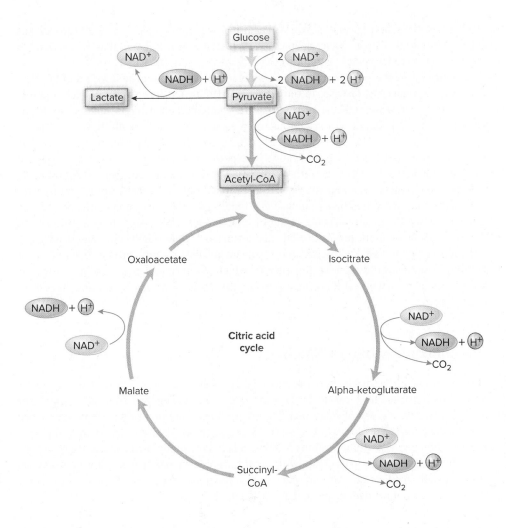

**Figure 13-10** The coenzyme form of niacin, NAD⁺, is required for glycolysis and the citric acid cycle. The NAD⁺ is reduced to NADH. When pyruvate is reduced to form lactate, NADH is converted to NAD⁺.

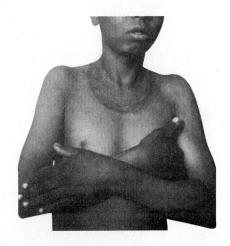

(a)

(b)

**Figure 13-11** The dermatitis of pellagra. (a) Dermatitis on both sides (bilateral) of the body is a typical symptom of pellagra. Sun exposure worsens the condition. (b) The rough skin around the neck is referred to as Casal's necklace.

(a) ©Dr. M.A. Ansary/Science Source; (b) Centers for Disease Control

pellagra develops from a poor diet, rather than a bacterial infection (as most believed until the early 1900s), is a fascinating story.

The first official record of pellagra, made in 1735 by Spanish physician Gaspar Casal, called this disease *mal de la rosa,* or "red sickness." This name referred to the rough, red rash that appears on skin exposed to sunlight, such as the forearms, backs of the hands, face, and neck (called Casal's necklace). The name *pellagra* comes from the Italian *pelle,* meaning "skin," and *agra,* meaning "rough" (Fig. 13-11). Other symptoms of pellagra include diarrhea and a variety of mental status changes, including headache, irritability, confusion, depression, and delirium. Thus, pellagra is identified by the 3 *D*s: dermatitis, diarrhea, and dementia. Death, the fourth *D*, can result if the disease is not treated.

Pellagra has long been associated with corn-based diets. Although there is no evidence of pellagra in Central and South America, where corn (also known as maize) has been the staple food in the diet for thousands of years, pellagra outbreaks followed the introduction of corn into Europe and Africa. As mentioned previously, the main reason for this was that in Latin America the traditional culinary practice was to treat corn with alkali (from lime water or wood ashes) which liberates niacin from protein. Unfortunately, this practice was not adopted in Europe, Africa, or the U.S. When maize, often prepared as a porridge or grits, became a staple food for poor people with limited diets, the result was a very low niacin intake and pellagra. Nutrition scientists have since discovered another reason that corn-based diets can lead to pellagra—corn contains little of the amino acid tryptophan.

During the early 1900s, pellagra was rampant in the southeastern U.S., where corn was a staple food for impoverished persons. More than 10,000 Americans died of pellagra in

## Critical Thinking

Both the vitamin niacin and protein rich foods can cure pellagra. Why are both effective?

1915. From 1918 until the end of World War II in 1945, approximately 200,000 Americans suffered from this disease. Many people had such severe dementia that they were forced to live out their lives in mental institutions.

One reason that pellagra remained a problem for so long was the false belief that pellagra was an infectious disease. In the 1910s and 1920s, Dr. Joseph Goldberger, a public health specialist, observed that institutionalized patients had pellagra but the better-fed staff did not—if pellagra was infectious, he reasoned, the staff should have "caught" it from their patients.[21] He, his wife, and his colleagues proved that pellagra is not caused by an infectious pathogen by participating in experiments that exposed them to biological samples, such as skin, feces, and scabs, from pellagra patients. Goldberger also induced pellagra in volunteer prisoners by serving a cornmeal-only diet, then cured them by adding meat, milk, and vegetables to their diets. Finally, in 1937, researchers discovered that nicotinic acid dramatically cures a similar disease in dogs, called black tongue. Soon after, the enrichment of grain products with niacin in the U.S. virtually eliminated pellagra, although isolated cases still occur due to severe malabsorption, chronic alcoholism, or Hartnup disease (a rare genetic disorder in which the tryptophan to niacin pathway is blocked). Today, pellagra still occurs rarely in individuals with poor diets or malabsorptive conditions.[22]

### Pharmacologic Use of Niacin

High-dose niacin has been used for many years to increase HDL-cholesterol and lower LDL-cholesterol and triglyceride levels with the aim of reducing the risk of stroke and heart attack. However, a study of more than 25,000 persons found that those people prescribed niacin-containing medications did not have fewer heart attacks and stroke, but they did experience serious adverse complications.[23] Thus, niacin-containing medications are no longer recommended for most individuals with cardiovascular disease, particularly persons also taking a statin medication. An important way to promote lifelong heart health is to follow the lifestyle recommendations presented in Chapter 6.

### Knowledge Check

1. What are 2 metabolic pathways that require a niacin coenzyme?
2. Describe why niacin intakes in the U.S. are generally very good for many individuals.
3. Why are populations in Latin America not afflicted with pellagra, despite relying heavily on a corn-based diet?

## 13.5 Pantothenic Acid

The name *pantothenic acid* was taken from the Greek word *pantothen,* meaning "from every side," because it is present in all body cells and is supplied by a wide variety of foods. Pantothenic acid is part of acyl carrier protein (ACP), a compound used in many biosynthetic reactions, and coenzyme A (CoA), which is used throughout the body in energy metabolism. CoA forms when pantothenic acid combines with a derivative of adenosine diphosphate (ADP) and part of the amino acid cysteine. Cysteine provides the sulfur atom, which is the functional end of the coenzyme.[24]

One whole avocado provides more than 50% of the AI for pantothenic acid.

*A Biochemist's View*

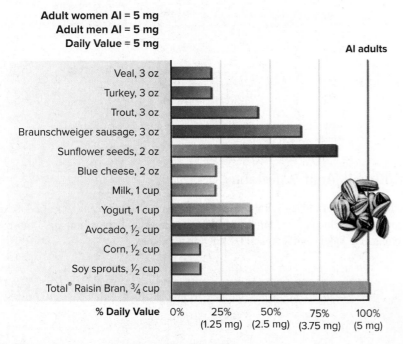

**Pantothenic acid**

**Coenzyme A (CoA)**

Pantothenic acid is part of the coenzyme A (CoA) molecule
*R* = Derivative of adenosine diphosphate (ADP)
Boxed area = Part of the amino acid cysteine

## Pantothenic Acid in Foods

Our food supply provides ample amounts of pantothenic acid. Common sources include meat, milk, and many vegetables (Fig. 13-12). Other foods rich in pantothenic acid include mushrooms, peanuts, egg yolks, yeast, broccoli, and soy milk. In general, unprocessed foods are better sources of pantothenic acid than processed foods because milling, refining, freezing, heating, and canning can reduce pantothenic acid in foods.[2]

**Adult women AI = 5 mg**
**Adult men AI = 5 mg**
**Daily Value = 5 mg**

AI adults

Veal, 3 oz
Turkey, 3 oz
Trout, 3 oz
Braunschweiger sausage, 3 oz
Sunflower seeds, 2 oz
Blue cheese, 2 oz
Milk, 1 cup
Yogurt, 1 cup
Avocado, ½ cup
Corn, ½ cup
Soy sprouts, ½ cup
Total® Raisin Bran, ¾ cup

**% Daily Value** 0%    25%    50%    75%    100%
(1.25 mg) (2.5 mg) (3.75 mg) (5 mg)

*Figure 13-12* Food sources of pantothenic acid.

©McGraw-Hill Education

## Pantothenic Acid Needs and Upper Level

For adults, the Adequate Intake for pantothenic acid is 5 mg/day.[2] Adults generally consume the Adequate Intake or more. The Daily Value on food and supplement labels is 5 mg. There is no known toxicity for pantothenic acid, so no Upper Level has been set.[2]

## Absorption, Transport, Storage, and Excretion of Pantothenic Acid

The pantothenic acid portion of any coenzyme A in the diet is released during digestion in the small intestine. It is then absorbed and transported throughout the body bound to red blood cells. Storage is minimal and is in the coenzyme form. Excretion of pantothenic acid is via the urine.[24]

## Functions of Pantothenic Acid

Coenzyme A is essential for the formation of acetyl-CoA from the breakdown of carbohydrate, protein, alcohol, and fat.[24] Acetyl-CoA molecules most often enter the citric acid cycle (with eventual ATP production). Beta-oxidation of fatty acids also requires coenzyme A. However, acetyl-CoA also is an important biosynthetic building block used to build fatty acids, cholesterol, bile acids, and steroid hormones. As coenzyme A, pantothenic acid also donates fatty acids to proteins in a process that can determine their location and function within a cell.

As a part of acyl carrier protein, pantothenic acid helps fatty acid biosynthesis. ACP attaches to fatty acids and shuttles them through the metabolic pathway that increases their chain length.

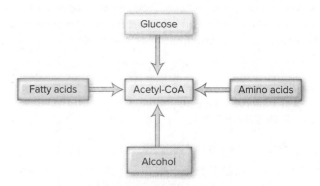

Pantothenic acid is required for the formation of coenzyme A, part of the structure of acetyl-CoA.

## Pantothenic Acid Deficiency

Pantothenic acid deficiency is very rare and has been observed only when a deficiency was experimentally induced.[2] Its symptoms include headache, fatigue, impaired muscle coordination, burning hands and feet, and GI tract disturbances.

*Knowledge Check*

1. What is the coenzyme formed from pantothenic acid?
2. How is pantothenic acid related to the formation of ATP?
3. What are 3 good food sources of pantothenic acid?

# 13.6 Biotin

Biotin's discovery was linked to what researchers in the 1920s called "egg-white injury." Rats fed large amounts of raw egg whites developed severe rashes, lost their fur, and became paralyzed. These symptoms were reversed when the rats were fed yeast, liver, and other foods. These observations led to the discovery of this B-vitamin. Biotin is a coenzyme that participates in reactions, known as **carboxylations,** that add carbon dioxide to compounds.[2,25]

## Sources of Biotin: Food and Microbial Synthesis

Biotin is widely distributed in very low amounts in foods and is found as the free vitamin and **biocytin**—biotin bound to the amino acid lysine in proteins. Sources include whole grains, mushrooms, egg yolks, nuts, and legumes (Fig. 13-13). The biotin content of food has been determined for only a small number of foods, so nutrient databases are incomplete.

We excrete more biotin than we consume; thus, it appears that bacteria in the large intestine synthesize biotin. However, it is not yet known how this contributes to overall biotin status, especially because biotin is absorbed most efficiently from the small intestine.

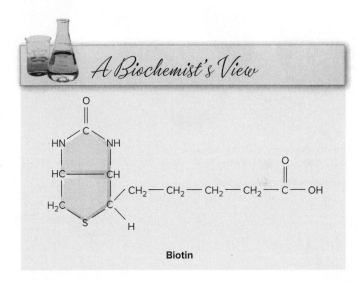

*A Biochemist's View*

Biotin

## Biotin Needs and Upper Level

The Adequate Intake for biotin for adults is 30 μg/day.[2] This amount is significantly less than the requirements for most of the water-soluble vitamins; remember, 1000 μg equal 1 mg. The diets of adults generally meet the Adequate Intake level. The Daily Value on food and supplement labels is 30 μg. There is no Upper Level for biotin.[2]

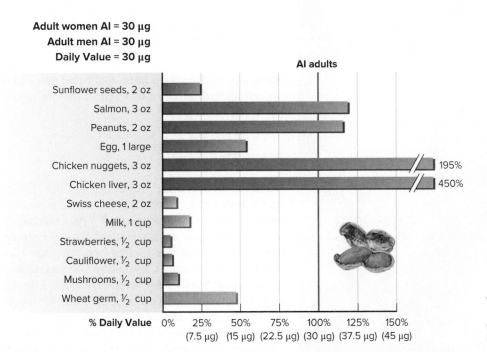

Adult women AI = 30 μg
Adult men AI = 30 μg
Daily Value = 30 μg

AI adults

Sunflower seeds, 2 oz
Salmon, 3 oz
Peanuts, 2 oz
Egg, 1 large
Chicken nuggets, 3 oz — 195%
Chicken liver, 3 oz — 450%
Swiss cheese, 2 oz
Milk, 1 cup
Strawberries, ½ cup
Cauliflower, ½ cup
Mushrooms, ½ cup
Wheat germ, ½ cup

% Daily Value    0%    25%    50%    75%    100%    125%    150%
            (7.5 μg)  (15 μg) (22.5 μg) (30 μg) (37.5 μg) (45 μg)

*Figure 13-13* Food sources of biotin.
©McGraw-Hill Education/Ken Cavanagh, Photographer

**carboxylation** Addition of a carboxyl group, COOH, into a compound or molecule.

**biocytin** Biotin bound to the amino acid lysine in food proteins.

The Adequate Intake level for biotin can be met with 3 tablespoons of peanuts.

©C Squared Stuidos/Getty Images RF

## Absorption, Transport, Storage, and Excretion of Biotin

In the small intestine, the enzyme biotinidase releases biotin from protein and lysine. Free biotin is absorbed in the small intestine via a sodium-dependent carrier. Biotin is stored in small amounts in the muscles, liver, and brain, and its excretion is mostly via the urine, although some is excreted in bile.[25]

## Functions of Biotin

Biotin functions as a coenzyme for several carboxylase enzymes that add carbon dioxide to various compounds. These enzymes are required for the metabolism of carbohydrates, proteins, and fats. Reactions that are dependent on biotin include the following.

- The carboxylation of pyruvate to form oxaloacetate, a citric acid cycle intermediate. Recall that, when glucose supplies run low, oxaloacetate serves as a starting point for gluconeogenesis (see Chapter 9).

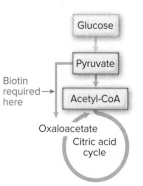

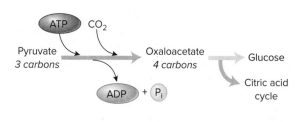

- The breakdown of the amino acids threonine, leucine, methionine, and isoleucine for use as energy
- The carboxylation of acetyl-CoA to form malonyl-CoA so that fatty acids can be synthesized (see Section 13.5)

Biotin also binds to the proteins that help DNA fold in the cell nucleus. In this role, biotin is thought to help maintain gene stability.[25]

## Biotin Deficiency

Overall, biotin deficiencies are rare. About 1 in 60,000 infants is born with a genetic defect that results in very low amounts of the enzyme biotinidase.[26,27] As a result, these infants cannot break down biocytin in foods for absorption. A biotin deficiency develops, and symptoms (skin rash, hair loss, convulsions, low muscle tone, and impaired growth) occur within a few weeks to months following birth. The affected individual is typically treated throughout life with regular doses of biotin supplements.[27]

Biotin deficiency also has resulted from the use of anticonvulsant medications, malabsorption in those with severe intestinal diseases, and the regular ingestion of large amounts (>12) of raw eggs each day. Raw egg

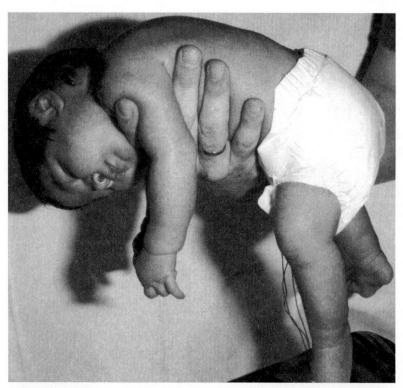

Babies with an untreated deficiency of biotinidase exhibit hypotonia (low muscle tone). Hypotonia also can be caused by other disorders that affect the nerves and muscles.

whites contain a protein, **avidin,** that binds biotin, limiting its absorption.[25] Cooking eggs denatures avidin, which prevents it from binding to biotin.

### Knowledge Check

1. Biotin is a coenzyme for several carboxylase enzymes. In general, what do these enzymes do?
2. How can a biotin deficiency occur?

## 13.7 Vitamin B-6

Nearly all amino acids require a vitamin B-6 coenzyme in their metabolism. Vitamin B-6 is a family of 3 compounds: pyridoxal, pyridoxine, and pyridoxamine. All 3 forms can be phosphorylated (have a phosphate group added) to become active vitamin B-6 coenzymes. The primary vitamin B-6 coenzyme is pyridoxal phosphate (PLP). Vitamin B-6 is converted to PLP by adding a phosphate group ($PO_4$) to its hydroxyl group. The generic name for this vitamin is B-6, or pyridoxine.[28]

### Vitamin B-6 in Foods

Vitamin B-6 is stored in the muscle tissues of animals; thus, meat, fish, and poultry are some of the richest sources. Although vitamin B-6 in foods of animal origin is often more readily absorbed than that in foods of plant origin, whole grains also are good sources of vitamin B-6. However, vitamin B-6 is lost during the refining of grains, and it is not added during enrichment. Most fruits and vegetables are not good vitamin B-6 sources, but there are some exceptions: carrots, potatoes, spinach, bananas, and avocados (Fig. 13-14). The leading contributors of vitamin B-6 in the U.S. diet are fortified ready-to-eat cereals, poultry, beef, potatoes, and bananas.[7] Like many other water-soluble vitamins, vitamin B-6 can be lost when foods are exposed to heat and other processing.

**avidin** Protein, found in raw egg whites, that can bind biotin and inhibit its absorption. Cooking destroys avidin.

▶ Biotin is a common ingredient in products marketed as treatment for thinning hair or brittle nails. The biotin doses in these products may be very high—as much as 300 times the AI. Most hair loss is hereditary, but in some cases diseases, nutrient deficiencies, and stress may contribute to hair loss. There are no scientific studies to support the claim that high doses of biotin can prevent or treat hair loss or brittle nails.

**Adult RDA = 1.3 mg**
**Daily Value = 1.7 mg**

| Food | % Daily Value |
|---|---|
| Tuna, 3 oz | |
| Turkey nuggets, 3 oz | |
| Beef, 3 oz | |
| Pistachios, 2 oz | |
| Pinto beans, 1 cup | |
| Feta cheese, 2 oz | |
| Banana, ½ cup | |
| Green peppers, ½ cup | |
| Sweet potato, ½ cup | |
| Potato, ½ cup | |
| Bagel, 4-inch | |
| Frosted Flakes® cereal, ¾ cup | |
| Oatmeal, ¾ cup | 190% |

% Daily Value: 0% — 25% (0.425 mg) — 50% (0.85 mg) — 75% (1.28 mg) — 100% (1.7 mg)

RDA adults

*Figure 13-14* Food sources of vitamin B-6. Note that, after age 51, the RDA increases (men: 1.7 mg; women: 1.5 mg).

©Stockdisc/PunchStock RF

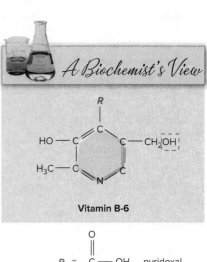

*A Biochemist's View*

**Vitamin B-6**

$R$ = C—OH   pyridoxal (with double-bonded O)

$R$ = $CH_2OH$   pyridoxine

$R$ = $CH_2NH_2$   pyridoxamine

Boxed area = Hydroxyl group where phosphate is added

A banana and 1 cup of spinach each provide nearly 25% of the vitamin B-6 RDA for adults.

Banana: ©Maksym Narodenko/123RF;
Spinach: ©Ingram Publishing RF

▶ Scientists are studying how high homocysteine blood concentrations may affect various disorders of the brain, bones, and cardiovascular system. Although evidence about the effects of too much homocysteine is still inconclusive, meeting B-vitamin (riboflavin, vitamin B-6, folate, and vitamin B-12) and choline requirements helps convert homocysteine to the amino acids methionine and cysteine. This keeps homocysteine levels in the body low.

**serotonin** Neurotransmitter that affects mood (sense of calmness), behavior, and appetite and induces sleep.

**dopamine (DOPA)** Neurotransmitter that leads to feelings of euphoria, among other functions; also forms norepinephrine.

**norepinephrine** Neurotransmitter released from nerve endings; also a hormone produced by the adrenal gland during stress. It causes vasoconstriction and increases blood pressure, heart rate, and blood sugar.

**gamma-aminobutyric acid (GABA)** Chief inhibitory neurotransmitter.

**histamine** Bioactive amine that participates in immune response, stimulates stomach acid secretion, and triggers inflammatory response. It regulates sleep and promotes smooth muscle contraction, increased nasal secretions, blood vessel relaxation, and airway constriction.

## Vitamin B-6 Needs and Upper Level

The RDA for vitamin B-6 is 1.3 mg/day for adult men and women up to age 50. For older adults, the RDA increases to 1.7 mg/day for men and 1.5 mg/day for women. The Daily Value on food and supplement labels is 1.7 mg. In the U.S., the average daily intake of vitamin B-6 from food is 2.7 mg in adult men, and for women it is 1.8 mg.[9]

Daily intakes of 2 to 6 g of vitamin B-6 (attainable only with a dietary supplement) have caused nerve damage so severe that the individuals were unable to walk. Less severe neurological symptoms, such as hand and foot tingling and numbness, may occur with lower amounts (200 to 500 mg) of vitamin B-6. The Upper Level for adults is set at 100 mg/day to avoid nerve problems.[2]

## Absorption, Transport, Storage, and Excretion of Vitamin B-6

The absorption of vitamin B-6 is by passive diffusion. The coenzyme form is normally converted to the free vitamin form for absorption, but at high concentrations some of the coenzyme may be absorbed as such. Vitamin B-6 is transported via the portal vein to the liver, where most of it is phosphorylated. From the liver, the phosphorylated forms (mainly PLP) are released for transport in the blood bound to the transport protein albumin. Muscle tissue is the main storage site for vitamin B-6. Excess vitamin B-6 is generally excreted in the urine.[2,28]

## Functions of Vitamin B-6

Vitamin B-6 coenzymes participate in numerous metabolic reactions. For example, PLP is a coenzyme in more than 100 enzymatic reactions, almost all of which involve nitrogen-containing compounds, such as amino groups ($NH_2$).[2,28,29]

### Metabolism

A major role of PLP is to participate in amino acid metabolism. An important function of PLP is as a coenzyme for transamination reactions that transfer amino groups to allow the synthesis of nonessential amino acids (see Chapter 7). Without PLP, every amino acid would be essential because it would have to be supplied by the diet (see Fig. 7-3). For example, Figure 7-3 shows an example of a transaminase enzyme pathway that utilizes vitamin B-6 to synthesize nonessential amino acids. In this example, pyruvic acid gains an amino group from glutamic acid to form the nonessential amino acid alanine. PLP also helps convert homocysteine to the amino acid cysteine, which occurs during the metabolism of the amino acid methionine (see Appendix C for details on methionine and homocysteine metabolism).

PLP also is required for folate metabolism. In addition, PLP is involved in the release of glucose from glycogen. In this way, PLP helps maintain blood glucose concentration.[28]

### Synthesis of Compounds

In the red blood cell, PLP catalyzes a step in the synthesis of heme, a nitrogen-containing ring that is inserted into certain proteins to hold iron in place. The best known of these proteins is hemoglobin, which uses iron to transport oxygen in the blood.

Amino acids are used not only to build proteins but also to make nonprotein nitrogen-containing compounds. Many of these compounds are neurotransmitters, which are important for brain function. PLP is required for the synthesis of several neurotransmitters: **serotonin** from tryptophan, **dopamine (DOPA)** and **norepinephrine** from tyrosine, and **gamma-aminobutyric acid (GABA)** from glutamic acid. The synthesis of **histamine** from the amino acid histidine also requires vitamin B-6.[28]

PLP participates in vitamin formation, too. It plays an important role in the synthesis of the B-vitamin niacin from the amino acid tryptophan.[28]

## Other Functions

Vitamin B-6 helps support normal immune function and the regulation of gene expression. It also may help prevent colon cancer—individuals with high blood concentrations of PLP have a reduced risk of colon cancer.[30] Low vitamin B-6 levels also have been observed in inflammatory diseases, such as cardiovascular disease, inflammatory bowel disease, diabetes, and rheumatoid arthritis.[31,32] It is not known whether improving vitamin B-6 status will reduce inflammation; some studies indicate vitamin B-6 supplementation alone or in combination with other B-vitamins, such as folic acid, does not alter biomarkers of inflammation.[33] More research is needed to clarify the effect of vitamin B-6 on inflammation.

## Vitamin B-6 Deficiency

Outright vitamin B-6 deficiency is rare in North America, but about 10% of the U.S. population has low vitamin B-6 blood concentrations.[34] When a deficiency does occur, the symptoms may include seborrheic dermatitis, **microcytic hypochromic anemia** (from decreased hemoglobin synthesis), convulsions, depression, and confusion due to altered tryptophan metabolism or neurotransmitter synthesis.[28,29] Women, older adults, Black individuals, smokers, users of oral contraceptive agents, and those who have alcoholism, are underweight, or consume poor diets are at risk of low blood vitamin B-6 concentrations.[29,34] Acetaldehyde, produced during alcohol metabolism, decreases the formation of PLP by cells and may reduce its biological activity. A number of medications can decrease the amount of PLP in the blood: l-DOPA, used to treat Parkinson disease; isoniazid, an antituberculosis medication; and theophylline, used to treat asthma. Individuals taking these medications may need vitamin B-6 supplementation.

## Pharmacologic Use of Vitamin B-6

Supplemental vitamin B-6 has a long history as a treatment for carpal tunnel syndrome, premenstrual syndrome (PMS), and nausea during pregnancy. Carpal tunnel syndrome is a common, painful disorder of the wrist and hand caused by nerve compression. PMS is a common, multisymptom disorder that occurs 1 to 2 weeks prior to menstruation. The symptoms include fluid retention, bloating and weight gain, breast tenderness, abdominal discomfort, headache, cravings for sugar and alcohol, mild depression, and anxiety. Nausea is experienced by 70 to 85% of women during the 1st trimester of pregnancy (see Chapter 16). In general, there is limited research supporting the use of vitamin B-6 in treating these conditions, but the FDA has approved a medication that contains vitamin B-6 and another compound for treating mild to moderate nausea in pregnancy.[35] Additionally, some physicians recommend single vitamin B-6 supplements to treat this common complaint. For each of these conditions, additional research is needed to determine if supplemental vitamin B-6 is beneficial.

**microcytic hypochromic anemia** Anemia characterized by small, pale red blood cells that lack sufficient hemoglobin and thus have reduced oxygen-carrying ability. It also can be caused by an iron deficiency.

▶ In the early 1950s, some infants were accidentally fed a commercial formula in which vitamin B-6 had been destroyed by oversterilization. The infants developed abnormal electroencephalogram (EEG) readings and experienced convulsions. The reason was probably a lack of neurotransmitter synthesis in the brain. The situation was successfully treated.

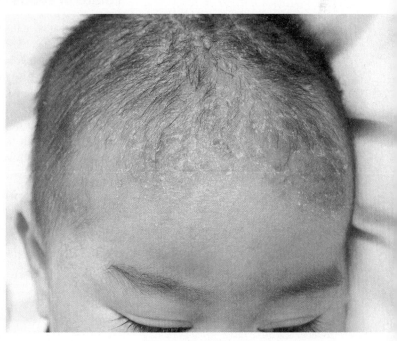

Seborrheic dermatitis can be caused by a vitamin B-6 deficiency.

Eaaw/Shutterstock

### Knowledge Check

1. How is the PLP coenzyme used in amino acid metabolism?
2. What are 3 good food sources of vitamin B-6?
3. Which individuals are at risk of having poor vitamin B-6 status?

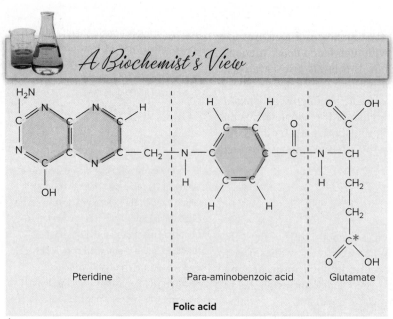

*A Biochemist's View*

Pteridine     Para-aminobenzoic acid     Glutamate

**Folic acid**

\* = In foods, additional glutamate molecules are usually linked here to the carboxyl group

# 13.8 Folate

The name of the B-vitamin folate comes from the Latin word *folium,* meaning "leaf." It was given this name because leafy green vegetables are excellent sources. *Folate* is the generic name, referring to the various forms of the vitamin found naturally in foods. The term **folic acid** refers to the synthetic form of the vitamin found in supplements and fortified foods.

Folate consists of 3 parts: pteridine, para-aminobenzoic acid (PABA), and 1 or more molecules of the amino acid glutamic acid (glutamate). If, as shown in its structure, only 1 glutamate molecule is present, it is designated folic acid (folate monoglutamate). In food, about 90% of the folate molecules have 3 or more glutamates attached to the carboxyl group (red asterisk in the structure) and are known as polyglutamates.[36]

## Folate in Foods

▶ PABA by itself is sometimes added to supplements, but there is no current scientific rationale for this.

The foods that have the largest amount of and most bioavailable folate are liver, legumes, and leafy green vegetables (Fig. 13-15). Other rich sources of folate include nuts, avocados, and citrus fruits. Mandatory fortification of enriched cereal grains with folic acid in the U.S. began in 1998, making these food products good sources of this vitamin, providing an additional 190 µg on average to the diet of adults. Foods made with flours fortified with folic acid, such as yeast bread, rolls, cakes, cookies, pasta, and pancakes, as well as ready-to-eat cereals and rice, are leading sources of this vitamin in the U.S. diet. Folate rich foods such as legumes, lentils, and vegetables contribute relatively small amounts to our diets.[7] The biological availability of naturally occurring folate in mixed diets is 50 to 80% of folic acid in fortified foods.[37]

▶ Many breakfast cereals provide 100% of the Daily Value of folic acid. They include

   Wheat Chex®
   Raisin bran
   Kashi® Honey Toasted Oats
   Kellogg's All-Bran®
   Cinnamon Life®

See the complete list at www.cdc.gov /ncbddd/folicacid.

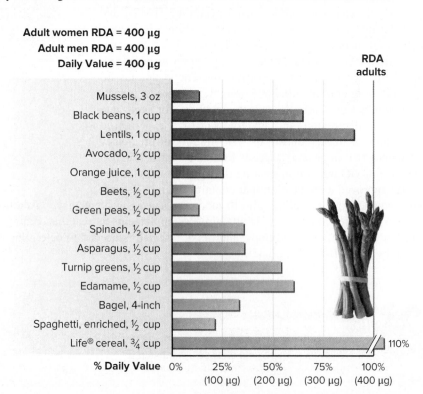

**Adult women RDA = 400 µg**
**Adult men RDA = 400 µg**
**Daily Value = 400 µg**

RDA adults

Mussels, 3 oz
Black beans, 1 cup
Lentils, 1 cup
Avocado, ½ cup
Orange juice, 1 cup
Beets, ½ cup
Green peas, ½ cup
Spinach, ½ cup
Asparagus, ½ cup
Turnip greens, ½ cup
Edamame, ½ cup
Bagel, 4-inch
Spaghetti, enriched, ½ cup
Life® cereal, ¾ cup — 110%

**% Daily Value**   0%    25%    50%    75%    100%
                  (100 µg) (200 µg) (300 µg) (400 µg)

*Figure 13-15* Food sources of folate.

©Rozenbaum/E Cirou/Getty Images RF

Food processing and preparation can destroy up to 90% of the folate in food. Folate is extremely susceptible to destruction by heat, oxidation, and ultraviolet light. (Vitamin C in foods helps protect folate from oxidative destruction.) The regular consumption of fresh or lightly cooked fruits and vegetables can help you gain the full benefits of their folate contents.

## Dietary Folate Equivalents

The RDA for folate is expressed as dietary folate equivalents (DFE).[2] DFE reflect the differences in the absorption of food folate and synthetic folic acid. The relationship among DFE, food folate, and folic acid is as follows.

$$1 \text{ DFE} = 1 \text{ μg food folate} = 0.6 \text{ μg folic acid} = 0.5 \text{ μg folic acid taken}$$
$$\text{taken with food} \quad \text{on an empty stomach}$$

DFE are calculated using this equation:

$$\text{DFE} = \text{μg food folate} + (\text{μg folic acid} \times 1.7)$$

For example, the Daily Value for a serving of ready-to-eat breakfast cereal is listed on the label as 50%, so the amount of folic acid is 200 μg per serving (Daily Value of 400 μg × 0.50). Because this folate is mainly synthetic folic acid, the 200 μg is multiplied by 1.7, yielding 340 μg DFE. If the diet also contains 300 μg of food folate, the total DFE intake is 640 μg DFE (300 μg + 340 μg), which exceeds the adult RDA. Nutrient databases typically report several values for folate. These values include μg food folate, μg folic acid, total folate (μg food folate + μg folic acid), and folate DFE (μg food folate + [μg folic acid × 1.7]).

## Folate Needs

The folate RDA for adults is 400 μg/day, expressed as DFE. For women of childbearing age, the RDA specifies that this amount be consumed as folic acid—that is, from fortified foods and supplements—in addition to folate from the diet. This recommendation is in response to evidence that supplemental folic acid protects against the development of neural tube birth defects that occur early in pregnancy (see Clinical Perspective: Neural Tube Defects).[38] Because of fortification of the U.S. food supply, average daily folate intake exceeds the RDA. The average intake in adult men is 601 DFE, and in adult women it is 459 DFE.[9] Despite this good intake, nearly 20% of women of childbearing age do not consume enough of this vitamin and are at risk of developing a folate deficiency.[39] The Daily Value on food and supplement labels is 400 μg/day.

## Upper Level for Folate

The Upper Level for synthetic folic acid is set at 1000 μg (1 mg); intakes above this level may mask a vitamin B-12 deficiency (see Section 13.9).[2] The Upper Level does not apply to folate in foods because absorption is limited. Usual daily intakes greater than the Upper Limit are reported primarily by those taking supplements. In response to the concern that high doses of folic acid might mask a vitamin B-12 deficiency, the FDA limits the amount of folic acid in nonprescription vitamin supplements. These levels are set at 400 μg for nonpregnant individuals when no statement of age is listed on the supplement label. When age-related doses are listed, there can be no more than 100 μg for infants, 300 μg for children, and 400 μg for adults. Over-the-counter prenatal supplements can contain 800 μg.

## Absorption, Transport, Storage, and Excretion of Folate

To be absorbed, folate polyglutamates must be broken down (hydrolyzed) in the GI tract to the monoglutamate form. Enzymes known as folate conjugases, produced by the absorptive cells, remove the additional glutamates. The monoglutamate form is then actively transported across the intestinal wall. Very large doses of folic acid from supplements are

## *Culinary Perspective*

### Beans, Lentils, and Dried Peas

The edible seeds of legumes, more commonly known as beans, lentils, and dried peas, have been cultivated and consumed across the globe for thousands of years. These common foods are nutritional powerhouses providing protein, complex carbohydrates, vitamins, minerals, and phytochemicals. For example, ½ cup of black beans provides 7 grams of protein and 7 grams of fiber, and it supplies enough folate to be considered an excellent source. Black beans also are a good source of thiamin, iron, and magnesium and provide a variety of phytochemicals, including flavonoids and antioxidants.

Today, there are dozens of varieties of beans and lentils available. They can be purchased in canned, dried, or frozen form. Canned beans, such as pinto, red, black, navy, and garbanzo beans, are convenient additions to soups, salads, and casseroles and can even be used in baking. Garbanzo beans, also known as chickpeas, mashed with other ingredients make the popular hummus spreads. Cooked dried beans are appreciated for their superior flavor and low cost. Cooking a pot of dry beans can take 2 or more hours, but lentils and dried peas (e.g., split green peas) can be cooked in 30 minutes or less due to their small size.

© Pixtal/AGE Fotostock RF

The old English expression "full of beans" is a happy one, meaning an energetic and cheerful mood. Are these healthy, versatile, and inexpensive foods part of your diet? If not, give them a try!

absorbed by passive diffusion. When synthetic folic acid is consumed as a supplement and without food, it is nearly 100% bioavailable. Consumed with food, as in fortified cereal grains, absorption is slightly reduced.[2]

The portal vein delivers the monoglutamate form of folate from the small intestine to the liver. Then, folate is either stored in the liver or released into the blood for delivery to other tissues in the body. Once folate is transported into a cell, it is converted to a polyglutamate form that "traps" the folate in the cell. Some liver folate is excreted in bile and reabsorbed by the enterohepatic circulation. Alcohol interferes with this process, which is a reason those with alcoholism often become folate deficient. Folate is excreted in both the urine and the feces.

### Functions of Folate

Folate coenzymes are required for the synthesis and maintenance of new cells. Folate coenzymes function in 1-carbon transfer reactions in the body. The folate coenzymes are formed from a central coenzyme form called tetrahydrofolic acid (THFA). Folate coenzymes are critical for DNA synthesis, DNA methylation, and amino acid metabolism.[5,36]

## DNA Synthesis

Normal cell division requires adequate folate because THFA is required for the synthesis of DNA. DNA contains 4 nitrogenous bases: cytosine and thymine (pyrimidines) and adenine and guanine (purines). The pyrimidine thymine is formed by the addition of a methylene group ($CH_2$) to the pyrimidine uracil. A folate coenzyme supplies the $CH_2$. THFA also is needed for the synthesis of the purines (adenine and guanine) in DNA. Thus, DNA synthesis and repair may decline in a folate shortage, impacting cell division.

THFA ($-CH_2-$)    THFA (free)

Uracil ▪▪▪▪▪▪▪▪▪▪▪▶ Thymine ▪▪▪▶ ▪▪▪▶ DNA

Folate and vitamin B-12 functions are closely linked. A vitamin B-12 coenzyme is required to recycle the folate coenzyme needed for DNA synthesis (see Section 13.9). Thus, folate and vitamin B-12 deficiencies can produce identical signs and symptoms.

▶THFA transfers these 1-carbon groups:

Methyl ($-CH_3$)

Formyl ($-CH=O$)

Methylene ($-CH_2-$)

Methenyl ($-CH=$)

## Epigenetic Modification of DNA

In addition to their role in DNA synthesis, folate coenzymes influence DNA function by the process of methylation of specific regions of DNA molecules. DNA methylation is a type of epigenetic modification that can affect our risk for cancer. Both hypomethylation and hypermethylation of DNA have been linked to the development of cancer and other diseases. Scientists now know that alterations in folate metabolism due to reductions in enzyme activities can influence DNA methylation.[5,41] Testing for this defect is not yet routine. Although some companies are already selling supplements intended to regulate methylation reactions, much more research is needed before these supplements can be recommended.

▶Although folate deficiency can be induced to treat cancer, folate deficiency also may cause concern with regard to promoting cancer. Because folate is required for DNA synthesis, even mild folate deficiency may contribute to DNA damage, which in turn affects cancer risk. A daily intake of 400 µg (the RDA) may protect against cancers such as colorectal cancer.

# CLINICAL PERSPECTIVE

## Folate and the Cancer Drug Methotrexate

The cancer drug methotrexate takes advantage of the key role of THFA in DNA synthesis. Methotrexate, referred to as a folate antagonist, interferes with THFA metabolism. This, in turn, reduces DNA synthesis throughout the body. The reduction in DNA synthesis can halt the growth of cancer cells, but it also affects other rapidly proliferating cells, such as intestinal and red blood cells. As a result, the typical side effects of methotrexate therapy are the same as for a folate deficiency (e.g., mouth sores, nausea, diarrhea, and anemia). Methotrexate also is used to treat several immunological disorders, such as rheumatoid arthritis, psoriasis, asthma, and inflammatory bowel disease. Individuals treated with methotrexate are sometimes advised by their physicians to take supplemental folic acid to reduce the drug's toxic side effects. This supplementation probably reduces side effects but does not influence the effectiveness of methotrexate.[40] Cancer and diet are discussed further in the Clinical Perspective in Chapter 15.

▶ The enzyme methylenetetrahydrofolate reductase (MTHFR) is critical for normal folate metabolism. Several MTHFR gene variants decrease the formation of 5-methyltetrahydrofolate (5-MeTHFA), an important methyl donor. The presence of an MTHFR gene variant is thought to be a factor in the increased risk of a pregnancy affected by a neural tube defect (see Clinical Perspective: Neural Tube Defects). However, it's important to know that MTHFR gene variants are relatively common in populations around the world and most people with these variants do not experience neural tube defects.

*Steps in Folate Deficiency*

1. A decrease in blood folate concentration
2. A decrease in red blood cell folate
3. Defective DNA synthesis
4. A change in the structure of certain white blood cells
5. An increase in blood concentration of homocysteine (and methylmalonic acid)
6. Megaloblastic changes in bone marrow and other rapidly dividing cells
7. An increase in the size of circulating red blood cells
8. Megaloblastic (macrocytic) anemia

**megaloblast** Large, nucleated, immature red blood cell in the bone marrow, which results from the inability of a precursor cell to divide when it normally should.

**macrocyte** Literally, "large cell," such as a large red blood cell.

**megaloblastic (macrocytic) anemia** Anemia characterized by abnormally large, nucleated, immature red blood cells, which result from the inability of a precursor cell to divide normally.

## Amino Acid Metabolism

THFA is important in amino acid metabolism, especially the interconversions of amino acids. It accepts 1-carbon groups from various amino acids and is responsible for converting the amino acid glycine to the amino acid serine (the main source of methyl groups for THFA) and converting the essential amino acid histidine to the amino acid glutamic acid. THFA, along with vitamin B-12, is involved in a pathway that converts the amino acid homocysteine to the amino acid methionine (described in Section 13.9).

## Other Functions

Another key function of folate is the formation of the neurotransmitters serotonin, norepinephrine, and dopamine in the brain. Because of this, some scientists have hypothesized that supplemental folic acid could be a useful addition to the treatment of depression. Some studies have supported this, but others do not.[42] Folate also may help maintain normal blood pressure and reduce the risk of stroke, especially in those with low blood folate levels—a condition that is much more common in countries that do not fortify foods with folic acid.[43]

## Folate Deficiency

Folate deficiency was once fairly common in the U.S. However, folic acid fortification of the food supply has dramatically decreased the number of persons with folate deficiency. Prior to fortification, 30% of the U.S. population had low red blood cell folate, a measure of long-term body stores, but this has dropped to less than 1% of the population.[44,45] Despite this improvement, folate deficiency still occurs.[36] It can result from low intake, malabsorption, and an increased requirement. Those at highest risk include impoverished individuals, those with alcoholism or chronic gastrointestinal diseases, and those who use certain medications, such as some anticonvulsants and methotrexate. In individuals with a vitamin B-12 deficiency, folate utilization and metabolism are abnormal (see Section 13.9), resulting in a "functional" folate deficiency. In addition, pregnancy greatly increases the need for this vitamin (the RDA is 600 μg DFE/day) because of the increased rate of cell division, and thus of DNA synthesis, in the mother's body and that of her developing offspring.[2] Prenatal care often includes prenatal multivitamin and mineral supplements fortified with folate to compensate for the extra needs associated with pregnancy.

A deficiency of folate first affects cell types that are actively synthesizing DNA because these cells have a short life span and rapid turnover rate. For example, red blood cells have a 120-day life span and are vulnerable to folate deficiency. Without folate, precursor cells in the bone marrow cannot form new DNA and therefore cannot divide normally to become mature red blood cells. The cells grow larger because there is continuous formation of RNA, leading to the increased synthesis of protein and other cell components. Hemoglobin synthesis also intensifies. However, when it is time for the cells to divide, they lack sufficient DNA for normal division.

Unlike normal mature red blood cells, these cells (called **megaloblasts**) retain their nuclei and remain in a large, immature form. Most megaloblasts do not make it out of the bone marrow. Any of these large cells that do enter the bloodstream are called **macrocytes**. Their presence results in a form of anemia called **megaloblastic, or macrocytic, anemia** (Fig. 13-16).

Large, immature cells also appear throughout the GI tract during chronic folate deficiency. This occurs because DNA synthesis is impaired, which hinders cell division in the GI tract. This change contributes to a decreased absorptive capacity of the GI tract and persistent diarrhea. Mouth sores and abnormal liver function are other signs of a folate deficiency. White blood cell synthesis also is disrupted by a folate deficiency because these cells are made in rapid bursts during immune challenges (e.g., infections). Thus, immune function can be diminished during a folate deficiency.

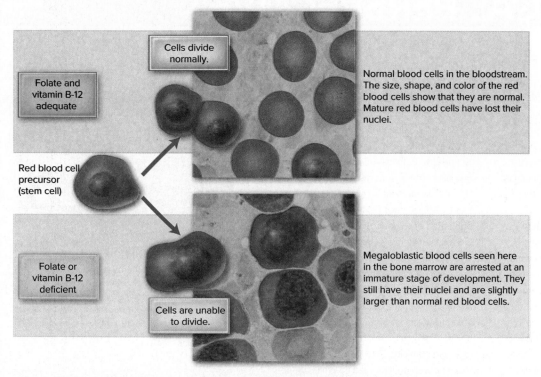

*Figure 13-16*  Megaloblastic anemia occurs when blood cells are unable to divide, leaving large, immature red blood cells. Either a folate or a vitamin B-12 deficiency may cause this condition. Measurements of blood concentrations of both vitamins are taken to help determine the cause of the anemia. Megaloblastic anemia due to folate deficiency has decreased substantially since folate fortification of grains began in 1998.

## CASE STUDY

©Corbis/VCG/Getty Images RF

Suzanne and Ted are planning to start a family. They are especially concerned because Suzanne's sister gave birth last year to a baby with spina bifida. In preparation for her pregnancy, Suzanne takes a multivitamin supplement and eats fortified breakfast cereal most days. She also tries to include oranges, orange juice, broccoli, and spinach salads in her diet. What is spina bifida? How do Suzanne's diet and multivitamin supplement help prevent this serious disorder?

### Knowledge Check

1. How do folate in food and synthetic folic acid differ? Which form is better absorbed?
2. What are 3 foods that are good sources of folate?
3. What are the 2 main ways that folate affects DNA function?
4. What type of anemia signifies a folate deficiency?

## Neural Tube Defects

A maternal deficiency of folate and a genetic predisposition have been linked to the development of **neural tube defects** (NTDs) in the fetus (Fig. 13-17). The 2 most common NTDs are spina bifida (a spinal cord or spinal fluid bulge through the back) and anencephaly (the absence of a brain). In both cases, there is a defect in the very early development of the neural tube, the structure that subsequently forms the brain, spinal cord, spinal nerves, and spinal column. Victims of spina bifida may exhibit paralysis, incontinence, hydrocephalus (the abnormal buildup of spinal fluid in the brain), and learning disabilities. Children born with anencephaly die shortly after birth.

Folate is critical to normal neural tube development. The neural tube forms and closes very early in pregnancy—the first 21 to 28 days after conception (see Chapter 16). During this time of critical development, many women are unaware that they are pregnant. Thus, ensuring good folate status for all women capable of becoming pregnant is crucial.

As discussed earlier in the chapter, the folic acid fortification of refined cereals and grains was begun in the U.S. in 1998, increasing folic acid intake on average by 190 µg/day. Before fortification, about 4000 pregnancies per year were affected by an NTD. Since fortification, scientists estimate that over 1300 cases of NTDs have been prevented each year. However, about 20% of U.S. women still have insufficient folate status, with red blood cell folate levels that may not be classified as deficient but that may increase the risk for NTDs.[46] Black women have the highest prevalence of insufficient folate levels at a rate of almost 35%. The likelihood of having a baby affected by an NTD varies by race and ethnicity. The highest rates are in Hispanic women, followed by White women. The lowest rates are among Black and Asian women.[47] Although the reasons for these differences are not known for certain, scientists hypothesize that several factors may place Hispanic women at higher risk.

Fumonisin mycotoxins may pose another risk factor for neural tube defects. Fumonisins are produced in mold-contaminated corn and are thought to disrupt cell metabolism in the embryo. Populations that rely heavily on corn as a dietary staple are most likely to be affected.[49]

©Wade Elmer

The recent occurrence of a cluster of anencephaly in 3 Washington State counties highlighted that scientists' knowledge of NTDs and their causes is incomplete. Between 2010 and 2014, 36 babies (or 8.6 in 10,000 live births) were delivered with anencephaly in these counties. Nationally, the rate for this period was 2.1 in 10,000 live births. The cause of this alarming NTD cluster is still unknown.[48]

It is recommended that all women capable of becoming pregnant consume 400 µg of folic acid daily from supplements or fortified foods, in addition to getting folate from a varied diet. Because many women do not follow these practices, some scientists are urging the government to double folic acid fortification levels to decrease further the incidence of NTDs. However, there are concerns that higher amounts of folic acid in foods may have unintended consequences, such as masking a vitamin B-12 deficiency (see Section 13.9).

- A lower intake of folic acid when foods (such as corn tortillas) made with corn masa flour are consumed in place of foods made with fortified wheat and other flours. Voluntary folic acid fortification of corn masa flour was not permitted until 2016.

- A higher incidence of the MTHFR genetic variant that decreases the ability to convert folic acid to its active form. Those who are of Hispanic heritage have an 18% incidence of this variant compared to 11% of non-Hispanic White individuals and 3% of Black individuals.

- Increased exposure to a variety of environmental factors, including ingestion of mycotoxins and exposure to pesticides, solvents, and nitrates

**Healthy spine** / **Spine affected by spina bifida**

Meninges — Skin on back — Spinal fluid — Spinal cord — Vertebra

*Figure 13-17* NTDs result from a developmental failure affecting the spinal cord or brain in the embryo. Very early in fetal development, a ridge of neural-like tissue forms along the back of the embryo. As the fetus develops, this material differentiates into the spinal cord and body nerves at the lower end and into the brain at the upper end. At the same time, the bones that make up the vertebrae gradually surround the spinal cord on all sides. If any part of this sequence of events goes awry, many defects can appear. The worst is total lack of a brain (anencephaly). Much more common is spina bifida, in which the backbones do not form a complete ring to protect the spinal cord. Deficient folate status in the mother during the beginning of pregnancy increases the risk of neural tube defects, as does a genetic predisposition.

©Biophoto Associates/Science Source

▶ Women who have had a child with a neural tube defect are advised to consume 4 mg/day of folic acid beginning at least 1 month before any future pregnancy. This must be done under strict physician supervision. For further information about neural tube defects, see the website www.spinabifidaassociation.org.

# 13.9 Vitamin B-12

Vitamin B-12, also known as cobalamin, is unique among the vitamins on 2 accounts. First, foods of animal origin, such as meat, poultry, fish, and dairy products, are the only reliable sources of vitamin B-12. Second, it is the only vitamin that contains the mineral cobalt as part of its structure.[47] Vitamin B-12 has a complex, multi-ring structure. The cyanocobalamin form of vitamin B-12 forms 2 active coenzymes (methylcobalamin and 5-deoxyadenosylcobalamin) by replacing the cyano group (shown in red in the Biochemist's View) with another group, such as a methyl group or a hydroxyl group. The discovery of vitamin B-12 and how it prevents the vitamin B-12 deficiency disease pernicious anemia was so significant that vitamin B-12 researchers were awarded 6 Nobel Prizes in the time period 1934–1965.

## Vitamin B-12 in Foods

Plants do not synthesize vitamin B-12. In fact, all vitamin B-12 compounds are synthesized exclusively by microorganisms, mainly bacteria. Animals acquire vitamin B-12 from soil ingested while eating and grazing. Ruminant animals, such as cows and sheep, also synthesize vitamin B-12 from bacteria in the multiple compartments of their stomachs.

   For humans, the sources of vitamin B-12 are foods of animal origin, such as meat, poultry, seafood, eggs, and dairy products. Especially rich sources of vitamin B-12 are organ meats, such as liver, kidneys, and heart, and fortified foods, such as ready-to-eat cereals (Fig. 13-18). Although algae and fermented soy products, such as tempeh and miso, are sometimes advertised as being good plant sources of vitamin B-12, vegans should not rely on them to meet vitamin B-12 requirements. These foods often contain vitamin B-12 analogs (compounds similar to vitamin B-12) that do not function as vitamin B-12 in the body.

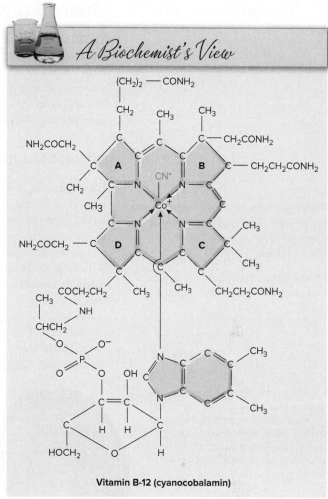

*A Biochemist's View*

**Vitamin B-12 (cyanocobalamin)**

*CN = cyano group

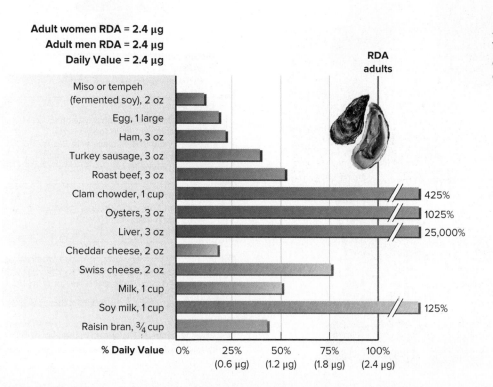

Adult women RDA = 2.4 µg
Adult men RDA = 2.4 µg
Daily Value = 2.4 µg

RDA adults

| Food | % Daily Value |
|---|---|
| Miso or tempeh (fermented soy), 2 oz | |
| Egg, 1 large | |
| Ham, 3 oz | |
| Turkey sausage, 3 oz | |
| Roast beef, 3 oz | |
| Clam chowder, 1 cup | 425% |
| Oysters, 3 oz | 1025% |
| Liver, 3 oz | 25,000% |
| Cheddar cheese, 2 oz | |
| Swiss cheese, 2 oz | |
| Milk, 1 cup | |
| Soy milk, 1 cup | 125% |
| Raisin bran, ¾ cup | |

% Daily Value    0%    25% (0.6 µg)    50% (1.2 µg)    75% (1.8 µg)    100% (2.4 µg)

*Figure 13-18* Food sources of vitamin B-12.

©Isabelle Rozenbaum & Frederic Cirou/ Getty Images RF

## Vitamin B-12 Needs and Upper Level

The RDA of vitamin B-12 for adults is 2.4 μg/day. The Daily Value on food and supplement labels also is 2.4 μg. Intake of vitamin B-12 from food in the U.S. typically exceeds the RDA. Adult men average 6.0 μg/day and adult women 3.9 μg/day.[9] This high intake provides the average meat-eating person with 2 to 3 years' storage of vitamin B-12 in the liver. No adverse effects have been observed with excess vitamin B-12 intake from food or from supplements, so there is no Upper Level for this vitamin.[2]

## Absorption, Transport, Storage, and Excretion of Vitamin B-12

Healthy adults absorb about 50% of the vitamin B-12 in the food they eat. As shown in Figure 13-19, the absorption of vitamin B-12 is quite complex. In food, vitamin B-12 is bound to protein. HCl and pepsin in gastric juice release vitamin B-12 from these proteins. In the stomach, the free vitamin B-12 binds to **R-protein,** which originates in the salivary glands. In the small intestine, pancreatic protease enzymes (e.g., trypsin) release vitamin B-12 from the R-protein/vitamin B-12 complex. The free vitamin B-12 then combines with **intrinsic factor,** a proteinlike compound produced by parietal cells in the stomach that enhances vitamin B-12 absorption. The vitamin B-12 intrinsic factor complex travels to the ileum, where vitamin B-12 is absorbed and transferred to the blood transport protein transcobalamin II. This vitamin B-12/transcobalamin II complex enters the portal vein and is delivered to the liver.[50] The liver can store enough vitamin B-12 to last several years; this is not the case with other water-soluble vitamins. Although vitamin B-12 is continually secreted

**R-protein** Protein produced by salivary glands that enhances absorption of vitamin B-12, possibly protecting the vitamin during its passage through the stomach.

**intrinsic factor** Substance in gastric juice that enhances vitamin B-12 absorption.

*Figure 13-19*  Absorption of vitamin B-12 requires several compounds produced in the mouth, stomach, and small intestine. Defects arising in the stomach or small intestine can interfere with vitamin B-12 absorption, in turn causing pernicious anemia.

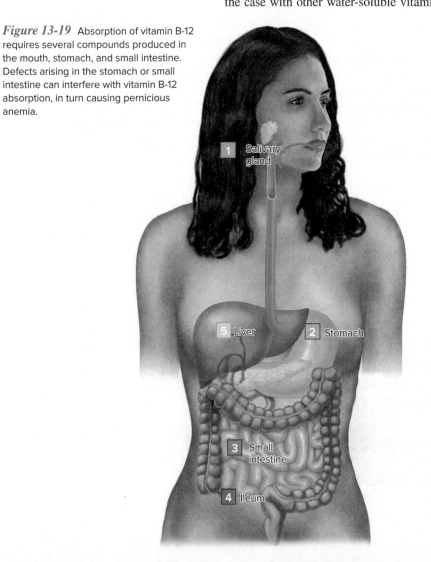

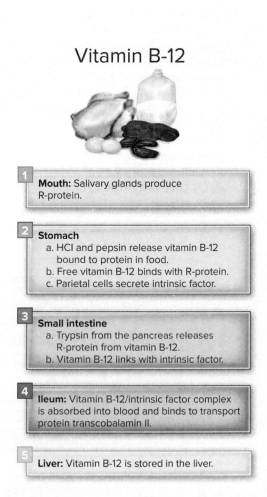

## Vitamin B-12

1
**Mouth:** Salivary glands produce R-protein.

2
**Stomach**
a. HCl and pepsin release vitamin B-12 bound to protein in food.
b. Free vitamin B-12 binds with R-protein.
c. Parietal cells secrete intrinsic factor.

3
**Small intestine**
a. Trypsin from the pancreas releases R-protein from vitamin B-12.
b. Vitamin B-12 links with intrinsic factor.

4
**Ileum:** Vitamin B-12/intrinsic factor complex is absorbed into blood and binds to transport protein transcobalamin II.

5
**Liver:** Vitamin B-12 is stored in the liver.

into the bile, most of it is reabsorbed by enterohepatic circulation, thereby efficiently "recycling" this vitamin. Little vitamin B-12 is excreted in the urine.[50]

## Functions of Vitamin B-12

Vitamin B-12 is required for 2 enzymatic reactions.[50] First, the formation of the amino acid methionine from the amino acid homocysteine is catalyzed by the enzyme methionine synthase, which requires the vitamin B-12 co-enzyme methylcobalamin (Fig. 13-20). Homocysteine accepts a methyl group from methylcobalamin, which forms methionine. Methionine, in turn, is the source of S-adenosyl methionine (SAM). In many reactions, SAM serves as a methyl donor. Methylation reactions are important for DNA and RNA regulation, myelin regulation, and the synthesis of many biochemical compounds. The methionine synthesis reaction also explains the close link between vitamin B-12 and folate: methylcobalamin obtains its methyl group from the folate coenzyme 5-methyltetrahydrofolate (5-MeTHFA). When the methyl group is donated to vitamin B-12, the folate coenzyme THFA is re-formed. When vitamin B-12 is lacking, THFA declines and the symptoms of folate deficiency can ensue. When either folate or vitamin B-12 is lacking, methionine and SAM synthesis decline, and the amount of homocysteine in the body increases. Choline (see Section 13.10) also can provide methyl groups to homocysteine.

The enzyme methylmalonyl **mutase** requires another vitamin B-12 coenzyme, 5-deoxyadenosylcobalamine. This enzyme is needed for the metabolism of fatty acids with an odd number of carbon molecules (most fatty acids have an even number of carbon molecules). It allows these fatty acids to be oxidized in the citric acid cycle and to provide energy.

## Vitamin B-12 Deficiency

Researchers in mid-19th-century England noted a form of anemia that causes death within 2 to 5 years of initial diagnosis. They called this disease **pernicious anemia** (*pernicious* means "leading to death"). It is now known that this disease can result from the inadequate production of the intrinsic factor required for vitamin B-12 absorption. As shown in Table 13-2, vitamin B-12 deficiency also results from malabsorption caused by various stomach and

Ground meat is an economical source of vitamin B-12.

©Comstock/Getty Images RF

**mutase** Enzyme that rearranges the functional groups on a molecule.

---

## CASE STUDY FOLLOW-UP

©Hill Street Studios/ Blend Images LLC RF

Mr. Martin's symptoms of weight loss, anemia, and neurological damage are most likely from a vitamin B-12 deficiency. Mr. Martin's vitamin B-12 intake has been very low for many years due to his long-standing vegan diet. Mr. Martin may also have low gastric acid production due to his age and his anti-reflux medication. This limits the release of any vitamin B-12 from food (although as a vegan he consumes little to no vitamin B-12). Mr. Martin will need to be treated by a physician for vitamin B-12 deficiency. In the future, if he remains a vegan, he will have to ensure a consistent and adequate intake of vitamin B-12. He will need a vitamin B-12 supplement and he can consume vitamin B-12–fortified foods, such as some breakfast cereals.

---

**pernicious anemia** Anemia that results from the inability to absorb sufficient vitamin B-12; is associated with nerve degeneration, which can result in eventual paralysis and death.

### Table 13-3 Causes of Vitamin B-12 Deficiency

| Impairment | Cause |
|---|---|
| Severe malabsorption: insufficient intrinsic factor production | Pernicious anemia (an autoimmune disorder) |
| | Gastrectomy (removal of all or part of the stomach) |
| | Gastric bypass surgery |
| Mild malabsorption: limited vitamin B-12 release from food protein | Atrophic gastritis (may be caused by *Helicobacter pylori* infection) limits HCl production and the activation of pepsin (enzyme that releases vitamin B-12 from food protein); occurs in 10 to 30% of older adults |
| | Gastric bypass surgery |
| | Metformin (drug used to lower blood sugar in type 2 diabetes)[52] |
| | Drugs that block stomach acid production |
| Mild or severe malabsorption: disorders of the ileum | Diseases of the small intestine, such as inflammatory bowel disease, celiac disease, intestinal overgrowth |
| | Surgical removal of the ileum |
| Low vitamin B-12 dietary intake | Long-term vegan or vegetarian diet |
| | Ingestion of breast milk from vitamin B-12–deficient mothers |

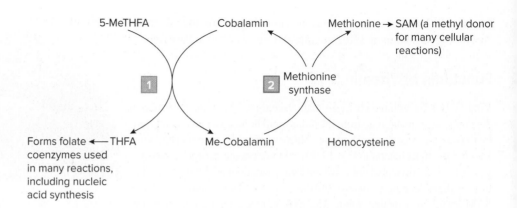

*Figure 13-20* Interrelationships of vitamin B-12 and folate with the amino acids homocysteine and methionine. **1** One methyl group from 5-methyltetrahydrofolate (5-MeTHFA) is donated to cobalamin (vitamin B-12). This reaction generates tetrahydrofolate (THFA) and methylcobalamin (Me-Cobalamin). **2** Me-Cobalamin donates the methyl group to homocysteine. This reaction is catalyzed by the enzyme methionine synthase. This reaction generates cobalamin and methionine. When vitamin B-12 is lacking, 5-MeTHFA builds up and THFA decreases, which slows or stops nucleic acid synthesis. Homocysteine levels also rise.

intestinal problems and the use of some medications.[51] Of course, insufficient vitamin B-12 intake due to very low intake of animal foods also can cause a deficiency.

### Megaloblastic (Macrocytic) Anemia

When a vitamin B-12 deficiency is severe enough that body stores are gone or almost gone, a megaloblastic (macrocytic) anemia results. The anemia produced in vitamin B-12 deficiency is identical to that produced in a folate deficiency. Because a lack of vitamin B-12 impairs folate metabolism, normal DNA and red blood cell synthesis is disrupted, resulting in megaloblastic anemia (see Fig. 13-16).

### Neurological Changes

A vitamin B-12 deficiency produces nerve degeneration in some individuals, which can be fatal. The neurological complications produce sensory disturbances in the legs, such as burning, tingling, prickling, and numbness (collectively referred to as paresthesia).[2,50] Walking is difficult and balance is seriously affected. Many mental problems exist as well, such as loss of concentration and memory, disorientation, and dementia. As the condition worsens, bowel and bladder control is lost. Visual disturbances are common, too. There also are numerous GI tract problems, ranging from a sore tongue to constipation. The neurological complications often precede the development of anemia.

### Elevated Plasma Homocysteine Concentrations

Poor vitamin B-12, folate, and vitamin B-6 status can each result in high circulating levels of the amino acid homocysteine (see Fig. 13-20). Some individuals with high homocysteine levels in the blood are at a higher risk for heart attack and stroke. This may be because homocysteine can damage the endothelial cells lining arteries. Other research has found that high levels of plasma homocysteine also are associated with cognitive dysfunction and osteoporotic fractures.

Researchers have long known that supplementing the diet with folate, vitamin B-12, and vitamin B-6 can reduce blood levels of homocysteine. However, the evidence that supplementing with these vitamins can reduce the diseases associated with high blood levels of homocysteine is not strong. Several large studies that compared vitamin supplements with placebo treatments found that the supplements did not prevent heart disease, even though homocysteine levels declined.[53] In addition, it appears that supplements of vitamin B-12, vitamin B-6, and folate generally do not improve cognitive function, although this is a complex research area.[54] Even though B-vitamin supplements have not been shown to decrease the risk of heart disease or improve cognition, ample B-vitamin intakes are vital for normal physiological function and good health.

### Persons at Risk of Vitamin B-12 Deficiency

Poor vitamin B-12 status affects about 20% of older Americans. The causes of vitamin B-12 deficiency are listed in Table 13-2. In the elderly, most cases are due to atrophic gastritis.[51] This deficiency usually is not severe enough to

Older adults have an increased risk of poor vitamin B-12 status.

produce anemia, but it can cause neurological problems and elevated blood homocysteine. The consumption of crystalline vitamin B-12, either as a dietary supplement or in fortified foods (many breakfast cereals are fortified with vitamin B-12), can improve vitamin B-12 status in older persons with or without gastric atrophy.[55]

As noted earlier, those with a malabsorption syndrome of any kind have an increased need for vitamin B-12. For those diagnosed with a vitamin B-12 deficiency due to impaired absorption, 3 options are available: (1) monthly injections of vitamin B-12 to bypass the GI tract, (2) the use of a vitamin B-12 nasal gel, which also bypasses the GI tract, and (3) very high oral doses (1 to 2 mg) daily of vitamin B-12. A very small amount of vitamin B-12 can be absorbed by passive diffusion and does not require the intrinsic factor system. When very large amounts of vitamin B-12 are ingested, enough can be absorbed to meet vitamin B-12 requirements and correct vitamin B-12 deficiency.[55]

Vegetarians also can become vitamin B-12 deficient; vegans are at highest risk for deficiency because they consume no animal foods. However, if a person becomes a vegetarian in adulthood, vitamin B-12 stores in the liver can delay signs of a deficiency for several years. Infants born to or breastfed by vegetarian or vegan mothers also can develop vitamin B-12 deficiency, accompanied by anemia and long-term neurological problems, such as diminished brain growth, degeneration of the spinal cord, and poor intellectual development. Vegetarians have several options for obtaining vitamin B-12. If they are not vegans, they can obtain vitamin B-12 from diary products and eggs. In addition, vitamin B-12 supplements and food products fortified with vitamin B-12 are available.

### Knowledge Check

1. What roles do HCl, pepsinogen, intrinsic factor, and the ileum play in vitamin B-12 absorption?
2. How are vitamin B-12 and folate metabolism related?
3. How is the condition pernicious anemia related to vitamin B-12 deficiency?
4. Which foods are good sources of vitamin B-12?

# 13.10 Choline

Choline, recognized by the Institute of Medicine as an essential nutrient in 1998, is a relative newcomer to the list of essential nutrients.[2] It cannot be considered a B-vitamin because it does not have a coenzyme function, and the amount in the body is much greater than that of the B-vitamins. Choline can be obtained from the diet and can be synthesized in cells. However, synthesis alone is not sufficient to meet choline requirements, and liver and muscle damage can occur in those with low dietary intakes.[2,56]

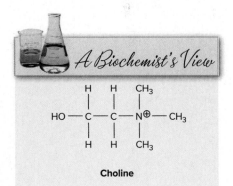

*A Biochemist's View*

**Choline**

## Choline in Foods

Choline is found in foods either as free choline or as part of another compound, such as phosphatidylcholine, also known as lecithin (see Chapter 6). Foods of animal origin (e.g., milk, eggs, chicken, beef, and pork) are large contributors of choline to the diet (Fig. 13-21). Grains, nuts, vegetables, and fruits also provide choline. Lecithins added to food during processing are yet another source.

## Choline Needs and Upper Level

The Adequate Intake for choline for adult men is 550 mg/day; for adult women, it is 425 mg/day. Although Adequate Intakes are set for choline, it may be that the choline requirement can be met by body synthesis at some or all stages of life. The *What We Eat in America* survey suggests that average choline intakes fall short—on average, adult males consume 402 mg per day and adult females consume 278 mg per day.[9] The Upper Level for adults is 3.5 g/day. Very high doses of choline have been associated with a fishy body odor (arising from a breakdown product), low blood pressure, vomiting, salivation, sweating, and GI tract effects.[2]

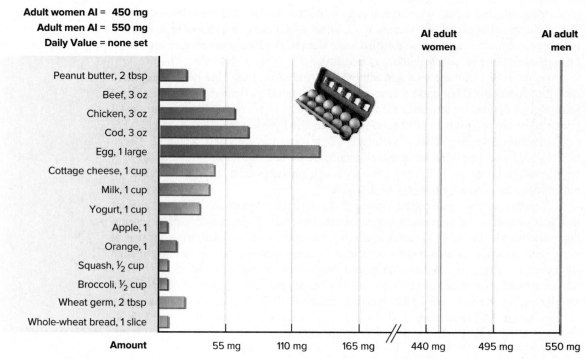

*Figure 13-21* Food sources of choline.

©Brand X Pictures/Getty Images RF

## Absorption, Transport, Storage, and Excretion of Choline

Choline is absorbed from the small intestine by way of transport proteins. The liver takes up choline rapidly from the blood delivered by the portal vein from the small intestine. All tissues contain some stores of choline. Choline is often oxidized to form the compound betaine, and both compounds are excreted in the urine.

## Functions of Choline

Choline is the precursor for a variety of compounds that enable cell membranes to form and that promote normal nervous system and liver function. It also is a source of methyl groups in the body.

- Choline is a component of the phospholipid molecules phosphatidylcholine and sphingomyelin. Phosphatidylcholine, also known as lecithin, is a component of cell membranes. Sphingomyelin is required for the synthesis of myelin, the substance that protects nerve fibers and facilitates transmission of electrical impulses. Because brain and nervous tissue growth is rapid during pregnancy and the 1st year of life, choline use and requirements are high during these periods.[57]
- Choline functions as a precursor for acetylcholine, a neurotransmitter associated with attention, learning, memory, muscle control, and many other functions.
- Choline helps prevent fat from accumulating in the liver because it is required for the synthesis and export of VLDL from the liver.
- Choline is an important source of methyl ($-CH_3$) groups required in many reactions in the body. Choline forms betaine, which can be used to form methionine from homocysteine. Supplementation with choline is another way to decrease homocysteine concentrations.

## Choline Deficiency

There is no deficiency disease associated with choline. However, liver and muscle damage has been observed in adults fed choline-deficient diets. Because of choline's critical roles in human metabolism, identifying and understanding the health effects of inadequate choline intake is an active area of research.

### Knowledge Check

1. Which foods are the best sources of choline?
2. Which organs are most affected by a choline-free diet?
3. Briefly describe 3 functions of choline.

# Take Action

## Energy Drinks with B-Vitamins

B-vitamins are sometimes marketed as a way to gain energy. Consider the following common scenario. Dan, age 20, is a junior in college with a full load of classes, a part-time job, and additional campus activities. He often feels tired and stressed and attributes this to his poor eating habits. Dan frequently misses meals and fills up on fast food and snacks. One of his friends suggests that he try a supplement drink to boost his energy, particularly when trying to study. Dan chooses an energy drink containing some B-vitamins and caffeine. According to the label, 1 container provides the following:

| | |
|---|---|
| Riboflavin | 100 mg |
| Niacin | 30 mg |
| Vitamin B-6 | 40 mg |
| Folic acid | 400 µg |
| Vitamin B-12 | 500 mg |
| Caffeine | 200 mg |

Compare the amounts provided in the supplement with the DRIs, including the UL for Dan. Do you think it will increase his energy? What if Dan decides to drink 2 supplements? Will the intake for any vitamin be over the UL? What other recommendations might you make to help Dan improve his energy and nutritional health? Learn more about caffeine and energy drinks at www.webmd.com/food-recipes /news/20121025/how-much-caffeine-energy-drink#1.

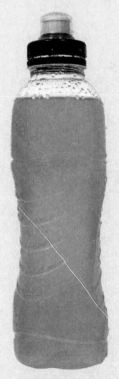

©Nitr/Shutterstock

Most animals can synthesize vitamin C from glucose. However, fruit bats, guinea pigs, gorillas, and humans do not have this ability and require a dietary source of vitamin C. Learn more at www.fao.org and www.ncbi.nlm.nih.gov/pmc/articles/PMC3145266.

Eric Gevaert/iStockphoto/Getty Images

# 13.11 Vitamin C

Most animals are able to synthesize vitamin C and, so, do not require a dietary source. However, humans and other primates, guinea pigs, fruit bats, and a few birds and fish are unable to synthesize this water-soluble vitamin. These animals rely on their diets for a vitamin C source.[58]

Vitamin C, also known as ascorbic acid, is involved in many processes in the human body, primarily as an electron donor. The term *vitamin C* actually refers not only to ascorbic acid but also to its oxidized form, dehydroascorbic acid. By adding or losing 2 hydrogens (in the boxed area of the structure), vitamin C undergoes reversible reduction and oxidation. Both forms of vitamin C are found in the foods we eat.

## Vitamin C in Foods

Most fruits and vegetables contain some vitamin C, but the richest sources are citrus fruits, peppers, and green vegetables (Fig. 13-22). Animal products and grains are generally not good sources. An intake of 5 servings per day of fruits and vegetables can provide ample vitamin C, depending on the foods chosen. However, according to the USDA, the leading fruits and vegetables consumed in the U.S. are white potatoes, tomatoes, lettuce, apples, and orange juice.[59] Although all of these foods contribute some vitamin C, only orange juice can be considered an excellent source of vitamin C.

Vitamin C, the least stable vitamin, is easily lost in storage, processing, and cooking. Normal cooking can decrease vitamin C content up to 40%, and potatoes stored for 5 months lose 50% of their vitamin C content.[3] This vitamin is very unstable when in contact with iron, copper, and oxygen. Juices are good foods to fortify with vitamin C because their acidity reduces vitamin C destruction.

## Vitamin C Needs

The RDA for vitamin C for adult men is 90 mg/day; for adult women, it is 75 mg/day;[3] and the Daily Value on food and supplement labels is 90 mg. For U.S. adults, average intakes

*A Biochemist's View*

**Ascorbic acid (reduced)**

**Dehydroascorbic acid (oxidized)**

Boxed area = Ascorbic acid (vitamin C) undergoes reversible oxidation and reduction by loss of 2 hydrogens (red) and 2 electrons (not shown).

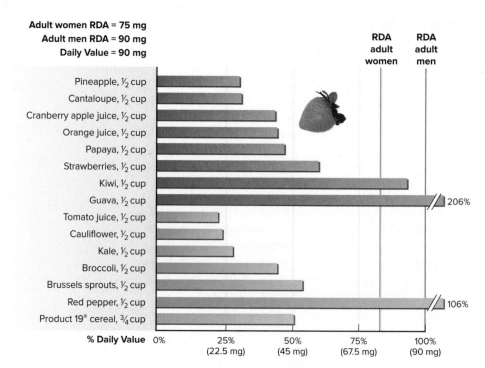

*Figure 13-22* Food sources of vitamin C. ©Ingram Publishing/Alamy RF

are close to the RDAs (83.3 mg for men and 75.1 mg for women),[9] but nearly 40% of adults consume less than the EAR for vitamin C. CDC data from 2003–2006 indicate that about 6% of persons ages 6 and older have vitamin C serum concentrations indicating a deficiency, and another 10 to 12% have concentrations that are low.[34] Vitamin C deficiency is more common in those with low incomes, as well as in young and middle-aged adults, compared with children, adolescents, and older adults, perhaps because many children and teens eat fortified cereals and many older adults take vitamin supplements.

Smokers have a higher requirement for vitamin C; the RDA increases by 35 mg/day for those who smoke. Smoking creates oxidative stress, which probably increases vitamin C turnover. Currently, about 17% of the U.S. adult population are smokers. More than 66% of adult smokers consume less than the EAR for vitamin C, and smokers' serum vitamin C levels are one-third lower than those of nonsmokers.[60] Thus, many Americans, especially smokers and those with reduced incomes, may need more vitamin C in their diets. Vitamin C needs rise in other situations, too. Women using oral contraceptive agents may require additional vitamin C. Individuals who have obesity may have low serum vitamin C concentrations; this may be due to a poor diet coupled with a more rapid use of the vitamin to battle the inflammation that can occur in obesity.[61] Vitamin C needs also increase in burn and trauma patients because collagen synthesis increases greatly when rebuilding new tissue. These patients often are provided an extra 500 to 1000 mg of vitamin C per day.

Citrus fruits are rich sources of vitamin C.

©D. Hurst/Alamy RF

## Upper Level for Vitamin C

The Upper Level for vitamin C is 2 g/day and is based on adverse gastrointestinal effects, such as bloating, stomach inflammation, and diarrhea.[3] High doses of vitamin C also can slightly increase the risk of kidney stone formation and excess iron absorption, but only in those who are predisposed to form kidney stones and who have preexisting iron absorption disorders. High doses of vitamin C can give false results in medical tests for blood in the stool. Persons taking large doses of vitamin C should discontinue the supplement before such tests. Informing your physician of your use of vitamin C and other nutrient supplements is a prudent practice.

## Absorption, Transport, Storage, and Excretion of Vitamin C

The absorption of vitamin C occurs in the small intestine by active transport (for ascorbic acid) and by facilitated diffusion (for dehydroascorbic acid). The efficiency of the absorptive mechanism decreases as intake increases. About 70 to 90% of vitamin C is absorbed at daily intakes between 30 and 200 mg, whereas the rate of absorption declines substantially with doses exceeding that amount. Excretion by the kidneys rises as dietary intake increases.[58]

The amount of vitamin C stored varies widely by tissue. High concentrations are found in the pituitary and adrenal glands, white blood cells, eyes, and brain. The lowest concentrations are in the blood and saliva.

## Functions of Vitamin C

Vitamin C performs a variety of important cell functions. It does so primarily by donating electrons in oxidation-reduction reactions.[58] As mentioned previously, substances that donate electrons become oxidized (lose electrons). As an electron donor, vitamin C has a cofactor role for several metalloenzymes and has antioxidant defense functions. **Metalloenzymes** contain metals, such as iron, copper, or zinc (usually as an ion), as a part of their structures. When a metalloenzyme catalyzes a reaction, the metal ion becomes oxidized. For example, reduced iron (ferrous iron, $Fe^{2+}$) is converted to its oxidized form (ferric iron, $Fe^{3+}$) during enzymatic activity. Ascorbic acid, by donating an electron to the oxidized iron, keeps the iron in its reduced ferrous form. This, in turn, allows enzymatic action to continue.

## HISTORICAL PERSPECTIVE

### A Treasure Chest Full of Vitamin C

Vitamin C, the substance that prevents scurvy, eluded us until about 90 years ago, when chemist Albert Szent-Györgyi was studying oxidation. He found a compound that loses and regains hydrogen atoms and later established that it prevents scurvy. To continue his work, he searched for vitamin C rich foods that could be easily purified. One evening while living in Hungary, the paprika capital of the world, Szent-Györgyi didn't feel hungry, so he took the fresh paprika he was served for dinner to his lab and within hours knew he had found "a treasure chest full of vitamin C." He also identified the proteins responsible for muscle contraction and demonstrated that ATP is the immediate source of energy for muscle contraction. Learn more about this Nobel Prize winner at www.nobelprize.org/nobel_prizes/medicine/laureates/1937/szent-gyorgyi-bio.html.

Both: ©Foodcollection RF

*Figure 13-23* Collagen synthesis requires vitamin C. **1** Vitamin C is needed for the addition of hydroxyl groups (–OH) to the amino acids proline and lysine in collagen molecules. **2** Hydroxyproline and hydroxylysine stablize the collagen fibers. **2** Without sufficient vitamin C available to perform this task, only weak connective tissue is formed.

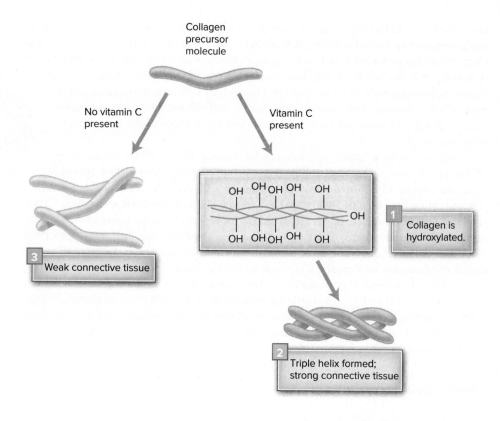

Collagen precursor molecule

No vitamin C present

Vitamin C present

OH OH OH OH OH
OH
OH OH OH OH OH

**1** Collagen is hydroxylated.

**3** Weak connective tissue

**2** Triple helix formed; strong connective tissue

**connective tissue** Cells and their protein products that hold different structures in the body together. Tendons and cartilage are composed largely of connective tissue. Connective tissue also forms part of bone and the nonmuscular structures of arteries and veins.

### Collagen Synthesis

**Collagen** is the major fibrous protein that holds together the various structures of the body. It is very strong and elastic. Collagen fibers are important constituents of **connective tissue,** such as tendons and ligaments, and they are found in bone, blood vessels, eyes, and skin. Collagen fibers also are essential in wound healing. A collagen molecule is like a 3-stranded rope—it consists of 3 polypeptide chains wound together to form a triple helix. Vitamin C is needed to get the 3 strands in the right shape to form the triple helix. Specifically, vitamin C works with the metalloenzymes prolyl and lysyl hydroxylase to convert the amino acids proline and lysine in collagen to hydroxyproline and hydroxylysine. These unusual amino acids (rarely found in other tissues) stabilize the collagen structure (Fig. 13-23). Vitamin C, as a reducing agent, acts to keep the iron in the metalloenzymes in the reduced $Fe^{2+}$ form required for enzyme action. Without vitamin C, hydroxyproline and hydroxylysine cannot be synthesized, and collagen is weak and fragile.[58]

### Synthesis of Other Vital Compounds

Vitamin C is required for the synthesis of many important biological compounds. These compounds include the amino acid tyrosine, the hormone thyroxine, the fatty acid transport compound carnitine, and the neurotransmitters norepinephrine, epinephrine, and serotonin. In addition, vitamin C is involved in the conversion of cholesterol to bile acids and biosynthesis of the hormones corticosteroids and aldosterone. In each case, vitamin C keeps the copper or iron in the metalloenzyme in the reduced state (as $Cu^+$ or $Fe^{2+}$).

### Antioxidant Activity

As an electron donor, vitamin C is an important water-soluble physiological antioxidant. By donating electrons to free radicals (molecules with unpaired electrons), vitamin C inactivates free radicals, thereby protecting various cellular compounds (including DNA, proteins, and lipids) from free radical–induced damage. For example, high concentrations of vitamin C in the eye may protect its tissues from free radicals produced by UV-light exposure.[62] Likewise, high concentrations in **neutrophils,** a type of white blood cell, may provide protection against the free radicals produced during immune functions.[58] Another important antioxidant function of vitamin C is to regenerate the active form of vitamin E and make it function more effectively. Many nutrition scientists believe that these antioxidant effects help prevent certain diseases, such as cancer and cardiovascular disease; however, some research suggests that vitamin C can increase oxidative stress, such as in people with diabetes.

### Iron Absorption

Vitamin C with meals modestly facilitates the intestinal absorption of nonheme iron (iron that is not in hemoglobin) because of the conversion of iron in the GI tract to ferrous iron ($Fe^{2+}$). Vitamin C also counters the action of certain food components that inhibit iron absorption.[3]

### Immune Function

White blood cells, part of the body's immune defenses, contain the highest vitamin C concentration of all body constituents. This may protect against the oxidative damage associated with cellular respiration. Free radicals generated during phagocytosis and neutrophil activation, although intended to kill bacteria or damaged tissue, also can damage the body's own immune cells. Vitamin C may reduce this self-destruction through its antioxidant defense actions. Vitamin C also may have other roles in immune function; however, supplemental vitamin C levels beyond the body's needs may not improve immune function.

## Vitamin C Deficiency

A deficiency of vitamin C prevents the normal synthesis of collagen, thus causing widespread, significant changes in connective tissue throughout the body. The first signs and symptoms of scurvy, the vitamin C deficiency disease, appear after about 20 to 40 days on a diet free of vitamin C and include fatigue and pinpoint hemorrhages around hair follicles (Fig. 13-24). These hemorrhages are the most characteristic sign of scurvy. In addition, the gums and joints bleed, a classic sign of connective tissue failure. Other effects of scurvy include impaired wound healing, bone pain, fractures, and diarrhea. Psychological problems, such as depression, are common in advanced scurvy. Scurvy is fatal if not treated.[3,58]

Worldwide, scurvy is associated with poverty. It is especially common in infants who are fed boiled milk (all forms of milk are poor sources of vitamin C) and not given a good food source of vitamin C or a supplement. Although scurvy is considered rare in North America, poor vitamin C status is relatively common, as mentioned previously. Those with poor vitamin C status may have nonspecific symptoms, including fatigue, irritability, gum bleeding, and joint and muscle pain.[3,58] Smokers and those with alcoholism or poor diets are at greatest risk.

▶ Nowadays, scurvy is a rare disease, but those who eat very limited diets can develop this condition. For example, an 8-year-old boy developed a rash, bleeding gums, bone pain, and weakness—all symptoms of scurvy—after consuming a low vitamin C diet consisting of spaghetti, meatballs, and potato chips. Two other young children, eating only yogurt in 1 case and sugar-sweetened cereal, ice cream, and crackers in the other case, also developed scurvy. These children were also diagnosed with autism or Asperger syndrome. Children with neurodevelopmental disorders such as these sometimes develop restrictive eating patterns that may put them at risk of scurvy and other nutrient deficiencies[63] (see Chapter 17).

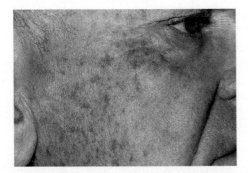
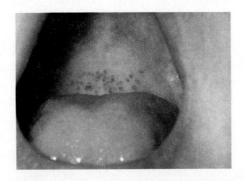

*Figure 13-24* Pinpoint hemorrhages of the skin—an early symptom of scurvy. The spots on the skin are caused by slight bleeding. The person may experience poor wound healing. These are all signs of defective collagen synthesis.

Both: ©Dr P. Marazzi/Science Source

Oranges are a rich source of vitamin C. The many phytochemicals they provide are an additional benefit and are not found in vitamin C supplements.

©Vanatchanan/Shutterstock, Inc.

### Vitamin C, Cancer, and Heart Disease

Because of its roles as an antioxidant and in promoting normal immune function, a great deal of research has examined vitamin C's ability to prevent both cancer and heart disease. For cancer, the evidence is best for cancers of the mouth, esophagus, stomach, and lung, but not all studies are positive and it is still not known if vitamin C, either in the diet or as a supplement, provides protection. For example, a study in men found that 500 mg of supplemental vitamin C per day did not reduce the risk of prostate cancer or overall cancer occurrence.[64] Many researchers believe that a healthy diet, along with a healthy lifestyle, affords the best cancer prevention. The value of using vitamin C as part of the treatment is not clear; some evidence suggests that its usefulness improves when administered into a vein (intravenously) rather than orally. However, in general, the benefits of high doses of vitamin C for cancer treatment are not proven.[65]

The situation is similar for vitamin C and cardiovascular disease. Many (but not all) studies suggest that good vitamin C status provides some protection against heart disease and stroke.[66] This protection may be the result of vitamin C's role as an antioxidant and the association between vitamin C rich diets and reduced blood pressure.[67] However, clinical

**Table 13-3  A Summary of Water-Soluble Vitamins and Choline**

| Vitamin | Major Functions | Deficiency Symptoms | People Most at Risk |
|---|---|---|---|
| Thiamin | Coenzyme in carbohydrate metabolism and energy release | Beriberi: anorexia, weight loss, weakness, peripheral neuropathy; Wernicke-Korsakoff syndrome | Those with alcoholism, people with very low incomes |
| Riboflavin | Coenzyme in numerous oxidation-reduction reactions, including those of energy release | Ariboflavinosis: inflammation of mouth and tongue, cracks at corner of mouth | People taking certain medications if no dairy products are consumed |
| Niacin | Coenzyme in numerous oxidation-reduction reactions in energy metabolism, synthesis and breakdown of fatty acids | Pellagra: diarrhea, dermatitis, dementia (death) | Those with alcoholism, people with very low incomes living in regions where corn is the dominant food |
| Pantothenic acid | Coenzyme in energy metabolism and fatty acid synthesis | Weakness, fatigue, impaired muscle function, GI tract disturbances; deficiency very rare | None |
| Biotin | Cofactor for carboxylase enzymes that participate in fatty acid, amino acid, and energy metabolism | Dermatitis, conjunctivitis, hair loss, nervous system abnormalities; deficiency very rare | Infants with a certain genetic defect |
| Vitamin B-6 (pyridoxine) | Coenzyme in amino acid metabolism, heme synthesis, lipid metabolism; homocysteine metabolism | Dermatitis, anemia, convulsions, depression, confusion | Those with alcoholism, those taking certain medications |
| Folate | Coenzyme in DNA synthesis, homocysteine metabolism | Megaloblastic (macrocytic) anemia, birth defects | Those with alcoholism, pregnant women, people on certain medications |
| Vitamin B-12 (cobalamin) | Coenzyme affecting folate metabolism, homocysteine metabolism | Megaloblastic (macrocytic) anemia, paresthesia, pernicious anemia | Older adults, vegans, patients with malabsorption syndromes |
| Vitamin C (ascorbic acid) | Collagen synthesis, some antioxidant capability, hormone and neurotransmitter synthesis, oxidation-reduction reactions | Scurvy: poor wound healing, pinpoint hemorrhages, bleeding gums | Those with alcoholism, smokers, individuals who eat few fruits and vegetables |
| Choline | Precursor for acetylcholine and phospholipids, homocysteine metabolism | Fatty liver, muscle damage | Older adults with very low choline intakes |

trials of supplements of vitamins C and E, another antioxidant, have been disappointing. Numerous scientific organizations, including the American Heart Association, have concluded that antioxidant supplements do not reduce the risk of heart disease.

## Vitamin C Intake above the RDA

Some popular authors and speakers advocate the consumption of vitamin C at amounts higher than the RDA. Surprisingly, there is little research comparing different vitamin C intake levels. If vitamin C intake is above about 100 mg/day, much of the additional vitamin C is excreted in the urine. Some research indicates that 200 mg/day is the highest amount needed to maximize the health benefits of vitamin C intake. Eating several vitamin C rich foods each day can boost intakes to 200 mg/day.

### Vitamin C and the Common Cold

One aspect of high vitamin C intake (up to 1000 mg/day) that has drawn a lot of attention is its possible use in preventing or treating the common cold. Almost 40 studies suggest that

*Critical* **Thinking**

Carlos just saw an advertisement claiming that vitamin C will cure just about everything from colds to heart disease. How would you explain to him vitamin C's main functions in the human body?

| Dietary Sources | RDA or Adequate Intake | Toxicity* |
|---|---|---|
| Pork and pork products, enriched and whole-grain cereals, eggs, nuts, legumes | Men: 1.2 mg/day; women: 1.1 mg/day | None recognized |
| Milk and milk products, mushrooms, eggs, liver, enriched grains | Men: 1.3 mg/day; women: 1.1 mg/day | None recognized |
| Meat, poultry, fish, enriched and whole-grain breads and cereals, tryptophan conversion to niacin | Men: 16 mg NE/day; women: 14 mg NE/day | Flushing of skin; Upper Level for adults is 35 mg/day from supplements, based on flushing of skin |
| Widely distributed in foods | Adequate Intake for adults: 5 mg/day | None recognized |
| Nuts, seeds, fish, whole grains, many other foods | Adequate Intake for adults: 30 µg/day | Unknown |
| Animal protein foods, potatoes, bananas, legumes, avocados | Adults 19–50: 1.3 mg/day; men over 50: 1.7 mg/day; women over 50: 1.5 mg/day | None from food but excess intake from supplements causes neuropathy, skin lesions; Upper Level is 100 mg/day, based on nerve destruction |
| Green vegetables, liver, enriched cereal products, legumes, oranges | 400 µg/day of dietary folate equivalents | None; Upper Level for adults set at 1000 µg/day for synthetic folic acid, exclusive of food folate, based on masking vitamin B-12 deficiency |
| Animal foods; fortified, ready-to-eat breakfast cereals | Adults 19–50: 2.4 µg/day; adults 51 and older: same, but use of fortified foods or supplements to meet needs is recommended | None recognized |
| Citrus fruits, papayas, strawberries, broccoli, potatoes, greens | Men: 90 mg/day; women: 75 mg/day; smokers need an additional 35 mg/day | Diarrhea and other GI tract problems; Upper Level is 2 g/day, based on development of diarrhea |
| Eggs, meat, fish, milk, wheat germ, self-synthesis | Adequate Intake for men: 550 mg/day; for women: 425 mg/day | Upper Level is 3.5 g/day, based on development of fishy body odor and reduced blood pressure |

*Toxicity arises only from supplement use.

any beneficial effect of vitamin C supplementation is very modest, perhaps reducing cold duration by 1 day per year in adults. For individuals exposed to extreme physical stress, such as marathon runners and cross-country skiers, vitamin C supplementation reduced the risk of contracting a cold after the physical stress. Despite its popularity, supplemental vitamin C cannot be recommended to prevent or treat common colds in most individuals.[68]

As you can see, the water-soluble vitamins and choline play important roles in the body. Table 13-3 provides a summary of the water-soluble vitamins and choline.

### Knowledge Check

1. How does vitamin C aid in the function of metalloenzymes in cells?
2. What are 4 good sources of vitamin C?
3. Impaired synthesis of what compound is responsible for the symptoms of scurvy, such as bleeding gums, easy bruising, and pinpoint hemorrhages?

## Take Action

### Spotting Fraudulent Claims for Vitamins and Vitamin-like Substances

Using your favorite search engine, search terms such as *energy boost, stress, mental acuity,* and *disease prevention* along with *vitamin supplement.* Identify several claims for vitamin supplements that you consider fraudulent or misleading.

1. What is the cost of these supplements?
2. How do the amounts of vitamins they contain compare with the DRIs?
3. What compounds are included that are not considered essential nutrients?
4. What disclaimers or warnings are made about the products?
5. How likely is the supplement to improve a person's health?

©Mike Kemp/Rubberball/Getty Images RF

### CASE STUDY FOLLOW-UP

©Corbis/VCG/Getty Images RF

Suzanne and Ted should remember that spina bifida is caused by a failure of the spinal cord to close during the first 28 days of pregnancy, a time when neither Ted nor Suzanne may realize that Suzanne is pregnant. The B-vitamin folate must be available at the time of conception to prevent spina bifida and other birth defects. The fact that a close relative of Suzanne's has already produced a child with this birth defect should be a warning sign. Suzanne and Ted would be wise to seek the advice of a registered dietitian nutritionist to ensure that Suzanne's prepregnancy diet provides enough synthetic folic acid.

# Chapter *Summary*

**13.1** **The water-soluble vitamins are the 8 B-vitamins and vitamin C.** A dietary source of choline also is required. The B-vitamins function as coenzymes in hundreds of metabolic reactions and are especially important in energy metabolism. The water-soluble vitamins are generally stored in the body to a lesser extent than the fat-soluble vitamins. Compared with fat-soluble vitamins, water-soluble vitamins are more easily destroyed during storage, processing, and cooking. Refined cereals and grains are enriched with thiamin, riboflavin, niacin, and folic acid.

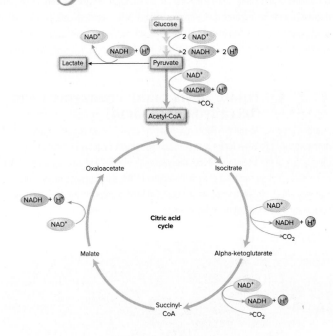

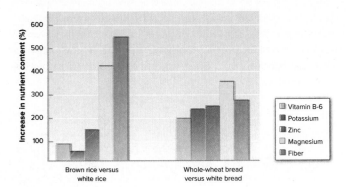

**13.2** **Thiamin in its functional form as TPP serves as a coenzyme in energy release.** Thiamin deficiency results in the disease beriberi. In North America, alcoholics are at risk of thiamin deficiency. Pork and enriched grains are reliable sources of thiamin.

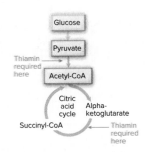

**13.3** **Riboflavin coenzymes, FAD, and FMN** participate in a wide variety of oxidation-reduction reactions, including those in numerous metabolic pathways that produce energy. A specific riboflavin deficiency is unlikely but can accompany other B-vitamin deficiencies. Dairy products and enriched grains are good dietary sources.

**13.4** **Niacin forms the coenzymes NAD⁺ and NADP⁺.** NAD⁺ is important in oxidation-reduction reactions in energy-yielding pathways. A deficiency of the vitamin produces the disease pellagra, which is seen most often in corn-based diets. Alcoholism can lead to a deficiency. Food sources of niacin are enriched cereal grains and protein foods. The body is able to synthesize the vitamin from the amino acid tryptophan. The UL for niacin is based on the flushing seen with high doses.

**13.5** **Among its functions, pantothenic acid in coenzyme form (CoA) shuttles 2 carbon fragments** from the metabolism of glucose, amino acids, fatty acids, and alcohol into the citric acid cycle during energy metabolism. A deficiency of pantothenic acid is unlikely because it is widely distributed in foods.

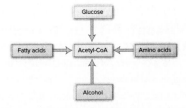

**13.6** **Biotin functions as a cofactor in enzymes that add carbon dioxide to a substance.** Biotin is widely distributed in foods. Intestinal bacteria also synthesize biotin. No deficiency exists in healthy people.

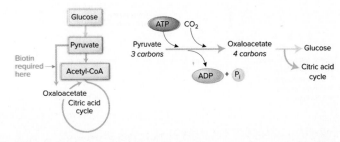

# 13.7 The vitamin B-6 coenzyme PLP participates in amino acid metabolism,
especially the synthesis of nonessential amino acids. It is essential in the synthesis of heme in hemoglobin, the formation of certain neurotransmitters, and the metabolism of homocysteine. Anemia, convulsions, and decreased immune response are symptoms of a deficiency, although deficiency is rare. Animal protein foods, a few fruits and vegetables, and whole-grain cereals are good sources of this vitamin. One of the toxic effects of excess consumption is nerve damage.

# 13.8 Folate in its many coenzyme forms (tetrahydrofolic acid)
accepts and donates 1-carbon groups. The most notable function performed by folate is DNA synthesis. Along with vitamin B-12, folate also participates in homocysteine metabolism. A dietary lack of this vitamin produces megaloblastic anemia and increases the risk of spina bifida. Folate contributes to methylation of DNA, an epigenetic modification that may affect cancer development. Folate is found in green vegetables, legumes, liver, oranges, and fortified cereal grains.

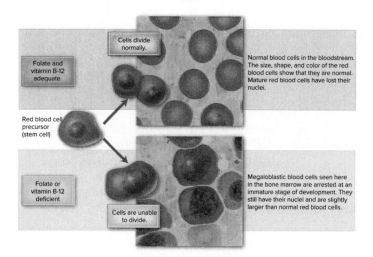

# 13.9 Vitamin B-12 in its coenzyme form transfers 1-carbon groups.
Because of its interaction with folate, a deficiency of vitamin B-12 results in the same type of megaloblastic anemia, as well as excess homocysteine in the blood. Pernicious anemia, an autoimmune disorder, is a cause of vitamin B-12 deficiency. Other causes are poor diet, malabsorption, and certain drugs. In such cases, injection of the vitamin or another pharmacologic approach is necessary. Vitamin B-12 is found in animal foods, but not in plant foods. Vegans need to look for foods fortified with the vitamin or take it as part of a multivitamin and mineral supplement.

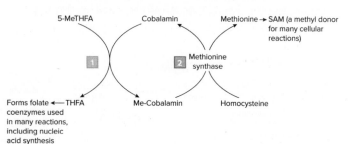

# 13.10 Eggs, meat, milk, and fish are good sources of choline,
but it also is found in other foods and is synthesized in the body. Choline provides methyl groups required in many reactions in the body. It is required for the formation of the neurotransmitter acetylcholine and is incorporated into phospholipids. Individuals fed choline-deficient diets may develop liver and muscle damage; however, choline deficiency is rare.

# 13.11 Vitamin C functions as an electron donor,
acting to keep metal cofactors such as iron and zinc in their reduced forms. As such, many compounds, including collagen, several neurotransmitters, and carnitine, require vitamin C for their synthesis. Vitamin C also is an antioxidant able to halt free radical formation. A deficiency of vitamin C causes the disease scurvy. Fresh fruits and vegetables are reliable sources of this vitamin. Like folate, vitamin C is destroyed by heat. Smokers have a higher vitamin C requirement. Among North Americans, those with alcoholism, smokers, and individuals who do not eat many fruits or vegetables are most likely to develop a deficiency.

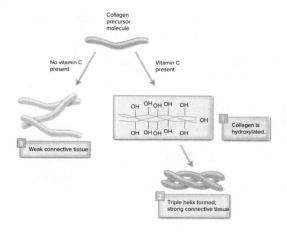

## Clinical Perspectives

The cancer drug methotrexate takes advantage of the key role of THFA in DNA synthesis. Methotrexate, referred to as a folate antagonist, interferes with THFA metabolism. This, in turn, reduces DNA synthesis throughout the body, which can halt the growth of cancer cells.

A maternal deficiency of folate and a genetic predisposition have been linked to the development of neural tube defects in the fetus. Folate is critical to normal neural tube development. All women capable of becoming pregnant should consume 400 μg of folic acid daily.

# Study Questions

1. When compared with whole-grain products, enriched cereals and grains provide the same or higher amounts of all the B-vitamins.

   a. true
   b. false

2. Thiamin pyrophosphate (TPP) is required for _____.

   a. protein synthesis
   b. fatty acid synthesis
   c. carbohydrate metabolism
   d. DNA synthesis

3. Thiamin deficiency can be found among _____.

   a. heavy users of alcohol
   b. poor people in developing regions reliant on white rice as a staple food
   c. poor people in low-income regions reliant on corn as a staple food
   d. both a and b
   e. both a and c

4. Which of the following water-soluble vitamins participate in oxidation-reduction reactions?

   a. thiamin, riboflavin, and niacin
   b. folate, vitamin B-12, and vitamin B-6
   c. biotin, pantothenic acid, and niacin
   d. vitamin C, riboflavin, and niacin

5. An individual who consumes no dairy products is most at risk of developing _____ deficiency.

   a. choline
   b. vitamin B-6
   c. riboflavin
   d. thiamin

6. Niacin can be synthesized in cells from _____.

   a. riboflavin
   b. fatty acids
   c. glucose
   d. tryptophan

7. Deficiencies of _____ and _____ are extremely rare.

   a. biotin; pantothenic acid
   b. vitamin C; niacin
   c. vitamin B-12; folate
   d. vitamin C; folate

8. Transamination reactions allow the formation of nonessential amino acids. Which vitamin is required for these reactions?

   a. folate
   b. vitamin B-12
   c. riboflavin
   d. vitamin B-6

9. The absorption of folic acid in supplements and fortified foods exceeds that of folate found in foods.

   a. true
   b. false

10. Good sources of folate include _____.

    a. lentils, spinach, asparagus, and fortified foods
    b. papayas, limes, oranges, and potatoes
    c. dairy products, fortified foods, and nuts
    d. tuna, chicken, beef, and dairy products

11. The prevention of neural tube defects is best achieved by _____.

    a. good folate status prior to becoming pregnant
    b. folic acid supplementation in the 2nd half of pregnancy
    c. folic acid supplementation during infancy
    d. all of the above

12. Macrocytic anemia, peripheral neuropathy, and impaired cognitive function are signs of _____ deficiency.

    a. ascorbic acid
    b. niacin
    c. vitamin B-12
    d. vitamin B-6

13. Sources of vitamin B-12 include _____.

    a. whole grains, tuna, and eggs
    b. dairy products, meat, and fish
    c. citrus fruits, papayas, and bananas
    d. dairy products, whole grains, and leafy green vegetables

14. Vitamin C is required for the formation of _____, required to synthesize collagen.

    a. tryptophan
    b. serotonin
    c. hydroxyproline
    d. acetyl-CoA

15. Ingesting 1000 mg of supplemental vitamin C per day has been proven to prevent common colds.

    a. true
    b. false

16. Define and explain the term *coenzyme*.

17. Explain why individual B-vitamin deficiencies are rare in the U.S. Which B-vitamins are added to enriched breads and cereals?

18. Draw a map of the main catabolic metabolic pathways described in the chapter, and identify where B-vitamins participate in energy metabolism.

19. Draw MyPlate, and place each B-vitamin, choline, and vitamin C into the food groups where they are most likely to be found.

**20.** Explain the relationships among folate, methotrexate, DNA synthesis, and cancer.

**21.** Describe how women capable of becoming pregnant can ensure they consume sufficient folic acid.

Banana: ©Maksym Narodenko/123RF; Spinach: ©Ingram Publishing RF

# References

1. Rosenfeld L. Vitamine-vitamin. The early years of discovery. *Clin Chem.* 1997;43:680.

2. Food and Nutrition Board, Institute of Medicine. *Dietary Reference Intakes for thiamin, riboflavin, niacin, vitamin B-6, folate, vitamin B-12, pantothenic acid, biotin, and choline.* Washington, DC: National Academies Press; 1998.

3. Food and Nutrition Board, Institute of Medicine. *Dietary Reference Intakes for vitamin C, vitamin E, selenium, and carotenoids.* Washington, DC: National Academies Press; 2000.

4. Lester GE, Makus DJ. Relationship between fresh-packaged spinach leaves exposed to continuous light or dark and bioactive contents: Effects of cultivar, leaf size, and storage duration. *J Agric Food Chem.* 2010;58:2980.

5. Shorter KR and others. Consequences of dietary methyl donor supplements: Is more always better. *Prog Biophy Molec Bio.* 2015;118:14.

6. Fulgoni III VL and others. Foods, fortificants and supplements: Where do Americans get their nutrients? *J Nutr.* 2011;141:1847.

7. O'Neil CE and others. Food sources of energy and nutrients among adults in the US: NHANES 2003–2006. *Nutrients.* 2012;4:2097.

8. Be'meur C, Butterworth F. Thiamin. In: Ross C and others, eds. *Modern nutrition in health and disease.* 11th ed. Philadelphia: Lippincott Williams & Wilkins; 2014.

9. U.S. Department of Agriculture, Agricultural Research Service. Nutrient intakes from food and beverages: Mean amounts consumed per individual, by gender and age. *What We Eat in America,* NHANES 2013–2014. 2016; www.ars.usda.gov/northeast-area/beltsville-md/beltsville-human-nutrition-research-center/food-surveys-research-group/docs/wweia-data-tables/.

10. Polegato BF and others. Role of thiamin in health and disease. *Nutr Clin Pract.* 2019;34:558.

11. Whitfield KC and others. Poor thiamin and riboflavin status is common among women of childbearing age in rural and urban Cambodia. *J Nutr.* 2015;145:628.

12. Densupsoontorn N and others. Prevalence of and factors associated with thiamin deficiency in obese Thai children. *J Clin Nutr.* 2019;1:116.

13. Pourhassan M and others. Prevalence of thiamine deficiency in older hospitalized patients. *Clin Inter Aging.* 2018;13:2247.

14. Donnelly A. Wernicke-Korsakoff syndrome: Recognition and treatment. *Nursing Standard.* 2017;31:46.

15. Dainty JR and others. Quantification of the bioavailability of riboflavin from foods by us of stable-isotope labels and kinetic modeling. *Am J Clin Nutri.* 2007;85:1557

16. Said HM, Ross CA. Riboflavin. In: Ross C and others, eds. Modern nutrition in health and disease. 11th ed. Philadelphia: Lippincott Williams & Wilkins; 2014.

17. Shaik MM, Gan SH. Vitamin supplementation as a possible prophylactic treatment against migraine with aura and menstrual migraine. *BioMed Res Int.* 2015;2015:Article ID 469529.

18. Aljaadi AM and others. Suboptimal biochemical riboflavin status is associated with lower hemoglobin and higher rates of anemia in a sample of Canadian and Malaysian women of reproductive age. *J Nutr.* 2019;00:1.

19. Park Y and others. Effectiveness of food fortification in the United States: The case of pellagra. *Am J Publ Health.* 2000;90:727.

20. Kirkland JB. Niacin. In: Ross C and others, eds. *Modern nutrition in health and disease.* 11th ed. Philadelphia: Lippincott Williams & Wilkins; 2014.

21. Kraut A. Dr. Joseph Goldberger & the war on pellagra. history.nih.gov/exhibits/goldberger.

22. Crook MA. The importance of recognizing pellagra (niacin deficiency) as it still occurs. *Nutr.* 2014;30:729.

23. The HPS2-Thrive Collaborative Group. Effects of extended release niacin with laropiprant in high risk patients. *N Engl J Med.* 2014;371:203.

24. Trumbo T. Pantothenic acid. In: Ross C and others, eds. *Modern nutrition in health and disease.* 11th ed. Philadelphia: Lippincott Williams & Wilkins; 2014.

25. Mock D. Biotin. In: Ross C and others, eds. *Modern nutrition in health and disease.* 11th ed. Philadelphia: Lippincott Williams & Wilkins; 2014.

26. Genetics Home Reference. Biotinidase deficiency. 2016; ghr.nlm.nih.gov/condition/biotinidase-deficiency.

27. Canda E and others. Single center experience of biotinidase deficiency: 259 patients and six novel mutations. *J Pediatr Endocrinol Metab.* 2018;31:917.

28. Da Silva VR and others. Vitamin B6. In: Ross C and others, eds. *Modern nutrition in health and disease.* 11th ed. Philadelphia: Lippincott Williams & Wilkins; 2014.

29. Stover PJ, Field MS. Vitamin B-6. *Adv Nutr.* 2015;6:132.

30. Gylling B and others. Vitamin B-6 and colorectal cancer risk: A prospective population-based study using 3 distinct plasma markers of vitamin B-6 status. *Am J Clin Nutr.* 2017;105:897.

31. Pusceddu I and others. Subclinical inflammation, telomere shortening, homocysteine, vitamin B6, and mortality: The Ludwigshafen risk and cardiovascular health study. 2019. https://doi.org/10.1007/s00394-019-01993-8.

32. Nix WA and others. Vitamin B status in patients with type 2 diabetes mellitus with and without incipient nephropathy. *Diabetes Res Clin Pract*. 2015;107:157.

33. Christen WG and others. Effect of combined treatment with folic acid, vitamin B6 and vitamin B12 on plasma biomarkers of inflammation and endothelial dysfunction in women. *J Am Heart Assoc*. 2018;7:1.

34. Centers for Disease Control and Prevention. Second national report on biochemical indicators of diet and nutrition in the U.S. population. 2012; www .cdc.gov/nutritionreport/.

35. McParlin C and others. Treatments for hyperemesis gravidarum and nausea and vomiting in pregnancy. A systematic review. *JAMA*. 2016;316:1392.

36. Stover PJ. Folic acid. In: Ross C and others, eds. *Modern nutrition in health and disease*. 11th ed. Philadelphia: Lippincott Williams & Wilkins; 2014.

37. Winkels R and others. Bioavailability of food folates is 80% of that of folic acid. *Am J Clin Nutr*. 2007;85:465.

38. Bibbins-Domingo K and others. Folic acid supplementation for the prevention of neural tube defects: US Preventive Services Task Force recommendation statement. *JAMA*. 2017;317:183.

39. Bailey R and others. Total folate and folic acid intake from foods and dietary supplements in the United States: 2003–2006. *Am J Clin Nutr*. 2010;91:231.

40. Shea B and others. Folic acid and folinic acid for reducing side effects in patients receiving methotrexate for rheumatoid arthritis. *Cochrane DB Syst Rev*. 2013;5:Art. No. CD000951. doi:10.1002/14651858.CD000951.pub2.

41. Naski FH and others. Folate: Metabolism, genes, polymorphisms and the associated diseases. *Gene*. 2014;533:11.

42. Sarris J and others. Adjunctive nutraceuticals for depression: A systematic review and meta-analyses. *Am J Psychiatry*. 2016;173:575.

43. Huo Y and others. Efficacy of folic acid therapy in primary prevention of stroke among adults with hypertension in China. The CSPPT randomized clinical trial. *JAMA*. 2015;313:1325.

44. Odewole O and others. Near-elimination of folate-deficiency anemia by mandatory folic acid fortification in older US adults: Reasons for Geographic and Racial Differences in Stroke Study 2003–2007. *Am J Clin Nutr*. 2013;98:1042.

45. Colapinto CK and others. Folate status of the population in the Canadian Health Measures Survey. *CMAJ*. 2011;183:E100.

46. Pfeiffer CM and others. Folate status in the US population 20 y after the introduction of folic acid fortification. *Am J Clin Nutr*. 2019;110:1088.

47. Williams J and others. Updated estimates of neural tube defects prevented by mandatory folic acid fortification—United States, 1995–2011. *MMWR*. 2015;64:1.

48. Barron S. Anecephaly: An ongoing investigation in Washington State. *Am J Nurs*. 2016;116:60.

49. Gelineau-van Waes J and others. Maternal fumonisin exposure as a risk factor for neural tube defects. *Adv Food Nutr Res*. 2009;56:145.

50. Carmel R. Cobalamin (vitamin B-12). In: Ross C and others, eds. *Modern nutrition in health and disease*. 11th ed. Philadelphia: Lippincott Williams & Wilkins; 2014.

51. Langan RC, Goodbred AJ. Vitamin B12 deficiency: Recognition and management. *Am Fam Physician*. 2017;96:384.

52. Lael R and others. Association of biochemical B12 deficiency with metformin therapy and vitamin B12 supplements. The National Health and Nutrition Examination Survey, 1999–2006. *Diabetes Care*. 2012;35:327.

53. Marti-Carvajal A and others. Homocysteine lowering interventions for preventing cardiovascular events. *Cochrane DB Syst Rev*. 2015;1:Art. No. CD006612.

54. Zhang C and others. Vitamin B12, B6, or folate and cognitive function in community-dwelling older adults: A systematic review and meta-analysis. *J Alz Dis*. 2020;77:781.

55. Shipton MJ, Thachil J. Vitamin B12 deficiency—A 21st century perspective. *Clin Med*. 2015;15:145.

56. Zeisel S. Choline. In: Ross C and others, eds. *Modern nutrition in health and disease*. 11th ed. Philadelphia: Lippincott Williams & Wilkins; 2014.

57. Wallace T and others. Choline: The underconsumed and underappreciated essential nutrient. *Nutrition Today*. 2018;53:240.

58. Levine M, Padayatty SY. Vitamin C. In: Ross C and others, eds. *Modern nutrition in health and disease*. 11th ed. Philadelphia: Lippincott Williams & Wilkins; 2014.

59. U.S. Department of Agriculture. Commodity consumption by population characteristics. 2016; www.ers.usda.gov/data-products/commodity-consumption -by-population-characteristics/.

60. Schleicher R. Serum vitamin C and the prevalence of vitamin C deficiency in the United States: 2003–2004 National Health and Nutrition Examination Survey (NHANES). *Am J Clin Nutr*. 2009;90:1252.

61. Hosseini B and others. Association between antioxidant intake/status and obesity: A systematic review of observational studies. *Biol Trace Elem Res*. 2016;175:287.

62. Mares J. Food antioxidants to prevent cataract. *JAMA*. 2015;313:1048.

63. Gulko E and others. MRI findings in pediatric patients with scurvy. *Skeletal Radiol*. 2015;44:291.

64. Wang L and others. Vitamin E and C supplementation and risk of cancer in men: Posttrial follow-up in the Physicians' Health Study II randomized trial. *Am J Clin Nutr*. 2014;100:915.

65. van Gorkom GNY and others. The effects of vitamin C (ascorbic acid) in the treatment of patients with cancer: A systematic review. *Nutrients*. 2019;11:977.

66. Aune D and others. Dietary intake and blood concentrations of antioxidants and the risk of cardiovascular disease, total cancer, and all-cause mortality: A systematic review and dose-response meta-analysis of prospective studies. *Am J Clin Nutr*. 2018;108:1069.

67. Buijsse B and others. Plasma ascorbic acid, a priori diet quality score, and incident hypertension: A prospective cohort study. *PLoS ONE*. 2016;10:e0144920.

68. Moore J and others. Dietary supplement use in the United States. Prevalence, trends, pros, and cons. *Nutr Today*. 2020;55:174.

©Ingram Publishing/Alamy RF

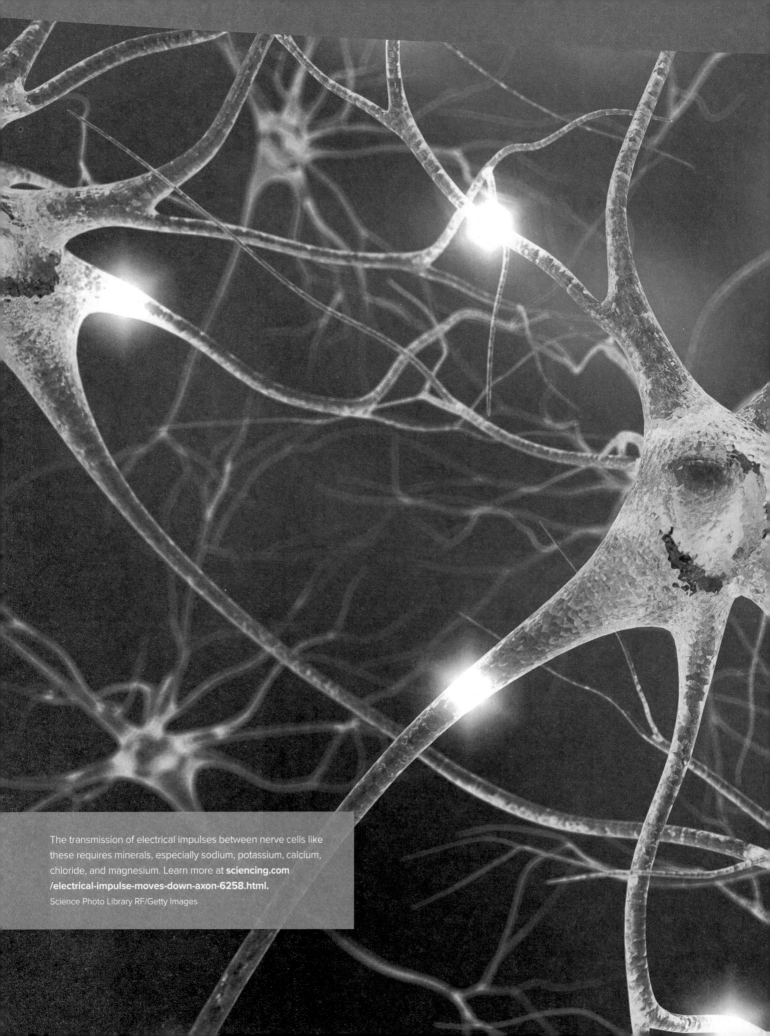

The transmission of electrical impulses between nerve cells like these requires minerals, especially sodium, potassium, calcium, chloride, and magnesium. Learn more at **sciencing.com/electrical-impulse-moves-down-axon-6258.html.**

# 14 Water and Major Minerals

## Learning Objectives

**After studying this chapter, you will be able to**

1. Describe the factors that influence water balance and how it is maintained in the body.
2. Discuss how both dehydration and water intoxication develop and how to prevent them.
3. Identify food sources of water and major minerals.
4. Explain the functions of water and major minerals in the body.
5. Discuss the problems with low and high intakes of major minerals and how to avoid inadequate or excessive intakes.
6. Explain the role of nutrition in the prevention and treatment of hypertension.
7. Estimate and evaluate the adequacy of dietary calcium intake.
8. Describe the role of nutrition in bone health and in the prevention of osteoporosis.

## Chapter Outline

**POUND FOR POUND, OUR BODIES** contain more water than any other component. After oxygen, water is the most important ingredient needed for life. Without water, biological processes—and life—cease within a matter of days. Because we cannot store this crucial nutrient, we must regularly replenish the water lost from the body. Water needs vary, depending on physical activity, environmental conditions (e.g., temperature and humidity), individual characteristics, and nutrient intake, especially protein and minerals.

Many minerals also are vital to health. These inorganic substances are critical to many body functions, including cell metabolism, nerve impulse transmission, and growth and development (Fig. 14-1).[1,2] Typical diets in high-income countries contain sufficient amounts of most minerals, either as natural components of foods or as additives through enrichment and fortification. Although severe deficiencies of minerals are rare in high-income countries, many people have lower than optimal intakes of some minerals, such as calcium, potassium, magnesium, iron, and iodine, and higher than recommended intakes of others, such as sodium. Deficiencies of certain minerals remain a major public health concern in low-income countries.

Minerals often are divided into 2 categories: major and trace minerals. Major minerals are those that are present and required in larger amounts in the body. They include sodium, potassium, and chloride, which are especially important in maintaining water and ion balance in cells. The other major minerals are calcium, phosphorus, magnesium, and sulfur. As you'll see in Chapter 15, the dietary requirements for trace minerals, such as iron and zinc, are small when compared with those for major minerals. This chapter will first explore water and its roles in the body, then will continue with the major minerals and their importance in human nutrition.

*Figure 14-1* Water and minerals are required for many processes in the body. Water is in all cells, carries nutrients around the body, and actively participates in many chemical reactions and physiological processes; thus, water is required for all of these functions.

©Ryan McVay/Getty Images

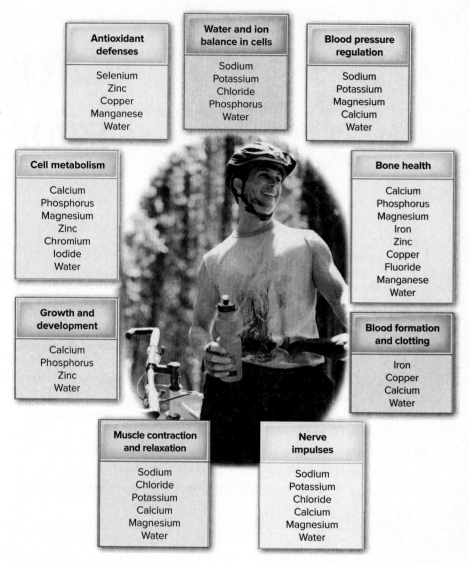

**Antioxidant defenses**

Selenium
Zinc
Copper
Manganese
Water

**Water and ion balance in cells**

Sodium
Potassium
Chloride
Phosphorus
Water

**Blood pressure regulation**

Sodium
Potassium
Magnesium
Calcium
Water

**Cell metabolism**

Calcium
Phosphorus
Magnesium
Zinc
Chromium
Iodide
Water

**Bone health**

Calcium
Phosphorus
Magnesium
Iron
Zinc
Copper
Fluoride
Manganese
Water

**Growth and development**

Calcium
Phosphorus
Zinc
Water

**Blood formation and clotting**

Iron
Copper
Calcium
Water

**Muscle contraction and relaxation**

Sodium
Chloride
Potassium
Calcium
Magnesium
Water

**Nerve impulses**

Sodium
Potassium
Chloride
Calcium
Magnesium
Water

# 14.1 Water

Each of the trillions of cells in the body contains and is surrounded by water. Thus, it is no surprise that maintaining the right amount and balance of water in the body is essential to life. An adult can survive for several weeks without food, but only several days without water. This difference in survival time between food and water occurs because the body has reserves of carbohydrate, fat, protein, vitamins, and minerals but no such reserves of water.

## Water in the Body: Intracellular and Extracellular Fluids

Water is the largest component of the human body, making up 50 to 75% of body weight, depending on age and body fat content. Water content is highest in infants and children and declines as we age. About 55% of an adult's body weight is water—that's about 10 gallons (40 liters) in a person weighing 160 pounds.[2,3] Lean individuals have a greater percentage of body water than those who are obese because lean tissue contains about 73% water, whereas adipose tissue is only 20% water.

Body water is found in 2 body compartments—the **intracellular fluid** compartment, or that inside cells, and the **extracellular** fluid compartment, or that outside cells (Fig. 14-2). Almost two-thirds of body water is found in the intracellular fluid compartment. The rest is in the extracellular fluid compartment, where it is divided into 2 additional compartments: **interstitial fluid,** the fluid between cells, and **intravascular fluid,** the fluid in the blood and lymph.

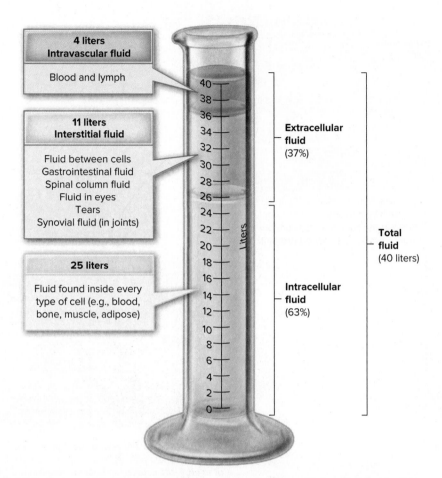

*Figure 14-2* Fluid compartments in the body. Total fluid volume is about 10 gallons (40 liters).

The fluid in these compartments is not pure water; it also contains dissolved substances known as **solutes.** The most abundant solutes are **electrolytes** that form when salts, such as sodium chloride or potassium phosphate, dissociate in solution and form **ions** (e.g., NaCl forms $Na^+$ and $Cl^-$). The major positively charged electrolytes (cations) and the negatively charged electrolytes (anions) found in each fluid compartment vary (Table 14-1). Intracellular fluids contain primarily potassium and magnesium cations, along with negatively charged phosphate anions. In extracellular fluids, positively charged sodium cations and the negatively charged chloride anions, along with bicarbonate ($HCO_3^-$), predominate.

### Maintenance of Intracellular and Extracellular Fluid Balance

The body controls the amount of water in each compartment mainly by controlling the electrolyte concentrations in the compartment. An extremely sophisticated gatekeeping system involving transmembrane pumps keeps intracellular and extracellular fluid volumes and electrolyte

**Table 14-1  Electrolytes in Intracellular and Extracellular Fluids**

| Intracellular Fluids | Extracellular Fluids |
|---|---|
| **Major Cations** | **Major Cations** |
| • Potassium ($K^+$) | • Sodium ($Na^+$) |
| • Magnesium ($Mg^{2+}$) | • Calcium ($Ca^{2+}$) |
| **Major Anions** | **Major Anions** |
| • Phosphate* | • Chloride ($Cl^-$) |
| • Sulfate ($SO^{4-}$) | • Bicarbonate ($HCO_3^-$) |

*Can ionize in different forms.
Note: Organic acids and proteins also contribute positive and negative charges in body fluids.

**solute** Substance dissolved in another substance (a solvent), forming a solution.

**electrolytes** Compounds that separate into ions in water and, in turn, are able to conduct an electrical current. These include sodium, chloride, and potassium.

**ion** Atom with an unequal number of electrons and protons. Negative ions (anions) have more electrons than protons and carry a negative charge; positive ions (cations) have more protons than electrons and carry a positive charge.

*Figure 14-3* Sodium-potassium pump. A sodium-potassium pump has a transmembrane transport protein that uses energy to transport Na and K ions through the membrane from a region of low concentration to a region of high concentration. This maintains a high concentration of $Na^+$ outside the cell and a high concentration of $K^+$ inside the cell. The continuous active transport can be broken into 4 steps.

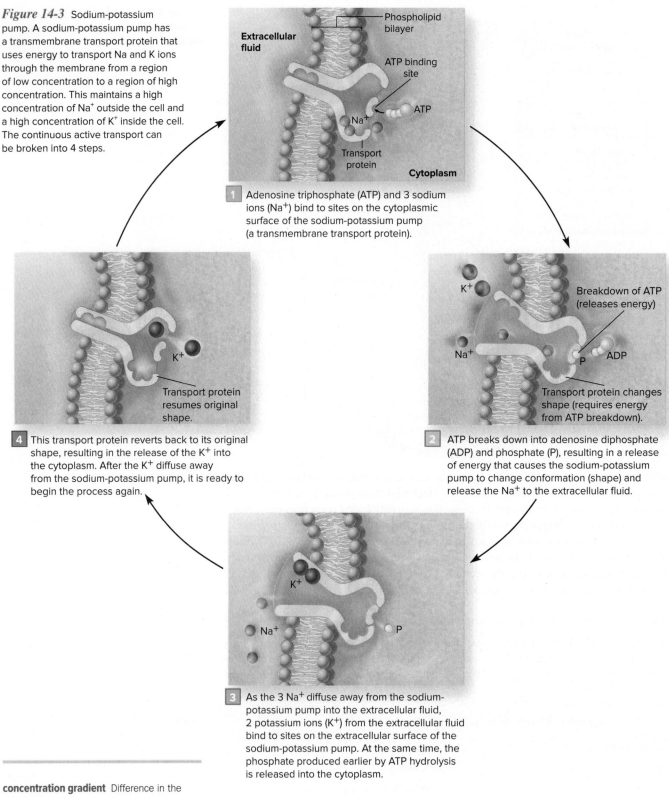

**1** Adenosine triphosphate (ATP) and 3 sodium ions ($Na^+$) bind to sites on the cytoplasmic surface of the sodium-potassium pump (a transmembrane transport protein).

**2** ATP breaks down into adenosine diphosphate (ADP) and phosphate (P), resulting in a release of energy that causes the sodium-potassium pump to change conformation (shape) and release the $Na^+$ to the extracellular fluid.

**3** As the 3 $Na^+$ diffuse away from the sodium-potassium pump into the extracellular fluid, 2 potassium ions ($K^+$) from the extracellular fluid bind to sites on the extracellular surface of the sodium-potassium pump. At the same time, the phosphate produced earlier by ATP hydrolysis is released into the cytoplasm.

**4** This transport protein reverts back to its original shape, resulting in the release of the $K^+$ into the cytoplasm. After the $K^+$ diffuse away from the sodium-potassium pump, it is ready to begin the process again.

**concentration gradient** Difference in the concentration of a solute from 1 area to another. Normally, a solute moves from where it is most concentrated to where it is least concentrated. When sodium is pumped outside the cell and potassium pumped inside the cell, they are moving instead to where each is most concentrated–that is, against the concentration gradient.

concentrations within quite narrow ranges. For example, a specific protein located in the cell membrane can pump potassium ions into and sodium ions out of a cell (Fig. 14-3). This sodium-potassium pump uses energy to move each ion against its **concentration gradient.**

Water is attracted to electrolytes and other ions and thus moves via osmosis from 1 fluid compartment to another as the concentration of solutes changes. **Osmosis** is the

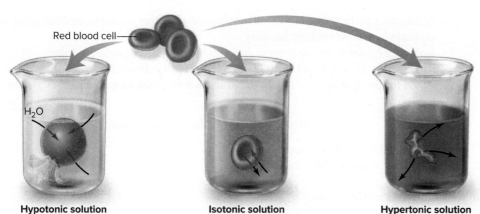

*Figure 14-4* Red blood cells affected by various ion concentrations. Osmosis causes fluid to shift into and out of the cells, depending on the ion concentration in the surrounding solution.

(a) A hypotonic (dilute) solution has a low ion concentration, which results in swelling *(black arrows)* and subsequent rupture (puff of red in the lower left part of the cell) of a red blood cell placed into the solution.

(b) An isotonic (normal) concentration (a concentration of ions outside the cell equal to that inside the cell) results in a typically shaped red blood cell. Water moves into and out of the cell in equilibrium *(black arrows)*, but there is no net water movement.

(c) A hypertonic solution has a high ion concentration, which causes shrinkage of the red blood cell as water moves out of the cell and into the concentrated solution *(black arrows)*.

passive diffusion of water across a semipermeable membrane—in the body, these are cell membranes. When the concentration of solutes (mainly electrolytes) differs on the 2 sides of a cell membrane, water will move from the side with a low solute concentration to the side with the higher solute concentration. Examples of osmosis that may be familiar to you are extracting water from cucumbers by sprinkling them with salt and crisping limp celery by placing it in water—water moves into the dehydrated celery cells.

Figure 14-4 illustrates how osmosis works. Figure 14-4*a* shows water moving from the hypotonic (dilute) solution into the more concentrated red blood cell, which causes the cell to swell and possibly burst. In Figure 14-4*b,* there is no net movement of water because solute concentrations are isotonic (equal) on both sides of the cell's semipermeable membrane. However, in Figure 14-4*c,* water is drawn across the red blood cell's membrane into the hypertonic (concentrated) solution surrounding it, causing the cell to shrink. In the body, the actual movement of water across the cell membrane is not quite as simple. Cell membranes are mainly lipid and poorly permeable to water; thus, in many cells, water moves through water channels made from proteins called aquaporins.

Adding water—instead of ions—to a fluid compartment dilutes its solute concentration, which causes water to move via osmosis to the more concentrated compartments nearby. This happens when we drink water—some of the absorbed water moves from the bloodstream into body cells, which equalizes the solute concentration in the cells with that in the bloodstream and in the interstitial fluid. Conversely, when blood loss occurs, plasma (the watery liquid part of blood) volume can be partially maintained by shifting fluid out of the intracellular compartment into the bloodstream. Tightly regulating the amount of water in each fluid compartment is critical because too little or too much water can severely disrupt cell and organ function.

▶ Osmotic pressure is the amount of force needed to prevent dilution of the compartment containing the higher particle concentration.

## Functions of Water

Because of its unique chemical and physical characteristics, many physiological functions depend on water:

- Maintaining blood volume that permits the transport of nutrients and oxygen throughout the body
- Forming specialized fluids throughout the body, such as saliva, tears, bile, and the amniotic fluid that surrounds a growing fetus in the uterus
- Helping form lubricants in the knees and other joints

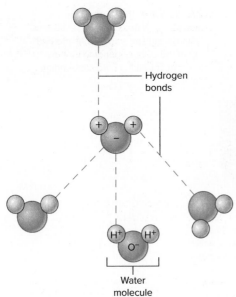

*Figure 14-5* At a molecular level, water is highly polar because it has regions of weak negative and positive charges. The positive charges are near the hydrogen molecules and the negative charges near the oxygen. The attraction of the negatively charged oxygen to the positively charged hydrogen results in weak bonds called hydrogen bonds. Each water molecule can form a maximum of 4 hydrogen bonds. The high specific heat of water results from hydrogen bonding.

**specific heat** Amount of heat required to raise the temperature of a gram of any substance 1°C. Water has a high specific heat, meaning that a relatively large amount of heat is required to raise its temperature; therefore, it tends to resist large temperature fluctuations.

▶ The simplest way to determine if water intake is adequate is to observe the color of one's urine. Water intake is adequate when urine is pale yellow and has little odor. Concentrated urine is very dark yellow and has a strong odor—it indicates that water intake needs to be increased (see Fig. 11-13, Chapter 11).

- Acting as a solvent, helping dissolve minerals and other nutrients, making them more available to cells throughout the body
- Keeping exposed tissues moist; this includes the eyes, nose, mouth, and skin
- Participating as a reactant in numerous chemical reactions; for example, water is required for the hydrolysis of the disaccharide sucrose to the monosaccharides glucose and fructose:

$$\text{Sucrose} + H_2O \longrightarrow \text{Glucose} + \text{Fructose}$$

- Regulating body temperature to keep it within a narrow range
- Removing waste products via urine

### Temperature Regulation

Keeping body temperature within a narrow range allows the body, especially enzymes, to function normally. Body temperatures just a few degrees higher or lower than normal can damage body systems and even lead to death. Water in the body helps maintain this range in 2 ways. First, water has a high heat capacity, or **specific heat.** That means water resists temperature changes, so its temperature rises slowly when it is heated. This occurs because water molecules form hydrogen bonds (Fig. 14-5) with each other, and a relatively large amount of heat is required to overcome this attraction. Think of heating equal amounts of oil and water in separate pans on a stove. The oil gets hot much faster than the water because the fat molecules are not strongly attracted to each other—oil has a lower specific heat.

Sweat, which is 99% water, is the other way water helps maintain normal body temperature. During exercise or hot weather, when body temperature increases, perspiration is secreted through the skin pores. Heat energy from the skin, transferred to the sweat (water on the skin), causes the sweat to evaporate. This loss of heat cools the skin. This is the main way in which the body cools itself.[2,3] For the most efficient cooling, perspiration must be allowed to evaporate—if it rolls off the skin or soaks into clothing, it doesn't cool us as much. Evaporation occurs most readily when humidity is low, which is why we feel more comfortable in hot, dry climates than in hot, humid ones. Of course, water lost through perspiration must be replaced, or dehydration and overheating will occur (see Chapter 11).

### Waste Product Removal

Water is an important vehicle for ridding the body of waste products. Most unwanted substances in the body are water soluble and can leave the body via the urine.[2] In addition, liver metabolism converts some fat-soluble compounds, such as certain medications and potentially cancer-causing substances, into water-soluble compounds that can be excreted in the urine. It is a common belief that increasing water intake helps the kidneys flush out "toxins"; however, little research evidence exists to support or refute this belief.[4]

A major body waste product is urea, the nitrogen-containing by-product of protein metabolism. As we eat more protein, more urea must be excreted in the urine. Likewise, the amount of sodium in the urine increases with higher dietary intakes of sodium.

A typical urine output of approximately 4¼ to 8½ cups (1 to 2 liters) daily can easily change in response to fluid, protein, and sodium intake.[2,3] The minimum urine output required to excrete usual amounts of urea and sodium waste is 2½ cups (600 ml) per day. If urine output is this low frequently, its heavy ion concentration may increase the risk of kidney stone formation in susceptible people, especially men.[5] Kidney stones are minerals and other substances that have precipitated out of the urine and accumulated in kidney tissues.

## Water in Beverages and Foods

### Beverages

Beverages, including water, milk, tea, juice, and soda, and liquid foods (e.g., soups, broths) provide the greatest amount of water in the diet. In addition to supplying water, many beverages also provide calories (Table 14-2) and may contribute to obesity and cardiovascular

**Table 14-2 Calorie Content of Popular Beverages***

| Beverage | Calories | Beverage | Calories |
|---|---|---|---|
| Mocha with whipped cream | 250 | Beer | 145 |
| Cranberry juice | 200 | Fat-free milk | 120 |
| Orange juice | 180 | Red Bull® energy drink, 8 oz | 105 |
| Regular soft drinks | 160 | White wine, 4 oz | 90 |

*Amounts are 12 oz unless otherwise noted.

## CASE STUDY

After graduating from college, Patrick, a 24-year-old accountant, noticed he had gained 8 pounds in the past year. His employer provides free soft drinks, juice, and bottled water. Patrick decided to keep track of his food intake for a couple of days to track his calories. Just his beverage intake is provided here. To estimate his energy intake from beverages, use Table 14-2 and an online nutrient database or nutrient analysis computer program, or visit this website: fdc.nal.usda.gov/. Do you think he obtains too many calories from these beverages? What are better alternatives?

©Paul Bradbury/
age fotostock RF

**Breakfast**
12 oz mocha with whipped cream
6 oz orange juice

**Morning Break**
12 oz cranberry juice
10 oz water

**Lunch**
12 oz regular cola

**Afternoon Break**
12 oz regular root beer or apple juice

**After Work**
1 or 2 beers (12 oz each)
10 oz fat-free milk

disease risk.[6] Sugar-sweetened beverages, such as soft drinks, fruit drinks, energy and sports drinks, and tea and coffee drinks, can contribute substantially to sugar and calorie intake. Recall from Chapter 2 that the Dietary Guidelines for Americans recommend that added sugars from foods and beverages not exceed 10% of daily calorie intake for those age 2 and older—that amounts to 50 g/day in a 2000 calorie diet. Infants and toddlers should avoid added sugars. A single 20 oz soft drink provides 65 g added sugar. Sugar-sweetened beverage intake is gradually decreasing, perhaps in response to recommendations to decrease sugar intake.

Micronutrient content is another consideration for beverage choice. Many sweetened beverages supply few micronutrients in contrast to micronutrient rich milk and fruit juice. When soft drinks are selected over milk, riboflavin, vitamin D, calcium, and phosphorus intakes drop. Similarly, replacing fruit juice with sweetened beverages decreases vitamin C, vitamin A, and folate intake. Although some sweetened beverages and bottled waters are fortified with certain vitamins and minerals, the array of micronutrients provided is limited.

Both coffee and tea are popular beverage choices. Many consumers believe that caffeinated beverages are dehydrating and should not "count" toward daily fluid intake. However, research studies do not support this common belief.[7] Although caffeine is a mild diuretic, intakes up to 500 mg per day (the amount in about 4.5 cups of brewed coffee) do not cause dehydration or water imbalance in most people. The caffeine content of foods is given in Appendix I.

Alcoholic beverages (wine, beer, spirits) are primarily water, too. However, alcohol (ethanol) increases urine output by inhibiting the action of antidiuretic hormone. This hormone helps control the amount of fluid lost in the urine. When antidiuretic hormone action is blocked, dehydration can occur.

▶ The concern over excessive soft drink intake has led many school districts to eliminate or reduce students' access to soft drinks, fruit drinks, and energy drinks and to promote water as the beverage of choice.

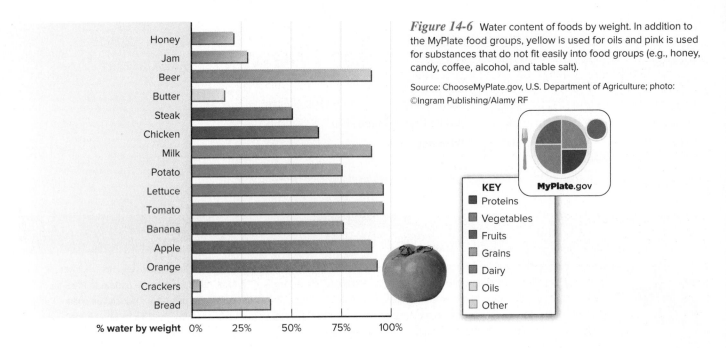

***Figure 14-6*** Water content of foods by weight. In addition to the MyPlate food groups, yellow is used for oils and pink is used for substances that do not fit easily into food groups (e.g., honey, candy, coffee, alcohol, and table salt).

Source: ChooseMyPlate.gov, U.S. Department of Agriculture; photo: ©Ingram Publishing/Alamy RF

**KEY**
- Proteins
- Vegetables
- Fruits
- Grains
- Dairy
- Oils
- Other

**antidiuretic hormone (ADH)** Hormone, secreted by the pituitary gland, that signals the kidney to cause a decrease in water excretion; also called arginine vasopressin.

**renin** Enzyme formed in the kidneys and released in response to low blood pressure. It acts on a blood protein called angiotensinogen to produce angiotensin I.

**angiotensin II** Compound, produced from angiotensin I, that increases blood vessel constriction and triggers production of the hormone aldosterone.

**aldosterone** Hormone, produced in the adrenal glands, that acts on the kidneys, causing them to retain sodium and, therefore, water.

Regular intake of fluid is essential to replace daily losses.

©Stockbyte/Getty Images RF

Although water is an excellent beverage choice for most of us, its source—the faucet or a bottle—deserves some thought. The U.S. is the world's leading consumer of bottled water, with an average annual consumption of 36.5 gallons (584 cups) per person.[8] Consumers may choose bottled water because they believe it tastes better, is more convenient, or is safer than tap water. Taste and convenience are personal views; however, in the U.S., bottled water usually is no safer than municipal tap water. Bottled water can originate from various sources, including springs, artesian wells, and public water sources (e.g., tap water). The label should state the water source. It also is important to consider that bottled water may not contain the fluoride needed to protect against dental caries (cavities). Additional considerations are the cost and environmental impact of producing and disposing of tons of plastic bottles each year.

## Foods

Water also is abundant in fruits and vegetables, which typically are 75 to 95% water by weight. Other sources that fall between 50 and 75% water are potatoes, chicken, and steak. Foods that are less than 35% water include jam, honey, crackers, and butter and margarine. Vegetable oils contain no water (Fig. 14-6).

## Water Needs

Water needs vary with factors such as body size, physical activity, environmental conditions, and dietary intake. Despite this variability, an Adequate Intake has been set to provide guidance to individuals for water intake. The Adequate Intake for total water intake per day is 15 cups (3.7 liters) for adult men and 11 cups (2.7 liters) for adult women.[2] This water can come from plain water, other beverages, and foods. The *What We Eat in America* survey indicates that 3 out of 4 people over the age of 2 years drink more than 4 cups of plain water daily.[9] Table 14-3 provides an example of a day's intake that meets the AI for an adult woman. Notice that, in this example, 35% of the water is provided by solid food. However, in typical U.S. diets, solid foods provide only about 20% of the water consumed each day.[2] Getting 80% of water from fluid translates to a daily fluid intake of about 13 cups (3 liters) for men and 9 cups (2.2 liters) for women. At a minimum, adults need 1 to 3 liters of fluid per day to replace daily water losses.

Water needs are met when water inputs and outputs are balanced. The inputs (Fig. 14-7) are the water in foods and beverages we consume and the water generated during metabolism (i.e., oxidation of carbohydrates, proteins, and fats). Water produced from metabolism is about 1 to 1½ ups (250 to 350 ml) per day.

Water output consists of sensible and insensible water losses. Sensible water losses, or those we notice, are urine output and heavy perspiration. Most of the sensible water loss is urine (typically, 1000 to 2000 ml/day). Insensible water losses, or those we do not normally notice, include water lost through the skin (450 to 1900 ml for normal perspiration), lungs (250 to 350 ml), and feces (100 to 200 ml).

The intestinal tract is efficient at recycling water—about 32 cups (8000 ml) of water enter the intestinal tract daily via secretions from the mouth, stomach, intestines, pancreas, and other organs, and the diet supplies an additional 8 to 13 cups (2 to 3 liters), but only about ½ to ¾ cup (100 to 200 ml) is lost in the feces. The kidneys also conserve water, reabsorbing about 97% of the water filtered from waste products. The kidneys' reabsorption of water is the primary means of regulating water balance.

## Regulation of Water Balance

The body has powerful mechanisms that regulate fluid balance and defend against both dehydration and overhydration. The kidneys are the main regulators, but the brain, lungs, and liver also have important roles in maintaining water balance. This regulation is so precise that, under normal conditions, a loss of 1% body water is usually compensated for within 24 hours.[3]

When water intake is too low to replace losses, the blood becomes more concentrated (the osmotic pressure of the extracellular fluid increases) and blood pressure falls. These changes signal the body that there is a shortage of water and trigger a series of fluid conservation measures. In response to increased osmotic pressure, the pituitary gland releases **antidiuretic hormone (ADH).** ADH signals the kidneys to retain water, thus reducing urine output (Fig. 14-8). At the same time, falling blood pressure initiates another sequence of events—highly sensitive pressure receptors in the kidneys trigger the release of the enzyme **renin** (Fig. 14-9). Renin, in turn, activates **angiotensinogen** (a circulating blood protein made in the liver), forming **angiotensin I.** In the lungs, angiotensin I is converted to **angiotensin II,** which, among other effects, causes the constriction of blood vessels and release of the hormone **aldosterone** by the adrenal glands. Aldosterone signals the kidneys to retain more sodium and chloride, and therefore more water. (Remember, water always follows electrolytes.) Thus, low blood pressure, through this roundabout sequence, causes the kidneys to increase water conservation in the body. There is, however, a limit to how

**Table 14-3  Water Content of a Typical Day's Diet**

| Meals | Beverage (fl oz) | Food (fl oz) |
|---|---|---|
| **Breakfast** | | |
| 1 cup fat-free milk | 7.5 | |
| 1 cup Cheerios® | | 0 |
| ½ cup strawberries | | 2.7 |
| 1½ cups coffee | 12.0 | |
| **Midmorning Snack** | | |
| 1½ cups diet soft drink | 12.0 | |
| 1 granola bar | | 0 |
| **Lunch** | | |
| 2 oz water-packed tuna | | 1.4 |
| 2 slices whole-wheat bread | | 0.7 |
| 2 tsp low-fat mayonnaise | | 0 |
| 3 large tomato slices | | 2.6 |
| 1 medium banana | | 3.0 |
| 1 cup water | 8.0 | |
| **Midafternoon Snack** | | |
| 1 cup low-fat yogurt | | 6.5 |
| 1 cup water | 8.0 | |
| **Dinner** | | |
| 3 oz chicken breast | | 1.8 |
| 1 medium baked potato | | 3.5 |
| 1 tsp butter | | 0 |
| 2 cups salad greens | | 3.5 |
| 1 tbsp salad dressing | | 0.3 |
| 1 cup raw carrots | | 3.6 |
| 1 cup fat-free milk | 7.5 | |
| **Evening Snack** | | |
| 1 cup tea | 8.0 | |
| 1 medium peach | | 4.5 |
| **Subtotal** | 63 (7.9 cups) | 34.1 (4.3 cups) |
| **Grand Total** | **97.1 fl oz (12.1 cups)\*** | |

*Meets the AI for water for a woman. Actual water requirements vary depending on body size, physical activity, and environmental conditions.

Consider the calorie and nutrient contents of beverages when selecting them. One ounce of orange juice supplies 14 calories. A frozen margarita has 23 calories per ounce (regular margaritas have 68 calories in an ounce). Milkshakes provide 40 calories or more per ounce.

margarita: ©John A. Rizzo/Getty Images RF; oj: ©Stockbyte/Getty Images RF; shake: ©Brand X Pictures/ Getty Images RF

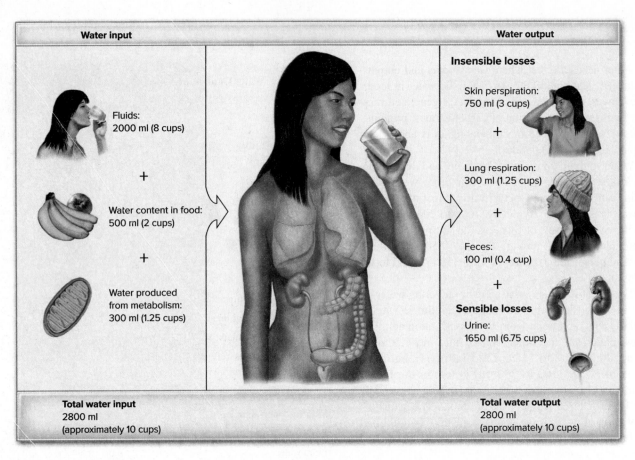

*Figure 14-7* Estimate of water input vs. water output in a woman. We primarily maintain our volume of body fluids by adjusting water output to input. As you can see, most water comes from the liquids we consume. Some comes from the moisture in solid foods, and the remainder is manufactured during metabolism. Water output includes insensible losses from the lungs, skin, and feces and sensible losses from urine and heavy perspiration.

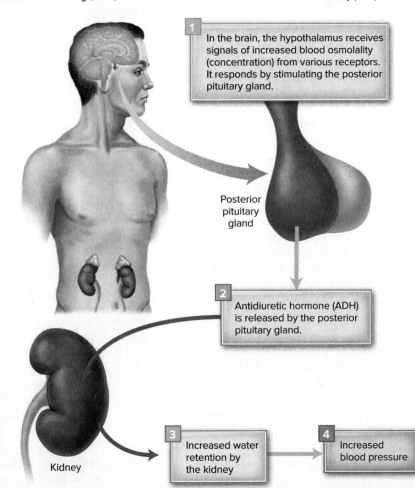

1 In the brain, the hypothalamus receives signals of increased blood osmolality (concentration) from various receptors. It responds by stimulating the posterior pituitary gland.

Posterior pituitary gland

2 Antidiuretic hormone (ADH) is released by the posterior pituitary gland.

Kidney

3 Increased water retention by the kidney

4 Increased blood pressure

*Figure 14-8* Antidiuretic hormone is released in response to an increased concentration of blood. Antidiuretic hormone acts on the kidney to increase water retention; therefore, blood volume and, in turn, blood pressure are restored to normal values. Alcohol (ethanol) inhibits the action of antidiuretic hormone.

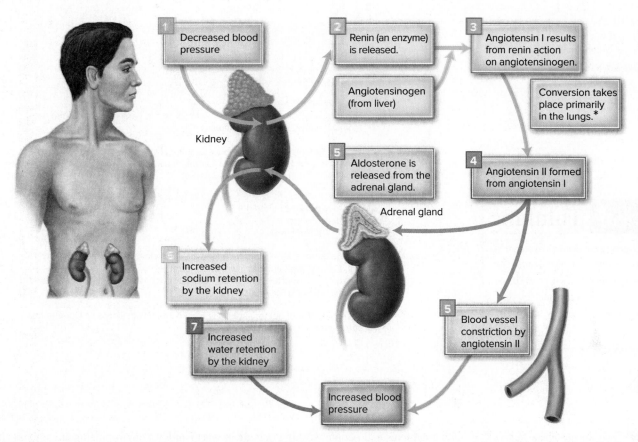

*Figure 14-9*  The renin-angiotensin system is 1 regulator of blood pressure and blood volume. A decrease in blood pressure ◼1 starts the cascade of reactions (◼2 – ◼7 ) that restore blood pressure to the normal range. This system functions with antidiuretic hormone to control blood pressure. ◼5 is listed twice because angiotensin II acts at both the adrenal gland and the blood vessels to help regulate blood pressure.

*The angiotensin-converting enzyme (ACE) inhibitors used to treat hypertension and other disorders act at this site. A new class of antihypertensive medications goes a step further to block the binding of angiotensin II to receptors in the body (e.g., in blood vessels). These are called angiotensin II receptor blockers (ARBs).

concentrated urine can become. Eventually, if sufficient fluid is not consumed, dehydration and ill effects ensue.[2]

When water intake exceeds that needed to excrete waste via the urine and replace insensible losses, the kidneys are able to decrease their reabsorption of water. With this, urine volume increases and the urine becomes more dilute. In rare cases, urinary excretion cannot match high water intake, and water intoxication occurs.

### Dehydration

Dehydration can result from a variety of conditions, including those related to medical conditions such as diarrhea, vomiting, fever, inability to consume sufficient fluid (e.g., when unconcious), poorly controlled diabetes mellitus, and burns. In healthy individuals, dehydration can occur as a result of heavy exercise, hot weather, dry environments, high altitudes (more rapid breathing causes increased fluid loss from the lungs), or even just ignoring feelings of thirst. In all cases of dehydration, fluid intake does not match fluid loss. Although thirst is a signal that more water is needed, the thirst mechanism does not always work well during intense exercise, illness, infancy, and old age. Sick children—especially those with fever, vomiting, diarrhea, and increased perspiration—and older persons often need to be reminded to drink plenty of fluids. As Chapter 17 discusses in further detail, infants can become dehydrated easily.

*Figure 14-10* The effects of dehydration range from thirst to death, depending on the extent of water weight loss.

©Brian Hagiwara/Getty Images RF

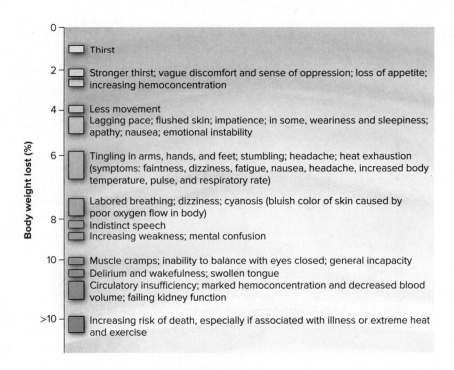

## Critical Thinking

Lily is suffering from diarrhea related to a viral infection. Why is it important for Lily to continue to drink fluids even while she is sick and has a poor appetite? Do you think that drinking fluids may help her feel better?

▶ It's easy to become dehydrated during air travel because the air on planes has very low humidity. This causes extra losses of water from the skin and lungs to the cabin air. Drinking extra fluid (but not alcoholic beverages), even though this means extra trips to the restroom, can prevent dehydration. Babies and children especially may need reminders to drink more.

Athletes and people working outside in warm and hot environments are at extra risk of dehydration and its effects (Fig. 14-10). During prolonged physical activity, sweat loss ranges from 3 to 8 cups (750 to 2000 ml) per hour. These individuals are advised to plan ahead to assure that plenty of water is available to maintain hydration. Specific recommendations for calculating fluid losses and fluid replacement in athletes are provided in Chapter 11.

The signs of mild to moderate dehydration include dry mouth and skin, fatigue and muscle weakness, decreased urine output, deep yellow (concentrated) urine, headache, and dizziness. As dehydration progresses, solute concentrations in the blood rise, blood pressure decreases, and heart rate increases due to low blood volume. If fluid losses continue, kidney failure, seizures, delirium, and coma can occur (see Fig. 14-10). The only treatment for dehydration is to replace the lost fluids. In cases of severe dehydration, fluid replacement should be done under medical supervision. Left unchecked, severe dehydration can progress to death.

### Water Toxicity

Drinking too much water too rapidly can be as dangerous as drinking too little. In rare instances, it can cause a condition called water intoxication. Water intoxication develops when the kidneys cannot remove water fast enough to keep pace with water intake. As a result, water accumulates in the blood and dilutes the sodium in the serum, a condition known as **hyponatremia.** To balance intracellular and extracellular electrolyte concentrations, water from the diluted blood is pulled by osmosis into the cells, causing them to swell. Adverse effects, such as headache, nausea, blurred vision, respiratory arrest, convulsion, and even death, occur when brain cells swell excessively. Fortunately, very few people drink water in excess of their kidneys' ability to excrete it.

Water intoxication has occurred in individuals with some mental disorders[2] and infants under 6 months of age given bottles of water or overdiluted formula,[10] in isolated incidences of forced water ingestion, and as part of the condition known as

# GL⬤BAL PERSPECTIVE

## Water for Everyone

In North America, we wash our bodies, clothes, dishes, and cars and water our plants and lawns, often with little thought to the source of the water, its safety, or how much we use. According to the United Nations, the average daily per capita water use in the U.S. and Europe ranges from 52 to 160 gallons (200 to 600 liters).[12] Contrast this with the approximately 5½ to 13 gallons (20 to 50 liters) of clean water that the United Nations suggests is needed per person daily for drinking, cooking, and cleaning. In reality, water needs greatly exceed this amount because water is required for agriculture, energy production, and industry. Agriculture, primarily for irrigation, accounts for about 70% of the world's water consumption. A region's dietary patterns are an important determinant of water needs—high meat diets require more water to produce than low meat diets. Producing a kilogram (2.2 lb) of grain requires nearly 400 gallons (1500 liters) of water, whereas producing the same amount of beef requires 8 to 10 times as much water.[12]

We all require water, but millions of people (farmers, herders, fishing people), especially in rural areas, rely on water for their incomes and food production. When water is in short supply, incomes and economic development fall, and poverty, malnutrition, and poor health increase. Water scarcity affects about 40% of the world's population at some point during the year.[13] In some places, there is inadequate water and, in others, there is inadequate infrastructure to move water from its sources to where it is needed. Some water sources are several miles away from homes. People, often women, may walk several miles daily to the water source and then carry the water back to their homes. When you consider that each gallon of water weighs more than 8 pounds, the energy needed just to carry water adds up to a huge calorie expenditure.

©Gavin Hellier/Alamy Stock Photo

Equitable water distribution is a major challenge. Many rivers and aquifers cut across political boundaries. Countries and regions may disagree on how to manage and share water resources. Droughts and floods caused by climate changes further increase the challenge of water management.

Equally important are basic sanitation and access to safe water. Although 9 out of 10 people now have access to an improved drinking water supply, a lack of basic sanitation is still a major problem for about one-third of the world's population.[14] According to the World Health Organization, 2 billion people use a drinking water source contaminated by feces.[14] Water contaminated with sewage, herbicides, pesticides, and toxins, such as arsenic and lead, can cause disease, and those who are weakened from malnutrition are most likely to succumb to waterborne diseases. Diarrheal diseases, linked to a variety of pathogens in water and poor sanitation, cause nearly 300,000 deaths each year in children under the age of 5 years.[14] Other water-related diseases, such as cholera and the parasitic disease schistosomiasis (which can damage organs), afflict millions more. Breaking the cycle of waterborne diseases depends on water purification and wastewater treatment.

exercise-associated hyponatremia (i.e., when athletes overconsume water while trying to prevent dehydration).[11] To combat this practice, sports medicine professionals no longer recommend that athletes drink predetermined amounts of water prior to, during, and after exercise or events. Instead, they recommend that athletes drink when they are thirsty (see Chapter 11).

## Knowledge Check

1. How is body water distributed?
2. What are the main functions of water in the body?
3. What factors affect water requirements in an individual?
4. How do the ADH and renin-angiotensin systems conserve water?
5. What are the signs of dehydration?

## CASE STUDY FOLLOW-UP

©Paul Bradbury/age fotostock RF

Beverages contribute 1100 to 1250 calories per day to Patrick's diet. This would be an easy place to reduce his calorie intake. He could replace the mocha with a non-fat latté, substitute water or a mix of half juice/half water in place of the full-strength juice, choose diet soda and lite beer, and/or decrease his alcohol intake.

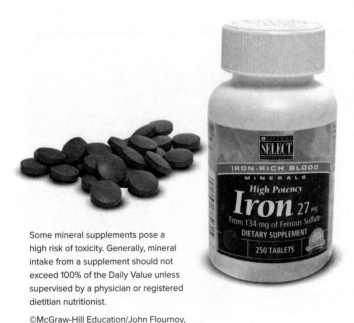

Some mineral supplements pose a high risk of toxicity. Generally, mineral intake from a supplement should not exceed 100% of the Daily Value unless supervised by a physician or registered dietitian nutritionist.

©McGraw-Hill Education/John Flournoy, photographer

## 14.2 Overview of Minerals

Minerals are naturally occurring, inorganic, solid substances. They cannot be synthesized in the body. There are close to 4000 recognized minerals, yet only 16 minerals and a few **ultratrace elements** (e.g., nickel and cobalt) are recognized as nutrients. These mineral nutrients are needed in small amounts in the diet for the normal function, growth, and maintenance of body tissues. As you know, to be considered a nutrient, biological function or health must decline when the substance is lacking in the diet. Restoring the substance to the diet should improve health unless permanent damage has occurred.

Mineral nutrients are divided into major minerals and trace minerals, depending on the amount needed each day. Generally, if we require 100 mg or more of a mineral daily, it is considered a major mineral, or macromineral; otherwise, it is considered a trace mineral, or micromineral. Major minerals are found in larger quantities in the body than trace minerals (Fig. 14-11). Using these criteria, calcium and phosphorus are examples of major minerals, and iron and copper are trace minerals.

### Food Sources of Minerals

**bioavailability** Degree to which the amount of an ingested nutrient is absorbed and is available to the body.

Minerals in the average North American's diet come from both plant and animal sources. For some minerals, animal-based foods are the richest source and have the best **bioavailability.** For instance, dairy products are rich sources of bioavailable calcium, whereas meat is a rich source of bioavailable iron and zinc. On the other hand, potassium, magnesium, and manganese are more plentiful in plant than animal foods, but compounds in plants may diminish their bioavailability.

*Figure 14-11* Approximate amounts of minerals present in the average human body. Some other trace minerals of nutritional importance not shown are selenium, zinc, chromium, fluoride, and molybdenum.

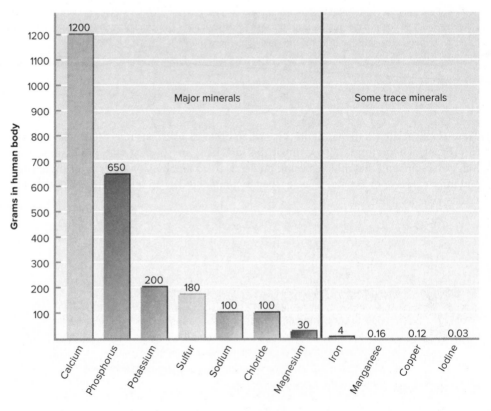

The quantity of minerals found in food is influenced by many agricultural factors, including genetic variations that affect the ability of plants and animals to absorb and store minerals, the mineral composition of animal feed and medications, and the mineral content of soil, water, fertilizers, and pesticides. Food mineral content also is affected by food processing. For example, iron from cooking equipment and food containers can migrate into food. Phosphorus, calcium, and other minerals are in additives used to enhance flavor, maintain texture, and preserve foods. Sanitizing solutions may leave mineral-containing residues, such as iodine, which are picked up by foods prepared in the cleaned equipment. Processing also can decrease mineral content. Typically, the more refined a plant food, the lower its mineral content. Milling grains, for instance, removes iron, selenium, zinc, copper, and other minerals. Boiling cubed potatoes reduces potassium, magnesium, manganese, sulfur, and zinc content by 50 to 75% because the minerals leach from the plant cells into the cooking water.[15]

More and more foods are being fortified with minerals. Iron has been added to milled grains since the enrichment program began in the 1940s. In the U.S., table salt fortified with iodine has been available for nearly 90 years. Many products, such as orange juice, are enriched with calcium. Most breakfast cereals are fortified with a wide array of minerals.

Spinach often is touted as a rich source of calcium, but little of the calcium present is bioavailable—that is, available to the body.

©Kuttelvaserova Stuchelova/Shutterstock, Inc.

## Absorption and Bioavailability of Minerals

Foods offer a plentiful supply of many minerals, but the body varies in its capacity to absorb and use them. The ability to absorb minerals from the diet depends on many factors. One significant factor is the physiological need for a mineral at the time of consumption. In general, when the need for a mineral is high, such as the need for iron in growing children, the absorption of that mineral increases. In contrast, absorption tends to decline when the body has adequate stores of the mineral.

Bioavailability is the other significant factor in the body's ability to absorb minerals (Table 14-4). The bioavailability of minerals can be greatly influenced by the amount of minerals consumed—that's because many minerals have similar molecular weights and charges (valences). For example, magnesium, calcium, iron, and copper can each exist in the $2^+$ valence state. These minerals can compete with each other for absorption, thereby affecting each other's bioavailability. As an example, an excess of zinc in the diet can decrease the absorption and metabolism of the mineral copper. This competition for absorption is of little concern when minerals are supplied by a varied diet; however, individual mineral supplements can create a serious imbalance. Thus, it is safest to choose mineral supplements that contain 100% or less of the Daily Value and use individual mineral supplements only under medical supervision.

Mineral bioavailability also is strongly affected by nonmineral substances in the diet. **Phytic acid (phytate)** in wheat grain and legume fiber can limit the absorption of some minerals by chemically binding to them and preventing their release during digestion. As noted in Chapter 5, fiber intake greatly above the Adequate Intake of 25 to 38 g/day can adversely affect mineral status. However, if grains are leavened with yeast, enzymes produced

**phytic acid (phytate)** Constituent of plant fibers that binds positive ions (e.g., zinc as $Zn^{2+}$) to its multiple phosphate groups and decreases their bioavailability.

**unleavened bread** Bread that does not contain leavening agents, such as yeast or baking powder. Leavening agents cause bread dough to rise. Flat breads, such as pita bread and tortillas, are unleavened. French and Italian bread, biscuits, and muffins are leavened breads.

**oxalic acid (oxalate)** Organic acid, found in spinach, rhubarb, and other leafy green vegetables, that can depress the absorption of certain minerals (e.g., calcium) present in the food.

**Table 14-4 Factors That Affect Mineral Bioavailability**

| Factors That Increase Bioavailability | Factors That Decrease Bioavailability |
| --- | --- |
| Normal gastric acidity | Reduced gastric acidity |
| Vitamin C intake | Phytic acid in whole grains and legumes |
| Good vitamin D status | Oxalic acid in leafy vegetables |
| | Polyphenols in tea, coffee, and red wine |
| | High-dose supplements of single minerals |

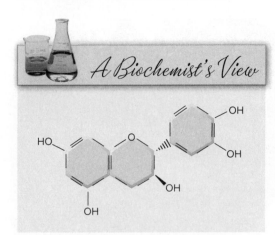

*A Biochemist's View*

There are many polyphenols in foods. This structure is catechin, found in the cacao bean used to make chocolate.

by the yeast can break some of the chemical bonds between phytic acid and minerals. Breaking these bonds increases the bioavailability of the minerals. The zinc deficiencies found among some Middle Eastern populations are attributed partly to their heavy reliance on **unleavened breads,** resulting in low bioavailability of dietary zinc.

**Oxalic acid (oxalate),** found in leafy green plants, also binds minerals and makes them less bioavailable. Spinach, for example, contains plenty of calcium, but only about 5% of it can be absorbed because of the vegetable's high concentration of oxalic acid.[16] On the other hand, about 32% of the dietary calcium is absorbed from milk and milk products.

**Polyphenols** are a group of compounds containing at least 2 ring structures that each have at least 1 hydroxyl group (OH) attached. Polyphenols also can lower the bioavailability of minerals, especially iron and calcium. Many polyphenols occur naturally in plants, such as tea, dark chocolate (cacao beans), and wine (grapes). Some types of polyphenols, such as flavonoids and tannins, may help prevent cancer and heart disease.

Mineral bioavailability can be enhanced by some vitamins. Vitamin C can improve iron absorption when both are consumed in the same meal. The vitamin D hormone calcitriol improves calcium, phosphorus, and magnesium absorption.

Gastric acidity also promotes the bioavailability of many minerals. Hydrochloric acid (HCl) in the stomach makes minerals more bioavailable by dissolving them and converting them to a form that can be more easily absorbed. For example, HCl provides an electron to ferric iron ($Fe^{3+}$) to yield ferrous iron ($Fe^{2+}$), which is better absorbed than ferric iron. Reduced stomach acid production, common in old age and with the use of antacids, can hinder mineral bioavailability.

## Transport and Storage of Minerals

Once absorbed, minerals travel in the blood, either in a free form or bound to proteins. For example, calcium ions can be found free in the blood or bound to the blood protein albumin. Trace minerals in their free form are often highly reactive and are toxic if not bound. Thus, many trace minerals have specific binding proteins that transport them in the bloodstream. Many also are bound by specific cellular proteins once they are taken up by cells.

Mineral stores provide a ready source of the mineral when dietary intake is lacking. Mineral storage in the body varies tremendously (see Fig. 14-11). Some minerals are stored in large quantities; for example, calcium and phosphorus are stockpiled in bones and teeth. Other minerals, such as iron stored in the liver and bone marrow, are stored in smaller amounts. The human body has extremely small stores of most microminerals and ultratrace elements.

## Excretion of Minerals

Excretion of the major minerals takes place primarily through the urine. However, some trace minerals, such as copper, are secreted by the liver into the bile for excretion in the feces. When kidney function fails, mineral intake must be controlled to avoid mineral toxicity, such as with phosphorus and potassium.

## Functions of Minerals

The metabolic roles of minerals vary considerably. Water balance requires sodium, potassium, calcium, and phosphorus. Sodium, potassium, and calcium aid in the transmission of nerve impulses throughout the body. Some minerals, such as iron, magnesium, copper, and selenium, function as cofactors and enable enzymes to carry out chemical reactions. Minerals also are components of many body compounds. For example, iron is a component of hemoglobin in red blood cells. Body growth and development also depend on certain minerals, such as calcium and phosphorus. At all levels—cellular, tissue, organ, and whole body—minerals play important roles in maintaining body functions (see Fig. 14-1).

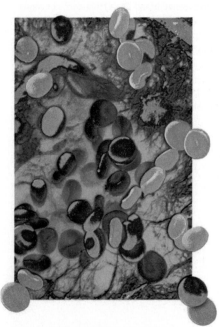

Red blood cells contain the mineral iron.

## Mineral Deficiencies

The Dietary Guidelines for Americans identify calcium and potassium as minerals likely to fall short of DRI recommendations in diets of all age groups in the U.S. Other minerals are underconsumed during specific life stages, such as iron intakes in teen-age girls. Because of their association with health concerns, low intakes of calcium (bone health), potassium (cardiovascular health), and iron (blood health, especially in children and during pregnancy) are considered to be public health concerns in the U.S.[17] Deficiencies of zinc and iodine also are common in some regions of the world. Mineral deficiencies are discussed further in this chapter and Chapter 15.

## Mineral Toxicity

Excess mineral intake can be toxic, particularly trace minerals, such as iron and zinc. The use of mineral supplements, especially if intake will exceed the Upper Level, is best considered after consultation with a registered dietitian nutritionist or physician. The potential for toxicity is not the only reason to carefully consider the use of mineral supplements. As discussed previously, high intakes of a mineral can hinder the absorption of others. Also, in the past, some mineral supplements were found to be contaminated with toxic compounds, such as with lead. Selecting brands approved by the U.S. Pharmacopeia (USP) lessens this risk. The USP-approved brands are tested to assure that contaminants are not present in harmful amounts, that the ingredients listed on the label are present and will dissolve in the body, and that the supplements were made under safe and sanitary conditions.

▶ Salt has a colorful history. It once was the custom to rub salt on newborn babies as a symbol of purity and to ensure their good health. Salt was once so scarce that it was used as money. Caesar's soldiers received part of their pay in salt. This part of their pay was known as their "salarium," and from this custom came today's word *salary*. The expression "not worth his salt" meant that a man did not earn his wages.

### Knowledge Check

1. How are the major minerals differentiated from microminerals?
2. What are 3 factors that can alter the bioavailability of a mineral?
3. What are 3 functions of minerals in the body?

# 14.3 Sodium (Na)

Salt—the most important source of the essential nutrient sodium—is produced by the evaporation of seawater. Inland salt deposits were created when ancient seas dried up; these deposits can be removed through mining or by forcing fresh water into them, pumping the salty brine that forms to the surface, and removing the water using a heat and vacuum process. Salt also is harvested by evaporating seawater and collecting the remaining salt crystals.

## Sodium in Foods

Salt, sodium chloride (NaCl), contributes most of the sodium to our diets (Fig. 14-12). Salt is 40% sodium and 60% chloride, which means that a teaspoon of salt (about 6 g) provides 2300 mg of sodium. However, most of the sodium we consume doesn't come from the salt shaker at home. The majority—75 to 80%—is added during food processing and at restaurants, as either salt or sodium-containing food additives. Sodium naturally present in foods provides about 10% of the sodium we consume, and the salt added in cooking and at the table provides another 10 to 15%. Other sodium sources are softened tap water and certain medicines.

Almost all unprocessed foods naturally contain little sodium; the higher amount found in milk (about 100 mg/cup) is an exception. Consider that, if we ate only unprocessed foods and added no salt, our daily sodium intake would be about 500 mg.[2] Comparing this with the average daily intake of 4094 mg by men and 2997 mg by women in the U.S.,[18] it is clear that food processing contributes most of our dietary sodium. Processed foods tend to be higher in sodium because this versatile mineral is part of many compounds—flavorings (table

Because bread and rolls are so widely consumed, they are leading contributors of sodium in the U.S. diet.

©McGraw-Hill Education/Mark Dierker, photographer

*Culinary Perspective*

## Sea and Specialty Salts

Salt has been highly valued as a food flavoring and preservative for thousands of years—an early reference can be found in the biblical Book of Job, written around 300 B.C. Today, many of us enjoy the flavor salt adds to food. Salt also is important in the production of many common food products, such as cheese, cured meats, pickled vegetables, and bread. Despite these attributes, many nutrition scientists believe that our intake of sodium is too high for good health.

Sea salt, made by evaporating seawater, is a popular ingredient because of its coarse, crunchy texture and stronger flavor than typical table salt. Specialty, or "gourmet," salt may be colored, flavored, or smoked. Many specialty salts originate in an exotic area of the world, such as the Himalaya mountains or Hawaii. Some individuals mistakenly believe that sea salt and specialty salt are healthier choices, with a lower sodium content and more trace minerals than typical table salt. However, whether chiseled from a mine or evaporated from the sea, table salt, sea salt, and specialty salt all have similar sodium contents—they all provide approximately 2300 mg of sodium per teaspoon. And none of them is a rich source of trace minerals.

Salt harvested after evaporating off seawater is a traditional source of sodium.

Pham Le Huong Son/Moment/Getty Images

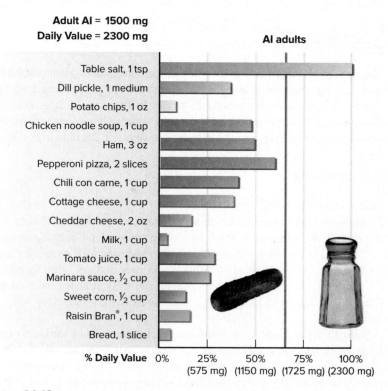

**Figure 14-12** Food sources of sodium.

pickle: ©Brand X Pictures/Getty Images RF; salt: ©C Squared Studios/Getty Images

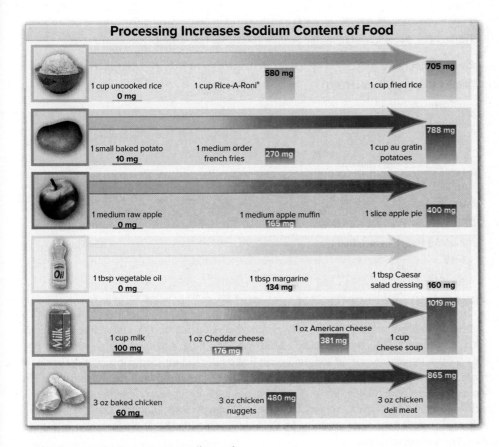

**Processing Increases Sodium Content of Food**

1 cup uncooked rice
**0 mg**

1 cup Rice-A-Roni®
**580 mg**

1 cup fried rice
**705 mg**

1 small baked potato
**10 mg**

1 medium order french fries
**270 mg**

1 cup au gratin potatoes
**788 mg**

1 medium raw apple
**0 mg**

1 medium apple muffin
**165 mg**

1 slice apple pie
**400 mg**

1 tbsp vegetable oil
**0 mg**

1 tbsp margarine
**134 mg**

1 tbsp Caesar salad dressing
**160 mg**

1 cup milk
**100 mg**

1 oz Cheddar cheese
**176 mg**

1 oz American cheese
**381 mg**

1 cup cheese soup
**1019 mg**

3 oz baked chicken
**60 mg**

3 oz chicken nuggets
**480 mg**

3 oz chicken deli meat
**865 mg**

*Figure 14-13* Processed foods generally contain more sodium than fresh foods. For example, a sodium free apple increases to 400 mg when that apple is incorporated into apple pie. The salt used to flavor the pie along with sodium-containing preservatives account for the increase. To keep sodium intake in check, consume foods that are minimally processed.

rice: ©Ryan McVay/Getty Images RF; potato: ©Stockbyte/Getty Images RF; apple: ©Ingram Publishing/Alamy RF; oil: ©McGraw-Hill Education/ Jacques Cornell photographer; milk: ©Ingram Publishing/Fotosearch RF; chicken: ©Elena Elisseeva/Shutterstock Inc.

salt), flavor enhancers (monosodium glutamate), preservatives (sodium benzoate), leavening agents (sodium bicarbonate, also called baking soda), curing agents (sodium nitrite), wetting agents for quick-cooking cereals (sodium phosphate), color preservatives (sodium bisulfite), anticaking agents that keep powdered foods from clumping (sodium aluminum silicate), and many others. Figure 14-13 shows how sodium levels change when food is processed.

The major contributors of sodium in the U.S. diet are mixed dishes (including burgers, pizza, pasta dishes, and sandwiches), protein foods (including cold cuts, cured meats, and processed meats), various grains, vegetables, and snacks and sweets.[17] Bottled sauces, condiments, spreads, and dips also contribute significantly to sodium intake. According to the CDC, more than 65% of our sodium intake is from retail processed foods.[19] Many of these foods are now available in reduced sodium versions. Restaurant and fast foods tend to contain excessive amounts of sodium. For example, a large order of cheese fries with ranch dressing contains more than 4000 mg of sodium, and a chicken fajita dinner may have more than 3500 mg of sodium.

**Sodium Sources in the U.S. Diet[17]**

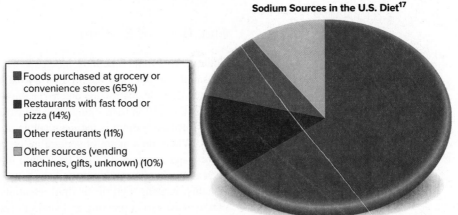

- ■ Foods purchased at grocery or convenience stores (65%)
- ■ Restaurants with fast food or pizza (14%)
- ■ Other restaurants (11%)
- □ Other sources (vending machines, gifts, unknown) (10%)

## Sodium Needs

Several recommendations for sodium intake for adults are published. As Table 14-5 indicates, the recommended sodium intakes range from 1200 mg/day to 2400 mg/day, with lower intakes recommended for older adults and those with medical conditions such as hypertension. Most healthy adults should strive to consume no more than 2300 mg per day. This is a generous recommendation; normal physiological functions can be met with much less.

▶ Many medical journals report sodium intake or excretion using the mmol unit instead of mg. To convert mmol Na to mg Na, multiply mmol by 23 (the molecular weight of sodium). Thus, a reported intake of 50 mmol Na equals 1150 mg Na:

$$50 \text{ mmol} \times 23 \text{ mg/mmol} = 1150 \text{ mg}$$

**Table 14-5  Recommendations for Sodium Intake for U.S. Adults**

| Recommendation | Amount Not to Exceed |
|---|---|
| **Dietary Reference Intake (DRI)** | |
| **Adequate Intake** | |
| Ages 20–51 years | 1500 mg |
| Ages 51–70 years | 1300 mg |
| Ages 71 years and over | 1200 mg |
| **Tolerable Upper Intake Level** | 2300 mg |
| **Daily Value** | 2300 mg |
| **Dietary Guidelines for Americans** | 2300 mg |
| **American Heart Association** | 1500 mg |

## Absorption, Transport, Storage, and Excretion of Sodium

Almost all the sodium consumed is absorbed in the intestinal tract. Sodium, like potassium and chloride ions, is absorbed by active transport in both the small and large intestines. Energy for the active transport of sodium is supplied by the sodium-potassium pump, shown in Figure 14-3. Most sodium in the body is found in the extracellular fluid compartment (ECF), where it is tightly regulated. When sodium intake is high, excess sodium is excreted by the kidneys. Conversely, when the concentration of sodium in the blood is low, the hormone aldosterone inhibits sodium excretion by the kidneys (see Fig. 14-9). Sodium also is lost via the feces and perspiration.

## Functions of Sodium

Sodium has 3 main functions: it helps in the absorption of glucose and some amino acids in the small intestine, it is required for normal muscle and nerve function, and it aids in water balance. Muscle contraction and nerve impulse conduction rely on the electrical charge created by the shift of both sodium and potassium ions across the cell membrane.

Because sodium is the main solute in the ECF, it regulates the ECF and plasma volumes. When the amount of sodium in the body increases (from more sodium in the diet), more water is retained in the body until the excess sodium is excreted. In some diseases, such as **nephrotic syndrome** and congestive heart failure, sodium excretion by the kidneys is faulty, causing significant fluid retention and edema. Such water and sodium retention can increase blood pressure and strain the cardiovascular system. Even in healthy persons, water retention can occur, especially when standing for long periods of time in hot weather. Consuming less sodium can improve this condition.

## Sodium Deficiency

Sodium deficiency is rare because of the abundance of sodium in the food supply, coupled with relatively low requirements for sodium. Nevertheless, sodium depletion can occur when losses exceed intake, such as in excessive perspiration. However, only when weight loss from perspiration exceeds about 2% of total body weight (or about 5 to 6 lb) should sodium losses be of concern.[3] Even then, merely salting foods is sufficient to restore body sodium levels for most people. Athletes, however, may need to consume sports drinks during competition to avoid the depletion of sodium and other electrolytes (see Chapter 11). Although perspiration tastes salty on the skin, sodium is not highly concentrated in perspiration. Rather, water evaporating from the skin leaves sodium behind. (Perspiration contains about two-thirds the sodium concentration found in blood.) Sodium depletion also can occur because of diarrhea or vomiting, especially in infants. Electrolyte drinks are used in such cases to replace sodium (see Chapter 17).

**nephrotic syndrome** Type of kidney disease that results from damage to the kidney, often caused by another disease, such as diabetes. The symptoms include fluid retention, weight gain, and high blood pressure.

Low blood sodium, hyponatremia, is a sign of sodium depletion. The symptoms of hyponatremia include headache, nausea, vomiting, fatigue, and muscle cramps. Seizures, coma, and death can occur in severe cases. Hyponatremia also results from the ingestion of excess water, as discussed in Section 14.1.

▶ Although the concentration of sodium in blood can be measured, it provides little information about an individual's sodium intake. Instead, it usually reflects hydration status. Low serum sodium (known as hyponatremia) suggests water excess, whereas high serum sodium (called hypernatremia) indicates dehydration or insufficient water intake.

## Excess Sodium Intake and Upper Level

The relationships among sodium intake, high blood pressure (hypertension), and cardiovascular disease have been widely studied and debated.[20,21,22] This body of research is complex, with many variables complicating its interpretation, including the following.

- Substantial differences in study designs, such as observational studies versus clinical trials
- A variety of methods used to measure sodium intake, such as dietary recall versus urinary sodium excretion (which can be measured in different ways, too)
- A diversity of study populations, such as normotensive versus hypertensive individuals, younger versus older adults, and healthy versus less healthy individuals

Although many studies find that, in hypertensive adults, blood pressure can be reduced by decreasing sodium intake, the effects of such reduction in normotensive individuals are less evident.[23,24,25] However, data from a few countries that have decreased sodium in the food supply show that blood pressure improves and deaths from cardiovascular disease decline.[26] More research is needed.

The Upper Level and Daily Value for sodium for adults are set at 2300 mg per day with the goal of reducing hypertension and cardiovascular disease. Fewer than one-third of adults consume less than this amount. As stated previously, the American Heart Association recommends that people eat no more than 1500 mg sodium/day but also acknowledges that, for many people, reducing to 2400 mg/day or less will bring cardiovascular benefits. For most North Americans, reducing sodium to the Daily Value will require that the amount of sodium in the food supply, especially in processed and restaurant foods, be reduced. Removing sodium from the food supply is challenging because many consumers have a longstanding preference for salty foods and diets. However, taste preferences can change, and the FDA is working with the food industry to develop voluntary sodium reduction targets.[27] Some public health experts estimate that reducing sodium intake to the Daily Value will reduce the prevalence of hypertension by at least 20% and will prevent thousands of heart attacks and strokes annually.[28]

Sodium intakes greater than 2 g per day also increase calcium loss in the urine and are of potential concern for the loss of calcium from bones. However, scientists have not determined that high sodium intakes cause or worsen osteoporosis.[29] Extra calcium in the urine

## Table 14-6 Guidelines for Decreasing the Amount of Salt (Sodium Chloride) in the Diet

| Choose These Foods More Often | Choose These Foods Less Often |
|---|---|
| Grains: whole grains or enriched breads and cereals, plain rice, pasta | Packaged rice and pasta mixes |
| Vegetables: fresh and frozen plain (no sauce) vegetables | Canned vegetables (read the labels), tomato and pasta sauces, frozen vegetables with sauce |
| Fruits: fresh, canned, and frozen fruits | Commercial fruit pies, turnovers |
| Dairy: low-fat milk, yogurt | Cheeses, especially processed |
| Meats and substitutes: fresh or frozen lean meats, poultry, fish, shellfish, unsalted lean pork, eggs, tuna or salmon canned without added salt, canned and drained meat and poultry, unsalted nuts or seeds, dried peas, beans, lentils | Salted nuts; frozen, breaded meat, fish, and poultry; processed meats, such as luncheon meats, bologna, salami, hot dogs, bacon, ham, beef jerky |
| Entrées: prepared from fresh ingredients or those labeled reduced sodium | Instant and canned soups, frozen entrées and dinners, pizza, many fast food items |
| Snack items: unsalted crackers, popcorn, pretzels, breadsticks | Salted crackers, popcorn, chips |
| Seasonings and condiments: fresh and dried herbs, lemon juice, low sodium seasoning products | Salt added during cooking, bouillon cubes, seasoning salts, soy sauce, teriyaki sauce, barbeque sauces, pickles, olives, bottled salad dressings |
| Hard tap water, low sodium bottled water | Softened water, sodium-containing well or bottled water |

*Critical* Thinking

Mrs. Massa has recently seen and heard a lot about the amount of salt in foods. She has been surprised by the number of articles that advise the public to decrease the amount of salt in their food. If sodium is such a bad thing, Mrs. Massa wonders, why do you need to have any at all? How would you explain to her the need for some sodium?

can lead to the formation of calcium oxalate kidney stones, the most common type; thus, reducing sodium levels is warranted for individuals prone to kidney stones.[30] In summary, reducing sodium intakes to below the Upper Level is likely to improve the health of many by lowering the risk of cardiovascular disease and kidney stone formation.

As discussed in Chapter 2, Nutrition Facts labels can help you become aware of the amount of sodium in your diet. Many food processors are responding to consumer desire for less sodium by offering modified foods. The terms that appear on food labels, such as *salt free, sodium free,* and *low sodium* (see Table 2-3), can help you quickly locate these foods. Table 14-6 provides suggestions for trimming dietary sodium intake. Note that the desire for salty taste is learned—that is, by eating salty food often, people gradually acquire a taste preference for salty foods. However, as sodium intake decreases, the preference for salty flavors declines.

### Knowledge Check

1. Which foods contribute high amounts of sodium to the diet?
2. How is excess sodium eliminated from the body?
3. What are the 3 main functions of sodium?
4. What is a good target for sodium intake for many healthy adults?
5. What are some strategies for decreasing sodium in the diet?

## 14.4 Potassium (K)

Potassium, a silvery gray metal, was discovered in the early 1800s. Its name comes from the word *potash,* which means "extracted in a pot from the ash of burnt trees." As you might guess from the origin of its name, plant-based foods are rich sources of this mineral. However, despite its abundance in the food supply, potassium intakes are low in many parts of the world, including the U.S. and Europe.

### Potassium in Foods

Potassium occurs naturally in many foods and, unlike sodium, unprocessed foods are the best sources (Fig. 14-14). Fruits, vegetables, milk, whole grains, dried beans, and

*Figure 14-14* Food sources of potassium.

©Brand X Pictures/Getty Images RF

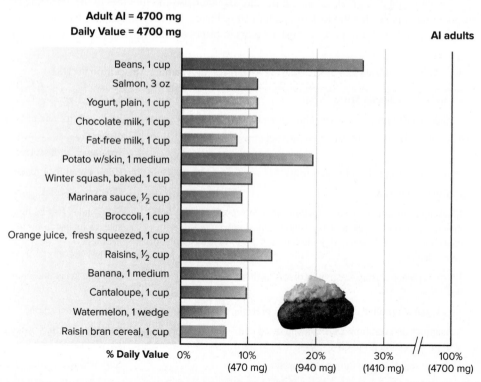

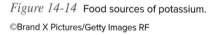

meats are all good sources. According to the national survey *What We Eat in America,* the major contributors to potassium in the U.S. are fruits and vegetables (20%), milk and milk drinks (11%), meats and poultry (10%), grain-based dishes (10%), coffee and tea (7%), and fruit and vegetable juices (5%).[31] Other sources of potassium are salt substitutes (potassium chloride) and a variety of food additives, such as acesulfame-K, an artificial sweetener, and potassium propionate, a preservative.

## Potassium Needs

The Adequate Intake for potassium for adults is 4700 mg per day.[2] The Daily Value used on food and supplement labels also is 4700 mg. Average daily potassium intakes for U.S. adults fall below both of these recommendations, ranging from 2320 mg in women to 3016 mg in men.[18] Men consume more potassium than women because they tend to eat more food overall. However, women tend to consume more potassium per calorie (higher potassium density) than men. Similarly, older adults have diets with higher potassium density than younger adults. Still, most adults need to boost their potassium intakes, preferably by eating more fruits, vegetables, whole-grain breads and cereals, and reduced fat milk and dairy products.

Vegetables are a rich source of potassium, as are fruits.

©James Gathany/CDC RF

## Absorption, Transport, Storage, and Excretion of Potassium

The body absorbs about 90% of the potassium consumed. Like sodium, potassium is absorbed in both the small and large intestines. The potassium ion ($K^+$) is transported to the body's cells, where 95% of the body's potassium is found. As with sodium, potassium balance is achieved primarily through kidney excretion or retention.

## Functions of Potassium

Potassium is the major cation inside the cell and performs many of the same functions as sodium. Both are involved in maintaining fluid balance, transmitting nerve impulses, and contracting muscle. Both muscle contraction and nerve impulse conduction rely on the electrical charge created by the shift of both potassium and sodium ions across the cell membrane. Like sodium, potassium influences the excretion of calcium, but in the opposite direction—when dietary potassium is high, the amount of calcium excreted in the urine declines.[30]

Potassium also is thought to blunt the effects of a high salt intake and help keep blood pressure normal. High dietary potassium intake suppresses the renin-angiotensin system (see Fig. 14-9) and promotes the excretion of excess sodium and water.[2] High rates of hypertension in the U.S. may be due, in part, to a high dietary sodium-to-potassium ratio.[32,33,34]

## Potassium Deficiency

Low blood potassium, known as **hypokalemia,** is a life-threatening problem. The symptoms include weakness, fatigue, constipation, and an irregular heartbeat (arrhythmia) that impairs the heart's ability to pump blood.[2] Consuming too little potassium also can raise blood pressure and the risk of stroke (see Clinical Perspective: Hypertension and Nutrition). Low potassium intakes also may increase the risk of kidney stones and bone loss.[35,36]

The depletion of potassium from the body and low blood potassium most often are caused by excessive potassium losses via the urine or the gastrointestinal

Potassium and sodium aid in muscle contraction.

©Fancy Collection/SuperStock RF

tract. Some diuretics used to treat hypertension deplete potassium from the body by increasing the amount excreted in the urine. For these people, high potassium foods are good additions to the diet, as are potassium chloride supplements if recommended by a physician. Not all diuretics increase potassium in the urine, however; some are formulated to spare potassium.

More rarely, very low dietary intake can cause low blood potassium. In persons with eating disorders, low food intake (along with vomiting and laxative use) makes potassium depletion and hypokalemia a common and very serious problem. Those with alcoholism also may have poor diets lacking in potassium. Athletes who exercise heavily may lose extra potassium in their sweat. These losses can be replaced with a healthy diet rich in high potassium foods.

## Excess Potassium and Upper Level

High blood potassium, known as **hyperkalemia,** is a life-threatening problem. Hyperkalemia almost never occurs in healthy persons. Even when dietary potassium intake is extremely high, the excess is readily excreted by the kidneys. However, when kidney function is poor, potassium quickly builds up in the blood and can cause an irregular heartbeat and even cardiac arrest. Potassium levels can be controlled in these cases by careful attention to the potassium content of the diet.

The use of potassium in supplement form to treat a deficiency or poor intake is harmless if the kidneys function normally. Thus, no Upper Level has been set.[2] However, taken in excessive amounts, potassium supplements can cause intestinal upset.

**hypokalemia**  Low potassium levels in the blood.

**hyperkalemia**  High potassium levels in the blood.

### Knowledge Check

1. Which food groups are generally good sources of potassium?
2. What conditions could predispose an individual to depletion of the body's potassium?
3. What are 2 chronic diseases that good potassium intake may protect against?
4. Why is hyperkalemia (high blood potassium) rare?

## 14.5 Chloride (Cl)

Chloride is an essential nutrient—it is the main anion ($Cl^-$) in the extracellular fluid. The chloride ion should not be confused with the element chlorine ($Cl_2$). Chlorine is a strong oxidant widely used to purify water, disinfect swimming pools, bleach fabrics, and produce many products (e.g., paper, plastics). Chlorine, along with chlorine gas, is toxic.

### Chloride in Foods

Almost all the chloride in the diet is from table salt—sodium chloride. Therefore, the same foods that provide sodium in the diet also provide most of the dietary chloride. Chloride also is found in seaweed, olives, rye, lettuce, a few fruits, and some vegetables. Salt substitutes usually contain potassium chloride.

Most chloride in the diet comes from table salt—sodium chloride.

©C Squared Studios/Getty Images

## Chloride Needs

The Adequate Intake for chloride for adults is 2300 mg per day. This amount is based on the 40:60 ratio of sodium to chloride in salt (the Adequate Intake of 1500 mg of sodium is accompanied by 2300 mg of chloride).[2] The Daily Value used on food and supplement labels also is 2300 mg. An average daily consumption of 9 g of salt yields 5400 mg of chloride.

## Absorption, Transport, Storage, and Excretion of Chloride

Chloride, like sodium and potassium, is almost completely absorbed in the small and large intestines. Chloride absorption follows right along with sodium absorption. This allows a balance of electrical charges between the negatively charged chloride ion ($Cl^-$) and the positively charged sodium ion ($Na^+$). Most chloride is found in the extracellular fluid, where it is associated with sodium. Like sodium and potassium, the excretion of chloride occurs mainly through the kidneys.

## Functions of Chloride

Chloride is the main anion ($Cl^-$) in the extracellular fluid, where its negative charge balances the positive charge from the sodium ion. Together, sodium and chloride help maintain extracellular fluid volume and balance. They also aid in the transmission of nerve impulses. In addition to its role as an electrolyte, chloride has other important functions. It is a component of the HCl produced in the stomach, and it is used during immune responses when white blood cells attack foreign cells. Finally, chloride helps maintain acid-base balance and dispose of carbon dioxide by way of exhaled air.

## Chloride Deficiency

A chloride deficiency is generally unlikely because our dietary sodium chloride (salt) intake is so high. Frequent and lengthy bouts of vomiting—if coupled with a nutrient poor diet—can cause a deficiency because of loss of HCl.[2] The symptoms include weakness, anorexia, and lethargy. The loss of HCl can disrupt the balance of acids and bases in the body.

▶ Much of what is known about chloride deficiency comes from studies of infants in the 1980s who were fed infant formula inadvertently manufactured with too little chloride. The infants suffered a variety of ill effects, including growth failure, lethargy, anorexia, and weakness. Some of these children were followed up 10 years later. The only problem noted in some children were deficits in language skills.[2]

## Upper Level for Chloride

The Upper Level for chloride is 3600 mg/day. This is based on the amount of chloride that parallels the Upper Level for sodium (2300 mg/day) using the 40:60 ratio of sodium to chloride in salt.[2] As just noted, the average adult typically consumes much more than this amount. Dietary chloride may contribute to the effects of sodium chloride on increasing blood pressure. Still, as one lowers sodium intake as part of hypertension therapy, chloride intake automatically falls as well.

### Knowledge Check

1. What is the source of most dietary chloride?
2. What are the functions of chloride in the body?
3. How were chloride needs and Upper Level set?

## Hypertension and Nutrition

Nearly 1 in 3 adults, or 75 million Americans, has high blood pressure (hypertension).[37,38] Blood pressure, measured in millimeters of mercury (mm Hg), is the force of the blood against artery walls. Simply put, it measures how hard the heart is working and what condition the arteries are in. It is expressed as 2 numbers. The higher number is the systolic blood pressure—the pressure in the arteries when the heart beats. The second number is the diastolic blood pressure—the pressure in the arteries between beats when the heart relaxes. Optimal blood pressure is less than 120 over 80 mm Hg. Table 14-7 gives the blood pressure categories currently in use for adults.

### Causes of Hypertension

Diseases such as kidney disease, liver disease, and diabetes can sometimes cause a condition known as secondary hypertension. This occurs in about 5 to 10% of individuals with hypertension and is due to the underlying disease. Most individuals with hypertension, however, are classified as having primary (essential) hypertension. Primary hypertension develops over a period of years in response to changes in the arteries, kidneys, and sodium/potassium balance.[32,38] As we age, our arteries tend to narrow and become more rigid through a process called arteriosclerosis. Additionally, endothelial cells that line the arteries often release vasoconstrictors, substances that cause the arteries to constrict, in response to arterial damage, poor blood flow, stress, and other factors. Although these events alone can result in increased blood pressure, their additive effects on kidney function increase arterial pressure.

In response to this, the kidney releases more renin, causing the formation of additional angiotensin II enzyme (see Fig. 14-9). Angiotensin II is a powerful vasoconstrictor that triggers the kidney's retention of sodium and water. Diets high in sodium and low in potassium worsen these physiological changes. Over time, the result is elevation of blood pressure.

### Risk Factors for Hypertension

Age, race, obesity, physical inactivity, poor diet, and diabetes all affect the risk of high blood pressure.[38,39]

- As people increase in age, so does blood pressure; over 90% of those over age 55 will develop high blood pressure in their lifetimes.
- Black Americans tend to develop hypertension more often and at a younger age than White Americans. Approximately 45% of Black American adults have hypertension—this is among the highest rates in the world.
- Obesity, especially abdominal obesity, also increases the risk of developing high blood pressure. An increase in fat mass adds extra blood vessels, which increases the heart's workload, as well as blood pressure.
- Physical inactivity causes blood vessels to become inelastic. This rigidity increases blood pressure.
- A poor diet with too much sodium, too many calories, and too little potassium contributes to high blood pressure. Poor diets also lead to obesity.
- Elevated blood insulin concentration, a hallmark of type 2 diabetes, is associated with

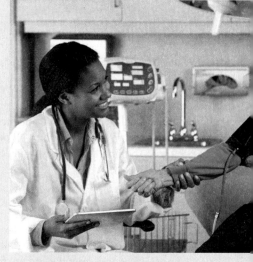

The risk of hypertension increases with age.

Monkey Business Images/Shutterstock

insulin-resistant adipose cells and is another reason for the link to obesity. Insulin increases sodium retention in the body and accelerates atherosclerosis. An estimated 65% of people with diabetes also have hypertension.

Hypertension is a serious chronic disease and health hazard. Recall from Chapter 6 that the high pressure in the arteries damages the arteries over time. Eventually, this high pressure damages the "target organs" of hypertension—the heart, brain, kidneys, and eyes. High blood pressure is considered the most significant risk factor for stroke and is a risk factor for heart attack, dementia, kidney disease, and vision

▶ Pregnancy-induced hypertension (PIH) is a type of hypertension that can occur during pregnancy. It can be very dangerous for the mother and growing fetus. (See Chapter 16.)

▶ The link between blood lead concentrations and hypertension risk appears to be mediated by biological factors, including race, ethnicity, and the form of the protein that carries lead in the blood. Lead, an environmental toxin, may damage the kidneys, cause arterial stiffness, or disrupt vascular function, eventually increasing blood pressure. Fortunately, lead exposure and blood levels are decreasing in the U.S.[40]

### Table 14-7  Classification of Blood Pressure*

| Category | SBP (mm Hg) | | DBP (mm Hg) |
|---|---|---|---|
| Normal | <120 | and | <80 |
| Prehypertension | 120–139 | or | 80–89 |
| Hypertension, stage 1 | 140–159 | or | 90–99 |
| Hypertension, stage 2 | ≥160 | or | ≥100 |

*SBP = systolic blood pressure; DBP = diastolic blood pressure.
Source: Reference Card from the *Seventh Report of the Joint National Committee on Prevention, Detection, Evaluation and Treatment of High Blood Pressure (JNC 7)*, NIH Publication Number 03-5231, May 2003. These classifications were not changed in the Eighth Joint National Committee (JNC 8), Evidence-based guideline for the management of high blood pressure in adults, 2014.

loss. Recall from Chapter 1 that cardiovascular disease, stroke, and kidney failure are leading causes of death in the U.S. Unfortunately, hypertension is often a silent disease, with no warning signs or symptoms. This is why it is important to get your blood pressure checked regularly. If hypertension is detected in its early stages, treatment is often more effective.

## Lifestyle Modifications to Prevent and Treat Hypertension

A healthy lifestyle is the cornerstone for the prevention and treatment of high blood pressure and its complications.[39] Maintaining a healthy weight and healthy eating patterns, engaging in regular physical activity, and limiting alcohol intake can help keep blood pressure normal throughout life. These modifications are presented in Tables 14-8, 14-9, and 14-10.

## The DASH Diet

The DASH (Dietary Approaches to Stop Hypertension) Diet was designed to test the effect of a diet low in saturated fat, total fat, and cholesterol and high in fruits, vegetables, and low-fat dairy products on blood pressure. The diet is rich in magnesium, potassium, calcium, protein, and fiber (see Table 14-9). In fact, the amounts of magnesium, potassium, calcium, and fiber greatly exceed typical intakes, but without using nutritional supplements. The DASH eating plan is shown in Table 14-10.

A careful test of the DASH Diet revealed that it significantly lowered blood pressure in those with normal blood pressure, prehypertension, and hypertension. The biggest blood pressure drop occurred in those with hypertension—the diet worked about as

Diets rich in fruits and vegetables help control blood pressure.
©C Squared Studios/Getty Images RF

Regular moderate exercise helps control blood pressure levels.
©Jeff Maloney/Getty Images RF

well as some common blood pressure drugs. The DASH Diet provides for 2 levels of sodium consumption: 2300 and 1500 mg/day. The DASH Diet combined with exercise and weight loss (for those who need them) may provide the greatest reduction in blood pressure.[41] The effectiveness of the DASH Diet tells us that dietary choices can be as important as medications in controlling high blood pressure.

The health benefits of the DASH Diet may extend beyond blood pressure reduction. In addition to nutrients, the fruits, vegetables, and whole grains in the diet contribute many other compounds to the diet. For instance, these foods are abundant in phytochemicals, such as polyphenols, antioxidants, and carotenoids, that can help prevent kidney stones, cancer, and heart disease.

**Table 14-8** Lifestyle Modifications to Lower Blood Pressure

| Modification | Recommendation | Expected Systolic Blood Pressure Reduction Range |
|---|---|---|
| Weight reduction | Maintain healthy body weight (BMI 18.5–24.9 kg/m²). | 5–20 mm Hg/10 kg weight reduction |
| DASH or similar eating plan | Consume a dietary pattern that emphasizes intake of vegetables, fruits, and whole grains; includes low-fat dairy products, poultry, fish, legumes, nontropical vegetable oils, and nuts; and limits intake of sweets, sugar-sweetened beverages, and red meats. | 6–7 mm Hg |
| Aerobic physical activity | Engage in moderate to vigorous aerobic physical activity at least 40 minutes per day, 3 to 4 days per week. | 2–5 mm Hg |
| Dietary sodium reduction | Consume no more than 2300 mg/day of sodium. Further reduction to 1500 mg/day will result in greater blood pressure reductions. Reduce sodium intake by at least 1000 mg per day. | 2–6 mm Hg |
| Limited alcohol consumption | Men should have no more than 2 drinks per day; women should have no more than 1 drink per day. | 2–4 mm Hg |

### Table 14-9 Nutrient Goals for the DASH Diet (for a 2100-Calorie Eating Plan)

| | | | |
|---|---|---|---|
| Total fat | 27% of calories | Sodium | 2300 mg* |
| Saturated fat | 6% of calories | Potassium | 4700 mg |
| Protein | 18% of calories | Calcium | 1250 mg |
| Carbohydrate | 55% of calories | Magnesium | 500 mg |
| Cholesterol | 150 mg | Fiber | 30 g |

*1500 mg sodium was a lower goal tested and found to be even better for lowering blood pressure. It was particularly effective for middle-aged and older individuals, Black individuals, and those who already had high blood pressure.
Source: *Your Guide to Lowering your Blood Pressure with DASH*, NIH Publication Number 06-4082, April 2006.

### Table 14-10 The DASH Eating Plan

| Food Group | Daily Servings | Serving Sizes |
|---|---|---|
| Grains* | 6–8 | 1 slice bread<br>1 oz dry cereal†<br>½ cup cooked rice, pasta, or cereal |
| Vegetables | 4–5 | 1 cup raw leafy vegetable<br>½ cup cut-up raw or cooked vegetable<br>½ cup vegetable juice |
| Fruits | 4–5 | 1 medium fruit<br>¼ cup dried fruit<br>½ cup fresh, frozen, or canned fruit<br>½ cup fruit juice |
| Fat free or low-fat milk and milk products | 2–3 | 1 cup milk or yogurt<br>1½ oz cheese |
| Lean meats, poultry, and fish | 6 or less | 1 oz cooked meats, poultry, or fish<br>1 egg |
| Nuts, seeds, and legumes | 4–5 per week | ⅓ cup or 1½ oz nuts<br>2 tbsp peanut butter<br>2 tbsp or ½ oz seeds<br>½ cup cooked legumes (dry beans and peas) |
| Fats and oils | 2–3 | 1 tsp soft margarine<br>1 tsp vegetable oil<br>1 tbsp mayonnaise<br>2 tbsp salad dressing |
| Sweets and added sugars | 5 or less per week | 1 tbsp sugar<br>1 tbsp jelly or jam<br>½ cup sorbet, gelatin<br>1 cup lemonade |

*Whole grains are recommended for most grain servings as a good source of fiber and nutrients.
†Serving sizes vary between ½ cup and 1¼ cups, depending on cereal type.
Source: In Brief: *Your Guide to Lowering your Blood Pressure with DASH*, NIH Publication Number 06-5834, August 2015.

This means that only about 25 to 50% of people experience high blood pressure with a high salt intake. Unfortunately, there is no easy way for a person or medical provider to determine if sodium sensitivity is a problem, but it is known that Black individuals, overweight persons, those with diabetes, and older adults are more likely to be sodium sensitive. These individuals should limit sodium intake to 1500 mg per day; others can aim for less than 2300 mg per day.

As described previously, the amount of potassium, compared with sodium, in the diet affects blood pressure. Diets high in potassium and low in sodium offer the most protection against high blood pressure. Such diets are consistent with the DASH Diet and recommendations from the Dietary Guidelines for Americans. Many individuals can approximate the AI for potassium (4700 mg) by adding an additional 1500 mg of potassium to their daily diets. One banana, 1/2 cup of beans, and 1 sweet or white baked potato easily provide this added potassium. On the other hand, low potassium and high sodium diets, which many people eat today, are more likely to result in high blood pressure. Diets rich in calcium, magnesium, and fiber also have been linked to lower blood pressure, but individual supplements of these nutrients generally do not show beneficial effects unless individuals have low intakes of these nutrients.[43]

### Chocolate, Cocoa, Caffeine, and Hypertension

Chocolate, cocoa, and caffeine consumption also has been linked to blood pressure. Cocoa is made from cacao beans, which contain the flavanols epicatechin, catechin, and procyanidins. These flavanols may reduce blood pressure by improving vascular function and insulin sensitivity, as well as inhibiting the renin-angiotensin system. Cocoa is used to make chocolate, with dark chocolate having more cocoa and a higher

▶ Dr. Kempner's Rice Diet was an early therapy for very severe hypertension. The diet, comprised mainly of rice and fruit, was low in calories, fat, protein, and sodium (<150 mg/day). Many people following this diet achieved normal blood pressure and improved their heart and kidney function. Kempner's work was not widely accepted by the medical establishment because the relationship between diet and blood pressure was not yet understood and because his diet was restrictive and nutritionally inadequate.[48]

### Minerals, Phytochemicals, and Hypertension

As noted previously, the issue of sodium restriction and its role in treating and preventing high blood pressure has been vigorously debated for years. The Intersalt study, conducted with over 10,000 people from 32 countries and often regarded as the "landmark"

sodium intake and hypertension study, found that, as urinary sodium excretion increased, so did blood pressure.[42] (Urinary sodium excretion is a much more sensitive indicator of sodium intake than estimating dietary intake of sodium.) However, with higher sodium intakes, a phenomenon called "salt sensitivity" occurs.

flavanol content than milk chocolate. However, even among dark chocolates with the same cocoa content, the flavanol content can vary significantly due to processing and cacao bean differences. Some studies have found that consuming flavanol rich cocoa and dark chocolate lowers blood pressure modestly in the short term.[44,45] However, more research on chocolate composition, dose, and long-term effects is needed before its consumption can be recommended as a reliable way to reduce blood pressure. Eating chocolate should not substitute for the lifestyle measures that have been proven to help reduce blood pressure! More is known about the effects of coffee and caffeine on blood pressure. Although they can temporarily increase blood pressure in some individuals, coffee and other caffeinated beverages are not thought to cause or worsen hypertension.[46,47]

© Isabelle Rozenbaum / Getty Images

### Drug Therapy for Hypertension

Medication usually isn't prescribed for hypertension until diastolic blood pressure measures at least 90 mm Hg and/or systolic blood pressure reaches 140 mm Hg on 3 or more occasions (older adults may be not treated until blood pressure reaches 150 mm Hg). The following are the general classes of hypertension medications.

- Diuretics are used most frequently. They increase water and salt excretion from the body. Some diuretics also increase potassium excretion, requiring individuals using these medications to monitor their potassium intake carefully. Examples of diuretics include furosemide (Lasix®) and hydrochlorothiazide (HydroDIURIL®).
- Beta-blockers (metropolol [Lopressor®]) slow the heart rate and force of heart contraction.
- Angiotensin-converting enzyme (ACE) inhibitors (captropril [Capoten®]) reduce the conversion of angiotensin I to angiotensin II in the lungs (see Fig. 14-9), which leads to vasodilation.
- Angiotensin II receptor blockers (ARBs) (losartan [Cozaar®]) prevent angiotensin II from binding on receptors on blood vessels, causing the blood vessels to dilate.
- Calcium channel blockers (nifedipine [Adalat®, Procardia®]) prevent calcium from entering the cells of the heart and blood vessels, which causes vasodilation.

Opting for fruits, vegetables, and low-fat foods recommended by the DASH Diet is a sound approach to nutrition for most people, regardless of hypertension risk.

©Rhoda Baer/National Cancer Institute RF

### Knowledge Check

1. What are the risk factors for developing high blood pressure?
2. Why is the periodic measurement of blood pressure important?
3. Why is hypertension a dangerous condition?
4. What lifestyle changes can help prevent and treat hypertension?

## 14.6 Calcium (Ca)

Calcium is an essential mineral for normal bone and tooth development. The bone disease osteoporosis, or "porous bone," has been known since early history. Archeologists have even discovered 4000-year-old Egyptian mummies with the classic sign of osteoporosis—the dowager's hump, or curved spine.[49] Calcium also has many industrial applications—including plaster of Paris, which was first used to set broken bones 1000 years ago.

Cottage cheese contains less calcium than other dairy products because most of its calcium is lost during production.

Didecs/Shutterstock

*Figure 14-15* Food sources of calcium.

©Stockbyte/Getty Images RF

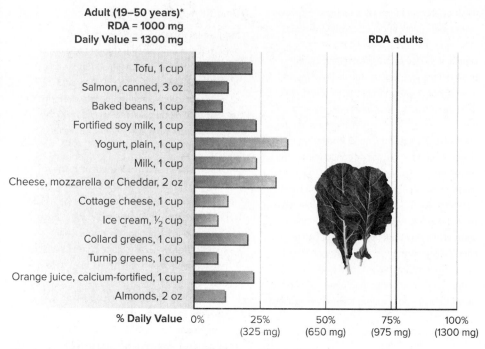

**Adult (19–50 years)***
**RDA = 1000 mg**
**Daily Value = 1300 mg**

RDA adults

| | |
|---|---|
| Tofu, 1 cup | |
| Salmon, canned, 3 oz | |
| Baked beans, 1 cup | |
| Fortified soy milk, 1 cup | |
| Yogurt, plain, 1 cup | |
| Milk, 1 cup | |
| Cheese, mozzarella or Cheddar, 2 oz | |
| Cottage cheese, 1 cup | |
| Ice cream, ½ cup | |
| Collard greens, 1 cup | |
| Turnip greens, 1 cup | |
| Orange juice, calcium-fortified, 1 cup | |
| Almonds, 2 oz | |

**% Daily Value**   0%    25%    50%    75%    100%
(325 mg)    (650 mg)    (975 mg)    (1300 mg)

*Beginning at age 51 for women and age 71 for men, RDA is 1200 mg.

## Calcium in Foods

Dairy products, such as milk and cheese, provide a rich supply of bioavailable calcium and make up nearly half of the calcium in U.S. diets (Fig. 14-15).[50] Yeast breads, rolls, crackers, and other foods made with dairy products are other contributors. Kale, collards, turnip greens, broccoli, almonds, and calcium-fortified fruit juice, soy milk, rice milk, and breakfast cereals supply this mineral. The tiny, soft bones present in canned fish, such as salmon and sardines, provide calcium, too. Tofu, when made by adding calcium carbonate to coagulate the soy protein, also provides calcium. The amount varies from brand to brand, so check Nutrition Facts labels for calcium content.

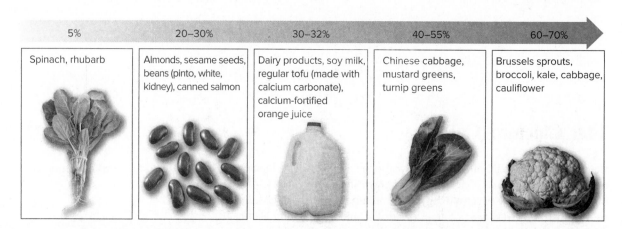

| 5% | 20–30% | 30–32% | 40–55% | 60–70% |
|---|---|---|---|---|
| Spinach, rhubarb | Almonds, sesame seeds, beans (pinto, white, kidney), canned salmon | Dairy products, soy milk, regular tofu (made with calcium carbonate), calcium-fortified orange juice | Chinese cabbage, mustard greens, turnip greens | Brussels sprouts, broccoli, kale, cabbage, cauliflower |

*Figure 14-16* The amount of calcium absorbed differs, depending on the food.[14,48] Even though only about 30% of the calcium in dairy products is absorbed, these foods are rich in calcium and provide a greater total quantity of calcium than many foods containing better-absorbed calcium.

spinach: ©Florea Marius Catalin/E+/Getty Images; beans: ©MRS.Siwaporn/Shutterstock RF; milk: ©Ryan McVay/Getty Images RF; greens: ©C Squared Studios/Getty Images RF; cauliflower: ©Brand X Pictures/Getty Images RF

## CASE STUDY

Your neighbor, Mrs. Leland, is a 50-year-old White high school English teacher. At a recent checkup, she learned that her blood pressure was 138 mm Hg/85 mm Hg and her BMI was 30. Mrs. Leland walks 3 miles twice a week. What she typically eats is described below. What advice would you give Mrs. Leland to help reduce her blood pressure? Use an online nutrient database or nutrient analysis computer program, or visit fdc.nal.usda.gov to determine her sodium intake.

©Thinkstock/Getty Images RF

Breakfast: 1 cup raisin bran, ¾ cup 1% milk, ¾ cup orange juice, 2 cups brewed coffee with 2 tablespoons flavored liquid coffee creamer

Morning break: 1 cup brewed coffee with 1 tablespoon flavored liquid coffee creamer, 2 small vanilla sandwich-type cookies

Lunch: ½ sandwich made with multigrain bread, 1½ slices deli ham, 1 slice American cheese, and ½ tablespoon mayonnaise; 1 small apple; ½ cup cheese crackers; 1 diet cola

Dinner: ½ cup 2% cottage cheese; ½ baked large chicken breast, with the skin, topped with ¼ cup marinara sauce and 2 tablespoons Parmesan cheese; ½ cup canned green beans; 1 cup garden salad with lettuce, tomatoes, carrots, and 2 tablespoons ranch dressing

Evening snack: 3/4 cup vanilla ice cream

Answer the following questions.

1. What is the classification for her blood pressure?
2. Does her physical activity meet recommendations for reducing blood pressure? If not, what would you recommend for her?
3. Compare her diet to the DASH eating plan. What could she do to improve her diet?
4. What is Mrs. Leland's sodium intake? How does it compare to recommendations?
5. What other recommendations can you make for Mrs. Leland?

When selecting foods as sources of calcium, both the amount of calcium per serving and the bioavailability of the calcium must be considered.[16,51] Recall that, in some plant foods, such as leafy greens, much of the calcium is bound to oxalic acid and is poorly absorbed. As shown in Figure 14-16, the amount of calcium absorbed varies widely. For example, as little as 5% of the 250 mg of calcium in a cup of spinach is bioavailable—about 13 mg. In contrast, a third of the 300 mg of calcium in a cup of milk is absorbable, yielding nearly 100 mg for absorption. Calcium bioavailability is especially important for vegans and non-dairy users to consider.

### Calcium Needs

The RDA for calcium was established in 2010 and is based on promoting bone growth and maintenance.[52] For most adults, the RDA is 1000 mg/day, but it increases to 1200 mg/day for women over age 50 and men over age 70. Adolescents need 1300 mg/day to allow for increases in bone mass during this time of rapid growth. The Daily Value used on food and supplement labels is 1300 mg.

In the U.S., average calcium intakes are approximately 842 mg/day and 852 mg/day for adolescent and adult females, respectively, and 1189 mg/day and 1086 mg/day for adolescent and adult males, respectively.[18] Females, especially those in the peak bone-building years from ages 9 through 18 years and those over the age of 71, have the lowest intakes. Only 15% of younger females and 39% of older women meet the RDA for calcium, even when both dietary and supplemental sources of calcium are considered. Older men, especially those over age 70, also do not fare well, with nearly 70% not reaching the RDA.[53]

*Critical* **Thinking**

Lieu is a vegetarian. She stopped eating meat and dairy products when she was 15 years old. She does eat fish and eggs, though, along with a variety of other foods. Lieu is now 28 years old and wants to start a family. She is concerned about obtaining enough calcium from her diet to ensure her baby's health. How can she consume enough calcium to meet her own and her baby's needs?

▶ No form of natural calcium, such as coral calcium or oyster shell calcium, is superior to typical supplement forms. Companies making such claims of superiority have been prosecuted by the U.S. Federal Trade Commission for false advertising.

▶ Some calcium supplements are poorly digested because they do not readily dissolve. To test for solubility, put a supplement in 6 oz of vinegar. Stir every 5 minutes. It should dissolve within 30 minutes.

▶ A 154 lb (70 kg) person has about 2.2 to 3 lb (1000 to 1400 g) of calcium stored in bones and teeth.

Calcium carbonate is found in many over-the-counter antacid tablets. Calcium in these chewable pills is best absorbed when taken with food.

PTZ Pictures/Shutterstock

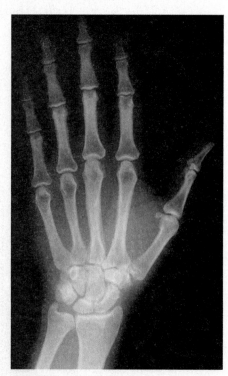

Ninety-nine percent of the calcium in the body is found in bones.

©Peter Miller/Getty Images RF

## Calcium Supplements

Calcium supplements are commonly used, especially by older women, to remedy dietary calcium shortfalls and to prevent bone loss and fracture.[53] How well calcium supplements (with or without vitamin D) protect bones in older individuals has come into question because many studies have found that calcium supplementation results in only very small increases in bone mineral density and has no effect on fractures.[54,55] Thus, the U.S. Preventive Services task force has recommended against daily supplementation for noninstitutionalized postmenopausal women (the primary users of these supplements).[56]

Calcium supplements are generally considered safe but should be used with care.[57] Adverse effects in some individuals include the following.

- Gastrointestinal symptoms, mainly constipation and flatulence
- Calcium-alkali syndrome characterized by hypercalcemia, or high blood levels of calcium. Hypercalcemia is a serious condition that can cause kidney stones, high blood pressure, and kidney failure if untreated. The risk of hypercalcemia is negligible when calcium supplement doses remain below 1500 mg/day.[58]
- Calcification of coronary arteries, which may increase risk of coronary heart disease. Some studies show a modest increase in risk of heart attack in those taking calcium supplements, but these studies cannot prove (or disprove) that supplemental calcium is harmful to cardiovascular health.[59,60] Calcium in food does not appear to increase this risk.
- Potential interactions with other minerals. As noted earlier, calcium supplements may decrease zinc, iron, and magnesium absorption, although such effects appear to be small. To minimize the potential for mineral interactions, calcium supplements should not be taken at the same time as other mineral supplements.
- Ingestion of lead in contaminated supplements.[61,62] Recall from Chapter 3 that lead ingestion causes an array of harmful effects on the body. Since 2008, the FDA has required companies to test the purity and composition of their dietary supplement products. Buying the brands approved by the U.S. Pharmacopeia (USP) lessens the risk of purchasing contaminated supplements.

The most widely used supplements are calcium carbonate (the form in calcium-based antacid tablets) and calcium citrate. The amount of calcium in a supplement depends on the form used. Calcium carbonate supplements have the highest proportion of calcium (40%), whereas those with calcium gluconate have the least (9%). The Supplement Facts label lists the amount of elemental calcium found in each pill. To boost absorption levels, calcium supplements may contain vitamin D. Absorption is best when doses do not exceed 500 mg and they are taken with or just after meals because the stomach acid produced during digestion improves absorption. Calcium citrate, which itself is acidic, is better absorbed by those with low stomach acid.

## Calcium Absorption, Transport, Storage, Regulation, and Excretion

Calcium absorption occurs along the length of the intestinal tract.[16] However, absorption is most efficient in the upper part of the small intestine because its slightly acidic pH helps keep the calcium dissolved in its ionic form ($Ca^{2+}$). Intestinal contents become more alkaline as they pass through the intestine; thus, calcium absorption decreases at the terminal end of the small intestine and in the colon, although some still occurs via passive diffusion. In addition, calcitriol (1,25 dihydroxy $D_3$) promotes and regulates the active transport of calcium that occurs in the upper intestinal tract (see Chapter 12). When an individual has poor vitamin D status, calcium absorption is reduced.

Adults absorb about 25 to 30% of the calcium present in the foods they eat, but during periods of growth when the body needs extra calcium—such as infancy and pregnancy—absorption levels might reach 75%.[52] Calcium absorption tends to decline with age, especially after age 70; postmenopausal women generally absorb the least calcium. Calcium absorption is enhanced by eating calcium rich foods with other foods, especially those that contain lactose, other sugars, and protein. Calcium absorption is reduced when there is

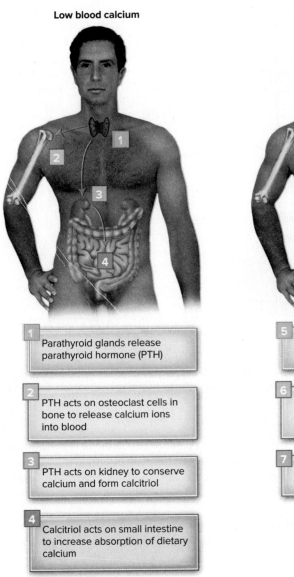

**Low blood calcium**

| | |
|---|---|
| **1** | Parathyroid glands release parathyroid hormone (PTH) |
| **2** | PTH acts on osteoclast cells in bone to release calcium ions into blood |
| **3** | PTH acts on kidney to conserve calcium and form calcitriol |
| **4** | Calcitriol acts on small intestine to increase absorption of dietary calcium |

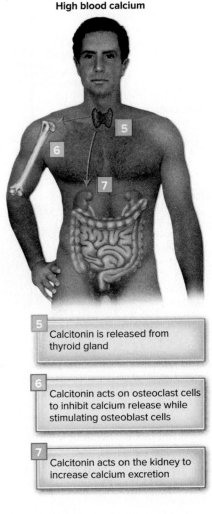

**High blood calcium**

| | |
|---|---|
| **5** | Calcitonin is released from thyroid gland |
| **6** | Calcitonin acts on osteoclast cells to inhibit calcium release while stimulating osteoblast cells |
| **7** | Calcitonin acts on the kidney to increase calcium excretion |

*Figure 14-17* Blood parathyroid hormone and calcitonin are key hormones for controlling blood calcium levels. When blood calcium level is low, the parathyroid glands release parathyroid hormone (PTH) **1**. PTH acts on bone **2** to release calcium ions into the bloodstream and kidneys **3** to conserve calcium and form calcitriol, which promotes absorption of calcium in the small intestine **4**. When blood calcium levels are too high, calcitonin is released from the thyroid gland **5**. Calcitonin acts on bone **6** to limit release of calcium to the blood and on the kidney **7** to increase calcium excretion in the urine. On a day-to-day basis, parathyroid hormone is the most important regulator of blood calcium levels.

reduced secretion of stomach acid, chronic diarrhea, or large intakes of phytic acid in fiber, oxalic acid, dietary phosphorus, and polyphenols, such as tannins in tea.[16,52] Fat malabsorption that occurs with some intestinal disorders also can reduce calcium absorption because calcium binds with fatty acids, forming unabsorbable soaps in the intestinal lumen.

Calcium in the bloodstream is transported to cells either as free ionized calcium or bound to proteins. The skeleton and teeth hold more than 99% of the body's calcium. However, all cells, not just those in bone, have a crucial need for calcium.

The concentration of calcium in the bloodstream is regulated by very tight hormonal control. This means that normal blood calcium can be maintained even when calcium intake is poor because calcium absorption and excretion and its release from bone are modified to keep blood and cellular concentrations normal. (This situation makes blood calcium a poor measure of calcium status.) As discussed in Chapter 12, when blood calcium falls, the parathyroid gland releases parathyroid hormone (Fig. 14-17). This hormone raises blood calcium levels by increasing the kidneys' reabsorption of calcium so that urinary calcium decreases. Parathyroid hormone also helps increase calcium absorption indirectly by promoting the synthesis of calcitriol in the kidney. In addition, parathyroid hormone often works in conjunction with calcitriol to release calcium from bones, a rich nutrient reserve for both calcium and phosphorus. A medical condition known as **hyperparathyroidism** can cause persistently elevated blood calcium. Hyperparathyroidism must be diagnosed and treated by a physician.

**hyperparathyroidism** Overproduction of parathyroid hormone by the parathyroid glands, usually caused by a nonmalignant tumor or abnormal growth of the glands. In most cases, there are no symptoms except hypercalcemia but, in more severe cases, weakness, confusion, nausea, and bone pain occur. Bone fractures and kidney stones also are problems.

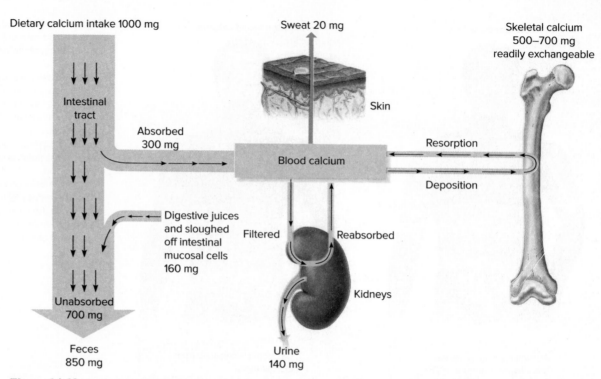

**Figure 14-18** Calcium balance in an adult. From a 1000 mg intake, only about 300 mg are absorbed. The remaining 700 mg are excreted in the feces. To maintain calcium balance, approximately 300 mg of calcium are excreted daily via the kidneys, skin, and secretions, as well as intestinal cells sloughed off in the feces.

When blood calcium levels rise too high, less parathyroid hormone is released. This causes urinary calcium excretion to increase. The synthesis of calcitriol also decreases, causing a drop in calcium absorption. In addition, the thyroid gland secretes the hormone calcitonin, which blocks calcium loss from bones. All these metabolic changes keep blood calcium within the normal range.

In addition to being excreted via the urine, other routes for dietary calcium excretion are the skin and the feces (Fig. 14-18). Calcium that is part of intestinal secretions also can pass out of the body in the feces.[52]

## Functions of Calcium

Developing and maintaining bones and teeth are calcium's major functions in the body. Calcium also is required for blood clotting, the transmission of nerve impulses, muscle contraction, and cell metabolism.

### Bone Development and Maintenance

Bone consists of a network of protein fibers, primarily collagen, and minerals. Calcium and phosphorus, obtained from the diet, are the main minerals in bone. The diet also supplies other nutrients needed for healthy bones, including protein, magnesium, potassium, sodium, fluoride, sulfur, and vitamins D and K. Calcium and phosphorus form the latticelike crystal **hydroxyapatite,** $Ca_{10}(PO_4)6OH_2$, which binds to the collagen fibers. This combination of materials allows bone to be resilient and strong. Collagen protein allows the skeleton to absorb impact (e.g., bone does not usually break when you jump), and the hydroxyapatite crystal makes bone strong (e.g., bone does not bend or collapse when you jump).

The outer, dense shell of bone is **compact bone** (also known as **cortical bone**); it makes up about 75% of the skeletal mass. The remainder is **spongy bone** (also known as **trabecular bone**), a hard, spongy network of rods, plates, and needlelike spines that adds strength without much weight (Fig. 14-19). Spongy bone is abundant at the ends of the long

**hydroxyapatite** Compound composed primarily of calcium and phosphate; it is deposited in bone protein matrix to give bone strength and rigidity ($Ca_{10}(PO_4)6OH_2$).

**bone remodeling** Lifelong process of building and breaking down of bone.

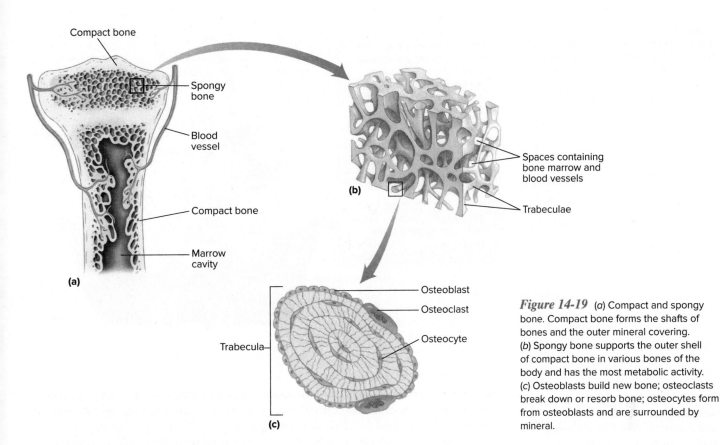

**Figure 14-19** (*a*) Compact and spongy bone. Compact bone forms the shafts of bones and the outer mineral covering. (*b*) Spongy bone supports the outer shell of compact bone in various bones of the body and has the most metabolic activity. (*c*) Osteoblasts build new bone; osteoclasts break down or resorb bone; osteocytes form from osteoblasts and are surrounded by mineral.

bones, inside the spinal vertebrae, and inside the flat bones of the pelvis. Spongy bone also is where most minerals move into and out of bone.

Bone is continually being built, broken down, and reshaped. This process of removing and replacing bone is referred to as **remodeling.** Remodeling is vital for bone health because it allows bones to grow normally and to repair and replace damaged (e.g., very small cracks) or brittle areas. Remodeling also permits calcium and phosphorus to be "withdrawn" and used for other functions when dietary intake is insufficient. Three main types of bone cells function in bone growth and remodeling: osteoblasts, osteocytes, and osteoclasts (see Fig. 14-19). **Osteoblasts** are bone-building cells that produce collagen and add minerals to form healthy bone. Some of the fully mineralized osteoblasts mature to form osteocytes, the most numerous cells in bones. **Osteocytes** are biochemically active; they can take up calcium from the blood and release it back into the blood, as well as help bone become more dense, if needed. In contrast, **osteoclasts** are cells on the bone surface that dissolve bone (termed **bone resorption**) by releasing acid and enzymes. Their activity is stimulated by parathyroid hormone, often in conjunction with 1,25(OH)$_2$ vitamin D. Osteoclasts are very active when a diet is deficient in calcium—they release calcium from the bone so that it can enter the blood. Remember, a supply of calcium is vital to all cells, not just to bone cells.

During times of growth, total osteoblast activity exceeds osteoclast activity, so we make more bone than we break down. This also can occur when bone is put under high stress—a right-handed tennis player, for example, builds more bone in that arm than in the left arm. Most bone is built from infancy through the late adolescent years. Calcium recommendations are set high during these times to support this bone-building activity. Small increases in bone mass continue between 20 and 30 years of age. Genetic background controls up to 80% of the variation in the peak bone mass ultimately built.[63]

Throughout adulthood, bone remodeling is ongoing. In fact, most of the adult skeleton is replaced about every 10 years. However, beginning in middle age, osteoclast activity generally becomes more dominant in both men and women. This can result in a total loss of about 25% of bone, depending on how long the person lives.[64] Women experience

▶ Milk provides good amounts of many of the nutrients needed for building bone. Drinking milk during childhood (the bone-building years) is associated with a lower risk of fracture later in life.

▶ The tooth consists of a hard, yellowish tissue called dentin, which is covered with enamel in the crown and cementum in the root. Enamel, the hardest substance in the body, is almost entirely hydroxyapatite crystals. Produced before the tooth erupts, enamel is not a living tissue and, unlike bone, does not undergo remodeling. Enamel can be easily damaged by acids produced by bacteria in the mouth from the metabolism of sugars and carbohydrates. This damage is known as dental caries, or cavities. Repairing dental caries requires the skills of a dentist. The number of dental caries in the U.S. population has declined because of the addition of fluoride (see Chapter 15) to some municipal water supplies, bottled water, and oral care products. Fluoride helps strengthen the enamel and makes it more resistant to acids. It also helps remineralize enamel in the earliest stages of damage.

*Figure 14-20* The release of neurotransmitters from nerve cells requires an influx of calcium through specialized calcium channels. This physiological function of calcium is vital for normal functioning of the nervous system.

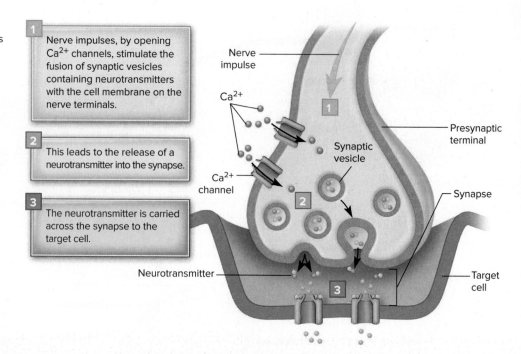

1 Nerve impulses, by opening $Ca^{2+}$ channels, stimulate the fusion of synaptic vesicles containing neurotransmitters with the cell membrane on the nerve terminals.

2 This leads to the release of a neurotransmitter into the synapse.

3 The neurotransmitter is carried across the synapse to the target cell.

Nerve impulse

$Ca^{2+}$

$Ca^{2+}$ channel

Synaptic vesicle

Neurotransmitter

Presynaptic terminal

Synapse

Target cell

**bone resorption** Process in which osteoclasts break down bone and release minerals, resulting in a transfer of calcium from bone to blood.

**tetany** Continuous, forceful muscle contraction without relaxation.

even greater bone loss when estrogen levels fall in menopause because estrogen inhibits bone breakdown by decreasing osteoclast activity. Women can experience an additional 20% loss of bone in the first 5 to 7 years following menopause.[64] Absent or irregular menses in younger women, which can occur because of very low body weight or eating disorders, signal low estrogen levels and a likelihood of substantial bone loss before middle age. When the rate of bone loss exceeds the rate at which bone is rebuilt, bone mass and bone strength decline and the risk of fracture rises greatly. Significant bone mass loss, known as osteoporosis, is discussed in Clinical Perspective: Osteoporosis.

### Blood Clotting

Calcium ions participate in several reactions in the cascade that leads to the formation of fibrin, the main protein component of a blood clot. (See Fig. 12-22 in Chapter 12 for more details on blood clotting.)

### Transmission of Nerve Impulses to Target Cells

When a nerve impulse reaches its target site—such as a muscle, other nerve cells, or a gland—the impulse is transmitted across the **synapse,** the junction between the nerve and its target cells. In many nerves, the arrival of the impulse at the target site causes calcium ions from the extracellular medium to flow into nerve cells. The rise in calcium ions in the nerve triggers synaptic vesicles to release their store of neurotransmitters. The released neurotransmitter then carries the impulse across the synapse to the target cells (Fig. 14-20).

In an entirely different process, nerve impulses develop spontaneously if insufficient calcium is available, leading to what is called hypocalcemic **tetany** (Fig. 14-21). This condition is characterized by muscle spasms because the muscles receive continual nerve stimulation. Inadequate parathyroid hormone release or action is the typical cause of hypocalcemia (low blood calcium).[16]

### Muscle Contraction

The critical role of calcium in muscle contraction is most easily understood in the context of skeletal muscles, but other types of muscles use calcium in a similar fashion. When a skeletal muscle fiber is stimulated by a nerve impulse from the brain, calcium ions are released from intracellular stores within the muscle cells. The resulting increase in the concentration

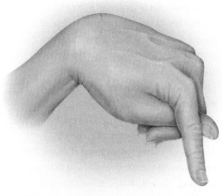

*Figure 14-21* Muscle tetany occurring in the hands and feet, such as carpopedal spasm, can be a sign of hypocalcemia.

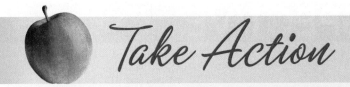

## Take Action

### Estimate Your Calcium Intake

| Dietary Sources of Calcium | Number of Servings per Day | Calcium Content (mg per Serving) | Calcium Intake (mg per Day) |
|---|---|---|---|
| Calcium-fortified juice, 6 oz<br>Yogurt, 8 oz | | × 350 | |
| Milk, any type, 1 cup<br>Calcium-fortified soy milk, 1 cup<br>Canned sardines, 3 oz<br>Parmesan cheese, grated, 1 oz<br>Lasagna or manicotti, 1 cup<br>Quiche, 1 piece<br>Tofu (processed with calcium), 4 oz | | × 300 | |
| Hard natural cheeses—Cheddar, mozzarella, Swiss, 1 oz<br>Cheese pizza, 1 slice of 12″ pie<br>Broccoli, 1 cup | | × 200 | |
| Soft cheeses—Ricotta or fortified cottage cheese (not regular cottage cheese), ¼ cup<br>Calcium-fortified cereal or granola bar, 1<br>Canned pink salmon, 3 oz<br>Meat-topped pizza, 1 slice of 12″ pie<br>Chinese cabbage and mustard greens, 1 cup | | × 150 | |
| Processed cheeses—American or Swiss, 1 slice<br>Macaroni and cheese, prepared from box, 1 cup<br>Almonds, ¼ cup | | × 100 | |
| Calcium intake from other foods | | + 250 | |
| Other calcium sources to consider: fortified cereals and vitamin and mineral supplements (check the labels) | | Varies | |
| 500 mg per serving calcium supplements: pills, calcium-fortified candies | | × 500 | |
| **Total calcium intake:** | | | |

aquariagirl1970 © 123RF.com

Compare your calcium intake with the RDA for calcium, found at the back of this book.

of calcium ions in a muscle cell, along with sufficient ATP, permits the contractile proteins to slide along each other.[16] This sliding leads to muscle contraction. Then, to allow for subsequent relaxation, the calcium ions are actively transported to the intracellular storage site, and the contractile proteins slide apart. (See Fig. A-4 in Appendix A.)

### Cell Metabolism

Calcium ions help regulate metabolism in the cell by participating in the **calmodulin** system. Each calmodulin binds 4 calcium ions. When calcium enters a cell (often because of hormone action) and binds to the protein calmodulin, the resulting calcium-calmodulin complex activates many intracellular enzymes, including an enzyme that initiates the breakdown of glycogen (Fig. 14-25).[16]

**calmodulin** Calcium-binding protein occurring in many tissues and participating in the regulation of many biochemical and physiological processes.

## Osteoporosis

Osteoporosis is the disease most often linked to low intakes of calcium. When calcium intake is insufficient, the body withdraws calcium from bone. This action preserves the indispensable functions of calcium, such as those that keep the heart and muscles working. However, bone health is related to many factors, including genetic background and several other nutritional and lifestyle factors.

Osteoporosis develops over a period of many years. A failure to maintain adequate bone mass in the body leads to a state of low bone mass called osteopenia. Factors associated with an increased risk of osteoporosis are listed in Table 14-11. Osteoporosis is diagnosed when bone loss and strength decline significantly and bones become fragile and likely to break (Fig. 14-22). The bones most likely to break are the hip, the wrist, and the vertebrae in the spine. Hip fractures are the most devastating—about 20% of those who suffer a hip fracture will die during the 1st year after the fracture, and those who do survive often can no longer walk without help and must move into long-term- care facilities. Loss of bone in the spine leads to compression fractures in the vertebrae, loss of height, and eventually **kyphosis** (dowager's hump).

Osteoporosis and low bone mass afflict many individuals, with most being over the age of 50. Approximately 10% of adults (10.2 million persons) over the age of 50 in the U.S. have osteoporosis and an additional 43 million (44%) have low bone mass.[65] Black individuals have the lowest rates of osteoporosis, followed by those of Hispanic/Latino heritage. The highest rates are seen in Caucasians and Asians. Half of all women and a quarter of all men over age 50 living in the U.S. are likely to suffer an osteoporosis-related bone fracture in their lifetimes. According to the National Osteoporosis Foundation, 2 million fractures per year are caused by osteoporosis.[64] The economic costs associated with osteoporosis treatment and fractures are significant, and these costs will only increase in future years as our population ages. Fortunately, there are now good ways to diagnose, treat, and prevent osteoporosis.

### Osteoporosis Diagnosis

A simple and very accurate test to identify osteoporosis is a dual energy X-ray absorptiometry (DEXA) bone scan.[64] DEXA scans measure bone mineral density in the spine, hip, and total body using a very low dose of X-ray radiation (about 1/10th of that received during a chest X-ray). For this test, a person lies on his or her back on a table for 10 to 20 minutes while an imaging arm glides over the length of the body (see Fig. 10-14 in Chapter 10). The ability of a bone to block the path of radiation is used as a measure of bone mineral density at each site. The DEXA measurements of the observed bone density are compared with those of a person at peak bone density (Fig. 14-23). The score resulting from these comparisons indicates whether bone mineral density is normal, low (osteopenia), or very low (osteoporosis).

Peripheral DEXA and ultrasound are other ways to measure the bone density of 1 part of the body, such as the wrist or heel. Peripheral methods are faster than DEXA but are not as accurate because the density of a single part of the body may not reflect the density of other areas susceptible to fractures, such as the spine.

---

**kyphosis** Abnormal convex curvature of the spine, resulting in a bulge at the upper back; often caused by osteoporosis of the spine.

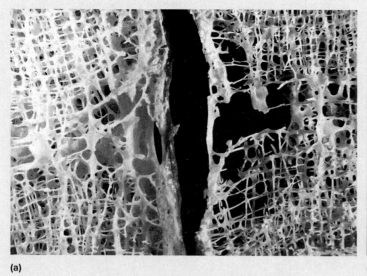

(a)

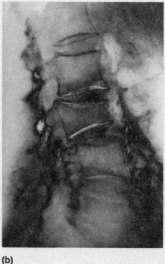

(b)

(c)

*Figure 14-22* (*a*) Normal and osteoporotic spongy bone. Note in the picture on the right how there is much less spongy bone. Breaks in horizontal or vertical spongy beams weaken a bone's support system and increase the risk of bone fracture. (*b*) Colorized X ray of vertebrae severely damaged by osteoporosis. (*c*) Abnormal curvature of the upper spine—kyphosis.

(a): ©Michael Klein/Photolibrary/Getty Images; (b): ©Dr. P. Marazzi/Science Source; (c): ©Yoav Levy/Phototake

## Table 14-11 Factors Associated with a Higher Risk of Osteoporosis

- History of osteoporotic fracture in a 1st-degree relative (i.e., parent or sibling)
- Small, thin skeletal frame
- Advancing age
- Too little estrogen in women, caused by
  - Menopause, especially in Caucasian and Asian women
  - Amenorrhea (cessation of normal menstrual periods), indicating low estrogen
  - Oophorectomy (removal of ovaries), resulting in low estrogen
- Smoking, which can have direct toxic effects on bone and impair absorption and metabolism of bone-forming nutrients; smokers also are likely to be less physically active
- Low intake of calcium, vitamin D, and other nutrients
- Excessive intake of coffee, cola, or other caffeinated beverages
- Low physical activity
- Three or more drinks (alcohol) per day
- Presence of other diseases (e.g., cystic fibrosis, anorexia nervosa, type 1 diabetes mellitus, inflammatory bowel disease, rheumatoid arthritis, celiac disease, multiple sclerosis, or epilepsy) that can impair absorption, metabolism, and utilization of bone-forming nutrients; in some cases, nutrient excretion may be increased
- Chronic use of some medications, including corticosteroids (used to treat many inflammatory and autoimmune diseases), anticonvulsants, and proton pump inhibitors (used to treat GERD; see Chapter 4)[64]

Source: National Osteoporosis Foundation, *Healthy Bones For Life: Patient's Guide*, 2014, www.nof.org.

### Osteoporosis Prevention

A healthy diet that provides adequate amounts of calcium, vitamin D, magnesium, phosphorus, potassium, vitamin K, and protein is key to bone health. Throughout the life span, the aim should be to meet the RDA or AI for these critical nutrients.

- A *calcium* rich diet, especially in the bone-building years, decreases bone turnover, increases bone density, and reduces the risk of fractures.[1,52,63,66,67] Calcium should come primarily from food sources, not supplements.
- *Vitamin D* allows us to use the calcium we consume. The optimal amount of vitamin D needed for bone health is an area of active research (see Chapter 12).
- *Magnesium*, like vitamin D, improves calcium utilization and the hormones that regulate calcium.
- *Potassium* plays a somewhat different role than the bone-forming minerals in osteoporosis prevention. By combining with

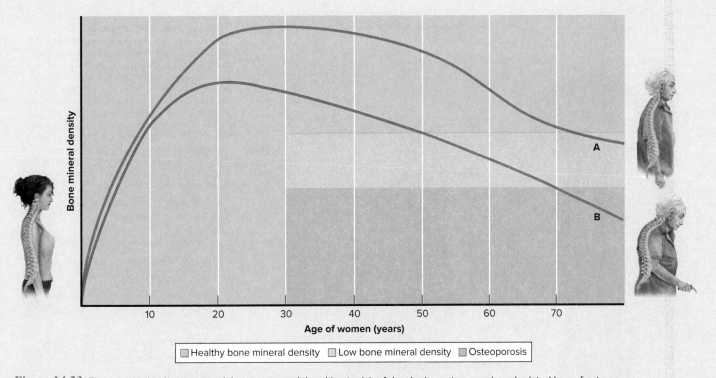

**Figure 14-23** The relationship between peak bone mass and the ultimate risk of developing osteoporosis and related bone fractures.

- *Woman A* had developed a high peak bone mass by age 30. Her bone loss was slow and steady between ages 30 and 50 and sped up somewhat after age 50 because of the effects of menopause. At age 75, the woman had a healthy bone mineral density value and did not show evidence of osteoporosis.
- *Woman B* with low peak bone mass experienced the same rate of bone loss as woman A. By age 50, she already had low mineral density, and by age 70 kyphosis and spinal fractures had occurred.

Given the low calcium intakes that are common among young women today, line B is a sobering reality. Following a diet and lifestyle pattern that contributes to maximal bone mineral density can help women follow line A and significantly reduce their risk of developing osteoporosis.

alkaline compounds, such as citrate and bicarbonate, potassium helps keep the acidity in body fluids in check.[35,36] Too much acidity can cause calcium to be leached from bones.

- *Vitamin K* helps form proteins in the bone matrix; thus, a deficiency may weaken bones (see Chapter 12).
- *Protein* also is important for healthy bones, but high protein diets increase calcium excretion and, thus, may negatively affect bone health.[63]
- Nutrient rich *dietary patterns*, like the DASH Diet, with more fruits, vegetables, whole grains, and low-fat dairy products (and less meat, desserts, fried foods, sodium, alcohol, and sugar-sweetened beverages), may also provide protection and nutrition for our bones.

In addition to dietary factors, other strategies can help build and maintain peak bone mass, thereby preventing osteoporosis. An active lifestyle that includes weight-bearing physical activity helps build and maintain muscle and bone mass. An added bonus is that exercise can help improve balance and strength, which reduces the risk of falling—a major cause of bone fractures in older and frail adults. Not smoking helps prevent osteoporosis, too.

At menopause, women should discuss osteoporosis prevention strategies with their health-care providers. They also need to keep accurate measures of their height—a decrease of more than 1½ inches from premenopausal height is a sign that significant bone loss is taking place (Fig. 14-24). The following medical therapies can be used to slow bone loss at menopause.

- Estrogen replacement slows osteoclast activity, which limits bone resorption.
- Bisphosphonates bind to hydroxyapatite crystals and osteoclasts, which slows bone resorption. (Bisphosphonates are often used in lieu of estrogen because of the increased risk of certain cancers and cardiovascular disease in some postmenopausal women.) Examples include alendronate (Fosamax®) and ibandronate (Boniva®).
- Selective estrogen receptor modulators (SERMs) (e.g., raloxifene [Evista®]) increase the utilization of existing estrogen in the body, which slows osteoclast activity.

To measure height accurately at home, tape a measuring tape (preferably made of metal) to a wall that does not have a baseboard. Be sure the measuring tape starts exactly at the floor and goes straight up the wall (perpendicular to the floor). With your shoes off and your back to the wall, stand up straight in front of the measuring tape. Your heels, buttocks, shoulders, and head should touch the wall. Keep eyes parallel with the floor. To determine your height, have a friend hold a piece of rigid cardboard above your head. The cardboard should be parallel to the floor with an edge touching the measuring tape. Then, slide the cardboard down the wall until it touches the highest point of your head—this is your maximum height.

©Don Mason/Blend Images LLC RF

- Calcitonin (Miacalcin®) inhibits osteoclast activity and bone resorption.
- RANK Ligand (RANKL) inhibitor (Prolia®) decreases osteoclast activity and bone resorption.
- Parathyroid hormone (PTH) (teriparatide [Forteo®]) stimulates osteoblast activity and new bone formation.

▶ To reduce the risk of falls, older people should exercise, limit medications and alcohol that affect physical coordination, and wear corrective lenses if their visual function is impaired.

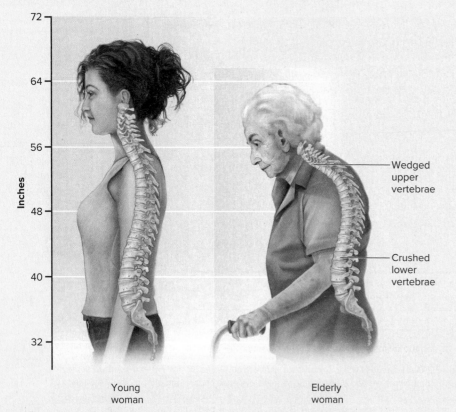

Young woman     Elderly woman

Wedged upper vertebrae

Crushed lower vertebrae

*Figure 14-24* Normal and osteoporotic women. Osteoporosis generally leads to loss of height, distorted body shape, fractures, and loss of teeth. Monitor your adult height changes to detect early osteoporosis.

## Take Action

### Bone Health

Jana, a senior in high school, recently stopped consuming dairy products. She also just started smoking, and her level of physical activity is low. Jana's diet on a recent day consisted of the following items. For breakfast, she had oatmeal made with water, a banana, and a cup of fruit juice. At midmorning, she bought a snack cake from the vending machine. At lunch, she had vegetable pasta, bread with olive oil, a small salad, 1 ounce of mixed nuts, and a soft drink. For dinner, she had a chicken burger, along with fries and water. As an evening snack, she had some cookies and hot tea.

1. What factors place Jana at risk of poor bone health?

2. What changes to her current lifestyle could reduce that risk?

3. Create an eating plan that could help Jana reduce her risk of osteoporosis.

©Stockbyte/PunchStock RF

## Potential Health Benefits of Calcium

Researchers have examined links between calcium intake and the risks of a wide array of diseases. Overall, the benefits of a diet providing adequate calcium extend beyond bone health. Both calcium and dairy products may protect against the development of colorectal cancer.[68,69,70] Dietary calcium (but not calcium from supplements) also may protect against the formation of calcium oxalate kidney stones.[71] Calcium binds oxalate in the small intestine, preventing its absorption and concentration in the kidneys. A calcium rich diet (800 to 1200 mg/day) may help decrease blood pressure, as discussed in Clinical Perspective: Hypertension and Nutrition.

## Upper Level for Calcium

The Upper Level for calcium is 2500 mg/day for adults ages 19 to 50 and 2000 mg/day for those over age 50. The UL is based on the potential for increased risk of developing kidney stones and hypercalcemia at higher intakes. Normally, the small intestine prevents the absorption of excess calcium. However, in some individuals, high amounts of calcium can cause hypercalcemia. Excessive dietary calcium also may cause irritability, headache, kidney failure, kidney stones, and decreased absorption of other minerals. Ordinarily, calcium in food and usual doses of calcium supplements do not pose a health threat because this mineral is present in relatively modest amounts.

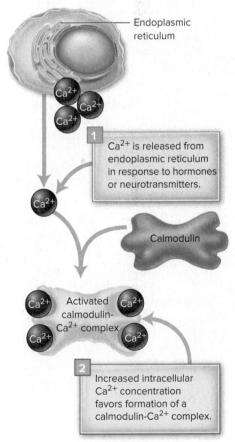

1 $Ca^{2+}$ is released from endoplasmic reticulum in response to hormones or neurotransmitters.

Calmodulin

2 Increased intracellular $Ca^{2+}$ concentration favors formation of a calmodulin-$Ca^{2+}$ complex.

Activated calmodulin-$Ca^{2+}$ complex

*Figure 14-25* Intracellular calcium ($Ca^{2+}$) binds to the protein calmodulin, activating it. Many intracellular enzymes require calmodulin for activity. These include enzymes required to contract smooth muscle and to initiate the breakdown of glycogen.

### Knowledge Check

1. Which foods provide the most bioavailable sources of calcium?
2. How do parathyroid hormone and vitamin D regulate serum calcium?
3. What are the functions of osteoblasts, osteocytes, and osteoclasts in bone?
4. Other than building bones and teeth, what are the functions of calcium in the body?

In addition to being a nutrient, phosphorus has many industrial uses, including in fireworks, matches, flares, fertilizer, and steel.

Smileus/Shutterstock

# 14.7 Phosphorus (P)

Phosphorus is an essential mineral required by every cell in the body. This phosphorescing (glowing) mineral wasn't discovered until the late 1600s, when Henning Brand, a merchant and amateur alchemist, distilled urine, hoping to manufacture gold! Instead, he discovered glowing white phosphorus in the bottom of a flask. Phosphorus was obtained in this way until researchers discovered that bones are another rich (and more pleasant!) source of this mineral. As you might guess from its history, this essential mineral is a major component of bone and teeth, and dietary excesses are removed by the kidneys.

## Phosphorus in Foods

Milk, cheese, meat, bakery products, and cereals provide most of the phosphorus in the adult diet. Bran, eggs, nuts, and fish also are sources (Fig. 14-26). Food additives, such as monosodium phosphate (emulsifier), monocalcium phosphate (jelling agent/dough conditioner), and nutrients used for fortification (e.g., iron phosphate), also contribute as much as 400 mg per day of phosphorus to the diet. Note that phosphorus from food additives may not be included in nutrient databases and in estimates of phosphorus intakes.[72]

## Phosphorus Needs

The RDA for both adult men and women is 700 mg/day.[1] Average intakes in adults, 1625 mg/day for men and 1187 mg/day for women, greatly exceed the RDA.[18] Thus, a phosphorus deficiency is unlikely in healthy adults, especially because it is so efficiently absorbed. The Daily Value for phosphorus used on food and supplement labels is 1250 mg.

## Absorption, Transport, Storage, and Excretion of Phosphorus

Adults absorb up to 70% of dietary phosphorus. Absorption occurs mainly in the upper small intestine by both active transport and diffusion. The active hormone $1,25(OH)_2$ vitamin D enhances absorption. Phosphorus in grains and legumes is mainly in the form of phytates and is poorly absorbed because we lack enzymes that release the phosphorus.[1] However, phosphorus in grain products prepared with yeast (e.g., bread) is better absorbed due to the breakdown of the phytate by the yeast.

Approximately 80% of the phosphorus in the body is found in bones and teeth as calcium phosphate (hydroxyapatite). The remainder of the phosphate is found in every cell in the body and in the extracellular fluid as $PO_4^{2-}$. Phosphorus is excreted by the kidneys. The

Meats are a rich source of phosphorus.

©Comstock/PunchStock RF

*Figure 14-26* Food sources of phosphorus.

©Brand X Pictures/Getty Images RF

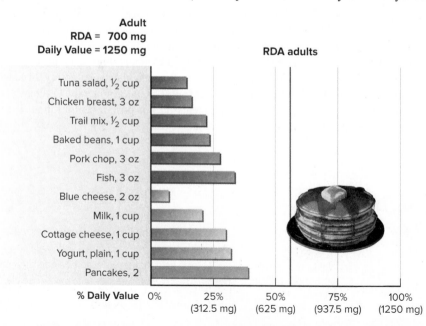

**Adult**
**RDA = 700 mg**
**Daily Value = 1250 mg**

**RDA adults**

|  | |
|---|---|
| Tuna salad, ½ cup | |
| Chicken breast, 3 oz | |
| Trail mix, ½ cup | |
| Baked beans, 1 cup | |
| Pork chop, 3 oz | |
| Fish, 3 oz | |
| Blue cheese, 2 oz | |
| Milk, 1 cup | |
| Cottage cheese, 1 cup | |
| Yogurt, plain, 1 cup | |
| Pancakes, 2 | |

**% Daily Value**  0%   25% (312.5 mg)   50% (625 mg)   75% (937.5 mg)   100% (1250 mg)

degree of excretion, regulated by PTH (like calcium) and another hormone, FGF23, produced in bone osteocytes, is the primary mechanism by which blood phosphorus level is regulated.[73]

## Functions of Phosphorus

In addition to being a major component of bones and teeth, phosphorus is critical to the function of every cell in the body. As $HPO_4^{2-}$ or $H_2PO_4^-$, phosphorus is the main intracellular anion, similar to chloride in the extracellular fluid. As a component of ATP and creatine phosphate, phosphorus is critical to energy production and storage. This mineral also is a part of DNA and RNA, phospholipids in cell membranes, and numerous enzyme and cellular message systems. Many hormones depend on phosphorylation for their activation. Phosphorus also helps regulate acid-base balance in the body.

## Phosphorus Deficiency

Phosphorus deficiency is rare, but a chronic deficiency can contribute to bone loss, decreased growth, and poor tooth development. The symptoms of rickets may occur in phosphorus-deficient children because of insufficient bone mineralization. Other symptoms of phosphorus deficiency include anorexia, weight loss, weakness, irritability, stiff joints, and bone pain. Marginal phosphorus status can be found in preterm infants, those with alcoholism, older people eating nutrient poor diets, those with long-term bouts of diarrhea and weight loss, and people who frequently use aluminum-containing antacids, which can bind phosphorus in the small intestine.[73]

## Toxicity and Upper Level for Phosphorus

Like phosphorus deficiency, toxicity is rare. As noted earlier, phosphorus intakes tend to be high; the long-term effects of high intake are unknown. However, high blood concentrations of phosphorus (hyperphosphatemia) can cause calcium-phosphorus precipitates to form in body tissues. Poor kidney function is the most common cause of hyperphosphatemia. Individuals with kidney failure must carefully monitor phosphorus intake and blood concentrations.

The typical American diet often has a low calcium intake coupled with a high phosphorus intake. For many decades, nutritionists thought this ratio of calcium to phosphorus contributed to mild hyperparathyroidism and bone loss. However, it is now thought that elevated PTH and bone loss occur in response to not meeting calcium needs (as can occur when soft drinks are substituted for milk and other calcium sources), rather than to the ratio of calcium to phosphorus in the diet. The Upper Level for phosphorus in adulthood is 3 to 4 g/day, based on the risk of developing a high blood concentration.[1]

▶ People who have experienced starvation and extreme weight loss are at risk of low blood phosphorus and a related condition called refeeding syndrome. If these individuals are aggressively refed, such as in a hospital or in a famine relief setting (in low-income countries), much of the small amount of phosphorus in the bloodstream will shift into cells in order to participate in essential metabolic pathways. This shift can cause blood phosphorus to be so low that respiratory failure, cardiac arrest, and other critical health conditions result. To avoid this problem, clinicians generally start by checking a starving person's blood phosphorus and correcting any deficiency, then gradually refeeding while monitoring and adjusting blood phosphorus to keep it in the normal range.[73]

### Knowledge Check

1. Which foods are high in phosphorus?
2. In addition to its structural roles for bones and teeth, what are critical roles of phosphorus in the body?
3. What are some of the symptoms of phosphorus deficiency?

# 14.8 Magnesium (Mg)

Magnesium was first discovered in Magnesia, a region of Greece. This silvery white metal is abundant in soil and in ocean water. It is an essential nutrient for both plants and animals—a wide array of foods supply us with magnesium.

## Magnesium in Foods

Magnesium is found in chlorophyll. Thus, some of the richest sources of magnesium are plant products, such as green leafy vegetables, broccoli, squash, beans, nuts, seeds, whole grains, and chocolate (Fig. 14-27). Animal products, such as milk and meats, supply some

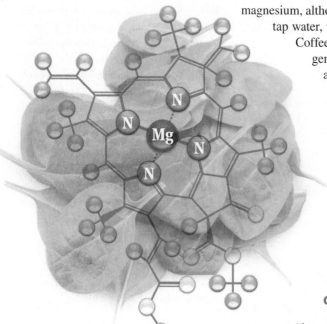

Chlorophyll has an atom of magnesium in the center of its structure. That is why plants are excellent sources of magnesium.

MysticaLink/Shutterstock

magnesium, although less than the plant-based foods. Another source of magnesium is hard tap water, which contains a high mineral content (hard water also contains calcium). Coffee and tea also contribute significantly to dietary magnesium. Refined foods generally are low in magnesium. The form of magnesium in multivitamins and mineral supplements (magnesium oxide) is not well absorbed.

### Magnesium Needs

The RDAs for magnesium are 400 mg/day for men 19 to 30 years of age and 310 mg/day for women 19 to 30 years of age.[1] Magnesium needs increase slightly (10 to 20 mg/day) beyond age 30. The Daily Value on food and supplement labels is 420 mg. Adults in the U.S. fall short on magnesium intake, with adult men and women averaging 345 and 268 mg per day, respectively. Fewer than 25% of adults meet the RDA.[18]

### Absorption, Transport, Storage, and Excretion of Magnesium

About 30 to 40% of the magnesium consumed is absorbed, but absorption efficiency can rise to about 80% when intakes are low.[74] Magnesium is absorbed in the small intestine by both passive and active absorption. The active vitamin D hormone [1,25(OH)$_2$ vitamin D] enhances magnesium absorption to a limited extent. About half of the magnesium is found in bones and the rest is stored in other tissues, such as muscles. The kidneys primarily regulate blood concentrations of magnesium and are able to reduce magnesium loss into the urine when blood magnesium is low.[74]

Nuts are a rich source of magnesium.

©C Squared Studios/Getty Images RF

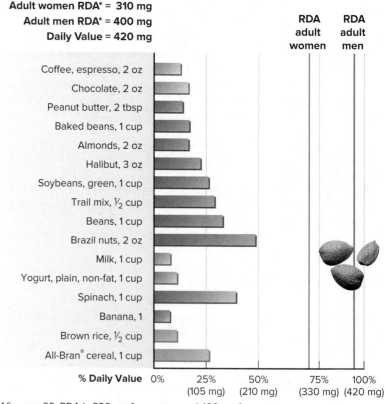

Adult women RDA* = **310 mg**
Adult men RDA* = **400 mg**
Daily Value = **420 mg**

| | RDA adult women | RDA adult men |
|---|---|---|

Coffee, espresso, 2 oz
Chocolate, 2 oz
Peanut butter, 2 tbsp
Baked beans, 1 cup
Almonds, 2 oz
Halibut, 3 oz
Soybeans, green, 1 cup
Trail mix, ½ cup
Beans, 1 cup
Brazil nuts, 2 oz
Milk, 1 cup
Yogurt, plain, non-fat, 1 cup
Spinach, 1 cup
Banana, 1
Brown rice, ½ cup
All-Bran® cereal, 1 cup

**% Daily Value**   0%   25% (105 mg)   50% (210 mg)   75% (330 mg)   100% (420 mg)

*After age 30, RDA is 320 mg for women and 420 mg for men.

*Figure 14-27* Food sources of magnesium.

©C Squared Studios/Getty Images RF

## CASE STUDY
## FOLLOW-UP

©Thinkstock/Getty Images RF

1. Mrs. Leland's blood pressure is classified as prehypertension.

2. Her physical activity level does not meet recommendations for reducing blood pressure. She needs to engage in moderate to vigorous physical activity for 40 minutes 3 to 4 times per week. She needs to walk more often to meet this guideline.

3. A comparison of her diet to the DASH recommendations indicates she needs to eat more whole grains, fruits, and vegetables and she consumes too many processed foods that contribute to empty calories and a high sodium intake. She does not appear to eat nuts, seeds, legumes, and fish. To improve her diet, she could use whole-grain bread for her sandwich; add a vegetable at lunch; replace deli meat and American cheese with tuna or sliced chicken breast; eat a piece of fruit instead of cookies at break; add a small handful of walnuts or almonds to her dinner salad; leave the chicken plain or use lemon or a low sodium seasoning mix to add flavor; and rinse the canned green beans to reduce sodium or choose fresh, frozen, or low sodium canned green beans.

4. Her sodium intake is about 3700 mg, which is much higher than the 2400 mg per day or less she should be getting.

5. Mrs. Leland's BMI indicates that she is obese. Losing weight may help reduce her blood pressure.

| Food Group | DASH Recommendation | Mrs. Leland's Diet |
|---|---|---|
| Grains | 6–8 servings, mostly whole grains | 4 servings, only 2 are whole grain |
| Vegetables | 4–5 servings, 1 cup leafy vegetable or ½ cup cooked vegetable | 2 cups |
| Fruits | 4–5 servings, 1 medium fruit or ½ cup | 2 cups fruit, which is close to the DASH recommendation |
| Fat-free or low-fat dairy milk and milk products | 2–3 servings (1 cup) | 1¼ cups |
| Lean meats, poultry, and fish | 6 ounces or less | 5½ ounces |
| Nuts, seeds, and legumes | 4–5 per week | None on this day |
| Fats and oils | 2–3 servings (1 tsp vegetable oil, 1 tbsp mayonnaise, 2 tbsp salad dressing) | 5 teaspoons |
| Sweets and added sugars | 5 or less per week | 3 in 1 day: 2 cookies, vanilla ice cream |

## Functions of Magnesium

Magnesium is the 2nd most abundant intracellular cation ($Mg^{2+}$) in the body and has a vital role in a range of biochemical and physiological processes. Magnesium helps stabilize ATP by binding to the phosphate groups of this molecule. In fact, magnesium is required by more than 300 enzymes that utilize ATP, including those required for energy metabolism, muscle contraction, and protein synthesis. A magnesium-dependent enzyme system pumps sodium out of cells and potassium into cells—this process seems especially sensitive to magnesium deficiency. Magnesium also is needed for DNA and RNA synthesis. Its role in calcium metabolism contributes to bone structure and mineralization. Magnesium also is important for nerve transmission, heart and smooth muscle contraction, and glucose and insulin metabolism.

Research suggests that diets high in magnesium rich foods may help protect against Metabolic Syndrome (see Chapter 5) and diabetes by improving insulin action.[75,76] Other research suggests that adequate magnesium in the diet decreases the risk of heart disease.[77,78] Magnesium has been shown to reduce inflammation, decrease blood pressure (and possibly stroke) by decreasing arterial stiffness,[79] and prevent heart rhythm abnormalities.

## Magnesium Deficiency

Magnesium deficiency causes an irregular heartbeat, sometimes accompanied by weakness, muscle spasms, disorientation, nausea, vomiting, and seizures. These symptoms may be related to abnormal nerve cell function caused by an impairment of sodium and potassium

▶ Magnesium sulfate is sometimes used to treat pregnancy-induced hypertension (see Chapter 16). Magnesium likely relaxes blood vessels, leading to a fall in blood pressure.

pumping. Magnesium deficiency also can cause a fall in PTH release, resulting in low blood calcium. The action of 1,25(OH)$_2$ vitamin D also is blunted during magnesium deficiency. Both of these factors may contribute to an increasing risk of osteoporosis in those with poor magnesium status. A magnesium deficiency develops very slowly because the body stores it readily.[1] Poor magnesium status has been linked to cardiovascular disease.

Excessive loss of magnesium from either the intestinal tract or the urine is responsible for most magnesium deficiencies.[74] People with GI disorders who experience prolonged bouts of diarrhea and vomiting are at risk. Those who have alcoholism, have poorly controlled diabetes, or take certain diuretics excrete more magnesium in their urine. Diets low in magnesium can worsen this situation. In addition, heavy perspiration for weeks in hot climates can increase magnesium requirements.

## Upper Level for Magnesium

The Upper Level of 350 mg per day for magnesium refers to supplement and other nonfood sources only, such as certain antacids and laxatives (e.g., Milk of Magnesia®). Intakes exceeding this amount from nonfood sources can lead to diarrhea.[1] Toxicity also can occur during kidney failure because the kidneys are the primary regulator of blood magnesium. High amounts of magnesium in the blood cause weakness, nausea, slowed breathing, eventual malaise, coma, and death. Older people are at particular risk of magnesium toxicity because kidney function tends to decline with age.

### Knowledge Check

1. Which foods are good sources of magnesium?
2. What are the functions of magnesium?
3. What are the symptoms of magnesium deficiency?
4. Which groups are most likely to be magnesium deficient?

Sulfur-containing compounds are what make rotten eggs, intestinal gas, and skunks smell bad.

Eric Issele/Shutterstock

## 14.9 Sulfur (S)

Sulfur, a bright yellow mineral, is provided primarily by the sulfur-containing amino acids methionine and cysteine. Inorganic sulfate also is found in water and food—for example, as a preservative that protects the color of dried fruits and white wines. There is no AI or RDA established for sulfur because we are able to obtain ample sulfur from protein-containing foods.[2] Sulfur is required for the synthesis of several sulfur-containing compounds, it helps stabilize the structure of proteins (e.g., collagen, hair, nails, skin), and it participates in regulating the acid-base balance in the body. No Upper Level for sulfur is established.

As you can see, the major minerals play important roles in the body. Table 14-12 provides a summary of the major minerals.

Protein rich foods supply sulfur in the diet.

©Comstock/Getty Images RF

**Table 14-12  Summary of the Major Minerals**

| Mineral | Major Functions | RDA or Adequate Intake | Dietary Sources | Deficiency Symptoms | Toxicity Symptoms |
|---|---|---|---|---|---|
| Sodium | Major positive ion of the extracellular fluid, aids nerve impulse transmission and muscle contraction, water balance, aids glucose and amino acid absorption | *Age 19–50 years:* 1500 mg *Age 51–70 years:* 1300 mg *Age over 70 years:* 1200 mg | Table salt, processed foods, condiments, sauces, soups, chips | Muscle cramps, headache, nausea, vomiting, fatigue | Contributes to hypertension in susceptible individuals, increases calcium loss in urine; Upper Level is 2300 mg |
| Potassium | Major positive ion of intracellular fluid, aids nerve impulse transmission and muscle contraction, water balance | 4700 mg | Many fruits and vegetables, milk and milk products, meat, legumes, whole grains | Irregular heartbeat, loss of appetite, muscle cramps, increased risk of hypertension and stroke | Slowing of the heartbeat, as seen in kidney failure |
| Chloride | Major negative ion of extracellular fluid, participates in acid production in stomach, aids nerve impulse transmission, water balance | 2300 mg | Table salt, some vegetables, processed foods | Convulsions in infants | Linked to hypertension in susceptible people when combined with sodium; Upper Level is 3600 mg |
| Calcium | Bone and tooth structure, blood clotting, aids in nerve impulse transmission, muscle contractions, enzyme regulation | *Age 9–18 years:* 1300 mg *Age over 18 years:* 1000–1200 mg | Milk and milk products, canned fish, leafy vegetables, tofu, fortified foods | Increased risk of osteoporosis | May cause kidney stones and other problems in susceptible people; Upper Level is 2500 mg |
| Phosphorus | Major ion of intracellular fluid, bone and tooth strength, part of ATP and other metabolic compounds, acid-base balance | *Age 9–18 years:* 1250 mg *Age over 18 years:* 700 mg | Milk and milk products, processed foods, fish, soft drinks, bakery products, meats | Possibility of poor bone maintenance | Impairs bone health in people with kidney failure, poor bone mineralization if calcium intakes are low; Upper Level is 3–4 g |
| Magnesium | Bone formation, aids enzyme function, aids nerve and heart function | *Men:* 400–420 mg *Women:* 310–320 mg | Wheat bran, green vegetables, nuts, chocolate, legumes | Weakness, muscle pain, poor heart function, seizures | Diarrhea, weakness, nausea, and malaise in people with kidney failure; Upper Level is 350 mg, but refers to nonfood sources (e.g., supplements) only |
| Sulfur | Part of vitamins and amino acids, aids in drug detoxification, acid-base balance | None | Protein foods | None observed | None likely |

# Chapter Summary

## 14.1 Water accounts for 50 to 75% of the weight of the human body.

The intracellular compartment holds two-thirds of the body's water, with the balance in the extracellular compartment. Electrolytes dissolved in the body's water help maintain fluid balance. Water's unique chemical properties enable it to dissolve substances and to serve as a medium for chemical reactions, temperature regulation, and lubrication. For adults, daily fluid needs are estimated at 9 cups (women) to 13 cups (men), but temperature, physical exertion, and other factors can greatly affect water requirements. Water balance is regulated by hormones that act on the kidneys. A water deficit results in dehydration, but too much water can cause water intoxication and hyponatremia.

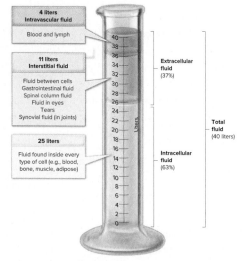

## 14.2 Minerals are divided into the major and trace minerals.

Animal foods are the best sources of calcium, iron, and zinc. Plant foods are good sources of potassium and magnesium. The absorption of minerals can be affected by the need for the minerals, the consumption of supplements, and the presence of phytic and oxalic acids. Minerals are needed for water balance, the transmission of nerve impulses, and muscle contraction. They function as enzyme cofactors and as components of body tissues. Intakes of calcium, potassium, magnesium, and iron are likely to fall short of DRI recommendations in the U.S. diet. Minerals taken in excess can be toxic.

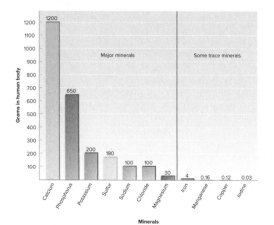

## 14.3 Sodium, the major positive ion (cation) found outside cells,

is vital in fluid balance and nerve impulse transmission. It also aids muscle contraction. One teaspoon of table salt provides 2300 mg of sodium. Most individuals in the U.S. consume more than the sodium AI (1500 mg) and UL (2300 mg). Most healthy adults should aim for no more than 2300 mg/day. The majority of dietary sodium comes from processed and restaurant foods. A high sodium intake is linked to hypertension and cardiovascular disease. Sodium deficiency is rare. Efforts to reduce the sodium content of processed and restaurant foods are under way.

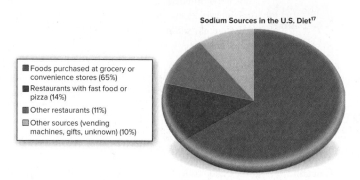

**Sodium Sources in the U.S. Diet[17]**

- Foods purchased at grocery or convenience stores (65%)
- Restaurants with fast food or pizza (14%)
- Other restaurants (11%)
- Other sources (vending machines, gifts, unknown) (10%)

## 14.4 Potassium, the major positive ion (cation) found inside cells, has functions similar to those of sodium.

Milk, fruits, and vegetables are good sources. Potassium intakes in the U.S. fall below the Adequate Intake. Low potassium diets increase the risk of hypertension and stroke. Too much dietary potassium is a problem only when kidney function is poor. Both high and low blood potassium concentrations are dangerous conditions.

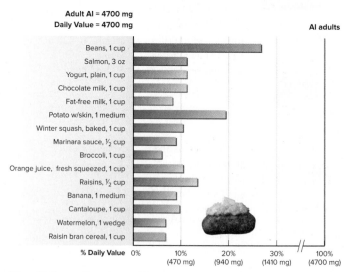

©Brand X Pictures/Getty Images RF

# 14.5 Chloride is the major negative ion (anion) found outside cells. It is important

in digestion as part of gastric hydrochloric acid and in immune and nerve functions. Table salt supplies most of the chloride in our diets.

# 14.6 Calcium forms a vital part of bone structure and is very important in blood clotting,

muscle contraction, nerve transmission, and cell metabolism. Calcium absorption is enhanced by stomach acid and the active vitamin D hormone. Blood calcium levels are tightly regulated by parathyroid hormone. Calcium is withdrawn from bone when intake is insufficient. Milk and milk products are rich calcium sources and many individuals use calcium supplements to help meet needs. Many girls, women, and older men consume inadequate calcium. Adequate calcium intake may help reduce the risk of hypertension, kidney stones, and colon cancer.

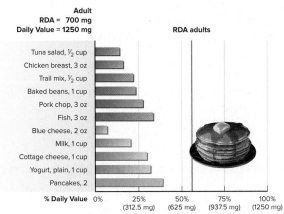

Spinach: ©Florea Marius Catalin/E+/Getty Images; Beans: ©MRS. Siwaporn/Shutterstock RF; Milk: ©Ryan McVay/Getty Images RF; Greens: ©C Squared Studios/Getty Images RF; Cauliflower: ©Brand X pictures/Getty Images RF

# 14.7 Phosphorus is a part of ATP and other key metabolic compounds,

aids the function of some enzymes, and forms part of cell membranes and bone. It is efficiently absorbed, and deficiencies are rare. Typical food sources are dairy products, bakery products, and meats.

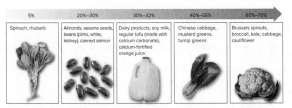

©Brand X Pictures/Getty Images RF

# 14.8 Magnesium, a mineral found mostly in plants,

is important for nerve and heart function and as a cofactor for many enzymes. Whole grains (bran portion), vegetables, nuts, seeds, milk, and meats are typical sources. Few adults meet the magnesium RDA. Magnesium deficiency causes irregular heartbeat, weakness, disorientation, nausea, vomiting, and seizures. This can be seen in those who abuse alcohol and take certain diuretics. Low magnesium intake is linked to Metabolic Syndrome, type 2 diabetes, and cardiovascular disease. The UL refers only to supplements. Intakes over the UL can cause diarrhea.

# 14.9 Sulfur is incorporated into certain vitamins and amino acids. Its ability to

bond with other sulfur atoms enables it to stabilize protein structure. There is no AI, RDA, or UL established for sulfur.

## Clinical Perspectives

Hypertension is a serious health problem that afflicts 1 in 3 adults. It increases the risk of cardiovascular disease, stroke, dementia, and kidney and eye disease. An impaired ability of the kidneys to excrete sodium along with increased arterial resistance are major contributors to hypertension. Hypertension is more prevalent in those who are older, Black individuals, overweight, insulin resistant, or diabetic. The DASH Diet, which is low in sodium and fat and high in fiber, potassium, magnesium, and calcium, is effective in treating hypertension. Reducing weight, limiting alcohol, and increasing physical activity also play important roles. Hypertension often is treated with medication when lifestyle measures are not adequate.

## Global Perspective

Per capita water use in the U.S. and Europe is much higher than the United Nations daily estimates for drinking, cooking, and cleaning. Agriculture accounts for most of the world's water consumption. Water scarcity affects more than half the world's population at some point during the year. Equitable water distribution is a major challenge. Worldwide, many people lack access to safe water supplies.

**Table 14-9 Nutrient Goals for the DASH Diet (for a 2100-Calorie Eating Plan)**

| | | | |
|---|---|---|---|
| **Total fat** | 27% of calories | **Sodium** | 2300 mg* |
| **Saturated fat** | 6% of calories | **Potassium** | 4700 mg |
| **Protein** | 18% of calories | **Calcium** | 1250 mg |
| **Carbohydrate** | 55% of calories | **Magnesium** | 500 mg |
| **Cholesterol** | 150 mg | **Fiber** | 30 g |

*1500 mg sodium was a lower goal tested and found to be even better for lowering blood pressure. It was particularly effective for middle-aged and older individuals, Black individuals, and those who already had high blood pressure.
Source: Your guide to lowering your blood pressure with DASH. NIH Publication Number 06-4082. April 2006; www.nhlbi.nih.gov/health/public/heart/hbp/dash/new_dash.pdf.

Osteoporosis is a common disease that develops over many years. Half of all women are likely to suffer an osteoporosis-related bone fracture. Osteoporosis represents an increase in bone resorption over bone building. Prevention and treatment of osteoporosis involve consuming adequate bone-building nutrients, engaging in weight-bearing physical activity, minimizing the risk of falls, and not smoking. Medications to treat osteopenia and osteoporosis are available.

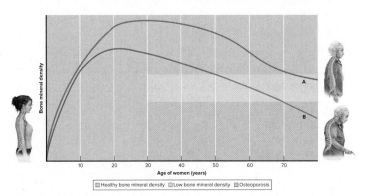

## Study Questions

1. The main electrolytes found inside the cell are _____.

   a. calcium, sodium, and chloride
   b. sulfate, potassium, and chloride
   c. potassium, magnesium, and phosphate
   d. sodium, chloride, and potassium

2. When the amount of water in the body is low, the hormones _____ and _____ signal the kidneys to conserve water and salt.

   a. insulin; calcitonin
   b. antidiuretic hormone; aldosterone
   c. estrogen; angiotensin II
   d. calmodulin; aldosterone

3. Both _____ and _____ can bind to minerals and limit their absorption.

   a. phytic acid; oxalic acid
   b. vitamin C; iron
   c. sodium; potassium
   d. protein; fiber

4. Most sodium in the diet comes from _____.

   a. table salt added at home
   b. processed foods
   c. restaurant foods
   d. both b and c

5. Sodium intake in the U.S. typically exceeds the Upper Level.

   a. true
   b. false

6. Low potassium intakes are associated with _____.

   a. eating disorders
   b. diets that contain few fresh foods and high amounts of processed foods
   c. excessive alcohol intake
   d. all of the above

7. The functions of the chloride ion include _____.

   a. serving as the main anion in the intracellular fluid
   b. aiding the transmission of nerve impulses
   c. forming acid in the stomach
   d. both b and c

8. _____ potassium intake and _____ sodium intake are associated with a higher risk of high blood pressure and stroke.

   a. High; low
   b. Low; high
   c. Low; low
   d. High; high

9. A blood pressure of 135/90 mm Hg is classified as _____.

   a. hypertension
   b. normal blood pressure
   c. prehypertension
   d. low blood pressure

10. Important lifestyle modifications for preventing hypertension include _____.

    a. maintaining a healthy BMI
    b. eating a diet rich in potassium, calcium, and magnesium
    c. not smoking
    d. all of the above

11. Significant non-dairy sources of calcium include _____.

    a. almonds, beans, mustard greens, broccoli, and tofu
    b. meat, eggs, and fish
    c. whole-grain breads and cereals, peanuts, bananas, and spinach
    d. all of the above

12. Calcium absorption is likely to be highest in _____.

    a. postmenopausal women
    b. adults in middle age
    c. adolescent males and females
    d. both a and b

13. Which of the following is *not* a risk factor for osteoporosis?

    a. low physical activity
    b. low calcium and vitamin D intakes
    c. obesity
    d. amenorrhea

14. Phosphorus deficiency is a common problem in North America.

    a. true
    b. false

15. Over half of the magnesium in the body is found in the _____.

    a. heart
    b. liver
    c. bones
    d. brain

16. Protein rich foods are the main source of the mineral sulfur in our diets.

    a. true
    b. false

17. Describe how sodium, potassium, and chloride are involved in maintaining water balance in cells.

18. What factors increase and decrease the bioavailability of minerals?

19. How are blood calcium levels maintained?

20. Why are phosphorus deficiencies rare?

21. Create a menu that is rich in magnesium.

22. Describe lifestyle modifications that can help reduce hypertension.

23. Discuss the factors that increase the risk of developing osteoporosis.

24. Describe factors contributing to water scarcity.

**Answer Key:** 1-c; 2-b; 3-a; 4-d; 5-a; 6-d; 7-d; 8-b; 9-c; 10-d; 11-a; 12-c; 13-c; 14-b; 15-c; 16-a; 17-refer to Sections 14.1, 14.3, 14.4, and 14.5; 18-refer to Section 14.2; 19-refer to Section 14.6; 20-refer to Section 14.7; 21-refer to Section 14.8; 22-refer to Clinical Perspective: Hypertension and Nutrition; 23-refer to Clinical Perspective: Osteoporosis; 24-refer to Global Perspective

# References

1. Food and Nutrition Board, Institute of Medicine. *Dietary Reference Intakes for calcium, phosphorus, magnesium, vitamin D and fluoride*. Washington, DC: National Academies Press; 1997.

2. Food and Nutrition Board, Institute of Medicine. *Dietary Reference Intakes for water, potassium, sodium, chloride and sulfate*. Washington, DC: National Academies Press; 2005.

3. Jéquier E, Constant F. Water as an essential nutrient: The physiological basis of hydration. *Eur J Clin Nutr*. 2010;64:115.

4. Negoianu D, Goldfarb S. Just add water. *J Am Soc Nephrol*. 2008;19:1.

5. Feehally J, Khosravi M. Effects of acute and chronic hypohydration on kidney health and function. *Nutr Rev*. 2015;73:110.

6. Ludwig DS and others. Dietary carbohydrates: Role of quality and quantity in health and disease. *BMJ*. 2018;361:k2340.

7. Mullens E, Jimenez F. Effects of caffeine on hydration status: Evidence analysis review. *J Acad Nutr Diet*. 2016;116:S1A78.

8. International Bottled Water Association. 2020; www.bottledwater.org.

9. Sebastian RS and others. Drinking water intake in the U.S.: What we eat in America, NHANES 2005–2008. Food Surveys Research Group Dietary Data Brief No. 7. September 2011; ars.usda.gov/Services/docs.htm? docide=19476.

10. Bruce RC, Kliegman RM. Hyponatremic seizures secondary to oral water intoxication in infancy: Association with commercial bottled drinking water. *Pediatr*. 1997;100:e4.

11. Bennett B and others. Wilderness Medical Society Clinical Practice Guidelines for the Management of Exercise-Associated Hyponatremia: 2019 Update. *Wild Environ Med*. 2020;31:50.

12. United Nations. UN water. 2012; www.unwater.org.

13. Mekonnen MM, Hoekstra AY. Four billion people experiencing water scarcity. *Sci Adv*. 2016;2:e1500323.

14. World Health Organization. Drinking water fact sheet. 2019; www.who.newsroom/fact-sheets/detail/drinking-water.

15. Bethke PC, Jansky SH. The effect of boiling and leaching on the content of potassium and other minerals in potatoes. *J Food Sci*. 2008;75:H80.

16. Weaver C, Heaney R. Calcium. In: Ross C and others, eds. *Modern nutrition in health and disease*. 11th ed. Philadelphia: Lippincott Williams & Wilkins; 2014.

17. U.S. Department of Health and Human Services and U.S. Department of Agriculture. Dietary Guidelines for Americans, 2020–2025, 9th ed. 2020. dietaryguidelines.gov.

18. U.S. Department of Agriculture, Agricultural Research Service. Nutrient intakes from food and beverages: Mean amounts consumed per individual, by gender and age. *What We Eat in America*, NHANES 2013–2014. 2016; www.ars.usda.gov/northeast-area/beltsville-md/beltsville-human-nutrition-research-center/food-surveys-research-group/docs/wweia-data-tables/.

19. Centers for Disease Control and Prevention. Vital signs: Food categories contributing the most to sodium consumption—United States, 2007–2008. *MMWR*. 2012;61(5):92.

20. Davy BM and others. Sodium intake and blood pressure: New controversies, new labels . . . new guidelines? *J Acad Nutr Diet*. 2015;115:200.

21. Trinquart L and others. Why do we think we know what we know? A metaknowledge analysis of the salt controversy. *Int J Epidemiol*. 2016;45:251.

22. Friedan TR. Sodium reduction—Saving lives by putting choice into consumers' hands. *JAMA*. 2016;316:579.

23. Mente A and others. Associations of urinary sodium excretion with cardiovascular events in individuals with and without hypertension: A pooled analysis of data from four studies. *The Lancet*. 2016;388:465.

24. Adler AJ and others. Reduced dietary salt for the prevention of cardiovascular disease. *Cochrane DB Syst Rev*. 2014;12:Art. No. CD009217.

25. Graudal NA and others. Effects of low sodium diet on blood pressure, renin, aldosterone, catecholamines, cholesterol, and triglyceride (review). *Cochrane DB Syst Rev*. 2017;4:Art. No. DD04022.

26. He FJ and others. Salt reduction in England from 2003 to 2011: Its relationship to blood pressure, stroke and ischaemic heart disease mortality. *BMJ Open*. 2014;4:e004549.

27. Food and Drug Administration. Draft guidance for industry: Voluntary sodium reduction goals. Target mean and upper bound concentrations for sodium in commercially processed, packaged, and prepared foods. 2016; www.fda.org.

28. Mozaffarian D and others. Global sodium consumption and death from cardiovascular disease. *N Engl J Med*. 2014;371:624.

29. Carbone L and others. Sodium intake and osteoporosis. Findings from the Women's Health Initiative. *J Clin Endocrinol Metab*. 2016;101:1414.

30. Heilberg IP, Goldfarb DS. Optimum nutrition for kidney stone disease. *Adv Chronic Kidney Dis*. 2013;20:165.

31. Hoy MK, Goldman JD. Potassium intake of the U.S. population: What we eat in America, NHANES 2009–2010. Food Surveys Research Group Dietary Data Brief No. 10. 2012; ars.usda.gov/Services/docs.htm?docid=19476.

32. Binia A and others. Daily potassium intake and sodium-to-potassium ratio in the reduction of blood pressure: A meta-analysis of randomized controlled trials. *J Hypertens*. 2015;33:1509.

33. Jackson SL and others. Association between urinary sodium and potassium excretion and blood pressure among adults in the United States National Health and Nutrition Examination Survey, 2014. *Circ*. 2018;137:237.

34. Levings JL, Gunn JP. The imbalance of sodium and potassium intake: Implications for dietetic practice. *J Acad Nutr Diet*. 2014;114:838.

35. Dawson-Hughes B and others. Potassium bicarbonate supplementation lowers bone turnover and calcium excretion in older men and women: A randomized dose-finding trial. *J Bone Min Res*. 2015;30:2103.

36. Granchi D and others. Potassium citrate supplementation decreases the biochemical markers of bone loss in a group of osteopenic women: The results of a randomized, double-blind, placebo-controlled pilot study. *Nutrients*. 2018;10:1293.

37. National Center for Health Statistics. *Health, United States, 2017: With special feature on mortality*. Hyattsville, MD: U.S. Government; 2018. www.cdc.gov/nchs/data/hus/hus17.pdf.

38. National Heart, Lung, and Blood Institute. 7th report of the Joint National Committee on Prevention, Detection, Evaluation and Treatment of High Blood Pressure. 2004; www.nhlbi.nih.gov/guidelines/hypertension/jnc7full.htm.

39. Whelton PK and others. 2017 ACC/AHA/AAPA/ABC/ACPM/AGS/APhA/ASH/ASPC/NMA/PCNA guideline for the prevention, detection, evaluation, and management of high blood pressure in adults. *J Am College Card*. 2018;71:127.

40. Centers for Disease Control and Prevention. Learn more about CDC's childhood lead poisoning data. 2019; www.cdc.gov/nceh/lead/data/learnmore.htm.

41. Blumenthal J and others. Effects of DASH Diet alone and in combination with exercise and weight loss on blood pressure and cardiovascular biomarkers in men and women with high blood pressure. *Arch Intern Med*. 2010;170:126.

42. Intersalt Cooperative Research Group. Intersalt: An international study of electrolyte excretion and blood pressure. Results for 24-hour urinary sodium and potassium excretion. *Brit Med J*. 1988;297:319.

43. Beyer FR and others. Combined calcium, magnesium and potassium supplementation for the management of primary hypertension in adults. *Cochrane DB Syst Rev*. 2006;3:Art. No. CD004805.

44. Ried K and others. Effect of cocoa on blood pressure. *Cochrane DB Syst Rev*. 2012;8:Art. No. CD008893.

45. Wang Y and others. Cocoa flavanols and blood pressure reduction: Is there enough evidence to support a health claim in the United States? *Trend Food Sci Technol*. 2019;83:203.

46. Steffen M and others. The effect of coffee consumption on blood pressure and the development of hypertension: A systematic review and meta-analysis. *J Hypertens*. 2012;30:2245.

47. Rhee JL and others. Coffee and caffeine consumption and the risk of hypertension in postmenopausal women. *Am J Clin Nutr*. 2016;103:210.

48. Kempner W. Treatment of hypertensive vascular disease with rice diet. *Am J Med*. 1948;4:545.

49. Patlak M. Bone builders: The discoveries behind preventing and treating osteoporosis. *FASEB J*. 2001;15:1677e.

50. O'Neil CE and others. Food sources of energy and nutrients among adults in the US: NHANES 2003–2006. *Nutrients.* 2012;19:2097.

51. Titchenal AC, Dobbs J. A system to assess the quality of food sources of calcium. *J Food Comp Analysis.* 2007;20:717.

52. Food and Nutrition Board, Institute of Medicine. *Dietary Reference Intakes for calcium and vitamin D.* Washington, DC: National Academies Press; 2011.

53. Bailey RL and others. Estimation of total usual calcium and vitamin D intakes in the United States. *J Nutr.* 2010;140:817.

54. Tai V and others. Calcium intake and bone mineral density: Systematic review and meta-analysis. *BMJ.* 2015;351:h4183.

55. Bolland MJ and others. Calcium intake and risk of fracture: Systematic review. *BMJ.* 2015;351:h4580.

56. US Preventive Services Task Force. Vitamin D, Calcium, or Combined Supplementation for the Primary Prevention of Fractures in Community-Dwelling Adults: US Preventive Services Task Force Recommendation Statement. *JAMA.* 2018;319:1592.

57. Reid IR and others. Calcium supplements: Benefits and risks. *J Intern Med.* 2015;278:354.

58. Patel AM, Goldfarb S. Got calcium? Welcome to the calcium-alkali syndrome. *J Am Soc Nephrol.* 2010;21:1440.

59. Bolland MJ and others. Calcium supplements and cardiovascular risk. 5 years on. *Ther Adv in Drug Safe.* 2013;4(5):199.

60. Lewis JR and others. The effects of calcium supplementation on verified coronary heart disease hospitalization and death in post-menopausal women: A collaborative meta-analysis of randomized controlled trials. *J Bone Min Res.* 2015;30:165.

61. Ross EA and others. Lead content of calcium supplements. *JAMA.* 2000;284:1425.

62. Rehman S and others. Calcium supplements: An additional source of lead contamination. *Trace Elem Res.* 2011;143:178.

63. Tucker KL, Rosen CJ. Prevention and management of osteoporosis. In: Ross C and others, eds. *Modern nutrition in health and disease.* 11th ed. Philadelphia: Lippincott Williams & Wilkins; 2014.

64. National Osteoporosis Foundation. Healthy bones for life—Clinicians guide. 2014; www.nof.org.

65. Wright NC. The recent prevalence of osteoporosis and low bone mass in the United States based on bone mineral density at the femoral neck or lumbar spine. *J Bone Miner Res.* 2014;29:2520.

66. Heaney RP. Dairy intake, dietary adequacy, and lactose intolerance. *Adv Nutr.* 2013;4:151.

67. Weaver CM and others. The National Osteoporosis Foundation's position statement on peak bone mass development and lifestyle factors: a systematic review and implementation recommendations. *Osteoporosis Int.* 2016;27:1281.

68. Protiva P and others. Calcium and 1,25-dihydroxy D3 modulate genes of immune and inflammatory pathways in the human colon: A human crossover trial. *Am J Clin Nutr.* 2016;103:1224.

69. Keum N and others. Calcium intake and colorectal cancer risk: Dose-response meta-analysis of prospective observational studies. *Int J Cancer.* 2014;135:1940.

70. Tantamango-Bartley Y and others. Independent associations of dairy and calcium intakes with colorectal cancers in the adventist health study-2 cohort. *Public Health Nutr.* 2017;20:2577.

71. Taylor EN, Curhan GS. Dietary calcium from dairy and nondairy sources, and the risk of symptomatic kidney stones. *J Urol.* 2013;190:1255.

72. Calvo MS, Uribarri J. Public health impact of dietary phosphorus excess on bone and cardiovascular health in the general population. *Am J Clin Nutr.* 2013;98:6.

73. O'Brien KO and others. Phosphorus. In: Ross C and others, eds. *Modern nutrition in health and disease.* 11th ed. Philadelphia: Lippincott Williams & Wilkins; 2014.

74. Rude RK. Magnesium. In: Ross C and others, eds. *Modern nutrition in health and disease.* 11th ed. Philadelphia: Lippincott Williams & Wilkins; 2014.

75. Dubey P and others. Role of minerals and trace elements in diabetes and insulin resistance. *Nutrients.* 2020;12:1864.

76. Sarrafzadegan MD and others. Magnesium status and the Metabolic Syndrome: A systematic review and meta-analysis. *Nutr.* 2016;32:409.

77. Rosique-Esteban N and others. Dietary magnesium and cardiovascular disease: A review with emphasis in epidemiological studies. *Nutrients.* 2018;10:168.

78. Kieboom BC and others. Serum magnesium and the risk of death from coronary heart disease and sudden cardiac death. *J Am Heart Assoc.* 2016;5:e002707.

79. Joris PJ and others. Long-term magnesium supplementation improves arterial stiffness in overweight and obese adults: Results of a randomized, double-blind, placebo-controlled intervention trial. *Am J Clin Nutr.* 2016;103:1260.

©C Squared Studios/Getty Images RF

Squid, as well as lobster, crayfish, crab, and snails, are great sources of copper. That's because hemocyanin, a copper-based compound, transports oxygen in their bodies, as hemoglobin does in ours. Learn more about these blue-blood animals at **ocean.si.edu/ocean-life/invertebrates/cephalopods.**

# 15

# Trace Minerals

## Learning Objectives

**After studying this chapter, you will be able to**

1. Discuss the major functions of each trace mineral.

2. List 3 important food sources for each trace mineral.

3. Describe how each trace mineral is absorbed, transported, stored, and excreted.

4. Describe the deficiency symptoms of trace minerals.

5. Describe the toxicity symptoms from the excess consumption of certain trace minerals.

6. Describe the development of cancer and the effects of genetic, environmental, and dietary factors on the risk of developing cancer.

## Chapter Outline

**15.1** Iron (Fe)

**15.2** Zinc (Zn)

**15.3** Copper (Cu)

**15.4** Manganese (Mn)

**15.5** Iodine (I)

**15.6** Selenium (Se)

**15.7** Chromium (Cr)

**15.8** Fluoride (F)

**15.9** Molybdenum (Mo) and Ultratrace Minerals

**Global Perspective: Global Nutrition Actions**

**Clinical Perspective: Nutrients, Diet, and Cancer**

TRACE MINERALS ARE ESSENTIAL, INORGANIC substances needed in small quantities in the diet (less than 100 mg daily) and are found in relatively minute amounts in the body (less than 5 g). In fact, all of them combined make up less than 1% of the minerals in the body. Nevertheless, trace minerals are essential for normal development, function, and overall health. For example, iron is needed to build red blood cells; zinc, copper, selenium, and manganese help protect us from the damaging effects of free radicals; and iodine is needed to maintain normal metabolism.

With the exception of iron and iodine, the essentiality of trace minerals in humans has been recognized only within the last 50 years. Intensive trace mineral research began in the early 1960s with the discovery of a group of Middle Eastern adolescent boys who had impaired growth and sexual development because they were zinc deficient.[1] Since then, scientists have discovered that many trace minerals are essential for normal body function and that physiological abnormalities occur when they are deficient in the diet.

The unique characteristics of trace minerals have presented many challenges to researchers. For example, because there are such small quantities in the body, measuring the amount in the body and changes in trace mineral concentrations is often difficult.[2] Therefore, it is hard to accurately assess trace mineral status in the body and set recommended intake levels. Evaluating the amount of trace minerals in foods also can be difficult. Many agricultural factors and food characteristics affect the amount of trace minerals in food, especially in plant-based foods. Thus, nutrient databases may not provide an accurate indication of the total amount and/or bioavailability of a trace mineral in a specific food. Despite these challenges, scientists' knowledge of trace minerals continues to expand as new technologies emerge and permit more precise measurement of their levels in the body and foods, as well as the study of their roles in physiological and cellular function.

The trace mineral content of plant-based foods depends, in part, on the mineral content of the soil in which they were grown.

©Pixtal/AGE Fotostock RF

Oysters, clams, and other shellfish are good sources of many trace minerals.

©C Squared Studios/Getty Images RF

*Figure 15-1* Food sources of iron. In addition to the MyPlate food groups, yellow is used for oils and pink is used for substances that do not fit easily into food groups (e.g., honey, candy, coffee, alcohol, and table salt)

©Comstock/Getty Images RF
Source: ChooseMyPlate.gov, U.S. Department of Agriculture

# 15.1 Iron (Fe)

The importance of iron for the maintenance of health has been recognized for centuries. In 4000 B.C., Persian physician Melampus gave iron supplements to sailors to compensate for the iron they lost from bleeding during battles. Maintaining iron status is still a problem throughout the world. Globally, 1 in 4 people has anemia, with children and women the most commonly affected.[3] *Anemia* means to have lacking or deficient (*an-*) blood (*-emia*). In the case of *iron deficiency* anemia, a lack of iron has reduced the number of red blood cells in the body. Iron deficiency is the cause of about half of all cases of anemia. Although anemia overall as well as iron deficiency anemia are less prevalent in the U.S. and other high-income countries, the World Health Organization considers anemia to be a leading global health concern.[3,4]

## Iron in Foods

Dietary iron occurs in several forms. In animal flesh (beef, pork, seafood, and poultry), most of the iron is present as hemoglobin and myoglobin, which collectively is called **heme iron**. The rest of the iron present in these foods, as well as all the iron in vegetables, grains, and supplements, is called **nonheme iron**.

The richest sources of iron in North American diets are meats and seafood (Fig. 15-1). Because iron is added to refined flour as a part of the enrichment process, bakery products (e.g., bread, rolls, and crackers) provide iron as well. Relatively high amounts of iron also are found in spinach and other dark, leafy greens and in kidney, garbanzo, and navy beans. However, many factors cause the iron in enriched and plant-based foods to be less bioavailable than iron from animal-based foods.

In addition to the iron in foods, iron cookware can contribute to iron intake. When foods are cooked in iron pans, small amounts of iron from the cookware are transferred to the food. Acidic food, such as tomato sauce, increases the amount of iron transferred from the cookware to the food and, in turn, increases the total iron content.

## Iron Needs

The RDA for adult women is 18 mg/day; it is 8 mg/day for adult men.[5] After age 51, the RDA for women drops to 8 mg/day because most women enter menopause and no longer lose iron via menstrual blood. Although iron absorption can vary considerably, the RDA

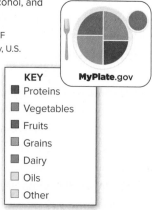

**KEY**
- Proteins
- Vegetables
- Fruits
- Grains
- Dairy
- Oils
- Other

MyPlate.gov

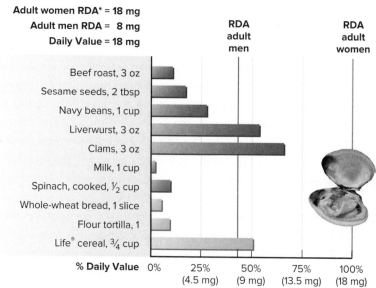

Adult women RDA* = 18 mg
Adult men RDA = 8 mg
Daily Value = 18 mg

RDA adult men

RDA adult women

- Beef roast, 3 oz
- Sesame seeds, 2 tbsp
- Navy beans, 1 cup
- Liverwurst, 3 oz
- Clams, 3 oz
- Milk, 1 cup
- Spinach, cooked, ½ cup
- Whole-wheat bread, 1 slice
- Flour tortilla, 1
- Life® cereal, ¾ cup

**% Daily Value** 0%  25% (4.5 mg)  50% (9 mg)  75% (13.5 mg)  100% (18 mg)

*After age 51, RDA is 8 mg.

values are based on the premise that approximately 18% of dietary iron is absorbed each day from typical Westernized diets, such as those eaten in most of North America.[5] The Daily Value for iron used on food labels and supplements is 18 mg.

Westernized diets typically contain about 6 mg of iron for every 1000 kcal.[5] Thus, the average 2000 kcal diet provides about 12 mg of iron. In North America, the average daily iron intake is approximately 17 mg for men and 12 mg for women.

## Absorption, Transport, Storage, and Excretion of Iron

Iron is absorbed across the brush border membrane into the small intestine by carrier-mediated mechanisms. The enterocytes (intestinal absorptive cells) produce different iron-binding carrier proteins, which play an important role in the absorptive process and overall regulation of iron status.[6] **Ferritin**, a key iron-binding protein produced in the enterocytes (and other tissues), binds and stores mucosal iron, thereby preventing it from entering the bloodstream (Fig. 15-2). The amount of mucosal ferritin produced is in proportion to body iron stores. Thus, when iron stores are low, very little ferritin is made,

Red meat is a major source of heme iron in the North American diet.

©Stockbyte/Getty Images RF

**ferritin** Iron-binding protein in the intestinal mucosa that binds iron and prevents it from entering the bloodstream; also the primary storage form of iron in liver and other tissues.

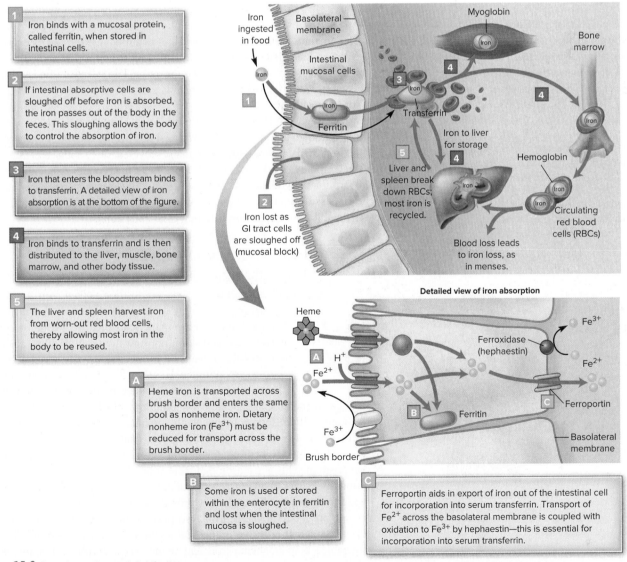

1   Iron binds with a mucosal protein, called ferritin, when stored in intestinal cells.

2   If intestinal absorptive cells are sloughed off before iron is absorbed, the iron passes out of the body in the feces. This sloughing allows the body to control the absorption of iron.

3   Iron that enters the bloodstream binds to transferrin. A detailed view of iron absorption is at the bottom of the figure.

4   Iron binds to transferrin and is then distributed to the liver, muscle, bone marrow, and other body tissue.

5   The liver and spleen harvest iron from worn-out red blood cells, thereby allowing most iron in the body to be reused.

A   Heme iron is transported across brush border and enters the same pool as nonheme iron. Dietary nonheme iron (Fe$^{3+}$) must be reduced for transport across the brush border.

B   Some iron is used or stored within the enterocyte in ferritin and lost when the intestinal mucosa is sloughed.

C   Ferroportin aids in export of iron out of the intestinal cell for incorporation into serum transferrin. Transport of Fe$^{2+}$ across the basolateral membrane is coupled with oxidation to Fe$^{3+}$ by hephaestin—this is essential for incorporation into serum transferrin.

*Figure 15-2* Iron absorption and distribution.

Iron in cookware can leach into food, increasing its iron content. The iron vats used by Bantu peoples in Africa to brew beer can increase iron intake to 100 mg/day. Learn more at **www.hematology.org/Patients/Blood-Disorders.aspx**.

©alisafarov/Shutterstock, Inc.

▶ The copper-containing proteins (hephaestin and ceruloplasmin) needed to oxidize iron from the ferrous ($Fe^{2+}$) to the ferric ($Fe^{3+}$) form demonstrate the close link between copper and iron metabolism.

**lysosome** Cell organelle that digests proteins, such as transferrin, and breaks down bacteria and old or damaged cell components.

**hemosiderin** Iron-binding protein in the liver that stores iron when iron levels in the body exceed the storage capacity of ferritin.

which allows greater amounts of iron to enter the mucosal iron pool for transport out of the enterocytes into the bloodstream. If iron stores are high or saturated, larger amounts of ferritin are made to bind iron as it enters the intestinal cells. Although a portion of this ferritin-bound iron remains in the intestinal iron pool, much of it is excreted when the intestinal cells are sloughed off after several days. This process is called a "mucosal block" because it prevents iron from entering the bloodstream and, in effect, blocks the excess accumulation of iron. Large doses of iron, however, can overtax the mucosal block's protective ability and increase the risk of toxicity.

When iron needs are high, most of the iron absorbed into enterocytes is released into an intestinal iron pool. This iron is then transported out of the enterocytes, by a protein called ferroportin, into the interstitial fluid for release into the bloodstream and distribution to body cells.[6] To transport absorbed iron to body cells, the iron is oxidized from the ferrous ($Fe^{2+}$) form to the ferric ($Fe^{3+}$) form by a copper-containing enzyme (either hephaestin in the enterocyte or ceruloplasmin in the blood) and bound to a serum protein called transferrin.[6] Each transferrin molecule can bind 2 molecules of ferric iron for transport through the blood to body cells.

All cells have transferrin receptors, located on their surface membrane, that allow them to take in the transferrin-iron complex.[7] Cells can control the amount of iron they take in by altering the synthesis of transferrin receptors. When more iron is needed, the cell increases the number of transferrin surface receptors to enhance iron uptake. Conversely, when cellular iron need is low, the number of receptors decreases.

After transferrin binds to its surface receptor, it is engulfed into the cell by endocytosis (see Chapter 4). Within the cell **lysosomes**, iron is released from transferrin and the receptor-protein complex is returned to the cell surface for reuse. The released iron is utilized for cellular functions or stored in the liver, bone marrow, and spleen with ferritin (the primary storage protein) or **hemosiderin**.

Intestinal absorption, cellular uptake, and the storage of iron are tightly regulated because the body has a limited ability to excrete iron that has been absorbed. In fact, approximately 90% of the iron used each day is recovered and recycled. Only about 10% is excreted, mainly via the small amount of bile lost in the feces. One of the proteins that aid in the regulation of iron absorption and balance is called hepcidin.[8] When iron stores are adequate or high, hepcidin is released from the liver and promotes the degradation of ferroportin, which in turn decreases the release of iron from enterocytes to body cells.

### Factors Affecting Iron Absorption

The amount of iron absorbed is affected by the body's iron needs and stores and by diet composition. When iron status is adequate, approximately 14 to 18% of dietary iron is absorbed from a typical North American diet.[5] However, when iron need is high and stores are low, the small intestine absorbs up to 35 to 40% of dietary iron. In contrast, when iron need is low and stores are saturated, less than 5% of dietary iron is absorbed.

Iron absorption also is affected by the form of iron in foods eaten (heme or nonheme), the total amount of iron present in the diet, diet composition (food components that increase or decrease the bioavailability of iron), and the acidity of the gastric contents. Heme iron is absorbed much more readily than nonheme iron and is not highly affected by dietary composition. This is a reason that meat is an efficient way to obtain iron. Additionally, meat provides greater amounts of iron than naturally occur in plant-based foods, so smaller quantities provide similar amounts of iron.

Plant-based nonheme iron absorption is hindered by several dietary factors. Phytic acid in whole grains and legumes and oxalic acid in leafy green vegetables bind nonheme iron and reduce its absorption. For this reason, whole grains and leafy green vegetables do not contribute significant amounts

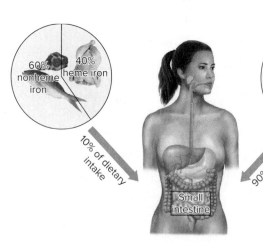

60% nonheme iron
40% heme iron

100% nonheme iron

10% of dietary intake

90% of dietary intake

Small intestine

17% of nonheme iron absorbed
25% of heme iron absorbed

Left: ©Comstock Images/Getty Images; Right: ©Pixtal/age fotostock

Overall iron absorption = 18%

**Table 15-1 Factors That Affect Iron Absorption**

| Factors That Increase Absorption | Factors That Decrease Absorption |
|---|---|
| High body demand for red blood cells (blood loss, high altitude, physical training, pregnancy, anemia) | Low need for iron (high level of storage iron) |
| Low body stores of iron | Phytic acid in whole grains and legumes |
| Heme iron in food | Oxalic acid in leafy vegetables |
| Meat protein factor (MPF) | Polyphenols in tea, coffee, red wine, and oregano |
| Vitamin C intake | Reduced gastric acidity |
| Gastric acidity | Excessive intake of other minerals (zinc, manganese, calcium) |

of iron, despite containing relatively high amounts of iron for a plant food. High intakes of dietary fiber from plant-based diets also bind nonheme iron and reduce its bioavailability.

Polyphenols, such as the tannins found in tea, and related substances in coffee are known to reduce nonheme iron absorption. Therefore, individuals trying to rebuild iron stores are advised to reduce coffee and tea consumption, particularly at mealtimes. Excessive intakes of other minerals, such as zinc, manganese, and calcium, may interfere with nonheme iron absorption. Because of these potential mineral-mineral interactions, individuals with high iron needs should avoid taking iron supplements with foods or supplements containing significant amounts of these minerals.

The absorption of nonheme iron can be enhanced by a component of meat called meat protein factor (MPF). Eating even a small amount of meat with nonheme iron–containing foods can be an effective means of boosting nonheme iron absorption. Vitamin C, or other organic acids, in the diet also increases nonheme iron absorption. Vitamin C provides an electron to $Fe^{3+}$ (ferric iron) to yield $Fe^{2+}$ (ferrous iron), which then forms a soluble complex with vitamin C. Ferrous iron is better absorbed than ferric iron because it more readily crosses the mucosal layer of the small intestine and reaches the brush border of the intestinal absorptive cells. (Heme iron does not need to undergo this reduction reaction because the iron in heme is already in the more soluble ferrous form.)

Although no iron absorption occurs in the stomach, gastric acid plays an important role in the absorption of nonheme iron by promoting the conversion of ferric ($Fe^{3+}$) iron to ferrous ($Fe^{2+}$) iron. If the amount of gastric acid produced is low, less ferric iron is converted to ferrous iron and overall nonheme iron absorption is decreased. This is of concern in individuals who regularly take antacids or other medications to reduce gastric acidity (and symptoms of reflux) as well as in older adults, many of whom have reduced gastric acid production. The factors that affect iron absorption are summarized in Table 15-1.

## Functions of Iron

Iron plays an important role in diverse functions in the body. Many of these functions are dependent on iron's ability to participate in oxidation and reduction (redox) reactions, changing $Fe^{2+}$ (ferrous) to $Fe^{3+}$ (ferric) iron and back. Although iron's ability to switch back and forth between $Fe^{2+}$ and $Fe^{3+}$ is vital, this reactivity also can be harmful because iron can form free radical compounds that damage cell membranes and DNA. To prevent these destructive effects and preserve iron for healthful uses, very little free iron is found in the body. Instead, iron is tightly bound to transport, functional, or storage proteins (Table 15-2).

Plant-based foods, such as beans, contain fiber and phytic acid, which can reduce the absorption of trace minerals.

©deepak bishnoi/Shutterstock, Inc.

**Table 15-2 Iron Proteins**

| Iron in Functional Proteins |
|---|
| Hemoglobin |
| Myoglobin |
| Iron-containing enzymes |
| **Iron in Transport Proteins** |
| Transferrin |
| Ferroportin |
| **Iron in Storage Proteins** |
| Ferritin |
| Hemosiderin |

*Figure 15-3*  Most iron in the body is present in heme within hemoglobin and myoglobin. Iron gives hemoglobin and myoglobin the ability to carry oxygen. Hemoglobin contains 4 heme compounds, whereas myoglobin contains 1 heme compound.

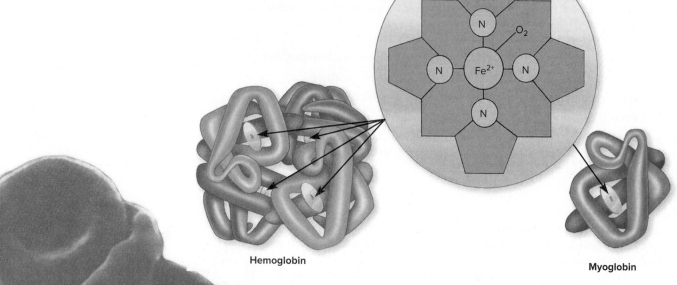

**Hemoglobin**

**Myoglobin**

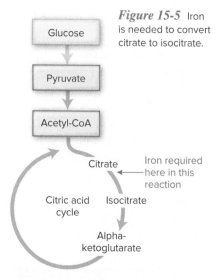

*Figure 15-4*  Red blood cells have a life span of about 120 days. As a red blood cell matures, its nucleus is expelled, along with its DNA. Thus, red blood cells cannot duplicate. The body recycles iron from worn-out red blood cells to help produce new red blood cells.

©SPL/Science Source

*Figure 15-5*  Iron is needed to convert citrate to isocitrate.

Glucose

↓

Pyruvate

↓

Acetyl-CoA

Citrate ← Iron required here in this reaction

Citric acid cycle    Isocitrate

↓

Alpha-ketoglutarate

Iron is an essential part of 2 proteins, hemoglobin and myoglobin, that are involved in the transport and metabolism of oxygen. Hemoglobin, which is found in erythrocytes (red blood cells), is composed of 4 iron-containing heme compounds that each bind 1 molecule of oxygen (Fig. 15-3). As a component of hemoglobin, iron carries oxygen in the blood from the lungs to all tissues of the body. It also transports carbon dioxide back to the lungs for expiration. Because the body produces approximately 200 billion erythrocytes each day, much of the body's iron is contained in hemoglobin. If the oxygen-carrying capacity of erythrocytes declines, the kidneys produce the hormone erythropoietin, which stimulates the bone marrow to produce more red blood cells (Fig. 15-4). This, in turn, increases the body's need for iron to support hemoglobin production.

Iron has a similar oxygen-carrying role in myoglobin, a protein in muscle cells. Each myoglobin contains 1 iron molecule (attached to heme) that transports oxygen from red blood cells to skeletal and heart muscle cells. During iron deficiency, when the delivery of oxygen to these cells becomes limited, individuals often develop shortness of breath and fatigue, especially with physical activity or exertion. Thus, maintaining adequate iron status is critical to myoglobin function and, in turn, to exercise and work performance.

Iron-containing enzymes play a vital role in functions such as energy metabolism, drug and alcohol transformation, and the excretion of organic compounds. Within the mitochondria, iron is a component of cytochromes that carry electrons from $NADH + H^+$ and $FADH_2$ to molecular oxygen in the electron transport chain. Iron also is required in the 1st step of the citric acid cycle for the conversion of citrate to subsequent compounds (Fig. 15-5). Alcohol and many drugs are metabolized in the liver by the iron-containing P-450 enzymes prior to excretion.

Iron is a cofactor for enzymes involved in the synthesis of neurotransmitters (dopamine, epinephrine, norepinephrine, and serotonin). These are important for normal early cognitive development and lifelong brain function.

The immune system requires iron for the production of lymphocyte and natural killer (NK) cells that help prevent infections. If iron status is low, the effectiveness of these cells is impaired and the likelihood of infections increases. Although iron deficiency increases infection risk, iron overload also can increase the incidence of infections because bacteria require iron to grow and proliferate. Thus, iron status must be maintained within a defined range to prevent deficiency as well as overload.

**Table 15-3 Stages of Iron Deficiency**

| Marginal or Early Iron Deficiency → | Moderate Iron Deficiency → | Severe Iron Deficiency, Iron Deficiency Anemia |
|---|---|---|
| • ↓ Iron intake<br>• ↑ Iron losses<br>• ↓ Iron stores<br>• ↑ Transferrin receptors<br>• No apparent symptoms | • Depleted iron stores<br>• ↓ Work/exercise capacity<br>• ↓ Immune function<br>• ↓ Iron transport | • ↓ Hemoglobin and red blood cell synthesis (anemia)<br>• ↓ Oxygen transport<br>• ↑ Fatigue<br>• Poor work/exercise performance<br>• ↑ Incidence of infection<br>• ↓ Growth and cognitive development in children<br>• ↑ Mortality |

▶ Anemia develops when the number of red blood cells falls below normal levels. There are many causes of anemia. The lack of iron for hemoglobin synthesis and red blood cell production in severe iron deficiency is a cause of anemia.

▶ The consumption of soil and other nonfood substances (called pica) (see Chapter 16) can lead to iron deficiency anemia because these substances bind iron in the intestinal tract and prevent absorption. Blood loss caused by intestinal or blood-borne parasite infections is another common cause of anemia in many parts of the world.

## Iron Deficiency

Iron deficiency is the most widespread nutritional deficiency in the U.S.[9] In the early stages of iron deficiency (summarized in Table 15-3), symptoms may be minimal or unapparent because the body can mobilize stores of iron from ferritin (the primary iron storage protein). However, even a mild to moderate deficiency of iron can compromise immune function and work performance. As iron deficiency progresses and stores are depleted, the lack of iron for heme and hemoglobin synthesis results in the development of iron deficiency anemia. This impairs oxygen transport in the blood, causing fatigue and a decreased ability to perform normal activities. Iron deficiency anemia also impairs energy metabolism, compromises immune function, and delays cognitive development.[10] Iron deficiency anemia is of particular concern in young children because cognitive and developmental impairments may not be reversible.

Clinicians can, in part, diagnose anemia by viewing the erythrocytes (red blood cells) under a microscope. In iron deficiency anemia, the red blood cells are smaller than normal (called microcytic) and paler (called hypochromic) (Fig. 15-6). Clinicians also can measure blood hematocrit (percentage of total blood volume comprised of red blood cells) and blood hemoglobin, both of which decrease in iron deficiency anemia (Fig. 15-7). However, these are not sensitive measures of iron status because they remain unchanged in the earlier stages

(a)

(b)

**Figure 15-6** Iron deficiency anemia. (*a*) Red blood cells that are the normal size and color. (*b*) Iron-deficient red blood cells. They are smaller (microcytic) and paler (hypochromic) than normal. Color loss is a result of a decreased amount of the pigment hemoglobin. Hypochromic red blood cells also have a reduced capacity to carry oxygen. That is why people feel tired when they have iron deficiency anemia.

a: ©SPL/Science Source; b: ©Omikron/Science Source

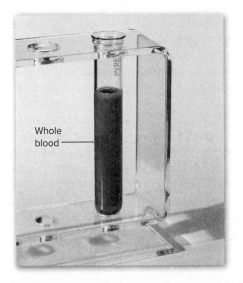

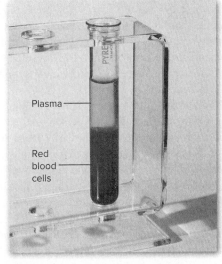

Whole blood

Plasma

Red blood cells

**Figure 15-7** Iron deficiency reduces the number of red blood cells produced. A hematocrit blood test measures the percentage of blood composed of red blood cells. The photo on the right is centrifuged whole blood. Note how red blood cells account for about 55% of the whole blood volume, which is the top of the normal range for adult males. The normal range for males is a hematocrit of 40 to 55% and for females is 37 to 47%.

Both: ©Martyn F. Chillmaid/Science Source

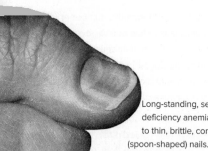

Long-standing, severe iron deficiency anemia can lead to thin, brittle, concave (spoon-shaped) nails.

©Dr P. Marazzi/Science Source

## CASE STUDY

©Image Source

Chloe is on the college cross-country team. She tries to eat plenty of grains and fruits to get the carbohydrates needed for her high level of physical activity. She eats 3 meals a day and a snack in the evenings. A typical day for Chloe would include eating cheese sticks and a glass of orange juice for breakfast; a peanut butter sandwich, an apple, a bag of pretzels, and iced tea for lunch; spaghetti with marinara sauce, garlic bread, salad, and a glass of milk for dinner; and an orange for an evening snack. This semester, Chloe feels tired more often and takes a sweatshirt to class to keep warm. She also is unable to concentrate well. What do you think might be causing Chloe to feel fatigued, feel cold more often, and have difficulty studying? What nutrient may be missing in her diet? What laboratory tests might help identify the dietary component of concern? What recommendations would you make to Chloe to improve her dietary intake?

▶ Hemosiderosis is the storage of excess iron in the form of hemosiderin. It does not cause organ damage the way hemochromatosis does because, in hemosiderosis, excess iron is stored in normal iron storage proteins. In hemochromatosis, however, the storage proteins are saturated with iron, and iron accumulates in the liver, the heart, and other organs.

of iron deficiency and are affected by many factors (e.g., disease, inflammation, blood loss, deficiencies of other nutrients) other than iron status. Thus, many experts recommend using transferrin receptor number to assess iron status because it reflects cellular iron need and is unaffected by the factors that limit the use of other iron biomarkers.[7] Some researchers assess body iron status by using lab data for transferrin receptors and ferritin.[9]

Many individuals are at risk of iron deficiency and iron deficiency anemia.[5] Premature infants (born before 37 weeks of pregnancy) are at increased risk because iron stores needed for the 1st few months of life accumulate during the last weeks of pregnancy. Thus, these infants are born with low stores, which can be depleted quickly by their high iron needs. Young children also are at risk because they are growing fast and typically have low intakes of iron rich meat and high intakes of iron poor cow's milk. One in 4 children in the U.S. ages 12 to 23 months has intakes below the RDA, increasing the risk of iron deficiency.[11] In the U.S., iron-fortified formulas and cereals provided by the Special Supplemental Nutrition Program for Women, Infants, and Children (WIC) to children from limited-resource families who are at nutritional risk have been instrumental in decreasing rates of iron deficiency anemia. However, in low-income countries, many young children are iron deficient because iron supplementation programs frequently are not available.

Teenage girls and women of childbearing age are at risk of iron deficiency because of monthly menstrual blood losses and low intake of iron rich foods. In the U.S. alone, approximately 10% of women ages 12 to 49 have iron deficiency anemia.[9] Even though menstruation stops during pregnancy, pregnant women need to be certain to consume sufficient iron because their own blood volume (and red blood cell number) is increasing, as is that of the fetus. Vegetarians and others who lack food sources of heme iron also are at increased risk of iron deficiency. Although they may have high intakes of plant-based iron foods, their diets contain many factors that decrease the bioavailability of this iron (see Table 15-1). These individuals need to include vitamin C rich foods in their diets and consider using iron-fortified foods as well as multivitamin-mineral supplements that contain iron.

Individuals who donate blood more than 2 to 4 times a year may be at increased risk of iron deficiency. The donation of 1 pint (0.5 L) of blood represents a loss of 200 to 250 mg of iron. For most healthy individuals, it takes several months to replace this iron, although women may need a longer interval between donations to rebuild their iron stores. As a health precaution, blood banks test potential donors' blood for anemia prior to allowing blood donations.

## Iron Overload and Toxicity

Although iron deficiency is a major public health concern, iron also poses a risk for toxicity. Thus, an Upper Level of 45 mg/day has been set for iron.[4] Intakes above this level, especially from supplements and highly fortified foods, can cause nausea and vomiting, stomach irritation, diarrhea, and impaired absorption of other trace minerals.

In the U.S., accidental iron overdose is the leading cause of poisoning in young children under the age of 6 years. Children are more vulnerable to iron poisoning than adults because their absorptive mechanisms cannot respond as rapidly as those of an adult. The primary cause of iron overload in children is the consumption of excess chewable iron-containing supplements. The FDA has ruled that all iron supplements must carry a warning about toxicity. As an added precaution, supplements that contain 30 mg or more per tablet must be individually wrapped.

In adults, iron overload is most commonly the result of **hemochromatosis**, a genetic condition that affects about 1 person in every 200 to 500. In hemochromatosis, the mucosal block that usually protects the body from excess iron absorption is ineffective.[12] Due to a deficiency of hepcidin, which prevents the normal degradation of the transport protein ferroportin, higher than normal amounts of iron are absorbed and transported across the enterocyte for binding to transferrin and distribution to tissues.[7,12,13] Because the body lacks a mechanism for eliminating this excess iron, iron accumulates in the body, leading to iron overload and tissue injury. This causes saturation of iron-binding proteins and, over time, results in iron deposits in the liver, the heart, and other organs. Left untreated, this eventually can lead to liver disease and heart failure.

Adult iron overload also may result from excess supplementation and frequent blood transfusions. The iron introduced into the body through repeated blood transfusion bypasses the protective mucosal block and can result in dangerously high iron stores in the body if not monitored closely.[5]

The treatment of iron toxicity depends on the underlying cause. For individuals using excess supplemental iron, supplement use should be stopped. In hemochromatosis, treatment consists of periodic blood removal (a procedure similar to that used for obtaining blood donations). Another treatment approach is to administer a chelator drug that binds iron and increases its excretion. However, chelator drugs also bind other trace minerals, possibly causing secondary trace mineral deficiencies.

▶ Several strains of healthy bacteria of the gut microbiome require little to no iron for their metabolism. However, certain disease-causing bacteria need iron and may increase in number and cause diarrhea in young children given iron supplements or foods fortified with iron.[14] Some reports suggest that the risk of diarrhea may be more likely in children who are not iron deficient. More research is needed to determine the effects of iron on the gut microbiota.

### Knowledge Check

1. How is iron involved in the metabolism of oxygen in the body?
2. Why are heme-containing foods a more efficient means of obtaining dietary iron than nonheme-containing foods?
3. What factors increase and decrease iron absorption?
4. What are the symptoms of iron deficiency and iron deficiency anemia?
5. What population groups are at greatest risk of iron deficiency anemia?
6. Why are children at increased risk of iron toxicity?

## 15.2 Zinc (Zn)

Zinc has been recognized as an essential nutrient in animals since the 1930s. However, it was 30 years later before it became apparent from studies in the Middle East that zinc is essential in humans for normal growth and development.[1] Since that time, scientists have learned that almost all the cells of the body contain zinc and use it for many different functions.

### Zinc in Foods

Protein rich meat and seafood are usually good sources of zinc (Fig. 15-8). North Americans obtain about 70% of their dietary zinc from animal-based foods, such as beef, lamb, and

Shellfish, such as lobster, crab, and oysters, are excellent zinc (and copper) sources.

©Ingram Publishing/SuperStock RF

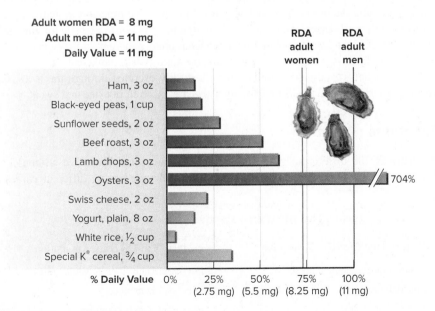

Adult women RDA = 8 mg
Adult men RDA = 11 mg
Daily Value = 11 mg

RDA adult women
RDA adult men

Ham, 3 oz
Black-eyed peas, 1 cup
Sunflower seeds, 2 oz
Beef roast, 3 oz
Lamb chops, 3 oz
Oysters, 3 oz — 704%
Swiss cheese, 2 oz
Yogurt, plain, 8 oz
White rice, ½ cup
Special K® cereal, ¾ cup

**% Daily Value**  0%   25% (2.75 mg)   50% (5.5 mg)   75% (8.25 mg)   100% (11 mg)

*Figure 15-8* Food sources of zinc.

©C Squared Studios/Getty Images RF

## HISTORICAL PERSPECTIVE

### Unleavened Breads and Zinc Deficiency

The high intake of unleavened breads in traditional Middle Eastern diets, combined with the relatively low zinc intakes from other dietary sources, resulted in the symptoms of zinc deficiency in adolescent boys (Fig. 15-9), lead scientists to recognize zinc as an essential nutrient in humans over 60 years ago.

*Figure 15-9* On the right, a 16-year-old Egyptian farm boy with delayed growth and sexual development associated with zinc deficiency.

Photo courtesy of Harold H. Sandstead, M.D.

**metallothionein** Protein involved in the binding and release of zinc and copper in intestinal and liver cells.

pork. Plant-based foods, such as nuts, beans, wheat germ, and whole grains, also can contribute significant amounts of zinc to our diets. Although some breakfast cereals are fortified with zinc, it is not included in the enrichment process of flour. Therefore, refined flour products are poor sources of zinc.

Whole-grain breads and cereals can contribute significant amounts of zinc to the diet; however, unleavened whole-grain bread is very high in phytic acid and other factors that decrease zinc bioavailability. Yeast fermentation (used in the preparation of yeast-leavened breads) reduces the effect of phytic acid up to 10-fold and thus increases zinc absorption.

## Dietary Needs for Zinc

The RDAs for zinc are 11 mg/day for adult men and 8 mg/day for adult women.[5] The RDAs are based on the average amount needed to replace the daily losses in feces, urine, and sweat and on an estimated dietary absorption of 40%. The Daily Value for zinc is 11 mg. Data from NHANES surveys indicate that mean zinc intakes for many adults in the U.S. currently meet RDA guidelines.[15]

## Absorption, Transport, Storage, and Excretion of Zinc

Zinc is absorbed into the cells of the small intestine by simple diffusion and active transport. When zinc is absorbed into enterocytes, it is exported to the bloodstream or bound to **metallothionein** in the intestinal epithelial cells. Metallothionein is a protein that binds zinc in much the same way that mucosal ferritin binds iron. In fact, the regulation of zinc absorption may be, in part, related to the synthesis of metallothionein because it hinders the movement of zinc from intestinal cells.[16] If zinc is not transported out of the intestinal absorptive cells into the bloodstream before the intestinal cells are sloughed off, it passes out of the body in the feces. Thus, a mucosal block, similar to that for iron, decreases excess absorption of zinc. However, large doses of zinc can override the mucosal block.

Like iron, the absorption of zinc is affected by diet composition and the body's need for the mineral. As summarized in Table 15-4, zinc absorption increases when zinc intake is low or marginal and when body needs for zinc are elevated. In contrast, zinc absorption decreases when zinc or nonheme iron intake is excessive, dietary fiber and phytic acid intake is high, and zinc status is adequate.

Zinc absorbed into the bloodstream binds to blood proteins, such as albumin, for transport to the liver. The liver repackages and releases zinc into the blood bound to alpha-2-macroglobulin, albumin, and other proteins. Multiple zinc transport systems appear to be involved in the cellular uptake of zinc.[17] Although there is no storage site for zinc, the body maintains an exchangeable pool of zinc in the liver, bone, pancreas, kidneys, and blood. This allows the body to recycle zinc and maintain zinc status when intake is low.

Fortunately, excess zinc (unlike iron) is readily excreted through the feces, thus decreasing the risk of toxicity. Small amounts also are excreted in urine and sweat.

## Functions of Zinc

As many as 300 different enzymes requiring zinc have been identified, although the actual number is lower when adjusted for the same enzyme having different names across

**Table 15-4  Factors That Affect Zinc Absorption**

| Factors That Increase Absorption | Factors That Decrease Absorption |
| --- | --- |
| Low to moderate zinc intake | Phytic acid and fiber in whole grains |
| Zinc deficiency | Excessive zinc intake |
| Certain amino acids | High nonheme iron intake |
| Increased need for zinc | Good zinc status |

species.[18] There are so many zinc enzymes, it is hard to name a body process or structure that isn't affected either directly or indirectly by zinc. Zinc contributes to DNA and RNA synthesis, heme synthesis, bone formation, taste acuity, immune function, reproduction, growth and development, and the antioxidant defense network (as a part of the Cu/Zn superoxide dismutase [SOD] enzyme). In addition, zinc stabilizes the structures of cell membrane proteins, gene transcription fingers (known as "zinc fingers"), and receptor proteins for vitamin A, vitamin D, and thyroid hormone.[18]

Zinc also may play a role in shortening the duration of common colds in healthy people, when taken as zinc lozenges within the first 24 hours of the onset of symptoms.[19] Doses of at least 75 mg per day been effective yet are above the Upper Level for zinc. More research is needed to determine optimal dosing and administration to achieve benefits while minimizing the potential side effects of nausea and bad taste.

## Zinc Deficiency

In many parts of the world where poverty limits food choices, zinc deficiency is a major health concern.[17] The symptoms include loss of appetite, delayed growth and sexual maturation, dermatitis, impaired vitamin A function, alopecia, decreased taste sensitivity, poor wound healing, immune dysfunction, severe diarrhea, birth defects, and increased infant mortality (see Figs. 15-9 and 15-10). Compromised zinc status also impairs the integrity of zinc-containing structural proteins in cell membranes, zinc fingers, and protein receptors. As a result, these proteins can no longer perform their functions.[18]

In North America, overt zinc deficiencies were not observed until the early 1970s, when hospitalized patients received total parenteral nutrition. Zinc was not added to the original parenteral solutions because the protein source in these solutions contained zinc. When the composition of these solutions was changed to provide isolated amino acids (lacking zinc) as the primary protein source, zinc deficiency symptoms quickly developed.

In North America, overt zinc deficiency is uncommon. Mild zinc deficiencies have been reported in young children, individuals with Crohn's disease and other malabsorptive diseases, those on kidney dialysis, and individuals who restrict their intake of animal-based foods.[18] Other individuals also may be at risk of mild or marginal deficiencies. However, it is difficult to detect marginal deficiencies because assessment measures that reflect changes in zinc status are lacking. In addition, marginal zinc deficiencies are likely to go undetected because they do not typically result in specific physical symptoms. More sensitive ways of assessing zinc status are needed.[20]

Severe zinc deficiency can result from a rare genetic condition called acrodermatitis enteropathica. This condition develops after weaning and results in impaired intestinal zinc absorption.[21] Treatment with supplemental zinc is effective in restoring zinc status.

## Zinc Toxicity

Signs of zinc toxicity have been reported with supplemental intakes of zinc at 5 or more times the RDA. Thus, the Upper Level is set at 40 mg/day.[5] The symptoms include loss of appetite, nausea, vomiting, intestinal cramps, and diarrhea. Toxicities also have been reported to impair immune function and reduce copper absorption and the activity of copper-containing enzymes.[18] Individuals taking zinc supplements and/or zinc lozenges (for the relief of cold symptoms) should do so cautiously and with the guidance of a registered dietitian nutritionist to avoid toxicity symptoms and potential mineral-mineral interactions.

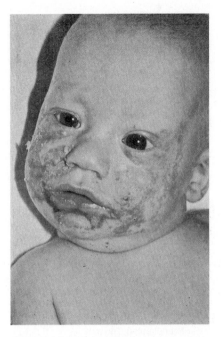

*Figure 15-10* This preterm infant's zinc deficiency was caused by low zinc stores at birth and below normal levels of zinc in his mother's breast milk.

Photo courtesy of Stephanie A. Atkinson, Ph.D.

### Knowledge Check

1. What functions in the body are dependent on zinc?
2. What are good dietary sources of zinc?
3. What are the symptoms of zinc deficiency?
4. Why are mild zinc deficiencies difficult to detect?

## Critical Thinking

Before class last week, Josh heard his friends talking about taking zinc lozenges to prevent colds and flu. He wasn't sure that this was effective but knew that his friends hadn't been sick lately, so he starting taking 50 mg daily. Is this a safe level of intake? What recommendations would you give Josh?

## Take Action

### Iron and Zinc Intake in a Sample Vegan Diet

Two months ago, Steve decided to become a vegan to improve his diet and overall health. He has asked you to evaluate his diet. His food and beverage intake yesterday is listed below. Does Steve's diet pose any nutritional risks? Is he meeting his needs for iron and zinc?

**Breakfast**
Soy milk, 1 cup
Raisin bran cereal, 1 cup
Orange, 1
Black coffee, 12 oz

**Lunch**
Sandwich:
　Whole-wheat bread, 2 slices
　Peanut butter, 2 tbsp
Granola bar, 1
Banana, 1
Water, 12 oz

**Snack**
Oatmeal cookies, 3 small
Apple juice, 12 oz

**Dinner**
Salad:
　Romaine lettuce, 1½ cups
　Tomato, 1 small
　Carrot, 1 shredded
　Cucumber, ½ sliced
　Mushrooms, ⅓ cup
　French dressing, 3 tbsp
White bean soup, 2 cups
Whole-wheat crackers, 8
Soy cheese, 1 oz
Iced tea, 12 oz

**Snack**
Popcorn, 3 cups
Root beer, 12 oz

©Sam Edwards/Getty Images

Start by analyzing Steve's iron and zinc intake using an online nutrient database, nutrient analysis software, or a website (fdc.nal.usda.gov/). What is your conclusion? Does Steve's diet appear to be a healthy way to eat? What other nutrient intakes may be of concern?

---

Nuts and legumes are rich sources of copper.

©I. Rozenbaum & F. Cirou /Getty images RF

## 15.3 Copper (Cu)

The use of copper to treat disease dates back to around 400 B.C. However, the essentiality of copper was not fully recognized until 1964, when conclusive evidence of human copper deficiency was reported. Copper has vital functions as part of many important proteins and enzymes in the body.

### Copper in Foods

Copper is found in a variety of foods (Fig. 15-11). Good sources of copper include liver, shellfish, nuts, seeds, lentils, soy products, and dark chocolate. Dried fruits, whole-grain products, and the tap water in many communities also are important sources. Although meat is only a marginal source of copper, it may promote copper absorption from other foods, as it does for iron.

### Dietary Needs for Copper

The adult RDA for copper is 900 µg/day.[5] This allowance is based on the amount needed for the normal activity of copper-containing enzymes and proteins in the body. The average adult intake in North America ranges from about 1000 to 1600 µg/day. The Daily Value on food and supplement labels is 0.9 mg (900 µg).

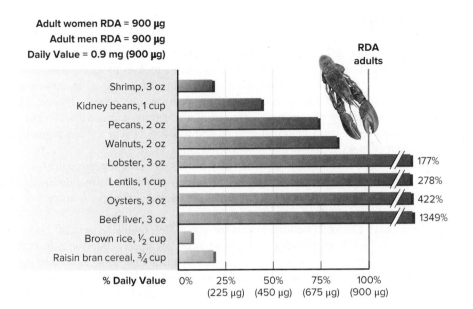

**Adult women RDA = 900 μg**
**Adult men RDA = 900 μg**
**Daily Value = 0.9 mg (900 μg)**

*Figure 15-11* Food sources of copper.
©C Squared Studios/Getty Images RF

## Absorption, Transport, Storage, and Excretion of Copper

Copper absorption occurs primarily in the small intestine. Like zinc, copper is absorbed by simple diffusion and active transport into intestinal absorptive cells and then transported out of the mucosal cells into the bloodstream.[22] In the blood, copper is bound to albumin and other proteins and moves rapidly to the liver (the main storage site) and kidneys. Copper is transported from the liver to other tissues bound primarily to the protein ceruloplasmin. Within the tissues, ceruloplasmin binds to specific receptors, which release copper to transporters within the cells.[22]

Very little copper is stored in the body. However, excess copper can bind to intestinal metallothionein, which may increase the short-term availability of copper.[16,22] Copper is excreted mainly through the bile into the intestinal tract for fecal elimination.

Copper absorption is the primary means of regulating copper balance. Thus, absorption can vary from approximately 12 to 70% of dietary intake. Copper absorption increases when dietary copper is low and decreases when intakes of copper, iron, and/or zinc are excessive. High phytate intake from cereals and legumes also decreases copper absorption.

## Functions of Copper

Like iron, copper is an important component of many enzymes because of its ability to alternate between 2 oxidation states ($Cu^{1+}$ and $Cu^{2+}$). Copper-containing enzymes have many functions in metabolism.[22] For example, the enzyme ceruloplasmin (also called ferroxidase I), is involved in oxidizing ferrous ($Fe^{2+}$) iron to ferric ($Fe^{3+}$) iron for incorporation into transferrin and subsequent transport from the liver to body cells. One effect of low ceruloplasmin levels is that little iron is transported from storage, resulting in decreased hemoglobin synthesis and the development of anemia. Ceruloplasmin also increases during inflammation and infection to prevent damage to body cells.

In combination with zinc, copper also functions as part of a family of enzymes known as **superoxide dismutase (SOD)** enzymes. These enzymes eliminate superoxide free radicals, which prevents oxidative damage to cell membranes. Another copper-containing enzyme, cytochrome C oxidase, catalyzes the last step of the electron transport chain in energy metabolism. Copper is involved in the regulation of neurotransmitters (serotonin, tyrosine, dopamine, and norepinephrine) via monoamine oxidase enzymes. In addition, copper also has an important role in connective tissue formation as a component of lysyl oxidase. Lysyl oxidase cross-links the strands in 2 structural proteins (elastin and collagen) that give tensile strength to connective tissues in the lungs, blood vessels, skin, teeth, and bones.

**superoxide dismutase (SOD)** Enzyme that deactivates a superoxide negative free radical ($O^{2-}$). SOD can contain the trace minerals copper and zinc or magnesium.

▶ Deficiencies of many micronutrients can cause anemia.

- Vitamin E deficiency can lead to hemolytic anemia.
- Vitamin K deficiency, especially coupled with the use of certain antibiotics, can lead to blood loss and hemorrhagic anemia.
- Vitamin B-6 deficiency can lead to microcytic hypochromic anemia.
- Folate deficiency can lead to megaloblastic (macrocytic) anemia.
- Vitamin B-12 malabsorption can lead to megaloblastic (macrocytic) anemia.
- Iron deficiency can lead to microcytic hypochromic anemia.
- Copper deficiency can lead to iron deficiency anemia

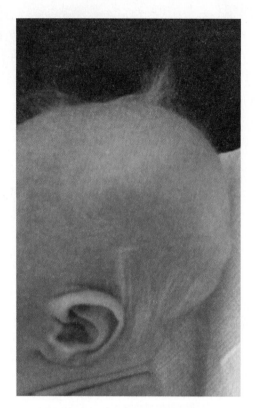

Menkes disease is a genetic condition that impairs the transport of copper and utilization by enzymes. Characteristics of this disease include fine, silvery, brittle hair; doughy, pale skin; below normal body temperature; low muscle tone; and neurological degeneration of the brain. Learn more at rarediseases.info.nih.gov/diseases/1521 /menkes-disease.

Reproduced with permission from Visual Diagnosis: 8-Day-Old Hypotonic Newborn With Sparse Hair, vol. 35, e53-e56, Copyright © 2014 by the AAP.

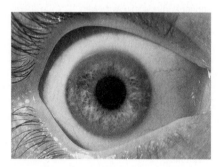

*Figure 15-12*  In individuals with Wilson disease, copper is deposited in the outer edges of the cornea. The rings (called Kayser-Fleischer rings) are golden to greenish-brown.

©Medical-on-Line/Alamy

## Copper Deficiency

Severe copper deficiency is relatively rare in humans. Deficiencies have been reported in premature infants fed milk-based formulas, in infants recovering from malnutrition, in patients on long-term total parenteral nutrition without added copper, in those consuming excessive amounts of zinc, and in individuals with the rare genetic disorder Menkes disease.[23] The symptoms of copper deficiency include anemia, decreased white blood cell counts (leukopenia), skeletal abnormalities (osteopenia), loss of skin and hair pigmentation, cardiovascular changes, and impaired immune function.

Studies suggest that copper deficiency may increase the risk of neurological disorders, such as amyotrophic lateral sclerosis (Lou Gehrig disease) and Alzheimer disease.[23] However, further studies are needed to determine the possible role of copper homeostasis in the etiology and treatment of these disorders.

Researchers also have become concerned about the possible effects of mild copper deficiencies resulting from prolonged marginal intakes of copper. Unfortunately, detecting marginal copper deficiency can be challenging due to a lack of reliable indicators. As with zinc, analyses of copper levels in tissues and blood may not fully reflect changes in copper status and can be affected by various diseases.[24] The symptoms associated with marginal intakes include suboptimal immune function, glucose intolerance, increased serum cholesterol, and cardiac abnormalities.[22]

## Copper Toxicity

Although copper toxicity is not common in humans, it has been reported in children taking accidental overdoses, in individuals consuming copper-contaminated food or water, and in Wilson disease (a genetic disorder resulting in excess copper storage) (Fig. 15-12). The symptoms of toxicity include abdominal pain, nausea, vomiting, and diarrhea. In severe cases, an accumulation of copper in the liver and the brain causes cirrhosis and neurological damage, respectively. The Upper Level for copper is set at 10 mg/day because higher intakes increase the risk of liver damage.[5]

> ### Knowledge Check
>
> 1. Which enzymes require copper for functioning?
> 2. What are 3 rich sources of copper?
> 3. Why do many individuals with copper deficiency develop iron deficiency anemia?

# 15.4 Manganese (Mn)

Manganese has been recognized as an essential trace mineral since the early 1930s. However, much less is known about manganese than iron, zinc, and copper. As scientists develop more sensitive measures of trace mineral status and function, they will have a clearer picture of the role of manganese in health and disease.

## Manganese in Foods

Whole-grain cereals, nuts, legumes, leafy greens, and tea are the best sources of manganese (Fig. 15-13). Meat and dairy products contribute very little manganese to the diet. Based on dietary surveys, the North American diet provides between 2 and 6 mg/day.

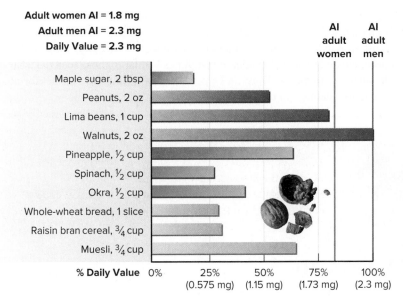

Adult women AI = 1.8 mg
Adult men AI = 2.3 mg
Daily Value = 2.3 mg

*Figure 15-13* Food sources of manganese.

©C Squared Studios/Getty Images RF

Manganese is found in peanuts and other legumes.

©C Squared Studios/Getty Images RF

## Dietary Needs for Manganese

Data are currently insufficient to determine the dietary requirement for manganese. Thus, an Adequate Intake (AI) for manganese was set at 2.3 mg/day for adult men and 1.8 mg/day for adult women.[5] The Daily Value for manganese on food and supplement labels is 2.3 mg.

## Absorption, Transport, Storage, and Excretion of Manganese

Manganese is absorbed in the small intestine via simple diffusion and active transport. Following absorption, most manganese is bound to alpha-2-macroglobulin for transport to the liver and then transported via transferrin, alpha-2-macroglobulin, or albumin to other tissues, such as the pancreas, kidneys, and bone. Excretion occurs mainly via bile, which is the primary means of regulating manganese levels in the body.[25]

Approximately 5 to 10% of the manganese in food is absorbed. Absorption is affected by the amount of manganese in the diet and by iron status. Absorption is increased by low manganese intake and iron deficiency (manganese and iron compete for transporters when iron intake is higher) and is decreased by high intakes of manganese and possibly copper, nonheme iron, fiber, phytates, and oxalates.

## Functions of Manganese

Manganese shares some functional similarities with zinc and copper.[25] For example, like zinc and copper, it serves as a cofactor for many enzymes in the body. Manganese-dependent enzymes have important functions in carbohydrate metabolism, gluconeogenesis, collagen formation, the antioxidant defense network (as Mn superoxide dismutase), and the nervous system.[25,26] In the body, manganese can alternate between different oxidation states ($Mn^{2+}$ and $Mn^{3+}$). This enables it to participate in various metabolic reactions.

## Manganese Deficiency and Toxicity

Manganese deficiency has not been well documented in humans. In fact, only a few cases have ever been reported. The symptoms associated with deficiency include skin rash, weight loss, poor growth, skeletal abnormalities, and impaired carbohydrate and lipid metabolism.[25]

©Image Source

## CASE STUDY FOLLOW-UP

Chloe decided to visit the campus health services clinic. She met with a physician and registered dietitian nutritionist. From an assessment of Chloe's dietary intake and laboratory data for hemoglobin and transferrin receptors, the health-care team determined that Chloe had iron deficiency anemia and an inadequate iron intake. The registered dietitian nutritionist worked with Chloe to help increase her intake of the better absorbed heme iron found in meat, fish, and poultry. For example, the marinara sauce could include ground beef. The peanut butter could be replaced with turkey or tuna. She also recommended that Chloe eat fortified breakfast cereals; have a vitamin C rich food or beverage, such as orange juice, when she eats iron-containing foods; and avoid coffee and tea.

Although manganese is considered to be less toxic than most trace minerals, toxicities have been reported in children receiving long-term parenteral nutrition and from the inhalation of airborne industrial and automobile emissions.[25,26] Toxicity causes severe neurological impairment and the development of symptoms similar to those seen in Parkinson disease (muscle stiffness, tremors). Thus, an Upper Level of 11 mg/day was set for manganese to prevent possible damage to the nervous system.[4]

### Knowledge Check

1. What are 3 rich sources of manganese?
2. How is manganese function similar to that of zinc and copper?
3. How is manganese status regulated in the body?

## 15.5 Iodine (I)

Iodine ($I_2$), present in food as iodide ($I^-$) and other nonelemental forms, is unique in many ways from other trace minerals. It is the heaviest element needed for human health and is responsible for only a single function in the body, the synthesis of thyroid hormones.[27]

### Iodine in Foods

The natural iodine content of most foods is relatively low. Saltwater seafood, seaweed, iodized salt, and dairy products are the best sources of iodine (Fig. 15-14). Dairy products are not naturally good sources of iodine, but they often provide significant amounts because iodide is added to cattle feeds and sanitizing solutions used in dairy processing. Breads and cereals also may contribute dietary iodine if they are prepared with iodized salt and/or dough conditioners (compounds that strengthen dough and improve bread volume and texture) because these contain iodates ($IO_3^-$).

Plant-based foods also may provide iodine if the soil in which they were grown was rich in this mineral. However, soil content can vary significantly from region to region, resulting in great differences in iodine content. Iodine rich soil tends to be near oceans because seawater is naturally high in iodine. When seawater evaporates, the iodine in the air falls on nearby soil.

For many Americans, iodized salt used in cooking and at the table provides most of their iodine. Iodized salt contains approximately 76 µg of iodine per gram of salt, but there

*Figure 15-14* Food sources of iodine.

©McGraw-Hill Education/Jacques Cornell, photographer

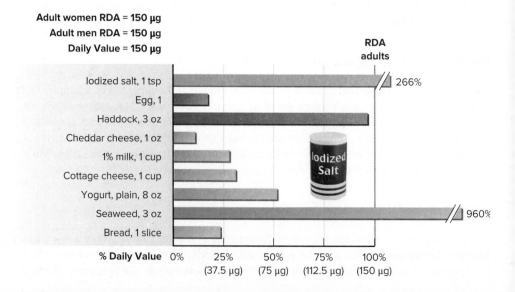

Adult women RDA = 150 µg
Adult men RDA = 150 µg
Daily Value = 150 µg

RDA adults

| | |
|---|---|
| Iodized salt, 1 tsp | 266% |
| Egg, 1 | |
| Haddock, 3 oz | |
| Cheddar cheese, 1 oz | |
| 1% milk, 1 cup | |
| Cottage cheese, 1 cup | |
| Yogurt, plain, 8 oz | |
| Seaweed, 3 oz | 960% |
| Bread, 1 slice | |

**% Daily Value**  0%   25%   50%   75%   100%
(37.5 µg)  (75 µg)  (112.5 µg)  (150 µg)

is variability among salt samples. In practical terms, this means that about a half teaspoon (about 2 g) of salt supplies the adult RDA for iodine. However, the salt used in processed foods and specialty salts, such as kosher and sea salts, are not iodized. It is important to read labels when purchasing different types of salt.

The bioavailability of iodine is decreased by compounds, called goitrogens, found in raw vegetables such as turnips, cabbage, Brussels sprouts, cauliflower, broccoli, rutabagas, potatoes, and cassava, as well as peanuts, soy, millet, peaches, and strawberries. **Goitrogens** decrease iodine absorption and inhibit iodine use by the thyroid gland. The risk of iodine deficiency is elevated in low-income parts of the world where raw vegetable intake is high and iodine intake low.[28] Goitrogens are of little concern in high-income countries because these foods are typically cooked (which helps destroy goitrogen activity) and iodine rich foods are widely available.

## Dietary Needs for Iodine

The adult RDA for iodine is 150 µg/day.[5] This recommendation is based on the amount needed to maintain adequate uptake and turnover by the thyroid gland. In the U.S., usual dietary iodine intakes range between 190 and 300 µg/day. This does not account for amounts provided by iodized salt sprinkled on foods at the table. The Daily Value for iodine used on food and supplement labels is 150 µg.

## Absorption, Transport, Storage, and Excretion of Iodine

Most of the iodine present in foods is in the form of iodide and, to a lesser extent, iodates. These forms are very efficiently absorbed throughout the small intestine. After absorption, most of the iodine (the general term used for the mineral) is transported to the thyroid gland. Using a sodium-dependent active transport system, the thyroid gland actively accumulates iodine—in essence, "trapping" it to support thyroid hormone synthesis. Excess iodine is excreted primarily via the kidneys into the urine. Very little iodine is excreted in the feces.

A small amount of iodized salt in the diet meets iodine needs.

©Tom Grill/Getty Images RF

## Functions of Iodine

Iodine is an essential component of the thyroid hormones thyroxine ($T_4$) and triiodothyronine ($T_3$). Most of the circulating thyroid hormone in the body is in the form of $T_4$. Within body cells, $T_4$ loses an iodine molecule and is converted to $T_3$, the active form of the hormone. The enzymes involved in this conversion (called deiodinase enzymes) require the trace mineral selenium. Thus, a selenium deficiency can limit the activity of deiodinase enzymes, resulting in decreased $T_3$ levels.

Although iodine has a singular function, this function has widespread consequences because of the vital role that thyroid hormones play in maintaining normal metabolism. As a component of $T_3$, iodine is involved in the regulation of many important metabolic and developmental functions. This includes the regulation of basal energy expenditure, macronutrient metabolism, brain and nervous system development, and overall growth.[27]

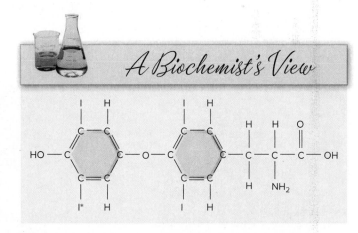

*A Biochemist's View*

Thyroxine ($T_4$). Triiodothyronine ($T_3$) has a similar structure but lacks 1 iodine (I), indicated in this figure with a red asterisk.

## Iodine Deficiency Disorders (IDD)

Iodine deficiency disorders, the collective name for **endemic** goiter and endemic cretinism, occur when dietary iodine intake is insufficient. When iodine availability decreases and plasma levels of $T_4$ hormone drop, the pituitary gland secretes thyroid-stimulating hormone

**endemic** Habitual presence within a given geographic area of a disease.

(TSH). In response to increased TSH levels, the thyroid gland enlarges in an attempt to increase its efficiency at trapping iodine. The characteristic enlargement of the thyroid gland that occurs is called a **goiter**. This early adaptive response allows the thyroid gland to temporarily maintain thyroid hormone synthesis. Although a goiter is a painless condition, it can cause pressure on the esophagus and trachea and impair their function. If the underlying iodine deficiency is not corrected, decreased $T_4$ synthesis, slowed metabolism, and more serious complications can develop.[28]

Iodine deficiency is of particular concern during pregnancy because of the adverse effects it can have on the developing offspring.[28,29] These include congenital abnormalities, low birth weight, neurological disorders, impaired mental function, poor physical development, and death. The resulting restriction of brain development and growth, called **cretinism** (Fig. 15-15), is characterized by severe mental retardation, loss of hearing and speech abilities, short stature, and muscle spasticity.

Descriptions of goiter have been documented for many centuries in China and other parts of the world. Prior to World War I, goiter was very common in the Great Lakes region of the U.S., where the soil and lake waters are very low in iodine. Cretinism also was common in parts of the U.S. The discovery in the early 1920s, by scientists in the U.S. and Switzerland, that iodine can prevent the development of goiter resulted in the fortification of table salt in the U.S. and many other countries. Using a small amount of iodized salt has eradicated endemic iodine deficiency in many countries. However, deficiencies continue to be prevalent in many low-income countries. In fact, the World Health Organization (WHO) believes that iodine deficiency is the "greatest single cause of preventable brain damage and mental retardation" and set a goal of eliminating this deficiency by increasing the use of fortified salt, oil, milk, and other food products.[28,29] This international focus on eliminating iodine deficiency has helped 9 out of 10 households worldwide gain access to iodized salt.[30,31] Nonetheless, nearly 20% of the world's countries remain at risk of iodine deficiency[31] (Fig. 15-16). The risk of inadequate iodine status is elevated in pregnancy, when needs are increased.

## Iodine Toxicity

The Upper Level is set at 1100 µg/day for adult men and women to prevent health-related risks.[5] Like iodine deficiency, iodine toxicity can cause enlargement of the thyroid gland

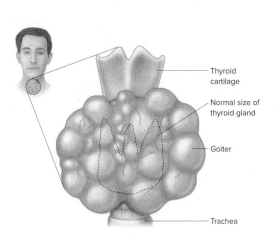

Thyroid cartilage

Normal size of thyroid gland

Goiter

Trachea

*Figure 15-15*  (*a*) This Nigerian woman has a goiter caused by iodine deficiency. (*b*) This 35-year-old Peruvian woman is 43 inches (110 cm) tall due to cretinism, which is caused by an iodine deficiency.

(a): ©Scott Camazine/Science Source; (b): ©Dr. Eduardo A. Pretell

(a)                              (b)

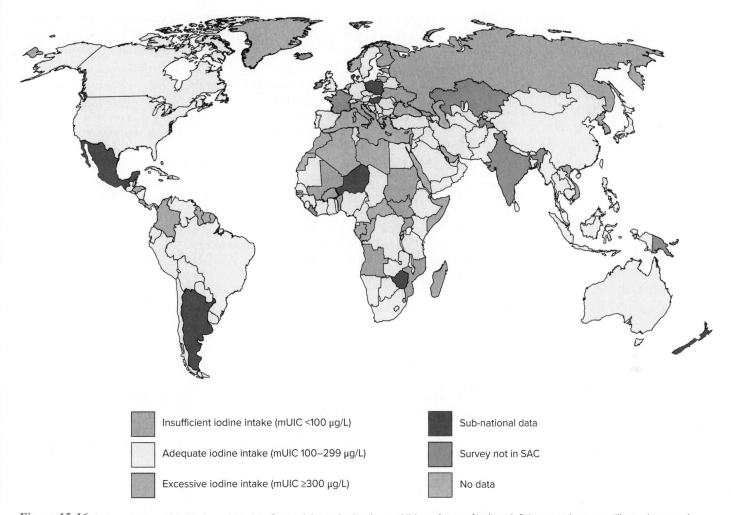

| | |
|---|---|
| ▨ Insufficient iodine intake (mUIC <100 μg/L) | ▨ Sub-national data |
| ☐ Adequate iodine intake (mUIC 100–299 μg/L) | ▨ Survey not in SAC |
| ▨ Excessive iodine intake (mUIC ≥300 μg/L) | ▨ No data |

***Figure 15-16***  Iodine status worldwide based on data from adults and school-age children. Areas of iodine deficiency and excess still remain around the world.

Source: Iodine Global Network. Global scorecard of iodine nutrition 2017. https://www.ign.org/cm_data/IGN_Global_Map_AllPop_30May2017.pdf, Accessed October 13, 2019.

and decreased thyroid hormone synthesis.[32] Toxicities have been reported in Japan, due to high intakes of iodine rich seaweed, and in Chile, due to high levels of environmental iodine, increased iodine use in water purification, and excessive fortification of salt. Although hypothyroidism is the most common result of iodine toxicity, excess iodine intakes also may increase the risk of hyperthyroidism, autoimmune thyroid disease, and thyroid cancer.

## Knowledge Check

1. What are 3 rich sources of iodine?
2. What is the main function of iodine?
3. Why is iodine deficiency still prevalent in many areas of the world?
4. What are the symptoms of iodine deficiency?

## CASE STUDY

©Ryan McVay/Getty Images RF

At a recent family reunion, Gina learned that an aunt was currently undergoing treatment for colon cancer. Her grandmother explained that 2 other family members had died of the disease before Gina was born. After the reunion, Gina decided to learn more about colon cancer and how her family history of the disease might affect her. Gina learned that 100,000 Americans are diagnosed with colon cancer each year and that her family history increases her chances of developing the disease. While searching online, she came across a site that recommended taking 200 mcg/day of the trace mineral selenium to prevent the disease. She then went to her local supermarket and found that 100 selenium tablets containing 200 mcg each cost only $7.50 a bottle. Gina figured this supplement was cheap "insurance" against developing the disease and, so, began taking 200 mcg of selenium a day. Is Gina's practice harmful? Are there other dietary practices she should consider to help protect her from developing colon cancer?

# 15.6 Selenium (Se)

The essentiality of selenium in human health was not recognized until 1979, when Chinese scientists noted that a cardiac condition in children and young women living in the Keshan province could be prevented by selenium. Scientists' understanding of the role of selenium in Keshan disease and other aspects of human health has developed rapidly since that time.[33]

## Selenium in Foods

The selenium content of food varies significantly in relation to the soil content where the plant was grown or the animal raised.[34] For example, the selenium content of grains grown in selenium rich soil will be much greater than that of grains grown in low selenium soil. In general, the best sources of selenium are seafood, meats, cereals, and grains (Fig. 15-17).

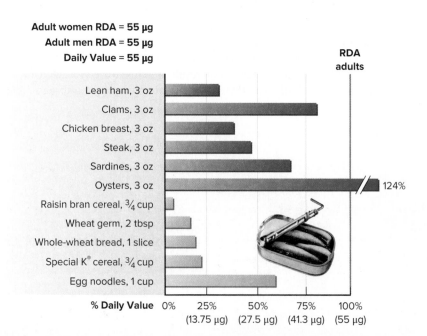

*Figure 15-17* Food sources of selenium.
©Brand X Pictures/Getty Images RF

## Dietary Needs for Selenium

The adult RDA for selenium is 55 µg/day.[35] The RDA is based on the amount of selenium needed to maximize glutathione peroxidase activity in the blood. In North America, daily intakes of selenium are typically above the RDA. The Daily Value on food and supplement labels is 55 µg.

## Absorption, Transport, Storage, and Excretion of Selenium

Most of the selenium in foods is bound to the amino acids methionine (as selenomethionine) and cysteine (as selenocysteine). Both forms are well absorbed in the small intestine, with absorption ranging from 50 to 100% of dietary intake. Unlike many of the other trace minerals, absorption is not affected by the body's selenium stores, and absorption does not play a role in maintaining homeostasis. Selenium balance is achieved primarily through urinary excretion, rather than intestinal absorption.

After absorption, selenium is bound to transport proteins in the blood and is distributed to target tissues and cells. Little is known about the transport of selenium across membranes into cells. The highest concentrations of selenium are found in the liver, pancreas, muscle, kidneys, and thyroid. Within tissues, selenomethionine provides a "storage pool" of selenium, whereas selenocysteine serves as the biologically active form of the mineral.

## Functions of Selenium

Selenium is a component of at least 25 different enzymes and proteins in the body.[33] One of its most recognized functions is in the antioxidant defense network, as a part of glutathione peroxidase (GPx) enzymes, thioredoxin reductase enzymes, and selenoprotein P (Fig. 15-18). As part of the antioxidant defense network, selenium helps prevent lipid

Pasta made from wheat grown in most parts of North America is a good source of selenium.

©Corbis Premium RF/Alamy RF

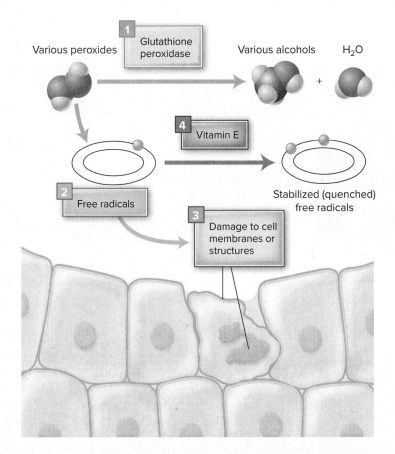

*Figure 15-18* Selenium is part of the glutathione peroxidase system, [1] which breaks down peroxides, such as hydrogen peroxide ($H_2O_2$), to water ($H_2O$) before they can form free radicals, [2] and damage cells, [3]. This breakdown of peroxides, in turn, helps spares vitamin E, [4], which is a major free radical scavenger.

peroxidation and cell membrane damage. Its ability to destroy highly reactive peroxyl free radicals spares vitamin E for use in other antioxidant functions. Another important function of selenium is in thyroid metabolism as a part of iodothyronine deiodinase enzymes. Recall from Section 15.5 that the deiodinase enzymes that convert thyroxine ($T_4$) to triiodothyronine ($T_3$) require selenium.

It is likely that selenium also plays an important role in immune function.[33] Scientists believe that selenium may prevent Keshan disease by inactivating a virus linked to its development. Additionally, selenium may decrease the risk of prostate, breast, lung, or other cancers, although studies have provided conflicting results.[33,36,37] Further research is needed to clarify the role of selenium in the prevention of chronic disease.

## Selenium Deficiency

Inadequate selenium intake is not known to cause a specific deficiency disease. As mentioned previously, selenium deficiency is associated with changes in thyroid hormone metabolism and a possible increased risk of certain cancers. Selenium deficiency also is associated with the development of Keshan disease, a disease characterized by insufficient cardiac function. As mentioned previously, this disease was 1st observed in the Keshan province of China, where the soil is almost devoid of selenium. It has since been diagnosed in other regions, including New Zealand and Finland. Although Keshan disease can be prevented through selenium supplementation, the related cardiac disorders are not corrected by selenium once the disease has developed.

## Selenium Toxicity

Excess supplementation of selenium can result in toxicity. In fact, selenium toxicity (selenosis) has been observed in intakes as low as 1 to 3 mg daily taken over many months. The symptoms of toxicity include nausea, diarrhea, fatigue, hair loss, changes in nails, and impairment of sulfur and protein metabolism. The Upper Level for selenium is 400 µg/day.[35]

### CASE STUDY FOLLOW-UP

©Ryan McVay/Getty Images RF

Gina's supplement dose of 200 µg/day, plus a typical dietary intake of 105 µg/day, is below the Upper Level of 400 µg/day for selenium. Therefore, her practice is probably safe. Whether it will be helpful in reducing colon cancer risk awaits further research. Until more information is available, the widespread use of such a high dose of selenium is not recommended. Dietary recommendations that Gina can follow to help decrease cancer risk are presented in this chapter's Clinical Perspective: Nutrients, Diet, and Cancer.

*Critical* Thinking

When Eric was shopping for antioxidant supplements, he noticed that many contain copper, zinc, manganese, and selenium. Why are these likely to be included in antioxidant supplements?

*Knowledge Check*

1. What are 3 rich sources of selenium?
2. How is selenium function linked to iodine metabolism?
3. What diseases are associated with selenium deficiency?

# 15.7 Chromium (Cr)

The importance of chromium in the diet has been recognized only in recent years. Like many trace minerals, the functions of this nutrient are emerging with advances in research technology.

Broccoli is a good source of chromium.

©C Squared Studios/Getty Images

## Chromium in Foods

Chromium is widely distributed in a variety of foods. However, information regarding the chromium content of foods is lacking. Thus, many nutrient databases do not yet include values for chromium. Meats, liver, fish, eggs, whole-grain products, broccoli, mushrooms, dried beans, nuts, and dark chocolate tend to be good sources of the mineral. Chromium is used to manufacture steel; thus, small amounts of chromium also can be transferred to food through food processing equipment.

## Dietary Needs for Chromium

The Adequate Intakes for chromium in adults ages 19 to 50 years are 35 µg/day for men and 25 µg/day for women.[5] After age 50, the Adequate Intake decreases to 30 µg/day for men and 20 µg/day for women. The Adequate Intake is based on the amounts typically found in nutritious diets. The average intake in North America generally meets the Adequate Intake level. The Daily Value for chromium is 35 µg.

## Absorption, Transport, Storage, and Excretion of Chromium

Very little chromium is absorbed from dietary sources. Absorption appears to increase when intakes are low and when consumed with vitamin C, although bioavailability of the mineral has been difficult to assess. It is likely that phytates in whole grains decrease absorption. Once absorbed, chromium is transported by transferrin via the bloodstream and accumulates in the bones, liver, kidneys, and spleen. Concentrations in human tissues are very low because most dietary chromium is excreted in the feces. Chromium also is excreted in the urine.

## Functions of Chromium

The functions of chromium are not fully known. Chromium may enhance insulin action, promote glucose uptake into cells, and normalize blood sugar levels.[38] However, chromium supplementation in patients with type 2 diabetes has not been shown to be effective in controlling blood glucose.[38,39] Many athletes use chromium supplements to enhance muscle mass and strength, despite a lack of research evidence supporting their effectiveness.

## Chromium Deficiency and Toxicity

Chromium deficiency has been difficult to assess due to the lack of sensitive measures of chromium status. Several cases of chromium deficiency have been reported in individuals receiving chromium-free **parenteral nutrition** solutions. The symptoms included weight loss, glucose intolerance, and nerve damage.

Few serious effects have been reported from excess dietary chromium intake. Thus, an Upper Level has not been set.[5] Nutritionists have expressed concern about the safety of high doses of chromium supplements (especially chromium picolinate) used by many athletes and have recommended continued monitoring for possible toxicities.

## 15.8  Fluoride (F)

Fluoride, the ionic form of fluorine, may not be an essential nutrient because all basic body functions can occur without it. However, in the early 1930s, it was observed that individuals living in the southwestern U.S., where the water naturally contained high concentrations of fluoride, had fewer dental caries (cavities). Many people in these areas also had small spots on their teeth (known as mottling or fluorosis) due to excess fluoride (Fig. 15-19). Although discolored, the mottled teeth were virtually free of dental caries. This discovery led to research studies confirming this mineral's ability to reduce cavities and the start of controlled water fluoridation in parts of the U.S.

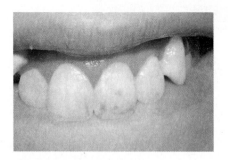

*Figure 15-19*  Fluorosis, or mottled enamel, caused by overexposure to fluoride.

©Paul Casamassimo, DDS, MS

### Fluoride in Foods

Today, in North America, the major source of fluoride is fluoridated water. Adding fluoride to the public water supply adjusts the fluoride level to 0.7 ppm or 0.7 mg/liter; this concentration is considered safe and optimal for the prevention of dental caries.[40] However, not all public or private water sources are fluoridated (Fig. 15-20).

In addition to fluoridated water, tea, seafood, and seaweed provide the greatest amounts of dietary fluoride. The use of fluoridated toothpastes and mouth rinses and fluoride treatments provides nondietary ways of obtaining fluoride. Most bottled waters do not contain fluoride.[40]

▶ To determine if your water supply is fluoridated, contact your local water supplier or have the water in your home analyzed for fluoride content. Your dentist can help you decide the best means for obtaining sufficient fluoride.

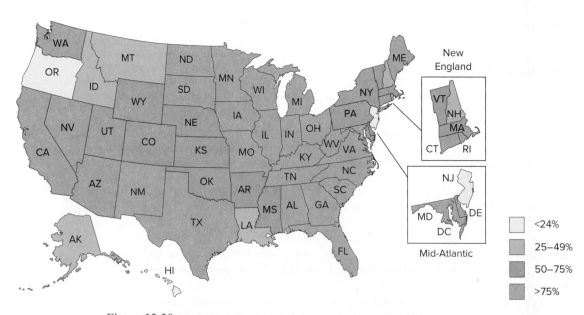

*Figure 15-20*  Percentage of state populations served by public water systems with fluoridated water. More information is available at www.cdc.gov/fluoridation/statistics.

Source: National Center for Chronic Disease Prevention and Health Promotion

## Dietary Needs for Fluoride

The Adequate Intakes for fluoride are 3 mg/day for adult women and 4 mg/day for adult men.[41] For infants up to 6 months of age, the Adequate Intake is 0.01 mg/day. This increases to 0.5 mg/day for infants 6 to 12 months of age and ranges from 0.7 to 3 mg/day for young children and adolescents. The Adequate Intake recommendations are based on the amount needed to provide resistance to dental caries without causing mottling of the tooth enamel.[41]

## Absorption, Transport, Storage, and Excretion of Fluoride

The absorption of dietary fluoride occurs rapidly in the stomach and small intestine via passive diffusion. Overall, approximately 80 to 90% of fluoride consumed is absorbed. Absorbed fluoride is transported in the bloodstream and concentrated in teeth and the skeleton. The amount of fluoride deposited in the teeth and bones is greatest during infancy, childhood, and adolescence. Calcified tissue deposition and urinary excretion are the major means for removing fluoride from the circulation.

Following the fluoridation of water in 1945 in many North American cities, the incidence of dental caries decreased dramatically.

©BananaStock/Getty Images RF

## Functions of Fluoride

Although a truly essential function for fluoride has not been described, fluoride is recognized for its beneficial role in supporting the deposition of calcium and phosphorus in teeth and bones and in protecting against the development of dental caries.[40] Fluoride works in several ways to prevent caries. During the development of teeth and bones, fluoride forms hydroxyfluorapatite crystals. These crystals provide greater resistance (than typical hydroxyapatite crystals) to bacteria and acids in the mouth that can erode tooth enamel. Fluoride in the blood contributes to fluoride in the saliva, which promotes the remineralization of enamel lesions and reduces the net loss of minerals from tooth enamel.

## Fluoride Deficiency and Toxicity

A lack of fluoride is associated with an increased incidence of dental caries. However, no specific deficiency disorder or disease appears to be caused by insufficient fluoride intake. In contrast, fluoride toxicity has been reported in young children who have swallowed fluoride tablets or solutions. Although rare, acute toxicity can occur rapidly and be life threatening. Thus, fluoridated toothpastes, mouth rinses, and supplements need to be kept out of the reach of children. The signs of toxicity include nausea, vomiting, diarrhea, sweating, spasms, convulsions, and coma.

Mottling, or fluorosis, of the enamel is the result of chronic intake of excess fluoride during tooth development. Dental fluorosis is not associated with any health risk but does result in discoloration and possible pitting of the enamel. To minimize the risk of fluorosis, an Upper Level has been set at 0.1 mg/kg body weight/day (0.7 to 2.2 mg/day) for infants and children up to 8 years of age. The Upper Level for children over age 8 and for adults is 10 mg/day.[41]

*Knowledge Check*

1. How does fluoride help prevent dental caries?
2. What are rich sources of fluoride?
3. Why should fluoridated toothpastes, mouth rinses, and supplements be kept out of the reach of children?

## Table 15-5  Key Trace Mineral Summary

| Mineral | Major Functions | Deficiency Symptoms | Individuals at Risk | Adult RDA or Adequate Intake | Good Dietary Sources | Toxicity Symptoms |
|---|---|---|---|---|---|---|
| Iron | Functional component of hemoglobin and other key compounds used in respiration; immune function; cognitive development; energy metabolism | Fatigue on exertion; poor immune function; anemia | Infants; preschool children; women in childbearing years | Males: 8 mg<br>Females: 18 mg | Meats; seafood; enriched breads; fortified cereals; eggs | Gastrointestinal upset; Upper Level is 45 mg/day |
| Zinc | Required for many enzymes; immune function; growth and development; stabilizes cell membranes and body proteins | Dermatitis; diarrhea; decreased appetite and sense of taste; infection; poor growth and development | Vegetarians; elderly people; people with alcoholism; malnourished populations | Males: 11 mg<br>Females: 8 mg | Seafoods; meats; whole grains | Decreased copper absorption, diarrhea, nausea, cramps, depressed immune function; Upper Level is 40 mg/day |
| Copper | Aids in iron metabolism; works in antioxidant enzymes and those involved in connective tissue metabolism | Anemia; low white blood cell count; poor growth | People using excessive zinc supplementation | 900 μg | Liver; cocoa; nuts; whole grains; shellfish; legumes | Excessive supplement use can cause vomiting, nausea, diarrhea, and nervous system and liver disorders; Upper Level is 10 mg/day |
| Manganese | Cofactor of several enzymes; involved in carbohydrate metabolism and antioxidant protection | Poor growth; skeletal abnormalities | Rare | Males: 2.3 mg<br>Females: 1.8 mg | Nuts; tea; legumes; whole-grain cereals | Nervous system disorders; Upper Level is 11 mg/day |
| Iodine | Component of thyroid hormones that regulate basal metabolism, growth, and development | Goiter; cretinism | Major problem in parts of the world where soil iodine is low and fortified foods are not available | 150 μg | Iodized salt; saltwater fish; dairy products | Inhibition of function of the thyroid gland; Upper Level is 1.1 mg/day |
| Selenium | Part of an antioxidant system as glutathione peroxidase; activates thyroid hormones | Keshan disease; reduced thyroid hormone | People in areas of the world with low selenium content in the soil | 55 μg | Meats; eggs; fish; seafoods; whole grains; nuts | Nausea, vomiting, hair loss, diarrhea, changes in nails; Upper Level is 400 mg/day |
| Chromium | Enhances insulin action | Glucose intolerance | Rare | Males: 30–35 μg<br>Females: 20–25 μg | Eggs; liver; whole grains; nuts; mushrooms; processed meats | No report of dietary toxicity; no Upper Level set |
| Fluoride | Increases resistance of tooth enamel to dental caries; mineralization of bones and teeth | Although not a true deficiency symptom, dental caries is a risk | Areas where water is not fluoridated | Males: 4 mg<br>Females: 3 mg | Fluoridated water; toothpaste; dental treatments; tea; seaweed | Fluorosis, or mottling (staining) of tooth enamel; acute toxicity can be fatal; Upper Level is 10 mg/day |
| Molybdenum* | Cofactor for several enzymes | Not known in humans | Rare | 45 μg | Grains; nuts; legumes | Poor growth in laboratory animals; Upper Level is 2 mg/day |

*Often classified as an ultratrace mineral, but unlike other ultratrace minerals it has an established RDA.

## Take Action

### Is Your Local Water Supply Fluoridated?

A goal of *Healthy People* is to increase the percentage of people in the U.S. served by community water systems who have an optimally fluoridated water supply. Today, about 66% of Americans have access to fluoridated water. For those receiving community water, 75% have a fluoridated supply. Do you know if your community water supply is fluoridated? To find the answer, check the My Water's Fluoride link of the website of the Centers for Disease Control and Prevention (nccd.cdc.gov/DOH_MWF). If your state or water system is not included at this link, or if you want additional information, you can check with your local water department. Is the water naturally rich in fluoride, or is this mineral added to the water supply? What amount of fluoride is in the drinking water? If it is added to the water supply, how long has this procedure been in operation? If the water does not supply fluoride, how can you obtain sufficient fluoride?

©Comstock/Getty Images RF

## 15.9 Molybdenum (Mo) and Ultratrace Minerals

Molybdenum is often classified as an ultratrace mineral. Although it is needed in extremely small, or "ultratrace," amounts, experts recognize that molybdenum, like many trace minerals, is essential for the activity of several enzymes. Molybdenum content is greatest in plant-based foods, such as grains, legumes, and nuts. Like iodine and selenium, the molybdenum content of foods can vary, depending on the soil in which the plant was grown. The RDA for molybdenum is very small (45 µg/day) and is based on its role as a cofactor required for the activity of several oxidase enzymes.[5] The dietary intakes in North America typically meet or exceed the RDA. The Daily Value used on food and supplement labels is 45 µg. Deficiencies of molybdenum are very rare, as are toxicities. However, an Upper Level of 2000 µg/day for adults was set to prevent the development of the goutlike symptoms of joint inflammation noted in several studies.[5] Table 15-5 summarizes the function of molybdenum and other trace minerals.

The body contains other minerals in ultratrace amounts that do not yet have clearly defined essential physiological functions. The minerals arsenic, boron, nickel, silicon, and vanadium fall within this classification (Table 15-6). It is likely that these ultratrace minerals

**Table 15-6  Ultratrace Mineral Summary**

| Mineral | Proposed Functions | Typical Intake by Adults | Estimated Daily Need | Upper Intake Level | Dietary Sources |
|---|---|---|---|---|---|
| Arsenic | Amino acid metabolism, DNA function | 30 µg | 12–25 µg | None set* | Fish, grains, cereals |
| Boron | Cell membrane function (ion transport), steroid hormone metabolism | 0.75–1.35 mg | 1–13 mg | 20 mg | Legumes, fruits, vegetables, potatoes, wine |
| Nickel | Metabolism of amino acids, fatty acids, vitamin B-12, and folic acid | 69–162 µg | 25–35 µg | 1 mg | Chocolate, nuts, legumes, whole grains |
| Silicon | Bone formation | 19–40 mg | 35–40 µg | None set | Root vegetables, whole grains |
| Vanadium | Mimics insulin action | 6–18 µg | 10 µg | 1.8 mg | Shellfish, mushrooms, parsley, dill |

*Although an Upper Level has not been set, the addition of this mineral to food is not recommended.

# GL🌐BAL PERSPECTIVE

## Global Nutrition Actions

Malnutrition affects several billion people worldwide. Both undernutrition and overweight are linked with serious health problems. Underweight, combined with suboptimal breastfeeding and micronutrient deficiencies, increases the risk of illness and death. Worldwide more than 149 million children under the age of 5 years have stunted growth and million are wasted, while over 40 million are overweight. [42] (low weight for height due to undernutrition). About one-third of females of reproductive age have anemia; iron deficiency increases the health risks that low-birth-weight babies face. Iodine deficiency is a preventable cause of brain damage in children. Zinc and vitamin A deficiencies increase the risks for infections and death in children.

The proportion of overweight children and adults worldwide raises concerns about the potential effects on the development of related disease, such as cardiovascular disease, diabetes, and cancer.[43] The majority of the world's population lives in countries where more deaths are attributable to overweight and obesity than to underweight.[43] Dietary intakes need improvement; for example, carbonated beverages are consumed more than any other non-alcoholic beverage across countries of all income levels.[42]

In an effort to decrease the number of deaths and morbidities associated with malnutrition, the World Health Organization (WHO) maintains the e-Library of Evidence for Nutrition Actions (eLENA). The online eLENA project provides case studies and easily accessed, evidence-informed information to help countries launch successful nutrition interventions. One project partner is the Cochrane Collaboration, an international network of people who prepare systematic reviews of health-care research. Interventions on eLENA are linked with WHO global nutrition targets.

The following are some of the interventions presented on eLENA.

- Promotion of breastfeeding
- Fortification of wheat and corn flours with iron, zinc, vitamin A, folic acid, and/or vitamin B-12
- Iron and folic acid supplementation for pregnant women
- Iodine fortification of salt
- Vitamin A supplementation for infants and children 6 to 59 months of age in settings where vitamin A deficiency is a public health problem
- Reduction of the impact of marketing foods and beverages that are high in fat, sugar, and/or salt
- Reduction of sodium intake
- Reduction of sugar-sweetened beverage intake

To learn more about the eLENA project and the global nutrition targets, visit the website www.who.int/elena/en.

©McGraw-Hill Education/Barry Barker, photographer

function as cofactors for specific enzymes or compounds, promote normal growth and development, and/or decrease the risk of certain diseases. However, further research is needed to define their specific roles in the body and to set RDAs or Adequate Intakes. Deficiency symptoms for these elements are not yet known. An Upper Level has been established for boron, nickel, and vanadium because of toxicity concerns associated with increased exposure to these minerals.[5]

### Knowledge Check

1. What are good sources of molybdenum?
2. What is the function of molybdenum?
3. Why are arsenic, boron, nickel, silicon, and vanadium classified as ultratrace minerals?
4. Which of the ultratrace minerals have an established Upper Level?

## Nutrients, Diet, and Cancer

Cancer is the 2nd leading cause of death for North American adults. Many nutrients, such as antioxidants (e.g., selenium and vitamin E) and calcium, have been studied for potential protective effects. At present, it is unclear how single nutrients or groups of nutrients may impact the risk for various cancers. Evidence supports a diet emphasizing plant-based foods to decrease cancer risk, whereas excessive calorie intake may increase cancer risk.

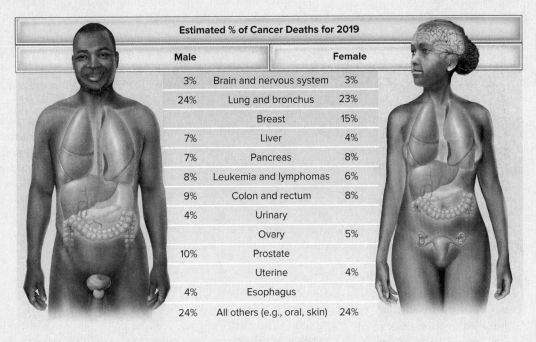

| | Estimated % of Cancer Deaths for 2019 | |
|---|---|---|
| Male | | Female |
| 3% | Brain and nervous system | 3% |
| 24% | Lung and bronchus | 23% |
| | Breast | 15% |
| 7% | Liver | 4% |
| 7% | Pancreas | 8% |
| 8% | Leukemia and lymphomas | 6% |
| 9% | Colon and rectum | 8% |
| 4% | Urinary | |
| | Ovary | 5% |
| 10% | Prostate | |
| | Uterine | 4% |
| 4% | Esophagus | |
| 24% | All others (e.g., oral, skin) | 24% |

**Figure 15-21** Cancer is many diseases. Numerous types of cells and organs are its target.

Source: American Cancer Society. Cancer facts & figures 2019. Atlanta: American Cancer Society; 2019.

### What Is Cancer?

Cancer is not a single disease. It occurs in many forms in different organs and types of cells in the body (Fig. 15-21). In fact, even in the same organ or tissue, cancers may be quite different. Lung, prostate, breast, and colorectal cancers account for over 45% of all cancers in the U.S. and are the leading cause of cancer-related death for nearly all racial and ethnic groups. Although these statistics are unsettling, the decline in cancer rates and cancer-related deaths over the past decade is encouraging. Early screening programs, new detection methods, and more effective cancer therapies have helped improve the prognosis for many cancers.

In general, cancer is characterized by the abnormal and uncontrolled division of altered cells. As these altered cells multiply, they often form tumors, which can be benign or malignant. **Benign** tumors are noncancerous because they are enclosed in a membrane that prevents them from spreading. They are harmful only if they interfere with normal function. For instance, a benign brain tumor can cause complications and death if it blocks blood flow in the brain. Unlike benign tumors, **malignant** tumors are capable of invading surrounding structures and spreading to other areas of the body. They can **metastasize** (spread) to distant parts of the body via the blood and/or lymph and can form invasive tumors in almost any area.

Most cancers are classified as carcinomas, sarcomas, lymphomas, or leukemias. Carcinomas constitute about 80 to 90% of all cancers. They develop from epithelial cells that cover external and internal areas (or surfaces) of the body and affect secretory organs, such as the breast. Sarcomas are cancers of connective tissues, such as in bone. Lymphomas, such as Hodgkin and non-Hodgkin lymphomas, are malignant tumors in the lymph nodes and lymphoid tissues. Leukemias are cancers of precursor white blood cells formed in the bone marrow. Although leukemias do not produce a tumor as other cancers do, they exhibit the basic characteristics of rapid and uncontrolled cell growth that can spread to other areas of the body (Fig. 15-22). Therefore, they are still classified as cancers.

### Development of Cancer

Most cells exist in a balanced cycle between turning cell replication on and turning it off. This cell cycle is controlled by the genes in DNA that promote cell replication (called protooncogenes) and prevent replication (called tumor suppressor genes). Cancer often results from a lack of suppressor genes or the overactivity of protooncogenes. The cancer gene (oncogene) is like an out-of-control protooncogene—it makes hundreds of copies of itself without any constraints. Normally, tumor suppressor genes act as braking mechanisms within the cell to prevent uncontrolled growth. When something goes wrong with these genes, the oncogenes are free to promote rapid cell growth. Within a cell, repair mechanisms constantly look for errors in DNA replication and make corrections. However,

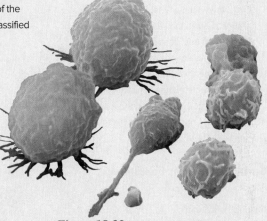

**Figure 15-22** Cancer cells are abnormal. The 2 cancerous leukemia cells on the left are larger than the normal marrow cells on the right.

©Eye of Science/Science Source

sometimes the repair mechanisms fail, resulting in an inherited defect, or a DNA mutation. Early defects that are not caught and repaired predispose the cell to additional errors, which increase the risk of cancer.

**Carcinogenesis**, the development of cancer, is a 3-step process (Fig. 15-23).

• Step 1 is the exposure of a cell to a carcinogen (cancer-causing agent) that triggers the initiation of cancer. Initiation can develop spontaneously or be induced by carcinogens, such as tobacco, radiation (e.g., sun exposure), alcohol, occupational toxins, viruses, food contaminants (e.g., aflatoxin mold), dietary factors, and drugs. The initiation stage of carcinogenesis, during which DNA is altered, is relatively short, ranging from minutes to days.

• Step 2 is the promotion state, which may last for months or even years. During this period, the mutation is locked into the genetic material of the cell. Compounds that increase cell division, called promoters, encourage uncontrolled replication of the altered DNA. Promoters may reduce the time available for repair mechanisms to enhance the replication of altered DNA cells. Excess alcohol, estrogen, and *Helicobacter pylori* bacteria in the stomach all may act as promoters.

• Step 3, cancer progression, begins with the appearance of cells that grow autonomously (out of control). During the progression phase, these malignant cells proliferate, invade surrounding tissue, and metastasize to other sites. Early in this stage, the immune system may find the altered cells and destroy them. Alternately, the cancer cells may become so defective that their DNA prevents their growth. However, if nothing stops cancer cell growth, 1 or more tumors eventually develop.

**Genetic, Environmental, and Dietary Factors**

Although our genetic makeup may contribute to a risk of certain cancers, such as colorectal, breast, and prostate cancer, inherited genetic alterations account for only about 5% of all cancers. This does not

explain the dramatic differences in cancer types and rates noted in different parts of the world. For example, in poorer countries, cancers of the stomach, liver, mouth, esophagus, and uterus are most common. In affluent countries, cancers of the lung, colon and rectum, breast, and prostate gland predominate. Because of these differences, many experts believe that environmental factors, such as exposure to radiation, chemicals, and air and water pollutants; smoking; lack of physical activity; obesity; and diet play a greater role in cancer initiation and development. For the U.S., estimates are that 1 in 5 cancer cases is due to poor diet, physical inactivity, overweight, obesity, and alcohol, and another one-third are caused by tobacco use.[44] The menu in Table 15-7 provides an example of a dietary plan to decrease cancer risk.

Various dietary factors have been associated with cancer. Evidence supports overall benefits of foods, not dietary supplements. The following list provides a summary of factors likely related to risk of cancer. Much of the evidence is based on epidemiological studies, which cannot show cause and effect. Further research is needed to confirm whether these factors increase or decrease cancer risk.[45–58]

• *Fruits and vegetables.* Diets low in fruits and vegetables (rich sources of phytochemicals and antioxidant nutrients, such as selenium, vitamin C, and vitamin E) have been associated with an increased incidence of certain cancers. Fruit and vegetable intake also may contribute to a decreased risk for some cancers by

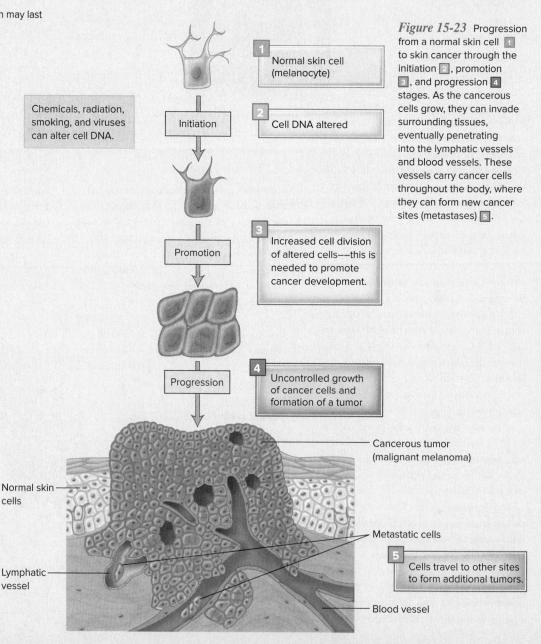

*Figure 15-23* Progression from a normal skin cell [1] to skin cancer through the initiation [2], promotion [3], and progression [4] stages. As the cancerous cells grow, they can invade surrounding tissues, eventually penetrating into the lymphatic vessels and blood vessels. These vessels carry cancer cells throughout the body, where they can form new cancer sites (metastases) [5].

Normal skin cell (melanocyte) [1]

Chemicals, radiation, smoking, and viruses can alter cell DNA.

Initiation — Cell DNA altered [2]

Promotion — [3] Increased cell division of altered cells—this is needed to promote cancer development.

Progression — [4] Uncontrolled growth of cancer cells and formation of a tumor

Cancerous tumor (malignant melanoma)

Normal skin cells

Lymphatic vessel

Metastatic cells

[5] Cells travel to other sites to form additional tumors.

Blood vessel

replacing energy-dense foods and helping maintain a healthy weight.

- *Excessive energy intake and obesity.* Excessive energy intake and obesity have been associated with an increased risk of breast, colorectal, and other cancers. This relationship may occur because obesity increases the production of hormones such as estrogen and insulin. Additionally, cancer cells may multiply more readily when excess energy is available to support their growth. Obesity also may contribute to cancer recurrence and reduce survival rates.
- *Meat.* High intakes of meats, especially red meat and meat cooked at high temperatures, such as when grilling, are associated with an increased risk of colorectal, kidney, pancreatic, and stomach cancers. The saturated fat content of red meat, and the production of polyaromatic hydrocarbons (e.g., benzopyrene) when meats are charbroiled, may increase the risk of cancers. Nitrosamine compounds found in processed meats such as sausage, hot dogs, and bacon also may increase cancer risk.
- *Fried foods.* In addition to increasing calories, fried foods also contain acrylamide, which can increase cancer risk. Acrylamide is not naturally found in these foods but is produced when potatoes and other starches are fried at high temperatures.
- *Whole grains and fiber.* Increased intake of dietary fiber from whole grains is associated with a decreased risk of developing colorectal cancer. Diets containing more

High intakes of grilled meats and seafood may increase the risk of some cancers.

© Alexander Raths/Shutterstock, Inc.

## Table 15-7 Example of a Diet Intended to Limit the Risk of Cancer—High in Fruits, Vegetables, and Antioxidants

**Breakfast**

| | |
|---|---|
| 4 oz calcium-fortified orange juice | 1 cup fresh blueberries |
| 1 cup ready-to-eat whole-grain breakfast cereal | 1 slice whole-wheat toast, jelly, soft margarine |
| 1 cup 1% milk | Hot green tea |

**Lunch**

| | |
|---|---|
| Sandwich: ½ cup curried chicken salad served on 1 whole-wheat bagel | 1 cup unsweetened ice tea |
| Assorted raw vegetables: carrots, celery, broccoli, cucumber, tomatoes | Fresh fruit cup with strawberries, melon, grapes, apple |
| | 1 cup yogurt |

**Snack**

¼ cup mixed nuts and raisins
4 oz apple-blueberry juice

**Dinner**

| | |
|---|---|
| 3 oz grilled salmon (or other fish) | Fresh garden salad with low-fat Italian dressing |
| Black beans (½ cup) with green peppers and onions | 1 whole-wheat dinner roll |
| Roasted corn on the cob, soft margarine | 1 scoop sherbet with fresh raspberries |
| | Hot raspberry tea |

**Snack**

6 oz cranberry juice
2 cups popcorn

whole grains and fewer refined grains improve weight control.

- *Alcohol.* Excess alcohol intake increases the risk of cancers of the mouth, throat, esophagus, breast, and colon. Alcohol abuse also severely damages the liver and other tissues in the body, which may predispose them to cancer development.
- *Dairy and calcium.* Increased intake of dairy, including milk and cheese, and dietary calcium are linked with a decreased risk of colorectal cancer. Calcium may bind free fatty acids and bile acids in the colon and prevent them from interacting with potential cancer cells. Vitamin D may inhibit the progression of cancer growth from malignant polyps in the colon, yet the evidence is inconsistent on whether adequate intakes

of vitamin D decrease the risk of colorectal cancer. Supplements of vitamin D were not linked with decreased risk of various cancers, including colorectal cancer, in a recent clinical study. More research is needed.

Based on current knowledge of diet and cancer risk, the American Institute for Cancer Research has provided guidelines for reducing cancer risk (Table 15-8). A recent study links following cancer prevention guidelines with a decreased risk of death from cancer and cardiovascular disease.[47] Additional information also can be found on the following websites.

www.cancer.org (American Cancer Society)
www.aicr.org (American Institute for Cancer Research)
cancer.gov (National Cancer Institute)

## Table 15-8 Diet and Health Guidelines for Cancer Prevention

1. Choose a diet rich in a variety of plant-based foods, eating plenty of vegetables, fruits, whole grains, and beans.
2. Do not rely on supplements to decrease cancer risk.
3. Maintain a healthy weight and be physically active.
4. Drink alcohol only in moderation, if at all.
5. Avoid processed foods high in fat.
6. Avoid sugary drinks and foods with added sugars.
7. Limit the intake of red meats and processed meats.
8. Exclusively breastfeed for 6 months.

*And always remember . . .* do not use tobacco in any form.

# Chapter Summary

## 15.1 The absorption of iron depends on the body's need for iron and on the form of iron in food (heme vs. nonheme).

Heme iron (from animal-based foods) is better absorbed than nonheme iron (primarily from plant-based foods). Liver, beef, and seafood are the best sources of dietary iron. The iron in plant foods is poorly absorbed. The body cannot readily excrete excess iron. Thus, the body regulates iron absorption and iron storage to maintain iron homeostasis. Iron is a critical component of hemoglobin, myoglobin, and many enzyme systems. Two-thirds of the body's iron is contained in hemoglobin, where it helps transport oxygen from the lungs to the tissues. In the early stages, iron deficiency does not result in obvious physical symptoms because iron stores can be mobilized to temporarily maintain many of iron's functions. As stores are depleted, iron deficiency anemia develops, causing fatigue, weakness, increased infection rate, delayed growth, and impaired brain development. Accidental iron overdose is the leading cause of poisoning in toddlers and young children. In adults, iron toxicity usually occurs from excess supplementation or from a genetic disorder, called hemochromatosis, that causes an overabsorption of iron.

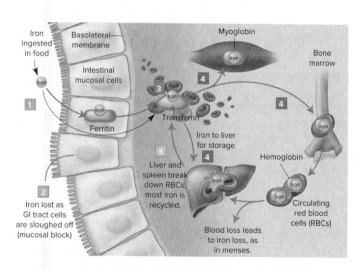

## 15.2 Like iron, zinc absorption is affected by the need for the mineral and the amount in the diet.

Absorption also plays a primary role in maintaining zinc balance. The best sources of zinc are oysters, meat, nuts, legumes, and whole grains. Zinc functions as a component of proteins and enzymes involved in stabilizing the structures of cell membrane proteins, reproduction, growth and development, immune function, and antioxidant defense (with copper as Cu/Zn SOD enzymes). Zinc deficiency causes impairments in taste, appetite, immune function, growth, sexual development, and reproduction. In many parts of the world, zinc deficiency (like iron deficiency) is still a serious public health concern. Zinc toxicity also is of concern in populations where zinc supplementation has become more common.

Photo courtesy of Harold H. Sandstead, M.D.

## 15.3 The best dietary sources of copper are liver, legumes, whole grains, and dark chocolate.

Copper is involved in the mobilization of iron from body stores, in the cross-linking of proteins in connective tissue formation, and as a part of antioxidant defense. Although copper deficiency and toxicity are rare in humans, RDA and UL guidelines have been established to promote intakes within safe ranges. In Wilson disease, copper is deposited in the edges of the cornea.

©Medical-on-Line/Alamy

## 15.4 Manganese is found in whole grains, nuts, legumes, and tea.

Manganese functions in antioxidant defense and as a component of different metabolic enzymes. Manganese deficiency and toxicity rarely have been reported in humans.

## 15.5 Iodine is an essential part of the thyroid hormones thyroxine ($T_4$) and triiodothyronine ($T_3$)

and thus plays a role in many metabolic and developmental functions in the body. The iodine content of the soil in which a plant was grown determines the iodine content of the plant food. The iodine content of most foods is low; thus, many populations obtain iodine from iodized salt. Iodine fortification programs have eradicated endemic iodine deficiency in these areas. In countries where foods fortified with iodine are not available, iodine deficiency is a serious health concern. Iodine deficiency results in enlargement of the thyroid gland (goiter) and severe growth and mental impairment (cretinism).

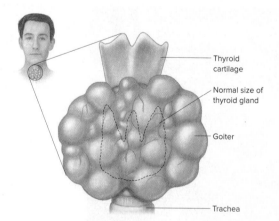

## 15.6 Selenium acts as a cofactor for glutathione peroxidase, which prevents the oxidative

destruction of cell membranes by hydrogen peroxide and free radicals. Selenium also aids in the conversion of the thyroid hormone $T_4$ to $T_3$. The selenium content of the soil in an area greatly affects the amount of selenium in the foods from that area. In general, meat, eggs, seafood, grains, and seeds are the best sources of selenium. In areas where the soil lacks selenium, deficiencies are more prevalent. Selenium deficiency increases the risk of Keshan disease and possibly certain cancers. Because selenium has a narrower range of safety for intake than most other trace minerals, selenium supplementation can cause symptoms of toxicity at lower levels of intake.

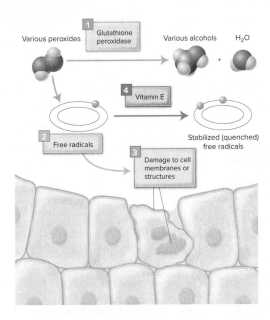

## 15.7 The functions of chromium are not fully known.
Chromium may enhance the action of insulin and promote the uptake of glucose into cells. Chromium is widely distributed in a variety of foods. Chromium deficiency and toxicity have not been well documented in humans.

## 15.8 Although fluoride is not truly classified as an essential trace mineral,
it is beneficial in decreasing the incidence of dental caries. Most North Americans obtain fluoride from fluoridated drinking water, toothpaste, mouth rinses, and dental treatments. Excess fluoride causes discoloration of the teeth, known as enamel mottling or fluorosis.

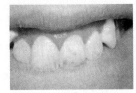

©Paul Casamassimo, DDS, MS

## 15.9 The best dietary sources of molybdenum are legumes, grains, and nuts.
Molybdenum is a cofactor required for the activity of several enzymes. Reports of molybdenum deficiency and toxicity are very rare. Arsenic, boron, nickel, silicon, and vanadium are classified as ultratrace minerals. Further research is needed to determine their functions in the body.

## Clinical Perspective

Cancer is a disease that develops from uncontrolled replication of mutated cells. It begins by exposure to cancer-causing agents (carcinogens), such as tobacco, radiation, viruses, and certain dietary factors. Low intakes of fruits, vegetables, and whole grains and excessive intakes of energy, sugary drinks, red meats, fried foods, and alcohol have been associated with increased risk of cancer.

## Global Perspective

The World Health Organization (WHO) launched the e-Library of Evidence for Nutrition Actions (eLENA) to help decrease the number of deaths and morbidities associated with malnutrition. The online eLENA project provides easily accessed, evidence-informed information to help countries launch successful nutrition interventions. The Cochrane Collaboration is an eLENA partner.

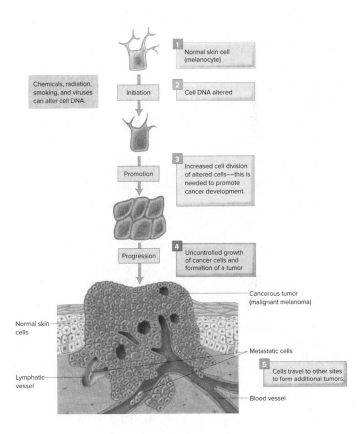

## Study Questions

1. Trace mineral status often is difficult to evaluate due to the relatively small amounts of minerals contained in blood and tissues and the lack of sensitive measures that reflect body mineral content.

   a. true               b. false

2. Which of the following decreases iron absorption?

   a. increased need        c. meat protein factor
   b. decreased intake      d. phytic acid

3. Which of the following minerals is a component of hemoglobin and myoglobin?

   a. zinc                  c. iron
   b. copper                d. manganese

4. Which of the following groups is at greatest risk of iron deficiency anemia?

   a. college students      c. well-trained athletes
   b. middle-aged males     d. adolescent girls

5. Accidental iron overdose is the leading cause of poisoning in young children.

   a. true               b. false

6. Which of the following is a good source of zinc?

   a. oysters               c. low-fat yogurt
   b. blueberries           d. lean bacon

7. Which of the following is a symptom of zinc deficiency?

   a. microcytic anemia     c. goiter
   b. poor growth           d. cardiomyopathy

8. Which of the following proteins/enzymes contains copper?

   a. glutathione peroxidase   c. thyroxine
   b. deiodinase               d. lysyl oxidase

9. Manganese is involved in the body's antioxidant defense network as a component of the enzyme manganese superoxide dismutase.

   a. true               b. false

10. Which of the following is a function of iodine?

    a. acts as a component of superoxide dismutase enzyme
    b. aids in thyroid hormone metabolism
    c. functions in antioxidant defense
    d. enhances insulin activity

11. Which of the following is associated with the development of cretinism?

    a. iron deficiency       c. iodine deficiency
    b. zinc deficiency       d. selenium deficiency

12. Selenium deficiency can cause impaired iodine metabolism.

    a. true               b. false

13. Which of the following helps protect against dental caries?

    a. iodine                c. fluoride
    b. manganese             d. chromium

14. Which of the following is *not* classified as an ultratrace mineral?

    a. arsenic               c. iodine
    b. boron                 d. vanadium

15. Low intakes of plant-based foods have been associated with increased risk of cancer.

    a. true               b. false

16. Describe factors that increase and decrease iron absorption.

17. Why does zinc affect so many body processes?

18. Describe the interrelationship between copper and iron storage in the body.

19. Which trace minerals function in antioxidant defense?

20. What regions of the world have adequate iodine intake?

21. Why are foods of animal origin better sources of selenium, iron, and zinc than foods of plant origin?

22. What are the main functions of chromium?

23. Why is fluoride not considered an essential nutrient?

24. Which nutrients are classified as ultratrace minerals?

25. What dietary practices are associated with an increased risk of cancer?

26. What is eLENA, and how can it help improve nutritional health?

©C Squared Studios/Getty Images RF

Answer Key: 1-a; 2-d; 3-c; 4-d; 5-a; 6-a; 7-b; 8-d; 9-a; 10-b; 11-c; 12-a; 13-c; 14-c; 15-a; 16-refer to Section 15.1; 17-refer to Section 15.2; 18-refer to Sections 15.1 and 15.3; 19-refer to Sections 15.2, 15.3, 15.4, and 15.6; 20-refer to Section 15.5; 21-refer to Sections 15.1, 15.2, and 15.6; 22-refer to Section 15.7; 23-refer to Section 15.8; 24-refer to Section 15.9; 25-refer to Section 15.8; 26-refer to Global Perspective; Clinical Perspective

# References

1. Prasad AS. Zinc in human health: Effect of zinc on immune cells. *Mol Med.* 2008;14:353.

2. Hooper L and others. Assessing potential biomarkers of micronutrient status by using a systematic review methodology: Methods. *Am J Clin Nutr.* 2009;89:1953S.

3. de Benoist B and others, eds. *Worldwide prevalence of anemia 1993–2005.* Geneva: World Health Organization; 2008.

4. World Health Organization. *Global nutrition targets 2025: Anaemia policy brief (WHO/NMH/NHD/14.4).* Geneva: WHO; 2014

5. Food and Nutrition Board, Institute of Medicine. *Dietary Reference Intakes for vitamin A, vitamin K, arsenic, boron, chromium, copper, iodine, iron, manganese, molybdenum, nickel, silicon, vanadium, and zinc.* Washington, DC: National Academies Press; 2001.

6. Anderson GJ, Frazer DM. Current understanding of iron homeostasis. *Am J Clin Nutr.* 2017;106:1559S.

7. Kawabata H. Transferrin and transferrin receptors update. *Free Rad Biol Med.* 2019;133:46.

8. Ganz T, Nemeth E. Iron homeostasis in host defence and inflammation. *Nature Rev Immunol.* 2015;15:500.

9. Cogswell ME and others. Assessment of iron deficiency in US preschool children and nonpregnant females of childbearing age: National Health and Nutrition Examination Survey 2003–2006. *Am J Clin Nutr.* 2009;89:1334.

10. World Health Organization. *The global prevalence of anaemia in 2011.* Geneva: WHO; 2015.

11. Hamner HC and others. Usual intake of key minerals among children in the second year of life, NHANES 2003–2012. *Nutrients.* 2016;8:468.

12. Bergamaschi G, Villani L. Serum hepcidin: A novel diagnostic tool in disorders of iron metabolism. *Haematologica.* 2009;94:1631.

13. Meynard D and others. The liver: Conductor of systemic iron balance. *Blood.* 2014;123:168.

14. Ghanchi A and others. Gut, germs, and iron: A systematic review on iron supplementation, iron fortification and diarrhea in children aged 4–59 months. *Curr Dev Nutr.* 2019;3:nzz005.

15. Food Surveys Research Group. What we eat in America, NHANES 2001–2002. 2007; www.ars.usda.gov/foodsurvey.

16. Kang YJ. Metallothionein redox cycle and function. *Exp Biol Med.* 2006;231:1459.

17. Hess SY and others. Recent advances in knowledge of zinc nutrition and human health. *Food Nutr Bulletin.* 2009;30:S5.

18. King JC, Cousins RJ. Zinc. In: Ross C and others, eds. *Modern nutrition in health and disease.* 11th ed. Philadelphia: Lippincott Williams & Wilkins; 2014.

19. Hemila H, Chalker E. The effectiveness of high dose zinc acetate lozenges on various common cold symptoms: A meta-analysis. *BMC Fam Pract.* 2015;16:24.

20. Hennigar SR and others. Serum zinc concentrations in the US population are related to sex, age, and time of blood draw but not dietary or supplementary zinc. *J Nutr.* 2018;148:1341.

21. Rashed AA and others. Acrodermatitis enteropathica in a pair of twins. *J Dermatol Case Rep.* 2016;10:65.

22. Collins JF. Copper. In Ross C and others, eds. *Modern nutrition in health and disease.* 11th ed. Philadelphia: Lippincott Williams & Wilkins; 2014.

23. Spain RI and others. When metals compete: A case of copper-deficiency myeloneuropathy and anemia. *Nature Clin Prac Neurol.* 2009;5:106.

24. Zatta P, Frank A. Copper deficiency and neurological disorders in man and animals. *Brain Res Rev.* 2007;54:19.

25. Buchman AL. Manganese. In: Ross C and others, eds. *Modern nutrition in health and disease.* 11th ed. Philadelphia: Lippincott Williams & Wilkins; 2014.

26. Livingstone C. Manganese provision in parenteral nutrition: An update. *Nutr Clin Prac.* 2018;33:404.

27. Freake HC. Iodine. In: Stipanuk MH, ed. *Biochemical, physiological, molecular aspects of human nutrition.* 2nd ed. St. Louis: Saunders; 2006.

28. Zimmermann MB and others. Iodine-deficiency disorders. *The Lancet.* 2008;372:1251.

29. World Health Organization. Micronutrient deficiencies: Iodine deficiency disorders. 2016; www.who.int/nutrition/topics/idd/.

30. Zimmermann MB, Andersson M. Update on iodine status worldwide. *Curr Opin Endocrinol Diabetes Obes.* 2012;19:382.

31. UNICEF. *NutriDash: Facts and figures. Nutrition programme data for the SDGs (2015–2030).* New York: UNICEF, 2017.

32. Teng W and others. Effect of iodine intake on thyroid diseases in China. *N Engl J Med.* 2006;354:2783.

33. Loscalzo J. Keshan disease, selenium deficiency and the selenoproteome. *N Engl J Med.* 2014;370:1756.

34. Ullah H and others. A comprehensive review on environmental transformation of selenium: Recent advances and research perspectives. *Environ Geochem Health.* 2019;41:1003.

35. Food and Nutrition Board, Institute of Medicine. *Dietary Reference Intakes for vitamin C, vitamin E, selenium, and carotenoids.* Washington, DC: National Academies Press; 2000.

36. Lippman SM and others. Effect of selenium and vitamin E on risk of prostate cancer and other cancers: The Selenium and Vitamin E Cancer Prevention Trial (SELECT). *JAMA.* 2009;301:39.

37. Cai X and others. Selenium exposure and cancer risk: An updated meta-analysis and meta-regression. *Sci Rep.* 2016;6:19213.

38. Wang ZQ, Cefalu WT. Current concepts about chromium supplementation in type 2 diabetes and insulin resistance. *Curr Diab Rep.* 2010;10:145.

39. Costello RB and others. Chromium supplements for glycemic control in type 2 diabetes: Limited evidence of effectiveness. *Nutr Rev.* 2016;74:455.

40. Position of the Academy of Nutrition and Dietetics: The impact of fluoride on health. *J Acad Nutr Diet.* 2012;112:1443.

41. Food and Nutrition Board, Institute of Medicine. *Dietary Reference Intakes for calcium, phosphorus, magnesium, vitamin D, and fluoride.* Washington, DC: National Academies Press; 1997.

42. 2020 Global Nutrition Report: Action on equity to end malnutrition. Bristol, UK: Development Initiatives. globalnutritionreport.org/reports/2020-global-nutrition-report.

43. World Health Organization. *Global health risks: Mortality and burden of disease attributable to selected major risks.* Geneva: World Health Organization; 2009.

44. American Cancer Society. *Cancer facts & figures 2019.* Atlanta: American Cancer Society; 2019.

45. Ruiz RB, Hernandez PS. Diet and cancer: Risk factors and epidemiological evidence. *Maturitas.* 2014;77:202.

46. World Cancer Research Fund/American Institute for Cancer Research. Diet, nutrition, physical activity and cancer: A global perspective. Continuous update project expert report 2018. 2018; www.dietandcancerreport.org.

47. Kabat G and others. Adherence to cancer prevention guidelines and cancer incidence, cancer mortality, and total mortality: A prospective study. *Am J Clin Nutr.* 2015;101:558.

48. Pierce JP and others. Influence of a diet very high in vegetables, fruit and fiber, and low in fat on prognosis following treatment for breast cancer. *JAMA.* 2007;298:289.

49. Yuan G and others. Dietary effects on breast cancer molecular subtypes, a 1:2 paired case–control study. *Food Sci Nutr.* 2020;8:5545.

50. van Duijnhoven FJB and others. Fruit, vegetables and colorectal cancer risk: The European Prospective Investigation into Cancer and Nutrition. *Am J Clin Nutr.* 2009;89:1441.

51. Alexander DD and others. Meta-analysis of prospective studies of red meat consumption and colorectal cancer. *FASEB J.* 2010;24:207.

52. Cross AJ and others. A large prospective study of meat consumption and colorectal cancer risk: An investigation of potential mechanisms underlying this association. *Cancer Res.* 2010;70:2406.

53. Chen P and others. Meta-analysis of vitamin D, calcium and the prevention of breast cancer. *Breast Cancer Res Treat.* 2010;121:469.

54. Newberry SJ and others. Vitamin D and calcium: A systematic review of health outcomes (update). Evidence report/technology assessment no. 217. *Agency for Healthcare Res Qual.* 2014; www.effectivehealthcare.ahrq.gov/reports/final.cfm.

55. Newmark HL, Heaney RP. Dairy products and prostate cancer risk. *Nutr Cancer.* 2010;62:297.

56. American Cancer Society. *Cancer prevention & early detection facts & figures 2019–2020.* Atlanta: ACA; 2019.

57. Hullings AG and others. Whole grain and dietary fiber intake and risk of colorectal cancer in the NIH-AARP diet and health study cohort. *Am J Clin Nutr.* 2020;112;603.

58. American Institute for Cancer Research. Recommendations for cancer prevention. 2019; www.aicr.org/reduce-your-cancer-risk/recommendations-for-cancer-prevention/.

©deepak bishnoi/Shutterstock, Inc.

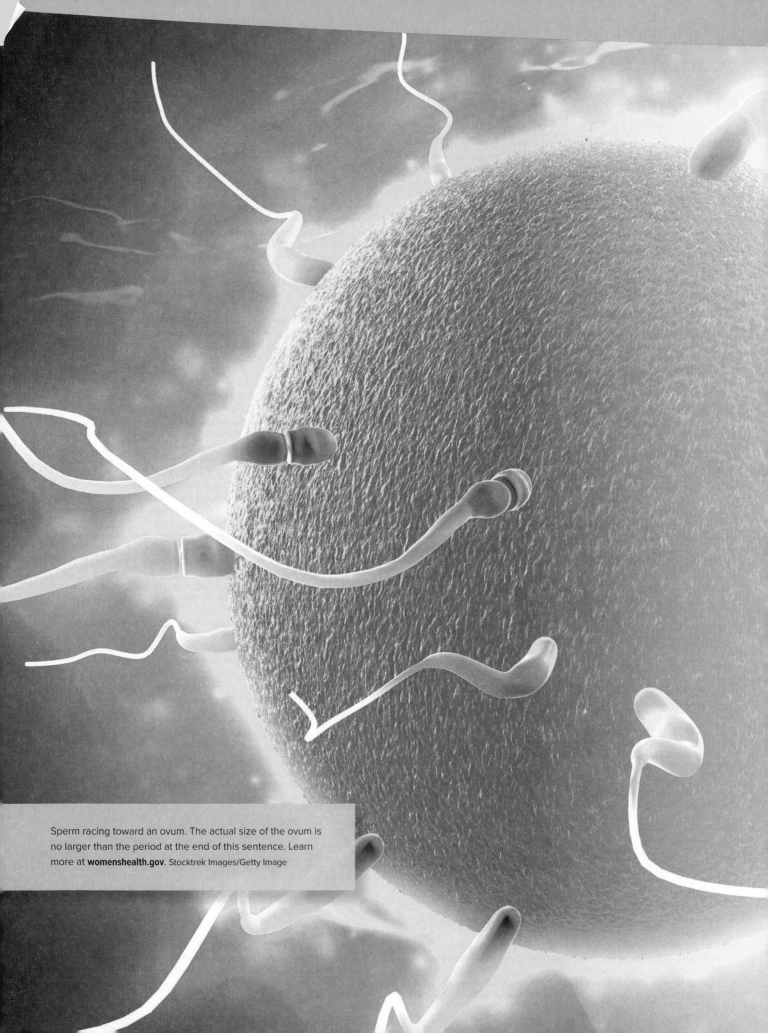

Sperm racing toward an ovum. The actual size of the ovum is no larger than the period at the end of this sentence. Learn more at **womenshealth.gov**. Stocktrek Images/Getty Image

# 16 Nutritional Aspects of Pregnancy and Breastfeeding

## Learning Objectives

**After studying this chapter, you will be able to**

1. Describe the factors that predict a successful pregnancy outcome.

2. List major physiological changes that occur in the body during pregnancy and describe how nutrient needs are altered.

3. Specify the optimal weight gain during pregnancy for adult women.

4. Describe the special nutritional needs of pregnant and lactating women, summarize factors that put them at risk for nutrient deficiencies, and plan a nutritious diet for them.

5. Identify nutrients that often need to be supplemented during pregnancy and lactation and explain the reason for each.

6. Discuss nutrition-related problems that can occur during pregnancy and suggest techniques for coping with these problems.

7. List substances and practices to avoid during pregnancy and lactation and describe why they are harmful.

8. Describe the physiological process of breastfeeding.

PREGNANCY IS A VERY SPECIAL time. Most parents-to-be want to do everything possible to maximize their chances of having a robust, lively newborn. Ideally, women begin improving their health and nutrient intake (especially folic acid) as well as eliminating potentially harmful habits long before becoming pregnant.

Despite these intentions, the number of infants who die in North America is higher than in many other industrialized nations. In Canada, about 4.5 of every 1000 infants per year die before their 1st birthday, and the rate is 5.8 in the U.S. Comparing these numbers with Monaco, which has a rate of about 1.8 of every 1000 infants, indicates there is room for improvement.

Although some genetic and environmental factors that affect fetal and newborn health are beyond a woman's control,[1] her conscious decisions about health and nutrition can significantly affect both her own and her baby's lifelong health.[2] This chapter will examine how a woman who eats nutritiously before and during pregnancy and while breastfeeding can help her baby have a healthy start in life.

A healthy full-term newborn usually weighs about 7.5 pounds and is 20 inches long. Photodisc Collection/Getty Images

▶ The American College of Obstetricians and Gynecologists and the Society for Maternal-Fetal Medicine have 4 definitions of "term" deliveries.[4]

*Early-term:* babies born between 37 weeks 0 days and 38 weeks 6 days of gestation

*Full-term:* babies born between 39 weeks 0 days and 40 weeks 6 days of gestation

*Late-term:* babies born between 41 weeks 0 days and 41 weeks 6 days

*Post-term:* babies born after 42 weeks 0 days of gestation

Full-term babies have the lowest risk for breathing and feeding problems, as well as death.

**preterm** Born before 37 weeks of gestation (also referred to as premature).

**low birth weight (LBW)** Infant weight of less than 5.5 pounds (2.5 kilograms) at birth; most commonly results from preterm birth.

**prenatal** During gestation (pregnancy); before birth.

**small for gestational age** Weighing less than the expected weight for length of gestation; typically defined as below the 10th percentile. This corresponds to less than 5.5 pounds (2.5 kilograms) in a full-term newborn; these infants are at increased risk of medical complications.

# 16.1 Pregnancy

Pregnancy, or **gestation**, may be the most sensitive stage of the life cycle—nutrient and calorie supplies at this time impact the final outcome.[3,4] To help ensure the optimal health of the mother and her offspring, adequate nutrition is vital both before and during pregnancy. Researchers often define a favorable pregnancy outcome as a full-term gestation period (39 or 40 weeks) that results in a live, healthy infant weighing more than 5.5 pounds (2.5 kg)[1,3] and permits the mother to return to her prepregnancy health status.

An infant's chances of survival depend greatly on both gestation length and birth weight.[3] The closer the gestation period comes to full-term, the greater the physical maturation and birth weight. Many **preterm** (born before 37 weeks) infants have numerous medical problems, including those that complicate nutritional care, such as poor sucking and swallowing abilities and **low birth weight (LBW)** (weighing less than 5.5 pounds, or 2500 grams). A healthy full-term newborn usually weighs about 7.5 lb. Low-birth-weight babies are much more likely to die during the 1st 4 weeks of life than are heavier infants. Those who survive are more likely to be ill and disabled than are heavier babies.

Suboptimal maternal nutrient and calorie intakes are linked with low birth weights. Although low birth weight is most commonly associated with being born preterm, infants who suffer growth retardation during **prenatal** development (also called intrauterine growth retardation) weigh less than expected for their gestational age—these infants are **small for gestational age**. Thus, a full-term infant weighing less than 5.5 lb at birth is small for gestational age. In contrast, a preterm infant is almost always low birth weight but may not be small for gestational age.

Low-birth-weight infants are more likely than normal-weight infants to have medical and nutritional complications, including problems with blood glucose control, temperature regulation, growth, and development in the early weeks after birth. These babies also are at risk of developing more body fat and less lean body mass in childhood, which increases their risk of chronic disease in adulthood.[5,6] The long-term consequences of prenatal growth retardation are much more profound than previously imagined—sometimes fetal health is harmed to the extent that its own future offspring also will be affected.[6] By eating a nutritious diet, pregnant women promote adequate growth and development throughout gestation and help ensure that the baby is born healthy, on time, and with the mental, physical, and physiological capabilities to grow and develop normally. A nutritious diet also helps protect the mother's health.

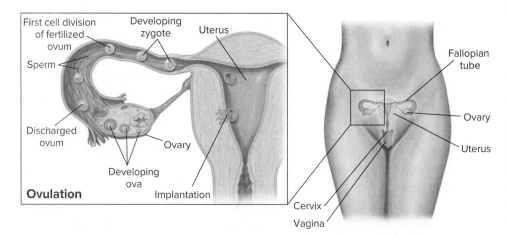

*Figure 16-1* After release from an ovary (ovulation), the ovum enters the abdominal cavity and finds its way into the fallopian tube. Receptor molecules on the ovum surface attract and "trap" sperm cells "swimming" up the fallopian tube. As soon as 1 sperm enters the ovum, complex mechanisms in the egg are activated to block the entry of another sperm. The 23 chromosomes from the sperm combine with the 23 chromosomes already in the ovum to make up the 46 chromosomes of the developing offspring.

## Prenatal Developmental Stages: Conception, Zygotic, Embryonic, and Fetal

Gestation begins at conception, when a sperm unites with an egg (ovum) (Fig. 16-1). About 30 hours after conception, the **zygote**, as the fertilized egg is called, begins the lifelong process of cell division. This cluster of cells drifts down the fallopian tube to the woman's uterus; within a week after conception, it has nestled deeply into the uterine lining and firmly attached itself there. The next stage begins 2 weeks after conception; the zygote is now called an **embryo**.

When the embryonic stage begins, the cells have already separated into a stack of 3 thin layers (Fig. 16-2). One layer of cells, called the endoderm, will develop into the digestive system, liver, and pancreas. Another layer, the mesoderm, provides the cells that become the skeleton, muscles, heart, and blood vessels. From the last cell layer, the ectoderm, will come the skin, nervous system, and sensory organs. When this stage ends after

▶ In the U.S., more than 8% of births are low-birth-weight babies—many of these are from multiple births. Most are premature and many experience prenatal growth retardation.

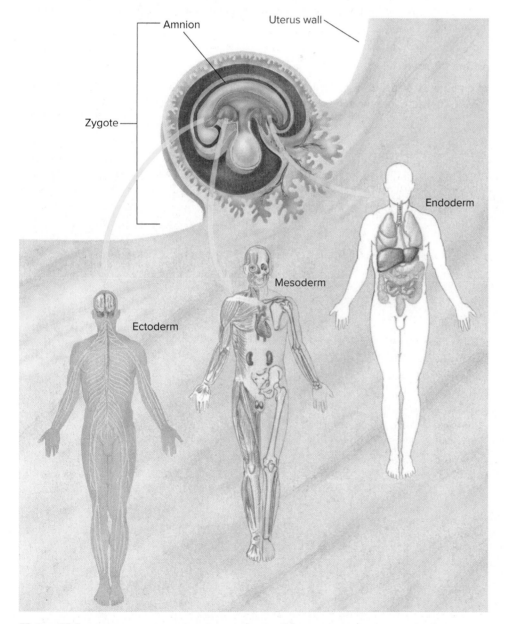

*Figure 16-2* The body systems that develop from the zygote's 3 cell layers.

**zygote**  Fertilized ovum; the cell resulting from the union of an egg cell (ovum) and a sperm until it divides.

**embryo**  In humans, the developing in utero offspring from about the beginning of the 3rd week to the end of the 8th week after conception.

**fetus** In humans, developing offspring from about the beginning of the 9th week after conception until birth.

**trimester** Three 13- to 14-week periods into which the normal pregnancy is somewhat arbitrarily divided for purposes of discussion and analysis (the length of a normal pregnancy is about 40 weeks, measured from the 1st day of the woman's last menstrual period).

week 8, the embryo is quite complex, yet it is no larger than a pea (approximately 3/8 inch, or 8 mm, long). The major organs are in place and some, such as the heart and liver, have begun to function.

From week 9 until birth, the developing offspring is known as a **fetus**. The fetal stage is when the most rapid growth occurs. In fact, about 90% of all fetal growth occurs in the last 20 weeks of gestation. During the fetal stage, length increases 20 times or more to an average of 20 to 22 inches (51–56 cm). Weight skyrockets, increasing about 3500 times, to an average weight of 7 to 8 pounds (3.2–3.6 kg). Body proportions change rapidly, too. When this stage begins, the head and the body are nearly the same size (Fig. 16-3). Between 21 and 30 weeks, body proportions shift to be similar to those of newborn infants, although the fetus is still quite slim and has loose, baggy skin. Body fat rises from less than 1% at 20 weeks to 16% by 38 weeks. The stored subcutaneous fat smoothes and tightens the skin and, most importantly, insulates the body. Infants born prematurely have difficulty regulating body temperature, which greatly affects their nutrient and calorie needs.

Premature infants also are at risk of nutrient deficiencies because nutrient stores do not accumulate appreciably until the last 4 to 6 weeks of gestation. By 38 to 40 weeks, a healthy fetus has built substantial nutrient stores and developed sufficiently to thrive outside the uterus.

### Critical Periods

For purposes of discussion, pregnancy is often divided into 3 periods called **trimesters**. Figure 16-4 describes the major developmental milestones of gestation during each trimester.

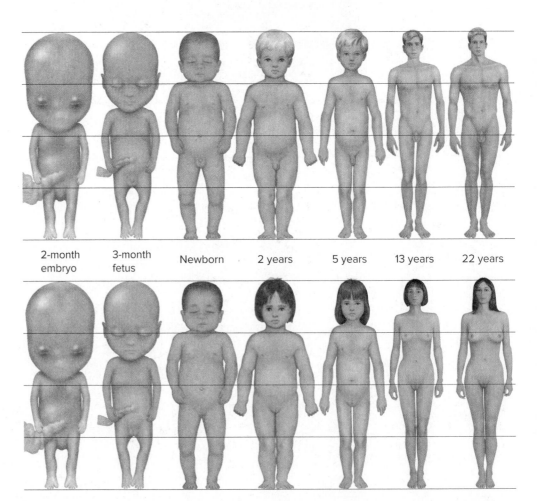

| 2-month embryo | 3-month fetus | Newborn | 2 years | 5 years | 13 years | 22 years |

*Figure 16-3* Body proportions change significantly during fetal development and continue until adulthood.

This complicated developmental process must occur precisely on schedule. That is, there is a finite window of opportunity, called a **critical period**, for cells to develop into a particular tissue or organ. As you can see, most critical periods occur during the 1st trimester. For instance, critical heart development occurs during the 3rd to 6th weeks. Critical tooth development happens during the 6th to 8th weeks of gestation. Nutrient deficiencies or excesses, certain pathogens, trauma, radiation, tobacco smoke, and toxins (e.g., drugs and alcohol) during a critical period can interfere with normal development, causing effects ranging from severe physical or mental abnormalities to spontaneous abortion (a naturally occurring premature termination of a pregnancy prior to 20 weeks of gestation) (Table 16-1).[7] Early spontaneous abortions, also called miscarriages, usually result from a genetic defect or fatal error in fetal development. About half or more of all pregnancies end in this way, often so early

**critical period** Finite period during embryonic and fetal development when cells for a particular tissue or organ can develop.

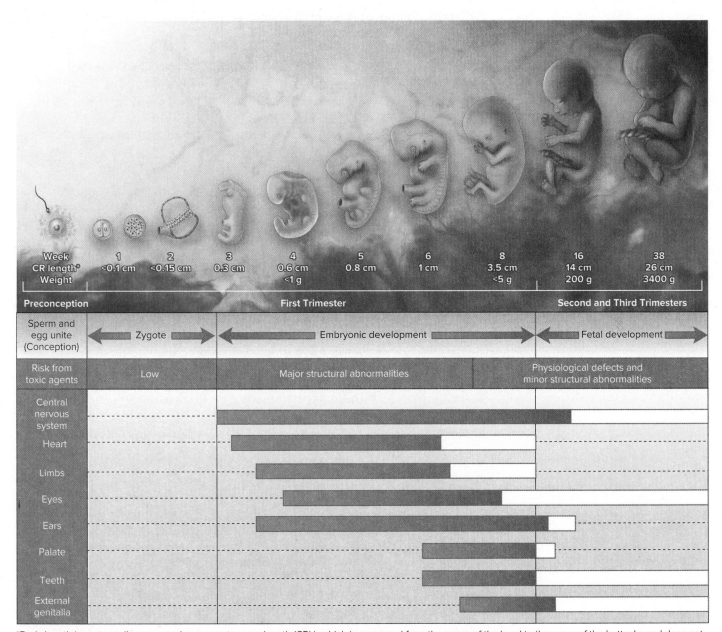

| Week | 1 | 2 | 3 | 4 | 5 | 6 | 8 | 16 | 38 |
|---|---|---|---|---|---|---|---|---|---|
| CR length* | <0.1 cm | <0.15 cm | 0.3 cm | 0.6 cm | 0.8 cm | 1 cm | 3.5 cm | 14 cm | 26 cm |
| Weight | | | | <1 g | | | <5 g | 200 g | 3400 g |

*Body length is customarily expressed as crown-to-rump length (CRL), which is measured from the crown of the head to the curve of the buttocks and does not include the lower limbs. Recall that 2.54 cm = 1 in.

*Figure 16-4* Critical periods of development are indicated with purple bars. The purple shading indicates when the effects of malnutrition and/or exposure to toxins, such as alcohol and drugs, are likely to be most severe. As the white bars in the chart show, damage to the eyes, brain, and genitals also can occur during the last months of pregnancy.

that a woman does not realize she was pregnant. An additional 15 to 20% are lost before the full-term gestation period ends.

Although the greatest risk is in the 1st trimester, a detrimental prenatal environment during the fetal stage also can adversely affect development; however, the damage often has less catastrophic consequences. Maternal nutrient deficiencies or exposure to toxins during the fetal stage can cause low nutrient stores, retarded growth, abnormal organ function, mental abnormalities, and/or a shorter than normal gestation period. Nutrient deficits during fetal life often can be partly reversed by adequate nutrition after birth.

*Perspective*
**on the Future**

About 3 decades ago, David Barker proposed that many adult diseases begin during fetal life. He observed that low birth weight, an indicator of poor nutrition and growth, increased the risk for heart disease, obesity, and type 2 diabetes in adulthood. Growing evidence indicates that both nutritional and non-nutritional stresses during critical growth periods affect lifelong health. Increased emphasis on fetal health care is growing as a preventive measure against chronic diseases.[8]

**Table 16-1   Potential Effects of Maternal Calorie and Nutrient Deficiencies and Excesses on Developing Offspring**

| Food Component | Potential Effect of a Deficiency | Potential Effect of an Excess |
|---|---|---|
| **Calories** | Growth retardation<br>Low birth weight | High birth weight<br>Complications during labor and delivery |
| **Protein** | Reduced head circumference<br>Fewer cells than normal, impact particularly severe in the brain | If high consumption is coupled with low carbohydrate intake, may lower glucose availability and restrict energy available to the fetus |
| **Vitamin C** | Premature birth | Sudden drop in vitamin C after birth may cause vitamin C deficiency symptoms |
| **Folate** | Spontaneous abortion<br>Fluid accumulation in the skull, leading to brain damage<br>Growth retardation<br>Premature birth<br>Neural tube defects | May inhibit maternal absorption of other nutrients<br>Hinders diagnosis of maternal vitamin B-12 deficiency |
| **Vitamin A** | Premature birth<br>Eye abnormalities and impaired vision<br>Maternal death | Birth defects that affect the nervous and cardiovascular systems<br>Facial deformities |
| **Vitamin D** | Low birth weight<br>Rickets<br>Lack of enamel on teeth | Calcification of soft tissues, such as the kidneys<br>Mental retardation<br>Growth retardation |
| **Calcium** | Decreased bone density | May hinder maternal absorption of minerals, such as iron and zinc |
| **Iron** | Low birth weight<br>Premature birth<br>Increased risk of fetal or infant death | May hinder maternal absorption of minerals, such as zinc and calcium |
| **Iodine** | Cretinism (mental and physical growth retardation) | Thyroid disorders |
| **Zinc** | Nervous system malformations<br>Growth retardation<br>Birth defects that affect the brain and bones | May hinder maternal absorption of minerals, such as copper and iron |

## Nourishing the Zygote, Embryo, and Fetus

A zygote nourishes itself by absorbing secretions from glands in the uterus and digesting some of the uterine lining. As a zygote develops into an embryo, the placenta begins to form inside the mother's uterus. The placenta takes over the role of delivering nourishment to the developing organism throughout the remainder of pregnancy. The placenta is a spongy, pancake-shaped, temporary organ that taps into the mother's blood supply.

The umbilical cord is the pipeline that connects the placenta to the fetus. Fetal blood travels from the fetal heart to the placenta by way of 2 umbilical arteries and returns (nutrient enriched and waste free) to the fetus by means of 1 umbilical vein. As shown in Figure 16-5, the placenta contains both maternal and fetal blood vessels. Although these vessels are not directly connected, they are so close together that nutrients and oxygen pass easily from the mother to the fetus, and fetal wastes are shuttled from the fetus to the mother for excretion. To accomplish these tasks, the placenta uses all the absorption mechanisms employed by the GI tract (see Chapter 4). The placenta also synthesizes fatty acids, cholesterol, and the fetus's main fuel—glycogen. In addition, the placenta produces hormones that help direct maternal nutrients to the fetus, control fetal metabolism, promote the changes in the mother's body that support pregnancy, and may cause maternal nausea.[9]

The umbilical cord is the pipeline that delivers nutrients and oxygen to and removes waste from the developing offspring. Steve Allen/ Getty Images

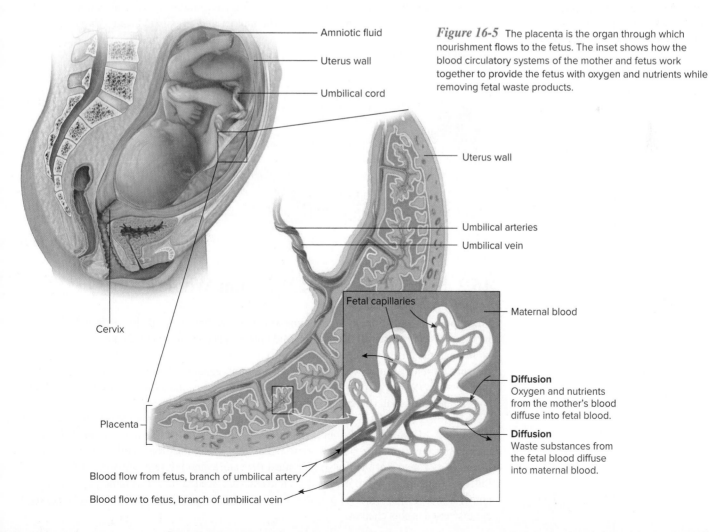

**Figure 16-5** The placenta is the organ through which nourishment flows to the fetus. The inset shows how the blood circulatory systems of the mother and fetus work together to provide the fetus with oxygen and nutrients while removing fetal waste products.

Amniotic fluid

Uterus wall

Umbilical cord

Cervix

Placenta

Blood flow from fetus, branch of umbilical artery

Blood flow to fetus, branch of umbilical vein

Uterus wall

Umbilical arteries

Umbilical vein

Fetal capillaries

Maternal blood

**Diffusion**
Oxygen and nutrients from the mother's blood diffuse into fetal blood.

**Diffusion**
Waste substances from the fetal blood diffuse into maternal blood.

## CASE STUDY

PhotoDisc/Getty
Images

Tracey and her husband have decided that they are ready to have a child. Tracey has been reading everything she can find on pregnancy because she knows that her prepregnancy health is important to the success of her pregnancy. She has just turned 25 and avoids alcohol, does not smoke, does not take any medications, and limits her coffee intake to 4 cups a day and her soft drink intake to 3 colas per day. She has decided to breastfeed her infant and has already inquired about childbirth classes. She has modified her diet to include some extra protein, along with more fruits and vegetables. Recently, she started swimming 5 days a week and plans to continue swimming throughout her pregnancy. She also has started taking an over-the-counter vitamin and mineral supplement. Tracey and her husband think that they have covered all the key areas of prepregnancy care. List a few positive attributes of her current practices. What are some potential problem areas and information they may have missed?

▶ The 13 square meters between maternal and fetal circulation are equal to the sleeping area of 3 king-size beds.

The placenta grows throughout pregnancy to keep up with the increasing demands of the developing fetus. At birth, a healthy placenta weighs about 1.5 lb (0.7 kg) and is about 6 to 8 inches in diameter and 1 inch thick. It has about 13 square meters of contact between maternal and fetal circulation.

The placenta's size and ability to support optimal fetal growth depend on the mother's nutritional status. Poorly nourished women tend to have smaller placentas with fewer blood vessels and smaller cells than those of well-nourished mothers. If the placenta is smaller than normal, the area of contact between the mother and fetus is reduced, which decreases the placenta's capacity to deliver nutrients and remove wastes. Small placentas may hinder optimal fetal growth and development.

### Knowledge Check

1. What is a favorable pregnancy outcome?
2. What are the risks associated with being born at a low birth weight?
3. What are the phases of prenatal growth and development?
4. What are critical periods?
5. What is the role of the placenta?

## 16.2 Nutrient Needs of Pregnant Women

Pregnancy is a nutritionally demanding stage of the life cycle. In just 9 months or so, a mother's body provides all the calories and nutrients needed to produce an infant 5000 times larger than the fertilized egg. To accomplish this, a pregnant woman needs additional calories and more of almost every nutrient than a nonpregnant woman. The extra calories and nutrients support the growth and development of the fetus, placenta, and mother's body, as well as increased maternal metabolism.

To meet her calorie and nutrient needs, a pregnant woman should eat more nutrient dense food. Metabolic adjustments that allow pregnant women to use some nutrients more efficiently (e.g., protein), absorb some better (e.g., calcium, iron), and/or excrete less of others (e.g., zinc, riboflavin) also help meet the calorie and nutrient demands of pregnancy. Sufficient quantities of calories and all nutrients are needed for a favorable pregnancy outcome; however, only calories and the nutrients of particular importance during pregnancy are discussed in this chapter.

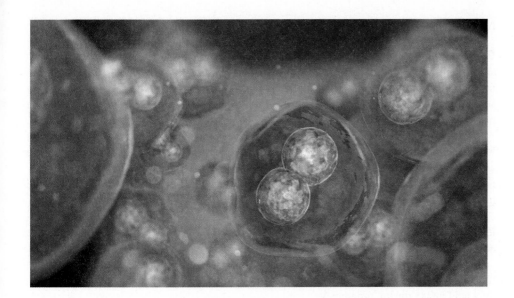

Soon after a woman's ovum is fertilized with a man's sperm, the fertilized ovum begins dividing. This 2-cell embryo will divide again and again until, by the time the offspring is born, there will be trillions of cells. Andrii Vodolazhskyi/Shutterstock

## Energy Needs

A pregnant woman needs extra calories to support the growth of her own tissues, as well as those of the fetus. Additional calories also are needed to fuel the extra metabolic workload pregnancy puts on a woman's heart, lungs, and other organs. Few, if any, of these extra calories are needed during the 1st trimester of pregnancy, when the developing offspring gains little weight. During the 2nd trimester, a daily increase of about 350 calories is recommended. And, in the 3rd trimester, a daily increase of approximately 450 calories is recommended.[10] Women who begin pregnancy overweight or obese should aim for a somewhat smaller increase in calories. Those who are in their teens, underweight, or physically active likely will need more calories. In fact, underweight women who increase their energy intake are more likely to give birth to healthier babies and experience fewer infant deaths than those who do not increase their energy intake. Women who are physically active during pregnancy may need to increase their intake by more than 350 to 450 calories because greater body weight requires more energy for activity.

Infants born to women who consumed insufficient calories are small and more likely to die soon after birth. Those who survive are likely to experience severe, lifelong consequences. There is substantial evidence that individuals who suffered calorie restrictions during fetal life develop the capacity to use calories in a "thrifty" manner. That is, throughout their lives they need fewer calories to maintain their bodies and support physical activity. Although this thrifty calorie usage promotes survival when the food supply is restricted, it elevates the risk of obesity and type 2 diabetes when food is abundant. Infants who are born small also have a greater risk of developing heart disease, high blood cholesterol levels, diabetes, and high blood pressure and experiencing impaired immune function.

The primary regulator of energy (calorie) use in the body is thyroid hormone. To ensure adequate production of thyroid hormone, pregnant women need to consume sufficient amounts of iodine. Using iodized salt can easily meet the need for this mineral. As you learned in Part 4, iodine deficiency during pregnancy can lead to severe birth defects.

To ensure optimal health and the rapid treatment of medical conditions that develop during pregnancy, a pregnant woman should consult with her health-care provider on a regular basis. Ideally, this consultation should begin before she becomes pregnant. JGI/Tom Grill/Blend Images/Corbis

▶ A goal of *Healthy People 2030* is to reduce the number of preterm births.

## Critical Thinking

Alexandra wants to have a baby. She has read that it is very important for women to be healthy during pregnancy. However, Jane, her sister, tells her that the time to begin to assess nutritional and health status is before she becomes pregnant. What additional information should Jane give Alexandra?

▶ A goal of *Healthy People 2030* is to increase the proportion of pregnancies that begin with optimal folate status in an effort to reduce the occurrence of neural tube defects.

▶ Congenital heart defects affect nearly 1 in 100 newborns and are a leading cause of infant illness and deaths due to birth defects. Taking multivitamins containing folic acid during early pregnancy is associated with a significant reduction in risk of heart defects.

▶ Women who have previously given birth to an infant with a neural tube defect, such as spina bifida, should consult their physician about the need for folic acid supplementation. A daily intake of 400 to 500 μg of synthetic folic acid at least 1 month prior to conception is recommended, but it must be taken under a physician's supervision. It is important to note that not all neural tube defects are caused by insufficient folic acid intake.[13]

## Nutrients Needed for Building New Cells

Cells grow and develop at a rapid rate during fetal life. Over the course of gestation, a single cell zygote will divide millions of times, creating trillions of cells. Although every nutrient plays an important role in manufacturing these new cells, the roles of protein, essential fatty acids, zinc, folate, vitamin B-12, and iron are especially noteworthy during pregnancy.

*Protein* intake recommendations for pregnant women are more than 50% above those for nonpregnant women. Even with this substantial increase, insufficient intakes of protein are uncommon in the U.S. and Canada because this nutrient is so plentiful that women, pregnant or not, consume amounts exceeding the RDA. Protein supplements are neither needed nor recommended during pregnancy.

*Essential fatty acids* are required for normal fetal growth and development, particularly of the brain and eyes. Sufficient intakes, especially of the omega-3 fat docosahexaenoic acid (DHA), may improve gestation duration and infant birth weight, length, and head circumference. Many women need to increase the amount of omega-3 fats in comparison with the amount of omega-6 fats they consume. In addition, they should minimize their *trans* fatty acid intake during pregnancy.

*Zinc* intake by both pregnant and nonpregnant women often is less than the RDA; however, deficiencies are uncommon. Severe zinc deficiency may cause birth defects, fetal growth retardation, premature birth, and spontaneous abortion.[11] Pregnant women may experience prolonged labor, bleeding, infections, and serious complications, such as pregnancy-induced hypertension and preeclampsia, discussed later in this chapter. Those who eat zinc poor diets and have a high fiber or iron intake (which impairs zinc absorption) need to pay particular attention to zinc intake. Certain medications, cigarette smoking, alcohol abuse, and strenuous exercise may prevent the placenta from transferring adequate zinc to the fetus.

*Folate and vitamin B-12* are critical for the synthesis of DNA and fetal and maternal cells. For instance, red blood cell formation, which requires folate, increases during pregnancy. When folate intake is insufficient, fewer red blood cells are synthesized, causing folate-related anemia. Insufficient folate intake also may cause premature birth, low birth weight, fetal growth retardation, spontaneous abortion, poor placenta development, and other pregnancy complications. An insufficient supply of folate may be the result of a low dietary intake. In addition, some women have a genetically determined inefficient use of folic acid and, thus, have an increased need for this vitamin.

Folate deficiencies in the very early stages of pregnancy also can cause neural tube defects (see Chapter 13). The neural tube is tissue that develops into the brain and spinal cord (Fig. 16-6). It starts as a shallow groove running down the back of the embryo that, at 28 days of gestation, folds in on itself to create a tube. If it fails to close at the top, the brain does not develop fully and death occurs soon after birth. The spine is incompletely formed when the tube does not seal at the bottom, causing spina bifida. Depending on the severity of spina bifida, children may be paralyzed or have curvature of the spine, dislocated hips, or other physical handicaps. There is some evidence that poor folate status also may play a role in the development of heart defects, other birth defects, and autism.[1,12]

Folic acid, the synthetic form of folate used in supplements and fortified foods, can prevent half or more of all neural tube defects. Folic acid is absorbed almost twice as well as the folate that naturally occurs in foods. Thus, experts recommend that all women who have the potential to become pregnant take folate supplements or eat folate-fortified foods in addition to eating a folate rich diet. This recommendation extends to all these women because a significant proportion of pregnancies are unplanned and neural tube defects occur in the 1st few weeks of pregnancy—before many realize they are pregnant. Women who have previously delivered an infant with a neural tube defect may need to consume more folate than the RDA. However, before exceeding the Upper Level of folate, women should consult a health-care provider because large doses of folate can make it difficult to diagnose a vitamin B-12 deficiency.

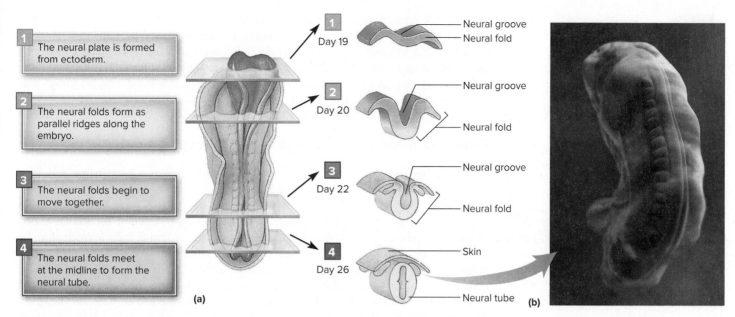

*Figure 16-6* (*a*) The formation of the neural tube (longitudinal view is on left; cross-sectional view is on right). (*b*) The neural tube of a 23-day-old embryo; the neural tube is beginning to fuse. b: Medical.com

Insufficient intake of choline and/or vitamin B-12 also may contribute to the development of neural tube defects.[14] Women who eat animal products usually consume sufficient amounts of vitamin B-12; however, vegans need a vitamin B-12 supplement.

Prior to 1998, when the FDA began requiring folate fortification of grains and cereals, only 10% of women met the folate RDA. Almost 1 in every 1000 babies born in the U.S. had a neural tube defect. Since folate fortification began, the percentage of women meeting folate needs has increased and the number of babies born with neural tube defects has declined.[13,16]

*Iron* needs rise significantly during pregnancy, mainly because of the increasing number of maternal red blood cells and accumulating fetal iron stores. The increased demand for iron puts many pregnant women at a greater risk of iron deficiency anemia than individuals in any other stage of the life cycle. Iron stored prior to pregnancy can supply some of the extra iron; however, most women enter pregnancy with poor iron stores. To help meet iron demands, maternal iron absorption increases up to 3 times and this mineral is conserved because menstruation stops during pregnancy. Even with these adaptations and a carefully planned diet, it is very difficult to meet the need for iron.

Many experts recommend that pregnant women take a low-dose iron supplement (30 mg/daily). However, because iron can interfere with zinc and copper absorption and utilization, women taking iron supplements also may need a zinc and copper supplement.[3] Iron supplements can decrease appetite and cause nausea and constipation; taking them between meals or just before going to bed can minimize these problems. Coffee or tea should not be consumed with an iron supplement because these beverages contain substances that interfere with iron absorption. Eating foods rich in vitamin C along with iron supplements or foods that contain heme iron helps boost iron absorption.

Recall that individuals with iron deficiency anemia have fewer red blood cells than normal; consequently, they have a reduced capacity to deliver oxygen to their body cells. In pregnancy, iron deficiency anemia means that less than optimal amounts of oxygen may reach the fetus. Iron deficiency anemia can cause low birth weight, premature birth, and infant death, and it may result in low iron stores in the infant. In addition, pregnant women with this condition may experience preeclampsia, labor and delivery complications, and an increased risk of death.[1]

▶ The following are maternal factors that increase the risk for neural tube defects.[15]

- Low folate intake
- Lack of folic acid supplementation
- Family history
- Hispanic ethnicity
- Antiseizure medication use
- Pre-pregnancy BMI >25
- Diabetes
- Opioid use early in pregnancy
- Elevated body temperature in early pregnancy (e.g., caused by fever, hot tub, sauna)

Iron deficiency anemia, which is a dangerous condition, should not be confused with a normal change called the anemia of pregnancy. During pregnancy, the number of maternal red blood cells increases 20 to 30%, but the liquid portion of the blood (plasma) expands 50%. Thus, there is a lower ratio of red blood cells to total blood volume. This hemodilution, known as physiological anemia, is a common, expected condition during pregnancy and does not pose a danger to the health of the mother or fetus.

## Nutrients Needed for Bone and Tooth Development

The fetus needs large quantities of vitamin D, calcium, phosphorus, magnesium, and fluoride for normal bone and tooth development. Of these, calcium and vitamin D need particular attention.

Even though a full-term fetus stores about 30,000 mg of calcium, the recommended intake of this mineral does not rise during pregnancy. That's because, early in pregnancy, the mother's body adjusts to absorb calcium much more efficiently. The calcium is stockpiled in her bones, to be drawn on during late pregnancy and lactation.[17] However, many women, pregnant or not, fail to consume the RDA for calcium, which puts them at increased risk of osteoporosis later in life. Too little vitamin D may cause mothers to develop the vitamin D deficiency disease osteomalacia and their fetuses to develop rickets, grow poorly, and inadequately calcify bones and teeth.

Vegans, pregnant teens, women at risk of pregnancy-induced hypertension, and those who do not consume dairy products are at risk of consuming insufficient calcium. Those who do not consume milk, which is fortified with vitamin D, or have limited exposure to sunlight also risk getting too little vitamin D. These women should choose foods fortified with calcium and vitamin D and discuss using a supplement with their health-care provider. A vitamin D supplement is particularly important during the winter in northern latitudes (see Part 4).

## Pregnant Women Do Not Have an Instinctive Drive to Consume More Nutrients

It is a common myth that women instinctively know what to eat during pregnancy and that, by responding to "cravings," they get the nutrients they need. During pregnancy, many women report craving some foods or beverages, ranging from the unusual (clay and rubber bands) to the ordinary (ice cream, pickles, and chocolate). In addition to cravings, many pregnant women experience aversions to certain odors and flavors, such as alcohol, eggs, coffee, fried foods, meat, and tomato sauce. The cause of cravings and aversions remains a mystery; they may be related to hormonal changes in the mother or family traditions. There is no evidence that cravings and aversions are the result of nutrient deficiencies.

The desire to eat or avoid certain foods or combinations of foods won't affect a woman's health as long as her overall diet provides adequate nutrients and calories. To cope with strong cravings for ordinary food items, a pregnant woman should eat small amounts of the desired food along with regular meals or snacks. When craved foods dominate the diet to the extent that variety is limited or when aversions severely inhibit food intake, malnutrition may occur.

Some women also practice **pica**, the eating of nonfood substances, such as laundry starch, coal, clay, and tire inner tubes, over a sustained period of time. Abnormal cravings for certain food items, such as baking soda and cornstarch, also are considered pica.[1,18] Pica is not unique to any group. It is practiced by both sexes and many racial and ethnic groups. Pica is especially common among women who, during childhood, observed family members eating nonfood items. It seems to be more frequently practiced during pregnancy than other times.

The potential dangers posed by pica far outweigh any perceived benefits. Some nonfood substances, such as mothballs, toilet bowl freshener blocks, and chalk, may contain

Milk and other dairy products supply the calcium pregnant women need to support optimal fetal development as well as protect their own bones. vadimgozhda/123

The time to begin thinking about prenatal nutrition is before becoming pregnant. This includes making sure folic acid intake is adequate (400 μg of synthetic folic acid per day) and any supplemental use of preformed vitamin A does not exceed 100% of the Daily Value. Shutterstock/ Gratsias Adhi Hermawan

▶ To learn more about folic acid and neural tube defects, visit www.marchofdimes.org.

**pica** Practice of eating nonfood items, such as dirt, laundry starch, or clay.

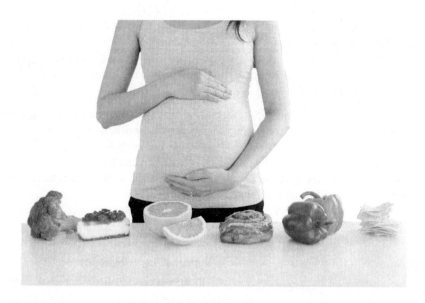

Humans cannot rely on cravings to meet nutrient needs. Nutrition advice from experts is much more reliable. JGI/Tom Grill/Blend Images LLC

dangerously high levels of toxins, such as lead. Clay can block the intestines and may contain parasites and pathogens that impair mineral absorption. Laundry starch is high in calories (1800 calories per pound) but lacks all other nutrients, except carbohydrate, and can lead to malnutrition and obesity. Pica can have dire consequences for a pregnant woman (e.g., malnutrition, anemia, and death) and her fetus (e.g., premature birth, low birth weight, poor nutrient stores, and death).

### Knowledge Check

1. What effect can eating too few calories during pregnancy have on offspring?
2. What nutrients are needed to build cells during fetal growth? What effect will consuming insufficient amounts of these nutrients during pregnancy have on offspring?
3. Why is it important to ensure an adequate folate intake before pregnancy?
4. How do iron deficiency anemia and physiological anemia differ?
5. How might practicing pica during pregnancy affect the pregnancy outcome?

## 16.3 Diet and Exercise Plan for Pregnancy

A pregnant woman's nutrient needs increase dramatically, whereas her calorie needs increase only a small amount. Selecting low-fat foods helps keep calories in line and increases the nutrient density of her diet. By choosing carefully, the woman can meet the increased nutrient demands of pregnancy and facilitate optimal fetal growth without consuming excessive calories.

The recommended daily increase of 350 to 450 calories doesn't sound like much, but the usefulness of this increase depends on how the woman decides to "spend" those calories. Two soft drinks or a chocolate candy bar supplies nearly the number of extra calories daily a woman needs in the last 2 trimesters—and provides almost no nutrients. On the other hand, 2 glasses of fat-free milk, a small spinach salad, and a bowl of strawberries will supply

Pregnancy leads to increased nutrient needs for the mother. Meeting these needs is an important step toward a successful pregnancy. Plush Studios/Brand X Pictures/Getty Images

▶ To help pregnant and lactating women personalize MyPlate, the USDA has created this website: www.choosemyplate.gov.

two-thirds of the needed calcium and vitamin A and one-fourth of the folate and will exceed the vitamin C needs for 1 day, all for less than 350 calories.

One approach to a diet plan that supports a successful pregnancy outcome is based on MyPlate. For an active adult woman in the 1st trimester, about 2200 kcal are recommended. The plan should include the following.

- Dairy Group: 3 cups of calcium rich, low-fat or fat-free foods to supply protein, calcium, and carbohydrate, as well as other nutrients; calcium-fortified foods from other food groups can make up for gaps between calcium intake and need
- Protein Group: 6 ounce-equivalents to deliver needed iron and zinc
- Vegetables Group: 3 cups to provide vitamins and minerals; 1 cup should be rich in vitamin C and 1 cup should be rich in folate
- Fruit Group: 2 cups to supply vitamins and minerals
- Grains Group: 7 ounce-equivalents emphasizing whole-grain and enriched foods

In the 2nd and 3rd trimesters, the plan recommends slight increases in almost every food group. Specifically, the plan should include about 2600 kcal, divided like this:

- Dairy Group: 3 cups
- Protein Group: 6½ ounce-equivalents
- Vegetables Group: 3½ cups
- Fruit Group: 2 cups
- Grains Group: 8 ounce-equivalents

A salad each day provides many nutrients for the prenatal diet.

Brand X Pictures/Getty Images

**Table 16-2 Sample 2600 kcal Daily Menu That Meets the Nutritional Needs of Most Pregnant and Breastfeeding Women**

Spinach salad with pepper rings adds important nutrients to a pregnant woman's diet.
Nancy R. Cohen/Getty Images

|  | Amount |
|---|---|
| **Breakfast** | |
| Fortified whole-grain breakfast cereal | 1 cup |
| Orange juice | 1 cup |
| Fat-free milk | 1 cup |
| **Calories** | **420** |
| **Morning Snack** | |
| Whole-wheat toast | 1 slice |
| Plain low-fat yogurt | 1 cup |
| Tangerine and blueberries | ½ cup |
| **Calories** | **240** |
| **Lunch** | |
| Spinach salad | 2 cups |
| Italian dressing | 2 tablespoons |
| Sweet pepper rings | ½ pepper |
| Hard-cooked egg | 1 egg |
| Whole-wheat bread | 1 slice |
| Peanut butter | 2 tablespoons |
| **Calories** | **450** |
| **Afternoon Snack** | |
| Whole-wheat crackers | 6 crackers |
| Tomato juice | 1 cup |
| Cheddar cheese | 1½ ounces |
| **Calories** | **290** |
| **Dinner** | |
| Grilled chicken breast, skinless[†] | 3½ ounces |
| Brown rice | 1 cup |
| Broccoli, carrots, and bok choy | 1 cup |
| **Calories** | **425** |
| **Evening Snack** | |
| Granola bar | 1-ounce bar |
| Raisins | 1 ounce |
| Sunflower seeds | 1 ounce |
| **Calories** | **375** |
| Additional calories | Up to 400 kcal* |
| **Total calories for the day** | **2600** |

Total intake of fluids, such as water, should be 10 cups or so per day.
*Amount of additional calories will vary based on the actual food choices made within each MyPlate group.
[†]Substituting a veggie burger for the chicken breast makes this menu a practical guide for a vegetarian.

Pregnant women also should include about 6 to 7 teaspoons of vegetable oil that is rich in essential fatty acids.

Table 16-2 illustrates a daily menu based on the basic diet plan for women in the last 2 trimesters. This menu meets the extra nutrient needs associated with pregnancy. Women who need to consume more than 2600 kcal—and some do for various reasons—should incorporate into their diets additional fruits, vegetables, and whole-grain breads and cereals, not poor nutrient sources such as desserts and regular soft drinks.

## Prenatal Vitamin and Mineral Supplements

Special supplements formulated for pregnancy are prescribed routinely for pregnant women by most physicians. Some are sold over the counter, whereas others are dispensed by prescription because of their high synthetic folic acid content (1000 µg), which could pose problems for others, such as older people (see Part 4). Prenatal supplements also are high in iron. It is important not to exceed the recommended dose because megadoses of vitamin and mineral supplements can be dangerous for both the pregnant woman and the fetus. For example, iron, zinc, selenium, and vitamins A, B-6, C, and D can exert toxic effects when taken in large doses.[3] Supplemental preformed vitamin A is especially important to keep under control; it should not exceed 3000 µg RAE/day (15,000 IUs per day) because higher levels are linked with teratogenic birth defects (see Part 4), mainly during the 1st trimester.

Many health professionals believe a pregnant woman should take nutrient supplements only when there is evidence that her usual diet is likely to limit maternal or fetal growth and development. For many pregnant women, the only supplement needed is iron during the last 2 trimesters. A multivitamin and mineral supplement is recommended for women who have a history of frequent dieting; are teenagers or vegans; have a low income; are underweight; smoke or abuse alcohol or illegal drugs; are carrying multiple fetuses; and/or are eating a diet restricted in variety. Supplements of specific nutrients may be recommended when these nutrients are inadequate; for instance, vegans usually need a vitamin B-12 supplement.

Pregnancy, in particular, is not a time to self-prescribe medications or vitamin and mineral supplements. For example, although vitamin A is a routine component of prenatal vitamins, intakes over 3 times the RDA can have toxic effects on the fetus. Andersen Ross/Getty Images

## Physical Activity during Pregnancy

A low- or moderate-intensity exercise program can offer physical and psychological benefits to a woman experiencing a normal, healthy pregnancy. Benefits include improved cardiovascular function, an easier and less complicated labor, and an improved attitude and mental state. Exercise also can help prevent or treat gestational diabetes. Infants born to women who exercise tend to be leaner and more neurologically mature than babies born to nonexercisers. Women with high-risk pregnancies, such as those experiencing premature labor contractions, may need to restrict their physical activity.

To ensure optimal health for both herself and her infant, a pregnant woman should consult her health-care provider before beginning or continuing with an exercise program. The following recommendations for exercise during pregnancy can be used to plan a safe exercise program.

- Exercise moderately for about 30 minutes daily on most days of the week.
- Drink plenty of liquids to maintain normal fluid and electrolyte balance and avoid dehydration.
- Keep heart rate below 140 beats per minute to maintain adequate blood (and oxygen) flow to the fetus.
- Include a cooldown period at the end of exercise sessions so that heart rate can gradually return to normal.

*Critical* Thinking

Hannah, a 16-year-old high school student, has just discovered that she is pregnant. At 5 feet 3 inches tall and 105 lb, she is underweight and her typical diet lacks many essential nutrients. For breakfast, she has coffee and a croissant, if anything at all. She often skips lunch or eats chips from the vending machine. She then eats a well-rounded dinner with her family. What risks do you see in this situation for Hannah and her baby?

# GL⊕BAL PERSPECTIVE

## Pregnancy and Malnutrition

Prolonged undernutrition is detrimental at any life stage, but its effects are particularly profound during pregnancy and fetal life. More than 303,000 women worldwide die each year from complications of pregnancy and childbirth—most are in low-income countries.[19] In Africa, for example, women give birth, on average, to more than 5 live babies. Coupled with chronic undernutrition, these high birth rates result in a 1 in 180 chance that a woman will die from pregnancy-related causes—chances rise to 1 in 54 in impoverished and war-torn areas. In contrast, women in high-income countries face a risk of only 1 death from pregnancy-related causes in about 4900 births. Pregnancy-related death is the social indicator that differs most between the low-income and high-income countries.

Connie Coleman/Getty Images

The fetus also faces major health risks from undernutrition during gestation. When nutrient needs are not met, the infant is often born prematurely and, as a result, has reduced lung function and a weakened immune system. These conditions not only compromise health but also increase the likelihood of premature death. Long-term problems in growth and development can result if the infant survives. In extreme cases, low-birth-weight infants face 15 to 20 times the normal risk of dying before the age of 1 year. Worldwide, more than 20 million infants are born each year with low birth weights.

Although it is difficult to specify the extent to which poor nutrition will affect each pregnancy, a daily diet containing only 1000 kcal has been shown to greatly restrict fetal growth and development.[20] Increased maternal and infant death rates seen in famine-stricken areas of Africa and wartime observations provide further evidence. For example, during World War II, food supplies in Holland were extremely restricted for about 6 months. Babies conceived before the food shortage and delivered during or after the shortage were only slightly shorter and lighter in weight than normal. They were less likely to be spontaneously aborted, stillborn (fetal death before onset of labor), or premature; to have birth defects; or to develop mental disorders later in life than were babies conceived near the end of the food shortage, when maternal nutrient stores and body weight were depleted. Similar outcomes were seen among women who became pregnant during the siege of Leningrad, when food was extremely limited.

---

Health-care providers typically encourage healthy, well-nourished pregnant women to engage in moderate exercise as long as increased energy and nutrient needs are met. JGI/Blend Images/ Alamy Stock Photo

- After about the 4th month, avoid exercises that are done while lying down because the enlarged uterus compresses blood vessels and cuts down blood flow to the fetus.
- Avoid deep flexing (e.g., deep knee bends), joint extensions (e.g., leg stretches), and activities that jar the joints (e.g., jumping) because connective tissue that has become more elastic to facilitate normal childbirth can be damaged if overstressed in exercise.
- Prevent increases in body temperature by not exercising in hot, humid weather or engaging in strenuous activities for more than 15 minutes. High body temperatures can damage enzymes that regulate fetal development.
- Stop exercising immediately if discomfort occurs—aches and pains are a warning that something may be wrong.
- Avoid strenuous or endurance activities because they can cause lower than normal birth weights.
- Avoid activities that could cause abdominal trauma (e.g., horseback riding, martial arts), that involve rapid shifts in balance or body position that may cause accidental falls (e.g., basketball, skiing, hockey), or that compress the uterus (e.g., scuba diving).
- Avoid becoming overly tired.

 ## Knowledge Check

1. How many servings from each food group should pregnant women consume in the last 2 trimesters?
2. Nutrient supplements during pregnancy are recommended for which groups of women?
3. What types of physical activities should pregnant women avoid?

# 16.4 Maternal Weight and Pregnancy Outcome

An infant's birth weight is closely related not only to the length of gestation but also to the woman's prepregnancy weight and the amount of weight she gains during pregnancy.[1] Starting pregnancy at a healthy weight and following weight gain recommendations increase the likelihood of a successful pregnancy outcome.

## Maternal Prepregnancy Weight

Infants born to women who began pregnancy substantially above or below a healthy weight are more likely to experience problems than are those born to women who begin pregnancy at a healthy weight (Table 16-3). For instance, babies born to women with obesity are at increased risk of having birth defects such as neural tube defects and oral clefts, being stillborn or dying in the 1st few weeks after birth, having high birth weights, and having obesity in childhood.[21,22] Many pregnant women with obesity experience high blood pressure, gestational diabetes, and Cesarean (also called surgical) deliveries.[22] Women can reduce the risks associated with obesity by losing weight *before* becoming pregnant.

Women who begin pregnancy underweight are more likely to have infants who are low birth weight and premature than are women at a healthy weight.[1] These differences may occur because women who are underweight tend to have lighter placentas and lower nutrient stores, especially iron, than women who weigh more, which can affect fetal growth negatively. Women who are underweight can improve nutrient stores and pregnancy outcome by gaining weight before pregnancy or gaining extra weight during pregnancy.

Prepregnancy weight and nutrient stores affect not only the pregnancy outcome but also the woman's ability to become pregnant.[21] Many women who are underweight experience amenorrhea, which may reduce their ability to ovulate. The chances of ovulating and becoming pregnant improve when body fat increases to a healthy level. Women with obesity also may have difficulty becoming pregnant. Both the additional estrogen produced by their extra fat stores (see Chapter 10) and insulin resistance (related to type 2 diabetes) (see Chapter 5) reduce fertility. Weight loss helps increase fertility in women with obesity.[22] On the other hand, obesity, inactivity, high-fat diets, and low nutrient intakes affect a man's ability to produce enough viable sperm. When the number of sperm is low and/or the sperm's ability to propel itself is reduced, fertilization is unlikely to occur. Zinc, folate, and vitamin C affect the quality of sperm; when these nutrients are in short supply or fat or alcohol intake is high, men may experience fertility problems.[23,24,25]

## Recommendations for Maternal Weight Gain

The weight a woman gains during pregnancy allows the fetus to grow and her body to accommodate it, as well as prepare for lactation. Figure 16-7 illustrates how the weight gained during pregnancy is divided. The recommendations for prenatal weight gain have steadily increased in the last

**Table 16-3  Maternal Factors That Increase Risk of Nutrient Deficiencies and Poor Pregnancy Outcome**

| Factor | Condition Increasing Risk of Nutrient Deficiencies |
|---|---|
| Body weight | |
|   Prepregnancy | BMI less than 19.8 or greater than 26 |
|   During pregnancy | Inadequate or excessive weight gain<br>Inappropriate pattern of weight gain |
| Age | Young age |
| Eating patterns | Regular omission of foods from 1 or more major food groups (e.g., vegan diets) |
| | Excessive consumption of a single type of food |
| | Fasting and weight loss diets |
| | Special diets to control maternal health conditions, such as heart disease, kidney disease, diabetes, and genetic disorders (e.g., phenylketonuria) |
| | Eating disorders |
| | Food cravings, aversions, and pica |
| Health | Pregnancies spaced less than 12 to 18 months apart |
| | More than 3 previous pregnancies if under age 20, more than 4 if over age 20 |
| | Carrying more than 1 fetus |
| | Inadequate prenatal health care |
| | Diseases such as HIV/AIDS and diabetes |
| | Onset of gestational diabetes mellitus |
| | Onset of pregnancy-induced hypertension, preeclampsia, and/or eclampsia |
| Sociocultural factors | Low income |
| | Limited educational achievement |
| | Lack of family or social support |
| Food supply safety | Food contaminants (e.g., mercury, lead, PCBs, pesticides) |
| | Foodborne illness pathogens |
| Lifestyle choices | High caffeine intake |
| | Use of alcohol, drugs, tobacco, or herbal and botanical products |

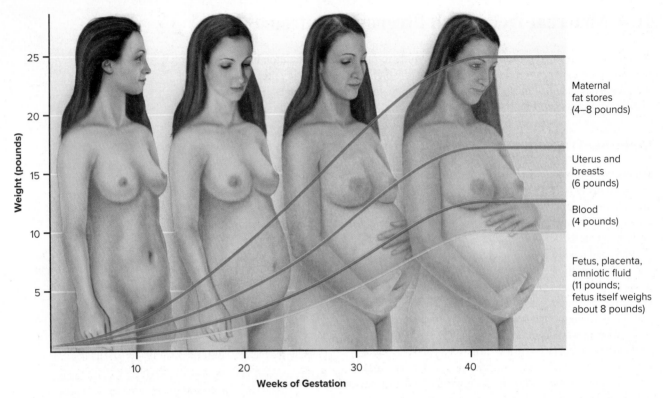

*Figure 16-7* Components of weight gain in pregnancy. A weight gain of 25 to 35 lb is recommended for most women. Note that the various components total about 25 lb.

**Table 16-4  Recommended Weight Gain in Pregnancy Based on Prepregnancy Body Mass Index (BMI)[23]**

| Prepregnancy BMI Category | Total Weight Gain* | |
| --- | --- | --- |
| | **(lb)** | **(kg)** |
| Underweight (BMI less than 18.5) | 28 to 40 | 12.5 to 18 |
| Healthy (BMI 18.5 to 24.9) | 25 to 35 | 11.5 to 16 |
| Overweight (BMI 25 to 29.9) | 15 to 25 | 7 to 11.5 |
| Obese (BMI 30 and higher) | 11 to 20 | 5 to 9 |

*The listed values are for pregnancies with 1 fetus. Weight gain recommendations for twins are 37 to 45 lb (17 to 20.5 kg) for women at a healthy weight, 31 to 50 lb (14 to 23 kg) for those in the overweight category, and 25 to 42 lb (11 to 19 kg) for those in the obese weight category.

▶ Before becoming pregnant, women who have had gastroplasty for weight loss (see Chapter 10) are advised to wait until their weight stabilizes (after approximately 12 to 18 months) and seek care from an obstetrician specializing in high-risk pregnancies.[21]

60 years from 15 to 16 pounds in the 1950s to 25 to 35 pounds, now recommended for women of healthy prepregnancy weight. Studies have shown repeatedly that gaining the amount of weight recommended by the National Academy of Science's Institute of Medicine improves the chances of optimal health for both mother and fetus if gestation lasts at least 38 weeks.

As shown in Table 16-4, a healthy goal for total weight gain for a woman of healthy BMI averages about 25 to 35 lb (11.5 to 16 kg).[23] For women with a low BMI (<18.5), the goal increases to 28 to 40 lb (12.5 to 18 kg). The goal decreases to 15 to 25 lb (7 to 11.5 kg) for those with BMIs ranging from 25 to 29.9 and 11 to 20 lb (5 to 9 kg) for a woman with obesity (BMI >30). The weight gain recommendations are given as ranges to allow for differences within each BMI group. Weight gain recommendations from the Institute of Medicine indicate that women who are short, are from various racial and ethnic groups, and/or are teenagers should follow the recommendations in Table 16-4 until more research is done to determine if special recommendations are needed for them.[23] Weight gain recommendations for healthy-BMI women carrying twins are 37 to 45 lb (17 to 20.5 kg). Women with BMIs ranging from 25 to 29.9 who are carrying twins should aim for a weight gain of 31 to 50 pounds (14 to 23 kg), and those who have obesity should aim to gain 25 to 42 pounds (11 to 19 kg).

The weight gain recommendations in Table 16-4 promote optimal fetal growth while minimizing the risks of complications at delivery, postpartum maternal weight retention, and the infant's chances of developing chronic disease later in life. Unfortunately, many women do not gain the recommended amounts. Women who gain less than recommended amounts have an increased risk of giving birth to a baby that is premature, is small for gestational age, and/or dies soon after birth.[20] In contrast, women who gain excessively higher amounts than the recommendations typically give birth to very large babies and experience an increase in complications at delivery, infant mortality, and weight retention postpartum. Only about one-third of women gain the recommended weight during pregnancy, with another 20% gaining less and about 50% gaining more than the recommendations.[26]

## Pattern of Maternal Weight Gain

The pattern the woman's weight gain follows is as important as the amount she gains. During the 1st trimester, most women should gain about 1.1 to 4.4 lb (0.5 to 2 kg), most of which is accounted for by the enlarging breasts and uterus. After that, a woman who starts at a healthy weight should gain about 0.8 to 1 lb (0.35 to 0.45 kg) a week at a slow, steady rate. Women with BMIs <18.5 should gain about 1 to 1.3 lb (0.45 to 0.6 kg) per week. Women with weights in the overweight category should gain about 0.5 to 0.7 lb (0.23 to 0.32 kg), and women with obesity should gain 0.4 to 0.6 lb (0.18 to 0.27 kg) each week.[27] In the 2nd trimester, weight gain is divided between mother and fetus. In the 3rd trimester, weight gain is almost all fetal tissue. Low weight gain in either of the last 2 trimesters increases the chances of prenatal growth retardation. Low weight gain in the last trimester raises the risk of premature birth.

If a woman deviates from the desirable weight gain pattern, she should work with her health-care provider to make appropriate adjustments. For example, if a woman begins to gain too much weight during her pregnancy, she should not try to lose weight. Instead, she should minimize her intake of foods that provide unnecessary calories. Even if a woman gains 35 lb in the first 7 months of pregnancy, she must still gain more during the last 2 months to have an optimal outcome. However, she should slow the increase in weight to parallel the weekly recommended rise in prenatal weight gain. Alternately, if a woman has not gained the desired weight by a given point in pregnancy, she should not try to gain the needed weight rapidly. Instead, she should slowly gain a little more weight than the typical pattern to meet the goal by the end of the pregnancy.

A sudden change in weight (up or down) may signal health problems for the mother. For instance, a sudden, erratic gain may be caused by fluid retention. This may indicate that the mother has pregnancy-induced hypertension, a potentially life-threatening condition.

## Postpartum Weight Loss

Weight gain recommendations also consider the importance of losing weight after pregnancy. Women lose some weight immediately after giving birth. Within a few weeks, the new mother's blood volume returns to normal. The time it takes to lose the extra fat stores will depend on the amount gained during pregnancy, as well as her postpartum diet and exercise pattern. For lactating mothers, some of the fat can be used as the energy source needed to produce breast milk. Many women retain some weight after each pregnancy, with those who began pregnancy with BMIs >25 retaining more weight.[28]

Following weight gain recommendations during pregnancy helps increase the chances of a successful outcome. Andersen Ross/Getty Images

*Knowledge Check*

1. How do maternal prepregnancy weight and weight gain during pregnancy affect pregnancy outcome?
2. What are the maternal weight gain recommendations for underweight, healthy-weight, overweight, and obese BMI categories?

▶ The following maternal characteristics affect pregnancy outcome.

- Age
- Eating patterns
- Health
- Sociocultural factors
- Safety of food supply
- Lifestyle choices

Very young mothers often are ill prepared to cope with the physiological, emotional, and societal pressures of pregnancy. These factors contribute to the added risks they face. Purestock/Getty Images

# 16.5 Nutrition-Related Factors Affecting Pregnancy Outcome

Evidence shows that a nutritious diet before pregnancy as well as during pregnancy can have a major effect on both the mother's health and that of her child. In addition to maternal body weight, the other factors listed in Table 16-3 are associated with less than optimal nutritional status and put women at great risk of a poor pregnancy outcome.

## Maternal Age

Teenage pregnancy poses special nutrition and health challenges for both the mother and the developing offspring. Young women continue maturing physically for about 5 years after the onset of menstruation (menarche). Because the average age for menarche is 13 years in the U.S., most women younger than 18 years are not as physically ready to be pregnant as they will be later. In addition, the teen years are a period of high nutrient needs, and teens' nutrient intakes frequently are below recommended amounts. It is difficult for a pregnant adolescent to consume adequate amounts of nutrients to support her own growth. It is even more difficult to meet the additional needs to support fetal growth. Many teens enter pregnancy underweight and gain too little weight during pregnancy. In addition, many teens do not receive adequate prenatal care. For all these reasons, babies born to teens are at greater risk of premature birth, prenatal growth retardation, and death soon after birth. Teens also have more stillbirths and spontaneous abortions than women in their 20s and early 30s.

Older first-time mothers and their offspring also face health challenges. Women who delay childbearing into their mid-30s to 50s have higher rates of complication than do women in their 20s and early 30s. These complications often are related to chronic diseases, such as diabetes and cardiovascular disease. Although it is not clear why, older mothers have higher rates of Cesarean deliveries, low-birth-weight infants, preterm births, fetal deaths, and infants born with birth defects.

## Maternal Eating Patterns

Pregnant women with diets that deviate greatly from the recommended diet by either restricting dietary intake or eating excessive quantities of a limited number of foods risk getting too few nutrients and a poor pregnancy outcome. For example, ketosis can result from restricting carbohydrate intake or fasting and is not desirable for the growing fetus. Ketone bodies are thought to be poorly used by the fetal brain and may slow its development. Pregnant women can develop significant amounts of ketones after only 20 hours of fasting, so eating regular meals and avoiding fasting for more than 12 hours is important. In addition, weight loss diets should never be attempted during pregnancy, regardless of prepregnancy weight or weight gain during pregnancy, because of the rise in ketones that can occur. Including at least 175 g of carbohydrate in the diet every day helps prevent ketosis.

Women who have eating disorders or other nutrition-related health conditions, such as diabetes or phenylketonuria, need to work with health-care providers to be certain their diets meet both their own and fetal needs. As mentioned previously, food cravings, aversions, and pica also can affect dietary quality and pregnancy outcome.

Lacto-ovo-vegetarians and lacto-vegetarians generally do not face special difficulties in meeting nutritional needs during pregnancy. On the other hand, for a vegan, careful diet planning before and during pregnancy is crucial to ensure sufficient intakes of protein, vitamin D (or sufficient sun exposure), vitamin B-6, iron, calcium, zinc, and especially vitamin B-12. The basic vegan diet listed in Chapter 7 should be modified to include more grains, beans, nuts, and seeds to supply the necessary extra amounts of some of these nutrients. The use of a prenatal multivitamin and mineral supplement is generally advocated to help fill micronutrient gaps. However, supplements are not high in calcium (200 mg per pill). If iron and calcium supplements are used, they should not be taken together to avoid possible competition for absorption.

## Maternal Health

Women with conditions that affect their psychological or physical health may not consume an adequate diet and have less than optimal nutrient stores. These conditions also may increase the chances that the pregnancy will result in preterm birth, intrauterine growth retardation, birth defects, stillbirth, trauma to the mother and infant at delivery, and/or maternal and/or infant death.

### Pregnancy History

Numerous previous pregnancies and/or closely spaced pregnancies (less than 1 year apart) may deplete a woman's nutrient stores, which decreases fertility and, should she become pregnant again, increases the risk that a subsequent pregnancy will result in a preterm birth, low-birth-weight infant, or small for gestational age infant. Carrying more than 1 fetus also increases nutrient needs and makes it difficult to consume sufficient calories and nutrients to support optimal growth of the fetuses.

Mothers who have previously experienced problems during pregnancy or have given birth to small or large for gestational age infants are at increased risk for having difficult future pregnancies. Eating a healthy diet before and during pregnancy as well as getting adequate medical care can improve a woman's chances for having healthier future pregnancies.

### Prenatal Care

If prenatal care is inadequate, delayed, or absent, untreated maternal nutritional deficiencies can deprive a developing fetus of needed nutrients. In addition, untreated chronic diseases, such as hypertension or diabetes, increase the risk of fetal damage. Without prenatal care, a woman is 3 times more likely to deliver a low-birth-weight baby. Although the ideal time to start prenatal care is before conception, about 20% of women in the U.S. receive no prenatal care throughout the 1st trimester—a critical time to positively influence pregnancy outcome.

### Phenylketonuria (PKU)

For a favorable pregnancy outcome, women with the hereditary condition PKU must consume a low phenylalanine diet at least 3 months before and throughout pregnancy. Uncontrolled PKU results in poor pregnancy outcomes, including delayed fetal growth and development, intellectual disabilities, and heart and other birth defects. Risks can be reduced by weekly monitoring of phenylalanine and tyrosine during pregnancy and adjusting dietary intake to keep the levels of these amino acids within the normal range. Mothers with PKU typically do not give birth to babies with PKU. Having PKU does not seem to affect male fertility.

### Acquired Immune Deficiency Syndrome

Individuals infected with human immunodeficiency virus (HIV) or who have acquired immune deficiency syndrome (AIDS) have increased needs for energy and some vitamins and minerals. Supplements of some nutrients, such as vitamin A, zinc, and iron, may adversely affect those who are HIV positive. Thus, careful dietary planning is needed by women with HIV or AIDS.

### Hypertension

The offspring of women who begin pregnancy with hypertension have a greater risk of experiencing intrauterine growth retardation and preterm birth. High blood pressure may cause the placenta to separate from the uterus before birth, which can lead to stillbirth. Controlling hypertension before and during pregnancy reduces the risk of complications.

**HISTORICAL PERSPECTIVE**

**The First 10 Minutes of Life**

Since the early 1950s, almost every baby has benefited from Virginia Apgar's work. Based on her observations, this anesthesiologist created a standardized evaluation of newborns—the Apgar Score. This score indicates the quality of an infant's heart rate, respiration, muscle tone, reflexes, and color at 1 minute and 5 minutes after birth. This early evaluation has allowed doctors to quickly identify infants in distress and has saved millions of lives. Learn more at www.nlm.nih.gov/ changingthefaceofmedicine/physicians/ biography_12.html.

Comstock/Getty Images

▶ The following maternal factors increase the risk for pregnancy-induced hypertension.

- First pregnancy
- History of preeclampsia, diabetes, kidney disease, lupus, or rheumatoid arthritis
- Chronic or gestational hypertension
- Family history of preeclampsia
- BMI >25
- >1 fetus
- Black ethnicity
- Age <20 or >35 years

▶ The following maternal factors increase the risk for gestational diabetes.

- Age >25
- Prepregnancy BMI >25
- Excessive weight gain during pregnancy
- Diabetes or pre-diabetes
- Family diabetes history
- Pregnancy history (gestational diabetes, large for gestational age baby)
- Polycystic ovarian syndrome
- Glycosuria
- Hispanic, Black, Asian, Native American, Pacific Islander ethnicity/race

**large for gestational age** Weighing more than the expected weight for length of gestation; typically defined as above the 90th percentile.

Closely spaced pregnancies increase the risk that the woman's nutrient stores will become depleted.
Fuse/Corbis/Getty Images

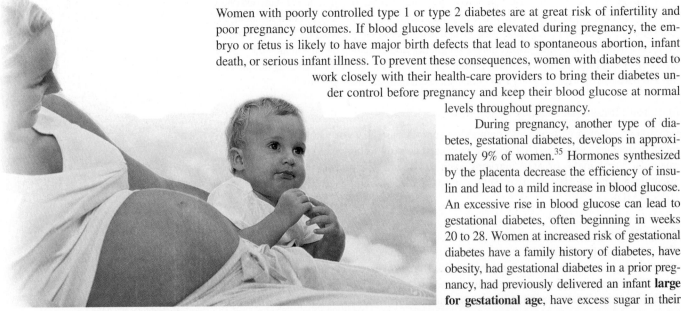

During pregnancy, it is normal to see an elevation in blood pressure; however, pregnancy-induced hypertension (gestational hypertension) causes it to rise abnormally high. Pregnancy-induced hypertension impairs the delivery of oxygen and nutrients to the fetus; thus, the fetus may experience retarded growth and premature birth. Pregnancy-induced hypertension can escalate into potentially deadly conditions called preeclampsia and eclampsia (formerly called toxemia). Preeclampsia is high blood pressure (see Part 4) accompanied by protein in the urine, decreased urine output, headache and dizziness, blurred vision, changes in blood clotting, nervous system disorders, abdominal pain, nausea, vomiting, and edema throughout the body. It can progress to eclampsia, which causes symptoms to become more severe and may lead to maternal convulsions, stroke, and coma. Blood pressure may climb so high that the kidneys and liver are damaged, and both the mother and her fetus may die. In fact, eclampsia is the leading cause of maternal and newborn death in the U.S.

Preeclampsia seems to be related to abnormal implantation of the zygote, which reduces the function of the placenta. Certain substances produced by the poorly performing placenta also appear to be involved in the development of preeclampsia.[1,29] Preeclampsia and eclampsia are most common among women who have a high BMI, are pregnant for the 1st time or have multiple-birth pregnancies, are older than 35 years, are African-American, or had diabetes and/or hypertension before pregnancy.[1] In addition, women who experienced pregnancy-induced hypertension, preeclampsia, and/or eclampsia in a previous pregnancy or who have a family history of these conditions are at increased risk of developing these conditions.[1]

Some evidence indicates that inadequate nutrient intakes may contribute to the development of preeclampsia. For instance, adequate antioxidant status may reduce the risk of preeclampsia. In women with low calcium intakes, large doses of calcium supplements may reduce the risk of preeclampsia.[29] However, in well-nourished women, calcium supplements probably do not help with this problem,[30,31] and insufficient evidence exists to conclude that vitamin E and fish oil supplements are effective in reducing the risk of preeclampsia.[32] Restricting sodium also does not reduce preeclampsia risk,[33] although moderate exercise throughout pregnancy may reduce the risk.

Pregnancy-induced hypertension usually resolves itself within about 2 days after the pregnancy ends, although it takes longer in some cases. However, because the problem often begins well before the normal end of gestation (usually around 20 weeks for preeclampsia and 34 weeks for eclampsia), treatment may be necessary to prevent the worsening of the disorder. Bed rest and magnesium sulfate are the most effective treatment methods, although their effectiveness varies.[34] Magnesium likely relaxes blood vessels and, so, leads to a fall in blood pressure. Several other treatments, such as various antiseizure and antihypertensive medications, are under study.

### Diabetes Mellitus

Women with poorly controlled type 1 or type 2 diabetes are at great risk of infertility and poor pregnancy outcomes. If blood glucose levels are elevated during pregnancy, the embryo or fetus is likely to have major birth defects that lead to spontaneous abortion, infant death, or serious infant illness. To prevent these consequences, women with diabetes need to work closely with their health-care providers to bring their diabetes under control before pregnancy and keep their blood glucose at normal levels throughout pregnancy.

During pregnancy, another type of diabetes, gestational diabetes, develops in approximately 9% of women.[35] Hormones synthesized by the placenta decrease the efficiency of insulin and lead to a mild increase in blood glucose. An excessive rise in blood glucose can lead to gestational diabetes, often beginning in weeks 20 to 28. Women at increased risk of gestational diabetes have a family history of diabetes, have obesity, had gestational diabetes in a prior pregnancy, had previously delivered an infant **large for gestational age**, have excess sugar in their urine (glycosuria), have polycystic ovarian syndrome, or are from a high-risk group (e.g., Hispanic, Asian, Black, and Native American individuals).[1] High-risk women should be screened for gestational diabetes as soon as possible, with repeated screenings if earlier tests were negative. Average-risk women are usually screened for gestational diabetes between 24 and 28 weeks of pregnancy.

Exercise and a diet that distributes low glycemic load carbohydrates throughout the day are important for keeping gestational diabetes under control. Some women also may need insulin therapy. Untreated gestational diabetes can severely deplete fetal iron stores. In addition, uncontrolled diabetes can cause the fetus to grow quite large.[36] The oversupply of glucose from maternal circulation signals the fetus to increase insulin production, which causes fetal tissues to readily use glucose for growth. Another threat is that the infant may have low blood glucose at birth because of the tendency to produce extra insulin, which began during gestation. Other concerns are the potential for preterm delivery and an increased risk of birth trauma and malformations to the heart, spine, brain, kidneys, and digestive system. The abnormally high blood glucose levels caused by gestational diabetes often return to normal after giving birth; however, the mother's risk of developing diabetes later in life rises, especially if she has obesity.[35] Infants born to mothers with gestational diabetes also may have lung and breathing problems as well as higher risks of developing obesity and type 2 diabetes as they grow to adulthood.

## Maternal Sociocultural Factors

Women with limited income or educational achievement and those who lack social support networks tend to have inadequate diets. Thus, to help women of low socioeconomic status procure the foods and nutrition education they need, the U.S. Department of Agriculture funds the Expanded Food and Nutrition Education Program (EFNEP), the Supplemental Nutrition Assistance Program (SNAP), and the Women, Infants, and Children (WIC) program.

EFNEP and SNAP provide nutrition education for those with limited resources. Program participants learn about good nutrition, meal planning, and techniques for stretching food dollars. SNAP also provides foods for those with limited financial resources. WIC provides nutritious foods specifically to low-income pregnant, postpartum, and breastfeeding women, as well as infants and children up to age 5 who are at nutritional risk. WIC participants receive milk, cheese, eggs, fruits, vegetables, juices, iron-fortified cereal, whole grains, soy beverages, tofu, beans, peanut butter, and other nutritious foods that supply nutrients often lacking in their diets, such as vitamins A and C, folate, iron, potassium, calcium, and protein. They attend nutrition education classes and are encouraged to seek and maintain appropriate medical care. This combination of supplemental foods, nutrition education, and health-care referrals has helped improve the nutrition-related health conditions of WIC participants. In addition, participation in WIC is linked to higher birth weights, fewer infant deaths, and lower medical costs after birth.

## Maternal Food Supply

The food supply in the U.S. and Canada can provide all the nutrients needed for a successful pregnancy outcome. However, intakes of the compounds found in some foods should be limited during pregnancy.

### Environmental Contaminants

Contaminants from the environment can enter food through food containers, polluted water, and farming and food preparation practices. As described in Chapter 3, common food contaminants that pose dangers to a pregnant woman and her fetus include lead, mercury, polychlorinated biphenyls (PCBs),

**Dietary Recommendations for Gestational Diabetes**

1. Consume adequate calories to allow for appropriate weight gain.
2. Consume a minimum of 175 grams of carbohydrate daily.
3. Avoid ketosis.
4. Divide carbohydrate intake among 3 meals and 2 to 4 snacks. The bedtime snack is especially important because it helps prevent low blood sugar levels during the night. This snack should include both protein and complex carbohydrate.

Healthy nutrition, prenatal care starting before pregnancy, and the avoidance of unsafe behaviors, such as smoking and taking drugs, are key factors in promoting a successful pregnancy outcome.
JGI/Tom Grill/Blend Images LLC

and pesticides. Lead can leach into food from lead crystal glasses, some dishes, and the solder used to seal copper water pipes. Fish are the food source most likely to be contaminated with mercury, PCBs, and other pollutants that were dumped into waterways and accumulated in the fish living there. To minimize pregnant women's exposure to mercury, the FDA and Environmental Protection Agency (EPA) advise pregnant women to avoid swordfish, shark, king mackerel, and tilefish and keep their intake of other fish and shellfish to 12 ounces (no more than 6 ounces from albacore tuna) per week. The effect of pesticide residues on humans is largely unknown. However, everyone, including pregnant women, can minimize exposure to pesticides by thoroughly washing all fruits and vegetables.

### Foodborne Pathogens

The risk of foodborne illness increases during pregnancy. As noted in Chapter 3, pregnant women should avoid foods frequently found to be contaminated with pathogens, such as raw sprouts, unpasteurized milk and juices, and raw or undercooked meat, fish, shellfish, and eggs. Exposure to the bacterium that causes the foodborne illness listeriosis (*Listeria monocytogenes*) is especially dangerous during pregnancy; it can cause spontaneous abortion, premature delivery, stillbirth, and infections in the newborn. Contaminated soft cheeses, such as Mexican-style cheese, feta, Brie, Camembert, and blue-veined cheeses (e.g., Roquefort), are often the cause of listeriosis. Thus, pregnant women are advised to avoid these cheeses. In addition, they are advised to cook leftovers and processed meats (e.g., hot dogs, deli meats) until steaming hot.

Maternal exposure to the parasite that causes toxoplasmosis can lead to infant death or brain damage. This parasite is found in cat feces, bird feces, and the soil. Raw meat can be contaminated with this parasite. To avoid toxoplasmosis, pregnant women should avoid litter boxes, kittens, and birds; carefully wash all produce to remove any soil; and thoroughly cook all meat.

Pregnant women should moderate their caffeine intake during pregnancy. Ken Karp/McGraw-Hill Education

### Caffeine

Caffeine is a stimulant and diuretic found in coffee, tea, some soft drinks and energy drinks, and chocolate products. It also is a common additive in many medications, including headache and cold remedies. High caffeine intake (more than 500 mg daily) may hinder a woman's ability to become pregnant and increase the risk of having a spontaneous abortion or a low-birth-weight infant.[37] During pregnancy, caffeine can affect fetal heart rate and breathing and decrease blood flow through the placenta. In addition, the fetus is unable to detoxify caffeine. Caffeine also may decrease the absorption of certain nutrients, such as calcium, iron, and zinc.[3] Heavy caffeine use during pregnancy may lead to caffeine withdrawal symptoms in the newborn. However, the evidence that caffeine consumption during pregnancy exerts a lasting adverse effect on the fetus is limited and very unclear. Nevertheless, many experts agree that it is sensible to limit caffeine intake during pregnancy to around 200 mg daily,[38] which is about 2 cups of coffee or 3 cups of caffeinated soft drinks (see Appendix I).

### Food Additives

Food additives, such as phenylalanine (a component of the noncaloric sweetener aspartame in NutraSweet® and Equal®), may be a cause for concern for some pregnant women. High amounts of phenylalanine in maternal blood disrupt fetal brain development if the mother has the genetic disease phenylketonuria (see Chapter 9). If the mother does not have this condition, it is unlikely that the baby will be affected by moderate aspartame use.

## Maternal Lifestyle

▶ Other lifestyle choices unrelated to nutrition that can harm the developing fetus include

- Exposure to hot temperatures in hot tubs and saunas
- X-ray exposure, including dental X rays
- Job- and hobby-related hazards, such as chemicals and toxins used in manufacturing, hairdressing, and artwork
- Exposure to Zika-infected mosquitoes

Lifestyle choices can have an important impact on pregnancy outcome. Alcohol, drugs, herbal and botanical products, and smoking are lifestyle choices that can alter genes or the expression of genes and increase the risk of a poor pregnancy outcome. A woman should avoid substances that can harm the developing offspring, especially during the 1st trimester. This caution holds true, as well, for the time a woman is trying to become pregnant.

## Alcohol

More than half of all women in the U.S. of childbearing age consume alcohol. Alcohol intake can impair the ability to become pregnant. During pregnancy, alcohol may displace nutrient dense foods in the mother's diet. Alcohol also slows nutrient and oxygen delivery to the embryo or fetus, retarding growth and development. Although the most severe damage occurs during the embryo stage, a time with many critical periods, consuming alcohol at any time during pregnancy can cause damage to the embryo or fetus that lasts a lifetime.

Of every 1000 babies born in the U.S. each year, as many as 2 have fetal alcohol spectrum disorders (FASD). The rate is much higher in some areas of the world. In South Africa, for example, the rate is more than 60 per 1000 births.[39] Recall from Chapter 8 that children born to women who drank alcohol during pregnancy may have facial malformations, growth retardation, and central nervous system defects, including profound mental retardation and a small brain size. These children are among the smallest in height, weight, and head circumference for their age. Prenatal exposure to alcohol also may cause somewhat lesser effects, such as lifelong learning difficulties, short attention spans, poor coordination, and hyperactivity. Growing evidence indicates that alcohol alters the genes of both the mother and the developing offspring.

It is not clear how alcohol causes these malformations and disabilities. However, alcohol freely crosses the placenta. Within minutes of its consumption, alcohol is present in the amniotic fluid, where the alcohol's intensity is magnified by the small size of the embryo or fetus. In addition, the effects of the alcohol are prolonged because the developing offspring is unable to metabolize the alcohol. It must wait for maternal blood to slowly carry the alcohol away.

No one is sure how much alcohol it takes to cause developmental problems. However, consuming as little as 1 ounce per day has resulted in mental and physical defects. The more alcohol consumed during pregnancy, the worse the effects are likely to be. Until a safe level of alcohol consumption during pregnancy is known, experts recommend that pregnant women drink no alcohol. Because of the potentially deleterious effect on the embryo early in pregnancy, experts recommend that women planning a pregnancy avoid alcohol in case they do become pregnant and that women who drink alcohol avoid becoming pregnant.

## Drugs

Over-the-counter, prescription, and illegal drugs all have the potential to cause detrimental effects on a pregnant woman's nutrition status and pregnancy outcome. Common drugs that can cause problems during pregnancy include aspirin (especially when used heavily), hormone ointments, nose drops, cold medications, rectal suppositories, weight control pills, medications prescribed for previous illnesses, marijuana, and cocaine. They may deplete nutrient stores, alter nutrient absorption, and decrease the desire to eat. In addition, drugs may reduce blood flow to the fetus, which deprives it of oxygen and nutrients. Opioids and illegal drugs can impair placental and fetal growth and development and frequently result in preterm births and low-birth-weight infants, many of whom have birth defects and damage to their nervous systems. A prescription drug that deserves special attention is Accutane® (isotretenoin), a form of vitamin A commonly used to treat severe acne. Babies born to mothers who take high doses of vitamin A, whether it is from isotretenoin or supplements, may experience severe birth defects, including nervous system abnormalities and facial and cardiovascular deformities. Prior to using any drug, a pregnant woman should discuss its use with her health-care provider.

## Herbals and Botanicals

Herbal and botanical products can exert potent, druglike effects on both the mother and the fetus. Until the safety of these products can be verified, experts recommend that pregnant women use all herbal and botanical products—including herbal tea—with caution and under the guidance of a health-care professional.

## Smoking and Nicotine

Nicotine and carbon monoxide affect the fetus when a pregnant woman smokes or is exposed to secondhand cigarette or cigar smoke. Smoking restricts blood flow, reducing the amount of oxygen and nutrients, especially zinc, that reaches the fetus, which in turn impairs

▶ Alcohol is found in beer, wine, and hard liquor (distilled spirits), as well as in many mouthwashes and cough syrups.

▶ The head circumference of a child from birth until about age 3 years is an indicator of brain growth and development. The smaller the head in proportion to the body, the more likely the brain has suffered irreversible growth retardation.

▶ For more information about fetal alcohol spectrum disorders, visit the website www.cdc.gov/ncbddd/fasd.

Rocketclips, Inc./Shutterstock

## Take Action

### Healthy Diets for Pregnant Women

A college friend, Angie, tells you that she is newly pregnant. You are aware that she usually likes to eat the following foods for her meals.

**Breakfast**

Skips this meal or eats a granola bar
Coffee

**Lunch**

Sweetened yogurt, 1 cup
Small bagel with ¼ cup cream cheese
Occasional piece of fruit
Regular caffeinated soda, 12 oz

**Afternoon Snack**

Chocolate candy bar or 2 slices of bologna

**Dinner**

2 slices of pizza, fried shrimp dinner, or 2 eggs with 2 slices of toast
Seldom eats a salad or vegetable
Regular caffeinated soda, 12 oz

**Evening Snack**

Pretzels or chips, 1 oz
Regular caffeinated soda, 12 oz

1. Evaluate Angie's diet for protein, carbohydrate, iron, vitamin B-6, folate, zinc, calcium, and caffeine. How does her intake compare with the recommended amounts for pregnancy?

2. Now redesign Angie's diet and make sure that her intake meets pregnancy needs for protein, carbohydrate, iron, vitamin B-6, folate, zinc, and calcium. (Hint: Fortified foods, such as breakfast cereals, are usually nutrient rich foods that can help individuals meet their nutrient needs.)

bhofack2/123

▶ A goal of *Healthy People 2030* is abstinence from alcohol, cigarettes, and illicit drugs by pregnant women.

growth. Smokers have an increased risk of spontaneous abortion and placenta problems. Smokers' babies are more likely to be preterm, experience prenatal growth retardation, and die soon after birth than are nonsmokers' babies. They also have an increased risk of birth defects and sudden infant death. Contributing to the smaller birth size is the fact that smokers are likely to have a lower prepregnancy weight and to gain less weight during pregnancy. Plus, many smokers' diets are less nutritious than those of nonsmokers. Nicotine alone, delivered by cigarettes, electronic cigarettes, chewing tobacco, patches, or exposure to electronic cigarette aerosol, has multiple health consequences, such as fetal brain, lung, and development abnormalities. Thus, experts recommend minimizing exposure to nicotine during pregnancy.[40,41]

After giving birth, mothers should be encouraged to maintain the healthy lifestyle changes they made during pregnancy, such as increased intake of fruits and vegetables, smoking cessation, and exercising. After pregnancy, women need to eat a healthy diet to replenish their nutrient stores and gradually return to a healthy weight. Women who experienced gestational diabetes should have a blood glucose screening 6 weeks after giving birth and then annually.[1] Those with elevated blood glucose levels should receive appropriate treatment to preserve their health (see Chapter 5).

### Knowledge Check

1. What are the symptoms of pregnancy-induced hypertension?
2. Why is it important to keep blood glucose levels under control during pregnancy?
3. What precautions should pregnant women take to avoid foods that might contain environmental contaminants or foodborne disease pathogens?
4. What effects can lifestyle choices related to alcohol, drugs, herbal and botanical products, smoking, and tobacco have on pregnancy outcome?

# CLINICAL PERSPECTIVE

## Nutrition-Related Physiological Changes of Concern during Pregnancy

The intense physical and hormonal changes that affect almost every aspect of a pregnant woman's body begin soon after the egg is fertilized. As the pregnancy progresses, the physical changes become more pronounced, and the growing fetus places an increasing burden on the mother. Although the woman's body adapts well to this burden, she may experience some nutrition-related problems, such as heartburn, constipation, morning sickness, and edema. Usually, these difficulties are relatively minor and easy to remedy with diet and lifestyle changes. However, if the problems continue and prevent the woman from eating a nutritious diet, she should contact her health-care provider.

### Heartburn

During pregnancy, the expanding uterus crowds abdominal organs and compresses them (Fig. 16-8). Crowding the stomach reduces its ability to stretch enough to accommodate a normal-size meal. Consequently, stomach acid and partially digested food may be pushed upward out of the stomach into the esophagus, causing a burning sensation commonly called heartburn (see Chapter 4). Hormones (e.g., progesterone) that slow the speed of digestion

and relax sphincters in the gastrointestinal tract also can contribute to heartburn.

Women usually can ease heartburn by eating several small meals instead of a few large ones, avoiding spicy and fatty foods, and limiting caffeine and chocolate intake. Women also should consume liquids mostly between meals to decrease the volume of food in the stomach after meals and thus relieve some of the pressure that encourages reflux. To keep acid from backing up into the esophagus, women should wait a couple of hours after eating to lie down and should sleep with their heads elevated slightly. Antacids contain certain minerals that may cause constipation, excessive sodium intake, and other problems, so they should be used only on the recommendation of a health-care provider.

### Constipation

Hormones relax the intestinal muscles, which slows digestion and permits increased nutrient absorption. This slowdown also causes more water than usual to be reabsorbed, which leads to hard, dry stools. In addition, iron supplements may cause hard, dry stools. Such stools are difficult to excrete, especially late in pregnancy as the fetus compresses the GI tract.

Straining to excrete hard, dry stools may cause hemorrhoids (see Chapter 4).

Pregnant women often can avoid both constipation and hemorrhoids by consuming high fiber foods, drinking more fluids, and exercising. The Adequate Intake for fiber in pregnancy is 28 g/day, slightly more than for a nonpregnant woman. Fluid needs are 10 cups/day. Iron supplement dosage also may need adjustment. Stool softeners and laxatives may contain substances that can harm the fetus or cause dehydration; thus, they should be used only on the advice of a health-care provider.

### Nausea and Vomiting

During the 1st few months of pregnancy, about 70 to 85% of women experience frequent bouts of nausea and vomiting. This nausea may be related to the increased sense of smell induced by pregnancy-related hormones circulating in the bloodstream. This is commonly called morning sickness, although it can occur at any time of the day and persist all day. It is often the 1st signal to a woman that she is pregnant. Mild cases often can be treated by breathing cool, fresh air; avoiding offensive odors; avoiding large fluid intakes early in the morning; avoiding an empty stomach; and eating specific foods

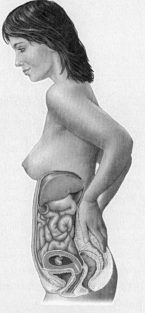
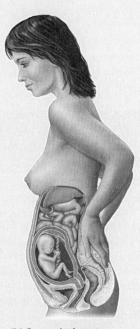
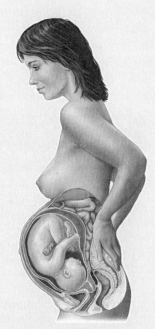

*Figure 16-8* The growing fetus crowds and compresses abdominal organs.

**(a) First trimester**      **(b) Second trimester**      **(c) Third trimester**

that ease symptoms. The iron in prenatal supplements may trigger nausea in some women; changing the type of supplement used or postponing iron supplements until the 2nd trimester may provide relief. If a woman thinks her prenatal supplement is related to morning sickness, she should talk to her physician about switching to another brand. Some women find that starchy, bland foods, such as dry toast, crackers, or cereal, relieve feelings of nausea better than high fat or high protein foods. Sweets, such as hard candy, popsicles, and carbonated beverages, help other women cope. Women need to learn from experience which foods help quell queasy feelings and have them readily available.[9,42,43]

Usually, nausea stops after the 1st trimester; however, in about 10 to 20% of cases, it continues throughout the pregnancy. Severe nausea and vomiting that continue beyond 14 weeks of gestation, known as hyperemesis gravidarium, is a serious condition that may require hospitalization. Left untreated, it can result in dehydration, malnutrition, and electrolyte or metabolic disturbances that adversely affect both maternal and fetal health.

### Edema

Placental hormones cause various body tissues to retain fluid during pregnancy. Blood volume also greatly expands during pregnancy. The extra fluid normally causes some swelling (edema). Plus, the enlarging uterus compresses blood vessels in the legs, which slows down blood circulation and the removal of waste

A few crackers on waking or between meals can help relieve the nausea of morning sickness.

mikeledray/Shutterstock

products (including water). As a result, edema in the lower legs, especially around the ankles, is expected to a certain degree in late pregnancy. Edema in the lower legs is of concern only when accompanied by high blood pressure, protein in the urine, or failure of the edema to subside when a woman elevates her feet. The presence of any of these factors may signal pregnancy-induced hypertension.

▶ To prevent and treat nausea and vomiting during pregnancy, the American College of Obstetricians and Gynecologists recommends that women begin taking supplements prior to conception and use vitamin B-6 alone (see Chapter 13) or with the antihistamine doxylamine. Ginger also may help relieve nausea.

### Critical | Thinking

Sandy, who is 6 months pregnant, has been having heartburn after meals, constipation, and difficult bowel movements. What remedies might you suggest to Sandy to relieve her problems?

### Knowledge Check

1. Why are heartburn and constipation common in pregnancy?
2. What are some practices that can help women cope with nausea during pregnancy?
3. Edema accompanied by which factors may signal pregnancy-induced hypertension?

## CASE STUDY
## FOLLOW-UP

PhotoDisc/Getty Images

From a dietary standpoint, Tracey is smart to take a close look at her protein intake because her needs will increase somewhat during pregnancy. More fruits and vegetables supply extra folate, and her use of an over-the-counter vitamin and mineral supplement helps ensure that she will have an ample amount. Still, she should schedule an appointment with her physician to discuss her plans to become pregnant and learn more about vitamin and mineral supplements. She would probably eventually benefit more from a prenatal supplement because it will have more iron and folic acid than over-the-counter multivitamin and mineral supplements. Her diet may not have enough calcium, so she should pay as much attention to consuming extra calcium as she does to consuming protein. Avoiding alcohol and tobacco is a smart move. Many experts would say that Tracey is consuming too much caffeine and would be wise to cut back on coffee and caffeine-containing soft drinks to a total of 3 or fewer servings per day. Swimming is an excellent choice for exercise, as long as it is not too vigorous. Brisk walking and stationary biking (spinning) are also good choices.

# 16.6 Lactation

Lactation (breastfeeding) is a natural physiological process of female mammals that occurs in the postpartum period when the mother's breast secretes milk and suckles the offspring. Preparation for lactation begins when a young girl enters puberty. Hormones, particularly estrogen, secreted during puberty stimulate the growth and development of the mammary glands. The most obvious change in the breasts is their size, which is due to the deposition of fat (Table 16-5). Less visible changes include the growth and development of the milk-producing/storage cells (lobules) and the network of ducts that connect the glands to the nipple (Fig. 16-9). Malnutrition during adolescence can impair breast development and limit a woman's future ability to provide adequate nourishment for a growing baby.

Once developed, the mammary glands are fairly inactive until early in pregnancy. At that time, hormones secreted by the placenta cause the milk-producing glands to mature and the lactiferous ducts to branch more.

Breastfeeding is the preferred way to feed an infant. JGI/Blend Images LLC

## Milk Production

The manufacture and secretion of **prolactin**, the principal hormone that promotes milk production, are stimulated by the birth of the baby and by suckling. When the baby suckles at the breast, nerve signals stimulate the pituitary gland in the brain, causing it to release prolactin. The prolactin travels in the blood to the milk-producing glands in the breast and stimulates them to synthesize milk. If suckling stops (or never occurred after delivery), prolactin is not released and milk secretion usually stops in a few days. The breasts gradually return to their prepregnancy state.

During the earliest days of life, sucking is strongest in the 1st hour or so after birth; thus, this is the best time to begin feeding. In the days right after delivery, the baby should suckle often (at least every 2 or 3 hours for 15 to 20 minutes on each breast) to promote the establishment of lactation. After a few weeks, nursing sessions can become less frequent. As long as the mother breastfeeds her baby regularly, milk production will continue and can theoretically go on for years.

Throughout lactation, the amount of milk produced closely parallels infant demand. That is, the more an infant suckles, the more milk that is produced. This is what makes it possible for a woman to adequately breastfeed even twins and triplets. To **wean** a child, it is best to gradually stop breastfeeding, usually by eliminating 1 daily feeding each week. Abruptly stopping causes the breast to become painfully engorged with milk for several days.

### Release of Milk from the Breast

An important brain-breast connection—commonly called the **let-down reflex**—is necessary for breastfeeding (Fig. 16-10). The pituitary gland releases the hormone **oxytocin**, which causes the musclelike cells in the breast tissues to contract and release (let down) the milk from the lobules. The milk then travels through the ducts to the nipple area. During the 1st few weeks after delivery, this reflex must be triggered by the infant suckling the mother's breasts. Once lactation is established, the let-down reflex becomes automatic. It can be triggered just by thinking about her infant or hearing a baby cry. It generally takes 2 to 3 weeks to fully establish a feeding routine in which both the infant and the mother feel comfortable,

**prolactin** Hormone secreted by a mother's pituitary gland that stimulates the synthesis of milk in her breast.

**wean** To accustom an infant to a diet containing foods other than just milk.

**let-down reflex** Reflex stimulated by infant suckling; causes the release (ejection) of milk from milk ducts in the mother's breasts; also called *milk ejection reflex*.

**oxytocin** Hormone secreted by a mother's pituitary gland that causes musclelike cells surrounding the ducts in the breasts and the smooth muscle in the uterus to contract.

### Table 16-5  Mammary Gland Development

*Puberty:* Ovarian hormones stimulate the development of the milk-producing cells (lobules) and ducts.

*During pregnancy:* Placental hormones stimulate lobules to mature and ducts to grow and branch.

*Following childbirth:* Pituitary hormones cause the breasts to begin producing milk (the hormone prolactin) and releasing milk (the hormone oxytocin). Suckling stimulates the pituitary to continue releasing prolactin. If the nipple is not suckled regularly, milk production ceases.

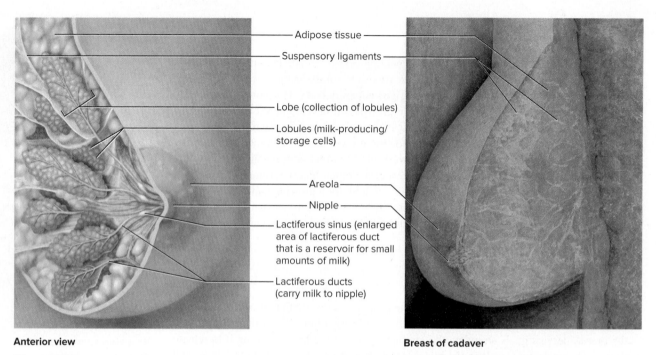

Anterior view                                                          Breast of cadaver

*Figure 16-9*  Breast anatomy. Many types of cells form a coordinated network to produce and secrete human milk.

From Anatomy & Physiology Revealed, McGraw-Hill Education/The University of Toledo, photography and dissection

the milk supply meets infant demand, and initial nipple soreness disappears. Establishing the breastfeeding routine requires patience, but the rewards are great.

The let-down reflex is easily inhibited by nervous tension, a lack of confidence, and fatigue. If the let-down reflex does not occur, the baby will be able to obtain only the small amount of milk that trickles out of the lobules between feedings. This will lead to a hungry, fussy baby, which likely will increase the mother's stressful, tense feelings—a vicious cycle. Women have a much better chance of successfully establishing lactation if they are in a relaxed, supportive environment and understand the process of lactation.[44] Careful monitoring during the 1st week of lactation by a physician or lactation consultant also helps ensure that breastfeeding and infant weight gain are proceeding normally.

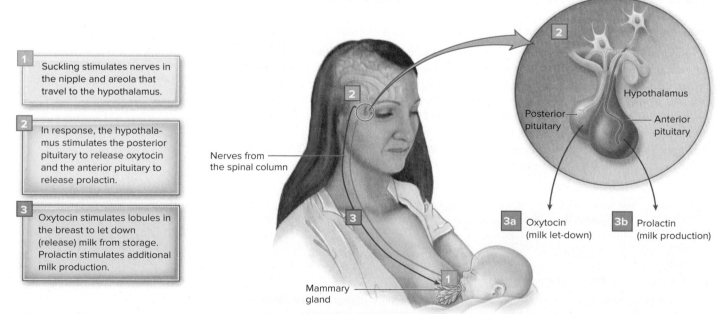

1  Suckling stimulates nerves in the nipple and areola that travel to the hypothalamus.

2  In response, the hypothalamus stimulates the posterior pituitary to release oxytocin and the anterior pituitary to release prolactin.

3  Oxytocin stimulates lobules in the breast to let down (release) milk from storage. Prolactin stimulates additional milk production.

*Figure 16-10*  Suckling sets in motion the sequence of events that triggers milk production and the let-down reflex. This reflex releases milk from the milk-producing lobules into the ducts that carry the milk to the nipple.

## Milk Types and Composition

Colostrum, transitional milk, and mature milk are the successive types of milk produced while lactation is being established. **Colostrum**—thin, yellowish, "immature" milk—may appear anytime from late pregnancy to several days postpartum. This early milk is richer in protein, minerals, and vitamin A than later milk, but it has less carbohydrate (lactose) and slightly fewer calories. It also contains antibodies and immune system cells, some of which pass unaltered through the infant's immature GI tract into the bloodstream and provide the infant with a defense against some diseases during the 1st few months of life when the immune system is very immature.[44] (The early few months of life are the only time when we can readily absorb whole proteins across the GI tract.) One component of colostrum, the *Lactobacillus bifidus* factor, encourages the growth of *Lactobacillus bifidus* bacteria, which limits the growth of potentially toxic bacteria in the infant's large intestine. In addition, colostrum contains a potent laxative, which helps the infant excrete the fecal matter that collected in the intestinal tract during fetal life.

The colostrum gradually changes in the early days after delivery to become transitional milk. Transitional milk contains more fat, lactose, water-soluble vitamins, and calories than colostrum. Within a week or so, transitional milk is replaced by mature milk. Mature human milk is thin and almost watery in appearance and often has a slightly bluish tinge. It provides about 20 calories per ounce and, with the possible exceptions of vitamin D and iron,[44] supplies all the nutrients the growing baby needs (see Chapter 17).

Before the let-down reflex becomes automatic, the infant must suck hard to cause the milk to be released. This vigorous suckling may make nipples sore. Health-care providers can help pregnant women prepare their nipples to minimize soreness. Most women find that nipple soreness is not a problem after about 2 or 3 weeks of breastfeeding. Westend61/Getty Images

### Knowledge Check

1. What is the role of prolactin?
2. How is oxytocin involved in the let-down reflex?
3. How does colostrum differ from mature milk?

# 16.7 Nutrient Needs of Breastfeeding Women

The nutrient and calorie demands of lactation on the mother are tremendous—in some cases, exceeding those of pregnancy. The quantities of nutrients needed during lactation are not so surprising when you consider that the milk secreted by a lactating woman can fulfill her infant's entire nutrient and calorie needs until the infant is approximately 6 months old—the time when the infant's growth rate and nutrient needs per pound of body weight are at a lifetime high.

The lactating woman's Dietary Reference Intakes (DRIs) are based on the quantity of milk produced by the average mother, its nutrient content, and the amounts of her nutrient stores used to produce milk. The RDA value for iron is lower than that of both a pregnant and a nonpregnant woman because human milk contains only a small amount of iron and because women who exclusively breastfeed their infants often do not resume menstruating for 6 months or so, thereby conserving iron. However, when a lactating woman begins menstruating again, her iron RDA returns to that of nonpregnant women.

## Maternal Nutritional Status

The "recipe" for human milk is set by nature. The "ingredients" (nutrients) are drawn from the mother's diet, with deficits of some nutrients being made up by the mother's stores. These 2 sources of nutrients keep the composition and volume of the milk at fairly constant levels. If the mother's diet is poor for a prolonged period and she has depleted nutrient stores, the quantity of milk may decline. Maternal malnutrition must be extremely severe before the quality of milk produced drops.

At the other end of the spectrum lie maternal diets that contain nutrients in amounts exceeding recommended intakes. In the case of macronutrients and water, excess amounts usually have no effect on milk composition or volume, although the proportions of different fatty acids in the milk vary with maternal intake.[45] If vitamin and mineral intakes exceed the recommendations, increased levels of these nutrients may appear in the milk.

An adequate intake of all nutrients by the mother is vital during lactation. Many lactating women consume calcium, magnesium, zinc, folate, and vitamin B-6 in less than recommended amounts. Adequate calcium intake is especially important because women who breastfeed their babies for 6 months or more lose significant amounts of calcium from their bones. Fortunately, with adequate calcium intake, bone density begins to return to normal within a few months after the baby is weaned. Adequate water and calorie intakes are particularly important.

*Water* is the main component of human milk, like other milk. Inadequate water intake can alter milk composition, decrease the amount of milk produced, and lead to maternal dehydration. To keep herself well hydrated, a breastfeeding woman should drink to satisfy her thirst. Most need an extra 32 ounces of fluid daily, in addition to the 72 ounces recommended for nonpregnant women. Some women, especially those breastfeeding more than 1 baby, need to consume even greater quantities.

*Calorie* needs can be met by eating the same number of servings from each food group as recommended for a pregnant woman in the last 2 trimesters (see Table 16-2). The average breastfeeding woman uses about 800 calories per day during the 1st 6 months of lactation to produce 750 ml (3 cups) of milk daily. Approximately 400 to 500 calories should come from her diet, with the remainder supplied by the fat stored during pregnancy.[9] Relying on stored fat to meet part of her calorie need helps the lactating woman gradually lose the extra body fat accumulated during pregnancy, especially if breastfeeding is continued for 6 months or more and the woman performs some physical activity. Overweight women appear to be able to rely on stored fat to meet their entire daily calorie need for lactation without adversely affecting the growth of their infants. However, severe calorie restrictions and weight loss greater than 1 to 4 pounds per month likely will reduce the total amount of milk produced.

After 6 months of breastfeeding, all these extra calories likely will need to come from the woman's diet if she has returned to her prepregnancy weight. Inadequate calorie intakes will cause her body weight to continue to drop and, if BMI is substantially below 18.5, she may become unable to produce enough milk for her growing child.

## Food Choices during Lactation

Lactating women can enjoy all foods in moderation. Although single food items that a mother eats have very little bearing on her milk's nutritional quality or the amount she manufactures, many cultures believe that certain foods, such as garlic, oatmeal, or ginger, boost the amount of milk produced. There is no scientific evidence that any food in particular promotes milk production or alters its nutrient composition but, as discussed later in this chapter, there is some very good evidence that certain substances in food may affect the infant.

As in pregnancy, a serving of a fortified, ready-to-eat breakfast cereal or a multivitamin and mineral supplement can help meet extra nutrient needs. Like pregnant women, breastfeeding mothers should consume sufficient amounts of omega-3 fatty acids because they are secreted into breast milk and are important for the development of the infant's nervous system. Fish and supplements can provide these fatty acids. Fish intake recommendations during pregnancy also apply during breastfeeding (see Chapter 3). Experts do not recommend that breastfeeding women restrict their diets as a strategy for preventing food allergies in children.[46]

Water needs rise during lactation.
Ingram Publishing

## Knowledge Check

1. What effect does maternal nutritional status have on the quality and quantity of the breast milk a woman produces?
2. What are the calorie intake recommendations for lactating women?
3. What steps can lactating women take to ensure they consume an adequate amount of nutrients?
4. What is the basic food plan suitable for breastfeeding women?

# 16.8 Factors Affecting Lactation

Breastfeeding offers nutritional, immunological, and psychological benefits to the infant.[47] Breastfeeding mothers, too, may gain health benefits, including a lower risk of ovarian and premenopausal breast cancer, bone remineralization to levels exceeding those before lactation, weight loss, quicker return of the uterus to its prepregnancy state, and less postpartum bleeding (Table 16-6). Although breastfeeding confers many advantages to both the mother and the child, several factors may affect milk quality and/or safety: maternal weight, age, and eating patterns; maternal and infant health; sociocultural factors; and maternal food supply and lifestyle choices.

## Maternal Weight

Women who were obese prior to pregnancy often have greater difficulty initiating and continuing breastfeeding and producing sufficient milk.[21] Obstacles these women may face include difficulty getting the infant to latch on adequately to their breasts and positioning the infant appropriately for breastfeeding. Women with obesity may have a low milk supply because they release less prolactin and are more likely to have had Cesarean deliveries, which delay the first suckling. These women may need to supplement human milk with infant formula until they develop a sufficient milk supply. Weight loss prior to pregnancy may improve the breastfeeding success rates of women with obesity.

## Maternal Age

Infants breastfed by adolescent mothers may grow more slowly than infants of older mothers. Teens can successfully breastfeed their offspring, but they need to work with their health-care providers to ensure that their own as well as their infant's nutrient needs are adequately met.

## Maternal Eating Patterns

A nutritious diet is the best choice for both the breastfeeding mother and her infant. However, an occasional day of poor intake is not a cause for concern—nutrient and/or calorie inadequacies usually can be made up using the mother's nutrient stores. It is only when a mother's diet is poor for several weeks or months that maternal nutrient stores may become depleted and her milk supply negatively affected.

## Maternal and Infant Health

Breastfeeding may be ruled out by certain medical conditions in either the infant or the mother. For example, breastfeeding may be detrimental to infants with the inborn error of metabolism phenylketonuria or galactosemia (see Chapter 9). The high concentration of phenylalanine and galactose in human milk may overwhelm the impaired ability of these infants to metabolize these nutrients, leading to serious illness and even death.

Infectious or chronic diseases and their compatibility with breastfeeding need to be carefully considered by the mother and her health-care provider because medications and infectious agents can be transmitted to the baby via the mother's milk. A minor disease, such as a cold, is not a reason to stop breastfeeding. The baby is likely to catch the disease, anyway, because he or she was exposed before the mother's

Eating fish twice a week will help breastfeeding women ensure that their infants receive important omega-3 fatty acids. It is important, however, to avoid fish that are likely contaminated with mercury (see Chapter 3). David Papazian/Getty Images

**Table 16-6 Advantages of Breastfeeding for Mothers***

- Earlier recovery from pregnancy due to the action of hormones that promote a quicker return of the uterus to its prepregnancy state
- Decreased risk of ovarian and premenopausal breast cancer
- Potential for quicker return to prepregnancy weight
- Potential for delayed ovulation and therefore reduced chances of pregnancy (a short-term benefit, however)
- Potential bone remineralization to levels exceeding those before lactation
- Less postpartum bleeding
- Reduced risk of type 2 diabetes among women with no history of gestational diabetes,[48] delayed onset of type 2 diabetes among women who had gestational diabetes.
- Reduced risk of Metabolic Syndrome later in life[49]
- Reduced risk of rheumatoid arthritis if the mother breastfeeds for more than 12 months during her lifetime
- Decreased feelings of depression in postpartum period after childbirth[46]
- Lower household food costs (extra food for a breastfeeding mother is less expensive than infant formula)

*See Table 17-1 (Chapter 17) for a summary of advantages for infants provided by human milk.

symptoms were apparent. Also, the immune factors in human milk may provide some protection for the breastfed baby against the minor disease. However, serious infectious diseases, such as tuberculosis and hepatitis C, can be life threatening, so it may be safer for women with such diseases to feed their babies infant formula.

Medical advances have made it possible for women with several serious chronic diseases, such as diabetes, multiple sclerosis, lupus, phenylketonuria, and cystic fibrosis, to safely breastfeed their infants. Chronic diseases that are incompatible with breastfeeding are cancer being treated with chemotherapy medications and psychiatric conditions treated with lithium. Human immunodeficiency virus (HIV) can be transmitted to the baby via human milk; however, transmission of HIV to infants is greatly reduced when HIV-exposed infants or HIV-infected mothers receive antiretroviral (ARV) interventions. The U.S. Centers for Disease Control and Prevention advises against breastfeeding by women who have HIV or are taking ARVs.[50] However, in regions of the world where infectious disease and malnutrition are the primary causes of infant death, the risk of not breastfeeding may outweigh the risk of possible transmission of HIV infection.[51] Widespread lack of basic sanitation (e.g., lack of clean water and soap) may make it more dangerous to feed the baby infant formula because formula prepared with polluted water and placed in a dirty bottle frequently causes diarrhea—the number 1 killer of infants in low-income countries (see Chapter 4).

Breast surgery may affect a woman's ability to produce milk and/or secrete it. Many women who have had breast implants are able to breastfeed. Women who have had breast reduction surgery may not be able to breastfeed if the milk-producing glands have been removed or their connection to the lactiferous ducts has been disrupted.

Planning ahead makes it possible for women to continue breastfeeding a baby after returning to work. Many state laws, as well as the federal Fairness for Breastfeeding Mothers Act and Patient Protection & Affordable Care Act in the U.S., support breastfeeding by requiring most employers to provide mothers with time and space to pump breast milk at work. runzelkorn/Shutterstock

## Sociocultural Factors

Overall, breastfeeding is a learned skill, and mothers need knowledge to breastfeed safely, especially with their 1st child. Women are more likely to decide to breastfeed and continue to do so if they know the advantages of breastfeeding, how to breastfeed their babies, what problems to expect, and how to cope with those problems; receive support from their partners; deliver their babies in a hospital that supports breastfeeding; and have health-care providers and experienced friends who are knowledgeable and able to lend needed support. Lack of information, low self-confidence, a lack of role models, and/or an inadequate support system may limit breastfeeding success.

The widespread increase in availability of lactation consultants over the past several years has helped increase access to accurate information, role models, and social support. In almost every community, a group called La Leche League offers classes in breastfeeding and advises women who have difficulties with it.

Misinformation or a lack of information deters many women from attempting breastfeeding and causes many others to abandon it after only a few weeks. The following are some points that are important to know about breastfeeding.

- Practically all women are physically able to breastfeed their children.
- Women with anatomical problems in the breast (e.g., flat or inverted nipples) usually can have those problems corrected prior to pregnancy and subsequently breastfeed successfully.
- Women with small breasts can produce all the milk their babies need; there is no relationship between breast size and quantity of milk produced.
- Even after returning to work or school, many women can continue to breastfeed. They may alternate breastfeeding with bottle-feeding by other caregivers, express breast milk into a bottle that can be given by others, or breastfeed during breaks if the baby is nearby. Mothers can use a breast pump to express milk into a sterile bottle and then store it in a refrigerator or freezer. A schedule of expressing milk and using supplemental formula feedings is most successful if begun after 1 to 2 months of exclusive breastfeeding.

- Many modest women find that breastfeeding in public can be done discreetly with the breast covered by a loose-fitting blouse or baby blanket. To the casual observer, it appears that the woman is simply holding her infant closely.
- In the U.S., no state or territory has a law prohibiting breastfeeding. However, indecent exposure (including the exposure of women's breasts) has long been a common law or statutory offense. Starting in the 1990s, nearly all states clarified the right to breastfeed and decriminalized public breastfeeding.
- A breastfeeding mother can tell whether her baby is getting enough milk if the baby has 6 or more wet diapers each day and shows normal growth.
- Breastfeeding women can become pregnant. It's true that, when babies are exclusively breastfed, ovulation is delayed and the chances of becoming pregnant during the 6-month period after a birth are low. However, there is no guarantee that a lactating woman won't ovulate and get pregnant, so a reliable birth control method should be used. Oral contraceptives may reduce the protein content and total quantity of breast milk.
- In many cases, mothers can breastfeed a premature and/or low-birth-weight infant. If the infant is unable to suckle, the mother can express milk and feed it to the infant. This type of feeding demands great maternal dedication; however, evidence indicates that human milk aids mental development in premature infants.[52] The fortification of human milk with certain nutrients (calcium, phosphorus, sodium, and protein) is often necessary to match the preterm infant's rapid growth.

▶ For more information on breastfeeding, visit the following websites.

www.llli.org

wicworks.fns.usda.gov

▶ A proposed goal of *Healthy People 2030* is to increase the proportion of infants who are breastfed exclusively through age 6 months.

▶ Partners shouldn't be left out of breastfeeding educational opportunities. This is because the single most important influence on a woman's decision to breastfeed or bottle-feed her baby is her partner's attitude.

## Maternal Food Supply

Compounds in the food the mother eats can be secreted into her milk. Environmental contaminants, such as pollutants and pesticides, can appear in human milk. The risks of these substances are largely unknown. However, the benefits of human milk are well established, and the effects of environmental contaminants on the breastfed infant have been seen mostly when the mother herself was suffering from the effects of the contaminants. A few measures a woman can take to counteract some known contaminants are to avoid freshwater fish from polluted waters, to carefully wash and peel fruits and vegetables, and to remove the fatty parts of meat because pesticides concentrate in fat. In addition, a woman should not try to lose weight rapidly while nursing (no more than ¾ to 1 lb/week) because contaminants stored in her fat tissue might then enter her bloodstream and affect her milk. If a woman questions whether her milk is safe, especially if she has lived in an area known to have a high concentration of toxic wastes or environmental pollutants, she should consult her local health department.

Caffeine may make infants fussy and unable to sleep. Lactating women can prevent these adverse effects by avoiding caffeine-containing foods or medications or limiting intake to 1 or 2 cups of coffee, tea, or cola daily.

Certain foods and compounds in foods (e.g., cabbage, chocolate) may cause some infants to become upset or not want to eat. Some foods can impart flavors to milk that may affect an infant's desire to eat. For instance, when milk smells of garlic, babies spend more time suckling. When it smells of alcohol, they suckle less. If a woman notices a connection between her diet and the baby's fussiness or unwillingness to eat, she should avoid the offending food. It's a good idea to experiment to see if eating the food causes feeding problems again because the irritability or unwillingness may not be related to the mother's diet at all.

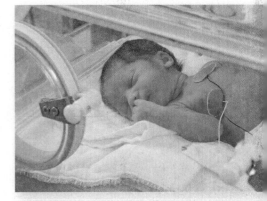

Human milk fed to preterm infants often must be fortified with certain nutrients to match their unique nutrient needs. Larry Mulvehill/Corbis

Tetra Images/Getty Images

## Maternal Lifestyle Choices

Substances related to lifestyle choices (e.g., alcohol, drugs, herbal and botanical products, nicotine) also are secreted into breast milk. The amount secreted is usually small (1 to 2% of the mother's dose) but may be a potent and harmful dose for a baby.

Alcohol, especially beer or wine, was recommended for centuries as a way for new mothers to relax and allow babies to nurse longer. However, research shows that consuming alcohol actually reduces milk output and causes babies to drink less and have disrupted sleep patterns. And infants are not as efficient as adults at breaking down alcohol, so the effects

## Investigating Breastfeeding

Margarite is 8 months pregnant and considering breastfeeding her baby. She doesn't know anyone who has breastfed and isn't sure that it is the best choice.

1. What additional information would you provide Margarite to help her make the most informed decision? (Note that the next chapter explores the benefits of breast milk for infants.)

2. Where can Margarite get additional accurate information to help her decide?

3. If Margarite decides to breastfeed, what can she do now to improve her chances of success?

Oh Baby Productions/Alamy Stock Photo

linger longer. The safest route for mothers and babies is to avoid alcohol altogether. However, lactating women who want to drink alcohol (beer, wine, or liquor) are advised to limit their intake and wait 3 to 4 hours before nursing again. The amount of alcohol in breast milk peaks about 30 to 60 minutes after the mother ingests it, then declines.

Medication use should be carefully considered by the lactating woman and her health-care provider. If a drug is necessary, she and the health-care provider need to decide whether the drug and lactation are compatible. If so, they can work out a plan for taking the drug and observing the infant's response. If the medication will be taken for a short time, the woman can express her milk by hand or with a breast pump and discard it. The baby can be fed formula until the mother is finished taking the medication. All illegal drugs should be avoided during lactation. The most commonly used illegal drugs, marijuana and cocaine, depress milk production and pass easily into breast milk. Marijuana in human milk slows infant development, and cocaine in milk causes babies to vomit and have tremors, breathing difficulties, and convulsions. Women addicted to drugs should not breastfeed their infants.

Many herbal and botanical products contain druglike compounds, which may appear in milk. All herbal and botanical products should be used cautiously during lactation.

Smoking also affects breastfeeding success and safety. In comparison with nonsmokers, women who smoke or breathe secondhand smoke produce less milk, and their babies gain less weight. Mothers who use cigarettes, electronic cigarettes, or nicotine patches also secrete nicotine in their milk, which can cause vomiting, slow breathing, increased blood pressure, and apathy in the infant. Lactating women, like pregnant women, are advised not to smoke. However, those who are smokers are advised to breastfeed because the benefits of breast milk outweigh the risk of nicotine exposure.

Breastfeeding may not be the optimal choice for some women. Alternatives to breastfeeding, along with the nutritional needs of infants, children, and adolescents are explored in Chapter 17.

### Knowledge Check

1. How do maternal prepregnancy weight and age affect breastfeeding success?
2. What diseases are incompatible with breastfeeding?
3. What steps can pregnant and lactating women take to overcome common sociocultural factors that affect breastfeeding success?
4. What effects can lifestyle choices related to alcohol, drugs, herbal and botanical products, cigarette, and nicotine use have on breastfed infants?

# Chapter *Summary*

## 16.1 A favorable pregnancy outcome lasts the full-term gestation period; results in a live,

healthy infant weighing more than 5.5 pounds; and permits the mother to return to her prepregnancy health status. Low-birth-weight babies weigh less than 5.5 pounds and have a high infant mortality rate. Low birth weight often is associated with premature birth. Full-term and preterm infants who weigh less than the expected weight for their duration of gestation are described as small for gestational age. A critical period is the specific time during embryonic or fetal life when cells develop into a particular tissue or organ. Nutrient deficiencies or excesses can interfere with normal development and cause physical or mental abnormalities. The zygote nourishes itself by absorbing secretions from glands in the uterus and digesting some of the uterine lining. Then, the placenta takes over the role of delivering nourishment to the developing offspring. The placenta contains both maternal and fetal blood vessels, which are so close together that nutrients and oxygen pass easily from the mother to the fetus and fetal wastes are shuttled from the fetus to the mother for excretion.

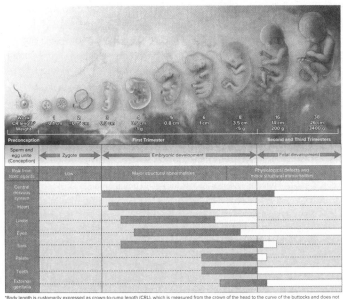

*Body length is customarily expressed as crown-to-rump length (CRL), which is measured from the crown of the head to the curve of the buttocks and does not include the lower limbs. Recall that 2.54 cm = 1 in.

## 16.2 Pregnancy is a nutritionally demanding stage of the life cycle.

A pregnant woman needs additional calories (about 350 in the 2nd trimester and 450 in the 3rd trimester) and more of almost every nutrient than a nonpregnant woman to support the growth and development of the fetus, placenta, and mother's body. The roles of protein, essential fatty acids, zinc, folate, vitamin B-12, iron, calcium, and vitamin D are particularly noteworthy during pregnancy. Pica, the eating of nonfood substances, can have dire consequences for a pregnant woman and her fetus.

## 16.3 A pregnant woman's meal plan should include nutrient dense foods from every food group.

MyPlate can help pregnant women plan nutrient dense diets. Special vitamin and mineral supplements formulated for pregnancy are prescribed routinely for pregnant women by most physicians. A low- or moderate-intensity exercise program offers physical and psychological benefits to a woman experiencing a normal, healthy pregnancy.

## 16.4 An infant's birth weight is closely related to the length of gestation,

the mother's prepregnancy weight, and the amount of weight she gains during pregnancy. Starting pregnancy at a healthy weight and following weight gain recommendations increase the likelihood of a successful pregnancy outcome. A weight gain of 25 to 35 lb is recommended for most women.

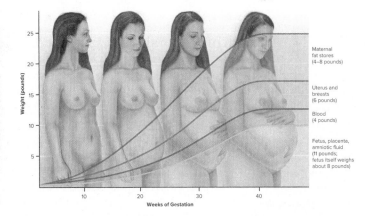

## 16.5 Numerous nutrition-related factors—including maternal age,

eating patterns, and health; sociocultural factors; the safety of the food supply; and maternal lifestyle choices—affect pregnancy outcome. Teenage pregnancy poses special health challenges because young women are not physically mature, have high nutrient needs, frequently do not consume a healthy diet, are underweight at the beginning of pregnancy and gain too little weight during pregnancy, and often do not receive adequate prenatal care. Pregnant women with eating patterns that deviate greatly from the recommended diet or who have eating disorders or other nutrition-related health conditions risk getting too few nutrients. Women with conditions that affect their health, such as numerous and/or closely spaced pregnancies, multiple fetuses, HIV or AIDS, or diabetes mellitus, may not consume an adequate diet and have less than optimal nutrient stores. Inadequate prenatal care and untreated diabetes or pregnancy-induced hypertension can lead to poor pregnancy outcomes. Those with limited income or educational achievement and those who lack social support networks tend to have inadequate diets. Environmental contaminants, foodborne illness pathogens, and certain compounds in food (e.g., caffeine) should be limited during pregnancy. Alcohol, drugs, herbal and botanical products, cigarettes, and nicotine intake are lifestyle choices that increase the risk of a poor pregnancy outcome.

# 16.6 Lactation (breastfeeding) is a natural physiological process that occurs in the postpartum period,

when the mother's breast secretes milk and suckles the offspring. Prolactin is the principal hormone that promotes milk production. The let-down reflex occurs when oxytocin causes milk to be released from the milk-producing lobules. The let-down reflex is easily inhibited by nervous tension, a lack of confidence, and fatigue. Colostrum is a thin, yellowish, "immature" milk that appears anytime from late pregnancy to several days postpartum. Mature human milk is thin and almost watery in appearance; with the possible exceptions of vitamin D and iron, it supplies all the nutrients the growing baby needs.

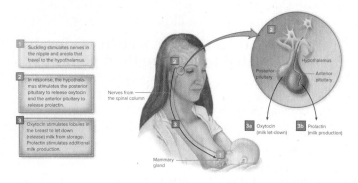

# 16.8 Maternal factors that can negatively affect breastfeeding

**success** include high prepregnancy weight, young age, long-term poor nutrient intake, certain diseases, lack of information and role models, and inadequate support system. Compounds in the food the mother eats, including contaminants, caffeine, alcohol, drugs, and nicotine, can be secreted into her milk.

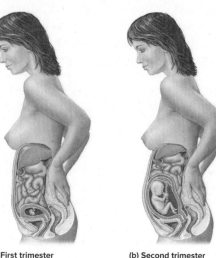

**(a) First trimester**     **(b) Second trimester**     **(c) Third trimester**

# 16.7 A lactating woman's DRI is based on the quantity of milk produced by the average mother,

its nutrient content, and the amounts of her nutrient stores used to produce milk. If the mother's diet is poor for a prolonged period and she has depleted nutrient stores, the quantity of milk produced may decline. Maternal malnutrition must be extremely severe before the quality of milk produced drops. The average breastfeeding woman uses about 800 calories per day during the 1st 6 months of lactation to produce 750 ml (3 cups) of milk daily. Approximately 400 to 500 calories should come from her diet, with the remainder supplied by the fat stored during pregnancy. Breastfeeding offers benefits to the infant and mother.

## Clinical Perspective

The physical and hormonal changes that affect almost every aspect of a woman's body begin soon after the egg is fertilized and may cause nutrition-related problems, such as heartburn, constipation, morning sickness, and edema. The growing fetus crowds and compresses abdominal organs, which contributes to heartburn and constipation.

## Global Perspective

Worldwide, thousands of women die each year from pregnancy and childbirth complications. Their fetuses also face major health problems and risk of premature death. Most of these deaths are in low-income countries. Undernutrition and closely spaced pregnancies are major contributors to these deaths.

| Maternal Characteristics Affecting Pregnancy Outcome | |
|---|---|
| • Age | • Eating patterns |
| • Health | • Sociocultural factors |
| • Safety of food supply | • Lifestyle choices |

**Table 16-6  Advantages of Breastfeeding for Mothers***

- Earlier recovery from pregnancy due to the action of hormones that promote a quicker return of the uterus to its prepregnancy state
- Decreased risk of ovarian and premenopausal breast cancer
- Potential for quicker return to prepregnancy weight
- Potential for delayed ovulation and therefore reduced chances of pregnancy (a short-term benefit, however)
- Potential bone remineralization to levels exceeding those before lactation
- Less postpartum bleeding
- Reduced risk of type 2 diabetes among women with no history of gestational diabetes[39]
- Reduced risk of Metabolic Syndrome later in life[41]
- Reduced risk of rheumatoid arthritis if the mother breastfeeds for more than 12 months during her lifetime
- Decreased feelings of depression in postpartum period after childbirth[39]
- Lower household food costs (extra food for a breastfeeding mother is less expensive than infant formula)

*See Table 17-1 (Chapter 17) for a summary of advantages for infants provided by human milk.

# Study Questions

1. Which of the following is true of a favorable pregnancy outcome?

   a. lasts longer than 37 weeks
   b. results in an infant weighing more than 5.5 pounds
   c. permits the mother to return to her prepregnancy health status
   d. all of the above

2. Iron needs rise significantly during pregnancy because _____.

   a. the fetus is building iron stores
   b. the number of red blood cells increases in the mother
   c. the fetus breaks down red blood cells rapidly
   d. all of the above
   e. both a and b

3. A woman who begins pregnancy at a healthy weight should gain _____.

   a. 15 pounds or less
   b. 25 to 35 pounds
   c. 28 to 40 pounds
   d. none of the above

4. If development does not occur during a critical period, the embryo can make up for this development later when more nutrients are available.

   a. true
   b. false

5. The placenta delivers nutrients and oxygen to the fetus.

   a. true
   b. false

6. Compared with nonpregnant women, during the 3rd trimester pregnant women need to increase calorie intake by about _____.

   a. 200 kcal
   b. 250 kcal
   c. 450 kcal
   d. 800 kcal

7. Prescription prenatal supplements often contain higher levels of which nutrient than those that can be purchased over the counter?

   a. vitamin A
   b. vitamin B-6
   c. calcium
   d. folic acid

8. Which of the following is *not* likely to increase the risk of poor pregnancy outcome?

   a. weight gain of less than 15 pounds
   b. pregnancy within 12 months of a previous pregnancy
   c. maternal age of 20 to 25 years
   d. limited educational achievement
   e. both c and d

9. Hormonal changes during pregnancy may cause _____.

   a. heartburn
   b. diarrhea
   c. edema
   d. both a and c

10. To cope with morning sickness, many women find relief by drinking a large glass of water on awakening in the morning.

    a. true
    b. false

11. Which of the following hormones promotes milk production?

    a. oxytocin
    b. prolactin
    c. estrogen
    d. placental hormone

12. The let-down reflex is triggered by _____.

    a. a baby suckling at the breast
    b. a baby crying
    c. thoughts about a baby
    d. all of the above

13. If a lactating woman's vitamin and mineral intakes exceed the RDA, increased levels of these nutrients may appear in the milk.

    a. true
    b. false

14. The thin, yellowish milk secreted in the few days right after birth is rich in antibodies and immune system cells.

    a. true
    b. false

15. Which of the following compounds can be secreted into human milk?

    a. alcohol
    b. food flavors
    c. drugs
    d. caffeine
    e. all of the above

16. Describe the role of the placenta. What factors affect its development?

17. How are fetal growth and development affected when calorie intake is insufficient?

18. Why is following MyPlate advocated to meet the increased nutrient needs of pregnancy?

**19.** What is the recommended weight gain for pregnancy? What is the basis for these recommendations?

**20.** Discuss how maternal age, eating patterns, health, and lifestyle choices affect pregnancy outcome.

**21.** Describe the physiological mechanisms that stimulate milk production and release. How does knowing about these mechanisms help mothers breastfeed successfully?

**22.** How should the basic food plan for pregnancy be modified during lactation?

**23.** Provide 5 tips for mothers who want to maximize their chances of breastfeeding successfully. Why did you choose these factors?

**24.** Discuss the implications of maternal malnutrition on fetal growth and development.

**25.** Explain why heartburn, constipation, nausea and vomiting, and edema are common during pregnancy, and describe how pregnant women can safely manage or avoid these conditions.

Answer Key: 1-d; 2-e; 3-b; 4-b; 5-a; 6-c; 7-d; 8-c; 9-d; 10-b; 11-b; 12-d; 13-a; 14-a; 15-e; 16-refer to Section 16.1; 17-refer to Section 16.2; 18-refer to Section 16.3; 19-refer to Section 16.4; 20-refer to Section 16.5; 21-refer to Section 16.6; 22-refer to Section 16.7; 23-refer to Section 16.8; 24-refer to Global Perspective; 25-refer to Clinical Perspective

# References

1. Procter SB, Campbell CG. Position of the Academy of Nutrition and Dietetics. Nutrition and lifestyle for a healthy pregnancy outcome. *J Acad Nutr Diet.* 2014;114:1099.

2. Procter SB, Campbell CG. Practice paper of the Academy of Nutrition and Dietetics abstract: Nutrition and lifestyle for a healthy pregnancy outcome. *J Acad Nutr Diet.* 2014;114:1099.

3. Institute of Medicine, Food and Nutrition Board. *Nutrition during pregnancy.* Washington, DC: National Academies Press; 1990.

4. Berti C and others. Pregnancy and infants' outcome: Nutritional and metabolic implications. *Crit Rev Food Sci Nutr.* 2016;56:82.

5. American College of Obstetricians and Gynecologics Committee on Obstetric Practice, Society for Maternal-Fetal Medicine. Definition of term pregnancy. *Obstet Gyn.* 2018;122:1139.

6. Prentice Hyatt MA and others. Suboptimal maternal nutrition, during early fetal liver development, promotes lipid accumulation in the liver of obese offspring. *Reproduc.* 2011;141:119.

7. Lee HS. Impact of maternal diet on the epigenome during in utero life and the developmental programming of diseases in childhood and adulthood. *Nutrients.* 2015;7:9492.

8. BMA Board of Science. *Growing up in the UK: Ensuring a healthy future for our children.* London: British Medical Association; 2013.

9. Herrell HE. Nausea and vomiting of pregnancy. *Am Family Physician.* 2014;89:965.

10. Institute of Medicine, Food and Nutrition Board. *Dietary Reference Intakes for energy, carbohydrate, fiber, fat, fatty acids, cholesterol, protein, and amino acids.* Washington, DC: National Academies Press; 2002.

11. Institute of Medicine, Food and Nutrition Board. *Dietary Reference Intakes for vitamin A, vitamin K, arsenic, boron, chromium, copper, iodine, iron, manganese, molybdenum, nickel, silicon, vanadium, and zinc.* Washington, DC: National Academies Press; 2001.

12. Wang M and others. The association between maternal use of folic acid supplements during pregnancy and risk of autism spectrum disorders in children: A meta-analysis. *Molec Autism.* 2017;8:51.

13. Centers for Disease Control and Prevention. Folic acid. 2020; www.cdc.gov /ncbddd/folicacid/index.html.

14. Caudill MA. Pre- and postnatal health: Evidence of increased choline need. *J Am Diet Assoc.* 2010;110:1198.

15. Agopian AJ and others. Proportion of neural tube defects attributable to known risk factors. *Birth Defects Res A Clin Mol Teratol.* 2013;97:42.

16. Pfeiffer CM and others. Folate status in the U.S. population 20 y after the introduction of folic acid fortification. *Am J Clin Nutr.* 2019;110:1088.

17. Institute of Medicine, Food and Nutrition Board. *Dietary Reference Intakes for calcium, phosphorus, magnesium, vitamin D, and fluoride.* Washington, DC: National Academies Press; 1997.

18. Miao D and others. A meta-analysis of pica and micronutrient status. *Am J Human Bio.* 2015;27:84.

19. World Health Organization. *Maternal mortality database.* Geneva: World Health Organization; 2019.

20. Stamnes Koepp U and others. Maternal pre-pregnant body mass index, maternal weight change and offspring birthweight. *Acta Obstet Gyn Scan.* 2012;91:243.

21. Stang J, Huffman LG. Position of the Academy of Nutrition and Dietetics. Obesity, reproduction, and pregnancy outcomes. *J Acad Nutr Diet.* 2016;116:667.

22. Kim SY and others. Percentage of gestational diabetes mellitus attributable in overweight and obesity. *Am J Public Health.* 2010;100:1047.

23. Giahi L and others. Nutritional modifications in male infertility: A systematic review covering 2 decades. *Nutr Rev.* 2016;74:118.

24. Campbell J and others. Paternal obesity negatively affects male fertility and assisted reproduction outcomes: A systematic review and meta-analysis. *Repro Biomed Online.* 2015;31:593.

25. McPherson NO and others. Paternal obesity, interventions, and mechanistic pathways to impaired health in offspring. *Ann Nutr Metab.* 2014;64:231.

26. Deputy NP and others. Gestational weight gain—United States, 2012–2013. *MMWR.* 2015;64:1215.

27. Rasmussen KM, Yaktine AL, eds. *Weight gain during pregnancy: Reexamining the guidelines.* Washington, DC: National Academies Press; 2009.

28. Ketterl TG and others. Association of pre-pregnancy BMI and postpartum weight retention before second pregnancy, Washington State, 2003–2013. *Matern Child Health J.* 2018;22:1339.

29. Perni U and others. Angiogenic factors in superimposed preeclampsia: A longitudinal study of women with chronic hypertension during pregnancy. *Hyperten.* 2012;59:740.

30. Roberts JM and others. Vitamins C and E to prevent complications of pregnancy-associated hypertension. *New Engl J Med.* 2010;362:1282.

31. Hofmeyr GJ and others. Prepregnancy and early pregnancy calcium supplementation among women at high risk of pre-eclampsia: A multicentre, double-blind, randomized, placebo-controlled trial. *The Lancet.* 2019;393:330.

32. Rumbold A and others. Vitamin E supplementation in pregnancy. *Cochrane DB Syst Rev.* 2015;Art. No. CD004069.

33. Duley L and others. Altered dietary salt for preventing pre-eclampsia, and its complications. *Cochrane DB Syst Rev.* 2015;9:Art. No. CD004069.pub3.

34. Kenny L. Improving diagnosis and clinical management of pre-eclampsia. *Med Lab Observ.* 2012;4:12.

35. DeSisto CL and others. Prevalence estimates of gestational diabetes mellitus in the United States, pregnancy risk assessment monitoring system (PRAMS), 2007–2010. *Prev Chronic Dis.* 2014;11:130415.

36. American Diabetes Association. Standards of medical care in diabetes—2019. *Diabetes Care.* 2019;42(Suppl 1).

37. Peacock A and others. Adherence to the caffeine intake guideline during pregnancy and birth outcomes: A prospective cohort study. *Nutrients.* 1018;10:319.

38. Committee on Obstetric Practice, American College of Obstetricians and Gynecologists. Moderate caffeine consumption during pregnancy. *Obstet Gyn.* 2010;116:467.

39. NIH, National Institute on Alcohol Abuse and Alcoholism. Fetal alcohol exposure. 2020; www.niaaa.nih.gov/publications/brochures-and-fact-sheets/fetal-alcohol-exposure.

40. England JL and others. Nicotine and the developing human: A neglected element in the electronic cigarette debate. *Am J Prev Med.* 2015;49:286.

41. Wong MK and others. Adverse effects of perinatal nicotine exposure on reproductive outcomes. *Reproduction.* 2015;150:R185.

42. Matthews A and others. Interventions for nausea and vomiting in early pregnancy. Update. *Cochrane DB Syst Rev.* 2014;3:Art. No. CD007575.

43. Marx W and others. Is ginger beneficial for nausea and vomiting. An update of the literature. *Curr Opin Support Pall Care.* 2015;9:189.

44. Section on Breastfeeding, American Academy of Pediatrics. Breastfeeding and the use of human milk. *Pediatrics.* 2012;129:e827.

45. Andreas NJ and others. Human breast milk: A review on its composition and bioactivity. *Early Hum Dev.* 2015;91:629.

46. Togias A and others. Addendum guidelines for the prevention of peanut allergy in the United States: Report of the National Institute of Allergy and Infectious Diseases-sponsored expert panel. *J Allergy Clin Immunol.* 2017;139:29.

47. Lessen R, Kavanaugh K. Position of the Academy of Nutrition and Dietetics. Promoting and supporting breastfeeding. *J Acad Nutr Diet.* 2015;115:444.

48. Ziegler AG and others. Long-term protective effect of lactation on the development of type 2 diabetes in women with recent gestational diabetes mellitus. *Diabetes.* 2012;61:3167.

49. Choi SR and others. Association between duration of breast feeding and Metabolic Syndrome: The Korean national health and nutrition examination surveys. *J Women Health.* 2017;26:361.

50. Centers for Disease Control and Prevention. HIV: Is it safe for a mother infected with HIV to breastfeed her infant? 2020; www.cdc.gov/hiv/group/gender/pregnantwomen.

51. World Health Organization. *Guideline updates on HIV and infant feeding.* Geneva: World Health Organization; 2016.

52. Lok KYW and others. Increase in weight in low birth weight and very low birth weight infants fed fortified breast milk versus formula milk: A retrospective cohort study. *Nutrients.* 2017;9:520.

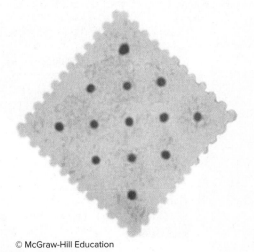

© McGraw-Hill Education

A nutritious diet supports normal growth and development throughout the growing years. Learn more at **www.cdc.gov**.
wavebreakmedia/Shutterstock

# 17 Nutrition during the Growing Years

## Learning Objectives

**After studying this chapter, you will be able to**

1. Describe normal growth and development during infancy, childhood, and adolescence and the effect of nutrition on growth and development.

2. Describe the calorie and nutrient needs of infants, children, and adolescents.

3. Compare the nutritional qualities of human milk and infant formula.

4. Explain the rationale—from the standpoints of both nutrition and physical development—for the delay in feeding infants solid foods until 4 to 6 months of age.

5. Describe the recommended rate and sequence for introducing solid foods into an infant's diet.

6. Discuss the factors that affect the food intake of children and adolescents.

7. Plan nutritious diets for infants, children, and adolescents using MyPlate.

8. Describe the nutrition-related problems that may occur during the growing years and their impact on future health.

## Chapter Outline

**IN THE SPAN OF LESS THAN 2 DECADES,** a helpless human newborn grows and develops into an independent, physically mature adult. This transformation is guided by genetic endowment and is highly dependent on an adequate supply of energy and nutrients. More than 14 million calories, 430 pounds of protein, 14 pounds of calcium, and vast quantities of every other nutrient are needed over the course of the growing years to develop into a healthy adult.[1,2]

Current trends in nutrition and overall health among children and adolescents in North America indicate that more children are receiving vaccinations than ever before and fewer teenagers are giving birth. In contrast to this good news, the percentages of children and teenagers who are obese, have Metabolic Syndrome and type 2 diabetes, and are not getting enough sleep are rising. In addition, physical activity is dropping as "screen time" (use of computers, television, and videos) increases. Milk intake is down and soft drink consumption is up. Whole grains, fruits, and vegetables also are in short supply in the diets of children and teens.

This chapter explores the effect of nutrition, from the beginning of infancy to the end of adolescence, on growth, development, and health. Because of the critical importance of adequate nutrition in infancy and the difficulties encountered in feeding some infants, a special emphasis is placed on this developmental stage.

# 17.1 Growing Up

As humans move from infancy to adulthood, height and weight increase. Body composition changes and organs mature. Normal growth and development are highly dependent on calorie and nutrient intake. Insufficient calories and nutrients, along with too little sleep and a lack of loving care, can impair the ability to thrive (grow and develop to the fullest physical and mental genetic potential).

When nutrients are missing at critical developmental phases, growth slows and may even stop. As with the fetus, the effects of poor dietary intake and other health choices in infancy, childhood, and adolescence depend on its severity, timing, and duration. Overall, eating a poor diet during the growing years hampers the cell division that occurs at critical stages. Consuming an adequate diet later usually cannot compensate for lost growth because the hormonal and other conditions needed for growth likely will not be present. Once the time for growth ceases, a sufficient nutrient intake helps maintain health and weight but cannot make up for lost growth.

## Height and Weight

Physical growth rate is at its peak velocity during infancy, which causes nutrient needs, per unit of body weight, to be at their lifetime highest level. Most babies have doubled their birth weight by 4 to 6 months of age. By their 1st birthday, they have tripled their birth weight and increased in length by 50%.[3]

In contrast to the rapid and usually smooth increases in height and weight of the 1st year, the physical growth rate of childhood is much slower and occurs in bursts.[2] In fact, it is normal for weight and height to remain unchanged for weeks, then suddenly spurt up. Nutrient and calorie needs, as well as the child's appetite, tend to rise and fall in response to these normal growth fluctuations. Healthy children grow a few inches taller each year. As you can see in Figure 17-1, they gain 4 to 6 pounds (2 to 3 kg) or so yearly until age 8 or 9; then the rate of weight gain increases to about 8 to 10 pounds annually until just before puberty, when many children normally store a few pounds of extra body fat.

Adolescence, the transition from childhood to adulthood, is a rapid phase of physical growth. One-third or more of all the growth in a lifetime occurs during this stage. Adolescence starts with the onset of **puberty**, which on average begins between the ages of 10 and 13 in girls and approximately 2 years later in boys. Puberty ends about 8 to 10 years after it

**puberty**  Period when a child physically matures into an adult capable of reproduction. Puberty is initiated by the secretion of sex hormones: primarily estrogen in females and testosterone in males.

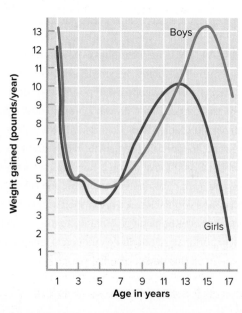
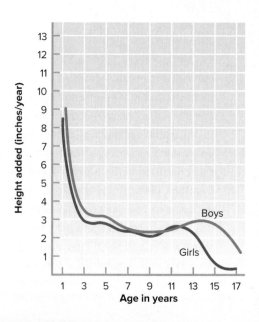

*Figure 17-1*  Growth rates for height and weight. The higher the line in any year, the greater the amount of annual gain compared with other years. For example, boys gain about 13 pounds in their 1st year of life, 5 pounds when they are age 7, and about 13 when they are 15 years old. Large weight gains occur in both infancy and puberty, whereas the very high length gain in infancy is never reached again. If graphs such as these were plotted in smaller time segments, they would appear as zigzag lines, rather than smooth lines, reflecting short, periodic spurts and plateaus in growth that occur from infancy to the end of adolescence.

starts, when the person is physically mature and capable of reproduction. Early-maturing girls may begin their growth spurt as early as age 7 to 8, whereas early-maturing boys may begin their growth spurt by age 9 to 10.[4]

When puberty begins, height and weight increase rapidly and the extra fat stored just before puberty usually decreases if the child did not enter puberty obese. The rate of growth in height peaks about 18 months after puberty begins, and then it slows down. For most females, height increases at its fastest rate at age 11, then slowly decelerates until they reach their adult height at about age 14 or 15. Girls usually begin menstruating during this growth spurt, and they gain little additional height 2 years after menarche. Most males experience peak velocity in height increases at age 13 to 14 years and attain their adult height around age 18. Both males and females may continue growing taller into their 20s. During adolescence, females grow about 10 inches (25 cm) in height and boys gain about 12 inches (30 cm). When they finish growing, teens will have gained about 15% of their adult height and weigh 45 to 85% more than when they entered this stage.

The term *length* is used in place of *height* when referring to children less than 2 to 3 years old because young children are measured lying on their backs with their body straight and legs extended. ©SECA

## Body Composition

As males and females move through the growing years, body composition changes, too. As you'll see later in this chapter, these changes affect nutrient needs in many ways. The proportion of body water declines during the 1st 2 or 3 years of life, at which point it achieves levels similar to those of adults.

The proportion of lean body tissue increases as infants and children grow older. Males and females enter adolescence with similar percentages of lean body tissue; however, during adolescence, males secrete testosterone, which causes them to gain more muscle mass, develop a heavier skeleton, and build a greater quantity of red blood cells than females. By the end of adolescence, females have two-thirds as much lean body tissue as males.

The proportion of body fat rises from birth to age 1, then declines slowly until age 7, at which point it begins gradually to increase again. During adolescence, body fat continues to rise in females, but it declines in males. This is because females secrete estrogen, which stimulates the accumulation of subcutaneous fat. This body fat is essential for sexual maturation.[5] When a female's body fat equals about 16 to 17% of body weight and her body weight reaches about 100 pounds (46 kg), menstrual periods begin and become regular once body fat reaches approximately 22%. The increasing regularity of the menstrual periods indicates that a female is nearing adult levels of body fat and is ending her growth. When adolescence ends, females have twice as much fat (as a percentage of body weight) as males.

Researchers once speculated that overfeeding during infancy may increase adipose tissue cell numbers. Today, it is known that the number of adipose cells can increase in adulthood as well. Still, if energy intake is limited during infancy to keep down the number of adipose cells, the growth and development of other organ systems, especially the brain and nervous system, also may be severely restricted.[6] In addition, most infants with obesity become normal-weight preschoolers without excessive diet restrictions. The risk of stunted growth and development makes it unwise to greatly restrict the dietary intakes of infants, as well as that of children and teens.

▶ The rapid weight gain during adolescence is difficult for many adolescent girls to accept because they fear they are becoming fat. Consequently, they may restrict calories and impair their bodies' chances to "grow up."

## Body Organs and Systems

In addition to the obvious outward changes in weight and height, babies, children, and teens are maturing inside. For instance, during infancy the kidneys double in size and begin to eliminate waste more efficiently. The stomach gradually increases its capacity and begins

Comstock Images

secreting digestive enzymes. By about age 4 to 6 months, the digestive tract has matured greatly, too. These changes enable infants to eat larger amounts of food at mealtime and use nutrients from a greater variety of foods than just human milk or infant formula.

Organs continue to grow and develop throughout childhood, with many reaching their full adult size and function during this stage. For instance, brain growth is three-quarters complete by age 2 and is finished by age 6 to 10 years. By age 9, the heart is nearly the size of an adult's and the respiratory system is approaching the functioning level of adulthood. By late childhood, the digestive tract has reached its full adult functional maturity. The growth and maturation of these systems permit them to meet the needs of the child's growing body. For example, a mature digestive system absorbs nutrients more completely, and a mature circulatory system efficiently delivers the nutrients and oxygen a growing body needs. More complete nutrient absorption also means that children are able to begin building nutrient stores that can help them meet the high nutrient demands of adolescence. During adolescence, any remaining growth and maturation are completed. Half of peak adult bone mass is accrued during adolescence. The most visible changes during adolescence are the maturation of the reproductive system and the development of secondary sexual characteristics.

### Knowledge Check

1. Approximately how much would you expect a baby born weighing 8 pounds to weigh by age 1 year?
2. At which stage of the growing years (infancy, childhood, adolescence) is growth the fastest?
3. How does body composition differ between males and females at the beginning and end of adolescence?
4. How does a mature digestive system support growth?

# 17.2 Physical Growth

The single best indicator of a child's nutritional status is growth. Thus, health-care professionals use growth charts matched to the child's biological sex to determine if growth is progressing normally. Figure 17-2 shows sample growth charts. They have several **percentile** curves (shown in blue) because children grow at different rates—many of which can be considered normal. The percentile curve a specific child follows depends on the child's dietary intake and genetic potential. For example, a child who is adequately nourished and has tall parents may fall in the 90th percentile for stature-for-age and 75th percentile for weight-for-age. By showing several percentile curves on growth charts, health-care professionals can compare a child's growth with the growth of others of the same age. For instance, if a boy's stature-for-age falls at the 25th percentile, he is shorter than 75% of the other boys of the same age. If a girl's weight-for-age is at the 95th percentile, she is heavier than all but 5% of girls of the same age. If a child's BMI-for-age is at the 50th percentile, the child has a lower BMI than half of all children of the same age.

## Tracking Growth

Growth should be tracked over time to identify a child's growth percentile and determine if growth is progressing normally. It takes 1 to 3 years for an infant to establish his or her own percentile. (Preterm infants may move up several percentiles—especially in length-for-age—because many experience "catch-up" growth and grow to the size they would have been if they had been full term.) Once growth percentiles are established, a healthy child

▶ For ages ranging from birth to 36 months, growth chart options include weight-for-age, length-for-age, weight-for-length, and head circumference-for-age. For ages 2 to 20 years, growth charts are available for weight-for-age, stature-for-age, and body mass index (BMI)-for-age. See Appendix K.

▶ BMI has fixed cutoff points for adults (e.g., a BMI of 25 for an adult is considered overweight). As Figure 17-2 shows, this is not true for children, for whom BMI is both biological sex- and age-specific.

**percentile** Classification of a measurement of a unit into divisions of 100 units.

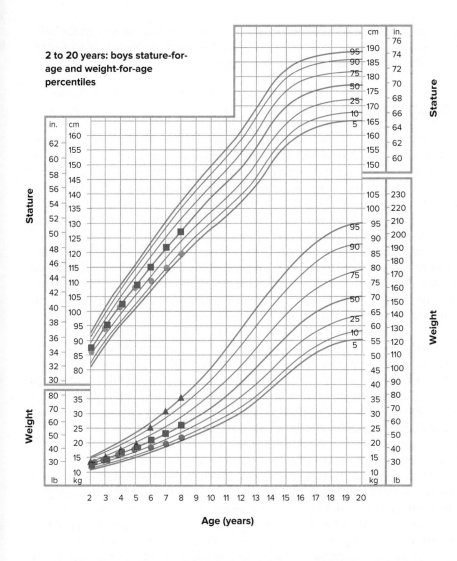

**2 to 20 years: boys stature-for-age and weight-for-age percentiles**

Stature

Weight

Age (years)

*Figure 17-2*  These growth charts are used to track the growth of males between the ages of 2 and 20. This growth chart depicts the gains in weight and stature that are expected as males increase in age. A certain weight and stature (height) correspond to a percentile value, which is a ranking of the person among 100 peers. The green squares represent a boy who is adequately nourished and growing normally. Notice how this boy continues along the same percentile curve from year to year. He is in the 50th percentile for weight from ages 2 to 8. The purple triangles show what might happen to his weight if the boy begins overeating. The red dots illustrate how weight is affected if the boy's access to food is restricted. The orange dots indicate how height can be affected if food restrictions are prolonged and severe. Growth charts for all age groups and biological sexes are available at www.cdc.gov/growthcharts and in Appendix K.

Source: www.cdc.gov/growthcharts

who is eating a nutritious diet will maintain about the same percentile curves from year to year (shown in green in Fig. 17-2). Small growth spurts and lags are expected, but a sudden jump up or down 2 or more percentile curves may signal that a child is experiencing growth problems caused by calorie excesses or deficits, nutrient inadequacies, illness, or psychosocial problems.[4] If, for example, a child begins overeating, BMI-for-age may jump several percentile curves (shown in purple in Fig. 17-2). When BMI-for-age reaches the 85th percentile, the child is considered overweight; when it reaches the 95th percentile, children have obesity.[7,8] At the 95th percentile, the diagnosis of obesity can be established if a physical exam indicates that the child is truly overfat, which is generally the case at this percentile.

If a child's access to food is restricted, BMI-for-age may drop to a lower percentile (shown in red in Fig. 17-2). A child is underweight when BMI-for-age drops below the 5th percentile. If food restriction is prolonged and severe, stature-for-age also may drop to a lower percentile (shown in orange in Fig. 17-2). Unless the child has short parents, he or she is likely experiencing growth stunting if stature-for-age drops below the 5th percentile. Special growth charts that include the 3rd and 97th percentiles are available for medical specialists to use when caring for children growing at the outer percentiles.

In early physical checkups, a health professional usually measures the head circumference as another means of assessing growth, especially brain growth. The brain grows faster in infancy than at any other time of life, with this rapid growth ending at about 18 months of age. How nutritional status affects brain development and intelligence quotient (IQ) is difficult to measure because scientists haven't figured out how to separate the effects of nature from those of nurture. However, several studies have determined that breastfed infants have

**Indicators of Nutritional Status**

**At Risk of Developmental Problems**

Birth to 2 years: head circumference-for-age <5th percentile or >95th percentile

**Stunted Growth**

Birth to 2 years: length-for-age <5th percentile

2 to 20 years: stature-for-age <5th percentile

**Underweight**

Birth to 2 years: weight-for-length <5th percentile

2 to 20 years: BMI-for-age <5th percentile

**Overweight**

Birth to 2 years: weight-for-length >95th percentile

2 to 20 years: BMI-for-age ≥85th and <95th percentile

**Obese**

2 to 20 years: BMI-for-age ≥95th percentile or BMI ≥30, whichever is smaller

higher IQs than infants fed with infant formula.[4,9] At the same time, studies from Central America suggest that IQ after age 5 years relates more closely to the amount of schooling a child receives than to nutritional intake during childhood.

## Using Growth Chart Information

A child's growth rate reflects calorie and nutrient intake.[2] BMI-for-age (or weight-for-length for children younger than 2 years old) is a good indicator of recent nutritional status. An insufficient diet in the immediate past causes a drop in BMI because weight decreases while height stays relatively the same. Excessive calorie intake results in a rise in BMI because weight increases without a corresponding increase in height. Stature-for-age is a good indicator of long-term nutritional status because undernutrition usually must be prolonged before stature is affected noticeably.

Infants and young children who do not grow at the expected rate for several months and are dramatically smaller or shorter than other children the same age, especially those who fall below the 5th percentile, are said to experience failure to thrive. Although physical abnormalities (e.g., heart defects, cleft palate), infections, intestinal problems, and hard-to-diagnose inborn errors of metabolism cause some cases of failure to thrive, nutrition or feeding problems are the cause of many cases. Nutrition problems include a lack of access to sufficient and/or appropriate food caused by limited income as well as parents who do not know how to meet a child's nutrient needs. Feeding problems may occur as a result of physical problems (e.g., the child has a weak sucking ability), poor feeding techniques (e.g., the parents limit feeding time), mental depression in the mother, and/or negative socialization factors (e.g., poor parent-infant relations).[7,9] Infants not only need food; they also need to be cuddled, hear voices, and have eye contact, especially at feeding times. In all cases, a physician should determine the cause of failure to thrive and work with parents to treat the problem.[4]

The long-term effects of failure to thrive caused by nutrition and food problems depend on the severity and length of time the child is malnourished. Continually receiving minimal quantities of food may permanently and irreversibly stunt growth and development. However, if dietary restrictions are followed by an adequate diet and positive social stimulation, children of all ages are likely to experience a faster than expected growth rate and "catch up" to where they would have been if malnutrition had not occurred.[8]

Growth in height ceases when the growth plates at the ends of the bones, called **epiphyses**, fuse. This process begins at around 14 years of age in girls and 15 years of age in boys, and it ends about 5 years later. For these reasons, a 16-year-old undernourished girl who is 4 feet 8 inches tall cannot attain her full adult height simply by eating better. She will be able to increase the diameter of her muscles, but overall muscle growth will be limited by the length of her bones.[4] Catch-up growth is possible even when growth retardation has been severe and prolonged, if the epiphyses have not closed.

Brain growth is faster in infancy than in any other stage of life. Head circumference measurements can help determine if growth is proceeding as expected.
RUTH JENKINSON/Getty Images

**epiphyses**   Ends of long bones.

## CASE STUDY

Damon is a 7-month-old boy who was taken to a clinic for a routine checkup. On examination, he seemed thin, and he plotted on the growth chart at the 25th percentile for weight and the 50th percentile for length. His physician scheduled a follow-up appointment in 3 months. At the 10-month visit, Damon appeared sluggish. He was again plotted on the growth chart and was now at the 5th percentile for weight but still at the 50th percentile for length. A registered dietitian nutritionist interviewed Damon's 16-year-old mother to collect information on his dietary intake. A 24-hour diet recall consisted of 2 bottles of infant formula, 3 bottles of Kool-Aid®, and a hot dog. However, Damon's mother was still in school, and at night she often left Damon with a neighbor so that she could go out for a few hours. Thus, she was not aware of all that he ate. What problems do you think are present in Damon's diet? What potential dangers await Damon if his health status continues along this current growth trend?

Muralinath/Getty Images

*Knowledge Check*

1. What factors affect the percentile growth curve a child follows?
2. Why are growth charts used to track growth over time?
3. What factors contribute to failure to thrive?
4. When is it no longer possible to grow taller?

## 17.3 Nutrient Needs

All of the changes that occur during the growing years influence energy and nutrient intakes and needs, but growth rate has the greatest effect. The faster the growth rate, the greater the nutrient and calorie needs per pound of body weight. Thus, the greatest needs, pound for pound, occur during infancy, when growth is at its peak velocity. Although calorie and nutrient needs *per pound* of body weight steadily decline after infancy, the total quantity of calories and nutrients needed rises throughout childhood because the body grows larger. During puberty, nutrient needs increase sharply because the growth rate is so rapid. In addition, biological sex differences in nutrient needs become more obvious during puberty (Fig. 17-3). The total quantities of nutrients and calories required are greater during adolescence than any other time except pregnancy and lactation. Males need more of many nutrients than females do because males are larger, develop more muscle mass and bone density, and have a longer, more intense growth period.

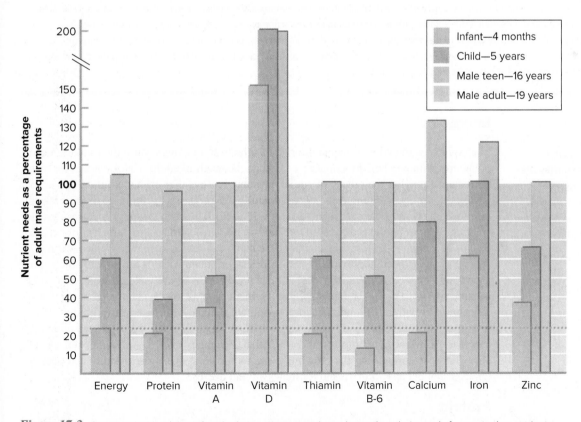

*Figure 17-3* Compared with adults, infants' relative energy needs are lower than their needs for most other nutrients, as illustrated by the different heights of the orange bars. Thus, infants need to obtain relatively larger amounts of nutrients from a smaller intake of food than adults. This also is true for young children (blue bars), but to a lesser extent.

# GL🌐BAL PERSPECTIVE

**Autism**

Autism spectrum disorder (ASD) is a complex, lifelong developmental disorder that affects millions around the world. Autism tends to be diagnosed during early childhood. It affects brain function, especially social and communication skills. Symptoms include delayed talking, unusual speech patterns, lack of eye contact, repetitive behaviors, short attention span, and sensory problems. Sensory problems range from hypersensitivity (overly responsive to smells, flavors, colors, light, and other stimuli) to hyposensitivity (no response to stimuli).

Chris Bernard/Getty Images

Many children with autism have unusual eating behaviors that can cause health concerns. In fact, they are 5 times more likely than their peers to have a feeding problem.[10] Food taste, smell, color, and/ or texture may cause hypersensitive children to restrict or avoid certain foods or groups of foods and be unwilling to try new foods. Some children with autism restrict themselves to 5 foods or less.[10] Insufficient food intake may result from hypersensitivity, an inability to stay focused long enough to finish a meal, and/ or extreme tantrums or ritualistic eating behaviors that interrupt meals. Feeding problems like these cause many children with autism to consume insufficient calcium, protein, and other nutrients. Some medications used to treat autism can add to feeding problems by depressing appetite and reducing food intake. Other medications may enhance appetite and lead to excessive calorie intake and/or reduce the absorption of certain vitamins and minerals. Constipation, another common difficulty, may be caused by eating a restricted diet. Feeding problems can lead to growth retardation, obesity, and heart disease.

Certain genetic and environmental factors are thought to increase the risk of having autism. Adequate intake of folic acid before and throughout pregnancy may reduce the risk of having children with autism.[11] Treatment tends to include educational programs, behavioral treatments, developmental therapies, and medications. Although there is little scientific evidence to indicate they are effective, complementary and alternative therapies are commonly used. These therapies include using vitamin, mineral, and omega-3 supplements as well as restricting food allergens, yeast, gluten, and casein.

Many children with autism are given a diet free of gluten (a protein in wheat, rye, and barley) and casein (a protein in milk); however, research evidence does not support gluten-free and casein-free diets as a primary treatment for autism.[12] These diets are commonly used because some children with autism have digestive problems. Removing casein and gluten from the diet is hypothesized to reduce symptoms. However, a diet free of dairy and wheat can be difficult to follow and, without careful planning, may provide insufficient nutrients, especially essential amino acids and calcium. Given that children with autism tend to have thinner bones than those without autism,[13] restricting calcium rich dairy sources can further impair bone health.

Studies are under way to determine the usefulness of diet therapies in the treatment of autism. Until then, those considering a gluten-free, casein-free diet or other diet therapies should work closely with health-care providers, including a registered dietitian nutritionist, to be certain the diet includes adequate amounts of all nutrients. Learn more at nih.gov/health-information.

## Energy

▶ From birth until age 2 years, basal metabolic rate continues to rise rapidly. After age 2, it rises slowly until puberty, when it increases dramatically. After puberty, basal metabolic rate rises slowly until about age 30; then it gradually decreases throughout life.

The rapid growth and high metabolic rate of infants and toddlers cause their calorie needs, pound for pound, to be 2 to 4 times greater than the needs of adults. Calorie needs for metabolic rate are higher than those of adults for many reasons. For example, infant hearts beat 120 to 140 times each minute compared to adults' 70 to 80 beats per minute. Babies take 20 to 40 breaths per minute but adults take only 15 to 20. The main reason babies have greater calorie needs per pound is that an infant's body has a large surface area, which allows a great deal of body heat to be lost. Thus, infants use many calories to keep their bodies warm.

Newborns need approximately 50 calories per pound each day to support their rapid growth and high basal metabolic rate. After 2 to 3 months of age, calorie needs drop to approximately 40 calories per pound daily and remain at about this level until age 3 years or so. The slower growth of childhood in comparison with infancy translates into a gradual reduction in calorie needs per pound of body weight. For example, calorie needs per pound decrease to 32 by age 5 and drop to about half that amount by age 15. Although calorie needs per pound decrease during the growing years, total calorie needs rise steadily and peak for females at about age 15 or 16 and for males around age 18.[2]

## Protein

Daily protein needs in infancy are roughly 1.5 g/kg (0.7 g/lb) of body weight daily to support their rapid rate of tissue synthesis—that is nearly twice as much protein per pound of body weight as adults. However, extra protein can be a problem for babies because it stresses the liver and kidneys, key organs in protein metabolism. Recall that the nitrogen from protein consumed in excess of need must be removed, then excreted via the kidneys. Large intakes of protein, as well as minerals, may overtax infants' immature kidneys and cause dehydration. Young infants need all the essential amino acids that adults do, as well as some others that are considered essential for infants.

During childhood and early adolescence, protein needs per pound of body weight are lower than during infancy but are 20 to 40% higher than in adulthood.[2] Children's protein needs are affected greatly by the growth and maturation of body organs. The protein needs for children 1 to 3 years is 1.1 g/kg (0.5 g/lb) of body weight/day, dropping to 0.95 g/kg (0.4 g/lb) of body weight/day for older children. Protein needs in adolescence are slightly higher than in adulthood because teens are increasing their lean body mass.

Protein malnutrition during the growing years can have profound, lifelong consequences. Physical development will be impaired if protein intake is inadequate or if calories are restricted to the point that protein must be used for energy. Inadequate protein intakes are uncommon in technologically developed countries, such as the U.S. and Canada; only excessive dilution of infant formula with water or severely restricted food intake is likely to lead to low protein intake. Dietary protein inadequacies, however, are a leading contributor to childhood illness, delayed or stunted growth, and death in low-income countries.

## Fat

Fat is an important part of infants' diets. It provides constituents such as cholesterol and essential fatty acids. In addition, because it contains many calories in a small volume, fat can meet infants' high calorie demands without overfilling their small stomachs.

Total fat should account for about 40 to 55% of a baby's calorie intake.[2] Infants, as well as children and teens, need at least 5 grams of essential fatty acids each day. In infancy, the fatty acids arachidonic acid and docosahexaenoic acid have especially important roles—the eyes and nervous system, especially the brain, depend on them for normal development.

Heart disease has its roots in childhood. However, dietary recommendations meant to reduce the risk of heart disease do *not* apply to children younger than 2 years old unless children are at increased risk of heart disease. In fact, many health experts, including those from the American Academy of Pediatrics and the National Cholesterol Education Program, warn against low-fat diets before age 2 because they can deprive young children of nutrients and calories and impair growth. Most health experts believe it is wise to reduce fat intake gradually between ages 2 and 5 until children are getting an average of 30 to 35% of their calories from fat. As fat intake declines, children should replace fat calories with nutrient rich foods, such as fruits, vegetables, lean meats, and low-fat dairy products.

## Carbohydrate

Lactose is the primary carbohydrate in the diets of most infants.[2] Starch intake increases when solid foods, such as baby cereals and vegetables, are added to an infant's diet. Once the diet includes foods other than human milk or infant formula, children should slowly increase their starch intake to equal about half their total calorie intake and limit simple carbohydrate intake. Many children and teens consume much larger amounts of simple carbohydrates than recommended.

Fiber intake recommendations for children less than age 1 are not yet set. However, after age 1, the daily Adequate Intake is 14 grams of total fiber per 1000 calories eaten.[2] Many children in the U.S. do not consume sufficient fiber and frequently experience constipation.[2] High fiber diets are not recommended for children because they tend to be high in bulk and inadequate in calories. In addition, they may bind minerals and block absorption to the extent that deficiencies occur.

## Water

Water is critical throughout the life cycle. But this nutrient is of special importance for babies because their need for water, per pound of body weight, is greater than that of older

▶ The American Academy of Pediatrics has indicated that reduced fat milk is appropriate for children ages 12 months to 2 years for whom overweight is a concern or who have a family history of obesity, high blood cholesterol levels, or cardiovascular disease. Parents are advised to consult with a registered dietitian nutritionist to ensure that reduced fat diets for children at risk of heart disease adequately meet their dietary needs.

▶ The National Cholesterol Education Program (NCEP) is designed to decrease heart disease by reducing the number of Americans with high blood cholesterol. To achieve this goal, the NCEP and many health experts recommend screening children over the age of 2 years who have a family history of premature heart disease. Children and teens with other risk factors, such as high blood cholesterol levels, obesity, high blood pressure, diabetes mellitus, a sedentary lifestyle, or cigarette smoking, also should be screened. Those with elevated total cholesterol and/or low-density lipoprotein (LDL) cholesterol levels should reduce dietary saturated fat and cholesterol. If, after dietary intervention, LDL levels remain very high in children over age 8, medication should be considered.[14,15] Visit www.nhlbi.nih.gov/files /docs/guidelines/peds_guidelines_full.pdf to learn more about the NCEP.

Candy, cookies, and soft drinks supply many of the carbohydrates children and teens eat.
Mike Kemp/Getty Images

individuals. One reason babies need more water is that their body surface area per pound of weight is about 3 times greater than adults—thus, babies lose much more of their body water through the skin. Another reason is that infants have proportionately more body water than adults and turn over body water 7 times faster than adults. Yet another reason is that the high metabolic rate of infants (about twice that of adults) produces a greater quantity of wastes that need to be excreted by the kidneys and lungs. A final reason is that a newborn's kidneys are only about half as efficient as adult kidneys. As a result, infants use much more water than adults use to wash away the same amount of waste in the urine.

A young baby's need for water is usually met by human milk or formula. Infants need supplementary water once they begin eating foods other than human milk and formula because of the increased amount of waste that must be filtered out by the kidneys and excreted in the urine. Feeding cow milk to infants who are less than 1 year old is not recommended for many reasons, a major reason being that it produces a greater amount of waste than either human milk or infant formula.[16]

Giving infants too much water can be harmful as well because it can lead to water intoxication. Overdiluting infant formula and feeding babies water instead of formula or human milk are the most common causes of water intoxication. Overall, it is best to limit supplemental fluids to about 4 ounces (120 ml) per day, unless a physician thinks a greater need exists because of disease or other conditions, such as high environmental temperatures. Fever, diarrhea, and vomiting can cause infants to quickly experience life-threatening dehydration. In cases of vomiting, diarrhea, or fever, a physician likely will recommend special fluid-replacement formulas containing electrolytes, such as sodium and potassium. Fluid-replacement formulas are available in supermarkets and pharmacies to treat dehydration; they should not be confused with bottled water.

## Vitamins and Minerals

All vitamins and minerals play an important role in supporting normal growth, but some are of particular concern during the growing years. For example, throughout the growing years, iron deficiency anemia is common. Children and adolescents also consume too little calcium, zinc, folate, and vitamins A and C. Many newborns have low stores of fluoride, vitamin K, and vitamin D.

### Iron

Healthy full-term infants are born with internal iron stores. However, by the time birth weight has doubled, usually by 4 to 6 months of age, iron stores are depleted. If the mother is iron deficient during pregnancy, these iron stores will be exhausted even sooner. To maintain a desirable iron status, the American Academy of Pediatrics recommends that breastfed infants receive iron supplementation and formula-fed infants be given an iron-fortified formula starting at birth.[4] Low iron infant formulas are sometimes prescribed to treat infants with various intestinal problems, but their use is discouraged. In addition, infants need solid foods to supply extra iron by about 6 months of age. In fact, this need for iron is a major consideration in deciding when to introduce solid foods.

Children 1 to 2 years old are particularly vulnerable to iron deficiency anemia because their diets often are dominated by cow milk, a low iron food, and most are no longer receiving iron-fortified formula. In addition, they typically do not like meat or have difficulty chewing it. In some cases, intestinal parasites contribute to iron deficiency. Iron-fortified cereal and easily chewed meats (e.g., ground beef) can help children boost iron intake; serving iron rich food with vitamin C rich food increases the absorption of this mineral.

Teens also are vulnerable to iron deficiency anemia because their need increases by about 40% for males and nearly 90% for females during this life stage.[17] Much of the increase, especially in males, results from expanding lean body mass, which directly incorporates substantial quantities of iron. As the body grows, blood volume also increases, including red blood cells, which contain iron. In addition, with the onset of menstrual periods, adolescent females need additional iron to replace the blood lost. About 10% of teenagers have low iron stores or iron deficiency anemia. Iron deficiency anemia sometimes occurs in girls after they start menstruating (particularly those with heavy menstrual flows) and in boys during their growth spurt when they are building large amounts of lean body tissue.

Iron deficiency anemia is of particular concern during the growing years. Low iron intakes adversely and irreversibly affect brain development, making it hard for children to learn and remember. Low intakes also can cause children to have disruptive behavioral patterns. The changes in the brain begin occurring even before blood tests detect low blood iron levels; thus, ensuring that children eat an iron rich diet is critical to lifelong intellectual ability. If anemia does develop, iron supplements should be used under a physician's guidance. Fortunately, childhood anemia is less common today in North America, probably because of children's consumption of iron-fortified breakfast cereals and the Special Supplemental Nutrition Program for Women, Infants, and Children (WIC). WIC emphasizes the importance of iron-fortified formulas and cereals and distributes them to low-income parents of infants and preschool children considered to be at nutritional risk.

### Potassium

Potassium is a nutrient of concern for many children as well as adults. A diet rich in fruits and vegetables can help children and teens meet their potassium needs.

### Calcium

Optimum calcium intake and weight-bearing exercise throughout the growing years are important to forming strong bones. Calcium needs rise sharply starting around age 9 years and remain high until the end of adolescence, largely because bones are growing longer and denser. In fact, most bone formation occurs between the ages of 9 and 18. Less than optimal calcium intakes may lead to decreased bone density and a greater risk of osteoporosis later in life. Unfortunately, teens' calcium intake usually falls well below the RDA, although males are more likely to have adequate calcium intake than females. Many teenage females replace calcium rich milk, which they perceive to be "fattening," with calcium poor soft drinks.

### Fluoride

Both the American Dental Association and the American Academy of Pediatrics recommend fluoride supplements for those between the ages of 6 months and 16 years whose drinking water is low in fluoride.[1] Parents should consult their dentist for advice on meeting their children's need for fluoride.

### Zinc

Although zinc deficiency does not appear to be a problem, some children and teens in the U.S. may be consuming too little.[19] Low intakes occur because children consume small portions of rich sources, such as meat. Low zinc intake during the growing years may impair growth. Breakfast cereal fortified with zinc can help provide this mineral.

### Folate

The diets of many older children and teenagers are low in folate. This is probably because their vegetable intake tends to be less than recommended. Also, during the teenage years, meals frequently are eaten away from home—compared with foods prepared at home, cafeteria and restaurant foods tend to be lower in folate, as well as vitamins A and C. Insufficient intakes of these vitamins may impair normal growth. The low folate intake of teenage girls is of particular concern as they reach childbearing age because folate deficiency can lead to neural tube defects in their offspring. (See Part 4.) Leafy green vegetables and enriched grains can supply ample folate.

### Vitamin D

Vitamin D is of special importance during the growing years because of its role in normal bone development. A lack of vitamin D will lead to the bone deformations seen in rickets (see Part 4). Ample vitamin D is provided by brief exposure to sunlight daily. When sunlight exposure is limited, as is often the case in northern latitudes, dietary vitamin D is needed. Fortified milk can supply sufficient vitamin D for children and teens. Infant formula supplies vitamin D to infants. The American Academy of Pediatrics recommends that infants receive a vitamin D supplement.

Getting sufficient calcium during childhood is difficult if milk is excluded from the diet. Both plain and flavored milk improve intakes of calcium and other nutrients and do not have adverse effects on BMI.[18] However, overconsumption of milk can displace other foods and lead to nutrition-related problems, such as iron deficiency anemia.

Ingram Publishing/SuperStock

► The warning given to parents to keep medicines (including vitamin and mineral supplements) out of the reach of children and not treat them as candy cannot be overemphasized. The most common cause of poisoning in children is overdose of supplements, especially those containing iron. Just 6 high-potency iron pills can be fatal for a 1-year-old. Iron poisoning causes bloody diarrhea, shock, liver damage, coma, and even death. Immediate medical care is essential because, once the iron is absorbed into the body, it is very difficult (if not impossible) to remove.

### Vitamin K

Infants are at risk of vitamin K deficiency because they are born with little or no vitamin K stores. The vitamin K–producing bacteria that thrive in the intestines begin to grow when a baby is fed for the 1st time. However, it may take several weeks for the bacteria to multiply to a level that can provide the infant with adequate vitamin K. In the meantime, infants have low vitamin K levels and may experience slowed blood clotting and unchecked bleeding.[16] To prevent these problems, the American Academy of Pediatrics recommends that all infants receive a dose of vitamin K at birth, and some state laws require it.

### Vitamin and Mineral Supplements

With the exception of a vitamin K supplement for newborns, an iron supplement for breastfed infants, a vitamin B-12 supplement for breastfed infants of vegan mothers, a vitamin D supplement for infants,[20] and a fluoride supplement for infants, children, and teens with an unfluoridated water supply, routine nutrient supplementation is not needed by healthy children and teens.[21] However, supplements may be recommended for children and teens who are poor eaters, vegans, pregnant, on programs to manage obesity, and/or deprived, neglected, or abused. The American Academy of Pediatrics suggests that these children and teens may benefit from a children's multivitamin and mineral supplement not exceeding 100% of the RDA or Adequate Intakes. Still, supplements are not a substitute for a healthy diet.

### Knowledge Check

1. Why do adolescent males need more of many nutrients than females?
2. Why are the calorie and protein needs of infants, per pound of body weight, higher than those of adults?
3. What are the recommended fat intake levels for infants and young children?
4. Why is the need for water of critical importance during infancy?
5. Why are children 1 to 2 years old vulnerable to iron deficiency anemia?
6. Why do calcium needs rise sharply during adolescence?
7. What vitamin and mineral supplements are recommended for infants, children, and teens?

## 17.4 Feeding Babies: Human Milk and Formula

Parents of a new baby have a thousand things to do, but menu planning isn't among them. With few exceptions, human milk or iron-fortified infant formula, coupled with the internal nutrient stores the baby built during fetal life, should meet an infant's nutrient needs at least until age 4 to 6 months.[21] The decision to breastfeed, formula-feed, or feed a baby a combination of these is personal. There are many valid reasons for selecting either human milk or formula, and both enable babies to grow normally.

### Nutritional Qualities of Human Milk

According to the American Academy of Pediatrics and Academy of Nutrition and Dietetics, human milk is the most ideal and desirable source of nutrients for infants, including premature and sick newborns.[22,23] Both of these organizations recommend breastfeeding exclusively from birth until 6 months of age, with the continued combination of breastfeeding and infant foods until age 1 year. The World Health Organization goes beyond that to recommend breastfeeding (with appropriate solid food introduction) for at least 2 years. However, surveys show that only about 83% of North American mothers begin to breastfeed their infants after birth, 47% are exclusively breastfeeding at 3 months, and at 12 months only 36% are still breastfeeding their infants.[24] Breastfeeding for the recommended length of time is best, but breastfeeding for even just a few weeks is beneficial.

Human milk may have many similarities with the milk from other mammals, but its nutrient composition and bioavailability are uniquely engineered by nature for human babies, just as cow milk is designed for calves and sheep milk is designed for lambs (Table 17-1). Unless altered, milk from cows or other animals should never be fed to infants younger than 12 months old. Except in cases of severe maternal malnutrition or other special cases, human milk contains ample supplies of all the nutrients needed until age 6 months, except for vitamin D, iron, and fluoride.[20,22]

The American Academy of Pediatrics recommends a vitamin D supplement for infants who are exclusively breastfed or those who receive less than 1 liter (or 1 quart) of vitamin D–fortified formula daily. Starting at age 4 months, exclusively breastfed infants and those receiving half their feedings as breast milk and no foods that contain iron usually need supplemental iron (1 mg/kg body weight daily). Supplemental iron may not be needed after infants begin eating iron-containing foods. At age 6 months, infants not receiving fluoridated water may need fluoride supplements.[25]

### Protein

The protein in human milk is mostly alpha-lactalbumin, which is synthesized by breast tissue. Some proteins, such as immune factors (e.g., antibodies) and enzymes, enter the milk directly from the mother's bloodstream. The major proteins in human milk are easier for the infant to digest and less stressful to the immature kidneys than is the protein in cow milk. Also, the proteins in human milk are not likely to cause allergic reactions in infants; thus, breastfed babies are less likely than formula-fed babies to develop allergies and food intolerances.

Another human milk protein, lactoferrin, increases the rate of iron absorption by the infant. As a result, more iron is absorbed from human milk than infant formula or milk from other animals, even though human milk contains less iron than these foods. Lactoferrin also protects against harmful bacteria. Immune factor proteins and other compounds, such as bifidus factor, work against harmful viruses, bacteria, and parasites. Because the immune system is not fully mature until age 2, these proteins and compounds offer a distinct advantage not found in infant formula. In fact, breastfed babies have fewer infections, fewer and less severe bouts with diarrhea, and better survival rates than those fed formula.

### Fat

The fats in human milk come from the mother's diet and are synthesized by breast tissue. The mother's current and long-term dietary intake of fat affects the profile of fatty acids of her breast milk. Human milk is high in cholesterol and linoleic acid, both of which are required for normal brain growth and development. It also contains omega-3 fatty acids, such as docosahexaenoic acid, which are needed for normal development of the retina in the eye and nervous system tissue. The lipids in human milk also promote efficient digestion.

The amount of fat in human milk changes during a feeding session. In the early part of a feeding session (1st 5 to 10 minutes on a breast), the baby receives fore milk. Fore milk is watery and contains less fat and fewer calories than hind milk, which is released after the fore milk. Infants need to breastfeed long enough (a total of 20 or more minutes) to get the calories in the fat rich hind milk to be satisfied between feedings and to grow well.

### Carbohydrate

Lactose is the main carbohydrate in human milk. Recall that lactose is comprised of 2 monosaccharides: galactose and glucose. Galactose is synthesized in the breast, and glucose enters from the mother's bloodstream.[4] Human milk, like milk from other animals, is mostly water. Thus, it provides adequate water for the infant when the baby is exclusively breastfed.

Human milk oligosaccharides are another type of carbohydrate found in human milk. These nondigestible complex carbohydrates function as prebiotics—they prevent pathogens from binding to intestinal

▶ There are substantial differences in nutrient composition between the milk of mothers who deliver prematurely and the milk of those who deliver at term. Researchers don't know why these differences exist, but they hypothesize that the preterm milk may be designed to meet the specific nutrient and immunological needs of the premature infant, which are different from the needs of a baby born at term.

**Table 17-1 Comparison of Human Milk and Cow Milk (per Liter)**

| Nutrient | Human Milk | Cow Milk |
|---|---|---|
| Total protein | 10.6 g | 30.9 g |
| Fat | 45.4 g | 38 g |
| Lactose | 71 g | 47 g |
| Calcium | 344 mg | 1370 mg |
| Phosphorus | 141 mg | 910 mg |

Formula-fed infants should remain on infant formula until age 1 year.

Ariel Skelley/Blend Images LLC

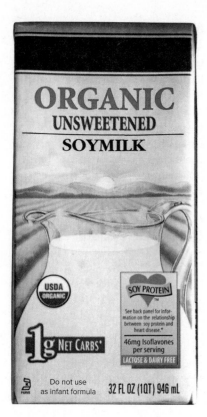

*Figure 17-4* Soy, rice, barley, oat, and nut beverages should not be confused with infant formula—they do not meet an infant's nutrient needs. Using these beverages in place of infant formula or human milk can cause serious health problems. To help prevent confusion, many soy beverage manufacturers print a warning on their beverages similar to this: "Do Not Use as Infant Formula." Andrew Resek/McGraw-Hill Education

cells and support healthy bacteria in the GI tract and immune system.[26] There are over 150 different human milk oligosaccharides; together, they make up 10% of human breast milk.

## Nutritional Qualities of Infant Formula

Commercial iron-fortified infant formulas provide a safe, nutritious alternative to human milk in areas of the world where high standards for water purity and cleanliness are common. Only formulas that are commercially made specifically for infants are safe to use. It is important to note that milk (e.g., cow, goat, soy), sweetened condensed milk, evaporated milk, homemade formulas, and plant-based "milks" (e.g., cashew, peanut, coconut, rice, oat, barley, soy) are inappropriate choices for infants—they do not conform to strict federal guidelines for calorie content, nutrient composition, and sanitation and, as a result, can cause life-threatening problems (Fig. 17-4). For example, goat milk is too low in folate, iron, and vitamin C. Cow milk is a poor source of vitamin C, vitamin E, copper, iron, and linoleic acid. Its high protein and mineral content overtaxes an infant's immature kidneys and increases the risk of dehydration. Its calcium is so high that it can cause bleeding in the stomach and intestine. The protein in cow milk is difficult for infants to digest and absorb.

In the U.S., the nutrient content of commercial infant formulas is regulated by the Infant Formula Act of 1980, which requires that they meet standards set by the American Academy of Pediatrics. These standards are set to match the nutrient composition of human milk as closely as possible. Although laboratory analysis procedures are quite sophisticated, the exact composition of human milk is not totally known. Thus, formulas only closely approximate the nutrient composition of human milk. Formulas, however, do not duplicate the immunological protection of human milk.

### Infant Formula

Infant formulas generally contain lactose and/or sucrose for carbohydrate, vegetable oils for fat, and modified proteins from cow milk, soy, or meat. Formula based on cow milk is recommended excepted in special circumstances.[21] A wide variety of commercial infant formulas are sold to meet the various health needs of infants. For example, soy-based infant formulas are available for infants who can't tolerate the lactose or proteins in cow milk–based infant formulas. Infant formula made with protein that has been "predigested" (broken down into peptides and amino acids) can help infants with digestive problems. Formulas that address the special nutrient needs of preterm infants are available. Other, more specialized formulas also are made for specific medical conditions, such as phenylketonuria. It is important to use an iron-fortified formula unless a physician recommends otherwise.

Sometimes infant formulas must be switched several times before the best choice for the infant is found. Parents should consult a physician when choosing or changing their infant's formula.

### Transitional Formula

Transitional formulas/beverages have been introduced by some manufacturers for older infants and toddlers. Some of these products are for use after 6 months of age, whereas others are intended only for toddlers. These transitional products have less fat than human milk or standard infant formulas and a mineral content more like human milk than cow milk. According to the manufacturers, the advantages of these transitional formulas/beverages over standard infant formulas include lower cost and better flavor. As stated previously, before making any formula change, parents should seek advice from their child's physician.

### Comparing Human Milk and Infant Formula

In addition to the nutritional and immunological differences described previously, human milk and commercial formula also differ in terms of health benefits, cost, convenience, and possibly mother-child bonding (Table 17-2). The health benefits of human milk for infants are not limited to its nutritional contributions.[23] Factors in human milk promote the maturation of the immune system and intestinal tract. Breastfed infants have a reduced risk of childhood asthma, obesity, diabetes, intestinal infections, misaligned teeth, ear infections,

### Table 17-2 Advantages to Infants Provided by Human Milk*

- Provides nutrients that are easily digestible and highly bioavailable and in amounts matched to needs
- May reduce risk of food allergies and intolerances, as well as some other allergies
- Provides immune factor proteins (antibodies) and other compounds (*Lactobacillus bifidus,* human milk oligosaccharides, lactoferrin) that reduce the risk of infections and diarrhea while the immune system is still immature
- Contributes to maturation of the GI tract and immune system
- Reduces risk of childhood asthma, obesity, diabetes, intestinal infections, misaligned teeth, ear infections, respiratory infections, and sudden infant death syndrome (SIDS)
- May enhance visual acuity, nervous system development, and learning ability by providing the fatty acid docosahexaenoic acid
- Establishes the habit of eating in moderation, thus decreasing the possibility of obesity later in life by about 20%
- Contributes to normal development of jaws and teeth for better speech development
- Is bacteriologically safe
- Is always fresh and ready

*For a summary of the advantages of breastfeeding for mothers, see Table 16-6 (Chapter 16).

respiratory infections, and sudden infant death syndrome (SIDS). They also are less likely to be obese in childhood and adolescence than formula-fed peers. In addition, there is evidence that children who were breastfed have significantly higher visual acuity and cognitive development scores than those who received no maternal milk. As the amount of human milk an infant receives during the 1st 6 months of life increases, the risk of developing health problems decreases. In addition, the longer children are breastfed, the more their cognitive development scores rise.

Human milk is almost always less expensive than formula, even after accounting for the costs of breast pump equipment and the slight increase in food required by the mother. Whether breastfeeding is more or less convenient than formula-feeding depends greatly on the circumstances. Preparing formula requires considerable time, careful sanitation, and exact measurements. Thus, human milk, which requires no preparation, is much more convenient to prepare. However, if the mother wants to feed her child human milk during times she is away from the child, she will need to express her milk and store it in bottles. The preparation time and sanitation measures needed for this are similar to those needed to prepare formula.

Mother-child bonding is a benefit frequently attributed to breastfeeding. It is true that breastfeeding requires an intimate physical relationship between mother and child that helps form a strong emotional bond. However, bonding depends more on close physical contact than on method of feeding. Formula-fed babies and their parents also can develop a strong bond if the babies are held during feeding.

## Feeding Technique

Newborn infants usually need 2 or 3 ounces of human milk or commercial formula every 2 to 4 hours. They need to be fed often because their stomachs hold only about 3 ounces, so they fill up and empty rapidly. As infants mature, the frequency of feeding decreases because the quantity of milk consumed at a time increases.

Infants should be followed closely over the 1st week of life to ensure that feeding and weight gain are proceeding normally. Monitoring by a physician or lactation consultant is especially important with mothers who have not breastfed before.

Breastfed infants tend to have fewer ear infections (otitis media) because they do not sleep with a bottle in their mouths.

Diane Mcdonald/Getty Images

Bottle-feeding allows caregivers other than the mother to participate and may promote closer bonding with the father.

Ariel Skelley/Getty Images

▶ Bisphenol A (BPA) is a compound used to make some plastics. Concerns about the effects of BPA on the development of infants and children led the FDA to ban its use in baby bottles and sippy cups. However, many other food containers, such as water bottles and linings in some food and beverage cans, may still contain BPA. To learn more, visit www.hhs.gov/safety/bpa.

The Safe to Sleep (formerly called Back to Sleep) campaign advises that infants be placed on their backs for sleeping.

Lucy von Held/Blend Images/Getty Images

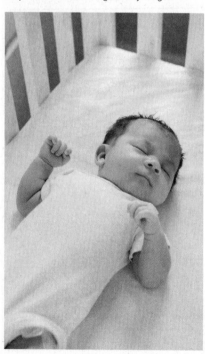

Parents may erroneously believe that encouraging a baby to eat more than desired will lengthen the time between feedings. This encouragement teaches them to overeat and causes physical pain when the stomach is overfilled. It is better to feed babies amounts they can easily accommodate often, rather than feeding larger amounts less frequently. Overfeeding can be avoided by watching for signs that the infant is full and terminating feeding at that time, even if some milk is left in the bottle (Table 17-3). Common signals that a bottle-feeding or breastfeeding infant has had enough include turning the head away, being inattentive, falling asleep, and becoming playful. Generally, the infant's appetite is a better guide than standardized recommendations concerning feeding amounts. By carefully observing infants while feeding them and responding to their cues appropriately, caregivers can be assured that the infants' energy needs are being met and can foster trust and responsiveness.[4]

Because a mother cannot measure the amount of milk a breastfed infant takes in, the mother may fear that she is not adequately nourishing the infant. As a rule, a well-nourished breastfed infant should (1) have 6 or more wet diapers per day after the 2nd day of life, (2) show a normal weight gain, and (3) pass at least 1 or 2 stools per day that look like lumpy mustard. In addition, softening of the breast during a feeding session helps indicate that enough milk is being consumed. Parents who sense that their infant is not consuming enough milk should consult a physician immediately because dehydration can develop rapidly.

Infants swallow a lot of air as they ingest either formula or human milk, so it's important to burp them after either 10 minutes of feeding or 1 to 2 oz (30 to 60 ml) from a bottle and again at the end of feeding. Spitting up a bit of milk is normal at this time. Once fed, infants should be placed on their backs. Infants should not be placed on their stomachs because this sleeping position has been linked to SIDS. The Safe to Sleep campaign, started in 1994 in the U.S., has reduced SIDS by 50%; however, plagiocephaly (flat-head syndrome) has increased as a result. To avoid flat-head syndrome, the American Academy of Pediatrics recommends periodically repositioning an infant's head while asleep and providing time on his or her stomach while awake.

## Preparing Bottles

Stored human milk, as well as infant formula, is fed to infants using bottles. All equipment and utensils used to prepare, store, and/or feed infant formula or human milk should be thoroughly washed and rinsed. This includes breast pumps, bottles, nipples, measuring spoons, and the like.

It is important to prepare formula by exactly following instructions on the label of powdered and concentrated formulas—adding too much or too little water can be very dangerous for infants. Only clean, cold water should be used—hot water from faucets in older homes pose a risk of high lead content (see Chapter 3). If well water is used, it should be boiled before making formula for at least the infant's 1st 3 months of life. In addition, it should be analyzed for naturally occurring nitrates—excessive nitrates can cause a severe form of anemia. If municipal water systems are high in nitrates, consumers will be warned (e.g., in local news outlets) not to use the water in infant formula. If water contaminants are a concern, formula can be mixed with bottled nursery water, sold in most supermarkets. Nursery water is purified water processed using steam distillation.

Prepared formula and expressed human milk can be fed immediately or refrigerated for up to 1 day. Human milk can be frozen for several weeks. Most infants accept room temperature formula. To warm a cold bottle of formula or stored human milk, run hot water over the bottle or place it briefly in a pan of simmering water. Infant formula and human milk fed from a bottle should not be heated in a microwave oven because hot spots may develop, which can burn the infant's mouth and esophagus. Discard any leftovers in the bottle—they are contaminated by bacteria and enzymes from the infant's saliva.

**Table 17-3** Summary of Physical and Eating Skills, Hunger and Fullness Cues, and Appropriate Food for Children 0 to 24 Months of Age

| Age | Skills and Developmental Signs* | Hunger Signs | Satiety Signals | Age-Appropriate New Foods to Introduce |
|---|---|---|---|---|
| **Birth to 4 months** | • Finds nipple through rooting reflex<br>• Tongue moves up and down<br>• Needs burping<br>• Little head and neck control<br>• Strong extrusion reflex (tongue thrusting) | • Cries until fed<br>• Hands form fists<br>• Body is tense<br>• Roots and sucks until fed<br>• Needs 8 to 10 feedings daily | • Removes mouth from nipple<br>• Falls asleep<br>• Relief of body tension | • Human milk or commercial infant formula |
| **4 to 6 months** | • Can swallow nonliquids<br>• Tongue protrudes in anticipation of nipple<br>• Tongue moves back and forth<br>• Learns to move food from the front of the tongue to the back<br>• Extrusion reflex diminishes and disappears<br>• Gaining control of head and neck<br>• Grasps items with entire hand and brings them to the mouth<br>• Teeth begin to erupt<br>• Near 6 months, sits with support | • Eagerly anticipates eating<br>• Opens mouth when sees bottle or breast<br>• Grasps and draws bottle or breast to mouth<br>• Needs 5 or 6 feedings daily | • Tosses head back or turns away<br>• Covers mouth with hands<br>• Spits out food<br>• Becomes playful or interested in surroundings<br>• Protests (fusses or cries) | • Iron-fortified infant cereal<br>• Puréed fruits and vegetables |
| **7 to 9 months** | • Tries to grasp feeding spoon<br>• Can form lips to rim of a cup and drink from a cup with help<br>• Holds bottle alone<br>• Develops pincher grasp (can pick items up with the thumb and index finger)<br>• Brings fist to mouth and feeds self finger foods<br>• Sits up alone, rolls over from back to front<br>• Jaw begins to move up and down<br>• Begins to chew and bite | • Reaches for food<br>• Looks for food when dish is removed<br>• Reacts to food preparation sounds<br>• Vocalizes hunger (cries or babbles) | • Changes body position<br>• Clamps mouth shut or puts hands in mouth<br>• Shakes head<br>• Says "no"<br>• Becomes playful (throws or plays with utensils or food)<br>• Pushes utensils or food away | • Fruit juice†<br>• Puréed protein rich foods (meat, fish, poultry, egg yolk, yogurt, cheese, tofu, beans)<br>• Finger foods (teething biscuits, crackers, toast, fruit slices, thin vegetable strips) |
| **10 to 12 months** | • Is able to chew<br>• Increases skill in biting, chewing, and swallowing<br>• Tongue is used to lick lips<br>• Demands to self-feed | • Grasps eating utensils<br>• Vocalizes hunger rather than cries<br>• Points to food | • Shakes head<br>• Says "no"<br>• Pushes food away<br>• Fidgets<br>• Grasps feeder's hand to control food intake | • Chopped, mashed table foods or commercial junior foods<br>• Grains (pasta, rice, etc.) |
| **12 to 24 months** | • Holds cup and drinks unassisted<br>• Uses spoon to feed self<br>• Becomes skilled self-feeder<br>• Food patterns become more individualized | • Vocalizes hunger, perhaps by asking for food or banging, waving, or dropping spoon<br>• Points to food or leads adult to the refrigerator | • Shakes head<br>• Says "no"<br>• Fidgets | • Same as for 10- to 12-month-old child with amounts determined by the child's appetite<br>• Whole milk, egg white, orange juice, table food |

*This timeline is just an estimate; skills/developmental signs of individual infants may vary by several months from the ages given. A pediatrician should be consulted if caregivers are concerned about an infant's development.

†Experts recommend trying to wait until the child is 12 to 24 months old before introducing juice and limit it to 4 to 6 oz daily.

(birth to 4 months): ©Barbara Penoyar/Getty Images RF; (4 to 6 months): Kwame Zikomo/Purestock/SuperStock; (7 to 9 months): ©Stockbyte/Getty Images RF; (10 to 12 months): ©Arthur Tilley/Getty Images RF; (12 to 24 months): ©lostinbids/E+/Getty Images RF

*Knowledge Check*

1. Why is human milk better suited to infants than cow or goat milk?
2. How do fore milk and hind milk differ?
3. What are 5 benefits of breastfeeding?
4. How might an infant signal he or she has had enough to eat?
5. How can a breastfeeding mother judge whether her infant is receiving enough nourishment?

# 17.5 Feeding Babies: Adding Solid Foods

As babies grow, the nutrient reserves they had at birth become depleted and human milk alone can no longer meet all their nutrient needs. Slowly, they need to begin getting some of their nutrients from solid foods (any food other than human milk or infant formula). Although iron-fortified commercial infant formulas can meet the nutrient needs of infants 6 to 12 months old, infants older than 6 months of age rarely are fed just infant formulas. Plus, adding solid foods helps ensure that any as yet unknown nutrient needs are met. An added bonus to exposing babies and children to a wide variety of food is helping them develop a willingness to taste new foods and learn to eat a widely varied diet. The more varied one's diet is, the more nutritious it is likely to be. The Dietary Guidelines for Americans 2020-2025 has issued guidelines for children from birth to age 23 months (Figure 17-5). These are congruent with the guidelines from the American Academy of Pediatrics (Fig 17-6).

## Deciding When to Introduce Solid Foods

The recommended age for introducing solid foods has changed as scientific knowledge has grown. The American Academy of Pediatrics and World Health Organization agree that *most* babies are not ready for solid food until they are 6 months old, although a few may be ready at 4 or 5 months. The time to introduce solid foods into an infant's diet hinges on these important factors:[4]

- *Nutritional need.* The nutrient stores a baby had at birth are exhausted by the time an infant has doubled his or her birth weight and weighs at least 13 pounds. A breastfed infant needs solid foods when he or she demands to be fed more than 8 to 10 times each day. A formula-fed baby needs solids when he or she drinks 8 ounces of formula and is hungry in less than 4 hours or consumes more than a quart of formula each day and still seems hungry. This description applies to most 6-month-old infants and a few 4-month-old infants. If solid foods are delayed much past the point a baby has a nutritional need for them, growth will slow.
- *Physiological capabilities.* Kidney function is quite limited until about 4 to 6 weeks of age. Until then, waste products from excessive amounts of dietary protein or minerals can cause so much urine output that dehydration occurs. An infant's intestinal tract is immature and cannot readily digest starch before 3 months. Because infants can easily absorb whole proteins until 4 to 5 months of age, exposing them to many different proteins before then—especially those in unaltered cow milk and egg whites—may predispose a child to future allergies and other health problems. For this reason, it's best to minimize the types of proteins in a young infant's diet by focusing exclusively on human milk or infant formula as a nutrient source.
- *Physical ability.* Young babies do not have the coordination to swallow solid food safely. These foods may cause young babies to choke or inhale food into the lungs (aspiration).[28] Several signs indicate that babies are developmentally ready for solid

In the early stages of solid food introduction, these foods complement rather than replace human milk or infant formula.

RuslanDashinsky/Getty Images

*Figure 17-5* The Dietary Guidelines for Americans from Birth to 23 Months of Age

*Figure 17-6* American Academy of Pediatrics guidelines for feeding infants.

**Dietary Guidelines: Birth to 23 Months**

**For about the first 6 months of life:**
- Exclusively feed infants human milk. Continue to feed infants human milk through at least age 1 year, and longer if desired. Feed infants iron-fortified infant formula during the 1st year of life when human milk is unavailable.
- Provide infants with supplemental vitamin D beginning soon after birth.

**At about 6 months:**
- Introduce infants to nutrient-dense complementary foods.
- Introduce infants to potentially allergenic foods along with other complementary foods.
- Encourage infants and toddlers to consume a variety of foods from all food groups. Include foods rich in iron and zinc, particularly for infants fed human milk.
- Avoid foods and beverages with added sugars.
- Limit foods and beverages higher in sodium.
- As infants wean from human milk or infant formula, transition to a healthy dietary pattern.

**The American Academy of Pediatrics has issued the following guidelines.[27]**

**Breastfeeding**
- Exclusive breastfeeding for the 1st year

**Bottle Feeding**
- Use appropriate bottle feeding practices:
  - *Avoid bottle propping.*
  - *Feed only breastmilk or formula in bottle unless otherwise directed by physician.*

**Food Introduction**
- Introduce solid foods around 6 months of age.
- Expose baby to a wide variety of healthy foods.
- Offer a variety of textures.

**Healthy Snacking**
- After 9 months, offer 2 to 3 healthy and nutritious snacks per day.
- Maintain fruit and vegetable intake after finger foods are introduced.

**Foster Self-feeding**
- Encourage babies to use spoons and fingers to feed themselves.
- Encourage babies to drink from a cup starting at 6 months of age.
- Parents recognize baby's hunger and satiety cues.

**Healthy Drinks**
- Babies should drink breastmilk or formula for the 1st year of life.
- Try to avoid introducing juice until child is a toddler. If juice is introduced, wait until 6–9 months and limit consumption to 4–6 ounces.
- Avoid introduction of sugar-sweetened beverages.

foods. These abilities usually occur around 4 to 6 months of age, but they vary with each infant.

- They can push up from lying face down and keep their elbows straight. They control head movements well and sit alone with support. These skills enable them to show interest in food by moving forward and to indicate satiety by pushing or turning away. A sitting position also provides a clear passageway for food to travel from the mouth to the stomach.
- They show readiness for different food textures by putting their hands or toys in their mouths.
- The extrusion reflex weakens and they are able to move food from the tip of the tongue to the back of the mouth. The **extrusion reflex** (also called tongue thrusting) helps a baby express milk from the nipple, but it also causes an infant to push objects placed on the tip of the tongue, such as a spoon or food, out of the mouth.
- They can make a chewing motion. At birth, babies have a poorly developed lower jaw and large fat pads in the cheeks that enable them to suck well. As the infant matures, the jaw line changes and the fat pads diminish—these changes allow the infant to chew and swallow, rather than just suck.

There is no advantage to introducing solid foods before an infant needs them or is ready for them. In addition to the strain that solid food puts on a young infant's organs, introducing food too early may lead to feeding problems and food dislikes. Also, the child is likely to eat far more calories than needed. If solid food takes the place of human milk or formula, reduced nutrient intake (especially calcium) may occur. Some mistakenly believe that adding solid food early will help infants sleep through the night. Actually, this achievement is a developmental milestone not affected by the amount of food consumed.

**extrusion reflex** Normal response present in 1st few months of life that causes the tongue to thrust outward when touched or depressed. It helps a baby express milk from a nipple, but it also causes an infant to push objects placed on the tip of the tongue, such as a spoon or food, out of the mouth; also called tongue thrusting.

▶ Many parents feed babies solid food before the age of 4 months. Getting solid foods too early may increase infants' risk for diabetes, obesity, celiac disease, and eczema.[29]

Infants show signs that they are ready for solid foods: they are able to sit up with support, they can push themselves up from a face-down position and keep their elbows straight, the extrusion reflex weakens, and they can make a chewing motion.

E. Dygas/Photodisc/Getty Images

### Rate for Introducing Solid Foods

Between 6 and 12 months of age, human milk or formula intake gradually decreases while solid food intake slowly increases. Initially, solid foods are a very small addition (1 to 2 teaspoons) to the diet (Table 17-4). As the baby's 1st birthday approaches, calories should be evenly divided between human milk or formula and a variety of foods from all the major food groups.

Experts recommend slowly adding foods with just 1 ingredient and waiting several days before offering another new food. This method makes it easy to identify food sensitivities and allergies by watching for reactions, such as gas, diarrhea, vomiting, rash, or breathing problems (e.g., wheezing). If the baby is fed new foods too often or is fed a food with several ingredients before each ingredient has been offered alone, there is no way to tell which has caused the problem. If a symptom appears, the suspected problem food should be avoided for several weeks and then reintroduced in a small quantity. If the problem continues, a physician should be consulted. Many babies outgrow food sensitivities in childhood. Some foods that commonly cause an allergic response in infants are egg whites, chocolate, nuts, and cow milk.

### Sequence for Introducing Solid Foods

The recommended sequence for introducing solid foods is designed to respond to the physical maturation and increasing nutrient needs of infants. Iron and vitamin C are the 1st nutrients needed by infants in quantities larger than those supplied by human milk. Thus, the 1st "solid" food generally recommended is iron-fortified baby cereal. Puréed fruits and vegetables are the next foods usually added. In fact, fruit is often introduced at about the same time as cereal.

Cereals and fruit juices marketed especially for babies are the preferred choice because they have no added salt, sugar, or monosodium glutamate (a flavor enhancer). Plus, the iron in infant cereals is much more absorbable than that found in adult cereals. Baby juices are fortified with vitamin C, which promotes iron absorption. Rice usually is the 1st cereal introduced be-

---

**Table 17-4  Tips for Introducing Solid Foods**

1. Start with teaspoon amounts of a single-ingredient food item, such as rice cereal, and increase the serving size gradually.

2. Offer solid foods after some breastfeeding or formula-feeding, when the edge has been taken off the infant's hunger.

3. Always feed solid foods from a spoon (a baby spoon [small spoon with a long handle] is best).

   - Spoon feeding is a skill babies need to learn.

   - Do not mix solid foods with a liquid and feed from a bottle because babies may choke. Also, eating baby foods from a bottle leads to poor eating habits and may cause children to be unwilling to try new foods and accept new textures.

4. Hold the infant comfortably on the lap, as for breastfeeding or bottle-feeding, but a little more upright to ease swallowing.

   - Put a small dab of food on the spoon tip and gently place it on the infant's tongue.

   - Be calm and go slowly enough to give the infant time to get used to food.

   - Expect the infant to take only 2 or 3 bites of the 1st meals.

   - Let infants decide when they are hungry and when they have had enough to eat.

---

## Critical | Thinking

Irena and Chris had a baby 11 months ago. At the last checkup, the doctor told them to start feeding the baby some new solid foods. After 5 days of eating a new food, the baby woke up with a runny nose and vomiting. The doctor told them to stop giving the baby that food. How can the doctor justify her recommendation?

cause it is least likely to cause allergies. Wheat, the cereal most likely to cause allergies, usually is introduced last. Orange juice may be too acidic for infants and is not recommended until age 1 year.

Protein rich foods, such as puréed meat, fish, chicken, beans, yogurt, egg yolk, and tofu, are introduced around age 6 to 8 months. These foods help supply the increasing amount of protein needed by the rapidly growing infant. It's usually best to wait until the child is a year old before serving egg whites and unaltered cow milk because they frequently cause allergic reactions in younger babies. Once cow milk is added to a child's diet, the American Academy of Pediatrics recommends that the child receive whole milk until age 2 years or older. Children who receive reduced fat milk before age 2 years have difficulty

There is considerable concern that some young children are consuming more fruit juice than is healthful. Excessive juice intake can cause diarrhea, gas, abdominal bloating, and tooth decay. Large amounts of juice also displace formula or human milk in the diet, causing the infant to receive inadequate amounts of calcium and other nutrients. Excessive fruit juice intake is associated with failure to thrive, GI tract complications, obesity, and short stature. The American Academy of Pediatrics advises parents not to introduce juices before age 12 months. Then, they should limit daily fruit juice intake to 4 ounces daily for children ages 1 to 3 years old, 4 to 6 ounces for children ages 4 to 6, and 8 ounces for children ages 7 to 18.[30]

lostinbids/Getty Images

meeting their calorie needs without exceeding their protein needs. After age 2 years, parents should consult with the child's health-care provider before switching to reduced fat milk.

As teeth begin to appear, babies are ready for foods with more texture, such as lumpy or chopped foods (e.g., cottage cheese, cooked vegetables). By about 9 months, many babies can pick up finger foods (e.g., crackers) and begin feeding themselves. By 1 year, most can eat table foods that were cooked until tender. Parents should aim to introduce infants to a variety of foods so that, by age 1 year, the infant is consuming many different foods from all food groups and the diet begins to resemble a balanced diet (Table 17-5).[4] Presenting each new food for several consecutive days can aid in an infant's acceptance of that food.

### Table 17-5 Sample Daily Menu for a 1-Year-Old Child*

| **Breakfast** | **Snack** |
|---|---|
| Applesauce, 1–2 tbsp | Cheddar cheese, ½ oz |
| Cheerios®, ¼ cup | Wheat crackers, 4 |
| Whole milk, ½ cup | Whole milk, ½ cup |
| **Lunch** | **Snack** |
| Roasted chicken, minced, 1 oz | Hard-cooked egg, ½ |
| Rice, 1–2 tbsp with ½ tsp soft margarine | Wheat toast, ½ slice, with ½ tsp soft margarine |
| Cooked peas, 1–2 tbsp | Orange juice, ½ cup |
| Whole milk, ½ cup | |
| **Dinner** | **Snack** |
| Ground beef, 1 oz | Banana, ½ |
| Mashed potatoes, 1–2 tbsp with ½ tsp soft margarine | Oatmeal cookies (no raisins), 2 |
| Cooked carrots (cut in strips, not coins), 1–2 tbsp | Whole milk, ½ cup |
| Whole milk, ½ cup | |

| **Nutritional Analysis** | |
|---|---|
| Total energy (kcal) | 1100 |
| % energy from | |
| Carbohydrate | 40% |
| Protein | 19% |
| Fat | 41% |

41%  40%  19%

*This diet is just a start. A 1-year-old may need more or less food. In those cases, serving sizes should be adjusted. The milk can be fed by cup; some can be put into a bottle if the child has not been fully weaned from the bottle. The juice should be fed in a cup.

▶ Some experts recommend introducing vegetables before fruits because, if fruits are offered 1st, the infant may prefer their sweet taste and resist vegetables.

**Typical Solid Food Progression Starting at 6 Months of Age***

Week 1: Rice cereal

Week 2: Add strained carrots.

Week 3: Add applesauce.

Week 4: Add oat cereal.

Week 5: Add cooked egg yolk.

Week 6: Add strained chicken.

Week 7: Add strained peas.

Week 8: Add strained plums.

*Extending the rice cereal step for a month or so is advised if solid food introduction begins at 4 months of age. If at any point signs of allergy or intolerance develop, substitute another, similar food item.

Older infants enjoy finger-feeding.

Corbis VCG/Getty Images

**Infant Feeding Summary**

*Breastfed Infants*

- Breastfeed for 6 months or longer, if possible. Then introduce iron-fortified infant formula when breastfeeding declines or ceases.
- Provide 400 IU/day of vitamin D.
- Consult a physician about the need for fluoride, vitamin B-12, and iron supplements.

*Formula-Fed Infants*

- Use an iron-fortified infant formula for the 1st year of life.
- Consult a dentist about the need for a fluoride supplement.

*All Infants*

- Provide a variety of basic soft foods after 6 months of age, progressing slowly to a varied diet.
- Add iron-fortified infant cereal at about 6 months of age.

*What Not to Feed Infants*

- Honey
- Very salty or very sweet foods
- Excessive amounts of infant formula or human milk; about 24 to 32 oz (750 to 1000 ml) of human milk or formula daily is ideal after 6 months, with food supplying the rest of the infant's energy needs
- Cow milk before age 1 year and reduced fat cow milk before age 2 years
- Large amounts of fruit juice

A wide variety of infant foods are sold in supermarkets. Single-food items are more desirable than mixed dinners and desserts, which are less nutrient dense. Grinding plain, unseasoned cooked foods and serving them fresh or freezing in ice-cube-size portions for serving later is an alternative to feeding commercial infant food. Careful attention to cleanliness is necessary when making baby food at home.

## Weaning from the Breast or Bottle

To prevent feeding problems and the overconsumption of milk, juice, or other sweetened beverages, babies should be completely weaned off bottles by age 18 months. Drinking from a sippy (spill-proof) cup helps begin weaning. Using a cup also helps prevent early childhood dental caries. If an infant drinks continually from a bottle, the carbohydrate rich fluid bathes the teeth, providing an ideal growth medium for bacteria. Bacteria on the teeth then make acids, which dissolve tooth enamel. Infants should never be put to bed with a bottle or placed in an infant seat with a bottle propped up because fluid (even milk) pools around the teeth, increasing the likelihood of dental caries and ear infections.

Honey should not be given to children until they are a year old because honey often contains *Clostridium botulinum* spores, which can grow and produce the deadly botulism toxin in an infant's digestive tract.
©D. Hurst/Alamy RF

## Learning to Self-feed

Infants begin to learn to feed themselves in late infancy and continue to develop these skills into the preschool years. Self-feeding skills require coordination and can develop only if the infant is allowed to practice and experiment. It's important that parents be patient and supportive, even though self-feeding is inefficient and very messy. Infants are solely finger feeders and don't always hit the "target" of their mouths. By the end of the 1st year, finger-feeding becomes more efficient and chewing is easier as more teeth erupt. In addition, by age 1 year, children drink from a cup without help, which makes fewer bottle- and/or breastfeedings necessary. By age 2, children can manage lifting, tilting, and lowering a cup quite well. In addition, they are fairly accomplished spoon users. Young children continue to be interested in the feel of foods, so they often engage in finger-feeding or filling the spoon with their fingers. They begin using a fork by age 3 or 4 and a knife to cut soft foods by age 4 or 5.

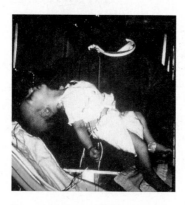

This young baby has infant botulism.

Source: Centers for Disease Control and Prevention

*Knowledge Check*

1. What physical ability signs indicate babies are developmentally ready for solid foods?
2. Why do babies need solid foods at around 6 months of age?
3. How can you determine whether a particular food has caused a sensitivity or an allergy in an infant?
4. Why should juice be served in a cup?

# Homemade Baby Food Made Easy

When babies are ready for solid foods, there are many options. Commercial baby food includes jarred strained fruits, vegetables, and meats; pouches of puréed fruit, vegetables, and grains; mini prepackaged meals; and puffs/crackers and cereals. There are organic and natural food options and foods with limited or no additives. Another choice is to make baby food at home.

New Africa/Shutterstock

Here is how to make baby food easily and safely at home.

1. Gather tools—be sure they are clean.

| Food Preparation | Food Storage |
|---|---|
| Pot and steamer basket (or electric steamer) | Ice cube tray or small freezer-friendly cups |
| Sharp knife and cutting board used only to make baby food | Freezer bags |
| Blender (immersion or standard) or a potato masher, ricer, or fork to break cooked food into small pieces | Permanent marker to label the bag with the date frozen |

2. Prepare and serve safely.

   a. Wash fresh fruits or vegetables well. Steam fresh (or frozen) fruits and vegetables until soft enough to easily pierce with a fork. Remove peels that are tough. Always steam fruits and vegetable for babies younger than 6 months old. After 6 months, soft, ripe fruits can be mashed with a fork—no steaming necessary.

   b. Break steamed food into small pieces. (Fruits canned in water or juice, canned low-sodium plain vegetables, or drained and rinsed canned plain vegetables can be used, too.) Add 1 tablespoon of water at a time to thin blended food to the desired consistency.

   - For babies between 4 and 7 months, blend food until very smooth.
   - As infants get older, food can have small, soft lumps. When babies reach 9 to 10 months old, soft, bite-size chunks can be served. Keep chunks small to prevent choking.

3. Make baby food ahead of time and store safely.

   a. Place thinned, blended food in ice cube trays and freeze. Each cube equals about 1 ounce. Remove frozen cubes from the tray and store in freezer bags labeled with the date frozen. Thaw and use within 6 months.

   b. Once children moved beyond purées, add frozen cubes to soups, casseroles, or smoothies to boost nutrient content.

Learn more at www.eatright.org/food/planning-and-prep/snack-and-meal-ideas/how-to-make-homemade-baby-food.

## Potential Nutrition-Related Problems of Infancy

Parents, other caregivers, and clinicians should be on the alert for a variety of health problems related to nutrition that can occur as children move through the growing years. Typically, these problems can be prevented or, in most cases, treated effectively with dietary modifications and/or medical intervention. Iron deficiency anemia, a common nutrition-related problem, was described earlier in this chapter. Parents and other caregivers usually need to consult a physician when dealing with many nutrition-related health conditions. The website of the American Academy of Pediatrics (www .aap.org) also provides useful information.

### Colic

**Colic** is sharp abdominal pain in otherwise healthy infants. The infants have repeated crying episodes, lasting 3 or more hours, that don't respond to typical remedies—such as feeding, holding, or changing diapers. Colic affects about 10 to 30% of all infants, starting at about 2 to 6 weeks of age and lasting until about 3 months of age. Crying episodes typically occur in the late afternoon and early evening. Nighttime sleeping also is usually disturbed by crying spells. Many colicky infants have flatulence, clench their fists, hold their bodies straight, draw up their legs, and want to be held.

The cause of colic is not known. It generally occurs in the absence of any physical problem in the infant. It tends to be most common in "temperamental" infants—those who are more sensitive, irritable, and

Colic causes inconsolable crying and can make parents feel frustrated and helpless.

EMPPhotography/Getty Images

intense and less adaptable and consolable than average for their age. Some researchers have speculated that an immature nervous system may cause colic. In addition, a lack of harmonious interaction between parents and the infant may contribute to the problem.

To help reduce excessive crying, parents should check to see whether the infant is tired, is bored, or wants to suckle. Holding the infant snugly to the shoulder causes many babies to become quiet and alert. Some infants can be calmed with pacifiers or by rhythmic sounds or movement.

Breastfeeding of colicky infants should continue. Mothers can try decreasing or stopping their consumption of milk and milk products, caffeine, chocolate, and strongly flavored vegetables to see if it helps reduce colic. Formula-fed infants with severe colic are sometimes helped by changing to soy-based or predigested protein formula. In addition, physicians may prescribe medication to calm colicky infants and reduce gas buildup.[4]

Coping with repeated crying spells can be challenging. Most parents benefit from the support of other adults and advice from those who have been through similar experiences. To optimize their ability to be sensitive and responsive to their infants, parents need to be well rested and set aside some time for themselves.

### Gastroesophageal Reflux

Many infants develop gastroesophageal reflux (GER), more commonly known as "spitting up," during their 1st year of life. In most cases, GER develops before age 2 to 3 months and usually resolves on its own by the time an infant is 12 months old. The problem occurs because the lower esophageal sphincter does not close completely, which allows milk or food in the infant's stomach to move back up into the esophagus. This can cause a painful burning sensation. In most cases, GER poses no serious medical concerns. In very rare cases, surgery may be required to remedy the problem.[4]

### Milk Allergy

Cow milk contains more than 40 proteins that can cause allergic reactions in infants.

Although some of these proteins are inactivated by heating (scalding) milk, others are not. A true milk allergy develops in about 1 to 3% of formula-fed infants. These infants may experience vomiting, diarrhea, blood in the stool, constipation, and other symptoms. If milk allergy is suspected, a formula-fed infant can be switched to a soy-based formula. However, in 20 to 50% of cases, infants also develop a soy protein allergy. In such cases, a predigested protein formula is necessary. A breastfeeding mother can experiment by eliminating cow milk from her diet. Fortunately, milk allergies seldom last beyond 3 years of age.[4]

### Food Allergies

More than 15 million Americans have food allergies. These allergies often begin in childhood. Although it is not known why some develop food allergies, research suggests that both genetic and environmental factors are involved. To prevent the development of food allergies, the American Academy of Pediatrics recommends exclusive breastfeeding for the 1st 4 to 6 months of life, feeding high-risk infants (those who have eczema or parents or siblings with food allergies) partially hydrolyzed formula if they are not breastfed, and introducing solid foods between 4 and 6 months.[31] Then, once children who are not at high risk for food allergies show they are able to tolerate some solid foods, parents should introduce small amounts of potentially allergenic foods (e.g., thinned peanut butter, eggs, milk).[32] Parents of children at high risk for food allergies should consult a pediatrician before giving these children potentially allergenic foods. If a food allergy is diagnosed, children must avoid the problematic foods. Dietary restrictions can make it difficult for children to meet their nutrient and calorie needs; thus, parents would be wise to work with a registered dietitian nutritionist to ensure their children have optimal diets.

### Ear Infections (Otitis Media)

When a baby sleeps with a bottle filled with formula, milk, juice, or a sweetened drink, liquids dripping from the nipple pool in the mouth and back up in the throat and the tubes leading to the ears. This backup can promote bacterial growth, which can result in painful ear infections and possibly hearing loss. Many ear infections can be prevented by never allowing a child to take a bottle to bed.

## Dental Caries

After the common cold, dental caries (cavities) are the most common childhood disease. Cavities form when bacteria in the mouth metabolize sugars and starch and form acids that erode tooth enamel. **Early childhood caries** (formerly called nursing bottle syndrome or baby bottle tooth decay) is a likely consequence when babies and children sleep with a bottle in the mouth (Fig. 17-7). The liquid from the dripping nipple spreads over the top teeth and rear bottom teeth. (The tongue covers the front bottom teeth.) Saliva normally washes the liquid from the bottle off the teeth, but during sleep the saliva flow stops. This gives bacteria in the mouth the

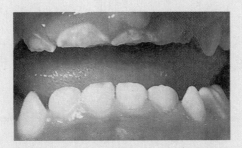

*Figure 17-7* Early childhood dental caries are the result of sleeping with a bottle in the mouth.

Paul Casamassimo, DD, MS

opportunity to metabolize the sugar in the liquid and produce acids that eat away tooth enamel.

The prevalence of dental caries in the U.S. has dropped significantly because of the increased use of fluoride-containing toothpaste, school-based dental care programs, fluoride in community water, professional fluoride treatments, and tooth sealant (a plasticlike material applied to teeth). Restricting carbohydrate intake is not an appropriate method for preventing tooth decay.

However, brushing after sticky, high sugar snacks and, if gum is chewed, selecting sugarless gum help prevent caries.

## Diarrhea

In the U.S., about 400 infants die each year of simple dehydration resulting from diarrhea, and about 210,000 are hospitalized for this problem. Typical symptoms of dehydration include dry mouth or tongue, few or no tears when crying, no wet diapers for 3 hours or more, irritability and listlessness, and sunken eyes and cheeks. To prevent dehydration, infants with diarrhea should be given plenty of fluids, under the advice of a physician. Specialized electrolyte-replacement fluids, such as Pedialyte®, may be recommended.[4] Once diarrhea subsides, a bottle-fed infant may be switched to a soy-based, lactose-free formula for a few days to allow time for the small intestine to produce sufficient lactase enzyme. A breastfed infant should continue to be breastfed for the duration of the diarrhea (see Chapter 4).

## CASE STUDY FOLLOW-UP

Muralinath/Getty Images

Damon's diet is inadequate for a 10-month-old infant because it lacks enough of the nutritious foods his growing body needs to support weight gain. These foods include iron-fortified cereal, puréed infant foods, and appropriate table foods. Damon should stay on infant formula until 1 year of age and should not be given sugary drinks, nor should these drinks be fed by bottle, if used. Damon needs a more nutrient dense diet containing a healthful variety of solid foods to provide him with enough energy and essential nutrients to grow and develop. Damon's mother would benefit from nutrition education classes taught at her school or at the local WIC program. The WIC program also can help her obtain healthy foods for Damon and herself.

## 17.6 Children as Eaters

The preschool years are the best time for children to start a healthful pattern of living and eating, focusing on regular physical activity and nutritious food. In the U.S. and Canada, children and teens tend to be fairly well nourished. However, there is room for improvement. For example, few meet the MyPlate recommendations (Fig. 17-8).[33]

As noted earlier in the chapter, children at greatest risk of nutrient inadequacies are those who have poor eating habits, are vegans, and/or are from families with limited resources. Vegan diets must be planned carefully, especially during the growing years, to ensure that they provide sufficient amounts of calories and nutrients, especially protein, omega-3 fatty acids, calcium, iron, zinc, iodine, vitamin B-12, and vitamin D (if sun exposure is limited) to support normal growth.[34]

During infancy, a child's attitudes toward foods and the whole eating process begin to take shape. If parents and other caregivers practice good nutrition and are flexible, they can lead a child into lifelong healthful food habits.

Imcsike/Shutterstock

# Healthy Eating for Preschoolers Daily Food Plan

| Food group | 2 year olds | 3 year olds | 4 and 5 year olds | What counts as: |
|---|---|---|---|---|
| **Fruits** | 1 cup | 1–1½ cups | 1–1½ cups | **½ cup of fruit?**<br>½ cup mashed, sliced, or chopped fruit<br>½ cup 100% fruit juice<br>½ medium banana<br>4–5 large strawberries |
| **Vegetables** | 1 cup | 1½ cups | 1½–2 cups | **½ cup of veggies?**<br>½ cup mashed, sliced, or chopped vegetables<br>1 cup raw leafy greens<br>½ cup vegetable juice<br>1 small ear of corn |
| **Grains**<br>Make half your grains whole | 3 ounces | 4–5 ounces | 4–5 ounces | **1 ounce of grains?**<br>1 slice bread<br>1 cup ready-to-eat cereal flakes<br>½ cup cooked rice or pasta<br>1 tortilla (6" across) |
| **Protein Foods** | 2 ounces | 3–4 ounces | 3–5 ounces | **1 ounce of protein foods?**<br>1 ounce cooked meat, poultry, or seafood<br>1 egg<br>1 Tablespoon peanut butter<br>¼ cup cooked beans or peas(kidney, pinto, lentils) |
| **Dairy**<br>Choose low fat or fat-free | 2 cups | 2 cups | 2½ cups | **½ cup of dairy?**<br>½ cup milk<br>4 ounces yogurt<br>¾ ounce cheese<br>1 string cheese |

*Figure 17-8* The Daily Food Plan for preschoolers is based on MyPlate.

Source: www.choosemyplate.gov

▶ **USDA School Nutrition Programs**

- School Breakfast Program
- National School Lunch Program
- Special Milk Program
- Fresh Fruit and Vegetable Program
- After-School Snack Program
- After-School Supper Program
- Summer Food Service Program

▶ The American Academy of Pediatrics recommends that families and schools take a whole diet approach when it comes to children's diets. Instead of focusing just on sugar, fat, or specific nutrients, parents and caregivers should be sure kids get a variety of vegetables, fruits, grains, low-fat dairy, and quality protein (e.g., lean meats, fish, nuts, seeds, eggs); avoid highly processed foods; use small  amounts of sugar, salt, fats, and oils to boost kids' enjoyment and intake of nutritious foods; serve appropriate portion sizes; and give kids diverse experiences with foods.[36]

The Special Supplemental Nutrition Program for Women, Infants, and Children (WIC) can help limited-resource children up to age 5 years who are at nutritional risk get the nutrients they need. In addition, encouraging school-age children and teens, especially those from limited-resource families, to participate in the U.S. Department of Agriculture–administered School Breakfast Program, National School Lunch Program, and other programs can help them improve their nutrient intake.[35] School breakfasts include 3 food items (milk, grain, and vegetable or fruit), and lunch includes 5 items (milk, protein food, grain, vegetables, and fruit). School breakfast and lunch menus must follow USDA Nutrition Standards, which are based on the Dietary Guidelines for Americans. School breakfast participants tend to perform better in school than those who don't participate. School lunch participants can obtain about one-third of their total daily nutrient intake from their lunch meal.

## Appetites

Compared with babies, children tend to have erratic appetites. Before and during a period of rapid growth, children have good appetites. When growth slows or plateaus, appetite drops off significantly. Appetite fluctuations are considered a problem only when low or high intakes occur for extended periods or the child exhibits signs of undernutrition (e.g., fatigue, increased susceptibility to infection, underweight, or failure to thrive) or overnutrition (e.g., obesity).

Caregivers who do not understand that appetite lulls are to be expected and that healthy, normal-weight children have built-in feeding mechanisms that regulate food intake to match needs may resort to bribes, forcing, teasing, or trickery to get children to eat. Bribing children to eat a new food (e.g., "Eat 3 bites of carrots and you can have dessert") may achieve the parent's immediate goal, but it often has negative results in the long run. Research studies show that, when children must clear an arbitrary hurdle (e.g., 3 bites of carrots) to receive a reward (e.g., dessert), preference for the reward rises, whereas preference for the hurdle decreases. In subsequent meals when the reward is removed, children eat less of the hurdle food. Bribing children with food, like dessert, also teaches them that food is an appropriate reward. At the other end of the spectrum lie parents who are so excessively concerned that their children will become obese that they restrict their children's food intake. In some cases, caregivers restrict food intake so severely that the children become underweight and fail to thrive.

FamVeld/Shutterstock

*Critical* **Thinking**

Tim refuses to eat breakfast before school. He doesn't like cereal, toast, or any of the other usual breakfast foods. What can Tim's parents do to ensure that he eats nutritious foods before leaving for school?

Pressuring a child to eat more or less than is desired tells the child not to trust his or her own hunger and satiety signals—this can lead to a lifetime battle with weight problems. Teaching a child to use food to fulfill emotional needs (e.g., as a reward) can have the same effect. Successful child feeding depends on a division of responsibility between parents and children. Parents are responsible for providing appropriate, nutritious, appealing, regular meals and snacks. Children are responsible for deciding how much to eat. If the division of responsibility is respected, a normal, healthy child will eat adequate amounts with minimal fuss. Problems often arise when parents don't fulfill their responsibilities and/or take over the child's responsibilities. A goal should be to make mealtime a happy, social time, sharing enjoyment of healthful foods.

▶ Choking is a possible but very preventable hazard for infants and young children. Avoid choking hazards by

- Setting a good example at the table by taking small bites and chewing foods thoroughly
- Insisting that children sit at the table, take their time, and focus on the food during meals and snacks
- Never giving infants and children any foods that are round, firm, sticky (gummy candy, peanut butter), or cut into large chunks; foods that can cause choking include hot dogs (unless finely cut into sticks, not coin shapes), candy, nuts, grapes, cherries, beans, popcorn, coarsely cut meats, and hard pieces of fruits or vegetables (e.g., raw carrots, apples)

## When, What, and How Much to Serve

Children have small stomachs. Offering them 6 or so small meals succeeds better than limiting them to 3 meals each day. Sticking to 3 meals a day offers no special nutritional advantages; it's just a social custom. When we eat isn't nearly as important as what we eat. Frequent meals and snacks help children meet their nutrient and calorie needs, as well as keep their blood glucose levels high enough to support the activity of their rapidly developing brain and nervous system. Breakfast and snacks are especially important. Children who eat breakfast or a nourishing morning snack have a greater daily intake of several vitamins and minerals than children who skip breakfast. Also, breakfast eaters may perform better in school and have a longer attention span than breakfast skippers.

Snacks provide about 25% of the total calorie intake of many children. Nutritious snacks can add significant amounts of nutrients to the diet. However, many children eat high calorie, high fat snacks instead of healthy snacks.

From the 1st birthday onward, children, like adults, need several servings from each of the major food groups every day. However, the size of each serving and the number of servings are different. A serving size for children is equal to about 1 tablespoon per year of age, depending on the child's appetite. For example, a 2-year-old's dinner might contain 2 tablespoons each of ground beef, pasta, and peas. Serving sizes increase gradually until late childhood, when they become the same as those for adults.

MyPlate for children is a useful meal planning tool. Table 2-6 (see Chapter 2) shows the recommendations for daily amounts of food to consume from the food groups based on energy needs. Young children (ages 2 and 3) typically need between 1000 and 1400 calories daily. Preschool children (ages 4 and 5) usually need 1200 to 1600 calories each day. During the early school years (ages 6 to 8), daily calorie needs tend

Sweets should be consumed in moderation during childhood, but they do not have to be avoided completely.

Rita Maas/Getty Images

▶ Schools play an important role in improving children's health and diet and combatting childhood obesity. Federal legislation requires schools participating in USDA Child Nutrition Programs, such as the School Breakfast and National School Lunch programs, to have a local school wellness policy.[37] This legislation mandates that school wellness policies be written documents with the following.

- Specific goals for nutrition promotion and education, physical activity, and other school activities that   promote wellness
- Standards and nutrition guidelines for all foods and beverages sold to students at school during school hours that are consistent with federal regulations for school meals and smart snacks nutrition standards
- Standards for all foods provided (but not sold) to students at school during the school day (e.g., foods at class parties, class snacks, foods used as incentives)
- Policy that allows marketing and advertising of only foods and beverages that meet the smart snacks nutrition standards
- Description of public involvement, public updates, policy leaders, and evaluation plan related to the wellness policy

Schools also must identify wellness policy leaders (e.g., school officials) who are responsible for   ensuring that the school complies with its wellness policy, allow the general public   (e.g., parents, students, teachers) to participate in developing the policy, and assess compliance with the wellness policy every 3 years.

Interest in food starts early in life.

Westend61 GmbH/Alamy Stock Photo

to range from 1200 to 2000 calories. Older children's calorie needs are higher, ranging from 1600 to 3200 calories. Visit choosemyplate.gov to learn more about food plans for children.

Because of the reduced appetite of younger children, providing nutrient dense foods is particularly important. There is no need to decrease fat or simple sugar intake severely, but fatty and sweet food choices should not overwhelm more nutritious choices. An overemphasis on fat reduced diets during childhood has been linked to an increase in eating disorders and encourages an inappropriate "good food, bad food" attitude. When planning menus, also keep in mind that children tend to like foods with crisp textures and mild flavors, as well as familiar foods. Because their taste buds are more sensitive than those of adults, preschool children often refuse to eat strongly flavored foods. In addition, young children are especially sensitive to hot-temperature foods and tend to reject them. Children also may object to having foods mixed, as in stews and casseroles, even if they like the ingredients separately.

## Food Preferences

Food preferences begin to be established during fetal life and continue to develop in the years spanning infancy and adolescence. In the early childhood years, the family usually has the greatest influence on the development of food preferences and habits. As children grow older, peers and teachers begin to provide new ideas about food, eating, and nutrition. Television programming and advertising also present strong lessons about food and eating.

Steering children toward healthful foods is likely to be more successful if parents understand the eating behaviors of young children (Table 17-6) and expose them to nutrition education. Because children spend much of their younger years in school, it is a great place to learn about positive, healthy eating habits.[37] This education can help children understand why eating a nutritious diet will make them feel more energetic, look better, and work more efficiently. Regular family meals daily—whether breakfast, lunch, or dinner—help children apply what they learn via nutrition education and build good eating habits.

The most important nutrition lessons for children of all ages involve expanding their familiarity with new foods and helping them develop a willingness to accept new foods. Teaching these lessons is challenging because children tend to reject unfamiliar foods, but the nutritional rewards of a varied diet are worth the effort. Children become familiar with new foods and are more willing to try them when they look attractive and are served in a social setting by calm, supportive, approving adults who are eating the foods. If a child observes adults and older children eating and enjoying a food, there's a good chance that he or she eventually will accept it. One possible policy is the 1-bite rule: within reason, children should take at least 1 bite or taste of the foods presented to them.

Giving children opportunities to make food decisions also can encourage them to try new foods. For example, parents can select several acceptable snack choices and let children choose the snack they want. Involving children in preparing a new food and serving small amounts of a new food at the beginning of the meal, when the child is most hungry, helps increase his or her acceptance of the food, too. As many as 15 to 20 exposures to a new food may be needed before the child accepts it. There may be some foods that a child will never accept—keep in mind that no single food is a dietary "essential" and children are entitled to their own likes and dislikes.

**Table 17-6  Food-Related Traits, Behaviors, and Skills of Preschool Children**

| | 1 to 2 Years Old | 3 Years Old | 4 Years Old | 5 Years Old |
|---|---|---|---|---|
| **Emotional Traits** | • Fears new things, including<br>• foods<br>• Shares with difficulty<br>• Likes to help<br>• Needs close supervision<br>• Inquisitive<br>• Often rebellious<br>• Seeks attention | • Wants to be included<br>• Shares with difficulty<br>• Fairly inflexible about the "right" way to do<br>• Reacts well to options instead of demands | • Seeks adult approval and attention<br>• Shares well<br>• Still inflexible about the "right" way to do something<br>• Requires and understands limits<br>• Willing to adhere to rules most of the time<br>• Shows off | • Cooperative with family routines and activities<br>• Attached to home and family<br>• Still inflexible about the "right" way to do something |
| **Eating Behaviors** | • "Picky" eater<br>• May insist on eating the same food at every meal (food jag)<br>• May not swallow food in the mouth | • Eats a wide variety of foods, except those that are strongly flavored<br>• Dawdles over eating when not hungry<br>• Talks about foods served at meals and snacks | • Likes to talk, which can get in the way of eating<br>• Strong food preferences<br>• Expresses independence by refusing to eat | • Wants to eat preferred foods<br>• Prefers most vegetables raw<br>• "Copies" food dislikes of family members |
| **Eating and Food Preparation Skills** | • Uses spoon fairly well<br>• Can tear, snap, and dip foods<br>• Can lift, drink from, and set down a cup<br>• Can scrub firm fruits and vegetables and pat them dry<br>• Can carry ingredients from 1 place to another | • Uses spoon well and is learning to use a fork<br>• Has moderate hand muscle development<br>• Feeds self independently<br>• Can pour milk, juice, and premeasured ingredients<br>• Can serve portions from a serving dish<br>• Can spread butter and dips<br>• Can knead and shape dough | • Uses all eating utensils<br>• Has small finger coordination<br>• Can peel, spread, cut, roll, and mash food; cracks eggs<br>• Can wipe, wash, and set table | • Well-coordinated hands and fingers<br>• Makes simple breakfast, lunch, and snacks<br>• Can measure, cut, grind, and grate with some supervision<br>• Can help clear the table and put away leftovers |

RubberBall Productions/Getty Images

Source: Sigman-Grant, M. "Feeding Preschoolers: Balancing Nutritional and Developmental Needs," *Nutrition Today*, vol. 27, issue no. 4, July/August 1992, p. 13.

1 to 2 Years Old: Amos Morgan/Getty Images; 3 Years Old: Sappington Todd/Getty Images; 4 Years Old: RubberBall Productions/Getty Images
5 Years Old: Mike Kemp /Getty Images

## Mealtime Challenges

Many preschool children go through periods of unpredictable and unusual eating behavior. One of the most common is food jags—demanding the same meal 3 times a day for a week or more. Food jags can be monotonous (e.g., eating only green foods or only peanut butter and jelly sandwiches) but generally don't present a nutritional problem unless the food demanded is excessively high in sugar, fat, or sodium or if the jag lasts for more than a few weeks. The best way to handle most food jags is to serve the food demanded while keeping in mind that the food jag will soon pass.

Sometimes children refuse to eat. When they do, it's best not to overreact. Doing so may give the child the idea that eating is a means of getting attention or manipulating a situation. Children rarely starve themselves to any point approaching physical harm. When children refuse to eat, parents should have them sit at the table for a while; if they still aren't interested in eating, the parents can remove the food and wait until the next scheduled meal or snack. Hunger is still the best means for getting a child

Getting children involved with food preparation can help them accept new foods.

White Rock/Getty Images

## Getting Young Bill to Eat

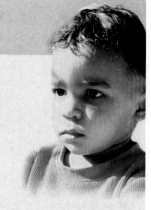

Bill is 3 years old, and his mother is worried about his eating habits. He absolutely refuses to eat vegetables, meat, and dinner in general. Some days he eats very little food. He wants to eat snacks most of the time. His mother wants him to eat a sit-down lunch and dinner to make sure he gets all the nutrients he needs. Mealtime is a battle because Bill says he isn't hungry, but his mother wants him to eat everything served on his plate. He drinks 5 or 6 glasses of whole milk per day because that is the only food he likes.

When his mother prepares dinner, she makes plenty of vegetables, boiling them until they are soft, hoping this will appeal to Bill. Bill's dad waits to eat his vegetables last, regularly telling the family that he eats them only because he has to. He also regularly complains about how dinner has been prepared.

Bill saves his vegetables until last and usually gags when his mother orders him to eat them. Bill has been known to sit at the dinner table for an hour until the war of wills ends. Bill's mother serves casseroles and stews

regularly because they are convenient. Bill likes to eat breakfast cereal, fruit, and cheese and regularly requests these foods for snacks. However, his mother tries to deny his requests so that he will have an appetite for dinner. Bill's mother asks you what she should do to get Bill to eat.

Jill Braaten/McGraw-Hill Education

### Analysis

1. List 4 mistakes Bill's parents are making that contribute to Bill's poor eating habits.

2. List 4 strategies Bill's parents might try to promote good eating habits for Bill.

---

▶ **The American Academy of Pediatrics Guidelines for Parents of Toddlers**[38]

**Healthy Beverages**

- Choose milk or water for your child's beverage.

**Healthy Snacking**

- Offer 2 to 3 healthy, nutritious snacks as a part of the daily routine.

**Picky Eaters**

- Introduce a variety of foods multiple times and in multiple ways.

**Parent Provides, Child Decides**

- Offer healthy food in age-appropriate portions at meals and snacks.
- Let children decide what and how much to eat.

to eat. A sudden loss of appetite, however, is reason for concern because it may indicate the child is ill.

Many parents describe their preschoolers as "picky" eaters. Picky eating is usually just another method children use to express their strong desire for independence. Nagging, forcing, or bribing children to eat reinforces picky eating behaviors because of the extra attention. Overall, parents should focus on offering a variety of healthy foods and let the child exert some autonomy over specific types of food and the amounts eaten.

Tensions between parents, or between parents and children, especially during mealtime, often contribute to eating problems. Getting to the root of family problems and creating a more harmonious family atmosphere are important steps toward resolving many childhood feeding issues. In addition, many parents need to learn what to expect of children and appropriate food-related goals to set.

### Knowledge Check

1. When are appetite fluctuations during childhood considered a problem?
2. Why is bribing a child to eat a food not recommended?
3. What are some steps caregivers can take to help children accept new foods?
4. What is the best way to handle children who go on food jags or refuse to eat?

## Potential Nutrition-Related Problems of Childhood

During the childhood years, common nutrition-related problems are constipation and obesity. Some people consider hyperactivity to be diet-related, even though scientific evidence indicates otherwise.

### Constipation

Many children have bouts with constipation; for some, it is a chronic problem. The most common cause is eating too little fiber, drinking too little water, drinking too much milk, and/or not responding promptly to urges to defecate. Children who have been constipated may hold in feces because they fear they will have a painful bowel movement—this can create a cycle that makes constipation worse. To prevent constipation, children over the age of 1 year need to consume 14 grams of fiber per 1000 kcal, drink plenty of water, keep their milk intake to about 16 to 24 ounces daily, and get 60 minutes of active play each day. The treatment of constipation may involve the use of stool softeners, laxatives, or enemas; parents should consult a physician before using these. (Chapter 4 covers constipation in more detail.)

### Obesity

In the U.S., about one-third of school-age children have weights that are too high for good health. The number of cases of overweight is higher in minority populations.[39] As indicated earlier in this chapter, obesity is generally diagnosed when a child reaches the 95th percentile for BMI and a physical exam indicates the child is truly overfat. The main consequences of obesity include ridicule, embarrassment, possible depression, and, in females, short stature linked to early puberty.[40] These early maturers tend to be shorter and fatter throughout life than later maturers. Other significant health problems associated with obesity are cardiovascular disease, type 2 diabetes, and hypertension. Obesity, glucose intolerance, and hypertension in childhood increase the risk of premature death.

The likelihood that obesity and its problems will follow children and teens into adulthood is great.[41] Although overweight infants seldom become obese children, approximately 40% of all children with obesity and 80% of teens with obesity will be obese in adulthood and face an increased risk of several chronic diseases. Significant weight gain generally begins between ages 5 and 7, during puberty, or during the teenage years.[41]

Numerous factors, including heredity, environment, and dietary and activity behaviors, influence the development of obesity in children. In terms of heredity, the risk of obesity increases with parental obesity. Although the genetic potential of children born to 1 or both obese parents is high, the actual development is largely determined by whether the young person's environment promotes overeating and inactivity. Diet is an important factor, but inactivity is a key contributor to excess weight gain. One environmental factor, in particular, is of concern: inactivity caused by frequent television watching and computer and video games. Children and teens are getting more than 7 hours of screen time daily.[42] The American Academy of Pediatrics recommends a limit of 7 hours of TV and video time per week.[4] In addition, excessive snacking, overreliance on fast food restaurants, parental neglect, the media, a lack of safe areas to play, and easy availability of high fat/high energy food choices also contribute to childhood obesity.

The best approach is to prevent obesity from occurring by offering children a nutritious diet and plenty of opportunities for physical exercise (Fig. 17-9).[41] However, if the precursors to obesity (inactivity and overeating) begin to appear, prevention efforts are needed immediately (more activity and fewer high calorie foods, such as soft drinks, chips, and whole milk). If a child has obesity, the 1st step is to assess how much physical activity he or she engages in. If a child spends much free time in sedentary activities, more physical activities should be encouraged. Both the U.S. government and health professionals recommend 60 minutes or more of moderate to intense physical activity per day for children and adolescents. An overall active lifestyle will help children not only attain a healthy body weight but also keep a healthy body weight later in life and reduce the risk of obesity-related diseases. An increase in physical activity won't just happen; parents and other caregivers need to plan for it. Two good ideas are getting the family together for a brisk walk after dinner and finding an after-school sport the child enjoys.

Children with obesity often need to find a new way to relate to foods, especially snack foods. An important family rule might be that children are allowed to eat only while sitting at the dining table or in the kitchen. This rule can stop mindless snacking in front of the television and make all family members more conscious of when they are eating. It also might be helpful to put portions of snack foods on plates rather than to allow unlimited amounts of snacks, as often happens when children eat directly from a full box of crackers or cookies.

Children with obesity also need support, admiration, and encouragement to bring their weight under control. A child's self-esteem is extremely fragile. Obesity itself often affects the child's psyche and mental outlook (e.g.,

Children and teens should get 60 minutes or more of moderate to intense physical activity every day.

Bananastock/Getty Images

# Enjoy Moving

### Be physically active every day

**Plenty** — Moving Whenever You Can

**More** — Making Your Heart Work Harder

**Enough** — Stretching and Building Your Muscles

**Less** — Sitting Around

*Figure 17-9* For optimal health, children and teens should engage in a variety of physical activities and minimize sedentary activity. Most of the recommended 60 minutes per day of exercise should be devoted to aerobic activity, such as brisk walking or running. Muscle strengthening activities, such as gymnastics or push-ups, should be included at least 3 days per week as part of the daily 60 minutes of exercise. Bone strengthening activities, such as jumping rope or running, should be included at least 3 days each week as part of the daily 60 minutes of exercise.[43]

depression). Humiliation doesn't work; it only makes the child feel worse.

Resorting to a weight loss diet is usually not necessary. Children have an advantage over adults in dealing with obesity; their bodies can use stored energy for growth. Thus, if weight gain can be moderated, their normal increases in height can help bring weight under control. That is, growing taller while controlling weight gain lets the child "grow into" his or her weight. Low calorie diets that result in rapid weight loss are not recommended for children and teens because they can stunt growth, cause micronutrient deficiencies, and adversely affect eating behaviors and parent-child relationships. In children younger than 2 years of age, energy-restricted diets are not recommended because insufficient fatty acid and micronutrient intake may impair brain development.[2]

Teens who have attained their full adult height and have obesity may need a weight loss program. Still, weight loss should be gradual, perhaps 1 lb/week, and should generally follow the advice in Chapter 10. Weight loss medications and/or surgery may be indicated for some older children and teens.

Using a weight loss diet at any time during childhood or adolescence requires careful supervision by physicians and registered dietitian nutritionists.

Finally, it is important to remember that not all children will be tall and slender. In other words, some children simply weigh more than others. A healthy lifestyle with plenty of physical activity and nutritious foods remains the goal. (For a complete discussion of obesity, see Chapter 10.)

## Hyperactivity

Hyperactivity is a medical condition characterized by distractibility, impulsiveness, disruptive behavior, and overactivity. The specific cause of hyperactivity is not known; however, a variety of diet-related causes have been proposed, such as food allergies, food additives, and/or consumption of large amounts of sugar. Numerous carefully controlled research studies offer little scientific evidence to support claims that dietary factors cause hyperactivity or that eliminating certain dietary constituents will "cure" hyperactivity, although a healthy diet low in sugar, fat, and sodium with

adequate amounts of vitamins, minerals, fatty acids, fiber, whole grains, fruits, vegetables, fish, and legumes may offer some relief.[44] Providing megadoses of vitamins will not "cure" hyperactivity or attention disorders and can cause liver damage and intestinal upset.

## Modifications of Child and Teen Diets to Reduce Future Disease Risk

Parents sometimes wonder whether dietary modifications during the growing years can help reduce future risks of cardiovascular disease, hypertension, and type 2 diabetes. Recall that the development of atherosclerosis often begins in childhood and that many experts recommend screening for children whose families have histories of early development of cardiovascular disease or high blood cholesterol and treating children with high blood cholesterol with appropriate dietary changes and possibly medication (see Chapter 6).[45] In general, it's unnecessary to discourage children and teens from consuming nutrient dense foods, such as milk and animal proteins, just because they contain some animal fat. The overriding message is moderation in these and other fat sources.

Scientific data neither confirm nor refute the idea that eating less salt (sodium) reduces the risk of future hypertension. However, moderation in salt intake does help build good health habits for the future. If children become accustomed to less salt, they'll be less inclined to eat very salty foods as adults. Children with hypertension can benefit from dietary modifications, such as the DASH Diet (see Part 4), that lower sodium intake and increase fruit, vegetable, and reduced fat dairy food. However, if a child with hypertension does not respond to diet and lifestyle therapy, medications may be prescribed.

Type 2 diabetes has become more common in children and teenagers, primarily due to the rise in obesity in this age group. Up to 90% of children with type 2 diabetes are overweight at diagnosis. Experts are currently recommending that at-risk children be screened for this disease every 3 years, starting at age 10 or at the onset of puberty.[46] Besides obesity and a sedentary lifestyle, other risk factors include having a close relative with the disease and having a non-White heritage.[46] Appropriate diet and lifestyle intervention should be implemented, along with the use of medications when necessary. A focus on low glycemic load fruits, vegetables, and whole-grain breads and cereals is especially important (see Chapter 5).

# 17.7 Teenage Eating Patterns

As the teenage growth spurt begins, teenagers begin to eat more. The types and amounts of foods recommended for teenagers are the same as those for adults, except that teens have a greater need for calcium rich foods. Adolescents in the U.S. and Canada are relatively well nourished, but they are more likely to have poorer nutrient intake than those in other life stages. Teens generally eat less than the recommended amounts of fruits and vegetables. They also tend to choose foods that are higher in cholesterol, sugar, fat, protein, and sodium than recommended. Some consume excess amounts of alcohol, too.

In comparison with teenage females, the diets of teenage males are better (but not totally adequate), mainly because boys eat 700 to 1000 calories more each day than girls.[19] (The more calories you eat, the more opportunities you have to increase your nutrient intake.) A particular problem with teenage girls is that many replace milk with soft drinks, so they may not consume enough calcium to allow for maximum mineralization of bones, which increases their risk of osteoporosis later on. The RDA for calcium for both males and females between ages 9 and 18 years is 1300 mg per day, compared with only 1000 mg per day for children ages 4 to 8 years.

The diets of teens are less than optimal because they frequently eat out, skip meals, and snack. As teens get older, they tend to eat more and more meals and snacks away from home. Many of these meals and snacks are purchased from vending machines, convenience stores, and fast food restaurants. The main concern with these foods is that they often are higher in calories, fat, sugar, and sodium and lower in vitamin A, vitamin C, folate, calcium, iron, and zinc than foods served at home.

Meal skipping, especially by females and older teens, is common. Breakfast is the meal skipped most often, with more than half of teens skipping it most of the time. Snacking is very common, too—9 out of 10 teens report snacking. Overall, snacks account for about 25% of a teen's total calorie intake and, depending on the foods selected, can provide substantial nutrient contributions and even make up for nutrients missed when meals are skipped. However, most of the snacks popular among teens (e.g., potato and corn chips, ice cream, candy, cookies, cake, crackers, popcorn, presweetened breakfast drinks) are of limited nutritional value and are high in calories, fat, and/or sugar. Snacks as well as meals purchased away from home, with careful selection, can be rich in nutrients while keeping calories under control.

Nutrition experts agree that a way to get teens to eat more nutritiously is to make sure that, when teens do eat at home, they find nutritious foods, such as fruits, vegetables, milk, and fruit juice, instead of soft drinks, high fat snacks, and sugary foods.[47] Another technique for helping teens eat nutritiously is to have family meals at least a few times a week.

Drinking soft drinks in place of milk causes many teenagers to have inadequate calcium intake. This practice is linked to decreased bone mass and increased bone fractures in this age group. Figure 2-3 in Chapter 2 shows the stark contrast between milk and soft drinks with respect to calcium and other nutrients. If dairy products are not consumed, alternative calcium sources need to be included in the diet.

Fancy Collection/SuperStock

## Factors Affecting Teens' Food Choices

Teenagers face a variety of challenges. The struggle to establish independence and individual identity, gain peer acceptance, and cope with their heightened concern about physical appearance has profound effects on almost every aspect of teenagers' lives, including their food choices.[48] For instance, as they become more independent, teens begin making more decisions on their own, especially decisions regarding what, when, where, and with whom they will eat. As they forge their individual identities, they may participate in activities (e.g., clubs, sports, and part-time jobs) that change their lifestyles and cause them to miss many meals at home. Other factors that affect teens' food choices are their perceived and desired body images, their participation in athletics, and substance use.

### Body Image

Teens are very concerned and sensitive about their appearance. The physical changes of adolescence, peer pressure, messages from the media, and the struggle to develop self-identity cause many teens to be dissatisfied with their bodies and become vulnerable to body image problems. Boys report wanting to gain weight to look stronger

As teens become more independent, they make more and more decisions on their own, especially decisions about what, when, where, and with whom they will eat.

Sarah Casillas/DigitalVision/Getty Images

Females most at risk for irregular menstrual periods or amenorrhea are those who train many hours each day and those involved in endurance sports (e.g., marathon running) or sports that emphasize slimness (e.g., ballet, gymnastics).

fizkes/Getty Images

**Parent Responsibilities for Child and Teen Nutrition**

- Have regular family meals to promote social interaction.
- Eat and enjoy a wide variety of nutritious foods in healthy amounts.
- Provide a wide variety of nutrient dense foods, such as fruits and vegetables.
- Limit high energy density/nutrient poor foods, such as salty snacks, fried foods, and cookies.
- Limit sweet beverages, such as juice, soda, and sports drinks.
- Serve non-fat or low-fat dairy products after age 2 years.
- Serve portions appropriate for the child's size and age.
- Allow children with normal BMI or weight-for-height to self-regulate total calorie intake.
- Teach children about food and nutrition at the grocery store and when preparing meals.
- Limit snacking during sedentary behavior or in response to boredom.
- Promote and participate in regular daily physical activity.
- Limit sedentary behaviors, with no more than 2 hours of screen time daily and no TVs in children's bedrooms.
- Ensure that children and teens get sufficient restful sleep to lower obesity risk.
- Counteract inaccurate information from the media and other influences.
- Teach care providers (e.g., baby-sitters, coaches) what parents want their children to eat and how much they want them to engage in active play.
- Work with health-care providers to develop a plan for children with weight that is too low or too high for their height.

Adapted from Rattay JS and others. Dietary recommendations for children and adolescents: A guide for practitioners. *Pediatrics.* 2006;117:544.

and more muscular. Thus, they may opt for high protein diets or supplements, both of which can impair their health without producing the desired physique. Girls frequently perceive themselves as being larger than they really are and want to lose weight, even when they are within or below the average range. The fear of obesity and a distorted body image can cause children and teens to adopt fad diets, use diet pills, set weight goals that are unrealistic and unhealthy, severely restrict food intake, and develop eating disorders[49] (see Chapter 10).

### Athletics and Physical Performance

Many teenagers become involved in athletic activities. Often, their diets do not include the additional calories and nutrients they need to maintain normal growth and maturation and support their physical activity. Some athletes turn to restricted or unbalanced diets in the hope of improving performance or adhering to weight limits for their sport. For example, as described in Chapters 10 and 11, some participants in sports such as gymnastics, cheerleading, wrestling, and dancing dangerously restrict calories and/or water intake to keep their weight down. Football players and bodybuilders may overemphasize dietary protein and use protein supplements in hopes of building muscle. Endurance athletes may practice carbohydrate loading.

Diets that restrict or overemphasize any nutrient to the exclusion of other nutrients can detrimentally affect physical growth. For example, physical activity can help young people increase bone density and build a stronger skeleton. However, if calorie intake is so low that it leads to diminished body fat, which causes hormonal changes that result in irregular menstrual periods or amenorrhea, the effect on bone density is negative. The effect of these hormonal changes, which are frequently coupled with low calcium intakes at a time of increased need, can leave teenage women more susceptible to stress fractures now and osteoporosis later in life. Female athletes who experience irregular menstrual periods or amenorrhea because they are underweight need to gain weight and perhaps take a vitamin and mineral supplement (see Chapter 11). Coaches, with the help of dietitians, can encourage and teach teenage athletes to choose diets that help them grow, develop, and perform optimally.

### Substance Use

Substance use among adolescents is common. Substance users tend to have poor diets, a reduced interest in food and eating, and poor nutritional status. The diets of alcohol abusers are poor because alcohol, an empty calorie food, displaces nutritious foods (see Chapter 8). Alcohol also alters metabolism and increases the need for and/or excretion of nutrients. Malnutrition and growth stunting can result from excessive alcohol use. Cocaine abusers tend to lose interest in food because their desire for cocaine overrides hunger sensations. As a result, they frequently lose significant amounts of weight, are malnourished, and develop eating disorders. (When rats are given unlimited cocaine and food, they choose the cocaine over food until they starve to death.) Nicotine suppresses hunger, which can lead to inadequate food intake and weight loss. In addition, the diets of smokers tend to be lower in fiber, vitamins, and minerals than those of nonsmokers, and their need for vitamin C is nearly twice that of nonsmokers. Marijuana distorts the appetite and creates a desire in many users to snack on foods that are high in carbohydrates, fat, and calories.

## Helping Teens Eat More Nutritious Foods

Because psychological, social, and physical changes occur so rapidly during adolescence, it may be difficult to help teenagers see the value of healthy diets and exercise. In addition, many teens have a hard time relating today's actions to tomorrow's health outcomes.

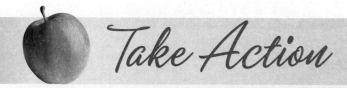

## Evaluating a Teen Lunch

These are 2 typical teen lunches and nutritional information for each.

| | Meal 1 | Meal 2 | Nutrient Needs for Teens |
|---|---|---|---|
| | 2 slices pepperoni pizza<br>1 chocolate candy bar<br>20 oz cola | 1 large hamburger<br>30 french fries<br>20 oz cola | |
| | | | Brand X Pictures |
| **Energy (kcal)** | 1100 | 1000 | Males: 3000 |
| | | | Females: 2200 |
| **Protein** | 32 | 20 | Males: 59 |
| | | | Females: 44 |
| **Vitamin C (mg)** | 5 | 18 | All sexes 9–13: 45<br>Males 14–18: 75<br>Females 14–18: 65 |
| **Vitamin A (g RAE)** | 300 | 10 | Males: 900 |
| | | | Females: 700 |
| **Iron (mg)** | 3 | 4 | Males: 11 |
| | | | Females: 15 |
| **Calcium (mg)** | 545 | 100 | All sexes: 1300 |

Burke/Triolo/Brand X Pictures

1. Keeping in mind that meals should meet about one-third of nutrient needs, what are the shortcomings and excesses of these meals? That is, given the nutritional information, compare these meals with one-third the RDA for protein, vitamin C, vitamin A, iron, and calcium.

2. How would you change these meals to improve balance and meet nutrient needs?

3. Reflect on your food choices as a teenager. Do you think your meal choices were balanced and varied? Why or why not? What could you have done to improve your nutritional habits at that time? How do your nutritional habits differ now?

Therefore, it's more effective to focus on the benefits they can reap right now than to talk about health hazards that may or may not happen later. One strategy for working with teenage boys is to stress the importance of nutrition and physical activity for physical development—especially muscular development—and for fitness, vigor, and health. One approach with teenage girls is to help them understand how choosing nutrient dense foods and enjoyable physical activities will lead to better health and a healthy weight.

Teens also need to know that healthful food habits don't mean giving up favorite foods. Small portions of fatty or sweet foods can complement larger portions of reduced fat dairy products, lean meats, fruits, vegetables, and whole-grain products. MyPlate is a useful tool for teenage meal plans. Teens typically need 1600 to 3200 calories per day. Table 2-6 (see Chapter 2) shows the amounts needed daily from each food group to meet energy needs.

### Knowledge Check

1. Which factors affect the quality of teenagers' diets as they get older?
2. What steps can caregivers take to help teens eat more nutritiously?
3. What effects can diminished body fat have on the bones of teenage girls?
4. What effects can substance use have on nutritional status?

# CLINICAL PERSPECTIVE

## Potential Nutrition-Related Problems of Adolescence

During adolescence, common nutrition-related challenges are teenage pregnancy, inadequate nutrition knowledge, and obesity (see Clinical Perspective: Potential Nutrition-Related Problems of Childhood). Some also consider acne to be a diet-related problem of adolescence, although research indicates otherwise.

### Teenage Pregnancy

The nutritional demands of pregnancy superimposed on those of adolescence can make it difficult for teens to consume a diet that can support their own growth and fetal growth. In an attempt to hide their pregnancy or retain a slim figure, pregnant teens may resort to excessive dieting. This practice, coupled with typical teenage eating patterns, lack of early prenatal care, and socioeconomic problems (e.g., incomplete education, lack of financial and emotional support), makes it even more difficult to meet nutritional demands.

A female who becomes pregnant within 2 years of menarche is at greatest nutritional risk because she is physically immature and her chances of developing complications (e.g., anemia, pregnancy-induced hypertension, spontaneous abortion, premature birth, a low-birth-weight infant) are much greater than for a female who becomes pregnant at a later point in her physical development. In addition, mothers less than 15 years old are much more likely to die as a direct result of pregnancy complications than are those who are older. (For a more complete discussion of pregnancy, refer to Chapter 16.)

Opportunities to learn about good nutrition at home and at school help prepare children and teens to make lifelong healthy food choices.

Fuse/Getty Images

An active lifestyle coupled with a healthy diet should be part of the growing years. Lack of exercise can lead to weight problems and high blood cholesterol.

Creatas/Jupiter Images

### Inadequate Nutrition Knowledge

Although children and teenagers are increasingly knowledgeable in many health areas, including nutrition, a large proportion are not exposed to nutrition education in any depth. Teaching children and teens about nutrition is important because those who do not have accurate nutrition knowledge cannot make informed food choices. In addition, learning how to select a nutritious diet can help them establish healthy eating habits early in life, and these eating habits tend to carry over into the adult years.

School-based nutrition education programs can help children and teens become more informed decision makers. Informed decision makers are able to make the best food choices for themselves at present, as well as safeguard their own health and the health of those for whom they are responsible (e.g., children, aging parents) throughout the adult years.

### Acne

Acne is a common teen concern—about 80% of teens experience it.[50] Although it's popularly believed that eating nuts, chocolate, and pizza can make acne worse, scientific studies have failed to show a strong link between any dietary factor and acne. Actually, acne develops when excess oily secretions block glands in the skin. Large doses of vitamins or minerals will not cure acne and can be toxic. As described in Part 4, physicians can prescribe a form of vitamin A (isotretenoin [Accutane® and Retin-A®]) to treat severe cases of acne. Although these treatments can be quite effective, close supervision by a physician is crucial because these vitamin A analogs can be toxic. Vitamin A itself is no help in treating acne, and excess amounts of vitamin A or related analogs can cause birth defects. Thus, females taking these vitamin A medications must not become pregnant.

678

# Chapter *Summary*

## 17.1 Normal growth and development are highly dependent on calorie and nutrient intake.

Physical growth rate is at its peak velocity during infancy, slows during childhood, then increases again during adolescence. During the growing years, height and weight increase, body composition changes, and body organs and systems mature.

inefficient kidneys. A young baby's need for water is usually met by human milk or formula. However, giving infants too much water can lead to water intoxication. Newborns frequently have low stores of fluoride, vitamin K, and vitamin D. Many children and adolescents consume too little calcium, zinc, folate, and vitamins A and C.

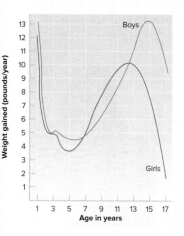

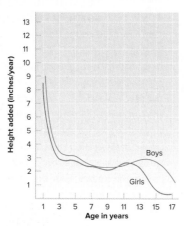

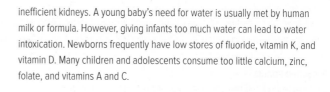

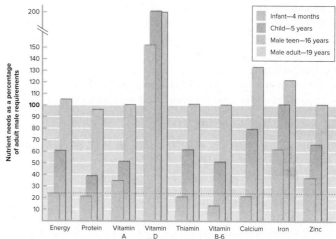

## 17.2 The single best indicator of a child's nutritional status is growth.

Growth charts matched to a child's biological sex are used to track physical growth over time to identify a child's growth percentile curves and determine if growth is progressing normally. BMI-for-age (or weight-for-length for children younger than 2 years old) is a good indicator of recent nutritional status. Stature-for-age is a good indicator of long-term nutritional status. Infants and young children who do not grow at the expected rate for several months and are dramatically smaller or shorter than other children the same age, especially those who fall below the 5th percentile, experience failure to thrive. Continually receiving minimal quantities of food may permanently and irreversibly stunt growth and development. Growth in height ceases when the growth plates at the ends of the bones fuse.

## 17.4 With few exceptions, human milk or iron-fortified infant formula and the baby's internal nutrient

stores meet an infant's nutrient needs at least until age 4 to 6 months. Human milk is the most ideal and desirable source of nutrients for infants. Commercial iron-fortified infant formulas provide a safe, nutritious alternative to human milk. Breastfed infants must be followed closely over the 1st week of life to ensure that feeding and weight gain are proceeding normally. All equipment and utensils used to prepare, store, and/or feed infant formula or human milk should be thoroughly washed and rinsed.

## 17.3 The greatest calorie and nutrient needs, pound for pound, occur during infancy.

The total quantity of calories and nutrients needed rises throughout childhood. The total quantities of nutrients and calories needed are greater during adolescence than at any other time, except pregnancy and lactation. Adolescent males need more of many nutrients than females because males are larger, develop more muscle mass and bone density, and have a longer, more intense growth period. Total fat should account for about 40 to 55% of a baby's calorie intake. It is wise to reduce fat intake gradually between ages 2 and 5 years until children are getting an average of 30 to 35% of their calories from fat. Dietary recommendations meant to reduce the risk of heart disease do *not* apply to children younger than 2 years old. Water is of special importance for babies because they have a large body surface area per pound of weight, turn over body water quickly, produce a large quantity of wastes, and have

**Table 17-2  Advantages to Infants Provided by Human Milk***

- Provides nutrients that are easily digestible and highly bioavailable and in amounts matched to needs
- Reduce risk of food allergies and intolerances, as well as some other allergies
- Provides immune factor proteins (antibodies) and other compounds (*Lactobacillus bifidus*) that reduce the risk of infections and diarrhea while the immune system is still immature
- Contributes to maturation of the GI tract and immune system
- Reduces risk of childhood asthma, obesity, diabetes, intestinal infections, misaligned teeth, ear infections, respiratory infections, and sudden infant death syndrome (SIDS)
- May enhance visual acuity, nervous system development, and learning ability by providing the fatty acid docosahexaenoic acid
- Establishes the habit of eating in moderation, thus decreasing the possibility of obesity later in life by about 20%
- Contributes to normal development of jaws and teeth for better speech development
- Is bacteriologically safe
- Is always fresh and ready

*For a summary of the advantages of breastfeeding for mothers, see Table 16-6 (Chapter 16).

# 17.5 When to introduce solid foods into an infant's diet hinges on the infant's nutritional needs, physiological capabilities, and

physical ability. Between 6 and 12 months of age, human milk or formula intake gradually decreases while solid food intake slowly increases. The recommended sequence for introducing solid foods is iron-fortified baby cereal, puréed fruits and vegetables, juices, protein rich foods, finger foods, and table foods. Infants should never be put to bed with a bottle or placed in an infant seat with a bottle propped up.

### Breastfed Infants

- Breastfeed for 6 months or longer, if possible. Then introduce iron-fortified infant formula when breastfeeding declines or ceases.
- Provide 400 IU/day of vitamin D.
- Consult a physician about the need for fluoride, vitamin B-12, and iron supplements.

### Formula-Fed Infants

- Use an iron-fortified infant formula for the 1st year of life.
- Consult a physician about the need for a fluoride supplement.

### All Infants

- Provide a variety of basic soft foods after 6 months of age, progressing slowly to a varied diet.
- Add iron-fortified infant cereal at about 6 months of age.

| **Table 17-4 Tips for Introducing Solid Foods** |
|---|
| 1. Start with teaspoon amounts of a single-ingredient food item, such as rice cereal, and increase the serving size gradually. |
| 2. Offer solid foods after some breastfeeding or formula-feeding, when the edge has been taken off the infant's hunger. |
| 3. Always feed solid foods from a spoon (a baby spoon [small spoon with a long handle] is best). |
|  • Spoon feeding is a skill babies need to learn. |
|  • Do not mix solid foods with a liquid and feed from a bottle because babies may choke. Also, eating baby foods from a bottle leads to poor eating habits and may cause children to be unwilling to try new foods and accept new textures. |
| 4. Hold the infant comfortably on the lap, as for breastfeeding or bottle-feeding, but a little more upright to ease swallowing. |
|  • Put a small dab of food on the spoon tip and gently place it on the infant's tongue. |
|  • Be calm and go slowly enough to give the infant time to get used to food. |
|  • Expect the infant to take only 2 or 3 bites of the 1st meals. |
|  • Let infants decide when they are hungry and when they have had enough to eat. |

# 17.6 Compared with babies, children tend to have erratic appetites. Pressuring a child

to eat more or less than desired tells the child not to trust his or her own hunger and satiety signals—this can lead to a lifetime battle with weight problems. Offering children 6 or so small meals succeeds better than limiting them to 3 meals daily. A serving size for children is equal to about 1 tablespoon per year of age, depending on the child's appetite. MyPlate for children is a useful meal planning tool. The most important nutrition lessons for children of all ages involve expanding their familiarity with new foods and helping them develop a willingness to accept new foods. Many preschool children go through periods of unpredictable and unusual eating behavior, such as going on food jags, refusing to eat, and being a picky eater. The best way to handle most of these behaviors is to not overreact, offer a variety of healthy foods, and let the child exert some autonomy over the specific types of food and amounts eaten.

# 17.7 The types and amounts of foods recommended for teenagers are the same as those for adults, except that

teens have a greater need for calcium. The diets of teens are less than optimal because teens frequently eat out, skip meals, and snack. The struggle to establish independence and individual identity, gain peer acceptance, and cope with their heightened concern about physical appearance affects teens' food choices. Other factors that affect their food choices are perceived and desired body image, participation in athletics, and substance use. MyPlate is a helpful tool for teenage meal plans.

©Fancy Collection/SuperStock RF

## Clinical Perspectives

### Infants
Common nutrition-related health problems during infancy are iron deficiency anemia, colic, gastroesophageal reflux, milk allergy, constipation, ear infections, dental caries, and diarrhea. Parents usually need to consult a physician when dealing with many nutrition-related conditions of infancy.

### Childhood
Potential nutrition-related problems of childhood are constipation and obesity. Some people consider hyperactivity to be diet-related, even though limited scientific evidence supports this.

### Adolescence
Common nutrition-related challenges of the adolescence life stage are teenage pregnancy, inadequate nutrition knowledge, and obesity. Children and teens from families with histories of early development of cardiovascular disease or high blood cholesterol should be screened for high blood cholesterol levels and, if necessary, treated with appropriate dietary changes and possibly medication.

## Global Perspective

Many children with autism have unusual eating patterns that can cause malnutrition, growth retardation, obesity, and heart disease. The usefulness of diet therapies in treating autism is under investigation.

# Study Questions

1. A 7-pound, 20-inch-long newborn who is growing normally will likely be _____ at 1 year old.

   a. 14 pounds, 25 inches
   b. 18 pounds, 30 inches
   c. 21 pounds, 30 inches
   d. 28 pounds, 40 inches

2. When adolescence ends, females have _____.

   a. twice as much body fat as males
   b. two-thirds as much lean body mass as males
   c. one-tenth more body water than males
   d. all of the above
   e. both a and b

3. The percentile growth curve a child follows depends mostly on _____.

   a. dietary intake
   b. genetic endowment
   c. biological sex
   d. age
   e. both a and b

4. Children are likely experiencing growth stunting if their _____.

   a. stature-for-age falls below the 5th percentile
   b. BMI-for-age falls below the 25th percentile
   c. weight-for-length rises above the 75th percentile
   d. head circumference-for-age exceeds the 95th percentile

5. A good indicator of long-term nutritional status is stature-for-age.

   a. true
   b. false

6. Pound for pound, which age group needs the most calories?

   a. newborns
   b. 6-month-olds
   c. 1-year-olds
   d. 3-year-olds

7. One reason water is of critical importance to infants is that they have _____.

   a. less body surface area per pound of body weight than adults
   b. a slow metabolic rate
   c. very efficient kidneys
   d. proportionately more body water than adults

8. To maintain a desirable iron status, breastfed infants should receive iron supplementation.

   a. true
   b. false

9. More iron is absorbed from human milk than from infant formula.

   a. true
   b. false

10. Which of the following is *not* true of proteins in human milk?

    a. They are easier to digest than those in cow milk.
    b. They increase the rate of iron absorption by an infant.
    c. They protect an infant from harmful pathogens.
    d. They often lead to allergies.

11. One sign that babies are developmentally ready for solid foods is that they _____.

    a. are 3 months old
    b. can control head movements
    c. have a strong extrusion reflex
    d. have more than 6 wet diapers per day

12. Which of the following commonly causes an allergic response in infants?

    a. egg whites
    b. peaches
    c. carrots
    d. egg yolk

13. Bribing a child to eat a food increases the child's desire for the food.

    a. true
    b. false

14. To help children accept a new food, caregivers should _____.

    a. serve the new food at the end of the meal
    b. give the child a reward for trying it
    c. make the child stay at the table until he or she tries it
    d. involve the child in preparing the food

15. Which of the following is true of infants with colic?

    a. They should not be breastfed.
    b. They cry for several hours each day.
    c. They stop crying after being fed or having their diapers changed.
    d. They sleep without waking during the night.
    e. All of the above are true.

16. Describe how height and body composition change during the growing years.

17. What are the advantages of regularly plotting a child's growth on a growth chart?

18. List 3 nutrients of concern during infancy, and describe the factors causing the concern.

**19.** Compare the advantages and disadvantages of feeding infants breast milk or infant formula.

**20.** Describe 3 factors that help determine when to introduce solid foods to an infant's diet.

**21.** List 3 reasons preschoolers are noted for "picky" eating. For each, describe an appropriate parental response.

**22.** Outline steps parents can take to help teens eat a nutrient dense diet.

**23.** Describe the common nutrition-related problems of infancy, childhood, and adolescence.

**24.** Explain how the eating patterns of some children with autism may put them at risk for malnutrition.

Answer Key: 1-c; 2-e; 3-e; 4-a; 5-a; 6-a; 7-d; 8-b; 9-a; 10-d; 11-b; 12-a; 13-b; 14-d; 15-b; 16-refer to Section 17.1; 17-refer to Section 17.2; 18-refer to Section 17.3; 19-refer to Section 17.4; 20-refer to Section 17.5; 21-refer to Section 17.6; 22-refer to Section 17.7; 23-refer to Clinical Perspectives; 24-refer to Global Perspective

# References

1. Institute of Medicine, Food and Nutrition Board. *Dietary Reference Intakes for calcium and vitamin D.* Washington, DC: National Academies Press; 2011.

2. Institute of Medicine, Food and Nutrition Board. *Dietary Reference Intakes for energy, carbohydrate, fiber, fat, fatty acids, cholesterol, protein, and amino acids.* Washington, DC: National Academies Press; 2002.

3. Butte NF and others. Body composition during the first two years of life: An updated reference. *Pediatr Res.* 2000;47:578.

4. U.S. Department of Agriculture, U.S. Department of Health and Human Services. Dietary Guidelines for Americans, 2020–2025, 9th Ed. 2020. dietaryguidelines.gov.

5. Lin-Su K and others. Body mass index and age at menarche in an adolescent clinic population. *Clin Pediatr.* 2002;41:501.

6. Kulkami B and others. Nutritional influences over the life course on lean body mass of individuals in developing countries. *Nutr Rev.* 2014;72:190.

7. Barlow SE and others. Expert committee recommendations regarding the prevention, assessment, and treatment of child and adolescent overweight and obesity: Summary report. *Pediatrics.* 2007;120:S164.

8. Gaffney KF and others. Baby steps in the prevention of childhood obesity. IOM guidelines for pediatric practice. *J Pediatr Nurs.* 2014;29:108.

9. Horta BL and others. Breastfeeding and intelligence: A systematic review and meta-analysis. *Acta Paediatr.* 2015;104:13.

10. Sharp WG and others. Feeding problems and nutrient intake in children with autism spectrum disorders: A meta-analysis and comprehensive review of the literature. *J Autism Dev Disord.* 2013;43:2159.

11. Suren P and others. Association between maternal use of folic acid supplements and risk of autism spectrum disorders in children. *JAMA.* 2013:309:570.

12. Buie T and others. Evaluation, diagnosis, and treatment of gastrointestinal disorders in individuals with ASDs: A consensus report. *Pediatrics.* 2010;125:S1.

13. Hediger ML and others. Reduced bone cortical thickness in boys with autism or autism spectrum disorder. *J Autism Dev Disord.* 2008;38:848.

14. Urbina EM, de Ferranti SD. Lipid screening in children and adolescents. *JAMA.* 2016;316:589.

15. National Heart, Lung, and Blood Institute. Expert panel on integrated guidelines for cardiovascular health and risk reduction in children and adolescents. NIH Pub #12-7486; 2012.

16. Heird WC. Nutritional requirements during infancy. In: Erdman IA and others, eds. *Present knowledge in nutrition.* 10th ed. Washington, DC: ILSI Press; 2012.

17. Institute of Medicine, Food and Nutrition Board. *Dietary Reference Intakes for vitamin A, vitamin K, arsenic, boron, chromium, copper, iodine, iron, manganese, molybdenum, nickel, silicon, vanadium, and zinc.* Washington, DC: National Academies Press; 2001.

18. Fayet-Moore F. Effect of flavored milk vs plain milk on total milk intake and nutrient provision in children. *Nutr Rev.* 2016;74:1.

19. Butte NF and others. Nutrient intakes of US infants, toddlers, and preschoolers meet or exceed Dietary Reference Intakes. *J Am Diet Assoc.* 2010;110:S27.

20. Dwyer J T. Feeding infants and toddlers study (FITS) 2016: moving forward. *J Nutr.* 2018;148:1575S.

21. National Academies of Sciences, Engineering, and Medicine 2020. Feeding Infants and Children from Birth to 24 Months: Summarizing Existing Guidance. Washington, DC: The National Academies Press.

22. American Academy of Pediatrics. Breastfeeding and the use of human milk. *Pediatrics.* 2012;115:496.

23. Lessen R, Kavanaugh K. Position of the Academy of Nutrition and Dietetics. Promoting and supporting breastfeeding. *J Acad Nutr Diet.* 2015;115:444.

24. Centers for Disease Control and Prevention. Breastfeeding report card, United States. 2018; www.cdc.gov/media/releases/2018/p0820-breastfeeding-report -card.html.

25. American Academy of Pediatrics. *Pediatric nutrition.* 8th ed. Chicago: AAP; 2019.

26. Bode L. The functional biology of human milk oligosaccharides. *Early Hum Dev.* 2015;91:619.

27. American Academy of Pediatrics. Infant food; www.uptodate.com/contents /starting-solid-foods-during-infancy-beyond-the-basics. *Infant Food and Feeding.* 2019. www.aap.org/en-us/advocacy-and-policy/aap-health-initiatives /HALF-Implementation-Guide/Age-Specific-Content/pages/infant-food-and -feeding.aspx.

28. Duryea TK and others. Patient education: Starting solid foods during infancy. 2020; www.uptodate.com/contents/starting-solid-foods-during-infancy-beyond -the-basics.

29. Clayton HB and others. Prevalence and reasons for introducing infants early to solid foods: Variations by milk feeding type. *Pediatr.* 2013;131:e1108.

30. Heyman MB and others. Fruit juice in infants, children, and adolescents: Current recommendations. *Pediatrics.* 2017;139:e20170967.

31. Greer FR and others. Effects of early nutritional interventions on the development of atopic disease in infants and children: The role of maternal dietary restriction, breastfeeding, time of complementary foods, and hydrolyzed formulas. *Pediatrics.* 2008;121:183.

32. Togias A and others. Addendum guidelines for the prevention of peanut allergy in the United States. *J Allergy Clin Immunol.* 2017;139:29.

33. Ogata BN, Hayes D. Position of the Academy of Nutrition and Dietetics: Nutrition guidance for children ages 2 to 11 years. *J Acad Nutr Diet.* 2014;114:1257.

34. Melina V, Craig W. Position of the Academy of Nutrition and Dietetics. Vegetarian diets. *J Acad Nutr Diet.* 2016;116:1970.

35. Roy PG, Stretch T. Position of the Academy of Nutrition and Dietetics: Child and adolescent federally funded nutrition assistance programs. *J Nutr Diet.* 2018;118:1490.

36. Council on School Health, Committee on Nutrition. Snacks, sweetened beverages, added sugars, and schools. *Pediatr.* 2015;135:1.

37. USDA, FNS. Team Nutrition local school wellness policy. 2016; www.fns.usda.gov /tn/local-school-wellness-policy.

38. American Academy of Pediatrics. Toddler food and feeding children. 2018; www.heart.org/en/healthy-living/healthy-eating/eat-smart/nutrition-basics/dietary-recommendations-for-healthy-children.

39. Hales CM and others. Prevalence of childhood and adult obesity in the U.S., 2015–2016. National Center for Health Statistics, Data Brief 288; 2017.

40. Centers for Disease Control and Prevention. *Childhood obesity causes & consequences*. Atlanta: CDC; 2016.

41. Witlock EP and others. Effectiveness of weight management interventions in children: A targeted systematic review for the USPSTF. *Pediatrics*. 2010;125:e396.

42. Chen W, Adler JL. Assessment of screen exposure in young children,1997–2014. *JAMA Pediatr*. 2019;173:391.

43. U.S. Department of Health and Human Services. *Physical activity guidelines for Americans*. 2nd ed. Washington, DC: U.S. DHHS; 2018.

44. Millichap JG, Yee MM. The diet factor in attention deficit/hyperactivity disorder. *Pediatrics*. 2012;129:300.

45. American Heart Association. AHA scientific statement. Dietary recommendations for healthy children. www.heart.org/en/healthy-living/healthy-eating/eat-smart/nutrition-basics/aha-diet-and-lifestyle-recommendations.

46. American Diabetes Association. Standards of medical care in diabetes—2019. *Diabetes Care*. 2019;42(Suppl 1).

47. Zabinski M and others. Psychosocial correlates of fruit, vegetable, and dietary fat intake among adolescent boys and girls. *J Am Diet Assoc*. 2006;106:814.

48. Henry B. Importance of the where as well as what and how much in food patterns of adolescents. *J Am Diet Assoc*. 2006;106:373.

49. U.S. Department of Health and Human Services, National Institute of Mental Health. Eating disorders. NIH publication no. 11-4901. 2014.

50. Kucharska A and others. Significance of diet in treated and untreated acne vulgaris. *Postepy Dermatol Alergol*. 2016;33:81.

©Ingram Publishing/Alamy RF

Living a long, healthy life depends on eating a nutritious diet and getting ample physical exercise. Learn more at **www.aoa. gov**. mavo/Shutterstock.

# 18 Nutrition during the Adult Years

## Learning Objectives

**After studying this chapter, you will be able to**

1. Describe the hypotheses about the causes of aging.
2. Discuss the factors that affect the rate of aging.
3. Explain how the basic concepts that underlie the Dietary Guidelines for Americans relate to adult health.
4. Compare the dietary intake of adults with the current recommendations.
5. Discuss the effects of physical, physiological, psychosocial, and economic factors on the food intake and nutrient needs of adults.
6. Describe community nutrition services for older persons.
7. Identify nutrition-related health issues of the adult years and describe the prevention and treatment of these health problems.
8. List the potential benefits and risks associated with the use of complementary and alternative medicine practices.

A LONG AND HEALTHY ADULT LIFE—that's what most people wish for. Thanks to a more abundant and nutritious food supply, a higher standard of living, and advances in medical technology, this wish can come true. In fact, it's already coming true. In the last century, the life expectancy for individuals born in the U.S. jumped more than 30 years. Babies born today can expect to live into their 80s—many will live even longer, and some will live the full life span possible for humans. The maximum life span for humans is generally accepted to be about 115 to 120 years.

More Americans may be living longer than before, but many are not living as healthfully (and perhaps not as long) as they could be. That is, their span of "healthy years" doesn't always correlate with the revolutionary increase in life expectancy that has occurred. Fewer than 1 in 5 people who live to age 65 or beyond are fully functional in their last year of life. For example, many suffer from diet-related chronic diseases, such as coronary heart disease, certain cancers, and osteoporosis, that impair their health. Older adults have less lean body mass and more body fat than when they were young, their body systems have slowed appreciably, and it takes longer for their body systems to restore balance. Even though old age is often equated with "going downhill," it doesn't have to be that way.[1]

This chapter will describe the physical and physiological changes that occur during adulthood and their impact on nutrient needs and food intake. It also will explore how dietary choices can help people keep their bodies performing efficiently and delay, prevent, or manage the declines associated with aging. That is, you'll discover how dietary practices, as well as exercise habits, can help you control the rate of aging and make the most of all the years of your life. Age quickly or slowly—it is partly your choice.

The increase in life expectancy has resulted in a phenomenon that is popularly referred to as "the graying of North America." That is, the proportion of people over age 65 is growing faster than any other segment of the population. A hundred years ago, only 1 person in 25 was 65 years or older. Today, this proportion has risen to 1 person in 6 and, in less than 40 years, people 65 years or older will account for 1 in every 4 people.

©George Doyle/Stockbyte/Getty Images RF

# 18.1 Physical and Physiological Changes during Adulthood

Adulthood, the longest stage of the life cycle, begins when an adolescent completes his or her physical growth. Unlike in earlier stages of the life cycle, nutrients are used primarily to maintain the body rather than support physical growth. (Pregnancy is the only time during adulthood when substantial amounts of nutrients are used for growth.) As adults get older, their nutrient needs change. For example, vitamin D needs are higher for persons in older stages of adulthood. Based on the needs for various nutrients, the Food and Nutrition Board divided the adult years into 4 stages: ages 19 to 30, 31 to 50, 51 to 70, and beyond 70 years. The intervals encompassing ages 19 through 50 are often referred to as young adulthood, 51 to 70 as middle adulthood, and beyond 70 as older adulthood.

Adulthood is characterized by body maintenance and gradual physical and physiological transitions, often referred to as aging. **Aging** can be defined as the time-dependent physical and physiological changes in body structure and function that occur normally and progressively throughout adulthood as humans mature and become older (Fig. 18-1). From the beginning of adulthood until age 30 or so, body systems are at their peak efficiency rate. Stature, stamina, strength, endurance, efficiency, and health are at their lifetime highs. The rates of cell synthesis and breakdown are balanced in most tissues. Then, after about age 30, the rate of cell breakdown slowly begins to exceed the rate of cell renewal, leading to a gradual decline in organ size and efficiency.

As the years progress, the cumulative effects of tissue breakdown lead to an erosion of the body's efficiency in functioning—these losses occur so slowly that it usually is decades before any great differences are obvious (Fig. 18-2). Even then, body systems and organs usually retain enough **reserve capacity** to handle normal, everyday demands throughout one's lifetime. Problems caused by diminished capacity typically don't arise unless severe demands are placed on the aging body. For example, alcohol intake can overtax an aging liver. The stress of shoveling a snow-covered sidewalk can exceed the capacity of the heart and lungs. Coping with an illness also can push an older body beyond its capacity.

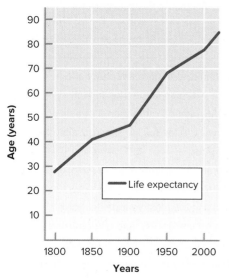

*Figure 18-1* Revolutionary advances that we now take for granted—more nutritious diets, antibiotics, vaccines, sewer systems, and women giving birth in hospitals—were responsible for increasing life expectancy in the 20th century. Unfortunately, in the past few years, life expectancy in the U.S. has dropped slightly due to high rates of drug overdose and suicide.

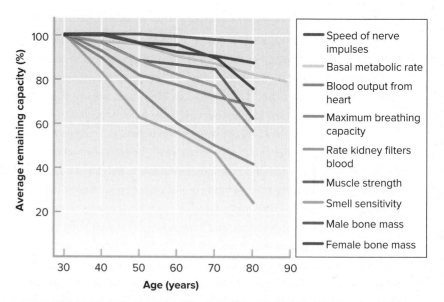

*Figure 18-2* Physiological functions typically decline with age. However, when compared with the average of younger adults, some older people show minimal or no losses. The physiological changes shown here likely are more reflective of those experiencing usual aging instead of successful aging.

**Table 18-1  Current Hypotheses about the Causes of Aging**

**Errors occur in copying the genetic blueprint (DNA).**

Once sufficient errors in DNA copying accumulate, a cell can no longer synthesize the major proteins needed to function and it dies. Damage to DNA in the mitochondria also contributes to the aging process.

**Free radicals damage cell parts.**

Electron-seeking free radicals can break down cell membranes and proteins. DNA in mitochondria typically shows this type of damage, which is linked to the aging process. One way to prevent free radical damage is to consume adequate amounts of antioxidants.

**Neuroendocrine communication and coordination diminish.**

Neurological and endocrine systems work together to regulate hormone secretions. With age, the communication and coordination between these systems decline. The blood concentration of many hormones, such as testosterone in men, falls during the aging process. Replacement of this and other hormones is possible, but the resulting risks and benefits are largely unknown.

**The immune system loses some efficiency.**

The immune system is most efficient during childhood and young adulthood but, with advancing age, it is less able to recognize and counteract foreign substances, such as viruses, that enter the body. Nutrient deficiencies can impair immune function.

**Autoimmunity develops.**

Autoimmune reactions occur when white blood cells and other immune system components begin to attack body tissues. Many diseases, including some forms of arthritis, involve this autoimmune response.

**Cross-linking, or glycosylation, of proteins occurs.**

Body proteins develop attachments or unnecessary cross-links that damage proteins or affect their function. For example, blood glucose, when chronically elevated, attaches to (glycates) various blood and body proteins. This action decreases protein function and can encourage the immune system to attack these altered proteins. Cross-linking also may decrease flexibility in key body components, such as connective tissue.

**Death is programmed into the cell.**

Each human cell can divide about 50 to 70 times. Once this total number of divisions has occurred, the cell dies (cellular senescence) and the contents of the cell spill out and damage neighboring cells. Considerable evidence is accumulating to indicate that cellular senescence is a key factor in aging and age-related diseases.[3]

**Excess energy intake speeds body breakdown.**

Underfed animals, such as spiders, mice, and rats, live longer than those that are well fed. Usual energy intake must be reduced by about 30% to see this effect. This approach may delay aging and the onset of age-related diseases, but nutritious diets without food restriction also lengthen life.[4]

---

*Critical* **Thinking**

Alexis has read several books supporting the idea that reducing typical energy intake by 30% can significantly extend one's life. Because she wants to do what is best for her 2 preschool children, she is thinking about adjusting her family's dietary habits to match this calorie restriction. What should you discuss with Alexis before she proceeds?

---

The causes of aging remain a mystery. Most likely, the physiological changes of aging are the sum of automatic cellular changes, lifestyle practices, and environmental influences, as listed in Table 18-1.[2] Even with the most supportive environment and healthy lifestyle, cell structure and function inevitably change over time. Eventually, cells lose their ability to regenerate vital internal parts and they die. This unavoidable dying of deteriorating cells is actually beneficial because it likely prevents diseases such as cancer. Unfortunately, there are negative consequences to this natural cell progression because, as more and more cells in an organ system die, organ function decreases. For example, kidney structures (nephrons) that filter blood to remove waste and reabsorb water, glucose, amino acids, and other nutrients are continually lost as we age. In some people, this loss exhausts the kidneys' reserve capacity and ultimately leads to kidney failure.

**reserve capacity** Extent to which an organ can preserve essentially normal function despite decreasing cell number or cell activity.

Even very healthy people have a shortened life expectancy if they are exposed to sufficient environmental stress, such as radiation and certain chemical agents (e.g., industrial solvents). Because cell aging and diseases such as cancer are aggravated by environmental factors, it makes good sense to avoid risks such as excessive sunlight exposure, hazardous chemicals, and environmental pollutants.

Chris Knorr/Designpics/PunchStock/Getty Images

The diseases and physical and physiological degeneration commonly observed in older people have long been assumed to be unavoidable consequences of aging. Certainly, some of the declines we blame on aging may be inevitable, such as gradual reductions in tissue and organ cell numbers, graying hair, and reduced lung capacity. However, many of the so-called usual or degenerative age-related changes can, in fact, be minimized, prevented, and/or reversed by healthy lifestyles (e.g., eating nutritious diets, exercising regularly, getting enough sleep) and avoiding adverse environmental factors (e.g., avoiding excessive exposure to sunlight and cigarette smoke). These discoveries have led researchers to introduce the concepts of "usual aging" and "successful aging."

## Usual and Successful Aging

Body cells age, no matter what health practices we follow. However, to a considerable extent, we can choose how quickly we age throughout our adult years. **Usual aging** refers to the age-related physical and physiological changes commonly thought to be a typical or expected part of aging, such as increasing body fatness, decreasing lean body mass, rising blood pressure, declining bone mass, and increasingly poor health. Many of these changes really represent the aging process accelerated by unhealthy lifestyle choices, adverse environmental exposures, and/or chronic disease. For instance, blood pressure does not tend to rise with age among people whose diets are traditionally low in sodium. Also, lean body mass is maintained much better in older people who exercise than in those who don't.

**Successful aging** (also called delayed aging), on the other hand, describes physical and physiological function declines that occur only because we grow older, not because lifestyle choices, environmental exposures, and chronic disease have aggravated or sped up the rate of aging.[5] Those who are successful agers experience age-related declines at a slower rate and the onset of chronic disease symptoms at a later age than usual agers. Striving to have the greatest number of healthy years and the fewest years of illness is often referred to as **compression of morbidity.** In other words, a person tries to delay the onset of disabilities caused by chronic disease and to compress significant sickness related to aging into the last few years—or months—of life. An example of this concept is illustrated for cardiovascular disease in Figure 18-3. Scientists don't yet know all of the factors that promote successful aging but, as you'll see later in this chapter, numerous nutrition-related and lifestyle factors are known or thought to play a role in slowing the aging process and increasing health span and life expectancy.

## Factors Affecting the Rate of Aging

The rate at which a person ages is individual; it is determined by heredity, lifestyle, and environment. Most of the factors that influence the rate of aging are directly linked to choices that are under our control, with the exception of heredity.

### Heredity

Heredity defines who you are biochemically and, to some extent, how long you will live (longevity).[1] Living to an old age tends to run in some families. If your parents and grandparents lived a long time, you are likely to have the potential to live to an old age, too. One of the most obvious genetic characteristics influencing longevity is biological sex. In the case of humans, as well as most other species, females tend to live longer than males.

Another genetic characteristic that can influence longevity is metabolic efficiency. Individuals with a "thrifty metabolism" require fewer calories for metabolic processes and are able to

A healthy diet throughout life promotes successful aging.

©Monkey Business Images/Shutterstock, Inc.

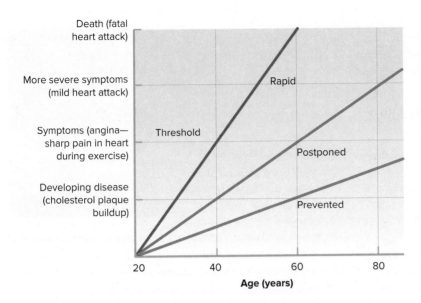

*Figure 18-3* Compression of morbidity, with cardiovascular disease as an example. The purple line depicts rapid deterioration in health; symptoms of cardiovascular disease appear by about age 40, symptoms and disability occur between ages 40 and 60, and death occurs at about age 60. A healthier lifestyle follows the blue line. Here, cardiovascular disease is postponed so that the symptoms are not apparent until age 60, severe symptoms are delayed until age 80, and death follows a few years later. The red line is ideal. Disease progresses so slowly that symptoms do not appear during a person's lifetime, so the disease never impedes activities.

store body fat more easily than those with faster metabolic rates. Throughout history, it was the individuals with thrifty metabolism who tended to live the longest because they efficiently stored fat during times of plenty and thus had the energy stores needed to survive frequent periods of food scarcity. However, today people with a thrifty metabolism living in technologically advanced countries where food is abundant and periods of scarcity are virtually nonexistent may find that their thrifty metabolism actually reduces longevity. Thrifty metabolism may enable them to store excessive amounts of body fat, which increases their risk of developing health problems that can shorten their lives (e.g., heart disease, hypertension, certain cancers). Yet another example of a genetic characteristic that may influence longevity is the rate of HDL-cholesterol production. Individuals who inherit an increased ability to produce abundant HDL-cholesterol may have a decreased risk of cardiovascular disease and a longer life. In contrast, those who inherit a reduced ability to produce HDL-cholesterol have a greater risk of premature heart disease and a shorter life.[1]

As you know, heredity is largely unchangeable, but it is not necessarily destiny—individuals can exert some control over the expression of their genetic potential. Both lifestyle and environment can modify the expression of genetic potential.

### Lifestyle

Lifestyle is one's pattern of living—it includes food choices, exercise patterns, and substance use (e.g., alcohol, drugs, tobacco). Lifestyle choices can have a major impact on health and longevity,[1,5,6] as well as on the expression of genetic potential. If individuals have a family history of premature heart disease, they can adjust their diets, exercise, and tobacco use patterns to slow the progression of the disease, get the medical care needed, and possibly extend their lifetime. The converse is true, too; that is, lifestyle choices (e.g., a high-fat diet and couch potato attitude) can increase susceptibility to diseases that hasten the rate of aging, ultimately shortening life expectancy, even if a person's genetic potential is for a very long life.

### Environment

Some aspects of the environment that exert a powerful influence on the rate of aging are income, education level, health care, shelter, and psychosocial factors. For instance, incomes that enable individuals to purchase nutritious foods, quality health care, and safe housing help decrease the rate of aging. Having the education needed to earn a sufficient income, as well as the knowledge required to select a nutritious diet and make wise lifestyle choices, also can slow the aging process. In addition, the willingness to seek health

▶ Besides having other long-lived family members, people who live to 100 years generally

- Do not smoke and do not drink heavily
- Gain little weight in adulthood
- Eat many fruits and vegetables
- Perform daily physical activity
- Challenge their minds
- Have a positive outlook
- Maintain close friendships
- Are (or were) married (especially true for men)
- Have a healthy rate of HDL-cholesterol production

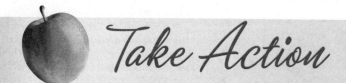

## Take Action

### Stop the Clock! Are You Aging Healthfully?

Americans spend several billion dollars every year on potions, gadgets, and books that are purported to extend life expectancy and stave off aging. None of these products live up to their promises; consequently, the quest for the fountain of youth remains alive. There is no surefire elixir that will stop aging in its tracks, but there are many things you can do to control the rate of aging.

If you want to stay younger longer, the following tips can help. Studies have found that adults who follow all of these suggestions have a physical health status comparable to people 30 years younger who follow few of these tips. How closely are you (or a parent or another older relative) following such a plan?

- ☐ Eat regular meals that, in combination, provide all the calories and nutrients you need in adequate, but not excessive, amounts.
- ☐ Limit alcohol consumption to a maximum of 1 or 2 drinks daily.
- ☐ Keep your body weight at a desirable level. Severe underweight and extreme overweight greatly reduce life expectancy.
- ☐ Exercise at least 3 or 4 times each week for an hour or more. But don't overdo exercise to the point that you become too thin, you are physically injured, or menstruation ceases.
- ☐ Sleep about 7 or 8 hours each day.
- ☐ Do not use any tobacco products or street drugs.
- ☐ Limit exposure to direct sunlight to no more than 15 minutes each day.
- ☐ Have regular medical and dental checkups. Seek health care as soon as it is needed and follow the instructions of your health-care provider.
- ☐ Take responsibility for maintaining your own health— don't leave everything in the hands of your health-care provider.
- ☐ Protect yourself from environmental pollutants.

- ☐ Minimize emotional stress, adjust to the causes of stress, or learn constructive techniques for dealing with, relieving, or managing it (e.g., meditation, massage, relaxation techniques, time management, or exercise).

©Nick Koudis/Getty Images RF

- ☐ Develop close, sustained, supportive friendships.
- ☐ Have an optimistic outlook and find ways to add meaning to your life. Laugh and relax regularly.
- ☐ Continue to learn and challenge your mind.

How many of the tips do you follow? If there are any you don't follow, it's probably a good idea to find ways to make them part of your lifestyle. Remember, you may not be able to turn back the clock, but you can keep it from ticking faster than it should.

Strength training helps older adults maintain muscle mass.
©Ronnie Kaufman/Blend Images LLC RF

### Critical | Thinking

The "fountain of youth" remains a mystery. Many people believe a source exists that can stop the aging process, allowing youth to remain. However, your friend asserts that the fountain of youth is not a place or a particular thing but, rather, a combination of diet and lifestyle. How can she justify this claim?

care promptly when it is needed, the capacity to follow the instructions of a health-care provider, and the desire to accept the responsibility for maintaining one's own health can slow the rate of aging. Likewise, shelter that protects individuals from physical danger, climatic extremes, and solar radiation helps slow the aging process. Allowing people to make at least some decisions for themselves and control their own activities (autonomy) and providing psychosocial support (informational and emotional resources)[7] promote successful aging and psychological well-being. In contrast, the body may be physically stressed and aging accelerated if any or all of the converse (i.e., insufficient income, low education level, lack of health care, inadequate shelter, and/or lack of autonomy and psychosocial support) are true.

Regular exercise can lead to a healthier, more energetic, and perhaps longer life. When regular exercisers and nonexercisers of the same age are compared, death rates are significantly higher among those who don't exercise. One reason for this difference in death rate may be that inactivity increases the risk of chronic diseases. Another reason may be that inactivity is responsible for many of the physiological changes seen in usual aging that gradually weaken the body.

©Keith Brofsky/Getty Images RF

Close, supportive relationships help promote successful aging.

©Kristy-Anne Glubish/Design Pics RF

## Knowledge Check

1. Why is an organ's reserve capacity important as a person ages?
2. What factors contribute to successful aging?
3. What does *compression of morbidity* mean?
4. What effects can lifestyle have on heredity?

## 18.2 Nutrient Needs during Adulthood

The challenge of the adult years is to maintain the body, preserve its function, and avoid chronic disease—that is, to age successfully. A healthy diet can help achieve this goal. One blueprint for a healthy diet comes from the Dietary Guidelines for Americans, discussed in Chapter 2. Its advice can be summarized into these main points:

1. Follow a healthy dietary pattern at every life stage to meet nutrient needs, achieve a healthy body weight, and reduce risk of chronic disease.
2. Customize and enjoy nutrient-dense food and beverages to reflect personal preferences, cultural traditions, and your food budget.
3. Focus on meeting food group needs with nutrient-dense foods and beverages and stay within calorie limits.
4. Limit foods and beverages higher in added sugars, saturated fat, and sodium, and limit alcoholic beverages.

Researchers have adapted MyPlate to acknowledge the unique needs of older adults. Appendix D reviews diet planning guidelines issued by Health Canada for Canadians. In addition, Chapter 1 discusses *Healthy People,* a U.S. federal agenda aimed at disease prevention and health promotion.

Overall, good nutrition benefits adults in many ways. Meeting nutrient needs delays the onset of certain diseases; improves the management of some existing diseases; speeds

**menopause** Cessation of menses in women, usually beginning at about age 50.

As we age, our nutrient needs change, but not the need to follow a healthy diet.

Lane Oatey/Getty Images

recovery from many illnesses; increases mental, physical, and social well-being; and often decreases the need for and length of hospitalization.[1] As you know, American adults are fairly well nourished, although some dietary excesses and inadequacies do exist. For instance, common dietary excesses are calories, fat, sodium, and, for some, alcohol. The diets of adult women tend to fall short of the recommended amounts of vitamins D and E, folate, magnesium, potassium, calcium, zinc, and fiber.[8] The diets of adult men tend to be low in the same nutrients, except vitamin D, which does not become problematic until age 50. The iron intake of most women during their childbearing years (ages 19 to 50) is insufficient to meet their needs; however, due to a reduced iron need after **menopause,** older women do get enough iron.[6]

People ages 65 and up, particularly those in long-term care facilities and hospitals, are the single largest group at risk of malnutrition. They may become underweight and show signs of numerous micronutrient deficiencies (e.g., vitamins B-6 and B-12 and folate). Friends, relatives, and health-care personnel should look for poor nutrient intake in all older people, including those who live in nursing home settings. Family members have a unique opportunity to make sure nutrient needs are met by looking for weight maintenance based on regular, healthful meal patterns.

To pinpoint those over age 65 at risk of nutrient deficiencies, the American Academy of Family Physicians, the Academy of Nutrition and Dietetics, and the National Council on Aging developed the Nutrition Screening Initiative checklist (Fig. 18-4). Older Americans, family members, and health-care providers can use the checklist to identify those at nutritional risk *before* health deteriorates significantly. If problems arise in consuming a healthful diet, a registered dietitian nutritionist can offer professional and personalized advice.

## Defining Nutrient Needs

The DRIs for adults are divided by biological sex into 4 age groups to reflect how nutrient needs change as adults grow older. These changes in nutrient needs take into consideration aging-related physiological alterations in body composition, metabolism, and organ function. For example, the calcium and vitamin D recommendations for older adult age groups exceed those for the youngest group to help offset changes such as reductions in the ability to absorb calcium and synthesize vitamin D in the skin.[9] In contrast to the rising need for calcium and vitamin D, the iron RDA for women declines in older age groups to reflect the decrease after menopause.[1]

### Calories

After age 30 or so, total calorie needs of physically inactive adults fall steadily throughout adulthood. This is caused by a gradual decline in basal metabolism. To a great extent, adults can exert considerable control over this reduction in calorie need by exercising. Exercise can halt, slow, and even reverse reductions in lean body mass and subsequent declines in calorie need. Also, keeping calorie needs high makes it easier to meet one's nutrient needs and avoid becoming overweight.

### Protein

The protein intake of adults of all ages in the U.S. and Canada tends to exceed current recommended levels. However, some recent studies indicate that consuming protein in amounts slightly higher than the RDA, but within the AMDR, may help preserve muscle and bone mass. Adults who have limited food budgets, have difficulty chewing meat, or are lactose intolerant may not get enough protein. Recall that any protein consumed in excess of that needed for the maintenance of body tissue will be broken down and used as energy or stored as fat. The waste products produced when protein is used for energy or stored as fat must be removed by the kidneys; excessive protein intake may accelerate kidney function decline.

▶ The Nutrition Screening Initiative uses the acronym "DETERMINE" to help identify older people whose health needs require extra attention.

- **D**isease
- **E**ating poorly
- **T**ooth loss or mouth pain
- **E**conomic hardship
- **R**educed social contact and interaction
- **M**ultiple medications
- **I**nvoluntary weight loss or gain
- **N**eed assistance with self-care
- **E**lder years above age 80

## A Nutrition Test for Older Adults

Here's a nutrition check for anyone over age 65. Circle the number of points for each statement that applies. Then compute the total and check it against the nutritional score.

| | Points | |
|---|---|---|
| | 2 | 1. The person has a chronic illness or current condition that has changed the kind or amount of food eaten. |
| | 3 | 2. The person eats fewer than 2 full meals per day. |
| | 2 | 3. The person eats few fruits, vegetables, or milk products. |
| | 2 | 4. The person drinks 3 or more servings of beer, liquor, or wine almost every day. |
| | 2 | 5. The person has tooth or mouth problems that make eating difficult. |
| | 4 | 6. The person does not have enough money for food. |
| | 1 | 7. The person eats alone most of the time. |
| | 1 | 8. The person takes 3 or more different prescription or over-the-counter drugs each day. |
| | 2 | 9. The person has unintentionally lost or gained 10 pounds within the last 6 months. |
| | 2 | 10. The person cannot always shop, cook, or feed himself or herself. |
| **Total** | | |

### Nutritional Score

**0–2: Good.** Recheck in 6 months.

**3–5 Moderate nutritional risk.** A local agency on aging has information about nutrition programs to help the elderly improve eating habits and lifestyle. Visit eldercare.acl.gov to find a local agency. Recheck in 6 months.

**6 or more: High nutritional risk.** A doctor or registered dietitian nutritionist should review this test and suggest how to improve nutritional health.

*Figure 18-4* There is more variation in health status among adults over age 50 than in any other age group. Among people age 70 and over, some are independent, healthy people, whereas others are frail and require almost total care. This means that chronological age is not useful in predicting physical health status (physiological age) or nutritional risk. This checklist can be used by older Americans themselves, family members, and health-care providers to identify those at nutritional risk before health deteriorates significantly. People at marginal risk should look for ways to avoid reaching the high-risk category. Those at high risk should discuss their responses to the checklist with a health professional to identify ways to improve their nutritional health.

Elderly individuals living in long-term care facilities are at increased risk for malnutrition.

©Thinkstock/Getty Images RF

A daily serving of a whole-grain breakfast cereal is a rich source of vitamins, minerals, and fiber and, so, contributes to healthy aging.

©C Squared Studios/Getty Images RF

**ostomy** Surgically created opening in the intestinal tract. The end point usually opens from the abdominal cavity rather than the anus (e.g., a colostomy).

### Fat

The total fat intake of adults of all ages is often at or above the recommendations. It's a good idea for almost all adults to reduce their total fat intake because of the strong link between high-fat diets and obesity, heart disease, and certain cancers. In addition, reducing fat intake "frees up" some calories that can be better "spent" on complex carbohydrates.

Intake of omega-3 fatty acids is often below recommendations. Eating fish twice weekly, as well as flaxseed and canola oil, can help adults meet omega-3 fatty acid needs and reduce the risk for age-related macular degeneration.

### Carbohydrates

The total carbohydrate intake of adults of all ages in the U.S. and Canada is often lower than recommended. In addition, many adults need to shift the carbohydrate composition of their diets to emphasize complex carbohydrates more and sugary, simple carbohydrates less. A diet richer in complex carbohydrates makes it easier to meet nutrient needs and stay within calorie bounds because many highly sweetened foods are low in nutrients and high in calories. Substituting foods rich in complex carbohydrates for sweets also makes it easier for the body to control blood glucose levels—a function that becomes less efficient as the increases in body fatness and inactivity associated with usual aging occur.[10] Declines in carbohydrate metabolism are so common that 20% of those age 65 years or older have diabetes. A diet rich in fiber helps adults reduce their risk of colon cancer and heart disease, lower their blood cholesterol levels, avoid constipation, and age successfully.[6] The typical American adult gets slightly more than half the recommended amount of dietary fiber.[11,12]

### Water

Many adults, especially those in the later years, fail to consume adequate quantities of water. In fact, many are in a constant state of mild dehydration and at risk of electrolyte imbalances. Low fluid intakes in older adults may be caused by a fading sensitivity to thirst sensations, chronic diseases, and/or conscious reductions in fluid intake in order to reduce the frequency of urination.[1] Some also may have increased fluid output because they are taking certain medications (i.e., diuretics and laxatives), have an **ostomy,** and/or experience an age-related decline in the kidneys' ability to concentrate urine. Dehydration is very dangerous and, among other symptoms, can cause disorientation and mental confusion, constipation, impacted fecal matter, and death.

### Minerals and Vitamins

Adequate intake of all vitamins and minerals is important throughout the adult years. The micronutrients that need special attention because they tend to be present in less than optimal amounts in the diets of many adults are calcium, vitamin D, iron, zinc, magnesium, folate, and vitamins B-6, B-12, and E. Adults with impaired absorption or who are unable to consume a nutritious diet may benefit from mineral or vitamin supplements matched with their needs. In fact, many nutrition experts recommend a daily balanced multivitamin and mineral supplement for older adults, especially for those 70 years of age and older. Supplements or fortified foods can be especially helpful when it comes to meeting vitamin D and vitamin B-12 needs.

Calcium and Vitamin D These bone-building nutrients tend to be low in the diets of all adults. They become particularly problematic after age 50. Inadequate intake of these nutrients coupled with their reduced absorption, the reduced synthesis of vitamin D in the skin, and the kidneys' decreased ability to put vitamin D in its active form greatly contribute to the development of osteoporosis.[9] Too little vitamin D may increase the risk of Alzheimer disease or other types of dementia.[13,14] Getting enough of these nutrients is a challenge for many older adults because food sources of vitamin D are limited in the North American diet, and the major sources—fatty fish and fortified milk—are not widely consumed by older adults. Plus, with increasing age, lactase production frequently decreases. As you know, 1 of the richest and most absorbable sources of these nutrients, milk, contains lactose. To get the vitamin D and calcium needed, many with lactose intolerance can consume small amounts of milk at mealtime with no ill effects. Calcium-fortified foods, cheese, yogurt, fish eaten with bones (e.g., canned sardines or salmon), and dark green leafy vegetables can help those with lactose intolerance meet calcium needs—but these foods often do not provide vitamin D. Just 10 to 15 minutes per day of sunlight can make a large difference in vitamin D status.

Iron   Iron deficiency anemia, the most common type of malnutrition during the adult years, is found most frequently in women in their reproductive years because their diets do not provide enough iron to compensate for the iron lost monthly during menstruation. Other common causes of iron deficiency in adults of all ages include digestive tract injuries that cause bleeding (e.g., bleeding ulcers or hemorrhoids) and the use of medicines, such as aspirin, that cause blood loss. Impaired iron absorption due to age-related declines in stomach acid production may contribute to iron deficiency in older adults.

Zinc   In addition to less than optimal dietary zinc intake during adulthood, zinc absorption declines as stomach acid production diminishes with age. Poor zinc status may contribute to the taste sensation losses, mental lethargy, and delayed wound healing many elderly adults experience.[11]

Sun exposure can help older adults meet their increased vitamin D need.

©Steve Mason/Getty Images RF

Magnesium   This mineral tends to be low in adults' diets. Inadequate magnesium intakes may contribute to the loss of bone strength, muscular weakness, and mental confusion seen in some elderly adults. It also can lead to sudden death from poor heart rhythm and is linked to cardiovascular disease, osteoporosis, and diabetes. The best source of magnesium is the diet; supplements can cause loose stools and diarrhea.

Potassium   Potassium is a nutrient of concern for most adults. Adequate intakes of this mineral are needed for maintaining fluid balance, transmitting nerve impulses, and contracting muscle. Adequate intakes also may help keep blood pressure normal. Fruits, vegetables, whole grains, legumes, and meats are good sources of potassium.

Folate and Vitamins B-6 and B-12   Sufficient quantities of folate, because of its neural tube defect prevention qualities, are very important to women during the childbearing years. In later years, folate and vitamins B-6 and B-12 are especially important because they are required to clear homocysteine from the bloodstream; elevated blood concentrations of homocysteine are associated with increased risk of the cardiovascular disease, stroke, bone fracture, and neurological decline seen in some elderly people.[15] Vitamin B-12 is particularly a problem for the older population because a deficiency may exist even when intake appears to be adequate. As people age, the stomach slows its production of acid and intrinsic factor, which leads to poor absorption of vitamin B-12 and eventually to pernicious anemia. Adults ages 51 and older need to meet vitamin B-12 needs with foods or supplements fortified with synthetic vitamin B-12.[16]

Vitamin E   The dietary intake of most of the population falls short of recommendations for vitamin E. Low vitamin E intake means the body has a reduced supply of antioxidants, which may increase the degree of cell damage caused by free radicals, promote the progression of chronic diseases and cataracts, and speed aging. In addition, low vitamin E levels can lead to declines in physical abilities.[17]

▶ The relationship between nutrition and chronic disease and/or the need for medications is a 2-way street. That is, food intake and nutrient needs are influenced by chronic diseases and the need for medications. However, food and nutrient intake influences the onset of chronic diseases and reduces the amount of medication needed in the early stages of chronic disease. A nutritious diet is a key factor in helping delay the onset of chronic disease and need for medication.

### Carotenoids

Dietary intakes of certain carotenoids have been shown to have a variety of important anti-aging and health protective effects. Specifically, beta-carotene (lutein and zeaxanthin) along with vitamin C, vitamin E, and zinc have been linked with the prevention of cataracts and age-related macular degeneration (see Part 4).[18] Diets high in fruit and vegetables, the major sources of carotenoids and other beneficial phytochemicals, are consistently shown to protect the body from a wide variety of age-related conditions.

### Knowledge Check

1. What are some examples of how a healthy diet can benefit adults?
2. Which nutrients tend to be too low in the diets of adults?
3. What are 3 signs that an older person's health needs extra attention?

## 18.3 Factors Influencing Food Intake and Nutrient Needs

Like those of other age groups, the food choices and nutritional adequacy of adults' diets depend on the interplay of physical, physiological, psychosocial, and economic factors. Alterations in any of these factors, such as age-related changes in body organs and systems or diminished psychological well-being, social interaction, or financial status, can result in deteriorations in the quality of dietary intake, nutritional status, and health.

### Physical and Physiological Factors

The implications of many of the physical and physiological changes that occur during adulthood for dietary intake and nutrient needs are summarized in Table 18-2. Some of the changes listed (e.g., tooth loss, loss in taste and smell perceptions) can influence dietary intake. Other changes (e.g., menopause, loss of lean body tissue) can alter nutrient and/or calorie needs. Still other changes (e.g., reduced stomach acidity, diminished kidney function) can cause changes in nutrient utilization. Chronic diseases and the need for medications are additional physiological changes many adults experience that can influence food intake and nutrient needs.

### Body Composition

The primary changes in body composition that occur as the adult years progress are diminished lean body mass (Fig.18-5), increased fat stores, and decreased body water.[21,22] Some muscle cells shrink and others are lost as muscles age; some muscles lose their elasticity as they accumulate fat and collagen. Loss of muscle mass leads to a decrease in basal metabolism, muscle strength, and energy needs. Less muscle mass also leads to lower physical activity, which makes the prognosis for maintaining muscle even worse. Clearly, it is best to avoid this vicious cycle.

Lifestyle greatly determines the rate of muscle mass deterioration. As you might predict, an active lifestyle helps maintain muscle mass, whereas an inactive lifestyle encourages its loss. In fact, much of what we associate with old age is due to a lifetime of physical inactivity. Ideally, an active lifestyle should be maintained throughout life and

The goal of adulthood is to preserve health as late into life as possible.

©David Sucsy/Getty Images RF

**Table 18-2** Nutritional Implications of Physical and Physiological Changes That Occur after about Age 30

| Usual Changes | Nutritional Implications of the Changes | Ways to Minimize the Changes and Promote Successful Aging |
|---|---|---|
| **Body Composition**<br><br>Gradual, steady decline in lean body mass (sarcopenia) and body water; slow increase in fatty tissue and redistribution of body fat from the limbs to the torso<br><br>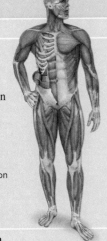<br>©McGraw-Hill Education<br> | Loss of lean body mass decreases metabolic rate, causing calorie needs to drop. Adequate fluid intake is important because a decrease in total body water elevates the risk of dehydration and decreases the body's ability to regulate its internal temperature. Excessive increases in fatty tissue raise the risk of developing conditions (e.g., high blood pressure, high blood glucose levels, type 2 diabetes, heart disease, and certain cancers) that may alter nutrient needs. | • Eating a nutritious diet that meets but does not exceed calorie needs, coupled with getting regular exercise (including strength training), helps minimize increases in body fat and maintains, or even rebuilds, lean body tissue and muscular strength, which keeps basal metabolic rate up. Eating high quality protein at each meal, increasing intake to be above the RDA but within the AMDR for protein, may help reduce the risk of sarcopenia. |
| **Skeletal System**<br><br>Slow, steady loss of bone minerals; in women, loss rises greatly in the 1st 5 to 10 years after menopause; may lead to osteoporosis<br><br>©McGraw-Hill Education | Adequate calcium and vitamin D during young adulthood helps build bone density and in the remainder of adulthood maintain and perhaps even increase bone mass. If osteoporosis causes adults to limit physical exercise, calorie needs will drop. | • Eating a diet rich in calcium and vitamin D, combined with participating in weight-bearing exercises, can help build bone mass until bones stop increasing in density (usually around age 35) and then can preserve bone minerals. Some older adults may benefit from medications that help preserve and rebuild bone.<br>• Keeping weight at a healthy level can help preserve bone mass.<br>• Avoiding smoking and chronic alcohol intake can help preserve bone mass because engaging in these behaviors increases the risk of osteoporosis. |
| **Cardiovascular and Respiratory Systems**<br><br>Gradual decrease in the ability of the heart and lungs to deliver oxygen and nutrient rich blood to body cells (aerobic capacity) and remove metabolic wastes; rise in blood pressure<br><br>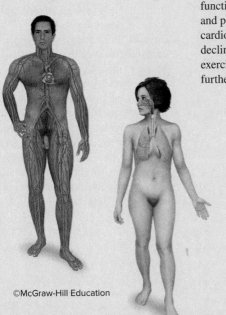<br>©McGraw-Hill Education | Reductions in cardiovascular and respiratory systems negatively affect the function of other organs (e.g., kidney, brain) and decrease their function, thus lowering calorie needs and possibly altering nutrient needs. If cardiovascular and respiratory system declines cause adults to limit physical exercise, calorie needs will drop further. | • Eating a low-fat diet rich in antioxidant nutrients, maintaining a healthy weight, avoiding cigarette smoke, and exercising regularly help minimize atherosclerotic plaque and reduce the risk of heart disease, which is responsible for many of the common age-related changes in the cardiovascular and respiratory systems.<br>• Exercising regularly helps maintain a high level of heart and lung fitness (aerobic capacity), raises blood levels of HDL-cholesterol, and keeps blood pressure under control.<br>• Lowering sodium intake, eating less animal protein, and maintaining a healthy weight may delay the onset of age-related rises in blood pressure.<br>• Monitoring and treating high blood lipids and hypertension help minimize damage to the cardiovascular system.<br>• Eating an antioxidant rich diet and avoiding polluted air and cigarette smoke help protect the lungs.<br>• A change that probably cannot be avoided is the decrease in lung capacity that occurs with aging: lungs shrink about 40% between the ages of 20 and 80. |

*(continued)*

**Table 18-2  Continued**

| Usual Changes | Nutritional Implications of the Changes | Ways to Minimize the Changes and Promote Successful Aging |
|---|---|---|
| **Digestive System** | | |
| Diminished chewing ability if gum disease occurs and leads to tooth loss or poorly fitting dentures; decline in efficiency of digestion and nutrient absorption due to reduced secretions of HCl and gastric, pancreatic, and intestinal digestive enzymes; decline in vitamin B-12 absorption due to decreased secretion of intrinsic factor; decline in the liver's ability to metabolize alcohol and drugs; slowdown in the movement of chyme through the intestines  ©McGraw-Hill Education | Chewing problems may result in reduced intake of crisp or chewy foods, such as raw fruits and vegetables, whole grains, and meats. Low HCl levels may impair absorption of iron, calcium, folate, vitamin B-6, and protein. Low HCl levels also may allow larger than normal numbers of bacteria to survive and establish colonies in the small intestine, where they may impair fat and fat-soluble vitamin absorption, compete for B-vitamins, and lead to weight loss and vitamin deficiencies. Diminished secretions of HCl and intrinsic factor virtually halt vitamin B-12 absorption. Reduced secretions of digestive enzymes may impair digestion and absorption of macronutrients; however, digestion is relatively complete and malabsorption does not seem to be a problem in most older adults. Decline in liver function slows detoxification of alcohol and drugs; thus, safe intake levels of these may drop. | • Consuming a diet rich in vitamin C (to maintain gums), calcium, and vitamin D (to maintain bone surrounding teeth) and practicing good dental health habits help prevent gum disease.<br>• Eating several smaller meals each day instead of a few larger ones may maximize digestion and absorption.<br>• Making dietary modifications and/or getting treatment to improve malabsorption problems caused by conditions such as celiac disease, diverticulitis, and excessive bacterial growth in the small intestine helps ensure nutrient needs are met.<br>• Consuming alcohol in moderation, if at all, helps avoid overtaxing the liver's capacity for detoxifying alcohol.<br>• Avoiding megadoses of vitamins and minerals helps prevent imbalances in nutrient absorption and the possibility of nutrient toxicities.<br>• Eating a fiber rich diet, drinking plenty of fluids, and exercising regularly help prevent constipation. |
| **Urinary System** | | |
| Decreased efficiency of kidneys in filtering out metabolic wastes, concentrating urine, and putting vitamin D synthesized in the skin in its active form; progressive weakening of the muscles that control urination <br><br>©McGraw-Hill Education | Diminished kidney function may impair reabsorption of glucose, amino acids, and vitamin C and impair vitamin D status. Excessive intakes of protein, electrolytes, water-soluble vitamins, and other substances that must be filtered out by the kidneys should be avoided. Vitamin D rich foods need to be emphasized or supplements may be needed. Reductions in the ability to concentrate urine increase the need for fluid. | Reductions in kidney filtration efficiency are not necessarily an inevitable part of aging; however, the factors that work to preserve filtration efficiency are not yet known.<br><br>• Avoiding excessive intakes of nutrients and other substances that must be filtered out by the kidneys throughout life may help preserve kidney function.<br>• Engaging in practices that maintain cardiovascular health (so that the blood supply to the kidneys is sufficient) and prevent hypertension (which can damage the kidneys) can help preserve kidney function.<br>• In those experiencing reduced kidney function, limiting protein and electrolyte intake may restore some kidney function.<br>• Doing muscular exercises and using behavior modification and medications can help improve functioning of the muscles that control urination. |

**Table 18-2 Continued**

| Usual Changes | Nutritional Implications of the Changes | Ways to Minimize the Changes and Promote Successful Aging |
|---|---|---|
| **Nervous System** | | |
| Gradual decline in number of cells that transmit nerve signals, which may result in decreased sensory perceptions (e.g., taste and smell), slowed reaction times, and impaired neuromuscular coordination, reasoning, and memory 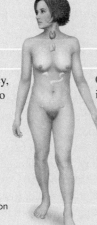 ©McGraw-Hill Education | Loss in taste and smell may reduce desire to eat, leading to weight loss. Diminished sensory perceptions may decrease secretions from the salivary glands, stomach, and pancreas and result in impaired digestion and blood glucose regulation. Neuromuscular coordination losses may make it difficult to cook or feed oneself. Reduced reasoning abilities can result in an inability to choose a nutritious diet, and memory losses may result in forgetting to eat altogether. | At present, there is no known way to prevent reductions in nerve cells. Studies suggest that smell and taste perception losses and decreases in intellectual performance are small in healthy older people. Losses appear greater in those with arteriosclerosis, inflammation, or obesity.<br><br>• Engaging in practices that promote a healthy cardiovascular system (exercise, low-fat diet) may help preserve nerve function.<br>• Experimenting with herbs, spices, and flavorings can boost the taste and smell of foods.<br>• Drinking enough fluids to prevent dehydration, engaging in lifelong learning, and getting enough sleep can help avoid mental confusion.<br>• Keeping blood pressure under control and consuming a nutrient dense diet may help preserve mental functioning.[19,20] |
| **Immune System** | | |
| Progressive decline in efficiency, which increases susceptibility to infection and disease 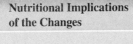 ©McGraw-Hill Education | Calorie and nutrient needs rise during infection and disease. | Reductions in immune system functioning may not be an inevitable part of aging; however, not all of the factors that preserve functioning are known.<br><br>• Eating a diet that meets nutrient needs and prevents obesity can help lower the risk of immune dysfunction.<br>• Exercising regularly may improve immune function.<br>• Avoiding prolonged emotional stress helps preserve immune function. |
| **Endocrine System** | | |
| Gradual decrease in hormone synthesis, hormone release, or sensitivity to hormones  ©McGraw-Hill Education | Decrease in sensitivity to insulin means that it takes longer for blood glucose levels to return to normal after a meal. Reduction in thyroid hormone slows metabolic rate and decreases calorie need. Decline in growth hormone leads to loss of lean body tissue and an increase in adipose tissue, both of which decrease metabolic rate and calorie need. Growth hormone reductions also cause the thinning of skin. | Eating a nutritious diet may influence endocrine activity by providing ample quantities of the compounds necessary for hormone synthesis and transport.<br><br>• Maintaining a healthy weight, getting physical exercise, and eating a low-fat, high fiber diet can slow, prevent, and perhaps even reverse decreased sensitivity to insulin.<br>• Maintaining lean body mass helps keep thyroid hormone levels steady.<br>• Getting injections of growth hormone may lead to increases in lean body mass, decreases in body fat, and thicker skin; however, long-term safety and usefulness are not known. |

*(continued)*

**Table 18-2  Continued**

| Usual Changes | Nutritional Implications of the Changes | Ways to Minimize the Changes and Promote Successful Aging |
|---|---|---|
| **Reproductive System** | | |
| Females: few changes until menopause (menopause is characterized by diminishing estrogen secretions and cessation of ovulation); males: slow decline in testosterone after age 60 | In females, iron needs drop when menopause occurs. Healthy diets and exercise are important after menopause because the decline in estrogen causes the risk of heart disease and osteoporosis to soar. In males, the reduction in testosterone may contribute to the loss of lean body tissue, which diminishes calorie needs. | Age-related changes in the reproductive system currently appear to be unalterable. |

©McGraw-Hill Education

include both aerobic and strength training (Table 18-3). Physical activity increases muscle strength and mobility, improves balance and decreases the risk of falling, eases daily tasks that require some strength, improves sleep, slows bone loss, and increases joint movement, thus reducing injuries. It also has a positive impact on a person's mental outlook. Strength training (resistance) helps reverse some of the decline in daily function associated with the muscle loss typically seen in older adulthood.

As lean tissue declines with age, body fat often increases. Much of this increase results from overeating and limited physical activity, although even athletic men and lean women typically gain some degree of midsection fat after age 50. A small fat gain in adulthood may not compromise health, but large gains are problematic. Recall that obesity can raise blood pressure and blood glucose and make walking and performing daily tasks more difficult.

Decreases in body weight are common in adults age 70 and older. Weight loss is a problem for older people in particular because it increases the risk of nutrition-related

▶ Researchers believe that maintaining lean muscle mass may be the most important strategy for successful aging. This is because maintaining lean muscle mass

- Maintains basal metabolic rate, which helps decrease the risk of obesity
- Keeps body fat low, which helps control blood cholesterol levels and helps avoid the onset of type 2 diabetes
- Maintains body water, which decreases the risk of dehydration and improves body temperature regulation

Loss of muscle mass (sarcopenia) is very prevalent in elderly individuals and greatly increases their risk of illness and death.

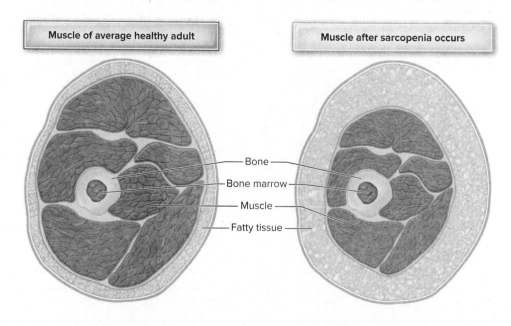

*Figure 18-5* Body composition changes as we age. Loss of muscle mass causes a condition called sarcopenia. Exercise coupled with a healthy diet can maintain muscle mass and reduce the risk of sarcopenia.

**Table 18-3 Strength Training Recommendations for Older Adults**

- Exercise at least 2 days per week in 30-minute sessions.
- Exercise different muscle groups in each session.
- Perform exercises that involve the major muscle groups (e.g., arms, shoulders, chest, abdomen, back, hips, and legs) and exercises that build grip strength.
- If weights are used, start with a weight that can be lifted only 8 times. Use it until able to lift it easily 10 to 15 times. When able to perform 2 repetitions, add more weight until it can be lifted only 8 times. Continue this process until reaching the goal weight.
- Rest between sets of exercises.
- Avoid locking the joints in the arms and legs.
- Stretch after completing all exercises.
- Stop exercising if pain begins.
- Breathe during strength exercises.

Source: National Institute on Aging; www.nia.nih.gov/health/exercise-physical-activity

illness and death. It may indicate illness, reduced tolerance to medication, or withdrawal from life.[23] Weight loss also may indicate that nervous system and hormonal factors are depressing feelings of hunger (see Chapter 10). The effects of current medication, as well as changes in taste and smell, may inhibit appetite, too. In addition, many older people live alone, a circumstance associated with depressed appetite.

## Skeletal System

Recall from Chapter 15 that bone loss in women occurs primarily after menopause. Bone loss in men is slow and steady from middle age throughout later life. Many older people may suffer from undiagnosed osteomalacia, a condition primarily caused by insufficient vitamin D. Osteoporosis can limit the ability of older people to move about, shop, prepare food, and live normally. Consuming adequate vitamin D, calcium, and protein and not smoking, drinking alcohol moderately or not at all, and engaging in weight-bearing exercises can help preserve bone mass. Medications also can help lessen bone loss.

## Cardiovascular System

The heart pumps blood less efficiently in some older people, usually because of insufficient physical activity. However, the decline in the heart's ability to pump blood is not inevitable with aging and does not occur among older people who remain physically active. In fact, it is thought that inactive lifestyles may contribute as much to the risk of cardiovascular disease as does smoking a pack of cigarettes per day. Exercising regularly, not smoking, and eating a low-fat diet rich in nutrients and moderate in sodium can help protect the cardiovascular system (see Chapter 6).

## Respiratory System

Lung efficiency declines somewhat with age and is especially pronounced in older people who have smoked or continue to smoke tobacco products. Breathing becomes shallower, faster, and more difficult as the amount of active lung tissue decreases. Smoking often leads to emphysema and lung cancer. The decrease in lung efficiency contributes to a general downward spiral in body function; breathing difficulties limit physical activity and endurance and frequently discourage eating. Besides not smoking, eating an antioxidant rich diet and being physically active help preserve lung function.

## Digestive System

As mentioned previously, the production of HCl, intrinsic factor, and lactase declines with advancing age and, as a result, impairs the absorption of several nutrients.[24] Constipation is the main intestinal problem for older people. To prevent constipation, older people should

**Exercise Guidelines for Adults**

**Aerobic (endurance) exercise:** moderate intensity exercise for 30 minutes daily on 5 days weekly or vigorous intensity exercise for 20 minutes daily for 3 days weekly. Brisk walking, yard work, and dancing are examples of endurance exercises.

**Strength (resistance) exercise:** 2 or 3 days per week, exercise each major muscle group with 2 to 4 sets of exercise, 8 to 20 repetitions each. Lifting weights and using resistance bands are examples of strength exercises.

**Flexibility exercise:** 2 or 3 days per week, stretch muscles 10 to 30 seconds, repeating 2 to 4 times for a total of 60 seconds per stretch. To improve flexibility, try yoga or stretching exercises.

**Neuromotor (balance) exercise:** 20 to 30 minutes on 2 or 3 days per week, exercise to improve balance, agility, coordination, and walking and to prevent falls. Practicing Tai Chi, standing on 1 foot, and walking heel-to-toe are ways to increase balance. Learn more at **go4life.nia.nih.gov/how-exercise-helps/.**

One in 5 older people has trouble walking, shopping, and cooking food.

©Brand X Pictures/Getty Images RF

▶ Older people should work with their physicians to develop a plan for limiting falls. Many falls are caused by neuromuscular coordination losses, impaired vision, walking and balance disorders, lack of regular physical activity, side effects of medication, and environmental hazards.

meet fiber needs, drink enough fluids, and exercise. Fiber medications are generally unnecessary but may be useful when total energy consumption does not allow for enough fiber intake. Because some medications can cause constipation, a physician should be consulted to determine if a laxative or stool softener is needed.

In addition to changes in the GI tract, the functions of the accessory organs decline as we age. For instance, the liver functions less efficiently. A history of significant alcohol consumption or liver disease will cause the liver to function even less efficiently. As liver efficiency declines, its ability to detoxify many substances, including medications, alcohol, and vitamin and mineral supplements, drops (see Chapter 8). The possibility for vitamin toxicity increases.

The gallbladder also functions less efficiently in later years. Gallstones can block the flow of bile out of the gallbladder into the small intestine, thereby interfering with fat digestion. Obesity is a major risk factor for gallbladder disease, especially in older women. A low-fat diet or surgery to remove the gallbladder may be necessary.

Although pancreatic function may decline with age, this organ has a large reserve capacity. One sign of a failing pancreas is high blood glucose, although this can occur as a result of several conditions. The pancreas may be secreting less insulin, or cells may be resisting insulin action (as is commonly seen in obese people with upper body fat storage). Where appropriate, improving nutrient intake, engaging in regular physical activity, and achieving and maintaining a healthy weight can improve insulin action and blood glucose regulation.[1]

Poor dental health contributes to digestive and food intake problems. About 19% or more of older people in North America have lost all their teeth.[25] Attention to dental hygiene and care throughout life greatly lowers this risk. Periodontal (gum) disease commonly causes tooth loss. Replacement dentures enable some people to chew normally, but many older adults—especially men—have denture problems. Some older adults also have trouble swallowing foods and/or liquids. This problem, called dysphagia, may decrease food intake and lead to weight loss, dehydration, and malnutrition. When people have problems chewing or swallowing, serving softer foods and allowing extra time for chewing and swallowing encourages eating.

## Urinary System

Over time, the kidneys filter wastes more slowly as they lose nephrons. The deterioration significantly decreases the kidneys' ability to excrete the products of protein breakdown, such as urea. As a result, individuals often need to avoid excess protein and keep intake at or slightly below the RDA.

Incontinence, the inability to control the muscle responsible for holding urine in the bladder, affects up to 20% of older adults living at home and about 55% of those in nursing homes. The fear of being unable to control one's bladder or the embarrassment of having to wear leak-proof, protective underwear causes many to restrict fluid intake (resulting in dehydration and constipation) and become socially isolated.

## Nervous System

A gradual loss of the nerve cells that transmit signals may decrease taste and smell perceptions and impair neuromuscular coordination, reasoning, and memory. Both hearing and vision decline with age. Hearing impairment is greatest in those who have been exposed constantly to loud noises, such as urban traffic, aircraft noise, and music. Because they cannot hear well, older people may avoid social contacts, which increases their risk of poor dietary intake.

Declining eyesight, frequently caused by retina degeneration and cataracts, can affect a person's abilities to grocery shop, locate desired foods, read labels for nutritional content, and prepare foods at home. Vision losses also may cause people to curtail social contacts, reduce physical activity, and not practice daily personal health and grooming routines. Macular degeneration, a form of failing eyesight in old age, is quite common, affecting about 1.75 million adults in the U.S. A

Grocery shopping can become more difficult in one's older years. Assistance from others is often very helpful.

©Steve Debenport/Getty Images RF

major risk factor is cigarette smoking. Diets rich in carotenoids help reduce the risk of macular degeneration. The risk of developing cataracts is decreased by consuming a diet rich in fruits and vegetables.

Neuromuscular coordination losses may make it difficult to shop for and prepare food. Physical tasks as simple as opening food packages can become so difficult that individuals restrict their dietary intake to foods that require little preparation and depend on others to provide food that is ready to eat. Eating may become difficult, too. Loss of coordination makes it a challenge to grasp cup handles and manipulate eating utensils. As a result, older adults may avoid foods that can be easily spilled (e.g., soups and juices) or that need to be cut (e.g., meats, large vegetable pieces) and restrict food intake to easy-to-eat finger foods. Some may even withdraw from social interaction and eat alone, which can lead to inadequate nutrient intake.

Changes in the immune system make handling food safely even more important as adults get older.

©Stockbyte/Getty Images RF

### Immune System

With age, the immune system often operates less efficiently. Consuming adequate protein, vitamins (especially folate and vitamins A, D, and E), iron, and zinc helps maximize immune system function. Recurrent sicknesses and poor wound healing are warning signs that a deficient diet (especially protein and zinc) may be hindering the function of the immune system. On the other hand, overnutrition appears to be equally harmful to the immune system. For example, obesity and excessive fat, iron, and zinc intake can suppress immune function.

### Endocrine System

As adulthood progresses, the rate of hormone synthesis and release can slow. A decrease in insulin release or sensitivity to insulin, for instance, means that it takes longer for blood glucose levels to return to normal after a meal. Maintaining a healthy weight, exercising regularly, eating a diet low in fat and high in fiber, and avoiding foods with a high glycemic index can enhance the body's ability to use insulin and restore elevated blood glucose levels to normal after a meal.

### Reproductive System

When menopause occurs, iron needs decline. A diet rich in vitamin D and calcium can help stave off the rapid loss of bone minerals that occurs after menopause. Testosterone production may decline as men age, leading to a loss of lean body mass, which results in a decreased metabolic rate and lowered calorie needs.

### Chronic Disease

The prevalence of obesity, heart disease, osteoporosis, cancer, hypertension, and diabetes rises with age.[1] Half of all adults have a chronic and potentially debilitating disease. One out of every 4 adults has at least 2 chronic conditions—this rises to 3 out of every 4 people over age 65. Chronic diseases may have a strong impact on dietary intake. For instance, excessive fatness, heart disease, and osteoporosis may impair physical mobility to the extent that victims are unable to shop for and prepare food. Chronic disease also can influence nutrient and calorie needs. Cancer, for example, boosts both nutrient and calorie needs. Hypertension may indicate a need to lower sodium intake. Nutrient utilization can be affected by chronic disease, too. For instance, diabetes alters the body's ability to utilize glucose. In addition, the effects of heart disease on the kidneys may impair their ability to reabsorb glucose, amino acids, and vitamin C.

Keeping blood pressure under control helps preserve the function of the circulatory system, which in turn helps keep other body systems working well.

©Ariel Skelley/Getty Images RF

# Drug-Nutrient Interactions

Many drugs can interact with food and/or nutrients. Food-drug or drug-nutrient interactions can change drug absorption, breakdown, or excretion or the way the drugs work. For instance, nutrients or compounds in food may increase or decrease the rate or amount of the drug that is absorbed. Food or nutrients may alter the rate or amount of the drug excreted or reabsorbed by the kidneys. Food or nutrients also can boost or limit the metabolism of drugs.

Drugs also affect food and nutrients. Drugs may increase, decrease, or block nutrient absorption. They can raise or lower the rate of nutrient excretion. Drugs also can accelerate the rate at which nutrients are metabolized, which results in increased need for the nutrient. The interactions that can occur between drugs and food or nutrients vary by drug. Table 18-4 provides examples of commonly used drugs that interact with food and nutrients.

Preventing interactions usually means avoiding certain foods completely or eating them at times well before or after taking a drug. To prevent interactions between drugs and food or nutrients, know which drugs you are taking and why. Also, carefully follow the instructions for taking each drug. For instance, should you take it on an empty stomach, wait an hour before taking calcium supplements, or avoid drinking grapefruit juice when taking the drug? Review the information sheet that comes with prescriptions or appears on the drug manufacturer's website to learn about how to take the drug—or ask a pharmacist to summarize the interactions that may occur with certain foods. Be sure to tell health-care providers and pharmacists about nutrient and herbal supplements you take—these can interact with drugs, too. It also is a good idea to fill all prescriptions at the same pharmacy—having a complete record of prescriptions from all prescribers helps pharmacists provide accurate advice. Learn more about herb-drug interactions in Clinical Perspective: Complementary and Alternative Health Approaches.

## Drugs and Supplements

According to the Food, Drug, and Cosmetic Act and FDA regulations, a drug is a substance that is used to diagnose, cure, mitigate, treat, or prevent disease and articles (other than food) intended to affect the structure or function of the body. Older adults are major consumers of drugs (prescription and over-the-counter) and nutrient and herbal supplements. Half of all people over age 65 take several medicines each day. The rate of supplement use increases throughout adulthood, until by age 50 approximately half of all adults are using supplements daily.[26] Physiological declines that occur during aging (e.g., reduced body water, reduced liver and kidney function) cause the effects of medications and nutrient supplements to be exaggerated and persist longer in older adults.

Drugs can improve health and quality of life, but some also adversely affect nutritional status, particularly of those who are older and/or take many different medications. For instance, some drugs depress taste and smell acuity or cause anorexia or nausea that can blunt interest in eating and lead to reduced dietary intake. Some drugs alter nutrient needs. Aspirin, for example, increases the likelihood of stomach bleeding, so long-term use may elevate the need for iron, as well as other nutrients. Antibiotics may deplete the body of vitamin K. Some drugs may impair nutrient utilization—diuretics and laxatives may cause excessive excretion of water and minerals. Even vitamin and mineral supplements can affect nutritional status. Iron supplements taken in large doses can interfere with the functioning of zinc and copper. Folate supplements can mask vitamin B-12 deficiencies.

People who must take drugs should eat nutrient dense foods and avoid any specific food or supplement that interferes with the function of the drug. For example, vitamin K can reduce the action of oral anticoagulants, aged cheese can interfere with certain drugs used to treat hypertension and depression, and grapefruit can interfere with tranquilizers and drugs that lower cholesterol levels (see Clinical Perspective: Drug-Nutrient Interactions). A physician, pharmacist, or registered dietitian nutritionist should be consulted about any restrictions on food and/or supplements.

## Psychosocial Factors

A positive outlook on life and intact support networks help make food and eating interesting and satisfying. In contrast, apathy and depression caused by feelings of social

▶ Obtaining enough food may be difficult for some older persons, especially if they are unable to drive and do not have friends or relatives close by to help with cooking or shopping. Older people may equate a request for help with a loss of independence. Pride or the fear of being victimized by those they hire may stand in the way of much needed help.

▶ Pharmaceutical companies market liquid meal replacement formulas to older adults. Previously, these products were primarily used in hospitals and nursing homes. Many of these products have an unusual taste because of the vitamins and types of proteins that have been added. Older adults can decide if the convenience, cost, and taste make this a wise diet choice.

**Table 18-4 Potential Drug-Nutrient Interactions for Commonly Used Drugs**

| Drugs (Examples) | Use | Nutrients Affected | Potential Mechanism |
|---|---|---|---|
| Antacids (Maalox®) | Reduce stomach acidity | Calcium, vitamin B-12, and iron | Decreased absorption due to altered gastrointestinal pH |
| Anticoagulants (Coumadin®) | Prevent blood clots | Vitamin K | Interference with utilization |
| Aspirin | Is an anti-inflammatory; reduces pain | Iron | Anemia from blood loss |
| Cathartics (laxatives) | Induce bowel movement | Calcium and potassium | Poor absorption |
| Cholestyramine | Reduces blood cholesterol | Vitamins A, D, E, and K | Poor absorption |
| Cimetidine (Tagamet®) | Treats ulcers | Vitamin B-12 | Poor absorption |
| Colchicine | Treats gout | Vitamin B-12, carotenoids, and magnesium | Decreased absorption due to damaged intestinal mucosa |
| Corticosteroids (prednisone) | Are an anti-inflammatory | Zinc and calcium | Poor absorption of zinc and poor utilization of calcium |
| Furosemide (Lasix®) | Decreases blood pressure; is a potassium-wasting diuretic | Potassium and sodium | Increased loss |
| Hydrochlorothiazide | Decreases blood pressure; is a diuretic | Potassium and magnesium | Increased loss and decreased absorption |
| MAO inhibitors (Parnate®) | Are an antidepressant | Tyramine (in cheese, wine, and other aged foods) | High blood pressure caused by limited tyramine metabolism |
| Tricyclic antidepressants (Elavil®) | Are an antidepressant | — | Weight gain from appetite stimulation |

isolation, grief, or change in lifestyle can lead to losses in appetite and interest in food, as well as disability. About 5 to 15% of persons 65 years old or older experience depression. The ratio is even higher for those with chronic disease and among those who need home health care or are in hospitals and nursing homes. Left untreated, depression can lead to a continual decline in appetite, which results in weakness, poor nutrition, mental confusion, and increased feelings of isolation and loneliness (Fig. 18-6). Sometimes people cope by overeating, which can lead to obesity and its associated problems. As many as 5 to 8% of depression cases may end in suicide. Depression is often treatable with medication, social support, and psychological intervention.[27]

Social isolation; perhaps spouse has died.

Loses interest in food: diet deteriorates.

Poor diet leads to weakness; this increases a feeling of isolation and abandonment.

Further isolation can then decrease desire for self-care.

Health declines visibly; weakness remains.

Self-care is seriously hampered.

*Figure 18-6* The decline in health often seen in older adults needs to be prevented whenever possible.

Worries about possible embarrassment caused by deteriorating physical capabilities may cause older adults to withdraw from social interaction and eat alone rather than with others. Those who eat alone, regardless of the reason, seldom eat as much or as nutritiously as they should. Both young and old people who eat without companionship tend to feel unmotivated to shop for or prepare foods. Many develop an apathetic attitude toward life, which over time causes health and nutritional status to decline.[1] As you'll see later in this chapter, in the U.S., several nutrition assistance programs can help older people obtain the food and social support needed for good health. The guidelines in Table 18-5 provide some practical suggestions to help older adults eat nutritiously.

©Image Source/Alamy Stock Photo

**Table 18-5  Guidelines for Healthful Eating in Later Years**

- Eat regularly; small, frequent meals may be best. Use nutrient dense foods as a basis for menus.
- Eat often with friends, with relatives, or at a senior center.
- Eat in a well-lit or sunny area; serve meals attractively; use foods with different flavors, colors, shapes, textures, and smells.
- If possible, take a walk before eating to stimulate appetite.
- Arrange kitchen and eating area so that food preparation and cleanup are easier.
- Use labor-saving equipment and foods (e.g., precut salad mix, canned beans, and frozen fruit).
- Try new foods, new seasonings, and new ways of preparing foods.
- Keep easy-to-prepare foods on hand for times when you feel tired.
- Share cooking responsibilities with a neighbor.
- Keep a box of dry milk handy to add nutrients to recipes for baked foods, casseroles, and meat loaf.
- If you have a freezer, cook larger amounts, divide them into small portions, and freeze.
- When necessary, chop, grind, or blend hard-to-chew foods. Softer protein rich foods can be substituted for meat when poor dental health limits normal food intake. Prepare soups, stews, cooked whole-grain cereals, and casseroles.
- If arm or hand movements are limited, cut the food before serving it, use utensils with deep sides or handles, and obtain specialized utensils if needed.
- Use community resources for help in obtaining meals (e.g., the Congregate Meal Program), shopping for groceries, and meeting other daily personal or household care needs.
- Buy only what you can use; small containers may be expensive, but letting food spoil also is costly.
- Ask the grocer to break family-sized packages of wrapped meat or fresh vegetables into smaller units.
- Buy several pieces of fruit—in various stages of ripeness—so that the fruit can be eaten over a period of several days.
- Consider buying meal replacement bars or liquid formulas for occasional snacks or meals.
- Stay physically active.

## Economic Factors

The amount of money available for purchasing food can have a great impact on the types and amounts of food a person eats. Unemployment, underemployment, retirement, or anything else that limits income makes it difficult to get the best, most healthful foods and can diminish nutritional status and health. Insufficient income is particularly a problem among those ages 65 and up; as a result, they frequently have trouble making sure they remain well nourished. The Commodity Supplemental Foods Program and Supplemental Nutrition Assistance Program (SNAP) are 2 federal U.S. programs that can help low-income individuals of all ages procure the foods they need.

### Knowledge Check

1. How does body composition tend to change as people age?
2. What body systems are adversely affected by physical inactivity?
3. How might the nervous system changes caused by aging affect dietary intake?
4. What are the risks of eating alone?

# 18.4 Nutrition Assistance Programs

The U.S. Department of Agriculture (USDA) and Administration on Aging programs provide food and nutrition services for adults. The USDA administers the following programs. Each program functions independently, has its own eligibility requirements and target audience, and may not be available in all states.

In many communities, programs assist older adults with daily tasks, which helps them meet their nutrient needs.

©Steve Mason/Getty Images RF

- The *Commodity Supplemental Food Program* distributes, free of charge, surplus agricultural products (e.g., cheese, peanut butter, canned foods) produced by U.S. farmers to low-income elderly persons and eligible children. The foods provided vary, depending on the farm products available at the time.
- The *Supplemental Nutrition Assistance Program* (*SNAP*, formerly the Food Stamp Program) provides nutrition education and supplements the food budgets of low-income households so that participants can purchase a greater quantity of food than they could afford to buy on their food budgets alone. The budget supplement, delivered in the form of electronic benefit transfer cards (like debit cards) can be used like money to purchase foods, food-producing plants and seeds, hot meals in group homes and shelters, and, in some areas, hot meals in certain authorized restaurants. SNAP benefits cannot be used to purchase alcohol or nonfood items (e.g., soap, paper goods).
- The *Child and Adult Care Food Program* provides nutritious meals and snacks to low-income children enrolled in child-care centers or residing in emergency shelters, as well as adults who are functionally impaired or ages 60 and older in nonresidential adult day-care centers.
- The *Senior Farmers' Market Nutrition Program* helps provide low-income older adults with coupons that can be exchanged for fresh fruits, vegetables, and herbs at farmers' markets, roadside stands, and other community-support agriculture programs.

The Administration for Community Living (formerly Agency on Aging) administers the Older Americans Act. This act is designed to help adults ages 60 years and older remain living independently in their homes and communities. Each community decides which programs and services will be provided. Community-based nutrition, health, and supportive services may include adult day care, senior center activities, transportation, information and counseling services, and health and physical activity programs. In-home care can include health and personal care, home maintenance assistance, and caregiver support services, such as nutrition advice and help with coordinating care needs.

The Elderly Nutrition Program, an important aspect of the Older Americans Act, provides nutrition services through the Congregate Meal Program and Home Delivered Meal Program (often referred to as Meals on Wheels). Anyone age 60 or older can participate; however, priority usually is given to those with the greatest economic, social, and health needs. Both meal programs can help older people obtain some of the food needed for good health. Each meal meets at least one-third of the daily recommendations. The social aspect of the Congregate Meal Program provides opportunities for socialization that often improve appetite and general outlook on life. These programs also may provide shopping assistance, counseling, nutrition education, and referral to other social, rehabilitative, and transportation services.

Opportunities to socialize at mealtime help improve appetite and nutritional status.

©Thinkstock/Getty Images RF

To learn about nutrition assistance programs, visit www.fns.usda.gov/programs.

## CASE STUDY

©Keith Brofsky/Getty Images RF

Frances is an 84-year-old woman who suffers from macular degeneration, osteoporosis, and arthritis. Since her husband died a year ago, she has moved from their family house to a small apartment. Her eyesight is progressively getting worse, making it hard to go to the grocery store or even to cook (for fear of burning herself). She is often lonely. Her only son lives 1 hour away and works 2 jobs, but he visits her as often as he can. Frances has lost her appetite and, as a result, often skips meals during the week. She has resorted to eating mostly cold foods, which are simple to prepare but are seriously limiting variety and palatability in her diet. She is slowly losing weight as a result of her dietary changes and loss of appetite. Her typical diet usually consists of a breakfast that may include a slice of wheat toast with margarine, honey, and cinnamon and a cup of hot tea. If she has lunch, she normally has a small can of peaches, half a turkey and cheese sandwich, and a small glass of water. For dinner, she might have half a tuna salad sandwich made with mayonnaise and a glass of iced tea. She usually eats 2 cookies at bedtime. What are the potential consequences of this dietary pattern? What services are available that could help Frances improve her diet and increase her appetite? What other easy-to-prepare foods could be included in her diet to make it more healthful and more varied?

### Knowledge Check

1. Which agencies provide food and nutrition services for adults?
2. How do the Commodity Food Program and the Supplemental Nutrition Assistance Program differ?
3. What can and cannot be purchased with SNAP benefits?
4. What types of services may be provided under the Older Americans Act?

## 18.5 Nutrition-Related Health Issues of the Adult Years

Diet is a key factor directly involved in the development of several health conditions during adulthood. The effects of diet and nutrients on many of these conditions, including atherosclerosis, cancer, constipation, diabetes, diverticular disease, heartburn, hypertension, obesity, and osteoporosis, were discussed in earlier chapters. In addition to these conditions, those discussed in this section are important to consider during adulthood. A slowing in the restoration of the internal balance of the body (homeostasis) is at least partially diet-related. Other health problems, such as arthritis and Alzheimer disease, are considered by some to be diet-related, even though scientific evidence currently indicates otherwise.

### Alcohol Use

The consequences of alcohol use, especially alcohol abuse, rise with advancing age. Older adults become intoxicated on a smaller amount of alcohol than when younger because they metabolize alcohol more slowly and have lower amounts of body water to dilute the alcohol. Both men and women over the age of 65 should limit alcohol consumption to no more than 1 standard drink (alcohol drink-equivalent) per day. (Recall from Chapter 8 that 1 drink per day is 5 ounces of wine, 12 ounces of beer, or 1.5 ounces of 80-proof liquor.) Even small amounts of alcohol can react negatively with common medications many older persons take. In addition to having adverse effects on the liver, large amounts of alcohol increase the risk of stroke and may aggravate hypertension in older adults.

Alcohol abuse is a problem among a small but significant group of older individuals who may continue this pattern from earlier in life or develop heavy drinking patterns and alcoholism in later life. Later development of this problem sometimes arises from the loneliness and social isolation caused by retirement or the loss of a spouse. Some symptoms of alcoholism in older persons include trembling hands, sleep problems, memory loss, and unsteady gait; these symptoms might be easily disregarded because they also are common symptoms of old age in general.

## Slowed Restoration of Homeostasis

As body tissues and systems age and their functioning diminishes, the body takes longer and longer to restore homeostasis. For example, it takes twice as long for the kidneys to remove wastes and restore blood levels to normal after eating excess protein in a person age 80 than in a person age 30. Similarly, it takes older people longer to break down alcohol, drugs, and nutrient supplements. Consequently, blood levels of these substances rise higher and have a stronger and longer effect in older adults than in younger people.

Even though the return to homeostasis is slowed, this slowdown usually is not a major problem unless disease strikes and stresses the body's capabilities. This slowdown makes an elderly person more vulnerable to illness and death. In the absence of disease, the slowed restoration of homeostasis usually is not a major problem until the end of the life span approaches. Prudent lifestyle choices may make it possible to keep the rate at which the body restores homeostasis high. Taking steps to avoid stressing the body's capabilities (e.g., getting flu shots, avoiding excessive protein intake) also can help preserve the optimal function of body tissues and systems. Getting prompt medical attention when the body's capabilities are stressed by illness helps preserve optimal functioning, too. In the case of already damaged tissues and systems, avoiding stressful practices helps prevent pushing an aging body beyond its capabilities.

Learning new skills throughout life helps preserve cognitive function.

Hello Lovely/Blend Images/Getty Images

## Alzheimer Disease

Alzheimer disease is an irreversible, abnormal, progressive deterioration of the brain. Healthy human brains contain billions of neurons—cells that process and transmit information using chemical and electrical signals. Alzheimer disease disrupts this communication of information by damaging connections between neurons. This disruption is caused by changes in normal brain proteins, beta-amyloid and tau, that damage neurons. Beta-amyloid protein levels rise abnormally high and clump together to form plaques that block neuron functions. Abnormal chemical changes cause tau molecules to stick together and form threads, which get tangled inside neurons and impair their functioning. As neurons are injured and die, the brain shrinks in size (Fig. 18-7). Memory, language, reasoning, and social behaviors are impaired. Chronic inflammation and circulatory system problems seem to promote the development and progression of Alzheimer disease.[28]

Alzheimer disease often takes a terrible toll on the mental and eventual physical health of older people. It is the 6th leading cause of disease in the U.S. About 5.8 million adults in the U.S. have the disease—5.6 million of these are over 65. Two-thirds of the cases are women. Compared to White populations, Alzheimer disease is about twice as common in Black and Hispanic populations.

No one is sure what causes this disease, but scientists have proposed various causes, including alterations in cell development or protein production in the brain, strokes, altered blood lipoprotein composition, obesity, poor blood glucose regulation (e.g., diabetes), high blood pressure, high blood cholesterol, and high free radical levels.

Preventive measures for Alzheimer disease focus on maintaining brain activity through lifelong learning, a healthy diet rich in fruits and vegetables and lower in fat, and exercise. The role of nutrition in preventing or minimizing the risk of this disease is being investigated. Getting enough antioxidant nutrients, such as vitamins C and E, helps protect the body from the damaging effects of free radicals. Adequate intakes of folate and vitamins B-6 and B-12 are especially important because elevated blood homocysteine and related inflammation also are risk factors. Dietary fats, too, may play a role in keeping this disease at bay. Individuals with diets rich in omega-3 and omega-6 foods and low in saturated and *trans* fatty acids seem to have a reduced risk of Alzheimer disease.[27] The Mediterranean Diet and DASH and other heart-healthy diet plans may help decrease Alzheimer disease risk.

*Figure 18-7* Damaged brain neurons impair communication in Alzheimer disease.

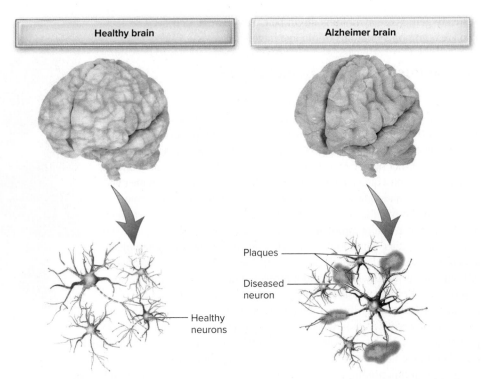

**Healthy brain**

**Alzheimer brain**

Healthy neurons

Plaques

Diseased neuron

**10 Warning Signs of Alzheimer Disease**

1. Memory loss that disrupts daily life
2. Difficulty completing familiar tasks
3. Problems with language in writing or speaking
4. Disorientation to time and place
5. Faulty, decreased, or poor judgment
6. Challenges with planning and problem solving
7. Misplacing things and inability to retrace steps
8. Changes in mood or personality
9. Difficulty understanding visual images and spatial relationships
10. Withdrawal from work or social activities

▶ To find out more about Alzheimer disease, visit www.alz.org or www.nia.nih.gov/alzheimers.

▶ Glucosamine and chondroitin sulfate may stimulate the production of new cartilage in humans, so they may help the body rebuild damaged cartilage. Also, individuals who take supplements may slow the progression of osteoarthritis, but more research is needed to confirm the effectiveness of these supplements.[31]

The dietary intakes of those with Alzheimer disease are poor compared with those of a similar age without this disease. The caregivers of those who have Alzheimer disease need to monitor their patients to ensure that they maintain a healthy weight and nutritional state. Other tips are to serve fish rich in omega-3s twice per week, serve a variety of foods, keep mealtimes calm, offer bite-size foods, and make sure eating habits do not pose a health risk (e.g., holding food in one's mouth or forgetting to swallow). Regular physical activity also has been shown to improve mental status in people afflicted with this disease.[30]

## Arthritis

There are over 100 forms of arthritis, a disease that causes the degeneration and roughening of the once smooth cartilage that covers and cushions the bone joints and/or the formation of calcium deposits (spurs). These changes in the joints cause them to ache and become inflamed and painful to move. Osteoarthritis becomes more common with age and, by age 80, almost everyone has this condition—it is the leading cause of disability among older persons. Rheumatoid arthritis, which is not as common, is more prevalent in younger adults.

Although it is not known what causes or cures arthritis, many unproven "remedies" have been publicized. Unusual diets, food restrictions, and nutrient supplements are some of the more popular "remedies." However, no special diet, food, or nutrient ever has been proven to prevent, relieve, or cure arthritis in humans.[26] The only diet-related treatment known to offer some relief is to maintain a healthy weight. This is because excess body weight adds extra stress to already painful arthritic joints. Although those with arthritis are not likely to benefit from dietary changes, they can experiment with their diets, provided they avoid practices that will result in inadequate, unbalanced, or excessive nutrient intakes, to see if such changes provide relief. It is important to keep in mind, though, that altering dietary practices may work for some, but no dietary changes have been found to be of help to all arthritis sufferers.[31]

*Critical* **Thinking**

Jamilla is 68 years old. She has heart disease and takes several medications. She went to her local pharmacy yesterday to look for a product to help her sleep through the night. On the shelves, she found an herbal supplement claiming to be a remedy for sleeplessness. She thought that, because a pharmacy carries the product, it should be safe and should work as indicated on the label. Is she correct in these assumptions? Are there specific risks associated with taking such herbal remedies? What should Jamilla do before taking the supplement?

*Knowledge Check*

1. Why might even small amounts of alcohol be problematic for older adults?
2. What are some steps an older adult can take to minimize the effects of slowed restoration of homeostasis?
3. Which nutrients may provide protection against Alzheimer disease?
4. How does maintaining a healthy weight help those with arthritis?

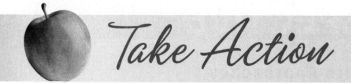

*Take Action*

## Helping Older Adults Eat Better

During their lifetimes, most people usually eat meals with families or loved ones. As people reach older ages, many are faced with living and eating alone. In a study of the diets of 4400 older adults in the U.S., 1 man in every 5 living alone who was over age 55 ate poorly. One in 4 women between the ages of 55 and 64 years had a low-quality diet. These poor diets can contribute to deteriorating mental and physical health. Consider the following example of the living situation of an older adult.

Neal, a 70-year-old man, lives alone at his home in a local suburban area. His wife died a year ago. He doesn't have many friends; his wife was his primary confidante. His neighbors across the street and next door are friendly, and Neal used to help them with yard projects in his spare time. Neal's health has been good, but he has had trouble with his teeth recently. His diet has been poor, and in the past 3 months his physical and mental vigor has deteriorated. He has been slowly lapsing into depression, so he keeps the shades drawn and rarely leaves his house. Neal keeps very little food in the house because his wife did most of the cooking and shopping, and he just isn't that interested in food.

©Keith Brofsky/Getty Images RF

If you were Neal's friend or relative and learned of his situation, what are 6 things you could do or suggest to help improve his nutritional status and mental outlook?

1. _____

2. _____

3. _____

4. _____

5. _____

6. _____

## CASE STUDY
## FOLLOW-UP

©Keith Brofsky/Getty Images RF

Frances or her son could contact a local government office that offers Congregate Meal Programs and inquire about where meals are served and the transportation available to the site. This meal program would give her social contact with other older persons, which is probably an important element that is missing in her life and could help alleviate her loneliness. She also could request home-delivered meals (if available in her area) to provide at least 1 meal each day, which may be just what she needs to help stimulate her appetite. She also could have groceries delivered to her home if her budget could handle the extra cost. Convenience foods that could improve her diet are milk, peanut butter, breakfast cereals, chicken or tuna in a can or pouch, yogurt, sliced cheese, cottage cheese, calcium-fortified orange juice, canned or frozen fruits and vegetables, and some fresh fruits and vegetables that do not require preparation, such as bananas and pre-washed salad greens. A further possibility is eating a nutrition bar or a liquid nutritional supplement each day. The resulting increase in her nutrient intake would help prevent disease in the future and increase her sense of well-being.

## Complementary and Alternative Health Approaches

Many people, especially older adults and those with chronic diseases, use complementary and alternative approaches to health (also called complementary care and integrative medicine).[32] "Complementary" or "integrative" health is using a nonmainstream approach to health management in addition to conventional medicine. "Alternative" means using a nonmainstream approach instead of conventional medicine.[33] Complementary approaches are most common. An example is using acupuncture along with medications to control pain.

Complementary health approaches typically fall into 2 categories.[33]

1. *Natural products:* using substances found in nature, such as vitamins, minerals, probiotics, fish oil, and herbs to treat or prevent disease. An herb is any plant or part of a plant used primarily for medicinal purposes.
2. *Mind-body practices:* using techniques such as yoga, Tai Chi, qi gong, chiropractic and osteopathic manipulation, meditation, massage, hypnotherapy, healing touch, and movement therapies.

Other complementary approaches outside the 2 main categories include practices such as Ayurvedic medicine (a natural healing process from India), naturopathy, homeopathy, traditional Chinese medicine, and traditional healers.

Some complementary health approaches show promise in the treatment of certain conditions.[33,34] For example, research indicates that acupuncture is effective for the treatment of chronic pain.[35] However, unlike conventional medicine, most of these approaches have little or no scientific evidence to support them, either because they do not work or because few studies have been conducted.[36]

If a complementary health approach doesn't work or hasn't been proven to work, why do people use them? One reason is that some may not have heard research results that products or approaches they use are ineffective.[36] Others may be confused by advertising claims. Still others assume that natural substances are gentler forms of therapy than pharmacologic medicines. Others use complementary health approaches because standard medical treatments didn't work or had too many adverse effects. Many who use alternative medicines have illnesses for which conventional medicine cannot offer a cure, such as arthritis, terminal stages of AIDS or cancer, and stress-related conditions. To cope with health conditions such as these, sufferers may resort to almost anything. Another reason unproven remedies remain popular is that they appear to relieve symptoms in a few people. This relief may be caused by a placebo effect, or it may be the result of the natural ups and downs of symptoms, remission of disease, or the possibility that the remedy contains a medicine not listed on the label. If improvement happens to coincide with the use of a remedy, a person may mistakenly believe that the remedy works.

The possibility that an unproven remedy itself actually causes the relief does exist. For example, a person with arthritis may find that restricting a certain food eliminates an unknown food allergy that irritates the joints. Another possibility is that dietary changes may improve the nutritional status of a person who is in such poor nutritional health that the immune system is unable to respond effectively to the joint inflammation causing arthritis pain. Experimenting with complementary therapies may help some individuals find relief from health conditions.

A rational approach for someone wanting to try complementary approaches to health is to keep a diary of symptoms, follow only a single therapy at a time, check with his or her physician first before discontinuing a medication or medical treatment, and find out if the alternative practitioner has experience with the medical problem to be treated. Interested consumers also might see if they can enroll in a study investigating the effectiveness of the substance or procedure in question. If adverse side effects occur from a complementary or alternative therapy, consumers should contact a physician right

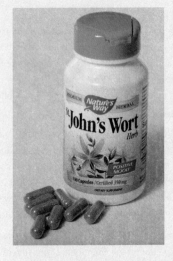

Some herbal products are effective for treating specific medical problems. To be safe, follow label instructions carefully, note potential side effects, and use them under the supervision of a physician.

©Science History Images/Alamy Stock Photo

▶ Studies indicate that the following herbs may be especially dangerous: germander, pokeroot, sassafras, mandrake, pennyroyal, comfrey, chaparral, yohimbe, lobelia, jin bu huan, kava kava, products containing stephanie and magnolia, senna, hai gen fen, paraguay tea, kombucha tea, tung shueh (Chinese black balls), and willow bark.

away. Physicians are encouraged to report such adverse effects to the FDA, state and local health departments, and consumer protection agencies.

When considering whether to use a complementary health approach, keep these points in mind:

- We often tend to believe what we hear or what close acquaintances tell us. This well-meaning advice is not a substitute for scientific proof of safety and effectiveness when it comes to health practices.

- The U.S. government provides little regulation of nutrient or herbal supplements or remedies (see Chapter 1). Another concern is that the actual content of the active ingredients in herbal supplements may be less or more than stated on the label. "Let the buyer beware" is prudent advice when using these products.

- Fraudulent claims for diet- and health-related remedies always have been a part of our culture. Carefully scrutinize the credentials and motives of anyone providing medical or health advice. Phony credentials and bogus practitioners are common.

- If it sounds too good to be true, it probably is. The medical community gains nothing by holding back effective cures from the public, despite what the alternative practitioners may say.

**A Closer Look at Herbal Therapy**

Throughout history, healers have gone to the garden, forest, and sea to seek herbal remedies. Some natural products may be harmless, others are potentially toxic, and still others may be effective for some problems but dangerous when taken in the wrong dose or by people with certain medical conditions (Table 18-6). Interactions between complementary and alternative therapies and pharmaceutical drugs can be drastic and include complications such as delirium, clotting abnormalities, rapid heartbeat, and even death. Table 18-6 describes some of the most commonly used herbal remedies. Learn more about herbs at nccih.nih.gov/health/herbsataglance.htm.

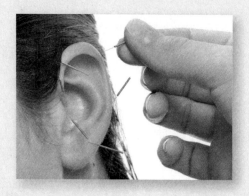

Acupuncture may offer relief from nausea after surgery or chemotherapy.

tankist276/Shutterstock

**Table 18-6  Popular Herbal Remedies, Nutrient Supplements, and Hormones**

| Substances | Potential Effects | Side Effects | Who Should Avoid Them |
|---|---|---|---|
| Black cohosh<br>Manfred Ruckszio/Shutterstock | May reduce postmenopausal symptoms | Nausea, fall in blood pressure | Women taking estrogen, hypertension medications, or aspirin and related drugs |
| Echinacea<br>©Brand X/Getty Images RF | May stimulate the immune system and shorten the duration of flulike illnesses; current studies show little or no effect | Nausea, skin irritation, allergic reactions | Anyone with an autoimmune disease or allergic reactions to daisies |
| Feverfew<br>©Bildagentur-online/Alamy | May reduce the pain and frequency of migraines | Abdominal pain, mouth sores, skin rash | Anyone allergic to ragweed or taking anti-inflammatory drugs, such as aspirin |
| Garlic<br>©Tetra Images/Getty Images RF | May have antibiotic properties and slightly lower blood cholesterol and blood pressure | In large amounts, burning of the mouth, nausea, sweating, stomach irritation, light-headedness, reduced blood clotting | Anyone taking anticoagulant medications, such as warfarin, for cardiovascular disease or AIDS medicines |
| Ginger<br>©Foodcollection RF | May prevent motion sickness and nausea related to surgery and pregnancy | Gastrointestinal (GI) tract discomfort with high doses on an empty stomach | People with a history of gallstones |

**Table 18-6 Continued**

| Substances | Potential Effects | Side Effects | Who Should Avoid Them |
|---|---|---|---|
| Ginkgo biloba<br><br>©Photographer's Choice/<br>Getty Images RF | May increase the circulation of blood in the body, especially to the brain and lower extremities; evidence is very weak | GI tract upset, headache, irritability, reduced blood clotting | Anyone taking anti-inflammatory or anticoagulant medications, including vitamin E and aspirin, and anyone who has had a stroke or is prone to them |
| Glucosamine<br><br>Carlos Yudica/Shutterstock | May decrease joint inflammation and pain associated with osteoarthritis | GI tract discomfort, which may disappear after 2 weeks | May disrupt blood glucose regulation in people with diabetes |
| Milk thistle<br><br>©Pixtal/SuperStock RF | May have a protective effect on the liver, thought to be due in part to its ability to prevent toxins from contaminating liver cell membranes | Diarrhea | No specific persons are at risk. |
| St. John's wort<br><br>©Pixtal/SuperStock RF | Mild antidepressant effect that may work by inhibiting monoamine oxidase (an enzyme in the brain that destroys "feel-good" hormones, such as serotonin, epinephrine, and dopamine) | Nausea, fatigue, dry mouth, dizziness, photosensitivity; increases metabolism and removal of many prescription drugs from the body | People taking medications to control depression, HIV, epilepsy, cardiovascular disease, and asthma or to suppress the immune system to keep the body from rejecting a transplanted organ |
| Saw palmetto<br><br>©Ingram Publishing/<br>SuperStock | May reduce symptoms of enlarged prostate gland by increasing urinary flow and easing urination; some studies show moderate evidence of effectiveness, whereas other studies have shown little or no benefit | Generally uncommon; when taken in large doses: headache, GI tract upset | People taking medication to treat enlarged prostate or BPH and anyone with a chronic GI tract disease |
| Valerian<br><br>©Pixtal/SuperStock RF | May alleviate restlessness and other sleeping disorders caused by nervous conditions | Headache, morning grogginess, irregular heartbeat, GI tract upset (also has a disagreeable odor) | Anyone taking central nervous system depressants, such as sedatives, and anyone who drinks alcohol |

Note that pregnant or lactating women, children under 2 years old, those over 65 years old, and anyone with a chronic disease should never take herbal remedies or supplements unless under the guidance of a physician.

Dosage forms of herbs include capsules, tablets, extracts or tinctures, powders, dried plant parts (herbs), teas, creams, ointments, and vapors (aromatherapy). Some herbal products are effective for treating specific medical problems. The best advice is to use these substances only under strict supervision of a physician. The FDA requires herbal supplement labels to include the herb name, quantity, dosage per day, and ingredients (Fig. 18-8).

Another important risk to consider is that traditional herbal products can be mislabeled, be adulterated with prescription drugs or contaminants (e.g., lead), or vary greatly in potency. Combination herbs always should be avoided because of the reported cases of adverse health effects and adulteration.

▶ Supplement labels can claim a benefit related to a classic nutrient deficiency disease, describe how the supplement affects body structure or function (i.e., structure/function claim), and state that general well-being results from consumption of the ingredient or ingredients. A supplement label cannot claim that the product treats, cures, or prevents a disease not completely related to a given nutrient deficiency. For example, an herbal product label can claim that it may help brain function, but not that it cures Alzheimer disease. The latter would constitute a drug claim.

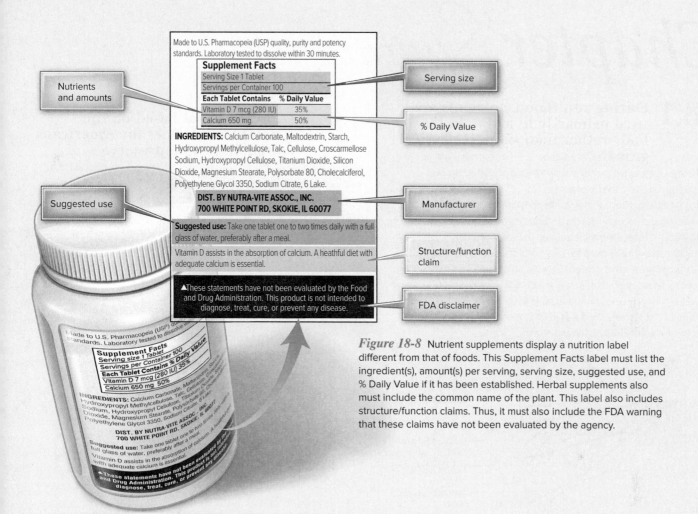

Made to U.S. Pharmacopeia (USP) quality, purity and potency standards. Laboratory tested to dissolve within 30 minutes.

**Supplement Facts**

Serving Size 1 Tablet
Servings per Container 100

| Each Tablet Contains | % Daily Value |
|---|---|
| Vitamin D 7 mcg (280 IU) | 35% |
| Calcium 650 mg | 50% |

**INGREDIENTS:** Calcium Carbonate, Maltodextrin, Starch, Hydroxypropyl Methylcellulose, Talc, Cellulose, Croscarmellose Sodium, Hydroxypropyl Cellulose, Titanium Dioxide, Silicon Dioxide, Magnesium Stearate, Polysorbate 80, Cholecalciferol, Polyethylene Glycol 3350, Sodium Citrate, 6 Lake.

**DIST. BY NUTRA-VITE ASSOC., INC.
700 WHITE POINT RD, SKOKIE, IL 60077**

**Suggested use:** Take one tablet one to two times daily with a full glass of water, preferably after a meal.

Vitamin D assists in the absorption of calcium. A healthful diet with adequate calcium is essential.

▲These statements have not been evaluated by the Food and Drug Administration. This product is not intended to diagnose, treat, cure, or prevent any disease.

Nutrients and amounts

Suggested use

Serving size

% Daily Value

Manufacturer

Structure/function claim

FDA disclaimer

*Figure 18-8* Nutrient supplements display a nutrition label different from that of foods. This Supplement Facts label must list the ingredient(s), amount(s) per serving, serving size, suggested use, and % Daily Value if it has been established. Herbal supplements also must include the common name of the plant. This label also includes structure/function claims. Thus, it must also include the FDA warning that these claims have not been evaluated by the agency.

These questions can help you evaluate a company's herbal products:

- What analyses are done to ensure quality, quantity, and reproducibility of the ingredients in individual doses as labeled?
- Is the product labeled with Latin botanical names?
- Does the label have expiration dates and lot numbers? How are expiration dates determined?
- Does the manufacturer offer a certificate of analysis for each product?
- Has the manufacturer been in business for at least 5 years? Does it nationally distribute the product?

Overall, herbal products should be used with great caution and in consultation with a person's primary physician. Otherwise, potential side effects may go undiagnosed, dangerous herb-medicine interactions may occur, or serious complications during surgery may develop. Pregnant and nursing women, children under 2 years of age, those over age 65, and anyone with a chronic disease should not take herbal supplements unless his or her physician consents to the practice and monitors the individual for potential complications.

For a balanced discussion of herbal medicine, visit

**National Center for Complementary and Integrative Health**
nccih.nih.gov

**American Botanical Council**
www.abc.herbalgram.org

**Natural Medicine Comprehensive Database**
naturalmedicines.therapeuticresearch.com/

Simply because herbal remedies come from natural sources does not mean they are without health risks. Even those that have been used for centuries may cause harmful effects.

©Comstock/Getty Images RF

# Chapter *Summary*

## 18.1 During adulthood, nutrients are used primarily to maintain the body rather than support physical growth.

As adults get older, nutrient needs change. Adulthood is characterized by body maintenance and gradual physical and physiological transitions, often referred to as "aging." The physiological changes of aging are the sum of cellular changes, lifestyle practices, and environmental influences. Many of these changes can be minimized, prevented, and/or reversed by healthy lifestyles. *Usual aging* refers to the age-related physical and physiological changes that are commonly thought to be a typical or expected part of aging. *Successful aging* (also called delayed aging) describes the declines in physical and physiological function that occur because an individual grows older. Striving to have the greatest number of healthy years and the fewest years of illness is referred to as compression of morbidity. The rate at which someone ages is individual; it is determined by heredity, lifestyle, and environment.

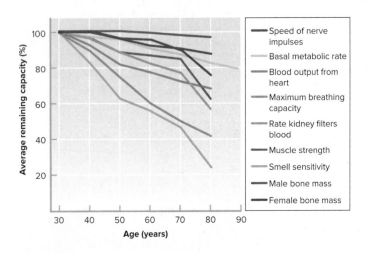

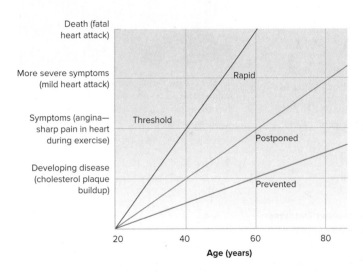

## 18.2 A healthy diet based on the Dietary Guidelines for Americans can help people preserve the body's

function, avoid chronic disease, and age successfully. American adults are fairly well nourished, although common dietary excesses are calories, fat, sodium, and, for some, alcohol. Common dietary inadequacies include vitamins D and E, folate, magnesium, potassium, calcium, zinc, and fiber. People ages 65 and up, particularly those in long-term care facilities and hospitals, are the single largest group at risk of malnutrition. The Nutrition Screening Initiative checklist can help identify older adults at risk of nutrient deficiencies. The DRIs for adults are divided by biological sex and age to reflect how nutrient needs change as adults grow older. These changes in nutrient needs take into consideration aging-related physiological alterations in body composition, metabolism, and organ function.

### A Nutrition Test for Older Adults

Here's a nutrition check for anyone over age 65. Circle the number of points for each statement that applies. Then compute the total and check it against the nutritional score.

| | Points | |
|---|---|---|
| | 2 | 1. The person has a chronic illness or current condition that has changed the kind or amount of food eaten. |
| | 3 | 2. The person eats fewer than 2 full meals per day. |
| | 2 | 3. The person eats few fruits, vegetables, or milk products. |
| | 2 | 4. The person drinks 3 or more servings of beer, liquor, or wine almost every day. |
| | 2 | 5. The person has tooth or mouth problems that make eating difficult. |
| | 4 | 6. The person does not have enough money for food. |
| | 1 | 7. The person eats alone most of the time. |
| | 1 | 8. The person takes 3 or more different prescription or over-the-counter drugs each day. |
| | 2 | 9. The person has unintentionally lost or gained 10 pounds within the last 6 months. |
| | 2 | 10. The person cannot always shop, cook, or feed himself or herself. |
| Total | | |

#### Nutritional Score

**0–2: Good.** Recheck in 6 months.

**3–5 Moderate nutritional risk.** A local agency on aging has information about nutrition programs to help the elderly improve eating habits and lifestyle. Visit eldercare.acl.gov to find a local agency. Recheck in 6 months.

**6 or more: High nutritional risk.** A doctor or registered dietitian nutritionist should review this test and suggest how to improve nutritional health.

## 18.3 The food choices and nutritional adequacy of adults' diets depend on physical,

physiological, psychosocial, and economic factors. Alterations in any of these factors can result in deteriorations in the quality of dietary intake, nutritional status, and health. The physical and physiological changes in body composition and body systems that occur during adulthood can influence dietary intake, alter nutrient and/or calorie needs, and/or alter nutrient utilization. The use of medications and supplements can improve health and quality of life, but some also can adversely affect nutritional status. Psychosocial status can affect food intake and health. Economic factors can have a great impact on the types and amounts of food someone eats.

©David Sucsy/Getty Images RF

# 18.4 Programs publicly funded by the U.S. Department of Agriculture (USDA) and

Administration on Community Living provide food and nutrition services for adults. The USDA administers food and nutrition assistance programs, including the Commodity Supplemental Foods Program, Supplemental Nutrition Assistance Program, Child and Adult Care Food Program, and Senior Farmers' Market Nutrition Program. The Administration on Community Living administers the Older Americans Act, which provides community-based nutrition, health, and supportive services and may include adult day care, senior center activities, transportation, information and counseling services, and health and physical activity programs. The Congregate Meal Program and Home-Delivered Meal Program can help older people obtain some of the food needed for good health.

©Thinkstock/Getty Images RF

# 18.5 Diet is a primary factor directly involved in the development of several

health conditions during adulthood, including atherosclerosis, cancer, constipation, diabetes, diverticular disease, heartburn, hypertension, obesity, osteoporosis, and periodontal disease. The consequences of alcohol use, especially alcohol abuse, rise with advancing age. A slowdown in restoration of the internal balance of the body (homeostasis) is at least partially diet-related. Arthritis and Alzheimer disease are considered by some to be diet-related, even though scientific evidence currently indicates otherwise.

©Steve Cole/Getty Images RF

## Clinical Perspectives

Food-drug or drug-nutrient interactions can change drug absorption, breakdown, or excretion or the way the drugs work. Drugs may increase, decrease, or block nutrient absorption and affect the rate of nutrient excretion and metabolism. The interactions that may occur between drugs and food or nutrients vary by drug. Preventing interactions usually means avoiding certain foods completely or eating them at times well before or after taking the drug.

"Complementary" or "integrative" health is using a nonmainstream approach to health management in addition to conventional medicine. "Alternative" means using a nonmainstream approach instead of conventional medicine. The 2 main categories are natural products and mind-body practices. Most alternative methods lack scientific evidence to support them. Some herbal products may be harmless, others are potentially toxic, and still others may be effective for some problems but dangerous when taken in the wrong dose or by people with certain medical conditions. Herbal products should be used with great caution.

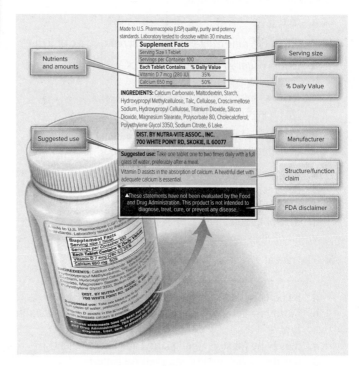

## Study Questions

1. During adulthood, nutrients are used primarily to maintain the body.

   a. true                    b. false

2. _____ is the time-dependent physical and physiological changes in body structure and function that occur normally and progressively throughout adulthood as humans mature and become older.

   a. Aging                   c. Usual aging
   b. Successful aging        d. Graying

3. Compression of mortality is the extent to which an organ can preserve essentially normal function despite decreasing cell number or cell activity.

   a. true                    b. false

4. Which of the following is *not* considered a cause of aging?

   a. Errors occur in the copying of DNA.
   b. Free radicals damage cell parts.
   c. Body system reserve capacity declines.
   d. Death is programmed into cells.

5. Physical and physiological changes associated with usual aging include _____.

   a. increasing body fatness
   b. decreasing lean body mass
   c. rising blood pressure
   d. declining bone mass
   e. all of the above

6. The rate at which a person ages is determined _____.

   a. by heredity, lifestyle choices, and environment
   b. mostly by heredity, education level, and access to health care
   c. mostly by lifestyle, diet quality, and environment
   d. mostly by diet quality and exercise pattern
   e. by heredity only

7. The impact lifestyle choices can have on the expression of genetic potential is _____.

   a. none
   b. minor
   c. major

8. The diets of adults tend to be low in _____.

   a. vitamin E
   b. calcium
   c. zinc
   d. fiber
   e. all of the above

9. The DRIs for adults do not take into account aging-related changes in body composition.

   a. true                    b. false

10. Which of the following is a sign that an older person's health needs extra attention?

    a. The person eats fewer than 2 meals daily.
    b. The person drinks 3 or more servings of alcohol often.
    c. The person eats alone often.
    d. All of the above are signs.
    e. Only a and b are signs.

11. Which of the following changes commonly occurs as someone ages but does not influence nutrient utilization?

    a. loss in taste and smell perceptions
    b. reduced stomach acidity
    c. diminished kidney function
    d. presence of chronic disease

12. A change that tends to occur as the adult years progress is _____.

    a. increased body water
    b. decreased lung efficiency
    c. increased intrinsic factor
    d. decreased hormone synthesis and release
    e. both b and d

13. Which publicly funded program distributes, free of charge, surplus agricultural products to low-income households?

    a. Supplemental Nutrition Assistance Program
    b. Commodity Foods Program
    c. Child and Adult Care Food Program
    d. Congregate Meal Program

14. Diet is *not* directly involved in the development of _____.

    a. atherosclerosis
    b. cancer
    c. diverticular disease
    d. arthritis

15. A potential risk associated with herbal products is that _____.

    a. they can interact with medicines
    b. they may vary in potency
    c. they may be contaminated
    d. all of the above
    e. none of the above

16. Describe 2 hypotheses proposed to explain the causes of aging, and note evidence for each in your daily life experiences.

17. How might the nutritional needs of older adults differ from those of young adults? How are their needs similar?

18. List 4 organ systems that can decline in function in later years, and describe a diet/lifestyle response to help cope with the decline.

19. Describe 3 community resources that can assist older adults in maintaining their nutritional status.

**20.** Defend the recommendation for regular physical activity during older adulthood.

**21.** Explain why food, nutrient, and supplement intake should be considered when taking drugs.

**22.** Describe a rational approach for trying complementary approaches to health.

Perspective: Complementary and Alternative Health Approaches
Perspective: Drug-Nutrient Interactions; 22-Refer to Clinical
18.4; 20-refer to Sections 18.3 and 18.5; 21-refer to Clinical
to Section 18.2; 18-refer to Section 18.3; 19-refer to Section
11-a; 12-e; 13-b; 14-d; 15-d; 16-refer to Section 18.1; 17-refer
Answer Key: 1-a; 2-a; 3-b; 4-c; 5-e; 6-a; 7-c; 8-e; 9-b; 10-d;

©C Squared Studios/Getty Images RF

# References

1. Bernstein M, Munoz N. Position of the Academy of Nutrition and Dietetics: Food and nutrition for older adults: Promoting health and wellness. *J Acad Nutr Diet.* 2012;112:1255.

2. Woo J. Nutritional strategies for successful aging. *Med Clin North Am.* 2011;95:477.

3. LaBrasseur NK and others. Cellular senescence and the biology of aging, disease, and frailty. *Nestle Nutrition Institute Workshop Series.* 2015;83:11.

4. Santos J and others. Dietary restriction and nutrient balance in aging. *Oxid Med Cell Longev.* 2016:Art. No. #4010357.

5. Beltran-Sanchez H and others. Past, present, and future of healthy life expectancy. *Perspect Med.* 2015;November 2:5.

6. Gopinath B and others. Adherence to dietary guidelines positively affects quality of life and functional status of older adults. *J Acad Nutr Diet.* 2014;114:220.

7. Campbell AD and others. Does participation in home-delivered meals programs improve outcomes for older adults? Results of a systematic review. *J Nutr Gerontol Geriatr.* 2015;34:124.

8. Centers for Disease Control and Prevention, National Center for Health Statistics. *Dietary intake of macronutrients, micronutrients, and other dietary constituents: United States, 1988–1994.* Washington, DC: U.S. Department of Health and Human Services; 2002.

9. Institute of Medicine, Food and Nutrition Board. *Dietary Reference Intakes for calcium and vitamin D.* Washington, DC: National Academies Press; 2011.

10. Huang ES. Management of diabetes mellitus in older people with comorbidities. *BMJ.* 2016;353:i2200.

11. Institute of Medicine, Food and Nutrition Board. *Dietary Reference Intakes for vitamin A, vitamin K, arsenic, boron, chromium, copper, iodine, iron, manganese, molybdenum, nickel, silicon, vanadium, and zinc.* Washington, DC: National Academies Press; 2001.

12. Gopinath B and others. Adherence to dietary guidelines and successful aging over 10 years. *J Geron Ser A.* 2016;71:349.

13. Littlejohns TJ and others. Vitamin D and dementia. *J Prev Alzheimers Dis.* 2016;3:43.

14. Yang K and others. Vitamin D concentration and risk of Alzheimer disease: A meta-analysis of prospective cohort studies. *Medicine.* 2019;98:e16804.

15. Bailey RL, vanWijngaarden JP. The role of B-vitamins in bone health and disease in older adults. *Curr Osteo Rep.* 2015;13:256.

16. Institute of Medicine, Food and Nutrition Board. *Dietary Reference Intakes for thiamin, riboflavin, niacin, vitamin B6, folate, vitamin B12, pantothenic acid, biotin, and choline.* Washington, DC: National Academies Press; 1998.

17. Bertali B and others. Serum micronutrient concentrations and decline in physical function among older adults. *JAMA.* 2008;299:308.

18. Lawrenson JG, Grzybowski A. Controversies in the use of nutritional supplements in ophthalmology. *Curr Pharma Des.* 2015;21:4667.

19. Kuczmarski MF and others. The association of healthful diets and cognitive function: A review. *J Nutr Gerontol Geriatr.* 2014;33:69.

20. Lehert P and others. Individually modifiable risk factors to ameliorate cognitive aging: A systematic review and meta-analysis. *Climacteric.* 2015;18:678.

21. Flicker L and others. Body mass index and survival in men and women aged 70 to 75. *J Am Geriatr Soc.* 2010;58:234.

22. Jura M, Kozak LP. Obesity and related consequences to ageing. *Age.* 2016;38:23.

23. McNaughton SA and others. Diet quality is associated with all-cause mortality in adults aged 65 years and older. *J Nutr.* 2012;142:320.

24. Soenen S and others. The ageing gastrointestinal tract. *Curr Opin Clin Nutr Metab Care.* 2016;19:12.

25. Dye BA and others. Dental caries and tooth loss in adults in the United States, 2011–2012. NCHS Data Brief #197. 2015; www.cdc.gov/nchs/data/databriefs/db197.htm.

26. Office of Dietary Supplements, USDHHS. Multivitamin/mineral supplements. 2015; ods.od.nih.gov/factsheets/MVMS-HealthProfessional/#h2.

27. Unwin BK and others. Nursing home care. Part II. Clinical aspects. *Am Fam Phys.* 2010;81:10.

28. National Institute on Aging. Alzheimer's disease fact sheet. 2019; www.nia.nih.gov/health/alzheimers-disease-fact-sheet.

29. Swaminathan A, Jicha AS. Nutrition and prevention of Alzheimer's dementia. *Front Aging Neurosci.* 2014;6:282.

30. Portugal EM and others. Aging process, cognitive decline and Alzheimer's disease: Can strength training modulate these responses? *CNS & Neuro Disor Drug Targ.* 2015;14:1209.

31. National Institute of Arthritis and Musculoskeletal and Skin Diseases. Osteoarthritis. 2016; www.niams.nih.gov/health_info/osteoarthritis/.

32. Kantor ED and others. Trends in dietary supplement use among US adults from 1999–2012. *JAMA.* 2016;316:1464.

33. National Center for Complementary and Alternative Health. Complementary, alternative, or integrative health: What's in a name? 2018; nccih.nih.gov/health/integrative-health.

34. Sargent PD and others. Integrative medical practices for combat-related posttraumatic stress disorder. *Psychiatr Ann.* 2013;43:181.

35. Vickers AJ and others. Acupuncture for chronic pain: Individual patient data meta-analysis. *Arch Intern Med.* 2012;179:1444.

36. Cohen PA. The supplement paradox. Negligible benefits, robust consumption. *JAMA.* 2016;316:1453.

# Appendix A

## HUMAN PHYSIOLOGY: A TOOL FOR UNDERSTANDING NUTRITION

This appendix explores the various systems in the body beyond the digestive system, focusing specifically on how these systems relate to the study of human nutrition. This focus will set the stage for investigating the various nutrients associated with human nutrition. Before that process can begin, however, it is important to review the processes taking place in a human cell.

## The Cell: Structure and Function

The cell is the basic structural and functional unit of life. Living organisms are made of many different kinds of cells specialized to perform particular functions, and all cells are derived from preexisting cells. In the human body, the trillions of cells all have certain basic characteristics that are alike. All cells have compartments, particles, or filaments that perform specialized functions; these structures are called organelles. There are at least 15 different organelles, but this section discusses only 8. The numbers preceding the names of the cell structures correspond to the structures illustrated in Figure A-1.

### 1 Cell (Plasma) Membrane

There is an outside and inside to every cell, as defined by the cell (plasma) membrane. This very thin and elastic membrane holds in the cellular contents and regulates the direction and flow of substances into and out of the cell. Cell-to-cell communication also occurs by way of this membrane. Some cells can even penetrate another cell membrane and, so, invade that cell.

The cell membrane is a lipid bilayer (double membrane) of **phospholipids** with their water-soluble (polar) heads facing into the interior of the cell or out to the exterior of the cell. The water-insoluble (nonpolar) tails are tucked into the interior of the cell membrane. (Chapter 6 reviews phospholipids in detail, and Appendix B reviews the concept of polar and nonpolar compounds)

Cholesterol is a fat-soluble component of the membrane, so it is embedded in the bilayer. This cholesterol provides rigidity, and thus stability, to the membrane.

There also are various proteins embedded in the membrane. Proteins provide structural support, act as transport vehicles, and function as enzymes that affect chemical processes within the membrane. Some proteins are open channels that allow water-soluble substances to pass into and out of the cell. Proteins on the outside surface of the membrane act as receptors, snagging essential substances the cell needs and drawing them into the cell. Other proteins act as gates, opening and closing to control the flow of various particles into and out of the cell.

In addition to the lipid and protein, the membrane also contains carbohydrates that mark the exterior of the cell. Carbohydrates combined with protein (known as glycoproteins) and lipids (known as glycolipids) make up a portion of the plasma membrane. Collectively, the glycoproteins, glycolipids, and carbohydrates on the outer surface of the plasma

**phospholipid** Class of fat-related substances that contain phosphorus, fatty acids, and a nitrogen-containing base. Phospholipids are an essential part of every cell.

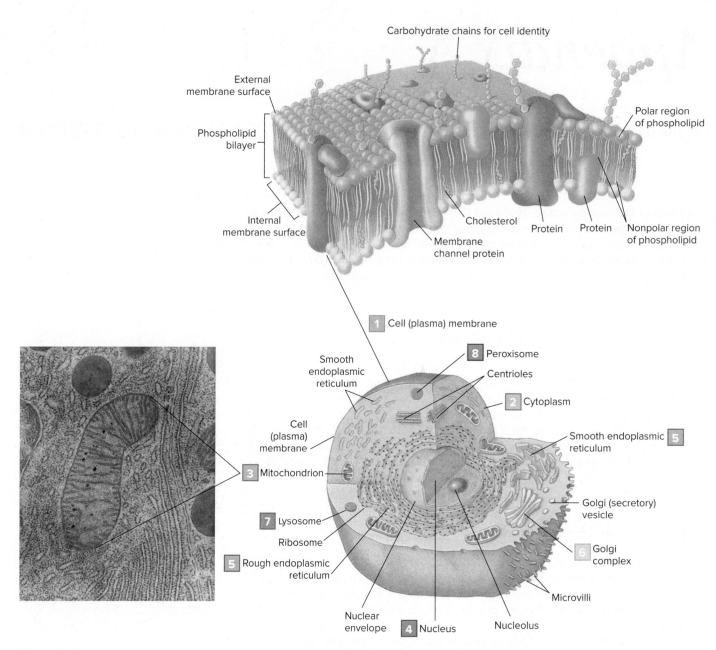

*Figure A-1*  An animal cell. Almost all human cells contain these various organelles. Shown in greater detail are mitochondria and the cell membrane. Note: not all cells have microvilli. The nuclear envelope encloses the nucleus. The centrioles participate in cell division.

**organelle**  Compartment, particle, or filament that performs specialized functions in a cell.

**cytoplasm**  Fluid and organelles (except the nucleus) in a cell.

**cytosol**  Water-based phase of a cell's cytoplasm; excludes organelles, such as mitochondria.

membrane make up the glycocalyx. The glycocalyx has multiple functions, including cell identification, protection against invaders, defense against cancer, and allowing cells to bind together.

Included within the cell membrane are **organelles.** They carry out vital roles in cell functions. Some structures allow the cell to replicate itself, others provide energy, and others destroy the cell when it is worn out. Still other organelles produce and secrete products destined for other cells.

## 2  Cytoplasm

The **cytoplasm** is the fluid material and organelles within the cell, not including the nucleus. (The **cytosol** is the fluid surrounding the organelles.) It also contains the many protein rods and tubes that form the cytoskeleton. A small amount of ATP energy for use by the cell can be produced by glycolysis reactions that occur in the cytoplasm. This contributes to our survival because it is the key process in red blood cell energy metabolism; it is called anaerobic metabolism because it doesn't require oxygen.

## 3  Mitochondria

**Mitochondria** are sometimes called the "power plants," or powerhouses, of cells. These organelles are capable of converting the energy in our energy-yielding nutrients (carbohydrate, protein, and fat) to a form that cells can use, ATP. This is an aerobic process that uses the oxygen we inhale, as well as water, enzymes, and other compounds (see Chapter 9 for details). With the exception of red blood cells, all cells contain mitochondria; only the sizes, shapes, and numbers vary.

Mitochondria have a double membrane, and this characteristic is key to overall mitochondrial function. Within the inner membrane, the electron transport chain and ATP synthesis take place. In the inner matrix of the mitochondria, the citric acid cycle, the beta-oxidation of fatty acids, and the transition reaction involving pyruvate take place.

The biochemical pathways that operate in the mitochondrial matrix also are capable of synthesizing cell components, such as the **carbon skeletons** needed to produce amino acids. These will eventually become cellular protein.

## 4  Cell Nucleus

The **cell nucleus** is surrounded by its own double membrane. The nucleus controls actions that occur in the cell, using the hereditary material deoxyribonucleic acid (DNA). DNA is the "code book" that contains directions for making substances the cell needs. It consists of genes on **chromosomes.** This code book remains in the nucleus of the cell, but it conveys its information to other cell organelles by way of a similar molecule called **ribonucleic acid (RNA).** RNA is responsible for *transcribing* the information of the DNA and moving out through pores in the nuclear membrane to the cytoplasm. The RNA then carries the code to protein-synthesizing sites called **ribosomes.** There, the RNA code is *translated* into a specific protein (see Chapter 7 for details on protein synthesis). With the exception of the red blood cell, all cells have 1 or more nuclei.

The **nucleoli** are areas within the nucleus of the cell containing a combination of protein and RNA. This is where RNA is produced for export to the cytoplasm.

DNA has the secondary task of cell replication. DNA is a double-stranded molecule; when the cell begins to divide, each strand is separated and an identical copy of each is made. Thus, each new DNA molecule contains 1 new strand of DNA and 1 strand from the original DNA. In this way, the genetic code is preserved from 1 cell generation to the next. The mitochondria contain their own DNA, so they reproduce themselves independently of the nucleus.

The transport of proteins, vitamins, and other material from the cytoplasm to the nucleus also occurs through pores in the nuclear membrane, as just mentioned. These small molecules serve a variety of functions, including the activation (or inactivation) of certain parts of the DNA.

## 5  Endoplasmic Reticulum (ER)

The outer membrane of the cell nucleus is continuous with a network of tubes called the **endoplasmic reticulum (ER).** The ER is found in 2 types: rough and smooth. The rough endoplasmic reticulum has ribosomes bound to it, whereas the smooth does not. As noted earlier, ribosomes are sites of protein synthesis. Many of these proteins play a central role in human nutrition. Smooth ER is involved in lipid synthesis, the detoxification of toxic substances, and calcium storage and release in the cell.

**mitochondria**  Main sites of energy production in a cell. They also contain the pathway for oxidizing fat for fuel, among other metabolic pathways.

**carbon skeleton**  Amino acid without the amino group ($—NH_2$).

**cell nucleus**  Organelle bound by its own double membrane and containing chromosomes that hold the genetic information for cell protein synthesis.

**chromosome**  Complex of DNA and protein containing the genetic material of a cell's nucleus. There are 46 chromosomes in the nucleus of each cell, except in germ cells.

**ribonucleic acid (RNA)**  Single-stranded nucleic acid involved in the transcription of genetic information and the translation of that information into protein structure.

**ribosomes**  Cytoplasmic particles that mediate the linking together of amino acids to form proteins; attached to endoplasmic reticulum as bound ribosomes or suspended in cytoplasm as free ribosomes.

**nucleolus**  Center for production of ribosomes within the cell nucleus. Nucleoli is the plural for nucleolus.

**endoplasmic reticulum (ER)**  Organelle in the cytoplasm composed of a network of canals running through the cytoplasm. Rough ER contains ribosomes. Smooth ER contains no ribosomes.

### 6 Golgi Complex

The **Golgi complex** is a packaging site for proteins and lipids that are used in the cytoplasm or exported from the cell. The Golgi complex consists of stacks of sacs within the cytoplasm in which products of the rough endoplasmic reticulum are received, processed, separated according to function and destination, and "packaged" in **secretory vesicles** for secretion by the cell.

### 7 Lysosomes

**Lysosomes** are the cell's digestive system. They are sacs that contain enzymes for the digestion of foreign material. Sometimes known as "suicide bags," they are responsible for digesting worn-out or damaged cells. They carry out **apoptosis,** or programmed cell death, which occurs naturally or is associated with illness or infection. Certain cells that are associated with immunity contain many lysosomes.

### 8 Peroxisomes

**Peroxisomes** are membrane-bound vesicles that contain enzymes that detoxify harmful chemicals. **Hydrogen peroxide** ($H_2O_2$) is formed as a result of such enzyme action. Peroxisomes contain a protective enzyme called *catalase,* which prevents excessive accumulation of hydrogen peroxide in the cell, which would be very damaging. Peroxisomes also play a minor role in metabolizing a possible source of energy for cells—alcohol.

The remainder of this appendix looks at the body systems. Keep in mind that these systems depend on the cell functions just discussed.

## Integumentary System

The system you probably are most familiar with is the **integumentary system,** which is made up of dissimilar elements, such as the skin, hair, various glands, and nails. The largest organ in the body, the skin, consists of 2 principal layers, the **epidermis** and the **dermis** (Fig. A-2). The epidermis is the layer of skin composed largely of dead cells, which are used for protection from environmental pathogens, ultraviolet light, toxins, injury, and water. We don't want to absorb water through the skin, nor do we want water to readily escape the body.

The dermis is a deeper and thicker layer of skin, with an extensive network of blood vessels, sweat glands, oil-secreting glands, nerve endings, and hair follicles. It confers strength and flexibility to the skin. When people are confined to bed for long periods of time, **decubitus ulcers,** also called bedsores, may develop because of restricted blood flow to the dermis. This lack of blood causes cells to die and open wounds to develop—a potentially life-threatening situation. Adequate intakes of protein, vitamin A, vitamin C, and zinc may help prevent this problem.

The appearance of the skin, hair, and nails is clinically important because it can indicate nutritional deficiencies. For instance, hot, dry skin is an obvious sign of dehydration due to inadequate water intake. (Other signs and symptoms of nutrient deficiencies, as manifested by the skin, are described in Chapters 5, 6, 7, 12, 13, 14, and 15 as the functions of individual nutrients are explained.)

The skin plays a vital role in temperature regulation. Heat produced by the body's metabolic processes, especially the processes that occur in muscle, must be removed before cells are damaged. Heat is removed from the body through the skin. When we are cold, we warm ourselves by shivering because muscle contractions generate heat.

---

**Golgi complex** Cell organelle near the nucleus; processes newly synthesized protein for secretion or distribution to other organelles.

**secretory vesicles** Membrane-bound vesicles produced by the Golgi apparatus; contain proteins and other compounds to be secreted by the cell.

**lysosome** Cell organelle that digests proteins, such as transferrin, and breaks down bacteria and old or damaged cell components.

**apoptosis** Process that occurs over time in which enzymes in a cell set off a series of events that disable numerous cell functions, eventually leading to cell death.

**peroxisome** Cell organelle that uses oxygen to remove hydrogens from compounds. This produces hydrogen peroxide ($H_2O_2$), which breaks down into $O_2$ and $H_2O$.

**integumentary system** Having to do with the skin, hair, glands, and nails.

**epidermis** Outermost layer of the skin; composed of epithelial layers.

**dermis** Second, or deep, layer of the skin under the epidermis.

**decubitus ulcer** Chronic ulcer (also called bedsore) that appears in pressure areas of the skin over a body prominence. These sores develop when people are confined to bed or otherwise immobilized.

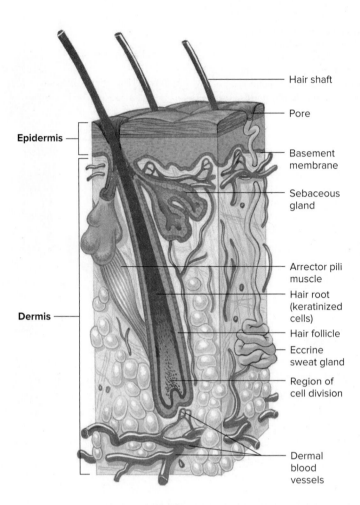

*Figure A-2*  Cross section of the skin. This is the major organ of the integumentary system.

Hair shaft

Pore

Epidermis

Basement membrane

Sebaceous gland

Arrector pili muscle

Hair root (keratinized cells)

Dermis

Hair follicle

Eccrine sweat gland

Region of cell division

Dermal blood vessels

An important nutrient, vitamin D, can be obtained from our diets, but the skin also can make it from a cholesterol derivative in the skin. There is more detail about this process in Chapter 15.

The sweat glands produce perspiration, or sweat, which helps evaporate fluids to cool the body and excrete certain wastes. Mammary glands within the breasts are modified sweat glands designed to secrete milk to feed a newborn.

## Skeletal System

The skeletal system includes bones, cartilage, and ligaments. The 206 bones that make up the skeletal system provide the rigid framework for the body. Joints connect many bones to one another. Bones that make up the skull and vertebral column protect the brain and spinal cord from injury. Likewise, the rib cage protects the heart, lungs, liver, and spleen from external damage. Bones have attachment sites for skeletal muscles, ligaments, and tendons. (Bones attached to muscles allow body movement when muscles contract.) Blood cell formation, known as **hematopoiesis,** takes place within the marrow of some bones. Bones also are a storehouse for minerals, such as calcium, phosphorus, magnesium, sodium, and fluoride. Bones are metabolically active and constantly adapting to a changing environment.

**hematopoiesis** Production of blood cells.

*Figure A-3* Diagram of a long bone. The epiphysis, consisting of spongy bone, is surrounded by a layer of compact bone. The epiphyseal line indicates that the bone has completed growth. The production of blood cells occurs in the porous chambers of spongy bone. The collagen material, the structural material of bone, is observed by the open flap. The skeletal system provides a reserve of calcium and phosphorus for day-to-day needs when dietary intake is inadequate.

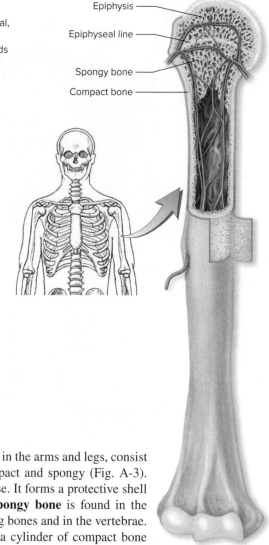

Epiphysis

Epiphyseal line

Spongy bone

Compact bone

**compact bone**  Dense, compact bone that constitutes the outer surface and shaft of bone; makes up about 75% of total bone mass; also called cortical bone.

**spongy bone**  Spongy, inner matrix of bone found primarily in the spine, pelvis, and ends of bones; makes up 20 to 25% of total bone mass; also called trabecular bone.

**epiphysis**  End of a long bone. The epiphyseal plate—sometimes referred to as the growth plate—is made of cartilage and allows bone to grow. During childhood, the cartilage cells multiply and absorb calcium to develop into bone.

**epiphyseal plate**  Cartilage-like layer in the long bone sometimes referred to as the growth plate. It functions in linear growth. During childhood, the cartilage cells multiply and absorb calcium to develop into bone.

**epiphyseal line**  Line that replaces the epiphyseal plate when bone growth is complete.

**collagen**  Major protein of the material that holds together the various structures of the body.

**hydroxyapatite**  Compound composed primarily of calcium and phosphate; it is deposited in bone protein matrix to give bone strength and rigidity ($Ca_{10}[PO_4]_6OH_2$).

**remodeling**  Lifelong process of the building and breaking down of bone.

**resorption**  Loss of a substance by physiological or pathological means.

**osteoblast**  Cell in bone; secretes mineral and bone matrix.

**osteoclast**  Bone cell that arises from a type of white blood cell; secretes substances that lead to bone erosion. This erosion can set the stage for subsequent bone mineralization.

Long bones, such as those in the arms and legs, consist of 2 types of body tissue: compact and spongy (Fig. A-3). **Compact bone** is hard and dense. It forms a protective shell on the exterior of the bone. **Spongy bone** is found in the compact bone at the ends of long bones and in the vertebrae. The shaft of the long bones is a cylinder of compact bone surrounding a central cavity containing the marrow (see Chapter 15).

At the end of the long bone is the **epiphysis,** consisting of spongy bone covered by compact bone. The epiphysis is strong and allows for the attachment of tendons and ligaments. Red bone marrow is made of spongy bone and is the source of red blood cells, as well as white blood cells and platelets. In children, just behind the epiphysis is the **epiphyseal plate.** This area of bone is responsible for linear growth. When linear growth is complete, an **epiphyseal line** replaces the plate.

Bones are constructed from several types of cells under the influence of a variety of growth factors. These factors stimulate the formation of **collagen,** a type of flexible protein matrix, which forms the basic shape of bone. Minerals—principally calcium and phosphorus—are embedded in the matrix, which give the bone strength. **Hydroxyapatite,** the calcium phosphorus salt deposited in the protein matrix, constitutes about 85% of the minerals in bone and makes it possible for the bone to resist compression and bending.

Calcification (also called ossification) of bone varies from bone to bone, but most bones mature (are ossified) by ages 17 to 25. However, some bones, such as the sternum (breastbone), may not complete growth until age 30-plus years.

Bone is constantly **remodeled** throughout life. Formation and **resorption** of bone occur because of the continual activity of **osteoblasts** and **osteoclasts.** Osteoblasts are bone-building cells, and osteoclasts are bone-resorbing cells. In the first 20 or so years of life, bone formation is greater than resorption. By age 40 or so, resorption generally exceeds deposition and we start to lose bone mass. Bone diseases are more likely to occur after age 50. Exercise promotes bone deposition, whereas a lack of exercise results in bone loss.

Bone deposition (ossification) and bone resorption (dissolution) also maintain homeostasis of calcium and phosphorus in the blood. Three hormones control the process: the vitamin D hormone calcitriol, calcitonin, and parathyroid hormone.

Other hormones are involved in bone maintenance, such as growth hormone; thyroid hormones; sex hormones, especially estrogen; and adrenocorticoid hormones. In addition, vitamins A, K, and C perform important jobs in bone metabolism. More information about bones can be found in Chapters 11 through 15, which cover exercise, vitamins, and minerals.

## Muscular System

The functions of the muscular system are to provide movement and to generate body heat. Most of the energy released by a muscle cell during physical exercise is in the form of heat. **Muscle fibers** respond when stimulated by motor **neurons** (nerve cells). A muscle cell converts the chemical energy in ATP into the mechanical energy of muscle contraction.

There are 3 types of muscle tissue: **smooth, cardiac,** and **skeletal muscle.** Smooth muscle fibers have a single nucleus and function in involuntary movements within internal organs. Cardiac muscle fiber is striated (striped) with a single nucleus. The stripes in muscle fibers are caused by the arrangement of alternating dark and light contractile proteins (**myosin** and **actin**). This type of muscle performs the involuntary, rhythmic contractions of the heart. Skeletal muscle, also containing striated muscle fibers, has several nuclei and is involved in voluntary movements. Skeletal muscle is attached to bone by **tendons.** (Note that Chapter 11 also discusses some specific muscle fiber types.) By age 70 or 80, significant loss of muscle mass is evident. Exercise can help maintain muscle mass.

### Skeletal Muscle

Skeletal muscle fibers are actually long cells with the same organelles that are in other cells. However, unlike most other cells, skeletal muscle cells possess an excellent supply of fuel in the form of glycogen, the body's storage form of the sugar glucose.

Skeletal muscles contract when stimulated by motor neurons. Motor neurons can stimulate several muscle fibers simultaneously. A single muscle fiber is not very efficient. The activation of numerous muscle fibers by multiple motor neurons results in increased muscle strength as the number of fibers stimulated by neurons increases.

### Muscle Contraction

As previously mentioned, within muscle fibers are dark and light stripes called striations. Each muscle cell, when viewed in the electron microscope, contains subunits called **myofibrils.** The myofibrils are the source of the light and dark bands, or stripes. The importance of these structures is the presence of the unique proteins actin and myosin. The functioning structure of the myofibril is the **sarcomere,** the contracting unit.

When a muscle fiber is stimulated by a neuron to contract, among the 1st events to occur is the release of large amounts of calcium from storage in the smooth endoplasmic reticulum (also called the sarcoplasmic reticulum). This is the "on" switch. The presence of calcium allows the 2 main proteins, myosin and actin, to get ready to slide into each other

**muscle fiber** Essentially, a single muscle cell; an elongated cell, with contractile properties, that forms the muscles in the body.

**neuron** Structural and functional unit of the nervous system, consisting of cell body, dendrites, and axon.

**smooth muscle** Muscle tissue under involuntary control; found in the GI tract, artery walls, respiratory passages, urinary tract, and reproductive tract.

**cardiac muscle** Muscle that makes up the walls of the heart; produces rhythmic, involuntary contractions.

**skeletal muscle** Muscle tissue responsible for voluntary body movements.

**myosin** Thick filament protein that connects with actin to cause a muscle contraction.

**actin** Protein in muscle fiber that, together with myosin, is responsible for contraction.

**tendon** Dense connective tissue that attaches a muscle to a bone.

**myofibril** Bundle of contractile fibers within a muscle cell.

**sarcomere** Portion of a muscle fiber that is considered the functional unit of a myofibril.

**power stroke** Movement of the thick filament alongside the thin filament in a muscle cell, causing muscle contraction.

and set the **power stroke** in motion. Of course, energy is required to carry out the muscle contraction. Here is where ATP plays the key role (Fig. A-4).

Another ATP is needed to release the actin from the myosin. This is the end of the contraction. As the muscle moves to the "off" position, the calcium is transported back to storage, the muscle fiber relaxes, and it gets ready for another contraction.

The factor that allows muscle action to occur at all is the essential nutrient calcium. When the muscle is relaxed, there is very little calcium in the cytoplasm of the muscle cell because calcium is in storage. However, when the muscle is ready to go to work, as directed by the motor neuron, calcium is moved out of storage, which sets the stage for the power stroke. When the contraction ends, the calcium is released from the muscle fibers and goes back into storage.

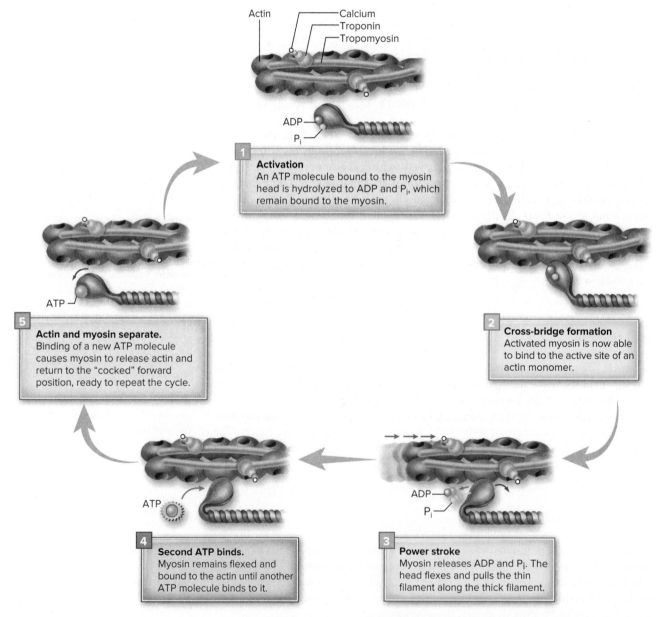

**1  Activation**
An ATP molecule bound to the myosin head is hydrolyzed to ADP and P$_i$, which remain bound to the myosin.

**2  Cross-bridge formation**
Activated myosin is now able to bind to the active site of an actin monomer.

**3  Power stroke**
Myosin releases ADP and P$_i$. The head flexes and pulls the thin filament along the thick filament.

**4  Second ATP binds.**
Myosin remains flexed and bound to the actin until another ATP molecule binds to it.

**5  Actin and myosin separate.**
Binding of a new ATP molecule causes myosin to release actin and return to the "cocked" forward position, ready to repeat the cycle.

*Figure A-4* Muscle contraction. In Step 1, activation, an ATP is bound to the myosin head (purple) and is split into ADP and P$_i$. Troponin and tropomyosin are proteins that participate in this process. During Step 2, the activated myosin can now bind to the actin (the red beads). In Step 3, the myosin head releases the ADP and P$_i$. The head flexes and pulls the thin filament along the thick filament. This is the *power stroke*. In Step 4, the myosin remains bound to the actin until another ATP binds to the myosin. The new ATP causes the myosin to release the actin so that it can get ready for another cycle, Step 5. Actin and myosin separate, which leads to muscle relaxation. The white dots in this figure represent calcium, a nutrient required for muscle action.

## Cardiac and Smooth Muscle

Cardiac muscle and smooth muscle, although similar in many ways to skeletal muscle in their use of calcium as an off/on switch, operate under involuntary control. In cardiac muscle, the stimulation occurs automatically by a group of muscle cells. These cells initiate the heartbeat and set the heart rate under control of the brain and the influence of certain hormones.

Smooth muscles are found in the lungs, blood vessels, GI tract, and other internal organs. In the GI tract, they produce important contractions in peristalsis (see Chapter 4 for details).

## Circulatory System

The circulatory system is made up of 2 separate systems: the cardiovascular system and the lymphatic system. The cardiovascular system consists of the heart and blood vessels. The lymphatic system consists of lymphatic vessels, lymph, and a number of lymphatic tissues.

One organ vital to our existence is the heart, a 4-chambered pump that keeps blood continuously circulating around the body. It takes about 1 minute for blood to leave the heart, circulate to all tissues in the body, and return to the heart. When we are exercising strenuously, the blood can circulate at a rate of 6 times per minute.

The cells that make up the tissues of the body need a constant supply of water, oxygen, and nutrients. In addition, the body needs ATP energy, which comes from the breakdown of energy nutrients within the cells. The blood carries oxygen from the lungs to all organs in the body. The blood also carries nutrients from the digestive tract to all tissues and to storage sites when nutrients are not needed immediately for energy, growth, or repair. Waste materials produced by cells must be removed by way of the skin, lungs, kidneys, and digestive tract. This, too, is a function of the cardiovascular system. The delivery of hormones to their target cells, the maintenance of a constant body temperature, and the distribution of white blood cells to protect against invading pathogens are all performed by the blood and circulatory systems without our being aware of any specific action. The circulatory system has chemical means to prevent excessive blood loss from damaged vessels using the clotting process (see Chapter 14).

## Blood Constituents

Red blood cells, known as **erythrocytes,** carry oxygen to all tissues and play a role in the return of carbon dioxide to the lungs. The white blood cells, known as **leukocytes,** function as part of the immune system. They protect the body from invading pathogens. The blood is able to clot because of platelets and other clotting factors. The liquid part of blood is known as **plasma.** In contrast, **serum** is the fluid that results after the blood is first allowed to clot before being centrifuged; it does not contain the blood clotting factors.

## Heart Structure

The heart has 2 sides, left and right. The right side is closest to your right arm; likewise, the left side is closest to your left arm. The upper part of the heart has left and right **atria,** which empty simultaneously into the lower part of the heart, consisting of the left and right **ventricles.**

Blood travels from the left side of the heart to the body through the **aorta** to major **arteries.** Arteries become smaller and smaller until they are so tiny they are classified as **arterioles.** The blood flows from the arterioles into microscopic, weblike structures called **capillaries.** Capillaries are just 1 cell layer thick; they have pores, which allow oxygen, water, and other nutrients to leave the blood for surrounding cells and allow waste and other products

▶ A unique feature of smooth muscle is its ability to stretch. By the end of pregnancy, the smooth muscle in the uterus can be stretched up to 8 times its prepregnant length.

**erythrocyte**  Mature red blood cell. It has no nucleus and a life span of about 120 days. It contains hemoglobin, which transports oxygen and carbon dioxide.

**leukocyte**  White blood cell.

**plasma**  Fluid, noncellular portion of the circulating blood. This includes the blood serum plus all blood clotting factors. In contrast, serum is the fluid that results after the blood is first allowed to clot before being centrifuged; does not contain the blood clotting factors.

**serum**  Portion of the blood fluid remaining after the blood is allowed to clot and the red and white blood cells and other solid matter are removed by centrifugation.

**atria**  Plural of *atrium*. Two upper chambers of the heart that receive venous blood.

**ventricles**  Two lower chambers of the heart, which contain blood to be pumped out of the heart.

**aorta**  Major blood vessel of the body, leaving from the left ventricle of the heart.

**artery**  Blood vessel that carries blood away from the heart.

**arteriole**  Tiny artery branch that ends in capillaries.

**capillary**  Smallest blood vessel; the major site for the exchange of substances between the blood and the tissues.

**venule**  Tiny vessel that carries blood from the capillary to a vein.

**vein**  Blood vessel that conveys blood to the heart.

**systemic circuit**  Part of the circulatory system concerned with the flow of blood from the heart's left ventricle to the body and back to the heart's right atrium.

of cellular metabolism to enter the blood. There are few cells in the body that aren't close to a capillary. Larger blood vessels are not porous, so blood cannot escape these vessels. Only in the capillaries can the blood discharge and recover substances associated with nearby cells.

As the blood exits the capillaries, it flows into tiny **venules,** which enlarge and become **veins,** returning the blood to the right side of the heart. The route from the left side of the heart to the capillaries and then back to the right side of the heart is called the **systemic circuit** of blood (Figs. A-5, A-6, and A-7).

The flow of blood through the circulatory system is measured by pressure in millimeters of mercury. The average arterial (artery) pressure is about 100 mm Hg, whereas the average venous pressure is only 2 mm Hg. To guarantee return flow to the heart, blood is moved through the veins by the contraction of skeletal muscles. Also, valves in the veins prevent a backflow of blood.

*Figure A-5*  Blood circulation through the body. This figure shows the paths that blood takes from the heart to the lungs (Steps 1 – 3 ), back to the heart (Step 4 ), and through the rest of the body (Steps 5 – 9 ). The red color indicates blood that is richer in oxygen; blue is for blood carrying more carbon dioxide. Arteries and veins go to all parts of the body.

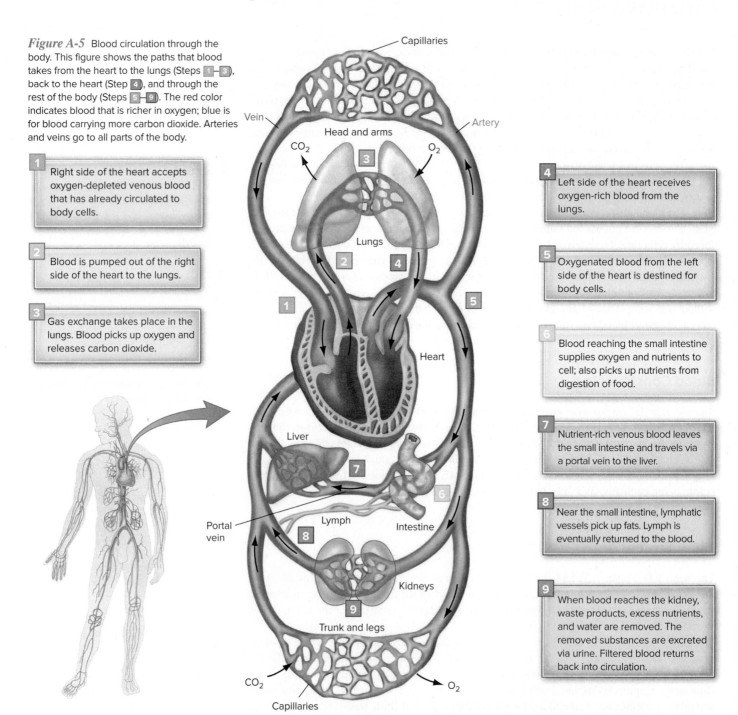

**1** Right side of the heart accepts oxygen-depleted venous blood that has already circulated to body cells.

**2** Blood is pumped out of the right side of the heart to the lungs.

**3** Gas exchange takes place in the lungs. Blood picks up oxygen and releases carbon dioxide.

**4** Left side of the heart receives oxygen-rich blood from the lungs.

**5** Oxygenated blood from the left side of the heart is destined for body cells.

**6** Blood reaching the small intestine supplies oxygen and nutrients to cell; also picks up nutrients from digestion of food.

**7** Nutrient-rich venous blood leaves the small intestine and travels via a portal vein to the liver.

**8** Near the small intestine, lymphatic vessels pick up fats. Lymph is eventually returned to the blood.

**9** When blood reaches the kidney, waste products, excess nutrients, and water are removed. The removed substances are excreted via urine. Filtered blood returns back into circulation.

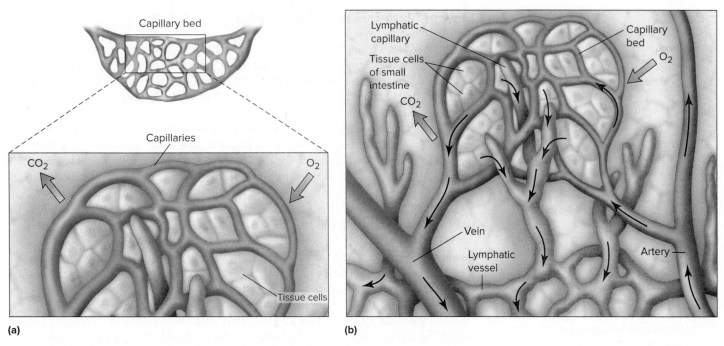

**(a)**                                                        **(b)**

*Figure A-6* Capillary and lymphatic vessels. (*a*) Exchange of oxygen and nutrients for carbon dioxide and waste products occurs between the capillaries and the surrounding tissue cells. (*b*) Lymphatic vessels also are present in capillary beds, such as in the small intestine. Lymphatic vessels in the small intestine are also called *lacteals*. Note that the lymphatic vessels are blind-ended.

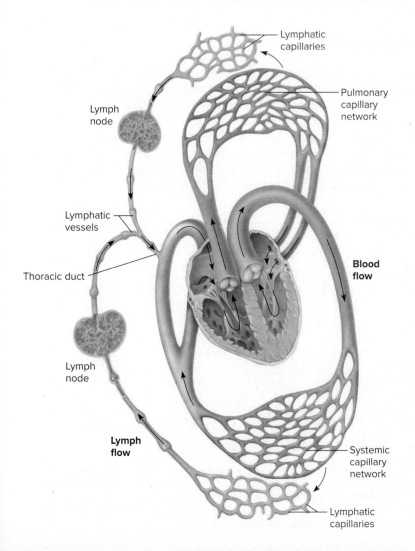

*Figure A-7*  Lymph. As lymph moves through the lymphatic system, it encounters lymph nodes containing immune cells that destroy invading pathogens. Lymph also carries dietary fat and fat-soluble nutrients from the digestive tract to the blood, using the thoracic duct.

## Flow of Materials between Capillaries and Cells

As the blood flows from the arterioles into the capillaries, the hydrostatic pressure generated by the force of the heart's contraction causes fluid to flow into spaces around the surrounding cells, called the **extracellular fluid (ECF)** (Fig. A-8). Some of this fluid returns to the capillaries and some enters another nearby vessel called a **lymphatic vessel.**

Oxygen and nutrients leave the capillaries, enter the ECF, and are then delivered to cells by 1 of the mechanisms mentioned in Chapter 4: passive and facilitated diffusion, active transport, and pinocytosis. Cellular products plus waste substances are collected in the ECF and are either released to the capillaries that connect to the venules or channeled into the lymphatic vessels. Oxygen travels to the cell by diffusing from the blood into the extracellular fluid and then in through the cell membrane. Carbon dioxide exits the cell and goes to the blood in the same way. This is 1 of 2 important gas exchange activities in the body and is often referred to as internal respiration.

The right atrium of the heart receives dark red venous blood from the body, which is then pumped into the right ventricle. The right ventricle pumps blood through the pulmonary arteries to the capillaries in the lungs. The lungs then return the freshly oxygenated blood to the left atrium of the heart via the pulmonary veins. This route is known as **pulmonary circulation.**

As the blood moves through the pulmonary capillaries, carbon dioxide is released for expiration, and the inhaled oxygen is taken up by the blood. This is the other site for gas exchange in the body, often referred to as external respiration. The oxygenated blood (now bright red) in the atrium is pumped to the left ventricle. The blood is pumped out of the left ventricle and through the aorta to the systemic circuit.

## Other Circulatory Systems

One capillary bed does not send blood back to the heart but, rather, directs it toward the liver. This is the **portal system** of the GI tract, composed of veins draining blood from the capillaries of the intestines and stomach. These veins empty into a large **portal vein,** which acts as a direct pipeline to the liver. (The brain also has a portal system.)

The heart has its own circulatory system, too. Coronary vessels supply blood to meet cardiac needs. These arteries are particularly susceptible to damage by deposits of cholesterol and other lipids in the artery wall. This accumulation of cholesterol can lead to coronary heart disease. (There is more about this disease in Chapter 6.)

**extracellular fluid (ECF)**  Fluid present outside the cells. It includes intravascular and interstitial fluids and represents one-third of all body fluid.

**lymphatic vessel**  Vessel that carries lymph.

**pulmonary circulation**  System of blood vessels from the right ventricle of the heart to the lungs and back to the left atrium of the heart.

**portal system**  Veins in the GI tract that convey blood from capillaries in the intestines and portions of the stomach to the liver.

**portal vein**  Large vein leaving from the intestine and stomach and connecting to the liver.

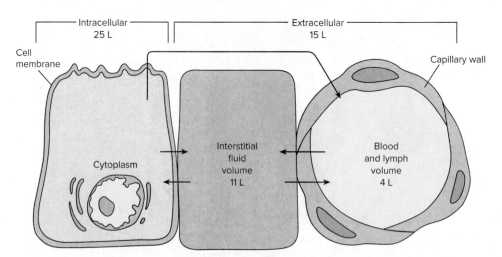

*Figure A-8*  Distribution of body fluids. The intracellular compartment contains fluid in the cell, which is free to move into the extracellular compartment. The extracellular compartment contains the fluid between the cell and the capillary, called interstitial fluid. The extracellular fluid also includes the fluid within blood and lymphatic vessels. This figure shows the fluid (plasma) from the blood moving freely between cells and capillaries through the interstitial fluid.

# Lymphatic System

The lymphatic system is closely related to the immune system in that both defend us against pathogenic invaders. As the lymphatic system collects fluid from tissues, it picks up microorganisms as well. The fluid passes through many **lymph nodes** as it makes its way back to the bloodstream. In the nodes is an abundant collection of white blood cells ready to detect pathogens in the lymph fluid and quickly destroy them. The lymphatic system consists of lymphatic vessels, lymph fluid, lymph nodes, and lymphatic tissue, with its population of immune cells.

The interstitial, or extracellular, fluid (fluid surrounding the cell) contains many components that are too large to pass through holes in the capillaries, so they are blocked from returning directly to the bloodstream. Therefore, they take an indirect route via the lymphatic system back to general circulation (see Fig. A-7).

Lymph also serves as the passageway by which fat-soluble nutrients are absorbed from the GI tract and carried into the bloodstream. Lymph also contains bacteria, viruses, cellular trash, and cancer cells on their way to invade some distant site. Lymph generates immune cells, called **lymphocytes,** that combat these invaders (see the section Immune System).

At the terminal end of the capillaries, the amount of fluid released from the capillaries into the venules is less than the amount of fluid entering the capillaries from the arterioles. The missing 15% of fluid represents the extracellular fluid that is returned to the vascular system via the lymphatic system. This fluid is subsequently delivered to the lymphatic system by way of specialized capillaries called lymphatic capillaries. Blood plasma and fluid in the tissues are constantly being interchanged. The fluid, which is now called **lymph,** enters these porous vessels and consists of extracellular fluid and proteins too large to squeeze back into the capillaries.

In addition to microorganisms, the lymph contains absorbed dietary fat. The absorption of fats occurs only in the **lacteals,** which are lymphatic capillaries of the small intestine, not the portal vein. From the lacteals, lymph is directed into larger vessels, called **lymph ducts,** and is moved toward the heart by the action of skeletal muscle contractions and other body movements.

As the lymph makes its way back to the heart, it encounters clusters of lymph nodes containing phagocytic cells, lymphocytes, and mobile **macrophages,** which help destroy invading pathogens and filter the lymph. **T lymphocytes** and **B lymphocytes** are found in these nodes and are major players in immunity (see the section Immune System). When you are ill and seek medical attention, do you ever wonder why your physician checks the lymph glands in your neck for swelling? Swelling means the lymph nodes are in combat against an invading pathogen.

The spleen, thymus, and tonsils are considered lymphoid organs. The spleen contains phagocytes, which filter out foreign substances and destroy worn-out red blood cells. The thymus is important in immunity during childhood. Tonsils protect against invaders that are inhaled or eaten.

Eventually, the lymph empties into the thoracic duct and the right lymph duct, then into veins that enter the right atrium of the heart, and finally into general circulation (see Fig. A-7). (There is further discussion of the transport of lipid substances in the lymphatic system in Chapters 4 and 6.)

# Immune System

The cells that carry out immune functions are known collectively as the immune system. Unlike other systems in the body, they do not exist as anatomically connected organs but, rather, as separate collections of cells throughout the body. They defend against invading pathogens—microorganisms or substances capable of producing disease. They discriminate between "self" and "non-self." They are very sensitive indicators of the body's nutritional status. The most numerous of the immune system cells are the leukocytes.

**lymph node** Small structure located along the course of the lymphatic vessels.

**lymphocyte** Class of white blood cells involved in the immune system, generally comprising about 25% of all white blood cells. There are several types of lymphocytes with diverse functions, including antibody production, allergic reactions, graft rejections, tumor control, and regulation of the immune system.

**lymph** Clear, plasmalike fluid that flows through lymphatic vessels.

**lacteal** Tiny vessels in the small intestine villi that absorb dietary fat.

**lymph duct** Large lymphatic vessel that empties lymph into the circulatory system.

**macrophage** Large, mononuclear, phagocytic cell derived from a monocyte in the blood and found in body tissues. Besides functioning as phagocytes, macrophages secrete numerous cytokines and act as antigen-presenting cells.

**T lymphocyte** Type of white blood cell that recognizes intracellular antigens (e.g., viral antigens in infected cells), fragments of which move to the cell surface. T lymphocytes originate in the bone marrow but must mature in the thymus gland.

**B lymphocyte** Type of white blood cell that recognizes antigens (e.g., bacteria) present in extracellular sites in the body and is responsible for antibody-mediated immunity. B lymphocytes originate and mature in the bone marrow and are released into the blood and lymph.

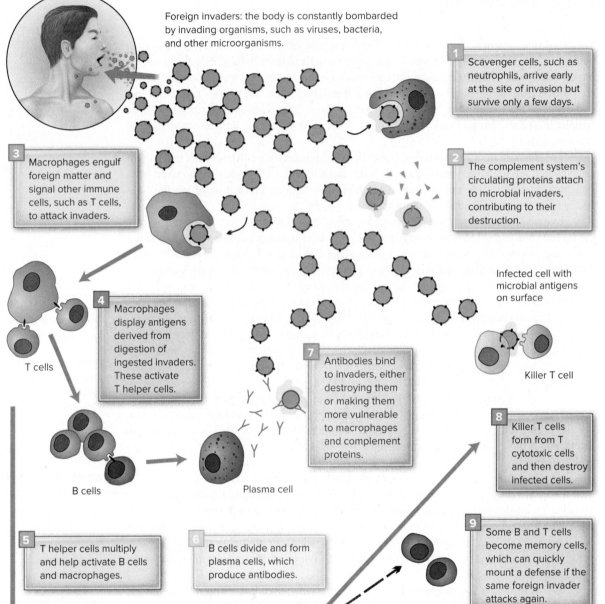

Foreign invaders: the body is constantly bombarded by invading organisms, such as viruses, bacteria, and other microorganisms.

**1** Scavenger cells, such as neutrophils, arrive early at the site of invasion but survive only a few days.

**2** The complement system's circulating proteins attach to microbial invaders, contributing to their destruction.

**3** Macrophages engulf foreign matter and signal other immune cells, such as T cells, to attack invaders.

Infected cell with microbial antigens on surface

**4** Macrophages display antigens derived from digestion of ingested invaders. These activate T helper cells.

T cells

Killer T cell

**7** Antibodies bind to invaders, either destroying them or making them more vulnerable to macrophages and complement proteins.

**8** Killer T cells form from T cytotoxic cells and then destroy infected cells.

B cells

Plasma cell

**5** T helper cells multiply and help activate B cells and macrophages.

**6** B cells divide and form plasma cells, which produce antibodies.

**9** Some B and T cells become memory cells, which can quickly mount a defense if the same foreign invader attacks again.

*Figure A-9* Biological warfare. The body commands a wide assortment of defenders to reduce the danger of infection and help guard against repeated microbial infections. The ultimate target of the immune response is an antigen—commonly, a foreign protein from a bacterium or another microbe.

Our bodies constantly wage war against disease-producing microorganisms, such as bacteria, viruses, fungi, and parasites; against substances capable of producing disease, such as toxins from snake venom; and against allergens, which trigger allergic reactions via **antigen** release (see Chapter 7), and cancer cells (Fig. A-9). The most common invaders are bacteria, which are 1-celled organisms with a cell wall in addition to a cell membrane, and viruses, which are nucleic acids surrounded by a protein coat. Viruses can't multiply by themselves because they lack ribosomes for protein synthesis, so they survive by taking over host cells and instructing them to produce the proteins and energy the viruses need for survival.

## Leukocytes and Macrophages

Leukocytes, also known as white blood cells, are produced in the bone marrow and may undergo further development in tissues outside the marrow. They travel via the blood and enter tissues, where they function. They are classified by their structure and their affinity for certain types of dye. For example, a monocyte has a single, prominent nucleus. Another

**antigen** Foreign substance, generally large in size, that is capable of inducing a specific immune response. Often binds with an antibody.

**Table A-1 Types and Functions of Leukocytes**

| Leukocyte | Functions |
|---|---|
| Neutrophil | Phagocytizes bacteria; forms highly toxic compounds that destroy bacteria |
| Eosinophil | Phagocytizes antigen-antibody complex, allergy-causing antigens, and inflammatory chemicals; attacks parasites, such as worms |
| Basophil | Secretes histamine, a vasodilator, thus increasing blood flow to tissues; secretes heparin, which prevents blood clotting |
| Lymphocyte | Natural killer cell that attacks cells infected with viruses or that have turned cancerous; B lymphocytes present antigens and activate other cells of the immune system; can become plasma cells that secrete antibodies; serves as a memory cell in humoral immunity; T lymphocytes destroy foreign cells, regulate the immune response, and serve as memory cells in cellular immunity |
| Monocyte | Differentiates into numerous types of macrophages; macrophages phagocytize pathogens, dead neutrophils, and cellular debris; macrophages present antigens and activate other cells of the immune system |

type of immune cell takes up the red dye eosin and, so, is called an eosinophil. There are 5 general types of leukocytes, which are listed in Table A-1, along with a brief description of their functions.

Macrophages are found in almost all tissues of the body. They are derived from a single kind of leukocyte, the monocyte. When a monocyte leaves the blood and enters a tissue, it is transformed into a macrophage. At birth, a baby is already supplied with macrophages, which continue to develop throughout life. They are strategically located throughout the body to phagocytize foreign material.

**Mast cells** are produced in the bone marrow and exist in almost all tissues and organs. They release histamine and the other chemicals involved in inflammation.

Other participants in the immune system are **cytokines,** a complicated group of protein messengers produced by various cells throughout the body. They regulate the host cells' function and growth.

There are 2 types of immunity: nonspecific, or natural, immunity and specific, or acquired, immunity. Nonspecific immunity protects against foreign invaders without having to recognize the specific appearance of the invaders, whereas specific immunity is acquired.

## Nonspecific Immunity

**Nonspecific immunity** is an array of mechanisms that are present at birth and do not require any activation. They are barriers such as the skin and the **mucous membranes** of the GI tract, reproductive system, urinary tract, and respiratory tract. The **mucus** produced by these tissues traps invaders. Internally, another form of nonspecific immunity is phagocytic cells, which can swallow bacteria and other harmful substances and ultimately destroy them. Acid produced by the stomach (HCl) can destroy ingested pathogens. Inflammation is a local response to infection or injury. The purpose is to destroy or inactivate foreign invaders and begin the process of repair. Fever also is an internal defense mechanism. It seems to aid in the recovery process by reducing the amount of iron in the blood, which reduces bacterial activity. Fever also seems to be associated with an increase in **interferons.** Viral infections are subject to short-term control by this group of proteins released by infected cells. Interferons are receiving a lot of attention as potent weapons against cancers, hepatitis C, and other diseases.

## Specific Immunity

**Antibody-mediated immunity,** or **specific immunity,** is directed at specific molecules. When nonspecific immunological defenses fail to halt an invasion by pathogens or by toxins produced by pathogens, the antibody-mediated immunity mechanism goes into action. This mechanism is based on the action of antibodies, lymphocytes, and other cells of the immune system.

Recall that antigens are molecules that are generally large and foreign to the body. A molecule can have a number of antigenic determinant sites that stimulate the production

**mast cell** Tissue cell that releases histamine and other chemicals involved in inflammation.

**cytokine** Protein, secreted by a cell, that regulates the activity of neighboring cells.

**nonspecific immunity** Defenses that stop the invasion of pathogens; requires no previous encounter with a pathogen.

**mucous membranes** Membranes that line passageways open to the exterior environment; also called mucosae.

**mucus** Thick fluid, secreted by glands throughout the body, that lubricates and protects cells; contains a compound that has both a carbohydrate and a protein nature.

**interferon** Protein released by virus-infected cells that bind to other cells, stimulating the synthesis of antiviral proteins, which in turn inhibit viral multiplication.

**antibody-mediated immunity** Specific immunity provided by B lymphocytes; also known as humoral immunity.

**specific immunity** Function of lymphocytes directed at specific antigens.

of various antibodies. When we successfully fight off an invader, chemicals called **antibodies** are in action. Antibodies are highly specific proteins produced by B lymphocytes in response to antigens. Antigens are detected as dangerous intruders. They are detected because the immune system can identify "self" from "non-self" molecules. (Recall that a role of the carbohydrates on the cell membrane is to identify "self.")

The lymphocytes that produce antibodies, designated B lymphocytes, are produced in the bone marrow. These B lymphocytes wage war against bacterial infections, as well as some viral infections and even a few parasites. B lymphocytes (B cells) and antibodies, also known as **immunoglobulins,** come in 5 major classifications. These bind to the antigen on the invader and begin a process of attack. This antibody-antigen interaction soon produces **plasma cells,** resulting in the production of more antibody proteins to continue the attack. A person can produce as many different antibodies as there are exposures to specific antigens. It is estimated that there are 100 million trillion antibody molecules per person, representing a few million species of antigens.

**Memory cells** are then produced by B cells and provide active immunity. Once you have been exposed to an antigen, you develop active immunity. Obviously, this is the basis of vaccinations; an inactivated pathogen is injected and the body develops immunity to that pathogen.

The blood also contains a group of proteins called **complement** proteins. Complement proteins are released into the area of infection and attach to the target pathogen to be destroyed. The antibody-antigen combination does not cause the destruction of the pathogenic invaders, but it does identify them so that they can be attacked by nonspecific immune processes, such as the complement proteins. Complement proteins attach to the pathogenic invader and drill holes in its membrane, thus leading to its destruction. (The hole in the wall allows water to flow into the cell, causing the cell to burst.)

T lymphocytes (T cells) directly attack and destroy specific cells, which are identified by specific antigens on the cell surface. T lymphocytes produce **cell-mediated immunity** because they are in contact with the enemy cell. T cells must be first activated in the thymus.

The T lymphocytes that are killers are known as **cytotoxic T cells.** They recognize the infected cell and attach themselves through a CD8 receptor. There also are **helper T cells,** which attach to an infected cell through a CD4 receptor. They promote phagocytic activity. Together, the cytotoxic and helper T cells bind to the infected cell and lead to the cell's destruction. You may have heard of CD4 cells because they are markers for AIDS. When the disease progresses, the CD4 count decreases as the virus attacks helper T cells (and macrophages).

Most information concerning the relationship of nutrition to immunity comes from studies in low-income countries, where children die of infectious diseases secondary to malnutrition. Protein-energy malnutrition deficiencies of vitamins and minerals and an inadequate intake of certain fatty acids seriously alter immune function. (There is more information about how individual nutrients make it possible to support an immune response in Chapters 12 through 15.)

Allergies are types of immune responses. One type of allergic response is almost immediate. The symptoms are produced by B lymphocytes exposed to an allergen, as demonstrated by a runny nose, red eyes, and itchy skin (dermatitis). The culprit is **histamine,** an altered form of the common amino acid histidine. This type of immune response can be treated by antihistamine drugs. (Allergies are further discussed in Chapter 7.)

Delayed hypersensitivity, an abnormal T cell response, can occur as late as 72 hours after exposure. The best-known example of this type of immune response is contact dermatitis caused by contact with poison ivy, poison oak, or poison sumac.

A final type of immunity is known as autoimmunity. Here the immune system fails to recognize "self," thinking a normal cell is an antigen. The immune system then goes on the attack by activating T lymphocytes and the production of antibodies by B lymphocytes, which kills the cell. In other words, the defense mechanisms are confused and attack the body rather than invaders. There are at least 40 autoimmune diseases. Some well-known examples are rheumatoid arthritis, type 1 diabetes, and multiple sclerosis.

**antibody** Blood protein that inactivates foreign proteins found in the body; helps prevent and control infections.

**immunoglobulins** Proteins (also called antibodies) in the blood that are responsible for identifying and neutralizing antigens, as well as pathogens that bind specifically to antigens.

**plasma cells** B lymphocytes that produce about 2000 antibodies per second.

**memory cells** B lymphocytes that remain after an infection and provide long-lasting or permanent immunity.

**complement** Series of blood proteins that participate in a complex reaction cascade following stimulation by an antigen-antibody complex on the surface of a bacterial cell. Various activated complement proteins can enhance phagocytosis, contribute to inflammation, and destroy bacteria.

**cell-mediated immunity** Process in which T lymphocytes come in contact with invading cells in order to destroy them.

**cytotoxic T cell** Type of T cell that interacts with an infected host cell through special receptor sites on the T cell surface.

**helper T cell** Type of T cell that interacts with macrophages and secretes substances to signal an invading pathogen; stimulates B lymphocytes to proliferate.

**histamine** Bioactive amine that participates in immune response, stimulates stomach acid secretion, and triggers inflammatory response. It regulates sleep and promotes smooth muscle contraction, increased nasal secretions, blood vessel relaxation, and airway constriction.

# Respiratory System

To produce sufficient energy to meet body needs, there must be oxygen present to help convert food energy into ATP. When oxygen is supplied to the tissues, carbon dioxide is produced and removed from the body by the combined actions of the cardiovascular and respiratory systems.

The organs of the respiratory system are the nose, pharynx, larynx, trachea, bronchi, and lungs. *Respiration* refers to breathing and to the exchange of gases between the blood and other tissues. The respiratory tract features the **alveoli** (plural) in the lungs. These are tiny structures where a form of gas exchange takes place, known as external respiration (Fig. A-10). The **alveolus** (singular), the basic functional unit of respiration, recovers oxygen from inhaled air and loads it onto red blood cells for transport to target tissues throughout the body. Simultaneously, carbon dioxide in the blood is released into the lungs and ultimately is exhaled into the air.

Air reaches the lungs from the nasal cavity and the mouth by initially passing through the **pharynx** to the **larynx.** The larynx is open to the trachea during breathing but closes during swallowing. The **trachea** is the tube that connects the larynx to the **bronchial tree.** The bronchial tree, located in the lungs, looks like a tree with branches. The branches on this tree get smaller and smaller the farther out they go from the tree trunk (the trachea) into lung tissue, until finally they turn into **bronchioles,** the location of the pulmonary alveoli.

The distance across the alveoli is 2 cells thick—1 cell for the alveoli plus 1 cell for the pulmonary capillaries. Gas exchange allows $CO_2$ and $O_2$ to diffuse easily between the blood and lungs. There are approximately 300 million alveoli in the lungs, providing a tremendous surface area for the diffusion of gases.

Another aspect of respiration is the discharge of water through the lungs. This process is obvious on a cold day when the breath we exhale turns to ice crystals and we can see vapor forming around our mouth and nose. Of course, such water loss is much more extensive during hot, humid weather when the body loses heat via the lungs.

**alveoli, alveolus** Basic functional units of the lungs where respiratory gases are exchanged.

**pharynx** Organ of the digestive tract and respiratory tract located at the back of the oral and nasal cavities.

**larynx** Structure located between the pharynx and trachea; contains the vocal cords.

**trachea** Airway leading from the larynx to the bronchi.

**bronchial tree** Bronchi and the branches that stem out to bronchioles.

**bronchioles** Smallest division of the bronchi.

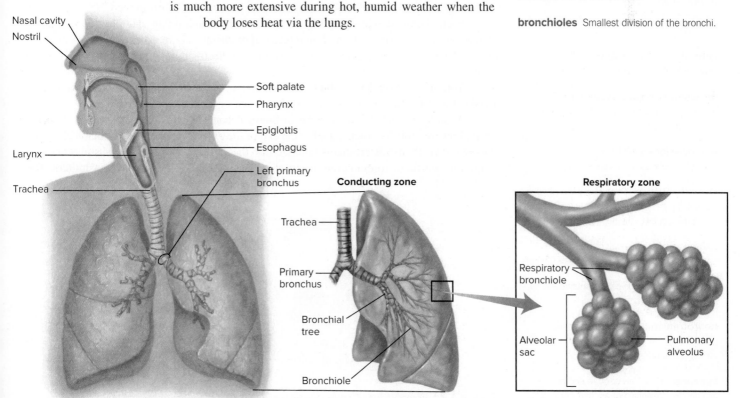

*Figure A-10* Anatomy of the respiratory system. Air enters through the nose and mouth and is conducted into the bronchioles of the lungs. Gas exchange occurs in the alveoli.

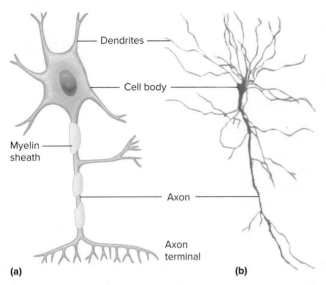

**(a)**                                                                                                    **(b)**

*Figure A-11* (*a*) An illustration of a neuron, or nerve cell, showing the cell body with dendrites and the axon. The axon releases the neurotransmitters. (*b*) How a neuron looks under a light microscope.

**homeostasis** Series of adjustments that prevent change in the internal environment in the body.

**central nervous system (CNS)** Brain and spinal cord portions of the nervous system.

**peripheral nervous system (PNS)** System of nerves that lie outside the brain and spinal cord.

**neuroglia (glial cells)** Specialized support cells of the central nervous system.

**dendrite** Relatively short, highly branched nerve cell process that carries electrical activity to the main body of the nerve cell.

**axon** Part of a nerve cell that conducts impulses away from the main body of the cell.

**nerve** Bundle of nerve cells outside the central nervous system.

**neurotransmitter** Compound made by a nerve cell that allows for communication between it and other cells.

**synapse** Space between the end of 1 nerve cell and the beginning of another nerve cell.

**dopamine** Neurotransmitter that leads to feelings of euphoria, among other functions; also forms norepinephrine.

## Nervous System

The nervous system is a regulatory system controlling a variety of body functions. It can detect changes in various organs and take corrective action when needed to maintain the constancy of the internal environment, **homeostasis.** The nervous system regulates activities that occur almost instantaneously, such as muscle contractions and perception of danger.

The nervous system consists of the **central nervous system (CNS)** and the **peripheral nervous system (PNS).** The central nervous system contains the brain and spinal cord. The peripheral nervous system, with its nerves coming from the central nervous system, branch out to all organs of the body.

The basic structural and functional unit of the nervous system is the neuron—a cell that responds to electrical and chemical signals, conducts electrical impulses, and releases chemical regulators (Fig. A-11).

Neurons allow us to perceive what is occurring in our environment, engage in learning, store vital information in memory, and control the body's voluntary actions. Incoming information depends on sensory receptors, such as visual, auditory, smell, and tactile receptors.

Neurons can't produce new cells, although some can regenerate parts of their structures. Loss of nerve tissue causes loss of important functions. A spinal cord injury is likely to cause permanent paralysis.

**Neuroglia (glial cells)** protect neurons and aid in their function. They are far more abundant than neurons. For example, 1 group of neuroglia wraps nerves in a protective myelin sheath, a job associated with vitamin B-12 (see Fig. A-11). This sheath acts as an insulating material, isolating each nerve conduction pathway from the others. Another group of neuroglia phagocytizes pathogens and disposes of cellular debris in the CNS.

Each neuron contains a cell body with a nucleus and rough endoplasmic reticulum, **dendrites,** and an **axon.** Information (electrical or chemical stimuli) enters the cell through the dendrites and/or the cell body, and the output of electrical impulses leaves by way of the axon.

You may be wondering about the term *nerve.* A **nerve** is a bundle of axons located outside the CNS. Nerves contain axons of both sensory and motor neurons.

Axons end close to, or may be in physical contact with, the next neuron. In most cases, however, the electrical signal at the end of the axon is converted to a chemical signal, called a **neurotransmitter,** that is released into the gap (Fig. A-12). The transmission from neuron to neuron or from neuron to muscle cell is by way of these neurotransmitters. The

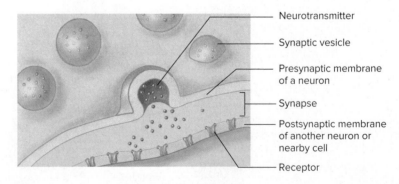

*Figure A-12* Transmission of a message from a neuron to another neuron or another cell relies on neurotransmitters. Vesicles containing neurotransmitters fuse with the membrane of the neuron, and the neurotransmitter is released into the synapse. The neurotransmitter then binds to the receptors on the nearby neuron (or cell). In this way, the message is sent from neuron to neuron, or to the cell that ultimately performs the action directed by the message.

space between 1 neuron and the next is known as a **synapse.** Neurotransmitters that bridge the gap are derived from common nutrients found in foods (see Chapter 7 for more details). There are a variety of neurotransmitters—**dopamine, norepinephrine, acetylcholine,** and **serotonin** are just a few.

The body's fight-or-flight mechanism—the ability to survive a threat—depends on the **adrenergic** effect provided by adrenergic neurons secreting **epinephrine** and norepinephrine. The adrenergic effect stimulates the heart to beat faster, constricts blood vessels to raise blood pressure, increases breathing, and promotes the breakdown of glycogen in the liver. These changes are essential to survival because they make it possible to provide plenty of glucose, our basic muscle fuel, instantly when there is an emergency and muscles need to respond quickly. **Cholinergic** effects usually produce the opposite response of adrenergic effects.

The brain has a tremendous metabolic rate; the blood it requires accounts for 20% of the total cardiac output. This translates into 750 ml of blood per minute being pumped through the brain, yielding a steady supply of oxygen and glucose. Any interruption in the supply of these 2 substances is life threatening. The brain also generates waste materials, which are promptly removed by this high blood flow rate.

All the structures that make up the nervous system are related to nutritional status. For example, most of the axons of the CNS and PNS are covered by myelin. Vitamin B-12 plays a key role in the formation of myelin.

The transmission of information through the nervous system depends on nutrients obtained from the diet: calcium, sodium, and potassium. The sodium ion ($Na^+$) (mostly extracellular) and the potassium ion ($K^+$) (mostly intracellular) located on either side of the axon membrane exchange places as they flow through ion channels in response to electrical stimulation. This is how an electrical signal is transmitted. They are later pumped back to their previous location.

Calcium allows the release of neurotransmitters from the axon of a neuron. As we have seen, the neurotransmitter carries the signal to the next neuron as it jumps the synapse. Fortunately, a calcium-deficient diet will never have a major effect on nerve transmission; the body can always find enough calcium to keep the nervous system functioning. However, in rare instances, a deficiency of calcium causes tetany. (More about tetany appears in Chapter 15.)

Other nutrients required for the nervous system are various amino acids. One amino acid we obtain from dietary protein, tryptophan, is converted to serotonin by neurons. This neurotransmitter has a variety of behavioral effects. Varying the amount of dietary tryptophan controls the amount of serotonin produced by neurons. The amino acid tyrosine can be converted to dopamine and norepinephrine.

Nerves help regulate the digestive system. The sight or smell of food, or one's emotions, can signal muscle cells and glands to prepare the way for food and turn on digestive processes. (Chapter 4 has more details.)

## Endocrine System and Hormones

Endocrine glands secrete regulatory substances, hormones, into the blood for distribution to target tissues or organs. The endocrine gland that secretes a hormone is responding to the need to restore homeostasis. This section is not a complete exploration of all the body's hormones; it concentrates on those that are associated with nutrition (Fig. A-13).

**norepinephrine** Neurotransmitter released from nerve endings; also a hormone produced by the adrenal gland during stress. It causes vasoconstriction and increases blood pressure, heart rate, and blood sugar.

**acetylcholine** Neurotransmitter, formed from choline, that is associated with attention, learning, memory, muscle control, and other functions.

**serotonin** Neurotransmitter that affects mood (sense of calmness), behavior, and appetite and induces sleep.

**adrenergic** Relating to the actions of epinephrine and norepinephrine.

**epinephrine** Hormone produced by the adrenal gland in times of stress. It also may have neurotransmitter functions, such as in the brain.

**cholinergic** Relating to the actions of acetylcholine.

▶ The most important nutrient for continued efficient brain function is carbohydrate in the form of glucose. Should the diet fail to deliver enough carbohydrate that can form glucose, the body will synthesize it in sufficient amounts to provide for the needs of the brain. Alternately, the brain will use an alternative fuel called ketone bodies, but this is not healthy for the body over the long term (see Chapter 9).

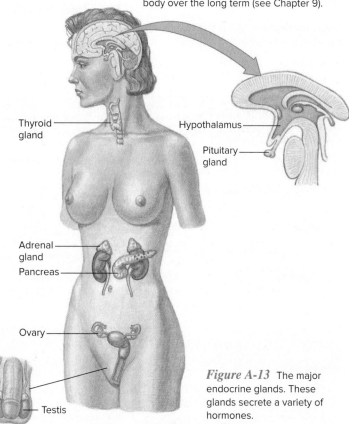

*Figure A-13* The major endocrine glands. These glands secrete a variety of hormones.

Some hormones control metabolic functions, such as appetite, and the transport of substances through cell membranes. Others control growth, and still others are responsible for sex and reproduction. Of all these hormones, some are described as local, in that they function in the immediate vicinity of their production. There also are many interrelationships between hormones and the nervous system. For example, the adrenal gland and pituitary gland respond to neural stimuli.

Some hormones from the pituitary gland control the secretion of other endocrine glands. And, as mentioned in the previous section, a substance such as norepinephrine secreted as a neurotransmitter can act as a hormone.

## Chemical Classification of Hormones

General hormones are classified according to chemical categories: **steroid hormones, glycoproteins, polypeptides,** and amines.

Steroid hormones are lipid substances synthesized from cholesterol (Table A-2). The glycoproteins are long chains of amino acids (100 or more) bound to carbohydrate (Table A-3). Follicle-stimulating hormone (FSH), luteinizing hormone (LH), thyroid-stimulating hormone (TSH), and several other pituitary hormones are such hormones and are referred to as **tropic hormones** because they stimulate the secretion of another hormone and usually stimulate the growth of the associated gland. For example, TSH stimulates the production of the thyroid hormone. Another group of hormones consists of polypeptide chains made of fewer than 100 amino acids per chain (Table A-4). Amines are hormones synthesized from the amino acids tyrosine and tryptophan (Table A-5).

There also are hormones that regulate the digestive tract. (These are discussed in Chapter 4.)

## Interesting Features of Hormones

Steroid and thyroid hormones can be taken in pill form because they are not digested in the GI tract; thus, they can be absorbed into the body in their active state. All the other hormones are deactivated when taken by mouth because their biological activity is destroyed

**steroid hormones** Group of hormones and related compounds that are derivatives of cholesterol.

**glycoprotein** Protein containing a carbohydrate group.

**polypeptide** Ten or more amino acids bonded together.

**tropic hormone** Hormone that stimulates the secretion of another secreting gland.

**Table A-2  Steroid Hormones**

| Hormone | Gland | Target | Effect | Role in Nutrition |
|---------|-------|--------|--------|-------------------|
| Testosterone | Testes, adrenal glands | Reproductive organs | Reproduction, secondary sexual development | Muscle growth |
| Estrogens, progesterone | Ovaries, adrenal glands | Reproductive organs | Reproduction, secondary sexual characteristics | Maintenance of bone |
| Cortisol | Adrenal glands | Liver | Glucocorticoid activity | Metabolism of protein, carbohydrate, fat |
| Aldosterone | Adrenal glands | Kidneys | Mineral-corticoid activity | Electrolyte balance |

**Table A-3  Glycoprotein Hormones**

| Hormone | Gland | Target | Effect | Role in Nutrition |
|---------|-------|--------|--------|-------------------|
| FSH, LH, TSH | Pituitary gland | Variety of organs | Stimulation of target organ to produce its own hormone | None directly |

by digestive enzymes. That is why the hormone insulin must be taken by injection to bypass the digestive tract.

Some hormones must undergo chemical changes before they can function. For example, vitamin D synthesized in the skin and/or obtained from food is converted to an active hormone by the kidneys and liver.

▶ In most cases, a single gland secretes a single hormone but, in a few cases, a gland secretes multiple hormones. In addition, sometimes a hormone is produced by more than 1 gland.

### Table A-4  Polypeptide Hormones

| Hormone | Gland | Target | Effect | Role in Nutrition |
|---|---|---|---|---|
| Antidiuretic hormone | Pituitary gland | Kidneys | Water retention, vasoconstriction | Maintenance of normal blood volume |
| Prolactin | Pituitary gland | Mammary glands | Milk production; in males, indirect enhancement of testosterone secretions | Nourishment of newborn |
| Oxytocin | Pituitary gland | Uterus, mammary glands | Contraction of uterus, mammary secretions | Milk production |
| Insulin | Pancreas | Fat and muscle cells | Decreased blood glucose concentration | Storage of glucose as glycogen, increased fat storage, increased amino acid uptake by cells |
| Glucagon | Pancreas | Liver | Increased blood glucose concentration | Release of glucose from liver stores, increased fat mobilization |
| ACTH (adrenocorticotropic hormone) | Pituitary gland | Adrenal glands | Secretion of glucocorticoids | Secretion of adrenal cortical hormones |
| Growth hormone | Pituitary gland | Most cells | Promotion of amino acid uptake by cells | Promotion of protein synthesis and growth, increased fat utilization for energy |
| Parathyroid hormone | Parathyroid glands | Intestinal tract, kidneys | Increased blood calcium | Release of calcium from bone into blood |
| Calcitonin | Thyroid gland | Bone | Inhibition of breakdown of bone, stimulation of calcium excretion by kidneys | Reduced blood calcium concentration |
| Leptin | No gland, just adipose tissue | Hypothalamus | Targeting of satiety center | Decreased appetite |

### Table A-5  Amine Hormones

| Hormone | Gland | Target | Effect | Role in Nutrition |
|---|---|---|---|---|
| Epinephrine, norepinephrine* | Adrenal glands | Heart, blood vessels, brain, lungs | Increased metabolic rate | Release of glucose into the blood, fat mobilization |
| Thyroid hormone | Thyroid gland | Most organs | Increased oxygen consumption, growth, brain development, development of CNS in fetus | Protein synthesis, increased metabolic rate |
| Melatonin | Pineal gland | Specific neurons | Maintenance of body (circadian) rhythms, sleep | Scavenging of atoms and molecules that are highly reactive and dangerous |

*Norepinephrine also functions as a neurotransmitter, depending on location in the body. Epinephrine is suspected of doing the same, such as in the brain.

## Neural and Endocrine Regulation

Whether a chemical is acting as a hormone or a neurotransmitter, the target cell must have a receptor protein to combine with it. This causes a change in the target cell. (Chapter 4 provides a fuller discussion of this concept.) This also means that there must be a mechanism to turn off the action. Hormones are subject to control by an "off" switch. For example, when the blood glucose concentration has been returned to normal by the action of the hormone insulin, insulin production is turned off. If it were not, the person would experience decreasing glucose concentrations until the concentration dropped so low that the person would go into shock and die.

## Urinary System

**ureter**  Tube that transports urine from the kidney to the urinary bladder.

**urethra**  Tube that transports urine from the urinary bladder to the outside of the body.

The urinary system is composed of 2 kidneys located on the back of the abdominal wall, 1 on each side of the vertebral column (Fig. A-14). Each is connected to the urinary bladder by a **ureter.** The bladder is emptied by way of the **urethra.**

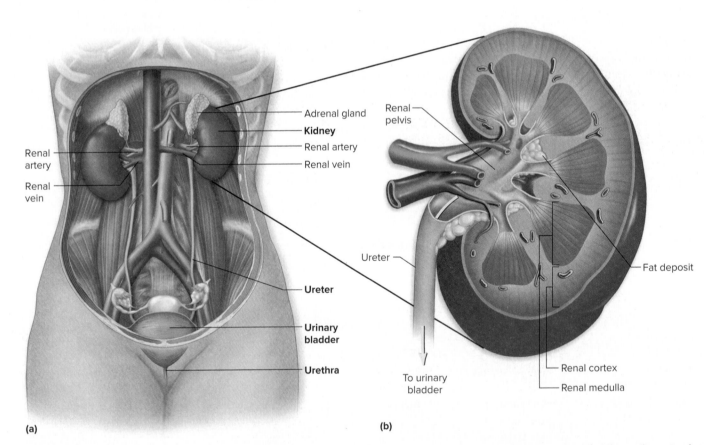

(a)

(b)

*Figure A-14* Organs of the urinary system. (*a*) The urinary system of the female. The male's urinary system is the same, except that the urethra extends through the penis. (*b*) A cross section of the kidney. The kidneys are bean-shaped organs located on each side of the spinal column. They filter waste from the blood, which is then stored in the bladder as urine. The kidneys are connected to the urinary bladder by ureters. The outer section of the kidney is the cortex; the inner section is the medulla. The functional unit of the kidney, the nephron, loops through the cortex and medulla, and the fluid that flows through these tiny structures is separated so that the waste is removed from the blood into collecting ducts and drains into the renal pelvis. Thus, the urine exits by way of the ureter to the bladder. The remaining fluid is returned to the circulatory system to maintain the normal composition of the blood.

Each bean-shaped kidney has an outer section called the cortex and an inner section called the medulla. The medulla is composed of cone-shaped pyramid structures, which empty waste materials into a funnel-shaped tube ending in the ureter. Ureters carry urine from the kidneys to the bladder for temporary storage. Blood flows through the kidneys at a rate of about 120 ml/minute.

▶ Together with the lungs, the kidneys maintain the pH of the blood.

## Kidney Functions

The kidneys regulate the composition of the blood (plasma) and the interstitial fluid, known together as the extracellular fluid. This regulation is accomplished by filtering the blood and forming urine, which is basically the filtrate. As a result of kidney action and the formation of urine, the volume of blood plasma is controlled and blood pressure is maintained. The kidneys remove metabolic waste and foreign chemicals from the blood, and they maintain a certain concentration of electrolytes, such as $Na^+$, $K^+$, and $HCO_3^-$ (bicarbonate) in the plasma. The kidneys constantly monitor the composition of the blood and produce hormones to maintain homeostasis. For example, the kidneys produce the hormone **erythropoietin,** which is responsible for the synthesis of red blood cells. The kidneys convert a form of vitamin D into its active hormone form. During times of fasting, the kidneys can produce glucose from amino acids.

## Kidney Structure

Each kidney is enclosed in a fatty, fibrous sac that protects it from external physical damage. Examined microscopically, the functional unit of the kidney, the **nephron,** is disclosed. Nephrons extend through the renal cortex and the renal medulla. Each kidney has more than 1 million nephrons. A nephron consists of small tubules allied with small blood vessels. The tiny capillary filtration unit, the **glomerulus,** is held in a small capsule (Bowman's capsule). The glomerulus filters large amounts of fluid from the blood, removing the dissolved waste and excess fluid to form urine, which leaves by way of the tubules. The remaining fluid is returned to the blood.

This ingenious mechanism constantly adjusts the composition of the blood. In this process, the essential components are recovered and returned to general circulation, waste products and excess water are removed, and unneeded nutrients (those in which storage compartments are full or there are no storage facilities) are flushed away by the urine.

# Reproductive System

Reproduction is a fundamental property of all living things. We die, but our genes live on in our progeny. In humans, both ova and sperm, called gametes (sex cells), contain 23 chromosomes. The fertilized egg contains 46 chromosomes (23 from each parent) and is programmed to produce a new human. At conception, the instructions for the developing embryo are all present. Through the actions of the female reproductive organs, supported by hormonal secretions, a full-term infant develops in the 37 to 40 weeks after conception, if essential nutrients are present and no genetic defects are encountered. The most precarious time of pregnancy is during the initial 13 weeks, when a woman is least likely to know she is pregnant.

The male reproductive organs consist of the scrotum (containing the testes), penis, urethra, seminal vesicles, and prostate. The female reproductive organs consist of the ovaries, uterus, and vagina.

**erythropoietin** Hormone, secreted mostly by the kidneys, that enhances red blood cell synthesis and stimulates red blood cell release from bone marrow.

**nephron** Functional unit of the kidney.

**glomerulus** Capillaries in the kidney that filter the blood.

**menarche** Onset of menstruation. Menarche usually occurs around age 13, 2 or 3 years after the 1st signs of puberty start to appear.

In addition to reproduction, the sex hormones stimulate bone growth and the closure of the epiphyseal plate, thus ending bone growth. Estrogens protect against bone loss. The sex hormone testosterone stimulates protein synthesis, such as muscle growth and bone growth.

Puberty, or the period of sexual maturation, takes place during early adolescence. **Menarche,** the onset of menstruation, occurs usually between the ages of 11 and 14 in females. In the male, sexual maturation occurs somewhat later and is initiated by hormonal secretions from the brain. (The female reproductive system is discussed in Chapter 16 in more detail.)

# Appendix B

## CHEMISTRY: A TOOL FOR UNDERSTANDING NUTRITION

The study of human nutrition requires a basic awareness of and familiarity with general chemistry, organic chemistry, and biochemistry. This appendix provides only a review of key chemistry principles and fundamental concepts regarding atoms, molecules, chemical bonds, pH, organic compounds, and biochemical structures that may come up as you study nutrition. An understanding of basic chemistry makes the study of nutrition easier and more interesting. It helps connect nutrient characteristics with the structural and chemical attributes of the individual components of food (Table B-1).

▶ The physical and chemical properties of almost anything—whether atoms, molecules, or organisms—are intimately related to its structure. A basic knowledge of chemical structures can help you visualize important fundamental concepts in nutrition.

## Properties of Matter and Mass

All living and nonliving things are composed of matter. Matter exists in 3 states: solid, liquid, or gas. An example of a solid is ice, a liquid is water, and a gas is steam. Two characteristics of matter are that it has mass and it occupies space (volume). Mass is related to the amount of force it takes to move an object—it takes less force to move a paper clip than a pencil; therefore, the clip has less mass. Volume is related to the amount of space an object occupies—a pint of water occupies less space than a gallon; therefore, a pint has a smaller volume. Both these properties depend on how much of the substance there is.

Another property of matter is density. Density is defined as the mass of an object divided by its volume:

$$\text{Density} = \frac{\text{Mass}}{\text{Volume}}$$

Density is independent of how much matter is available. The density of water in a lake is the same as in a cup. Density is commonly expressed in units of grams per cubic centimeter ($g/cm^3$).

You can use density to compare objects. Using the density of pure water as a comparison ($1.0 \ g/cm^3$), lean body tissue has a density of about $1.1 \ g/cm^3$. The density of body fat, in comparison, is about $0.9 \ g/cm^3$. Substances that are less dense than water are buoyant (they tend to float), whereas substances that are more dense than water sink. The next time you are in a swimming pool, note the density of men and women. Women tend to have more body fat, so they float; men are generally more muscular (have more lean tissue), so they tend to sink deeper in the water. This physical property is used to determine the amount of body fat stored in a person (see Chapter 10).

### Physical and Chemical Properties of Substances

Every substance has a characteristic set of physical and chemical properties. Physical properties can be determined without altering the chemical composition of the substance. Ice melts at 1°C. Sugar melts at 186°C. Melting and boiling points are common examples of physical properties.

**Table B-1 Periodic Table of the Elements**

**Legend:**
- Nonmetals
- Alkali metals
- Alkaline Earth metals
- Transition elements
- Other metals
- Metalloids
- Halogenes
- Noble gases
- Lanthanides
- Actinides

| Period | Group 1 | 2 | 3 | 4 | 5 | 6 | 7 | 8 | 9 | 10 | 11 | 12 | 13 | 14 | 15 | 16 | 17 | 18 |
|---|---|---|---|---|---|---|---|---|---|---|---|---|---|---|---|---|---|---|
| 1 | 1 H 1.008 | | | | | | | | | | | | | | | | | 2 He 4.003 |
| 2 | 3 Li 6.941 | 4 Be 9.012 | | | | | | | | | | | 5 B 10.81 | 6 C 12.01 | 7 N 14.01 | 8 O 16 | 9 F 19 | 10 Ne 20.18 |
| 3 | 11 Na 22.99 | 12 Mg 24.31 | | | | | | | | | | | 13 Al 26.98 | 14 Si 28.09 | 15 P 30.97 | 16 S 32.07 | 17 Cl 35.45 | 18 Ar 39.95 |
| 4 | 19 K 39.10 | 20 Ca 40.08 | 21 Sc 44.96 | 22 Ti 47.88 | 23 V 50.94 | 24 Cr 52 | 25 Mn 54.94 | 26 Fe 55.85 | 27 Co 58.47 | 28 Ni 58.69 | 29 Cu 63.55 | 30 Zn 65.39 | 31 Ga 69.72 | 32 Ge 72.59 | 33 As 74.92 | 34 Se 78.96 | 35 Br 79.9 | 36 Kr 83.8 |
| 5 | 37 Rb 85.47 | 38 Sr 87.62 | 39 Y 88.91 | 40 Zr 91.22 | 41 Nb 92.91 | 42 Mo 95.94 | 43 Tc (98) | 44 Ru 101.1 | 45 Rh 102.9 | 46 Pd 106.4 | 47 Ag 107.9 | 48 Cd 112.4 | 49 In 114.8 | 50 Sn 118.7 | 51 Sb 121.8 | 52 Te 127.6 | 53 I 126.9 | 54 Xe 131.3 |
| 6 | 55 Cs 132.9 | 56 Ba 137.3 | 57 La 138.9 | 72 Hf 178.5 | 73 Ta 180.9 | 74 W 183.9 | 75 Re 186.2 | 76 Os 190.2 | 77 Ir 192.2 | 78 Pt 195.1 | 79 Au 197 | 80 Hg 200.5 | 81 Tl 204.4 | 82 Pb 207.2 | 83 Bi 209 | 84 Po (210) | 85 At (210) | 86 Rn (222) |
| 7 | 87 Fr (223) | 88 Ra (226) | 89 Ac (227) | 104 Rf (257) | 105 Db (260) | 106 Sg (263) | 107 Bh (262) | 108 Hs (265) | 109 Mt (266) | 110 Ds (271) | 111 Rq (272) | 112 Uub (285) | 113 Uut (284) | 114 Uuq (289) | 115 Uup (288) | 116 Uuh (292) | 117 Uus 0 | 118 Uuo 0 |

**Lanthanides (6):**

| 58 Ce 140.1 | 59 Pr 140.9 | 60 Nd 144.2 | 61 Pm (147) | 62 Sm 150.4 | 63 Eu 152 | 64 Gd 157.3 | 65 Tb 158.9 | 66 Dy 162.5 | 67 Ho 164.9 | 68 Er 167.3 | 69 Tm 168.9 | 70 Yb 173 | 71 Lu 175 |
|---|---|---|---|---|---|---|---|---|---|---|---|---|---|

**Actinides (7):**

| 90 Th 232 | 91 Pa (231) | 92 U (238) | 93 Np (237) | 94 Pu (242) | 95 Am (243) | 96 Cm (247) | 97 Bk (247) | 98 Cf (249) | 99 Es (254) | 100 Fm (253) | 101 Md (256) | 102 No (254) | 103 Lr (257) |
|---|---|---|---|---|---|---|---|---|---|---|---|---|---|

**Key to Abbreviations**

| Name | Symbol | Name | Symbol | Name | Symbol | Name | Symbol |
|---|---|---|---|---|---|---|---|
| Actinium | Ac | Cerium | Ce | Gold | Au | Meitnerium | Mt |
| Aluminum | Al | Cesium | Cs | Hafnium | Hf | Mendelevium | Md |
| Americium | Am | Chlorine | Cl | Hassium | Hs | Mercury | Hg |
| Antimony | Sb | Chromium | Cr | Helium | He | Molybdenum | Mo |
| Argon | Ar | Cobalt | Co | Holmium | Ho | Neodymium | Nd |
| Arsenic | As | Copper | Cu | Hydrogen | H | Neon | Ne |
| Astatine | At | Curium | Cm | Indium | In | Neptunium | Np |
| Barium | Ba | Dubnium | Db | Iodine | I | Nickel | Ni |
| Berkelium | Bk | Dysprosium | Dy | Iridium | Ir | Niobium | Nb |
| Beryllium | Be | Einsteinium | Es | Iron | Fe | Nitrogen | N |
| Bismuth | Bi | Erbium | Er | Krypton | Kr | Nobelium | No |
| Bohrium | Bh | Europium | Eu | Lanthanum | La | Osmium | Os |
| Boron | B | Fermium | Fm | Lawrencium | Lr | Oxygen | O |
| Bromine | Br | Fluorine | F | Lead | Pb | Palladium | Pd |
| Cadmium | Cd | Francium | Fr | Lithium | Li | Phosphorus | P |
| Calcium | Ca | Gadolinium | Gd | Lutetium | Lu | Platinum | Pt |
| Californium | Cf | Gallium | Ga | Magnesium | Mg | Plutonium | Pu |
| Carbon | C | Germanium | Ge | Manganese | Mn | Polonium | Po |

**Table B-1** Periodic Table of the Elements, Concluded

### Key to Abbreviations

| Name | Symbol | Name | Symbol | Name | Symbol | Name | Symbol |
|---|---|---|---|---|---|---|---|
| Potassium | K | Rutherfordium | Rf | Tantalum | Ta | Uranium | U |
| Praseodymium | Pr | Samarium | Sm | Technetium | Tc | Vanadium | V |
| Promethium | Pm | Scandium | Sc | Tellurium | Te | Xenon | Xe |
| Protactinium | Pa | Seaborgium | Sg | Terbium | Tb | Ytterbium | Yb |
| Radium | Ra | Selenium | Se | Thallium | Tl | Yttrium | Y |
| Radon | Rn | Silicon | Si | Thorium | Th | Zinc | Zn |
| Rhenium | Re | Silver | Ag | Thulium | Tm | Zirconium | Zr |
| Rhodium | Rh | Sodium | Na | Tin | Sn | | |
| Rubidium | Rb | Strontium | Sr | Titanium | Ti | | |
| Ruthenium | Ru | Sulfur | S | Tungsten | W | | |

Chemical properties, such as whether a compound is an acid or a base, determine the changes that a substance undergoes in **chemical reactions.** Other substances affect the chemical properties of a substance. A chemical reaction is a process whereby the composition of 1 or more substances is changed. What actually takes place is affected by the chemical properties of the participants. For example, given the right conditions, exposing glucose to oxygen causes it to break down to carbon dioxide and water:

$$C_6H_{12}O_6 \quad + \quad 6O_2 \quad \rightarrow \quad 6CO_2 \quad + \quad 6H_2O$$

| Glucose | Oxygen | Carbon dioxide | Water |

**chemical reaction** Interaction between 2 or more chemicals that changes both chemicals.

**Celsius** Centigrade measure of temperature; for conversion: (degrees in Fahrenheit − 32) × 5/9 = °C; (degrees in Celsius × 9/5) + 32 = °F.

## Units

The SI (*Système International d'Unités*) units used for scientific measurements designate specific metric units. The units used most frequently in nutrition are mass (kilogram), length (meter), temperature, and amount of substance. Prefixes indicate decimal fractions or multiples of the various units. For example, *kilo* means $1 \times 10^3$, and 1 *milli* is $1 \times 10^{-3}$.

The temperature scale commonly used in scientific studies is the **Celsius** scale. On this scale, water freezes at 0°Celsius (32°Fahrenheit). Water boils at 100°C (212°F). Normal body temperature is 37.0°C (98.6°F). For English-to-metric conversions for length, weight, temperature, and volume (amount), see Appendix H.

## Calories and Joules

Energy is measured in calories or joules. A calorie is the amount of energy required to raise the temperature of 1 gram of water 1 degree C. The SI unit of energy is the joule (J). A mass of 1 gram moving at a velocity of 1 meter per second possesses the energy equivalent of 1 J. A calorie or joule is not a large amount of energy, so kilocalories (kcal) and kilojoules (kJ) are widely used in nutrition chemistry, biology, and biochemistry. In terms of the joule, 1 kcal = 4.184 kJ.

## Scientific Notation

In science, very large and very small numbers frequently must be used, but they are awkward because large numbers have a long string of trailing zeros, and small numbers have a long string of leading zeros. A more convenient way to express these numbers is to use the power of 10, or scientific notation.

▶ To change a number greater than 1 into scientific notation, move the decimal point to the left until the number is greater than 1 but less than 10. This number is the coefficient. The number of places that the decimal is moved becomes the exponent of 10. To change a number less than 1 into scientific notation, move the decimal point to the right until the number is greater than 1 but less than 10. This number is the coefficient. The number of places that the decimal is moved is again the exponent of 10, but this time a negative sign is placed in front of it.

**atom** Smallest unit of an element that still has all the properties of the element. An atom contains protons, neutrons, and electrons.

**proton** Part of an atom that is positively charged.

**neutron** Part of an atom that has no charge.

**electron** Part of an atom that is negatively charged. Electrons orbit the nucleus.

In scientific notation, a number is expressed as a product of a coefficient multiplied by a power of 10. The coefficient is a number equal to or greater than 1 but less than 10. The power of 10 is the exponent. In other words,

$$a \times 10^b$$

where $a$ is the coefficient and $b$ is the exponent.

$$6.02217 \times 10^{23} = 602,217,000,000,000,000,000,000$$

$$2.99161 \times 10^{-23} = 0.0000000000000000000000299161$$

In the previous examples, the positive exponent for the number indicates that the number is very large, whereas the negative exponent indicates a very small number.

## Atoms

The smallest unit of matter that can undergo a chemical change is called an **atom.** An element is composed of atoms of only 1 kind. For example, the element carbon is composed of just carbon atoms. There are more than 100 different elements.

### Atomic Structure

The center of an atom is, for the most part, a nucleus containing 2 (subatomic) particles: **protons,** which carry a positive charge, and **neutrons,** with no charge. Usually, the mass of the proton equals the mass of the neutron. Adding the number of protons and number of neutrons together yields the atomic mass of the atom. An atom of carbon containing 6 protons and 6 neutrons has an atomic mass of 12. The atomic mass of nitrogen is 14, and the atomic mass of oxygen is 16.

The atomic number is equal to the number of protons in the nucleus. What are the atomic numbers of hydrogen, carbon, nitrogen, and oxygen?

Surrounding the nucleus of the atom are negatively charged subatomic particles called **electrons.** The nucleus is actually surrounded by an electron cloud. Electrons have about 2000 times less mass than the mass of protons or neutrons. Thus, all the mass of an atom essentially is in the nucleus. The structure of an atom can therefore be pictured as a very tiny, highly dense nuclear core surrounded by a cloud of electrons. The number of electrons in an atom equals the number of protons, so the net charge is 0.

Electrons surrounding the nucleus have a somewhat peculiar, nonintuitive (contrary to what would be expected) behavior. For instance, it's impossible to know precisely where any given electron is located at any given moment. It is only possible to define a volume of space where the electron is most likely to be found. This volume has a specific distribution of electron density in space and is called an orbital. An orbital is a volume of space. Each orbital has its own characteristic energy and shape.

▶ Hydrogen has an atomic mass of 1 because it has 1 proton and no neutrons.

Orbitals of similar energy are grouped together into energy levels. The energy levels are assigned coordinate numbers—1, 2, 3, etc.—that increase as you move away from the nucleus. Energy level 1 contains only 1 orbital, an "s" orbital. This orbital can hold a maximum of 2 electrons. Energy level 2 contains an "s" orbital and a "p" orbital; the "s" orbital can contain up to 2 electrons, the "p" orbital up to 6. Energy level 3 contains an "s" orbital, a "p" orbital, and a "d" orbital. As before, the "s" orbital and "p" orbital can hold up to 2 and 6 electrons, respectively, whereas the "d" orbital can contain as many as 10. Thus, each energy level can hold a maximum of 2, 8, or 18 electrons, depending on the number of orbitals, and any energy level can hold less than the maximum number of electrons.

Atoms tend to exist in the lowest possible energy state. Thus, electrons tend to occupy orbitals at low energy levels before filling orbitals at higher energy levels. The 1st energy level outside the nucleus has room for just 2 electrons. When that is full, the next energy level away from the nucleus is available for electrons, and there is room for 8 electrons. For

**Table B-2** Atoms Commonly Present in Organic Molecules

| Atom | Symbol | Atomic Number | Atomic Mass | Energy Level 1 | Energy Level 2 | Energy Level 3 | Number of Chemical Bonds to Attain Electron Stability |
|------|--------|---------------|-------------|---------------|---------------|---------------|------------------------------------------------------|
| Hydrogen | H | 1 | 1 | 1 | 0 | 0 | 1 |
| Carbon | C | 6 | 12 | 2 | 4 | 0 | 4 |
| Nitrogen | N | 7 | 14 | 2 | 5 | 0 | 3 |
| Oxygen | O | 8 | 16 | 2 | 6 | 0 | 2 |
| Sulfur | S | 16 | 32 | 2 | 8 | 6 | 2 |

example, hydrogen has 1 electron in energy level 1. Carbon has 2 electrons in energy level 1 and 4 in energy level 2. In energy level 3, there is room for 8 electrons. Sulfur, with an atomic number of 16, has 2 electrons in the 1st energy level, 8 in the 2nd, and 6 in the 3rd (Table B-2).

An atom tends to bond with other atoms that will fill its outermost energy level and produce a number of valence electrons equal to the noble gas that is the farthest to the right in its row in the periodic table (e.g., helium and argon). For instance, a hydrogen atom, with only a single electron, will react with other atoms that provide another electron and fill the energy level with 2 electrons, the same number of electrons as in the noble gas helium.

▶ Only the electrons in the outermost energy level (if it is incomplete) can participate in chemical reactions to form chemical bonds. The outermost electrons of an atom are known as its valence electrons.

## Isotopes and Atomic Weight

All the atoms of an element have the same number of protons in the nucleus, but the number of neutrons in the nuclei of elements such as carbon, nitrogen, and oxygen may vary. All elements have such varieties, called **isotopes,** that differ from each other only in the number of neutrons and, consequently, atomic mass. Most hydrogen atoms have only 1 proton, but isotopic forms can have 1 or 2 neutrons. Some isotopes are radioactive, but most are not. Tritium, a radioactive isotope of hydrogen, has 1 proton and 2 neutrons. Carbon nuclei can contain 5, 6, 7, or 8 neutrons.

Isotopes are distinguished by adding the number of protons and neutrons together and writing the resultant sum as a superscript to the left of the symbol for the element. For example, a carbon nuclei with 6 protons and 6 neutrons is written as $^{12}$C. The isotope containing 7 neutrons is labeled $^{13}$C, and the isotope containing 8 neutrons is labeled $^{14}$C. Note that, because all these atoms have 6 protons, they are all carbon atoms. However, because they possess different numbers of neutrons, they represent isotopes of carbon. All isotopes of an element behave the same way chemically.

Atomic weight takes into account that an element is a mixture of isotopes. If all carbon were $^{12}$C, the atomic weight would be the same as its atomic mass, 12. But because some carbon exists as $^{13}$C and $^{14}$C, the atomic weight is slightly higher, 12.011. The atomic weight is based on the relative abundance of the various isotopes.

Although the ordinary chemical behaviors of different isotopes of the same element are virtually identical, the radiochemical behavior is sometimes different. Isotopes exhibit such differences in physical behavior because they decay (break down) to more stable isotopes by giving off nuclear particles of ionizing radiation. Certain unstable isotopes (radioisotopes) are in an obvious process of decay. Every element has at least 1 such radioisotope. These radioisotopes have a physical half-life, which is the time required for 50% of its atoms to decay to a more stable state. Isotopes such as $^{32}$P (phosphorus) emit radiation that can be measured by instruments, such as Geiger counters and scintillation counters. The isotope $^{14}$C decays more rapidly than other isotopes of carbon.

Other isotopes are not radioactive but still can be traced in body fluids or tissues using other types of instruments. Examples include $^{13}$C and $^{15}$N; these are called stable isotopes because they decay very slowly and do not emit radiation.

$^{12}$Carbon
6 Protons
6 Neutrons
6 Electrons

$^{13}$Carbon
6 Protons
7 Neutrons
6 Electrons

$^{14}$Carbon
6 Protons
8 Neutrons
6 Electrons

**isotope** Alternate form of a chemical element. It differs from other atoms of the same element in the number of neutrons in its nucleus.

Isotope "markers," such as $^{32}P$ and $^{13}C$, have a practical use; they can be used to trace nutrients as they follow various chemical pathways in the body. For example, researchers can "mark" a glucose molecule with a radioactive carbon atom ($^{14}C$). This marking allows the researchers to see where the carbons of glucose are distributed in the body, and it helps indicate what chemical transformations glucose undergoes when metabolized. Such studies have demonstrated that glucose can become part of the lipid stored in adipose cells and can form $CO_2$ (detected as $^{14}CO_2$) that is exhaled. Isotope techniques are widely used in nutrition research.

## Atomic and Molar Mass

▶ *Dalton* is another term used to indicate atomic mass, such as for proteins, DNA, and RNA. One Dalton is equivalent to 1 atomic mass unit (amu).

Atoms are very small. One $^{12}C$ atom has a mass of $1.993 \times 10^{-23}$ g. The units used to quantify atomic mass are called atomic mass units (amu). The carbon amu is calculated by dividing the mass of a carbon atom by $1.6605 \times 10^{-24}$, which is essentially the mass of 1 proton or neutron. By performing this calculation on $^{12}C$, you will find that the mass of a carbon atom is 12 amu. The amu for each element is listed in the bottom portion of each entry in the periodic table. Each amu is based on comparing the element's mass with that of $^{12}C$.

You are familiar with counting units, such as the number of sticks in a package of chewing gum. In chemistry, the unit for counting atoms, ions (electrically charged atoms), and molecules (a combination of atoms) is the mole (mol). A mole is defined as the amount of matter that contains as many objects as the number of atoms in 12 g of $^{12}C$. The number of atoms in 12 g of $^{12}C$ is

$$12 \text{ g } ^{12}C \times \frac{1 \text{ atom}}{1.993 \times 10^{-23} \text{ g } ^{12}C} = 6.023 \times 10^{23} \text{ atoms}$$

It is not the weight but the *number* of molecules that determines the physiological effect of a substance. Therefore, the number of "objects" in a mole of carbon (or any other substance) is $6.02 \times 10^{23}$, which is called Avogadro's number—for example,

$$1 \text{ mol } ^{12}C \text{ atoms} = 6.02 \times 10^{23} \, ^{12}C \text{ atoms}$$

$$1 \text{ mol of water molecules} = 6.02 \times 10^{23} \, H_2O \text{ molecules}$$

$$1 \text{ mol } NO_3^- \text{ ions} = 6.02 \times 10^{23} \, NO_3^- \text{ ions}$$

A single $^{12}C$ atom has a mass of 12 amu, but a single $^{24}Mg$ is twice as massive, 24 amu. Because a mole always has the same number of particles, a mole of Mg is twice as massive as a mole of $^{12}C$ atoms. A mole of carbon weighs 12 g; a mole of Mg weighs 24 g. The same number that refers to the mass of a single atom of an element (in amu) also represents the mass (in g) of 1 mol of atoms of that element. For example, 1 $^{12}C$ atom weighs 12 amu. One mol $^{12}C$ weighs 12 g. One $^{24}Mg$ atom weighs 24 amu, and 1 mol $^{24}Mg$ weighs 24 g.

The mass in g of 1 mole of a substance is called its molar mass. The molar mass (in g) of any substance is always numerically equal to its formula weight (in amu). For example, 1 $H_2O$ molecule weighs 18.0 amu, and 1 mol of $H_2O$ weighs 18.0 g. One NaCl molecule weighs 58.5 amu, and 1 mol of NaCl weighs 58.5 g.

**molecule**  Atoms linked (bonded) together; the smallest part of a compound that still has all the properties of a compound (see also *compound*).

**bond**  Link between 2 atoms by the sharing of electrons, charges, or attractions.

**compound**  Atoms of 2 or more elements bonded together in specific proportions. Not all chemical compounds exist as molecules. Some compounds are made up of ions attracted to each other, such as Na⁺Cl⁻ (table salt).

## Molecules, Covalent Bonds, Hydrogen Bonds, and Ions and Ionic Compounds

### Molecules

**Molecules** are formed through the interaction of the electrons in the outermost orbitals (valence electrons) of 2 or more atoms. When electrons are shared, chemical **bonds** are formed. The term **compound** refers to molecules composed of more than 1 element. Water

**Table B-3  Important Ions in the Human Body**

| Common Ion | Symbol | Some Functions |
|---|---|---|
| Calcium | $Ca^{2+}$ | Component of bones and teeth; necessary for blood clotting, muscle contraction, and nerve transmission |
| Sodium | $Na^{2+}$ | Helps maintain membrane potentials (electrical charge differences across a membrane) and water balance |
| Potassium | $K^+$ | Helps maintain membrane potentials |
| Hydrogen | $H^+$ | Helps maintain acid-base balance |
| Hydroxide | $OH^-$ | Helps maintain acid-base balance |
| Chloride | $Cl^-$ | Helps maintain acid-base balance |
| Bicarbonate | $HCO^{3-}$ | Helps maintain acid-base balance |
| Ammonium | $NH_4^+$ | Helps maintain acid-base balance |
| Phosphate | $PO_4^{3-}$ | Component of bones and teeth; involved in energy exchange and acid-base balance |
| Iron | $Fe^{2+}$ | Necessary for red blood cell formation and function |
| Magnesium | $Mg^{2+}$ | Necessary for enzyme function |
| Iodide | $I^-$ | Part of the thyroid hormones |
| Fluoride | $F^-$ | Strengthens bones and teeth |

is a compound. Each molecule (or compound) possesses its own properties, such as color, taste, and density.

Hydrogen can form just 1 chemical bond because it has room for just 1 electron in its orbital of 2 electrons, in turn yielding a noble gas electron configuration. Carbon can form 4 chemical bonds, nitrogen 3, and oxygen 2 (Table B-3).

A molecular formula gives the elemental composition of a molecule or compound. This formula consists of the symbols of the atoms in the molecule plus a subscript denoting the number of each type of atom.

A structural formula shows how the atoms are arranged with respect to each other. As an extension, molecular space-filling and molecular ball-and-stick models approximate the space and shape of the molecule (Fig. B-1).

When molecules combine with each other, atoms do not increase or decrease in number. Atoms present in starting materials must be present in the products. For example, compare the number of oxygen atoms in glucose and the oxygen itself with the number in the products of the reaction (18 vs. 18). This example also illustrates the process of conservation of mass.

$$C_6H_{12}O_6 + 6\ O_2 \rightarrow 6\ CO_2 + 6\ H_2O$$

## Covalent Bonds

When atoms share their valence electrons, a **covalent bond** is formed (Fig. B-2). The electrons shared between atoms are bonding electrons; these represent the adhesive that holds the atoms together in molecular form.

When 2 identical atoms share electrons, as in the formation of hydrogen gas ($H_2$) or oxygen gas ($O_2$), the covalent bond is very strong because the electrons are shared equally. This equal distribution between the atoms makes the molecule nonpolar. Consider the simple compound methane ($CH_4$). Hydrogen has 1 valence electron, and its outermost (only) orbital can hold a maximum of 2 electrons. Carbon has 4 electrons in its outermost energy level, or valence shell, and that shell can hold a maximum of 8 electrons. Both carbon and

**covalent bond** Union of 2 atoms formed by the sharing of electrons.

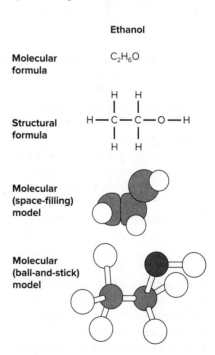

**Ethanol**

Molecular formula          $C_2H_6O$

Structural formula

Molecular (space-filling) model

Molecular (ball-and-stick) model

*Figure B-1* Examples of the molecular and structural formulas and the molecular models of ethanol. The space-filling model gives a more realistic feeling of the space occupied by the atoms. On the other hand, the ball-and-stick type shows the bonds and bond angles more clearly.

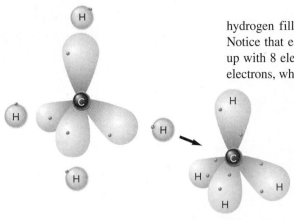

*Figure B-2*  Covalent bonds. In each of the 4 bonds, 1 electron of the carbon is shared with the electron of a hydrogen atom in a single, sausage-shaped molecular orbital encompassing the 2 nuclei. Methane is the simplest organic molecule. Even the largest organic molecules are held together by strong covalent bonds like these.

Methane ($CH_4$)

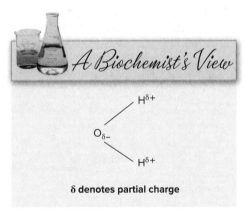

*A Biochemist's View*

$H^{\delta+}$

$O_{\delta-}$

$H^{\delta+}$

**δ denotes partial charge**

▶ Water is a good example of a dipole compound. The oxygen atom pulls electrons from the 2 hydrogen atoms toward its side of the water molecule so that the oxygen side is more negatively charged than the hydrogen side of the molecule. Water, the most abundant molecule in the body, is a good solvent because of the nature of its basic structure.

▶ Polar molecules are weakly attracted both to ions and to other polar molecules. The positive end of the molecule can align itself with an anion or with the negative end of another molecule. These attractive forces—called, respectively, ion-dipole and dipole-dipole forces—are much weaker than covalent bonds individually, but, when there are many of them, they make a significant contribution to the total energy of a collection of molecules. Water, for instance, has a much higher boiling point than expected because the molecules are held together by such forces.

hydrogen fill their valence shells to the maximum by sharing electrons with each other. Notice that each hydrogen in methane contains 2 electrons and that the carbon atom ends up with 8 electrons. A good way to look at this is that the hydrogen atoms share 1 pair of electrons, whereas the carbon atoms share 4 pairs of electrons (see Fig. B-2).

Guidelines that govern the formation of covalent bonds are as follows:

1.  The valence shell of each element must have room to accommodate additional electrons.
2.  Second-row nonmetallic elements of the periodic table (e.g., carbon, nitrogen, and oxygen) and hydrogen typically fill their outermost energy levels by sharing the necessary number of electrons with another element.
3.  Third-row nonmetals and those beyond this point in the periodic table (e.g., phosphorus and sulfur) frequently attain stability by giving up electrons in the outermost energy level rather than adding them. Phosphorus, for example, typically makes 5 bonds to attain stability instead of the 3 that are needed to have the electron configuration of the noble gas argon (18 electrons).

A single covalent bond forms when 2 atoms share 1 electron pair. A double covalent bond forms when 2 atoms share 2 electron pairs.

When electrons spend approximately equal time around each atom nucleus, the bond is called a nonpolar covalent bond. These are the strongest covalent bonds. If the 2 nuclei are not equally attractive to electrons, their atoms can form a polar covalent bond in which the electrons spend more time orbiting the more attractive nucleus. For example, when hydrogen bonds with oxygen, the electrons are more attracted to the oxygen nucleus and orbit that nucleus more than they do the hydrogen nucleus. Electrons carry a negative charge, which makes the oxygen region of the molecule slightly negative and the hydrogen region slightly positive. The Greek letter delta (δ) is used to symbolize a charge less than that of 1 electron or proton. A slightly negative region of a molecule is shown as $\delta^-$ and a slightly positive region is shown as $\delta^+$. A molecule such as this is called a dipole because it has 2 charged ends.

When 2 different atoms form a covalent bond, the bonding electrons are never shared equally. Consider again the H—O bond in water. It is unreasonable to expect that the hydrogen nucleus (containing 1 proton) and the oxygen nucleus (containing 8 protons) have identical forces of attraction for the shared electron pair. In addition, other factors come into play, such as how many energy levels each atom has, how many electrons are in each, and the distance the shared electrons are from each nucleus. All these factors lead to an unequal sharing of electrons in a covalent bond between different atoms.

The ability of an atom in a molecule to attract electrons is called electronegativity. Elements toward the top right corner of the periodic table have the highest electronegativity, and those toward the bottom left have the lowest (electronegativity generally increases from left to right in a row of the periodic table, and it decreases going down a column; the difference in the electronegativities of bonded atoms can be used to determine the polarity of a bond). Metals have low electronegativity, whereas nonmetals have relatively high electronegativity. Oxygen and nitrogen have the highest electronegativities of the elements typically found in compounds important to nutrition. The electronegativity values of atoms determine the type of chemical bond formed. If the electronegativity values are not very different, a covalent bond is formed. If the electronegativities of 2 bonding atoms differ greatly, electron transfer occurs to yield an ionic bond, as in $Na^+Cl^-$ (see the section Ions and Ionic Compounds).

## Hydrogen Bonds

Water, and most other molecules containing an O—H or N—H bond, exhibits a particularly strong interaction called hydrogen bonding (Fig. B-3). In this case, the hydrogen atom of 1 molecule is attracted to a nonbonded electron pair (called a lone pair) of a highly electronegative atom on a neighboring molecule, such as oxygen. Water molecules are attracted to

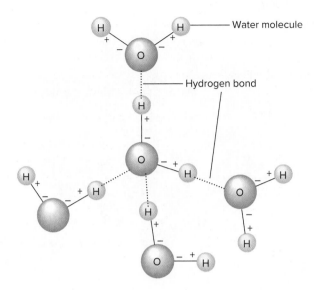

*Figure B-3*  Hydrogen bonds between water molecules. The oxygen atoms of water molecules are weakly joined together by the attraction of the electronegative oxygen for the positively charged hydrogen. These weak bonds are called hydrogen bonds.

each other by hydrogen bonds. This attraction is responsible for many of the biologically important properties of water. Hydrogen bonds, such as those found in large proteins and DNA, help hold the molecule together. These molecules fold or twist into 3-dimensional shapes due in part to the action of hydrogen bonds. Hydrogen bonds are usually symbolized by a dotted line between the atoms: —C—O · · · H—N—. Hydrogen bonds are the weakest of all chemical bonds.

**ionic bond**  Union between 2 atoms formed by an attraction of a positive ion to a negative ion, as in table salt ($Na^+Cl^-$).

## Ions and Ionic Compounds

Atoms that have an equal number of positively charged protons and negatively charged electrons are electrically neutral. Atoms or molecules that have positive or negative charges are called ions. **Ionic bonds** result when 1 or more valence electrons from 1 atom are completely transferred to another atom or molecule. Elements that have 1 to 3 valence electrons have a tendency to give up electrons, and those with 4 to 7 valence electrons have a tendency to accept electrons. The electrons are not shared. In both cases, the elements are giving up or taking on electrons to achieve the electron configuration of the closest noble gas. Consider sodium chloride. One atom loses electrons, so its number of electrons becomes smaller than its number of protons; thus, it becomes positively charged as $Na^+$ in sodium chloride. Now, sodium has the same number of electrons as neon. The other atom gains electrons, so its number of electrons is greater than its number of protons; it becomes negatively charged as $Cl^-$ in sodium chloride. Now, chloride has the same number of electrons as argon.

Positively charged ions are called cations; they move toward the negative pole in an electric field. An atom with more electrons than protons is negatively charged and is known as an anion; it moves to the positive pole. NaCl is an example of an ionic compound. Note the name change that occurs when an element gains an electron to become a negative ion; the suffix becomes -*ide*.

These charged atoms, where electrons have been added or removed, are collectively known as ions. Sodium ($Na^+$), potassium ($K^+$), and calcium ($Ca^{2+}$) are found in the body as cations. Chloride ($Cl^-$) is a common anion in the body. See Table B-3 for a more complete list of common ions found in the body.

Ionic bonds are weaker than polar covalent bonds. Ionic compounds easily separate when dissolved in water. Table salt (NaCl) is obvious when poured out of the salt shaker, but when the salt is stirred into a cup of water it disappears. It *dissociates*. The polar water's negative side (oxygen) is attracted to the $Na^+$, and the positive side (hydrogen) is attracted to $Cl^-$.

▶ Water molecules that surround ions attract other water molecules to form hydration spheres around each ion. This mechanism makes ions and numerous molecules soluble in water.

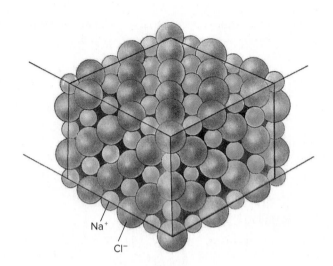

*Figure B-4*  Molecules of sodium chloride (table salt) in typical cube-shape formation.

▶ Salts separate to form positively and negatively charged ions when dissolved in water. Substances that dissolve in water and conduct electricity are called electrolytes. (A solute that produces ions in solution forms an electrolytic solution that conducts an electrical current.) Sodium ($Na^+$), potassium ($K^+$), calcium ($Ca^{2+}$), chloride ($Cl^-$), magnesium ($Mg^{2+}$), phosphate ($PO_4^{3-}$), and bicarbonate ($HCO_3^-$) are electrolytes commonly found in the body.

## Salts

Salts are substances composed of cations and anions. Table salt is NaCl. The $Na^+$ and $Cl^-$ are attracted to each other by electrostatic force, and the resulting ionic compound is known chemically as sodium chloride. Salts are formed by the interaction of acids and bases in a neutralization reaction. Water also is formed in such a reaction. In this type of reaction, hydrogen ions of an acid are replaced by the positive ions of a base, and a salt forms. For example, when hydrochloric acid reacts with sodium hydroxide, table salt is produced:

$$\text{HCl} \quad + \quad \text{NaOH} \quad \rightarrow \quad \text{NaCl} \quad + \quad \text{H}_2\text{O}$$

**Hydrochloric acid**  **Sodium hydroxide**  **Salt**  **Water**
**(neutralization reaction)**

The formula for salts can be misleading. For example, NaCl suggests that table salt exists as a discrete entity containing 1 sodium ion and 1 chloride ion. An inspection of the chemical structure of table salt shows that it is actually a 3-dimensional stack of layers—much like a ream of paper with all the pages glued together (Fig. B-4).

## Acids, Bases, and the pH Scales

You have a pretty good idea of what acids and bases are. You know that lemon juice is an acid and drain cleaners are strong bases.

A solution that has a higher concentration of protons ($H^+$) is said to be acidic. One that is lower is basic, or alkaline. An acid is a substance that can ionize and release protons ($H^+$) into solution. It is a proton donor.

Any substance that releases protons (hydrogen ions) when in water is an acid. For example, hydrogen chloride (HCl) forms hydrogen and chloride ions ($H^+$ and $Cl^-$) in solution and therefore is an acid:

$$\text{HCl} \rightarrow \text{H}^+ + \text{Cl}^-$$

A base is a negatively charged ion or a molecule that ionizes to produce an anion. This then can combine with a proton ($H^+$), removing it from solution. This base is a proton acceptor. Any substance that can accept hydrogen ions while in water is a base. Figure B-5 lists several common acids and bases.

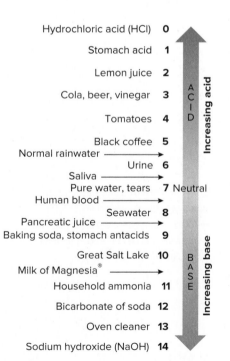

| | |
|---|---|
| Hydrochloric acid (HCl) | 0 |
| Stomach acid | 1 |
| Lemon juice | 2 |
| Cola, beer, vinegar | 3 |
| Tomatoes | 4 |
| Black coffee | 5 |
| Normal rainwater | |
| Urine | 6 |
| Saliva | |
| Pure water, tears | 7  Neutral |
| Human blood | |
| Seawater | 8 |
| Pancreatic juice | |
| Baking soda, stomach antacids | 9 |
| Great Salt Lake | 10 |
| Milk of Magnesia® | |
| Household ammonia | 11 |
| Bicarbonate of soda | 12 |
| Oven cleaner | 13 |
| Sodium hydroxide (NaOH) | 14 |

ACID — Increasing acid

BASE — Increasing base

*Figure B-5*  pH of various substances. Any pH value above 7 is basic, and any pH value below 7 is acidic.

Many bases can function as proton acceptors by releasing hydroxide ions ($OH^-$) when dissolved in water. Most strong bases release $OH^-$ into solution. The $OH^-$ combines with $H^+$ to form water:

$$\text{NaOH} \rightarrow \text{Na}^+ + \text{OH}^-$$

**Sodium hydroxide**       **Sodium ion**       **Hydroxide ion**

$$\text{OH}^- + \text{H}^+ \rightarrow \text{H}_2\text{O}$$

**Hydroxide ion**       **Hydrogen ion**       **Water**

## pH

Acidity is expressed in terms of pH, a measure of the molarity (the ratio of solute per liter of solution) of $H^+$. Molarity is expressed by square brackets, so the molarity of $H^+$ is symbolized as $[H^+]$. pH is defined as the negative logarithm of the hydrogen ion molarity (concentration), or $pH = -\log [H^+]$. The pH unit is the $H^+$ concentration of a solution. Pure water has a neutral pH because it contains equal amounts of hydrogen (hydronium) and hydroxyl ions. The pH scale runs from 0 to 14 (see Fig. B-5).

Because pH is a negative logarithmic scale, a solution with a pH of 4 has an **acidic pH** that is 10 times greater than that of a solution with a pH of 5, and it is 100 times more acidic than a solution with a pH of 6. These numbers may be confusing because they are inversely related to the hydrogen ion concentration: a solution with a high hydrogen ion concentration has a low pH number. A solution with a low hydrogen concentration has a high pH number. Acidic solutions have a pH of less than 7. Basic, or **alkaline pH,** solutions have a pH greater than 7.

A slight disruption of pH can seriously disturb normal physiological functions, so it is important for the body to be able to control pH. Blood normally has a pH range from 7.35 to 7.45. Any deviations from this range can cause dizziness, fainting, coma, paralysis, or death.

Acids and bases are classified as strong or weak. Strong acids and strong bases dissociate completely when dissolved in water. Consequently, they release all their hydrogen ions or hydroxide ions when dissolved. In general, the more completely an acid or a base dissociates, the stronger it is. Hydrochloric acid, for example, is a strong acid because it completely dissociates in water.

Weak acids only partially dissociate in water. Consequently, they release only some of their acidic hydrogens. For example, when acetic acid ($CH_3C{-}OH$, with a double-bonded O, the principal component of vinegar) dissolves in water, it dissociates only partially:

$$\underset{\text{Acetic acid}}{CH_3\overset{\overset{\text{O}}{\|}}{C}{-}OH} \longleftrightarrow \underset{\text{Acetate ion}}{CH_3\overset{\overset{\text{O}}{\|}}{C}{-}O^-} + \underset{\text{Proton}}{H^+}$$

The equilibrium lies far to the left so that only a small fraction of the acetic acid in the vinegar is dissociated into acetate ions and protons.

Most weak bases release hydroxide into solution by reacting with the water itself. For example, ammonia ($NH_3$) reacts with water to form $NH_4^+$ and $OH^-$:

$$\underset{\text{Ammonia}}{NH_3} + \underset{\text{Water}}{H_2O} \longleftrightarrow \underset{\text{Ammonium ion}}{NH_4^+} + \underset{\text{Hydroxide ion}}{OH^-}$$

## Buffers

Many of the biochemical reactions that occur in living tissues require tight control of pH. To prevent changes in the $H^+$ concentration in the body and to control the pH, a system of

▶ Water molecules of 2 hydrogens and 1 oxygen are held together by polar covalent bonds. Although these are strong bonds, a *small* proportion of them break, releasing a hydrogen ion and a hydroxide ion. The hydrogen ion (a proton) is transferred to another oxygen in a water molecule, forming a *hydronium ion.* This means that a pair of water molecules can act as an acid and a base because water self-ionizes, forming hydronium ions and hydroxide ions:

$$2\,H_2O \longleftrightarrow H_3O^+ + OH^-$$
**Water       Hydronium ion   Hydroxide ion**

For simplicity, ionized water will be represented by $H^+$ and $OH^-$.

**acidic pH**  pH less than 7; for example, lemon juice has an acidic pH.

**alkaline pH**  pH greater than 7. Baking soda in water yields an alkaline pH.

▶ The kidneys play a buffering role in the body by absorbing or releasing $H^+$ or $HCO_3^-$, depending on the acid-base balance in the person. In fact, much of the excess acid leaves the body via the urine (urine has an acidic pH). Thus, the kidneys and lungs keep this buffering system functioning and, in turn, are key to acid-base balance in the body.

buffers is maintained. These buffers are ions and molecules that stabilize the pH of a solution. In the blood (plasma), the pH is maintained by the carbonic acid–bicarbonate buffer system. The acid is formed by the combination of water and carbon dioxide. Carbonic acid separates into bicarbonate ion ($HCO_3^-$) and the hydrogen ion ($H^+$).

$$HCO_3^- \;+\; H^+ \;\longleftrightarrow\; H_2CO_3 \;\longleftrightarrow\; H_2O \;+\; CO_2$$

**Bicarbonate**  **Hydrogen ion**  **Carbonic acid**  **Water**  **Carbon dioxide**

The reaction can go either way. The direction depends on the concentration of ions on either side of the arrows. For example, if an acid were released into the blood plasma (more $H^+$ in solution), the reaction would be driven to the right. The carbon dioxide produced could then be exhaled via the lungs. Acids that are present in the plasma come from cellular activities but, despite the increase in $H^+$ by these activities, the blood plasma pH hardly changes; it is essentially constant. The buffer, bicarbonate, accomplishes this. It is constantly formed to maintain normal pH.

## Free Radicals

You are aware that atoms tend to share electron pairs when forming chemical bonds, and they tend to share enough electrons to completely fill the valence shell to form a noble gas configuration. A consequence is that atoms or elements are rarely found with unpaired electrons. When a molecule with an extra electron does arise, however, it is called a **free radical.** An example is the superoxide anion. Oxygen is composed of 2 oxygen atoms ($O_2$); if an electron is added, it becomes superoxide, or $O_2^{\cdot\,-}$. The dot signifies an unpaired electron.

Superoxide and other free radicals are reactive, primarily because they contain an unpaired electron. Free radicals seek an electron by attacking and removing electrons from other compounds, such as at the location where hydrogens are attached to carbon. This not only damages the other molecule but also transforms it into a free radical:

$$R^{\cdot} + {-}CH_2 \rightarrow RH^+{-}CH^{-\bullet}$$

Free radicals also are formed when a covalent bond breaks and each atom or molecule fragment recovers the electron originally used to make the bond. In this case, energy—usually in the form of sunlight, ultraviolet radiation, or heat—is used to break the bond:

$$A{-}B + energy \rightarrow A^{\bullet} + B^{\bullet}$$

Because free radicals are reactive, they can generate thousands of other free radicals within minutes in a chain-reaction process. The reactivity of free radicals sometimes produces detrimental effects in living systems. For instance, the development of cardiovascular disease and some types of cancer, such as skin and lung cancer, is probably promoted by free radicals. However, some normal physiological functions in the body involve free radical formation; for example, various white blood cells use free radicals to kill invading bacteria.

The body has a number of mechanisms, such as antioxidants, for neutralizing free radicals. Antioxidants are substances that react with and neutralize free radical forms of oxygen and nitrogen. The enzyme superoxide dismutase (SOD) converts superoxide into oxygen and hydrogen peroxide. One form of SOD contains the minerals copper and zinc, whereas another form contains manganese. Other antioxidants obtained from the diet are vitamin E and various phytochemicals.

Some substances are used extensively in the food industry to trap free radicals or prevent their formation. This use allows for increased food storage time by decreasing chemical breakdown. These substances are part of a class of food additives called preservatives (see Chapter 3). Vitamin E added to cooking oils protects C—C bonds by trapping free radicals.

**free radical** Compound with an unpaired electron, which causes it to seek an electron from another compound. Free radicals are strong oxidizing agents and can be very destructive to electron-dense cell components, such as DNA and cell membranes.

# Organic Chemistry

**Organic compounds** contain carbon in combination with other elements, such as hydrogen, oxygen, and nitrogen. Carbon compounds are associated with living things, but why carbon? It is because carbon forms very stable covalent bonds, such as single, double, and even triple bonds. Carbon also forms these bonds with many other atoms. Carbon atoms can even form rings and chains by bonding to other carbons. Variation in the length of the chains, and their atomic configurations, allows the formation of a wide variety of molecules. Organic molecules generally also contain hydrogen.

**organic compound** Substance that contains carbon atoms bonded to hydrogen atoms in the chemical structure.

## Cyclic and Chain Compounds

Cyclic organic compounds are common forms of hydrocarbons. Note the diagram of butyric acid (a chain) and compare that with the structure of glucose, which is a ring. Even though the 2 compounds are only carbon, oxygen, and hydrogen, their structures each confer a very different property. Some ring structures are referred to as **aromatic compounds**.

Hydrocarbons as chains or rings provide the backbone of many groups of compounds that make up important organic nutrients. Other groups are attached to these backbones. They usually contain atoms of oxygen, nitrogen, phosphorus, and sulfur. The functional or reactive groups provide the unique chemical properties of organic molecules. Classes of organic molecules are known by their functional groups.

**Glucose**

**Butyric acid**

## A Closer Look at Functional Groups

Several important organic compounds contain a functional group called a carbonyl group ($C = O$). The **carbonyl group** is the parent compound for ketones, aldehydes, and many related groups. Table B-4 lists all these compounds that are important to nutrition.

**Ketones** are organic compounds in which the carbonyl group occurs in the interior of a carbon chain and is therefore flanked by carbon atoms. Body fat that is breaking

$$\overset{\displaystyle O}{\overset{\displaystyle \|}{\phantom{.}}}$$

down at a rapid rate produces ketones (C—C—C), some of which are removed from the body by way of the urine (see Chapter 9).

$$\overset{\displaystyle O}{\overset{\displaystyle \|}{\phantom{.}}}$$

**Aldehydes** (—C—H) are organic compounds that contain a carbonyl group to which at least 1 hydrogen atom is attached. This active group is found in an important form of vitamin A. As an aldehyde, it plays a central role in vision.

Many of the most common substances in both foods and the body contain

$$\overset{\displaystyle O}{\overset{\displaystyle \|}{\phantom{.}}}$$

carboxylic acids. A **carboxylic acid** (—C—OH) contains the carbonyl group with an OH group attached. These acids are widely distributed in tissues and natural products. Vinegar contains acetic acid. Citrus fruits contain citric acid, and vitamin C is ascorbic acid.

The **carboxyl group** is an acid because it can donate a $H^+$ (proton) to a solution. A very common acid formed in muscle cells is lactic acid. When lactic acid ionizes, it releases the $H^+$ and becomes lactate. Because both forms of the acid (ionized and non-ionized) are in solution, the proportion depends on the pH of the solution.

An **alcohol** has the carbon-oxygen bond, but the O also is bonded to a single hydrogen. This leaves only a single bond between the carbon and oxygen, forming an —OH or hydroxide group (ROH).

**Table B-4  Typical Chemical Groups Found in Nutrients**

| Functional Group | Name | Typically Found In | Example |
|---|---|---|---|
| $-OH$ | Hydroxide | Alcohols | $CH_3-OH$ |
| $-C=O$ with $H$ below | Aldehyde | Sugars | $CH_3C=O$ with $H$ below |
| $C-C=O$ with $C$ below | Ketone | Ketones | $CH_3C=O$ with $CH_3$ below |
| $-C=O$ with $OH$ below | Carboxyl | Acids | $CH_3C=O$ with $OH$ below |
| $-S-S-$ | Disulfide | Proteins | $-CH_2-S-S-CH_2-$ |
| $-C=O$ | Carbonyl | Aldehydes, ketones, carboxylic acids, amides | $(CH_3)_2C=O$ |
| $-C-NH_2$ | Amine | Proteins | $CH_3-NH_2$ |
| $-C=O$ with $NH_2$ below | Amide | Vitamins | $-CH_2C=O$ with $NH_2$ below |
| $2O-P=O$ with $O-$ above and $O-$ below | Phosphate | High-energy compounds | $-CH_2-O-P=O$ with $O-$ above and $O-CH_2-$ below |
| $-C=O$ with $O-C$ below | Ester | | $-CH_2-C(=O)-O-C$ linked to $C-O-C(=O)-CH_2-$ and $C-O-C(=O)-CH_2-$ |
| $-O-C(=O)-CH_2-$ | Acyl | Triglycerides | $-CH_2-C(=O)-O-C$ linked to $C-O-C(=O)-CH_2-$ and $C-O-C(=O)-CH_2-$ |

An **ester** $(R-\overset{O}{\overset{\|}{C}}-O-C)$ is an organic compound that has an O—C group attached to a carbonyl group. An ester is the product of a reaction between a carboxylic acid and an alcohol. The formation of lipids called triglycerides involves the formation of ester bonds.

The carbonyl portion of a compound such as an ester is called an **acyl group.** Thus, removal of the hydroxyl group (OH) from an organic acid forms an acyl group.

Two sulfur atoms (S—S), each attached to a carbon, produce a **disulfide group.** This group is important to the structural characteristics of certain proteins.

A single carbon with an **amine** (also called **amino**) **group** attached (—NH₂) is a component of all amino acids.

# Isomerism

Molecules that have identical chemical formulas but different structures are called **isomers.** A simple example is 2 compounds with the formula $C_2H_6O$:

**CH₃CH₂OH**       **CH₃OCH₃**

**Ethanol**       **Methyl ether**

Both of these compounds can be harmful. However, there are intake levels at which ethanol produces no toxic symptoms (e.g., the amount in a small glass of wine) but at which methyl ether would cause very toxic effects. This fact illustrates an important point about isomers: because they have different structures, they can have different *chemical* properties.

The difference in properties between 2 isomers can be great (as in the preceding example) or very subtle, but the differences are detectable. There are different types of isomerism, but only 2 of the common types will be briefly reviewed in this section: structural isomers and stereoisomers.

## Structural Isomers

Isomers in which the number and kinds of bonds differ are called **structural isomers.** Molecules containing chains of carbon atoms typically have many structural isomers. Any variation in the way the chain is branched gives rise to a new isomer. For example, pentane ($C_5H_{12}$) has 3 isomers.

## Stereoisomers

**Stereoisomers** have the same number and types of chemical bonds but with different spatial arrangements (different configurations in space). Molecules containing double bonds illustrate stereoisomers. Because there is no freedom to rotate around a C—C bond, molecules containing such bonds frequently exhibit stereoisomerism. For example, hydrogens or various chemical compounds can be located on the same side of the bond (*cis* **configuration**) or on opposite sides of the double bond (*trans* **configuration**).

Consider oleic acid and its isomer elaidic acid (Fig. B-6a). Oleic acid is a *cis* isomer, or the form found naturally in food. With food processing technology, such as hydrogenation, some *cis* bonds of fatty acids are converted to *trans* bonds. When vegetable oils are converted to vegetable fats, as in margarine or shortening, some of the *trans* isomers are formed. The *trans* isomer elaidic acid is not the natural form. Isomers of these types (*cis* and *trans*) are called geometric isomers.

Describing each stereoisomerism depends on which way the functional groups are arranged with respect to each other. If there are 2 isomers, *D* stands for dextro or right-handed, and *L* stands for levo or left-handed, such as alanine in D-alanine and L-alanine (see Fig. B-6b). Stereoisomers that can't be superimposed on their mirror images are called optical isomers. Optical isomers can be identified from each other by their reaction to polarized light. One solution of an isomer that rotates the plane of polarized light to the right is dextrorotary. And the solution of its optical isomer rotates the plane of light to the left and, so, is levorotary.

The difference between 2 stereoisomers is "fit." This difference is important because the molecule has to fit an enzyme to make the chemical reaction proceed. For example, human enzymes use only L-amino acids (building blocks of protein) and D-sugars to build compounds. D-amino acids and L-sugars just won't function as such in the body. It is rather like trying to wear a left-hand glove on the right hand and do anything that requires manual skill.

**isomers** Different chemical structures for compounds that share the same chemical formula.

$\delta$ denotes partial change

Isomers of pentane

*cis* **configuration** Form seen in compounds with double bonds, such as fatty acids, in which the hydrogens on both ends of the double bond lie on the same side of the plane of that bond in the cell mitochondria.

*trans* **configuration** Compound that has hydrogens located opposite each other across a carbon-carbon double bond.

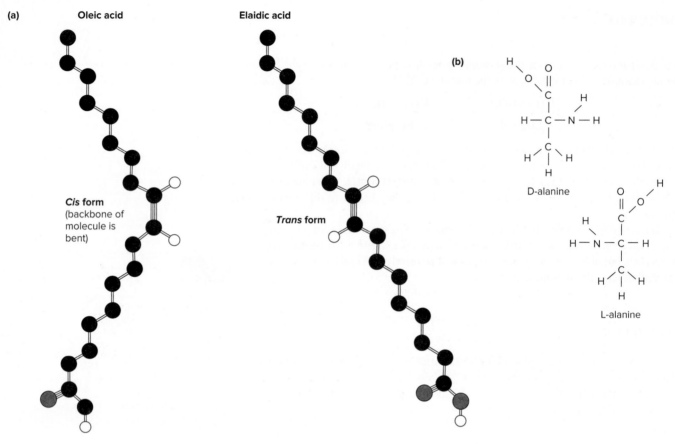

*Figure B-6*   (*a*) *Cis* and *trans* isomers of fatty acids. *Cis* forms are the most common forms in unprocessed foods (see Chapter 6). (*b*) Optical isomers of alanine—an amino acid. The L isomer is the most common amino acid in nature.

▶ *Trans* isomers of fatty acids are associated with an increased risk of cardiovascular disease (see Chapter 6).

A carbon atom with 4 different atoms or groups of atoms attached is described as being **chiral** (also called asymmetric). In living organisms, many molecules are chiral. A molecule with 1 chiral carbon can have 2 stereoisomers, such as alanine (see Fig. B-6*b*). When 2 or more (n) chiral carbons are present, there can be $2^n$ stereoisomers. Some stereoisomers are mirror images of each other; others are not:

When compounds have more than 1 chiral center, the "RS" system of naming is used, rather than the D and L system. Every chiral carbon is designated either *R* or *S,* based on specific rules.

This RS terminology is important to understanding vitamin E chemistry. It is now known that vitamin E as alpha-tocopherol has 3 chiral centers and, so, has 8 different stereoisomers ($2^3 = 8$). All 3 are found in synthetic preparations. The 3 chiral centers are identified as 2, 4, and 8 as related to the position on the tail of the molecule (see Chapter 15). The RRR isomer (R form at all of the 3 chiral centers on the tail) is the natural form. A transfer protein in the liver only recognizes the R form of the chiral center at the 2 position. Of all the 8 combinations of R and S in the tail of synthetic vitamin E, the only biologically active ones are RRR, RSR, RSS, RRS because they all have the R form in the 2 position.

# Biochemistry

The study of the chemistry or molecular basis of life and the reactions, structures, and composition of living materials is known as biochemistry. Biochemical reactions are possible because of enzymes. Living organisms convert the energy they extract from food into energy for growth, maintenance, and reproduction. Energy can be stored for future use. The energy in food is converted and used in the form of chemical energy contained in adenosine triphosphate (ATP). The fact that living organisms can self-replicate depends on deoxyribonucleic acid (DNA) and the genetic code. All forms of life store and transmit genetic information in the form of DNA.

Approximately 98.5% of the body's weight is composed of the elements oxygen, carbon, hydrogen, nitrogen, calcium, and phosphorus. Elements such as iron, zinc, and copper are present in trace amounts in the body, but that doesn't mean they are unimportant. For instance, iron combines with a blood protein to form hemoglobin, an oxygen carrier. Hemoglobin transports oxygen from the lungs to the tissues and assists in returning carbon dioxide from the tissues to the lungs for removal.

Water is the most abundant chemical in the body, making up to about 70% of human tissue. Other important classes of compounds in the body are the proteins, carbohydrates, lipids, and nucleic acids.

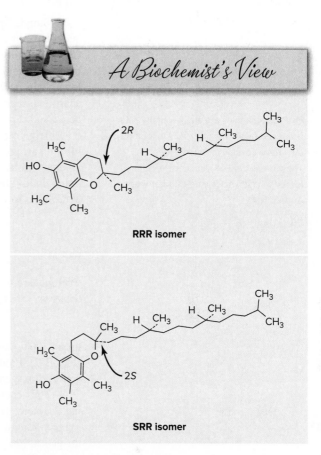

*A Biochemist's View*

RRR isomer

SRR isomer

RRR and SRR isomers of vitamin E. Of the 2, only the RRR isomer contributes to vitamin E needs.

## Biochemical Reactions

All the biochemical reactions that occur in the body are described as **metabolism.** The intermediate compounds in metabolism are termed metabolites. Metabolic reactions that build (synthesize) complex molecules are described as **anabolic.** An example is the synthesis of protein from amino acids. The reactions that break down (degrade) larger molecules into smaller ones are described as **catabolic.** An example is starch breaking down to glucose molecules.

## Carbohydrates

**Carbohydrates** are aldehydes with hydroxyl groups and ketones, containing carbon, hydrogen, and oxygen with the general formula $CH_2O$. (There are twice as many hydrogen atoms as carbon and oxygen atoms.) The suffix *-ose* indicates a sugar. *Hexose* refers to a 6-carbon monosaccharide. There are 3 structural isomers of hexose: galactose, glucose, and fructose. All have the same formula, $C_6H_{12}O_6$, but the arrangements of their individual atoms differ slightly.

The simplest carbohydrates are **monosaccharides.** When 2 monosaccharides are chemically bonded, they form a **disaccharide,** or double sugar. The table sugar sucrose is an example of a disaccharide, formed from glucose and fructose.

Polysaccharides are many monosaccharides joined by covalent bonds. Plant starch and cellulose are polysaccharides. Some starches have thousands of glucose subunits. In animals, carbohydrate is stored as an animal starch called glycogen, found in liver and muscle tissue.

▶ *Di-* and *poly*saccharides are assembled by a condensation reaction. Water is a by-product of the reaction. In contrast, hydrolysis, or splitting by the addition of water, digests di- and polysaccharides to smaller sugar units (for details, see the section Important Chemical Reactions Related to the Study of Nutrition).

## Lipids

**Lipids** are a class of nonpolar compounds that are grouped according to solubility in organic solvents. They don't readily dissolve in water because most are nonpolar or hydrophobic.

Simple lipids include fatty acids and steroids. The lipid cholesterol serves as the precursor (parent) for the steroid hormones, such as testosterone, estrogen, and progesterone. Complex lipids include triglycerides (often referred to as *triacylglycerols*), which are esters of glycerol and fatty acids. Phospholipids are composed of glycerol, phosphoric acid, and long-chain fatty acids; sphingolipids are composed of sphingosine, phosphoric acid, long-chain fatty acids, and choline; and glycosphingolipids are composed of sphingosine, fatty acids, and carbohydrates.

Triglycerides represent fuel found in food and stored in adipose tissues. Phospholipids are part polar and part nonpolar, which allows them to interact with water and function as emulsifiers. Sphingophospholipids make up the material surrounding nerves. Glycosphingolipids are structural material for brain and nerve tissue. These complex lipids can be hydrolyzed to yield fatty acids.

▶ Prostaglandins are a type of fatty acid produced by almost all organs in the body; they have specific regulatory functions. They are all derived from certain dietary (essential) fatty acids (see Chapter 6 and Appendix A).

## Proteins

**Proteins** are polymers of amino acids. Twenty common amino acids are incorporated into the great variety of body proteins. Although all the amino acids contain an amine (amino)

group ($NH_2$) and a carboxylic acid group ($—C—OH$), each has a distinctive structure (Fig. B-7). Proteins typically contain many atoms, such as carbon, nitrogen, sulfur, hydrogen, and oxygen.

The genetic information in the DNA in a cell's nucleus is the code book for constructing a protein. The sequence of amino acids in a protein follows the DNA code for synthesizing the protein. This protein can be made over and over again because of the code carried in the genes.

## Nucleic Acids (DNA and RNA)

Nucleic acids include DNA (deoxyribonucleic acid), RNA (ribonucleic acid), and the subunits from which they are formed, called nucleotides. A nucleotide is made of 3 components: a 5-carbon pentose sugar, a phosphate group, and a nitrogenous base (Fig. B-8). There are 2 kinds of nitrogenous base: purines (double ring) and pyrimidines (single ring).

The sugar contained in RNA is ribose. The pyrimidine bases in ribonucleic acids are uracil and cytosine, and the purine bases are guanine and adenine. RNA is a single polynucleotide strand, not a double strand like DNA.

DNA in the nucleus of the cell is the basis of the genetic code. The sugar in DNA, deoxyribose, can be covalently bonded to the purine bases adenine and guanine and to the pyrimidine bases cytosine and thymine (Fig. B-9). These 4 types of nucleotides can produce the long chain that makes up a single strand of DNA. The DNA is a 2-stranded sugar phosphate chain that twists around in such a way to form a helix. The bases project into the center of the helix, forming a staircase structure. The 2 strands are held together by hydrogen bonds (Fig. B-10).

DNA always contains an equal number of purine and pyrimidine bases. And there is a relationship called complementary base pairing—adenine pairs only with thymine, and guanine pairs only with cytosine. (In RNA, adenine pairs with uracil.)

Although there are only 4 bases, the number of sequences of bases is endless. The total human genome consists of billions of base pairs, making up thousands of genes. The applications of this knowledge can lead to genetic screening for breast cancer and, in the future, are likely to help produce drugs to treat obesity and inborn errors of metabolism.

During replication, the helix uncoils and separates so that each chain or strand serves as a template for the synthesis of its complementary chain. This step is important for cell division. Each daughter cell receives DNA containing 1 strand of the original molecule and 1 new strand.

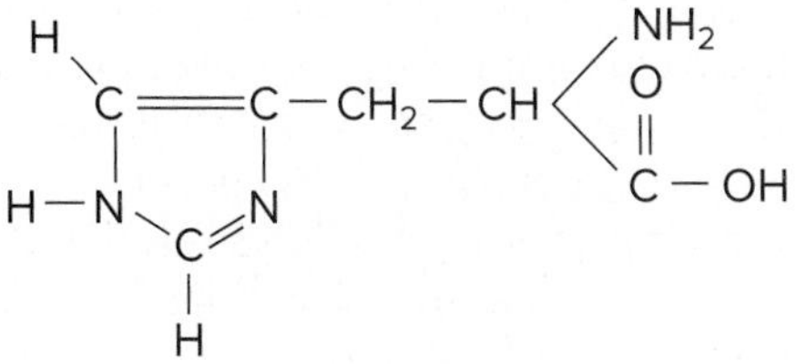

**Histidine (His)**
**(essential)**

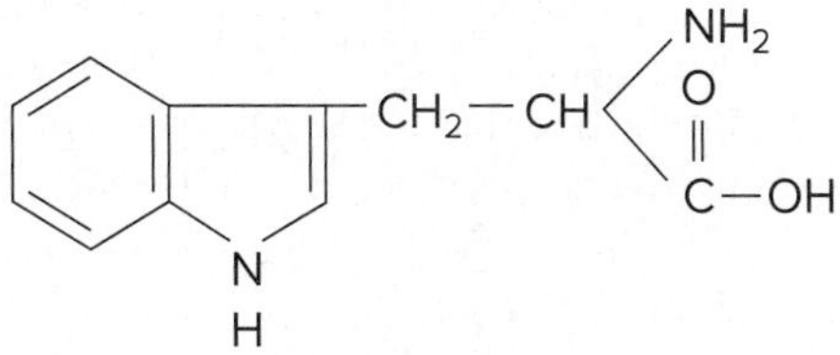

**Tryptophan (Trp)**
**(essential)**

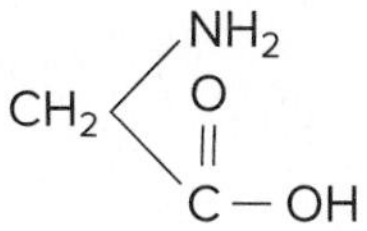

**Glycine (Gly)**

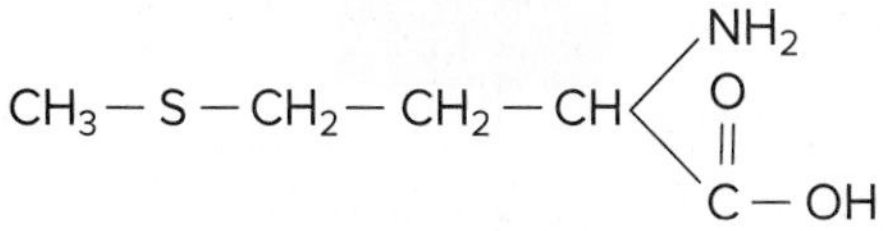

**Methionine (Met)**
**(essential)**

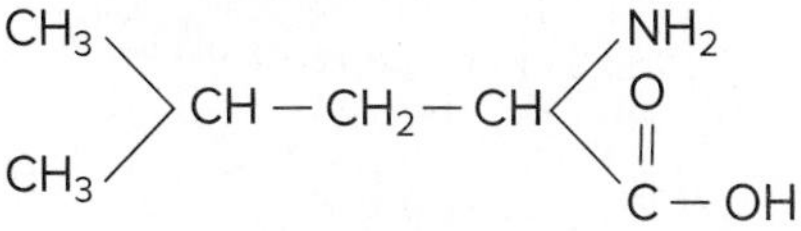

**Leucine (Leu)**
**(essential)**

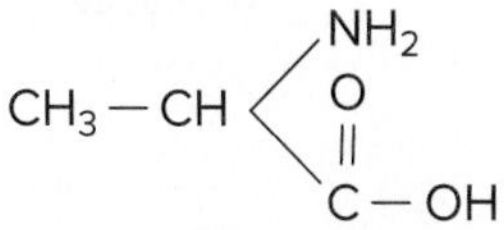

**Alanine (Ala)**

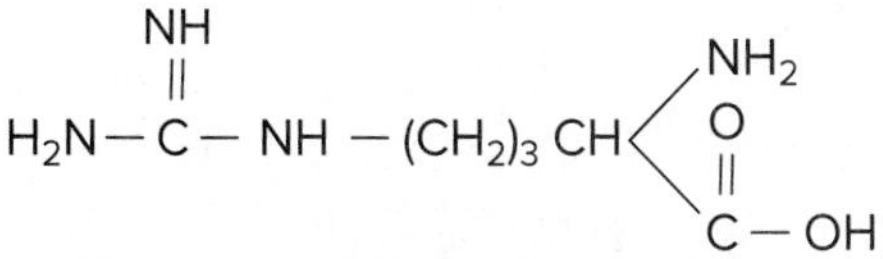

**Arginine (Arg)**
**(essential)**

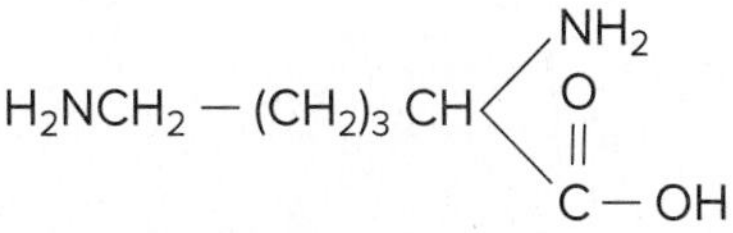

**Lysine (Lys)**
**(essential)**

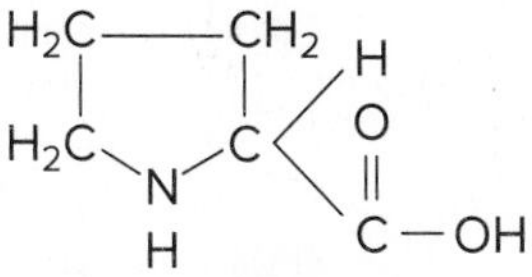

**Proline (Pro)**

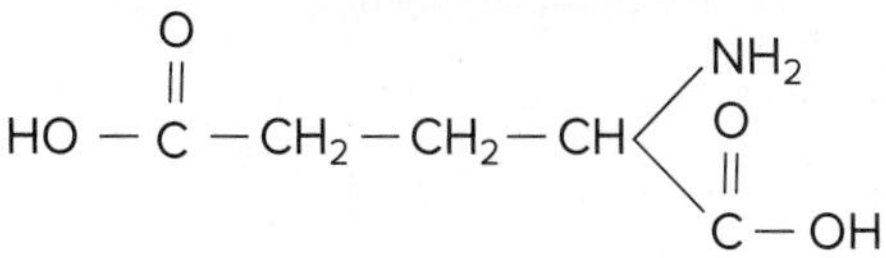

**Glutamic acid (Glu)**

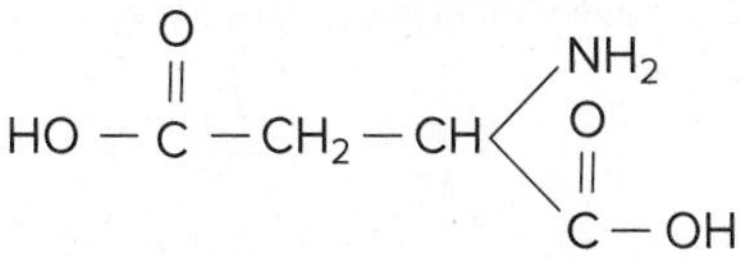

**Aspartic acid (Asp)**

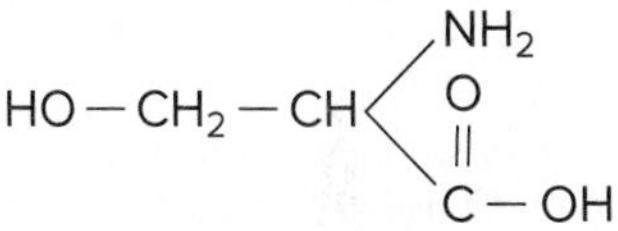

**Serine (Ser)**

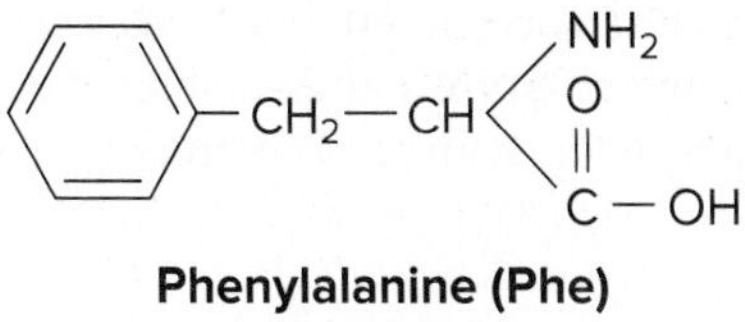

**Phenylalanine (Phe)**
**(essential)**

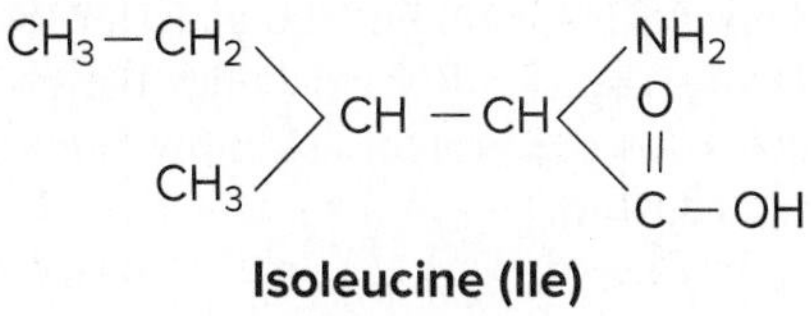

**Isoleucine (Ile)**
**(essential)**

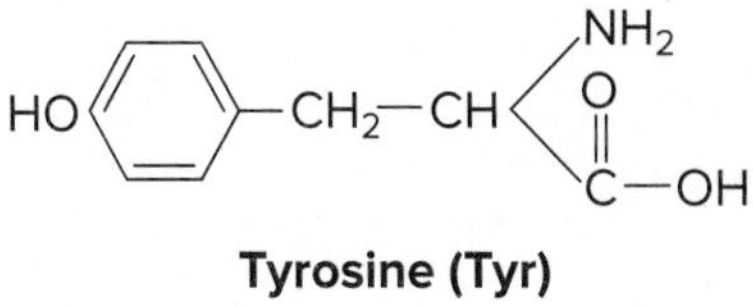

**Tyrosine (Tyr)**

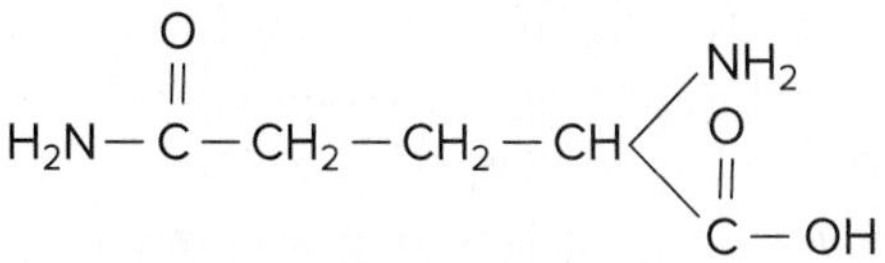

**Glutamine (Gln)**

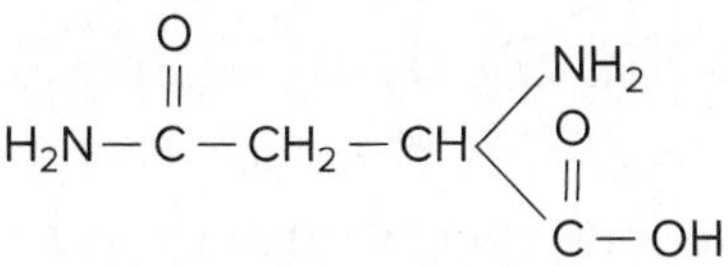

**Asparagine (Asn)**

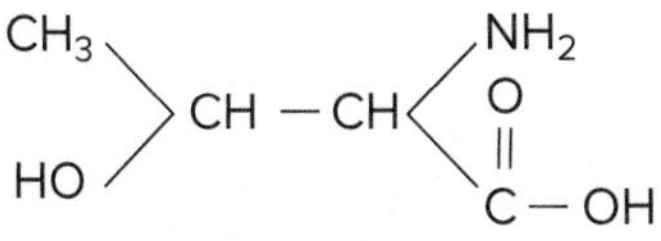

**Threonine (Thr)**
**(essential)**

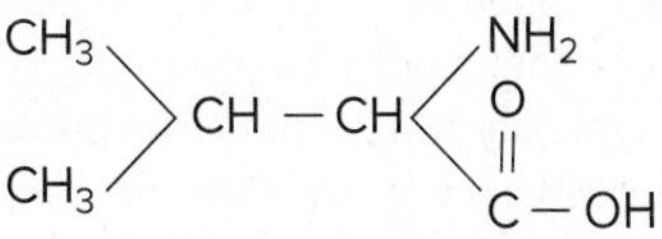

**Valine (Val)**
**(essential)**

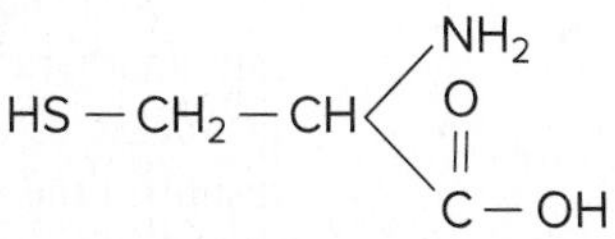

**Cysteine (Cys)**

*Figure B-7* The 20 common amino acids in foods.

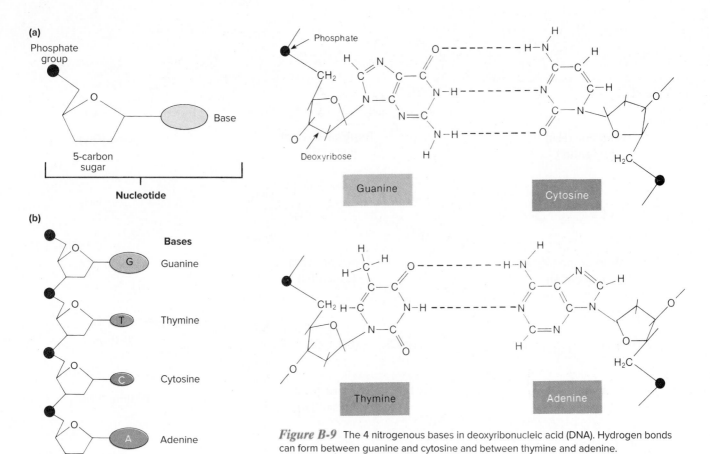

**(a)**

Phosphate group

Base

5-carbon sugar

**Nucleotide**

**(b)**

**Bases**

G — Guanine

T — Thymine

C — Cytosine

A — Adenine

*Figure B-8* (*a*) General structure of a nucleotide. (*b*) A polymer of nucleotides, or polynucleotide, is formed by sugar-phosphate bonds between nucleotides.

Phosphate

Deoxyribose

Guanine    Cytosine

Thymine    Adenine

*Figure B-9* The 4 nitrogenous bases in deoxyribonucleic acid (DNA). Hydrogen bonds can form between guanine and cytosine and between thymine and adenine.

RNA, another nucleic acid, takes its instructions from DNA. There are 3 types of RNA: ribosomal RNA, transfer RNA, and messenger RNA. Ribosomal RNA forms part of the structure of ribosomes in the cell; this is where proteins are synthesized. Messenger RNA contains the code for the synthesis of a specific protein transcribed from DNA. Transfer RNA decodes the genetic message in RNA and assembles the amino acids for the protein assembly line (see Chapter 7 for details). The process is called **translation.**

## Important Chemical Reactions Related to the Study of Nutrition

One of the most important properties of chemical compounds is the type of reactions they undergo. Chemical reactions are responsible for vision, thinking, movement, and everything else that occurs in the human body.

In a chemical reaction, a compound or set of compounds (the reactants) is converted into another compound or set of compounds (the products), accompanied by the absorption or release of energy, which is typically heat in biological processes. In effect, the reactants reshuffle their atoms to form products. Clearly, then, no atoms lose their identity during a chemical reaction and no atoms are gained, lost, or converted to another kind of atom during the course of chemical activity.

Chemists have grouped reactions according to their similarities in chemical behavior. Some of these reactions are performed over and over in each cell. Following is a brief overview of some important reaction types.

## Condensation Reactions

A **condensation reaction** occurs when 2 molecules join together to form a larger molecule and water is released. The 2 reactant molecules typically contain hydroxyl groups, meaning that there are 2 OH groups. A simple example is the condensation of glucose and galactose to make lactose and water:

$$C_6H_{12}O_6 \quad + \quad C_6H_{12}O_6 \quad \rightarrow \quad C_{12}H_{22}O_{11} \quad + \quad H_2O$$

**Glucose**  **Galactose**  **Lactose**  **Water**

One —OH group on the single sugar gains a proton and forms a water molecule. The OH group on the other single sugar loses a proton and forms a bond with the other molecule—in exactly the same place that the water molecule leaves. Note that this is an overall description of what happens, not how it happens. In addition, although it is typical for both molecules to contain an —OH group in a condensation reaction, it is not a requirement for the reaction. A condensation reaction can occur in which only 1 of the reactants contains an —OH group.

## Hydrolysis Reactions

**Hydrolysis reactions** are reactions that occur when water is added to a compound. In biological systems, hydrolysis reactions are very frequently the reverse of condensation reactions. That is, water is added to a large molecule, which results in the formation of 2 smaller molecules. This can be illustrated by the hydrolysis of lactose:

$$C_{12}H_{22}O_{11} \quad + \quad H_2O \quad \rightarrow \quad C_6H_{12}O_6 \quad + \quad C_6H_{12}O_6$$

**Lactose**  **Water**  **Glucose**  **Galactose**

## Oxidation-Reduction Reactions

**Oxidation-reduction (redox) reactions** are important in nutrition science because they release energy from food during oxidation and synthesize carbohydrates, fatty acids, and other organic compounds during reduction. Redox reactions follow 3 rules.

4. No oxidation reaction takes place without something being reduced at the same time, and no reduction takes place without something being oxidized.
5. Oxidation is the loss of electrons.
6. Reduction is the gain in electrons.

A simple redox reaction involving iron is as follows:

$$Fe^{3+} + e^- \longleftrightarrow Fe^{2+}$$

A biochemical redox reaction involving the coenzyme form of riboflavin occurs as follows:

$$\begin{array}{c} +2H \\ \searrow \\ FAD \longleftrightarrow FADH_2 \\ \swarrow \\ -2H \end{array}$$

(Chapters 9, 10, and 13 provide more information about coenzymes, cofactors, and oxidation-reduction reactions.)

## Energy and Enzymatic Reactions

Enzymes are large proteins with varying amino acid composition that behave as organic **catalysts.** They are highly specific. Enzymes help a reaction proceed by lowering the

▶ Many important compounds in cells are formed through condensation reactions, and the breakdown of many compounds into smaller fragments occurs via hydrolysis reactions.

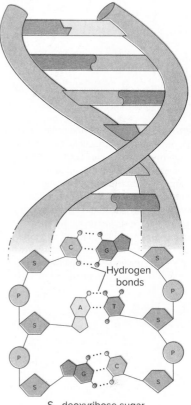

S—deoxyribose sugar
P—phosphate group

*Figure B-10* The double-helix structure of DNA. The 2 strands are held together by hydrogen bonds between complementary bases in each strand.

▶ In organic chemistry, oxidation is the loss of hydrogen (or gain of oxygen).

▶ In organic chemistry, reduction is the gain of hydrogen (or loss of oxygen).

**catalyst** Compound that speeds reaction rates but is not altered by the reaction.

Energy of activation

Reactants

Uncatalyzed path

Catalyzed path

Energy

Products

Course of the reaction in time

*Figure B-11* Enzymes and other catalysts accelerate chemical reactions by reducing the energy barrier to the reactions. Reactant molecules free in solution can react only if they meet in just the right orientation and with enough energy. An enzyme holds its substrate molecules in the right orientation to react and exerts forces on them that cause chemical bonds to break and form. In this way, an enzyme lowers the energy barrier that substrates must pass and, so, increases their reaction rates.

"energy of activation" so that the reaction can go faster (Fig. B-11). Enzymes lower this energy barrier between the reactants and the products. Some of the enzyme reactions that occur in the cell require coenzymes (vitamins) at the active site to make the reaction go, whereas many others don't. Fortunately, an enzyme isn't consumed by the reaction, so it can be used over and over.

## Common Chemical Structures

Most compounds in the body are composed of carbon, hydrogen, and oxygen, with carbon often being the predominant atom. Some commonly encountered combinations of atoms, called functional groups, have been given specific names because they appear in many molecules. You need to be familiar with them, for they are the most important features in many nutrients. The important functional groups were listed in Table B-4.

## The Drawing of Chemical Structures

Chemists have developed a shorthand notation for writing chemical formulas, called **skeletal structures.** In skeletal structures, neither carbon atoms nor the hydrogens bonded to the carbon atoms are expressly shown. What are shown are the bonds between the carbon atoms and the position of all atoms other than carbon and hydrogen. Keep in mind that there are carbon atoms at the apices (corners) of every angle in the structure (with the appropriate number of hydrogens attached to the carbon) and at the terminal end of the sticks. By way of illustration, look at a skeletal structure of propane ($CH_3CH_2CH_3$).

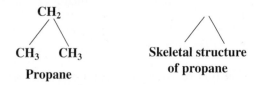

$CH_2$

$CH_3$    $CH_3$

**Propane**

**Skeletal structure of propane**

The advantage of using skeletal structures is that they allow for a clear representation of complex molecules without cluttering up the picture. They are handy when large structures, such as fatty acids, have to be represented. This notation is used throughout the text.

# *Appendix C*

## DETAILED DEPICTIONS OF GLYCOLYSIS, CITRIC ACID CYCLE, ELECTRON TRANSPORT CHAIN, CLASSES OF EICOSANOIDS, AND HOMOCYSTEINE METABOLISM

The following illustrations are provided to help you better visualize the changes in chemical structures throughout the metabolic processes described. These figures reflect greater scientific detail than the more simplified versions in Chapters 6, 9, and 13.

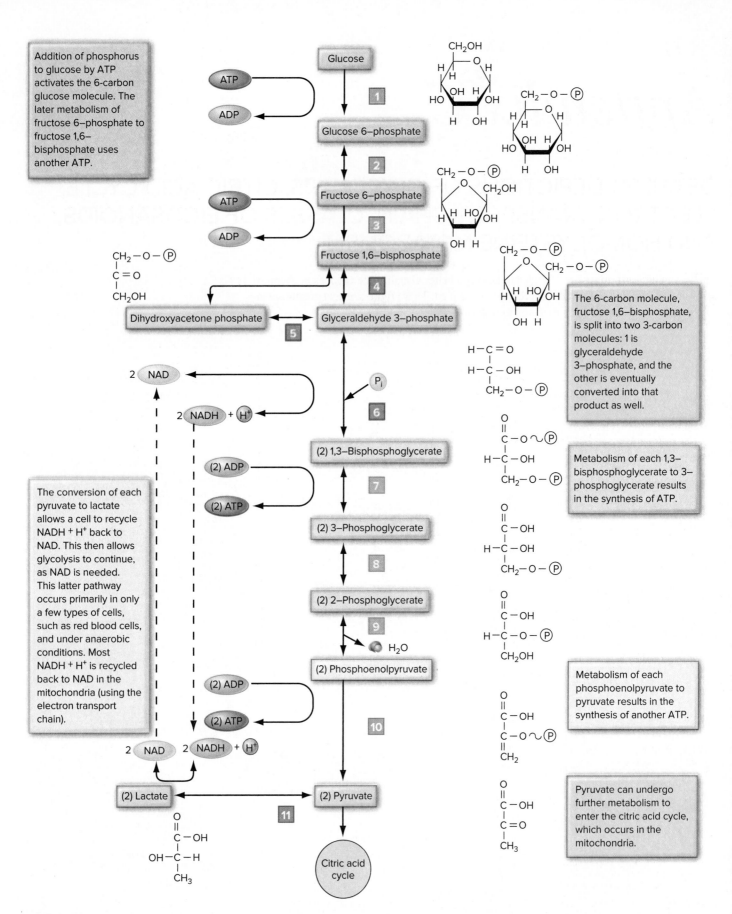

**Figure C-1** Detailed depiction of the individual chemical reactions that constitute glycolysis—glucose to pyruvate. Glycolysis takes place in the cytosol of the cell. The enzymes in the cytosol that participate at the following steps are [1] hexokinase, [2] phosphohexose isomerase, [3] phosphofructokinase, [4] aldolase, [5] phosphotriose isomerase, [6] glyceraldehyde-3-phosphate dehydrogenase, [7] phosphoglycerate kinase, [8] phosphoglycerate mutase, [9] enolase, and [10] pyruvate kinase. Sometimes [11] lactate dehydrogenase is used to recycle NADH + H⁺ back to NAD (anaerobic glycolysis). $P_i$ represents a phosphate group.

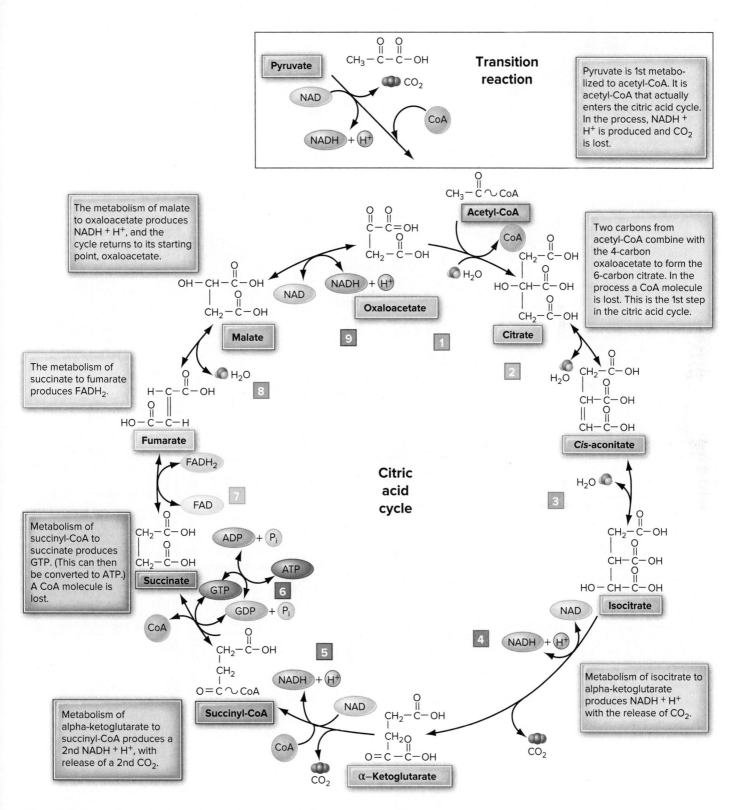

*Figure C-2* Detailed depiction of conversion of pyruvate to acetyl-CoA in the transition reaction and the individual chemical reactions of the citric acid cycle. Conversion of pyruvate to acetyl-CoA uses an enzyme complex that includes pyruvate dehydrogenase. The enzymes used in the citric acid cycle at the following steps are ▇ citrate synthase, ▇ aconitase, ▇ aconitase, ▇ isocitrate dehydrogenase, ▇ alpha-ketoglutarate dehydrogenase, ▇ succinate thiokinase, ▇ succinate dehydrogenase, ▇ fumarase, and ▇ malate dehydrogenase. *CoA* stands for coenzyme A, which is made from the vitamin pantothenic acid (see Chapter 13 for the chemical structure). Note that the $CO_2$ molecules lost during 1 turn of the citric acid cycle are not those from the carbons donated by acetyl-CoA. Instead, the carbons are broken off the portion of the citrate molecule derived from oxaloacetate.

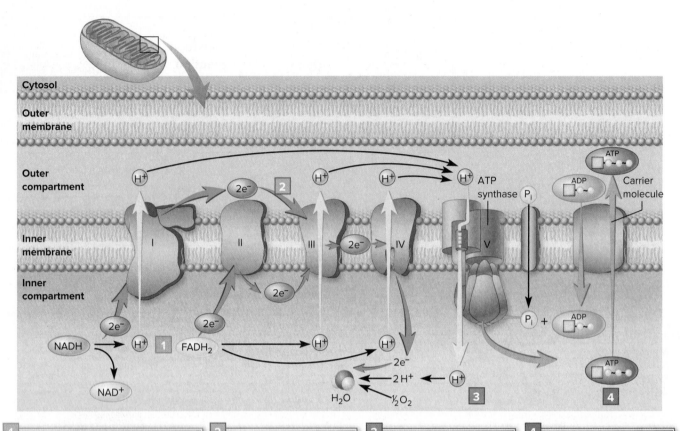

**1**
NADH + H$^+$ and FADH$_2$ transfer their hydrogen ions and electrons to the electron carriers located on the inner mitochondrial membrane. Although NADH + H$^+$ and FADH$_2$ transfer their hydrogens to the electron transport chain, the hydrogen ions (H$^+$), having been separated from their electron (H→H$^+$ + e$^-$), are not carried down the chain with the electrons. Instead, the hydrogen ions are pumped into the outer compartment (located between the inner and outer membranes of a mitochondrion). The NAD$^+$ and FAD regenerated from the oxidation of the NADH + H$^+$ and FADH$_2$ are now ready to function in glycolysis, the transition reaction, and the citric acid cycle.

**2**
Pairs of electrons are then separated by coenzyme Q (CoQ) and each electron is then passed along a group of iron-containing cytochromes. At each transfer from 1 cytochrome to the next, energy is released. Some of this energy is used to pump hydrogen ions into the outer compartment. A portion of the energy is eventually used to generate ATP from ADP and P$_i$, but much is simply released as heat.

**3**
As hydrogen ions diffuse back into the inner compartment through special channels, ATP is produced by the enzyme ATP synthase. At the end of the chain of cytochromes, the electrons, hydrogen ions, and oxygen combine to form water. Oxygen is the final electron acceptor and is reduced to form water.

**4**
One carrier molecule moves ADP into the inner compartment and a different carrier molecule moves phosphate (P$_i$) into the inner compartment. In the inner compartment, the energy generated by the electron transport chain unites ADP to P$_i$ to form ATP. ATP is transported out of the inner compartment by a carrier protein molecule that exchanges ATP for ADP.

*Figure C-3*   The electron transport chain.

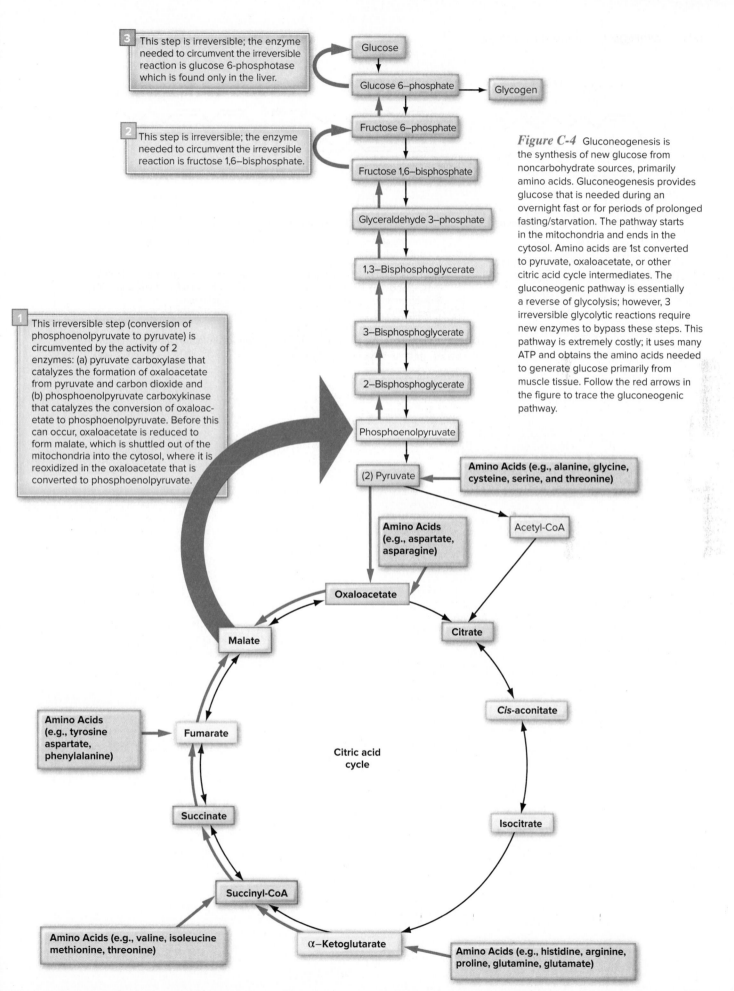

**3** This step is irreversible; the enzyme needed to circumvent the irreversible reaction is glucose 6-phosphotase which is found only in the liver.

Glucose

Glucose 6–phosphate → Glycogen

**2** This step is irreversible; the enzyme needed to circumvent the irreversible reaction is fructose 1,6–bisphosphate.

Fructose 6–phosphate

Fructose 1,6–bisphosphate

Glyceraldehyde 3–phosphate

1,3–Bisphosphoglycerate

3–Bisphosphoglycerate

2–Bisphosphoglycerate

Phosphoenolpyruvate

**1** This irreversible step (conversion of phosphoenolpyruvate to pyruvate) is circumvented by the activity of 2 enzymes: (a) pyruvate carboxylase that catalyzes the formation of oxaloacetate from pyruvate and carbon dioxide and (b) phosphoenolpyruvate carboxykinase that catalyzes the conversion of oxaloacetate to phosphoenolpyruvate. Before this can occur, oxaloacetate is reduced to form malate, which is shuttled out of the mitochondria into the cytosol, where it is reoxidized in the oxaloacetate that is converted to phosphoenolpyruvate.

*Figure C-4* Gluconeogenesis is the synthesis of new glucose from noncarbohydrate sources, primarily amino acids. Gluconeogenesis provides glucose that is needed during an overnight fast or for periods of prolonged fasting/starvation. The pathway starts in the mitochondria and ends in the cytosol. Amino acids are 1st converted to pyruvate, oxaloacetate, or other citric acid cycle intermediates. The gluconeogenic pathway is essentially a reverse of glycolysis; however, 3 irreversible glycolytic reactions require new enzymes to bypass these steps. This pathway is extremely costly; it uses many ATP and obtains the amino acids needed to generate glucose primarily from muscle tissue. Follow the red arrows in the figure to trace the gluconeogenic pathway.

(2) Pyruvate

**Amino Acids (e.g., alanine, glycine, cysteine, serine, and threonine)**

Acetyl-CoA

**Amino Acids (e.g., aspartate, asparagine)**

Oxaloacetate

Citrate

Malate

*Cis*-aconitate

**Amino Acids (e.g., tyrosine aspartate, phenylalanine)**

Fumarate

**Citric acid cycle**

Isocitrate

Succinate

Succinyl-CoA

**Amino Acids (e.g., valine, isoleucine methionine, threonine)**

α–Ketoglutarate

**Amino Acids (e.g., histidine, arginine, proline, glutamine, glutamate)**

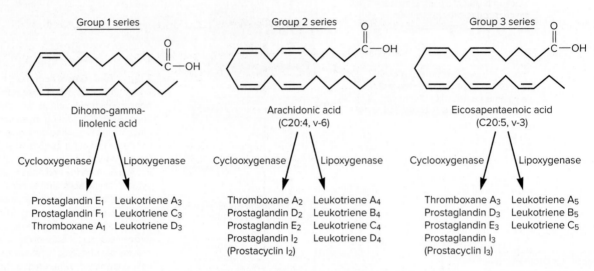

| Group 1 series | | Group 2 series | | Group 3 series | |
|---|---|---|---|---|---|
| Dihomo-gamma-linolenic acid | | Arachidonic acid (C20:4, v-6) | | Eicosapentaenoic acid (C20:5, v-3) | |
| Cyclooxygenase | Lipoxygenase | Cyclooxygenase | Lipoxygenase | Cyclooxygenase | Lipoxygenase |
| Prostaglandin $E_1$ | Leukotriene $A_3$ | Thromboxane $A_2$ | Leukotriene $A_4$ | Thromboxane $A_3$ | Leukotriene $A_5$ |
| Prostaglandin $F_1$ | Leukotriene $C_3$ | Prostaglandin $D_2$ | Leukotriene $B_4$ | Prostaglandin $D_3$ | Leukotriene $B_5$ |
| Thromboxane $A_1$ | Leukotriene $D_3$ | Prostaglandin $E_2$ | Leukotriene $C_4$ | Prostaglandin $E_3$ | Leukotriene $C_5$ |
| | | Prostaglandin $I_2$ | Leukotriene $D_4$ | Prostaglandin $I_3$ | |
| | | (Prostacyclin $I_2$) | | (Prostacyclin $I_3$) | |

*Figure C-5* Examples of eicosanoids from the 3 major groups. The parent fatty acid produces profound differences in how eicosanoids across the 3 groups act in the body (e.g., thromboxane $A_1$ vs. $A_2$ vs. $A_3$).

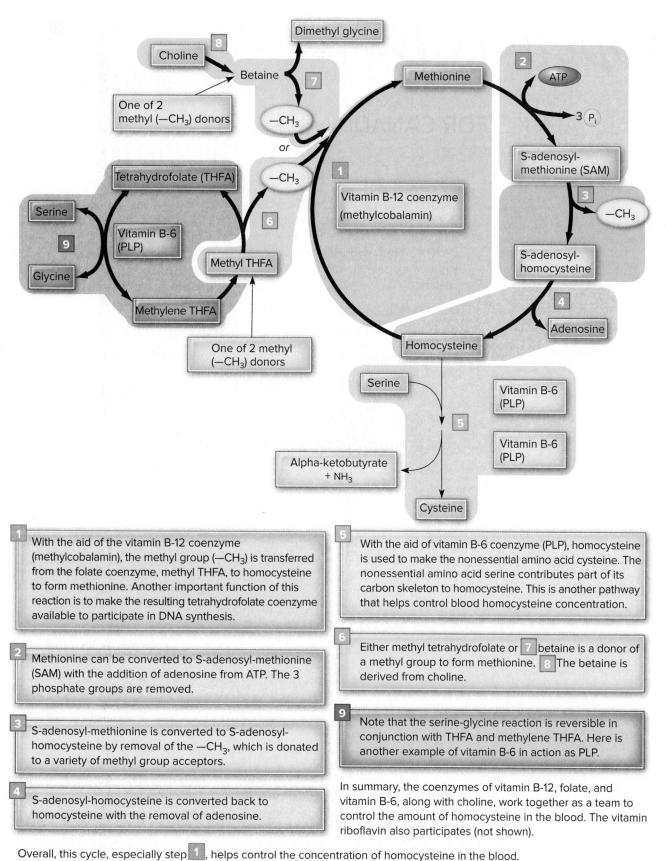

**1** With the aid of the vitamin B-12 coenzyme (methylcobalamin), the methyl group (—CH₃) is transferred from the folate coenzyme, methyl THFA, to homocysteine to form methionine. Another important function of this reaction is to make the resulting tetrahydrofolate coenzyme available to participate in DNA synthesis.

**2** Methionine can be converted to S-adenosyl-methionine (SAM) with the addition of adenosine from ATP. The 3 phosphate groups are removed.

**3** S-adenosyl-methionine is converted to S-adenosyl-homocysteine by removal of the —CH₃, which is donated to a variety of methyl group acceptors.

**4** S-adenosyl-homocysteine is converted back to homocysteine with the removal of adenosine.

**5** With the aid of vitamin B-6 coenzyme (PLP), homocysteine is used to make the nonessential amino acid cysteine. The nonessential amino acid serine contributes part of its carbon skeleton to homocysteine. This is another pathway that helps control blood homocysteine concentration.

**6** Either methyl tetrahydrofolate or **7** betaine is a donor of a methyl group to form methionine. **8** The betaine is derived from choline.

**9** Note that the serine-glycine reaction is reversible in conjunction with THFA and methylene THFA. Here is another example of vitamin B-6 in action as PLP.

In summary, the coenzymes of vitamin B-12, folate, and vitamin B-6, along with choline, work together as a team to control the amount of homocysteine in the blood. The vitamin riboflavin also participates (not shown).

Overall, this cycle, especially step **1** , helps control the concentration of homocysteine in the blood.

*Figure C-6*  Detailed diagram of folate, vitamin B-12, vitamin B-6, and choline metabolism in relation to homocysteine metabolism. Step **9** (serine → glycine) is the major source of methyl groups for this overall pathway.

# Appendix D

## DIETARY ADVICE FOR CANADIANS

**Recommended Nutrient Intake (RNI)**
Canadian version of RDA, published in 1990.

The information in this appendix includes advice on dietary patterns as well as regulations that apply to food labeling. Previous **Recommended Nutrient Intakes (RNIs)** for nutrients have been replaced by the Dietary Reference Intakes (DRIs) that apply to Canadian and U.S. citizens. These are listed on the last pages of this book. Both Canadian and American scientists worked on the various DRI committees, creating a set of harmonized DRIs for both countries.

Learn more at

food-guide.canada.ca/en/guidelines/

www.dietitians.ca/Downloads/Public/Senior-Friendly-collection.aspx

## Summary of the Nutrition Guidelines and Recommendations for Canadians

Figure D-1 Canada's food guide. ©All rights reserved. Eating Well with Canada's Food Guide. Health Canada. 2011. Reproduced with permission from the Minister of Health, 2014.

(continued)

**Canada's food guide**

# Eat well. Live well.

## Healthy eating is more than the foods you eat

**Be mindful of your eating habits**

**Cook more often**

**Enjoy your food**

**Eat meals with others**

**Use food labels**

**Limit foods high in sodium, sugars or saturated fat**

**Be aware of food marketing**

Discover your food guide at

## Canada.ca/FoodGuide

Health Canada　Santé Canada

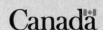
Canada

| Guidelines | Considerations |
|---|---|
| **Guideline 1**<br><br>Nutritious foods are the foundation for healthy eating.<br><br>• Vegetables, fruit, whole grains, and protein foods should be consumed regularly. Among protein foods, consume plant-based more often.<br><br>   • Protein foods include legumes, nuts, seeds, tofu, fortified soy beverage, fish, shellfish, eggs, poultry, lean red meat including wild game, lower fat milk, lower fat yogurts, lower fat kefir, and cheeses lower in fat and sodium.<br><br>• Foods that contain mostly unsaturated fat should replace foods that contain mostly saturated fat.<br><br>• Water should be the beverage of choice. | **Nutritious foods to encourage**<br>• Nutritious foods to consume regularly can be fresh, frozen, canned, or dried.<br><br>**Cultural preferences and food traditions**<br>• Nutritious foods can reflect cultural preferences and food traditions.<br>• Eating with others can bring enjoyment to healthy eating and can foster connections between generations and cultures.<br>• **Traditional food** improves diet quality among Indigenous Peoples.<br><br>**Energy balance**<br>• Energy needs are individual and depend on a number of factors, including levels of physical activity.<br>• Some **fad diets** can be restrictive and pose nutritional risks.<br><br>**Environmental impact**<br>• Food choices can have an impact on the environment. |

| Guidelines | Considerations |
|---|---|
| **Guideline 2**<br><br>Processed or prepared foods and beverages that contribute to excess sodium, free sugars, or saturated fat undermine healthy eating and should not be consumed regularly. | **Sugary drinks, confectioneries and sugar substitutes**<br>• **Sugary drinks** and **confectioneries** should not be consumed regularly.<br>• Sugar substitutes do not need to be consumed to reduce the intake of free sugars.<br><br>**Publically funded institutions**<br>• Foods and beverages offered in publically funded institutions should align with Canada's Dietary Guidelines.<br><br>**Alcohol**<br>• There are health risks associated with alcohol consumption. |
| **Guideline 3**<br><br>Food skills are needed to navigate the complex food environment and support healthy eating.<br><br>• Cooking and food preparation using nutritious foods should be promoted as a practical way to support healthy eating.<br><br>• Food labels should be promoted as a tool to help Canadians make informed food choices. | **Food skills and food literacy**<br>• Food skills are important life skills.<br>• Food literacy includes food skills and the broader environmental context.<br>• Cultural food practices should be celebrated.<br>• Food skills should be considered within the social, cultural, and historical context of Indigenous Peoples.<br><br>**Food skills and opportunities to learn and share**<br>• Food skills can be taught, learned, and shared in a variety of settings.<br><br>**Food skills and food waste**<br>• Food skills may help decrease household food waste. |

# Appendix E

## THE FOOD LISTS FOR DIABETES: A HELPFUL MENU PLANNING TOOL

The Food Lists are a valuable means for roughly estimating the energy, protein, carbohydrate, and fat content of food, meals, and diets. The food choices allow individuals to use the macronutrient composition of foods to plan and balance their daily intake. By using the food choices, you can plan menus to fall within dietary recommendations without having to look up or memorize the nutrient values of numerous foods.

In the Food Lists, individual foods are grouped with foods of similar macronutrient composition into the following lists: carbohydrates, proteins, fats, and alcohol. The carbohydrates are further divided into the groups of starch; fruits; milk and milk substitutes; nonstarchy vegetables; and sweets, desserts, and other carbohydrates. Proteins are subdivided into lean, medium-fat, high-fat, and plant-based categories. These lists are designed so that, when the recommended serving size is used, each food on a list provides about the same amount of carbohydrate, protein, fat, and energy. This equality allows food choices from each list.

The Food Lists concept was originally referred to as the Exchange System and was developed for planning diabetic diets. Diabetes is easier to control if the person's diet has a consistent macronutrient composition each day. Thus, the same number of foods from each of the various lists is recommended each day to maintain dietary balance and blood glucose control. Because the lists provide a quick way to estimate the energy, carbohydrate, protein, and fat content in a food or meal, they also are a valuable menu planning tool for everyone.

### Becoming Familiar with the Food Lists for Diabetes

To use the Food Lists, you must know which foods are on each list and the serving sizes for each food. Table E-1 gives the serving sizes for foods on each Food List, as well as the carbohydrate, protein, fat, and energy content per list. Note that the proteins and milk lists are divided into subclasses, which vary in fat content and, hence, in the amount of energy they provide. Except for the plant-based proteins, foods on the proteins and fats lists contain no carbohydrate; those on the fruit and fat lists lack protein; and those on the nonstarchy vegetable and fruit lists contain essentially no fat.

Note that the starch list includes not only bread, dry cereal, cooked cereal, rice, and pasta but also baked beans, corn on the cob, and potatoes. Although these foods are not included in the grain group in MyPlate, their macronutrient composition is actually very similar to grains because of their high starch content.

Additional information about the **Food Lists for Diabetes** is available in Table E-1. To learn more information about food choices and meal planning, visit the website of the American Diabetes Association, www.diabetes.org/nutrition.

The Food Lists are based on macronutrient composition rather than food origin (Fig. E-1). For example, plant-based proteins (i.e., soy-based meatless burgers, tempeh,

**Food Lists for Diabetes** System for classifying foods into numerous lists based on the foods' macronutrient composition and establishing serving sizes so that 1 serving of each food on a list contains about the same amount of carbohydrate, protein, fat, and energy content.

### Table E-1  Nutrient Composition of Food Lists

| Lists | Household Measures* | Carbohydrate (g) | Protein (g) | Fat (g) | Energy (kcal) |
|---|---|---|---|---|---|
| **Starch** | 1 slice bread, 1 cup cereal, or ½ cup cooked pasta | 15 | 3 | 1 | 80 |
| **Fruits** | 1 small piece, ¾ to 1 cup raw, ½ cup juice or canned | 15 | — | — | 60 |
| **Milk and Milk Substitutes** | 1 cup | | | | |
| Fat-free/low-fat (1%) | | 12 | 8 | 0–3† | 100 |
| Reduced fat (2%) | | 12 | 8 | 5 | 120 |
| Whole | | 12 | 8 | 8 | 160 |
| **Nonstarchy Vegetables** | 1 cup raw or ½ cup cooked or juice | 5 | 2 | — | 25 |
| **Sweets and Other Carbohydrates** | Varies | 15 | Varies | Varies | Varies |
| **Proteins** | 1 oz | | | | |
| Lean | | — | 7 | 2 | 45 |
| Medium-fat | | — | 7 | 5 | 75 |
| High-fat | | — | 7 | 8 | 100 |
| Plant-based | | Varies | 7 | Varies | Varies |
| **Fats** | 1 tsp | — | — | 5 | 45 |
| **Alcohol** | 5 oz wine 12 oz beer, 1½ oz hard liquor | Varies | — | — | Varies |

Source: Reproduction of the Food Lists in whole or in part, without permission of the Academy of Nutrition and Dietetics) or the American Diabetes Association, Inc. is a violation of federal law. This material has been modified from *Choose Your Foods: Food Lists for Diabetes,* which is the basis of a meal planning system designed by a committee of the American Diabetes Association and Academy of Nutrition and Dietetics. While designed primarily for people with diabetes and others who must follow special diets, the Food Lists are based on principles of good nutrition that apply to everyone. Copyright © 2019 by the American Diabetes Association and the American Dietetic Association.

*Just an estimate; see Food Lists for actual amounts.

†Calculated as 1 g for purposes of energy contribution.

*Figure E-1*  The Food Lists promote use of a variety of foods to plan a healthful diet. ©aluxum/Getty Images

and tofu) are included in the protein list because their composition is comparable to animal-based proteins. Some plant-based proteins, such as baked beans and peas, are counted as both a protein and a starch. Note that bacon appears in both the high-fat protein and fats Food Lists. Some foods that are essentially calorie free include bouillon, diet soda, coffee, tea, dill pickles, and vinegar, as well as herbs and spices.

## Using the Food Lists to Develop Daily Menus

Use the Food Lists to plan a 1-day menu. Target an energy content of 2000 kcal, with 55% derived from carbohydrates (1100 kcal), 15% from protein (300 kcal), and 30% from fat (600 kcal). These specifications can be translated into a meal plan that consists of 2 low-fat milk choices, 3 nonstarchy vegetable choices, 5 fruit choices, 11 starch choices, 4 lean protein choices, and 7 fat choices (Table E-2). Note that this example is only 1 of many possible combinations.

Table E-3 separates these choices into breakfast, lunch, dinner, and a snack. In this example, breakfast includes 1 reduced fat milk choice, 2 fruit choices, 2 starch choices, and 1 fat choice. This can be met by a meal comprised of ¾ cup of a ready-to-eat breakfast cereal, 1 cup of reduced fat milk, 1 slice of bread with 1 tsp of margarine, and 1 cup of orange juice.

Our lunch example consists of 2 fat choices, 4 starch choices, 1 nonstarchy vegetable choice, 1 reduced fat milk choice, and 2 fruit choices. This total can be met by having 1 slice of bacon with 1 teaspoon of mayonnaise on 2 slices of bread, with tomato—in other words, a bacon and tomato sandwich. You also can add lettuce to the sandwich. Lettuce can be considered a free vegetable choice. Add to this meal a medium banana (= 2 fruit choices), 1 cup of reduced fat milk, and 6 graham cracker squares. Later, a snack of ¾ oz of pretzels can be added to provide another starch choice.

In our example, dinner consists of 4 lean protein choices, 1 fruit choice, 2 vegetable choices, 2 fat choices, and 2 starch choices. This total could correspond to a dinner of 4 oz

**Table E-2  Possible Food List Patterns That Yield 55% of Energy as Carbohydrate, 30% as Fat, and 15% as Protein for Energy Intakes Greater Than 2000 kcal**

| kcal/Day<br>Food List | 1200* | 1600* | 2000 | 2400 | 2800 | 3200 | 3600 |
|---|---|---|---|---|---|---|---|
| Milk (reduced fat) | 2 | 2 | 2 | 2 | 2 | 2 | 2 |
| Nonstarchy vegetable | 3 | 3 | 3 | 4 | 4 | 4 | 5 |
| Fruit | 3 | 4 | 5 | 6 | 8 | 9 | 9 |
| Starch | 5 | 8 | 11 | 13 | 15 | 18 | 21 |
| Protein (lean) | 4 | 4 | 4 | 5 | 6 | 7 | 8 |
| Fat | 3 | 5 | 7 | 9 | 11 | 13 | 14 |

This is just 1 set of options. More meat could be included if less milk were used, for example.
*Energy intakes of 1200 and 1600 kcal contain 20% of energy as protein and 50% of energy as carbohydrate to allow for greater flexibility in diet planning.

**Table E-3 Sample 1-Day 2000 kcal Menu Based on the Food Lists**

**Breakfast**

| | |
|---|---|
| 1 reduced fat milk choice | 1 cup reduced fat milk (some on cereal) |
| 2 fruit choices | 1 cup orange juice |
| 2 starch choices | ¾ cup ready-to-eat breakfast cereal (bran), 1 piece whole-wheat toast |
| 1 fat choice | 1 tsp soft margarine on toast |

**Lunch**

| | |
|---|---|
| 4 starch choices | 2 slices whole-wheat bread, 6 graham crackers (2½ inches by 2½ inches) |
| 2 fat choices | 1 slice bacon, 1 tsp mayonnaise |
| 1 nonstarchy vegetable choice | 1 sliced tomato |
| 2 fruit choices | 1 banana (8 inches) |
| 1 reduced fat milk choice | 1 cup reduced fat milk |

**Snack**

| | |
|---|---|
| 1 starch choice | ¾ oz pretzels |

**Dinner**

| | |
|---|---|
| 4 lean protein choices | 4 oz lean steak (well trimmed) |
| 2 starch choices | 1 medium baked potato |
| 2 fat choices | 2 tsp soft margarine |
| 2 nonstarchy vegetable choices | 1 cup cooked broccoli |
| 1 fruit choice | 1 kiwi fruit |
| | Coffee (if desired) |

**Snack**

| | |
|---|---|
| 2 starch choices | ½ large bagel |
| 2 fat choices | 2 tbsp regular cream cheese |

broiled steak (meat only, no bone), 1 medium baked potato (1 choice = 3 ounces) with 2 tsp of margarine, 1 cup of broccoli, and 1 kiwi fruit. Coffee (if desired) is not counted because it contains no appreciable energy as long as no sugar or cream is added.

Finally, we have added a snack containing 2 starch choices and 2 fat choices to complete the menu. This can be provided by having ½ of a large bagel with 2 tbsp of regular cream cheese.

This menu is only 1 of many that are possible with the Food Lists. For example, apple juice could replace the orange juice; 2 small apples could be exchanged for the banana. The choices are endless. Notice that a diet with the Food Lists is often easier to plan if you use individual foods, as was done here. However, the Food List tables have some combination foods to help you plan more varied menus. For instance, lasagna typically has protein choices, nonstarchy vegetable choices, and starch choices. With practice, you will be able to break down combination foods into the corresponding Food List groups to plan a variety of meals (Fig. E-2).

*Figure E-2* Record the Food List pattern you have chosen in the left column. Then distribute the choices throughout the day, noting the food to be used and the serving size.

| Food List | Total Exchanges to Be Consumed Daily | Food Choices Consumed at Each Meal | | |
|---|---|---|---|---|
| | | Breakfast | Lunch | Dinner |
| **MILK AND MILK SUBSTITUTES** | | | | |
| **NONSTARCHY VEGETABLES** | | | | |
| **FRUIT** | | | | |
| **STARCH** | | | | |
| **PROTEINS** | | | | |
| **FAT** | | | | |

## Food Lists*

### Milk List

Fat-Free and Low-Fat Milk

(12 g carbohydrate, 8 g protein, 0–3 g fat, 100 kcal)

| | |
|---|---|
| 1 cup | fat-free and 1% milk and buttermilk |
| ⅓ cup | powdered (fat-free dry, before adding liquid) |
| ½ cup | canned, evaporated fat-free milk |
| 1 cup | buttermilk made from fat-free or low-fat milk |
| 1 cup | lactose free milk (low-fat or fat-free) |
| ⅔ cup (6 oz) | yogurt made from fat-free milk (plain or Greek) |
| ⅔ cup (6 oz) | yogurt, fat-free, flavored, sweetened with non-nutritive sweetener and fructose |
| 1 cup | chocolate milk (1 fat-free milk, 1 carbohydrate) |

*The Food Lists are the basis of a meal planning system designed by a committee of the American Diabetes Association and the Academy of Nutrition and Dietetics. While designed primarily for people with diabetes and others who must follow special diets, the lists are based on principles of good nutrition that apply to everyone. Copyright © 2014 by the American Diabetes Association and the Academy of Nutrition and Dietetics.

Reduced Fat Milk

(12 g carbohydrate, 8 g protein, 5 g fat, 120 kcal)
| | |
|---|---|
| 1 cup | 2% milk |
| 1 cup | lactose free milk (2%) |
| ⅔ cup | yogurt plain, low-fat (added milk solids) |
| 1 cup | sweet acidophilus milk |

Whole Milk

(12 g carbohydrate, 8 g protein, 8 g fat, 160 kcal)
| | |
|---|---|
| 1 cup | whole milk |
| ½ cup | evaporated whole milk |
| 1 cup | goat milk |
| 1 cup | kefir |
| 1 cup | yogurt, plain (made from whole milk) |
| 1 cup | chocolate milk (1 whole milk, 1 carbohydrate) |

Other Milk Items and Milk Substitutes

| | | |
|---|---|---|
| ⅓ cup | eggnog, fat-free | 1 carbohydrate |
| ⅓ cup | eggnog, low-fat | 1 carbohydrate, ½ fat |
| ⅓ cup | eggnog, whole milk | 1 carbohydrate, 1 fat |
| 1 cup | rice drink, plain fat-free | 1 carbohydrate |
| 1 cup | rice drink, flavored low-fat | 2 carbohydrates |
| 1 cup | soy milk, light or low-fat, plain | ½ carbohydrate, ½ fat |
| 1 cup | soy milk, regular, plain | ½ carbohydrate, 1 fat |
| ⅔ cup | yogurt, with fruit, low-fat | 1 fat-free milk, 1 carbohydrate |

## Nonstarchy Vegetable List

(5 g carbohydrate, 2 g protein, 0 g fat, 25 kcal)
1 vegetable choice equals:
½ cup cooked vegetables or vegetable juice
1 cup raw vegetables

| | | | |
|---|---|---|---|
| amaranth leaves | Brussels sprouts | hearts of palm | radishes |
| artichoke | cabbage | kohlrabi | sauerkraut |
| artichoke hearts | carrots | leeks | spinach |
| asparagus | cauliflower | mixed vegetables (without starchy | squash (summer) |
| baby corn | celery | vegetables, legumes, or pasta) | sugar snap peas |
| bamboo shoots | cucumber | mushrooms | tomato (fresh, canned, sauce) |
| beans (green, wax, Italian) | eggplant | okra | tomato/vegetable juice |
| bean sprouts | fennel | onions | turnips |
| beets | green onions or scallions | pea pods | water chestnuts |
| broccoli | greens (e.g., collard) | peppers (all varieties) | zucchini |

## Fruit List

Fruit

(15 g carbohydrate, 0 g protein, 0 g fat, 60 kcal)
1 fruit choice equals:

| | | | | |
|---|---|---|---|---|
| ½ cup | canned or frozen fruit or fruit juice | | ½ cup | cherries, canned |
| ¾ to 1 cup | fresh (1 small piece) | | 3 small | dates |
| 2 tbsp | dried fruit | | 1½ (3½ oz) | figs, fresh (large) |
| 1 (4 oz) | apple, unpeeled (small) | | 3 small | figs, dried |
| 4 rings | apple, dried | | ½ cup | fruit cocktail |
| ½ cup | applesauce (unsweetened) | | ½ (5½ oz) | grapefruit (large) |
| 4 (5½ oz) | apricots, fresh | | ¾ cup | grapefruit sections, canned |
| 8 halves | apricots, dried | | 17 (3 oz) | grapes (small) |
| ½ cup | apricots, canned | | 2 small (2½ oz) | guava |
| 1 (4 oz) | banana (extra small, 4 inches) | | 1 slice (10 oz) | honeydew melon |
| 1 cup | blackberries | | 1 (3½ oz) | kiwi |
| ¾ cup | blueberries | | ¾ cup | mandarin orange sections |
| 1 cup cubes | cantaloupe | | ½ (5½ oz) | mango (or ½ cup) |
| 12 (3½ oz) | cherries | | 1 (5½ oz) | nectarine (medium) |

| | | | |
|---|---|---|---|
| 1 (6½ oz) | orange (medium) | ½ cup | plums, canned |
| ½ (8 oz) | papaya (or 1 cup cubes) | 3 | plums, dried (prunes) |
| 1 (6 oz) | peach, fresh (medium) | ½ cup | pomegranate seeds |
| ½ cup | peaches, canned | 2 tbsp | dried fruits (raisins, cranberries, cherries, blueberries, mixed fruit) |
| ½ (4 oz) | pear, fresh | | |
| ½ cup | pear, canned | 1 cup | raspberries |
| ¾ cup | pineapple, fresh | 1¼ cups | strawberries (raw, whole) |
| ½ cup | pineapple, canned | 1 (6 oz) | tangerines (small) |
| ¼ (2½ oz) | plantain (raw) | 1¼ cups diced | watermelon |
| 2 (5 oz) | plums (small) | | |

### Fruit Juice

| | | | |
|---|---|---|---|
| ½ cup | apple juice/cider | ½ cup | orange juice |
| ⅓ cup | fruit juice blends, 100% juice | ½ cup | pineapple juice |
| ⅓ cup | grape juice | ½ cup | pomegranate juice |
| ½ cup | grapefruit juice | ⅓ cup | prune juice |

## Starch List

(15 g carbohydrate, 3 g protein, 1 g fat, 80 kcal)
1 starch choice equals:

### Bread

| | | | |
|---|---|---|---|
| ¼ (1 oz) | bagel | ¼ | naan, 8-inch × 2-inch |
| 2 slices (1½ oz) | bread, loaf-type, reduced calorie | 1 | pancake, 4 inches across × ¼ inch thick |
| 1 slice (1 oz) | bread, loaf-type (white, whole-wheat, pumpernickel, rye, sourdough, unfrosted raisin or cinnamon) | ½ | pita, 6 inches across |
| | | 1 (1 oz) | roll, plain (small) |
| 1 oz | bread, flat-type (chapatti, ciabatta, naan, roti) | 1 bun (1½ oz) | sandwich flat buns, whole-wheat |
| | | 1 | tortilla, corn, 6 inches across |
| | | 1 | tortilla, flour, 6 inches across |
| ½ | English muffin | ⅓ | tortilla, flour, 10 inches across |
| ½ (1 oz) | hot dog or hamburger bun | 1 | waffle, 4 inches square or across |

### Cereals and Grains

| | | | |
|---|---|---|---|
| ½ cup | bran cereal | ½ cup | kasha |
| ½ cup | bulgur | ⅓ cup | millet |
| ½ cup | cereal, cooked (oats, oatmeal) | ¼ cup | muesli |
| ¾ cup | cereal, unsweetened, ready-to-eat | ½ cup | oats |
| 3 tbsp | cornmeal (dry) | ⅓ cup | pasta |
| ⅓ cup | couscous | 1½ cups | puffed cereal |
| 3 tbsp | flour (dry) | ⅓ cup | rice, white or brown |
| ¼ cup | granola, low fat | ½ cup | Shredded Wheat, plain™ |
| ¼ cup | Grape-Nuts™ | ½ cup | sugar-frosted cereal |
| ½ cup | grits | 3 tbsp | wheat germ |

### Starchy Vegetables

| | | | |
|---|---|---|---|
| ⅓ cup | cassava | ½ cup or ½ medium (3 oz) | potato, boiled |
| ¼ cup | breadfruit | | |
| ½ cup | corn | ¼ large (3 oz) | potato, baked with skin |
| ½ (5 oz) | corn on the cob (large) | ¾ cup | pumpkin |
| ½ cup | marinara, pasta, or spaghetti sauce | ½ cup | succotash |
| 1 cup | mixed vegetables with corn or peas | 1 cup | squash, winter (acorn, butternut) |
| ½ cup | peas, green; parsnips | ½ cup | yam, sweet potato, plain |
| ⅓ cup | plantain | | |

## Crackers and Snacks

| | | | |
|---|---|---|---|
| 8 | animal crackers | ¾ oz | pretzels |
| 3 | graham crackers, 2½-inch square | 2 | rice cakes, 4 inches across |
| ¾ oz | matzoh | 6 | saltine-type crackers |
| 4 slices | melba toast | 8 (¾ oz) | snack chips, baked (pita, potato) |
| 20 | oyster crackers | 5 regular 1½-inch squares (¾ oz) | whole-wheat crackers, baked (10 thins) |
| 3 cups | popcorn (popped, no fat added or low-fat microwave) | | |

## Dried Beans, Peas, and Lentils

(counts as 1 starch choice plus 1 lean protein choice)

| | | | |
|---|---|---|---|
| ⅓ cup | baked beans | ½ cup | peas (black eyed or split) |
| ½ cup | beans (garbanzo, pinto, kidney, lima, white) | ½ cup | lentils |
| | | 3 tbsp | miso |

## Starchy Foods Prepared with Fat

(counts as 1 starch choice plus 1 fat choice)

| | | | |
|---|---|---|---|
| 1 | biscuit, 2½ inches across | ½ cup | potato, mashed |
| ½ cup | chow mein noodles | ⅓ (1 oz) | muffin, 5 oz |
| 1 (2 oz) | corn bread, 2-inch cube | 3 cups | popcorn, with butter |
| 6 | crackers, round butter type | 3 | sandwich crackers, cheese or peanut butter filling |
| 1 cup | croutons | | |
| 1 cup (2 oz) | french-fried potatoes (oven-baked) (see also the fast foods list) | ⅓ cup | stuffing, bread (prepared) |
| | | 2 | taco shell, 5 inches across |
| ¼ cup | granola | 1 | waffle, 4-inch square or across |
| ⅕ cup | hummus | 4–6 (1 oz) | whole-wheat crackers, fat added |

# Sweets, Desserts, and Other Carbohydrates

## Choices per Serving

| | | |
|---|---|---|
| 1/12 cake (about 2 oz) | angel food cake, unfrosted | 2 carbohydrates |
| 1 oz | biscotti | 1 carbohydrate, 1 fat |
| 2-inch square (about 1 oz) | brownie, unfrosted (small) | 1 carbohydrate, 1 fat |
| 2-inch square (about 1 oz) | cake, unfrosted | 1 carbohydrate, 1 fat |
| 2-inch square (about 2 oz) | cake, frosted | 2 carbohydrates, 1 fat |
| 1 oz | candy, chocolate | 1 carbohydrate, 2 fats |
| 5 pieces | candy, chocolate kisses | 1 carbohydrate, 1 fat |
| 3 pieces | candy, hard | 1 carbohydrate |
| 4 tsp | coffee creamer, non-dairy, powdered | ½ carbohydrate, ½ fat |
| 2 tbsp | coffee creamer, non-dairy, liquid | 1 carbohydrate |
| 2, 2¼ inches across | cookies, chocolate chip | 1 carbohydrate, 2 fats |
| 2 | cookies, fat-free (small) | 1 carbohydrate |
| 2 (about ⅔ oz) | cookies or sandwich cookies with creme filling (small) | 1 carbohydrate, 1 fat |
| ½ cup | cranberry juice cocktail | 1 carbohydrate |
| ¼ cup | cranberry sauce, jellied | 1½ carbohydrates |
| 1 (about 1¾ oz) | cupcake, frosted (small) | 2 carbohydrates, 1–1½ fats |
| 1 (1½ oz) | donut, plain cake (medium) | 1½ carbohydrates, 2 fats |
| 3¾ inches across (2 oz) | donuts, glazed | 2 carbohydrates, 2 fats |
| ½ cup | flan | 2½ carbohydrates, 1 fat |
| 1 | frozen pop | ½ carbohydrate |
| 1 cup | fruit drink or lemonade | 2 carbohydrates |
| ½ cup (3½ oz) | fruit cobbler | 3 carbohydrates, 1 fat |
| 1 bar (3 oz) | fruit juice bars, frozen, 100% juice | 1 carbohydrate |
| 1 roll (¾ oz) | fruit snacks, chewy (puréed fruit concentrate) | 1 carbohydrate |

| | | |
|---|---|---|
| 1½ tbsp | fruit spread, 100% fruit | 1 carbohydrate |
| ½ cup | gelatin, regular | 1 carbohydrate |
| 3 | gingersnaps | 1 carbohydrate |
| 1 bar (1 oz) | granola or snack bar (regular and low-fat) | 1½ carbohydrates |
| ½ cup | gravy | ½ carbohydrate, ½ fat |
| 1 tbsp | honey | 1 carbohydrate |
| 1 envelope | hot chocolate | 1 carbohydrate |
| ½ cup | ice cream | 1 carbohydrate, 2 fats |
| ½ cup | ice cream, fat-free | 1 carbohydrate |
| ½ cup | ice cream, light or no sugar added | 1 carbohydrate, 1 fat |
| ½ cup | ice cream, regular | 1 carbohydrate, 2 fats |
| 1 tbsp | jam or jelly, regular | 1 carbohdrate |
| 1 (4 oz) | muffin, low-fat | 4 carbohydrates, ½ fat |
| 1 (4 oz) | muffin, regular | 4 carbohydrates, 2½ fats |
| ⅙ pie | pie, fruit, 2 crusts (8 inches across) | 3 carbohydrates, 2 fats |
| ⅛ pie | pie, pumpkin or custard (8 inches across) | 1½ carbohydrates, 1½ fats |
| ½ cup | pudding, regular (made with reduced fat milk) | 2 carbohydrates |
| ½ cup | pudding, sugar-free (made with fat-free milk) | 1 carbohydrate |
| ¼ cup | salad dressing, fat-free | 1 carbohydrate |
| 1 (4 oz) | scone | 4 carbohydrates, 3 fats |
| ½ cup | sherbet, sorbet | 2 carbohydrates |
| 1 can (12 oz) | soft drink, regular | 2½ carbohydrates |
| 1 cup (8 oz) | sports drinks | 1 carbohydrate |
| 1 tbsp | sugar | 1 carbohydrate |
| 1½ tbsp | sweetener, mixture artificial sweetener and sugar | 1 carbohydrate |
| 1 (2½ oz) | sweet roll or Danish | 2½ carbohydrates, 2 fats |
| 2 tbsp | sweet and sour sauce | 1 carbohydrate |
| 2 tbsp | syrup, chocolate | 2 carbohydrates |
| 2 tbsp | syrup, light | 1 carbohydrate |
| 1 tbsp | syrup, regular | 1 carbohydrate |
| 5 | vanilla wafers | 1 carbohydrate, 1 fat |
| ½ cup | yogurt, frozen, fat-free | 1 carbohydrate |
| ½ cup | yogurt, frozen, regular | 1 carbohydrate, 0–1 fat |
| ½ cup | yogurt, frozen, Greek, low-fat or fat-free | 1½ carbohydrates |

## Protein List

Lean Protein List

(0 g carbohydrate, 7 g protein, 2 g fat, and 45 kcal)
1 lean protein choice equals:

***Beef***

1 oz    Select or Choice grades of beef trimmed of fat, such as round, sirloin, and flank steak; tenderloin; roast (chuck, rump); steak (T-bone, porterhouse, cubed), ground (90% or higher lean)

***Pork***

1 oz    lean pork, such as ham, Canadian bacon; tenderloin, rib or loin chop or roast

***Lamb***

1 oz    roast, chop, leg

***Veal***

1 oz    lean chop, roast, cutlet without breading

***Poultry***

1 oz    no skin: chicken, turkey, Cornish hen, duck, goose (drained of fat); lean ground turkey or chicken

***Fish***

1 oz    fresh or frozen cod, catfish, orange roughy, tilapia, flounder, haddock, halibut, or trout; tuna, fresh or canned in water or oil (drained); salmon, herring

***Shellfish***

1 oz    clams, crab, lobster, scallops, shrimp, imitation shellfish

***Game***

1 oz    venison, buffalo, ostrich, rabbit

### Cheese with 3 g or less fat per oz

| | |
|---|---|
| ¼ cup | cottage cheese (all types); ricotta (fat-free or light) |
| 1 oz | cheese, 3 g fat or less |

### Other

| | |
|---|---|
| ½ oz | beef jerky |
| 1 oz | processed sandwich meats with 3 g or less fat per oz, such as deli thin shaved meats, chipped beef, turkey ham, turkey pastrami |
| 2 | egg whites |

| | |
|---|---|
| ¼ cup | egg substitute, plain |
| 1 oz | hot dog with 3 g or less fat per oz |
| 1 oz | organ meats, such as, liver, kidney, heart (may be high in cholesterol) |
| 6 medium | oysters |
| 2 small | sardines |
| 1 oz | sausage with 3 g or less fat per oz |

***Counts as 1 lean protein and 1 starch exchange:***

| | |
|---|---|
| ½ cup | dried beans, peas, lentils (cooked) |

## Medium-Fat Protein List

(0 g carbohydrate, 7 g protein, 5 g fat, and 75 kcal)
1 medium-fat protein choice equals:

### Beef

| | |
|---|---|
| 1 oz | most beef products (ground beef of 85% or lower lean, meatloaf, corned beef, short ribs, tongue, prime grades of meat trimmed of fat, such as prime rib) |

### Pork

| | |
|---|---|
| 1 oz | cutlet, ground, shoulder roast |

### Lamb

| | |
|---|---|
| 1 oz | rib roast, ground |

### Poultry

| | |
|---|---|
| 1 oz | with skin: chicken, turkey, pheasant, dove, duck, goose; fried chicken |

### Fish

| | |
|---|---|
| 1 oz | any fried fish product |

### Cheese (with 4–7 g or less fat per oz)

| | |
|---|---|
| 1 oz | feta |
| 1 oz | mozzarella |
| 1 oz | processed cheese spread |
| 1 oz | cheeses with 4–7 g of fat per oz |
| ¼ cup (2 oz) | ricotta (regular or part skim) |

### Other

| | |
|---|---|
| 1 | egg |
| 1 oz | sausage with 4–7 g or less fat per oz |

## High-Fat Protein List

(0 g carbohydrate, 7 g protein, 8 g fat, and 100 kcal)
1 high-fat protein choice equals:

### Pork

| | |
|---|---|
| 1 oz | spareribs, pork sausage |

### Cheese

| | |
|---|---|
| 1 oz | all regular cheeses, such as American, Cheddar, Monterey Jack, Swiss, Parmesan, queso, goat, brie |

### Other

| | |
|---|---|
| 1 oz | processed sandwich meats with 8 g or less fat per oz, such as bologna, pastrami, salami |

| | |
|---|---|
| 1 oz | sausage, such as bratwurst, chorizo, Italian, knockwurst, Polish, smoked, summer |
| 1 | hot dog (turkey or chicken) (10 per pound) |
| 2 slices | pork bacon (1½ oz total before cooking) |
| 3 slices | turkey bacon (1½ oz total before cooking) |

***Counts as 1 high-fat meat plus 1 fat:***

| | |
|---|---|
| 1 | hot dog (beef, pork, or combination) (10 per pound) |

## Plant-Based Proteins

| | | |
|---|---|---|
| 2 strips | bacon, soy | 1 lean protein |
| ⅓ cup | baked beans | 1 lean protein, 1 starch |
| ½ cup | beans (black, garbanzo, kidney, lima, pinto, white; refried) | 1 lean protein, 1 starch |
| 1 oz | crumbles, no beef or sausage | 1 lean protein |
| 1 oz | deli slices, meatless | 1 lean protein |
| ½ cup | edamame, shelled | 1 lean protein, ½ carbohydrate |
| 3 patties | (2 inches across) falafel | 1 high-fat protein, 1 carbohydrate |
| ⅓ cup | hummus | 1 medium-fat protein, 1 carbohydrate |
| ½ cup | lentils and peas (black-eyed and split) | 1 lean protein, 1 starch |
| 2 oz | mycoprotein (no-chicken tenders and crumbles) | 1 lean protein, ½ carbohydrate |

| | | |
|---|---|---|
| 1 tbsp | nut spreads, such as peanut butter or almond butter | 1 high-fat protein |
| 1 (1½ oz) | sausage type patty, meatless | 1 medium-fat protein |
| 3 oz | soy-based burger, meatless | 2 lean proteins, ½ carbohydrate |
| 2 (1½ oz) | soy-based nuggets, no chicken | 2 medium-fat proteins, ½ carbohydrate |
| 1 (1½ oz) | soy-based hot dog, meatless | 1 lean protein |
| ¾ oz | soy nuts, unsalted | 1 medium-fat protein, ½ carbohydrate |
| ¼ cup | tempeh, plain | 1 medium-fat protein |
| ½ cup | tofu, regular | 1 medium-fat protein |
| ½ cup | tofu, light | 1 lean protein |
| 1 (2½ oz) | vegetable and starch based patty | 1 lean protein, ½ carbohydrate |

## Fats List

### Monounsaturated Fats List

(5 g fat and 45 kcal)
1 choice equals:

| | | | | |
|---|---|---|---|---|
| 1 cup | almond milk, unsweetened | 3 nuts | macadamia |
| 2 tbsp (1 oz) | avocado (medium) | 6 nuts | mixed (50% peanuts) |
| 1 tsp | oil (canola, olive, peanut) | 10 nuts | peanuts |
| | olives: | 4 halves | pecans |
| 8 | ripe, black (large) | 16 nuts | pistachios |
| 10 | green, stuffed (large) | ½ tbsp | peanut butter, almond butter, cashew butter |
| 2 tsp | plant stanol ester spread, regular | | (no *trans* fat nut butter products) |
| 1 tbsp | plant stanol ester spread, light | 1 tbsp | sesame seeds |
| 6 nuts | almonds, cashews | 2 tsp | tahini or sesame paste |
| 5 nuts | hazelnuts | | |

### Polyunsaturated Fats List

(5 g fat and 45 kcal)
1 choice equals:

| | | | | |
|---|---|---|---|---|
| | margarine: | | salad dressing: |
| 1 tsp | stick, tub, or squeeze (no *trans* fat) | 1 tbsp | regular |
| 1 tbsp | lower fat (30 to 50% vegetable oil, no *trans* fat) | 2 tbsp | reduced fat |
| | mayonnaise: | | mayonnaise-like salad dressing: |
| 1 tsp | regular | 2 tsp | regular |
| 1 tbsp | reduced fat | 1 tbsp | reduced fat |
| | nuts: | 1 tbsp | seeds: pumpkin, sunflower |
| 4 halves | English walnuts | 2 tsp | sesame or tahini paste |
| 1 tbsp | pine nuts | | |
| 1 tsp | oil (corn, safflower, soybean, sunflower, flaxseed, grapeseed) | | |

### Saturated Fats List

(5 g fat and 45 kcal)
1 choice equals:

| | | | | |
|---|---|---|---|---|
| 1 slice | bacon, cooked (pork or turkey) | | cream cheese: |
| 1 tsp | bacon, grease | | |
| 2 tbsp (½ oz) | chitterlings, boiled | 1 tbsp (½ oz) | regular |
| | coconut: | 1½ tbsp (¾ oz) | reduced fat |
| 2 tbsp | shredded, sweet | 1 tsp | shortening, lard, coconut oil, palm oil, palm kernel oil |
| 1½ tbsp | milk: canned, thick, regular | | |
| ⅓ cup | milk: canned, thick, light | ¼ oz | salt pork |
| 1 cup | milk: beverage, thin, unsweetened | | |

|        | butter:        |         | cream:         |
|--------|----------------|---------|----------------|
| 1 tsp  | stick          | 2 tbsp  | half and half  |
| 2 tsp  | whipped        | 1 tbsp  | heavy cream    |
| 1 tbsp | reduced fat    | 1½ tbsp | light cream    |
|        | butter blends, with oil: | 2 tbsp | whipped cream |
| ½ tbsp | regular        |         | sour cream:    |
| 1 tbsp | light or reduced fat | 2 tbsp | regular  |
|        |                | 3 tbsp  | reduced fat    |

## Snacks/Condiments/Sauces

Some foods and drinks contains less than 20 kcal or less than 5 g of carbohydrate per serving. Small portions of nonstarchy vegetables, fruits, starch, and fats (e.g., nuts) can be such snacks.

### Drinks

| | |
|---|---|
| bouillon, broth, consommé | diet soft drinks, sugar-free |
| bouillon or broth, low-sodium | drink mixes, sugar-free |
| carbonated or mineral water | tea, unsweetened or sweetened with sugar substitute |
| club soda | |
| coffee, unsweetened or sweetened with sugar substitute | tonic water, sugar-free |
| 1 tbsp   cocoa powder, unsweetened | water, plain or sugar-free flavored |

### Condiments/Sauces

| 3 tbsp | catsup                        | 1 carboydrate        |
|--------|-------------------------------|----------------------|
| 1 oz   | curry sauce                   | 1 carbohydrate, 1 fat |
| 3 tbsp | salad dressing, creamy, fat-free | 1 carbohydrate    |
| 3 tbsp | barbecue sauce; chili sauce, sweet and sour sauce | 1 carbohdrate |

### Seasonings

flavoring extracts
garlic
herbs, fresh or dried
spices
nonstick cooking spray

### Combination Foods List

| 1 cup (8 oz) | beef stew, other meat stews and vegetables | 1 carbohydrate, 1 medium-fat protein, 0–3 fats |
|---|---|---|
| 1 cup (8 oz) | tuna noodle casserole, lasagna, spaghetti with meatballs, chili with beans, macaroni and cheese | 2 carbohydrates, 2 medium-fat proteins |
| ¼ of 12-inch (4½–5 oz) | pizza, cheese or vegetarian, thin crust | 2 carbohydrates, 2 medium-fat proteins |
| ¼ of 12-inch (5 oz) | pizza, meat topping, thin crust | 2 carbohydrates, 2 medium-fat proteins, 1½ fats |
| ⅙ of 12-inch (4 oz) | cheese/vegetarian or meat, rising crust | 2½ carbohydrates, 2 medium-fat proteins |
| 1 (4½ oz) | pocket sandwich | 3 carbohydrates, 1 lean protein, 1–2 fats |
| 1 (5 oz) | burrito, beef and bean | 3 carbohydrates, 1 lean protein, 2 fats |
| 1 (7 oz) | pot pie | 3 carbohydrates, 1 medium-fat protein, 3 fats |
| 1 (about 7 oz) | chicken breast, breaded and fried | 1 carbohydrate, 6 medium-fat proteins |
| 1 (about 7 oz) | chicken breast, no skin or breading | 4 lean proteins |
| 1 (about 2½ oz) | chicken drumstick, breaded and fried | ½ carbohydrate, 2 medium-fat proteins |
| 1 (about 2½ oz) | chicken drumstick, no skin or breading | 1 lean protein, ½ fat |
| 6 (about 3½ oz) | chicken nuggets or tenders | 1 carbohydrate, 2 medium-fat proteins, 1 fat |

| | | |
|---|---|---|
| 1 (about 7½ oz) | chicken sandwich, grilled | 3 carbohydrates, 4 lean proteins |
| 1 (about 6 oz) | chicken sandwich, fried | 3 carbohydrates, 2 lean proteins, 3½ fats |
| 1 (about 2 oz) | chicken wing, fried, breaded | ½ carbohydrate, 2 medium-fat proteins |
| 1 (5 oz) | fish sandwich/tartar sauce | 2½ carbohydrates, 2 medium-fat proteins, 1½ fats |
| 1 regular (3½ oz) | hamburger | 2 carbohydrates, 1 medium-fat proteins, 1 fat |
| 1 large (8½ oz) | hamburger | 3 carbohydrates, 4 medium-fat proteins, 2½ fats |
| 1 | hot dog with bun | 1 carbohydrate, 1 high-fat protein, 2 fats |
| ⅛ 14-inch | pizza, cheese, meat, and vegetables, regular crust | 2½ carbohydrates, 2 high-fat proteins |
| ⅛ 14-inch | pizza, cheese, pepperoni, or sausage, thin crust | 1½ carbohydrates, 1 high-fat protein, 1 fat |
| ⅛ 14-inch | pizza, cheese, pepperoni, or sausage, thick or regular crust | 2½ carbohydrates, 1 high-fat protein, 1 fat |
| 1 sub (6 inches) | submarine sandwich, regular | 3 carbohydrates, 2 lean proteins, 1 fat |
| 1 sub (6 inches) | submarine sandwich, under 6 g fat | 3 carbohydrates, 2 lean proteins |
| 1 small (4–5 oz) | wrap, grilled chicken, cheese, vegetables, condiment | 2 carbohydrates, 2 lean proteins, 1½ fats |

### *Mexican and Asian*

| | | |
|---|---|---|
| 1 | egg roll, meat | 1½ carbohydrates, 1 lean protein, 1½ fats |
| 1 | fortune cookie | ½ carbohydrate |
| 1 cup | meat with vegetables in sauce | 1 carbohydrate, 2 lean proteins, 1 fat |
| 1 cup | meat with sweet sauce | 3½ carbohydrates, 3 medium fat proteins, 3 fats |
| 1 cup | noodles with vegetables in sauce, such as chow mein or lo mein | 2 carbohydrates, 2 fats |
| 1 cup | rice, fried, no meat | 2½ carbohydrates, 2 fats |
| 1 cup | soup, hot and sour | ½ carbohydrate, ½ fat |
| 1 small (6 oz) | burrito, beans and cheese | 3½ carbohydrates, 1 medium-fat protein, 1 fat |
| 8 (1 small order) | nachos with cheese | 2½ carbohydrates, 1 high-fat protein, 2 fats |
| 1 small (about 3 oz) | taco, crisp, meat and cheese | 1 carbohydrate, 1 medium-fat protein, ½ fat |
| 1 salad (1 lb) | taco salad, chicken, tortilla bowl | 3½ carbohydrates, 4 medium-fat proteins, 3 fats |
| 1 small (about 5 oz) | tostada with beans and cheese | 2 carbohydrates, 1 high-fat protein |

### *Soups*

| | | |
|---|---|---|
| 1 cup (8 oz) | bean, split pea, or lentil | 1½ carbohydrates, 1 lean protein |
| 1 cup (8 oz) | cream (made with water) | 1 carbohydrate, 1 fat |
| 1 cup (8 oz) | chowder (made with milk) | 1 caroboydrate, 1 lean protein, 1½ fats |
| 1 cup (8 oz) | tomato (made with water), borscht, or rice soup | 1 carbohydrate |
| 1 cup (8 oz) | vegetable beef, chicken noodle, or other broth-type | 1 carbohydrate, 1 lean protein |
| 1 cup (8 oz) | noodle soup, Ramen | 2 carbohydrates, 2 fats |
| 1 cup (8 oz) | miso soup | ½ carbohydrate, 1 lean protein |
| ½ cup | coleslaw | 1 carbohydrate, 1½ fats |
| ½ cup | pasta, macaroni salad | 2 carbohydrates, 3 fats |
| ½ cup | potato salad | 1½–2 carbohydrates, 1–2 fats |
| ½ cup (3½ oz) | chicken or tuna salad | ½ carbohydrate, 2 lean proteins, 1 fat |
| 1 small serving (3½ oz) | french fries | 2½ carbohydrates, 2 fats |
| 1 medium serving (5 oz) | french fries | 3½ carbohydrates, 3 fats |
| 1 large serving (6 oz) | french fries | 4½ carbohydrates, 4 fats |
| 8–9 rings (4 oz) | onion rings | 3½ carbohydrates, 4 fats |

# Appendix F

## FATTY ACIDS, INCLUDING OMEGA-3 FATTY ACIDS, IN FOODS

| Chain Length, Number, and Site of Double Bonds for Common Fatty Acids | |
| --- | --- |
| **Common Name of Fatty Acid** | **Number of Carbon Atoms and Number and Site of Double Bond(s), Counting from Methyl End (–CH$_3$) If Appropriate** |
| **Saturated Fatty Acids (No Double Bonds)** | |
| Formic | 1 |
| Acetic | 2 |
| Propionic | 3 |
| Butyric | 4 |
| Valeric | 5 |
| Caproic | 6 |
| Caprylic | 8 |
| Capric | 10 |
| Lauric | 12 |
| Myristic | 14 |
| Palmitic | 16 |
| Stearic | 18 |
| **Unsaturated Fatty Acids** | |
| Oleic | 18:1 (9-10) ω-9 |
| Linoleic | 18:2 (6-7, 9-10) ω-6 |
| Alpha-linolenic | 18:3 (3-4, 6-7, 9-10) ω-3 |
| Arachidonic | 20:4 (6-7, 9-10, 12-13, 15-16) ω-6 |
| Eicosapentaenoic | 20:5 (3-4, 6-7, 9-10, 12-13, 15-16) ω-3 |
| Docosahexaenoic | 22:6 (3-4, 6-7, 9-10, 12-13, 15-16, 18-19) ω-3 |

**Fatty Acid Composition of Selected Foods***

| Food Item | Saturated | | | | | Unsaturated | | | | |
|---|---|---|---|---|---|---|---|---|---|---|
| | <C12:0 | C12:0 | 14:0 | C16:0 | C18:0 | C18:1 ω-9 | C18:2 ω-6 | C18:3 ω-3 | C20:5 ω-3 | C22:6 ω-3 |
| | | Lauric Acid | Myristic Acid | Palmitic Acid | Stearic Acid | Oleic Acid | Linoleic Acid | Alpha-Linolenic Acid | EPA‡ | DHA‡ |
| **Fats and Oils** | | | | | | | | | | |
| Beef tallow | 0.0 | 0.90 | 3.70 | 24.9 | 18.9 | 36.0 | 3.1 | 0.60 | 0.00 | 0.00 |
| Butter | 8.9 | 2.6 | 7.4 | 21.7 | 10.0 | 20.0 | 2.2 | 0.3 | 0.0 | 0.0 |
| Canola oil | 0.0 | 0.0 | 0.0 | 4.3 | 2.1 | 61.7 | 18.6 | 9.1 | 0.0 | 0.0 |
| Cocoa butter | 0.0 | 0.0 | 0.10 | 25.4 | 33.2 | 32.6 | 2.8 | 0.10 | 0.0 | 0.0 |
| Coconut oil | 14.1 | 44.6 | 16.8 | 8.2 | 2.8 | 5.8 | 1.8 | 0.0 | 0.0 | 0.0 |
| Cod liver oil | — | 0.0 | 3.6 | 10.6 | 2.8 | 20.7 | 0.9 | 0.9 | 6.9 | 11.0 |
| Corn oil | 0.0 | 0.0 | 0.02 | 10.6 | 1.8 | 27.3 | 53.2 | 1.2 | 0.0 | 0.0 |
| Cottonseed oil | 0.0 | 0.0 | 0.80 | 22.7 | 2.3 | 17.0 | 51.5 | 0.20 | 0.0 | 0.0 |
| Lard | 0.1 | 0.20 | 1.30 | 23.8 | 13.5 | 41.2 | 10.2 | 1.0 | 0.0 | 0.0 |
| Margarine, stick | 0.0 | 0.0 | 0.05 | 8.4 | 6.2 | 38.7 | 22.3 | 2.0 | 0.0 | 0.0 |
| Margarine, tub | 0.04 | 0.4 | 0.2 | 7.1 | 5.8 | 35.5 | 21.0 | 4.7 | 0.0 | 0.0 |
| Menhaden oil | — | — | 8.0 | 15.1 | 3.8 | 14.5 | 2.2 | 1.5 | 13.2 | 8.6 |
| Olive oil | 0.0 | 0.0 | 0.0 | 11.3 | 2.0 | 71.3 | 9.8 | 0.8 | 0.0 | 0.0 |
| Palm kernel oil | 7.2 | 47.00 | 16.40 | 8.1 | 2.8 | 11.4 | 1.6 | 0.0 | 0.0 | 0.0 |
| Palm oil | 0.0 | 0.10 | 1.00 | 43.5 | 4.3 | 36.6 | 9.1 | 0.20 | 0.0 | 0.0 |
| Peanut oil | 0.0 | 0.0 | 0.10 | 9.5 | 2.2 | 44.8 | 32.0 | 0.0 | 0.0 | 0.0 |
| Safflower oil, high oleic | — | 0.0 | 0.0 | 4.9 | 1.9 | 74.7 | 12.7 | 0.1 | 0.0 | 0.0 |
| Shortenings | 0.0 | 0.0 | 0.1 | 12.5 | 11.4 | 41.0 | 26.2 | 1.9 | 0.0 | 0.0 |
| Soybean oil | 0.0 | 0.0 | 0.04 | 10.5 | 4.4 | 22.5 | 50.4 | 6.8 | 0.0 | 0.0 |
| **Meat, Fish, and Poultry** | | | | | | | | | | |
| Beef, lean only, cooked | 0.01 | 0.01 | 0.2 | 1.8 | 1.3 | 3.9 | 0.3 | 0.02 | 0.0 | 0.0 |
| Chicken, white meat, cooked | 0.0 | 0.01 | 0.03 | 0.7 | 0.3 | 1.0 | 0.6 | 0.03 | 0.01 | 0.02 |
| Salmon, coho, wild, cooked | 0.0 | 0.0 | 0.15 | 0.7 | 0.2 | 0.9 | 0.06 | 0.06 | 0.40 | 0.7 |
| Tuna, light, canned in water | 0.0 | 0.0 | 0.02 | 0.2 | 0.06 | 0.09 | 0.0 | 0.0 | 0.05 | 0.22 |
| **Nuts and Seeds** | | | | | | | | | | |
| Flaxseeds | 0.0 | 0.0 | 0.0 | 2.2 | 1.3 | 7.4 | 5.9 | 22.8 | 0.0 | 0.0 |
| Walnuts, English | 0.0 | 0.0 | 0.0 | 4.4 | 1.7 | 8.8 | 38 | 9 | 0.0 | 0.0 |

From USDA Nutrient Database for Standard Reference, Release 23.

*Only major fatty acids are presented.

†All values represent grams per 100 g edible portion.

‡EPA eicosapentaenoic acid ⎫
DHA docosahexaenoic acid ⎭ fish oil fatty acids

# Appendix G

## METROPOLITAN LIFE INSURANCE COMPANY HEIGHT-WEIGHT TABLE AND DETERMINATION OF FRAME SIZE

**1983 Metropolitan Life Insurance Company Height-Weight Table\*†**

| Women | | | | | Men | | | | |
|---|---|---|---|---|---|---|---|---|---|
| Height | | Frame | | | Height | | Frame | | |
| Feet | Inches | Small | Medium | Large | Feet | Inches | Small | Medium | Large |
| 4 | 10 | 102–111 | 109–121 | 118–131 | 5 | 2 | 128–134 | 131–141 | 138–150 |
| 4 | 11 | 103–113 | 111–123 | 120–134 | 5 | 3 | 130–136 | 133–143 | 140–153 |
| 5 | 0 | 104–115 | 113–126 | 122–137 | 5 | 4 | 132–138 | 135–145 | 142–156 |
| 5 | 1 | 106–118 | 115–129 | 125–140 | 5 | 5 | 134–140 | 137–148 | 144–160 |
| 5 | 2 | 108–121 | 118–132 | 128–143 | 5 | 6 | 136–142 | 139–151 | 146–164 |
| 5 | 3 | 111–124 | 121–135 | 131–147 | 5 | 7 | 138–145 | 142–154 | 149–168 |
| 5 | 4 | 114–127 | 124–138 | 134–151 | 5 | 8 | 140–148 | 145–157 | 152–172 |
| 5 | 5 | 117–130 | 127–141 | 137–155 | 5 | 9 | 142–151 | 148–160 | 155–176 |
| 5 | 6 | 120–133 | 130–144 | 140–159 | 5 | 10 | 144–154 | 151–163 | 158–180 |
| 5 | 7 | 123–136 | 133–147 | 143–163 | 5 | 11 | 146–157 | 154–166 | 161–184 |
| 5 | 8 | 126–139 | 136–150 | 146–167 | 6 | 0 | 149–160 | 157–170 | 164–188 |
| 5 | 9 | 129–142 | 139–153 | 149–170 | 6 | 1 | 152–164 | 160–174 | 168–192 |
| 5 | 10 | 132–145 | 142–156 | 152–173 | 6 | 2 | 155–168 | 164–178 | 172–197 |
| 5 | 11 | 135–148 | 145–159 | 155–176 | 6 | 3 | 158–172 | 167–182 | 176–202 |
| 6 | 0 | 138–151 | 148–162 | 158–179 | 6 | 4 | 162–176 | 171–187 | 181–207 |

\*Based on a weight-height mortality study conducted by the Society of Actuaries and the Association of Life Insurance Medical Directors of America, Metropolitan Life Insurance Medical Directors of America, Metropolitan Life Insurance Company, 1983 (latest revision).

†Weights at ages 25 to 59 based on lowest mortality. Height includes 1 in. heel. Weight for women includes 3 lb for indoor clothing. Weight for men includes 5 lb for indoor clothing.

Source: Permission granted courtesy of Metropolitan Life Insurance Company, *Statistical Bulletin*.

## Using the Metropolitan Life Insurance Table to Estimate Healthy Weight

The Metropolitan Life Insurance table was a common method for estimating healthy weight (body mass index is more commonly used now). The table lists for any height the weight that is associated with a maximum life span. The table does not tell the healthiest weight for a living person; it simply lists the weight associated with longevity.

Criticisms of this table stem from the inclusion of some people and the exclusion of others. For example, only policyholders of life insurance are included. In addition, smokers are included, but anyone over the age of 60 is excluded. Weight is measured only at the time of purchase of insurance, and there is no follow-up. All these factors contribute to the fact that this table should be used only as a rough screening tool; not meeting the exact recommendations should not be cause for alarm.

To diagnose overweight or obesity using the table, calculate the percentage of the Metropolitan Life Insurance table weight. Use the midpoint of a weight range for a specific height.

$$\frac{(\text{Current wt.} - \text{wt. from table})}{\text{Weight from table}} \times 100$$

Example:

$$\frac{140 - 120}{120} \times 100 = 17\% \text{ over standard}$$

Overweight can be defined as weighing at least 10% more than the weight listed on the table. Obesity weighs in at 20% more than that listed on the table. Moreover, this measure of obesity comes in degrees. Whereas mild obesity carries little health risk, severe obesity raises overall health risk 12-fold.

### Degrees of Obesity

| % over Healthy Body Weight | Form of Obesity |
|---|---|
| 20–40% | Mild |
| 41–99% | Moderate |
| 100%+ | Severe |

## Determining Frame Size

### Method 1

Height is recorded without shoes. Wrist circumference is measured just beyond the bony (styloid) process at the wrist joint on the right arm, using a tape measure. The following formula is used:

$$r = \frac{\text{Height (cm)}}{\text{Wrist circumference (cm)}}$$

Frame size can be determined as follows:[†]

| *Males* | *Females* |
|---|---|
| $r > 10.4$ small | $r > 11$ small |
| $r = 9.6–10.4$ medium | $r = 10.1–11$ medium |
| $r < 9.6$ large | $r < 10.1$ large |

[†]From Grant JP. *Handbook of total parenteral nutrition.* Philadelphia: WB Saunders; 1980.

### Method 2

The patient's right arm is extended forward, perpendicular to the body, with the arm bent so that the angle at the elbow forms 90 degrees, with the fingers pointing up and the palm turned away from the body. The greatest breadth across the elbow joint is measured with a sliding caliper along the axis of the upper arm, on the 2 prominent bones on either side of the elbow. This measurement is recorded as the elbow breadth. The following table gives elbow breadth measurements for medium-framed men and women of various heights. Measurements lower than those listed indicate a small frame size; higher measurements indicate a large frame size.

| Men | | Women | |
|---|---|---|---|
| **Height in 1″ Heels** | **Elbow Breadth** | **Height in 1″ Heels** | **Elbow Breadth** |
| 5′2″–5′3″ | 2½–2⅞″ | 4′10″–4′11″ | 2¼–2½″ |
| 5′4″–5′7″ | 2⅝–2⅞″ | 5′0″–5′3″ | 2¼–2½″ |
| 5′8″–5′11″ | 2¾–3″ | 5′4″–5′7″ | 2⅜–2⅝″ |
| 6′0″–6′3″ | 2¾–3¼″ | 5′8″–5′11″ | 2⅜–2⅝″ |
| 6′4″ and over | 2⅞–3¾″ | 6′0″ and over | 2½–2¾″ |

# Appendix H

## ENGLISH-METRIC CONVERSIONS AND NUTRITION CALCULATIONS

## English-Metric Conversions

### Length

| English (USA) | Metric |
| --- | --- |
| inch (in.) | = 2.54 cm, 25.4 mm |
| foot (ft) | = 0.30 m, 30.48 cm |
| yard (yd) | = 0.91 m, 91.4 cm |
| mile (statute) (5280 ft) | = 1.61 km, 1609 m |
| mile (nautical) | |
| (6077 ft, 1.15 statute mi) | = 1.85 km, 1850 m |

| Metric | English (USA) |
| --- | --- |
| millimeter (mm) | = 0.039 in. (thickness of a dime) |
| centimeter (cm) | = 0.39 in. |
| meter (m) | = 3.28 ft, 39.37 in. |
| kilometer (km) | = 0.62 mi, 1091 yd, 3273 ft |

### Weight

| English (USA) | Metric |
| --- | --- |
| grain | = 64.80 mg |
| ounce (oz) | = 28.35 g |
| pound (lb) | = 453.60 g, 0.45 kg |
| ton (short—2000 lb) | = 0.91 metric ton (907 kg) |

| Metric | English (USA) |
| --- | --- |
| milligram (mg) | = 0.002 grain (0.000035 oz) |
| gram (g) | = 0.04 oz ($\frac{1}{28}$ oz) |
| kilogram (kg) | = 35.27 oz, 2.20 lb |
| metric ton (1000 kg) | = 1.10 tons |

### Volume

| English (USA) | Metric |
| --- | --- |
| cubic inch | = 16.39 cc |
| cubic foot | = 0.03 m$^3$ |
| cubic yard | = 0.765 m$^3$ |
| teaspoon (tsp) | = 5 ml |
| tablespoon (tbsp) | = 15 ml |
| fluid ounce | = 0.03 liter (30 ml)* |
| cup (c) | = 237 ml |
| pint (pt) | = 0.47 liter |
| quart (qt) | = 0.95 liter |
| gallon (gal) | = 3.79 liters |

| Metric | English (USA) |
| --- | --- |
| milliliter (ml) | = 0.03 oz |
| liter (L) | = 2.12 pt |
| liter | = 1.06 qt |
| liter | = 0.27 gal |

## Metric and Other Common Units

| Unit/Abbreviation | Other Equivalent Measure |
| --- | --- |
| milligram/mg | $\frac{1}{1000}$ of a gram |
| microgram/µg | $\frac{1}{1,000,000}$ of a gram |
| deciliter/dl | $\frac{1}{10}$ of a liter (about ½ cup) |
| milliliter/ml | $\frac{1}{1000}$ of a liter (5 ml is about 1 tsp) |
| International Unit/IU | Crude measure of vitamin activity generally based on growth rate seen in animals |

1 liter ÷ 1000 = 1 milliliter or 1 cubic centimeter ($10^{-3}$ liter).

1 liter ÷ 1,000,000 = 1 microliter ($10^{-6}$ liter).

*Note: 1 ml = 1 cc.

## Fahrenheit-Celsius Conversion Scale

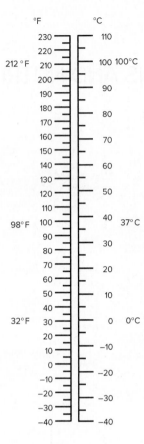

To convert temperature scales:
Fahrenheit to Celsius: $°C = (°F - 32) \times \frac{5}{9}$
Celsius to Fahrenheit: $°F = \frac{9}{5} (°C) + 32$

## Household Units

| | |
|---|---|
| 3 teaspoons | = 1 tablespoon |
| 4 tablespoons | = ¼ cup |
| 5⅓ tablespoons | = ⅓ cup |
| 8 tablespoons | = ½ cup |
| 10⅔ tablespoons | = ⅔ cup |
| 16 tablespoons | = 1 cup |
| 1 tablespoon | = ½ fluid ounce |
| 1 cup | = 8 fluid ounces |
| 1 cup | = ½ pint |
| 2 cups | = 1 pint |
| 4 cups | = 1 quart |
| 2 pints | = 1 quart |
| 4 quarts | = 1 gallon |

## Nutrition Calculations

Conversions are mathematical techniques for expressing the same quantity in different measurements. This section will walk you step by step through a few basic conversions that are important to understand when studying human nutrition.

**Example 1: Converting Pounds to Kilograms.** Begin with 2.2 pounds and 1 kilogram; they are equivalent. Each represents the same weight but is expressed in different units. The following is the conversion factor to change pounds to kilograms and kilograms to pounds.

$$\frac{2.2 \text{ lb}}{1 \text{ kg}} \text{ or } \frac{1 \text{ kg}}{2.2 \text{ lb}}$$

Because these factors equal 1, they can be multiplied by a number without changing the measurement value. This allows the units to be changed.

Convert the weight of 150 lb to kg:

Step 1: Choose the conversion factor in which the unit you are seeking is on top.

$$\frac{1 \text{ kg}}{2.2 \text{ lb}}$$

Step 2: Multiply 150 pounds by the factor.

$$150 \text{ lb} \times \frac{1 \text{ kg}}{2.2 \text{ lb}} = \frac{150 \text{ kg}}{2.2} = 59 \text{ kg}$$

**Example 2: Converting Cups to Milliliters.** Convert ½ cup to an approximate number of milliliters for use in a recipe:

Step 1: The conversion factor is

$$\frac{1 \text{ cup}}{240 \text{ ml}} \text{ or } \frac{240 \text{ ml}}{1 \text{ cup}}$$

Step 2: Multiply ½ cup by the conversion factor.

$$\frac{1}{2} \text{cup} \times \frac{240 \text{ ml}}{1 \text{ cup}} = 120 \text{ ml}$$

**Example 3: Calculating Percentage of Calories Supplied by Carbohydrate, Fat, Protein, and Alcohol.** Suppose that for 1 day in your diet you consumed 290 g of carbohydrate, 60 g of fat, 70 g of protein, and 15 g of alcohol. Calculate the energy intake for the day, as well as the percentage of carbohydrate, fat, protein, and alcohol in the day's diet.

Step 1: Calculate the total energy intake. Begin by multiplying the grams of carbohydrate, fat, protein, and alcohol by the number of kcal that each gram yields.

$$\text{Carbohydrate: } 290 \text{ g} \times \frac{4 \text{ kcal}}{\text{g}} = 1160 \text{ kcal}$$

$$\text{Fat: } 60 \text{ g} \times \frac{9 \text{ kcal}}{\text{g}} = 540 \text{ kcal}$$

$$\text{Protein: } 70 \text{ g} \times \frac{4 \text{ kcal}}{\text{g}} = 208 \text{ kcal}$$

$$\text{Alcohol: } 15 \text{ g} \times \frac{7 \text{ kcal}}{\text{g}} = 105 \text{ kcal}$$

Step 2: Add all the values together for total energy intake.

$$1160 + 540 + 280 + 105 = 2085 \text{ total kcal}$$

Step 3: Calculate the percentage of total carbohydrate by dividing the kcal from carbohydrate by the total energy intake.

$$\frac{1160 \text{ kcal from carbohydrate}}{2085 \text{ total kcal}} = 56\% \text{ carbohydrate}$$

Step 4: Calculate the percentage of total fat by dividing the grams of fat by the energy yield and total energy factor.

$$\frac{540 \text{ kcal from fat}}{2085 \text{ total kcal}} = 26\% \text{ fat}$$

Step 5: Calculate the percentage of total protein by dividing the grams of protein by the energy yield and total energy factor.

$$\frac{280 \text{ kcal from protein}}{2085 \text{ total kcal}} = 13\% \text{ protein}$$

Step 6: Calculate the percentage of total alcohol by dividing the grams of alcohol by the energy yield and total energy factor.

$$\frac{105 \text{ kcal from alcohol}}{2085 \text{ total kcal}} = 5\% \text{ alcohol}$$

**Example 4: Converting Amount of a Nutrient in a Quantity of Food to Another Quantity.** How many grams of saturated fat are contained in a 3 oz hamburger? A 5 oz hamburger contains 8.5 g of saturated fat.

Step 1: The conversion factor for grams of saturated fat is

$$\frac{8.5 \text{ g saturated fat}}{5 \text{ oz hamburger}}$$

Step 2: Multiply 3 oz hamburger by the conversion factor.

$$3 \text{ oz hamburger} \times \frac{8.5 \text{ g saturated fat}}{5 \text{ oz hamburger}} = \frac{3 \times 8.5 \text{ g}}{5} = \frac{25.5 \text{ g}}{5}$$

**Example 5: Converting Sodium to Salt.** The conversion factor to change milligrams of sodium to milligrams of salt, and milligrams of salt to milligrams of sodium, is

$$\frac{1000 \text{ mg sodium}}{2500 \text{ mg salt}} \text{ or } \frac{2500 \text{ mg salt}}{1000 \text{ mg sodium}}$$

A frozen pepperoni pizza contains 2200 mg of salt. How much sodium is in the pizza?

Step 1: Choose the conversion factor with the unit you are seeking on top.

$$\frac{1000 \text{ mg sodium}}{2500 \text{ mg salt}}$$

Step 2: Multiply 2200 mg of salt by the conversion factor.

$$2200 \text{ mg salt} \times \frac{1000 \text{ mg sodium}}{2500 \text{ mg salt}} = 880 \text{ mg sodium}$$

**Example 6: Converting Folate to Dietary Folate Equivalents.** To convert micrograms of synthetic folate in supplements and enriched foods to Dietary Folate Equivalents (micrograms DFEs), use this conversion factor:

$$\frac{1 \mu g \text{ synthetic folic acid}}{1.7 \mu g \text{ DFE}} \text{ or } \frac{1.7 \mu g \text{ DFE}}{1 \mu g \text{ synthetic folic acid}}$$

If a ready-to-eat breakfast cereal contains 200 μg of synthetic folic acid, how many μg of folate in the product are in DFE units?

Step 1: Choose the conversion factor with the unit you are seeking on top.

$$\frac{1.7\,\mu g\ DFE}{1\,\mu g\ synthetic\ folic\ acid}$$

Step 2: Multiply 200 μg of synthetic folic acid by the conversion factor.

$$200\,\mu g\ synthetic\ folic\ acid \times \frac{1.7\,\mu g\ DFE}{1\,\mu g\ synthetic\ folic\ acid}$$

For naturally occurring folate, assign each microgram of food folate a value of 1 microgram DFE:

$$\frac{1\,\mu g\ food\ folate}{1\,\mu g\ DFE}\ or\ \frac{1\,\mu g\ DFE}{1\,\mu g\ synthetic\ folic\ acid}$$

An orange has 50 μg of food folate. How many μg of folate in DFE does the orange contain?

Step 1: Choose the conversion factor with the unit you are seeking on top.

$$\frac{1\,\mu g\ DFE}{1\,\mu g\ food\ folate}$$

Step 2: Multiply 50 μg of food folate by the conversion factor.

$$50\,\mu g\ food\ folate \times \frac{1\,\mu g\ DFE}{1\,\mu g\ food\ folate} = 50\,\mu g\ DFE$$

**Example 7: Converting substances to milliequivalents from metric weight and vice versa.** Substances such as sodium, potassium, and calcium are sometimes reported in milliequivalents (mEq) instead of the metric weight (mg or g). A milliequivalent takes into account both the molecular or formula weight and the charge or valence of the substance.

To convert from mEq to mg, use the following formula:
mg = mEq × molecular weight/valence.

$$25\ mEq\ calcium = 500\ mg$$

The molecular weight of calcium is 40 and its valence is 2, so the formula is

$$25 \times 40/2 = 500\ mg$$

To convert from mg to mEq, use the following formula: mEq = mg/molecular wt × valence.

$$2000\ mg\ sodium = 87\ mEq$$

The molecular weight of sodium is 23 and its valence is 1, so the formula is

$$2000/23 \times 1 = 87\ mEq$$

# Appendix I

## CAFFEINE CONTENT OF BEVERAGES, FOODS, AND OVER-THE-COUNTER DRUGS

| Beverage or Food | Serving Size | Caffeine (mg)* |
|---|---|---|
| **Coffee Drinks** | | |
| Brewed, generic | 8 oz | 95–200 |
| Brewed, Starbucks® | 12 oz | 260 |
| Brewed, decaffeinated | 8 oz | 2–12 |
| Caffé latté | 12 oz | 75–233 |
| Decaf ground coffee | 1 tbsp for 6 oz | 1–5 |
| Dunkin Donuts® Iced Coffee, bottle | 12 oz | 327 |
| Espresso | 1 oz | 63–75 |
| Instant coffee | 8 oz | 69–135 |
| Keurig, k-cup, most varieties | 1 k-cup | 75–150 |
| Nespresso® capsule | 1 capsule | 55–65 |
| Starbucks® decaf | 16 oz | 25 |
| Starbucks® frappuccino blended coffee beverage, average | 16 oz | 95 |
| **Tea** | | |
| Brewed, black | 8 oz | 14–70 |
| Brewed, black decaf | 8 oz | <5 |
| Brewed, green | 8 oz | 24–45 |
| Brewed, herbal | 8 oz | 0 |
| KeVita Master Brew Kombucha | 15.2 oz | 36–68 |
| Starbucks® Tazo Chai Tea Latté | 16 oz | 95 |
| Bottled iced teas | 16 oz | 10–76 |
| **Soda**† | | |
| FDA limit | 12 oz | 72 |
| PepsiMax™ | 12 oz | 69 |
| Citrus-type (e.g., Mountain Dew®, Mello Yello®) | 12 oz | 42–55 |
| Cola-type | 12 oz | 34–57 |
| Pepper-type | 12 oz | 37–41 |
| Miscellaneous: Barq's Root Beer®, A & W® cream soda, Sunkist®, Big Red® | 12 oz | 22–42 |
| **Energy Drinks** | | |
| 5-Hour Energy™ | 2 oz | 200–207 |
| AMP Tall Boy® energy drink | 16 oz | 143 |

(continued)

| Beverage or Food | Serving Size | Caffeine (mg)* |
|---|---|---|
| Bai Antioxidants Infusion® | 8 oz | 35 |
| Bang Energy® | 16 oz | 300 |
| Crystal Light Energy® | 8 oz | 60 |
| Full Throttle®, Monster Energy®, Rockstar® | 16 oz | 160 |
| HiBall®, Venom Energy®, NOS Energy® | 16 oz | 160 |
| Red Bull® | 8.3 oz | 80 |
| SoBe® Energy Citrus | 20 oz | 80 |
| SoBe® No Fear | 16 oz | 174 |
| Starbucks Refresher® | 12 oz | 50 |
| **Shakes, Water, and Chocolate Drinks** | | |
| Ensure® Max Protein | 11 oz | 100 |
| Hershey's Chocolate Lowfat Milk | 12 oz | 2 |
| Kellogg's® Special K Vanilla Cappuccino Protein Shake | 10 oz | 65 |
| Odwalla® Chai Vanilla Protein Shake | 15.2 oz | 29–48 |
| Silk® Chocolate Soymilk or Almond Milk | 8 oz | <5 |
| Sparkling Ice® plus Caffeine | 8 oz | 70 |
| Water Joe™ | 23.7 oz | 85 |
| **Chocolate, Candies, and Snack Foods** | | |
| Awake Energy Chocolate | 1.5 oz | 100 |
| Bang!! Caffeinated Ice Cream | 4 oz | 125 |
| Clif Shot® Energy Gels | 1.2 oz | 50–100 |
| Coffee ice cream | ½ cup | 24–42 |
| Energy mints | 2 mints | 95–200 mg |
| Jelly Belly Extreme Sport Beans® | 1 oz (4 chews) | 20–40 |
| Hershey's® milk chocolate | 1.45 oz | 8 |
| Hershey's® special dark chocolate | 1.45 oz | 18–27 |
| Hot chocolate from mix | 8 oz | 3–13 |
| **Over-the-Counter Drugs** | | |
| Anacin® (maximum strength) | 2 tablets | 64 |
| Excedrin® (extra strength) | 2 tablets | 130 |
| NoDoz® (maximum strength) | 1 tablet | 200 |
| Vivarin® | 1 tablet | 200 |
| Zantrex Weight Loss | 2 tablets | 200–300 |

*Each of the following ingredients listed on food, supplement, or drug labels contributes to caffeine: caffeine; coffee or coffee beans; cocoa or cacao, Theobroma cacao; guarana or paullinia cupana; kola nuts or cola seeds, cola nitida; green tea, black tea, or Camellia Sinesis, Thea Sinesis, Camellia; yerba mate or Mate, Ilex paraguariensis.

†Caffeine levels can vary considerably from brand to brand. Many soft drinks contain no caffeine.

Source: USDA National Nutrient Database for Standard Reference 20; *Caffeine content of food & drugs*, Center for Science in the Public Interest; 2019 https://cspinet.org/eating-healthy/ingredients-of-concern/caffeine-chart; Caffeine content for coffee, tea, soda and more. 2014; www.mayoclinic.org/healthy-living/nutrition-and-healthyeating/in-depth/caffeine/art-20049372; Chou K-H, Bell LN. Caffeine content of prepackaged national-brand and private-label carbonated beverages. *J Food Sci*. 2007;72:C337. Dietary Supplement Ingredient Database, Agricultural Research Service, U.S. Department of Agriculture.

# Appendix J

## DIETARY REFERENCE INTAKES (DRIs)

**Estimated Average Requirements Set by the Food and Nutrition Board, Institute of Medicine, National Academies**

| Life-Stage Group | Calcium (g/d) | Carbohydrate (mg/d) | Protein (g/kg/d) | Vitamin A (µg/d) | Vitamin C (mg/d) | Vitamin D (mg/d) | Vitamin E (mg/d) | Thiamin (mg/d) | Riboflavin (mg/d) | Niacin (mg/d) |
|---|---|---|---|---|---|---|---|---|---|---|
| **Children** | | | | | | | | | | |
| 1–3 y | 500 | 100 | 0.88 | 210 | 13 | 10 | 5 | .4 | .4 | 5 |
| 4–8 y | 800 | 100 | 0.76 | 275 | 22 | 10 | 6 | .5 | .5 | 6 |
| **Males** | | | | | | | | | | |
| 9–13 y | 1100 | 100 | 0.76 | 445 | 39 | 10 | 9 | .7 | .8 | 9 |
| 14–18 y | 1100 | 100 | 0.73 | 630 | 63 | 10 | 12 | 1.0 | 1.1 | 12 |
| 19–30 y | 800 | 100 | 0.66 | 625 | 75 | 10 | 12 | 1.0 | 1.1 | 12 |
| 31–50 y | 800 | 100 | 0.66 | 625 | 75 | 10 | 12 | 1.0 | 1.1 | 12 |
| 51–70 y | 800 | 100 | 0.66 | 625 | 75 | 10 | 12 | 1.0 | 1.1 | 12 |
| >70 y | 1000 | 100 | 0.66 | 625 | 75 | 10 | 12 | 1.0 | 1.1 | 12 |
| **Females** | | | | | | | | | | |
| 9–13 y | 1100 | 100 | 0.76 | 420 | 39 | 10 | 9 | .7 | .8 | 9 |
| 14–18 y | 1100 | 100 | 0.71 | 485 | 56 | 10 | 12 | .9 | .9 | 11 |
| 19–30 y | 800 | 100 | 0.66 | 500 | 60 | 10 | 12 | .9 | .9 | 11 |
| 31–50 y | 800 | 100 | 0.66 | 500 | 60 | 10 | 12 | .9 | .9 | 11 |
| 51–70 y | 1000 | 100 | 0.66 | 500 | 60 | 10 | 12 | .9 | .9 | 11 |
| >70 y | 1000 | 100 | 0.66 | 500 | 60 | 10 | 12 | .9 | .9 | 11 |
| **Pregnancy** | | | | | | | | | | |
| ≤18 y | 1000 | 135 | 0.88 | 530 | 66 | 10 | 12 | 1.2 | 1.2 | 14 |
| 19–30 y | 800 | 135 | 0.88 | 550 | 70 | 10 | 12 | 1.2 | 1.2 | 14 |
| 31–50 y | 800 | 135 | 0.88 | 550 | 70 | 10 | 12 | 1.2 | 1.2 | 14 |
| **Lactation** | | | | | | | | | | |
| ≤18 y | 1000 | 160 | 1.05 | 880 | 96 | 10 | 16 | 1.2 | 1.3 | 13 |
| 19–30 y | 800 | 160 | 1.05 | 900 | 100 | 10 | 16 | 1.2 | 1.3 | 13 |
| 31–50 y | 800 | 160 | 1.05 | 900 | 100 | 10 | 16 | 1.2 | 1.3 | 13 |

This information taken from the various DRI reports (see www.nap.edu).

| Vitamin B-6 (mg/d) | Folate (µg/d) | Vitamin B-12 (µg/d) | Copper (µg/d) | Iodine (µg/d) | Iron (mg/d) | Magnesium (mg/d) | Molybdenum (µg/d) | Phosphorus (mg/d) | Selenium (µg/d) | Zinc (mg/d) |
|---|---|---|---|---|---|---|---|---|---|---|
| .4 | 120 | .7 | 260 | 65 | 3.0 | 65 | 13 | 380 | 17 | 2.2 |
| .5 | 160 | 1.0 | 340 | 65 | 4.1 | 110 | 17 | 405 | 23 | 4 |
| | | | | | | | | | | |
| .8 | 250 | 1.5 | 540 | 73 | 5.9 | 200 | 26 | 1055 | 35 | 7 |
| 1.1 | 330 | 2.0 | 685 | 95 | 7.7 | 340 | 33 | 1055 | 45 | 8.5 |
| 1.1 | 320 | 2.0 | 700 | 95 | 6 | 330 | 34 | 580 | 45 | 9.4 |
| 1.1 | 320 | 2.0 | 700 | 95 | 6 | 350 | 34 | 580 | 45 | 9.4 |
| 1.4 | 320 | 2.0 | 700 | 95 | 6 | 350 | 34 | 580 | 45 | 9.4 |
| 1.4 | 320 | 2.0 | 700 | 95 | 6 | 350 | 34 | 580 | 45 | 9.4 |
| | | | | | | | | | | |
| .8 | 250 | 1.5 | 540 | 73 | 5.7 | 200 | 26 | 1055 | 35 | 7 |
| 1.0 | 330 | 2.0 | 685 | 95 | 7.9 | 300 | 33 | 1055 | 45 | 7.5 |
| 1.1 | 320 | 2.0 | 700 | 95 | 8.1 | 255 | 34 | 580 | 45 | 6.8 |
| 1.1 | 320 | 2.0 | 700 | 95 | 8.1 | 265 | 34 | 580 | 45 | 6.8 |
| 1.3 | 320 | 2.0 | 700 | 95 | 5 | 265 | 34 | 580 | 45 | 6.8 |
| 1.3 | 320 | 2.0 | 700 | 95 | 5 | 265 | 34 | 580 | 45 | 6.8 |
| | | | | | | | | | | |
| 1.6 | 520 | 2.2 | 785 | 160 | 23 | 355 | 40 | 1055 | 49 | 10.5 |
| 1.6 | 520 | 2.2 | 800 | 160 | 22 | 290 | 40 | 580 | 49 | 9.5 |
| 1.6 | 520 | 2.2 | 800 | 160 | 22 | 300 | 40 | 580 | 49 | 9.5 |
| | | | | | | | | | | |
| 1.7 | 450 | 2.4 | 985 | 209 | 7 | 300 | 50 | 1055 | 59 | 11.6 |
| 1.7 | 450 | 2.4 | 1000 | 209 | 6.5 | 255 | 50 | 580 | 59 | 10.4 |
| 1.7 | 450 | 2.4 | 1000 | 209 | 6.5 | 265 | 50 | 580 | 59 | 10.4 |

## Dietary Reference Intakes for Individuals, Vitamins
### Food and Nutrition Board, Institute of Medicine, National Academies

| Life-Stage Group | Vitamin A (µg/d)[a] | Vitamin C (mg/d) | Vitamin D (µg/d)[b,c] | Vitamin E (mg/d)[d] | Vitamin K (µg/d) | Thiamin (mg/d) |
|---|---|---|---|---|---|---|
| **Infants** | | | | | | |
| 0–6 mo | 400* | 40* | **10** | 4* | 2.0* | 0.2* |
| 7–12 mo | 500* | 50* | **10** | 5* | 2.5* | 0.3* |
| **Children** | | | | | | |
| 1–3 y | **300** | **15** | **15** | **6** | 30* | **0.5** |
| 4–8 y | **400** | **25** | **15** | **7** | 55* | **0.6** |
| **Males** | | | | | | |
| 9–13 y | **600** | **45** | **15** | **11** | 60* | **0.9** |
| 14–18 y | **900** | **75** | **15** | **15** | 75* | **1.2** |
| 19–30 y | **900** | **90** | **15** | **15** | 120* | **1.2** |
| 31–50 y | **900** | **90** | **15** | **15** | 120* | **1.2** |
| 51–70 y | **900** | **90** | **15** | **15** | 120* | **1.2** |
| >70 y | **900** | **90** | **20** | **15** | 120* | **1.2** |
| **Females** | | | | | | |
| 9–13 y | **600** | **45** | **15** | **11** | 60* | **0.9** |
| 14–18 y | **700** | **65** | **15** | **15** | 75* | **1.0** |
| 19–30 y | **700** | **75** | **15** | **15** | 90* | **1.1** |
| 31–50 y | **700** | **75** | **15** | **15** | 90* | **1.1** |
| 51–70 y | **700** | **75** | **15** | **15** | 90* | **1.1** |
| >70 y | **700** | **75** | **20** | **15** | 90* | **1.1** |
| **Pregnancy** | | | | | | |
| ≤18 y | **750** | **80** | **15** | **15** | 75* | **1.4** |
| 19–30 y | **770** | **85** | **15** | **15** | 90* | **1.4** |
| 31–50 y | **770** | **85** | **15** | **15** | 90* | **1.4** |
| **Lactation** | | | | | | |
| ≤18 y | **1200** | **115** | **15** | **19** | 75* | **1.4** |
| 19–30 y | **1300** | **120** | **15** | **19** | 90* | **1.4** |
| 31–50 y | **1300** | **120** | **15** | **19** | 90* | **1.4** |

mg = milligram, µ = microgram

NOTE: This table (taken from the DRI reports; see www.nap.edu) presents Recommended Dietary Allowances (RDAs) in **bold type** and Adequate Intakes (AIs) in ordinary type followed by an asterisk (*). RDAs and AIs may both be used as goals for individual intake. RDAs are set to meet the needs of almost all (97 to 98%) individuals in a group. For healthy breastfed infants, the AI is the mean intake. The AI for other life stage and gender groups is believed to cover needs of all individuals in the group, but lack of data or uncertainty in the data prevents being able to specify with confidence the percentage of individuals covered by this intake.

[a]As retinol activity equivalents (RAEs). 1 RAE = 1 µg retinol, 12 µg β-carotene, 24 µg α-carotene, or 24 µg β-cryptoxanthin. To calculate RAEs from REs of provitamin A carotenoids in foods, divide the REs by 2. For preformed vitamin A in foods or supplements and for provitamin A carotenoids in supplements, 1 RE = 1 RAE.

[b]cholecalciferol. 1 µg cholecalciferol = 40 IU vitamin D.

[c]In the absence of adequate exposure to sunlight.

[d]As α-tocopherol. α-Tocopherol includes RRR-α-tocopherol, the only form of α-tocopherol that occurs naturally in foods, and the 2R-stereoisomeric forms of α-tocopherol (RRR-, RSR-, RRS-, and RSS-α-tocopherol) that occur in fortified foods and supplements. It does not include the 2S-stereoisomeric forms of α-tocopherol (SRR-, SSR-, SRS-, and SSS-α-tocopherol), also found in fortified foods and supplements.

[e]As niacin equivalents (NE). 1 mg of niacin = 60 mg of tryptophan; 0–6 months = preformed niacin (not NE).

[f]As dietary folate equivalents (DFE). 1 DFE = 1 µg food folate = 0.6 µg of folic acid from fortified food or as a supplement consumed with food = 0.5 µg of a supplement taken on an empty stomach.

[g]Although AIs have been set for choline, there are few data to assess whether a dietary supply of choline is needed at all stages of the life cycle, and it may be that the choline requirement can be met by endogenous synthesis at some of these stages.

[h]Because 10 to 30% of older people may malabsorb food-bound B-12, it is advisable for those older than 50 years to meet their RDA mainly by consuming foods fortified with B-12 or a supplement containing B-12.

[i]In view of evidence linking folate intake with neural tube defects in the fetus, it is recommended that all women capable of becoming pregnant consume 400 µg from supplements or fortified foods in addition to intake of food folate from a varied diet.

[j]It is assumed that women will continue consuming 400 µg from supplements or fortified food until their pregnancy is confirmed and they enter prenatal care, which ordinarily occurs after the end of the periconceptional period—the critical time for formation of the neural tube.

Adapted from the Dietary Reference Intakes series, National Academies Press. Copyright 1997, 1998, 2000, 2001, 2011, by the National Academy of Sciences. The full reports are available from the National Academies Press at www.nap.edu.

| Riboflavin (mg/d) | Niacin (mg/d)[e] | Vitamin B-6 (mg/d) | Folate (µg/d)[f] | Vitamin B-12 (µg/d) | Pantothenic Acid (mg/d) | Biotin (µg/d) | Choline (mg/d)[g] |
|---|---|---|---|---|---|---|---|
| 0.3* | 2* | 0.1* | 65* | 0.4* | 1.7* | 5* | 125* |
| 0.4* | 4* | 0.3* | 80* | 0.5* | 1.8* | 6* | 150* |
| | | | | | | | |
| 0.5 | 6 | 0.5 | 150 | 0.9 | 2* | 8* | 200* |
| 0.6 | 8 | 0.6 | 200 | 1.2 | 3* | 12* | 250* |
| | | | | | | | |
| 0.9 | 12 | 1.0 | 300 | 1.8 | 4* | 20* | 375* |
| 1.3 | 16 | 1.3 | 400 | 2.4 | 5* | 25* | 550* |
| 1.3 | 16 | 1.3 | 400 | 2.4 | 5* | 30* | 550* |
| 1.3 | 16 | 1.3 | 400 | 2.4 | 5* | 30* | 550* |
| 1.3 | 16 | 1.7 | 400 | 2.4[h] | 5* | 30* | 550* |
| 1.3 | 16 | 1.7 | 400 | 2.4[h] | 5* | 30* | 550* |
| | | | | | | | |
| 0.9 | 12 | 1.0 | 300 | 1.8 | 4* | 20* | 375* |
| 1.0 | 14 | 1.2 | 400[i] | 2.4 | 5* | 25* | 400* |
| 1.1 | 14 | 1.3 | 400[i] | 2.4 | 5* | 30* | 425* |
| 1.1 | 14 | 1.3 | 400[i] | 2.4 | 5* | 30* | 425* |
| 1.1 | 14 | 1.5 | 400 | 2.4[h] | 5* | 30* | 425* |
| 1.1 | 14 | 1.5 | 400 | 2.4[h] | 5* | 30* | 425* |
| | | | | | | | |
| 1.4 | 18 | 1.9 | 600[j] | 2.6 | 6* | 30* | 450* |
| 1.4 | 18 | 1.9 | 600[j] | 2.6 | 6* | 30* | 450* |
| 1.4 | 18 | 1.9 | 600[j] | 2.6 | 6* | 30* | 450* |
| | | | | | | | |
| 1.6 | 17 | 2.0 | 500 | 2.8 | 7* | 35* | 550* |
| 1.6 | 17 | 2.0 | 500 | 2.8 | 7* | 35* | 550* |
| 1.6 | 17 | 2.0 | 500 | 2.8 | 7* | 35* | 550* |

## Dietary Reference Intakes for Individuals, Elements
**Food and Nutrition Board, Institute of Medicine, National Academies**

| Life-Stage Group | Calcium (mg/d) | Chromium (µg/d) | Copper (µg/d) | Fluoride (mg/d) | Iodine (µg/d) | Iron (mg/d) |
|---|---|---|---|---|---|---|
| **Infants** | | | | | | |
| 0–6 mo | 200* | 0.2* | 200* | 0.01* | 110* | 0.27* |
| 7–12 mo | 260* | 5.5* | 220* | 0.5* | 130* | **11** |
| **Children** | | | | | | |
| 1–3 y | **700** | 11* | **340** | 0.7* | **90** | **7** |
| 4–8 y | **1000** | 15* | **440** | 1* | **90** | **10** |
| **Males** | | | | | | |
| 9–13 y | **1300** | 25* | **700** | 2* | **120** | **8** |
| 14–18 y | **1300** | 35* | **890** | 3* | **150** | **11** |
| 19–30 y | **1000** | 35* | **900** | 4* | **150** | **8** |
| 31–50 y | **1000** | 35* | **900** | 4* | **150** | **8** |
| 51–70 y | **1000** | 30* | **900** | 4* | **150** | **8** |
| >70 y | **1200** | 30* | **900** | 4* | **150** | **8** |
| **Females** | | | | | | |
| 9–13 y | **1300** | 21* | **700** | 2* | **120** | **8** |
| 14–18 y | **1300** | 24* | **890** | 3* | **150** | **15** |
| 19–30 y | **1000** | 25* | **900** | 3* | **150** | **18** |
| 31–50 y | **1000** | 25* | **900** | 3* | **150** | **18** |
| 51–70 y | **1200** | 20* | **900** | 3* | **150** | **8** |
| >70 y | **1200** | 20* | **900** | 3* | **150** | **8** |
| **Pregnancy** | | | | | | |
| ≤18 y | **1300** | 29* | **1000** | 3* | **220** | **27** |
| 19–30 y | **1000** | 30* | **1000** | 3* | **220** | **27** |
| 31–50 y | **1000** | 30* | **1000** | 3* | **220** | **27** |
| **Lactation** | | | | | | |
| ≤18 y | **1300** | 44* | **1300** | 3* | **290** | **10** |
| 19–30 y | **1000** | 45* | **1300** | 3* | **290** | **9** |
| 31–50 y | **1000** | 45* | **1300** | 3* | **290** | **9** |

NOTE: This table presents Recommended Dietary Allowances (RDAs) in bold type and Adequate Intakes (AIs) in ordinary type followed by an asterisk (*). RDAs and AIs may both be used as goals for individual intake. RDAs are set to meet the needs of almost all (97 to 98%) individuals in a group. For healthy breastfed infants, the AI is the mean intake. The AI for other life stage and gender groups is believed to cover needs of all individuals in the group, but lack of data or uncertainty in the data prevents being able to specify with confidence the percentage of individuals covered by this intake.

Source: Dietary Reference Intakes for Calcium, Phosphorus, Magnesium, Vitamin D, and Fluoride (1997); Dietary Reference Intakes for Thiamin, Riboflavin, Niacin, Vitamin B-6, Folate, Vitamin B-12, Pantothenic Acid, Biotin, and Choline (1998); Dietary Reference Intakes for Vitamin C, Vitamin E, Selenium, and Carotenoids (2000); Dietary Reference Intakes for Vitamin A, Vitamin K, Arsenic, Boron, Chromium, Copper, Iodine, Iron, Manganese, Molybdenum, Nickel, Silicon, Vanadium, and Zinc (2001); and Dietary Reference Intakes for Calcium and Vitamin D (2011). These reports may be accessed via www.nap.edu. Adapted from the Dietary Reference Intake series, National Academies Press. Copyright 1997, 1998, 2000, 2001, and 2011 by the National Academy of Sciences. The full reports are available from the National Academies Press at www.nap.edu.

| Magnesium (mg/d) | Manganese (mg/d) | Molybdenum (µg/d) | Phosphorus (mg/d) | Selenium (µg/d) | Zinc (mg/d) |
|---|---|---|---|---|---|
| 30* | 0.003* | 2* | 100* | 15* | 2* |
| 75* | 0.6* | 3* | 275* | 20* | 3 |
| | | | | | |
| 80 | 1.2* | 17 | 460 | 20 | 3 |
| 130 | 1.5* | 22 | 500 | 30 | 5 |
| | | | | | |
| 240 | 1.9* | 34 | 1250 | 40 | 8 |
| 410 | 2.2* | 43 | 1250 | 55 | 11 |
| 400 | 2.3* | 45 | 700 | 55 | 11 |
| 420 | 2.3* | 45 | 700 | 55 | 11 |
| 420 | 2.3* | 45 | 700 | 55 | 11 |
| 420 | 2.3* | 45 | 700 | 55 | 11 |
| | | | | | |
| 240 | 1.6* | 34 | 1250 | 40 | 8 |
| 360 | 1.6* | 43 | 1250 | 55 | 9 |
| 310 | 1.8* | 45 | 700 | 55 | 8 |
| 320 | 1.8* | 45 | 700 | 55 | 8 |
| 320 | 1.8* | 45 | 700 | 55 | 8 |
| 320 | 1.8* | 45 | 700 | 55 | 8 |
| | | | | | |
| 400 | 2.0* | 50 | 1250 | 60 | 12 |
| 350 | 2.0* | 50 | 700 | 60 | 11 |
| 360 | 2.0* | 50 | 700 | 60 | 11 |
| | | | | | |
| 360 | 2.6* | 50 | 1250 | 70 | 13 |
| 310 | 2.6* | 50 | 700 | 70 | 12 |
| 320 | 2.6* | 50 | 700 | 70 | 12 |

## Dietary Reference Intakes for Individuals, Macronutrients
### Food and Nutrition Board, Institute of Medicine, National Academies

| Life-Stage Group | Carbohydrate (g/d) | Total Fiber (g/d) | Fat (g/d) | Linoleic Acid (g/d) | α-Linolenic Acid (g/d) | Protein[a] (g/d) |
|---|---|---|---|---|---|---|
| **Infants** | | | | | | |
| 0–6 mo | 60* | ND | 31* | 4.4* | 0.5* | 9.1* |
| 7–12 mo | 95* | ND | 30* | 4.6* | 0.5* | **11.0** |
| **Children** | | | | | | |
| 1–3 y | **130** | 19* | ND[b] | 7* | 0.7* | **13** |
| 4–8 y | **130** | 25* | ND | 10* | 0.9* | **19** |
| **Males** | | | | | | |
| 9–13 y | **130** | 31* | ND | 12* | 1.2* | **34** |
| 14–18 y | **130** | 38* | ND | 16* | 1.6* | **52** |
| 19–30 y | **130** | 38* | ND | 17* | 1.6* | **56** |
| 31–50 y | **130** | 38* | ND | 17* | 1.6* | **56** |
| 51–70 y | **130** | 30* | ND | 14* | 1.6* | **56** |
| >70 y | **130** | 30* | ND | 14* | 1.6* | **56** |
| **Females** | | | | | | |
| 9–13 y | **130** | 26* | ND | 10* | 1.0* | **34** |
| 14–18 y | **130** | 26* | ND | 11* | 1.1* | **46** |
| 19–30 y | **130** | 25* | ND | 12* | 1.1* | **46** |
| 31–50 y | **130** | 25* | ND | 12* | 1.1* | **46** |
| 51–70 y | **130** | 21* | ND | 11* | 1.1* | **46** |
| >70 y | **130** | 21* | ND | 11* | 1.1* | **46** |
| **Pregnancy** | | | | | | |
| 14–18 y | **175** | 28* | ND | 13* | 1.4* | **71** |
| 19–30 y | **175** | 28* | ND | 13* | 1.4* | **71** |
| 31–50 y | **175** | 28* | ND | 13* | 1.4* | **71** |
| **Lactation** | | | | | | |
| 14–18 y | **210** | 29* | ND | 13* | 1.3* | **71** |
| 19–30 y | **210** | 29* | ND | 13* | 1.3* | **71** |
| 31–50 y | **210** | 29* | ND | 13* | 1.3* | **71** |

NOTE: This table presents Recommended Dietary Allowances (RDAs) in **bold type** and Adequate Intakes (AIs) in ordinary type followed by an asterisk (*). RDAs and AIs may both be used as goals for individual intake. RDAs are set to meet the needs of almost all (97 to 98%) individuals in a group. For healthy breastfed infants, the AI is the mean intake. The AI for other life stage and gender groups is believed to cover needs of all individuals in the group, but lack of data or uncertainty in the data prevents being able to specify with confidence the percentage of individuals covered by this intake.

[a]Based on 0.8g protein/kg body weight for reference body weight.

[b]ND = not determinable at this time.

Source: Dietary Reference Intakes for Energy, Carbohydrate, Fiber, Fat, Fatty Acids, Cholesterol, Protein, and Amino Acids (2002). This report may be accessed via www.nap.edu. Adapted from the Dietary Reference Intake series, National Academies Press. Copyright 1997, 1998, 2000, 2001, by the National Academy of Sciences. The full reports are available from the National Academies Press at www.nap.edu.

## Dietary Reference Intakes for Individuals, Electrolytes and Water
### Food and Nutrition Board, Institute of Medicine, National Academies

| Life-Stage Group | Sodium (mg/d) | Potassium (mg/d) | Chloride (mg/d) | Water (L/d) |
|---|---|---|---|---|
| **Infants** | | | | |
| 0–6 mo | 110* | 400* | 180* | 0.7* |
| 7–12 mo | 370* | 860* | 570* | 0.8* |
| **Children** | | | | |
| 1–3 y | 800* | 2000* | 1500* | 1.3* |
| 4–8 y | 1000* | 2300* | 1900* | 1.7* |
| **Males** | | | | |
| 9–13 y | 1200* | 2500* | 2300* | 2.4* |
| 14–18 y | 1500* | 3000* | 2300* | 3.3* |
| 19–30 y | 1500* | 3400* | 2300* | 3.7* |
| 31–50 y | 1500* | 3400* | 2300* | 3.7* |
| 51–70 y | 1500* | 3400* | 2000* | 3.7* |
| >70 y | 1500* | 3400* | 1800* | 3.7* |
| **Females** | | | | |
| 9–13 y | 1200* | 2300* | 2300* | 2.1* |
| 14–18 y | 1500* | 2300* | 2300* | 2.3* |
| 19–30 y | 1500* | 2600* | 2300* | 2.7* |
| 31–50 y | 1500* | 2600* | 2300* | 2.7* |
| 51–70 y | 1500* | 2600* | 2000* | 2.7* |
| >70 y | 1500* | 2600* | 1800* | 2.7* |
| **Pregnancy** | | | | |
| 14–18 y | 1500* | 2600* | 2300* | 3.0* |
| 19–50 y | 1500* | 2900* | 2300* | 3.0* |
| **Lactation** | | | | |
| 14–18 y | 1500* | 2500* | 2300* | 3.8* |
| 19–50 y | 1500* | 2800* | 2300* | 3.8* |

NOTE: The table is adapted from the DRI reports. See www.nap.edu. Adequate Intakes (AIs) are followed by an asterisk (*). These may be used as a goal for individual intake. For healthy breastfed infants, the AI is the average intake. The AI for other life stage and gender groups is believed to cover the needs of all individuals in the group, but lack of data prevent being able to specify with confidence the percentage of individuals covered by this intake; therefore, no Recommended Dietary Allowance (RDA) was set.

Source: Dietary Reference Intakes for Water, Potassium, Sodium, Chloride, and Sulfate (2005). Dietary Reference Intakes (DRIs) for Sodium and Potassium (2019). These reports may be accessed via www.nap.edu.

## Acceptable Macronutrient Distribution Ranges

| Macronutrient | Range (percent of energy) | | |
|---|---|---|---|
| | Children, 1–3 y | Children, 4–18 y | Adults |
| Fat | 30–40 | 25–35 | 20–35 |
| *omega*-6 polyunsaturated fats (linoleic acid) | 5–10 | 5–10 | 5–10 |
| *omega*-3 polyunsaturated fats[a] (α-linolenic acid) | 0.6–1.2 | 0.6–1.2 | 0.6–1.2 |
| Carbohydrate | 45–65 | 45–65 | 45–65 |
| Protein | 5–20 | 10–30 | 10–35 |

[a]Approximately 10% of the total can come from longer-chain n-3 fatty acids.

Source: Dietary Reference Intakes for Energy, Carbohydrate, Fiber, Fat, Fatty Acids, Cholesterol, Protein, and Amino Acids (2002). The report may be accessed via www.nap.edu. Adapted from the Dietary Reference Intakes series, National Academies Press. Copyright 1997, 1998, 2000, 2001, 2011, by the National Academy of Sciences. The full reports are available from the National Academies Press at www.nap.edu.

## Tolerable Upper Intake Levels (UL[a]), Vitamins
**Food and Nutrition Board, Institute of Medicine, National Academies**

| Life-Stage Group | Vitamin A (μg/d)[b] | Vitamin C (mg/d) | Vitamin D (μg/d) | Vitamin E (mg/d)[c,d] | Vitamin K | Thiamin | Riboflavin | Niacin (mg/d)[d] |
|---|---|---|---|---|---|---|---|---|
| **Infants** | | | | | | | | |
| 0–6 mo | 600 | ND | 25 | ND | ND | ND | ND | ND |
| 7–12 mo | 600 | ND | 38 | ND | ND | ND | ND | ND |
| **Children** | | | | | | | | |
| 1–3 y | 600 | 400 | 63 | 200 | ND | ND | ND | 10 |
| 4–8 y | 900 | 650 | 75 | 300 | ND | ND | ND | 15 |
| **Males, Females** | | | | | | | | |
| 9–13 y | 1700 | 1200 | 100 | 600 | ND | ND | ND | 20 |
| 14–18 y | 2800 | 1800 | 100 | 800 | ND | ND | ND | 30 |
| 19–70 y | 3000 | 2000 | 100 | 1000 | ND | ND | ND | 35 |
| >70 y | 3000 | 2000 | 100 | 1000 | ND | ND | ND | 35 |
| **Pregnancy** | | | | | | | | |
| ≤18 y | 2800 | 1800 | 100 | 800 | ND | ND | ND | 30 |
| 19–50 y | 3000 | 2000 | 100 | 1000 | ND | ND | ND | 35 |
| **Lactation** | | | | | | | | |
| ≤18 y | 2800 | 1800 | 100 | 800 | ND | ND | ND | 30 |
| 19–50 y | 3000 | 2000 | 100 | 1000 | ND | ND | ND | 35 |

[a]UL = The maximum level of daily nutrient intake likely to pose no risk of adverse effects. Unless otherwise specified, the UL represents total intake from food, water, and supplements. Due to lack of suitable data, ULs could not be established for vitamin K, thiamin, riboflavin, vitamin B-12, pantothenic acid, biotin, or carotenoids. In the absence of ULs, extra caution may be warranted in consuming levels above recommended intakes.

[b]As preformed vitamin A only.

[c]As α-tocopherol; applies to any form of supplemental α-tocopherol.

[d]The ULs for vitamin E, niacin, and folate apply to synthetic forms obtained from supplements, fortified foods, or a combination of the two.

[e]β-Carotene supplements are advised only to serve as a provitamin A source for individuals at risk of vitamin A deficiency.

[f]ND = Not determinable due to lack of data of adverse effects in this age group and concern with regard to lack of ability to handle excess amounts. Source of intake should be from food only to prevent high levels of intake.

Source: Dietary Reference Intakes for Calcium and Vitamin D (2011); Dietary Reference Intakes for Calcium, Phosphorus, Magnesium, Vitamin D, and Fluoride (1997); Dietary Reference Intakes for Thiamin, Riboflavin, Niacin, Vitamin B-6, Folate, Vitamin B-12, Pantothenic Acid, Biotin, and Choline (1998); Dietary Reference Intakes for Vitamin C, Vitamin E, Selenium, and Carotenoids (2000); Dietary Reference Intakes for Vitamin A, Vitamin K, Arsenic, Boron, Chromium, Copper, Iodine, Iron, Manganese, Molybdenum, Nickel, Silicon, Vanadium, and Zinc (2001); and Dietary Reference Intakes for Water, Potassium, Sodium, Chloride, and Sulfate (2004). These reports may be accessed via www.nap.edu. Adapted from the Dietary Reference Intakes series, National Academies Press. Copyright 1997, 1998, 2000, 2001, 2011, by the National Academy of Sciences. The full reports are available from the National Academies Press at www.nap.edu.

| Vitamin B-6 (mg/d) | Folate (μg/d)[d] | Vitamin B-12 | Pantothenic Acid | Biotin | Choline (g/d) | Carotenoids[e] |
|---|---|---|---|---|---|---|
| ND | ND | ND | ND | ND | ND | ND |
| ND | ND | ND | ND | ND | ND | ND |
| 30 | 300 | ND | ND | ND | 1.0 | ND |
| 40 | 400 | ND | ND | ND | 1.0 | ND |
| 60 | 600 | ND | ND | ND | 2.0 | ND |
| 80 | 800 | ND | ND | ND | 3.0 | ND |
| 100 | 1000 | ND | ND | ND | 3.5 | ND |
| 100 | 1000 | ND | ND | ND | 3.5 | ND |
| 80 | 800 | ND | ND | ND | 3.0 | ND |
| 100 | 1000 | ND | ND | ND | 3.5 | ND |
| 80 | 800 | ND | ND | ND | 3.0 | ND |
| 100 | 1000 | ND | ND | ND | 3.5 | ND |

## Tolerable Upper Intake Levels (UL[a]), Elements and Electrolytes[b,c]
**Food and Nutrition Board, Institute of Medicine, National Academies**

| Life-Stage Group | Arsenic[b] | Boron (mg/d) | Calcium (g/d) | Copper (µg/d) | Fluoride (mg/d) | Iodine (µg/d) | Iron (mg/d) | Magnesium (mg/d)[d] |
|---|---|---|---|---|---|---|---|---|
| **Infants** | | | | | | | | |
| 0–6 mo | ND[f] | ND | 1 | ND | 0.7 | ND | 40 | ND |
| 7–12 mo | ND | ND | 1.5 | ND | 0.9 | ND | 40 | ND |
| **Children** | | | | | | | | |
| 1–3 y | ND | 3 | 2.5 | 1000 | 1.3 | 200 | 40 | 65 |
| 4–8 y | ND | 6 | 2.5 | 3000 | 2.2 | 300 | 40 | 110 |
| **Males, Females** | | | | | | | | |
| 9–13 y | ND | 11 | 3 | 5000 | 10 | 600 | 40 | 350 |
| 14–18 y | ND | 17 | 3 | 8000 | 10 | 900 | 45 | 350 |
| 19–70 y | ND | 20 | 2.5[g] | 10,000 | 10 | 1100 | 45 | 350 |
| >70 y | ND | 20 | 2 | 10,000 | 10 | 1100 | 45 | 350 |
| **Pregnancy** | | | | | | | | |
| ≤18 y | ND | 17 | 3 | 8000 | 10 | 900 | 45 | 350 |
| 19–50 y | ND | 20 | 2.5 | 10,000 | 10 | 1100 | 45 | 350 |
| **Lactation** | | | | | | | | |
| ≤18 y | ND | 17 | 3 | 8000 | 10 | 900 | 45 | 350 |
| 19–50 y | ND | 20 | 2.5 | 10,000 | 10 | 1100 | 45 | 350 |

[a]UL = The maximum level of daily nutrient intake that is likely to pose no risk of adverse effects. Unless otherwise specified, the UL represents total intake from food, water, and supplements. Due to lack of suitable data, ULs could not be established for arsenic, chromium, and silicon. In the absence of ULs, extra caution may be warranted in consuming levels above recommended intakes.

[b]Although a UL was not determined for arsenic, there is no justification for adding arsenic to food or supplements.

[c]Although silicon has not been shown to cause adverse effects in humans, there is no justification for adding silicon to supplements.

[d]The ULs for magnesium represent intake from a pharmacological agent only and do not include intake from food and water.

[e]Although vanadium in food has not been shown to cause adverse effects in humans, there is no justification for adding vanadium to food and vanadium supplements should be used with caution. The UL is based on adverse effects in laboratory animals and this data could be used to set a UL for adults but not children and adolescents.

[f]ND = Not determinable due to lack of data of adverse effects in this age group and concern with regard to lack of ability to handle excess amounts. Source of intake should be from food only to prevent high levels of intake.

[g]Upper Limit declines to 2 after age 50.

Source: Dietary Reference Intakes for Calcium and Vitamin D (2011); Dietary Reference Intakes for Calcium, Phosphorus, Magnesium, Vitamin D, and Fluoride (1997); Dietary Reference Intakes for Thiamin, Riboflavin, Niacin, Vitamin B-6, Folate, Vitamin B-12, Pantothenic Acid, Biotin, and Choline (1998); Dietary Reference Intakes for Vitamin C, Vitamin E, Selenium, and Carotenoids (2000); Dietary Reference Intakes for Vitamin A, Vitamin K, Arsenic, Boron, Chromium, Copper, Iodine, Iron, Manganese, Molybdenum, Nickel, Silicon, Vanadium, and Zinc (2001); and Dietary Reference Intakes for Water, Potassium, Sodium, Chloride, and Sulfate (2004). Dietary Reference Intakes for Sodium and Potassium (2019). These reports may be accessed via www.nap.edu. Adapted from the Dietary Reference Intakes series, National Academies Press. Copyright 1997, 1998, 2000, 2001, 2011, 2019 by the National Academy of Sciences. The full reports are available from the National Academies Press at www.nap.edu.

| Manganese (mg/d) | Molybdenum (µg/d) | Nickel (mg/d) | Phosphorus (g/d) | Selenium (µg/d) | Vanadium (mg/d)[e] | Zinc (mg/d) | Sodium (mg/d) | Chloride (mg/d) |
|---|---|---|---|---|---|---|---|---|
| ND | ND | ND | ND | 45 | ND | 4 | ND | ND |
| ND | ND | ND | ND | 60 | ND | 5 | ND | ND |
| 2 | 300 | 0.2 | 3 | 90 | ND | 7 | 1200 | 2300 |
| 3 | 600 | 0.3 | 3 | 150 | ND | 12 | 1500 | 2900 |
| 6 | 1100 | 0.6 | 4 | 280 | ND | 23 | 1800 | 3400 |
| 9 | 1700 | 1.0 | 4 | 400 | ND | 34 | 2300 | 3600 |
| 11 | 2000 | 1.0 | 4 | 400 | 1.8 | 40 | 2300 | 3600 |
| 11 | 2000 | 1.0 | 3 | 400 | 1.8 | 40 | 2300 | 3600 |
| 9 | 1700 | 1.0 | 3.5 | 400 | ND | 34 | 2300 | 3600 |
| 11 | 2000 | 1.0 | 3.5 | 400 | ND | 40 | 2300 | 3600 |
| 9 | 1700 | 1.0 | 4 | 400 | ND | 34 | 2300 | 3600 |
| 11 | 2000 | 1.0 | 4 | 400 | ND | 40 | 2300 | 3600 |

# Appendix K

## CDC GROWTH CHARTS

# Birth to 24 months: Boys
## Length-for-age and Weight-for-age percentiles

NAME _____

RECORD # _____

AGE (MONTHS)

Birth   3   6   9   12   15   18   21   24

**LENGTH** — percentiles: 98, 95, 85, 75, 50, 25, 10, 5, 2

**WEIGHT** — percentiles: 98, 95, 90, 75, 50, 25, 10, 5, 2

| Mother's Stature _____ | | | Gestational | | |
|---|---|---|---|---|---|
| Father's Stature _____ | | | Age: _____ Weeks | | Comment |
| Date | Age | Weight | Length | Head Circ. | |
| | Birth | | | | |
| | | | | | |
| | | | | | |
| | | | | | |
| | | | | | |
| | | | | | |

Published by the Centers for Disease Control and Prevention, November 1, 2009
SOURCE: WHO Child Growth Standards (http://www.who.int/childgrowth/en)

# Birth to 24 months: Boys
## Head circumference-for-age and
## Weight-for-length percentiles

NAME _____

RECORD # _____

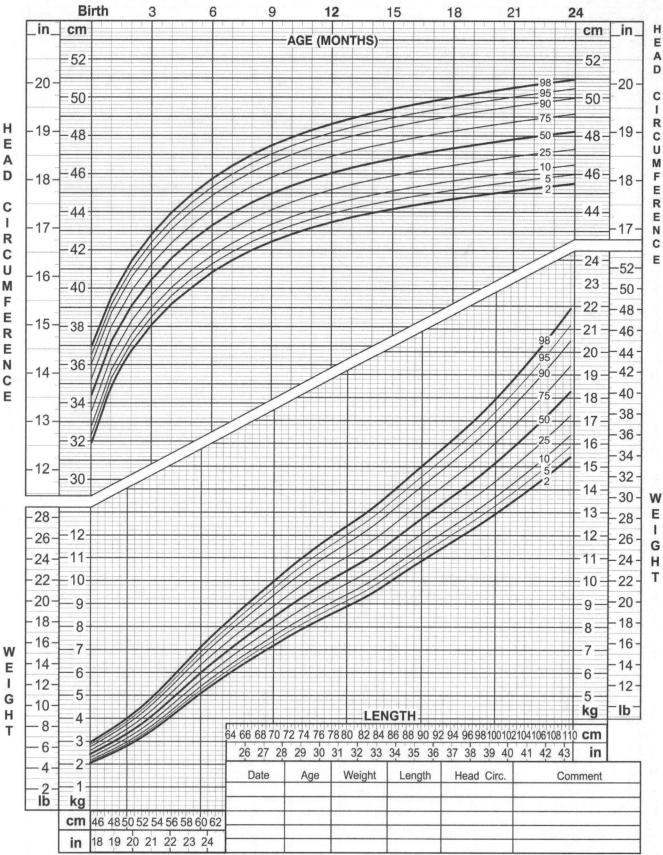

AGE (MONTHS)

| Date | Age | Weight | Length | Head Circ. | Comment |
|------|-----|--------|--------|-----------|---------|
| | | | | | |
| | | | | | |
| | | | | | |
| | | | | | |

Published by the Centers for Disease Control and Prevention, November 1, 2009
SOURCE: WHO Child Growth Standards (http://www.who.int/childgrowth/en)

# Birth to 24 months: Girls
## Length-for-age and Weight-for-age percentiles

NAME _____

RECORD # _____

AGE (MONTHS)

Birth    3    6    9    12    15    18    21    24    41

LENGTH

| Mother's Stature _____ | | Gestational | |  |
| Father's Stature _____ | | Age: _____ Weeks | | Comment |
| Date | Age | Weight | Length | Head Circ. | |
| | Birth | | | | |
| | | | | | |
| | | | | | |
| | | | | | |
| | | | | | |
| | | | | | |

AGE (MONTHS)

WEIGHT

Length percentile lines: 98, 95, 90, 75, 50, 25, 10, 5, 2

Weight percentile lines: 98, 95, 90, 75, 50, 25, 10, 5, 2

Published by the Centers for Disease Control and Prevention, November 1, 2009
SOURCE: WHO Child Growth Standards (http://www.who.int/childgrowth/en)

# Birth to 24 months: Girls
## Head circumference-for-age and
## Weight-for-length percentiles

NAME _____

RECORD # _____

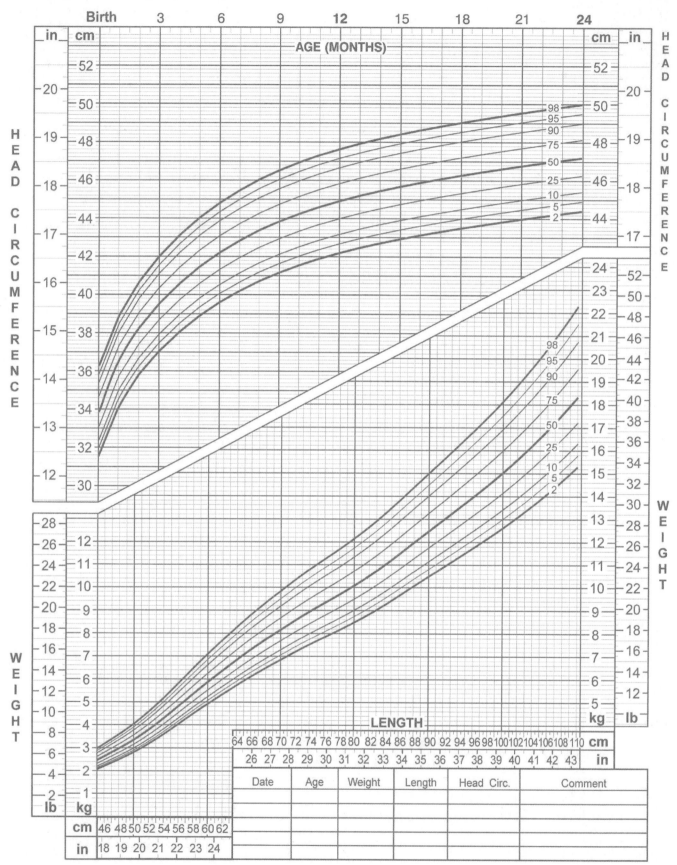

| Date | Age | Weight | Length | Head Circ. | Comment |
|------|-----|--------|--------|------------|---------|
|  |  |  |  |  |  |
|  |  |  |  |  |  |
|  |  |  |  |  |  |
|  |  |  |  |  |  |

Published by the Centers for Disease Control and Prevention, November 1, 2009
SOURCE: WHO Child Growth Standards (http://www.who.int/childgrowth/en)

# 2 to 20 years: Boys
## Stature-for-age and Weight-for-age percentiles

NAME _____

RECORD # _____

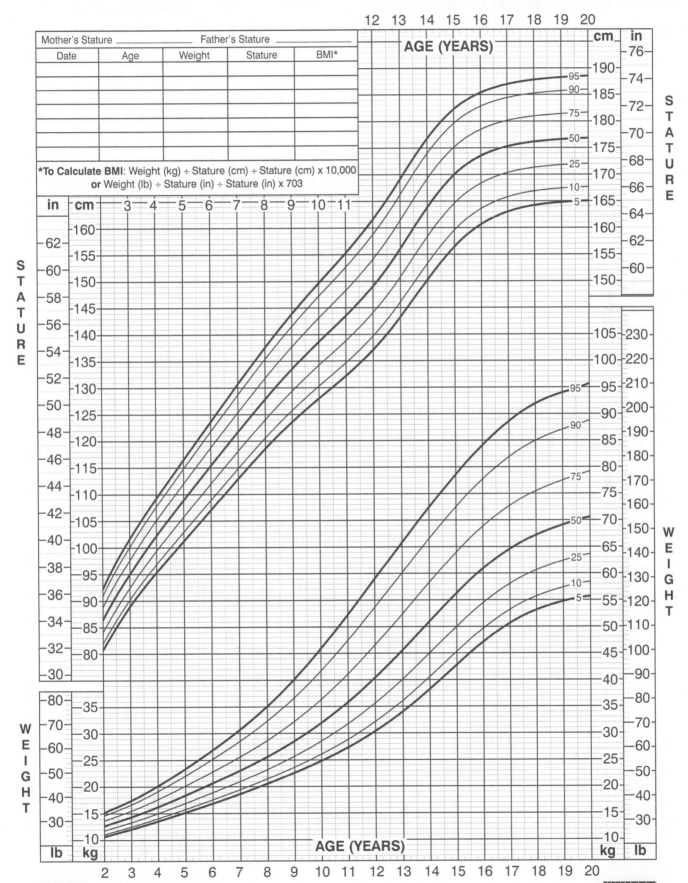

*To Calculate BMI: Weight (kg) ÷ Stature (cm) ÷ Stature (cm) x 10,000
or Weight (lb) ÷ Stature (in) ÷ Stature (in) x 703

Published May 30, 2000 (modified 11/21/00).
SOURCE: Developed by the National Center for Health Statistics in collaboration with
the National Center for Chronic Disease Prevention and Health Promotion (2000).
http://www.cdc.gov/growthcharts

SAFER · HEALTHIER · PEOPLE™

# 2 to 20 years: Boys
## Body mass index-for-age percentiles

NAME _____

RECORD # _____

| Date | Age | Weight | Stature | BMI* | Comments |
|------|-----|--------|---------|------|----------|
|      |     |        |         |      |          |
|      |     |        |         |      |          |
|      |     |        |         |      |          |
|      |     |        |         |      |          |
|      |     |        |         |      |          |
|      |     |        |         |      |          |
|      |     |        |         |      |          |
|      |     |        |         |      |          |
|      |     |        |         |      |          |

*To Calculate BMI: Weight (kg) ÷ Stature (cm) ÷ Stature (cm) x 10,000
or Weight (lb) ÷ Stature (in) ÷ Stature (in) x 703

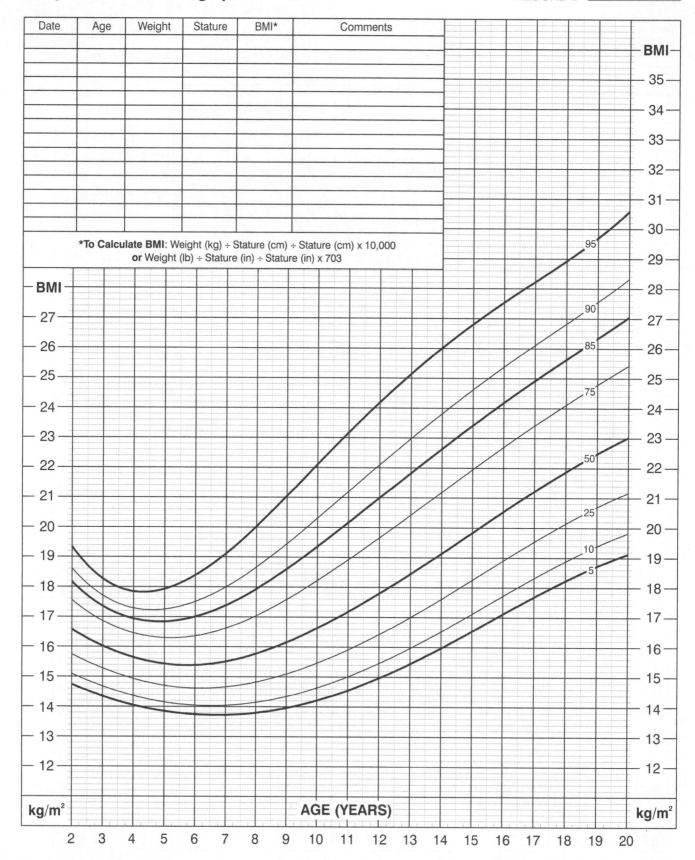

AGE (YEARS)

kg/m²

BMI

Published May 30, 2000 (modified 10/16/00).

SOURCE: Developed by the National Center for Health Statistics in collaboration with
the National Center for Chronic Disease Prevention and Health Promotion (2000).
http://www.cdc.gov/growthcharts

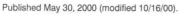

SAFER • HEALTHIER • PEOPLE™

# 2 to 20 years: Girls
## Stature-for-age and Weight-for-age percentiles

NAME _____

RECORD # _____

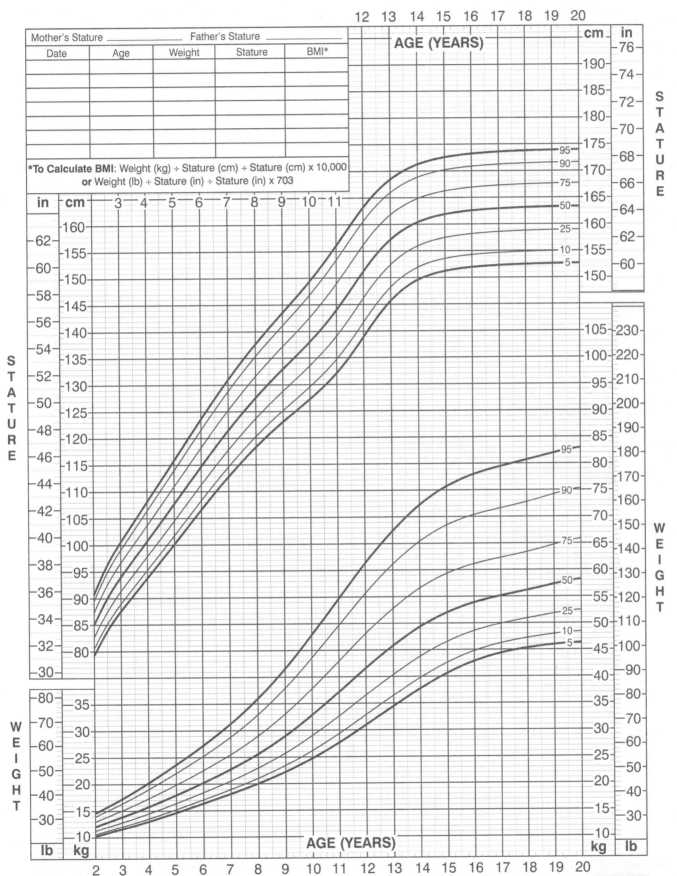

*To Calculate BMI: Weight (kg) ÷ Stature (cm) ÷ Stature (cm) x 10,000
or Weight (lb) ÷ Stature (in) ÷ Stature (in) x 703

Published May 30, 2000 (modified 11/21/00).
SOURCE: Developed by the National Center for Health Statistics in collaboration with
the National Center for Chronic Disease Prevention and Health Promotion (2000).
http://www.cdc.gov/growthcharts

SAFER • HEALTHIER • PEOPLE™

# 2 to 20 years: Girls
## Body mass index-for-age percentiles

NAME _____

RECORD # _____

| Date | Age | Weight | Stature | BMI* | Comments |
|------|-----|--------|---------|------|----------|
|      |     |        |         |      |          |
|      |     |        |         |      |          |
|      |     |        |         |      |          |
|      |     |        |         |      |          |
|      |     |        |         |      |          |
|      |     |        |         |      |          |
|      |     |        |         |      |          |

*To Calculate BMI: Weight (kg) ÷ Stature (cm) ÷ Stature (cm) x 10,000
or Weight (lb) ÷ Stature (in) ÷ Stature (in) x 703

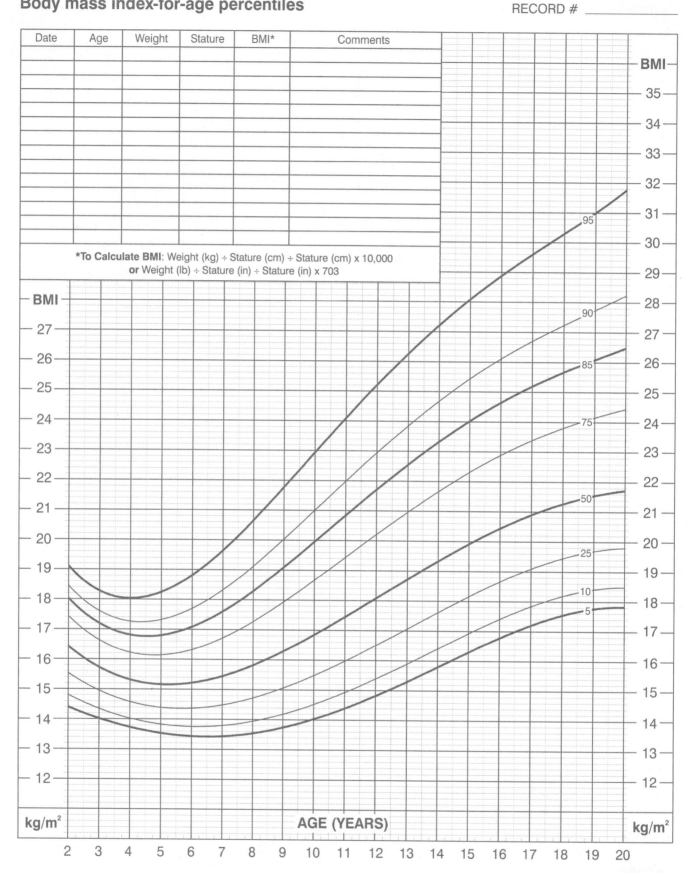

AGE (YEARS)

Published May 30, 2000 (modified 10/16/00).
SOURCE: Developed by the National Center for Health Statistics in collaboration with
the National Center for Chronic Disease Prevention and Health Promotion (2000).
http://www.cdc.gov/growthcharts

SAFER·HEALTHIER·PEOPLE™

CDC

# Appendix L

## SOURCES OF NUTRITION INFORMATION

Consider the following reliable sources of food and nutrition information.

**Journals That Regularly Cover Nutrition Topics**

*Amber Waves Magazine**
*American College of Sports Medicine's (ASCM's) Health & Fitness Journal*
*American Family Physician**
*American Journal of Clinical Nutrition*
*American Journal of Epidemiology*
*American Journal of Medicine*
*American Journal of Nursing*
*American Journal of Obstetrics and Gynecology*
*American Journal of Physiology*
*American Journal of Public Health*
*American Scientist*
*Annals of Internal Medicine*
*Annals of Nutrition and Metabolism*
*Annual Reviews of Medicine*
*Annual Reviews of Nutrition*
*Archives of Disease in Childhood*
*The BMJ*
*British Journal of Nutrition*
*Cancer*
*Cancer Research*
*Circulation*
*Critical Reviews in Food Science & Nutrition*
*Diabetes*
*Diabetes Care*
*Disease-a-Month*

*European Journal of Clinical Nutrition*
*FASEB Journal*
*FDA Consumer**
*Food Chemical Toxicology*
*Food Engineering*
*Food and Nutrition Bulletin*
*Food and Nutrition Research*
*Food Technology Magazine*
*Gastroenterology*
*Geriatrics*
*Gut*
*International Journal of Behavioral Nutrition and Physical Activity*
*International Journal of Environmental Research and Public Health*
*International Journal of Obesity*
*JAMA (Journal of the American Medical Association)*
*JAMA Internal Medicine*
*JNCI (Journal of the National Cancer Institute)*
*Journal of the Academy of Nutrition and Dietetics*
*Journal of Applied Physiology*
*Journal of Clinical Investigation*
*Journal of Foodservice*
*Journal of Hunger and Environmental Nutrition*
*Journal of Human Nutrition and Dietetics*
*Journal of Nutrition*

*Journal of Nutrition Education and Behavior**
*Journal of Nutrition for the Elderly*
*Journal of Nutrition, Health & Aging*
*Journal of Nutritional Biochemistry*
*Journal of Pediatrics*
*Journal of the American College of Nutrition*
*Journal of the American Geriatric Society*
*Journal of the Canadian Dietetic Association**
*Lancet*
*Mayo Clinic Proceedings*
*Medicine & Science in Sports and Exercise*
*Nature*
*New England Journal of Medicine Nutrition*
*Nutrients*
*Nutrition Reviews*
*Nutrition Today**
*Obesity Research*
*Pediatrics*
*The Physician and Sports Medicine*
*Postgraduate Medicine**
*Proceedings of the Nutrition Society*
*Progress in Lipid Research*
*Science*
*Science News**
*Scientific American**

The majority of these journals are available in college and university libraries. As indicated, a few journals will be filed under their abbreviations rather than the first word in their full name. A reference librarian can help you locate any of these sources. The journals with an asterisk (*) are those you may find especially interesting and useful because of the number of nutrition articles presented each month or the less technical nature of the presentation. To search for specific nutrition-related topic(s), use the free digital archive of biomedical and life science journals, PubMedCentral (PMC) from the U.S. National Institutes of Health at www.ncbi.nlm.nih.gov/pmc/.

## Textbooks and Other Sources for Advanced Study of Nutrition Topics

Groff JL and others. *Advanced human nutrition and metabolism.* 7th ed. Belmont, CA: Cengage Learning; 2017.

International Life Sciences Institute. *Present knowledge in nutrition.* 10th ed. Washington, DC: Nutrition Foundation; 2012.

Mahan LK and others. *Krause's food, and the nutrition care process.* 14th ed. Philadelphia: WB Saunders; 2016.

Rodwell V and others. *Harper's illustrated biochemistry.* 31st ed. New York: McGraw-Hill; 2018.

Ross AC and others. *Modern nutrition in health and disease.* 11th ed. Philadelphia: Lippincott Williams & Wilkins; 2012.

Stipanuk MH, Caudill MA. *Biochemical, physiological and molecular aspects of human nutrition.* 4th ed. St. Louis: Elsevier; 2019.

## Newsletters That Cover Nutrition Issues on a Regular Basis

*American Institute for Cancer Research Newsletter*
www.aicr.org

*Consumer Health Digest*
www.quackwatch.com

*Dairy Council Digest*
www.nationaldairycouncil.org

*Environmental Nutrition*
www.environmentalnutrition.com

*Harvard Health Letter*
www.health.harvard.edu/newsletters

*Nutrition Action Healthletter*
www.cspinet.org

*Tufts University Health & Nutrition Letter*
www.healthletter.tufts.edu

*University of California at Berkeley Wellness Letter*
www.wellnessletter.com

## Professional Organizations with a Commitment to Nutrition Issues

Academy of Nutrition and Dietetics
www.eatright.org

American Academy of Pediatrics
www.aap.org

American Cancer Society
www.cancer.org

American College of Sports Medicine
www.acsm.org

American Council on Health and Science
www.acsh.org

American Council on Fitness
www.acefitness.org

American Dental Association
www.ada.org

American Diabetes Association
www.diabetes.org

American Geriatrics Society
www.americangeriatrics.org

American Heart Association
www.heart.org

American Institute for Cancer Research
www.aicr.org

American Medical Association
www.ama-assn.org

American Public Health Association
www.apha.org

American Society for Nutrition
www.nutrition.org

The Calorie Control Council
www.caloriecontrol.org

Canadian Diabetes Association
www.diabetes.ca

Canadian Dietetic Association
www.dietitians.ca

Canadian Nutrition Society
www.cns-scn.ca

Environmental Working Group
www.ewg.org

Food Allergy and Anaphylaxis Network
www.foodallergy.org

Food Marketing Institute
www.fmi.org

Food and Nutrition Board
nationalacademies.org/

Institute of Food Technologies
www.ift.org

National Council on the Aging
www.ncoa.org

National Osteoporosis Foundation
www.nof.org

The Obesity Society
www.obesity.org

Oldways
www.oldwayspt.org

Society for Nutrition Education
www.sne.org

The Vegetarian Resource Group
www.vrg.org

## Professional or Lay Organizations Concerned with Nutrition Issues

Bread for the World Institute
www.bread.org

Food Research and Action Center
www.frac.org

Institute for Food and Development Policy
www.foodfirst.org

International Food Information Council
foodinsight.org/

La Leche League International, Inc.
www.lalecheleague.org

March of Dimes Birth Defects Foundation
www.marchofdimes.com

National Council Against Health Fraud, Inc.
www.ncahf.org

National Eating Disorders Association
www.nationaleatingdisorders.org

National WIC Association
www.nwica.org

Overeaters Anonymous
www.oa.org

Oxfam America
www.oxfamamerica.org

## Local Resources for Advice on Nutrition Issues

Registered dietitian or nutritionists in health-care, city, county, or state agencies, as well as in private practice

Cooperative extension agents in county extension offices and specialists at land-grant universities

Nutrition faculty affiliated with departments of food and nutrition, human ecology, family and consumer sciences, and dietetics

**Government Agencies Concerned with Nutrition Issues or That Distribute Nutrition Information**

*United States*

Centers for Disease Control and Prevention
www.cdc.org

Consumer Information Center
www.pueblo.gsa.gov

Food and Drug Administration (FDA)
www.fda.gov

Food and Nutrition Information and Education Resources Center
www.nal.usda.gov/fnic/education-resources

Health and Human Services
www.hhs.gov

*Healthy People*
healthypeople.gov

Human Nutrition Research Division Agricultural Research Center
www.ars.usda.gov

MyPlate
www.choosemyplate.gov

National Agriculture Library and USDA
www.nutrition.gov

National Cancer Institute
www.cancer.gov

National Center for Health Statistics
www.cdc.gov/nchs

National Heart, Lung, and Blood Institute
www.nhlbi.nih.gov

National Institute on Aging
www.nia.nih.gov

National Institutes of Health
www.health.nih.gov

National Institutes of Health Office of Dietary Supplements
www.ods.od.nih.gov

USDA, Center for Nutrition Policy and Promotion
www.fns.usda.gov/cnpp

USDA, Food Safety & Inspection Service
www.fsis.usda.gov

U.S. Government Printing Office
www.gpo.gov

*Canada*
Canadian Food Inspection Agency
www.inspection.gc.ca

Health Canada
www.hc-sc.gc.ca

*United Nations*
Food and Agriculture Organization (FAO)
www.fao.org

World Health Organization (WHO)
www.who.org

# Appendix M

## DIETARY INTAKE AND ENERGY EXPENDITURE ASSESSMENT

Although it may seem overwhelming at first, it is actually very easy to track the foods you eat. One tip is to record foods and beverages consumed as soon as possible after the actual time of consumption.

**I. Fill in the food record form that follows.** This appendix contains a blank copy (see the completed example in Table M-1). Then, to estimate the nutrient values of the foods you are eating, consult food labels and a nutrient database. If these resources do not have the serving size you need, adjust the value (see Appendix H). If you drink ½ cup of orange juice, for example, but a table has values only for 1 cup, halve all values before you record them. Then, consider grouping all servings of the same food to save time; if you drink a cup of 1% milk 3 times throughout the day, enter milk consumption only once as 3 cups. As you record your intake for use on the nutrient analysis form that follows, consider the following tips.

- Measure and record the amounts of foods eaten in portion sizes of cups, teaspoons, tablespoons, ounces, slices, or inches (or convert metric units to these units).
- Record brand names of all food products.
- Measure and record all those little extras, such as gravies, salad dressings, taco sauces, pickles, jelly, sugar, catsup, and margarine.
- For beverages
  —List the type of milk, such as whole, fat-free, 1%, evaporated, or chocolate.
  —Indicate whether fruit juice is fresh, frozen, or canned.
  —Indicate type for other beverages, such as fruit drink, fruit-flavored drink, and hot chocolate made with water or milk.
- For fruits
  —Indicate whether fresh, frozen, dried, or canned and whether processed in water, light syrup, or heavy syrup.
  —If whole, record number eaten and size with approximate measurements (e.g., 1 apple—3 in. in diameter).
- For vegetables
  —Indicate whether fresh, frozen, dried, or canned.
  —Record as portion of cup, teaspoon, or tablespoon or as pieces (e.g., carrot sticks—4 in. long, ½ in. thick).
  —Record preparation method.
- For cereals
  —Record cooked cereals in portions of tablespoon or cup (a level measurement after cooking).
  —Record dry cereal in level portions of tablespoon or cup.
  —If margarine, milk, sugar, fruit, or another ingredient is added, measure and record amount and type.
- For breads
  —Indicate whether whole-wheat, rye, white, and so on.
  —Measure and record number and size of portion (biscuit—2 in. across, 1 in. thick; slice of homemade rye bread—3 in. by 4 in., ¼ in. thick).
  —Sandwiches: list *all* ingredients (lettuce, mayonnaise, tomato, and so on).

**Table M-1 Food Record Example**

| Time | Minutes Spent Eating | M or S* | H† (0–3) | Activity While Eating | Place of Eating | Food Eaten and Amount | Others Present | Reason for Choosing Food |
|---|---|---|---|---|---|---|---|---|
| 7:10 A.M. | 15 | M | 2 | Standing, fixing lunch | Kitchen | Orange juice, 1 cup<br>Crispix®, 1 cup<br>Fat-free milk, ½ cup<br>Sugar, 2 tsp<br>Black coffee | No one | Health<br>Habit<br>Health<br>Taste<br>Habit |
| 10:00 A.M. | 4 | S | 1 | Sitting, taking notes | Classroom | Diet cola, 12 oz | Class | Weight control |
| 12:15 P.M. | 40 | M | 2 | Sitting, talking | Student center | Chicken sandwich with lettuce and mayonnaise (3 oz chicken, 2 slices bread, 2 tsp mayonnaise)<br>Pear, 1 medium<br>1% milk, 1 cup | Friends | Taste<br><br><br><br><br>Health<br>Health |
| 2:30 P.M. | 10 | S | 1 | Sitting, studying | Library | Regular cola, 12 oz | Friend | Hunger |
| 6:30 P.M. | 35 | M | 3 | Sitting, talking | Kitchen | Pork chop, 1<br>Baked potato, 1<br>Margarine, 2 tbsp<br>Lettuce and tomato salad, 1½ cups<br>Ranch dressing, 2 tbsp<br>Peas, ½ cup<br>Whole milk, 1 cup<br>Cherry pie, 1 small piece<br>Iced tea, 12 oz | Boyfriend | Convenience<br>Health<br>Taste<br>Health<br><br>Taste<br>Health<br>Habit<br>Taste<br>Health |
| 9:10 P.M. | 10 | S | 2 | Sitting, studying | Living room | Apple, 1<br>Mineral water, 12 oz | — | Weight control<br>Weight control |

*M or S: Meal or snack.
†H: Degree of hunger (0 = none; 3 = maximum).

- For meat, fish, poultry, and cheese
  —Give size (length, width, thickness) in inches or weight in ounces after cooking for meat, fish, and poultry (e.g., cooked hamburger patty—3 in. across, ½ in. thick).
  —Give size (length, width, thickness) in inches or weight in ounces for cheese.
  —Record measurements only for the cooked, edible part—without bone or fat that is left on the plate.
  —Describe how meat, poultry, or fish was prepared.
- For eggs
  —Record as soft- or hard-cooked, fried, scrambled, poached, or omelet.
  —If milk, butter, or other ingredients are used, specify kinds and amount.
- For desserts
  —List commercial brand or "homemade" or "bakery" under brand.
  —Specify kind and size of purchased candies, cookies, and cakes.
  —Measure and record portion size of cakes, pies, and cookies by specifying thickness, diameter, and width or length, depending on the item.

**II. Now complete the nutrient analysis form as shown, using your food record.**
A blank copy of this form is printed in this appendix for your use. See the example in Table M-2.

**Table M-2  Nutrient Analysis Example**

| Name | Quantity | kcal | Protein (g) | Carbohydrates (g) | Fiber (g) | Total Fat (g) | Monounsaturated Fat (g) | Polyunsaturated Fat (g) | Saturated Fat (g) | Cholesterol (mg) | Calcium (mg) | Iron (mg) |
|---|---|---|---|---|---|---|---|---|---|---|---|---|
| Egg bagel, 4-inch | 1 | 180 | 7.45 | 34.7 | 0.748 | 1.00 | 0.286 | 0.400 | 0.171 | 44.0 | 20.0 | 2.10 |
| Jelly | 1 tbsp | 49.0 | 0.018 | 12.7 | — | 0.018 | 0.005 | 0.005 | 0.005 | — | 2.00 | 0.120 |
| Orange juice, prepared fresh or frozen | 1½ cups | 165 | 2.52 | 40.2 | 1.49 | 0.210 | 0.037 | 0.045 | 0.025 | — | 33.0 | 0.411 |
| Cheeseburger, McDonald's® | 2 | 636 | 30.2 | 57.0 | 0.460 | 32.0 | 12.2 | 2.18 | 13.3 | 80.0 | 338 | 5.68 |
| French fries, McDonald's® | 1 order | 220 | 3.00 | 26.1 | 4.19 | 11.5 | 4.37 | 0.570 | 4.61 | 8.57 | 9.10 | 0.605 |
| Cola beverage, regular | 1½ cups | 151 | — | 38.5 | — | — | — | — | — | — | 9.00 | 0.120 |
| Pork loin chop, broiled, lean | 4 oz | 261 | 36.2 | — | — | 11.9 | 5.35 | 1.43 | 4.09 | 112 | 5.67 | 1.04 |
| Baked potato with skin | 1 | 220 | 4.65 | 51.0 | 3.90 | 0.200 | 0.004 | 0.087 | 0.052 | — | 20.0 | 2.75 |
| Peas, frozen, cooked | ½ cup | 63.0 | 4.12 | 11.4 | 3.61 | 0.220 | 0.019 | 0.103 | 0.039 | — | 19.0 | 1.25 |
| Margarine, regular or soft, 80% fat | 20 g | 143 | 0.160 | 0.100 | — | 16.1 | 5.70 | 6.92 | 2.76 | — | 5.29 | — |
| Iceberg lettuce, chopped | 2 cups | 14.6 | 1.13 | 2.34 | 1.68 | 0.212 | 0.008 | 0.112 | 0.028 | — | 21.2 | 0.560 |
| French dressing | 2 oz | 300 | 0.318 | 3.63 | 0.431 | 32.0 | 14.2 | 12.4 | 4.94 | — | 7.10 | 0.227 |
| Reduced fat milk | 1 cup | 121 | 8.12 | 11.7 | — | 4.78 | 1.35 | 0.170 | 2.92 | 22.0 | 297 | 0.120 |
| Graham crackers | 2 | 60.0 | 1.04 | 10.8 | 1.40 | 1.46 | 0.600 | 0.400 | 0.400 | — | 6.00 | 0.367 |
| Totals | | 2584 | 99.0 | 300 | 17.9 | 112 | 44.1 | 24.8 | 33.4 | 266 | 792 | 15.4 |
| DRI/nutrient standard* | | 2900 | 58 | 130 | 38 | | | | | | 1000 | 8 |
| % of nutrient needs | | 89 | 170 | 230 | 47 | | | | | | 79 | 193 |

Abbreviations: g = grams, mg = milligrams, µg = micrograms.

*Values from the last pages of this book. The values listed are for a male age 19 years. Note that number of kcal is just a rough estimate. It is better to base energy needs on actual energy output.

†In RAE units. Table values generally are in RE units today because the food values have not been updated to reflect the latest vitamin A standards. RAE equal RE for foods with preformed vitamin A, such as for the pork chop, but RAE are only about half the RE listed for foods with provitamin A carotenoids, such as for the peas (see Chapter 15 for details).

‡Amounts refer to actual folate content rather than dietary folate equivalents (DFEs). This difference is important to consider if the food contains added synthetic folic acid as part of enrichment or fortification. Any such folic acid is absorbed about twice as much as the folate present naturally in foods. Thus, the total contribution of folate in the food in comparison with human needs will be greater than if all the folate were naturally in the food product. Nutrient analysis tables have yet to be updated to reflect the dietary folate equivalents of products (see Chapter 13 for more details).

| Magnesium (mg) | Phosphorus (mg) | Potassium (mg) | Sodium (mg) | Zinc (mg) | Vitamin A (RE)† | Vitamin C (mg) | Vitamin E (mg) | Thiamin (mg) | Riboflavin (mg) | Niacin (mg) | Vitamin B-6 (mg) | Folate (µg)‡ | Vitamin B-12 (µg) |
|---|---|---|---|---|---|---|---|---|---|---|---|---|---|
| 18.0 | 61.0 | 65.0 | 300 | 0.612 | 7.00 | — | 1.80 | 2.58 | 0.197 | 2.40 | 0.030 | 16.3 | 0.065 |
| 0.72 | 1.00 | 16.0 | 4.00 | — | 0.200 | 0.710 | 0.016 | 0.002 | 0.005 | 0.036 | 0.005 | 2.00 | — |
| | | | | | | | | | | | | | |
| 36.0 | 60.0 | 711 | 3.00 | 0.192 | 28.5 | 145 | 0.714 | 0.300 | 0.060 | 0.750 | 0.165 | 163 | |
| 45.8 | 410 | 314 | 1460 | 5.20 | 134 | 4.10 | 0.560 | 0.600 | 0.480 | 8.66 | 0.230 | 42.0 | 1.82 |
| 26.7 | 101 | 564 | 109 | 0.320 | 5.00 | 12.5 | 0.203 | 0.122 | 0.020 | 2.26 | 0.218 | 19.0 | 0.027 |
| 3.00 | 46.0 | 4.00 | 15.0 | 0.049 | — | — | — | — | — | — | — | — | — |
| 34.0 | 277 | 476 | 88.2 | 2.54 | 3.15 | 0.454 | 0.405 | 1.30 | 0.350 | 6.28 | 0.535 | 6.77 | 0.839 |
| 55.0 | 115 | 844 | 16.0 | 0.650 | — | 26.1 | 0.100 | 0.216 | 0.067 | 3.32 | 0.701 | 22.2 | — |
| 23.0 | 72.0 | 134 | 70.0 | 0.750 | 53.4 | 7.90 | 0.400 | 0.226 | 0.140 | 1.18 | 0.090 | 46.9 | — |
| | | | | | | | | | | | | | |
| 0.467 | 4.06 | 7.54 | 216 | 0.041 | 199 | 0.028 | 2.19 | 0.002 | 0.006 | 0.004 | 0.002 | 0.211 | 0.017 |
| 10.1 | 22.4 | 177 | 10.1 | 0.246 | 37.0 | 4.36 | 0.120 | 0.052 | 0.034 | 0.210 | 0.044 | 62.8 | — |
| 5.81 | 3.63 | 7.03 | 666 | 0.045 | 0.023 | — | 15.9 | — | — | — | 0.006 | — | — |
| 33.0 | 232 | 377 | 122 | 0.963 | 140 | 2.32 | 0.080 | 0.095 | 0.403 | 0.210 | 0.105 | 12.0 | 0.89 |
| 6.00 | 20.0 | 36.0 | 86.0 | 0.113 | — | — | — | 0.020 | 0.030 | 0.600 | 0.011 | 1.80 | — |
| 298 | 1425 | 3732 | 3165 | 11.7 | 607 | 204 | 22.5 | 5.52 | 1.79 | 25.9 | 2.14 | 395 | 3.65 |
| 400 | 700 | 4700 | 1500 | 11 | 900† | 90 | 15 | 1.2 | 1.3 | 16 | 1.3 | 400‡ | 2.4 |
| 75 | 204 | 80 | 210 | 106 | 67 | 226 | 150 | 450 | 138 | 162 | 160 | 99 | 152 |

| Time | Minutes Spent Eating | M or S* | H† (0–3) | Activity While Eating | Place of Eating | Food Eaten and Amount | Others Present | Reason for Choosing Food |
|------|------|------|------|------|------|------|------|------|
|  |  |  |  |  |  |  |  |  |
|  |  |  |  |  |  |  |  |  |
|  |  |  |  |  |  |  |  |  |
|  |  |  |  |  |  |  |  |  |
|  |  |  |  |  |  |  |  |  |
|  |  |  |  |  |  |  |  |  |
|  |  |  |  |  |  |  |  |  |
|  |  |  |  |  |  |  |  |  |
|  |  |  |  |  |  |  |  |  |
|  |  |  |  |  |  |  |  |  |
|  |  |  |  |  |  |  |  |  |
|  |  |  |  |  |  |  |  |  |
|  |  |  |  |  |  |  |  |  |
|  |  |  |  |  |  |  |  |  |
|  |  |  |  |  |  |  |  |  |
|  |  |  |  |  |  |  |  |  |
|  |  |  |  |  |  |  |  |  |
|  |  |  |  |  |  |  |  |  |
|  |  |  |  |  |  |  |  |  |
|  |  |  |  |  |  |  |  |  |
|  |  |  |  |  |  |  |  |  |

*M or S: Meal or snack.
†H: Degree of hunger (0 = none; 3 = maximum).

**III. Complete the following table to summarize dietary intake.**

---

**Percent of kcal from Protein, Fat, Carbohydrate, and Alcohol**

**Intake**

| | | | |
|---|---|---|---|
| Protein (P): | ____ g/day × 4 kcal/g | = | (P) ____ kcal/day |
| Fat (F): | ____ g/day × 9 kcal/g | = | (F) ____ kcal/day |
| Carbohydrate (C): | ____ g/day × 4 kcal/g | = | (C) ____ kcal/day |
| Alcohol (A): | | = | (A) ____ kcal/day |
| | Total kcal (T)/day | = | (T) ____ kcal/day |

Percent of kcal from protein:

$\dfrac{(P)}{(T)} \times 100 = \underline{\quad}\%$

Percent of kcal from fat:

$\dfrac{(F)}{(T)} \times 100 = \underline{\quad}\%$

Percent of kcal from carbohydrate:

$\dfrac{(C)}{(T)} \times 100 = \underline{\quad}\%$

Percent of kcal from alcohol:

$\dfrac{(A)}{(T)} \times 100 = \underline{\quad}\%$

Note: The 4 percentages can total slightly more or less than 100% due to rounding. To calculate kcal provided by alcohol, subtract kcal from carbohydrate, fat, and protein from total kcal. The remaining kcal are from alcohol.

**Nutrient Analysis Form**

| Name | Quantity | kcal | Protein (g) | Carbohydrates (g) | Fiber (g) | Total Fat (g) | Monounsaturated Fat (g) | Polyunsaturated Fat (g) | Saturated Fat (g) | Cholesterol (mg) | Calcium (mg) | Iron (mg) |
|------|----------|------|-------------|-------------------|-----------|---------------|-------------------------|-------------------------|-------------------|------------------|--------------|-----------|
|  |  |  |  |  |  |  |  |  |  |  |  |  |
|  |  |  |  |  |  |  |  |  |  |  |  |  |
|  |  |  |  |  |  |  |  |  |  |  |  |  |
|  |  |  |  |  |  |  |  |  |  |  |  |  |
|  |  |  |  |  |  |  |  |  |  |  |  |  |
|  |  |  |  |  |  |  |  |  |  |  |  |  |
|  |  |  |  |  |  |  |  |  |  |  |  |  |
|  |  |  |  |  |  |  |  |  |  |  |  |  |
|  |  |  |  |  |  |  |  |  |  |  |  |  |
|  |  |  |  |  |  |  |  |  |  |  |  |  |
|  |  |  |  |  |  |  |  |  |  |  |  |  |
|  |  |  |  |  |  |  |  |  |  |  |  |  |
|  |  |  |  |  |  |  |  |  |  |  |  |  |
|  |  |  |  |  |  |  |  |  |  |  |  |  |
|  |  |  |  |  |  |  |  |  |  |  |  |  |
| Totals |  |  |  |  |  |  |  |  |  |  |  |  |
| DRI/nutrient standard* |  |  |  |  |  |  |  |  |  |  |  |  |
| % of nutrient needs |  |  |  |  |  |  |  |  |  |  |  |  |

*Values from the last pages of this book. Note that number of kcal is just a rough estimate. It is better to base energy neeeds on actual energy input.

Note that a food such as toast with soft margarine contributes to 2 categories—the grain group and the oils group. You can expect that many food choices will contribute to more than 1 group.

| Magnesium (mg) | Phosphorus (mg) | Potassium (mg) | Sodium (mg) | Zinc (mg) | Vitamin A (RE) | Vitamin C (mg) | Vitamin E (mg) | Thiamin (mg) | Riboflavin (mg) | Niacin (mg) | Vitamin B-6 (mg) | Folate (µg) | Vitamin B-12 (µg) |
|---|---|---|---|---|---|---|---|---|---|---|---|---|---|
| | | | | | | | | | | | | | |
| | | | | | | | | | | | | | |
| | | | | | | | | | | | | | |
| | | | | | | | | | | | | | |
| | | | | | | | | | | | | | |
| | | | | | | | | | | | | | |
| | | | | | | | | | | | | | |
| | | | | | | | | | | | | | |
| | | | | | | | | | | | | | |
| | | | | | | | | | | | | | |
| | | | | | | | | | | | | | |
| | | | | | | | | | | | | | |
| | | | | | | | | | | | | | |
| | | | | | | | | | | | | | |
| | | | | | | | | | | | | | |
| | | | | | | | | | | | | | |
| | | | | | | | | | | | | | |
| | | | | | | | | | | | | | |
| | | | | | | | | | | | | | |

**IV. Evaluation.** Are there weaknesses suggested in your nutrient intake that correspond to missing servings in MyPlate? Consider replacing the missing servings to improve your nutrient intake.

**V. For the same day you keep your food record, also keep a 24-hour record of your activities.** Include sleeping, sitting, and walking, as well as the obvious forms of exercise. Calculate your energy expenditure for these activities using Table 10-5 in Chapter 10 or the software available with this book. Try to substitute a similar activity if your particular activity is not listed. Calculate the total kcal you used for the day. Following is an example of an activity record and a blank form for your use.

**Weight (kg)\*: 70 kg**

| | | | Energy Cost | |
| Activity | Time (Minutes): Convert to Hours | Column 1 kcal/kg/hr (from Table 10-5) | Column 2 (Column 1 × Time) | Column 3 (Column 2 × Weight in kg) |
| --- | --- | --- | --- | --- |
| Brisk walking | (60 min) 1 hr | 4.4 | (× 1) = 4.4 | (× 70) = 308 |

*lb/2.2.

**Weight (kg)\*:**

| | | | Energy Cost | |
| Activity | Time (Minutes): Convert to Hours | Column 1 kcal/kg/hr (from Table 10-5) | Column 2 (Column 1 × Time) | Column 3 (Column 2 × Weight in kg) |
| --- | --- | --- | --- | --- |
| | | | | |
| | | | | |
| | | | | |
| | | | | |
| | | | | |
| | | | | |
| | | | | |
| | | | | |
| | | | | |
| | | | | |
| | | | | |
| | | | | |
| | | | | |
| | | | | |
| | | | | |

Total kcal used (add all items listed in column 3):

*lb/2.2.

# Glossary Terms

## A

**absorption** Process by which nutrient molecules are absorbed by the GI tract and enter the bloodstream.

**absorptive cells** Class of cells, also called enterocytes, that cover the surface of the villi (fingerlike projections in the small intestine) and participate in nutrient absorption.

**Acceptable Daily Intake (ADI)** Amount of a food additive considered safe for daily consumption over one's lifetime.

**Acceptable Macronutrient Distribution Range (AMDR)** Range of macronutrient intake, as percent of energy, associated with reduced risk of chronic diseases while providing for recommended intake of essential nutrients.

**acesulfame K (ay-SUL-fame)** Alternative sweetener that yields no energy to the body; 200 times sweeter than sucrose.

**acetaldehyde** A colorless, volatile liquid with the formula $C_2H_4O$ formed by the oxidation of ethanol.

**acetic acid (a-SEE-tic)** Two-carbon fatty acid used in the synthesis of lipids.

$$CH_3-\overset{\overset{\displaystyle O}{\|}}{C}-OH$$

**acetylcholine (a-SEE-tul-coal-ene)** Neurotransmitter, formed from choline, that is associated with attention, learning, memory, muscle control, and other functions.

**acetyl-CoA (acetyl coenzyme A) (a-SEE-tul)** An important metabolic intermediate formed by the breakdown of glucose, fatty acids, and some amino acids. Its synthesis requires coenzyme A derived from panthothenic acid and acetic acid.

**achlorhydria (ay-clor-HIGH-dre-ah)** Decrease in stomach acid primarily due to age-associated loss of acid-producing gastric cells.

**acidic pH** pH less than 7; for example, lemon juice has an acidic pH.

**acquired (specific) immunity** Immune response that develops over a lifetime; it is initiated by the recognition of a specific antigen and release of antibodies and other specialized immune cells that destroy the antigen. Also called adaptive immunity because after exposure to an antigen the immune system recognizes the antigen and adapts its response to it.

**acquired immune deficiency syndrome (AIDS)** Disorder in which a virus (human immunodeficiency virus [HIV]) infects specific types of immune system cells. This leaves the person with reduced immune function and, in turn, defenseless against numerous infectious agents.

**acrodermatitis enteropathica (Ak-roh-der-MAH-tight-is INN-teer-oh-PATH-ih-cah)** Rare inherited childhood disorder that results in the inability to absorb adequate amounts of zinc from the diet. Symptoms include skin lesions, hair loss, and diarrhea. If untreated, the condition can result in death during infancy or early childhood. Management of this condition is with zinc supplements.

**actin (AK-tin)** Protein in muscle fiber that, together with myosin, is responsible for contraction.

**active absorption** Absorption using a carrier and expending ATP energy. In this way, the absorptive cell can absorb nutrients, such as glucose, against a concentration gradient.

**acute alcohol intoxication** Temporary deterioration in mental and physical function, arising from drinking alcoholic beverages too rapidly. An intoxicated person may be confused and disoriented, lack coordination, and have increasing lethargy. Coma and death can occur.

**acyl carrier protein** Protein, formed from the vitamin pantothenic acid, that attaches to fatty acids and shuttles them through the metabolic pathway that increases their chain length.

**acyl group** Carbonyl portion of a compound, such as an ester.

**ad libitum (ad LIB-itum)** At one's desire or pleasure.

**added sugars** Any sugars added during food manufacturing, such as honey, syrups, and concentrated fruit and vegetables juices. Added sugars do not include sugars found naturally in foods.

**adenine** Nitrogenous base that forms part of the structure of DNA and RNA; a purine.

**adenosine diphosphate (ADP) (ah-DEN-o-scene di-FOS-fate)** Breakdown product of ATP. ADP is synthesized into ATP using energy from foods and a phosphate group (abbreviated $P_i$).

**adenosine monophosphate (AMP) (ah-DEN-o-scene mono-FOS-fate)** Breakdown product of ADP when a phosphate group is removed. AMP is produced when ATP is in short supply.

**adenosine triphosphate (ATP) (ah-DEN-o-scene tri-FOS-fate)** Chemical that supplies energy for many cellular processes and reactions.

**Adequate Intake (AI)** Nutrient intake amount set for any nutrient for which insufficient research is available to establish an RDA. AIs are based on estimates of intakes that appear to maintain a defined nutritional state in a specific life stage.

**adipose tissue (ad-i-POSE)** Group of fat-storing cells.

**ADP** See *adenosine diphosphate.*

**adrenergic (ADD-ren-er-gic)** Relating to the actions of epinephrine and norepinephrine.

**aerobic (air-ROW-bic)** Requiring oxygen.

**aerobic exercise** Physical activity that uses large muscle groups and aerobic respiration.

**aflatoxin** Mycotoxin found in peanuts, corn, tree nuts, and oilseeds that can cause liver cancer.

**aging** Time-dependent physical and physiological changes in body structure and function that occur normally and progressively throughout adulthood as humans mature and become older.

**agrobiodiversity** Variety and variability of animals, plants, and microorganisms that are used directly or indirectly for food and agriculture.

**air displacement** Method for estimating body composition based on the volume of space taken up by a body inside a small chamber.

**alcohol** Compound with a carbon-oxygen bond with the oxygen also bonded to a single hydrogen; also the type of alcohol consumed, ethyl alcohol or ethanol ($CH_3CH_2OH$).

**alcohol abuse** Alcohol consumption that results in severe physical, psychological, or social problems.

**alcohol dehydrogenase (dee-high-DRO-jen-ase)** Enzyme used in alcohol (ethanol) metabolism; the major enzyme used in the liver when alcohol is in low concentration.

**alcohol dependence** Chronic disease that includes the following symptoms: craving, loss of control, withdrawal symptoms, tolerance, and unsuccessful attempts to cut down on use.

**aldehyde (AL-dah-hide)** Organic compound that contains a carbonyl group to which at least 1 hydrogen atom is attached; found in 1 form of vitamin A.

$$\begin{array}{c} O \\ \| \\ -C-H \end{array}$$

**aldosterone (al-DOS-ter-own)** Hormone produced in the adrenal glands that acts on the kidneys, causing them to retain sodium and, therefore, water.

**alimentary canal** Tubular portion of the digestive tract that extends from the mouth to the anus. The word *alimentary* means "relating to nourishment or nutrition."

**alkaline pH** pH greater than 7. Baking soda in water yields an alkaline pH.

**allergen** Substance (e.g., a protein in food) that induces a hypersensitive response, with excess production of certain immune system antibodies. Subsequent exposure to the same protein leads to allergic symptoms.

**allergy** Hypersensitive immune response that occurs when immune bodies produced by us react with a protein we sense as foreign (an antigen).

**alpha (α) bond** Type of chemical bond that can be broken by human intestinal enzymes in digestion; drawn as C⌐o⌐C.

**alpha-linolenic acid (AL-fah-lin-oh-LE-nik)** Essential omega-3 fatty acid with 18 carbons and 3 double bonds (C18:3, omega-3).

**alpha-tocopherol (to-ca-FUR-all)** Most potent form of vitamin E for antioxidant function in humans.

**alveoli (al-VE-o-lye), alveolus** Basic functional units of the lungs where respiratory gases are exchanged.

**Alzheimer disease** Irreversible, abnormal, progressive deterioration of the brain that causes victims to steadily lose the ability to remember, reason, and comprehend.

**amenorrhea (A-men-or-ee-a)** Absence of 3 or more consecutive menstrual cycles; absence of menses in a female.

**amine (amino) group** Nitrogen-containing chemical group (—NH$_2$ derived from ammonia); attached to a single carbon in all amino acids. Examples include tyramine and histamine.

**amino acid (ah-MEE-noh)** Building block for proteins, containing a central carbon atom, an amino group (NH$_2$), a carboxylic acid group (COOH), and a side group.

**amniotic fluid (am-nee-OTT-ik)** Fluid contained in a sac within the uterus. This fluid surrounds and protects the fetus during development.

**AMP** See *adenosine monophosphate*.

**amphetamine (am-FET-ah-mean)** Group of medications that stimulate the central nervous system and have other effects in the body. Abuse is linked to physical and psychological dependence.

**amylase (AM-uh-lace)** Starch-digesting enzyme from the salivary glands or pancreas.

**amylopectin (AM-uh-low-pek-tin)** Digestible branched-chain type of starch composed of multiple glucose units.

**amylose (AM-uh-los)** Digestible straight-chain type of starch made of multiple glucose units.

**anabolic steroids** Hormones that increase strength and muscle mass; known to have severe and sometimes deadly side effects. The use of anabolic steroids is illegal.

**anabolic** Pathways that use small, simple compounds to build larger, more complex compounds.

**anaerobic (AN-ah-ROW-bic)** Not requiring oxygen.

**anaerobic exercise** Physical activity, such as sprinting, that uses anaerobic metabolism.

**anal sphincters** Group of 2 sphincters (inner and outer) that help control expulsion of feces from the body.

**analog (AN-a-log)** Chemical compound that differs slightly from another naturally occurring compound. Analogs generally contain extra or altered chemical groups and may have similar or opposite metabolic effects compared with the native compound; also spelled *analogue*.

**anaphylaxis** Severe allergic response that results in lowered blood pressure and respiratory and gastrointestinal distress. This reaction can be fatal.

**androgenic (AN-dro-jenic)** Hormones that stimulate development in male sex organs—for example, testosterone.

**android obesity (AN-droyd)** Type of obesity in which fat is stored primarily in the abdominal area; defined as a waist circumference greater than 40 inches (102 centimeters) in men and greater than 35 inches (89 centimeters) in women; closely associated with a high risk of cardiovascular disease, hypertension, and type 2 diabetes.

**anemia (ah-NEM-ee-a)** Decreased oxygen-carrying capacity of the blood. Can be caused by many factors, such as iron deficiency or blood loss.

**anencephaly (an-en-SEF-ah-lee)** Fatal birth defect in which parts of the brain and skull are missing.

**anergy (AN-er-jee)** Lack of an immune response to foreign compounds entering the body.

**angiotensinogen** Blood protein, synthesized in the liver, that forms angiotensin I.

**angiotensin I (an-jee-oh-TEN-sin)** Intermediary compound produced during the body's attempt to conserve water and sodium. It is converted in the lungs to angiotensin II.

**angiotensin II** Compound, produced from angiotensin I, that increases blood vessel constriction and triggers production of the hormone aldosterone.

**angular cheilitis (kee-LIE-tis)** Deep cracks at the corners of the mouth; may result from a B-vitamin deficiency.

**animal model** Laboratory animal useful in medical research because it can develop a health condition (e.g., disease or disorder) that closely mimics a human disease and, thus, can be utilized to learn more about causes of a condition and its diagnosis in humans, as well as assess the usefulness and safety of new treatments or preventive actions.

**anion** Negatively charged ion.

**anorexia nervosa (an-oh-REX-ee-uh ner-VOH-sah)** Eating disorder involving a psychological loss or denial of appetite followed by self-starvation; it is related, in part, to a distorted body image and to social pressures.

**anthropometric assessment (an-throw-PO-met-rick)** Pertaining to the measurement of body weight and the lengths, circumferences, and thicknesses of parts of the body.

**antibody (AN-tih-bod-ee)** Blood protein that inactivates foreign proteins found in the body; helps prevent and control infections.

**antibody-mediated immunity** Specific immunity provided by B lymphocytes; also known as humoral immunity.

**antidiuretic hormone (ADH) (an-tie-dye-u-RET-ik)** Hormone, secreted by the pituitary gland that signals the kidney to cause a decrease in water excretion; also called arginine vasopressin.

**antigen (AN-ti-jen)** Foreign substance, generally large in size, that is capable of inducing a specific immune response. Often binds with an antibody.

**antioxidant (an-tie-OX-ih-dant)** Compound that stops the damaging effects of reactive substances seeking an electron (i.e., oxidizing agent). This compound prevents the breakdown of substances in food or the body, particularly lipids. An antioxidant is able to donate electrons to electron-seeking compounds, which helps prevent the breakdown of unsaturated fatty acids and other cell (and food) components by oxidizing agents. Some compounds have antioxidant capabilities (i.e., stop oxidation) but are not electron donors per se.

**anus (A-nus)** Last portion of the GI tract; serves as an outlet for that organ through which feces are expelled.

**aorta (a-ORT-ah)** Major blood vessel of the body, leaving from the left ventricle of the heart.

**apoenzyme (ape-oh-EN-zime)** Inactive enzyme without its cofactor.

**apoferritin (ape-oh-FERR-ih-tin)** Protein in the intestinal cell and liver that binds with the ferric form of iron ($Fe^3$) to form ferritin.

**apolipoprotein (ape-oh-LIP-oh-pro-teen)** Protein attached to the surface of a lipoprotein or embedded in its outer shell. Apolipoproteins can help enzymes function, act as lipid-transfer proteins, or assist in the binding of a lipoprotein to a cell-surface receptor.

**apoptosis (ah-pop-TOE-sis)** Process that occurs over time in which enzymes in a cell set off a series of events that disable numerous cell functions, eventually leading to cell death.

**appetite** Primarily psychological (external) influences that encourage us to find and eat food, often in the absence of obvious hunger.

**arachidonic acid (ar-a-kih-DON-ik)** Omega-6 fatty acid with 20 carbon atoms and 4 carbon-carbon double bonds (C20:4, omega-6); a precursor to some eicosanoids.

**areola (ah-REE-oh-lah)** Circular, dark area of skin surrounding the nipple of the breast.

**ariboflavinosis (ah-rih-bo-flay-vih-NOH-sis)** Condition resulting from a lack of riboflavin; *a* means "without," and *osis* means "a condition of."

**aromatherapy** Use of the vapors of essential oils extracted from flowers, leaves, stalks, fruits, and roots for therapeutic purposes.

**arrhythmias (ah-RITH-me-ahs)** Abnormal heart rhythms that may be too slow, too early, too rapid, or irregular.

**arteriole (ar-TEAR-e-ol)** Tiny artery branch that ends in capillaries.

**arteriosclerosis (ar-TEAR-e-o-scle-ROH-sis)** Chronic disease characterized by abnormal thickening and hardening of the arterial walls, resulting in a lack of elasticity.

**artery** Blood vessel that carries blood away from the heart.

**arthritis** Inflammation at a point where bones join together. The disease has many possible causes.

**ascending colon** First part of the large intestine, between the ileocecal valve and the transverse colon.

**ascites (a-SITE-ease)** Fluid produced by the liver, accumulating in the abdomen, that is a sign of liver failure associated with cirrhosis.

**ascorbic acid** Water-soluble vitamin; also known as vitamin C.

**aseptic processing (ah-SEP-tik)** Method by which a food and a container are separately and simultaneously sterilized, allowing manufacturers to produce boxes of milk and other foods that can be stored at room temperature.

**aspartame (AH-spar-tame)** Alternative sweetener made of 2 amino acids and methanol; about 200 times sweeter than sucrose.

**ataxia (a-TAX-ee-a)** Inability to coordinate muscle activity during voluntary movement; incoordination.

**atherosclerosis (ath-e-roh-scle-ROH-sis)** Buildup of fatty material (plaque) in the arteries, including those surrounding the heart.

**atom** Smallest unit of an element that still has all the properties of the element. An atom contains protons, neutrons, and electrons.

**ATP** See *adenosine triphosphate.*

**atria (A-tree-a)** Plural of *atrium.* Two upper chambers of the heart that receive venous blood.

**atrophic gastritis (A-troh-fik)** Chronic inflammation of the stomach, in which the stomach glands and lining atrophy.

**atrophy (AT-row-fee)** Wasting away of tissue or organs.

**autodigestion** Literally, "self-digestion." The stomach limits autodigestion by covering itself with a thick layer of mucus and producing enzymes and acid only when needed for digestion of foodstuff.

**autoimmune** Immune reactions against normal body cells; self against self.

**avidin (AV-ih-din)** Protein, found in raw egg whites, that can bind biotin and inhibit its absorption. Cooking destroys avidin.

**axon (Ax-on)** Part of a nerve cell that conducts impulses away from the main body of the cell.

# B

**B lymphocyte (LIM-fo-site)** Type of white blood cell that recognizes antigens (e.g., bacteria) present in extracellular sites in the body and is responsible for antibody-mediated immunity. B lymphocytes originate and mature in the bone marrow and are released into the blood and lymph.

**bacteria** Single-cell microorganisms; some produce poisonous substances that cause illness in humans. They contain only 1 chromosome and lack many of the organelles found in human cells. Some can live without oxygen and survive harsh conditions by means of spore formation.

**basal metabolic rate (BMR)** Rate of energy use (e.g., kcal/min) by the body when at rest, fasting, and awake in a warm, quiet environment.

**basal metabolism** Minimum amount of energy the body uses to support itself when fasting, resting, and awake in a warm, quiet environment. It amounts to roughly 1 kcal per kilogram per hour for men and 0.9 kcal per kilogram per hour for women.

**benign** Noncancerous; describes tumors that do not spread.

**beriberi (BEAR-ee-BEAR-ee)** Thiamin deficiency disorder characterized by muscle weakness, loss of appetite, nerve degeneration, and sometimes edema.

**beta (β) bond** Type of chemical bond that cannot be broken by human intestinal enzymes during digestion when it is part of a long chain of glucose molecules (e.g., cellulose); drawn as $C^\cap O_\cup C$.

**betaine (bee-TAINE)** Product of choline metabolism and a methyl (—$CH_3$) donor in methionine metabolism.

**beta-oxidation** Breakdown of a fatty acid into numerous acetyl-CoA molecules; also known as fatty acid oxidation.

**BHA** Butylated hydroxyanisole; a synthetic antioxidant added to food.

**BHT** Butylated hydroxytoluene; a synthetic antioxidant added to food.

**bile** Liver secretion that is stored in the gallbladder and released through the common bile duct into the duodenum. It is essential for the digestion and absorption of fat.

**bile acids** Emulsifiers synthesized by the liver and released by the gallbladder during digestion.

**bilirubin (bi-li-RUBE-in)** Bile pigment derived from hemoglobin during the destruction of red blood cells; excess in the blood causes skin and eyes to become yellow (jaundiced).

**binge drinking** Consumption of 5 or more drinks by men or 4 or more drinks by women at a single occasion.

**binge eating** Eating episode, usually occurring in a discrete period of time (e.g., 2 hours), in which the amount of food consumed is larger than what most individuals would eat in a similar time frame and under comparable circumstances, accompanied by feelings of lack of control over eating during the episodes.

**binge-eating disorder** Eating disorder characterized by recurrent binge eating and feelings of loss of control over eating.

**bioavailability** Degree to which the amount of an ingested nutrient is absorbed and is available to the body.

**biochemical assessment** Assessment focusing on biochemical functions (e.g., concentrations of nutrient by-products or enzyme activities in the blood or urine) related to a nutrient's function.

**biochemical lesion** Indication of reduced biochemical function (e.g., low concentrations of nutrient by-products or enzyme activities in the blood or urine) resulting from a nutritional deficiency.

**biocytin (BY-oh-si-tin)** Biotin bound to the amino acid lysine in food proteins.

**bioelectrical impedance** Method to estimate total body fat that uses a low-energy electrical current. The more fat storage a person has, the more impedance (resistance) to electrical flow will be exhibited.

**biological pest management** Control of agricultural pests by using natural predators, parasites, or pathogens. For example, ladybugs can be used to control an aphid infestation.

**biological value (BV)** Measure of how efficiently food protein, once absorbed from the gastrointestinal tract, can be turned into body tissues.

**biopharming** Use of genetically engineered crops and livestock animals to produce pharmaceutical agents.

**biotechnology** Collection of processes that involve the use of biological systems for altering and, ideally, improving the characteristics of plants, animals, and other forms of life.

**biotin (BY-oh-tin)** Water-soluble vitamin that, in coenzyme form, participates in reactions where carbon dioxide is added to a compound. It is an essential cofactor for enzymes involved in energy and amino acid metabolism and in fatty acid synthesis. Peanuts, liver, and eggs are rich sources. It also can be synthesized by intestinal bacteria.

**bisphosphonates (bis-FOS-foh-nates)** Medications, composed primarily of carbon and phosphorus, that bind to bone mineral and in turn reduce bone breakdown.

**bleaching process** Process by which light depletes the rhodopsin concentration in the eye by separating opsin from all-*trans*-retinal. This fall in rhodopsin concentration allows the eye to become adapted to bright light.

**blind study** Experiment in which the participants, the researchers, or both are unaware of each participant's assignment (test or placebo) or the outcome of the study until it is completed. See also *double-blind study*.

**blood doping** Technique by which an athlete's red blood cell count is increased. Blood is taken from the athlete, the red blood cells are concentrated, and then later they are reinjected into the athlete. Alternately, a hormone may be injected to increase red blood cell synthesis (erythropoetin [Epogen®]).

**body mass index (BMI)** Weight (in kilograms) divided by height (in meters) squared. A normal value is 18.5 to 24.9. A value of 25 or greater indicates a risk of body weight–related health disorders, such as type 2 diabetes and cardiovascular disease, especially when it is 30 or greater. One BMI unit equals 6 to 7 lb.

**bolus (BOWL-us)** Mass of food that is swallowed.

**bomb calorimeter (kal-oh-RIM-eh-ter)** Instrument used to determine the energy content of a food.

**bond** Link between 2 atoms by the sharing of electrons, charges, or attractions.

**bone mass** Total mineral substance (e.g., calcium or phosphorus) in a cross section of bone, generally expressed as grams per centimeter of length.

**bone mineral density** Total mineral content of bone at a specific bone site divided by the width of the bone at that site, generally expressed as grams per cubic centimeter. Bone mineral density tests are used to diagnose osteopenia and osteoporosis.

**bone remodeling** Lifelong process of building and breaking down of bone.

**bone resorption** Process in which osteoclasts break down bone and release minerals, resulting in a transfer of calcium from bone to blood.

**bonking** State of exercise when the muscles and liver have run out of glycogen, characterized by extreme fatigue, confusion, anxiety, and sweating; sometimes referred to as "hitting the wall."

**botulism** Foodborne illness caused by the bacterium *Clostridium botulinum*.

**bran** Outer layer of grains, such as wheat; rich source of dietary fiber.

**branched-chain amino acids** Amino acids that contain branched methyl groups in their side chains; the essential amino acids valine, leucine, and isoleucine.

**bronchial tree (BRON-key-al)** Bronchi and the branches that stem out to bronchioles.

**bronchioles** Smallest division of the bronchi.

**brown adipose tissue (ADD-ih-pose)** Specialized form of adipose tissue that produces large amounts of heat by metabolizing energy-yielding nutrients without synthesizing much useful energy for the body. The unused energy is released as heat.

**brush border** Densely packed microvilli on the intestinal epithelial cells.

**buffer** Compound that helps maintain acid-base balance within a narrow range.

**bulimia nervosa (boo-LEEM-ee-uh)** Eating disorder in which large quantities of food are eaten on a single occasion (binge eating) and counteracted by purging food from the body, fasting, and/or excessive exercise.

**B-vitamins** Group of several water-soluble vitamins that includes thiamin, riboflavin, niacin, pantothenic acid, biotin, vitamin B-6, vitamin B-12, and folate. All B-vitamins function as coenzymes.

# C

**cachexia (ka-KEX-ee-a)** Widespread wasting of the body due to undernutrition and usually associated with chronic disease.

**calcitonin (kal-sih-TONE-in)** Thyroid gland hormone that inhibits bone resorption and lowers blood calcium.

**calcitriol (kal-sih-TRIH-ol)** Name sometimes given to the active hormone form of vitamin D [1,25(OH)$_2$ vitamin D].

**calcium** Major mineral component of bones and teeth. Calcium also aids in nerve impulse transmission, blood clotting, muscle contractions, and other cell functions. Milk and milk products, leafy vegetables, and tofu are good sources.

**calmodulin (kal-MOD-ju-lyn)** Calcium-binding protein occurring in many tissues and participating in the regulation of many biochemical and physiological processes.

**calorie** See *kilocalorie*.

**Campylobacter jejuni (kam-PILE-o-bak-ter je-JUNE-ee)** Bacterium that produces a toxin that destroys the mucosal surfaces of the small and large intestines. *Campylobacter* is a leading cause of bacterial foodborne illness. The chief food sources are raw poultry and meat and unpasteurized milk. It is easily destroyed by cooking.

**cancer** Condition characterized by uncontrolled growth of abnormal body cells.

**cancer initiation** Stage in the process of cancer development that begins with the exposure of a cell to a carcinogen and that results in alterations in DNA. These alterations may cause the cell to no longer respond to normal physiological controls.

**cancer progression** Final stage in the cancer process, during which the cancer cells proliferate, invade surrounding tissue, and metastasize to other sites.

**cancer promotion** Stage in the cancer process during which cell division increases, in turn decreasing the time available for repair enzymes to act on altered DNA and encouraging cells with altered DNA to develop and grow.

**capillary (KAP-ill-air-ee)** Smallest blood vessel; the major site for the exchange of substances between the blood and the tissues.

**capillary beds** Minute blood vessels, 1 cell thick, that create a junction between arterial and venous circulation. Gas and nutrient exchange occurs here between body cells and the bloodstream. **Figure A-6** in Appendix A provides a detailed view of a capillary bed.

**carbamates** Compounds that, like organophosphates, are toxic to the nervous system of insects and animals. Carbamates are less toxic than the organophosphates.

**carbohydrate (kar-bow-HIGH-drate)** Compound containing carbon, hydrogen, and oxygen atoms; most are known as sugars, starches, and fibers; supplies 4 kcal/gram.

**carbohydrate (glycogen) loading** Exercise and eating regimen that increases the amount of glycogen stored in muscles to levels higher than normal.

**carbohydrate counting** Diet method that assigns points (1 point = 15 g of carbohydrate) to each meal and snack.

**carbon skeleton** Amino acid without the amino group (—NH$_2$).

**carbonyl group (KAR-bow-neel)** Parent compound for ketones, aldehydes, and many related groups; (C•O).

**carboxyl group (KAR-BOX-ill)** The COOH group in an organic acid.

**carboxylation** Addition of a carboxyl group, COOH, into a compound or molecule.

**carboxylic acid (KAR-BOX-ih-lik)** Organic molecule with the carboxyl group. Examples include acetic acid and citric acid.

$$
\begin{array}{c}
O \\
\parallel \\
(\text{—C—OH})
\end{array}
$$

**carcinogenesis** Development of cancer.

**carcinogenic (Kar-sin-oh-JEN-ik)** Having the potential to cause cancer.

**carcinoma (Kar-sih-NOH-mah)** Invasive malignant tumor derived from epithelial tissues that cover external and internal areas of the body.

**cardiac muscle** Muscle that makes up the walls of the heart; produces rhythmic, involuntary contractions.

**cardiac output** Amount of blood pumped by the heart.

**cardiomyopathy (Kar-dee-oh-my-OP-ah-thee)** Disease in which the heart muscle is damaged and cannot pump blood efficiently.

**cardiovascular (heart) disease** Disease of the heart and circulatory system, characterized by the deposition of fatty material in the blood vessels (hardening of the arteries), which can lead to organ damage and death; also called coronary heart disease (CHD) because the vessels of the heart are the primary sites of the disease.

**cardiovascular system** Body system consisting of the heart, blood vessels, and blood. This system transports nutrients, waste products, gases, and hormones throughout the body and plays an important role in immune responses and body temperature regulation.

**cariogenic (CARE-ee-oh-jen-ik)** Literally "caries producing"; a substance, often carbohydrate rich (e.g., caramel), that promotes dental caries.

**carnitine (CAR-nih-teen)** Compound used to shuttle fatty acids from the cytosol of the cell into mitochondria.

**carotenoids (kah-ROT-en-oyds)** Pigmented materials in fruits and vegetables that range in color from yellow to orange to red (e.g., beta-carotene); 3 types yield vitamin A activity in humans and thus are called provitamin A. Many have antioxidant properties.

**carpal tunnel syndrome (CAR-pull)** Disease in which nerves that travel to the wrist are pinched as they pass through a narrow opening in a bone in the wrist.

**cartilage** Connective tissue, usually part of the skeleton, composed of cells in a flexible network.

**case-control study** Study in which individuals who have the condition in question, such as lung cancer, are compared with individuals who do not have the condition.

**casein (KAY-seen)** Protein, found in milk, that forms curds when exposed to acid and is difficult for infants to digest.

**catabolic (cat-ah-BOL-ik)** Pathways that break down large compounds into smaller compounds. Energy is usually released.

**catalase pathway** Alternative enzyme pathway to alcohol metabolism. Alcohol is broken down in conjunction with the breakdown of hydrogen peroxide (H$_2$O$_2$) by this enzyme.

**catalase** Enzyme that breaks down hydrogen peroxide (H$_2$O$_2$) to water.

**catalyst (CAT-ul-ist)** Compound that speeds reaction rates but is not altered by the reaction.

**cation** Positively charged ion.

**cecum (SEE-come)** First portion of the large intestine, which connects to the ileum.

**celiac disease (SEE-lee-ak)** Immunological or allergic reaction to the protein gluten in certain grains, such as wheat and rye. The effect is to destroy the intestinal enterocytes, resulting in a much reduced surface area due to flattening of the villi. Elimination of wheat, rye, and certain other grains from the diet restores the intestinal surface.

**cell** Minute structure; the living basis of plant and animal organization. In animals, the cell is bounded by a cell membrane. Cells contain both genetic material and systems for synthesizing energy-yielding compounds. Cells have the ability to take up compounds from and excrete compounds into their surroundings.

**cell differentiation** Process of transforming an unspecialized cell into a specialized cell.

**cell nucleus** Organelle bound by its own double membrane and containing chromosomes that hold the genetic information for cell protein synthesis.

**cell-mediated immunity** Process in which T lymphocytes come in contact with invading cells in order to destroy them.

**cellular respiration** Oxidation (electron removal) of food molecules resulting in the eventual release of energy, CO$_2$, and water. See *respiration*.

**cellulose (SELL-you-lows)** Straight-chain polysaccharide of glucose molecules that is indigestible because of the presence of beta bonds; part of insoluble fiber.

**Celsius (SEL-see-us)** Centigrade measure of temperature; for conversion: (degrees in Fahrenheit – 32) × 5/9 = °C; (degrees in Celsius × 9/5) + 32 = °F.

**central nervous system (CNS)** Brain and spinal cord portions of the nervous system.

**cerebrovascular accident (CVA) (se-REE-bro-VAS-cue-lar)** Death of part of the brain tissue due typically to a blood clot; also called stroke *or brain attack*.

**ceruloplasmin (se-RUE-low-PLAS-min)** Blue, copper-containing protein in the blood that can remove an electron from Fe$^{2+}$ (ferrous form) to yield Fe$^{3+}$ (ferric form). The Fe$^{3+}$ form can bind with iron transport and storage proteins, such as transferrin.

**chain-breaking** Breaking the link between 2 or more behaviors that encourage overeating, such as snacking while watching television.

**chelates (KEY-lates)** Complexes formed between metal ions and substances with polar groups, such as proteins. The polar groups form 2 or more attachments with the metal ions, forming a ringed structure.

The metal ion is then firmly bound and sequestered.

**chelation (key-LAY-shun)** Use of medicinal compounds, such as ethylene-diamine-tetra-acetic acid (EDTA), to bind metals and other constituents in the blood.

**chemical reaction** Interaction between 2 or more chemicals that changes both chemicals.

**chemical score** Ratio comparing the essential amino acid content of the protein in a food with the essential amino acid content in a reference protein. The lowest amino acid ratio calculated for any essential amino acid is the chemical score.

**chief cell** Gastric gland cell that secretes pepsinogen.

**Child and Adult Care Food Program** U.S. government program that provides nutritious meals and snacks to low-income children enrolled in child-care centers or residing in emergency shelters, as well as adults who are functionally impaired or age 60 and older in nonresidential adult day-care centers.

**chiral (KI-rell)** Carbon atom with 4 different atoms or groups of atoms attached.

**chloride** Major negative ion of extracellular fluid; aids in nerve impulse transmission and fluid balance in conjunction with sodium and potassium. It contributes to the function of white blood cells, aids in the transport of carbon dioxide from cells to the lungs, and is a component of hydro-chloric acid production in the stomach. Salt supplies most of the chloride in the diet.

**cholecystokinin (CCK) (ko-la-sis-toe-KY-nin)** Hormone that stimulates enzyme release from the pancreas and bile release from the gallbladder.

**cholera (KOL-er-a)** See *Vibrio cholerae*.

**cholesterol (ko-LES-te-rol)** Waxy lipid found in all body cells. It has a structure containing multiple chemical rings. It is an important component of cell membranes and serves as a precursor to many important biological compounds. Dietary cholesterol is found only in foods that contain animal products.

**choline (COAL-ene)** Water-soluble vitamin-like compound that functions as a precursor for acetylcholine, a neurotransmitter associated with attention, learning and memory, muscle control, and many other functions. Protein foods, especially eggs, are rich in choline.

**cholinergic (coal-in-NER-jic)** Relating to the actions of acetylcholine.

**chromium** Trace mineral that enhances the action of insulin. Egg yolks, whole grains, pork, nuts, and mushrooms are good sources.

**chromosome** Complex of DNA and protein containing the genetic material of a cell's nucleus. There are 46 chromosomes in the nucleus of each cell, except in germ cells.

**chronic (KRON-ik)** Long-standing, developing over time. When referring to disease, this term indicates that the disease tends to progress slowly; an example is cardio-vascular disease.

**chylomicron (kye-lo-MY-kron)** Lipoprotein made of dietary fats that are surrounded by a shell of cholesterol, phospholipids, and protein. Chylomicrons are formed in the absorptive cells (enterocytes) in the small intestine after fat absorption and travel through the lymphatic system to the bloodstream.

**chyme (KIME)** Liquid mixture of stomach secretions and partially digested food.

**ciguatera toxin (see-gwah-TER-ah)** Seafood toxin that causes gastrointestinal, neuro-muscular, and respiratory symptoms. It is most common in large fish from tropical waters.

**circular folds** Numerous folds of the mucous membrane of the small intestine.

**cirrhosis (see-ROH-sis)** Chronic degenerative disease, caused by poisons (e.g., alcohol) that damage liver cells, that results in a reduced ability to synthesize proteins and metabolize nutrients, drugs, and poisons.

*cis* **configuration (SIS)** Form seen in compounds with double bonds, such as fatty acids, in which the hydrogens on both ends of the double bond lie on the same side of the plane of that bond in the cell mitochondria.

**citric acid cycle** Pathway that breaks down acetyl-CoA, yielding carbon dioxide, $FADH_2$, $NADH^+ H^+$, and GTP. The pathway also can be used to synthesize compounds; also known as the tricarboxylic acid cycle (TCA cycle) and the Krebs cycle.

**clinical assessment** Physical evidence of diet-related disease. This type of assessment focuses on the general appearance of skin, eyes, and tongue; evidence of rapid hair loss; loss of sense of touch; and loss of ability to cough and walk.

**clinical lesion** Sign seen on physical examination or a symptom perceived by the patient resulting from a nutritional deficiency.

**clinical symptoms** Changes in health status noted by the individual (e.g., stomach pain) or clinician during physical examination (the latter is technically called a clinical sign).

**cloning** Creating genetically identical animals by nonsexual reproduction.

**Clostridium botulinum (closs-TRID-ee-um bot-u-LYE-num)** Bacterium in soil and possibly in food in the form of bacteria or spores. This bacterium multiplies in the absence of air and produces a deadly toxin. *C. botulinum* thrives primarily in canned food, especially incorrectly home-canned, low acid foods, such as string beans, corn, mushrooms, beets, asparagus, and garlic.

**Clostridium perfringens (per-FRING-ens)** Toxin-producing bacterium living throughout the environment, especially in soil, the intestinal tract of humans and animals, and sewage. It is often referred to as the "cafeteria germ" because most out-breaks of foodborne illness caused by it are associated with the food service industry or with events where large quantities of food are prepared and served. *Clostridium* thrives in an oxygen-free environment and forms heat-resistant spores.

**coagulation** Formation of blood clot.

**cobalamin (koh-BAL-ah-meen)** Vitamin B-12, a water-soluble vitamin.

**Cochrane Collaboration** Source of systematic reviews to inform health-care decisions.

**codon (KOH-don)** Specific sequence of 3 nucleotide units within DNA that codes particular amino acids needed for protein synthesis.

**coenzyme** Compound that combines with an inactive protein, called an apoenzyme, to form a catalytically active enzyme, called a holoenzyme. In this manner, coenzymes aid in enzyme function.

**cofactors** Organic or inorganic substance that binds to a specific region on an enzyme and is necessary for the enzyme's activity.

**cognitive behavior therapy** Psychological therapy in which a person's assumptions about dieting, body weight, and related issues are challenged. New ways of thinking are explored and then practiced by the person. In this way, the person can learn new ways to control disordered eating behaviors and related life stress.

**cognitive restructuring** Changing one's frame of mind regarding eating—for example, instead of using a difficult day as an excuse to overeat, substituting other pleasures or rewards, such as a relaxing walk with a friend.

**cohort study** Research that follows a healthy population over time, looking for indicators of the development of disease.

**colic (KOL-ik)** Sharp abdominal pain that generally occurs in otherwise healthy infants and is associated with periodic spells of inconsolable crying.

**colipase (co-LIE-pace)** Protein, secreted by the pancreas, that changes the shape of pancreatic lipase, facilitating its action.

**colitis (koh-LIE-tis)** Inflammation of the colon that can lead to ulcers (ulcerative colitis).

**collagen (KOL-ah-jen)** Major protein of the material that holds together the various structures of the body.

**colostrum (ko-LAHS-trum)** First fluid secreted by the breast during late pregnancy and the 1st few days after birth. This thick fluid is rich in immune factors and protein.

**Commodity Foods Program** U.S. government program that distributes, free of charge, surplus agricultural products (e.g., cheese, peanut butter, canned foods), produced by U.S. farmers, to low-income households.

**comorbid** Disease process that accompanies another disease. For example, if hypertension develops as obesity is established, hypertension is said to be a comorbid condition accompanying the obesity.

**compact bone** Dense, compact bone that constitutes the outer surface and shaft of bone; makes up 75 to 80% of total bone mass; also called cortical bone.

**complement** Series of blood proteins that participate in a complex reaction cascade following stimulation by an antigen-antibody complex on the surface of a bacterial cell. Various activated complement proteins can enhance phagocytosis, contribute to inflammation, and destroy bacteria.

**Complementary and Alternative Medicine (CAM)** Medical or health-care system, practice, or product not presently part of conventional medicine; also called complementary care and integrative medicine.

**complementary proteins** Two food protein sources that make up for each other's inadequate supply of specific essential amino acids. Together, they yield a sufficient amount of all 9 and, so, provide high quality (complete) protein for the diet.

**complete proteins** Proteins that contain ample amounts of all 9 essential amino acids.

**complex carbohydrate** Carbohydrate composed of many monosaccharide molecules. Examples include glycogen, starch, and fiber.

**compound** Atoms of 2 or more elements bonded together in specific proportions. Not all chemical compounds exist as molecules. Some compounds are made up of ions attracted to each other, such as $Na^+Cl^-$ (table salt).

**compression of morbidity** Delay of the onset of disabilities caused by chronic disease.

**concentration gradient** Difference in the concentration of a solute from 1 area to another.

Normally, a solute moves from where it is most concentrated to where it is least concentrated. When sodium is pumped outside the cell and potassium pumped inside the cell, they are moving instead to where each is most concentrated—that is, against the concentration gradient.

**conceptus (kon-SEP-tus)** Developmental stage derived from the fertilized ovum (zygote) until birth. The conceptus includes the extra-embryonic membranes, as well as the embryo or fetus.

**condensation reaction** Chemical reaction in which a bond is formed between 2 molecules by the elimination of a small molecule, such as water.

**cones** Sensory elements in the retina of the eye responsible for visual processes that occur under bright light, translating objects into color images.

**confabulation** Replacing a gap in a person's memory with a fabricated memory that he or she believes to be true.

**congenital (con-JEN-i-tal)** Literally, "present at birth." A congenital abnormality is a defect that has been present since birth. These defects may be inherited from the parents, may occur as a result of damage or infection while in the uterus, or may occur at the time of birth.

**congestive heart failure** Condition resulting from severely weakened heart muscle, resulting in ineffective pumping of blood. This leads to fluid retention, especially in the lungs. Symptoms include fatigue, difficulty breathing, and leg and ankle swelling.

**conjugase (KON-ju-gase)** Enzyme systems in the intestine that enhance folate absorption; they remove glutamate molecules from polyglutamate forms of folate.

**conjunctiva (kon-junk-TEA-vah)** Mucous membrane covering the front surface of the eye and the lining of the eyelids.

**connective tissue** Cells and their protein products that hold different structures in the body together. Tendons and cartilage are composed largely of connective tissue. Connective tissue also forms part of bone and the nonmuscular structures of arteries and veins.

**constipation** Condition characterized by infrequent and often painful bowel movements.

**contingency management** Forming a plan of action to respond to a situation in which overeating is likely, such as when snacks are within arm's reach at a party.

**control group** Participants in an experiment who are not given the treatment being tested.

**copper** Trace mineral that aids in iron metabolism, functions in antioxidant enzyme systems and with enzymes involved in connective tissue metabolism, and is used

in hormone synthesis. Liver, cocoa, beans, nuts, and whole grains are good sources.

**cortical bone (KORT-ih-kal)** See *compact bone*.

**corticosteroid (kor-ti-ko-STARE-oyd)** Steroid produced by the adrenal gland (e.g., cortisol).

**cortisol (KORT-ih-sol)** Hormone made by the adrenal glands that, among other functions, stimulates the production of glucose from amino acids and increases the desire to eat.

**covalent bond (ko-VAY-lent)** Union of 2 atoms formed by the sharing of electrons.

**creatine (CR) (CREE-a-tin)** Organic molecule in muscle cells that serves as a part of the high-energy compound creatine phosphate, or phosphocreatine.

**creatinine (cree-A-tin-in)** Nitrogenous waste product of the compound creatine found in muscles.

**cretinism (KREET-in-ism)** Stunting of body growth and mental development during fetal and later development that results from inadequate maternal intake of iodine during pregnancy.

**CRISPR-Cas9** A gene-editing technology that permits scientists to precisely cut out, modify, or add to DNA in human, animal, plant, and other cells. *CRISPR* refers to a segment of regularly repeating DNA found in some bacteria. *Cas9* refers to enzymes and RNA associated with the CRISPR.

**critical period** Finite period during embryonic and fetal development when cells for a particular tissue or organ can develop.

**Crohn's disease** Inflammatory disease of the GI tract that often limits the absorptive capacity of the small intestine. Family history is a major risk factor.

**crude fiber** Outdated term for what remains of fiber after extended acid and alkaline treatment. Crude fiber consists primarily of cellulose and lignins.

**cryptosporidiosis (krip-toe-spore-id-ee-O-sis)** Intestinal disease, characterized by diarrhea, that originates from a protozoan parasite of the genus *Cryptosporidium*.

**Cushing disease** Endocrine disorder characterized by elevated blood levels of the hormone cortisol. High cortisol levels can lead to the breakdown of body proteins, such as those in the skin and muscle.

**cyclamate (sigh-cla-MATE)** Alternative sweetener that yields no energy to the body; 30 times sweeter than sucrose. Not a legal food additive in the U.S.

**cyclooxygenase (COX) (sigh-clo-OXY-jen-ase)** Enzyme used to synthesize prostaglandins, thromboxanes, and other eicosanoids.

**cystic fibrosis (SIS-tik figh-BRO-sis)** Inherited disease that can cause overproduction of mucus. Mucus can block the pancreatic duct, decreasing enzyme output.

**cytochrome (SITE-o-krome)** Electron-transfer compound that participates in the electron transport chain.

**cytochrome P450** Set of enzymes in cells, especially in the liver, that act on compounds foreign to the body. This action aids in their excretion but also creates short-lived, highly reactive forms.

**cytokine (SITE-o-kine)** Protein, secreted by a cell, that regulates the activity of neighboring cells.

**cytoplasm (SITE-o-plas-um)** Fluid and organelles (except the nucleus) in a cell.

**cytosine (SIH-toe-zeen)** Nitrogenous base that forms part of the structure of DNA and RNA; a pyrimidine.

**cytosol (SHI-tae-sall)** Water-based phase of a cell's cytoplasm; excludes organelles, such as mitochondria.

**cytotoxic T cell (cite-o-TOX-ik)** Type of T cell that interacts with an infected host cell through special receptor sites on the T cell surface.

**cytotoxic test** Unreliable test to diagnose food allergies; it involves mixing white blood cells with food proteins.

# D

**Daily Reference Values (DRVs)** Nutrient-intake standards established for protein, carbohydrate, and some dietary components lacking an RDA or a related nutrient standard, such as total fat intake. The DRVs for sodium and potassium are constant; those for the other nutrients increase as energy intake increases. The DRVs constitute part of the Daily Values used in food labeling.

**Daily Values** Standard nutrient-intake values developed by the FDA and used as a reference for expressing nutrient content on nutrition labels. The Daily Values include 2 types of standards—RDIs and DRVs.

**danger zone** Temperature range of 41°F to 135°F, which supports the growth of pathogenic bacteria.

**dark adaptation** Process by which the rhodopsin concentration in the eye increases in dark conditions, allowing improved vision in the dark.

**deamination (dee-am-ih-NA-shun)** Removal of an amino group from an amino acid.

**decarboxylation (dee-car-box-ih-LAY-shun)** Removal of 1 molecule of carbon dioxide from a compound.

**decubitus ulcer (dee-CUBE-ih-tus)** Chronic ulcer (also called bedsore) that appears in pressure areas of the skin over a body prominence. These sores develop when people are confined to bed or otherwise immobilized.

**de-esterification** Process of removing a fatty acid from a glycerol molecule.

**defecation** Expulsion of feces from the rectum.

**dehydroascorbic acid (DEE-hy-dro-ah-scor-bik)** Oxidized form of ascorbic acid (vitamin C).

**Delaney Clause** Clause in the 1958 Food Additives Amendment of the Pure Food and Drug Act in the U.S. that prevents the intentional (direct) addition to foods of a compound that has been shown to cause cancer in laboratory animals or humans.

**dementia (de-MEN-sha)** General, persistent loss of or decrease in mental function.

**denaturation (dee-NAY-ture-a-shun)** Alteration of a protein's 3-dimensional structure, usually because of treatment by heat, enzymes, acidic or alkaline solutions, or agitation.

**dendrite (DEN-drite)** Relatively short, highly branched nerve cell process that carries electrical activity to the main body of the nerve cell.

**dental caries (KARE-ees)** Erosions in the surface of a tooth caused by acids made by bacteria as they metabolize sugars.

**deoxyribonucleic acid (DNA) (DEE-awks-ee-ry-boh-noo-KLAY-ik)** Site of hereditary information in cells. DNA directs the synthesis of cell proteins.

**depolarization** Reversal of membrane potential, which triggers generation of the nerve impulse in nerve cells.

**dermatitis (dur-ma-TIE-tis)** Inflammation of the skin.

**dermis (DUR-miss)** Second, or deep, layer of the skin under the epidermis.

**descending colon** Part of the large intestine between the transverse colon and the sigmoid colon.

**desirable nutritional status** State in which body tissues have enough of a nutrient to support normal functions and to build and maintain surplus stores.

**DEXA** See *dual energy X-ray absorptiometry.*

**dextrin** Partial breakdown product of starch that contains few to many glucose molecules. These appear when starch is being digested into many units of maltose by salivary and pancreatic amylase.

**diabetes (DYE-uh-BEET-eez)** Disease characterized by high blood glucose, resulting from either insufficient or no release of the hormone insulin by the pancreas or the general inability of insulin to act on certain body cells, such as muscle cells. The 2 major forms are type 1 (requires daily insulin therapy) and type 2 (may or may not require insulin therapy).

**diabetic ketoacidosis** Condition in which high amounts of ketones are in the blood, usually due to uncontrolled type 1 diabetes.

**diarrhea** Loose, watery stools occurring more than 3 times per day.

**diastolic blood pressure (dye-ah-STOL-ik)** Pressure in the arterial blood vessels when the heart is between beats.

**dietary assessment** Assessment that focuses on one's typical food choices, relying mostly on the recounting of one's usual intake or a record of one's intake of the previous day.

**dietary fiber** Fiber in food.

**Dietary Guidelines for Americans** General goals for nutrient intakes and diet composition set by the USDA and the U.S. Department of Health and Human Services.

**Dietary Reference Intakes (DRIs)** Nutrient recommendations made by the Food and Nutrition Board, a part of the Institute of Medicine, and the National Academy of Science. These include EARs, RDAs, AIs, EERs, and ULs.

**dietitian** See *registered dietitian.*

**diffusion** Net movement of molecules or ions from regions of higher concentration to regions of lower concentration.

**digestibility (dye-JES-tih-bil-i-tee)** Proportion of food substances eaten that can be broken down into individual nutrients in the intestinal tract for absorption into the body.

**digestion** Process by which large ingested molecules are mechanically and chemically broken down to produce smaller molecules that can be absorbed across the wall of the GI tract.

**digestive enzymes** Compounds that aid in the breakdown of carbohydrates, fats, and proteins.

**digestive system** Body system consisting of the gastrointestinal tract and accessory structures, such as the liver, gallbladder, and pancreas. This system performs the mechanical and chemical processes of digestion, absorption of nutrients, and formation and elimination of feces.

**diglyceride (dye-GLISS-er-ide)** Breakdown product of a triglyceride; consists of 2 fatty acids bonded to a glycerol backbone.

**dihomo-gamma-linolenic acid (dye-homo-gama-linoh-lenik)** Omega-6 fatty acid with 20 carbons and 3 double bonds; the precursor to some eicosanoids.

**direct calorimetry (kal-oh-RIM-eh-tree)** Method of determining a body's energy use by measuring heat that is released from the body, usually using an insulated chamber.

**disaccharide (dye-SACK-uh-ride)** Class of sugars formed by the chemical bonding of 2 monosaccharides.

**discretionary calories** Amount of energy theoretically allowed in a diet after a person has met overall nutrition needs. This generally small amount of energy gives individuals the flexibility to consume some foods and beverages that contain alcohol, added sugars, or added fats (e.g., many snack foods).

**disordered eating** Mild to short-term abnormal changes in eating patterns that occur in relation to a stressful event, an illness, or a desire to modify one's diet for a variety of health and personal appearance reasons.

**distill** To separate 2 or more liquids that have 2 different boiling points. Alcohol is boiled off and the vapors are collected and condensed. Distillation produces a high alcohol content in hard liquor.

**disulfide group (dye-SUL-fide)** Two sulfur atoms (S—S), each attached to a carbon; important structural characteristic of some proteins.

**diuretic (dye-u-RET-ik)** Substance that, when ingested, increases the flow of urine.

**diverticula (DYE-ver-TIK-you-luh)** Pouches that protrude through the exterior wall of the large intestine.

**diverticulitis (DYE-ver-tik-you-LITE-us)** Inflammation of the diverticula caused by acids produced by bacterial metabolism inside the diverticula.

**diverticulosis (DYE-ver-tik-you-LOW-sus)** Condition of having many diverticula in the large intestine.

**DNA transcription** Process of forming messenger RNA (mRNA) from a portion of DNA.

**DNA** See *deoxyribonucleic acid.*

**docosahexaenoic acid (DHA) (DOE-co-sa-hex-ee-noik)** Omega-3 fatty acid with 22 carbons and 6 carbon-carbon double bonds (C22:6, omega-3). It is present in large amounts in fatty fish and is slowly synthesized in the body from alpha-linolenic acid. DHA is concentrated in the retina and brain.

**dopamine (DOPA) (DOE-pah-mean)** Neurotransmitter that leads to feelings of euphoria, among other functions; also forms norepinephrine.

**double-blind study** Experiment in which neither the participants nor the researchers are aware of each participant's assignment (test or placebo) or the outcome of the study until it is completed. An independent 3rd party holds the code and the data until the study has been completed.

**dual energy X-ray absorptiometry (DXA)** Highly accurate method of measuring body composition and bone mass and density using multiple low-energy X-rays.

**duodenum (doo-oh-DEE-num, or doo-ODD-num)** First portion of the small intestine; leads from the pyloric sphincter to the jejunum.

**duration** Length of time (e.g., duration of an exercise session).

**dysbiosis** An imbalance in the intestinal bacteria that precipitates changes in the normal activities of the gastrointestinal tract, possibly resulting in health problems.

**dyslipidemia (DIS-lip-ah-DEEM-E-ah)** State in which various blood lipids, such as LDL or triglycerides, are greatly elevated or, in the case of HDL, are very low.

# E

**E. coli** See *Escherichia coli.*

**early childhood caries** Tooth decay that results from formula or juice (and even human milk) bathing the teeth as the child sleeps with a bottle in his or her mouth. Upper teeth are affected mostly because the lower teeth are protected by the tongue; formerly called nursing bottle syndrome and baby bottle tooth decay.

**early term birth** Baby born between 37 weeks 0 days and 38 weeks 6 days of gestation.

**eating disorder** Severe alterations in eating patterns linked to physiological changes; the alterations include food restricting, binge eating, purging, weight fluctuations, and emotional and cognitive changes in perceptions of one's body.

**eclampsia (ee-KLAMP-see-ah)** See *pregnancy-induced hypertension.*

**ecosystem (ek-OH-sis-tum)** "Community" in nature that includes plants, animals, and the environment.

**edamame (ed-a-MOM-ee)** Fresh green soybeans.

**edema (uh-DEE-muh)** Buildup of excess fluid in extracellular spaces.

**EFNEP** See *Expanded Food and Nutrition Education Program.*

**eicosanoids (eye-KOH-san-oyds)** Hormone-like compounds synthesized from polyunsaturated fatty acids, such as omega-3 fatty acids and omega-6 fatty acids.

**eicosapentaenoic acid (EPA) (eye-KOH-sah-pen-tahee-NO-ik)** Omega-3 fatty acid with 20 carbons and 5 carbon-carbon double bonds (C20:5, omega-3). It is present in large amounts in fatty fish and slowly synthesized in the body from alpha-linolenic acid. EPA is a precursor to some eicosanoids.

**Elderly Nutrition Program** U.S. government program that provides nutrition services through the Congregate Meal Program and Home Delivered Meal Program (often referred to as Meals on Wheels) to anyone age 60 and older.

**electrolytes (ih-LEK-tro-lites)** Compounds that separate into ions in water and, in turn, are able to conduct an electrical current. These include sodium, chloride, and potassium.

**electron transport chain** Series of reactions using oxygen to convert NADH + H$^+$ and FADH$_2$ molecules to free NAD$^+$ and FAD molecules with the donation of electrons and hydrogen ions to oxygen, yielding water and ATP.

**electron** Part of an atom that is negatively charged. Electrons orbit the nucleus.

**element** Substance that cannot be separated into simpler substances by chemical processes. Common elements in nutrition include carbon, oxygen, hydrogen, nitrogen, calcium, phosphorus, and iron.

**e-Library of Evidence for Nutrition Actions (eLENA)** Source of systematic reviews to help countries implement effective nutrition interventions, policies, and programs.

**elimination diet** Restrictive diet that systematically tests foods that may cause an allergic response by first eliminating them for 1 to 2 weeks and then adding them back, 1 at a time.

**embryo (EM-bree-oh)** In humans, the developing in utero offspring from about the beginning of the 3rd week to the end of the 8th week after conception.

**empty calorie foods** Foods that tend to be high in sugar and/or fat and low in other nutrients; that is, the calories are "empty" of nutrients.

**emulsifier (ee-MULL-sih-fire)** Compound that can suspend fat in water by isolating individual fat droplets using a shell of water molecules or other substances to prevent the fat from coalescing.

**endemic (en-DEM-ik)** Habitual presence within a given geographic area of a disease.

**endocrine cells** Cells throughout the gastrointestinal tract that contain regulatory peptides and/or biogenic amines.

**endocrine disrupter** Substance that interferes with the normal function of hormones produced in the body.

**endocrine gland (EN-doh-krin)** Hormone-producing gland.

**endocrine system** Body system consisting of the various glands and the hormones these glands secrete. This system has major regulatory functions in the body, such as in reproduction and cell metabolism.

**endocytosis (phagocytosis/pinocytosis)** Active absorption in which the absorptive cell forms an indentation in its membrane, and then particles (in phagocytosis) or fluids (in pinocytosis) entering the indentation are engulfed by the cell.

**endometrium (en-doh-ME-tree-um)** Membrane that lines the inside of the uterus. It increases in thickness during the menstrual cycle until ovulation occurs. The surface layers are shed during menstruation if conception does not take place.

**endoplasmic reticulum (ER) (en-doh-PLAZ-mik re-TIK-u-lum)** Organelle in the cytoplasm composed of a network of canals running through the cytoplasm. Rough ER contains ribosomes. Smooth ER contains no ribosomes.

**endorphins (en-DOR-fins)** Natural body tranquilizers that may be involved in the feeding response and function in pain reduction.

**endoscope** Small, lighted, flexible medical instrument with a camera that is used to exam areas inside the body; usually inserted into a natural opening, such as using the nose or mouth to observe the esophagus.

**endosperm** Starch interior of a cereal grain.

**endothelial cells (en-doh-THEE-lee-al)** Flat cells lining the blood vessels and lymphatic vessels and the chambers of the heart.

**enema (EN-ah-mah)** Injection of liquid through the rectum to cause the elimination of fecal matter.

**energy balance** State in which energy intake, in the form of food and beverages, matches energy expended, primarily through basal metabolism and physical activity.

**energy density** Comparison of the energy content of a food with the weight of the food. An energy dense food is high in energy but weighs very little (e.g., many fried foods), whereas a food low in energy density (e.g., an orange) weighs a lot but is low in energy content.

**energy equilibrium** State in which energy intake equals energy use; thus, the body maintains a stable condition.

**enriched** Term generally meaning that the vitamins thiamin, niacin, riboflavin, and folate and the mineral iron have been added to a grain product to improve its nutritional quality.

**enterocytes (en-TER-oh-sites)** Specialized absorptive cells in the villi of the small intestine.

**enterohepatic circulation (EN-ter-oh-heh-PAT-ik)** Continual recycling of compounds between the small intestine and the liver. Bile is an example of a recycled compound.

**environmental assessment** Assessment that focuses on one's education and economic background and other factors that affect one's ability to purchase, transport, and cook food and follow instructions given by health-care providers.

**enzyme (EN-zime)** Compound that speeds the rate of a chemical process but is not altered by the process. Almost all enzymes are proteins (some are made of nucleic acids).

**epidemiology (ep-uh-dee-me-OLL-uh-gee)** Distribution and determinants of diseases in human populations.

**epidermis (ep-ih-DUR-miss)** Outermost layer of the skin; composed of epithelial layers.

**epigenetic carcinogens (promoters) (ep-ih-je-NET-ik car-SIN-oh-jens)** Compounds that increase cell division and thereby increase the chance that a cell with altered DNA will develop into cancer.

**epigenetics** Non-Mendelian inheritance of DNA modifications that may influence gene expression on 1 or more alleles. Methylation of DNA is an example.

**epiglottis (ep-ih-GLOT-iss)** Flap that folds down over the trachea during swallowing.

**epinephrine (ep-ih-NEF-rin)** Hormone produced by the adrenal gland in times of stress. It also may have neurotransmitter functions, such as in the brain.

**epiphyseal line (ep-ih-FEES-ee-al)** Line that replaces the epiphyseal plate when bone growth is complete.

**epiphyseal plate** Cartilage-like layer in the long bone sometimes referred to as the growth plate. It functions in linear growth. During childhood, the cartilage cells multiply and absorb calcium to develop into bone.

**epiphyses (e-PIF-ih-seas)** Ends of long bones. The epiphyseal plate—sometimes referred to as the growth plate—is made of cartilage and allows bone to grow. During childhood, the cartilage cells multiply and absorb calcium to develop into bone.

**epithelial tissue (ep-ih-THEE-lee-ul)** Surface cells that line the outside of the body and all passageways within it.

**epithelium (ep-ih-THEE-lee-um)** Covering of internal and external surfaces of the body, such as the lungs, GI tract, blood vessel linings, and skin.

**equilibrium (ee-kwih-LIB-ree-um)** In nutrition, a state in which nutrient intake equals nutrient losses. Thus, the body maintains a stable condition, such as energy equilibrium.

**ergogenic aid (ur-go-JEN-ic)** Substance or treatment intended to directly improve exercise performance.

**erythrocyte (eh-RITH-row-site)** Mature red blood cell. It has no nucleus and a life span of about 120 days. It contains hemoglobin, which transports oxygen and carbon dioxide.

**erythropoietin (eh-REE-throw-POY-eh-tin)** Hormone, secreted mostly by the kidneys, that enhances red blood cell synthesis and stimulates red blood cell release from bone marrow.

**Escherichia coli** Bacterium commonly found in the intestinal tract of humans and animals (commonly called *E. coli*). The especially virulent strains 0157:H7 and 0111:H8 have been found in undercooked beef, especially ground beef. Foods implicated in *E. coli* infection include unpasteurized milk, unpasteurized fresh apple cider, salad greens, cantaloupe, dry-cured salami, and many types of sprouts. Cooking destroys *E. coli*.

**esophagus (eh-SOF-ah-gus)** Tube in the GI tract that connects the pharynx with the stomach.

**essential amino acids** Amino acids that the human body cannot synthesize in sufficient amounts or at all and therefore must be included in the diet. There are 9 essential amino acids. They also are called indispensable amino acids.

**essential fatty acids (EFAs)** Fatty acids that must be supplied by the diet to maintain health. Currently, only linoleic acid and alpha-linolenic acid are classified as essential.

**essential nutrient** In nutritional terms, a substance that, when left out of a diet, leads to signs of poor health. The body either can't produce this nutrient or can't produce enough of it to meet its needs. Then, if added back to a diet before permanent damage occurs, the affected aspects of health are restored.

**ester (ES-ter)** Organic compound that has an O'C group attached to a carbonyl group; product of a reaction between a carboxylic acid and an alcohol; formation of triglycerides involves forming ester bonds.

**esterification (e-ster-ih-fih-KAY-shun)** Process of attaching fatty acids to a glycerol molecule, creating an ester bond and releasing water. Removing a fatty acid is called de-esterification; reattaching a fatty acid is called re-esterification.

**Estimated Average Requirement (EAR)** Nutrient intake amounts estimated to meet the needs of 50% of the individuals in a specific life stage.

**Estimated Energy Requirement (EER)** Estimate of the energy (kcal) intake needed to match the energy use of an average person in a specific life stage.

**ethanol** Chemical term for the form of alcohol found in alcoholic beverages.

**eustachian tubes (you-STAY-shun)** Thin tubes connected to the middle ear that open into the throat.

**Evidence Analysis Library** Source of systematic reviews conducted by the Academy of Nutrition and Dietetics to guide clinical decisions.

**Exchange System** System for classifying foods into numerous lists based on the foods' macronutrient composition and establishing serving sizes so that 1 serving of each food on a list contains the same amount of carbohydrate, protein, fat, and calories.

**exchange** Serving size of a food on a specific exchange list.

**exercise** Physical activity done with the intent of providing a health benefit, such as improved muscle tone or stamina.

**exocrine gland (EK-so-krin)** Cluster of epithelial cells specialized for secretion. They have ducts that lead to an epithelial surface.

**exocytosis (ek-so-sigh-TOE-sis)** Process of cellular secretion in which the secretory products are contained within a membrane-enclosed vesicle. The vesicle fuses with the cell membrane and is open to the extra-cellular environment.

**Expanded Food and Nutrition Education Program (EFNEP)** U.S. government program that provides nutrition education for families with limited resources.

**experiment** Test made to examine the validity of a hypothesis.

**extracellular** Outside cells.

**extracellular fluid (ECF)** Fluid present outside the cells. It includes intravascular and interstitial fluids and represents one-third of all body fluid.

**extrusion reflex** Normal response present in 1st few months of life that causes the tongue to thrust outward when touched or depressed. It helps a baby express milk from a nipple, but it also causes an infant to push objects placed on the tip of the tongue, such as a spoon or food, out of the mouth; also called tongue thrusting.

# F

**facilitated diffusion** Absorption in which a carrier shuttles substances into the absorptive cell but no energy is expended. Absorption is driven by a concentration gradient that is higher in the intestinal contents than in the absorptive cell.

**failure to thrive** Inadequate gains in height and weight in infancy, often due to an inadequate food intake.

**famine** Extreme shortage of food that leads to massive starvation in a population; often associated with crop failures, war, and political unrest.

**fasting blood glucose** Measurement of blood glucose levels after a period of 8 hours or more without food or beverages; also called fasting blood sugar.

**fasting hypoglycemia (HIGH-po-gly-SEE-meah)** Low blood glucose that follows after about a day of fasting.

**fat** Substance that dissolves in organic solvents, such as benzene and ether. Fats are mostly composed of carbon and hydrogen, with relatively small amounts of oxygen and other elements. Dietary fat supplies 9 kcal/gram.

**fat-soluble vitamins** Vitamins that dissolve in fat and such substances as ether and benzene, but not readily in water; vitamins A, D, E, and K.

**fatty acid** Chain of carbons chemically bonded together and surrounded by hydrogen molecules. These hydrocarbons are found in lipids and contain a carboxyl (acid) group at 1

$$O$$
$$|$$
$$(—C—OH)$$

end and a methyl group ($—CH_3$) at the other.

**fatty acid oxidation** Breakdown of fatty acids into compounds that enter the citric acid cycle.

**fatty liver** Accumulation of triglycerides and other lipids inside liver cells; most often caused by excessive alcohol intake. Other causes include malnutrition and obesity.

**favorable pregnancy outcome** In humans, a full-term gestation period (longer than 37 weeks) that results in a live, healthy infant weighing more than 5.5 pounds.

**fecal transplant** The transfer of fecal matter (stool) from a healthy donor to a recipient. It is most commonly used to treat *Clostridium difficile* infection.

**feces (FEE-seas)** Substances discharged from the bowel during defecation, including undigested food residue, dead GI tract cells, mucus, bacteria, and other waste material.

**feeding center** Group of cells in the hypothalamus that, when stimulated, causes hunger.

**female athlete triad** Condition characterized by low energy availability, menstrual disorders, and low bone mineral density.

**fermentation** Breakdown of large organic compounds into smaller compounds, especially organic acids. The breakdown is often by anaerobic bacteria.

**ferritin (FER-ih-tin)** Iron-binding protein in the intestinal mucosa that binds iron and prevents it from entering the bloodstream; also the primary storage form of iron in liver and other tissues.

**fetal alcohol spectrum disorder (FEET-al)** Hyperactivity, attention deficit disorder, poor judgment, sleep disorders, and delayed learning as a result of prenatal exposure to alcohol.

**fetal alcohol syndrome** Group of irreversible physical and mental abnormalities in an infant that results from the mother's consuming alcohol during pregnancy.

**fetus (FEET-us)** In humans, developing offspring from about the beginning of the 9th week after conception until birth.

**FGF23** Fibroblast growth factor 23 is a protein hormone secreted by osteoclasts. It acts on the kidney to regulate serum phosphorus concentration.

**fiber** Substance in plant foods that is not broken down by the digestive processes of the stomach or small intestine. Fiber adds bulk to feces. Fiber naturally found in foods is called dietary fiber.

**flatulence (FLAT-u-lens)** Intestinal gas.

**flatus (FLA-tus)** Gas generated in the intestinal tract that may be passed through the anus.

**flavin** Group of compounds that contains riboflavin or a related compound.

**flavin adenine dinucleotide (FAD)** Coenzyme that readily accepts and donates electrons and hydrogen ions; formed from the vitamin riboflavin.

**flavin mononucleotide (FMN)** Coenzyme, formed from the vitamin riboflavin, that participates in oxidation-reduction reactions.

**flexibility exercise** Ability to move a joint through its full range of motion.

**fluoride** Trace mineral that increases the resistance of tooth enamel to dental caries. Typical sources are fluoridated water and toothpaste.

**fluoroapatite (fleur-oh-APP-uh-tite)** Fluoride-containing, acid-resistant, crystalline substance produced during bone and tooth development. Its presence in teeth helps prevent dental caries.

**folate** Water-soluble vitamin that shares a close relationship with vitamin B-12. In its coenzyme form, folate is necessary for the synthesis of DNA and in the metabolism of various amino acids and their derivatives, such as homocysteine. It also functions in the formation of neurotransmitters in the brain. A maternal deficiency of folate can lead to neural tube defects in the very early development of the fetus.

Asparagus, spinach, fortified grain products, and legumes are good sources.

**folic acid** Form of folate found in supplements and fortified foods.

**folk medicine** Medical treatment based on the beliefs, traditions, or customs of a particular society or ethnic/cultural group.

**follicular hyperkeratosis (fo-LICK-you-lar high-per-ker-ah-TOE-sis)** Condition in which keratin, a protein, accumulates around hair follicles.

**food additive** Substance added to foods to produce a desired effect, such as preservation or nutritional fortification. Over 3000 food additives are regulated by the FDA.

**food desert** Geographic area where fresh, affordable, healthy foods cannot be purchased easily; usually located in impoverished urban neighborhoods and rural areas.

**food diary** Written record of sequential food intake for a period of time. Details associated with the food intake are often recorded as well.

**food insecure** Condition in which the quality, variety, and/or desirability of the diet is reduced and there is difficulty at times providing enough food for everyone in the household.

**food intolerance** Adverse reaction to food that does not involve an allergic reaction.

**Food Lists for Diabetes** System for classifying foods into numerous lists based on the foods' macronutrient composition and establishing serving sizes so that 1 serving of each food on a list contains about the same amount of carbohydrate, protein, fat, and energy content.

**food secure** Condition in which food needs are met all of the time.

**food sensitivity** Mild reaction to a substance in a food; might be expressed as light itching or redness of the skin.

**Food Stamp Program** U.S. government program that provides nutrition education and foods for those with limited financial resources. It is now more commonly known as the Supplemental Nutrition Assistance Program (SNAP).

**Food sustainability** Ability to produce enough food to maintain the human population.

**foodborne illness** Sickness caused by the ingestion of food containing pathogenic microorganisms or toxins made by these pathogens.

**fore milk** First breast milk delivered in a breastfeeding session.

**fortified** Term generally meaning that vitamins, minerals, or both have been added to a food product in excess of what was originally found in the product.

**fraternal twins** Offspring that develop from 2 separate ova and sperm and therefore have separate genetic identities but develop simultaneously in the mother.

**free fatty acid** Fatty acid that is not attached to a glycerol molecule.

**free radical** Compound with an unpaired electron, which causes it to seek an electron from another compound. Free radicals are strong oxidizing agents and can be very destructive to electron-dense cell components, such as DNA and cell membranes.

**free water** Water not bound to the compounds in a food. This water is available for microbial use.

**frequency** In terms of exercise, number of activity sessions performed per week.

**fructans (FROOK-tans)** Polysaccharides composed of fructose units.

**fructose (FROOK-tose)** Monosaccharide with 6 carbons that forms a 5-membered or 6-membered ring with oxygen in the ring; found in fruits and honey.

**fruitarian (froot-AIR-ee-un)** Person who eats primarily fruits, nuts, honey, and vegetable oils.

**functional fiber** Fiber added to foods that has shown to provide health benefits.

**functional foods** Foods that provide health benefits beyond those supplied by the traditional nutrients they contain. For example, a tomato contains the phytochemical lycopene, so it can be called a functional food.

**fungi** Simple parasitic life forms, including molds, mildews, yeasts, and mushrooms. They live on dead or decaying organic matter. Fungi can grow as single cells, such as yeast, or as multicellular colonies, as seen with molds.

# G

**galactose (gah-LAK-tos)** Six-carbon monosaccharide that forms a 6-membered ring with oxygen in the ring; an isomer of glucose.

**galactosemia (gah-LAK-toh-SEE-mee-ah)** Rare genetic disease characterized by the buildup of the single sugar galactose in the bloodstream, resulting from the liver's inability to metabolize it. If present at birth and left untreated, this disease can cause severe mental retardation and cataracts in the infant.

**gallbladder** Organ attached to the underside of the liver and in which bile is stored and secreted.

**gamma-aminobutyric acid (GABA) (ah-MEE-noh-bu-tir-ik)** Chief inhibitory neurotransmitter.

**gastric inhibitory peptide (GIP) (GAS-trik in-HIB-ih-tor-ee PEP-tide)** Hormone

that slows gastric motility and stimulates insulin release from the pancreas.

**gastrin (GAS-trin)** Hormone that stimulates HCl and pepsinogen secretion by the stomach.

**gastroesophageal reflux disease (GERD) (gas-troh-eh-SOF-ah-jee-al)** Disease that results from stomach acid backing up into the esophagus. The acid irritates the lining of the esophagus, causing pain.

**gastrointestinal (GI) tract** Comprises the main sites in the body used in digestion and absorption of nutrients. The GI tract consists of the mouth, esophagus, stomach, small intestine, large intestine, rectum, and anus.

**gastrointestinal distension (gas-troh-in-TEST-in-al)** Expansion of the wall of the stomach or intestines due to pressure caused by the presence of gases, food, drink, or other factors. This expansion contributes to a feeling of satiety brought on by food intake.

**gastroplasty (GAS-troh-plas-tee)** Surgery performed on the stomach to limit its volume to approximately 30 milliliters.

**gene expression (JEAN)** Activation of a specific site on DNA, which results in either the activation or the inhibition of the gene.

**generally recognized as safe (GRAS)** List of food additives that in 1958 were considered safe for consumption. Manufacturers were allowed to continue to use these additives, without special clearance, when needed for food products. The FDA bears responsibility for proving they are not safe; it can remove unsafe products from the list.

**genes (JEANS)** Hereditary material on chromosomes that makes up DNA. Genes provide the blueprint for the production of cell proteins. The nucleus of the cell contains about 30,000 genes.

**genetic engineering** Manipulation of the genetic makeup of any organism with recombinant DNA technology.

**genetically modified organism (GMO)** Organism created by genetic engineering.

**genotoxic carcinogen (initiator) (JEE-no-TOK-sik car-SIN-oh-jen)** Compound that directly alters DNA or is converted in cells to metabolites that alter DNA, thereby providing the potential for cancer to develop.

**germ** Vitamin and lipid rich core of the whole grain.

**gestation (jes-TAY-shun)** Period of intrauterine development of offspring, from conception to birth. In humans, gestation lasts for about 40 weeks after the woman's last menstrual period.

**gestational diabetes (jes-TAY-shun-al)** High blood glucose concentration that develops during pregnancy and returns to normal after birth.

**ghrelin (GREL-in)** Hormone, made by the stomach, that increases food intake.

**glomerulus (glo-MER-you-lus)** Capillaries in the kidney that filter the blood.

**glossitis (glah-SI-tis)** Inflammation of the tongue. It becomes red, smooth, shiny, and sore.

**glucagon (GLOO-kuh-gon)** Hormone, made by the pancreas, that stimulates the breakdown of glycogen in the liver into glucose. This breakdown increases blood glucose.

**glucogenic amino acid (gloo-ko-JEN-ik)** Amino acid that can be converted into glucose via gluconeogenesis.

**gluconeogenesis (gloo-ko-nee-oh-JEN-uh-sis)** Generation *(genesis)* of new *(neo)* glucose from certain (glucogenic) amino acids.

**glucose (GLOO-kos)** Monosaccharide with 6 carbons; also called dextrose; a primary source of energy in the body; found in table sugar (sucrose) bound to fructose.

**glucose polymer (PAH-lah-mer)** Carbohydrate, used in some sports drinks, that consists of a few glucose molecules bonded together.

**glucose-dependent insulinotropic peptide** Hormone that slows gastric motility and stimulates insulin release from the pancreas; formerly known as gastric inhibitory peptide.

**glutamine (GLOO-tah-meen)** Amino acid that enhances the immune system during trauma and illness.

**glutathione (gloo-tah-THIGH-on)** Reducing agent; can remove toxic peroxides that form in the cell during aerobic respiration.

**glutathione peroxidase (gloo-tah-THIGH-on per-OX-ih-dase)** Selenium-containing enzyme that can destroy peroxides; acts in conjunction with vitamin E to reduce free radical damage to cells.

**glycemic index (GI) (gli-SEA-mik)** Ratio of the blood glucose response to a given food, compared with a standard (typically, glucose or white bread).

**glycemic load (GL)** Amount of carbohydrate in a food multiplied by the glycemic index of that carbohydrate. The result is then divided by 100.

**glycerol (GLIS-er-ol)** Three-carbon alcohol that provides the backbone of triglycerides.

**glycocalyx (gli-ko-KAL-iks)** Projections of proteins on the microvilli. They contain enzymes to digest protein and carbohydrate.

**glycogen (GLI-ko-jen)** Carbohydrate made of multiple units of glucose with a highly branched structure; sometimes known as animal starch; the storage form of glucose in humans; is synthesized (and stored) in the liver and muscles.

**glycogen storage disease** Genetic defect that does not allow glycogen to be stored in the muscles or liver.

**glycolipid (gli-ko-LIP-id)** Lipid (fat) containing a carbohydrate group.

**glycolysis (gli-KOL-ih-sis)** Metabolic pathway that converts glucose into 2 molecules of pyruvic acid, with the net gain of 2 ATP and 2 NADH + 2 H$^+$.

**glycoprotein (gli-ko-PRO-teen)** Protein containing a carbohydrate group.

**glycosylation (gli-COS-ih-lay-shun)** Process by which glucose attaches to (glycates) other compounds, such as proteins.

**goiter (GOY-ter)** Enlargement of the thyroid gland; can be caused by a lack of iodide in the diet.

**goitrogens (GOY-troh-jens)** Substances in food and water that interfere with thyroid gland metabolism and thus may cause goiter if consumed in large amounts.

**Golgi complex (GOAL-jee)** Cell organelle near the nucleus; processes newly synthesized protein for secretion or distribution to other organelles.

**gout (gowt)** Joint inflammation caused by accumulation of uric acid. Obesity is a risk factor for developing gout.

**green revolution** Increases in crop yields accompanying the introduction of new agricultural technologies in less developed countries, beginning in the 1960s. The key technologies were high-yielding, disease-resistant strains of rice, wheat, and corn; greater use of fertilizer and water; and improved cultivation practices.

**growth hormone** Pituitary hormone that stimulates body growth and the release of fat from storage, as well as other effects.

**guanine (GWAH-neen)** Nitrogenous base that forms part of the structure of DNA and RNA; a purine.

**gum** Soluble fiber consisting of chains of galactose and other monosaccharides; characteristically found in exudates from plant stems.

**gynecoid obesity (GI-nih-coyd)** Excess fat storage located primarily in the buttocks and thigh area.

# H

**H₂ blocker** Medication, such as cimetidine (Tagamet®), that blocks the increase of stomach acid production caused by histamine.

**Harris-Benedict equation** Equation that predicts resting metabolic rate based on a person's weight, height, and age.

**health claim** Claim that describes a well-researched and documented relationship between a disease and a nutrient, food, or food constituent. See also *preliminary health claim.*

**heart attack** Rapid fall in heart function caused by obstructed blood flow through the heart's blood vessels. Often, part of the heart dies in the process. It is technically called a myocardial infarction.

**heart disease** See *cardiovascular disease.*

**heartburn** Pain caused by stomach acid backing up into the esophagus and irritating the tissue in that organ.

**heat cramps** Frequent complication of heat exhaustion. Cramps usually occur in individuals who have experienced large sweat losses from exercising for several hours in a hot climate and have consumed a large volume of water. The cramps occur in skeletal muscles and consist of contractions for 1 to 3 minutes at a time.

**heat exhaustion** First stage of heat-related illness that occurs because of depletion of blood volume from fluid loss by the body. This depletion may increase body temperature and can lead to headaches, dizziness, muscle weakness, and visual disturbances, among other effects.

**heatstroke** Condition in which the internal body temperature reaches 104°F or higher. Blood circulation is greatly reduced. Nervous system damage may ensue, and death is likely. Sweating generally ceases, which cause the skin of individuals who suffer heatstroke to feel hot and dry.

**helminth (HEL-menth)** Parasitic worm that can contaminate food, water, feces, animals, and other substances.

**helper T cell** Type of T cell that interacts with macrophages and secretes substances to signal an invading pathogen; stimulates B lymphocytes to proliferate.

**hematocrit (hee-MAT-oh-krit)** Percentage of total blood volume occupied by red blood cells.

**hematopoiesis (hee-mat-oh-po-EE-sis)** Production of blood cells.

**heme (HEEM)** Iron-containing structure in hemoglobin and myoglobin.

**heme iron** Iron provided from animal tissues primarily as a component of hemoglobin and myoglobin. Approximately 40% of the iron in meat is heme iron; it is readily absorbed.

**hemicellulose (hem-ih-SELL-you-los)** Mostly insoluble fiber containing galactose, glucose, and other monosaccharides bonded together.

**hemochromatosis (heem-oh-krom-ah-TOE-sis)** Genetic disorder characterized by increased absorption of iron, saturation of iron-binding proteins, and iron deposits in the liver, heart, pancreas, joints, and pituitary gland.

**hemoglobin (HEEM-oh-glow-bin)** Iron-containing protein in red blood cells that transports oxygen to the body tissues and some carbon dioxide away from the tissues. It also is responsible for the red color of blood.

**hemolysis (hee-MOL-ih-sis)** Destruction of red blood cells caused by the breakdown of the red blood cell membranes. This causes the cell contents to leak into the fluid portion (plasma) of the blood.

**hemolytic anemia (hee-moe-LIT-ik)** Disorder that causes red blood cells to break down faster than they can be replaced.

**hemorrhage (hem-OR-ij)** Bleeding.

**hemorrhagic stroke (hem-oh-RAJ-ik)** Damage to part of the brain resulting from rupture of a blood vessel and subsequent bleeding within or over the internal surface of the brain.

**hemorrhoid (HEM-or-oid)** Pronounced swelling in a large vein, particularly a vein in the anal region.

**hemosiderin (heem-oh-SID-er-in)** Iron-binding protein in the liver that stores iron when iron levels in the body exceed the storage capacity of ferritin.

**hepatic portal vein system (he-PAT-ik)** Veins leaving from the stomach, intestines, spleen, and pancreas that drain into the hepatic portal vein, which flows into the liver.

**hepatitis A (hep-ah-TIE-tis)** Virus found in the human intestinal tract and feces; causes inflammation and loss of function of the liver. It can contaminate many foods, especially shellfish and raw foods, and can endure significant heat, cold, and drying.

**herbicide (ERB-ih-side)** Compound that reduces the growth and reproduction of plants.

**hexose (HEK-sos)** Carbohydrate containing 6 carbons.

**hiatal hernia (high-AY-tal HUR-nee-ah)** Protrusion of part of the stomach upward through the diaphragm into the chest cavity.

**high fructose corn syrup** Corn syrup that has been manufactured to contain between 42 and 90% fructose.

**high-density lipoprotein (HDL)** Lipoprotein that picks up cholesterol from cells and transfers it in the bloodstream to the liver. A low blood HDL value increases the risk of cardiovascular disease.

**high-quality (complete) proteins** Dietary proteins that contain ample amounts of all 9 essential amino acids.

**hind milk (HYND)** Milk secreted at the end of a breastfeeding session. It is higher in fat than fore milk.

**histamine (HISS-tuh-meen)** Bioactive amine that participates in immune response, stimulates stomach acid secretion, and triggers inflammatory response. It regulates sleep and promotes smooth muscle contraction, increased nasal secretions, blood vessel relaxation, and airway constriction.

**holoenzyme** Active enzyme complex composed of the apoenzyme and the cofactor.

**homeostasis (home-ee-oh-STAY-sis)** Series of adjustments that prevent change in the internal environment in the body.

**homocysteine (homo-SIS-teen)** Amino acid not used in protein synthesis. Instead, it arises during metabolism of the amino acid methionine. Homocysteine likely is toxic to many cells, such as those lining the blood vessels.

**hormone** Chemical substance produced in the body that controls or regulates the activity of certain cells or organs. Hormones can be amino acid–like (epinephrine), proteinlike (insulin), or fatlike (estrogen).

**hormone-sensitive lipase** Hormone that is responsible for breaking down stored triglycerides in fat cells into free fatty acids and glycerol.

**hospice care (HAHS-pis)** Supportive care that emphasizes comfort and dignity in death.

**human immunodeficiency virus (HIV)** Virus that leads to acquired immune deficiency syndrome (AIDS).

**hunger** Primarily physiological (internal) drive for food.

**hydrogen peroxide (pur-OX-ide)** Chemically, $H_2O_2$.

**hydrogenation (high-dro-jen-AY-shun)** Addition of hydrogen to some carbon-carbon double bonds and producing some *trans* fatty acids. This process is used to convert liquid oils into more solid fats.

**hydrolysis reaction (high-DROL-ih-sis)** Chemical reaction that breaks down a compound by adding water. One product receives a hydrogen ion ($H^+$); the other product receives a hydroxyl ion (—OH). Hydrolytic enzymes break down compounds using water in this manner.

**hydrophilic (high-dro-FILL-ik)** Literally, "water-loving"; attracts water.

**hydrophobic (high-dro-FO-bik)** Literally, "water-fearing"; repels water.

**hydroxyapatite (high-drox-ee-APP-uh-tite)** Compound composed primarily of calcium and phosphate; it is deposited in bone protein matrix to give bone strength and rigidity ($Ca_{10}[PO_4]6OH_2$).

**hyperactivity** Inattention, irritability, and excessively active behavior in children; technically referred to as attention deficit hyperactivity disorder.

**hypercalcemia (high-per-kal-SEE-mee-ah)** High concentration of calcium in the bloodstream. This condition can lead to loss of appetite, calcium deposits in organs, and other health problems.

**hypercarotenemia (high-per-car-oh-teh-NEEM-ee-ah)** Elevated amounts of carotenoids in the bloodstream, usually caused by consuming a diet high in carrots or squash or by taking beta-carotene supplements.

**hyperemesis gravidarum (high-per-EM-eh-sis gra-va-DAR-um)** Severe nausea and vomiting experienced during pregnancy that continues beyond 14 weeks of gestation.

**hyperglycemia (HIGH-per-gly-SEE-me-uh)** High blood glucose, above 125 mg/100 ml (dl) of blood.

**hypergymnasia (high-per-jim-NAY-zee-ah)** Exercising more than is required for good physical fitness or maximum performance in a sport; excessive exercise.

**hyperkalemia (high-per-kah-LEE-mee-ah)** High potassium levels in the blood.

**hyperlipidemia (high-per-lip-ih-DEE-me-ah)** Presence of an abnormally large amount of lipids in the circulating blood.

**hypernatremia (high-per-nay-TREE-mee-ah)** High sodium levels in the blood.

**hyperparathyroidism (high-per-pair-ah-THY-royd-iz-um)** Overproduction of parathyroid hormone by the parathyroid glands, usually caused by a nonmalignant tumor or abnormal growth of the glands. In most cases, there are no symptoms except hypercalcemia but, in more severe cases, weakness, confusion, nausea, and bone pain occur. Bone fractures and kidney stones also are problems.

**hyperplasia (high-per-PLAY-zee-uh)** Increase in cell number.

**hypertension (high-per-TEN-shun)** Persistently elevated blood pressure. Obesity, inactivity, alcohol intake, and excess salt intake all can contribute to the problem.

**hyperthyroidism** Condition characterized by high blood levels of thyroid hormone.

**hypertriglyceridemia (high-PURR-tri-GLISS-uh-ride-ee-me-ah)** Condition in which there are excess triglycerides in the blood.

**hypertrophy (high-PURR-tro-fee)** Increased tissue or organ size.

**hypervitaminosis A (HIGH-per-vi-tah-mi-NO-sis)** Condition resulting from intake of excessive amounts of vitamin A.

**hypocalcemia (HIGH-po-kal-SEE-me-ah)** Low blood calcium, typically arising from inadequate parathyroid hormone release or action.

**hypochromic (high-po-KROM-ik)** Pale; red blood cells lacking sufficient hemoglobin. Hypochromic cells have a reduced oxygen-carrying ability.

**hypoglycemia (HIGH-po-gly-SEE-me-uh)** Low blood glucose, below 50 mg/100 ml (dl) of blood.

**hypokalemia (high-po-kah-LEE-me-ah)** Low potassium levels in the blood.

**hyponatremia (high-po-nay-TREE-me-ah)** Low sodium levels in the blood.

**hypothalamus (high-po-THALL-uh-mus)** Region at the base of the brain; contains cells that play a role in the regulation of hunger, respiration, body temperature, and other body functions.

**hypothesis (high-POTH-eh-sis)** Tentative explanation by scientists for a phenomenon.

**hysterectomy (hiss-te-RECK-toe-mee)** Surgical removal of the uterus.

# I

**identical twins** Two offspring that develop from a single ovum and sperm and, consequently, have the same genetic makeup.

**ileocecal valve (ill-ee-oh-SEE-kal)** Ring of smooth muscle between the ileum of the small intestine and the colon; also known as the ileocecal sphincter.

**ileum (ILL-ee-um)** Terminal portion of the small intestine.

**immune system** Body system consisting of white blood cells, lymph glands, lymphocytes, antibodies, and other body tissues and cells. The immune system defends against foreign invaders.

**immunoglobulins (em-you-no-GLOB-you-lins)** Proteins (also called antibodies) in the blood that are responsible for identifying and neutralizing antigens, as well as pathogens that bind specifically to antigens.

**in utero (in-YOU-ter-oh)** "In the uterus," or during pregnancy.

**in vitro (in-VEE-troh)** Literally, "in glass," such as in a test tube (e.g., experiments performed outside the body).

**in vivo (in-VEE-vo)** Within the living body.

**incidence** Number of new cases of a disease in a defined population over a specific period of time, such as 1 year.

**incidental food additives** Additives that appear in food products indirectly, from environmental contamination of food ingredients or during the manufacturing process.

**incomplete (lower-quality) protein** Food protein that lacks enough of 1 or more of the essential amino acids to support human protein needs.

**indirect calorimetry (kal-oh-RIM-eh-tree)** Method to measure energy use by the body by measuring oxygen uptake. Formulas are used to convert this gas exchange value into energy use.

**infancy** Earliest stage of childhood—from birth to 1 year of age.

**infectious disease (in-FEK-shus)** Disease caused by an invasion of the body by microorganisms, such as bacteria, fungi, or viruses.

**infrastructure** Basic framework of a system or an organization. For a society, this includes roads, bridges, telephones, and other basic technologies.

**innate (nonspecific) immunity** Immune response present at birth that provides the first barrier of protection against invading antigens; includes physical barriers that prevent access to the inside of the body, chemical secretions that destroy antigens, physiological barriers that prevent the growth of antigens, and phagocytic cells that engulf and destroy antigens.

**inorganic substance (in-or-GAN-ik)** Substance lacking carbon atoms bonded to hydrogen atoms in the chemical structure.

**insensible water losses** Water losses not readily perceived, such as water lost with each breath.

**insoluble fibers** Fibers that mostly do not dissolve in water and are not metabolized by bacteria in the large intestine. These include cellulose, some hemicelluloses, and lignins; more formally called nonfermentable fibers.

**insulin (IN-su-lynn)** Hormone produced by beta cells of the pancreas. Among other processes, insulin increases the synthesis of glycogen in the liver and the movement of glucose from the bloodstream into muscle and adipose cells.

**insulin resistance** Condition in which body tissues have a lowered level of response to insulin, the hormone secreted by the pancreas that helps regulate blood glucose (sugar). Persons with insulin resistance produce large quantities of insulin to maintain normal levels of glucose in the blood.

**integumentary system (in-teg-you-MEN-tah-ree)** Having to do with the skin, hair, glands, and nails.

**intensity** In terms of exercise, the amount of effort expended or how difficult the activity is to perform.

**intentional food additive** Additive knowingly (directly) incorporated into food products by manufacturers.

**interferon (in-ter-FEAR-on)** Protein released by virus-infected cells that bind to other cells, stimulating the synthesis of antiviral proteins, which in turn inhibit viral multiplication.

**intermediate** Chemical compound formed in 1 of many steps in a metabolic pathway.

**Intermediate Density Lipoprotein (IDL)** Lipoprotein created from VLDL and carries both cholesterol and protein. IDL's are quickly cleared from the blood and are formed into low-density lipoproteins (LDL).

**international unit (IU)** Crude measure of vitamin activity, often based on the growth rate of animals. Today, IUs generally have been replaced by precise measurements of actual quantities, such as milligrams or micrograms.

**interstitial fluid (in-ter-STISH-al)** Fluid between cells.

**interstitial spaces** Spaces between cells.

**intracellular fluid (in-tra-SELL-you-lar)** Fluid contained within a cell; represents about two-thirds of all body fluid.

**intravascular fluid (in-tra-VAS-kyu-lar)** Fluid within the bloodstream (in the arteries, veins, capillaries, and lymphatic vessels); represents about 25% of all body fluids.

**intrinsic factor (in-TRIN-zik)** Substance in gastric juice that enhances vitamin B-12 absorption.

**inulin (IN-u-lin)** Type of soluble fiber made up mainly of fructose molecules; found in foods such as onion and chicory; acts as a prebiotic.

**iodine** Trace mineral that is a component of thyroid hormones. A deficiency can result in goiter. Iodized salt, saltwater fish, and iodine-fortified foods are good sources.

**ion (EYE-on)** Atom with an unequal number of electrons and protons. Negative ions (anions) have more electrons than protons and carry a negative charge; positive ions (cations) have more protons than electrons and carry a positive charge.

**ionic bond (eye-ON-ik)** Union between 2 atoms formed by an attraction of a positive ion to a negative ion, as in table salt ($Na^+Cl^-$).

**iron** Trace mineral that functions as a component of hemoglobin and other key compounds used in respiration; also important in immune function and cognitive development. Meats, seafood, molasses, and fortified foods are good sources.

**irradiation (ir-RAY-dee-AY-shun)** Process in which radiation energy is applied to foods, creating compounds (free radicals) within the food that destroy microorganisms that can lead to food spoilage. This process does not make the food radioactive.

**irritable bowel syndrome** Bowel disease characterized by diarrhea, constipation, abdominal pain, and distension; believed to be caused by abnormal function of the muscles and nerves of the gastrointestinal tract. It is more common in women than men.

**ischemia (ih-SKEE-mee-ah)** Lack of blood flow due to mechanical obstruction of the blood supply, mainly from arterial narrowing.

**ischemic stroke (ih-SKEE-mik)** Stroke caused by the absence of blood flow to a part of the brain.

**isomers (EYE-so-merz)** Different chemical structures for compounds that share the same chemical formula.

**isotope (EYE-so-towp)** Alternate form of a chemical element. It differs from other atoms of the same element in the number of neutrons in its nucleus.

# J

**jaundice (JOHN-diss)** Yellowish staining of skin, sclerae of the eyes, and other tissues by bile pigments that build up in the blood.

**jejunum (je-JOO-num)** Middle section of the small intestine. (The first 12 inches is the duodenum.)

# K

**ketogenic amino acid (kee-toe-JEN-ik)** Amino acid that can be converted to acetyl-CoA and can form ketones.

**ketone (kee-tone)** Produced in the liver during the breakdown of fat when carbohydrate intake is very low.

**ketone bodies (KEE-tone)** Incomplete breakdown products of fat, containing 3 or 4 carbons. Most contain a chemical group called a ketone. An example is acetoacetic acid.

**ketosis (kee-TOE-sis)** Condition of having a high concentration of ketone bodies and related breakdown products in the bloodstream and tissues.

**kidney nephron (NEF-ron)** Functional unit of kidney cells that filters the blood for reabsorption of compounds and elimination of waste.

**kilocalorie (kcal) (kill-oh-KAL-oh-ree)** Heat energy needed to raise the temperature of 1000 grams (1 L) of water 1 degree Celsius; also written as *Calorie*.

**kilojoule (kJ) (KIL-oh-jool)** Measure of work. A mass of 1 kilogram moving at a velocity of 1 m/sec possesses the energy of 1 kJ; 1 kcal equals 4.18 kJ.

**kwashiorkor (kwash-ee-OR-core)** Condition occurring primarily in young children who have an existing disease and consume a marginal amount of energy and severely insufficient protein. It results in edema, poor growth, weakness, and an increased susceptibility to further infection and disease.

**kyphosis (ky-FOH-sis)** Abnormal convex curvature of the spine, resulting in a bulge at the upper back; often caused by osteoporosis of the spine.

# L

**lactase** Enzyme made by absorptive cells of the small intestine; digests lactose to glucose and galactose.

**lactate (LAK-tate)** Three-carbon acid formed during anaerobic cell respiration; a partial breakdown product of glucose; also called lactic acid.

**lactation** Period of milk secretion following pregnancy; typically called breastfeeding.

**lacteal (LACK-tee-al)** Tiny vessels in the small intestine villi that absorb dietary fat.

***Lactobacillus bifidus* factor (lak-toe-bah-SIL-us BIFF-id-us)** Protective factor secreted in colostrum; encourages the growth of beneficial bacteria in a newborn's intestines.

**lacto-ovo-pesco-vegetarian (lak-toe-o-vo-pes-co-vej-eh-TEAR-ree-an)** Person who consumes only plant products, dairy products, eggs, and fish.

**lacto-ovo-vegetarian (lak-toe-o-vo-vej-eh-TEAR-ree-an)** Person who consumes plant products, dairy products, and eggs.

**lactose (LAK-tose)** Glucose bonded to galactose.

**lactose intolerance** Condition caused by a lack of the enzyme that digests lactose (lactase); symptoms include abdominal gas, bloating, and diarrhea.

**lacto-vegetarian (lak-toe-vej-eh-TEAR-ree-an)** Person who consumes plant products and dairy products.

**lanugo (lah-NEW-go)** Downlike hair that appears after a person has lost much body fat through semistarvation. The hair stands erect and traps air, acting as insulation for the body to compensate for the relative lack of body fat, which usually functions as insulation.

**large for gestational age** Weighing more than the expected weight for length of gestation; typically defined as above the 90th percentile.

**larva (LAR-vah)** Early developmental stage in the life history of some organisms, such as parasites.

**larynx (LAYR-ingks)** Structure located between the pharynx and trachea; contains the vocal cords.

**laxative** Medication or other substance used to relieve constipation.

**lean body mass** Body weight after subtracting the weight of body fat. Lean body mass includes organs such as the brain, muscles, and liver, as well as blood and other body fluids.

**leavened bread (LEV-end)** Bread prepared using a leavening agent, such as yeast or baking powder. Leavening agents create gas, which causes bread dough to rise. Flat breads, such as pita bread, do not contain leavening.

**lecithin (LESS-uh-thin)** Group of phospholipids containing 2 fatty acids, a phosphate group, and a choline molecule. Lecithins differ based on the types of fatty acids found on each lecithin molecule.

**leptin (LEP-tin)** Hormone made by adipose tissue that influences long-term regulation of fat mass. Leptin also influences reproductive functions, as well as other body processes, such as insulin release.

**let-down reflex** Reflex stimulated by infant suckling; causes the release (ejection) of milk from milk ducts in the mother's breasts; also called *milk ejection reflex*.

**leukemia (loo-KEY-mee-ah)** Malignant neoplasm of blood-forming tissues, the bone marrow.

**leukocyte (LOO-ko-site)** White blood cell.

**leukotriene (LT) (loo-ko-TRY-een)** Eicosanoid involved in inflammatory or hypersensitivity reactions, such as asthma.

**life expectancy** Average length of life for a given group of people (usually determined by the year of birth).

**life span** Oldest age a person can reach.

**lignan (LIG-nan)** Phytochemical class that acts as a phytoestrogen in the body. Food sources are whole grains and flaxseeds.

**lignin (LIG-nin)** Insoluble fiber made of a multi-ringed alcohol (noncarbohydrate) structure.

**limiting amino acid** Essential amino acid in the lowest concentration in a food or diet relative to body needs.

**linoleic acid (lin-oh-LEE-ik)** Essential omega-6 fatty acid with 18 carbon and 2 double bonds (C18:2, omega-6).

**lipase (LYE-pace)** Fat-digesting enzyme; produced by the stomach, salivary glands, and pancreas.

**lipid peroxidation (per-OX-ih-day-shun)** Process initiated by an environmental component that induces the formation of an organic free radical, R·. In the formation of a fatty acid of this type, 1st a carbon-

carbon double bond is broken. The resulting breakdown products react with oxygen to form peroxides (a) or free radicals (b):

a. 
```
   H   H
   |   |
  —C — C — O — O — H
   |   |
   H   H
```

b. 
```
   H   H
   |   |
  —C — C — O — O
   |   |
   H   H
```

**lipid** Group of organic compounds that includes oils and fats; triglycerides, phospholipids, and sterols. All these compounds contain carbon, hydrogen, and oxygen. None dissolve in water but do dissolve in organic solvents, such as chloroform, benzene, and ether.

**lipogenesis (lye-poh-JEN-eh-sis)** Building of fatty acids using derivatives of acetyl-CoA.

**lipogenic (lye-poh-JEN-ik)** Creation of lipid. The liver is the major organ with lipogenic potential in the human body.

**lipolysis (lye-POL-ih-sis)** Breakdown of triglycerides to glycerol and fatty acids.

**lipoprotein (ly-poh-PRO-teen)** Compound, found in the bloodstream, containing a core of lipids with a shell composed of protein, phospholipid, and cholesterol.

**lipoprotein lipase (lye-poh-PRO-teen LYE-pace)** Enzyme attached to the outside of endothelial cells that line the capillaries in the blood vessels. It breaks down triglycerides into free fatty acids and glycerol.

**lipoxin (LX) (lih-POX-in)** Eicosanoid made by white blood cells that is involved in the immune system and allergic response.

**lipoxygenase (lih-POX-ih-jen-ace)** Enzyme used to synthesize leukotrienes and some other types of eicosanoids.

**Listeria monocytogenes (lis-TEER-i-a mono-sy-TODGE-en-ees)** Bacterium widely distributed in the environment, often entering food from contamination with animal or human feces. Unpasteurized milk and soft cheeses made with unpasteurized milk are most often implicated. *Listeria* is very hardy, resisting heat, salt, cold, nitrate, and acidity much better than any other bacterium. Thorough cooking and pasteurization destroy *Listeria*.

**liter (L) (LEE-ter)** Measure of volume in the metric system; 1 liter equals 0.96 quart.

**liver** Organ located in the abdominal cavity below the diaphragm; performs many vital functions that maintain balance in blood composition.

**lobules (LOB-you-elz)** Saclike structures in the breast that store milk.

**long-chain fatty acids** Fatty acids that contain 12 or more carbons.

**low birth weight (LBW)** Infant weight of less than 5.5 pounds (2.5 kilograms) at birth; most commonly results from preterm birth.

**low-density lipoprotein (LDL) (lye-po-PRO-teen)** Lipoprotein in the blood containing primarily cholesterol; elevated LDL-cholesterol is strongly linked to cardiovascular disease risk.

**lower esophageal sphincter (e-sof-ah-GEE-al SFINK-ter)** Circular muscle that constricts the opening of the esophagus to the stomach.

**lower-quality (incomplete) proteins** Dietary proteins that are low in or lack 1 or more essential amino acids.

**lumen (LOO-men)** Inside of a tube, such as the inside cavity of the GI tract.

**lymph (LIMF)** Clear, plasmalike fluid that flows through lymphatic vessels.

**lymph duct** Large lymphatic vessel that empties lymph into the circulatory system.

**lymph node** Small structure located along the course of the lymphatic vessels.

**lymphatic system (lim-FAT-ick)** System of vessels that can accept fluid surrounding cells and large particles, such as products of fat absorption. Lymph fluid eventually passes into the bloodstream via the lymphatic system.

**lymphatic vessel (lim-FAT-ick)** Vessel that carries lymph.

**lymphocyte (LIM-fo-site)** Class of white blood cells involved in the immune system, generally comprising about 25% of all white blood cells. There are several types of lymphocytes with diverse functions, including antibody production, allergic reactions, graft rejections, tumor control, and regulation of the immune system.

**lymphoma (lim-FO-ma)** Malignant tumor arising from lymph nodes or other lymphatic tissues.

**lysosome (LYE-so-som)** Cell organelle that digests proteins, such as transferrin, and breaks down bacteria and old or damaged cell components.

**lysozyme (LYE-so-zime)** Set of enzyme substances produced by a variety of cells; can destroy bacteria by rupturing cell membranes.

# M

**macrocyte (MACK-ro-site)** Literally, "large cell," such as a large red blood cell.

**macrocytic anemia (mack-ro-SIT-ik ah-NEM-ee-a)** Anemia characterized by the presence of abnormally large red blood cells in the bloodstream.

**macronutrient** Nutrient needed in gram quantities in the diet. Fat, protein, and carbohydrates are macronutrients.

**macrophage (MACK-ro-faj)** Large, mononuclear, phagocytic cell derived from a monocyte in the blood and found in body tissues. Besides functioning as phagocytes, macrophages secrete numerous cytokines and act as antigen-presenting cells.

**macular degeneration (MAK-u-lar)** Chronic eye disease that occurs when tissue in the macula (the part of the retina responsible for central vision) deteriorates. It causes a blind spot or blurred vision in the center of the visual field.

**magnesium (mag-NEE-zee-um)** Major mineral essential to many biochemical and physiological processes, including calcium metabolism, active ATP formation, enzyme function, DNA and RNA synthesis, nerve and heart function, and insulin function. Spinach, squash, and wheat bran are good sources.

**major mineral** Mineral vital to health; required in the diet in amounts greater than 100 mg/day; also called a macromineral.

**malignant (ma-LIG-nant)** In reference to a tumor, the property of spreading locally and to distant sites.

**malnutrition** Can refer to either undernutrition or overnutrition; eventually contributes to failing health.

**malonyl-CoA (MAL-o-kneel)** Building block in fatty acid synthesis.

**maltase (MALL-tace)** Enzyme made by absorptive cells of the small intestine; digests maltose to 2 glucoses.

**maltose (MALL-tos)** Disaccharide made of 2 glucose molecules.

**manganese** Trace mineral that functions as a cofactor of some enzymes, such as those involved in carbohydrate metabolism and antioxidant protection. Nuts, oats, beans, and tea are good sources.

**mannitol (MAN-ih-tahl)** Alcohol derivative of fructose.

**marasmus (ma-RAZ-mus)** Condition that results from a severe deficit of energy and protein, which causes extreme loss of fat stores, muscle mass, and body weight.

**Marfan syndrome** Genetic disorder affecting muscles and skeleton, characterized by tallness, long arms, and little subcutaneous fat. Some medical historians speculate that Abraham Lincoln suffered from Marfan syndrome.

**mass movement** Peristaltic wave that simultaneously coordinates contraction over a large area of the large intestine. Mass movements propel material from 1 portion of the large intestine to another and from the large intestine to the rectum.

**mast cell** Tissue cell that releases histamine and other chemicals involved in inflammation.

**meconium (me-KO-nee-um)** First thick, mucuslike stool passed by an infant after birth.

**Mediterranean Diet** Dietary pattern that includes large amounts of fruits, vegetables, and olive oil; associated with a low incidence of coronary heart disease.

**medium-chain fatty acid** Fatty acid that contains 6 to 10 carbons.

**megadose** Intake of a nutrient far beyond human needs.

**megaloblast (MEG-ah-low-blast)** Large, nucleated, immature red blood cell in the bone marrow, which results from the inability of a precursor cell to divide when it normally should.

**megaloblastic (macrocytic) anemia (MEG-ah-low-BLAST-ik)** Anemia characterized by abnormally large, nucleated, immature red blood cells, which result from the inability of a precursor cell to divide normally.

**memory cells** B lymphocytes that remain after an infection and provide long-lasting or permanent immunity.

**menaquinone (men-ah-KWIH-nohn)** Form of vitamin K found in fish oils and meats; also is synthesized by bacteria in the human intestine.

**menarche (men-AR-kee)** Onset of menstruation. Menarche usually occurs around age 13, 2 or 3 years after the 1st signs of puberty start to appear.

**menopause (MEN-oh-pawz)** Cessation of menses in women, usually beginning at about age 50.

**meta-analysis** Summary of several scientific studies grouped together.

**metabolic equivalent (MET)** Exercise intensity that is relative to a person's metabolic rate.

**metabolic pathway** Series of chemical reactions occurring in a cell, such as glycolysis, beta-oxidation, the citric acid cycle, and the electron transport chain.

**Metabolic Syndrome** Condition characterized by poor blood glucose regulation, hypertension, increased blood lipids, and abdominal obesity; usually accompanied by lack of physical activity; previously called Syndrome X.

**metabolism (meh-TAB-oh-liz-m)** Chemical processes in the body that provide energy in useful forms and sustain vital activities.

**metabolites** Intermediate compounds in metabolism.

**metalloenzyme (meh-tal-oh-EN-zyme)** Enzyme that contains 1 or more metal ions that are required for enzymatic activity.

**metallothionein (meh-TAL-oh-THIGH-oh-neen)** Protein involved in the binding and release of zinc and copper in intestinal and liver cells.

**metastasize (ma-TAS-tah-size)** Spread of disease from 1 part of the body to another, even to parts of the body that are remote from the site of the original tumor. Cancer cells can spread via blood vessels, the lymphatic system, or direct growth of the tumor.

**meter (mee-ter)** Measure of length in the metric system; 1 meter equals 39.4 inches.

**micelle (my-SELLS)** Water-soluble, spherical structure formed by lecithin and bile acids in which the hydrophobic parts of the molecules face inward and the hydrophilic parts face outward.

**microbiome** Microorganisms in a specific environment, such as our bodies. Learn more at **commonfund.nih.gov/hmp**.

**microbiota** The microorganisms that inhabit a particular region, such as the gastrointestinal tract.

**microcytic (my-kro-SIT-ik)** Literally, "small cell" (e.g., red blood cells that are smaller than normal).

**microcytic hypochromic anemia (high-po-KROME-ik)** Anemia characterized by small, pale red blood cells that lack sufficient hemoglobin and thus have reduced oxygen-carrying ability. It also can be caused by an iron deficiency.

**microfractures** Small fractures, undetectable by X-rays or other bone scans, that may develop constantly in bones.

**micronutrient** Nutrient needed in milligram or microgram quantities in a diet. Vitamins and minerals are micronutrients.

**microsomal ethanol oxidizing system (my-kro-SO-mol)** Alternative pathway for alcohol metabolism when alcohol is in high concentration in the liver; uses rather than yields energy for the body.

**microvilli (my-kro-VIL-eye)** Microscopic, hairlike projections of cell membranes of certain epithelial cells.

**migrant study** Research that examines the health of people who move from 1 country to another.

**mineral** Element used in the body to promote chemical reactions and to form body structures.

**miscarriage** Nonelective termination of pregnancy that occurs before the fetus can survive; typically called spontaneous abortion.

**mitochondria (my-toe-KON-dree-ah)** Main sites of energy production in a cell. They also contain the pathway for oxidizing fat for fuel, among other metabolic pathways.

**mode** Type, as in the type of exercise that is performed.

**modified food starch** Product consisting of chemically linked starch molecules; is more stable than normal, unmodified starches.

**mold** Type of fungus that grows best in warm, dark, moist environments. Some molds produce toxins (mycotoxins) that cause illness when ingested by humans.

**molecule** Atoms linked (bonded) together; the smallest part of a compound that still has all the properties of a compound (see also compound).

**molybdenum (mo-LIB-den-um)** Trace mineral that aids in the action of some enzymes in the body. Beans, whole grains, and nuts are good sources.

**monoamine (MON-oh-ah-MEAN)** Molecule containing 1 amide group.

**monoglyceride (mon-oh-GLIS-er-ide)** Breakdown product of a triglyceride consisting of 1 fatty acid bonded to a glycerol backbone.

**monosaccharide (mon-oh-SACK-uh-ride)** Simple sugar, such as glucose, that is not broken down further during digestion.

**monounsaturated fatty acid (MUFA) (mon-oh-un-SAT-urated)** Fatty acid containing 1 carbon-carbon double bond.

**morbidity** Disease condition or state; amount of illness present in a population.

**mortality** Death rate of a population.

**motility** Ability to move spontaneously; also movement of food through the GI tract.

**mottling (MOT-ling)** Discoloration or marking of the surface of teeth from exposure to excessive amounts of fluoride (also called enamel fluorosis).

**mRNA translation** Synthesis of polypeptide chains at the ribosome according to information contained in strands of messenger RNA (mRNA).

**mucilage (MYOO-sih-laj)** Soluble fiber consisting of chains of galactose and other monosaccharides; characteristically found in seaweed.

**mucopolysaccharide (MYOO-ko-POL-ee-SAK-ah-ride)** Substance containing protein and carbohydrate parts; found in bone and other organs.

**mucosa (MYOO-co-sa)** Mucous membrane consisting of cells and supporting connective tissue. Mucosa lines cavities that open to the outside of the body, such as the stomach and intestine, and generally contains glands that secrete mucus.

**mucous membranes (MYOO-cuss)** Membranes that line passageways open to the exterior environment; also called mucosae.

**mucus (MYOO-cuss)** Thick fluid, secreted by glands throughout the body, that lubricates and protects cells; contains a compound that has both a carbohydrate and a protein nature.

**muscle** Body tissue made of groups of muscle fibers that contract to allow movement. There are 3 muscle types: smooth, skeletal, and cardiac.

**muscle fiber** Essentially, a single muscle cell; an elongated cell, with contractile properties, that forms the muscles in the body.

**muscle tissue** Type of tissue adapted for contraction.

**muscular system** System consisting of smooth, skeletal, and cardiac muscle. This system produces body movement, maintains posture, and produces body heat.

**mutagen (MYOO-tah-jen)** Agent that promotes a mutation (e.g., radioactive substances, X-rays, or certain chemicals).

**mutagenicity** Agent that can induce or increase the frequency of mutation in an organism.

**mutase (MYOO-tace)** Enzyme that rearranges the functional groups on a molecule.

**mutation (myoo-TAY-shun)** Change in the chemistry of a gene that is perpetuated in subsequent divisions of the cell in which it occurred; a change in the sequence of the DNA.

**mycotoxin (MY-ko-tok-sin)** Toxic compound produced by molds, such as aflatoxin B-1, found on moldy grains.

**myelin sheath (MY-eh-lyn)** Combined lipid and protein structure (lipoprotein) that covers nerve fibers.

**myocardial depression** Decreased activity of the heart muscle.

**myocardial infarction (MY-oh-CARD-ee-ahl in-FARK-shun)** Death of part of the heart muscle.

**myofibril (my-oh-FIB-ril)** Bundle of contractile fibers within a muscle cell.

**myoglobin (my-oh-GLOW-bin)** Iron-containing protein that controls the rate of diffusion of oxygen ($O_2$) from red blood cells to muscle cells.

**myosin (MY-oh-sin)** Thick filament protein that connects with actin to cause a muscle contraction.

# N

**nanotechnology** Study of controlling matter at the atomic or molecular level.

**narcotic** Agent that reduces sensations and consciousness.

**natural food** Food that has undergone minimal processing and does not contain food additives.

**natural toxins** Naturally occurring toxins in foods, especially plants. They rarely cause disease in humans.

**negative energy balance** State in which energy intake is less than energy expended, resulting in weight loss.

**negative nitrogen balance** State in which nitrogen losses from the body exceed intake, as in starvation.

**neoplasm (NEE-oh-plaz-em)** New and abnormal growth of tissues, which may be benign or cancerous.

**neotame** General-purpose non-nutritive sweetener that is approximately 7000 to 13,000 times sweeter than table sugar. It has a chemical structure similar to aspartame's. Neotame is heat stable and can be used as a tabletop sweetener, as well as in cooking applications. It is not broken down to its amino acid components in the body after consumption.

**nephron (NEF-ron)** Functional unit of the kidney.

**nephrotic syndrome (NEF-rot-ick)** Type of kidney disease that results from damage to the kidney, often caused by another disease, such as diabetes. The symptoms include fluid retention, weight gain, and high blood pressure.

**nerve** Bundle of nerve cells outside the central nervous system.

**nervous system** Body system consisting of the brain, spinal cord, nerves, and sensory receptors. This system detects sensations and controls physiological and intellectual functions and movement.

**nervous tissue** Tissue composed of highly branched, elongated cells that transport nerve impulses from 1 part of the body to another.

**neural tube defect** Defect in the formation of the neural tube occurring during early fetal development. This type of defect results in various nervous system disorders, such as spina bifida. A very severe form is anencephaly. Folate deficiency in a pregnant woman increases the risk that the fetus will develop this disorder.

**neuroendocrine (NEW-row-EN-do-krin)** Linked to the combined action of the endocrine glands and the nervous

system. Examples include substances released from glands in response to nerve stimulation.

**neuroglia (glial cells) (NEW-row-GLEE-ah)** Specialized support cells of the central nervous system.

**Neuromotor exercise** Exercises that include motor skills (e.g., balance, coordination, gait, agility) and training to help the body control joints to be stable and react appropriately to becoming off balance to prevent falls.

**neuromuscular junction (NEW-row-MUS-kyo-lar)** Chemical synapse between a motor neuron and a muscle fiber.

**neuron (NEW-ron)** Structural and functional unit of the nervous system, consisting of cell body, dendrites, and axon.

**neuropeptide Y (NEW-row-PEP-tide)** Small protein (36 amino acids) that increases food intake and reduces energy expenditure when injected into the brains of experimental animals.

**neurotransmitter (NEW-row-TRANS-mit-er)** Compound made by a nerve cell that allows for communication between it and other cells.

**neutron (NEW-tron)** Part of an atom that has no charge.

**neutrophil (NEW-tro-fil)** Type of phagocytic white blood cell, normally constituting about 60 to 70% of the white blood cell count; forms highly toxic compounds, which destroy bacteria.

**neutrophil/activation** Type of white blood cell being prepared for immune response.

**niacin** Water-soluble vitamin that, in coenzyme form, participates in numerous oxidation-reduction reactions in cellular metabolic pathways, especially those used to produce ATP. Tuna, chicken, beef, peanuts, and salmon are good sources.

**nicotinamide (nick-ah-TIN-ah-mide)** One of the 2 forms of the vitamin niacin.

**nicotinamide adenine dinucleotide (NAD) (nick-ah-TIN-ah-mide AD-ah-neen di-NEW-klee-a-tide)** Coenzyme that readily accepts and donates electrons and hydrogen ions; formed from the vitamin niacin.

**nicotinamide adenine dinucleotide phosphate (NADP)** Coenzyme that readily accepts and donates electrons and hydrogen ions; formed from the vitamin niacin.

**nicotinic acid (nick-ah-TIN-ick)** One of the 2 forms of the vitamin niacin. Physicians sometimes prescribe it to lower LDL-cholesterol and increase HDL-cholesterol levels.

**night blindness** Vitamin A deficiency condition in which the retina in the eye cannot adjust to low amounts of light.

**nitrate (NIE-trate)** Nitrogen-containing compound used to cure meats. Its use contributes a pink color to meats and confers some resistance to bacterial growth.

**nitrosamine (ni-TROH-sa-mean)** Carcinogen formed from nitrates and breakdown products of amino acids; can lead to stomach cancer.

**nonalcoholic fatty liver disease** Accumulation of fat in the liver, diagnosed by liver biopsy or with an imaging procedure, in the absence of excess alcohol intake.

**nonalcoholic steatohepatitis** Inflammatory and fatty infiltration of the liver in people who do not abuse alcohol. It is a chronic condition that may cause progressive scarring and cirrhosis of the liver.

**nonessential amino acids** Amino acids that the human body can synthesize in sufficient amounts. There are 11 nonessential amino acids. These also are called dispensable amino acids.

**nonheme iron (non-HEEM)** Iron provided by plant sources and elemental iron components of animal tissues. Nonheme iron is less efficiently absorbed than heme iron, and absorption is more closely dependent on body needs.

**nonpolar** Neutral compound, no positive or negative poles present.

**nonspecific immunity** Defenses that stop the invasion of pathogens; requires no previous encounter with a pathogen.

**nonsteroidal anti-inflammatory drugs (NSAIDs)** Class of medications that reduce inflammation, fever, and pain but are not steroids. Aspirin, ibuprofen (Advil®), and naproxen (Aleve®) are some examples.

**no-observable-effect level (NOEL)** Highest dose of an additive that produces no deleterious health effects in animals.

**norepinephrine (nor-ep-ih-NEF-rin)** Neurotransmitter released from nerve endings; also a hormone produced by the adrenal gland during stress. It causes vasoconstriction and increases blood pressure, heart rate, and blood sugar.

**norovirus (NOR-oh-VIE-rus)** Virus in the human intestinal tract and feces. It contaminates food via direct hand-to-food contact, when sewage is used to enrich garden/farm soil, or when shellfish are harvested from waters contaminated by sewage. Cooking destroys the virus. Shellfish and salads are the foods most often implicated. Noroviruses cause more cases of foodborne illness than any other microorganism. They can survive chlorination, and a relatively small amount can cause illness.

**nuclear receptor (NEW-klee-er)** Site on the DNA in a cell where compounds (e.g., hormones) bind. Cells that contain DNA receptors for a specific compound are affected by that compound.

**nucleolus (NEW-klee-o-less)** Center for production of ribosomes within the cell nucleus.

**nucleus (NEW-klee-us)** In chemistry, the core of an atom; contains protons and neutrons.

**nutrient** Chemical substance in food that contributes to health. Nutrients nourish us by providing energy, materials for building body parts, and factors to regulate necessary chemical processes in the body.

**nutrient content claim** Claim that describes the nutrients in a food, such as "low in fat" and "calorie free."

**nutrient density** Ratio derived by dividing a food's contribution to nutrient needs by its contribution to energy needs. When its contribution to nutrient needs exceeds its energy contribution, the food is considered to have a favorable nutrient density.

**nutrient receptors** Proposed sites in the small intestine that contribute signals to the brain, which in turn elicit a feeling of satiety. These receptors are stimulated by nutrient exposure in the lumen of the small intestine.

**nutrient requirement** Amount of a nutrient required to maintain health. This varies between individuals.

**nutrition** Science of food; the nutrients and the substances therein; their action, interaction, and balance in relation to health and disease; and the process by which the organism (i.e., body) ingests, digests, absorbs, transports, utilizes, and excretes food substances.

**Nutrition Care Process** Systematic approach used by registered dietitians to ensure patients receive high-quality, individualized nutrition care; process involves nutrition assessment, diagnosis, intervention, and monitoring, and evaluation.

**nutrition-focused physical exam** Nutrition-focused physical exam clinical assessment for physical evidence of nutrition-related deficiencies and toxicities

**nutrition label** Label containing "Nutrition Facts"; must be included on most foods. It depicts nutrient content in comparison with the Daily Values set by the FDA. Canada has a separate set of nutrition labels.

**nutritional status** Nutritional health of a person as determined by anthropometric measures (e.g., height, weight, circumferences), biochemical measurements of nutrients or their by-products in blood and urine, a clinical (physical) examination, a dietary analysis, and an economic evaluation.

**nutritionist** Person who advises about nutrition and/or works in the field of food and nutrition. In many states in the U.S., a person does not need formal training to use this title. Some states reserve this title for registered dietitians.

# O

**obesity (oh-BEES-ih-tee)** Condition characterized by excess body weight and/or body fat; typically defined in clinical settings as a body mass index (BMI) of 30 or above, but this cutoff is not always appropriate.

**Older Americans Act** U.S. government legislation administered by Administration on Aging designed to help adults ages 60 years and older remain living independently in their homes and communities; community-based nutrition, health, and supportive services may include adult day care, senior center activities, transportation, information and counseling services, and health and physical activity programs. In-home care can include health and personal care, home maintenance assistance, and caregiver support services.

**oleic acid (oh-LAY-ik)** Omega-9 fatty acid with 18 carbons and 1 double bond (C18:1, omega-9).

**olfactory (ol-FAK-toe-ree)** Related to the sense of smell.

**olfactory cells** Cells in the nasal region that discriminate among numerous chemical molecules and transmit that information to the brain. This information represents a component of flavor.

**oligosaccharide (ol-ih-go-SAK-ah-ride)** Carbohydrate containing 3 to 10 single sugar units.

**omega-3 (n-3) fatty acid** Unsaturated fatty acid with the 1st double bond on the 3rd carbon from the methyl end ($-CH_3$).

**omega-6 (n-6) fatty acid** Unsaturated fatty acid with the 1st double bond on the 6th carbon from the methyl end ($-CH_3$).

**omnivore (AHM-nih-voor)** Person who consumes foods from both plant and animal sources.

**oncogene (AHN-ko-jeen)** Protooncogene out of control.

**oncotic force (ahn-KAH-tik)** Osmotic potential exerted by blood proteins in the bloodstream.

**opportunistic infection** Infection that arises primarily in people who are already ill because of another disease.

**opsin (AHP-sin)** Protein in the rods of the retina in the eye that binds to 11-*cis*-retinal to form the visual pigment rhodopsin.

**organ** Structure (e.g., heart, kidney, or eye) consisting of cells and tissues that perform a specific function in the organism.

**organ system** Group of organs classified as a unit because they work together to perform a function or set of functions.

**organelle (OAR-gan-ell)** Compartment, particle, or filament that performs specialized functions in a cell.

**organic** Substance that contains carbon atoms bonded to hydrogen atoms in the chemical structure.

**organic compound** Substance that contains carbon atoms bonded to hydrogen atoms in the chemical structure.

**organic food** Food grown with farming practices such as biological pest management, composting, manure application, and crop rotation and without the use of synthetic pesticides, fertilizers, antibiotics, sewage sludge, genetic engineering, irradiation, and hormones.

**organism** Living thing (e.g., the human body is an organism consisting of many organs, which act in a coordinated manner to support life).

**organochlorine insectides** Compounds commonly used in the past to kill insects, but many (e.g., DDT and chlordane) have been removed from the market due to their negative health and environmental effects and their persistence in the environment.

**organophosphates** Compounds toxic to the nervous system of insects and animals; some were used in World War II as nerve agents; they usually are not persistent in the environment.

**osmolality (oz-mo-LAL-ih-tee)** Measure of the total concentration of a solution; number of particles of solute per kilogram of solvent.

**osmosis (oz-MO-sis)** Passage of a solvent, such as water, through a semipermeable membrane from a less concentrated compartment to a more concentrated compartment.

**osmotic pressure (oz-MAH-tick)** Exerted pressure needed to keep particles in a solution from drawing liquid toward them across a semipermeable membrane.

**osteoblast (OS-tee-oh-blast)** Cell in bone; secretes mineral and bone matrix.

**osteocalcin (OS-tee-oh-KAL-sin)** Protein produced in bone that is thought to bind calcium. Synthesis of osteocalcin is aided by vitamin K.

**osteoclast (OS-tee-oh-klast)** Bone cell that arises from a type of white blood cell; secretes substances that lead to bone erosion. This erosion can set the stage for subsequent bone mineralization.

**osteocytes (OS-tee-oh-sites)** Bone cells formed from osteoblasts that have become embedded in bone matrix.

**osteomalacia (OS-tee-oh-mal-AY-shuh)** Weakening of bones that occurs in adults as a result of poor bone mineralization linked to inadequate vitamin D status.

**osteopenia (os-tee-oh-PEE-nee-ah)** Decreased bone mass caused by cancer, hyperthyroidism, or other health conditions.

**osteoporosis (os-tee-oh-po-ROH-sis)** Decreased bone mass leading to risk of bone fractures. This bone loss is related to the effects of aging, genetic background, poor diet, and hormonal changes occurring in postmenopausal women.

**ostomy (OS-toe-me)** Surgically created opening in the intestinal tract. The end point usually opens from the abdominal cavity rather than the anus (e.g., a colostomy).

**overnutrition** State in which nutritional intake greatly exceeds the body's needs.

**overweight** Body weight that is greater than an acceptable standard for a given height. Excess weight is usually due to excess body fat; however, in very muscular individuals, higher weight may be due to muscle.

**ovum (OH-vum)** Egg cell from which a fetus eventually develops if the egg is fertilized by a sperm cell.

**oxalic acid (oxalate) (ox-AL-ick, ox-AH-late)** Organic acid, found in spinach, rhubarb, and other leafy green vegetables, that, can depress the absorption of certain minerals (e.g., calcium) present in the food.

**oxidation (ox-ih-DAY-shun)** Loss of an electron by an atom or a molecule; in metabolism, often associated with a gain of oxygen or a loss of hydrogen. Oxidation (loss of an electron) and reduction (gain of an electron) take place simultaneously in metabolism because an electron that is lost by 1 atom is accepted by another.

**oxidation-reduction (redox) reaction** Chemical reaction that releases energy from food (oxidation) and synthesizes carbohydrates, fatty acids, and other organic compounds (reduction).

**oxidative phosphorylation** Process by which energy derived from the oxidation of $NADH + H^+$ and $FADH_2$ is transferred to $ADP + P_i$ to form ATP.

**oxidative stress** Damage to lipids, proteins, and DNA produced by excessive production of free radicals.

**oxidize (OX-ih-dize)** Loss of an electron or gain of an oxygen in a chemical substance. This change typically alters the shape and/or function of the substance. An oxidizing agent is a substance capable of capturing an electron from another source. That source is then "oxidized" when it loses the electron.

**oxidized LDL** LDL (low-density lipoprotein) that has been damaged by free radicals. This type of damage is seen in both the lipids and the proteins that make up LDL.

**oxidizing agent** Substance capable of capturing an electron from another compound. A compound is "oxidized" when it loses an electron.

**oxygenase (OK-si-jen-ace)** Enzyme that incorporates oxygen directly into a molecule.

**oxytocin (ok-si-TO-sin)** Hormone secreted by a mother's pituitary gland that causes musclelike cells surrounding the ducts in the breasts and the smooth muscle in the uterus to contract.

# P

**p⁵³ gene** Tumor suppressor gene that can prevent inappropriate cell division.

**palatable (PAL-it-ah-bull)** Pleasing to taste.

**pancreas (PAN-kree-us)** Endocrine organ, located near the stomach, that secretes digestive enzymes into the small intestine and produces hormones—notably, insulin.

**pantothenic acid** Water-soluble vitamin that functions as a component of coenzyme A (CoA), which itself plays a pivotal role in energy metabolism and fatty acid synthesis. Most foods are sources.

**para-aminobenzoic acid (ah-MEE-noh-ben-ZOH-ick)** Compound that is a part of the B-vitamin folate.

**parasite** Organism that lives in or on another organism and derives nourishment from it.

**parasthesia (par-a-STEE-zya)** Abnormal spontaneous sensation, such as burning, prickling, and numbness.

**parathyroid hormone (PTH)** Hormone made by the parathyroid glands; increases synthesis of the vitamin D hormone and aids calcium release from bone and calcium uptake by the kidneys, among other functions.

**parenteral nutrition** Method of providing nutrients via the veins; also called intravenous feeding.

**parietal cell (PAH-rye-ah-tahl)** Gastric gland cell that secretes hydrochloric acid and intrinsic factor.

**passive diffusion** Absorption that requires permeability of the substance through the wall of the small intestine and a concentration gradient higher in the intestinal contents than in the absorptive cell.

**pasteurizing (PAS-tur-eye-zing)** Heating food products to kill pathogenic microorganisms and reduce the total number of bacteria.

**pathogen** Disease-causing microorganism.

**pathway** Metabolic progression of individual steps from starting materials to ending products, such as $C_6H_{12}O_6$ (glucose) + $O_2$ eventually yielding $CO_2$ + $H_2O$.

**pectin (PEK-tin)** Soluble fiber containing chains of various monosaccharides; characteristically found between plant cell walls.

**peer-reviewed journal** Journal that publishes research only after researchers who were not part of the study agree that the study was carefully designed and executed and the results are presented in an unbiased, objective manner. Thus, the research has been approved by peers of the research team.

**pellagra (peh-LAHG-rah)** Disease characterized by inflammation of the skin, diarrhea, and eventual mental incapacity; results from an insufficient amount of the vitamin niacin in the diet.

**pentose phosphate pathway** Pathway for glucose breakdown that produces NADPH used in biosynthetic pathways and the 5-carbon sugar used to synthesize RNA and DNA.

**pepsin (PEP-sin)** Protein-digesting enzyme produced by the stomach.

**pepsinogen (PEP-sin-oh-jin)** Inactive protein precursor to the protein-digesting enzyme pepsin; produced in the stomach.

**peptic ulcer** Hole in the lining of the stomach or duodenum.

**peptide bond** Chemical bond formed between amino acids in a protein.

**peptide** Amino acids (usually 2 to 4) chemically bonded together.

**percentile** Classification of a measurement of a unit into divisions of 100 units.

**peripheral nervous system (PNS) (peh-RIF-er-al)** System of nerves that lie outside the brain and spinal cord.

**peripheral neuropathy (peh-RIF-er-al new-ROP-ah-thee)** Problem caused by damage to the nerves outside the spinal cord and brain. Symptoms include numbness, weakness, tingling, and burning pain, often in the hands, arms, legs, and feet.

**peristalsis (per-ih-STALL-sis)** Coordinated muscular contraction that propels food down the GI tract.

**pernicious anemia (per-NISH-us)** Anemia that results from the inability to absorb sufficient vitamin B-12; is associated with nerve degeneration, which can result in eventual paralysis and death.

**peroxisome (per-OK-si-som)** Cell organelle that uses oxygen to remove hydrogens from compounds. This produces hydrogen peroxide ($H_2O_2$), which breaks down into $O_2$ and $H_2O$.

**peroxyl radical (per-OK-syl)** Peroxide compound containing a free radical; designated R-O-O˙, where R is a carbon-hydrogen chain broken off a fatty acid and the dot is an unpaired electron.

**pesticide** Agent that can destroy bacteria, fungi, insects, rodents, or other pests.

**pH** Measure of relative acidity or alkalinity of a solution. The pH scale is 0 to 14. A pH below 7 is acidic; a pH above 7 is alkaline.

**phagocytic cells (fag-oh-SIT-ick)** Cells that engulf substances; include neutrophils and macrophages.

**phagocytosis (FAG-oh-sigh-TOW-sis)** Form of active absorption in which the absorptive cell forms an indentation, and particles or fluids entering the indentation are then engulfed by the cell.

**pharynx (FAIR-ingks)** Organ of the digestive tract and respiratory tract located at the back of the oral and nasal cavities.

**phenobarbital (fee-noe-BAR-bit-ahl)** Medication used to treat seizure disorders.

**phenylalanine (fen-ihl-AL-ah-neen)** Essential (indispensable) amino acid.

**phenylketonuria (PKU) (fen-ihl-kee-toh-NEW-ree-ah)** Disease caused by a defect in the liver's ability to metabolize the amino acid phenylalanine into the amino acid tyrosine; untreated, toxic by-products of phenylalanine build up in the body and lead to mental retardation.

**phosphocreatine (PCr) (fos-fo-CREE-a-tin)** High-energy compound that can be used to re-form ATP from ADP.

**phospholipase (fos-fo-LY-pase)** Enzyme that splits a fatty acid from a cell membrane phospholipid.

**phospholipid** Class of fat-related substances that contain phosphorus, fatty acids, and a nitrogen-containing base. Phospholipids are an essential part of every cell.

**phosphorus** Major ion of intracellular fluid. It contributes to acid-base balance, bone and tooth strength, and various metabolic processes. Milk, milk products, and nuts are good sources.

**photoisomerization (foto-eye-SOM-er-eye-zay-shun)** Molecular isomerization of a compound by the energy of light.

**photon (FO-ton)** Unit of light intensity at the retina having the brightness of 1 candle.

**photosynthesis (fo-to-SIN-tha-sis)** Process by which plants use energy from the sun to produce energy-yielding compounds, such as glucose.

**phylloquinone (fil-oh-KWIN-own)** Form of vitamin K that comes from plants; also called vitamin $K_1$.

**physical activity** Body movement caused by muscular contraction, resulting in the expenditure of energy.

**physiological anemia** Normal increase in blood volume in pregnancy that dilutes the concentration of red blood cells, resulting in anemia; also called hemodilution.

**physiological fuel value** Calories supplied by each macronutrient; equal to 4, 9, 4, and 7 kcal/g for carbohydrate, fat, protein, and alcohol, respectively.

**phytic acid (phytate) (FY-tick, FY-tate)** Constituent of plant fibers that binds positive ions (e.g., zinc as $Zn^{2+}$) to its multiple phosphate groups and decreases their bioavailability.

**phytobezoar (fy-tow-BEE-zor)** Pellet of fiber characteristically found in the stomach.

**phytochemical (fie-toe-KEM-i-kahl)** Physiologically active compound in plants that may provide health benefits.

**pica (PIE-kah)** Practice of eating nonfood items, such as dirt, laundry starch, or clay.

**pinocytosis (pee-no-sigh-TOE-sis)** Formation of a vesicle that brings molecules into a cell; also called cell drinking.

**placebo (plah-SEE-bo)** Fake treatment (e.g., a sham medicine, supplement, or procedure) that seems like the experimental treatment; used to disguise whether a study participant is in the experimental or control group.

**placebo effect** *Placebo* is derived from the Latin word that means "I shall please." The placebo effect occurs when control group participants experience changes that cannot be explained by the action of the placebo they received. These changes may be linked to a treatment that is working, or a desire to help researchers achieve their goals. Overall, it is critical for researchers to take the placebo effect into consideration when interpreting research results.

**placenta (plah-SEN-tah)** Organ that forms in the uterus in pregnant women. Through this organ, oxygen and nutrients from the mother's blood are transferred to the fetus and fetal wastes are removed. The placenta also releases hormones that maintain the state of pregnancy.

**plaque (PLACK)** Cholesterol rich substance deposited in the blood vessels; contains white blood cells, smooth muscle cells, connective tissue (collagen), cholesterol and other lipids, and eventually calcium.

**plasma** Fluid, noncellular portion of the circulating blood. This includes the blood serum plus all blood clotting factors. In contrast, serum is the fluid that results after the blood is first allowed to clot before being centrifuged; does not contain the blood clotting factors.

**plasma cells** B lymphocytes that produce about 2000 antibodies per second.

**platelets** Cells that circulate in large numbers the blood and help stop bleeding by clumping together to form clots; also called thrombocytes.

**polar** Compound with distinct positive and negative charges (poles) on it. These charges act as poles on a magnet.

**polyglutamate form of folate (POL-ee-GLOO-tah-mate)** Folate with more than 1 glutamate molecule attached.

**polyneuropathy (POL-ee-nyoo-ROP-ah-thee)** Disease process involving a number of peripheral nerves.

**polypeptide (POL-ee-PEP-tide)** Ten or more amino acids bonded together.

**polyphenol** Group of compounds containing at least 2 ring structures, each with at least 1 hydroxyl group (OH) attached. Polyphenols occur naturally in tea, dark chocolate, and wine and can lower the bioavailability of minerals, especially iron and calcium.

**polysaccharide (POL-ee-SACK-uh-ride)** Large carbohydrate containing from 10 to 1000 or more monosaccharide units; also known as complex carbohydrate.

**polyunsaturated fatty acid (PUFA)** Fatty acid containing 2 or more carbon-carbon double bonds.

**pool** Amount of a nutrient within the body that can be easily mobilized when needed.

**portal system** Veins in the GI tract that convey blood from capillaries in the intestines and portions of the stomach to the liver.

**portal vein** Large vein leaving from the intestine and stomach and connecting to the liver.

**positive energy balance** State in which energy intake is greater than energy expended, generally resulting in weight gain.

**positive nitrogen balance** State in which nitrogen intake exceeds related losses. This state causes a net gain of nitrogen in the body, such as when tissue protein is gained during growth.

**potassium** Major positive ion in intracellular fluid. It performs many of the same functions as sodium, such as fluid balance and nerve impulse transmission. Potassium also influences the contractility of smooth, skeletal, and cardiac muscle. Spinach, squash, and bananas are good sources.

**poverty guidelines** Federal poverty level; income level calculated each year by the U.S. Census Bureau. Guidelines are used to determine eligibility for many federal food and nutrition assistance programs.

**power stroke** Movement of the thick filament alongside the thin filament in a muscle cell, causing muscle contraction.

**Prader-Willi syndrome** Genetic disorder characterized by shortness, mental retardation, and uncontrolled appetite, caused by a dysfunction of the nervous system, leading to extreme obesity.

**prebiotic** Substance that stimulates bacterial growth in the large intestine.

**precursor** Compound that comes before; a precedent.

**preeclampsia (pre-ee-KLAMP-see-ah)** Part of the disease called pregnancy-induced hypertension. This serious disorder can include high blood pressure, kidney failure, convulsions, and even death of the mother and fetus. Mild cases are known as preeclampsia; more severe cases are called eclampsia or, formerly, toxemia.

**pregnancy-induced hypertension** Serious disorder that can include high blood pressure, kidney failure, convulsions, and even death of the mother and fetus. Although its exact cause is not known, an adequate diet (especially adequate calcium intake) and prenatal care may prevent this disorder or limit its severity. Mild cases are known as preeclampsia; more severe cases are called eclampsia (formerly called toxemia).

**preliminary health claim** Claim made about foods that is based on incomplete scientific evidence.

**premenstrual syndrome (PMS)** Disorder occurring in some women a few days before a menstrual period begins; characterized by depression, anxiety, headache, bloating, and mood swings. Severe cases are currently termed premenstrual dysphoric disorder (PDD).

**prenatal** During gestation (pregnancy); before birth.

**preservatives** Compounds that extend the shelf life of foods by inhibiting microbial growth or minimizing the destructive effect of oxygen and metals.

**preterm** Born before 37 weeks of gestation (also is referred to as premature).

**prevalence** Proportion of people in a population at a specific time who have a certain disease, such as obesity or cancer.

**previtamin D$_3$** Precursor of 1 form of vitamin D, produced as a result of sunlight opening a ring on 7-dehydrocholesterol in the skin.

**primary disease** Disease process that is not simply caused by another disease process.

**primary prevention** Attempt to prevent a disease from developing—for example, following a diet low in saturated fat and cholesterol in an attempt to prevent cardiovascular disease.

**primary structure of a protein** Order of amino acids in the protein molecule.

**prion (PRE-on)** Protein involved in maintaining nerve cell function. Prions can become infectious and lead to diseases, such as bovine spongiform encephalopathy, also known as mad cow disease.

**prior sanctioned substances** Food additives in use prior to 1958 that have been approved by the FDA or the USDA.

**probiotic (PRO-bye-ah-tic)** Live microorganisms that when ingested in adequate amounts confer a health benefit on the host. Prebiotic Substance that stimulates bacterial growth in the large intestine.

**progestin (pro-JES-tin)** Hormone, including progesterone, that is necessary for maintaining pregnancy and lactation.

**prognosis (prog-NO-sis)** Forecast of the course and end of a disease.

**progression** Increase in exercise frequency, duration, and intensity over time.

**prohormone** Precursor of a hormone.

**prolactin (pro-LACK-tin)** Hormone secreted by a mother's pituitary gland that stimulates the synthesis of milk in her breast.

**prospective** Type of research that follows individuals during a current course of treatment, in contrast with retrospective research, which examines the past habits of individuals.

**prostacyclin (PGI) (prost-ah-SIGH-klin)** Eicosanoid made by the blood vessel walls; a potent inhibitor of blood clotting.

**prostaglandins (pros-tah-GLAN-din)** Potent compounds that are synthesized from polyunsaturated fatty acid that produce diverse effects in the body.

**prostanoids (PROS-ta-noidz)** Group of prostaglandins, prostacyclins, and thromboxanes produced from 20 carbon (C:20) fatty acids; not as inclusive a term as *eicosanoids* because leukotrienes and lipoxins are not included.

**prostate gland (PROS-tait)** Solid, chestnut-shaped organ surrounding the first part of the urinary tract in males. The prostate gland secretes substances into the semen.

**protease (PRO-tea-ace)** Protein-digesting enzyme.

**protein** Food and body components made of amino acids. Proteins contain carbon, hydrogen, oxygen, nitrogen, and sometimes other atoms in a specific configuration. Proteins contain the form of nitrogen most easily used by the human body. Protein supplies 4 kcal/gram.

**Protein Digestibility Corrected Amino Acid Score (PDCAAS)** Chemical score of a food multiplied by its digestibility.

**protein efficiency ratio (PER)** Measure of protein quality in a food, determined by the ability of a protein to support the growth of a young animal.

**protein quality** Measure of the ability of a food protein to support body growth and maintenance.

**protein turnover** Process by which a cell breaks down existing proteins and then synthesizes new proteins.

**protein-energy malnutrition (PEM)** Condition resulting from insufficient amounts of energy and protein, which eventually results in body wasting and increased susceptibility to infections.

**prothrombin (pro-THROM-bin)** Protein that participates in the formation of blood clots. Conversion of its precursor protein to the active blood clotting factor in the liver requires vitamin K.

**proton (PRO-ton)** Part of an atom that is positively charged.

**proton pump inhibitor** Medication that inhibits the ability of gastric cells to secrete hydrogen ions (e.g., esomeprazole [Nexium®], lansoprazole [Prevacid®], and omeprazole [Prilosec®]).

**protooncogenes (pro-toe-ON-ko-jeans)** Genes that cause a resting cell to divide.

**protozoa (pro-tah-ZOE-ah)** One-celled animals that are more complex than bacteria. Disease-causing protozoa can be spread through food and water.

**provitamin** Substance that can be made into a vitamin.

**psoriasis (sah-RIE-ah-sis)** Immune system disorder that causes a chronic inflammatory skin condition (painful patches of red, scaly skin).

**psyllium (SIL-ee-um)** Mostly soluble type of dietary fiber found in the seeds of the plantago plant (native to India and Mediterranean countries).

**pteridine (TER-ih-deen)** Bicyclic compound that makes up part of the structure of folate.

**puberty** Period when a child physically matures into an adult capable of reproduction. Puberty is initiated by the secretion of sex hormones: primarily estrogen in females and testosterone in males.

**pulmonary circulation (pulmonary circuit)** System of blood vessels from the right ventricle of the heart to the lungs and back to the left atrium of the heart.

**purine (PURE-een)** Double-ringed compound that forms the nitrogenous bases adenine and guanine found in DNA and RNA.

**pyloric sphincter (pi-LOR-ik SFINK-ter)** Ring of smooth muscle between the stomach and the duodenum.

**pyrethroid pesticides (pi-REE-throyd)** Compounds that mimic naturally occurring pesticides found in chrysanthemums, some of which are toxic to the nervous system of insects.

**pyridoxal (pir-ih-DOX-ahl)** One of the 3 forms of vitamin B-6.

**pyridoxamine (pir-i-DOX-ah-meen)** One of the 3 forms of vitamin B-6.

**pyridoxine (pir-i-DOX-een)** One of the 3 forms of vitamin B-6.

**pyrimidine (pie-RIM-i-deen)** Six-membered ring compound that forms the nitrogenous bases thymine and cytosine found in DNA and RNA.

**pyruvate (pie-ROO-vate)** Three-carbon compound formed during glucose metabolism; also called pyruvic acid.

# R

**racemase (RAS-ih-mace)** Group of enzymes that catalyzes reactions involving structural rearrangement of a molecule (e.g., conversion of D-alanine isomer to L-alanine isomer).

**radiation** Energy that is emitted from a center in all directions. Various forms of radiation energy include X-rays and ultraviolet rays from the sun.

**raffinose (RAF-ih-nos)** Indigestible oligosaccharide made of 3 monosaccharides (galactose-glucose-fructose).

**rancid (RAN-sid)** Containing products of decomposed fatty acids; yield unpleasant flavors and odors.

**Rating of Perceived Exertion (RPE)** Scale that defines the difficulty level of any activity; can be used to determine intensity.

**reactive hypoglycemia (HIGH-po-gly-SEE-mee-uh)** Low blood glucose that may follow a meal high in simple sugars, with corresponding symptoms of irritability, headache, nervousness, sweating, and confusion; actually called postprandial hypoglycemia.

**reactive oxygen species (ROS)** Several oxygen derivatives produced during the formation of ATP; formed constantly in the human body and shown to kill bacteria and inactivate proteins; also are implicated in a number of diseases and inflammatory processes.

**receptive framework for learning** Process by which a person opens up to learning more about a problem. It usually involves seeking more information about the issue from books and people. In the case of seeking behavior changes, it involves examining background experience to evaluate whether a behavior change is feasible.

**receptor (ri-SEP-ter)** Site in a cell where compounds (e.g., hormones) bind. Cells that contain receptors for a specific compound are partially controlled by that compound.

**receptor pathway for cholesterol uptake** Process by which LDL is bound by cell receptors and incorporated into the cell.

**recombinant DNA (re-KOM-bih-nant)** Molecule composed of the DNA from 2 different species spliced together, such as a combination of bacterial and human DNA used to produce unique bacteria that can synthesize human proteins.

**recombinant DNA technology** Test tube technology that rearranges DNA sequences in an organism by cutting the DNA, adding or deleting a DNA sequence, and rejoining DNA molecules with a series of enzymes.

**Recommended Dietary Allowances (RDAs)** Nutrient intake amount sufficient to meet the needs of 97 to 98% of the individuals in a specific life stage.

**Recommended Nutrient Intake (RNI)** Canadian version of RDA, published in 1990.

**rectum** Terminal portion of the large intestine.

**redox agents (RE-doks)** Chemicals that can readily undergo both oxidation (loss of an electron) and reduction (gain of an electron).

**reducing agent** Compound capable of donating electrons (also hydrogen ions) to another compound.

**reduction** In chemical terms, the gain of an electron by an atom; takes place simultaneously with oxidation (loss of an electron by an atom) in metabolism because an electron that is lost by 1 atom is accepted by another. In metabolism, reduction often is associated with the gain of hydrogen.

**re-esterification** Process of reattaching a fatty acid to glycerol that has lost a fatty acid.

**Reference Daily Intakes (RDIs)** Nutrient-intake standards set by the FDA based on the 1968 RDAs for various vitamins and minerals. RDIs have been set for 4 categories of people: infants, toddlers, people over 4 years of age, and pregnant or lactating women. Generally, the highest RDA value out of all categories is used as the RDI. The RDIs constitute part of the Daily Values used in food labeling.

**registered dietitian (R.D.)** Person who has completed a baccalaureate degree program approved by the Academy of Nutrition and Dietetics (formerly the American Dietetic Association), performed at least 900 hours of supervised professional practice, and passed a registration examination.

**reinforcement** Reaction by others in response to a person's behavior. Positive reinforcement entails encouragement; negative reinforcement entails criticism or penalty.

**relapse prevention** Series of strategies used to help prevent and cope with weight control lapses, such as recognizing high-risk situations and deciding beforehand on appropriate responses.

**remodeling** Bone remodeling Lifelong process of the building and breaking down of bone.

**renin (REN-in)** Enzyme formed in the kidneys and released in response to low blood pressure. It acts on a blood protein called angiotensinogen to produce angiotensin I.

**reproductive system** Body system, consisting of the gonads, accessory structures, and genitals of males and females, that performs the process of reproduction and influences sexual functions and behaviors.

**reserve capacity** Extent to which an organ can preserve essentially normal function despite decreasing cell number or cell activity.

**resistance exercise** Physical activities that use muscular strength to move a weight or work against a resistant load.

**resistant starch** Starch, found in whole grains and some fruit, that resists the action of digestive enzymes.

**resorption (ree-ZORP-shun)** Loss of a substance by physiological or pathological means.

**respiration** Use of oxygen; in the human organism, the inhalation of oxygen and the exhalation of carbon dioxide; in cells, the oxidation (electron removal) of food molecules, resulting in the eventual release of energy, $CO_2$, and water.

**respiratory system** Body system, consisting of the lungs and associated organs (e.g., the nose and various conducting tubes), that transports oxygen from outside air to the lungs and allows carbon dioxide to be expelled from the body. Oxygen and carbon dioxide are exchanged with the blood in the lungs. This system also regulates acid-base balance in the body.

**resting metabolism** Amount of energy the body uses when the person has not eaten in 4 hours and is resting (e.g., 15 to 30 minutes) and awake in a warm, quiet environment. It is approximately 6% higher than basal metabolism because of the less strict criteria for the test; often referred to as resting metabolic rate (RMR).

**resveratrol** A polyphenolic compound produced in grapes and other plants. It functions as an antioxidant and helps to prevent against various plant diseases. It may health benefits for humans.

**retina** Light-sensitive layer at the back of the eye; contains the photoreceptors of the eye, called rods and cones.

**retinoids (RET-ih-noydz)** Biologically active forms of vitamin A, including retinol, retinal, and retinoic acid.

**reverse transport of cholesterol** Process by which cholesterol is picked up by HDL particles and transferred to the liver or to other lipoproteins that can dispose of it in the liver.

**rhodopsin (row-DOP-sin)** Photoreceptor in rod cells composed of 11-*cis*-retinal and opsin.

**riboflavin (RYE-bo-fla-vin)** Water-soluble vitamin that functions in coenzyme form in oxidation-reduction reactions, thereby playing a key role in energy metabolism. Milk and milk products, liver, mushrooms, and green leafy vegetables are rich sources of riboflavin.

**ribonucleic acid (RNA) (RI-bow-new-CLAY-ik)** Single-stranded nucleic acid involved in the transcription of genetic information and the translation of that information into protein structure.

**ribose (RIGH-bos)** Five-carbon sugar found in genetic material—specifically, RNA.

**ribosomes (RI-bow-somz)** Cytoplasmic particles that mediate the linking together of amino acids to form proteins; attached to endoplasmic reticulum as bound ribosomes or suspended in cytoplasm as free ribosomes.

**rickets (RIK-its)** Disease characterized by inadequate mineralization of the bones caused by poor calcium deposition during growth. This deficiency disease arises in infants and children with poor vitamin D status.

**risk factor** Term used frequently when discussing diseases and factors contributing to their development. Risk factors include inherited characteristics, lifestyle choices (e.g., smoking), and nutritional habits that affect the chances of developing a particular disease.

**rods** Sensory elements in the retina of the eye responsible for visual processes occurring in dim light, translating objects into black-and-white images, and detecting motion.

**rough endoplasmic reticulum (EN-doe-PLAZ-mik re-TIK-you-lum)** Portion of the endoplasmic reticulum that contains ribosomes; site of protein synthesis in a cell.

**R-protein** Protein produced by salivary glands that enhances absorption of vitamin B-12, possibly protecting the vitamin during its passage through the stomach.

**RXR, RAR** Abbreviations for retinoid X receptor and retinoic acid receptor. These 2 subfamilies of retinoid receptors in the nucleus interact with retinoic acid and bind with specific sites on DNA, allowing for gene expression.

# S

**saccharin (SACK-ah-rin)** Alternative sweetener that yields no energy to the body; 300 times sweeter than sucrose.

**S-adenosyl methione (SAM) (ES-ah-DEN-oh-sill meh-THI-ohn)** Compound formed from methionine; serves as a methyl donor in many biochemical reactions.

**saliva (sah-LIGH-vah)** Watery fluid, produced by the salivary glands in the mouth, that contains lubricants, enzymes, and other substances.

**salivary amylase (SAL-ih-var-ee AM-ih-lace)** Starch-digesting enzyme produced by salivary glands.

**salmonella (sal-mo-NELL-a)** Large class of bacteria, many strains of which are toxic, commonly found in animal and human feces. Salmonella can multiply in raw meats, poultry, eggs, fish, sprouts, unpasteurized milk, and foods made with these products. Cooking destroys salmonella.

**salt** Compound of sodium and chloride in a 40:60 ratio.

**sarcoma (sar-KO-mah)** Malignant tumor arising from connective tissues, such as bone.

**sarcomere (SAR-koe-mere)** Portion of a muscle fiber that is considered the functional unit of a myofibril.

**satiety (suh-TIE-uh-tee)** State in which there is no longer a desire to eat; a feeling of satisfaction.

**saturated fatty acid (SFA)** Fatty acid containing no carbon-carbon double bonds.

**scavenger pathway for cholesterol uptake** Process by which LDL is taken up by scavenger cells embedded in the blood vessels.

**schistosomiasis (shis-to-soh-MY-ah-sis)** Diseases of the liver, bladder, and GI tract caused by drinking water infested with a parasitic worm.

**scientific integrity Integrity** demonstrated by adhering to professional values and practices when conducting, interpreting, reporting, and using scientific research. These values and practices help ensure that research activities are objective, clear, reproducible, and useful. These values and practices also help prevent bias, fabrication or falsification of data, plagiarism, inappropriate interference by others who might want to influence the way the research is conducted or reported, censorship of scientific findings, and inadequate research activity security.

**scurvy (SKER-vee)** Deficiency disease that results after a few weeks to months of consuming a diet that lacks vitamin C. Pinpoint sites of bleeding on the skin are an early sign.

**seborrheic dermatitis (seh-bor-REE-ik der-mah-TITE-is)** Skin condition that results in scaly, flaky, itchy, and red skin; may result from a B-vitamin deficiency.

**secondary deficiency** Deficiency caused not by lack of the nutrient in question but by lack of a substance or process needed for that nutrient to function.

**secondary disease** Disease process that develops as a result of another disease.

**secondary prevention** Intervention to prevent further development of a disease so as to reduce the risk of further damage to health—for example, smoking cessation for a person who already has suffered a heart attack.

**secretin (SEE-kreh-tin)** Hormone that causes bicarbonate ion release from the pancreas.

**secretory vesicles (see-KRE-tor-ee VES-ih-kels)** Membrane-bound vesicles produced by the Golgi apparatus; contain proteins and other compounds to be secreted by the cell.

**sedentary lifestyle** Lifestyle that includes only the light physical activity associated with typical day-to-day life.

**segmentation** Contractions of the circular muscles in the intestines that lead to a dividing and mixing of the intestinal contents. This action aids digestion and absorption of nutrients.

**selenium** Trace mineral that functions as part of antioxidant enzyme systems and in thyroid hormone metabolism. Animal protein foods and whole grains are good sources.

**self-monitoring** Tracking behavior and conditions affecting that behavior; actions are usually recorded in a diary, along with location, time, and state of mind. This tool can help people understand more about their eating habits.

**semiessential amino acids** Amino acids that, when consumed, spare the need to use an essential amino acid for their synthesis. Tyrosine in the diet, for example, spares the need to use phenylalanine for tyrosine synthesis; also called conditionally essential amino acids.

**Senior Farmers' Market Nutrition Program** U.S. government program that provides low-income older adults with coupons that can be exchanged for fresh fruits, vegetables, and herbs at farmers' markets, roadside stands, and other community-support agriculture programs.

**sensible water losses** Water losses readily perceived, such as urine output and heavy perspiration.

**sequestrants (see-KWES-trants)** Compounds that bind free metal ions, thereby reducing the ability of ions to cause rancidity in foods containing fat.

**serotonin (ser-oh-TONE-in)** Neurotransmitter that affects mood (sense of calmness), behavior, and appetite and induces sleep.

**serum (SEER-um)** Portion of the blood fluid remaining after the blood is allowed to clot and the red and white blood cells and other solid matter are removed by centrifugation.

**set-point theory** Theory that humans have a genetically predetermined body weight, which is closely regulated. It is not known what cells control this set point or how it actually functions in weight regulation.

**sexually transmitted disease (STD)** Contagious disease usually acquired by sexual intercourse or genital contact. Common examples include AIDS, gonorrhea, and syphilis; also called venereal disease and sexually transmitted infection.

**shellfish poisoning** Paralysis caused by eating shellfish contaminated with dinoflagellates (algae); also called paralytic shellfish poisoning.

**short-chain fatty acids** Fatty acid that contains fewer than 6 carbon atoms.

**sickle-cell disease (sickle-cell anemia)** Genetic disease that creates red blood cells with incorrect primary structure in part of the hemoglobin protein chains. Malformed (sickle-shaped) red blood cells can lead to episodes of severe bone and joint pain, abdominal pain, headache, convulsions, paralysis, and even death.

**sideroblastic anemia (si-der-oh-BLA-stik)** Form of anemia characterized by red blood cells containing an internal ring of iron granules. This anemia may respond to vitamin B-6 treatment.

**sigmoid colon (SIG-moyd)** Part of the large intestine that connects the descending colon to the rectum.

**sign** Physical attribute that can be observed by others, such as bruises.

**simple carbohydrate** Carbohydrate composed of 1 or 2 sugars (e.g., glucose, fructose, galactose, sucrose, maltose, lactose).

**simple sugar** Monosaccharide or disaccharide in the diet.

**skeletal fluorosis (flo-ROW-sis)** Condition caused by a very high fluoride intake, characterized by bone pain and damage to skeletal structure.

**skeletal muscle** Muscle tissue responsible for voluntary body movements.

**skeletal structure** Shorthand notation for writing chemical formulas in which neither carbon atoms nor the hydrogens bonded to the carbon atoms are expressly shown, although the bonds between the carbon atoms and the position of all atoms other than carbon and hydrogen are shown.

**skeletal system** Body system, consisting of the bones, associated cartilage, and joints, that supports the body, allows for body movement, produces blood cells, and stores minerals.

**skinfold thickness** Method for estimating body composition by measuring fat just under the skin (subcutaneous) with skinfold calipers and then using the measurement in a mathematical formula to estimate body fat.

**slough (SLUF)** To shed or cast off.

**small for gestational age (jes-TAY-shun-al)** Weighing less than the expected weight for length of gestation; typically defined as below the 10th percentile. This corresponds to less than 5.5 pounds (2.5 kilograms) in a full-term newborn. SGA infants are at increased risk of medical complications.

**smooth endoplasmic reticulum (EN-doe-PLAZ-mik ri-TIK-you-lum)** Portion of the endoplasmic reticulum that does not contain ribosomes. This is the site of lipid synthesis in a cell.

**smooth muscle** Muscle tissue under involuntary control; found in the GI tract, artery walls, respiratory passages, urinary tract, and reproductive tract.

**sodium bicarbonate (SO-dee-um bi-KAR-bown-ait)** Alkaline substance made basically of sodium and carbon dioxide ($NaHCO_3$).

**sodium** Major positive ion in extracellular fluid; essential for maintaining fluid balance and conducting nerve impulses. Salt added to foods during their production supplies most of the sodium in the diet.

**soft palate (PAL-it)** Fleshy posterior portion of the roof of the mouth.

**solanine (sou-lah-neen)** Toxin, produced in potatoes, that increases when potatoes are stored in a brightly lit environment; causes gastrointestinal and neurological symptoms.

**soluble fibers (SOL-you-bull)** Fibers that either dissolve or swell in water and are metabolized (fermented) by bacteria in the large intestine; include pectins, gums, and mucilages; more formally called viscous fibers.

**solute** Substance dissolved in another substance (a solvent), forming a solution

**solvent** Liquid substance that other substances dissolve in.

**sorbitol (SOR-bih-tol)** Alcohol derivative of glucose that yields about 3 kcal/g but is slowly absorbed from the small intestine; used in some sugarless gums and dietetic foods.

**Special Supplemental Feeding Program for Women, Infants, and Children (WIC)** U.S. government program that provides nutritious foods and nutrition education to low-income pregnant, postpartum, and breastfeeding women, as well as infants and children up to age 5 years who are at nutritional risk.

**specific heat** Amount of heat required to raise the temperature of a gram of any substance 1°C. Water has a high specific heat, meaning that a relatively large amount of heat is required to raise its temperature; therefore, it tends to resist large temperature fluctuations.

**specific immunity** Function of lymphocytes directed at specific antigens.

**sphincter (SFINK-ter)** Muscular valve that controls the flow of foodstuffs in the GI tract.

**sphincter of Oddi (ODD-ee)** Ring of smooth muscle between the common bile duct and the upper part of the small intestine (duodenum); also called the hepatopancreatic sphincter.

**spina bifida (SPY-nah BIF-ih-dah)** Birth defect in which the backbone and spinal canal do not close.

**spongy bone** Spongy, inner matrix of bone found primarily in the spine, pelvis, and ends of bones; makes up 20 to 25% of total bone mass; also called cancellous bone or trabecular bone.

**spontaneous abortion** Cessation of pregnancy and expulsion of the embryo or nonviable fetus prior to 20 weeks' gestation; result of natural causes, such as a genetic defect or developmental problem; also called miscarriage.

**spore** Dormant reproductive cell capable of turning into adult organisms without the help of another cell. Certain fungi and bacteria form spores.

**sports anemia (ah-NEE-me-ah)** Decrease in the blood's ability to carry oxygen, found in athletes, which may be caused by iron loss through perspiration and feces or increased blood volume.

**stachyose (STACK-ee-os)** Indigestible oligosaccharide made of 4 monosaccharides (galactose-galactose-glucose-fructose).

**Staphylococcus aureus (staf-i-lo-COCK-us OR-ee-us)** Bacterium found in nasal passages and in cuts on skin; produces a toxin when contaminated food is left for an extended time at danger zone temperature. Meats, poultry, fish, dairy, and egg products pose the greatest risk. *Staphylococcus aureus* can withstand prolonged cooking.

**starch** Carbohydrate made of multiple units of glucose attached together in a form the body can digest; also known as complex carbohydrate.

**stem cell** Unspecialized cell that can be transformed into a specialized cell.

**stenosis (ste-NO-sis)** Narrowing, or stricture, of a duct or canal.

**stereoisomers (stare-ee-oh-EYE-soh-mirz)** Isomers with the same number and types of chemical bonds but with different spatial arrangements (different configurations in space).

**steroid hormones (STARE-oyd)** Group of hormones and related compounds that are derivatives of cholesterol.

**steroid** Group of hormones and related compounds that are derivatives of cholesterol.

**sterol (STARE-ol)** Compound containing a multi-ring (steroid) structure and a hydroxyl group (—OH).

**stimulus control** Alteration of the environment to minimize the stimuli for eating—for example, removing foods from sight and storing them in kitchen cabinets.

**stomatitis (stow-mah-TIE-tis)** Inflammation and soreness of the mouth and throat.

**stress fracture** Fracture that occurs from repeated jarring of a bone. Common sites include bones of the foot.

**striated muscle** Muscles showing a striped pattern when viewed under a microscope. Stripes are due to the presence and specific organization of the contractile proteins actin and myosin.

**stroke** Loss of body function that results from a blood clot or another change in arteries in the brain that affects blood flow and leads to death of brain tissue; also called a cerebrovascular accident.

**structural isomers (EYE-soh-mirz)** Isomers in which the number and kinds of chemical bonds differ.

**structure/function claim** Claim that describes how a nutrient affects human body structure or function such as "iron builds strong blood."

**subclinical** Present but not severe enough to produce signs and symptoms that can be detected or diagnosed (e.g., a subclinical disease or disorder).

**submucosal layer (sub-myoo-KO-sal)** Layer of blood and lymphatic vessels along with nerve fibers and connective tissue that stretch the whole length of the GI tract.

**subsistence farmer** Farmer who grows food only for the farm family rather than for sale.

**successful aging** Physical and physiological function decline that occurs only because a person grows older, not because lifestyle choices, environmental exposures, and chronic disease have aggravated or sped up the rate of aging.

**sucralose (SOO-kra-los)** Alternative sweetener that has chlorines in place of 3 hydroxyl (—OH) groups on sucrose; 600 times sweeter than sucrose.

**sucrase (SOO-krace)** Enzyme made by the absorptive cells of the small intestine; digests sucrose to glucose and galactose.

**sucrose (SOO-kros)** Disaccharide composed of fructose bonded with glucose; also known as table sugar.

**sugar** Simple carbohydrate form with the chemical composition $(CH_2O)_n$. Most sugars form ringed structures when in solution; monosaccharides and disaccharides.

**sulfur** Major mineral primarily functioning in the body in non-ionic form as part of vitamins and amino acids. In ionic form, such as sulfate, it participates in the acid-base balance in the body. Protein rich foods supply sulfur in the diet.

**superoxide dismutase (SOD) (soo-per-OX-ide DISS-myoo-tase)** Enzyme that deactivates a superoxide negative free radical ($O^{2-}$). SOD can contain the trace minerals copper and zinc or magnesium.

**sustainable agriculture** Agricultural system that provides a secure living for farm families; maintains the natural environment and resources; supports the rural community; and offers respect and fair treatment to all involved, from farmworkers to consumers to the animals raised for food.

**sympathetic nervous system** Part of the nervous system that regulates involuntary vital functions, including the activity of the heart muscle, smooth muscle, and adrenal glands.

**symptom** Change in health status noted by the person with the problem, such as a stomach pain.

**synapse (SIN-aps)** Space between the end of 1 nerve cell and the beginning of another nerve cell.

**system** Collection of organs that work together to perform an overall function.

**systematic review** Critical evaluation and synthesis of research studies focusing on a specific topic or research question.

**systemic circuit** Part of the circulatory system concerned with the flow of blood from the heart's left ventricle to the body and back to the heart's right atrium.

**systolic blood pressure (sis-TOL-lik)** Pressure in the arterial blood vessels associated with the pumping of blood from the heart.

# T

**T lymphocyte (tee LYMF-oh-site)** Type of white blood cell that recognizes intracellular antigens (e.g., viral antigens in infected cells), fragments of which move to the cell surface. T lymphocytes originate in the bone marrow but must mature in the thymus gland.

**tagatose (TAG-uh-tose)** Isomer of fructose that is poorly absorbed and, so, yields only 1.5 kcal/g to the body. Tagatose is 90% as sweet as sucrose.

**taurine (TAH-reen)** Nonessential sulfur-containing amino acid; has many vital functions and is included in some energy drinks.

**telomerase (teh-LO-mer-ace)** Enzyme that maintains length and completeness of chromosomes.

**telomeres (TELL-oh-meers)** Caps at the end of chromosomes.

**tendon** Dense connective tissue that attaches a muscle to a bone.

**teratogenic (ter-A-toe-jen-ic)** Tending to produce physical defects in a developing fetus (literally, "monster producing").

**tertiary structure of a protein (TER-she-air-ee)** Three-dimensional structure of a protein formed by interactions of amino acids placed far apart in the primary structure.

**tetany (TET-ah-nee)** Continuous, forceful muscle contraction without relaxation.

**tetrahydrofolic acid (tet-rah-high-dro-FOE-lick)** Central coenzyme formed from folic acid; participates in 1 carbon metabolism.

**theory** Explanation for a phenomenon that has numerous lines of evidence to support it.

**thermic effect of food (TEF)** Energy the body uses to digest, absorb, transport, store, and metabolize nutrients. TEF represents 5 to 10% of energy consumed.

**thermogenesis (ther-mo-JEN-ih-sis)** Heat production by humans.

**thiamin (THIGH-a-min)** Water-soluble B-vitamin that functions in coenzyme form to play a key role in energy metabolism. Pork is a good source of thiamin.

**thiamin pyrophosphate (TPP) (pye-row-FOS-fate)** Coenzyme form of thiamin.

**thioredoxin (THIGH-o-re-dock-sin)** Family of 3 selenium-dependent enzymes that have an antioxidant role and other roles in the body.

**thrifty metabolism** Metabolism that characteristically conserves more energy than normal, such that it increases risk of weight gain and obesity.

**thromboxane (TX) (throm-BOK-sane)** Eicosanoid made by blood platelets that is a stimulant of blood clotting.

**thymine (THIGH-meen)** Nitrogenous base that forms part of the structure of DNA and RNA; a pyrimidine.

**thyroid hormone** Hormone, produced by the thyroid gland, that increases the rate of overall metabolism in the body.

**thyroid-stimulating hormone (TSH)** Hormone that regulates the uptake of iodine by the thyroid gland and the release of thyroid hormone. TSH is secreted in response to a low concentration of circulating thyroid hormone (thyroxine).

**tissue (TISH-you)** Collection of cells adapted to perform a specific function.

**tocopherols (tuh-KOFF-er-allz)** Group of 4 structurally similar compounds that have vitamin E activity. The RRR ("d") isomer of alpha-tocopherol is the most active form.

**tocotrienols (toe-co-TRY-en-olz)** Group of 4 compounds with the same basic chemical structure as the tocopherols but containing slightly altered side chains. They exhibit much less vitamin E activity than the corresponding tocopherols.

**Tolerable Upper Intake Level (UL)** Maximum chronic daily intake of a nutrient that is unlikely to cause adverse health effects in almost all people in a specific life stage.

**total fiber** Combination of dietary fiber and functional fiber in a food; also called fiber.

**total parenteral nutrition** Intravenous provision of all necessary nutrients, including the most basic forms of protein, carbohydrates, lipids, vitamins, minerals, and electrolytes. This solution is generally infused for 12 to 24 hours a day in a volume of about 2 to 3 L.

**toxic** Poisonous; caused by a poison.

**toxicity** Capacity of a substance to produce injury or illness at some dosage.

**toxin** Poisonous compound that can cause disease.

**trabecular bone (trah-BEK-you-lar)** See *spongy bone*.

**trace mineral** Mineral vital to health that is required in the diet in amounts less than 100 mg/day; also called micromineral.

**trachea (TRAY-key-ah)** Airway leading from the larynx to the bronchi.

***trans* configuration** Compound that has hydrogens located opposite each other across a carbon-carbon double bond.

***trans* fatty acids** Form of unsaturated fatty acids, usually monounsaturated when found in food, in which the hydrogens on both carbons forming that double bond lie on opposite sides of that bond (*trans* configuration). Margarine, shortenings, and deep fat–fried foods are rich sources.

**transamination (trans-am-ih-NAY-shun)** Transfer of an amino group from an amino acid to a carbon skeleton to form a new amino acid.

**transcobalamin II (trans-koh-BAL-ah-meen)** Blood protein that transports vitamin B-12.

**transferrin (trans-FER-in)** Blood protein that transports iron in the blood.

**transgenic (trans-JEN-ik)** Organism that contains genes originally present in another, different organism.

**transketolase (trans-KEY-toe-lace)** Enzyme whose functional component is thiamin pyrophospate (TPP). It converts glucose to other sugars.

**translation** Decoding of genetic information stored in RNA by transfer RNA and assembly of amino acids for protein synthesis.

**transverse colon** Part of the large intestine between the ascending colon and descending colon.

***Trichinella spiralis* (trik-i-NELL-a)** Parasitic nematode worm, found in wild game and pork, that causes the flulike disease trichinosis; easily destroyed by cooking. Modern sanitary feeding practices have drastically reduced *Trichinella* in commercial pork.

**triglyceride (try-GLISS-uh-ride)** Major form of lipid in the body and in food;

composed of 3 fatty acids bonded to glycerol, an alcohol.

**trimesters** Three 13- to 14-week periods into which the normal pregnancy is somewhat arbitrarily divided for purposes of discussion and analysis (the length of a normal pregnancy is about 40 weeks, measured from the 1st day of the woman's last menstrual period).

**tropic hormone (TROW-pic)** Hormone that stimulates the secretion of another secreting gland.

**trypsin (TRIP-sin)** Protein-digesting enzyme secreted by the pancreas to act in the small intestine.

**tumor** Mass of cells; may be cancerous (malignant) or noncancerous (benign).

**tumor suppressor genes** Genes that prevent cells from dividing.

**type 1 diabetes** Form of diabetes in which the person is prone to ketosis and requires insulin therapy.

**type 2 diabetes** Most common form of diabetes, in which ketosis is not commonly seen. Insulin therapy may be used but often is not required. This form of the disease is often associated with obesity.

# U

**ulcer (UL-sir)** Erosion of the tissue lining, usually in the stomach (gastric ulcer) or the upper small intestine (duodenal ulcer). These are generally referred to as peptic ulcers.

**ultratrace elements** Elements with requirements estimated at <1 mg/day and that do not have an AI or RDA established. Examples include arsenic, boron, nickel, silicon, vanadium, and cobalt.

**umami (you-MA-mee)** Brothy, meaty, savory flavor in some foods (e.g., mushrooms, Parmesan cheese). Monosodium glutamate enhances this flavor when added to foods.

**undernutrition** Failing health that results from a longstanding dietary intake that does not meet nutritional needs.

**underwater weighing** Method of estimating total body fat by weighing the individual on a standard scale and then weighing him or her again submerged in water. The difference between the 2 weights is used to estimate total body volume.

**underweight** Body mass index below 18.5.

**unleavened bread** Bread that does not contain leavening agents, such as yeast or baking powder. Leavening agents cause bread dough to rise. Flat breads, such as pita bread and tortillas, are unleavened. French and Italian bread, biscuits, and muffins are leavened breads.

**unsaturated fatty acid** Fatty acid with 1 or more carbon-carbon double bonds in its chemical structure.

**upper body obesity** Type of obesity in which fat is stored primarily in the abdominal area; a waist circumference more than 40 inches (102 centimeters) in men and more than 35 inches (88 centimeters) in women; closely associated with a high risk of cardiovascular disease, hypertension, and type 2 diabetes.

**uracil (YUR-ah-sil)** Nitrogenous base that forms part of the structure of DNA and RNA; a pyrimidine.

**urea (yoo-REE-ah)** Nitrogenous waste product of protein metabolism and the major source of nitrogen in the urine; chemically,

$$H_2N-\overset{\overset{\displaystyle O}{\|}}{C}-NH_2$$

**ureter (YOUR-ih-ter)** Tube that transports urine from the kidney to the urinary bladder.

**urethra (yoo-REE-thra)** Tube that transports urine from the urinary bladder to the outside of the body.

**urinary system** Body system, consisting of the kidneys, urinary bladder, and ducts that carry urine, that removes waste products from the circulatory system and regulates blood acid-base balance, overall chemical balance, and water balance in the body.

**USDA Nutrition Evidence Library** Source of systematic reviews to guide U.S. nutrition programs and policies.

**usual aging** Age-related physical and physiological changes commonly thought to be a typical or expected part of aging.

# V

**vagus nerves (VAY-guss)** Nerves arising from the brain that branch off to other organs and are essential for control of speech, swallowing, and gastrointestinal function.

**vegan (VEE-gun)** Person who eats only plant foods.

**vegetarian** Person who avoids eating animal products to a varying degree, ranging from consuming no animal products to simply not consuming mammals.

**vein** Blood vessel that conveys blood to the heart.

**ventricles (VEN-tri-kelz)** Two lower chambers of the heart, which contain blood to be pumped out of the heart.

**venule (VEN-yool)** Tiny vessel that carries blood from the capillary to a vein.

**very-low-calorie diet (VLCD)** Diet that contains 400 to 800 kcal per day, often in liquid form. Of this, 120 to 480 kcal are

carbohydrate; the rest is mostly high-quality protein; also known as protein-sparing modified fast (PSMF).

**very-low-density lipoprotein (VLDL)** Lipoprotein, created in the liver, that carries both cholesterol and lipids taken up from the bloodstream by the liver and those that are newly synthesized by the liver.

**Vibrio cholerae** Bacterium found in human and animal feces; causes the illness cholera; most often transmitted via contaminated drinking water.

**villi (VIL-eye)** Fingerlike protrusions into the small intestine that participate in digestion and absorption of food components.

**virus** Smallest known type of infectious agent, many of which cause disease in humans. They do not metabolize, grow, or move by themselves. They reproduce by the aid of a living cellular host. A virus is essentially a piece of genetic material surrounded by a coat of protein.

**visual cycle** Chemical process in the eye that participates in vision. Forms of vitamin A participate in the process.

**vitamin** Compound needed in very small amounts in the diet to help regulate and support chemical reactions in the body.

**vitamin A** Fat-soluble vitamin that exists in retinoid and carotenoid forms and is crucial to vision in dim light, color vision, cell differentiation, growth, and immunity. Significant food sources are beef liver, sweet potato, spinach, and mangoes.

**vitamin B-12** Water-soluble vitamin. In coenzyme form, it participates in folate metabolism and in the metabolism of fatty acids. Meats and shellfish are good sources; plants do not synthesize vitamin B-12.

**vitamin B-6** Group of water-soluble vitamins that, in coenzyme form, play a role in more than 100 enzymatic reactions in the body, almost all of which involve nitrogen-containing compounds. These reactions include amino acid metabolism, heme synthesis, and homocysteine metabolism. Salmon, potatoes, and bananas are good sources.

**vitamin C** Water-soluble vitamin involved in many processes in the body, primarily as an electron donor; contributes to collagen synthesis, iron absorption, and immune function; likely has in vivo antioxidant capability; also known as ascorbic acid. Fruits and vegetables in general contain some vitamin C; citrus fruit, green vegetables, tomatoes, peppers, and potatoes are especially good sources.

**vitamin D** Fat-soluble vitamin crucial to maintenance of intracellular and extracellular calcium concentrations; exists in cholecalciferol and ergocalciferol forms. Vitamin D is abundant in fatty fish, such as herring, eel, salmon, and sardines, and, in North America, in fortified milk.

**vitamin E** Fat-soluble vitamin that functions in the body as an antioxidant, preventing the propagation of free radicals; exists as tocopherols or tocotrienols. Significant food sources are seeds, nuts, and plant oils.

**vitamin K** Fat-soluble vitamin that contributes to the liver's synthesis of blood clotting factors and the synthesis of bone proteins; exists as phylloquinone or menaquinone. Green leafy vegetables, such as Brussels sprouts, kale, and lettuce, are excellent sources.

**VO$_{2max}$** Maximum volume of oxygen that can be consumed per unit of time.

# W

**water** Universal solvent of life; chemically, $H_2O$. The body is composed of about 60% water. Water serves as a solvent for many chemical compounds, provides a medium in which many chemical reactions occur, and actively participates as a reactant or becomes a product in some reactions.

**water-soluble vitamin** Vitamin that dissolves in water; includes the B-vitamins and vitamin C.

**wean (WEEN)** To accustom an infant to a diet containing foods other than just milk.

**weight cycling** Successfully dieting to lose weight, regaining weight, and repeating the cycle.

**Wernicke-Korsakoff syndrome (ver-NIK-ee KOR-sah-koff)** Thiamin deficiency disease caused by excessive alcohol consumption. Symptoms include eye problems, difficulty walking, and deranged mental functions.

**whey (WAY)** Protein, such as lactalbumin, found in great amounts in human milk; easy to digest.

**white blood cell** Formed element of the circulating blood system; also called leukocyte. Leukocyte types are lymphocytes, monocytes, neutrophils, basophils, and eosinophils. White blood cells are able to squeeze through intracellular spaces and migrate. Leukocytes phagocytize bacteria, fungi, and viruses, as well as detoxify proteins that may result from allergic reactions, cellular injury, and other immune system cells.

**whole grains** Grains containing the entire seed of the plant, including the bran, germ, and endosperm (starchy interior). Examples are whole wheat and brown rice.

**WIC** See *Special Supplemental Feeding Program for Women, Infants, and Children (WIC).*

# X

**xanthine dehydrogenase (ZAN-thin de-HY-droj-ehn-ace)** Enzyme containing molybdenum and iron; functions in the formation of uric acid and the mobilization of iron from liver ferritin stores.

**xenobiotic (ZEE-no-bye-OT-ic)** Compound that is foreign to the body; principal classes are drugs, chemical carcinogens, and environmental substances, such as pesticides.

**xerophthalmia (zer-op-THAL-mee-uh)** Condition marked by dryness of the cornea and eye membranes that results from vitamin A deficiency and can lead to blindness; specific cause is a lack of mucus production by the eye, which then leaves it more vulnerable to surface dirt and bacterial infections.

**xylitol (ZY-lih-tol)** Alcohol derivative of the 5-carbon monosaccharide xylose.

# Y

**Yersinia enterocolitica (yer-SIN-ee-ya)** Bacterium found throughout the environment; present in feces; can contaminate food and water; multiplies rapidly at room and refrigerator temperatures and is destroyed by thorough cooking.

**yo-yo dieting** See *weight cycling.*

# Z

**zinc** Trace mineral required for many enzymes, including those that participate in antioxidant enzyme systems; stabilizes cell membranes and other body molecules. Seafood, meats, and whole grains are good sources.

**zoochemical (zoh-uh-KEM-i-kuhl)** Physiologically active compounds in foods of animal origin that may provide health benefits.

**zygote (ZY-goat)** Fertilized ovum; the cell resulting from the union of an egg cell (ovum) and a sperm until it divides.

**zymogen (ZY-mow-gin)** Inactive form of an enzyme that requires the removal of a minor part of the chemical structure for it to work. The zymogen is converted into an active enzyme at the appropriate time, such as when released into the stomach or small intestine.

# Index